# Winemaking

## Basics and Applied Aspects

## Books Published in *Food Biology* series

1. **Microorganisms and Fermentation of Traditional Foods**
   *Ramesh C. Ray, Montet Didier (eds.), 2014*
2. **Bread and Its Fortification: Nutrition and Health Benefits**
   *Cristina M. Rosell, Joanna Bajerska, Aly F. El Sheikha (eds.), 2015*
3. **Authenticity of Foods of Animal Origin**
   *Ioannis Sotirios Arvanitoyannis (ed.), 2015*
4. **Fermented Foods, Part I: Biochemistry and Biotechnology**
   *Didier Montet, Ramesh C. Ray (eds.), 2015*
5. **Foodborne Pathogens and Food Safety**
   *Md. Latiful Bari, Dike O. Ukuku (eds.), 2015*
6. **Fermented Meat Products: Health Aspects**
   *Nevijo Zdolec (ed.), 2016*
7. **Fermented Foods of Latin America: From Traditional Knowledge to Innovative Applications**
   *Ana Lucia Barretto Penna, Luis A. Nero, Svetoslav D. Todorov (eds.), 2016*
8. **Lactic Acid Fermentation of Fruits and Vegetables**
   *Spiros Paramithiotis (ed.), 2016*
9. **Microbial Enzyme Technology in Food Applications**
   *Ramesh C. Ray, Cristina M. Rosell (eds.), 2016*
10. **Acetic Acid Bacteria: Fundamentals and Food Applications**
    *Ilkin Yucel Sengun (ed.), 2017*
11. **Fermented Foods, Part II: Technological Interventions**
    *Ramesh C. Ray, Montet Didier (eds.), 2017*
12. **Food Traceability and Authenticity: Analytical Techniques**
    *Didier Montet and Ramesh C. Ray (eds.), 2018*
13. **Food Molecular Microbiology**
    *Spiros Paramithiotis (ed.), 2019*

*Food Biology Series*

# Winemaking
## Basics and Applied Aspects

*Editors*

**V.K. Joshi**

Adjunct Professor, Department of Bioengineering and Food Technology, Shoolini University of Biotechnology and Management, Solan (HP), India,
Editor-in-chief: IJFFT, Former Consultant: CSIR, IHBT
*Formerly* Professor and Head
Department of Food Science & Technology
Dr Y S Parmar University of Horticulture & Forestry
Nauni, HP, India

**Ramesh C. Ray**

*Formerly* Principal Scientist (Microbiology)
ICAR – Central Tuber Crops Research Institute
Bhubaneshwar, Odisha, India

CRC Press is an imprint of the
Taylor & Francis Group, an **informa** business
A SCIENCE PUBLISHERS BOOK

CRC Press
Taylor & Francis Group
6000 Broken Sound Parkway NW, Suite 300
Boca Raton, FL 33487-2742



International Standard Book Number: 978-1-138-49091-8 (Hardback)
International Standard Book Number: 978-0-367-71334-8 (Paperback)



**Library of Congress Cataloging-in-Publication Data**

Names: Joshi, V. K., 1955- editor. | Ray, R. C., editor.
Title: Winemaking : basics and applied aspects / editors, V.K. Joshi,
Formerly Professor and Head Department of Food Science & Technology,
Dr Y S Parmar University of Horticulture & Forestry, Nauni, HP, India,
R.C. Ray, Formerly Principal Scientist (Microbiology) ICAR-Central Tuber
Crops Research Institute Bhubaneshwar, Odisha, India.
Other titles: Wine making
Description: Boca Raton, FL : CRC Press, Taylor & Francis Group, 2019. |
Series: Food biology series | Includes bibliographical references and index.
Identifiers: LCCN 2019042511 | ISBN 9781138490918 (hardcover : acid-free paper)
Subjects: LCSH: Wine and wine making.
Classification: LCC TP548 .W7729 2019 | DDC 641.2/2--dc23
LC record available at https://lccn.loc.gov/2019042511

**Visit the Taylor & Francis Web site at**
**http://www.taylorandfrancis.com**

**CRC Press Web site at**
**http://www.routledge.com**

**Science Publishers Web site at**
**http://www.scipub.net**

Dedicated respectfully to

*Professor M.A. Amerine, the great Oenologist, whose contribution to technology of wine is immense and has been a source of inspiration to the Oenologists world over. Professor Amerine is considered as the most significant wine scientist ever produced and his contribution to the wine industry would always be remembered.*

# Preface to the Series

Food is the essential source of nutrients (such as carbohydrates, proteins, fats, vitamins, and minerals) for all living organisms to sustain life. A large part of daily human efforts is concentrated on food production, processing, packaging and marketing, product development, preservation, storage, and ensuring food safety and quality. It is obvious therefore, our food supply chain can contain microorganisms that interact with the food, thereby interfering in the ecology of food substrates. The microbe-food interaction can be mostly beneficial (as in the case of many fermented foods such as cheese, butter, sausage, etc.) or in some cases, it is detrimental (spoilage of food, mycotoxin, etc.). The *Food Biology* series aims at bringing all these aspects of microbe-food interactions in form of topical volumes, covering food microbiology, food mycology, biochemistry, microbial ecology, food biotechnology and bio-processing, new food product developments with microbial interventions, food nutrification with nutraceuticals, food authenticity, food origin traceability, and food science and technology. Special emphasis is laid on new molecular techniques relevant to food biology research or to monitoring and assessing food safety and quality, multiple hurdle food preservation techniques, as well as new interventions in biotechnological applications in food processing and development.

The series is broadly broken up into food fermentation, food safety and hygiene, food authenticity and traceability, microbial interventions in food bio-processing and food additive development, sensory science, molecular diagnostic methods in detecting food borne pathogens and food policy, etc. Leading international authorities with background in academia, research, industry and government have been drawn into the series either as authors or as editors. The series will be a useful reference resource base in food microbiology, biochemistry, biotechnology, food science and technology for researchers, teachers, students and food science and technology practitioners.

**Ramesh C Ray**
*Series Editor*

# Preface

Wine as a food, a product of immense economic significance, and a social lubricant, needs no introduction. It has earned a place in every society, which no other beverage except milk could attain. It has been an integral component of human diet – since its discovery by the caveman, around 6000 B.C. It has travelled a long distance with human civilization and has witnessed many ups and downs. It gave inspiration to the poets and artists, and marveled in the eyes of consumers of diverse origins, withstanding the onslaught of different opinions, movements and cultures, and at the same time was condemned also. Recent researches have added another feather to the already decorated cap of wine in the form of its therapeutic properties.

Our understanding of the technology of wine production includes the knowledge of fruit cultivation, biochemistry and microbiology, methods to produce different types of wines coupled with the advances made in the analytical techniques that have enabled us to produce this fascinating beverage of the best quality, made possible through slow and gradual improvements made by folk-fore, keen observations made and of course, sheer luck not withstanding failures. Nevertheless, the consumer equipped with greater knowledge continues to demand better product. Consequently, the wine makers endeavour to satisfy his quest by providing diverse and quality products by imbibing the latest technological, scientific innovations and marketing strategies. The availability of the appropriate literature in the presentable form comes here into play and also steers the course of future research and development in wine.

The wine preparation technically involves a series of steps right from harvesting the grapes, fermenting the must, maturing the wine, stabilizing and preserving the same and finally, providing the bottled wine ready to the consumers. The fruit growers, wine manufacturers and the consumers are thus, intimately connected with the wine, rather are the main pillars on which the wine rests. The anxiety of the fruit grower is for the fruit production of the specific quality needed for the wine preparation by alcoholic fermentation and maintaining its stability during maturation where a microbiologist and biochemist plays a very significant role. A source of sound and reliable scientific information naturally comes up as a helping hand for this purpose.

No doubt, the wine is a simple and the traditionally natural food but the topic is vast. A large number of cultivars, a number of methods and style of wine along with presentation and consumption pattern makes it more complex and comprehensive subject. May be a table wine, dessert wine or sparkling wine, the technological input is different that has significance to the consumer. The science dealing with wine is called oenology. During the last few decades, a number of developments have taken place in enology (wine microbiology, its therapeutic and medicinal role in human health) that demand collective documentation. The role played by the recombinant DNA technology, wine microbiology, wine microbiome, biochemistry of wine, malo-lactic acid fermentation, control of wine fermentation and bioreactor technology, utilisation of wine waste, application of enzyme technology, the new analytical methods of wine evaluation and sensory evaluation, call for a comprehensive review of the research gains made so far.

The present book has been divided into three sections viz., Section 1: General Aspects of Wine; Section 2: Basics of Winemaking; Section 3: Applied Aspects of Winemaking. The first section covers the introduction and general issues connected with wine including health benefits of wine, the second section dwells upon the various basic issues involved in wine making and section three describes the applied aspects viz., preparation of wine and brandy and, valourisation of waste from wine in part A while part B describes the techniques used in wine evaluation. The various techniques of wine making have been included as illustrations of basic aspects described earlier.

The readers would have ample opportunities to peep into the holistic view of wine, both from the basics and applied aspects in this manuscript, comprising of 27 chapters. Though it is difficult to cover the entire process of wine making especially the detailed treatment used for the preparation of wines of different styles in a text of modest size, various innovations made in the oenology have been included in the book. The entire text is based on the latest research findings dotted with illustrations and the latest trends/or the innovations in the fields. References cited at the end of each chapter would enable the readers to search for more information on specific aspect. Indirectly, it is a stimulation for more research in the field. The book draws upon the expertise of leading researchers in the wine making world-over. It is a useful text and a reference book too. Efforts have been made to eliminate the errors of omissions and commissions in the book. We would, however, welcome constructive criticism and suggestions for the improvement of the manuscript in the later editions. On the whole, the publication should be viewed as continuance of the text by the great peers who have set the stage for scientific and technological advancement of oenology.

Both the editors are thankful to their respective families, without who's support this volume would not have been possible.

**V.K. Joshi**
**R.C. Ray**

# Contents

# Section 1
# General Aspects of Wine

# 1 Wine and Winemaking: An Introduction

**V.K. Joshi[1*] and Ramesh C. Ray[2]**

[1] Department of Bioengineering and Food Science, Shoolini University of Biotechnology and Management, Solan (HP), India. (Former) Department of Food, Science and Technology, Dr. Y.S. Parmar University of Horticulture and Forestry, Nauni, Solan, HP – 173230, India

[2] Regional Centre, Central Tuber Crop Research Institute, Bhubaneswar, Orissa – 751013, India

## 1. Introduction

### 1.1. Wine – Introduction and Definition

Wine has been prepared and consumed by man since antiquity and has a rich history dating back thousands of years (Unwin,1991; Amerie *et al.*, 1980; Joshi, 1997; Robinson, 1986, 2016; Joshi *et al.*, 2011a). Thus, wine is the oldest fermented product known to man and the history of this alcoholic beverage is as old as that of humans (Unwin, 1991; Joshi *et al.*, 2012). Ever since his settlement in the Tigris-Euphrates basin, wine is known to human civilization. It has been regarded as a gift from God and described as a divine liquid in Indian mythology (Vyas and Chakravorty, 1971; Amerine *et al.*, 1980; Joshi *et al.*, 1999a, b). Thus, the knowledge of grape and wine is as old as antiquity (Feher *et al.*, 2007), for example in Hindu mythology, *somras* has been described while in Christianity also, it has been shown to have religious connections with wine (McDonald, 1986). Wine has captured the imagination of poets and philosophers. It had been and still is an integral component of the cultures of many countries.

Wine is a completely or partially fermented juice of grapes, but fruits other than grapes like apple, plum, peach, pear, berries, strawberries, cherries, currants, apricots, banana, etc. have also been utilised in the production of wines (Amerine *et al.*, 1967, 1980; Amerine *et al.*, 1972; MAFF, 1980; Jackson and Schuster, 1981; Jackson, 1994; Joshi, 1997; Sandhu and Joshi, 1995; Jackson, 1994, 1999; Joshi, 1997; Mohanty *et al.*, 2006; Joshi *et al.*, 1990, 1999b, 2000, 2004, 2005; Joshi *et al.*, 2012 Brand *et al.*, 2001). Wine, when distilled, is known as brandy(Voguel, 2003). It is one of the most highly acceptable fermented alcoholic beverages throughout the world, unless forbidden by religion.

There is a wonderful diversity in the styles and quality of wines produced throughout the world (Jackson,1999; Jackson, 2000).

### 1.2. Wine as a Food

Wine has been used as a food adjunct by man. The word 'food' has many definitions, but it is reasonable to quote the *Codex Alimentarius Commission*, which says, 'Food' means any substance, whether processed, semi-processed, or raw, which is intended for human consumption and this includes drink (Burlingame, 2008). Admittedly, except for water and milk, no other beverage has earned such universal acceptance and esteem throughout the ages as wine. No other beverage is discussed, adored or criticised in the same way as wine. It makes an excellent combination with the diet of the people of the Western world because of their high nutritive value, like containing calories, amino acids, vitamins, minerals, etc. Efforts have been directed to find out ways and means to improve the useful components and eliminate those that are toxic to human health. Similar efforts for non-grape wines have not been made. Nevertheless, fruit wine is popular throughout the world and in some parts, it makes a significant contribution to the diet and income of millions of individuals.

### 1.3. Wine Production and Consumer Acceptance

Wine production is both an art and a science besides being a blend of individual creativity and innovative technology. At the same time, it is a business with economic factors attached to it. Some of the leading

*Corresponding author: vkjoshipht@rediffmail.com

wine-producing countries are France, Italy, Spain, Argentina, Portugal, Germany, South Africa, and the United States. Cider and perry (alcoholic drink made from fermented pear juice) are important products in England and northern France; fortified cherry and black currant wines are produced in Denmark; and important American fruit wines, produced mainly on the eastern coast, include apple, cherry, blackberry, elderberry, and loganberry (Amerine *et al.*,1980; Joshi *et al.*, 2011a). Compared to the quantity of grape wine produced and consumed in the world, the quantity of wine produced from non-grape fruits is insignificant except for cider and perry, which are produced and consumed in significant amounts throughout the world. Plum wine is quite popular in many countries. In India, wine production is negligible but is gradually picking up now (Joshi and Parmar, 2004).

Most of the wines are produced between 30° and 50°N, and 30° and 50°S of the equator, with major countries being France, Italy, Spain, USA, Chile, Argentina, South Africa, Australia, New Zealand, etc. Areas within these bands of latitude have a combination of warm summers and relatively mild winters, making it an ideal weather to grow high quality wine grapes. Location and soil are also important. Valleys give protection against wind and frost and thus soil, with good fertility induces vines to root deeply, protecting them from excessive damp and letting them draw water from deep subsoil during droughts. Suitable varieties of grapes are indispensable to the success of making wine, especially distinctive, high quality table wine. When did wine originate, what varieties of grapes are used, which technology is employed, what are the factors involved and how they are relevant would be interesting to know. Some of these aspects have been introduced here in this chapter, while in the entire book, other aspects connected with winemaking are discussed here.

## 2. Origin and History of Wine

### 2.1. Origin of Wine

It can easily be presumed that the origin of wine might have been accidental, when the juice of some fruit might have got transformed by itself into such a beverage having exhilarating or stimulating properties (Vine, 1981; Joshi, 1997; Gasteineu *et al.*, 1979). Clearly, the first wine master must have been a caveman to discover the magic of fermentation. Because the consumption of such beverages might have induced euphoria and pleasing relaxation from the strains of life, it eventually gained social acceptance, with inclusion in religious feasting and celebration or when entertaining guests. A peep into the history of mankind would clearly reveal that the preparation of wine has been a means of preserving perishable commodities, like grape and juice (Unwin, 1991; Amerine *et al.*, 1980; Joshi, 1997; Joshi *et al.*, 2011a; Joshi and Kumar, 2014), and has become an important method for preservation and preparation of products with appealing qualities, even today (Joshi *et al.*, 1999a; Joshi *et al.*, 2011a; Joshi *et al.*, 2017).

The existence of alcoholic beverages, like wine in ancient times, has been amply proven by paintings, articles and writings of historic themes in various parts of the world. Evidence of winemaking first appeared in representations of wine presses (Fig. 1) that date back to the reign of Udimu in Egypt, some 5,000 years ago (Petrie, 1923). Archeological excavations have also uncovered many sites with sunken jars (Fig. 2), indicating the existence of wine for more than 7,500 years (McGovern *et al.*, 1996, 2004).

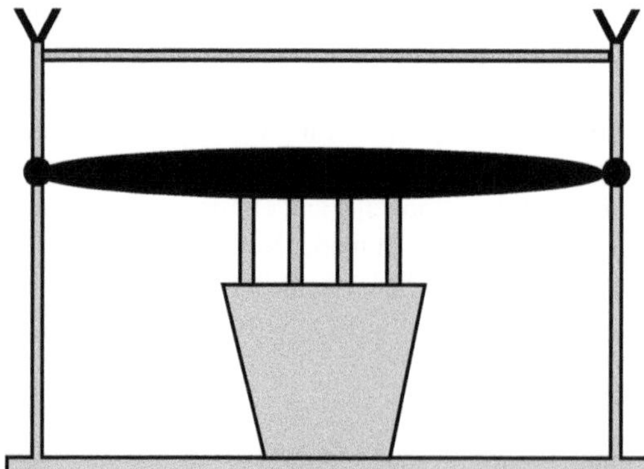

**Figure 1.** A diagrammatic view of a wine press (hieroglyph of Shem, God of the wine press) (Redrawn from Gasteineau *et al.*,1979)

Homer's *Odyssey* and *Iliad,* both have given an excellent description of wine. It was an important article of Greek commerce and Greek doctors, including Hippocrates, were among the first to prescribe wine for treating several ailments (Lucia, 1963). Most civilisations that took to drinking their characteristic wine or other alcoholic beverages believed in myths about the origin of winemaking and attributed its discovery to divine revelation. But the hypothesis of the near Eastern origin and spread of winemaking is also supported by a remarkable similarity between the words meaning 'wine' in most of the Indo-European languages (Renfrew, 1989).

Wine came to Europe with the spread of the Greek civilisation around 1600 BC. However, the beginning of the art of winemaking is shrouded in prehistoric darkness. Nevertheless, there is evidence to suggest that the process of winemaking existed long before the chronicles found in Egyptian hieroglyphics (Vine, 1981).

**Figure 2.** The oldest bottle of wine (Reproduced from ttps://www.google.co.in/search?q=oldest+bottle+of+wine&client=firefox-b&tbm =isch &imgil=BAiK19WA1QyPUM%253A%253BQ)

## 2.2. History of Wine Making

Some of the aspects related to wine history are summarized in Table 1. Starting around 1000 BC, the Romans made major contributions to the development of viticulture and oenology (the science of winemaking) by classifying grape varieties and observing the color and charting ripening characteristics, identifying diseases and recognising soil type preferences. Pruning skills, irrigation and fertilisation, which were also acquired by the Romans, increased the yields. The Greeks introduced viticulture

**Table 1.** Salient Features in the History of Wine

- The Romans organised grape and wine production during their era
- The Greeks stored wine in earthenware amphoras, while the Romans extended the life of their wines with improved oak cooperage
- Both Greek and Roman civilisations drank almost all of their wines within a year of vintage and disguised spoilage by adding flavouring agents
- After the collapse of the Western Roman Empire in the 5th century AD, the survival of viticulture depended on the symbolic role that wine played in Christianity
- The Muslims, however, destroyed the wine industry of the countries they conquered
- The need for wine at religious ceremonies led to the development of wine in Central Europe
- In Western Europe, vineyards were developed
- Early colonial fermented beverages were made from sugar-rich fruits
- The more aristocratic colonists preferred imported wines
- By the 17th century, coopers began building more and better casks and barrels for longer and safe aging of wine
- Wooden barrels remained the principal aging vessels until the 17th century, when mass production of glass bottles and the invention of cork stopper allowed wines to be aged for years in bottles
- The trend of wine consumption shifted towards distilled beverages in the 18th century, after the discovery of the process of distillation
- By the 19th century, the scientific work of Pasteur revolutionised the wine industry by recognising the roles of yeast and bacteria
- Pasteur also identified the bacteria that spoil wine and devised a heating method (called pasteurisation) to kill the bacteria
- In the 1960s, mechanisation (grape harvesters and field crushers) in the vineyards contributed to better quality control
- Advances in plant physiology and plant pathology led to better vine training and less mildew attack on grapes
- Stainless steel fermentation-and-storage tanks that could be easily cleaned and refrigerated at precise temperatures improved the quality of wine
- Automated, enclosed racking and filtration systems reduced the contact of bacteria in the air, thereby preventing spoilage of wine

*Source*: Adapted from Unwin, 1991; Joshi *et al*., 2011a; Vine, 1981

to France, northern Africa and Egypt, whereas the Romans exported vines to Bordeaux, to the valleys of the Rhone, Marne, Seine, etc., and to Hungary, Germany, England, Italy, and Spain. Evidence of the existence of grape farming in India has also been documented. The ancient Aryans are said to have possessed the knowledge of grape culture as well as preparation of beverages from it (Shanmugavelu, 2003). Information on the production of various fruit wines is totally lacking except for the fact that cider was made and consumed widely (Vine, 1981; Joshi *et al.*, 2017), especially in England and France, well before the 12th century.

The Roman technology of winemaking was highly developed though it lacked in the preparation of medicinal wines or in the methods of wine preservation. A breakthrough in the understanding of wine fermentation, however, came at the end of the 17$^{th}$ century, when Van Leeuwenhoek described the occurrence of microorganisms and Louis Pasteur who discovered yeast and how it works.

## 3. Classification of Wines

Wine is a non-distilled and fermented beverage produced by alcoholic fermentation of grapes (Voguel, 2003) or any other fruits. Thus, wine is a low-alcohol, fermented drink produced from grapes only though wine is given the prefix of the fruit from which it is prepared. However, in the European Union, the term 'wine' is exclusively reserved for fermented grape juice but in the USA, any fermented fruit or agricultural product containing 7-24 per cent alcohol is termed as wine (Harding, 2005).

Generally, there is no accepted system of classification of wines (Amerine *et al.*, 1980; Jackson, 2000, 2003), though they may be classified broadly based on their geographic origin, grape variety used, fermentation, or maturation process or may also bear generic names (Fig. 3). Wines can be classified based on alcohol content also, like table wines and fortified wines. Dessert wines are fortified sweet wines. The table wines generally have an alcohol content of 10-11 per cent, but can be as low as 7 per cent, whereas that of cider and perry is usually 2-8 per cent (Joshi, 1997; Joshi *et al.*, 1999, 2011a). Based on the carbon dioxide content, wines can be classified as those without it are known as still wines and those containing it are sparkling wines. Most of the wines are still wines, as they retain no carbon dioxide produced during the fermentation, in contrast to sparkling wines, which contain a considerable amount of carbon dioxide artificially induced. Champagne in France is the sparkling wine made in the Champagne region (Amerine *et al.*, 1980; Joshi *et al.*, 1995; Jeandet *et al.*, 2011; Joshi *et al.*, 2016). It is prepared either by bottle or tank fermentation for carbonation of the base wine through secondary fermentation.

For taxation purposes, wines are often divided into three basic categories, namely, still wines, sparkling wines and fortified wines. Still wines can be further classified into white, red and rosé groups, and into dry or sweet wines. Sparkling wines are often classified according to the method used for the production of high carbon dioxide content, i.e. those made by *methode champanoise*, tank method and *methode naturalis*.

### 3.1. Still Table Wines

These wines are grouped into white, rosé and red wines, depending upon the grapes used and the length of time, the skins have been left to ferment along with the juice (Joshi *et al.*, 2016). Further, these can be dry and sweet wines, depending on whether all the grape sugar has been fermented into alcohol or some residual sugar has been left out. However, the most acceptable division is based on wine colour that reflects the distinct differences viz., white, rose and red.

***Dry white wines*** are a large group of wines intended to be consumed with meals and designed to have an acidic taste. White winemaking demands immediate pressing of grapes, juice clarification, cold fermentation, racking and cold stabilisation before bottling.

***Sweet (not fortified) white wines*** have alcoholic strength varying from 6-14 per cent volume, sugar content varying from 25 to 45 g/l and ranging from aromatically simple to complexly fragrant. Sweet (not fortified) white wines can be classified into two main groups – botrytised and non-botrytised wines.

***Botrytised wines*** are derived from grapes infected by *Botrytis cinerea*. Typically, its effect on the quality of the resultant wine is negative, but under unique climatic conditions. the infected grapes develop the so-called noble rot.

***Non-botrytised sweet wines*** are produced in many winemaking regions of the world. Several vinification techniques have been developed in these regions, depending upon the climatic conditions and consumer demand.

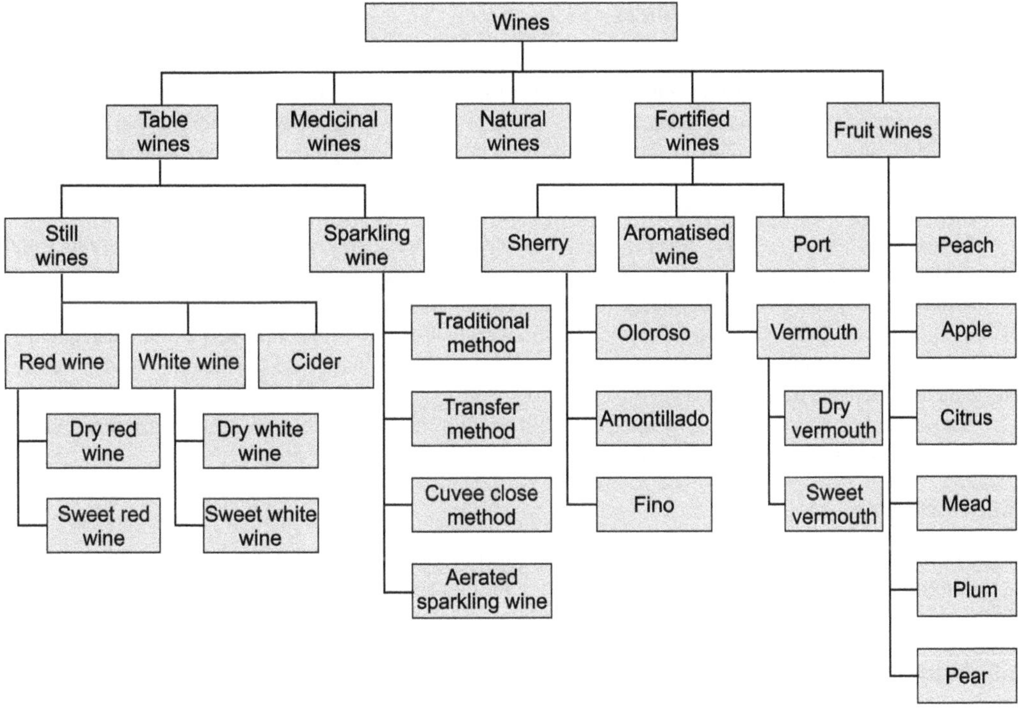


**Figure 3.** A generalised classification of wines of different types
*Source*: Based on Jackson, 2004; Joshi *et al.*, 2016

### 3.2. Rosé Wines

Still rosé wines are, generally classified according to their colour, regardless of their vinification technique. To achieve the desired rosé colour, the grape skins are removed from the juice shortly after fermentation has begun, resulting in lesser uptake of substances than that of red wines.

### 3.3. Red Wines

***Dry red wines,*** unlike white wines, still red wines are almost exclusively dry. The absence of a detectable sweet taste is consistent with their intended use as a food beverage. Rich in phenolic compounds and derived from grape skins, red wines possess a bitter and astringent sensory character, which perfectly combines with food proteins.

### 3.4. Sparkling Wines

Sparkling wines are the wines which have carbon dioxide derived exclusively from fermentation and have excessive pressure due to carbon dioxide in solution of not less than 3 bar when kept at temperature of 20°C, in closed containers. Sparkling wines can be white, rosé and red, and can be dry or sweet also.

### 3.5. Carbonated Wines

These are wines to which carbon dioxide is added exogenously.

### 3.6. Fortified Wines

These wines have high alcohol content due to the addition of alcohol or wine spirit at some stage of production characterising the large group of fortified wines (Amerine *et al.*,1980). Fortified wines can be made in a wide range of styles and could be white or red, dry or sweet, with or without added flavours. Classification of fortified wines may be done according to their sugar content, to the presence or not of added flavour or colour to the wine. The types are sherry, port, madeira, other liqueur wines and the aromatised wines (Karagiannins, 2011).

### 3.7. Sherry and Sherry-like Wines

The base wine for sherry production is commonly white wine fermented to dryness or to a low sugar content followed by addition of a small quantity of spirit to stabilise it while it matures in contact with air.

***Sherry (Jerez)*** is a protected designation of origin, restricted to special wines produced in and around *Jerez de la Frontera* (Spain). The grape varieties preferred for sherry production are the neutral flavored Palomino and Pedro Ximénez.

Three main groups of sherry are:

***Finos***, which is very pale, dry, light in body and of characteristic slightly pungent bouquet and flavour derived from the action of flor. The yeast used in secondary fermentation is called flor yeast.

***Amontillado*** sherry is a fino sherry that has been aged for a long period in wood and lost much of its original fino character. Its production is similar to that of fino at the first stages but subsequently, the frequency of transfers is reduced, leading to progressive elimination of the flor due to decrease of nutrients and increase of alcohol content.

***Olorosos*** sherry is a wine of deeper colour than fino, usually fairly sweet and sometimes called 'golden' or 'East India Sherry' in commerce.

### 3.8. Port Wine

Port (protected designation of origin) is a fortified wine produced in the demarcated area of Douro Valley, in northern Portugal. The Douro wine region is a delimited area mainly bordering the Douro river from Régua, some distance eastward (Karagiannins, 2011).

### 3.9. Madeira Wine

Madeira is fortified wine produced in the Portuguese island of the same name. It is a white wine of various type with naturally high acidity and fortified with alcohol, before the fermentation stops. Then, it is baked in cement tanks or wooden casks before aging. Addition of wine spirit supplies the alcohol that gets lost during heating and raises the alcohol content up to 20 per cent by volume.

### 3.10. Aromatised Wine

Aromatised wine is fortified wine containing alcohol ranging from 14.5-22 per cent by volume, is obtained from wine products to which ethyl alcohol is added and which has been flavoured with natural flavouring substances and/or aromatic herbs and/or spices (Amerine *et al.*, 1980). The main categories of aromatised wines are:

***Vermouth***, whose two classes are recognised in the trade – the sweet or Italian-type vermouth and the dry or French type (Panesar *et al.*, 2011). ***Sweet vermouth*** is produced in Italy, Spain, Argentina and the United States. The typical Italian vermouth is dark amber in colour, with a light musky, sweet nutty flavour and a well developed and pleasing fragrance. It has a generous and warming taste and a slightly bitter but agreeable aftertaste. It must contain at least 15.5 per cent by volume alcohol. ***Dry vermouth*** is mainly produced in France, the United States and Hungary. It is usually not only much lower in sugar content and lighter in colour than the sweet, but also is often higher in alcohol content and sometimes more bitter in flavour.

***Bitter aromatised wine*** possesses a characteristic bitter flavour (Quinquina wine, Bitter vino, Americano).

### 3.11. Other Wines

***Medicinal wine*** is flavoured wine containing extracts of several herbs of medicinal importance and may even include bitter compounds like quinine, for example. Such wine is usually sweet and has as high as 18-20 per cent by volume alcohol.

***Herbal wine*** is made by fermenting parts of herbal plants or by adding them as an additional substrate to conventional fruits. Aloe vera wine, sea buckthorn wine, aloe-mint wine, adean berry are a few examples of herbal wines (Negi *et al.*, 2013; Trivedi *et al.*, 2013; Trivedi *et al.*, 2012; Trivedi *et al.*, 2015a, b; Chauhan *et al.*, 2015; Vahos *et al.*, 2016).

***Natural wine*** is obtained through the use of approved formulas with a natural wine base, herbs, spices, fruit juices, aromatics, essences and other flavouring materials in order to attract the non-wine-drinking public. Many of these wines contained less than 14 per cent alcohol.

### 3.12. Fruit Wines

Fruit wine is made from fruits other than grapes, such as cherries, kiwifruit, apple, peach, berries, pear, muskmelon and *jamun* (Robinson, 2006; Joshi *et al.*, 2011a; Joshi *et al.*, 2017). The system already employed for the classification of wine from grapes is made use of for classification of non-grape wines. The broad overall classification of wine is depicted in Fig. 3 and is described here briefly; for more details, the reader may consult the literature cited (Karagiannins, 2011).

In Great Britain and France, the term 'cider' (or *cidre*) means apple wine, hard cider, or fermented apple juice, but in the United States, it may mean fermented or unfermented apple juice based on the definitions and available products. Broadly, it can be classified as soft cider, with 0.5 per cent to 5 per cent alcohol content, or hard cider, with 5-8 per cent alcohol content. Apple wine contains alcohol above 8 per cent, but may go up to 14 per cent. Similar to the wine obtained from grapes, table or fortified wine, sweet or dry wine and vermouth (Joshi *et al.*, 2011b; Joshi and Sandhu, 2000) are made.

*Boukha* is wine made from figs.

*Perry* is wine prepared from pear juice. It can be sweet or dry. Because the pear is more astringent, the same characteristic is imparted to the wine also.

*Mead* is a type of wine which was prepared by the ancient people of India and was known for its excellent digestive qualities.

## 4. Composition of Wine

A typical wine contains ethyl alcohol, sugar, acids, higher alcohols, tannins, aldehydes, esters, amino acids, minerals, vitamins, anthocyanins and flavouring compounds (Amerine *et al.*, 1980; Joshi, 1997 Soni *et al.*, 2011; Joshi and Kumar, 2011). Being fruit based, fermented and undistilled, wines retain most of the nutrients present in the original fruit juice (Joshi and Kumar, 2012; Swami *et al.*, 2014). Wine is a fruit product, but fermentation produces a variety of chemical changes in the must and so the wine is far from being juice with ethanol added (Saigal and Ray, 2007). It's a nutritionally rich and functionally active preparation. A typical 149 ml of wine is composed of 127 g water, 15.6 g alcohol, 3.8 g total carbohydrate, 0.1 g protein, 0.4 g ash and varying quantities of vitamins, minerals and phytochemicals (Gebhardt and Thomas, 2002). It has been noted that depending upon the substrate, the polyphenolic content of wine may range between 100-35,000 mg GAE/L (Rana and Singh, 2013; Di Lorenzo *et al.*, 2016). In Table 2, some physico-chemical characteristics (sugars, alcohol content, total acids and volatile acids) of a few wines are presented while in Table 3, mineral contents in different wines are shown. As an example of fruit wines, some physico-chemical characteristics of plum wine are shown in Table 4. Fermentation alters the composition of must by fermenting the sugars into ethanol, producing compounds like higher alcohols, lactic acid, acetic acid acetaldehyde, esters, especially ethyl acetate etc. modifying the conjugation of organic acids and phenolics, by extraction and formation of co-pigments and, therefore, improves the anti-oxidant potential. Since 19th century, our perception about wine, its structure and alterations has developed parallel with the advances made in microbiology, biochemistry and chemistry.

**Table 2.** Reducing Sugars, Titratable and Volatile Acid Contents of Various Wines

| *Wine* | *Reducing sugars (g/100 ml)* | *Total acids (g/100 ml)* | *Volatile acids (g/100 ml)* |
|---|---|---|---|
| Dry white | 0.134 | 0.586 | 0.101 |
| Dry red | 0.146 | 0.649 | 0.128 |
| Sweet white | 11.30 | 0.412 | 0.092 |
| Sweet red | 10.26 | 0.502 | 0.122 |
| Sparkling | 3.409 | 0.658 | 0.082 |

*Source*: Amerine *et al.*, 1972

**Table 3.** Mineral Contents of Different Types of Wines

| *Type of beverage* | *Minerals (mg/l)* | | | | | | | |
|---|---|---|---|---|---|---|---|---|
| | *Na* | *K* | *Ca* | *Mg* | *Cu* | *Fe* | *Mn* | *Zn* |
| **Wine** | | | | | | | | |
| Grape wine | 51 | 803 | 106 | 88 | 3.0 | 0.13 | 0.66 | 0.70 |
| Apricot wine (New Castle) | 11 | 1481 | 18 | 71 | 2.72 | 0.96 | 1.92 | 0.88 |
| Wild apricot wine (Chulli) | 43 | 2602 | 25 | 94 | 5.97 | 0.50 | 2.69 | 0.99 |
| Apple wine (Golden Delicious) | 18 | 1044 | 11 | 144 | 3.68 | 0.21 | 0.76 | 0.84 |
| Cider (hpmc apple juice conc.) | 61 | 1900 | 23 | 137 | 4.31 | 0.32 | 1.54 | 1.01 |
| Hard cider (Golden Delicious) | 19 | 1069 | 17 | 97 | 3.03 | 0.19 | 0.91 | 0.82 |
| Pear wine (Sand pear) | 87 | 1906 | 37 | 122 | 8.91 | 0.16 | 0.80 | 1.10 |
| Plum wine (Santa Rosa) | 20 | 1008 | 18 | 82 | 12.73 | 0.20 | 1.04 | 0.95 |
| **Vermouth** | | | | | | | | |
| Grape Vermouth (Quinas) | 111.64 | 735.64 | 89.25 | 62.18 | 0.53 | 6.95 | 0.58 | – |
| Aperitivos vinicos | 45.56 | 297.62 | 57.06 | 53.03 | 0.46 | 5.13 | 0.47 | – |
| Vermuts balancos | 58.70 | 225.00 | 59.50 | 37.17 | 0.48 | 4.31 | 0.38 | – |
| Vermuts royas | 73.65 | 524.72 | 54.96 | 57.57 | 0.42 | 7.13 | 0.34 | – |
| Plum Vermouth | 41.00 | 973 | 101 | 17.0 | 1.07 | 1.30 | 1.07 | 0.82 |
| Sand pear Vermouth | 45.0 | 967 | 43 | 15.0 | 1.23 | 7.11 | 1.23 | 2.39 |

*Source*: Amerine *et al.*, 1980; Joshi *et al.*, 1999; Bhutani *et al.*, 1989; Bhutani and Joshi, 1995

In addition to these components, wines contain a variety of phenolic components and a number of flavanoids (Mongas *et al.*, 2015). The components have a major role to play in prevention of CVD (cardio-vascular diseases) in wine consumers. Phenolic compounds known to have antioxidant activity have also been found in strawberry and wines made from apple, especially those treated with wood chips (Joshi *et al.*, 2009). The details of these aspects have been discussed in the literature cited (Muller, 1995; Stockley, 2011) and described extensively in Chapter 2 of this book.

Resveratrol is another compound found in grapes and the various wines made from them. The compound has been known to act as an anti-cancer agent. A lot of research on this aspect has been conducted and it has been extensively reviewed earlier (Okada and Yohotosuha, 1996). For more details, the reader can refer to the literature cited on this aspect (Joshi and Devi, 2009; Gusman, 2012; Gurbuz *et al.*, 2017).

**Table 4.** Physico-chemical Characteristics of Plum Wine

| *Characteristic* | *Range* |
|---|---|
| Ethanol (% v/v) | 8.5-11.0 |
| Total soluble solids (Brix) | 8.0-12.0 (Sweet) |
| Titratable acidity (% malic acid) | 0.62-0.68 |
| Volatile acidity (% acetic acid) | 0.028-0.040 |
| Total Esters (mg/litre) | 104.0-109.0 |
| Colour (tintometer colour units) | |
| Red | 6-10 |
| Yellow | 10 |
| Total colouring matter and tannins (mg/100 ml) | 119 |

*Source*: Joshi *et al.*, 1999a

**Table 5.** Concentration of Resveratrol in Wines Made from Grapes Grown in Different Areas of Japan

| *Wines* | *Resveratrol (μg/l)* |
|---|---|
| **White wines** | |
| Chardonnay | 34±20 |
| Koshu | 13±3 |
| Delaware | 1±1 |
| Niagara | 27±1 |
| Zalagyongye | 80±36 |
| Zalagyongye | 40±36 |
| Muller-Thurhau | 44±25 |
| Average ($n$ = 10) | 27 |
| **Red wines** | |
| Cabernet Sauvignon | 92±20 |
| Muscat Bailey A | 89±26 |
| Campbell Early | 24±12 |
| Concord | 71±25 |
| Merlet | 210±0 |
| Cabernet franc | 202±42 |
| Zweigelt rebe | 98±43 |
| Kiyomi | 82±70 |
| Average ($n$ = 9) | 157 |
| Pink wines | 6±5 |
| Kyoho | 3±0 |
| Average ($n$ = 2) | 5 |

The wine produced from tropical fruits is also a rich source of flavonoids (Saigal and Ray, 2007). Much has been said about the functionality of these compounds that have anti-carcinogenic, anti-atherogenic, anti-thrombotic, anti-microbial, analgesic vessel and show dilator effect (Joshi and Siby, 2002; Papadopoulou, 2005; Gusman *et al.*, 2012; Monagas *et al.*, 2015). Although, colourless or slightly yellowish, the flavonoids have an important role in strengthening and stabilising the colour of red wine. They contribute to the sensory characteristics, such as taste, astringency and harshness of wine and aid in the preservation of beverages and retaining the properties of wine during the aging process. Most of the phenols are in the form of tannins (molecules which preserve the wine by absorbing oxygen) or as flavonoid derivatives, such as flavan-3-ols and flavan-3, 4-iols which can polymerise tannins. The most important representatives of the group of flavan-3-ols are the catechin and epicatechin, epimers at carbon 3 (Boulton, 2011). According to Teissedre *et al.* (2016), among the various phenolic compounds present in wines, the greatest antioxidant activity in preventing LDL oxidation was impacted by flavan-3-ols, such as catechin, epicatechin, dimmers and trimers of procyanidins, which are responsible for the highest antioxidant activity (Teissedre *et al.*, 2016).

In young red grape wines, the concentrations of catechin and epicatechin vary between 4.96-7.14 mg $L^{-1}$ and 2.02-3.02 mg $L^{-1}$ and this is very similar to the data found in tropical fruit wines in this study, with the exception of mangaba wine, which contains higher values (14.01 mg $L^{-1}$ and 22.66 mg $L^{-1}$ of catechin and epicatechin, respectively) (Gomex Plaza *et al.*, 2015). In white wine, where there is a limited contact with the skin, the catechins are the main flavonoids. In winemaking, the catechins are transferred to the must in the soaking step. However, during the manufacture of white wine, since there is a quick press operation to separate the peel and pulp from the juice and this accounts for the low amount of phenolic compounds that are present in white wine (Auger *et al.*, 2014).

The jackfruit is a rich source of phenolics and flavonoids, which in turn have good antioxidant properties. Jagtap *et al.* (2014) prepared wine from jackfruit pulp and evaluated the total phenolic content, flavonoid contents and antioxidant properties of wine. Orange juice and wine are rich sources of phenolic compounds, hydroxyl benzoic acids, hydroxyl cinnamic acids and flavanones along with hesperidin, narirutin and ferulic acid.

## 5. Wine and Health

Wine has been espoused as a superior beverage, especially for health benefits (Waterhouse, 1995). It has been used as a medicine as has been mentioned in the *Rigveda*. It is without any dispute that moderate wine consumption is associated with lower mortality from coronary heart diseases (Delin and Lee, 1991, 1992; Stockley, 2011; Bission *et al.*, 1995; Joshi, 2012). Wine consumers wax lyrical about resveratrol and anthocyanins (Burlingame 2008; Joshi and Devi, 2009). The health benefits of wine are now also associated with the antimicrobial properties (Joshi and Siby, 2002). Wine is a nutraceutical product; when the fruit is consumed as table grapes or as a transformed product (juice, wine, etc.), it helps to reduce cardiovascular disease and has anti-carcinogenic properties. Both clinical and experimental evidences suggest that moderate consumption of red wine offers greater protection to health by reducing cardiovascular morbidity and mortality due to antioxidant polyphenolics which are found particularly in red grape wine (Halpern, 2008; Gresele *et al.*, 2011). The phenolic acids can scavenge free radicals and quench reactive oxygen species (ROS) and therefore, provide effective means of preventing and treating free radical-mediated diseases (Shahidi, 2009). Wine polyphenols can also lead to the modulation of both oral and gut microbiota (Requena *et al.*, 2015).

The beneficial effects of wine on health are associated with antimicrobial activities of ethanol and antioxidant properties of phenolic components and flavonoids (Nijveldt *et al.*, 2001; Tomaz and Maslov, 2015). In modern times, the role of wine has come into picture from France. Several observations regarding low mortality rates in people consuming moderate levels of red wine have led the investigators to scientifically evaluate the beneficial properties of wine. Since then, many case studies and experimental investigations have proved wine to be beneficial to health if consumed in moderate amounts (Drel and Sybrina, 2010; Creina Stockley, 2011). Both clinical and experimental evidences suggest that moderate consumption of red wine offers greater protection to health by reducing cardiovascular morbidity and mortality (Halpern, 2008; Gresele *et al.*, 2011).

Fruits containing a wide range of flavonoids and other phenolic compounds possess antioxidant activity. Phenolic compounds in red grape wine have been shown to inhibit *in vitro* oxidation of human low-density lipoprotein (LDL) (Treissedre *et al.*, 2016). The compounds of phenols and flavonoids present in wine are known for their positive effects on inflammation, cardiovascular diseases, besides having antibacterial and antioxidant effects. Some of the reported values of DPPH activity and phenolic composition of various tropical fruit wines are discussed earlier (Saigal and Ray, 2007).

Co-existence of ethanol and the antioxidant activity has a great implication in maintaining health as summarised in Table 6. In addition to these nutritional advantages, wine has safer psychotropic effects, that is why wines are preferable to distilled liquors (Jackson 1994; Soni *et al.*, 2011). Another compound, called salicylic acid, found in wine also plays a great role in maintaining health as summarised in Table 7. Glucose-tolerance factor, a chromium-containing compound synthesised by yeast, is considered beneficial in diabetes and is found in wine, as cited earlier (Sandhu and Joshi, 1995). Wine is considered an important adjunct to the human diet, as it increases satisfaction and contributes to relaxation necessary for proper digestion and absorption of food.

The jackfruit wine showed high radical scavenging capacity to the tune of 69 per cent as compared to that of wines prepared from lime (20.1±1.09 per cent), tamarind (15.7±0.63 per cent), garcinia (15.4±0.21 per cent), rambutan (15.1±0.26 per cent), star gooseberry (14.8±0.46 per cent) fruits, cashew apple 7.72 per cent, mango 42 per cent, *bael* wine 48 per cent, pineapple 36 per cent and *tendu* 52 per cent. As expected, wine from grapes has the highest antioxidant activity than any other fruit wine with the reported value of DPPH activity (93-95 per cent) (Jagtap *et al.*, 2014). The antioxidant capacity of orange juice however,has been found to be higher than that of orange wine (Kelebek et al., 2009).

*Jamun* wine also possesses medicinal properties, like antidiabetic properties and the ability to cure bleeding piles (Joshi *et al.*, 2012; Chowdhury and Ray, 2006). Mango wine along with aromatics is

**Table 6.** Health Benefits due to Co-existence of Antioxidants and Ethanol

- The co-existence of alcohol and antioxidants in beverages like wine have profound health implications.
- The antioxidants are protected by the processes used to detoxify the alcohol.
- In liver mostly, NAD[+] dependent oxidizing enzyme, alcohol dehydrogenase (ADH; converting acetaldehye into acetate) producing NADH, recycle the spent antioxidants by reducing and regenerating NAD.
- The alcohol content of about 12% might produce a lot of reducing equivalents of antioxidants during ethanol detoxification.
- Antioxidants are available in reduced state, excessive levels of NADH generated by detoxification of ethanol are lowered, thus, minimizing reductive damage to other systems.
- These properties are shared by wine or wine-like beverages.
- Neither alcohol as such nor dealcoholised wine exhibits such healthful properties (due to alcohol absorption antioxidants may not be facile in the gut).

*Source*: Muller, 1995; Joshi *et al.*, 2005; Kennedy and Tipton, 1990; Crouse and Grundy, 1984

**Table 7.** The Role of Salicylic Acid (SA) in Wine

- The ubiquitous compound has many health related characteristics, and pharmacological activity ranging from antipyretic and analgesic properties, to providing cardiovascular wellness due to fibrinolytic activity in whole blood, sharing this property with alcohol.
- Fibrin formation is important in the etiology of atherosclerotic plaque formation and lesions as LDL oxidation.
- Is a powerful antioxidant, capable of quenching the damaging hydroxyl radical, having antiatherogenic properties by ameliorating LDL oxidation.
- It is effective against some viral infections as common cold.
- Aspirin is deacetylated to yield salicylic acid having higher residence time in blood which indirectly indicates that many functions of aspirin are performed by salicylic acid.
- Aspirin (acetyl salicylic acid) has antiplatelet aggregation function due to its ability to inhibit thromboxane formation and subsequent release from platelets by acetylating a key serene residue on the enzyme cyto-oxygenase. Salicylic acid has little action on thromboxane and its release from platelets.
- Aspirin unless added (not permitted) is not detectable in wine. The total salicylic acid content of white and red wine are 11.00 mg/L and 18.5 mg/L, levels of 2-3 DHB 21.0 mg/L and 26.5 mg/L, respectively and SA levels are dependent upon the type of wine (red has more than white), oak wood aging and temperature.
- Salicylic acid is equally effective as aspirin in preventing chemotactic generation of 12-HETE from 12 HPETE in the lipo-oxygenase pathway of arachidonic acid metabolism by leucocytes—in prevention of inflammation.
- Wine due to SA and other compounds may be helpful in providing wellness due to serenogenic (relaxant) compounds of wine and of alcohol.
- SA is produced in plant, both constitutive and inductive, as a defence chemical.
- SA protects against the vehicular emission, smog and tobacco smoke containing oxides of nitrogen or oxidants, lung cancer, emphysema, respiratory tract infection and other pathological states to which smokers are at increased risk.

*Source*: Adapted from Joshi and Pandey, 1999b; Levy, 1979; Mahamy and Klessig, 1992; Muller and Fugelsang, 1993; Muller *et al.*, 1994; Yalpani and Raskin, 1993

recommended as a restorative tonic as it contains a good amount of vitamins A and C, which are useful in heat apoplexy (Reddy *et al.*, 2014, 2015). Mangoes with higher initial concentration of α-carotene are helpful as cancer-preventing agents.

The pomegranate wine is rich in anthocyanins, which are responsible for the antioxidant properties and colour of the wine. Similarly in another study, Mena *et al.* (2012) reported the presence of melatonin (N-acetyl-5-methoxytryptamine) in the wine made from pomegranate cultivars. Melatonin (N-acetyl-5-methoxytryptamine) is a neuro-hormone related to a broad array of physiological functions and has proven therapeutic properties.

Wine is one of the functional fermented foods having many health benefits, like anti-ageing effects, improvement of lung function (from antioxidants in white wine), reduction in coronary heart disease, development of healthier blood vessels and reduction in ulcer-causing bacteria. Many wines are made from fruits having medicinal value (Tapsell *et al.*, 2006). Research currently in progress continues to document the healthful properties of wine.

Ethanol consumption at moderate levels serves as a dietary source of energy. However, its excessive consumption is associated with several health-related problems, such as cirrhosis of lever, digestive tract cancer, hypertension, stroke, pancreatitis, etc. (Alves and Herbert, 2011). To cite an example, a diet high in alcohol causes decreased absorption of vitamins of B-group, such as $B_1$, $B_2$ and $B_{12}$, especially of cobalamine and folic acid. Heavy alcohol consumption thus, leads to the damage of liver and spleen.

## 6. Production of Wine

Winemaking is an attractive means of utilising surplus and over-ripe fruits (Verma and Joshi, 2000). Moreover, fermentation helps to preserve and enhance the nutritional value of foods and beverages (Dickinson, 2013). Winemaking is one of the most ancient arts and is now, one of the most common biotechnological processes (Moreno-Arribas and Polo, 2005; Ward and Ray, 2006). Wine production is a fine blend of art, skill and science emanating from individual creativity coupled with innovative technology evolved over thousands of years. The technique of winemaking is known since the dawn of civilisation and has followed human and agricultural progress (Chambers and Pretorius, 2010). The process of winemaking is unique in the sense that nearly all the physical, chemical and biological sciences, especially microbiology and biochemistry, contribute to its production (Joshi *et al.*, 2011a).

### 6.1. Wine Fermentation

Production of wine involves a fermentation process which is a product of high economic importance. It is a biotechnological process in which the desired microorganisms, either inoculated or selected spontaneously, are used in the production of value-added products of commercial importance. It occurs in Nature in any sugar-containing must made from fruit, juice, berries, honey, or sap tapped from palms. If left exposed in a warm atmosphere, airborne yeast acts on the sugar to convert it into alcohol and carbon dioxide. Over the years, it was established that industrial fermentation processes are conducted with the help of selected microorganisms under specified conditions with carefully adjusted nutrient concentrations (Rose and Harrison, 1971). The products of wine fermentation are many, such as alcohol, glycerol and carbon dioxide which are obtained from yeast fermentation of various sugars but in wine the focus is on ethyl alcohol and by-products of alcoholic fermentation.

### 6.2. Microbiology of Wine Production

In 1863, Louis Pasteur, the founder of modern oenology, revealed for the first time the involvement of microorganisms in winemaking. However, advances in the second half of the 20th century have shown that fermentation of grape must and the production of quality wine is a more complex subject than Pasteur had documented at that time. Considerable progress has since been made in the last few decades in controlling the winemaking process, right from vineyard to bottling plant, and in the understanding of the interactions of various groups of microorganisms, i.e. yeast, lactic acid bacteria and acetic acid bacteria involved in winemaking (Fleet, 2013 b).

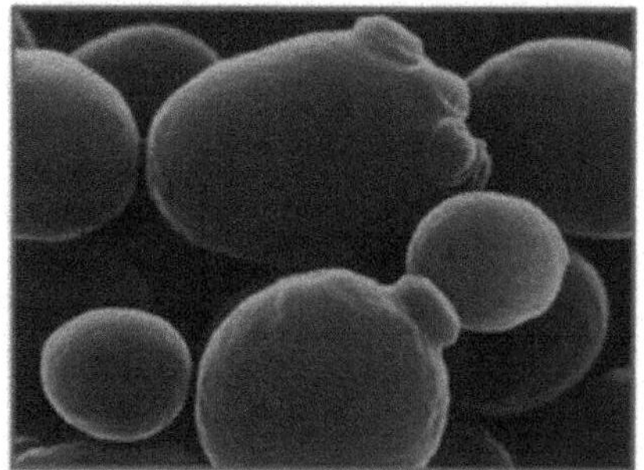

**Figure 4.** Electron micrograph of *Saccharomyces cerevisiae*, the wine yeast

In the preparation of wine and brandy, wine microbiology (*see* Chapter 9 of this text) plays a central and vital role (Riberau Gayon *et al.*, 2000). Different microorganisms involved in the process, mainly yeast *Saccharomyces cerevisiae*, produce normal transformation of juice into wine while malo-lactic acid bacteria are involved in reducing the acidity of wine from grapes produced in cold climate regions. But yeasts like wild yeast, lactic acid or acetic acid bacteria can cause spoilage (Fleet, 1993a; Fugelslang, 1992), so their growth must be controlled. In brief, the microorganisms are involved in winemaking in one or the other, as described in Table 8. The use of yeasts, like *Schizosaccharomyces pombe*, has been advocated for the deacidification of high acidic fruits containing a greater amount of malic acid to make palatable wine (Joshi *et al.*, 1991). Microbial aspects of wine production, such as malo-lactic fermentation, yeast cultures, genetic engineering (Dequin, 2001) and killer yeasts are, therefore of great significance and would continue to be focused in this section besides being described in detail in other chapters of this text too (*see* Chapters 9-12).

**Table 8.** Microorganism Associated with Wine Fermentation and their Role

- Grapes and other fruits are subjected to mould spoilage at different stages of harvesting, storage and transportation.
- Mould growth can also result in the production of toxic compounds like ochratoxin and patulin
- The moulds cannot grow in the wine mainly due to the inhibitory effect of ethanol and the anaerobic conditions, as most of the moulds are aerobic in nature
- Besides, damaging the fruits, the moulds can also affect the flavour of the wine by their growth on the cooperage
- The yeast *Saccharomyces cerevisiae* is involved in ethanol production, the major ingredient of alcoholic beverages
- The non-*Saccharomyces* yeast is involved in production of various chemical components, especially volatile compounds producing the mixture of different alcohols and compounds in the fermented broth to add the distinct aroma and flavour, having significance in wine quality, especially sensory quality
- Microorganisms also play an important role in winemaking by developing complex aroma and flavour due to the action of various enzymes
- Some microorganisms, like malo-lactic acid bacteria, are involved in malo-lactic fermentation that makes the high acidic wines palatable
- The microbial activities contribute in the development of a desirable quality wine
- Various microorganisms are involved in spoilage of wine, mainly the non-*Saccharomyces* yeasts/wild yeast, lactic acid bacteria and acetic acid bacteria

Further, today there is a new and controversial focal point for innovation in winemaking – the genetic modification of grape cultivars and wine yeast (*Saccharomyces cerevisiae*). The diversity in the yeast species associated with winemaking, the tailoring of wine yeast and the possible use of strains expressing novel designer genes make possible adoption of various exciting new approaches to winemaking in the 21$^{st}$ century (Pretorius, 1999, 2000).

### 6.3. Stuck Fermentation

Many a times, wine fermentation stops midway, that is, before the completion of the process and is called as 'stuck fermentation' (Bission, 1999). It refers to the premature termination of fermentation before all the sugars have been metabolised (Alexandre and Charpentier, 1998). Its occurrence has been associated with overheating during fermentation. In addition to this, several other reasons have been associated with stuck fermentation. The resulting wines are high in residual sugar concentration and thus, susceptible to microbial spoilage. In general, instability is increased if grapes are low in acidity and/or high in pH. Improvements in temperature control have eliminated overheating as an important cause of this fermentation.

The likelihood of stuck fermentation may be minimised by reduced pre-fermentation clarification, limited oxidation during crushing and pressing by the addition of ergosterol and/or long chain unsaturated fatty acids (i.e. oleic, linoleic acid) and the addition of ammonium salts, etc. (Bission *et al.*, 1995).

### 6.4. Malo-lactic Fermentation

Malo-lactic fermentation (MLF) has three distinct, but interrelated effects on wine quality. It reduces acidity (increases pH), influences microbial stability and may affect the sensory characteristics of wine.

Reduction in acidity increases the smoothness and drinkability of red wines (Kunkee, 1999), but excess reduction generates a 'flat' taste. In wines of excessive acidity, the reduction is desirable. Thus, winemakers in most cool wine-producing regions view MLF positively, especially for red wine. In contrast, wine produced in warm regions may be low in acidity or high in pH and it can leave the wine 'flat' tasting and microbiologically unstable (Liu, 2002); thus it is not encouraged. In addition to the decarboxylation of malic acid by lactic acid bacteria, wine also has the capacity to metabolise citric acid (Liu *et al.*, 1995a), which influences the formation of flavour in wine.

Depending on the genus or species, lactic acid bacteria (LAB) may ferment sugars solely to lactic acid or to lactic acid, ethanol and carbon dioxide. Bacteria capable of both types of fermentation (homo- and hetero-lactic fermentation) can grow in wine (Axelsson, 1993; Montet *et al.*, 2006). The most beneficial hetero-fermentative species is *Leuconostoc oenos* (Holzapfel and Schillinger, 1992) which is probably the most frequently occurring species of LAB in wine and is the only species inducing malo-lactic fermentation at low pH of 3.5. Although LAB grow under acidic conditions, their growth is poor in must and wine. Species of *Lactobacillus* and *Pediococcus* commonly cease growing below pH 3.5. Even *L. oenos,* the primary malo-lactic bacterium, is inhibited below pH 3.0 to 2.9 (Davis *et al.*, 1986a, b). *Leuconostoc* grows optimally within a pH range of 4.5-5.5. By metabolising a dicarboxylic acid (malic acid) to a monocarboxylic acid (lactic acid), acidity is reduced but the pH increases. *Leuconostoc* appears to show little ability to ferment sugar below pH 3.5 (Davis *et al.*, 1986b). As the pH of the wine changes, so does the relative proportion of various coloured and uncoloured forms of anthocyanin pigments. A sizeable loss in red colour (from full red to bluish hue) comes from an equilibrium shift in anthocyanin pigments configuration resulting from the increase in pH (Mazza, 1995). The metabolism of carbonyl compounds (notably acetaldehyde) by lactic acid bacteria and the accompanying release of $SO_2$ also may result in some pigment bleaching (Liu, 2002).

Microbial stability of wine is considered as one of the prime benefits of MLF as the consumption of malic acid and citric acid leaves behind only the more stable tartaric acid and lactic acids (Henick-Kling, 1993). The greatest merit of MLF, however, revolves around the flavour modification (de Revel *et al.*, 1999). Diacetyl (biacetyl, 2, 3-butanedione) commonly accumulates during MLF; however, at concentrations above 5 to 7 $mgL^{-1}$, its buttery character can become pronounced and undesirable (Martineau and Henick-Kling, 1995a, b). The other flavour compounds produced by lactic acid bacteria in sufficient amounts to affect sensory characters of wine include acetaldehyde, acetic acid and acetoin, 2-butanol, diethyl succinate, ethyl acetate and ethyl lactate (Ramos *et al.*, 1995; Rattray *et al.*, 2000). The undesirable buttery odour usually connected with MLF is induced by *Pediococci* or *Lactobacilli* above pH 3.5.

To ensure the development of MLF, wine-makers inoculate the wine with one or more strains of *L. oenos* by maintaining conditions favourable to bacterial activity. Simultaneous inoculation with both yeast and bacteria results in the concurrent development of alcoholic fermentation and MLF. As a result, the wine can be racked off the lees, cooled and sulphited immediately after the alcoholic fermentation. The bacteria suffer neither from a shortage of nutrients nor from the toxicity of alcohol. More details on malo-lactic acid fermentation can be found in chapter 12 of this book.

### 6.5. Biochemistry of Wine Fermentation

The main biochemical mechanisms of yeast metabolism during wine fermentation are shown in Fig. 5. When the concentration of sugars is high (22-24 per cent), such as in the grape must, *S. cerevisiae* can only be metabolized by the fermentation route. The transformation of glucose into ethanol occurs through glycolysis, resulting in the production of pyruvate, which is later transformed into ethanol and $CO_2$ by two additional enzymatic (pyruvate decarboxylase and alcohol dehydrogenase) reactions (Goyal, 1999). When the grape must is inoculated with *S. cerevisiae*, ethanol is not formed immediately. Due to the anaerobic conditions, a small fraction of the must sugars, approximately 6-8 per cent, is transformed through glycerol-pyruvic fermentation into glycerol and pyruvate. Further, the yeast produces from the pyruvate a low concentration of a range of volatile compounds that make up the so-called 'fermentation bouquet'. The main group of compounds, and hence, the best studied, are higher alcohols, fatty acids, aldehydes and esters (Moreno-Arribas and Polo, 2005). Simultaneously, there is also production of some

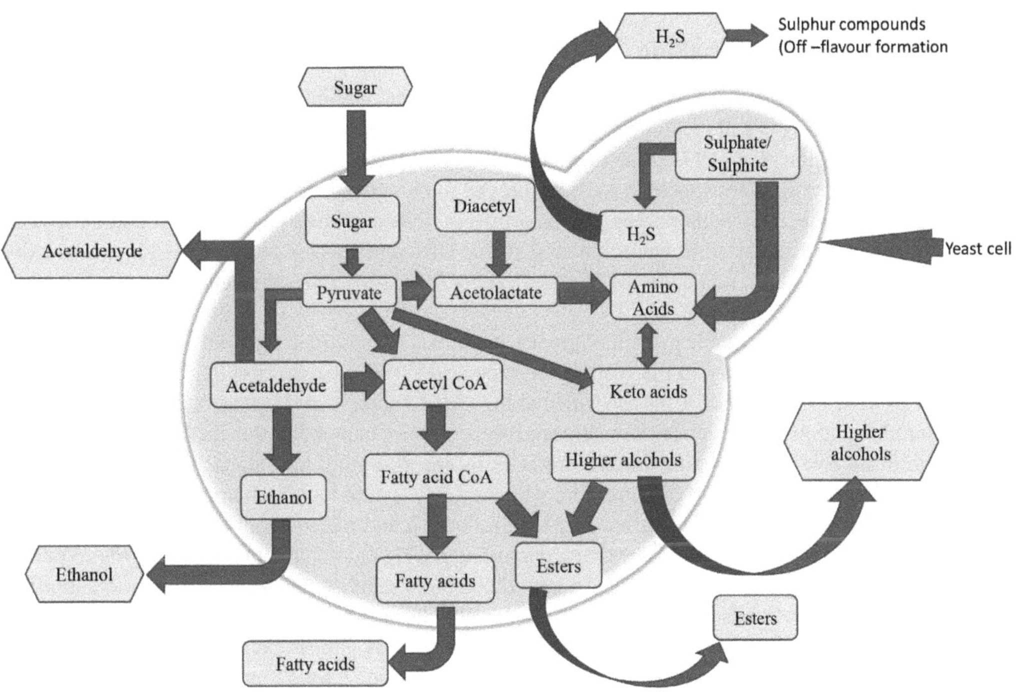


**Figure 5.** A schematic representation of the biochemical mechanisms of yeast metabolism during alcoholic fermentation (*Source*: Adapted from Pretorius (2000))

Color version at the end of the book

undesirable compounds, such as hydrogen sulphide and other volatile sulphur compounds (Spiropoulos *et al.*, 2000) regarded as 'negative' or 'unpleasant' aroma compounds (off-flavours) and, to a lesser extent, diacetyl and other related carbonyl compounds (Martineau and Henick-Kling, 1995a, b; Lavigne and Dubourdieu, 1996).

There are several other off-flavour compounds produced during winemaking, such as volatile phenols. The main volatile phenols of wine are 4-vinylphenol, 4-vinylguaicol, 4-ethylphenol and 4-ethylguaicol (Chatonnet *et al.*, 1993). These compounds are generated from non-oxidising decarboxylation of phenolic acids in the must by *S. cerevisiae*. Several lactic acid bacteria, i.e. *Pediococcus, Leuconostoc* and *Lactobacillus* and a few wine contaminant yeasts of the genus *Bretanomyces/Dekkera* have been reported to produce these ethyl phenols via enzymatic activities of cinnamate-decarboxylase and vinyl-phenol reductase (Edlin *et al.*, 1995; Chatonnet *et al.*, 1997).

Some strains of *S. cerevisiae* are marketed as 'enhancers of varietal/aroma expression', which have the ability to hydrolyse conjugated aroma precursors in juice, improving wine aroma (Lambrechts and Pretorius, 2000; Moreno-Arribas and Polo, 2005). The use of 'non-*Saccharomyces*' yeasts in winemaking is an excited field of future research and application. For example, this group of yeasts has been shown to contribute to the production of esters and other pleasant volatile compounds, albeit to a lower extent (Ciani and Maccarelli, 1998; Charoenchai *et al.*, 1997; Soden *et al.*, 2000). Some non-*Saccharomyces* strains of the genera *Kloeckera* and *Hanseniaspora* have been described to produce significant protease activity that influences protein profile of finished wines (Dizy and Bisson, 2000). Non-*Saccharomyces* yeasts have been shown to produce β-glycosidase enzymes involved in the flavour-releasing processes during the course of winemaking. Recently, the acetate esters, formed by enzymatic activities of yeast strains belonging to the genera *Hanseniaspora* and *Pichia*, have been studied in detail with very encouraging results (Rojas *et al.*, 2001, 2003).

# 7. Technology of Wine Production

## 7.1. Fruits used in Winemaking

Amongst the various fruits used to make wine, grapes are most technically and commercially used as substrates for winemaking and technology for wines and brandy has been well established and practiced the world over (Boulton *et al.*, 1995). The impact of the model plant, grape, is relevant, hence genetic and molecular studies on this plant species have been proved to be very successful in winemaking (Pretorius and Hoj, 2005). Grape, unlike other crops, can be grown in diverse climates and soils but similar to other fruit crops, grapes are subjected to environmental stress. Different species of grape used in winemaking are illustrated in Fig. 6. Different cultivars of grapes, other fruit crops and some cultivation conditions are described in Chapter 4 of this text.

Development and sustainable growth of grapes and other fruits is thus, the most essential component for growth of the wine industry in the world (Amerine *et al.*, 1980; Kosseva *et al.*, 2017). Selection of new varieties of grape and other fruits for winemaking (*see* Chapter 4 of this text) had been made in the past and would continue in future also. Genetic engineering (*see* Chapter 8 of this text) is expected to play a very significant role in developing various varieties by incorporating into grapes those characteristics which would improve the quality of wine. The wine industry would have to be mobilised against pests and diseases of grapes or other fruits used in wine production and has to adopt eco-friendly techniques to control the pests and diseases of fruit crops, including biological control methods. Efforts to protect the future crops have heightened the need for new disease-resistant grape cultivars. Because *V. vinifera* does not possess any notable disease resistant, new cultivars must incorporate resistance genes for a wide range of diseases in *V. vinifera* as the production of wine is almost exclusively made from these cultivars. Species and sub-genera of genus of Vitis are shown in Fig. 7. So, efforts towards mapping genes and genomics for better understanding of the mechanisms, genetics and expression of pest and disease resistance are being made that would enable the *V. vinifera* grapes to grow without pesticides in future.

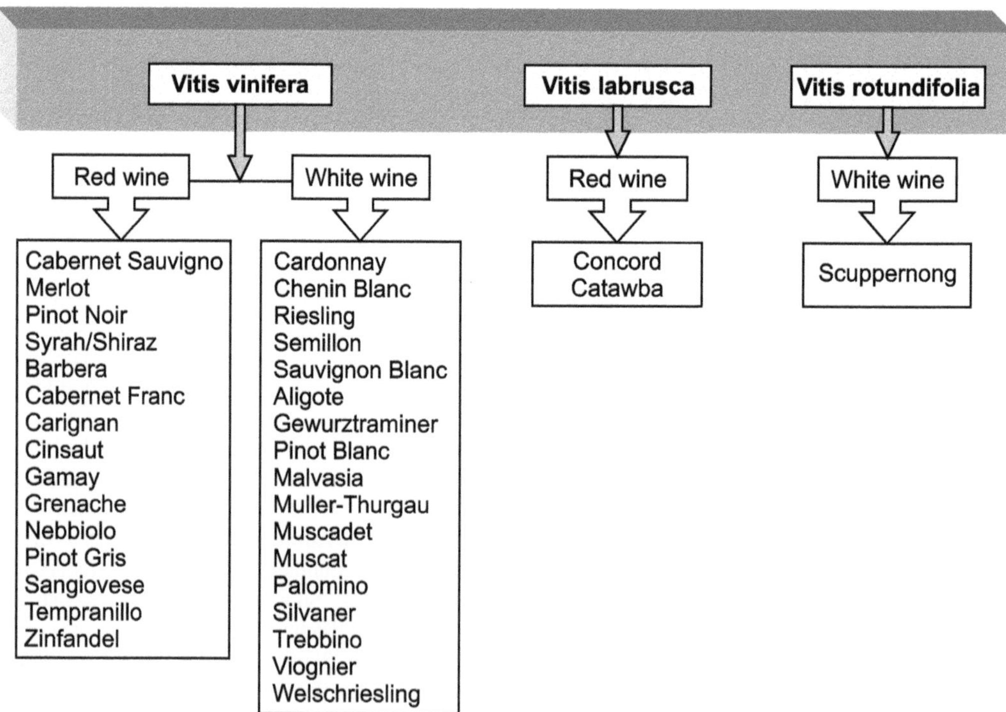


**Figure 6.** Main vine varieties used for wine production are classified by species (*Source*: Unwin, 1991)

In addition to grapes, several other fruits, including tropical fruits, are employed to produce wine. Major tropical fruits, like mango (Srisamatthakarn *et al.*, 2003), jackfruit (Panda *et al.*, 2016), *bael* (Panda *et al.*, 2013), *jamun* (Chowdhury and Ray, 2006), *tendu* (Sahoo *et al.*, 2012), *litchi* (Kumar *et al.*, 2008) and banana (Onwuka and Awan, 2012) could enhance local or international markets by appropriate utilisation processes where fermentation remains a technological attempt at such utilisation (Muniz *et al.*, 2008; Chowdhury and Ray, 2006). Similarly, wine has been prepared by the use of honey with fruits and wild fruits, like wild apricot (Joshi *et al.*,1990 a, b; Joshi and Attri, 2005).

**Figure 7.** Subgenera and main species of the genus Vitis (*Source*: Unwin, 1991)

## 7.2. Methods of Wine Production

The production of any wine basically involves fermentation with wine yeast, especially *Saccharomyces cerevisiae* var. *ellipsoideus* by engaging in several unit operations to produce wine (Amerine *et al.*, 1980; Boulton *et al.*, 1995; Ough, 1992; Jackson, 1999; Joshi *et al.*, 2011). Various unit operations in production of wine are depicted in Fig. 8. The fermentation of fruit must bring about many biochemical changes through the use of enzymes of fruits, yeasts and/or exogenously added enzymes for specific purposes (Canal-Llaub`eres, 1993; Bartowsky, 2003; Joshi and Bhutani, 1990) and, therefore, wine is not merely a fruit juice having ethanol (Jagtap and Bapat, 2015). Yeast, especially different strains of *Saccharomyces cerevisiae,* have long been used for the production of alcoholic beverages (Dunn *et al.*, 2015) and is the microorganism on whose activity the production of wine or any other alcoholic beverage depends upon (Rebordinos et al., 2011). An attempt to prepare wine by using the sequential culturing has also been made (Lee *et al.*, 2012). As a living organism, yeast primarily requires sugars, water and suitable temperature to stay alive. In addition, organic or inorganic nitrogenous material is also necessary for yeast to thrive (Amerine *et al.*, 1980; Joshi, 1997; Joshi *et al.*, 2011a). Where the nitrogenous material, be it organic or inorganic, is not sufficient, it is supplied in the form of salts, like ammonium hydrogen phosphate or ammonium sulphate. The organic source of nitrogen is yeast extract or amino acids. When needed, the vitamin requirement is met by addition of vitamin thiamin.

***Red and White Wine:*** In general, to produce red wine, the skin of coloured grapes is fermented, whereas for white wine, the fermentation of skin, pomace and free juice of white grape varieties is carried out (Fig. 9). Further, the fermentation of white grapes is generally conducted at low temperatures of 15-20°C, whereas that of red wine is done at 20-25°C (Amerine *et al.*, 1980; Vyas and Chakravorty, 1971; Kundu *et al.*, 1980). Generally, red winemaking comprises the following procedure: a) grape crushing and stem removal (not always), b) extraction of anthocyanins and other substances with sensory properties from skin to juice, c) fermentation at higher temperature than that on white winemaking, d) pressing by the time fermentation is completed or nearly complete, e) malolactic fermentation and f) aging (oxidative reductive) in most cases. The wines of other styles and types can be made from the base wine using different methods/modifications.

***Fortified Wines***

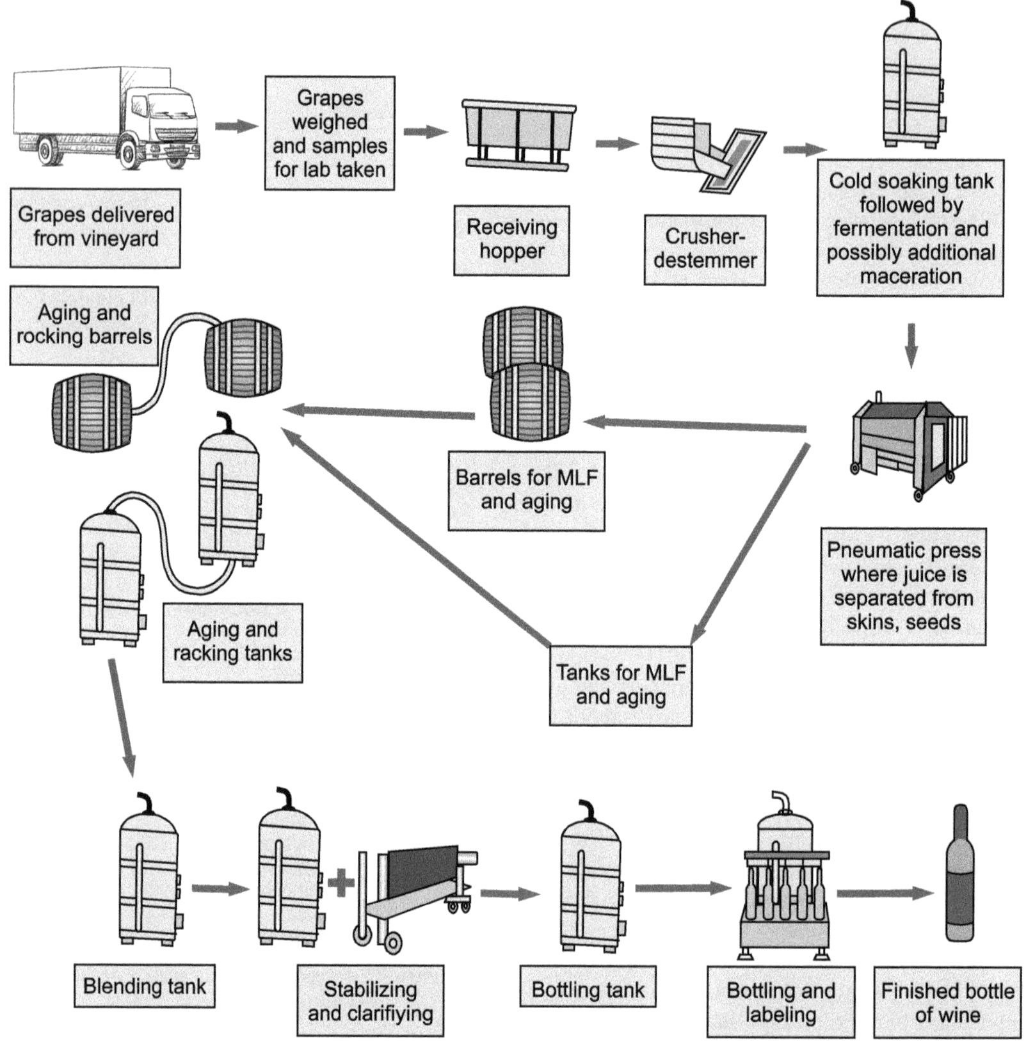


**Figure 8.** A generalised scheme of wine production
(*Source*: Adapted from http://encyclopediaZ.the free dictionary come./wine making">wine making/a)

***Sherry:*** Amongst the fortified wines sherry is basically a base wine fortified with brandy fermented with flor yeast. Vermouth is the aromatized fortified wine which makes use of spices and herbal extracts.

***Sparkling wine:*** In the preparation of sparkling wine, secondary fermentation of base wines is carried out either in bottles or in tanks. For more details, see chapter 21 on technology of wine production.

***Fruit Wine Production:*** Wine is also prepared from fruits like plum and the flow sheet of the process is given in Fig. 10. In preparation of cider (Fig. 11), either by natural fermentation or inoculated fermentation is carried out with an optimum level of initial sugar concentration using specific apple varieties so as to produce a low-alcohol beverage viz., soft or hard cider.

For more information on fruit wines, see the literature cited (Boulton *et al.*, 1995; Amerine *et al.*, 1980; Joshi, 1997; Joshi *et al.*, 1999a; Joshi and Attri Devender, 2005; Kosseva *et al.*, 2017 ).

## 7.3. Bioreactor Technology in Winemaking

A considerable amount of diversity is available to wine-makers in bioreactor technology, leading to a wide variety of wines. Winemaking involves two principal operations: (1) preparation of the grape must to tailor the composition and maintain the quantities of grapes at harvest and (2) implementation of microbial fermentation through biochemical activities of yeast and lactic acid bacteria. The equipment

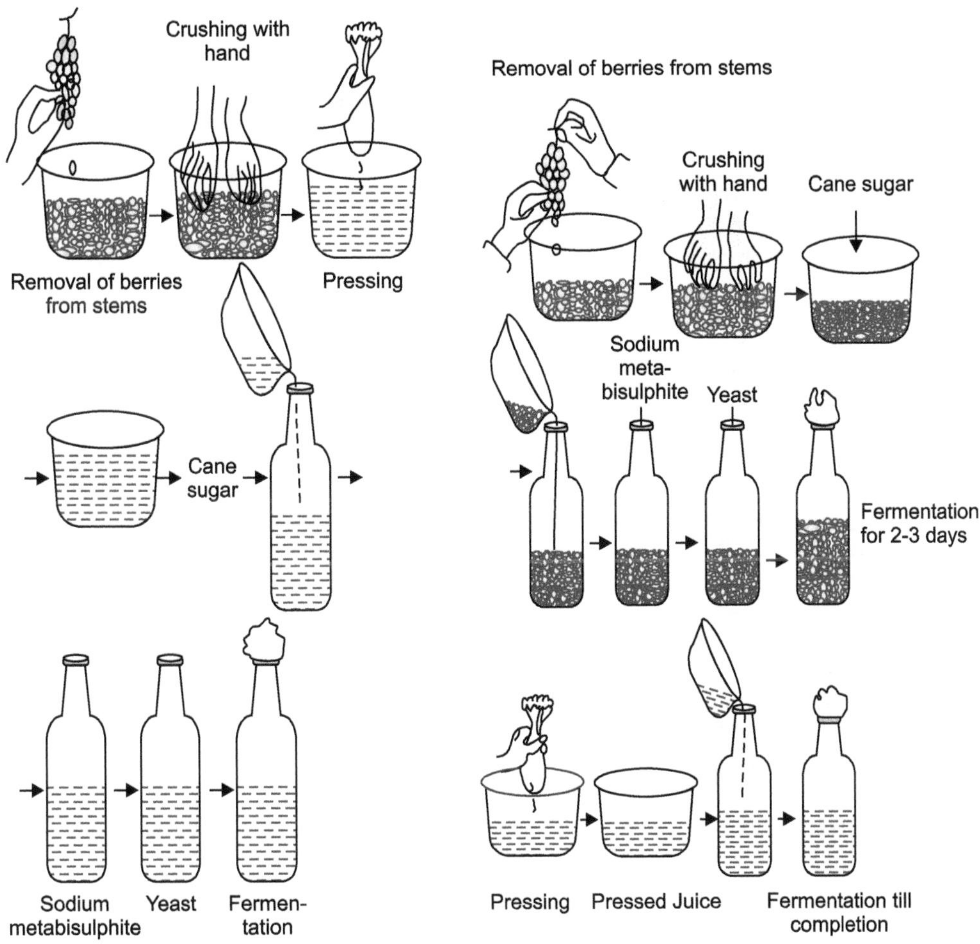


**Figure 9.** A diagrammatic representation of making a) white wine and b) red wine at home scale (*Source*: Vyas and Chakravorty, 1971)

used for fermentation has undergone extensive improvement. 'Autopiegeage' tanks (wine liquid is automatically recycled through the cap) are an improvement over the traditional methods in use, like wooden vats or steel tanks. These models also facilitate automation.

Most fermenters in winemaking are of the batch type. Attempts to make wine in continuous fermentation have also been made but with variable success. The substrate (must) is added continuously and an equivalent volume of the fermenting liquid (wine) is removed to maintain the cycle. Despite the potential advantages of continuous fermentation, it is rarely used in the wine industry, especially when preparing high quality wines owing to the subtle and complex association of hundreds of metabolites from the fruit used for fermentation and from fermentation itself. Moreover, the continuous fermenters are expensive, so they are economically feasible only if used the year round. More details including the immobilisation method and the control systems of wine fermentation can be seen in Chapters 18 and 19 of this text.

## 8. Quality and Safety of Wine

Quality and safety of wine is of utmost importance as it is a commodity consumed by a large number of human beings. It is essential that such a commodity should be free from any spoilage or contamination with microbial and non-microbial agents, especially the toxic components. Some of these aspects are described here while others are discussed in detail in Chapters 9 and 17 of this book.

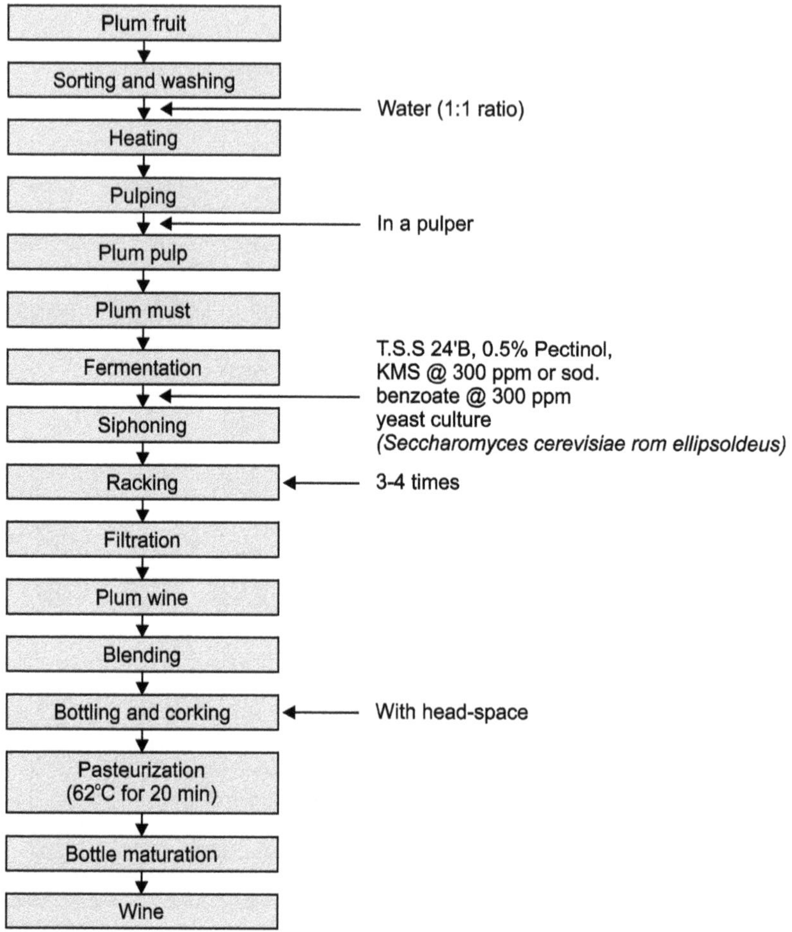


**Figure 10.** A diagrammatic representation of making plum wine (*Source*: Joshi *et al.*, 1999a)

## 8.1. Quality of Wine

Production of wine in different regions of the world is well established and regulated by various acts and regulations. Besides varietal differentiation, the regions where grapes are grown also influence the quality of the wine. Naturally, the consumers expect wine from a particular region to possess unique qualities that differentiate it from other regions (Vaidya *et al.*, 2011). Indeed, quality wines are currently being produced on all the six arable continents and emerging nations are active in the international wine trade. Wines of different styles and types also are produced and consumed the world over. Wines of different styles accompanying foods, method of serving and tasting, varieties of grapes used to make specific wine have also been described elsewhere.

Like any other food product, the quality of wine is of utmost importance. Chemical analysis of wine constitutes a significant component of quality measures (Zoecklein *et al.*, 1990; Zoecklein *et al.*, 2011; Plutowska and Wardencki, 2008), especially those prescribed by specific laws of nutritional significance and freedom from toxic compounds. To monitor and ensure the quality of wine, new techniques have emerged, including the use of molecular biological techniques like polymerase chain reaction (PCR) (Cocolin *et al.*, 2000). In future, these techniques are likely to be used more frequently, such as in direct profiling of yeast dynamics in wine fermentation. In the past few decades, significant advances have been made in the standardisation of wine quality and safety by introducing new techniques for the analysis of various types of phenolics and aromatics that affect the particular wine's characteristics and in the increased understanding of the factors influenced by vineyard practice and wine aging.

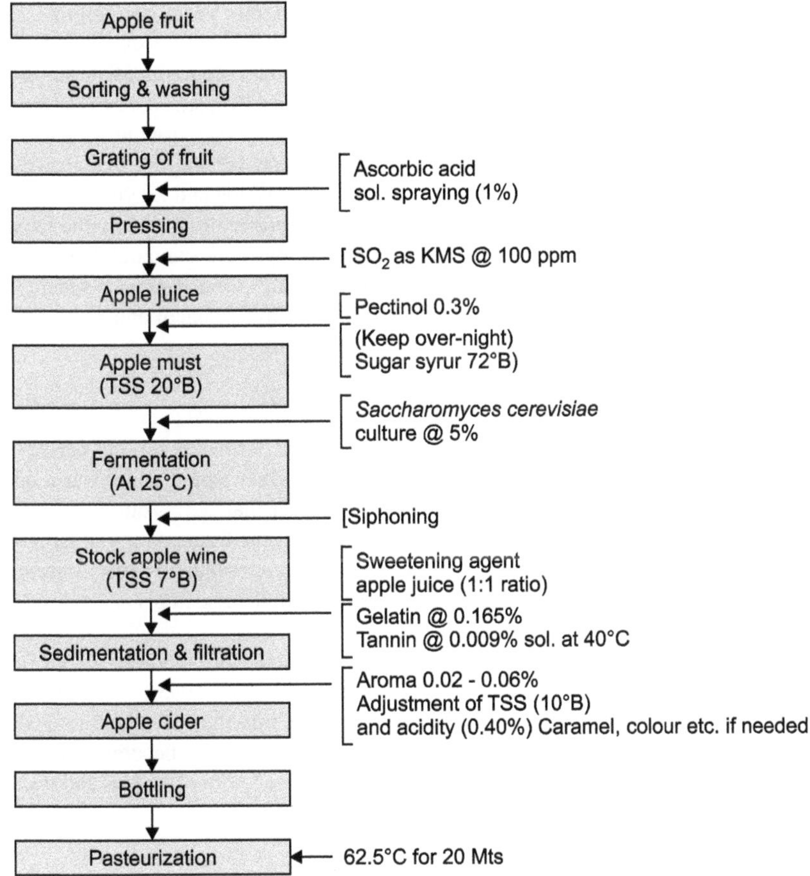


**Figure 11.** Flow sheet for making cider (*Source*: Joshi *et al.*,1999 a)

Out of the several factors, the sensory quality of wine is of great significance (Amerine *et al.*, 1980; Jackson, 2003; Joshi, 2006). Over the past few decades, advancements in the field of sensory science have enabled us to understand the variables that influence and contribute to the sensory perception of alcoholic beverages, like wine. The actual sensory quality characteristics desired in wine may differ but the pleasing sensory experience is always expected to be the same. The consumers are always willing to pay more for a product with a satisfying sensory experience (*see* Chapter 27 for details). To characterise wine flavour quantitatively, the application of descriptive analysis has led to the improvement of wine quality. Efforts have been directed at linking chemical and sensory measurements of flavour, i.e. a trained human subject sniffs (Acree, 1997) the effluent from a gas chromatogram. Gas chromatography olfactometry can link the detection and quantification of odourants to their sensory impact in wine (Brand *et al.*, 2001; Plutowska and Wardencki, 2008). The sub-set of compounds upon recombination that closely mimic the properties of the original aroma properties of the mixture can be chosen from such chromatograms. It is expected that application of such knowledge would be used in production of quality wine and the enologist must strive to achieve this excellence in the wine.

## 8.2. Microbial Spoilage of Wines

Different types of microorganisms are involved in the spoilage of wine. These include yeasts and bacteria.

### *8.2.1. Yeast*

A wide variety of yeast is involved in wine spoilage, such as haziness and deposition of sediment. The number of cells required to generate haziness, however, varies with species. With *Brettanomyces,* cloudiness is reported to develop at less than $10^2$ cells $ml^{-1}$. With most yeast, haze begins to develop only at $10^5$ cells $ml^{-1}$ (Hammond, 1976). *Zygosaccharomyces* can generate both flocculent and granular deposits (Rankine and Pilone, 1973) and can grow in bottling equipment, contaminating thousands of bottles. White and rose wines tend to be more susceptible than red wines. *Zygosaccharomyces bailii* is difficult to control because of its high resistance to yeast inhibitors. The yeast can grow in wine supplemented with 200 mg/litre of $SO_2$, sorbic acid or dimethylsulfide. Many yeasts form a film-like growth on the surface of wine under aerobic conditions, such as the strains of *S. cerevisiae* and *Zygosaccharomyces fermentii*. Out of the many spoilage-causing yeasts, species of *Brettanomyces* are probably the most serious. Both *B. intermedius* and *B. lambicus* produce 2-acetyltetrahydropyridine compounds that possess 'mousy odour' (Grbin and Henschke, 2000; Costelle and Henschke, 2002). *Brettanomyces* species also synthesise volatile phenolic compounds, including 4-ethyl guaiacol, 4-vinyl guaiacol and phenol that impact with smoky, spicy, medicinal and woody taints (Heresztyn, 1986a, b).

### *8.2.2. Lactic Acid Bacteria*

Lactic Acid Bacteria (LAB) are beneficial in malo-lactic acid fermentation, but certain strains are the cause of spoilage of wine, though the spoilage is limited to wines stored under warm conditions in the presence of insufficient sulphur dioxide and at pH value higher than 3.5. 'Tourne' is a spoilage problem caused primarily by *Lactobacillus brevis*. The primary action is the fermentation of tartaric acid to oxalic and acetic acid. Depending on the strain, the oxaloacetate is subsequently metabolised to lactic acid, succinic acid or acetic acid and carbon dioxide. Associated with the rise in pH is the development of 'flat' taste.

'Ropiness' is another type of spoilage of wine associated with synthesis of profuse amounts of mucilaginous polysaccharides (β-1,3-glucans) (Lonvaud-Funel *et al*., 1993). Typically *Leuconostoc* and *Pediococcus* are the responsible strains. The polysaccharides hold the bacteria in long silky chains that appear as floating threads in the affected wine. When dispersed, the wine becomes oily and has a viscous texture. Though visually unappealing, 'ropiness' is not associated with off-odour and taste.

### *8.2.3. Acetic Acid Bacteria*

The ability of acetic acid bacteria to oxidise ethanol to acetic acid induces wine spoilage, though it is vital for commercial vinegar production (Battcock and Azam Ali, 2001). Acetic acid bacteria can remain viable in wine for years together under anaerobic conditions. Thus, they are thought to be silent aerobes, which are unable to grow or survive in the absence of oxygen as they may grow in barrel- or bottled-wine if acceptable electron acceptors are present. It may be borne in mind that the spoilage can occur from the activity of bacteria at any stage of wine production (Joyeux *et al*., 1984). Mouldy grapes have a high population of acetic acid bacteria. The by-products of their metabolism, such as acetic acid and ethyl lactate, are retained throughout the fermentation and can taint the resultant wine. Spoilage of bottled wine by acetic acid bacteria, however, is limited to situations where failure of the closure permits seepage of oxygen into the bottle. Higher levels of acetic acid than 0.7 g/litre impart a vinegar-like odour and taste to the wine, thus making the wine unacceptable. Seriously spoiled wines are normally converted to wine vinegar.

## 8.3. Toxic Metabolites

A few toxic compounds have also been documented that need serious consideration from safety of consumers point of view. Currently, a major area of concern is the study of the metabolic role of LAB in the formation of compounds with undesirable implications for health. The two most toxic compounds that affect health are ethyl carbamate and biogenic amines.

### *8.3.1. Ethyl Carbamate*

Ethyl carbamate is a naturally-occurring component of fermented food and beverages, including wine (Liu *et al*., 1994). Due to its carcinogenic effect on health, considerable research has been carried out at exploring the origin of the compound in wine. Its formation in wine involves ethanol and carbamylic compounds, including urea produced by yeast during alcoholic fermentation and citruline and carbamyl phosphate, produced by LAB during malolactic fermentation (Liu *et al.*, 1994). In both the cases, these products are intermediates of arginine metabolism, one of the major amino acids of must and wine (Monteiro and Bisson, 1991).

### *8.3.2. Biogenic Amines*

In wine, as in other fermented products, biogenic amines are mainly derived from the presence of lactic acid bacteria that produce enzymes capable of decarboxylating the corresponding precursor amino acids (Montet *et al.*, 2006). Although 25 different biogenic amines have been described in wine, the most common ones are histamine, tyramine, putrescine and cadaverine; they are produced by the decarboxylation of histidine, tyrosine, ornithine/arginine and lysine, respectively (Liu *et al.*, 1995 b; Moreno-Arribas *et al.*, 2000; Lonvaud-Funel, 2001). Due to their undesirable effects on the health of the consumer, considerable efforts over the last few years to evaluate the biogenic amine content in wines from different origin areas, as well as the factors influencing biogenic amine production in wine have been made.

The formation of biogenic amines in wines mainly depends on the presence of lactic acid bacterial strains with the corresponding enzymatic decarboxylase activities. Other factors that may influence the abundance of amines are, for example, the presence of the amino acid precursor (which is influenced by the composition of grape must and by the type of winemaking involved (Lonvaud-Funel and Joyeux, 1994), since it affects the amino acid composition of wines before malolactic fermentation), pH and $SO_2$, which determine bacterial microflora and biological activity (Moreno-Arribas and Lonvaud-Funel, 1999, 2001). More details about the factors influencing amine formation by lactic acid bacteria in wines can be found in Chapter 17 of this book.

## 9. Waste from Wine Industry

Man is the most beautiful of all God's creation and like man, his environment too is beautiful with rivers to quench his thirst with sweet water, atmosphre to provide him pure oxygen, land and fire to provide him food. Primitive man ate uncooked food and lived in caves. Soon he learnt the use of firewood for cooking of raw meat and provided the first instance of pollution from food processing in history. However, as the man shifted from one place to another, this type of living never polluted the environment. But as man ascended the ladder of civilisation, his needs grew in arithmetic progression but corresponding pollution grew in geometric progression. With a huge increase in population and diverse employment opportunities in industries, a need was felt for processed food, giving rise to massive development of food processing industries all over the world (Joshi, 2011). Unfortunately, it also brought along with it a host of environmental problems, resulting in pollution of natural resources like water, air, land and vegetation (Joshi *et al.*, 1999c, Joshi *et al.*, 2011d).

Like other food industry, the waste from wine and brandy is highly polluting (Fig. 12), so measures need to be taken to control the same. At the same time, the waste is a rich source of nutrients and can be used in various ways to recover the value-added products or utilised as a ferment (*see* Chapter 24 of this text).

The waste from brandy production is more polluting than wine itself (Joshi, 2011). It is rich in several useful constituents, which are recoverable. Besides minimising the waste, the strategy should be for maximum use of the waste (*see* Chapter 24) for recovery of value-added products (Joshi *et al.*, 1999 c; Joshi *et al.*, 2011d). These include anthocyanins, flavour, fats, etc. Besides, the waste from winery can be used to produce ethanol, vinegar, single-cell proteins, animal feed, etc. More details can be found in a separate chapter 24 of this text and the literature cited.

## 10. Future Research and Development in Wine Industry

Obtaining continuous R&D input for the proper growth of the industry is desirable. The production of wines under defined conditions with proper quality control ensures the safety of product in the market. There are several issues for research and development in future. Some of these are discussed here and summarised in Table 9.

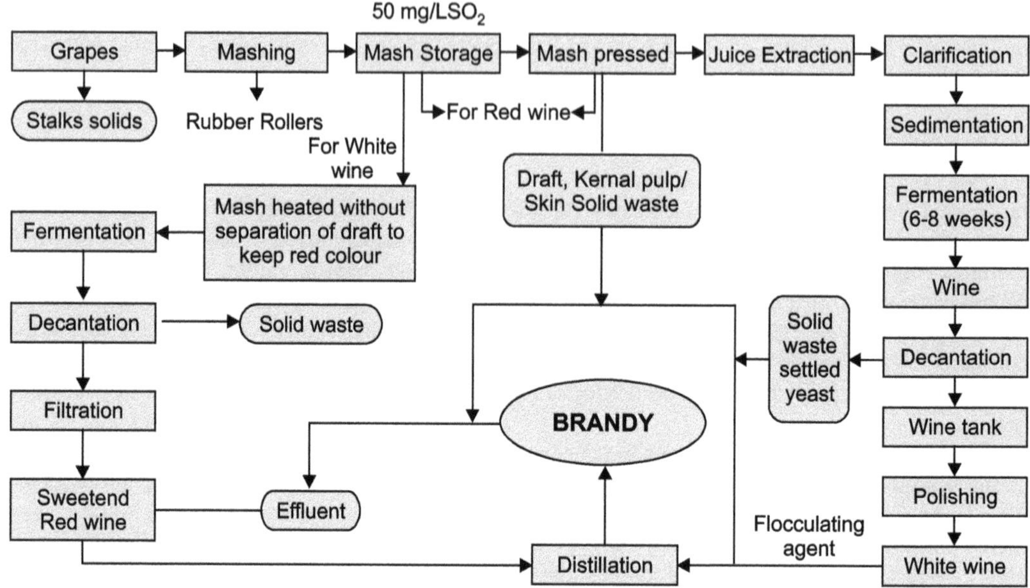


**Figure 12.** A diagrammatic view of the processes involved in wine and brandy along with the stages where waste is generated (*Source*: Joshi, 2011)

**Table 9.** Future Projections on Wine Fermentation Technology

- Covering more fruits, development of new fermented products, their evaluation as low alcoholic fruit beverages, including use of honey
- Improvement in quality of the already developed products through systematic process standardisation and evaluation
- In-depth investigation of fruit wines for antimicrobial or similar medicinal properties
- Therapeutic values of wines, though observed from ancient times, need more elaboration and scientific scrutiny
- Development of technology for continuous wine production by bioreactor technology using immobilised microorganisms or enzyme, especially β-glucosidase, to release flavour from fruits
- Development and evaluation of sparkling and other fortified wines from fruits
- Selection of suitable yeast strains or their manipulation for fruit fermentation application
- Use of yeasts other than *Saccharomyces cerevisiae* for specific purposes, such as more ester production, would offer an excited field of future research
- Use of enzyme to enhance flavour, yields and sensory qualities of wines
- Utilisation of wild fruits as such or in combination to produce acceptable and value-added fermented products
- Continuous monitoring and evaluation of locally made liquors for any toxic components and remedial measures for their improvement
- Preparation of medicinal, natural and nutritionally-fortified alcoholic beverages
- Development of suitable technology for waste treatment from wine fermentation industry as a profitable approach, such as animal feed, yeast recovery or other such products

There has been a considerable interest in the use of enzymes, notably glucosidases, to release flavour compounds bound in the fruit. There is also an increasing awareness of the disadvantages of overly protecting the juice from the minimal oxidation that occurs during crushing, racking and other winemaking practices. This is particularly reflected in the reduced use of sulphur dioxide before and after fermentation. In addition, there is growing interest in the use of several yeast strains to induce fermentation and to produce some of the perceived benefits of spontaneous fermentation while retaining the safety of induced fermentation. Inoculation with one or more strains of lactic acid bacteria is also becoming a common procedure to induce malo-lactic fermentation.

Considerable interest is being shown in the use of cell and enzyme immobilisation in wine production in batches as well as in continuous fermentation systems. Some of the fruits used in winemaking are highly acidic and the use of biological deacidification (using malo-lactic bacteria or deacidifying yeast, like *S. pombe*) is being made, but how these practices influence the composition and nutritive value of wines has not been evaluated so far and is an interesting and useful aspect of research in the near future.

Use of machinery and equipment being an inseparable component of wine would receive the attention of enologists in future also. Modelling and controlling of wine fermentation (Chapter 19) is comparatively a new aspect and is expected to be a hot area of future research. The production of wine and brandy of superior quality would also involve precise technology which could be applied directly by the industry and this aspect would continue to receive top priority in future too. The future outlook could foresee the usage of non-*Saccharomyces* yeasts for flavour improvement especially in biological de-acidification using malo-lactic bacteria or deacidifying yeasts, like *Schizosaccharomyces pombe*. Low or no alcoholic wines are also catching the fascination of consumers. It becomes all the more relevant in the present scenario to conduct in-depth research when the research by several well-established scientists have proved that wine possesses antioxidants properties and its limited consumption could be beneficial to prevent cardio-vascular diseases (CVD). Excess of everything is bad and wine is no exception to this rule. It is an admitted fact that great strides have been made on different aspects of fruit fermentation, but many indigenous fruit-based alcoholic beverages have never been investigated (Ray and Josh, 2014; Joshi, 2016). The therapeutic aspects of traditional beverages and their role in raising nutritional status of the consumers need to be stressed. Wine made from fruits other than grapes are just as diverse and their cascading flow of possibilities open up a whole new world to wine-drinkers. Fruit wine can be sweet and fruity or dry, more complex and nuanced. Of these, cider and other apple-based wines are produced and consumed in significant amounts throughout the world as is the case with plum wines. There are numerous varieties with a large range of flavours. The utility and scope of the wines from non-grape fruits reflect that systematic research on different facets of enology needs to be strengthened.

Needless to say, research currently in progress will continue to document the health-benefit properties of wine. With many tools in the hand of oenologist, it should not be a problem to evaluate the traditional fermented products also. The industry will however, need to play a highly visible role in the promotion of sound and sustainable environmental stewardship as a strong motivator in the purchase of wines. Further, since fermented products like wine are already accepted by the society, improvements in their production can be taken up easily by the industry immediately.

## References


Acree, T.E. (1997). GC/Olfactometry: GC with a sense of smell. *Anal. Chem.* 69: 170A.

Akubor, P.I., Obio, S.O., Nwadomere, K.A. and Obiomah, E. (2003). Production and quality evaluation of banana wine. *Plant Food Hum. Nutr.* 58: 1-6.

Alexándre, H. and Charpentier, C. (1998). Biochemical aspects of stuck and sluggish fermentation in grape must. *J. Ind. Microbiol. Biotechnol.* 20: 20-27.

Alexándre, H., Berlot, J.P. and Charpentier, C. (1992). Effect of ethanol on membrane fluidity of protoplats from *Saccharomyces cerevisiae* and *Kloeckera apiculata* grown with or without ethanol, measured by fluorescence anisotropy. *Biotechnol. Tech.* 5: 295-300.

Alves, A. and Herbert, P. (2011). Toxicological aspects related to wine consumption. pp. 209-233. *In*: V.K. Joshi (Ed.). Handbook of Enology: Principles, Practices and Recent Innovations, 3-vol. set. Asia-Tech Publisher and Distributors, New Delhi.

Amerine, M.A., Berg, H.W. and Cruess, W.V. (1972). The Technology of Wine Making. AVI Publishing Co. Inc., Westport, CT.

Amerine, M.A., Berg, H.W., Kunkee, R.E., Ough, C.S., Singleton, V.L. and Webb, A.D. (1980). The Technology of Wine Making, 4th ed. AVI Publishing, Westport, CT.

Auger, C., Al-Awwadi, N., Bornet, A., Rouanet, J.M., Gasc, F., Cros, G. and Teissedre, P.L. (2014). Catechins and procyanidins in Mediterranean diets. *Food Res. Int.* 37(3): 233-245.

Axelsson, L. (1993). Lactic acid bacteria: Classification and physiology. pp. 1-72. *In*: S. Salminen and A. von Wright (Eds.). Lactic Acid Bacteria: Microbiology and Functional Aspects, second ed. Marcel Dekker, New York, USA.

Bartowsky, E.J. (2003). Lysozime and winemaking. *Aus. J. Grape Wine Res.* 473a: 101-104.

Battcock, M. and Azam-Ali, S. (2001). Fermented fruits and vegetables, *FAO Agric. Ser. Bull.* 134: 96.

Bhutani, V.P., Joshi, V.K. and Chopra, S.K. (1989). Mineral contents of fruit wines produced experimentally. *J. Food Sci. Technol.* 26(6): 332-333.

Bhutani, V.P. and Joshi, V.K. (1996). Mineral composition of experimental sand pear and plum vermouth. *Alimentaria* 272: 99-103.

Bisson, L.F. (1999). Stuck and sluggish fermentations. *Am. J. Enol. Vitic.* 50: 107-119.

Bisson, L.F., Butzke, C.E. and Ebler, S.E. (1995). The role of moderate ethanol consumption in health and nutrition. *Am. J. Enol. Vitic.* 46: 449-457.

Boulton, R.B. (2011). The co-pigmentation of anthocyanins and its role in the colour of red wine: A critical review. *Amer. J. Enol. Viticul.* 52(2): 67-87.

Boulton, R.B., Singleton, V.L., Bision, L.F. and Kunkee, R.F. (1995). Principles and Practices of Wine Making. p. 13. Chapman and Hall, New York.

Brand, G., Millot, J.L. and Henquell, D. (2001). Complexity of olfactory lateralisation processes revealed by functional imaging: A review. *Neurosci. Biobehav. Rev.* 25: 159.

Burlingame, B. (2008). Wine: Food of poets and scientists. *J. Food Composition Analysis* 21: 587.

Canal-Llaubères, R.M. (1993). Enzymes in winemaking. pp. 477-506. *In*: G.H. Fleet (Ed.). Wine Microbiology and Biotechnology. Harwood Academic Publishers, Switzerland.

Chambers, P.J. and Pretorius, I.S. (2010). Fermenting knowledge: The history of winemaking, science and yeast research. *EMBO Reports* 11: 914-920.

Charoenchai, C., Fleet, G. and Henscke, P.A. (1998). Effects of temperature, pH and sugar concentration on the growth rates and cell biomass of wine yeasts. *Am. J. Enol. Vitic.* 49: 283-288.

Chatonnet, P., Dubourdieu, D., Boidron, J.-N. and Lavigne, V. (1993). Synthesis of volatile phenols by *Saccharomyces cerevisiae* in wines. *J. Sci. Food Agric.* 62: 191-202.

Chatonnet, P., Viala, C. and Dubourdieu, D. (1997). Influence of polyphenolic components of red wines on the microbial synthesis of volatile phenols. *Am. J. Enol. Vitic.* 48: 443-448.

Chauhan, A., Swami, U., Negi, B. and Soni, S.K. (2015). A valorised wine from *Aloe vera* and *Mentha arvensis* and its LC-Q-ToF-MS metabolic profiling. *Int. J. Food Ferment Technol.* 5: 183-190.

Chowdhury, P. and Ray, R.C. (2006). Fermentation of *Jamun* (*Syzygium cumini* L.) fruits to form red wine. *ASEAN Food J.* 14(1): 9-17.

Ciani, M. and Maccarelli, F. (1998). Oenological properties of non-*Saccharomyces* yeast associated with wine-making. *World J. Microbiol. Biotechnol.*, 14: 199-203.

Cocolin, L., Bission, L.F. and Mills, D.A. (2000). Direct profiling of the yeast dynamics in wine fermentations. *FEMS Microbiology Letters* 189: 81.

Costello, P.J. and Henschke, P. (2002). Mousy off-flavour of wine: Precursors and biosynthesis of the causative N-heterocycles 2-ethyltetrahydropyridine, 2-acetyltetrahydropyridine and 2-acetyl-1-pyrroline by *Lactobacillus hildargii* DSM 20176, *J. Agric. Food Chem.* 50: 7079-7087.

Crouse, J.R. and Grundy, S.M. (1984). Effect of alcohol on plasma lipoproteins and cholesterol and triglyceride metabolism in man. *J. Lipid Res.* 25: 486.

Davis, C.R., Wibowo, D., Fleet, G.H. and Lee, T.H. (1986a). Growth and metabolism of lactic acid bacteria during and after malo-lactic fermentation of wines at different pH. *Appl. Environ. Microbiol.* 51: 539-545.

Davis, C.R., Wibowo, D., Fleet, G.H. and Lee, T.H. (1986b). Growth and metabolism of lactic acid bacteria during fermentation and conser-vation of some Australian wines. *Food Technol. Aust.* 38: 35-40.

Davis, C.R., Fleet, G.H. and Lee, T.H. (1981). The microflora of wine corks. *The Australian Grapegrower and Winemaker* 208: 42-44.

de Revel, G., Martin, N., Pripis-Nicolau, L., Lonvaud-Funel, A. and Bertrand, A. (1999). Contribution to the knowledge of malo-lactic fermentation influence of wine aroma. *J. Agric. Food Chem.* 47: 4003-4008.

Delaquis, P., Cliff, M., King, M., Girrad, B., Hall, J. and Reynolds, A. (2000). Effect of two commercial malo-lactic cultures on the chemical and sensory properties of chancellor wines vinified with different yeasts and fermentation temperatures. *Am. J. Enol. Vitic.* 51: 42-48.

Delin, C.R. and Lee, T.H. (1991). The J-shaped curve revisited: Wine and cardiovascular health update. *Australian & New Zealand Wine Industry Journal* 6(1): 15-16.

Delin, C.R. and Lee, T.H. (1992). Psychological concomitant of the moderate consumption of alcohol. *J. Wine Res.* 3(5): 23.

Dequin, S. (2001). The potential of genetic engineering for improving brewing, wine-making and baking yeasts. *Appl. Microbiol. Biotechnol.* **56:** 577-588.

Di Lorenzo, A., Bloise, N., Meneghini, S., Sureda, A., Tenore, G.C., Visai, L., Arciola, C.R. and Daglia, M. (2016). Effect of winemaking on the composition of red wine as a source of polyphenols for anti-infective biomaterials. *Materials* 9: 316.

Dickinson, J.R. (2013). Carbon metabolism. pp. 591-595. *In*: J.R. Dickinson and M. Schweizer (Eds.). The Metabolism and Molecular Physiology of Saccharomyces cerevisiae. PA: CRC Press, Philadelphia.

Dizy, M. and Bisson, L.F. (2000). Proteolytic activity of yeast strains during grape juice fermentation. *Am. J. Enol. Vitic.*, 51: 155-167.

Drel, V.R. and Sybirna N. (2010). Protective effects of polyphenolics in red wine on diabetes associated oxidative/nitrative stress in streptozotocin-diabetic rats. *Cell Biol Int.* 34: 1147-1153.

Dunn, B., Levine, R.P. and Sherlock, G. (2015). Microarray karyotyping of commercial wine yeast strains reveals shared, as well as unique, genomic signatures. *BMC Genomics*, 6(1): 53.

Edlin, D.A.N., Narbad, A., Dickinson, J.R. and Lloyd, D. (1995). The biotransformation of simple phenolic compounds by *Brettanomyces anomalus*. *FEMS Microbiol. Lett.* 125: 311-316.

Feher, J., Lengyel, G. and Lugasi, A. (2007). The cultural history of wine: Theoretical background to wine therapy. *Cent Eur J Med.*, 2: 379-391.

Fleet, G.H. (1993a). The microorganisms of winemaking – isolation, enumeration and identification. pp. 1-46. *In*: G.H. Fleet (Ed.). Wine Microbiology and Biotechnology. Harwood Academic Publishers, GmbH, Switzerland.

Fleet, G.H. (2013). Yeast interaction and wine flavor. *Int. J. Food Microbiol.* 86(1-2): 11-22.

Fugelsang, K.C. (Ed.) (1997). Wine Microbiology. Chapman & Hall, New York, USA.

Gasteineau, F.C., Darby, J.W. and Turner, T.B. (1979). Fermented Food Beverages in Nutrition. Academic Press, New York.

Gomez Plaza, E., Gil Munoz, R., Lopez Roca, J.M., Martinez Cutillas, A., Fernandez, J. and Fernandez, J.I. (2015). Phenolic compounds and colour stability of red wines: Effect of skin maceration time. *Amer. J. Enol. Viticult.* 52(3): 266-270.

Goyal, R.K. (1999). Biochemistry of fermentation. pp. 87-129. *In*: V.K. Joshi and Ashok Pandey (Eds.). Biotechnology: Food Fermentation (Microbiology, Biochemistry and Technology). Educational Publishers and Distributors, New Delhi.

Grbin, P.R. and Henschke, P.A. (2000). Mousy off-flavour production in grape juice and wine. *Aus. J. Grape Wine Res.* 6: 255-262.

Gresele, P., Cerletti, C., Guglielmini, G., Pignatelli, P., Gaetano, G. and Violi, F. (2011). Effects of resveratrol and other wine polyphenols on vascular function: An update. *J. Nutr. Biochem.* 22: 201-211.

Gurbuz, O., Goçmen, D., Dagdelen, F., Gursoy, M., Aydin, S., Sahin, I., Buyukuysal, J. and Usta, M. (2017). Determination of flavan-3-ols and trans-resveratrol in grapes and wine using HPLC with fluorescence detection. *Food Chem.* 100(2): 518-525.

Gusman, J., Malonne, H. and Atassi, G. (2012). A repraisal of the potential chemo-preventive and chemo-therapeutic properties of resveratrol. *Carcinogenesis* 22(8): 1111-1117.

Halpern, R. (2008). A celebration of wine: Wine is medicine. *Inflammopharmacology* 16: 240- 244.

Hammond, S.M. (1976). Microbial spoilage of wines. pp. 38-44. *In*: F.W. Beech, A.G.H. Lea and C.F. Timeberlake (Eds.). Wine Quality – Current Problems and Future Trends. Long Asuton Research Station, University of Bristol, Bristol, UK.

Harding, G. (2005). A Wine Miscellany. Clarkson Potter Publishing, New York, USA.

Henick-Kling, T. (1993a). Malolactic fermentation. pp. 286-326. *In*: G.H. Fleet (Ed.). Wine Microbiology and Biotechnology. Harwood Academic, Amsterdam, The Netherlands.

Henick-Kling, T., Acree, T.E., Krieger, S.A., Laurent, M.-H. and Edinger, W.D. (1994). Modification of wine flavour by malo-lactic fermentation, *Wine East* 4: 8-15 and 29-30.

Henschke, P.A. (1993b). An overview of malo-lactic fermentation research. *Aust.–NewZealand Wine Ind. J.* 8: 69-79.

Henschke, P.A. and Dixon G.D. (1990). Effect of yeast strain on acetic acid accumulation during fermentation of Botrytis-affected grape juice. pp. 242-244. *In*: P.J. Williams, D. Davidson and T.H. Lee (Eds.). Proc. 7th Aust. Wine Ind. Tech. Conf. Australian Industrial Publishers, Adelaide, Australia.

Heresztyn, T. (1986a). Formation of substituted tetrahydropyridines by species of *Brettanomyces* and *Lactobacillus* isolated from mousy wines. *Am. J. Enol. Vitic.*, 37: 127-131.

Heresztyn, T. (1986b). Metabolism of volatile phenolic compounds from hydroxycinnamic acids by *Brettanomyces* yeast. *Arch. Microbiol.* 146: 96-98.

Holzapfel, W.H. and Scillinger, U. (1992). The genus *Leuconostoc*. pp. 1508-1534. *In*: A. Balow (Ed.). The Prokaryotes, second 2nd Edn. Springer Verlag, N.Y., USA.

Jackson, D. and Schuster, D. (1981). Grape Growing and Wine Making: A Handbook for Cool Climates. Martinborough, Alister Tagcor, New Zealand.

Jackson, R.S. (1994). Wine Science and Technology: Principles and Applications, Academic Press, San Diego, pp. 467.

Jackson, R.S. (1999). Grape-based fermentation products. pp. 647-744. *In*: V.K. Joshi and Ashok Pandey (Eds.). Biotechnology: Food Fermentation (Microbiology, Biochemistry and Technology), Vol. II. Educational Publishers and Distributors, New Delhi.

Jackson, R.S. (2000). Wine Science – Principles, Practices, Perception, 2nd edn. Academic Press, San Diego.

Jackson, R.S. (2003). Wines: Types of table wine. pp. 6, 342-372. *In*: B. Caballero, L. Trugo and P.N. Figlas (Eds.). Encylopedia of Food Sciences and Nutrition, 2nd Edn. Elsevier Science, UK.

Jagtap, U.B. and Bapat, V.A. (2015). Wines from fruits other than grapes: Current status and future prospectus. *Food Biosci.* 9: 80-96.

Jagtap, U.B. and Bapat, V.A. (2014). Phenolic composition and antioxidant capacity of wine prepared from custard apple (*Annona squamosa* L.) fruits. *J. Food Process. Preserv.* 122(9): 1595-1601.

Jagtap, U.B., Waghmare, S.R., Lokhande, V.H., Suprasanna, P. and Bapat, V.A. (2014). Preparation and evaluation of antioxidant capacity of jack fruit (*Artocarpus heterophyllus* L.) wine and its protective role against radiation induced DNA damage. *Ind. Crops Prod.* 34: 1595- 1601.

Jeandet, P., Vasserot, Y., Liger-Belair, G. and Richard, M. (2011). Sparkling wine production. pp. 1064-1115. *In*: V.K. Joshi (Ed.). Handbook of Enology: Principles, Practices and Recent Innovations, Vol. 3, Asia Tech Publisher, New Delhi.

Joshi, C. (2011). Waste management in winery and distillery. pp. 949-997. *In*: V.K. Joshi (Ed.). Handbook of Enology. 3 vol set. Asia Tech Publishers, Inc., New Delhi.

Joshi, V.K. (1997). Fruit Wines, second ed. Directorate of Extension Education, Dr. Y.S. Parmar, University of Horticulture and Forestry, Nauni, Solan, India, p. 255.

Joshi, V.K. (2006). Sensory Science: Principles and Application in Food Evaluation. Agro-Tech Academy, Udaipur.

Joshi, V.K. (2012). Health benefits and therapeutic value of fermented foods. pp. 183-235. *In*: R.K. Gupta, S. Bansal and M. Mangal (Eds.). Health Foods: Concept, Technology and Scope. Biotech, New Delhi.

Joshi, V.K. (Ed.). (2016). Indigenous Fermented Foods of South Asia. Rob Nout and Prabir Sarkar (Series Eds.). The Fermented Foods and Beverages Series. CRC Press, Roca, Florida, USA.

Joshi, V.K. and Attri, Devender (2005). A Panorama of research and development of wines in India. *J. Scientific and Industrial Research* 64(1): 9-18.

Joshi, V.K. and Bhutani, V.P. (1991). The influence of enzymatic clarification on the fermentation behaviour, composition and sensory qualities of apple wine. *Sciences Des Aliments* 11(3): 491-496.

Joshi, V.K. and Kumar, Vikas (2011). Importance, nutritive value, role, present status and future strategies in fruit wines in India. pp. 39-62. *In*: P.S. Panesar, H.K. Sharma and B.C. Sarkar (Eds.). Bio-Processing of Foods. Asia Tech Publisher, New Delhi.

Joshi, V.K. and Parmar, M. (2004). Present status, scope and future strategies of fruit wines production in India. *Indian Food Industry* 23(4): 48-52.

Joshi, V.K. and Sandhu, D.K. (2000). Influence of ethanol concentration, addition of spices extract and level of sweetness on physico-chemical characteristics and sensory quality of apple vermouth. *Braz. Arch. Biol. Technol.* 43(5): 537-545.

Joshi, V.K. and Siby John (2002). Antimicrobial activity of apple wine against some pathogenic and microbes of public health significance. *Alimentaria* 31(2): 67-72.

Joshi, V.K., Attri, Devender, Singh, Tuhin Kumar and Abrol, Ghanshyam (2011b). Fruit wines: Production technology. pp. 1177-1221. *In*: V.K. Joshi (Ed.). Handbook of Enology, Vol. III. Asia Tech Publishers, INC. New Delhi.

Joshi, V.K., Attri, B.L., Gupta, J.K. and Chopra, S.K. (1990a). Comparative fermentation behaviour, physico-chemical characteristics of fruit honey wines. *Indian J. Hort.* 47(1): 49-54.

Joshi, V.K., Bhutani, V.P. and Sharma, R.C. (1990b). Effect of dilution and addition of nitrogen source on chemical, mineral and sensory qualities of wild apricot wine. *Am. J. Enol. Vitic.* 41(3): 229-231.

Joshi, V.K., Bhutani, V.P. and Thakur, N.K. (1999a). Composition and nutrition of fermented products. pp. 259-320. *In*: V.K. Joshi and Ashok Pandey (Eds.). Biotechnology: Food Fermentation, Vol. I. Educational Publishers and Distributors, New Delhi.

Joshi, V.K., Sandhu, D.K. and Thakur, N.S. (1999b). Fruit-based alcoholic beverages. pp. 647-744. *In*: V.K. Joshi and Ashok Pandey (Eds.). Biotechnology: Food Fermentation (Microbiology, Biochemistry and Technology). Vol. II, Educational Publishers and Distributors, New Delhi.

Joshi, V.K., Pandey, A. and Sandhu, D.K. (1999c). Fermentation technology for food industry waste utilisation. pp. 1291-1348. *In*: V.K. Joshi and Ashok Pandey (Eds.). Biotechnology: Food Fermentation (Microbiology, Biochemistry and Technology), Vol. I. Educational Publishers and Distributors, New Delhi.

Joshi, V.K., Dev Raj and Joshi, C. (2011). Utilisation of wastes from food fermentation industry. pp. 295-356. *In*: Food Processing Wastes Management. New India Publishing Agency, New Delhi.

Joshi, V.K., Sharma, R., Girdhar, A. and Abrol, G.S. (2012). Effect of dilution and maturation on physico-chemical and sensory quality of jamun (Black plum) wine. *Indian J. Nat. Prod. Res.* 3(2): 222-227.

Joshi, V.K., Sharma, S. and Bhushan, S. (2005). Effect of method of preparation and cultivar on the quality of strawberry wine. *Acta Alimentaria* 34(3): 339-353.

Joshi, V.K., Sharma, S. and Sharma, A. (2016). Wines – white, red, sparkling, fortified, and Cider (Ch 12). pp. 353-407. *In*: Ashok Pandey (India), Guocheng Du (China), Maria Angeles Sanroman (Spain), Carlos Ricardo Soccol (Brazil), Claude-Gilles Dussap (France) (Eds.). Current Developments in Biotechnology and Bioengineering. Elsevier book series. Elsevier, London, UK.

Joshi, V.K., Sharma, Somesh and Rana, V.S. (2012). Wine and brandy. pp. 471-494. *In*: V.K. Joshi and R.S. Singh (Eds.). Food Biotechnology: Principles and Practices. I.K. International Publishing House, New Delhi.

Joshi, V.K., Sharma, Somesh, Bhushan, S. and Attri, D. (2004). Fruit-based alcoholic beverages. p. 335. *In*: Ashok Pandey (Ed.). Concise Encyclopedia of Bioresources Technology. Haworth Food Product Press, New York.

Joshi, V.K., Sharma, Somesh, John, Sibby, Kaushal, B.B.L. and Rana, Neerja (2009). Preparation of antioxidant rich apple and strawberry wine. *Proc. Nat. Acad. Sci. India* Sec. B, 79: 415-420.

Joshi, V.K., Thakur, N.S., Bhat, A. and Garg, C. (2011a). Wine and brandy: a perspective. pp. 3-45. *In*: V.K. Joshi (Ed.). Handbook of Enology: Principles, Practices and Recent Innovations, vol. 1, Asia Tech Publisher, New Delhi.

Joshi,V.K. and Devi, P.M. (2009). Resveratrol, role, contents in wine and factors influencing its production. *Proc. Nat. Acad. Sci. India* Sec. B, 79: 212-226.

Joshi, V.K. and Kumar, V. (2014). Research and development of fruit wines. *Food Marketing and Technology*, January, 39-41.

Joshi, V.K., Sharma, P.C. and Attri, B.L. (1991). A note on the deacidification activity of *Schizosaccharomyces pombe* in plum musts of variable composition. *J. Appl. Bacteriol.* 70: 386-390.

Joshi, V.K., Panesar, P.S., Rana, V.S. and Kaur, S. (2017). Science and technology of fruit wines: An overview. Chapter 1. pp. 1-72. *In*: Maria Kossovea, V.K. Joshi and P.S. Panesar (Eds.). Science and Technology of Fruit Wines. Elsevier, UK.

Joyeux, A., Lafon-la Fourcade, S. and Ribereau-gayon, P. (1984). Evolution of acetic acid bacteria during fermentation and storage of wine. *Appl. Environ. Microbiol.* 48: 153-156.

Karagiannins, S.D. (2011). Classification and characteristics of wines and brandies. pp. 46-65. *In*: V.K. Joshi (Ed.). Handbook of Enology: Principles, Practices and Recent Innovations, Vol. 3. Asia Tech Publisher, New Delhi.

Kelebek, H., Selli, S., Canbas, A. and Cabaroglu, T. (2009). HPLC determination of organic acids, sugars, phenolic compositions and antioxidant capacity of orange juice and orange wine made from a Turkish cv. Kozan. *Microchem J.* 9: 187-192.

Kennedy, N.P. and Tipton, K.F. (1990). Ethanol metabolism and alcoholic liver disease. *Essays Biochem* 25: 137.

Kinsella, J.E., Frankel, E.N., German, J.B. and Kanner, J. (1993). Possible mechanisms for the protective role of antioxidants in wine and plant foods. *Food Technol.* 47: 85.

Klatsky, A.L. and Armstrong, M.A. (1993). Alcoholic beverage choice and risk of coronary heart disease mortality: Do red wine drinkers fare best? *Am. J. Cardiol.* 71: 467-469.

Koehler, C. 1986. Handling of Greek Transport Amphoras. pp. 49-67. *In*: J.-Y. Empereur and Y. Garlan (Eds.). Recherches sur les amphores grecques. BCH Suppl. 13. Athens and Paris: École Française d'Athènes and de Boccard.

Kosseva, M.R., Joshi, V.K. and Panesar P.S. (Eds.) (2017). Science and Technology of Fruit Wine Production. Academic Press is an imprint of Elsevier, London, United Kingdom, pp. 705.

Kumar, K.K., Swain, M.R., Panda, S.H., Sahoo, U.C. and Ray, R.C. (2008). Fermentation of litchi (*Litchi chinensis* Sonn.) fruits into wine. *Food* 2: 43-47.

Kundu, B.S., Bardiya, M.C., Daulta, B.S. and Tauro, P. (1980). Evaluation of exotic grapes grown in Haryana for white table wines. *J. Food Sci. Technol.* 17: 221-224.

Kunkee, R.E. (1999). Some roles of malic acid in the malo-lactic fermentation in wine making. *FEMS Microbiol. Rev.* 88: 55-57.

Kunkee, R.E. and Bisson, L.F. (1993). Winemaking yeasts. *In*: A.H. Rose and J.S. Harrison (Eds.). The Yeast, 2nd edn. Academic Press, New York, USA.

Lavigne, V. and Dubourdieu, D. (1996). Demonstration and interpretation of the yeast's lees ability to adsorb certain volatile thiols contained in wine. *J. Int. Sci. Vigne Vin.* 30: 201-206.

Lee, P.R., Chong, I.S.M., Yu, B., Curran, P. and Liu, S.Q. (2012). Effect of sequentially inoculated *Williopsis saturnus* and *Saccharomyces cerevisiae* on volatile profiles of papaya wine. *Food Res. Int.* 45: 177-183.

Levy, G. (1979). Pharmacokinetics of salicylic acid in man. *Drug Metab. Rev.* 9: 379.

Liu, S.-Q. (2002). A review – Malolactic fermentation in wine beyond deacidification. *J. Appl. Microbiol.* 92: 589-601.

Liu, S.-Q. and Davis, C.R. (1994). Analysis of wine carbohydrates using capillary gas-liquid chromatography. *Am. J. Enol. Vitic.* 45: 229-234.

Liu, S.-Q., Pritchard, G.G., Hardman, M.J. and Pilone, G.J. (1994). Citrulline production and ethyl carbamate (urethane) precursor formation from arginine degradation by wine lactic acid bacteria *Leuconostoc oenos* and *Lactobacillus buchneri*. *Am. J. Enol. Vitic.* 45: 235-242.

Liu, S.-Q., Davis, C.R. and Brooks, J.D. (1995a). Growth and metabolism of selected lactic acid bacteria in synthetic wine. *Am. J. Enol. Vitic.* 46: 166-174.

Lonvaud-Funel, A. (2001). Biogenic amines in wines: Role of lactic acid bacteria. *FEMS Microbiol. Lett.* 199: 9-13.

Lonvaud-Funel, A. and Joyeux, A. (1994). Histamine production of wine by lactic acid bacteria: Isolation of a histamine producing strain of *Leuconostoc oenos*. *J. Appl. Bacteriol.* 77: 401-407.

Lonvaud-Funel, A., Guilloux, Y. and Joyeux, A. (1993). Isolation of a DNA probe for identification of glucan-producing *Pediococcus damnosus* in wines. *Appl. Microbiol. Biotechnol.* 56: 35-39.

Lucia, S.P. (1963). A History of Wine as Therapy. Lippincott, Philadelphia.

MAFF (1980). Ministry of Agriculture, Fisheries and Food. Grape for Wine, HMSO, London.

Mahamy, J. and Klessig, D.F. (1992). Salicylic acid and plant disease resistance. *Plant J.* 2: 643.

Martineau, B. and Henick-Kling, T. (1995a). Performances and diacetyl production of commercial strains of malolactic bacteria in wine. *J. Appl. Bacteriol.* 78: 526-536.

Martineau, B. and Henick-Kling, T. (1995b). Formation and degradation of diacetyl in wine during alcoholic fermentation with *Saccharomyces cerevisiae* strain EC1118 and malo-lactic fermentation with *Leuconostoc oenos* strain MCW. *Am. J. Enol. Vitic.* 46: 442-448.

Martinez, J., Millan, C. and Ortega, J.M. (1989). Growth of natural flora during the fermentation of inoculated musts from 'Pedro ximerez' grapes. *S. Afr. J. Enol. Vitic.* 10: 31-35.

Mazza, G. (1995). Anthocyanins in grapes and grape products. *Crit. Rev. Food Sci. Nutr.* 35: 341-371.

McDonald, J. (1986). A Symposium on Wine, Health and Society Nutrition. Wine Institute, Washington DC, USA.

McGovern, P.E., Glusker, D.L., Exner, L.J. and Voigt, M.M. (1996). Neolithic resinated wine. *Nature* 381: 480.

McGovern, P.E., Zhang, J., Tang, J., Zhang, Z., Hall, G.R. and Moreau, R.A. (2004). Fermented beverages of pre- and proto-historic China. *Proc. Natl. Acad. Sci.*, U.S.A. 101: 17593-17598.

Mohanty, S., Ray, P., Swain, M.R. and Ray, R.C. (2006). Fermentation of cashew (*Anacardium occidentale* L.) 'apple' into wine. *J. Food Process. Preserv.* 30: 314-322.

Monagas, M., Bartolome, B. and Gomez Cordoves, C. (2015). Update knowledge about the presence of phenolic compounds in wine. *Crit. Rev. Food Sci. Nutr.* 45: 85-118.

Monteiro, F. and Bisson, L.F. (1991). Amino acid utilisation and urea formation during vinification fermentations. *Am. J. Enol. Vitic.* 42: 199-208.

Montet, D., Loiseau, G. and Zakhia-Rozis, N. (2006). Microbial technology of fermented vegetables. pp. 309-344. *In*: R.C. Ray and O.P. Ward (Eds.). Microbial Biotechnology in Horticulture. Science Publishers, Enfield, New Hampshire, USA.

Moreno-Arribas, M.V. and Polo, M.C. (2005). Wine making biochemistry and microbiology: Current knowledge and future trends. *Crit. Rev. Food Sci. Nutr.* 45: 265-286.

Moreno-Arribas, V. and Lonvaud-Funel, A. (1999). Tyrosine decarboxylase activity of *Lactobacillus brevis* IOEB 9809 isolated from wine and *L. brevis* ATCC 367. *FEMS Microbiol. Lett.* 180: 55-60.

Moreno-Arribas, V. and Lonvaud-Funel, A. (2001). Purification and characterisation of tyrosine decarboxylase of *Lactobacillus brevis* IOEB 9809 isolated from wine. *FEMS Microbiol. Lett.* 195:103-107.

Moreno-Arribas, V., Torlois, S., Joyeux, A., Bertrand, A. and Lonvaud- Funel, A. (2000). Isolation, properties and behaviour, of tyramine producing lactic acid bacteria from wine. *J. Appl. Microbiol.* 88: 584-593.

Muller, C.J. and Fugelsang, K.C. (1993). Gentisic acid: An Aspirin like constituent of wine. *Pract Winery and Vineyard* Sept.-Oct.: 45.

Muller, C.J., Striegler, R.K., Fugelsang, K.C. and Wineman, D.R. (1994). Salicylic acid: Rootstock defence? *Pract Winery and Vineyard* March/April: 17.

Muller, C.J. (1995). Wine and Health – It is more than alcohol. pp. 15-29. *In*: B.W. Zoecklein (Ed.). Wine Analysis and Production. New York: Chapman and Hall.

Muniz, C.R., Borges, M.D.F. and Freire, F.D.C.O. (2008). Tropical and subtropical fruit fermented beverages. pp. 35-39. *In*: R.C. Ray and O.P. Ward (Eds.). Microbial Biotechnology in Horticulture, Vol. 2. Enfield, NH: Science Publishers

Negi, B., Kaur, R. and Dey, G. (2013). Protective effects of a novel sea buckthorn wine on oxidative stress and hypercholesterolemia. *Food Funct.* 4: 240-248.

Nijveldt, R.J., van Nood, E., van Hoorn. D.E., Boelens, P.G., van Norren, K. and van Leeuwen, P.A. (2001). Flavonoids: A review of probable mechanisms of action and potential applications. *Am. J. Clin. Nutr.* Oct. 74(4): 418-425.

Okada, T. and Yohotosuha, K. (1996). *Trans*-resveratrol concentrations in berry skins and wines from grapes grown in Japan. *Am. J. Enol. Vitic.* 47: 93-99.

Onwuka, U.N. and Awan, F.N. (2012). The potential for baker's yeast (*Saccharomyces cerevisiae*) in the production of wine from banana, cooking banana and plantain. *Food Sci. Technol.* 1: 127-132.

Ough, C.S. (1992). Winemaking Basic. Food Product Press: An Imprint of the Haworth Press, New York.

Panda, S.H., Parmanik, M., Sharma, P., Panda, S. and Ray, R.C. (2005). Microorganisms in food biotechnology. pp. 47-54. *In*: R.C. Mohanty and P.K. Chand (Eds.). Microbes in Our Lives. Department of Botany and Microbiology, Utkal University, Bhubaneswar, India.

Panda, S.K., Behera, S.K. and Ray, R.C. (2014). Fermentation of sapota (*Achras sapota* L.) fruits to functional wine. *Nutrafoods* 13(4): 179-186.

Panda, S.K., Behera, S.K., Sahu, U.C., Ray, R.C., Kayitesi, E. and Mulaba Bafubiandi, A.F. (2016). Bioprocessing of jackfruit (*Artocarpus heterophyllus* L.) pulp into wine: Technology, proximate composition and sensory evaluation. *Afr. J. Sci. Technol. Innov. Develop.* 8(1): 27-32.

Panda, S.K., Ray, R.C., Mishra, S.S. and Kayitesil1, K. (2017). Microbial processing of fruit and vegetable wastes into potential bio-commodities: A review. *Crit. Rev. Biotechnol.* 1-16.

Panda, S.K., Sahoo, U.C., Behera, S.K. and Ray, R.C. (2013a). Bio-processing of *bael* (*Aegle marmelos* L.) fruits into wine with antioxidants. *Food Biosci.* 5: 34-41.

Panda, S.K., Swain, M.R., Singh, S. and Ray, R.C. (2013b). Proximate compositions of herbal purple sweet potato (*Ipomoea batatas* L.) wine. *J. Food Process. Preserv.* 37: 596-604.

Panesar, P.S., Joshi, V.K., Panesar, R. and Abrol, G.S. (2011). Vermouth: Technology of production and quality characteristics. pp. 253-271. *In*: R.S. Jackson (Ed.). Advances in Food and Nutritional Research. 63, Elsevier, Inc., London, U.K.

Papadopoulou, C., Soulti, K. and Roussis, I.G. (2005). Potential antimicrobial activity of red and white wine phenolic extracts against strains of *Staphylococcus aureus*, *Escherichia coli* and *Candida albicans*. *Food Technol. Biotechnol.* 43: 41-46.

Petrie, W.M.E. (1923). Social Life in Ancient Egypt. Methuen, London.

Pino, J.A. and Queris, O. (2010). Analysis of volatile compounds of pineapple wine using solid-phase microextraction techniques. *Food Chem.* 122: 1241-1246.

Plutowska, B. and Wardencki, W. (2008). Application of gas chromatography eolfactometry (GCeO) in analysis and quality assessment of alcoholic beverages – A review. *Food Chem.* 107(1): 449-463.

Pretorius, I.S. (1999). Engineering designer genes for wine yeasts. *Aust. N.Z. Wine Indust. J.* 14: 42-47.

Pretorius, I.S. (2000). Tailoring wine yeast for the new millennium: Novel approaches to the ancient art of wine making. *Yeast* 16: 1-55.

Pretorius, I.S. (2001). Gene technology in wine making: New approaches to an ancient art. *Agric. Consp. Sci.* 66: 1-20.

Pretorius, I.S. and Hoj, P.B. (2005). Grape and wine biotechnology: Challenges, opportunities and potential benefits. *Aust. J. Grape Wine Res.* 11: 83-108.

Pretorius, I.S., Van der Westhuizen, T.J. and Augustyn, O.P.H. (1999). Yeast biodiversity in vineyards and wineries and its importance to the South African wine industries – A review. *S. Afr. J. Enol. Vitic.* 20: 61-74.

Ramos, S. (2008). Cancer chemoprevention and chemotherapy: Dietary polyphenols and signalling pathways. *Mol. Nutr. Food Res.* 52: 507-526.

Rana, A. and Singh, H.P. (2013). Bio-utilisation of wild berries for preparation of high valued herbal wines. *Ind. J. Nat. Prod. Res.* 4: 165-169.

Rankine, B.C. and Pilone, D.A. (1973). *Saccharomyces bailii*, a resistant yeast causing serious spoilage of bottled table wine. *Am. J. Enol. Vitic.* 24: 55-58.

Rattray, F.R., Walfridsson, M. and Nilsson, D. (2000). Purification and characterisation of a diacetyl reductase from *Leuconostoc pseudomesenteroides*. *Int. Dairy J.* 10: 781-789.

Ray, R.C. and Joshi, V.K. (2014). Fermented Foods: Past, present and future scenario. pp. 1-36. *In*: R.C. Ray and D. Montet (Eds.). Microorganisms and Fermentation of Traditional Foods. CRC Press, Boca Raton, Florida, USA.

Ray, R.C., Panda, S.K., Swain, M.R. and Sivakumar, P.S. (2012). Proximate composition and sensory evaluation of anthocyanin-rich purple sweet potato (*Ipomoea batatas* L.) wine. *Int. J. Food Sci. Tech.* 47: 452-458.

Rebordinas, L., Infante, J.J., Rodriguez, M.E., Vallejo, I. and Cantroal, J.M. (2011). Wine yeast growth and factors affecting. pp. 406-434. *In*: V.K. Joshi (Ed.). Handbook of Enology, Vol. 2. Asia Tech Publisher, New Delhi.

Reddy, A. and Reddy, S. (2005). Production and characterisation of wine from mango fruit (*Mangifera indica* L). *World J. Microbiol. Biotech.* 21: 345-350.

Reddy, L.V.A., Reddy, O.V.S. and Joshi, V.K. (2014). Production of wine from mango fruit: A review. *Int. J. of Food Ferment. Technol.* 4(1): 13-25.

Renfrew, C. (1989). The origins of Indo-European languages. *Sci. Am.* 261(4): 106-110.
Requena, T., Monagas, M., Pozo Bayo, M., Martin, P.J., Bartolome, B. and Campo, R. (2015). Perspectives of the potential implications of wine polyphenols on human oral and gut microbiota. *Trends Food Sci. Technol.* 21: 332-344.
Ribereau Gayon, P., Dubourdieu, D., Donèche, B. and Lonvaud, A. (2000). The microbiology of wine and vinifications. *Handbook of Enology*, Vol. 1, John Wiley and Sons, U.K., 358-405.
Robinson, J. (1986). Vines, Grape and Wines. Mitchell, Beazley, London.
Robinson, J. (2016). The Oxford Companion to Wine, 3rd Edn. Oxford University Press. 779- 787.
Rojas, V., Gil, J.V., Pinaga, F. and Manzanares, P. (2001). Studies on acetate ester production by non-*Saccharomyces* wine yeast. *Int. J. Food Microbiol.* 70: 283-289.
Rojas, V., Gil, J.V., Pinaga, F. and Manzanares, P. (2003). Acetate ester formation in wine by mixed cultures in laboratory fermentations. *Int. Food Microbiol.* 86: 181-188.
Rose, A.H. and Harrison, J.S. (Eds.) (1971). The Yeast. *Physiology and Biochemistry of Yeasts*, Vol. 2. Academic Press, New York.
Sahoo, U.C., Panda, S.K., Mohapatra, U.B. and Ray, R.C. (2012). Preparation and evaluation of wine from tendu (*Diospyros melanoxylon* L.) fruits with antioxidants. *Int. J. Food Ferment. Technol.* 2: 171-178.
Saigal, D. and Ray, R.C. (2007). Wine making: Microbiology, biochemistry and biotechnology. pp. 1-33. *In*: Ramesh C. Ray and O.P. Ward (Eds.). Microbial Biotechnology in Horticulture, Vol. 3. Science Publishers, New Hampshire, USA.
Sandhu, D.K. and Joshi, V.K. (1995). Technology, quality and scope of fruit wines with special reference to apple. *Indian Food Industry* 14(1): 24-34.
Shahidi, F. (2009). Nutraceuticals and functional foods: Whole versus processed foods. *Trends in Food Sci. Technol.* 20: 376-387.
Shanmugavelu, G.K. (2003). Grape Cultivation and Processing. Agrobios (India), Jodhpur.
Shikhamany, S.D. (2003). Grape Production in India. Website http//www.FAO.org/ DIEREP/03/X6897E/ X6897E05/HTM
Soden, A., Francis, I.L., Oakey, H. and Henschke, P.A. (2000). Effects of co-fermentation with *Candida stellata* and *Saccharomyces cerevisiae* on the aroma and composition of Chardonnay wine. *Aus. J. Grape Wine Res.*, 6: 21-30.
Soni, S.K., Marwaha, S.S., Marwaha, U. and Soni, R. (2011). Composition and nutritive value of wine. pp. 89-145. *In*: V.K. Joshi (Ed.). Handbook of Enology: Principles, Practices and Recent Innovations, Vol. 1. Asia Tech Publisher, New Delhi.
Spiropoulos, A., Tanaka, J., Flerianos, I. and Bisson, L.F. (2000). Characterisation of hydrogen sulfite formation in commercial and natural wine isolates of *Saccharomyces*. *Am. J. Enol. Vitic.* 51: 233-248.
Srisamatthakarn, P., Chanrittisen, T. and Buranawijarn, E. (2003). Effects of Sampee mango (*Mangifera indica* L.) ripening stage and flesh ratio on mango wine quality. Proceeding CD-ROM: The First International Symposium on Insight into the World of Indigenous Fermented Foods for Technology Development and Food Safety. Bangkok, Thailand, p. 11.
Stockley, C.S. (2011). Therapeutic value of wine: A clinical and scientific perspective. pp. 146-208. *In*: V.K. Joshi (Ed.). Handbook of Enology: Principles, Practices and Recent Innovations, Vol. 1. Asia Tech Publisher, New Delhi.
Swami, S.B., Thakor, N.J. and Divate, A.D. (2014). Fruit wine production: A review. *Journal of Food Research and Technology* 2: 93-94.
Tapsell, L.C., Hemphill, I., Cobiac, L., Patch, C.S., Sullivan, D.R., Fenech, M., Roodenrys, S., Keogh, J.B., Clifton, P.M., Williams, P.G., Fazio, V.A. and Inge, K.E. (2006). Health benefits of herbs and spices: The past, the present, the future. *Med. J. Aust.* 185: 4-24.
Tomaz, I. and Maslov, L. (2015). Simultaneous determination of phenolic compounds in different matrices using phenyl-hexyl stationary phase. *Food Anal. Methods*, DOI 10.1007/s12161-015-0206-7
Treissedre, P.L., Frankel, E.N., Waterhouse, A.L., Peleg, H. and German, J.B. (2016). Inhibition of *in vitro* human LDL oxidation by phenolic antioxidants from grapes and wines. *J. Sci. Food Agric.* 70(1): 55-61.

Trivedi, N. (2013). Process development for the production of herbal wines from *Aloe vera* and evaluation of their therapeutic potential. Thesis submitted to Panjab University, Chandigarh.

Trivedi, N., Rishi, P. and Soni, S.K. (2012). Production of a herbal wine from *Aloe vera* gel and evaluation of its effect against common food-borne pathogens and probiotics. *Int. J. Food for Fer Technol.* 2: 157-166.

Trivedi, N., Rishi, P. and Soni, S.K. (2015a). Protective role of *Aloe Vera* wine against oxidative stress induced by salmonella infection in a murine model. *IJFANS* 4: 64-76.

Trivedi, N., Rishi, P. and Soni, S.K. (2015b). Antibacterial activity of prepared *Aloe vera* based herbal wines against common food-borne pathogens and probiotic strains. *J Home Sci.*, 1: 91-99.

Unwin, T. (1991). Wine and the Vine: An Historical Geography of Viticulture and the Wine Trading. Routledge, London.

Vahos, I.C.Z., Ochoa, S., Maldonado, M.I.E., Zapata, A.D.Z. and Rojano, B.I. (2016). Cytotoxic effect and antioxidant activity of Andean berry (*Vaccinium meridionale* Sw) wine. *J. Med. Plant Res.* 10: 402-408.

Vaidya, Manoj Kumar, Vaidya, Devina and Joshi, V.K. (2011). Wine regions and status of wild wine production. pp. 66-88. *In*: V.K. Joshi (Ed.). Handbook of Enology: Principles, Practices and Recent Innovations, Vol. I. Asia Tech Publishers, Inc., New Delhi.

Verma, L.R. and Joshi, V.K. (2000). An overview of post-harvest technology. *In*: L.R. Verma and V.K. Joshi (Eds.). Postharvest Technology of Fruits and Vegetables. Vol. I. Indus Publishing Co., New Delhi, p. 1.

Vine, R.P. (1981). Wine and the history of western civilisation. *In*: Commercial Winemaking, Processing and Controls. The AVI Publishing Co., Westport, CT. *Vittic* 56: 139-147.

Voguel, W. (2003). Que es el vino? *In*: S.A. Acribia (Ed.). Elaboracio´n casera de vinos. Vinos de uvas, manzanas y bayas. Zaragoza, Spain.

Vyas, S.R. and Chakravorty, S.R. (1971). Wine Making at Home. Haryana Agricultural University Bulletin, Hissar.

Ward, O.P. and Ray, R.C. (2006). Microbial biotechnology in horticulture – An overview. p. 1. *In*: R.C. Ray and O.P. Ward (Eds.). Microbial Biotechnology in Horticulture, Vol. 1. Enfield, NH: Science Publishers.

Waterhouse, A.L. (1995). Wine and heart disease, *Chemistry Industry*, May, 338-341.

Yalpani, N. and Raskin, I. (1993). Salicylic acid: A systemic signal in induced plant disease. *Trends Microbiol.* 1: 88.

Zoecklein, B., Fugelsang, K.C., Gump, B.H. and Nury, F.S. (1990). Wine Analysis and Production. Van Nostrand-Rheinhold, New York.

# 2 Compositional, Nutritional and Therapeutic Values of Wine and Toxicology

**Hatice Kalkan Yıldırım**
Ege University, Department of Food Engineering, 35100 Bornova, Izmir, Turkey

## 1. Introduction

Wine is one of the oldest fermented foods that has been produced and consumed. For thousands of years, wine has always been important in human life and regarded as the gentlest light alcoholic drink throughout the ages. Consumption of wine at ceremonies, celebrations and banquets and other special occasions has always been at the forefront of drinks that accompanies food today as it did yesterday.

In the nineteenth century, wine was regarded as a basic food. With this view, wine was protected by governments by applying equal taxation of wine with all other basic foods. After emerging of phylloxera disease in Europe (after 1850), vine yield greatly reduced. In response to this situation, countries took urgent measures to meet the wine demand. In this sense, artificial wine production has begun. By special law, 'artificial wine' was allowed to be made by combining alcohol, sugar, glycerine, dyes and other substances in certain combinations (Penza and Cassano, 2004). In the meantime, phylloxera-resistant rootstocks were developed. Thus winemaking was revived again all over Europe. The production of artificial wine was banned by a special law called 'Fraud Prevention Law' or legal regulation in France. Later (1935) with the promulgation of a law concerning confirmation of grape origin and where grape varieties were used for wine production, the production was continued in a controlled and natural manner.

Even the fluctuations in its production, wine was always a point of interest as a medicine. Ancient Greek society used wine very often in herbal infusions. These applications were continued until the beginning of the twentieth century. The excessive abuse of spirits, religious and political approaches led to the prohibition of all foods containing alcohol. On realising alcoholism as a complex disease with genetic and environmental origin, wine components and their nutritional values, their therapeutic properties began to be investigated more seriously. Most of the results relate to the importance of moderate wine consumption, with 250-300 ml/day as volume enough for health benefit (food with antimicrobial, antioxidant, anticancer and anti-ageing properties) and control over excessive alcohol consumption (Fischer *et al.*, 2007).

Wine is not only an alcoholic beverage with special healthy properties but also an important product contributing to the economy with the added value provided by evaluating the most important fruit of agriculture. Today, wine-free vineyard economies are unthinkable in many vineyard countries, such as France, Italy, Spain and Portugal. In these countries, as much as 90 per cent of the grapes produced, sometimes even more, are converted into wine. Wine is also an important economic source in many new independent republics like Soviet Union, Hungary, Austria, Germany, Balkan countries, Algeria, Tunisia, Morocco, Chile, Argentina, South Africa, Australia, New Zealand and the USA. Grape productivity in these countries is expressed in hectolitres of wine rather than kilograms of grapes per acre, considering that most of the grapes will be processed for wine. In Anatolia (east of Turkey), the place where wine is born, wine is only processed at a very low rate of 2 per cent because grapes are given preference as table and dried grapes. The regions between 32 and 50 latitudes degrees of northern and southern hemispheres of the world are the most suitable places for viticulture. Anatolia, with its location in 34-42 latitude, of the northern hemisphere is one of the most suitable geographical positions to grow the vineyard (Fig. 1).

Wine production begins in the vineyard. A good winemaker should be from a good vineyard at the same time. Grapes create the wine when changed by fermentation and shaped during ageing. For this

E-mail: hatice.kalkan.yildirim@ege.edu.tr

reason, the winemaker must know the characteristics of each compounds taking place in the composition of grapes and respectively of wines.

The development of instrumental techniques, such as chromatography (thin-layer: TLC, gas: GC, liquid: HPLC), and spectroscopy (infrared: IRS, nuclear magnetic resonance: NMR) allow detection and evaluation of most compounds found in wines. Even with these useful techniques, the knowledge concerning possible changes in wine compounds during production could provide a real-time assessment. Generally, average concentrations of the major components of wine are water, 86 per cent; ethanol, 12 per cent; glycerol and polysaccharides or other trace elements, 1 per cent; different types of organic acids, 0.5 per cemt, and volatile compounds, 0.5 per cent (Sumbly *et al.*, 2010).

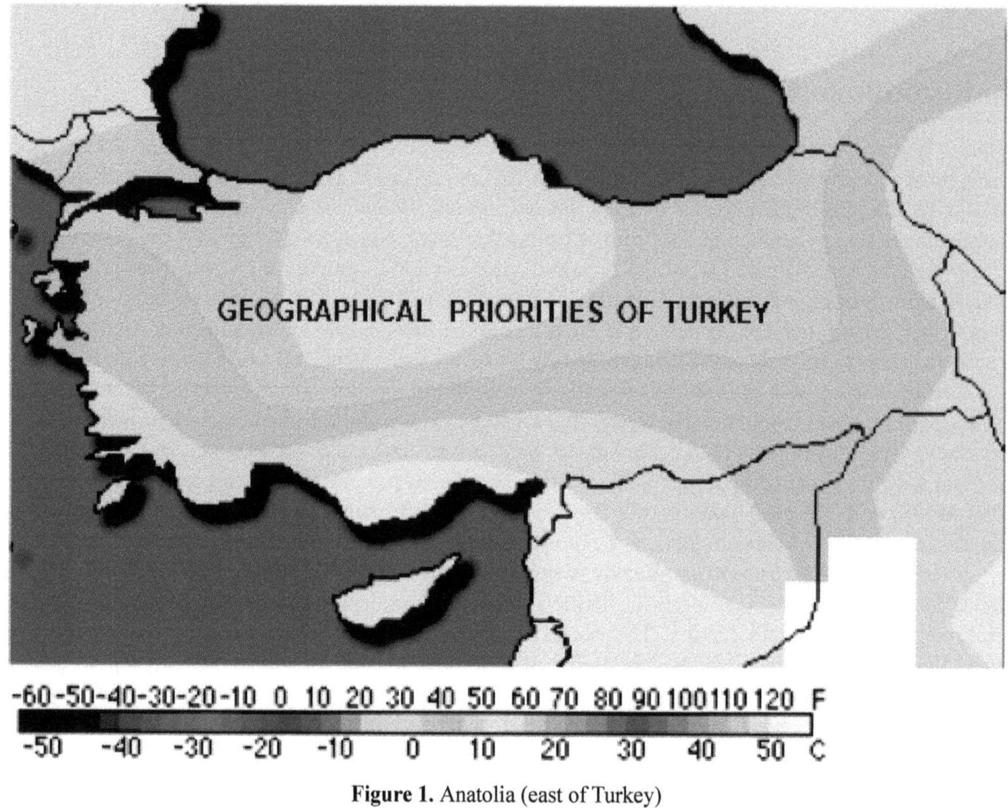


**Figure 1.** Anatolia (east of Turkey)

Color version at the end of the book

## 2. General Composition of Wine

According to the OIV, "wine is the beverage resulting exclusively from the partial or complete alcoholic fermentation of fresh grapes, whether crushed or not or of grape must. Its actual alcohol content shall not be less than 8.5 per cent vol. (Wright, 1975). Wines can be classified into five main groups – red, rose, white, sparkling and fortified wine. Wine compounds in wines is depending on grapes varieties and wine production techniques (Fig. 2) (Liu *et al.*, 2008).

### 2.1. Main Wine Components

#### *2.1.1. Water*

The water content of wine depends on the water level of grapes, but it is affected by production steps, such as maceration and fermentation, also. It changes around 80-85 per cent of total wine volumes (Wansbrough *et al.*, 1998). Additionally, in some cases depending on alchohol values, this range could be found as 85-90 per cent (Aktan and Kalkan, 2000). Vine water status has implications on yield and

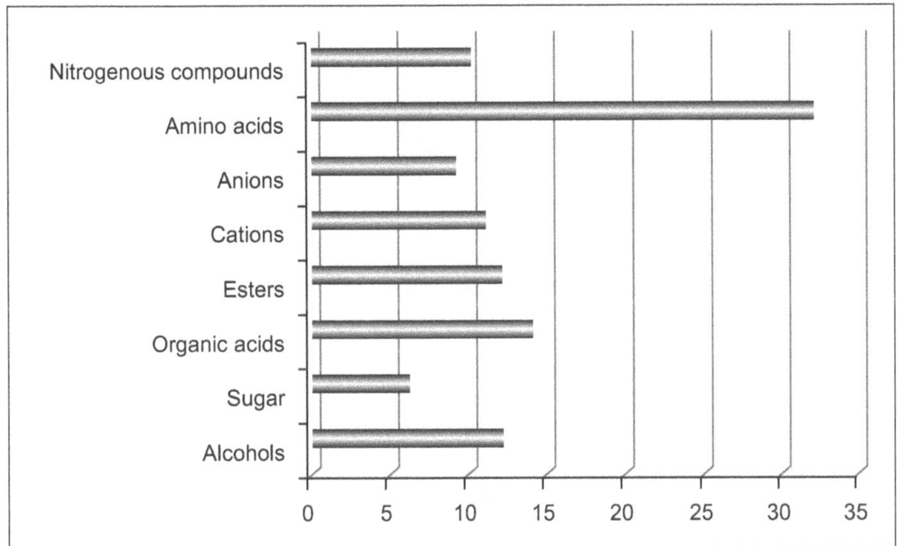


**Figure 2.** The number of wine compounds depending on grapes and production techniques

quality parameters and is therefore essential in the economics of vineyard management. Grape growers tend to focus on yields while grape buyers focus on quality parameters. Hence, an accurate and easy-to implement tool for assessing vine water status can clarify relations between grape growers and buyers in the wine industry (Leeuwen *et al.*, 2009).

### 2.1.2. Carbohydrates

Glucose and fructose are the main sugars in grapes juice (Fig. 3) and which are used by the yeast to produce alcohol and carbon dioxide. The pectins have no great importance in the juice itself, but if they are not broken down, they can create haziness in the wine. In wine, monomer sugars are around 2-2.7 per cent by weight fraction (Wansbrough *et al.*, 1998). Chardonnay and Pinot Blanc are classified as 'high fructose' varieties, while Chenin Blanc and Zinfandel are regarded as 'high glucose' varieties (Murli, 2010). Typically wine yeasts are capable of metabolising glucose and fructose. However, yeast produces enzymes that hydrolyses sucrose to its component glucose and fructose units, allowing it to be metabolised. This means that there is rarely more than trace amounts of sucrose left in the finished wine (with the exception of wines that have had sucrose deliberately added or where fermentation was stopped before completion). The pentose sugars (predominantly arabinose, xylose, ribose and rhamnose) are not metabolised by yeast and remain in the wine after fermentation at levels of 0.4-2 g/L. Thus in wine, the most significant sugars likely to be found are traces of unfermented glucose and fructose and small quantities of pentoses (Prenesti *et al.*, 2004).

### 2.1.3. Ethanol

Ethanol formation in wine is basically transformation (Fig. 4) of monomer sugars into carbon dioxide and ethanol by yeast strains for energy production. In wine, ethanol is present around 5.5-15.5 per cent by volume. Ethanol is metabolised in the liver and produces mainly acetaldehyde. The final product is acetic acid, which is either oxidised to provide energy or used by molecules, such as coenzyme A to create fatty acids, steroids and other useful biological molecules (Zakhari, 2006; https://pubs.niaaa.nih.gov/). Yeasts are chemoorganotrophic microfungi, which obtain their carbon and energy by metabolising organic substrates. Although glucose is commonly used as the sole carbon source for yeast growth in the laboratory, this sugar is usually not found freely in industrial fermentation media. In such an environment, the more common carbon sources are maltose (as for malt infusions), sucrose (such as those used in the production of molasses), lactose (such as whey-based drinks) and fructose (Walker and Stewart, 2016).

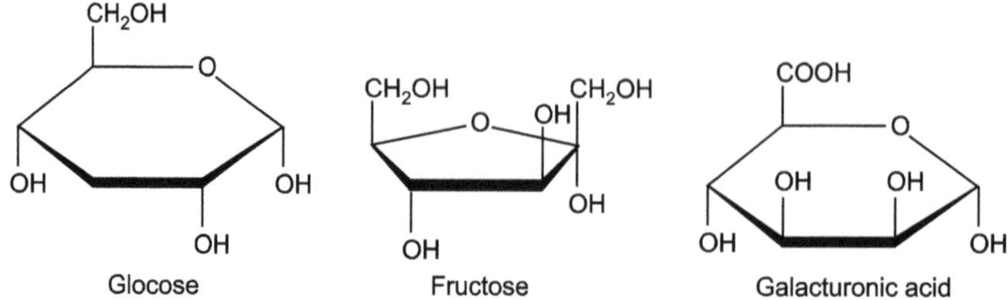


**Figure 3.** Sugars in wine (Wansbrough et al., 1998)

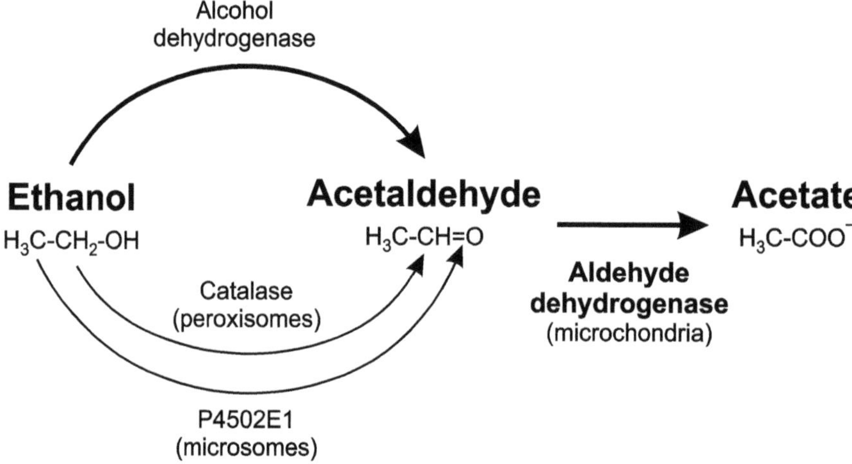


**Figure 4.** Metabolism of ethanol

### *2.1.4. Organic Acids*

Three main organic acids occur in grapes – malic, tartaric and citric acid. Yet there is one more acid in wine composition, named succinic acid, and it is formed from yeast metabolism. Studies show that wines made from grapes which are grown in cooler regions have more acidity than warmer regions. Additionally, acidity depends on cultivars which are basically grapes' acid level (Murli 2010; National Wine and Grape Industry Centre, 2017). Tartaric acid is most important in wine due to its prominent role in maintaining the chemical stability of wine and its colour and finally influencing the taste of the finished wine. The concentration varies depending on grape variety and the soil content of the vineyard. Some varieties, such as Palomino, are naturally disposed to high levels of tartaric acids, while Malbec and Pinot noir generally have lower levels (Kerem *et al.*, 2004). During flowering, high levels of tartaric acid are concentrated in the grape flowers and their young berries. As the vine progresses through ripening, tartaric does not get metabolised through respiration, like malic acid, so the levels of tartaric acid in the grape vines remain relatively consistent throughout the ripening process.

Malic acid is found in nearly every fruit and the berry plant but is most often associated with green (unripe) apples, the flavour it most readily projects in wine. Malic acid is involved in several processes which are essential for the health and sustainability of the vine (De *et al.*, 2011). Its chemical structure allows it to participate in enzymatic reactions that transport energy throughout the vine. Its concentration varies, depending on the grape variety, with some varieties, such as Barbera, Carignan and Sylvaner, being naturally disposed to high levels. Malic acid can be further reduced during the winemaking process through malo-lactic fermentation (MLF) (Palacios *et al.*, 2007). In this process, bacteria convert the stronger malic acid into the softer lactic acid, which is less acidic due to the fact that it exists in the monoprotic form. Thus, after MLF, wine has a higher pH (less acidic) and a different mouthfeel (Pickering *et al.*, 1999b). For some wines, the conversion of malic into lactic acid can be beneficial, especially if the wine has excessive levels of malic acid. For other wines, such as Chenin Blanc and Riesling, it produces off-flavours in the wine, such as the buttery smell of diacetyl that would not be appealing for that variety (Keller, 2010).

Lactic acid is produced during winemaking by lactic acid bacteria, which includes three genera – *Oenococcus, Pediococcus* and *Lactobacillus.* These bacteria convert both sugar and malic acid into lactic acid, the latter through MLF. This process can be beneficial for some wines, adding complexity and softening the harshness of malic acidity, but it can generate off-flavours and turbidity in others. While very common in citrus fruits, such as limes, citric acid is found only in very minute quantities in wine grapes. The citric acid most commonly found in wine is due to sucrose fermentation in grape juice. Also, it can be added into wine to boost total acidity. The acetic acid is most volatile of the primary acids associated with wine and is responsible for the sour taste of vinegar. During fermentation, a small amount of acetic acid is produced. If the wine is exposed to oxygen, *Acetobacter* bacteria convert ethanol into acetic acid. This process is known as the 'acetification' of wine and is the primary process behind wine degradation into vinegar. An excessive amount of acetic acid is also considered a wine fault. Succinic acid is most commonly found in wine, but can also be present in trace amounts in ripened grapes. While concentration varies among grape varieties, it is usually found in higher levels in red wine grapes. The acid is created as a byproduct of the metabolisation of nitrogen by yeast cells during fermentation. The combination of a succinic acid with one molecule of ethanol creates the ester monoethyl succinate responsible for a mild, fruity aroma in wines (Robinson, 2006).

### *2.1.5. Phenolic Compounds*

Flavan-3-ols, such as catechin, forms the most plentiful class and contains simple monomeric catechins, but most exist in oligomeric and polymeric proanthocyanid in forms which are derived from the skins and seeds of grapes. Flavonols, like quercetin, found in the berry skin, appear to function as sunscreen and their amount increases by being exposed to high sunlight. Anthocyanins, such as malvidin-3-glucoside, are the red-coloured phenols. The anthocyanins are converted to other coloured forms as the wine ages. The major flavonoid constituents of red wine are presented in Table 1 and the general composition of wine is presented in Table 2.

**Table 1.** The Major Flavonoid Constituents of Red Wine

| **Flavonoid** | **Quantities** |
|---|---|
| Catechin | 190 mg/L |
| Gallic Acid | 95 mg/L |
| Epicatechin | 82 mg/L |
| Malvidin 3-Glucoside | 24 mg/L |
| Rutin | 9 mg/L |
| Myricetin | 8 mg/L |
| Quercetin | 8 mg/L |
| Caffeic Acid | 7 mg/L |
| Cyanidin | 3 mg/L |
| Resveratrol | 1.5 mg/L |

Important stilbenoids are resveratrol and phenolic acids, such as benzoic and caffeic acid. Hydroxycinnamates are potent antioxidants but have no sensory impact, except that when oxidised, they may form brown pigments that eventually precipitate.

Tannat, Cabernet Sauvignon and Merlot wines are analysed for their phenolic richness, extractable anthocyanins contents and total potential. Tannat grapes present anthocyanins and total polyphenols contents significantly higher in both the studied years. Wines from this variety present colour intensity and phenolic contents higher than Cabernet Sauvignon and Merlot. The relationship between the phenolic contents of grapes, skins, musts and wines is very significant. Colour intensity and phenolic contents of

**Table 2.** Composition of Wine

| *Wine composition* | *Concentration (g/L)* |
|---|---|
| Water | 600-850 |
| Ethyl alcohol | 85-130 |
| Sugar | 1-230 |
| Non-sugar extract | 15-40 |
| Glycerine | 5-15 |
| Butylene glycol | 0.1-0.7 |
| Fusel oils | 0.1-0.15 |
| Tartaric and malic add | 3.0-7.0 |
| Lactic acid | 0.2-3.0 |
| Volatile acids (in terms of acetic acid) | 0.1-1.6 |
| Mineral compounds | 1.3-6.0 |
| Nitrogenous compounds | 0-0.9 |
| **Phenolic substances and pigments** | |
| For white wine | 0.1-1.0 |
| For red wine | 1.0-4.5 |
| Aroma components | 1.0-2.0 |
| Acetaldehyde | 0-0.15 |
| **Vitamins** | |
| For white wine | 0.04-0.1 |
| For red wine | Up to 0.2 |
| Sulphuric acid | 0.02-0.4 |

wines are highly correlated with the total polyphenols of the grapes and with anthocyanins of grapes, skins, musts and wines (González-Neves *et al.*, 2004). Another study performed with 24 Cabernet Sauvignon and 7 Merlot wines demonstrated the importance of vintage time and cultivar type on chemical and sensory properties of wine. The study indicated that phenolic compounds, tannins, anthocyanins, hue, astringency appear as relevant criteria for vintage differentiation for both cultivars (Chira *et al.*, 2011). Wines prepared from berries (Table 3), such as elderberry, blueberry, raspberry and cranberry have a greater antioxidant capacity than wine prepared from pome fruits, such as apple and pears (Rupasinghe and Clegg, 2006).

### *2.1.6. Aroma Compounds*

Wine has more than 1,000 aromatic compounds. The diversity of aromatic compounds in wine is immense and ranges in concentration from several mg $L^{-1}$ to a few ng $L^{-1}$. Wine flavour can be divided into classes – varietal aroma, typical of grape variety; pre-fermentative aroma, originated during grape processing; fermentative aroma, produced by yeast and bacteria during alcoholic and malo-lactic fermentations (MLFs) and post-fermentative aroma. The perception of wine flavour is the result of a multitude of interactions among a large number of chemical compounds and sensory receptors. Higher alcohols, acids and esters are quantitatively dominant in wine aroma and are important for sensory properties of wine (Stashenko *et al.*, 1992). Small amounts of higher alcohols contribute positively to wine quality, while excessive amounts of some compounds may detract wine quality (Rapp, 1986). Esters contribute to wine odour and relative concentrations of fatty acids give an appreciable strong odour character to wines (Gil *et al.*, 1996). The main compounds responsible for the most intense aromas in Sauvignon Blanc wines have been assumed to be methoxypyrazines and varietal thiols in the Marlborough region (Jouanneau *et*

**Table 3.** Phenolic Content and Antioxidant Capacity of Different Fruit Wines (Rupasinghe and Clegg, 2006)

| *Source of fruit wine* | *Sample size* | *Total antioxidant capacity (mg AAE/L)* | *Total phenolic content (mg GAE/L)* |
|---|---|---|---|
| Cabernet (grapes, red wine) | 6 | 2447 | 2005 |
| Elderberry | 5 | 1911 | 1753 |
| Blueberry | 6 | 1655 | 1676 |
| Black Currant | 6 | 1595 | 1509 |
| Cherry | 6 | 1102 | 991 |
| Raspberry | 6 | 1067 | 977 |
| Cranberry | 6 | 875 | 971 |
| Plum | 4 | 618 | 555 |
| Apple | 6 | 404 | 451 |
| Peach | 5 | 395 | 418 |
| Icewine (grapes) | 5 | 383 | 493 |
| Chardonnay (grapes, white wine) | 6 | 276 | 287 |
| Pear | 3 | 271 | 310 |
| Riesling (grapes) | 6 | 219 | 250 |
| **Mean** | | **944** | **903** |

*al.*, 2012). In Çalkarası rosé wines, the wine is dominated by fresh fruit, floral and red fruit characters. The aroma shows a complex profile with 28 compounds determined above their odour threshold (Darici *et al.*, 2014). Different studies show that the levels of volatile compounds are biochemically related to the yeast amino acid metabolism. Levels of isobutyric and isovaleric acids and their ethyl esters, isobutanol, isoamyl alcohol, â-phenylethanol, methionol and isoamyl and phenylethyl acetates are found to differ according to the variety of grape used in Spanish red wines (Murli, 2010; Wansbrough *et al.*, 1998).

### 2.1.7. Nitrogenous Compounds

The nitrogen content of the grape varies with variety, climate, soil, fertilisation and other cultural practices. The total nitrogen concentration of the fruit increases during the maturation period. Nitrogen influences biomass formation (cell population or cell yield), the rate of fermentation and production of various byproducts, which in turn affect the sensory attributes of wine (Murli, 2010; Wansbrough *et al.*, 1998). Quantitatively, next to sugars, nitrogenous compounds are the most important nutrient substances found in grape must. Ammonia, which exists as ammonium ($NH_4^+$) ions in must, and amino acids are the predominant nitrogen-containing compounds that are utilised by yeast. On a dry weight basis, about 10 per cent of yeast weight consists of nitrogen. All the nitrogen used in building cellular material (population 108 cells/ml) during fermentation is taken up from the must. It is therefore important that the must contains sufficient amounts of nitrogen to support a healthy yeast population during fermentation (Martin *et al.*, 2003).

### 2.1.8. Minerals

Minerals make up 0.4 per cent of the weight of grapes. The most important minerals are magnesium and potassium, which are important in fermentation and phosphate necessary for yeast growth. During ripening, the potassium content of the grape increases. Its movement into fruit leads to the formation of potassium bitartrate, which reduces the acidity and increases the juice pH. It should be noted that the tartaric acid salt of potassium is involved in wine instability problems (Murli, 2010; Rupasinge and Clegg, 2006; ,Wansbrough *et al.*, 1998). Each glass of red wine gives on average the following daily nutritional needs – 1 per cent vitamin K, 1 per cent thiamin, 2 per cent niacin, and 3 per cent riboflavin.

Each glass of wine gives close to the following percentages of one's daily adult requirement of minerals – 1 per cent calcium, 1 per cent copper, 1 per cent zinc, 3 per cent phosphorus, 4 per cent iron, 4 per cent magnesium, 5 per cent potassium and 10 per cent manganese. White wine is lower in carbohydrates, with only 2.6 carbohydrates on an average per serving. White wine provides of daily nutritional needs 3 per cent magnesium, 3 per cent vitamin $B_6$, 3 per cent vitamin $B_2$ and 3 per cent niacin, 1 per cent riboflavin along with trace elements of iron, calcium, potassium, phosphorus and zinc (Mouret *et al.*, 2004).

## 2.2. Changes of Wine Composition Depending on Different Factors

### *2.2.1. Effects of Grape Varieties*

In a study investigating the phenolic composition of commercial red wines in Brazil and Argentina, eight phenolic compounds were identified using the HPLC method. This work includes red wines from the São Francisco Valley (Pernambuco-Brazil) (32 bottles samples of different vintages between 2008 and 2015) and a sample of Cabernet Sauvignon (from Grande do Sul), four samples of Cabernet Sauvignon: 2013 (2) and 2014 (2); four samples of Syrah: 2012 (1), 2013 (2) and 2014 (1); one sample of Rubi Cabernet (2012) and one sample of Petite Syrah (2012) having different values of phenols, depending on grapes types and regions (Belmiro, Pereira and Paim, 2017). Another study concerning the mineral content of red wines (Nero d'Avola and Syrah cultivars of *Vitis vinifera*) emphasised the importance of grape varieties and regions for their contents like phenols (Potortí *et al.*, 2017). A study was performed with the purpose of evaluation of organic acid content, CIE $L^*C^*_{ab}h_{ab}$ chromaticity coordinates, phenolic compound contents (with spectrophotometric assays and HPLC-DAD) and nitrogen compounds (with HPLC-FLD) in 14 samples (vintage, 2013). Carignano wines showed a significant level of phenolic compounds (2023 ± 435 mg GAE/L) ($p < 0.001$). The nitrogen compounds found in samples were mainly amino acids and the content of essential amino acids was 61.4 ± 22.5 mg/L levels (Tuberoso *et al.*, 2017). The protein and polysaccharide contents differ in amounts in different wines. The protein content ranged between 212-253 mg/L. Polysaccharide and protein contents evolved differently throughout the winemaking process: a decrease of 18 per cent for polysaccharides and an increase of 19 per cent for proteins (Coelho *et al.*, 2017). Sparkling wines produced via Methode Traditionelle transfer and Charmat methods had similar free amino acid concentrations, on an average of 931-976 mg/L. In carbonated wines, amino acid levels ranged between 471-1924 mg/L. Ethyl hexanoate was approximately threefold more abundant than hexanoic acid for all the production methods. Decanoic acid was approximately twice as abundant as ethyl decanoate; ester to acid ratios were 0.50 for carbonation, 0.52 for Charmat, 0.47 for transfer, and 0.45 for Methode Traditionelle. Ethyl octanoate was approximately 10-15 per cent more abundant than octanoic acid (Culbert *et al.*, 2017). Higher concentrations of 1-octen-3-ol were noticed in selected wines that may be connected with the metabolic activity of *Botrytis cinerea*. Linalool, β-damascenone, α-ionone, β-ionone, γ-nonalactone, *cis*-rose oxide and 1-octen-3-ol had the greatest contribution to the overall aroma composition (odour activity value OAV >1). According to the results, linalool was the most abundant component ranging between 97.61-309.82 μg/L in ice wine samples. Norisoprenoids, such as β-damascenone and β-ionone, were also determined and ranged between 3.02-8.03 μg/L (Maslov *et al.*, 2017).

### *2.2.2. Effects of Wine Production Steps*

The combined effect of rainfed and cluster-thinning treatments increased the majority of individual aromatic compounds quantified in Tempranillo wines and also showed the highest total odour activity value (Talaverano *et al.*, 2017). Dehydrated wines samples contained greater amounts of volatile compounds. Because of fast dehydration, main detected compounds were free esters, linalool, rose oxide, citronellol and geraniol. The predominant glycosylated compounds were geraniol (43.1-45.0 per cent) and nerol (28.6-33.6 per cent), followed by 1-hexanol (7.5-8.4 per cent) and linalool (5.1-7.5 per cent) (Urcan *et al.*, 2017).

The malo-lactic fermentation (MLF) in barrels led to lower total proanthocyanidin (~5.4 g/L wine) and total anthocyanin (~288.0 mg Mlv/L wine) contents as compared to wines with MLF in the tank (~5.8 g/L wine, ~342.3 mg Mlv/L wine) ($p < 0.05$) (González-Centeno *et al.*, 2017).

Using high pressure during alcoholic fermentation, fermentation times are getting longer and at the end of the process, wine contains more residual sugars. Increases in temperature result in shorter fermentation time, higher ethanol productivity, but at higher temperatures concentration of some volatile compounds decreases (Galanakis *et al.*, 2012; Torija *et al.*, 2002).

Low-temperature alcoholic fermentations are becoming more frequent due to the desire to produce wines with more pronounced aromatic profiles. However, their biggest drawback is the high risk of stuck and sluggish fermentations. To see the effects of fermentation temperature and *Saccharomyces* species on the cell fatty acid composition and the presence of volatile compounds in wine, experiments were conducted. Tests were done with two strains of *Saccharomyces cerevisiae* and one strain of *Saccharomyces bayanus* and their performance at low temperatures 13°C taking 25°C as a reference were evaluated. Concentrations of volatile compounds, such as acetates, ethyls, 2-phenylethanol, hexanoic acid octanoic acid were higher in wines produced at lower temperatures, depending on used strains (Torija *et al.*, 2002).

Different maceration techniques affect the wine's chemical properties. For observing these changes, experiments were carried out at different temperatures, heating times and heating techniques. Microwave and thermal maceration resulted in higher alcohol content but lower aroma and colour compounds, while classical maceration had lower alcohol and higher colour and aroma compounds. During classical maceration, there was higher yeast activity due to lower temperature; thus more second metabolites (aroma compounds) were created by yeasts (Moldovan *et al.*, 2015).

The inoculation time affected colour, aroma, phenolic and volatile compounds of wines. Using co-inoculation, quick degradation of malic acid occurred. Additionally, co-inoculation acetate ester concentrations decreased while ethyl lactate concentrations increased. Ethyl esters of butanoate, hexanoate, decanoate and dodecanoate did not exhibit MLF-dependent concentration changes. Ethyl esters of propanoate, 2-methyl propanoate, 2-methyl butanoate and 3-methyl butanoate increased in concentration following MLF inoculation during alcohol fermentation. There was an overall decrease in wine colour densities with every MLF treatment compared to no-MLF. However, there were no significant differences in colour densities across MLF inoculation regimes. In wines obtained by co-inoculation treatment, concentrations of anthocyanins did not decrease as much as the other MLF inoculation regimes (Abrahamse and Bartowsky, 2012; Bauer and Dicks, 2004).

Chemical compositions of Tannat wines obtained by different treatment showed the expected differences in acidity, pH, malic and lactic acids contents. Esters, isoamyl, isobutyl and 2-phenylethyl acetate showed different behaviours related to the used strain during fermentations (Boido *et al.*, 2009; Soden *et al.*, 2000). Anthocyanin concentration generally decreased in both Cabernet Sauvignon and Merlot wine as storage time increased and small polymeric pigments and large polymeric pigments were formed. The anthocyanins were transformed into more stable oligomeric and polymeric pigments at the stage of maturation and ageing. According to the data, differences in anthocyanin levels was a function of storage temperature (Villamor *et al.*, 2009). A significant decrease of phenols was detected during storage, which results in a change in the colour of the wine from pale yellow to yellow-brown. However, it can be noted that their loss was significantly higher in the wine subjected to variable temperatures than in the one stored at a constant temperature after 12 months (Recalmes *et al.*, 2005). Although fermentation occurs under anaerobic conditions, oxidative processes, both enzymatic and non-enzymatic processes may play a significant role in phenolic composition changes. This is accompanied by the oligomerisation of the original phenolic compounds (Kallithraka *et al.*, 2008). Results showed that the contents of most of the phenols diminished with time, while the antioxidant activity increased with storage, depending on wine type, grape variety and wine quality, whereas reducing power remained significantly unaffected (Kallithraka *et al.*, 2007).

## 3. Nutritional Value of Wine

### 3.1. Mineral Compounds Present in Wine

Essential mineral compounds within wines are shown in Table 4.

**Potassium:** It constitutes 40 per cent of wine ash. This mineral provides the effective functionality to muscles, especially cardiac muscles. It is effective in impulses of muscles, nerves and glands.

**Table 4.** Essential Minerals Present in Wine

| *Mineral* | *Daily needs* | *The amount of mineral compounds in 1 L wine* |
|---|---|---|
| Potassium | 2-3 g | 0.5-5 g |
| Calcium | 0.8 g | 0.1-0.5 g |
| Magnesium | 0.3 g | 0.1-0.25 g |
| Sodium | 5 g | 0.01-2.3 g |
| Iron | 5.5 mg | 1-6 mg |
| Manganese | 3 mg | 1-3 mg |
| Zinc | 5-10 mg | 0.5-8.5 mg |
| Phosphorus | 1-2 g | 0.15-0.4 g |
| Fluorine | 1 mg | 0.04-1.75 mg |
| Copper | 1-2 mg | 1 mg and more |
| Iodine | 0.15-0.2 mg | 0.1-0.2 mg |
| Cobalt | 1-2 µg | 0.15-12 µg |

**Calcium:** In man 98 per cent of calcium is found in the skeleton and teeth combined with phosphorus as a building block of the skeleton. Calcium ions take part in neural system reactions and blood coagulation. It improves resistance to infections by fortifying the cell membrane.

**Magnesium:** 70 per cent of magnesium found in human body is present in teeth and bones as an important constituent. It is responsible for activities of a large number of enzymes. It helps to remove excess calcium from the body. In the case of magnesium deficiency, contraction in muscles, a disorder in the gastrointestinal tract, heart throbbing and signs of fatigue are noticeably main effects.

**Sodium:** It plays an important role in the production of urine and sweat. In the case of sodium deficiency, blood circulation disorders, dizziness and contraction in muscles is often seen.

**Iron:** It is an important factor in the formation of erythrocyte and leukocyte and is a building block of blood pigments. Since it has oxygen-carrying abilities, it takes place in cell respiration together with other heavy metals (manganese, zinc, etc.) and trace elements (nickel, cobalt, fluorine, copper, bromine, aluminium, etc). In its deficiency, headache, dizziness, prostration, changes in nail structure and gastroenterological disorders are observed.

**Manganese:** It is an important mineral taking place in exchange of carbohydrates in nerve cells and hematogenesis. It helps to remove toxins produced by the illness-causing microorganism. It is a building block of several enzymes and endocrine glands.

**Zinc:** It is an indispensable mineral taking place in blood sugar regulation. A number of enzymes need zinc for their activity.

**Phosphorus:** It is required for the formation of vital compounds, as nucleic acid, lecithin, nucleoprotein, phosphatide and nucleotide. It also plays a main role in the formation of the skeleton together with calcium.

**Trace elements:** Trace elements are required for the formation, maintenance and functions of cells, enzymes, hormones and vitamins. In wines, aluminium, fluorine, copper, vanadium, cobalt, molibden, iodine and bromine are found. In the case of lack of these compounds, the hormones, enzymes and vitamins are not able to perform their functions and some disorders in an organism occurs. Trace elements increase their activity together with proteins in the stomach and small intestine. For this reason, drinking of wine is suggested at a meal or after a meal.

**Aluminium** is found 0.5-1 mg/L in wine and researchers regard its physiological importance in the human body.

**Fluorine**: It is especially effective against tooth decay.

**Vanadium:** It inhibits cholesterol formation in body but accelerates the cholesterol break down in blood vessels. In wines, it is found in the range of 0.06-0.26 mg/L.

**Iodine:** It provides the functionality of the thyroid hormone (thyroxinin).

**Cobalt:** It takes place in the formation of $B_{12}$ vitamin, which is absolutely necessary for blood formation. Like copper it plays an important role in the absorption of iron in the blood structure.

**Molibden**: It takes place as a primary building block of some enzymes. Additionally, it increases teeth resistance against caries. Its quantity in wine is in the range of 0.008-0.15 mg/L.

## 3.2. Phenolic Compounds Present in Wine

Among the main components found in wine are alcohol, glycerine, aldehydes, esters, acids, volatile acids, nitrogen compounds, aromatic and phenolic substances. One of the most important groups are phenolic substances that impart specific characteristic to the wine. Phenolic compounds that are found in wine are important for human health due to their biological benefits (antioxidant, antimicrobial, anti-inflammatory). Red wines contain much more phenolic compounds than white wines.

The composition of phenolic compounds in the grape is given in Fig. 5.

Phenolic compounds found in grape are described as secondary metabolites of byproducts of carbohydrate metabolism. The formation of phenolic compounds starts from a single sugar unit, followed by erythrose-4-phosphate obtained as a result of phosphoenol-pyruvic acid condensation reactions during the pentose-phosphate cycle. A condensation reaction of 3 acetyl-Co-A molecules formed during the Krebs cycle causes the formation of a benzene ring. Condensation of the second benzene ring with cinnamic acid leads to the formation of flavonoids. The molecules of these compounds contain two benzene rings.

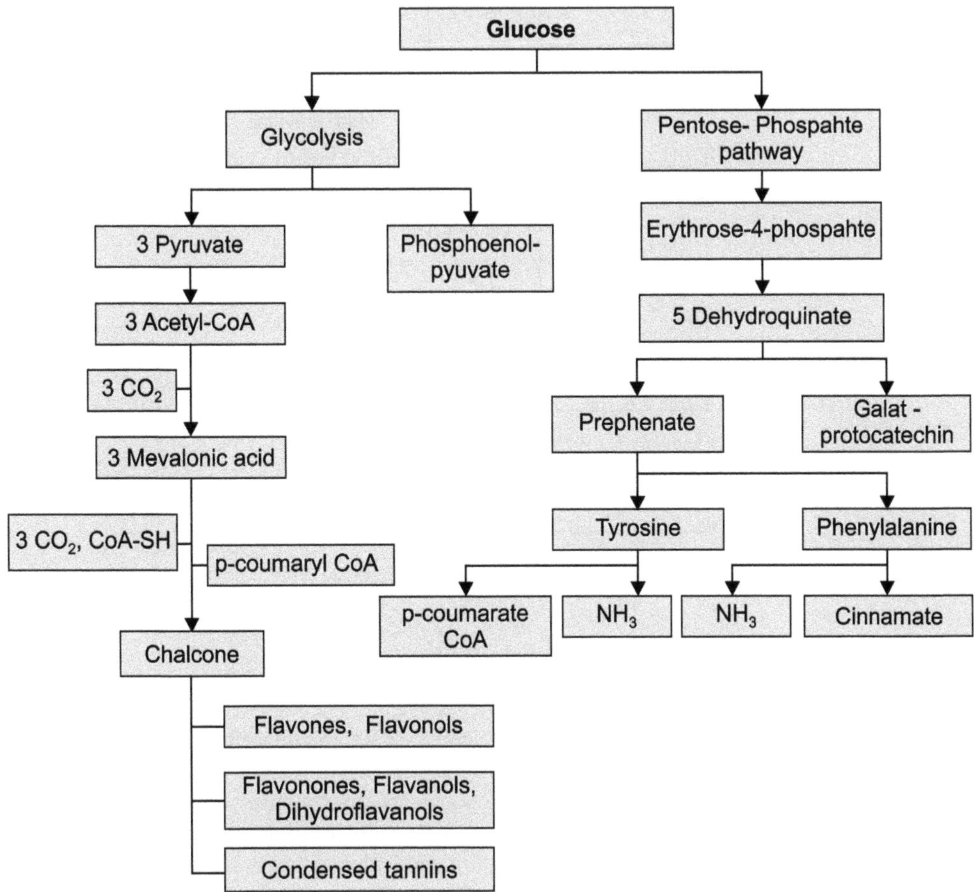


**Figure 5.** Composition of phenolic compounds in grape

After this step, the formation of different phenolic compounds occurs by different chemical changes (hydroxylation, methoxylation, and esterification). The phenylalanine ammonia provides generation of cinnamic acid and other phenols by eliminating $NH_3$ radical.

The phenolic compounds found in wines are divided into two groups – flavonoids and non-flavonoids compounds. Phenolic compounds containing a hydroxyl (-OH) group attached to a benzene ring in their structure are present in cells forming part of the grape, such as crust, seed and stem. The flavonoids are with $C_6$-$C_3$-$C_6$diphenylpropane structure and the triple carbon bridge between the phenyl groups forms a ring with oxygen. The differences among different flavonoids depend on a number of binding hydroxyl groups, the degree of unsaturation and the oxidation level of the triple carbon segment. Phenolic compounds found in wine are given in Fig. 6.

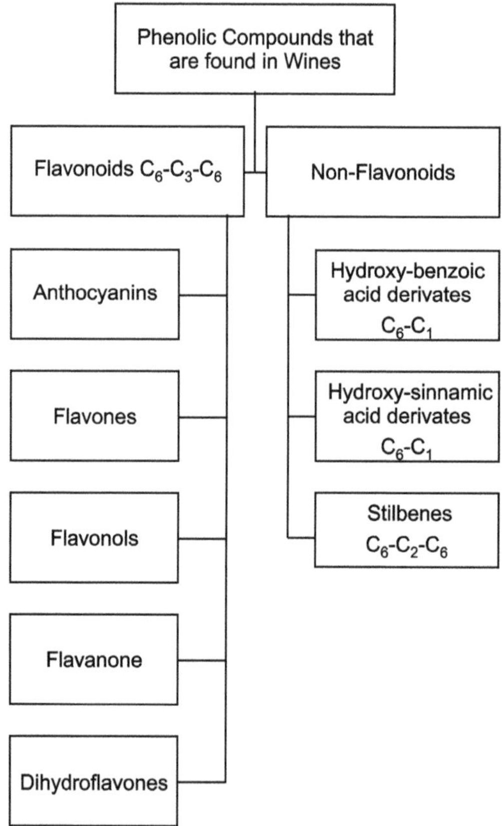


**Figure 6.** Phenolic compounds found in wine

## Flavonoids

### *Anthocyanins*

Anthocyanins are grouped under flavonoids. Anthocyanins are water-soluble and glycosidic compounds. They contain sugars and non-sugars constituents. Sugars are linked by the hydroxyl group of the third carbon atom. Anthocyanins are classified according to sugar molecules number. Single-sided anthocyanins (monoglycosides) contain only one sugar molecule in the third position; double-sided anthocyanins (diglycosides) contain two sugar molecules in the third and fifth position or third and seventh position. Diglycosides are unique for some species of *Vitis* (*V. riparia, V. rupestris*) and are not found in *V. vinifera* fruits. This difference is important for determination of grapes and wine origins. The anthocyanins structure of wines is presented in Fig. 7.

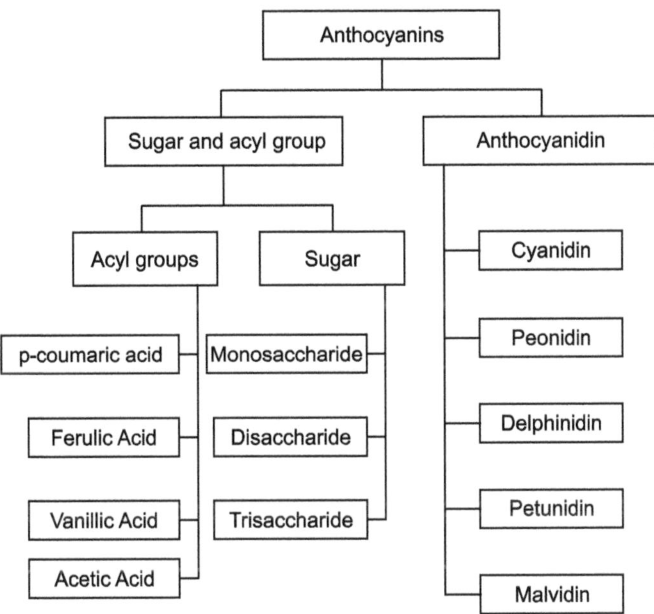


**Figure 7.** Structures of anthocyanins in wine

### *Flavones*

Another important group under flavonoids is flavanols. The most common compounds in this group are monomers containing one (OH) group in the $C_3$ attachment. They have four isomers due to two asymmetric carbon atoms present in their structure. Catechins generate proanthocyanidin as a result of condensation of oligomers and polymers obtained after their chemical and enzymatic reaction with oxygen.

### *Flavones, Flavanones and Dihydroflavones*

The other important groups of flavonoids are flavonols, flavonon and dihydroflavon. The main difference between flavones and flavonols is the absence of OH group in the middle ring, but the presence of H molecule.

## **Non-flavonoids Phenolic Compounds**

Non-flavonoid group includes mainly cinnamic acid, benzoic acid and their derivatives. The compounds are found in the form of their esters and glycosides as well as in their free forms. The main representative compounds of this group are cinnamic acid and its derivatives: o-coumaric acid, p-coumaric acid, caffeic acid, ferulic acid, isoferric acid and sinapic acid; and benzoic acid and its derivatives: m-hydroxybenzoic acid, p-hydroxybenzoic acid, gallic acid, vanillic acid, isovanillic acid, and syringic acid.

Resveratrol is mainly found in grapes and red wine. Resveratrol provides a wide range of benefits, including cardiovascular protective, antiplatelet, antioxidant, anti-inflammatory, blood glucose-lowering and anticancer activities; hence it exhibits a complex mode of action. Application of phytochemical substances, such as resveratrol, in therapy for malignant diseases in combination with conventional chemotherapeutic preparations can open new perspectives in this field. Resveratrol has also been entitled as a natural therapeutic agent with pharmacological potential in various neurodegenerative impairments, including Alzheimer's, Huntington's, Parkinson's diseases, amyotrophic lateral sclerosis and alcohol-induced neurodegenerative disorder (Drajen *et al.*, 2015; Kurvietien *et al.*, 2016; Tsai *et al.*, 2017).

### *3.2.1. Changes of Wine Phenols Depending on Different Factors*

The concentrations of phenolic compounds in wines change depending on grape variety, grape growing

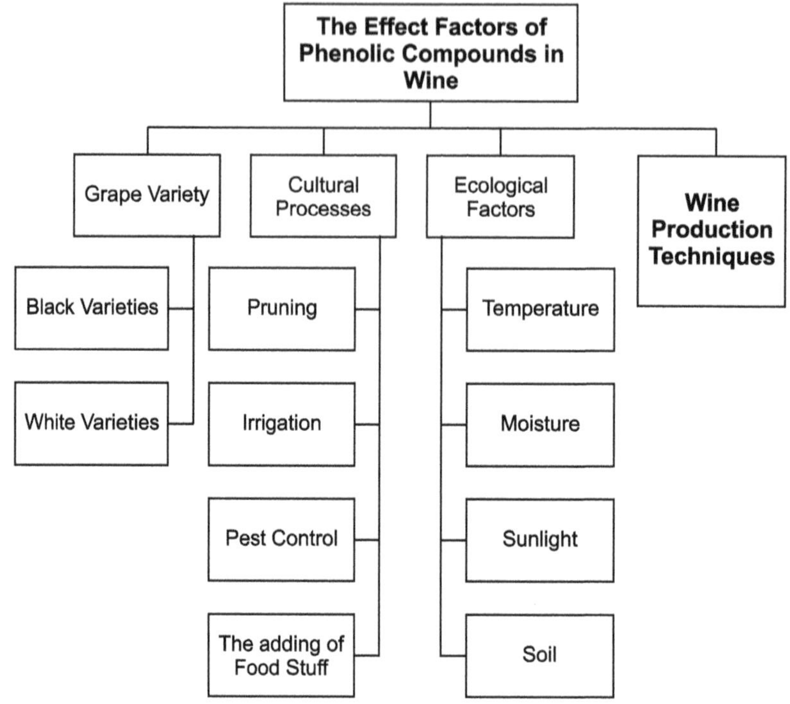


**Figure 8.** Factors affecting phenolic compounds in wine

and applied processes during wine production (Fig. 8). The amount for the non-flavonoid group is about 200 mg/L GAE and flavonoid group is about 1200 mg/L GAE. The concentration of anthocyanins, condensed tannins, other flavonoids and flavonols was determined as 150 mg/L GAE, 750 mg/L GAE, 250 mg/L GAE and 50 mg/L GAE, respectively. As the distribution within the fruit itself is different: fruit (1 per cent), fruit juice (5 per cent), skin (50 per cent), seed (55 per cent), the number and quantities of phenolic compounds vary, depending on grape varieties. Different phenolic compounds are located in different parts of the fruit. Anthocyanins are found in the husk, especially in hypodermal cells. Monomeric flavanols, such as catechins and polymeric tannins, are found in the stone, skin and in a very small amounts in the pulp. The cinnamic acid and its derivatives are mainly found in pulp and stone. Only in Alicante-Bouschet grapes, these compounds are present in a high amount in fruit flesh. Considering the distribution of these compounds in fruit, the importance of production techniques for extraction of phenolic compounds, which are useful for human health, cannot be denied.

While anthocyanins and polymeric pigments play a role in the colour formation of red wines, flavonols and polymers (tannins) are responsible for sensory properties, such as bitterness and bitterness of wines. The extraction of anthocyanins is very rapid in the initial stages of fermentation and reaches its maximum level in the first days. After this stage, their concentration drops. The reason for this is the beginning the formation of polymeric compounds. The effect of process and parameters on phenolic compounds during wine production is given in Fig. 9. Harvest in vineyard can be carried out at different brixes (percentage), pH values and acidity (g/L tartaric acid), depending on the grape variety and the intended wine to be produced. *V. vinifera* and *L. vidal* grapes were harvested at three different Brix values (17.5, 22.8, 37.2) and dry, semi-dry and icewine types were produced in order to determine the effect of different harvest times on phenolic compounds. According to results, the highest level of total phenol and phenolic acids content was detected in wines produced with the latest harvest date. Comparing the wine types, similar tendencies were observed. The pressing process is carried out at different periods, depending on the targeted wine types (red/white/rose). Traditionally, during the production of white wine from white grapes, the pressing process is carried out immediately after the crushing procedure. In the case of red wine produced from red grapes, pressing takes place after must fermentation. The number of phenols and

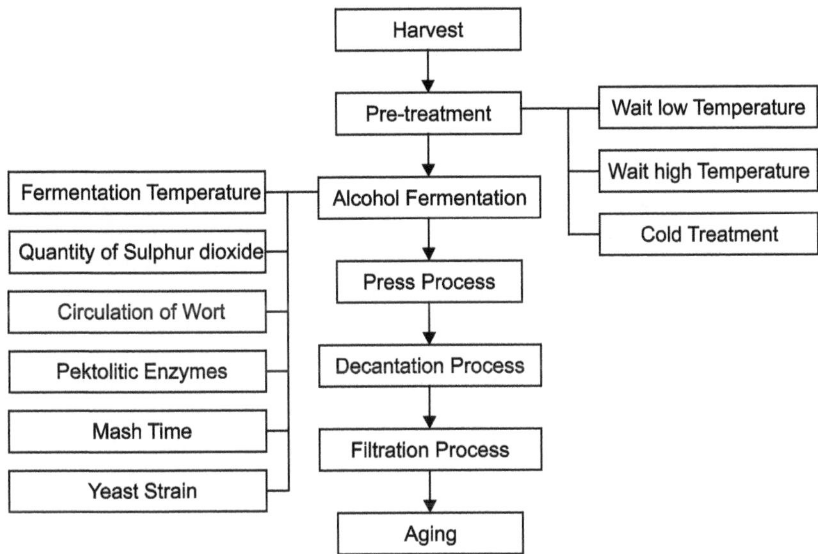


**Figure 9.** Factors affecting phenolic compounds formation during wine production

other substances obtained after pressing varies, depending on the type of pressing machine, pressure force and duration used. In a study performed by using Koshu grape, the press operation at high pressure caused the release of compounds, such as caffeic acid, catechin and epicatechin. In a study using white grapes, two different techniques were tried – pressing before and after the must fermentation (by hydraulic press). Results demonstrated that pressing after must fermentation caused more phenol (25-45 per cent) extraction. Grapes could be subjected to different pretreatments before fermentation. One of this treatment is cooling (a few days at 10-15°C). The phenol levels of wines produced from frozen grapes increased significantly. The cracks obtained in the cells lead to the breakdown of the cell membrane, which facilitates the release of phenolic compounds and especially of anthocyanins. In a study done with Merlot grapes, dry ice was used for the freezing process. The results demonstrated the production of wines with 50 per cent more tannins and 50 per cent more anthocyanins than control. Similar results were obtained with Cabernet Sauvignon and Cabernet Franc grapes. Another technique applied as a pretreatment was the preheating (60-70°C). Application of high temperature disintegrates hypodermal cells and facilitates the extraction of anthocyanins. Additionally, inactivation of polyphenol oxidase inhibits possible oxidation. The extraction of tannin does not take place due to the absence of ethanol in the medium at pretreatment stage. Application of hot treatment as pretreatment done during the study with Pinot noir grape demonstrated the increase of anthocyanin three times more than control ones (classical technique). One of the most important parameters during the fermentation stage is the applied fermentation temperature. It has been determined that high fermentation temperatures (>25°C) cause increase in the amount of extracted phenolic compounds. The effects of fermentation temperature on phenols were investigated by using Pinot Noir grape (15°C and 30°C). While the concentration of total phenols increased at higher temperatures, it was not the same in the case of anthocyanins concentration. The reason is that anthocyanins content in the first stage increases and their concentrations decrease proceeding fermentation. $SO_2$, added to wines as antimicrobial and antioxidant compounds, is still an indispensable additive used in wine production. Additionally, it causes quick ripening of the wines during the maturation process. At low fermentation temperatures (22°C) and high $SO_2$ (100 ppm) concentrations, increased extraction of phenolic compounds was reported. In a study conducted with white grapes (0; 75; 150 mg/L $SO_2$), the total phenols increased in the first stage but very little in later periods. In a study done with Cabernet Frank, two parallel samples were examined: containing 50 ppm $SO_2$ and sulphur dioxide-free samples. In sulphur dioxide-free wines, browning process begins in a short time. It has been determined that the amounts of other phenolic compounds except catechin vary, depending on the variety. The highest differences were detected for Pinot noir wines. Mechanical immersion and circulation processes were found to increase phenol contents by 100-200 per cent. Furthermore, it was found that mechanical submersion is more effective than circulation. The effects of manual immersion, circulation and rotating mixing tank techniques on the phenols were investigated by using Pinotage grapes. According to the results, the least values of total flavonoids, total tannins and anthocyanins were obtained by circulation technique. The order of maximum phenol extraction was as – rotating mixing scheme > manual immersion > circulation process. Pectolytic enzymes break down skin cell walls and allow extraction of wine pigments. The purity of used pectolytic enzymes is extremely important. If in the used enzyme remains beta-glucosidase, due to the anthocyanin hydrolysis, less stable pigments are obtained. It is well known that the usage of pectolytic enzymes increases the must yield. The effects of five different pectolytic enzyme preparations on the phenol amounts of Pinot Noir wines were investigated. The results demonstrated increase in tannin and polymeric pigments, and in anthocyanins. As the duration of must fermentation becomes longer, increase in the concentration of the tannin occurs, but changes in anthocyanins reverse. The maximum values of anthocyanins are reached on the fourth and fifth day of must fermentation. In the following periods, increases are at a very low level. It was demonstrated that the extended pomace fermentation leads to more anthocyanin formation in young wines and more tannin content after ageing. In a study conducted with Cabernet Sauvignon wine, changes in phenol contents during 7th, 13th and 21st days of must fermentation were investigated. Prolonged fermentation caused increase in the concentration of gallic acid, flavanols and total phenols. Total phenols and flavanols reached their maximum values on 13th and 21st days, respectively. Polymeric pigments were found to be present in large quantities in samples (Cabernet Sauvignon wines) that were subjected to 21 months of ageing after a long must fermentation (44 days). In a study done with Merlot wines, during 36 days of must fermentation, slow increase in total phenols up to 10 days was observed. The increase in polymeric pigments emerged within four days and afterwards began to decrease rapidly. According to the results of studies, prolonged mature fermentation time (>7) causes an increase in tannins and polymeric pigments. The selection of yeast strain is important for the phenolic content of red wines. Some studies demonstrate that dead yeast cells adsorb anthocyanins. The interaction between tannins and mannoproteins from different yeasts has been investigated and it was determined that the addition of mannoprotein, obtained from some special yeasts, leads to more anthocyanin-tannin condensation reactions and leads to decrease in astringency. The most commonly used clarifying agents are bentonite, egg white, gelatin, carbon and polyvinylpolypyrrolidone (PVPP). It was determined that antioxidant activities of wines decrease in case of PVPP usage. The betonite agent was found to cause the greatest reduction of polymeric anthocyanins. In a study done by using Cabernet Sauvignon grapes the effects of two different fermentation temperatures (15°C and 25°C), three different pressing time and six different fining agents (gelatin, kiselsol, egg white, isinglass, PVPP, bentonite) at three different concentrations on phenols and antioxidant activities was determined. The results of multivariate statistical analyses demonstrated that the parameters analysed by control and combination of gelatine and kiselsol were close to each other in the area of principal component distribution. The localisation of egg white and PVPP treated samples was different from the coordinates of total phenol and antioxidant activity, thus explaining the adverse effects of both fining agents. The highest values were obtained for gallic acid and p-hydroxybenzoic acid for both conditions by HPLC analysis. Concentrations of phenolic acids were found from highest to lowest order in the following manner: p-hydroxybenzoic acid > gallic acid > syringic acid > ferulic acid > p-coumaric acid > caffeic acid. Filtration is a process aimed at removing solid particles from wine. Cloth bag filter, cellulose plate filters and perlite and kieselguhr are used as auxiliary filter agents. Anthocyanins rapidly decrease during the ageing process. Factors that trigger the degradation reactions of anthocyanins include high temperature, oxidation of enzymes and secondary fermentation (Aktan and Kalkan, 2000; Auw *et al.*, 1996; Harbertson *et al.*, 2002; Yıldırım and Aktan, 2007; Sacchi, Bisson and Adams, 2005).

## 4. The Therapeutic Value of Wine

Wine is one of the most strong antioxidant food with special properties for human health (Aktan and Kalkan, 2000). The term antioxidant and its relation with wines is well defined. Oxidation caused by free radicals triggers the initiation of many diseases (coronary heart diseases, kidney inflammation) (Gaulejac, Glories and Vivas, 1999). Some free radicals may occur spontaneously during metabolism. Some environmental factors, such as pollution, radiation, cigarette smoke and herbicides could accelerate the formation of free radicals (Ames, 1983; Gaulejac, Glories and Vivas, 1999; Halliwell *et al.*, 1995). Antioxidants neutralise the free radicals by giving their own electrons. After this stage, antioxidants do not turn into free radicals since they are stable in both forms. These compounds play a protective role in food. Depending on the structure of the polyphenols, some compounds act as 'supplying $H^+$ ions antioxidants', while others are important as 'chelates forming' obtained by binding metal ions (Singleton and Esau, 1969). The degree of glycolisation influences the antioxidant activity of corresponding phenolic compounds. The stereochemical structure of these compounds also affects the antioxidant activities. The free form of quercetin is more effective than its glycoside form. Catechin and its isomers are accepted as powerful antioxidants due to their ability to bind the superoxide anion radicals. Tannins having larger molecular structures are reported to have the highest antioxidant activity (Acar and Gökmen, 2005; Kalkan and Aktan, 1998; Kalkan and Aktan, 1999a,b; Yıldırım *et al.*, 2005). The most important healthy properties of wine are connected with the antioxidant properties of its constituents. This feature also describes what is known as the 'French paradox' (low coronary heart disease despite high-fat consumption). It has been found that red wines prevent LDL oxidation in *in vitro* conditions. It has also been reported that this characteristic is caused by the phenol composition of wines (Kinsella *et al.*, 1993; Ole, Suadicani and Gyntenberg, 1996; Renaud and De Lorgeril, 1992).

The polyphenols compounds found in wine have the following properties:

*Biological Effects of Wines*

- Reduction of thrombus formation
- Inhibition of smooth muscle cells proliferation
- Prevention of free radicals formation
- Inhibition of LDL oxidation
- Prevention of cancer cell proliferation
- Relaxing effect
- Protecting against negative effects of nitrite
- Positive effects on aging
- Other positive effects

### 4.1. Powerful Antioxidants with Special Biological Effects

The resveratrol found in wines has higher antioxidant activity than C and E-vitamins. Additionally, it provides inhibition of lipoxygenase enzyme and binding of free radicals. Mechanisms of antioxidant action (Fig. 10) include removal of $O_2$, scavenging reactive oxygen/nitrogen species or their inhibiting reactive oxygen/nitrogen species formation, binding metal ions needed for catalysis of ROS generation and up-regulation of endogenous antioxidant defences (Halliwell, 1996). Studies show that anthocyanins present in wines exert a high protective effect as scavengers (hydroxyl and superoxide radicals) with significant capacity to transfer electrons (Rivero *et al.*, 2008). The phenolic structure of anthocyanins is responsible for their antioxidant ability: scavenging reactive oxygen species such as, superoxide ($O_2^{\cdot-}$), singlet oxygen, peroxide ($ROO^-$), hydrogen peroxide ($H_2O_2$) and hydroxyl radical ($OH^\cdot$). The antioxidant effects of anthocyanins *in vitro* demonstrate using several cell culture systems, including colon, endothelial, liver, breast and leukemic cells and keratinocytes. In these culture systems, anthocyanins exhibit multiple anti-toxic and anti-carcinogenic effects, such as directly scavenging reactive oxygen species, increasing the oxygen-radical absorbing capacity of cells, stimulating the expression of Phase II detoxification enzymes, reducing the formation of oxidative adducts in DNA, decreasing lipid peroxidation, inhibiting mutagenesis and carcinogens, reducing cellular proliferation by modulating signal transduction pathways. Although most of the protective effects of anthocyanins are attributed to their ability to scavenge ROS, they also play a role by chelating metals and direct binding to the proteins. The radical scavenging activity of anthocyanins is connected to the presence of hydroxyl groups in position 3 of ring C and also in the 3′, 4′ and 5′ positions in ring B of the molecule. In general, the radical scavenging activity of the anthocyanidins (aglycons) is superior to their respective anthocyanins (glycosides) (Wang and Stonder, 2008; Bagchi *et al.*, 2003). However to explain the antioxidant capacity of wines it is necessary to conjugate the effect of anthocyanins with the rest of the components of the wine matrix, which exert presumably certain synergic effects (Rivero *et al.*, 2008).

**Figure 10.** Mechanism for the antioxidant activity of phenolic

## 4.2. Reduction of Thrombus Formation

The mechanisms of thrombus reduction are given below:

Polyphenols ⟶ Thromboxane ⟶ Reduction of ⟶ Reduction of synthesis is inhibited aggregation thrombus formation

## 4.3. Inhibition of Smooth Muscle Cells Proliferation

The smooth muscle cells proliferation is given in the following picture:

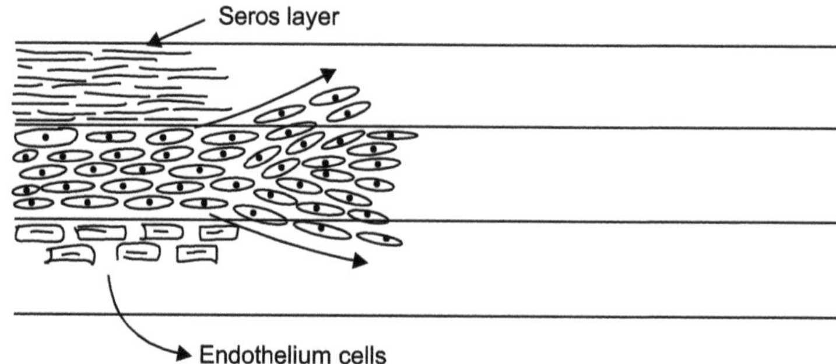


## 4.4. Prevention of Free Radicals Formation

The ways of prevention of free radicals formation are given below:

- The damage of phospholipid liposomes is prevented
- Damage to DNA structure is prevented
- Damage to the hemoprotein cytochrome C is prevented

## 4.5. Inhibition of LDL Oxidation

Phenolic compounds that are found in wines increase the affinity of the protein component of HDL (high-density lipoproteins) to cholesterol and in this case, an increase in the amount of HDL occurs (Fig. 11). Thus, cholesterol is pulled from the peripheral cells and sent to the liver cells for degradation. The LDL (lower density lipoproteins) level increases during cholesterol transport from the liver to the cells, 'foam cells' are formed by macrophages, causing the beginning of atheromatous plaque formation. As a result of this reduced perfusion, the initiation of many coronary heart diseases begins. In case of thrombosis' 'infarc.' formation at this stage occurs. These results have been supported by many types of research (Kontoudakis *et al.*, 2011).

Concentrations of some phenolic substances (gallic acid, catechin, caffeic acid, epicatechin, cyanidin, malvinidin, myricetin) and their inhibitory effects on LDL oxidation were determined in 20 wines. The amount of all polyphenols were expressed as GAE (gallic acid equivalent). The inhibitory effects of LDL oxidation were found to be 37-65 per cent for red wines and 27-37 per cent for white wines. A positive correlation was determined between total polyphenolic compounds present in wines and their inhibitory effects on LDL oxidation. A similar experiment was conducted with French wines to explore the effect of different periods of must fermentation on the oxidation process over LDL inhibition. The result emphasised the importance of must fermentation duration since a longer period (>5 days) caused higher inhibition of LDL oxidation (65-70 per cent). Additionally, the effects of used grape varieties were also determined as critical for such effects. Corresponding inhibition rates for different varieties is – Grenache wine (50 per cent), Merlot (60 per cent), Cabernet Sauvignon (44 per cent) (Frankel *et al.*, 1993; Akçay *et al.*, 2004, 2007; Teissedre, Waterhouse and Frankel, 1995; Yıldırım *et al.*, 2004, 2005). Resveratrol is an effective scavenger of hydroxyl, superoxide and metal-induced radicals. It inhibits two physiological oxidants: ferrylmyoglobin and peroxynitrite and formation of foam cells (Harborne and Willams, 2000; Bradamante *et al.*, 2004; Aviram and Fuhrman B., 2002; Kong *et al.*, 2003).

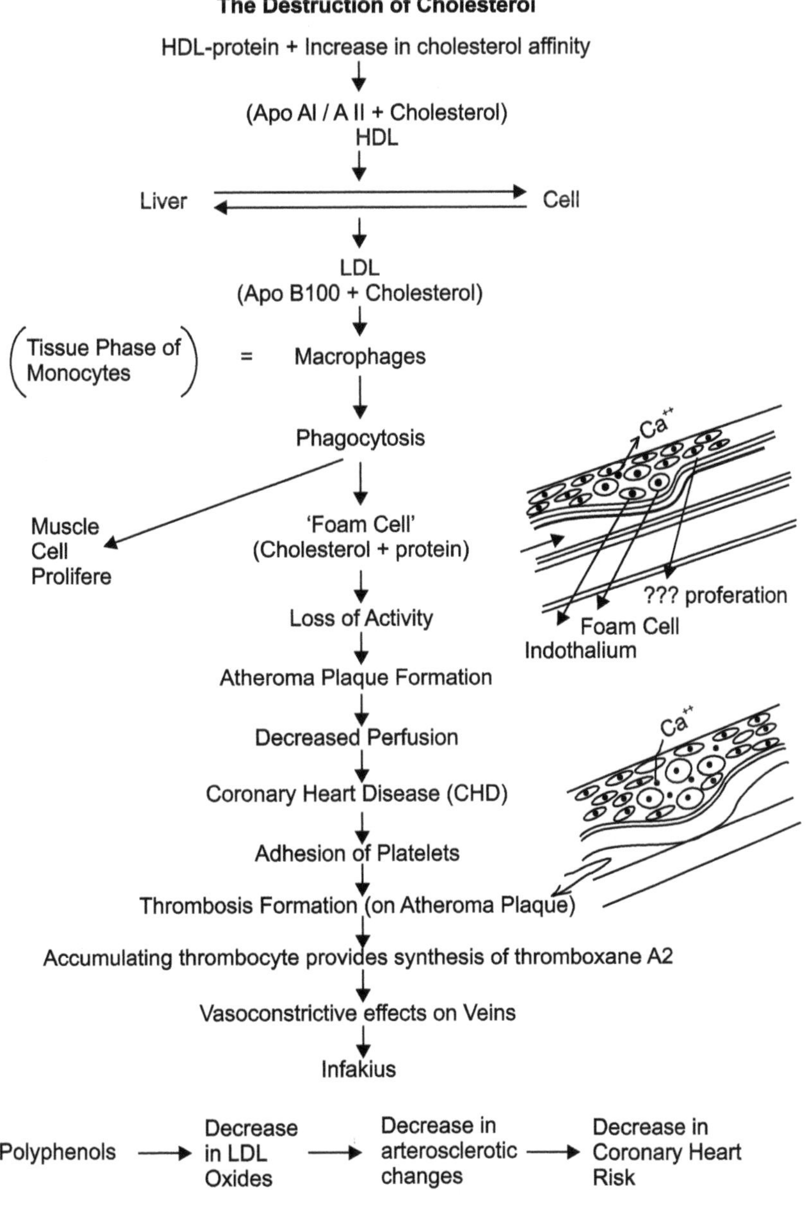


**Figure 11.** The effect of polyphenols on heart disease

### 4.6. Prevention of Cancer Cell Proliferation

Interesting results were obtained concerning the anti-tumoral effect of red wine. In mice consuming wine led to tumoral formation as much as 20 per cent after 95 days, whereas in the group consuming the standard diet without wine, this formation was higher and began after 65 days. Cancer cases have been examined in 28,200 people (15,117 male, 13,063 women) participating in the ages of 20-98 ranges. Monitoring and evaluation of results continued for 13 years. Wine consumption habits were grouped into four categories – 1. Group: 1 glass in the week; 2. Group: 6 glasses in the week; 3. Group: 7-21 glasses in the week; 4. Group: ≥ 22 glasses in the week. As a result of the survey, evaluation was concluded that moderate wine consumption patterns led to specific surviving patients (Clifford *et al.*, 1996). Another treatment with resveratrol on CaCo-2 (human colon cancer) cells resulted in the inhibition of these cells. Resveratrol causes a significant decrease in ornithine decarboxylase (ODC) activity, a key enzyme of polyamine biosynthesis, which enhances cancer cell growth. ODC inhibition results in the reduction of intracellular putrescine content. Resveratrol concentration of 25 μM exhibits anti-proliferative effect of CaCo-2 cells (Schneider *et al.*, 2000; Jackson and Lombard, 1993). Anti-cancer mechanism of resveratrol was explained by the damage of mitochondrial function that leads to increase in reactive oxygen species, apoptosis and possibly intracellular drug accumulation via inhibition of proteins involved in multi-drug resistance (Sun *et al.*, 2008).

### 4.7. Relaxing Effect

It was demonstrated that gamma-butyric acid found in wine has a positive effect on stress control. Additionally, this function was demonstrated to be related to neuromuscular activation and diuretic properties (Kinsella *et al.*, 1993).

### 4.8. Protection Against Negative Effects of Nitrite

Phenolic substances of wine eliminate the negative effects of nitrite if consumed in small quantities. However, overconsumption of nitrite with foods (pesticides and smoked meat/fish) cause the formation of diazophenols which trigger the beginning of mouth and gastric cancers.

### 4.9. Positive Effects on Aging

Phenolic compounds found in wines could activate proteins located in nerve cells, which facilitate learning. Liquid increases or bleeding in retinal vessels connected with age could be lowered by a small amount of wine consumption (Bianchini and Vainio, 2003).

### 4.10. Other Positive Effects

Ethyl alcohol in wine is absorbed quickly and provides energy for a very short time. Moderate wine consumption could suppress the growth of *Helicobacterium pylori*, providing prevention of ulcer formation. The antidiabetic properties of red wine from Portugal were studied *in vitro*. Four fractions from solid phase extraction with reversed-phase material C18 were subjected to the antidiabetic activity assay *in vitro*. Results showed that dealcoholised red wine and the four fractions exhibited strong inhibitory activities on α-glucosidase which breaks down long-chain carbohydrates. The results indicate that the main components responsible for such activities were monomeric and oligomeric flavan-3-of compounds. These data might provide further evidence that the prevention of hyperglycemia may be another beneficial effect of moderate consumption of red wine (Xia *et al.*, 2017).

## 5. Wine Toxicology

### 5.1. Ochratoxin Formation

Ochratoxin A (OTA) is a mycotoxin associated with outbreak of renal toxic disease and is classified as a possible human carcinogen. Its chemical structure consists of a chlorine-containing dihydro-isocoumarin linked through the 7-carboxyl group to l-â-phenylalanine. There are several OTA analogues, ochratoxins B, C, and alkyl esters of ochratoxins that having a similar structure but are less toxic (Fung and Clark,

2004). OTA is easily absorbed in the gastrointestinal tract, mainly in the duodenum and jejunum, based on animal studies (Kumagai, 1988). OTA was found in low concentrations in kidney, liver, fat and muscle tissues (Krogh *et al.*, 1974). Excretion is mainly via renal elimination (Chang *et al.*, 1977). The elimination half-life in an animal model was reported between 23.6 and 28.7 hours to as long as 35 days in monkeys (Fukui *et al.*, 1987). The LD 50 of OTA ranges from 0.5 mg/kg for dogs to over 50 mg/kg for mice (Kuiper-Goodman and Scot, 1989). The toxicity of OTA involves several mechanisms. OTA inhibits protein synthesis by competing with the phenylalanine aminoacylation reaction catalysed by Phe-tRNA synthase (Creppy *et al.*, 1984). This results in inhibition of protein as well as DNA and RNA synthesis. OTA also disrupts hepatic microsomal calcium homeostasis by impairing the endoplasmic reticulum membrane via lipid peroxidation (Omar and Rahimtula, 1991). OTA is the major ochratoxin component and is the most toxic among analogues. However, it has been estimated that an infant could eat up to 10 kg of food contaminated with 20 ppb without significant adverse health effects (Chu, 1974). It occurs in a variety of plant products, including wine, grape juice and dried vine fruits. The species responsible for OTA in cereals and other food products were not responsible for OTA contamination of grapes, wine and vine fruits. The common contaminant of these commodities was found to be *Aspergillus niger*. For processing of grapes, pH is also important. There have been some studies of the effect of water and temperature on germination and growth of *A. niger* (Magan and Aldred, 2005). Experiments with an *A. carbonarius* strain demonstrated that ochratoxin-A production differs significantly with the composition of the berries at different maturation stages. The study demonstrated that ochratoxin-A production was in the highest value when grapes are unripened and lowest at ripened grapes (3402, 1530 and 22 µg/kg, respectively) (Serra *et al.*, 2006).

## 5.2. Ethyl Alcohol and Its Metabolism in Wine

The human being burns 0.1 g of alcohol per kg of body weight in one hour. A human weighing 70 kg needs 25 grams of absolute alcohol to reach a concentration of 0.5 grams of alcohol concentration in blood. This amount of alcohol (percentage) is found in three bottles of beer in 0.3 litres, 0.25 litres of wine or 50-60 ml of spirits (Kallıthraka *et al.*, 2007). This alcohol is oxidised within three hours without affecting the central nervous system and provides a significant portion of the body's required calories. Occurring changes of ethyl alcohol in the body are given in Fig. 12. It is well known that blood contains 0.002 L of ethyl alcohol per litre even if it is not consumed. It is produced by microorganisms found in the body – *Escherichia coli, Lactobacillus* species and many other microorganisms. The alcohol that is taken in the body passing through the intestines is oxidised in the liver. The alcohol oxidation capacity of the liver is very high. However, if the blood alcohol concentration coming to the liver is higher than its absorption power, alcohol is spread to other parts of the body. If this level goes up too much, intoxication happens. 85 per cent of taken alcohol is metabolised in the liver. About 5-15 per cent of alcohol is excreted by breathing, sweat and urine. If 85 per cent is accepted as 100 per cent, 80 per cent of this amount is metabolised by alcohol dehydrogenase enzyme to acetaldehyde. The other 20 per cent is converted to acetaldehyde by the microsomal enzyme oxidation system of the liver cells. Acetaldehyde, which occurs both ways, is converted to acetate by a second enzyme named aldehyde dehydrogenase. Acetates found outside of the liver enter the citric acid cycle and are destroyed to $CO_2$ and $H_2O$ molecules. In the absence of aldehyde dehydrogenase enzyme system, the acetaldehyde takes place in fatty acids formation via acetyl Co-A molecules. So, in the case of taking a large amount of ethyl alcohol, the lipid layer around liver increases and predisposes to belly thickness. The alcohol dehydrogenase and aldehyde dehydrogenase (ALDH) enzymes are responsible for many more reactions in the body. Both enzymes have genetic origins. There are eight or more isoenzymes of alcohol dehydrogenase with different metabolic properties and different activities, depending on the human race. At least four of these ALDH isoenzymes are considered as clinically significant. In 50 per cent of the Oriental race, the ALDH is in very low concentration. The search for these enzymes and their genetic origin may illuminate the susceptibility to alcoholism. The oxidation of ethyl alcohol leads to the formation of acetaldehyde. Obtained acetaldehyde is unstable and is immediately converted to ethyl acetate. The reaction does not advance further. Once acetate gets out from the liver, it is rapidly oxidated. Ethanol oxidation is less concerned with oxygen consumption in the liver. Low ethanol concentrations slightly increase the oxygen consumption in the liver but decrease at high levels. As the alcohol oxidation increases in the liver, the oxidation of other substances decreases.

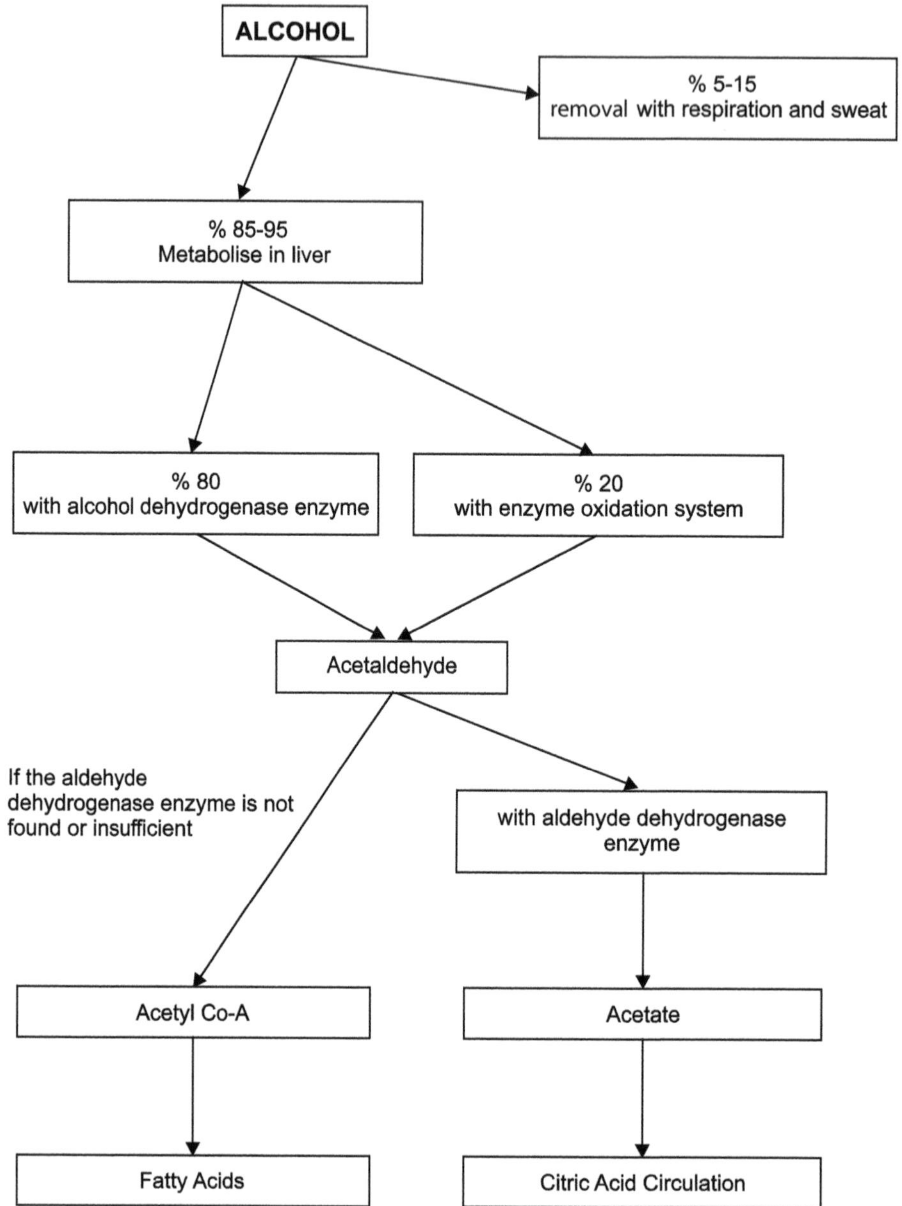


**Figure 12.** Changes of ethyl alcohol after its consumption

The ethyl alcohol metabolism in the body is rapid. In the presence of alcohol in the body, catabolytic reactions begin and lipid oxidation reactions in the liver are reduced.

### *Effects of Ethyl Alcohol on Human Health*

Ethanol (ethyl alcohol) has an important economic value but if taken in excess value, it causes paresthesia as in other foods containing tobacco, tea and coffee (Petrlungar *et al.*, 2002). Uncontrolled consumption leads to rude behaviour, loss of personality, moral crimes and traffic accidents. It disturbs health-related reactions (self-criticism = weakness of logic), prolongs the horror moment and increases the excitement (Reynolds *et al.*, 1996). The extent to how the behaviour is affected depends on the age, weight, sex, alcohol consumption habit and most importantly, the level of tolerance. The relationship between the alcohol level in the blood and motor functions are given in Table 5. In low doses, motor coordination, mild changes in perception and temperament occur. Amnesia and Stage 1 anesthesia develop at a level of 300 mg and above in 100 mg (1 dL) blood. The levels of 400-700 mg/dL in the blood may result in coma, respiratory insufficiency and death. A glass of wine contains approximately 8-10 grams of pure alcohol and an average of 70 calories. Higher alcohol concentration causes narcotic effects and leads to death. Lethal doses are 283 g for men, 193 g for women and 30-50 g for children (3 g/kg of body weight). Controlled consumption of ethyl alcohol encourages friendship in the community, increases appetite, and increases stomach fluid (especially $CO_2$ and sugar-containing sparkling wines).

**Table 5.** Relationship Between the Alcohol Level in the Blood and Motor Functions

| *Alcohol level in the blood (mg/dL)* | *Expected impact* |
|---|---|
| 20-99 | Coordination gets harmed |
| 100-199 | Mental capability decreases, judgment reduces, astatic-indecisive behaviour and ataxia are observed |
| 200-299 | Ataxia increases, speech becomes flattened, unsteady and nervous temperament; nausea and vomiting are possible |
| 300-399 | First-stage anesthesia develops, memory loss and unbalanced-unstable temperament are observed |
| >400 | Respiratory insufficiency, coma, death |

The value of the wine, rich in aromatic substances, becomes very important. Due to the regulatory feature, ethyl alcohol prevents circulatory disturbance. By drinking four glasses of wine, the higher alcohol level is reached in half an hour, the body veins become wider and the sensation of warmth increases. Diluted alcohol is not harmful to human health. On the contrary, it gives a stimulant and refreshing effect at a certain dose. The body temperature increases due to the loss of control in the nervous system, leading to accelerated pulsation. Actually, the role of ethyl alcohol has a more relaxing effect than stimulating. Ethyl alcohol warms up only the exterior surface of the body. Further, more energy is released from the body than the energy gained by ethyl alcohol. In fact, alcohol lowers the body temperature more effectively than ice. Alcohol dehydrogenase enzyme activity is higher in people who continuously drink wine. However, enzyme activity slows down in case of a large amount of alcohol consumption (Goel *et al.*, 1996).

## 5.3. Methanol Formation

Humans are uniquely sensitive to the toxicity of methanol as they have limited capacity to oxidise and detoxify formic acid. Thus, the toxicity of methanol in humans is characterised by formic acidemia, metabolic acidosis, blindness or serious visual impairment and even death (Skrzydlewska, 2003). Small amounts of methanol are a natural endogenous compound found in normal, healthy human individuals, as concluded by a study with a mean of 4.5 ppm in the exhaled breath of subjects (Turner *et al.*, 2006). The mean endogenous methanol in humans of 0.45 g/d may be obtained from pectin found in fruit. Additionally, one kilogram of apple produces up to 1.4 g methanol (Lindinger *et al.*, 1997). Methanol has a high toxicity in humans. Ten mL of pure methanol ingested is metabolised into formic acid, which can cause permanent blindness by destruction of the optic nerve. Thirty mL is potentially fatal (Vale, 2007), although the median lethal dose is typically 100 mL (3.4 fl oz) (i.e. 1-2 mL/kg body weight of pure methanol). The reference dose for methanol is 2 mg/kg/day. Toxic effects begin hours after ingestion and antidotes can often prevent permanent damage. Because of its similarity in both appearance and odour to ethanol (the alcohol in beverages), it is difficult to differentiate between the two. Such is also the case with denatured alcohol, adulterated liquors or very low-quality alcoholic beverages (Skrzydlewska, 2003).

Methanol is toxic by two mechanisms. Methanol enters the body through ingestion, inhalation and absorption through the skin and can be fatal due to its depressant properties in the same manner as seen in ethanol poisoning. Second, in a process of toxication, it is metabolised to formic acid, which is present as the formate ion via formaldehyde in a process initiated by the enzyme alcohol dehydrogenase in the liver (Schep *et al.*, 2009). Methanol is converted to formaldehyde via alcohol dehydrogenase (ADH) and formaldehyde is converted to formic acid (formate) via aldehyde dehydrogenase (ALDH). The conversion to formate via ALDH proceeds completely, with no detectable formaldehyde remaining (McMartin *et al.*, 1979). Liesivuori and Savolainen (1991) demonstrated that formate is a toxic substance since it inhibits mitochondrial cytochrome c oxidase, causing hypoxia at the cellular level and metabolic acidosis, among a variety of other metabolic disturbances. Methanol is a minor constituent of wine (0.1-0.2 g/L) and has no direct sensory effect. It is predominantly generated from the enzymatic breakdown of pectins. On degradation, methyl groups associated with pectins are released as methanol. Pectolytic enzymes added to juice or wine to aid clarification inadvertently increase the methanol content of wine. Wine has the lowest concentration of methanol of all the fermented beverages. A study with mango fruit (*Mangifera indica* L.) was performed to investigate wine capability of that fruit. Results demonstrated that mango fruit has high methanol concentration because of pectinolytic enzymes, which are commonly used in mango wine production and are responsible for the splitting of pectic substances to galacturonic acid and methanol (Reddy, 2005).

## 5.4. Ethyl Carbamate Formation

Ethyl carbamate (EC), or urethane is the ethyl ester of carbamic acid ($NH_2COOH$). Ethyl carbamate is a suspected carcinogen found in a variety of fermented foods and beverages as a natural consequence of the metabolic activity of microorganisms. Studies with rats, mice and hamsters have shown that ethyl carbamate will cause cancer when it is administered orally, injected or applied to the skin, but no adequate studies of cancer in humans has been reported. Acute toxicity studies show that the lowest fatal dose in rats, mice, and rabbits equals 1.2 grams/kg or more. When ethyl carbamate was used medicinally, about 50 per cent of the patients exhibited nausea and vomiting, and long-time use led to gastroenteric haemorrhages (Monteiro *et al.*, 1989). Its formation involves three main pathways – hydrolysis, N-hydroxylation or C-hydroxylation and side chain oxidation. The major pathway is through its hydrolysis by liver microsomal esterases to ethanol, ammonia and carbon dioxide. In rodents, approximately 5 per cent of it is excreted unchanged and more than 90 per cent is hydrolysed. Ethyl carbamate is also converted by cytochrome P-450 to form ethyl N-hydroxycarbamide (approximately 0.1 per cent), α-hydroxy ethyl carbamate and vinyl carbamate (approximately 0.5 per cent). This last compound, a more potent carcinogenic compound than ethyl carbamate, is converted via an epoxidation reaction to vinyl carbamate epoxide which can be covalently bound to DNA, RNA and proteins (Park *et al.*, 1993). Cytochrome P450, contained in the lung, lymph, liver and skin, catalyses the metabolism of urethane to *N*-hydroxyurethane. *N*-hydroxyurethane can be metabolised by esterase to hydroxylamine, which exerts its carcinogenic effect on multiple organs by generating $O_2$ and NO to cause oxidation and depurination of DNA (Sakano *et al.*, 2002). *Saccharomyces cerevisiae* metabolises arginine, one of the major amino acids in grape musts, to ornithine and urea during wine fermentation. Wine yeast strains of *S. cerevisiae* do not fully metabolise urea during grape must fermentation. Urea is secreted by yeast cells and it reacts spontaneously with ethanol in wine to form ethyl carbamate, a potential carcinogenic agent for humans (Monteiro and Bisson, 1991). More urea can be released from the cells when ethanol concentration gets elevated as a consequence of fortification of the must. There were significant turnover and degradation of arginine in cells incubated in the presence of ethanol, as would occur during commercial fortification of must. Thus, urea can be formed during vinification and if released into the medium, will yield ethyl carbamate through reaction with ethanol (Monteiro *et al.*, 1989). In case of controlled production conditions, most of these reverse effects could be overcome.

## 6. Conclusion

Wine is a food with complex structure having special properties:

- Wine composition could be modified by appropriate choice of grape varieties and production parameters.
- Nutritional values of wines depend on special compounds mainly phenolics which could be optimised by different production steps procedure.
- The therapeutic values of wines include a lot of properties such as antioxidant, antimicrobial and anti-cancerogenic. Some of them need further studies to be improved.
- The toxicity of some wine compounds could be considered as a potential risk in case of inappropriate wine production and consumption.

## References

Abrahamse, E. and Bartowsky, J. (2012). Timing of malo-lactic fermentation inoculation in Shiraz grape must and wine: Influence on chemical composition. *World J. Microbiol. Biotechnol.* 28: 255-265.

Acar, J. and Gökmen, V. (2005). Fenolik bileşikler ve doğal renk maddeleri. pp. 463-496. *In*: Saldamlı, İ. (Ed.). Gıda Kimyası, Hacettepe Üniversitesi Yayını, Ankara.

Akçay, Y.D., İlanbey, B., Yıldırım, H.K. and Sözmen, E.Y. (2007). Pomegranate wine has a greater protection capacity than red wine on LDL-oxidation. *Journal of Medicinal Food* 10(2): 371-374.

Akçay, Y.D., Yıldırım, H.K., Güvenç, U. and Sözmen, E.Y. (2004). The effects of consumption of organic and non-organic red wine on low-density lipoprotein oxidation and antioxidant capacity in humans. *Nutrition Research* 24: 541-554.

Aktan, N. and Kalkan, H. (2000). Şarap teknolojisi, Kavaklıdere Eğitim Yayınları. Ankara 4: 615.

Ames, B.N. (1983). Dietary carcinogens and anticarcinogens oxygen radicals and degenerative diseases. *Science* 221: 1256-1262.

Auw, J.M., Blanco, V., O'Keefe, S.F. and Sims, C.A. (1996). Effect of processing on the phenolics and colour of Cabernet Sauvignon, Chambourchin and Noble wines and juices. *Am. J. Enol. Vitic.* 47: 279-286.

Aviram, M. and Fuhrman, B. (2002). Wine flavonoids protect against LDL oxidation and atherosclerosis. The Lipid Research Laboratory. Technion Faculty of Medicine, The Rappaport Family Institute for Research in the Medical Sciences and Rambam Medical Centre, Haifa, Israel. *Ann. N.Y. Acad. Sci.* 957: 146-161.

Bagchi, D., Sen, C.K., Bagchi, M. and Atalay, M. (2003). Anti-angiogenic, antioxidant, and anti-carcinogenic properties of a novel anthocyanin-rich berry extract formula. *Biochemistry* (Moscow), 69(1): 75-80, translated from Biokhimiya 69(1): 95-102.

Bauer, R. and Dicks, T. (2004). Control of malo-lactic fermentation in wine: A review. Department of Microbiology, Stellenbosch University. *South African Journal of Enology & Viticulture* 25: 74-88.

Belmiro, T.M.C., Pereira, C.F. and Paim, A.P.S. (2017). Red wines from South America: Content of phenolic compounds and chemometric distinction by origin. *Microchemical Journal* 133: 114-120.

Bianchini, F. and Vainio, H. (2003). Wine and resveratrol: Mechanisms of cancer prevention. *Eur. J. of Cancer Prev.* 12: 417-425.

Boido, E., Medina, K., Farin, A., Carrau, F., Versini, G. and Dellacassa, E. (2009). The effect of bacterial strain and aging on the secondary volatile metabolites produced during malo-lactic fermentation of Tannat red wine. *J. Agric. Food Chem.* 57: 6271-6278.

Bradamante, S., Barenghi, L. and Villa, A. (2004). Cardiovascular protective effects of resveratrol. CNR – ISTM. Istituto di Scienze e Tecnologie Molecolari, Milan, Italy. *Cardiovascular Drug Reviews* 22(3): 169-188.

Chang, F.C. and Chu, F.S. (1977). The fate of ochratoxin A in rats. *Food Cosmet Toxicol.* 15(3): 199-204.

Chira, K., Pacella, N., Jourdes, M. and Teissedre, P.L. (2011). Chemical and sensory evaluation of Bordeaux wines (Cabernet-Sauvignon and Merlot) and correlation with wine age. *Food Chemistry* 126: 1971-1977.

Chu, F.S. (1974). Studies on ochratoxins. *CRC Crit. Rev. Toxicol.* 2(4): 499-524.

Clifford, A.J., Ebeler, S., Ebeler, J.D., Bills, N.D., Hiurichs, S.H., Teissedre, P.L. and Waterhouse, A.L. (1996). Delayed tumor onset in transgenic mice fed an amino acid-based in supplements. *The American Journal of Clinical Nutrition* 64(5): 748-756.

Coelho, C., Parot, J., Gonsior, M., Nikolantonaki, M., Schmitt-Kopplin, P., Parlanti, E. and Gougeon, R.D. (2017). Asymmetrical flow field-flow fractionation of white wine chromophoric colloidal matter. *Analytical and Bioanalytical Chemistry* 409(10): 2757-2766.

Creppy, E.E., Roschenthaler, R. and Dirheimer, G. (1984). Inhibition of protein synthesis in mice by ochratoxin A and its prevention by phenylalanine. *Food Chem. Toxicol.* 22(11): 883-886.

Culbert, J.A., McRae, J.M., Condé, B.C., Schmidtke, L.M., Nicholson, E.L., Smith, P.A. and Wilkinson, K.L. (2017). Influence of production method on the chemical composition, foaming properties and quality of Australian carbonated and sparkling white wines. *Journal of Agricultural and Food Chemistry* 65(7): 1378-1386.

Darici, M., Cabaroglu, T., Ferreira, V. and Lope, R. (2014). Chemical and sensory characterisation of the aroma of Çalkarası rosé wine. *Australian Journal of Grape and Wine Research* 20(3): 340-346.

De, E.D.C.N.R. and Vinho, O.D.E. (2011). Effect of cyclodextrins on off-odours removal of red wine: An innovative approach. *Ciência Téc. Vitiv.* 26: 63-68.

Draijer, R., de Graaf, Y., Slettenaar, M., de Groot, E. and Wright, C.I. (2015). Consumption of a polyphenol-rich grape-wine extract lowers ambulatory blood pressure in mildly hypertensive subjects. *Nutrients* 7(5): 3138-3153.

Fischer, U., Löchner, M. and Wolz, S. (2007). Red wine authenticity: Impact of technology on anthocyanin composition. pp. 239-253. *In*: Ebeler, S.E., Takeoka, G.R. and Winterhalter, P. (Eds.). Authentication of Food and Wine. ACS Symposium Series 952. Washington, DC.

Frankel, E.N., Kanner, J., German, J.B., Parks, E. and Kinsella, J.E. (1993). Inhibition of oxidation of human low-density lipoprotein by phenolic substances in red wine. *Lancet* 341: 454-457.

Fukui, Y., Hoshino, K., Kameyma, Y., Yasui, T., Toda, C. and Nagano, H. (1987). Placental transfer of ochratoxin A and its cytotoxic effect on the mouse embryonic brain. *Food Chem. Toxicol.* 25(1): 17-24.

Fung, F. and Clark, R.F. (2004). Health effects of mycotoxins: A toxicological overview. *Journal of Toxicology, Clinical Toxicology*, 42(2): 217-234.

Galanakis, C., Kordulisb, C., Kanellakib, M., Koutinasb, A., Bekatoroub, A. and Lycourghiotisb, A. (2012). Effect of pressure and temperature on alcoholic fermentation by *Saccharomyces cerevisiae* immobilised on γ-alumina pellets. *Bioresource Technology* 114: 492-498.

Gaulejac, N., Glories, Y. and Vivas, N. (1999). Free radical scavenging effect of anthocyanins in red wines. *Food Res. Int.* 32: 37-333.

Gil, J.V., Mateo, J.J., Jiménez, M., Pastor, A. and Huerta, T. (1996). Aroma compounds in wine as influenced by apiculate yeasts. *Journal of Food Science* 61(6): 1247-1250.

González-Centeno, M.R., Chira, K. and Teissedre, P.L. (2017). Comparison between malo-lactic fermentation container and barrel toasting effects on phenolic, volatile, and sensory profiles of red wines. *Journal of Agricultural and Food Chemistry* 65(16): 3320-3329.

González-Neves, G., Charamelo, D., Balado, J., Barreiro, L., Bochicchio, R., Gatto, G., Gil, G., Tessore, A., Carbonneau, A. and Moutounet, M. (2004). The phenolic potential of Tannat, Cabernet-Sauvignon and Merlot grapes and their correspondence with wine composition. *Analytica Chimica Acta* 513: 191-196.

Halliwell, B. (1996). Antioxidants: The basics – what they are and how to evaluate them. *Advances in Pharmacology* 38: 3-20.

Halliwell, B., Murcia, M.A., Chirico, S. and Auroma, O.I. (1995). Free radicals and antioxidants in food and *in vivo*: What they do and how they work? *Cri. Rev in Food Sci. and Nut.* 35: 7-20.

Harbertson, J.F., King, A., Block, D.E. and Adams, D.O. (2002). Factors that influence tannin extraction and formation of polymeric pigments during winemaking. *Am. J. Enol. Vitic.* 53: 245-259.

Harborne, J.B. and Williams, C.A. (2000). Advances in avonoid research since 1992. Department of Botany, School of Plant Sciences. The University of Reading, Reading RG6 6AS, UK. *Phytochemistry* 55: 481-504.

https://pubs.niaaa.nih.gov/publications/aa72/aa72.htm.

https://pubs.niaaa.nih.gov/

Hur, S., Lee, S., Kim, Y., Choi, I. and Kim, G. (2014). Effect of fermentation on the antioxidant activity in plant-based foods. *Food Chemistry* 160: 346-356.

Jackson, D.I. and Lombard, P.B. (1993). Environmental and management practices affecting grape composition and wine quality: A review. *American Journal of Enology and Viticulture* 44: 409-430.

Jouanneau, S., Weaver, R.J., Nicolau, L., Herbst-Johnstone, M., Benkwitz, F. and Kilmartin, P.A. (2012). Subregional survey of aroma compounds in Marlborough Sauvignon Blanc wines. *Australian Journal of Grape and Wine Research* 18(3): 329-343.

Kalkan, H. and Aktan, N. (1998). The importance of polyphenolic compounds in wines. Development of Food Science and Technology. *Scientific Works, HIFFI- Plovdiv* vol. XLIII.

Kalkan, H. and Aktan, N. (1999a). Kirmızi şarap üretim aşamasında değişik uygulamaların fenolik asit miktarlarına ilişkin etkisinin belirlenmesi üzerine bir araştırma. E.Ü. Müh. Fak., Gıda Müh. Bölümü, *2000'li Yıllarda Gıda Bilimi ve Teknolojisi Kongresi*, 18-20 October, İzmir.

Kalkan, H. and Aktan, N. (1999b). Carignane sek şarabında kateşin ve epikateşin konsantrasyonunu etkileyen faktörlerin belirlenmesi üzerine bir araştırma. E.Ü. Müh. Fak., Gıda Müh. Bölümü, *2000'li Yıllarda Gıda Bilimi ve Teknolojisi Kongresi*, 18-20 October, İzmir.

Kallithraka, S., Salacha, M. and Tzourou, I. (2007). Changes in phenolic composition and antioxidant activity of white wine during bottle storage: Accelerated browning test versus bottle storage. *Food Chemistry* 113: 500-505.

Kallithraka, S., Mamalos, A. and Makris, D.P. (2007). Differentiation of young red wines based on chemometrics of minor polyphenolic constituents. *J. Agric. Food Chem.* 55(9): 3233-3239.

Keller, M. (2010). Managing grapevines to optimise fruit development in a challenging environment: A climate change primer for viticulturists. *Australian Journal of Grape and Wine Research* 16: 56-69.

Kerem, Z., Bravdo, B., Shoseyov, O. and Tugendhaft, Y. (2004). Rapid liquid chromatography–ultraviolet determination of organic acids and phenolic compounds in red wine and must. *Journal of Chromatography A* 1052: 211-215.

Kinsella, J.E., Frankel, E., German, B. and Kanner, J. (1993). Possible mechanism and protective role of antioxidants in wine and plant foods. *Food Tech.* 47: 85-89.

Kong, J., Chia, L., Goh, N., Chia, T. and Brouillard, R. (2003). Analysis and biological activities of anthocyanins. National Institute of Education, Nanyang Technological University. 1 Nanyang Walk, Singapore 637616, Singapore. Institut de Chimie, Universite Louis Pasteur, 1 rue Blaise Pascal, Strasbourg 67008, France. *Phytochemistry* 64(5): 923-933.

Kontoudakis, N., Esteruelas, M., Fort, F., Canals, J.M. and Zamora, F. (2011). Use of unripe grapes harvested during cluster thinning as a method for reducing alcohol content and pH of wine. *Australian Journal of Grape and Wine Research* 17: 230-238.

Krogh, P., Axelsen, N.H., Elling, F., Gyrd-Hansen, N., Hald, B., Hyldgaard-Jensen, J., Larsen, A.E., Madsen, A., Mortensen, H.P., Moller, T., Petersen, O.K., Ravnkov, U., Rostgaard, M. and Aalund, O. (1974). Experimental porcine nephropathy: Changes of renal function and structure induced by ochratoxin A contaminated feed. *Acta Pathol. Microbio.l Scand.* 246: 1-21.

Kuiper-Goodman, T. and Scott, P.M. (1989). Risk assessment of the mycotoxin ochratoxin A. *Biomed. Environ. Sci.* 2(3): 179-248.

Kumagai, S. (1988). Effects of plasma ochratoxin toxin and luminal pH on the jejunal absorption of ochratoxin A in rats. *Food Chem. Toxicol.* 26(9): 753-758.

Kurvietien, L., Staneviien, I., Mongirdien, A. and Bernatonien, J. (2016). Multiplicity of effects and health benefits of resveratrol. *Medicina* (Lithuania) 52(3): 148-155.

Leeuwen, C.V., Tregoat, O., Chone, X., Bois, B., Pernet, D. and Gaudillere, J.P. (2009). Vine water status is a key factor in grape ripenıng and vintage quality for red bordeaux wine: How can it be assessed for vineyard management purposes? *Journal International des Sciences de la Vigne et du Vin.* 43(3): 121-134.

Liesivuori, J. and Savolainen, H. (1991). Methanol and formic acid toxicity: Biochemical mechanisms. *Pharmacol. Toxicol.* 69(3): 157-163.

Lindinger, W., Taucher, J., Jordan, A., Hansel, A. and Vogel, W. (1997). Endogenous production of methanol after the consumption of fruit. *Alcoholism, Clinical and Experimental Research* 1(5): 939-943.

Liu, L., Cozzolino, D., Cynkar, W.U. *et al.* (2008). Preliminary study on the application of visible-near infrared spectroscopy and chemometrics to classify Riesling wines from different countries. *Food Chemistry* 106: 781-786.

Magan, N. and Aldred, D. (2005). Conditions of formation of ochratoxin A in drying, transport and in different commodities. *Food Additives & Contaminants* 22(1): 10-16.

Maslov, L., Tomaz, I., Mihaljević Žulj, M. and Jeromel, A. (2017). Aroma characterisation of predicate wines from Croatia. *European Food Research and Technology* 243(2): 263-274.

Martin, O., Brandriss, M.C., Schneider, G. and Bakalinsky, A.T. (2003). Improved anaerobic use of arginine by *Saccharomyces cerevisiae*. *Appl. Environ. Microbiol.* 3: 1623-1628. doi: 10.1128/AEM.69.3.1623-1628.2003

McMartin, K.E., Martin-Amat, G., Noker, P.E. and Tephly, T.R. (1979). Lack of a role for formaldehyde in methanol poisoning in the monkey. *Biochem. Pharmacol.* 28(5): 645-649.

Moldovan, A., Mudura, E., Coldea, T., Rotar, A. and Pop, C. (2015). Effect of maceration conditions on chemical composition and colour characteristics of Merlot wines. Faculty of Food Science and Technology, University of Agricultural Sciences and Veterinary Medicine. *Food Science and Technology* 72(1): 104-108.

Monteiro, F.F. and Bisson, L.F. (1991). Amino acid utilisation and urea formation during vinification fermentations. *Am. J. Enol. Vitic.* 42: 199-208.

Monteiro, F.F., Trousdale, E. and Bisson, L.F. (1989). Ethyl carbamate formation in wine: Use of radioactively labeled precursors to demonstrate the involvement of urea. Department of Viticulture and Enology. University of California, Davis, CA95616-5270. *Am. J. Enol. Vitic.* 40(1): 1-8.

Mouret, J.R., Camarasa, C., Angenieux, M., Aguera, E., Perez, M., Farines, V. and Sablayrolles, J.M. (2014). Kinetic analysis and gas-liquid balances of the production of fermentative aromas during winemaking fermentations: Effect of assimilable nitrogen and temperature. *Food Res. Int.* 62: 1-10. doi:10.1016/j.foodres.2014.02.044

Murli, D. (2010). The Composition of Grapes. Iowa State University. Iowa State University Extension: Ames, Iowa, National Wine and Grape Industry Centre, 2017; web site: https://www.csu.edu.au/nwgic

Ole, H.H., Suadicani, P. and Gyntenberg, F. (1996). Alcohol consumption, serum low-density lipoprotein cholesterol concentration and risk of ischaemic heart disease: Six-year followup in the Copenhagen male study. *B.M.J.* 312: 736-741.

Omar, R.F. and Rahimtula, A.D. (1991). Role of cytrochrome P-450 and in ochratoxin A-stimulated lipid peroxidation. *J. Biochem. Toxicol.* 6(3): 203-209.

Palacios, G.A.T., Raginel, F. and Julien, A.O. (2007). Can the selection of *Saccharomyces cerevisiae* yeast lead to variations in the final alcohol degree of wines? *Australian and New Zealand Grapegrower and Winemaker* 527: 71-75.

Park, K.K., Liem, A., Stewart, B.C. and Miller, J.A. (1993). Vinyl carbamate epoxide, a major strong electrophilic, mutagenic and carcinogenic metabolite of vinyl carbamate and ethyl carbamate (urethane). *Carcinogenesis* 14: 441-450.

Penza, M. and Cassano, G. (2004). Recognition of adulteration of Italian wines by thin-film multisensory array and artificial neural networks. *Analytica Chimica Acta*. 509: 159-177.

Peterlunger, E., Celotti, E., Da Dalt, G., Stefanelli, S., Gollino, G. and Zironi, R. (2002). Effect of training system on Pinot noir grape and wine composition. *Am. J. Enol. Vitic.* 53(1): 14-18.

Pickering, G.J., Heatherbell, D.A. and Barnes, M.F. (1999b). The production of reduced-alcohol wine using glucose oxidase-treated juice. Part III: Sensory. *American Journal of Enology and Viticulture* 50: 307-316.

Prenesti, E., Toso, S., Daniele, P.G., Zelano, V. and Ginepro, M. (2004). Acidbase chemistry of red wine: Analytical multi-technique characterisation and equilibrium-based chemical modelling. *Analytica Chimica Acta* 507: 263-273.

Potortí, A.G., Lo Turco, V., Saitta, M., Bua, G.D., Tropea, A., Dugo, G. and Di Bella, G. (2017). Chemometric analysis of minerals and trace elements in Sicilian wines from two different grape cultivars. *Natural Product Research* 31(9): 1000-1005.

Rapp, A. and Mandery, H. (1986). Wine Aroma. Federal Research Station for Grapevine-Breeding. Geilweilerhof, D-6741 Siebeldingen (Federal Republic of Germany) and Institute for Food Chemistry. University of Karlsruhe, D-7500 Karlsruhe (Federal Republic of Germany). Birkh/iuser Verlag, CH 4010 Basel/Switzerland. *Experientia* 42: 873-884.

Recalmes, F., Sayago, A., Hernanz, D. and Gonzalez, M. (2005). The effect of time and storage conditions on the phenolic composition and colour of white wine. *Food Research International* 39: 220-229.

Reddy, L.V.A. and Reddy, O.V.S. (2005). Production and characterisation of wine from mango fruit (*Mangifera indica* L.). Department of Biochemistry. S.V. University, Tirupati, 517502, India. *World Journal of Microbiology & Biotechnology* 21: 1345-1350.

Renaud, S. and De Lorgeril, M. (1992). Wine, alcohol, platelets and the French paradox for coronary heart disease. *Lancet* 33: 1523-1526.

Reynolds, A.G., Wardle, D.A. and Dever, M. (1996). Vine performance, fruit composition and wine sensory attributes of Gewurztraminer in response to vineyard location and canopy manipulation. *Am. J. Enol. Vitic.* 47(1): 77-92.

Rivero, M.D., Muñiz, P. and González-Sanjosé, M.L. (2008). Contribution of anthocyanin fraction to the antioxidant properties of wine. Dpto. de Biotecnologı´a y Ciencia de los Alimentos, Facultad de Ciencias, Universidad de Burgos, Pza. Misael Banuelos s/n, Burgos 09001, Spain. *Food and Chemical Toxicology* 46(8): 2815-2822.

Robinson, J. (2006). The Oxford Companion to Wine, third edition. Oxford University Press, pp. 681.

Rupasinghe, H.P. and Clegg, S. (2006). Total antioxidant capacity, total phenolic content, mineral elements and histamine concentration in wines from different fruit sources. *Journal of Food Composition and Analysis* 20: 133-137.

Sacchi, L.K., Bisson, L.F. and Adams, D.O. (2005). A review of the effect of winemaking techniques on phenolic extraction in red wines. *Am. J. Enol. Vitic.* 56: 197-206.

Sakano, K., Oikawa, S., Hiraku, Y. and Kawanishi, S. (2002). Metabolism of carcinogenic urethane to nitric oxide is involved in oxidative DNA damage. *Free Radic. Biol. Med.*, 33(5): 703-714.

Schep, L.J., Slaughter, R.J., Vale, J.A. and Beasley, D.M. (2009). A seaman with blindness and confusion. *BMJ* (clinical research ed.) 339: b3929.

Schneider, Y., Vincent, F., Duranton, B., Badolo, L., Gosse, F., Bregman, C., Seiler, N. and Raul, F. (2000). Anti-proliferative effect of resveratrol, a natural component of grapes and wine, on human colonic cancer cells. ULP/CJF INSERM 95-09. Laboratory of Metabolic and Nutritional Control in Digestive Oncology, IRCAD, 1 Place de l'Ho Ãpital, 67091 Strasbourg. France Laboratory of Molecular Oncology, France. *Cancer Letters* 158: 85-91.

Serra, R., Mendonça, C. and Venâncio, A. (2006). Ochratoxin A occurrence and formation in Portuguese wine grapes at various stages of maturation. Centro de Engenharia Biológica. Universidade do Minho, Campus de Gualtar, 4710-057 Braga, Portugal. *International Journal of Food Microbiology* 111(1): 35-39.

Singleton, V.L. and Esau, P. (1969). Phenolic Substances in Grapes and Wine and Their Significance. Academic Press, New York.

Skrzydlewska, E. (2003). Toxicological and metabolic consequences of methanol poisoning. *Toxicology Mechanisms and Methods* 13(4): 277-293.

Soden, A., Francis, I.L., Oakey, H. and Henschke, P.A. (2000). Effects of co-fermentation with Candida stellata and *Saccharomyces cerevisiae* on the aroma and composition of Chardonnay wine. *Australian Journal of Grape and Wine Science* 6(1): 21-30.

Stashenko, H., Macku, C. and Shibamato, T. (1992). Monitoring volatile chemicals formed from must during yeast fermentation. *Journal of Agricultural and Food Chemistry* 40(11): 2257-2259.

Sun, W., Wang, W., Kim, J., Keng, P., Yang, S., Zhang, H., Liu, C., Okunieff, P. and Zhang, L. (2008). Anti-cancer effect of resveratrol is associated with induction of apoptosis via a mitochondrial pathway alignment. pp. XXIX. *In*: K.A. Kang *et al.* (Eds.). Oxygen Transport to Tissue. XXIX. Springer Science and Business Media LLC, USA, 179-186.

Talaverano, I., Valds, E., Moreno, D., Gamero, E., Mancha, L. and Vilanova, M. (2017). The combined effect of water status and crop level on Tempranillo wine volatiles. *Journal of the Science of Food and Agriculture* 97(5): 1533-1542.

Teissedre, P.L., Waterhouse, A.L. and Frankel, E.N. (1995). Principal phenolic chemicals in French syrah and grenache phones wines and their antioxidant activity in ınhibiting oxidation of human low-density lipoproteins. *J. Intern. Sciences de la vigne et du Vin*. 29: 205-212.

Torija, J., Beltran, G., Novo, M., Poblet, M., Guillamon, M., Mas, A. and Rozes, N. (2002). Effects of fermentation temperature and *Saccharomyces* species on the cell fatty acid composition and presence of volatile compounds in wine. Unitat d'Enologia del Centre de Referencia de Tecnologia d'Aliments. Dept. Bioquímica i Biotecnologia, Facultat d'Enologia de Tarragona. *International Journal of Food Microbiology* 85: 127-136.

Tsai, C.-C., Lee, M.-C., Tey, S.L., Liu, C.-W. and Huang, S.C. (2017). Mechanism of resveratrol-induced relaxation in the human gallbladder. *BMC Complementary and Alternative Medicine* 17(1): 254.

Tuberoso, C.I.G., Serreli, G., Congiu, F., Montoro, P. and Fenu, M.A. (2017). Characterisation, phenolic profile, nitrogen compounds and antioxidant activity of Carignano wines. *Journal of Food Composition and Analysis* 58: 60-68.

Turner, C., Španěl, P. and Smith, D. (2006). A longitudinal study of methanol in the exhaled breath of 30 healthy volunteers using selected ion flow tube mass spectrometry, SIFT-MS. *Physiological Measurement* 27(7): 637-648.

Urcan, D.E., Giacosa, S., Torchio, F., Río Segade, S., Raimondi, S., Bertolino, M. and Rolle, L. (2017). 'Fortified' wines volatile composition: Effect of different postharvest dehydration conditions of wine grapes cv. Malvasia moscata (*Vitis vinifera* L.). *Food Chemistry* 219: 346-356.

Vale, A. (2007). Methanol. *Medicine* 35(12): 633-634.

Villamor, R., Harbertson, J. and Ross, C. (2009). Influence of tannin concentration, storage temperature, and time on chemical and sensory properties of Cabernet Sauvignon and Merlot wines. *American Journal of Enology and Viticulture* 60: 442-449.

Walker, G.M. and Stewart, G.G. (2016) *Saccharomyces cerevisiae* in the production of fermented beverages. *Beverages* 2: 30. doi:10.3390/beverages2040030

Wang, L. and Stoner, G.D. (2008). Anthocyanins and their role in cancer prevention. Department of Internal Medicine and Comprehensive Cancer Center. Ohio State University College of Medicine, Columbus, OH 43210, USA. *Cancer Letters* 269(2): 281-290.

Wansbrough, H., Sherlock, R.S., Barnes, M. and Reeves, M. (1998). Chemistry in winemaking. pp. 1-12. *In*: John Packer, E. and Robertson, J. (Eds.). Chemical Processes in New Zealand. Vol 1, Chemistry of Wine Making 1-12.

Waterhouse, A. (2002). Wine phenolics. *Annals of the New York Academy of Sciences* 957(1): 21-36.

Wright, J. (1795). An essay on wines, especially on port wine; intended to instruct every person to distinguish that which is pure, and to guard against the frauds of adulteration. London.

Xia, X., Sun, B., Li, W., Zhang, X. and Zhao, Y. (2017). Anti-diabetic activity phenolic constituents from red wine against α-glucosidase and α-amylase. *Journal of Food Processing and Preservation* 41(3): 1-5.

Yildirim, H.K. and Aktan, N. (2007). Carignane şarabında polifenollerin stabilitesini etkileyen faktörelin belirlenmesi üzerine araştırmalar. *Hasat Gıda* 268: 37-43.

Yildirim, H.K., Akçay, Y.D., Güvenç, U. and Sözmen, E.Y. (2004). Protection capacity against low-density lipoprotein oxidation and antioxidant potential of some organic and non-organic wines. *International Journal of Food Sciences and Nutrition* 55(5): 351-362.

Yıldırım, H.K., Akçay, Y.D., Güvenç, U., Altındişli, A. and Sözmen, E.Y. (2005). Antioxidant activities of organic grape, pomace, juice, must, wine and their correlation with phenolic content. *International Journal of Food Science and Technology* 40(2): 133-142.

Zakhari, S. (2006). Overview: How is alcohol metabolised by the body? *Alcohol Research & Health* 29(4): 245-254.

# 3 Wine as a Complete Functional Beverage: A Perspective

**S.K. Soni[1*], Urvashi Swami[1], Neetika Trivedi[1] and Raman Soni[2]**

[1] Department of Microbiology, Panjab University, Chandigarh – 160014, India
[2] Department of Biotechnology, D.A.V. College, Chandigarh – 160011, India

## 1. Introduction

The knowledge of grape and wine is as old as traditional antiquity of mankind (Feher *et al.*, 2007). Humans have been consuming beer for about 6,000 years and wine for nearly 9,000 years. Wine, being man's oldest beverage, has been an integral part of the diet of the Mediterraneans for centuries. Wine has also been perceived to be the drink of gods. In Hindu religion, it is believed that wine was created by Lord Brahma himself and was named as *somras* and the process of its creation was revealed only to his seven students, the Saptarishis, who then revealed it to their followers. *Somras* was considered to reduce degeneration of cells and, therefore, was consumed to stay young for over 100 years (Shukla *et al.*, 2014). Among Christians also, it has religious connections (McDonald, 1986). It is rightly said by Martin Luther (1483-1556), "Beer is made by humans and wine is made by God." In modern times, wine therapy has come into picture from France. Several observations regarding low mortality rates in people consuming moderate levels of red wine led the investigators to scientifically evaluate the observation. Since then, many case studies and experimental investigations have proved wine to be beneficial to health when consumed in moderate quantities (Drel and Sybrina, 2010).

## 2. Wine: A Complete Functional Beverage

Wine is a non-distilled and fermented beverage produced by anaerobic fermentation of grapes. Other than grapes, there are many fruits that can be used for making wines. According to the definition, wine is a low-alcohol, fermented drink produced from grapes only; otherwise wine is given the prefix of the fruit (Voguel, 2003). In European Union, the term 'wine' is exclusively used for fermented grape juice. However, in US any fermented fruit or agricultural product containing 7-24 per cent alcohol is termed as wine (Harding, 2005). The fermentation of fruit must bring about many chemical changes and, therefore, wine is not merely a fruit juice to which ethanol has been added (Jagtap and Bapat, 2015). It's a nutritionally-rich and functionally-active preparation. A typical 149 mL of wine is composed of 127 g water, 15.6 g alcohol, 3.8 g total carbohydrates, 0.1 g protein, 0.4 g ash and varying quantities of vitamins, minerals and phytochemicals (Gebhardt and Thomas, 2002). It has been noted that depending upon the substrate, the polyphenolic content of wine may range from 100 to 35,000 mg GAE/L (Rana and Singh, 2013; Di Lorenzo *et al.*, 2016). Fermentation alters the must by modifying the conjugation of organic acids and phenolics, by extraction and formation of co-pigments and, therefore, improves the anti-oxidant potential. Since the 19th century, our perception about wine, its structure and alterations have impressively evolved along with the advances in pertinent technical fields of microbiology, biochemistry and chemistry. Every functional progress has led to improved regulation of winemaking, aging conditions and wine quality (Richter *et al.*, 2015).

## 3. Chemical Composition of Wines

Chemically, wine is a complex beverage. In addition to water and ethanol, it contains a variety of minerals, vitamins, esters, aldehydes and phenolic compounds. As the fermentation process includes crushing and

*Corresponding author: sonisk@pu.ac.in

pressing of the fruits, it enables the extraction of the fruit components into must, which thereby get imparted into wine (Geana *et al.*, 2016). Fruits and herbs contain various phyto-nutrients whose extraction and bio-availability in the wines depend upon the vinification process, ageing of wine and many other factors (Joshi, 2011). Apart from phenolics, the overall composition of wines remains similar. Average chemical composition of wines is shown in Fig. 1. The major constituents of wine are discussed at length in the following pages.

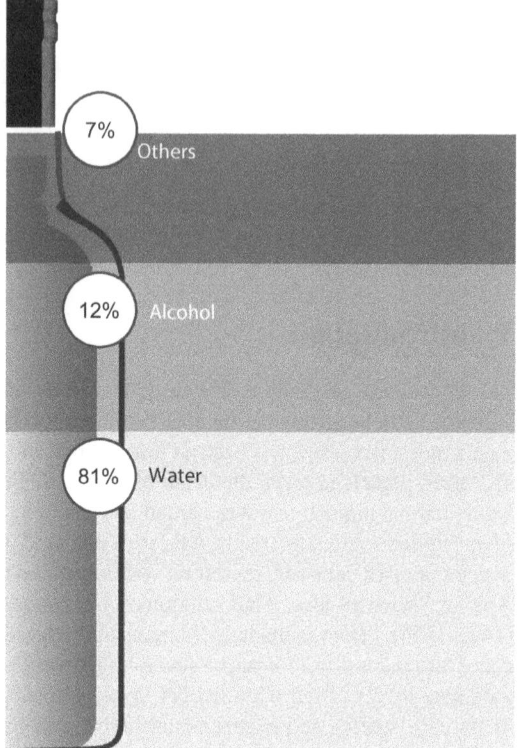


**Figure 1.** Chemical components of typical wine (Image adapted from benefits-of-resveratrol.com)

### 3.1. Water

Water constitutes a major part of wine and can be present up to 90 per cent v/v (Soleas *et al.*, 1997). Despite being a major ingredient, it is not given much importance. However, it is noteworthy that water plays a main role in extraction of bio-actives in wine and is necessary for the action of many enzymes. There are hardly any compounds in wine which are water insoluble. Therefore, it is an integral solvent in wine and is also an essential component of the fermentation process and ageing (Jackson, 2014).

### 3.2. Sugars

Sugars are a sub-type of carbohydrates, which possess the general formula $C_x(H_2O)_x$. Sugars are sweet in taste and high in energy. The most important and abundant sugars in grapes are hexoses, glucose and fructose. The sugar content in mature berries may range from 20 to 24° Brix (Crippen and Morrison, 1986). They are utilised by wine yeast and fermented to produce ethanol. Apart from hexoses and pentoses, other sugars are also present but in insignificant amounts. The wine yeast, *Saccharomyces cerevisiae* has a restricted ability to utilise the hexoses. Therefore, other sugars are of least importance in wine production. Some other carbohydrates can be problematic if present in abnormal quantities. For example, the presence of galacturonic acid and pectin, if not broken, can create an undesirable haziness in wine. Therefore, their quantity is required to be checked upon (Baskan *et al.*, 2016).

During fermentation, the carbohydrate content decreases as one gram of glucose with a calorific value of 3.86 Kcal is converted to about 0.51 g/0.64 mL of ethanol with a calorific value of 3.5 Kcal and roughly 20 per cent of the calories available from sugar are lost (Reed, 1981). Carbohydrates are important in wine production as they are not only the raw material for yeast, they also confer sweetness to wine, from crystal clear taste in sweet wines to background note in dry wines. In sparkling wines, sugars impart carbon dioxide to wine as a by-product of their breakdown. The unfermented sugars of wine are known as residual sugars. In dry wine, residual sugars are mainly pentose sugars, e.g. arabinose, rhamnose, xylose, etc., in addition to low quantities of hexoses. Sweetness of carbohydrates of wine in comparison to sucrose is given in Table 1. Sugars are generally higher in red wines than in white. The reason behind this may be lengthier aging which leads to sugar liberation from polyphenol conjugates and hydrolysis of hemicellulose during barrel ageing. The sugar quantity depends upon the type of oakwood used for maturation. Monosaccharides, especially fructose, galactose and xylose differ throughout the aging process and depend on the type of oakwood used (del Alamo *et al.*, 2000; Ribereau-Gayon *et al.*, 2000).

The residual sugar level is around 1.5 g/L in dry wines. The sweetness in wine is detected when sugar level reaches 1 per cent. The sweetness of wine is also affected by other components, like ethanol, tannins, aromatic compounds, acids, etc. (Deibner *et al.*, 1965). A sugar content of 1 per cent is generally insufficient for any microbial growth to occur and infect wine. But, an increase in the sugar levels may pose a threat to the integrity of wine by aiding microbial contamination. Contaminants like *Botrytis cinerea* affect the sugar content of wine through oxidation. Botrytised wines have 5-oxo-fructose which is formed by oxidation of fructose and xylosone (Ribereau-Gayon, 1976; Barbe *et al.*, 2000, 2002).

The sugar quantity in wine may increase during maturation due to the breakdown of oak glycosides and sugar released by yeast. Oak barrel ageing alters the percentage of sugar content of wine. The levels of fructose, galactose and xylose are greater than other monosaccharides during aging. Oak variety and manufacturing of oak barrel also affect the sugar structure of aged wine (del Alamo *et al.*, 2000).

The most significant product of sugar fermentation is ethanol. Other by-products, such as esters, fatty acids, higher alcohols and acids, etc. are also produced in minor quantities (Soni *et al.*, 2011). Wine sugars play a role during the ageing process. Brown melanoidin pigments are formed due to structural transformation of sugars and this leads to browning of wine. This occurs via a series of chemical reactions called Maillard reactions (Rivero-Pérez *et al.*, 2002). Mainly hexoses, along with amines, are involved in the process. Sugar dehydration and conversion of amines into imines occurs and their further chemical conversions and polymerisation causes the development of brown pigments. Metal ions, such as of iron and copper, act as catalysts and the reaction takes place at elevated temperatures. At lower temperatures, Maillard's reaction may occur but at comparatively slower rates (Li *et al.*, 2008; Jackson, 2014). Sugars have also been linked with the aroma of the wine. Sugars escalate the volatility of aromatic compounds (Sorrentino *et al.*, 1986; Robinson *et al.*, 2009).

**Table 1.** Sweetness of Carbohydrates of Wines (Relative to that of Sucrose)

| *Compound* | *Relative sweetness* |
|---|---|
| D-Fructose | 1.14 |
| D-Glucose | 0.69 |
| Sucrose | 1 |
| α,α-Trehalose | 0.45 |
| Xylitol | 1 |
| Sorbitol | 0.55 |
| Mannitol | 0.5 |

*Source*: Adapted from Sanz and Castro, 2008

### 3.3. Ethanol

The transformation of must into wine leads to the production of ethanol from the breakdown of sugars. This completely changes the chemical composition of wine. Fermentation is a catabolic process and for every one molecule of sugar metabolised, two molecules of ethanol are formed. Ethanol is the most important by-product of fermentation. Although it is also produced in grapes during carbonic maceration, the chief cause of ethanol is the fermentation itself. Generally, the level of ethanol in wine reaches 12-14 per cent v/v. However, it can be increased by addition of sugar exogenously into the medium or by fortification with brandy. It has also been found that excessive ethanol may lead to hotness or bitterness in the taste of wine. Therefore, adequate levels of ethanol should be maintained (Varela *et al.*, 2015).

Ethanol concentration can be expressed as percentage w/v or percentage v/v, specific gravity or proof. The generalised method of expression is percentage v/v. Ethanol, being the second most abundant component of wine, is an important co-solvent along with water. It aids the extraction of grape constituents into wine. The most important health beneficial constituents of wines, polyphenolics, are non-polar. Therefore, ethanol is a crucial factor for maintaining the phenolic content of wines. Not only this, ethanol also affects the volatile aromatic compounds and add to the beautiful aroma of wine. At low ethanolic concentrations, ethanol takes the form of a mono-dispersed, aqueous solution. Concentrations below seven per cent v/v favour the discharge of volatile compounds. This influences the sensorial attributes and the overall structure of wine (Williams and Rosser, 1981; Guth, 1998). However, with the increase in ethanol levels, the hydrophobic hydration gets reduced (D'Angelo *et al.*, 1994), leading to increase in the solubility of nonpolar compounds (Robinson *et al.*, 2009). High ethanol concentration also reduces tannin-protein interactions and leads to decreased astringency of wines and, therefore, moderate quantity of ethanol is desirable in wines as inadequate quantities hamper the overall quality of wine (McRae *et al.*, 2015). Ethanol content of different wines is shown in Table 2.

Ethanol also acts as a reactant and due to its chemical involvement, many important volatile compounds, like ethyl esters, are produced. They further impart the characteristic odour to the wine. Ethanol affects the type of compounds to be released by the action of yeasts by denaturing enzymes and altering the membrane permeability. The quality and quantity of phenolic to be produced and liberated is thus affected (Jackson, 2014).

Ethanol greatly affects the mouth-feel and taste of wine. It adds to the sweetness and makes the more acidic wines feel less sour and more balanced. However, when present in higher concentrations, ethanol causes a burning sensation (Jones *et al.*, 2008). It also affects the astringency of tannins and may increase the bitterness of wine (Lea and Arnold, 1978; Conner *et al.*, 1998). Due to the increased solubility of aromatics in ethanol, their loss is lowered due to solubility into ethanol (Williams and Rosser, 1981).

Apart from affecting the physico-chemical composition of wine, ethanol acts a protectant to the wine. It averts the spoilage of wine by progressively limiting the growth of other microbes that are potentially undesirable (Renouf *et al.*, 2007). The relative alcohol tolerance of *S. cerevisiae* allows the typical dominance of yeast in fermentation. Micro-organisms which produce off-odours and may prove damaging, are normally inhibited. The antimicrobial capacity of ethanol and wine's acidity provide stability to the wine for years in anaerobic environment (Jackson, 2014).

**Table 2.** Ethanol Contents of Different Wines

| *Beverages* | *Ethanol content (% v/v)* |
|---|---|
| Table wines | 11-14 |
| Dry red wines | 12.6 |
| Sweet white wines | 19.3 |
| Sweet red wines | 19.3 |
| Sparkling wines | 13.2 |
| Fortified wines | > 15 |
| Dry white wines | 12.4 |

Adapted and modified from Amerine *et al.*, 1972 and Soni *et al.*, 2011

## 3.4. Methanol

Methanol is present in wine in trace amounts (0.1 to 0.2 g/L) and has no virtual sensory traits. However, its conversion to formaldehyde and formic acid raises concern as both of them are toxic and affect the central nervous system and may cause blindness or death (Singkong *et al.*, 2012). Methanol never reaches lethal intensity in wine under appropriate winemaking techniques. The marginal quantity of methanol content in wine arises due to demethylation of pectins. Thus, methanol content depends on the pectin content of the must. As compared to other fruits, grapes have lower pectin content. Therefore, grape wine is composed of the least levels of methanol as compared to any other fruit-based beverage. However, it has been pointed out that pectolytic treatment of wine for its clarification can unintentionally escalate the methanol intensities (Jackson, 2014). Increased levels of methanol during fermentation have been observed due to pectin methyl esterase activity (Revilla and Gonzalez-SanJose, 1998). Addition of distilled spirits in wine can also add to the methanol concentration (Jackson, 2014). Other factors, like fermentation conditions, oenological practices and the yeast strain, can also affect the levels of methanol in wine (Van Rensburg and Pretorius, 2000; Cabaroglu, 2005). Paine and Devan (2001) have described 2 g of methanol as tolerable and 8 g as a toxic dose for a single day. For the production of a non-toxic wine, the methanol level is an important concern (Van Rensburg and Pretorius, 2000; Cabaroglu, 2005). Considering the serious impact of methanol on health, several methods have been optimised for decreasing the methanol content in wines, including utilisation of atmospheric-room temperature plasma method, use of malt and other adjuncts (Liang *et al.*, 2014), optimisation of the pectinase treatment conditions (Reddy and Reddy, 2005) by adding coumaric acid or gallic acid (Hou *et al.*, 2008), choosing the yeast strain which does not produce pectin esterase, applying demethanol columns (Nikicevic and Tesevic, 2005) and by using packed-bed columns for wine distillation (Carvallo et al., 2011).

### 3.5. Higher Alcohols

Higher alcohols are among the most vital volatile compounds produced during wine fermentation and are also known as fusel oils. They contain more than two carbon atoms and include the branched-chain alcohols, like 2-methylbutanol, 2-methylpropanol and 3-methylbutanol and the aromatic alcohols, like 2-phenylethanol and tyrosol (Ugliano and Henschke, 2009). In wines, the level of higher alcohol should be less than 300 mg/L as the higher quantities may affect the quality of wine (Rapp and Versini, 1996). However, some higher alcohols, like 2-phenylethanol, also contribute towards wine aroma and are desirable (Swiegers *et al.*, 2005). Major higher alcohols that are present in wines are listed in Table 3.

Most of the higher alcohols, excluding ethanol, are formed during the maceration process. They may generate from grape-derived aldehydes by reductive denitrification of amino acids, or via synthesis from sugars (Chen, 1978). Higher alcohols contribute to aged wine bouquet as they react with organic acids and lead to the production of esters (Rapp and Güntert, 1986). Other alcohols include diols, like 2,3-butanediol, polyols like glycerol and sugar alcohols. In dry wine, after water and ethanol, glycerol is the most abundant compound. It is present in higher quantities in red wine (~10 g/litre) as compared to the white ones (~7 g/litre) (Noble and Bursick, 1984).

**Table 3.** Higher Alcohols in Wines

| *Compound* | *mg/L (ppm)* |
|---|---|
| Propan-1-ol | 9-68 |
| Butan-1-ol | 1.4-8.5 |
| 2-methyl-Propan-1-ol | 6-174 |
| Pentan-1-ol | 0-0.4 |
| 3-methyl-Butan-1-ol | 83-490 |
| 2-methyl-Butan-1-ol | 17-150 |
| Hexan-1-ol | 0.5-12 |
| Hex-3-en-1-ol | 0.04-0.23 |
| Octan-1-ol | 0.2-1.5 |
| 2-phenyl-ethanol-1-ol | 0-50 |
| 3-thio-propen-1-ol | 46-96 |
| Glycerol | 5000-10000 |

*Source*: Adapted and modified from Bakker and Clarke, 2011

Glycerol possesses a moderately sweet taste and can add to the viscosity of wine. It may also attract spoilage bacteria (Calderone *et al.*, 2004). Sugar alcohols, such as arabitol, erythritol, myo-inositol, mannitol, alditol, and sorbitol, are also present in wine, but only in trace amounts. They are utilised by *Acetobacter* and converted to other compounds. Sugar alcohols have a small effect on the body of the wines (Jackson, 2014).

### 3.6. Acids

The range of desirable total acidity is 5.5-8.5 mg/L, for majority of wines. For white wines, a higher acidity is preferred with a pH of 3.1-3.4 as compared to red wines, which are acceptable within the pH range of 3.3-3.6. The two principal inorganic acids of wines, carbonic and sulphurous acid, are present as dissolved gases in the wine. The main organic acids found in wine are citric, malic, tartaric, succinic and acetic acid. These acids balance the flavour of the wine by providing sharpness to it (Dry and Coombe, 2004).

Wine acidity is recognised into two main types: volatile and fixed. Volatile acids can be evaporated by steam distillation; however, fixed acids are comparatively poorly volatile. Total acidity is deducted by combining them both and is generally expressed in terms of tartaric acid. Malic and tartaric acid constitute about 90 per cent of the fixed acidity of the wines. Fixed acidity may range between 4-7 g/L or even higher. If malo-lactic fermentation occurs in wine, malic acid gets processed into lactic acid, which is a smoother-tasting monocarboxylic acid (Jackson, 2014). However, it has been observed that malo-lactic fermentation does not lead to release of any foul compounds that may affect the acceptability of wines (Izquierdo-Cañas *et al.*, 2016). Various quantified acids, before and after malo-lactic fermentation, are shown in Table 4.

**Table 4.** Acidity of a Typical Wine Before and After Malo-lactic Fermentation

| *Acids (g/L)* | *Before* | *After* |
|---|---|---|
| Total | 4.9 | 3.8 |
| Volatile | 0.21 | 0.28 |
| Fixed | 4.7 | 3.6 |
| Malic | 3.2 | 0.5 |
| Lactic | 0.12 | 1.8 |

*Source*: Adapted from Ribereau-Gayon *et al.*, 1976

The increased chemical complexity of wine contributes to the progress of an aged bouquet. If the wine pH rises, then citric or tartaric acid is supplemented in wine for acidifying the undesirable high pH. Acids impart sensory characteristic to wines, enhance colour stability and limit the oxidation processes (Waterhouse and Ebeler, 1998; Villiers *et al.*, 2012). Succinate can impart a bitter salty facet; α-ketoglutarate can attach to sulphur dioxide and reduce its functional quantity in wine; and pyruvic acid can stabilise wine's red colour and may bind to sulphur dioxide (Jackson, 2014).

### 3.7. Phenolics

Wines contain a huge range of phenolic compounds, including flavonoids, non-flavonoids and phenolic-protein-polysaccharide complexes. These are predominantly found in grape skin, seeds and stems. Distribution of polyphenols in grape berry is shown in Fig. 2. Phenolics are higher in red wines than in white wines. Phenolic compounds influence taste, appearance, mouth-feel, fragrance, bitterness, astringency, colour and antimicrobial property of wine (Basha *et al.*, 2004; Lorenzo *et al.*, 2016). Phenolics are also highly responsible for the antioxidant capacity of wines (Lingua *et al.*, 2016). Although principally of grape derivation, smaller amounts are also extracted during wine maturation from oak and small quantities are produced via yeast metabolism (Jackson, 2014).

After harvesting, the phenolic content of wine depends on the processing in the winery. Majority of the phenolic compounds are derived from the skin and seed tissue, with stem-derived phenolics constituting only a minor component, if included and the exception of pulp derived hydroxycinnamic acids (Kennedy, 2008; El Darra *et al.*, 2016). The distribution of various phenolic compounds in a grape berry is shown in Fig. 2.

Phenols can be simple or complex; simple phenols contain a single aromatic ring attached to one or more hydroxyl groups; however, complex phenols, also called polyphenolic compounds, contain numerous phenol rings (Waterhouse, 2002). Because phenolic compounds possess both hydroxyl groups and benzene ring, they have both hydrophobic and hydrophilic attributes. Wine phenolics are clustered into two groups: non-flavonoids and flavonoids.

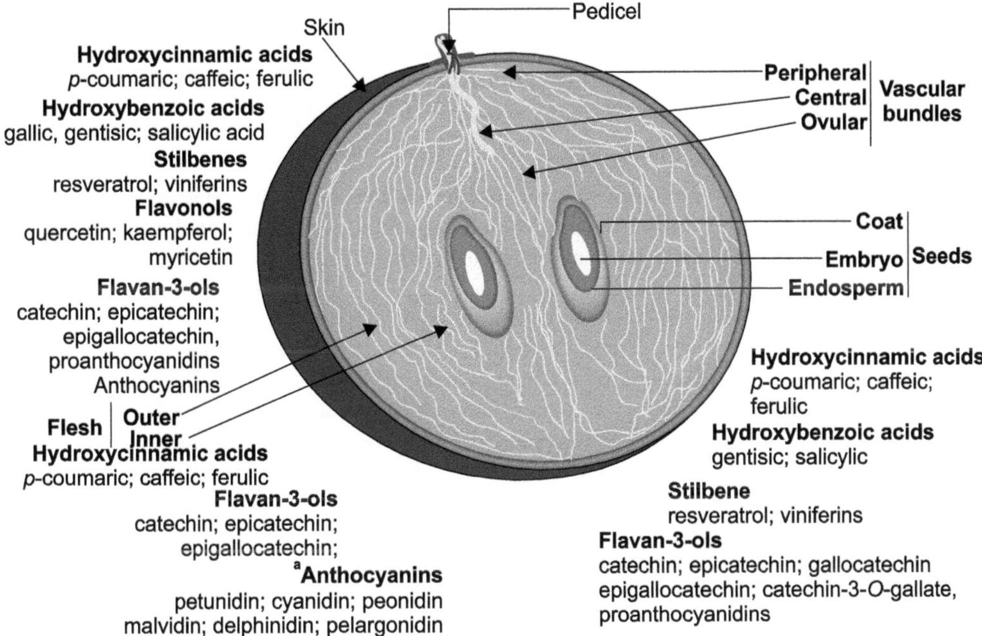


**Figure 2.** Distribution of important polyphenols in a grape berry (Image adapted from Teixeira, 2013)

### 3.7.1. Non-Flavonoids

The non-flavonoids include hydroxybenzoic acids, hydroxycinnamic acids and stilbenes (Cristino *et al.*, 2013). Hydroxybenzoic acids occurring in grapes and wine include gallic acid, protocatechuic acid, ellagic acid, syringic acid and vanillic acid (Macheix and Fleuriet, 1998). The occurrence of ellagic acid and its derivatives is mostly due to their extraction from wood during maturation in wooden barrels. The hydroxycinnamic acids, namely ferulic, *p*-coumaric, sinapic and caffeic acids, infrequently occur in the free form in fruits, but may be present in wine due to the vinification process (Macheix and Fleuriet, 1998). The soluble derivative of these compounds has one of the alcoholic groups esterified with tartaric acid and can also be glycosylated and acylated in different positions. Stilbenes have two ethane-bonded benzene rings. Among these trans-isomer compounds, resveratrol is the major stilbene, along with infrequent piceids, viniferins and pterostilbenes, respectively, dimethylated resveratrol derivatives, resveratrol glucosides and resveratrol oligomers (Jeandet *et al.*, 2002).

### 3.7.2. Flavonoids

Also known as diphenylpropanoids, flavonoids possess two phenyl groups and are derived from a reaction between a phenylpropanoid and three malonylCoA moieties. They contain 15 carbon atoms (Hossain *et al.*, 2016) and are further divided into the groups discussed below.

#### 3.7.2.1. Flavan-3-ols

Also known as proanthocyanidins, these are the most complex flavonoids, ranging from the simple monomers to oligomeric and polymeric, which are also called as condensed tannins (Rio *et al.*, 2013). They do not occur as glycosides. Catechin and epicatechin are the most common members of flavonoids present in wine and are found in greater quantities in red wine due to extraction from grape seeds and skins during vinification (Oszmianski *et al.*, 1986; Hossain *et al.*, 2016).

#### 3.7.2.2. Flavonols

These compounds are largely present in the skin of berry, the principal ones being quercetin, myricetin, kaempferol and their glycosides (Castillo-Munoz *et al.*, 2010).

*3.7.2.3. Proanthocyanidins*

Proanthocyanidins are complex flavonoids naturally present in legumes, cereals, some fruits, cocoa and beverages, such as wine and tea (Santos-Buelga and Scalbert, 2000). They yield anthocyanidins when heated in the presence of a mineral acid (Porter *et al.*, 1986).

*3.7.2.4. Anthocyanidins and Anthocyanins*

Anthocyanidins and anthocyanins are observable in fruits and flowers as they impart colour. Prominent anthocyanidin are aglycones, which include delphinidin, cyanidin, petunidin, peonidin, and malvidin, etc. which conjugate with sugars and organic acids to generate anthocyanins of different colours (Jaganath *et al.*, 2011; Ozeki *et al.*, 2011). Anthocyanins and their conjugates are responsible for the brilliant colour of red wine (Brouillard and Dangles, 1994; Darias-Martin *et al.*, 2001; Markovic *et al.*, 2000).

Polyphenolic content of wines varies with grape variety, origin, vinification process, maceration time period and maturation. Because of the high content of polyphenolics in the skin of red grapes and greater maceration period, red wine contains around tenfold higher polyphenolic content (Lingua *et al.*, 2016). Phenolic content is also influenced by the presence and concentration variations in ethanol content. The initial rate of flavonoid extraction from skins is not influenced by the alcohol content but as ethanol concentration increases, extraction of flavonoids from seeds increases simultaneously (Canals *et al.*, 2005). The phenolics in wine are also affected by the time and type of maturation process. Aged red wines have a different phenolic composition as compared to unmatured ones because of the development of polymeric compounds as well as oxidation, hydrolysis and other chemical transformations that take place in the already present phenolics (Soni *et al.*, 2011).

Apart from colour, aroma and mouth-feel related importance of the polyphenolic compounds, the one characteristic for which they have gained huge interest in the past century is their health benefits. Several in-vitro and in-vivo studies have established the medicinal effects of polyphenols. Because of their abundance in coloured fruits and vegetables, food products and beverages based on these phenolic-rich veggies have become tremendously popular. Most popular wines are grape wines and their polyphenols are accredited to the medicinal properties associated with wines. These polyphenols are located in different parts of the grapes, including the skin, pulp and seed. Anti-microbial, anti-cancer, hypoglycemic, hypolipidemic, anti-atherogenic, anti-oxidative, anti-carminative and many other health benefits of polyphenols have been scientifically evaluated in-vivo (Scalbert *et al.*, 2011). The molecular mechanism of their actions and genetic controls have also been studied (He *et al.*, 2008). Some of the disease-oriented genes, which are down-regulated by polyphenols, are depicted in Fig. 3.

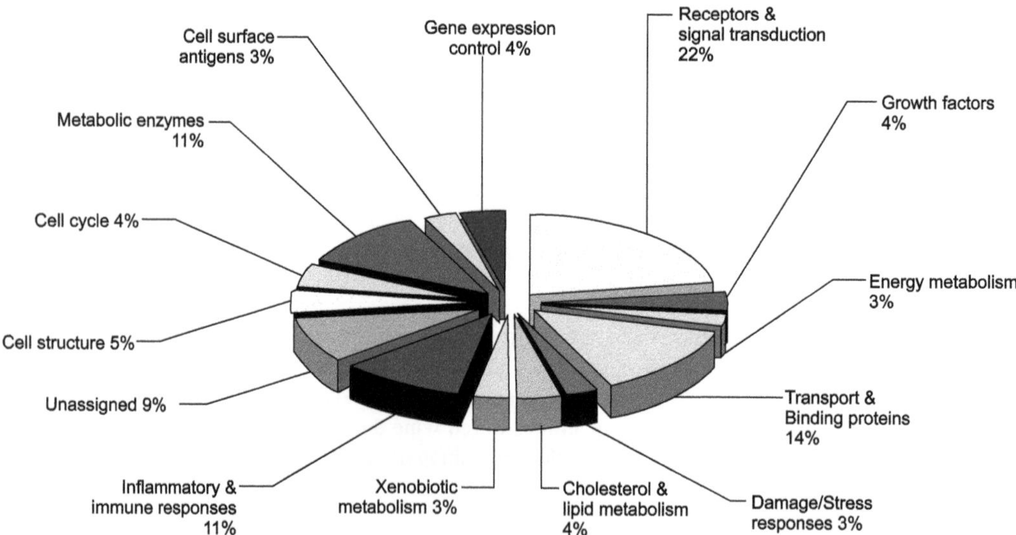


**Figure 3.** Per cent gene down-regulation by the action of polyphenols (Adapted from He *et al.*, 2008)

### 3.8. Esters

Esters are synthesised by the action of enzyme alcohol-acylcoA transferases on acyl-coA and alcohols. In wine, normally two types of esters are present: higher alcohols' acetate which impart diverse odours, such as rose (phenylethanol acetate), banana (isoamyl acetate), glue (ethyl acetate); and the second ones are the esters of ethanol and fatty acids which produce a fruity aroma. Apart from ethyl acetate, all esters contribute positively to the wine aroma (Sumby *et al.*, 2010).

Esters are produced in grapes, but rarely in significant quantities (Marais *et al.*, 1979). Most wine esters are the secondary products of yeast metabolism (Lee *et al.*, 2004). Over 160 wine esters have been identified till now (Marais and Pool, 1980). The range of major wine esters is described in Table 5. Out of all the esters, ethyl acetate is the most studied one. Its concentration is generally in a range of less than 50–100 mg/L in sound wine. At lower levels (<50 mg/L), it can enhance the complexity of the aroma, whereas at over 150 mg/L, it produces a sour-vinegary odour. The undesirable level of ethyl acetate is commonly associated with contamination with acetic acid bacteria (Amerine and Roessler, 1983; Sumby *et al.*, 2010).

**Table 5.** Range of Various Esters Identified in Wines

| *Ester* | *Range (mg/L)* |
|---|---|
| Ethyl acetate | 11-232 |
| n-Propyl acetate | 0.04-0.8 |
| Isobutyl acetate | 0-0.5 |
| Isoamyl acetate | Traces-9.3 |
| n-Hexyl acetate | 0-1.0 |
| Phenylethyl acetate | 0-1.14 |
| Ethyl butyrate | 0.2-0.44 |
| Ethyl caproate | Traces-1.8 |
| Ethyl caprylate | Traces-2.1 |
| Ethyl caprate | Traces-0.9 |
| Ethyl succinate | 0.2-6.3 |
| Mono-caffecyl tartarate | 70.9-233.8 |
| Mono-p-coumaroyl tartarate | 8.3-33.3 |
| Mono-feruloyl tartarate | 1.6-15.9 |
| Methyl anthranilate | 0.14-3.50 |

Adapted from Ough, 1991 and Soni *et al.*, 2011

### 3.9. Amino Acids

Amino acids are amino derivatives and contain a carboxyl group attached to the amine-linked carbon. They may originate from grapes and are also released into the wine by yeast autolysis (Martinez-Rodriguez *et al.*, 2004). They are critically significant components and are involved in biosynthesis of enzymes and other proteins. In addition, they are a major source of nitrogen and energy for metabolism of yeast. They constitute ≥90 per cent of the nitrogen content of musts. In addition to this, they also add to the flavour of wine after being metabolised to aldehydes, phenols, organic acids, lactones and higher alcohols. Few amino acids have undesirable odours and, therefore, above-threshold concentrations may lead to generation of bitter taste (Desportes *et al.*, 2001). The metabolism of amino acids does not affect the taste, but it may lead to production of biogenic amines and ethyl carbamate precursors, which are toxic and objectionable (Martinez-Rodriguez and Pueyo, 2008). In amino acids, proline is found in maximum quantities in both must and wine, as yeast does not metabolise it. Some amino acids, like ariginine (Callejon *et al.*, 2014), which are not present in the substrate, are found in the wines. These are formed due to metabolism of yeast strains.

### 3.10. Vitamins

Vitamins are involved in the regulation of cellular activity. They are found in minor quantities in wine and decrease during fermentation and aging due to oxidation by light, $SO_2$ or chemical reactions. Only *p*-aminobenzoic acid increases during fermentation (Jackson, 2014). Vitamins are added to the juice for stimulating vigorous fermentation and to reduce the use of $SO_2$. But when present in higher quantities, they may lead to generation of high levels of acetic acid which adds to volatile acidity of wine (Eglinton *et al.*, 1993; Ciani *et al.*, 2016). In wines, Vitamin A and C contents are insignificant and concentration of other vitamins may vary as thiamine (0-50), pyridoxine (70-100), riboflavin (5-120) and nicotinic acid (65-120) μg/100 mL respectively (Soni *et al.*, 2011).

### 3.11. Minerals

There are various mineral elements present in both grapes and wine. In adequate range, they act as important cofactors for vitamins and enzymes. However, some heavy metals, like cadmium, lead, selenium, and mercury are actually toxic and they usually precipitate during fermentation (von Hellmuth *et al.*, 1985; Cozzolino, 2015). The elemental composition of wine adds to the nutritional quality of the wine (Ibanez *et al.*, 2008). Metals affect the organoleptic characteristics of wine, including flavour, aroma, freshness, taste and colour (Gonzalez Hernandez *et al.*, 1996; Frias *et al.*, 2001; Galani-Nikolakaki *et al.*, 2002; Jos *et al.*, 2004). It has been revealed that a daily intake of wines in moderation, considerably adds to the necessity of humans for essential elements like Ca, Cr, Cu, Co, Fe, Mg, K, Mn, Mo, Zn or Ni (Lopez *et al.*, 1998; Teissedre *et al.*, 1998; Galani-Nikolakaki *et al.*, 2002; Lara *et al.*, 2005). Table 2.6 summarises the range of various minerals found in wines. Analysis of mineral content is also important for assessing the national safety of the wines as some of the heavy metals when present in higher quantities may cause damage to the body (Kartas *et al.*, 2015). Therefore, there are regulations in different countries to limit the heavy metal concentration in wines (Fiket *et al.*, 2011), as described in Table 6.

### 3.12. Proteins

Proteins in wine originate from yeast and grapes during vinification via skin/seed contact. The amount of contribution depends upon the grape diversity (Fukui and Yokotsuka, 2003). Molecular weight of wine proteins is in the range of 20-40 kD. Solubility of wine proteins depends on alcohol level, temperature and pH and abrupt alterations in any of these parameters may lead to protein precipitation. Precipitated proteins add to wine turbidity and interfere with organoleptic characteristics and foam formation (Boulton *et al.*, 1996). Yeast fermentation does not raise the protein or amino acids contents of the fermentation medium except where a nitrogen source is added. During fermentation, the amount of nitrogenous compounds including amino acid decreases because of their utilisation by yeast or precipitation of some proteins and their removal by centrifugation, decantation and filtration (Reed, 1981). Protein level of wines range from 10-300 mg/L. Agglomerated proteins may precipitate in the form of an evident unstructured fog and may lead to visual damage to the integrity of wine. This effect is augmented due to exposure to high temperatures.

## 4. Pharmacological Effects of Wines

There are numerous bioactive compounds in grape wines which are related the medicinal properties of the wines. The therapeutic capabilities of wines have been well investigated through animal and human trials. The starting point for wine and health studies was the 'French paradox' reported by Renaud and De Lorgeril (1992). Further clinical and epidemiological studies showed that regular and moderate wine consumption is associated with decreased incidence of diabetes and cardiovascular disease (CVD) (Rimm *et al.*, 1995; Leighton *et al.*, 1999; Mezzano *et al.*, 2001; Mukamal *et al.*, 2005; Avellone *et al.*, 2006; Djousse and Gaziano, 2008; Bertelli and Das, 2009; Sumpio *et al.*, 2016). Anti-inflammatory and antioxidant properties as well as actions on vascular function of wine inhibit atherosclerosis. It may be attributed to polyphenols present in wine which exhibit anti-carcinogenic (Ramos, 2008), antioxidant (Boban *et al.*, 2016; Vinson *et al.*, 2003; Pazzini *et al.*, 2015; del Pino-Gracia *et al.*, 2016), hypotensive (Bhatt *et al.*, 2011), anti-inflammatory (Calabriso *et al.*, 2016; Palmieri *et al.*, 2011), hepatoprotective (Silva *et al.*,

**Table 6.** Range of Various Metal Ions in Wines

| *Cation* | *Range (mg/L)* |
|---|---|
| Potassium | 90-2040 |
| Sodium | 3-320 |
| Calcium | 6-310 |
| Magnesium | 21-245 |
| Iron | 0.3-22 |
| Copper | Traces-2.4 |
| Lead | 0-1.26 |
| Zinc | Traces-11.7 |
| Aluminium | 0.67-5.4 |
| Arsenic | 0.001-0.53 |
| Antimony | 0.0012-0.009 |
| Barium | 0.02-0.22 |
| Boron | 3.9-80.3 |
| Cadmium | Traces-0.049 |
| Caesium | 0.0002-0.0047 |
| Chromium | 0.004-0.81 |
| Lithium | 0.005-0.11 |
| Manganese | 0.09-17.4 |
| Mercury | 0.02-0.65 |
| Molybdenum | 0.01-0.33 |
| Nickel | 0.01-0.40 |
| Rubidium | 0.58-4.99 |
| Silicon | 12.6-28.9 |
| Strontium | 0.14-3.13 |

Adapted and modified from Ough, 1991; Fiket *et al.*, 2011 and Soni *et al.*, 2011

2015) and even anticoagulant properties (Crescente *et al.*, 2009). The underlying mechanisms to explain these protective effects against CHD include a decrease in platelet aggregation, increase in high-density lipoprotein (HDL) cholesterol, increase in insulin sensitivity, restoration of endothelial dysfunction and a reduction in the levels of fibrinogen which are attributed to the ethanol and polyphenolic content in wine (Ruf, 2003; Lamuela-Raventos and Estruch, 2016; del Pino-Gracia *et al.*, 2016; Sumpio *et al.*, 2016). It is not clear as to which of the two factors, ethanol or polyphenolics, is more creditable for this activity and their synergism is ought to be of greatest significance for their reported medicinal effects (Lamuela-Raventós and Estruch, 2016).

Protective effects of moderate drinking have also been observed on kidney stones infection, hypertension, peptic ulcers and certain types of cancer, including basal cell, colon, ovarian and prostate carcinoma (Jindal, 1990; Arranz *et al.*, 2012). Prevention of cancer has been linked to the antioxidants present in wine (Bission *et al.*, 1995; Gronbaek *et al.*, 2000). Anti-carcinogenic activity of wine may be associated with its anti-inflammatory, antioxidant, anti-metastatic, anti-mutagen, anti-differentiation, anti-angiogenic, pro-apoptotic and anti-proliferative properties. It modulates the immune response, signal transduction, growth factors, transcription factors, cytokines, prostaglandin synthesis, interleukins and cell cycle-regulating proteins (Arranz *et al.*, 2012). Positive effects have also been stated for moderate wine consumption on cognitive function, cellular aging damage and dementia (Arranz *et al.*, 2012). Not only this, wine and wine phenolics lead to delay in degenerative diseases and exert protection against neurological disorders (Esteban-Fernández *et al.*, 2016). Several studies show that light-to-moderate alcohol consumers have an increased survival rate compared to abstainers (Di Castelnuovo *et al.*, 2006). Along with this, moderate wine consumption improves bone mass and structure (Kutlesa and Mrsic, 2016). Alcohol intake at regular intervals is observed to be beneficial. A summary of the biological events and triggering chemicals is presented in Fig. 4.

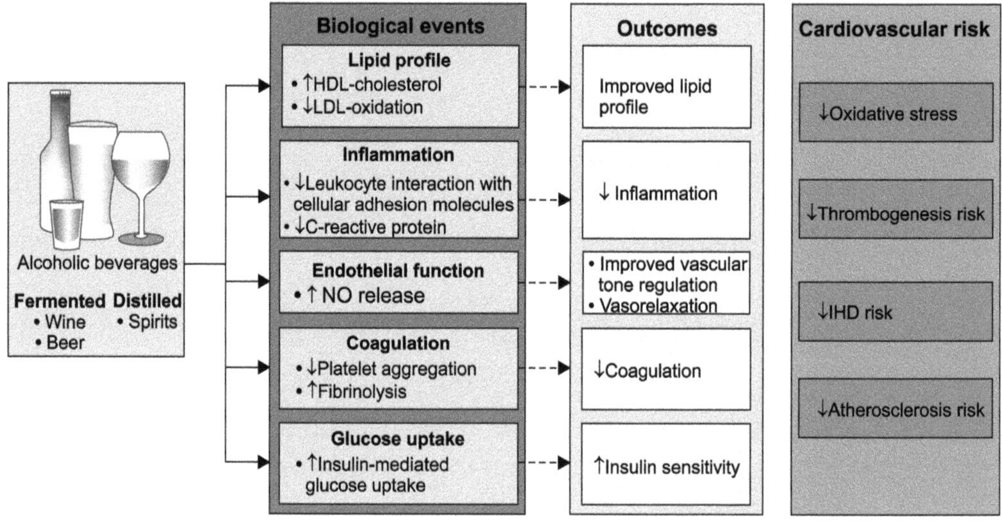


**Figure 4.** Schematic representation of the biological mechanisms of alcohol intake (HDL indicates high-density lipoprotein; IHD, ischemic heart disease; LDL, low-density lipoprotein; and NO, nitric oxide (*Source*: Adapted from Haseeb *et al.*, 2015)

All these studies confirm the potency of grape wine to combat various ailments. The establishment of relevant scientific data, proving the attributes of grape wine, has increased the popularity and consumption of this beverage. Over the past 20 years, wine consumption has increased steadily and is expected to continue to increase. According to Organisation Internationale (OIV), wine production and consumption has been estimated to be over 20,000 million litres per year worldwide (Grindlay *et al.*, 2011). Per capita, France consumes the maximum wine, followed by Italy, United States, Germany and Spain. Wine consumption in India is growing at the rate of 50 per cent per annum. One of the reasons for the consistent increase in annual sales is the health benefits provided by wine which further depends upon regular and moderate consumption (Clark, 2007). Although general recommendations are one drink (150 mL of wine or 10 g of alcohol) daily for women and two drinks (300 mL of wine or 20 g of alcohol) daily for men, individual ideals may vary based on age, gender, genetics, body type and drug/supplement use (Arranz *et al.*, 2012). A Chinese yellow wine has also been found to inhibit the onset and progression of atherosclerosis (Ji *et al.*, 2016). Thus, these new variants of the wines prove their medical potentials both in-vitro and in-vivo. These findings further widen the enthusiasm for search of other herbal substrates that can be explored for the production of wine and for evaluation of their benefits in animal models. It has been observed that red wines have a specifically beneficial effect in diabetes mellitus, which is a serious health burden and a prevalent syndrome. The detailed observed effects of wine on diabetes mellitus and the mechanism of action is discussed in the forthcoming section.

## 5. Wine as a Therapeutic Agent against Food-borne Pathogens

Wine consumption has been reported to have a protective effect as regards the susceptibility to develop bacterial food infection (Belido-Blasco *et al.*, 2002). Evidence exists that individuals who ingest wine during a meal become less susceptible to food-borne infections (Correia *et al.*, 2003). Thus, the potent antimicrobial activity exhibited by wine indicates a significant biological effect which has been demonstrated under various experimental conditions (Boban *et al.*, 2010a). The antibacterial efficacy of different variants of functional Aloe vera wine against *S. typhimurium, S. aureus* and *E. coli,* the common food-borne pathogens, was assessed by the presence or absence of inhibition zones and time-kill curves by Trivedi and co-workers (2013). It was observed that all the four functional wines showed remarkably higher efficacy of inhibition in comparison to respective controls, like 10 per cent ethanol and herbal extracts (Table 7). The average sizes of zones with *Aloe vera*, Aloe-*amla*, Aloe-ginger and Aloe-red grape wines against the three organisms measured about 10.0, 12.5, 11.0 and 10.5 mm, respectively making Aloe-*amla* wine proved slightly more efficacious in terms of its antibacterial activity as compared to its counterparts. Aloe-*amla* wine had the highest antibacterial activity against *S. typhimurium* and *E. coli* while Aloe-ginger wine worked best against *S. aureus*. Whereas, 10 per cent ethanol alone accounted for an average zone size of 2.7 mm, other components like *Aloe vera* extract, Aloe-*amla*, Aloe-ginger and Aloe-red grape extract measured 2.2, 3.5, 3.3 and 2.9 mm, respectively. The data signify that the synergistic effect of all components of wine results in a drastically improved antimicrobial potential as the unfermented herbal extracts are far weaker to inhibit bacterial growth.

**Table 7.** Size of Zone of Inhibition (mm)

| *Sample* | *Zone of Inhibition (mm)* | | |
|---|---|---|---|
| | *S. Typhimirium* | *S. aureus* | *E. coli* |
| Aloe vera gel | 2.4±0.03 | 2.0±0.04 | 2.2±0.04 |
| Aloe-*amla* extract | 4.0±0.06 | 3.0 ±0.06 | 4.1±0.07 |
| Aloe-ginger extract | 2.7±0.05 | 4.0±0.04 | 3.2±0.08 |
| Aloe-red grape juice | 2.6±0.03 | 3.0 ±0.04 | 3.1±0.04 |
| 10% ethanol (v/v) | 2.5±0.06 | 2.9±0.05 | 2.7±0.05 |
| Aloe vera wine | 10.1±0.12 | 9.0±0.18 | 11.0±0.08 |
| Aloe-*amla* wine | 14.0±0.09 | 11.0±0.07 | 12.0±0.07 |
| Aloe-ginger wine | 10.5±0.11 | 11.5±0.08 | 11.0±0.14 |
| Aloe-red grape wine | 11.0±0.08 | 10.5±0.06 | 10.0±0.10 |

Values are expressed as mean ± SD of the three different observations.

It is established that the wine has a greater antibacterial effect compared with the same concentration of diluted absolute ethanol and grape juice (Weisse *et al.*, 1995; Marimon *et al.*, 1998; Just and Daeschel, 2003). The antimicrobial activity has been studied for ethanol, methanol and acetone extracts of *Aloe vera* gel powder against four Gram-negative (*Escherichia coli*, *Salmonella typhi*, *Pseudomonas aeruginosa*, *Klebsiella pneumonia*) and Gram-positive (*Bacillus cereus*, *Bacillus subtilis*, *Streptococcus pyogenes*, *Staphylococcus aureus*) bacteria using the agar well diffusion method (Lawrence *et al.*, 2009). The methanol and ethanol extracts established higher activity than the acetone extract against most of the tested pathogens (the first two had an inhibition zone ranging between 12.6 and 23.3 mm and the maximum value was obtained for *Bacillus cereus*). However, the inhibition zone achieved on using the acetone extract ranged from 6.0 (for *Escherichia coli*) to 7.3 mm (for *Streptococcus pyogenes*) and no activity was detected for *Pseudomonas aeruginosa* and *Salmonella typhi*. Generally, these extracts showed better activity against Gram-positive bacteria (Lawrence *et al.*, 2009). The presence of high organic acids, high alcohol content and polyphenolic compounds contribute to the antibacterial property of wines.

The intrinsic antibacterial properties of wine have been verified by Weisse *et al.* (1995). They compared the antibacterial properties of red and white wines to bismuth subsalicylate, against pathogens generally responsible for traveller's diarrhoea. Red and white wines were found to cause the greatest reduction in number from $10^5$-$10^6$ cfu to none detected (<10 cfu). Bismuth subsalicylate was less effective than red and white wine, with a reduction in *E. coli* of $10^4$ cfu after a 20 min. exposure and *S. enteritidis* and *S. sonnei* requiring 60 and 120 min, respectively, for populations to become undetectable. Bismuth subsalicylate was more effective than the diluted tequila at reducing counts of *E. coli* and *S. enteritidis*, but only marginally more so with *S. sonnei*. Pure ethanol diluted to 10 per cent concentration did not exhibit any type of antibacterial activity.

Harding and Maidment (1996) inspected the antibacterial activity of fruit juices (grape juice, cider, and orange juice), red and white wines, pH buffers and industrial methylated spirits against *E. coli*, *S. sonnei* and *S. enteritidis*. They found the shortest survival time of bacteria when exposed to wine, whereas 10 per cent of methylated spirits and pH buffer had no effect on the bacteria, indicating that there were additional components that contribute to the wine's antibacterial activity beyond ethanol and pH. However, the addition of 10 per cent methylated spirit to grape juice was found to have antibacterial activity, greater than the two individually, suggesting a synergistic effect of ethanol and components in the wine. Fermentation process added considerably to the antibacterial activity as antibacterial action of the combined grape juice and methylated spirit was still less than that of a low-alcohol wine. Sugita-Konishi *et al.* (2001) also observed that red and white wines significantly decreased colony counts from $10^5$ to undetectable amounts within 30 min. It was concluded that neither ethanol nor sulphite was directly responsible for the wine's antibacterial activity.

Moretro and Daeschel (2004) took a different approach to investigate wine's antibacterial activity and examined it in conjunction with food-borne pathogens and mutants that lacked genes necessary to elicit a stress response. In the study, *E. coli* O157:H7 and an rpoS mutant, *Listeria monocytogenes* and a sigB mutant, *S. typhimurium* and an rpoS mutant, and *Staphylococcus aureus* and a sigB mutant were exposed to organic Chardonnay and Cabernet Sauvignon wine without added sulphites. They found the mutant strains lacking stress response genes had significantly less resistance to wine than the wild-type pathogens, except *S. typhimurium*, which was most susceptible and was not detectable after 5 min. of exposure. It was concluded that the genes encoding proteins necessary for stress response offer protection to ethanol, low pH, or organic acids. Red wine was also found to have a greater bactericidal effect than white wine on most strains tested. They also compared red and white wine to solutions containing a combination of organic acids, ethanol and low pH (0.15 per cent malic acid, 0.6 per cent tartaric acid, 15 per cent ethanol and pH 3). The synthetic mixture had significantly greater bactericidal activity than the red or white wine, or the acids, ethanol, or pH components tested individually, and the authors assumed that pH, organic acids and ethanol were largely responsible for the antibacterial effect found in wine.

A related finding was made by Trivedi *et al.* (2013) in a study that tested the antibacterial efficacy of functional Aloe vera wines against common food-borne pathogens i.e. *S. typhimurium, S. aureus* and *E. coli* by time-dependent bactericidal assay. Exposure to the prepared functional wines decreased the bacterial count of all the three pathogens from $10^7$ cfu/mL to undetectable levels within 20 min. as shown in Figs. 5a, b and c. Time-kill curve of the three pathogenic organisms was observed at five different time intervals, i.e. 0, 5, 10 and 15, 20 min. after exposure to all wine variants and corresponding unfermented herbal extracts, 10 per cent ethanol as controls separately. An initial count of nearly $10^7$ cfu/mL was used for the three pathogens. All the tested wines displayed a remarkable antimicrobial response against the pathogens. All three pathogens were completely inactivated after 20 min. of exposure to each wine.

*S. typhimurium* and *E. coli* were completely inactivated in 15 min. by Aloe-*amla* and Aloe-red grape wine while Aloe vera and Aloe-ginger wine took 20 min. for the same. Nearly 2 log reduction in growth was observed in the first five minutes of all the wine variants. Again, the respective controls showed relatively much weaker antibacterial activity with an average one log reduction in growth even after 20 min. of exposure.

In case of *S. aureus*, Aloe-ginger and Aloe-*amla* wines were most successful as these eliminated the pathogen after 15 min. of exposure. Here again, Aloe-ginger wine was slightly more effective in comparison to Aloe-*amla* wine. But, Aloe vera and Aloe-red grape wine were also fairly quick in action as they destroyed the pathogen within 20 min. of exposure. Synergism of all individual components in wines was probably responsible for the strong antimicrobial potential even in this case as the respective controls could only reduce the organism by one log after 20 min. of exposure.

Polyphenolic compounds in wine are also considered to be bactericidal. The antimicrobial activity of pure phenolic compounds, three flavonoids: rutin, catechin and quercetin, four phenolic acids: gallic, vanillic, protocatechuic and caffeic were tested against *Serratia marcescens*, *Flavobacterium* sp., *Escherichia coli*, *Proteus mirabilis*, *Klebsiella pneumoniae*, and *Staphylococcus aureus*. All of the bacteria showed different sensitivities to the phenolic compounds at different concentrations, with *Flavobacterium* being resistant to all compounds (Rodriguez Vaquero *et al.*, 2007a). Clarified wines were

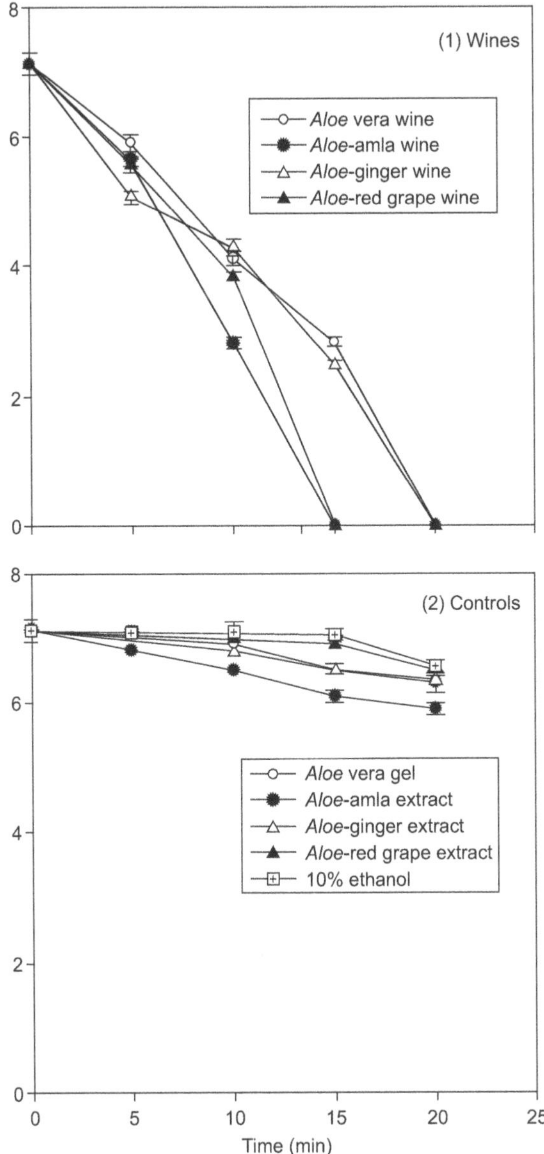


**Figure 5a.** The pattern of reduction of *S. typhimurium* counts in presence of different *Aloe vera* wines (1) and their respective unfermented extracts and 10 per cent (v/v) ethanol (2), indicating bactericidal activity. The data represents mean ± SD of three independent experiments

not found to be antimicrobial due to the lack of phenolic compounds. In the study by Radovanovic *et al.* (2009), antimicrobial activity of six red wines was found to be correlated with their total phenolic and monomeric anthocyanins content significantly. Likewise, Vaquero *et al.* (2007) in their study found that antimicrobial activity of tested wines increased with the increase in their polyphenols content, indicating an essential role of polyphenolic compounds present in red wines for the antimicrobial effects observed. It was also shown that antimicrobial activity of phenolic extracts from different red and white wines was related to their total phenolic content (Papadopoulou *et al.*, 2005). Nohynek *et al.* (2006) examined the mechanism of Nordic berry phenolics on the antibacterial activity against human pathogens. They found that *Bacillus cereus* was the only pathogen tested that was susceptible to all extracts, or strongly inhibited. *Candida albicans* was the most resistant, but was susceptible to three types of berries – raspberry, cloudberry and strawberry. Cloudberry and raspberry were found to be bactericidal rather than

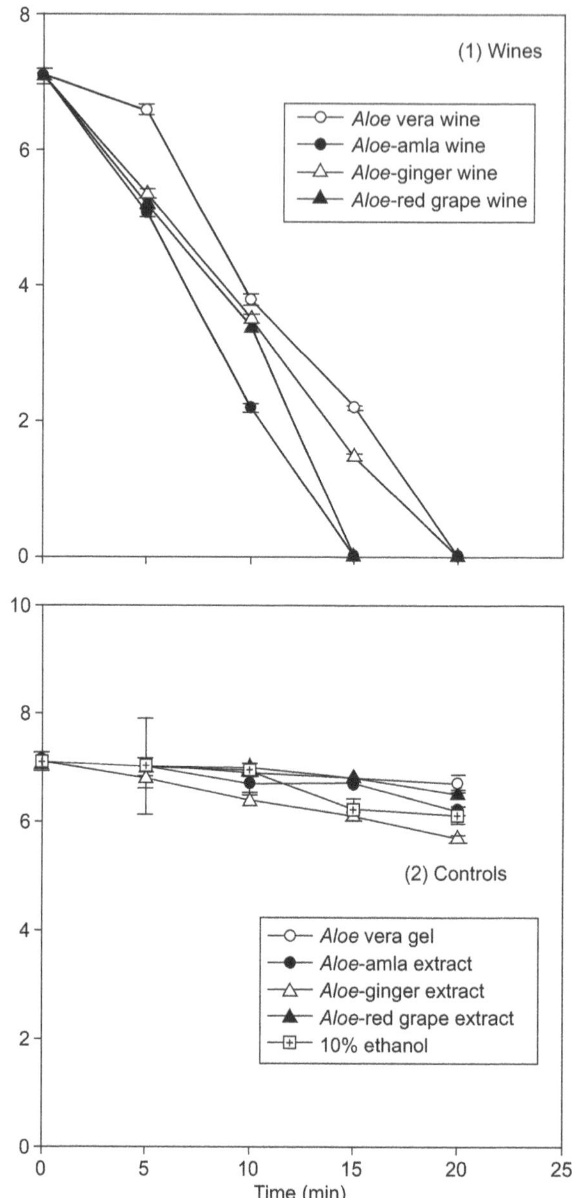


**Figure 5b.** The pattern of reduction of *S. aureus* counts in presence of different *Aloe vera* wines (1) and their respective unfermented extracts and 10 per cent (v/v) ethanol (2), indicating bactericidal activity. The data represents mean ± SD of three independent experiments

bacteriostatic and disrupted the outer membrane of *Salmonella*. Ellagic acid was found to be primarily responsible for the antimicrobial activity of the berries.

The effect of specific phenolic compounds and phenolic extracts of three red wines on the growth of *L. monocytogenes* was investigated by Rodriguez Vaquero *et al.* (2007b) for comparison to non-clarified concentrated wine. The wine concentrates were also clarified using activated charcoal to remove phenolic compounds. They found that all of the tested phenolic compounds inhibited the growth of *L. monocytogenes* and larger concentrations had a greater inhibitory effect. The wine extracts also showed greater effect from higher concentration, with a fourfold concentration showing the greatest inhibition. Polyphenolic compounds in wine were responsible for the inhibitory effect of wine as the clarified wines without phenolic compounds were found to be inactive against *L. monocytogenes*. Rodriguez Vaquero *et*

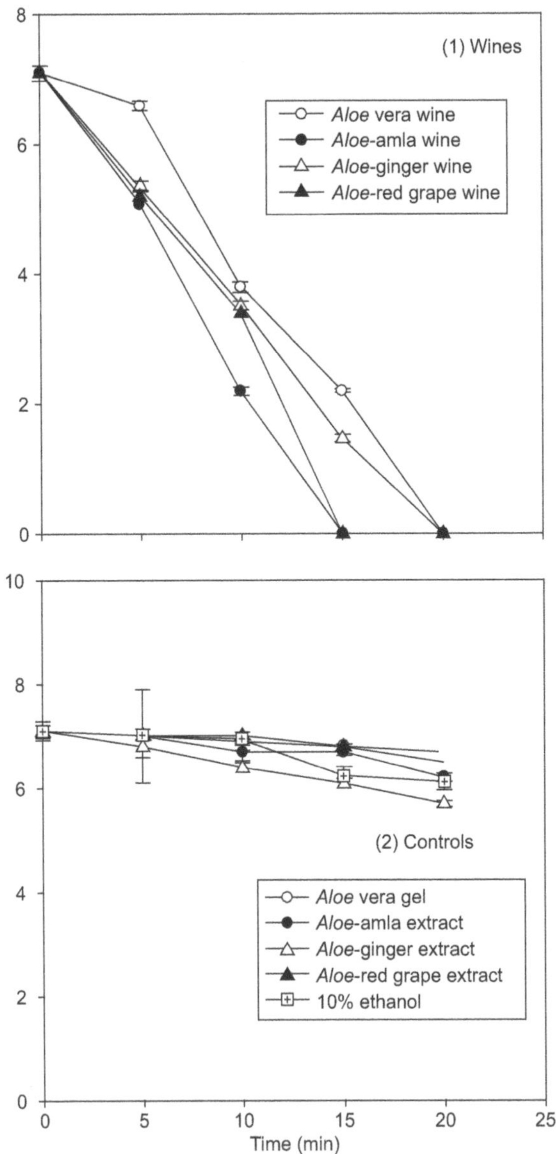


**Figure 5c.** The pattern of reduction of *E. coli* counts in presence of different *Aloe vera* wines (1) and their respective unfermented extracts and 10% (v/v) ethanol (2), indicating bactericidal activity. The data represents mean ± SD of three independent experiments

*al.* (2007a) made similar findings while examining the effect of phenolics and wine extracts on the growth of *S. marcescens, E. coli, K. pneumoniae, P. mirabilis,* and *Flavobacterium*. Effectiveness of the different phenolic compounds was different, with quercetin being inhibitorier than rutin; *E. coli* was found to be the most sensitive, while *Flavobacterium* was resistant to all phenolic compounds tested.

The combination of polyphenols can have a synergistic or an antagonistic effect on antibacterial properties. Arima *et al.* (2002) found that combinations of quercitin with either rutin or morin have significantly more activity than any flavonoid alone against *Bacillus cereus* and *S. enteritidis*. Related findings have been found with phenolic phytochemical-enriched alcoholic beverages on *Helicobacter pylori* (Lin *et al.*, 2005). Synergistic and antagonistic effects were also found on using a combination of polyphenols extracted from bergamot peel extracts (Mandalari *et al.*, 2007) and almond skins (Mandalari *et al.*, 2010).

Phenolic compounds in red, white and rose wine were found to possess antibacterial properties against *Campylobacter jejuni* (Ganan *et al.*, 2009). They found that red and rose wines were bactericidal at concentrations of 10 per cent or greater and white wine was bactericidal at 25 per cent and greater. Ethanol also contributes to the antibacterial activity as ethanol solutions were bactericidal at mixed concentrations of 25 per cent and greater as well. A study by Boban *et al.* (2010) examined the contributing factors of wine, specifically ethanol, phenolics and pH to the antibacterial activity of wine against *Salmonella enteric, S. enteritidis* and *Escherichia coli.* Their results showed that intact wine had the greatest antibacterial activity, followed by phenol-stripped wine, dealcoholised wine, a combination of low pH and ethanol, and finally low pH, followed by ethanol. Antioxidant capacity (as measured through a FRAP assay) and phenolic content of the solutions were closely related. The combination of ethanol and low pH was synergistic for antibacterial activity, having shown significant reduction in counts, where the ethanol and low pH treatments alone had a negligible effect.

The phenolic constituents of wine are judged as a 'protective factor' in regard to human health (Ghiselli *et al.*, 1995). These are important as they add to the sensory characteristics, principally astringency, colour and bitterness (Robichand *et al.*, 1990; Minussi *et al.*, 2003; Lesschaeve and Noble, 2005) and as they are also involved in pharmacological effects, including anti-microbial, anti-carcinogenic and anti-oxidant properties (Frankel *et al.*, 1993; Mazza *et al.*, 1993; Visioli *et al.*, 1998; Boselli *et al.*, 2009). Phenolic compounds in wine act as antioxidants through mechanisms involving both metal chelation and free-radical scavenging (Li *et al.*, 2009). The total phenolic content also defines the antioxidant competence of the wines (Lopez-Velez *et al.*, 2003). The quantity of phenolic material vary expansively in different types of wine, depending on the grape variety or the substrate used for wine production, environmental factors in the vineyard and the wine-processing techniques (Frankel *et al.*, 1995). Lopez-Velez *et al.* (2003) reported the content of total phenols, as determined by Folin-Ciocalteu method, varied from 1848 to 2315 mg GAE/L and these values were found to be in conformity with those reported by Rice-Evans (1996); Simonetti (1997) and Minnusi *et al.* (2003) Sato *et al.* (1996) reported 250-720 mg/L polyphenols in white wines and 735-2858 mg/L in red wines. Comparable levels were published by Frankel *et al.* (1995), while Singleton *et al.* (1980), in his publication, reported lower levels – 200 mg/L for white wines and 890–1200 mg/L for red wines. Kanner *et al.* (1994) found 1800–3200 mg/L polyphenols in selected red wines. Lugasi and Hovari (2003) informed the average concentration of polyphenols in white and red wines to be 392 and 1720 mg/L, respectively. Trivedi *et al.* (2013) reported the content of total phenols in the range of 1404.9 mg GAE/L-1797.05 mg GAE/L, averaging 1473.5 mgGAE/L, for the functional Aloe vera wine variants. The highest concentration was detected in Aloe-*amla* wine followed by Aloe-red grape wine, Aloe-ginger wine and Aloe vera wine (Table 8).

**Table 8.** Phenolic and Antioxidant Content of *Aloe* Functional Wines

| *Constituents* | *Aloe-amla wine* | *Aloe-ginger wine* | *Aloe-red grape wine* | *Aloe vera wine* |
|---|---|---|---|---|
| Phenolics (mgGAE/L) | 1797.05 ± 32.88 | 1218.6 ± 11.99 | 1404.9 ± 19.28 | 1065.0 ± 10.11* |
| Antioxidants (μmol/L) | 4550.0 ± 14.84 | 3040.0 ± 17.07 | 3270.0 ± 38.89 | 2950.0 ± 18.48* |

Values are expressed as mean±SD of the observations (in triplicate) from three independent experiments. *shows significant difference (*$p < 0.05$).

Phenol-rich beverages, like wines, contain a variety of low-molecular mass molecules and a lot of these have been credited with primary antioxidant role. In spite of much interest in the antioxidant activity of red wine, it is not yet known which of the phenols exhibit the greatest antioxidant effect. The high antioxidant activity of the wine sample could be accredited to the synergistic effects in a mixture of natural phenolic compounds. Loizzo *et al.* (2013) reported good antioxidant properties displayed by passito wine and the reported bioactivity could be related to the presence of flavonoids, anthocyanins and many other polyphenolic compounds that are known to be valuable to human health. The functional Aloe vera wine sample demonstrated significant antioxidant capacity with FRAP test as reported by Trivedi *et al.* (2013). The antioxidant activity assessed by determining the FRAP values varied from 3040.0±17.07 μmol/L to 4550.0±14.84 μmol/L averaging 3620.0 μmol/L for the functional variants of Aloe vera wine (Table 8). These values are quite similar to the commercial wine values reported in the literature (Cristino *et al.*, 2013; Loizzo *et al.*, 2013). The significant relationship between the antioxidant activity and total phenolic compounds suggests that phenolic compounds are the major contributors of antioxidant capacities of the functional wines. These results confirm that wine polyphenols are important antioxidants in vitro and may explain the beneficial effects of a moderate daily intake of wine. These results are in agreement with other reports in the literature (Sato *et al.*, 1996; Simonetti *et al.*, 1997; Henn and Stehle, 1998; Fogliano *et al.*, 1999; Sanchez-Moreno *et al.*, 1999; Lopez-Velez *et al.*, 2003; Minussi *et al.*, 2003; Katalinic *et al.*, 2004; Vichitphan and Vichitphan, 2007; Boban *et al.*, 2010a) as a very close relationship between total phenolic content and total antioxidant potential for wines was observed in these studies. Polyphenols can act as primary (chain breaking) and secondary (preventive) antioxidants and restrain the pathological lipid peroxidation processes. In addition, in-vitro FRAP values of wine might act as an empirical way of inferring the bio-availability of wine antioxidants which could serve to assess its potential biological benefit as antioxidant agent in-vivo, thereby protecting the organs against the oxidative challenges caused by ROS.

Phytochemical components, such as phenols, alkaloids, tannins, flavonoids, saponins and several other aromatic compounds are secondary metabolites of plants that serve a defence mechanism against many microorganisms, insects and other herbivores (Afolayan and Meyer, 1997; Lutterodt *et al.*, 1999; Marjorie, 1999; Bonjar *et al.*, 2004). Phytochemicals from fruits acting as quorum-sensing inhibitors could be another feature in wine which contributes to the health benefit against pathogenic bacteria. Quorum sensing is a chemical signalling system that transmits information about cell density and regulates the expression of genes in bacteria (Miller and Bassler, 2001). Auto-inducers are the signalling compounds which regulate a variety of cell behaviour, including antibiotic production, virulence, biofilm production and sporulation. Vattem *et al.* (2007) investigated the role of dietary phytochemicals in acting as quorum-sensing inhibitors, through the use of sub-lethal concentrations of herb extracts, fruit and spice. It was found that two different mechanisms were responsible for quorum-sensing inhibition, i.e. interference in the activity of the auto-inducer and the bacteria's ability to synthesise these compounds. Phytochemical profiling of functional Aloe vera wines revealed the presence of medicinally active constituents including tannins, flavonoids, glycosides, polysaccharides and free amino acids, as reported by Trivedi *et al.* (2013), which could be responsible for the antimicrobial properties of the functional wines (Table 9).

**Table 9.** Phytochemical Profile of Variants of *Aloe vera* Wine

| *Phytochemical analysis* | *Aloe vera wine* | *Aloe-ginger wine* | *Aloe-red grape wine* | *Aloe-amla wine* |
|---|---|---|---|---|
| Tannins | + | + | + | + |
| Flavonoids | + | + | + | + |
| Alkaloids | – | – | – | – |
| Saponins | – | – | – | – |
| Glycosides | + | + | + | + |
| Terpenoids | – | – | – | – |
| Polysaccharides | + | + | + | + |
| Free amino acids | + | + | + | + |

Trivedi and co-workers (2013) evaluated the therapeutic efficiency of different variants of functional Aloe vera wine prepared by supplementing the Aloe vera juice with additional herbs/fruits, including *amla*, ginger and red grapes. The bactericidal effect of the Aloe wines in murine infection model of *S. typhimurium* showed decreased bacterial infection in the homogenates of wine-fed *Salmonella*-challenged animals. A 2-3 logfold decrease in the bacterial load in liver, spleen and small intestine of the infected-wine-fed animals was observed relative to infected control, thus suggesting the therapeutic activity of wines against *S. typhimurium* infection as depicted in Fig. 6.

This describes the ability of Aloe-vera wine to arrest the bacteria in different organs of the infected mice and thus proving to be an alternate intervention for the same (Trivedi *et al.*, 2012; Trivedi *et al.*, 2015 a,b). Additionally, wine administration also had a restorative influence on generated oxidative stress as infected-wine-fed groups displayed reduced lipid peroxidation as compared to the infected counterpart. The maximal effect was observed with Aloe-*amla* wine, which showed 56.01 per cent decrease in MDA

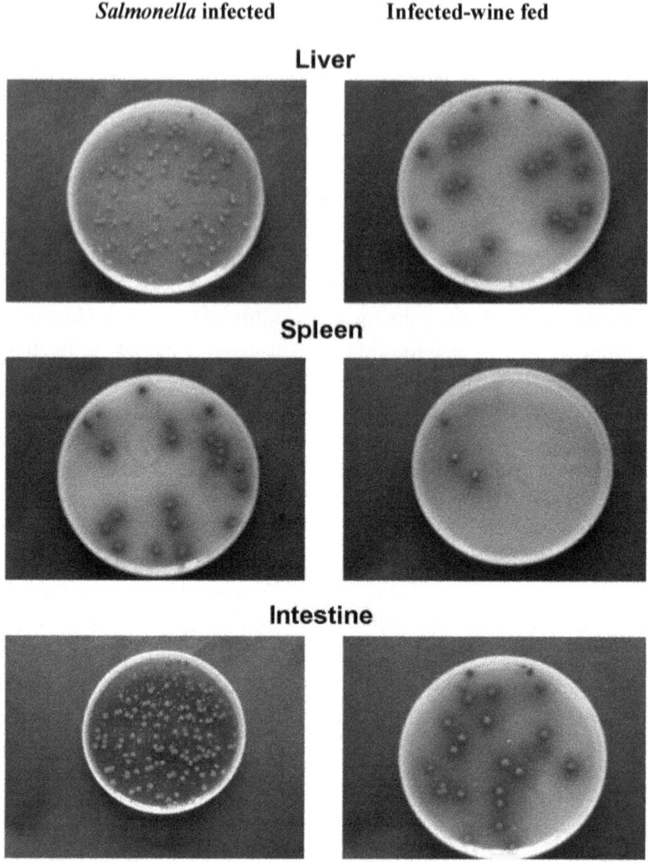


**Figure 6.** Representative *in-vivo* bacterial count plate of liver, spleen and intestine on BSA medium for the two mice groups: infected control group and infected-wine-fed group

levels as compared to infected control. All the other three wines, i.e. Aloe vera, Aloe-ginger, Aloe-red grape wine also significantly reduced the MDA levels ($p < 0.001$). Similarly, other markers of oxidative stress, like Superoxide Dismutase and Reduced Glutathione levels were also arrested by the administration of the wine variants (Trivedi, 2013). Another variant of Aloe wine was prepared by Chauhan (2015). *Mentha arvensis* and a probiotic strain, *Lactobacillus sporogenes* were the two additional components of this wine. The phenolic content of this wine was 1431 mgGAE/L and a FRAP value of 7.13 (AAmmol/ml). This wine was able to inhibit the common food-borne pathogens in in-vitro analysis. Minimum inhibitory concentration (MIC) for *S. typhimiurum* was found to be 24 per cent for Aloe-mentha wine. For *S. aureus,* MIC was recorded 24 per cent; MIC for *E. coli* was 16 per cent. Minimum bactericidal concentrations (MBCs) of wine against the three pathogens were found to be 32 per cent, 24 per cent and 32 per cent for *S. aureus, E. coli* and *S. typhimurium* respectively. The anti-bacterial effect in terms of MIC and MBC of different wine variants of Aloe vera is given in Table 10.

Seauckthorn is another berry whose wine has been validated in-vivo for its medicinal attributes. It has been found that the wine possesses a total phenolic content of 2182 mgGAE/L. when evaluated against phorone-induced hypercholesterolemia; seabuckthorn was able to arrest the disease and brought back the cholesterol to near normal value in mice. It was also observed that wine also reduced the oxidative stress markers in the mice and an overall positive influence was exerted on mice, when administered with seabuckthorn wine (Negi *et al.*, 2013). Apart from this, *Phyllanthus* wine was found to be quite rich in phenolics and inhibited cell proliferation and LDL-oxidation (Nambiar *et al.*, 2016). Olive wine too was found to reduce MDA levels in the liver of the mouse and improved carbonyl and SOD in liver and plasma, thereby exerting antioxidant effect (Yao *et al.*, 2016). Hibiscus wine has also been found to inhibit

**Table 10.** Efficacy of Some *Aloe vera*-based Herbal Wines against Common Food-borne Pathogens

| *Pathogen* | *Wine* | *MIC* | *MBC* |
|---|---|---|---|
| *S. Typhimurium* | Aloe vera wine | 35% | 40% |
| | Aloe-*amla* wine | 25% | 35% |
| | Aloe-ginger wine | 40% | > 50% |
| | Aloe-red grape wine | 35% | > 50% |
| | Aloe-mentha wine | 24% | 32% |
| *S. aureus* | Aloe vera wine | 45% | 50% |
| | Aloe-*amla* wine | 40% | 45% |
| | Aloe-ginger wine | 30% | 40% |
| | Aloe-red grape wine | 35% | 50% |
| | Aloe-mentha wine | 16% | 24% |
| *E. coli* | Aloe vera wine | 45% | > 50% |
| | Aloe-*amla* wine | 40% | 45% |
| | Aloe-ginger wine | 50% | > 50% |
| | Aloe-red grape wine | 50% | > 50% |
| | Aloe-*mentha* wine | 24% | 32% |

α-gluosidase in-vitro (Ifie *et al.*, 2016). A locally-prepared banana wine in Burundi, Africa has been found to eliminate excess uric acid from body when consumed in moderate quantities. Higher intake, however, was found to increase the storage of uric acid inside the body (Ndamanisha and Nkuririmana, 2016).

## 6. Wine Therapy for Management of Diabetes Mellitus

Since ancient times, wine consumption has been associated with health benefits. It's only since last two decades that the scientific basis of medicinal properties of wine is being technically confirmed. Numerous studies have demonstrated the antioxidant, anti-coagulant, anti-cancer and other pharmacological effects of wine. The greater remedial potential of red wine is due to the high polyphenolic content. Apart from bioactive components of red wine, the ethanol content has also been found to be effective medically. The mechanisms of health beneficial effects of wine are complex as various pathways are involved in the action. Some studies relate alcohol content with the positive effect of wine or cardiovascular diseases. Others describe polyphenols as the major contributing factor (Arranz *et al.*, 2012). Various effects of red wine and its active component, resveratrol, are described in Fig. 6.

Wine intake has beneficial effect on diabetes also. Safety studies showed that even a relatively large dose of ethanol like 400 ml of red wine with dinner had no negative effect on the glycemic control of insulin-dependent diabetes mellitus (IDDM) patients. It even intensified the meal-induced insulin secretion in non-insulin-dependent diabetes mellitus (NIDDM) patients, thus dropping blood glucose levels (German and Walzem, 2000). In human studies also, wine exertied no negative effect on diabetic patients (Gin *et al.*, 1992).

The polyphenolic extract of red wine at a dose of 200 mg/kg was found to maintain euglycemia in diabetic rats. It also reduced the food intake and absorption in streptozotocin-induced hyperglycemic rats (Al-Awwadi *et al.*, 2004). Even after treatment with polyphenol-enriched Chardonnay wines, the plasma antioxidant levels were restored in diabetic rats. The study confirmed the *in-vivo* transfer of antioxidant capacity from wine to organism (Landrault *et al.*, 2003). Oral administration of red wine was able to control even the post-prandial oxidative stress and thrombosis. It was observed that the absorptive phase-free radicals generated after meal intake in diabetics were arrested by red wine. Further, it also decreased the LDL oxidation and thrombosis and, therefore, its preventive effect in cardiovascular complications of diabetic was established (Ceriello *et al.*, 2001). Even in human studies, red wine, tannic acids and ethanol were found to have a positive effect on glucose levels of diabetic patients. However, this was ascribed to the non-alcoholic compounds of red wine (Gin *et al.*, 1999). Furthermore, in NIDDM, red wine intake improved whole body glucose disposal by 43 per cent. Concretely, it was established that red wine attenuates insulin resistance in patients in a span of two weeks only (Napoli *et al.*, 2005). Another study by Kwon and co-workers (2008) showed the α-amylase and β-glucosidase-inhibiting profiles of red wines in-vitro. Regular and moderate wine consumption has also been prescribed for ameliorating the diabetic-oxidative status. It has been pointed out that regular moderate intake could prevent the onset of diabetes and eliminate further diabetic complications (Caimi *et al.*, 2003). When taken with meals, significantly improved cardiac functions and reduced oxidative stress, pro-inflammatory cytokines (Marfella *et al.*, 2006) was seen. Not only this, wine phenolics also modulate gut microbiota of hyperglycemic subjects along with controlling their blood glucose levels (M1oreno-Indias *et al.*, 2016).

Resveratrol, a highly functional bioactive component of red wine, has also been investigated for its effects on diabetic animals. Its administration implicated a moderate yet persistent positive effect on hyperglycemia and hyperlipidemia. The activity is explained by reduced digestion, food intake and absorption of foods in the gastro-intestine. Other studies elucidated the flavonoid's hypoglycemic activity in pre-clinical studies. Epicatechin has the capability of inducing pancreatic β-cell regeneration; catechin hinders intestinal glucose absorption; epigallocatechin escalates hepatic glycogen synthesis. Further, it has been observed that oligomeric flavonoid procyanidins control glycemia by delaying glucose absorption in the intestine and exert an insulin-like effect on tissues (Su *et al.*, 2006). Not only in the gastric and hepatic system, but also red wine and resveratrol exhibit antioxidant capabilities in the nervous system

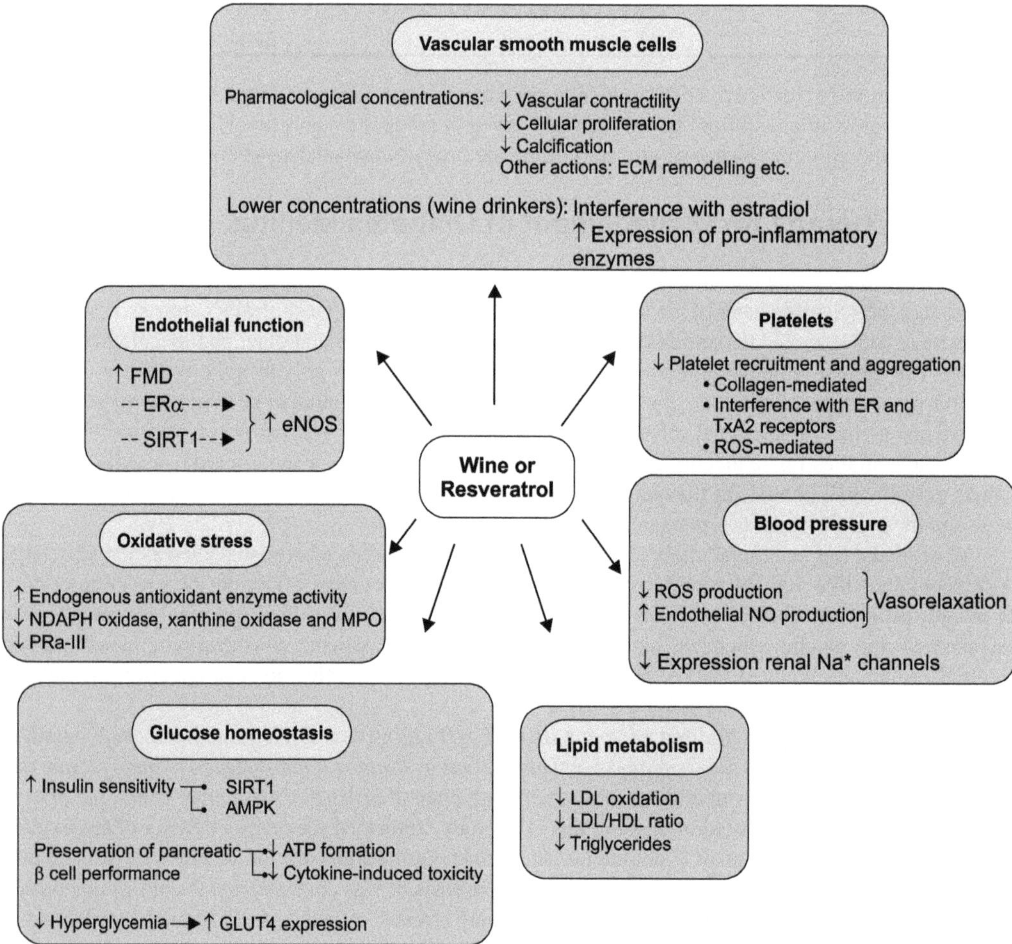


**Figure 7.** Biological effects associated with wines and resveratrol (*Source*: Adapted from Artero *et al.*, 2015) AMPK – 5'AMP-activated protein kinase; Erα – estrogen receptor alpha; eNOS – endothelialnitric oxide synthase; ECM – extracellular matrix; ER – estrogen receptor; FMD – flow-mediated dilation; GLUT4 – glucose transporter- 4; NO – nitric oxide; HDL – high-density lipoprotein cholesterol; LDL – low-density lipoprotein cholesterol; ROS – reactive oxygen species; TxA2 – thromboxane A2; SIRT1 – Sirtuin

of diabetic animals and do not cause any proliferation in the lower hippocampal cell (Venturini *et al.*, 2010). The biological effects of resveratrol and wines is also shown in Fig. 7. All these observations point towards the synergistic mode of the action of ethanol, phenolics and other biochemical constituents of wine against diabetes. This encourages the evaluation of other potent herbal candidates that can be used for preparation of wine and which may be even healthier and effective against DM.

Although wines have been prepared from various plants including some herbs, there are many more plants that can be investigated for their utilisation for winemaking or for making multi-substrate wines. Also, the pre-clinical studies show significant health benefits of wines on humans, clinical studies regarding the utilisation of wines as a functional food or medical aid are not up to par. The current human trials have investigated the effect of red wine on heart disease, kidney injury, obesity, etc. The effects of non-traditional wines need to be validated so as to increase their popularity in terms of therapeutic benefits. As the market of functional foods is increasing each day, wines can be targeted as a crucial component for a specific consumer population. However, the taste, disease-prevention abilities, price and moderate use of the wines are the criteria that should be emphasised for obtaining maximum advantage of the wine as a beverage as well as an aid.

## References

Afolayan, A.J. and Meyer, J.J. (1997). The antimicrobial activity of 3,5,7-trihydroxyflavone isolated from the shoots of *Helichrysum aureonitens*. *J. Ethnopharm.* 57: 177-181.

Al-Awwadi, N., Azay, J., Poucheret, P., Cassanas, G., Krosniak, M., Auger, C., Gasc, F., Rouanet, J.M., Cros, G. and Teissèdre, P.L. (2004). Antidiabetic activity of red wine polyphenolic extract, ethanol, or both in streptozotocin-treated rats. *J. Agric. Food Chem.* 52: 1008-1016.

Amerine, M.A., Berg, H.W. and Cruess, W.V. (1972). The Technology of Wine Making. AVI Publishing Co. Inc., Westport CT, USA.

Arima, H., Ashida, H. and Danno, G. (2002). Rutin-enhanced antibacterial activities of flavonoids against *Bacillus cereus* and *Salmonella enteritidis*. *Biosci. Biotechnol. Biochem.* 66: 1009-1014.

Arranz, S., Chiva-Blanch, G., Valderas-Martinez, P., Medina-Remon, A., Lamuela Raventos, R.M. and Estruch, R. (2012). Wine, beer, alcohol and polyphenols on cardiovascular disease and cancer. *Nutrients* 4: 759-781.

Artero, A., Artero, A., Tarín, J.J. and Cano, A. (2015). The impact of moderate wine consumption on health. *Maturitas* 80: 3-13.

Avellone, G., Di Garbo, V., Campisi, D. *et al.* (2006). Effects of moderate Sicilian red wine consumption on inflammatory biomarkers of atherosclerosis. *Eur. J. Clin. Nutr.* 60: 41-47.

Bakker, J. and Clarke, R.J. (2011). Volatile compounds. pp. 155-238. *In*: R.J. Clarke and J. Bakker (Eds.). Wine: Flavour Chemistry. John Wiley & Sons. Chichester, UK.

Barbe, J.C., de Revel, G. and Bertrand, A. (2002). Gluconic acid, its lactones and $SO_2$.binding phenomena in musts from botrytised grapes. *J. Agric. Food Chem.* 50: 6408-6412.

Barbe, J.C., de Revel, G., Joyeux, A., Lonvaud-Funel, A. and Bertrand, A. (2000). Role of carbonyl compounds in $SO_2$-binding phenomena in musts and wines from botrytised grapes. *J. Agric. Food Chem.* 48: 3413-3419.

Basha, S.M., Musingo, M. and Colova, V.S. (2004). Compositional differences in the phenolics compounds of muscadine and bunch grape wines. *Afr. J. Biotechnol.* 3: 523-528.

Başkan, K.S., Tütem, E., Akyüz, E., Özen, S. and Apak, R. (2016). Spectrophotometric total reducing sugars assay based on cupric reduction. *Talanta* 147: 162-168.

Belido-Blasco, J.B., Arnedo-Pena, A., Cordero-Cutillas, E., Canos-Cabedo, M., Herrero Carot, C. and Safont-Adsuara, L. (2002). The protective effect of alcoholic beverages on the occurrence of a *Salmonella* food-borne outbreak. *Epidemiol.* 13: 228-230.

Bertelli, A.A. and Das, D.K. (2009). Grapes, wines, resveratrol and heart health. *J. Cardiovasc. Pharmacol.* 54: 468-476.

Bhatt, S.R., Lokhandwala, M.F. and Banday, A.A. (2011). Resveratrol prevents endothelial nitric oxide synthase uncoupling and attenuates development of hypertension in spontaneously hypertensive rats. *Eur. J. Pharmacol.* 667: 258-264.

Bisson, L.F., Butzke, C.E. and Fbeler, S.F. (1995). The role of moderate ethanol consumption in health and human nutrition. *Am. J. Enol. Vitic.* 46: 449-462.

Boban, M., Stockley, C., Teissedre, P., Patrizia Restani, P., Fradera, U., Stein-Hammer, C. and Ruf, J. (2016). Drinking pattern of wine and effects on human health: Why should we drink moderately and with meals? *Food Func.* 7: 2937-2942.

Boban, N., Tonkic, M., Budimir, D., Modun, D., Sutlovic, D., Punda-Polic, V. and Boban, M. (2010a). Antimicrobial effects of wine: Separating the role of polyphenols, pH, ethanol, and other wine components. *J. Food Sci.* 75: 322-326.

Boban, N., Tonkic, M., Modun, D., Budimir, D., Mudnic, I., Sutlovic, D., Punda-Polic, V. and Boban, M. (2010b). Thermally-treated wine retains antibacterial effects to food-borne pathogens. *Food Control* 21: 1161-1165.

Bonjar, G.H.S., Nik, A.K. and Aghighi, S. (2004). Antibacterial and antifungal survey in plants used in indigenous herbal-medicine of southeast regions of Iran. *J. Biol. Sci.*, 4: 405-412.

Boselli, E., Bendia, E., Di Lecce, G., Benedetti, A. and Frega, N.G. (2009). Ethyl caffeate from Verdicchio wine: Chromatographic purification and in vivo evaluation of its antifibrotic activity. *J. Sep. Sci.*, 32: 1-6.

Boulton, R.B., Singleton, V.L., Bisson, L.F. and Kunkee, R.E. (1996). Principles and Practices of Winemaking. New York: Chapman & Hall.

Cabaroglu, T. (2005). Methanol contents of Turkish varietal wines and effect of processing. *Food Control* 16: 177-181.

Caimi, G., Carollo, C. and Lo Presti, R. (2003). Diabetes mellitus: Oxidative stress and wine. *Curr. Med. Res. Opin.* 19: 581-586.

Calabriso, N., Scoditti, E., Massaro, M., Pellegrino, M., Storelli, C., Ingrosso, I., Giovinazzo, G. and Carlucci, M.A. (2016). Multiple anti-inflammatory and anti-atherosclerotic properties of red wine polyphenolic extracts: Differential role of hydroxycinnamic acids, flavonols and stilbenes on endothelial inflammatory gene expression. *Eur. J. Nutr.* 55: 477-489.

Calderone, G., Naulet, N., Guillou, C. and Reneiro, F. (2004). Characterisation of European wine glycerol: Stable carbon isotope approach. *J. Agric. Food Chem.* 52: 5902-5906.

Callejon, S.R., Ferrer, S.S. and Pardo, I. (2014). Identification of a novel enzymatic activity from a lactic acid bacteria able to degrade biogenic amines in wine. *Appl. Microbiol. Biotechnol.* 98: 185-198.

Canals, R., Llaudy, M.C., Valls, J., Canals, J.M. and Zamora, F. (2005). Influence of ethanol concentration on the extraction of colour and phenolic compounds from the skin and seeds of Tempranillo grapes at different stages of ripening. *J. Agric. Food Chem.* **53**: 4019-4025.

Carvallo, J., Labbe, M., Perez-Correa, J.R., Zaror, C. and Wisniak, J. (2011). Modelling methanol recovery in wine distillation stills with packing columns. *Food Control*. 22: 1322-1332.

Castillo-Munoz, N., Gomez-Alonso, S., Garcia-Romero, E. and Hermosin-Gutierrez, I. (2010). Flavonol profiles of *Vitis vinifera* white grape cultivars. *J. Food Comp. Anal.* 23: 699-705.

Ceriello, A., Bortolotti, N., Motz, E., Lizzio, S., Catone, B., Assaloni, R., Tonutti, L. and Taboga, C. (2001). Red wine protects diabetic patients from meal-induced oxidative stress and thrombosis activation: A pleasant approach to the prevention of cardiovascular disease in diabetes. *Eur. J. Clin. Invest.* 31: 322-328.

Chauhan, A. (2015). Development of a non-traditional probiotic Aloe-mentha herbal wine and evaluation of its in-vitro efficacy against common food-borne pathogens. An M.Sc. thesis submitted to Panjab University, Chandigarh.

Chen, E.C.H. (1978). The relative contribution of Ehrlich and biosynthetic pathways to the formation of fusel alcohols. *J. Am. Soc. Brew. Chem.* 35: 39-43.

Ciani, M., Capece, A., Comitini, F., Canonic, L., Siesto, G. and Romano, P. (2016). Yeast interactions in inoculated wine fermentation. *Front Microbiol.* 7: 555. http://dx.Doi.org/10.3389/fmicb.2016.00555

Clark, D. (2007). California wine sales up 2 per cent in US California Wine and Food. http://californiawineandfood.com/articles/348/1/2007-California-Wine-Sales-Up-2-Percent-in-US/Page2.html

Conner, J.M., Birkmyre, L., Paterson, A. and Piggott, J.R. (1998). Headspace concentrations of ethyl esters at different alcoholic strengths. *J. Sci. Food Agric.* 77: 121-126.

Correia, A., Gomes, A., Oliveira, B., Gonçalves, G., Miranda, M. and Almeida, O. (2003). The protective effect of alcoholic beverages in food-borne outbreaks of *Salmonella enteritidis* PT1 in northern Portugal. *Eurosurveillance Weekly* 7: 1-13.

Cozzolino, D. (2015). Elemental composition in grapes and wine. *In*: M. de la Guardia and S. Garrigues (Eds.). Handbook of Mineral Elements in Food. John Wiley & Sons Ltd. Chichester, UK.

Crescente, M., Jessen, G., Momi, S., Holtje, H.D., Gresele, P., Cerletti, C. and de Gaetano, G. (2009). Interactions of gallic acid, resveratrol, quercetin and aspirin at the platelet cyclooxygenase-1 level: Functional and modelling studies. *Thromb. Haemost.*, 102: 336-346.

Crippen, D.D. and Morrison, J.C. (1986). The effects of sun exposure on the compositional development of Cabernet Sauvignon berries. *Am. J. Enol. Vitic.* 37: 235-242.

Cristino, R., Costa, E., Cosme, F. and Jordao, A.M. (2013). General phenolic characterisation, individual anthocyanin and antioxidant capacity of matured red wines from two Portuguese appellations of origins. *J. Sci. Food Agric.* 93: 2486-2493.

D'Angelo, M., Onori, G. and Santucci, A. (1994). Self-association of monohydric alcohols in water: Compressibility and infrared absorption measurements. *J. Chem. Phys.*, 100: 3107-3113.

Darias-Martin, J., Carillo, M., Diaz, E. and Boulton, R.B. (2001). Enhancement of red wine colour by pre-fermentation addition of co-pigments. *Food Chem.* 73: 217-220.

Deibner, L., Mourgues, L. and Cabibel-Hughes, M. (1965). Evolution d l'indice des substances aromatiques volatiles des raisins de deux cepages rouges au cours de leur maturation. *Ann. Technol. Agric.* 14: 5-14.

del Alamo, M., Bernal, J.L., del Nozal, M.J. and Gómez-Cordovés, C. (2000). Red wine aging in oak barrels: Evolution of the monosaccharides content. *Food Chem.* 71: 189-193.

del Pino-García, R., Gerardi, G., Rivero-Pérez, M.D., González-San José, M.L., García-Lomillo, J. and Muñiz, P. (2016). Wine pomace seasoning attenuates hyperglycaemia-induced endothelial dysfunction and oxidative damage in endothelial cells. *J. Func. Foods* 22: 431-445.

Desportes, C., Charpentier, M., Duteurtre, B., Maujean, A. and Duchiron, F. (2001). Isolation, identification and organoleptic characterisation of low-molecular-weight peptides from white wines. *Am. J. Enol. Vitic.* 52: 376-380.

Di Castelnuovo, A., Costanzo, S., Bagnardi, V., Donati, M.B., Iacoviello, L. and de Gaetano, G. (2006). Alcohol dosing and total mortality in men and women: An updated meta-analysis of 34 prospective studies. *Arch. Intern. Med.* 166: 2437-2445.

Di Lorenzo, A., Bloise, N., Meneghini, S., Sureda, A., Tenore, G.C., Visai, L., Arciola, C.R. and Daglia, M. (2016). Effect of winemaking on the composition of red wine as a source of polyphenols for anti-infective biomaterials. *Materials* 9: 316.

Djoussé, L. and Gaziano, J.M. (2008). Alcohol consumption and heart failure: A systematic review. *Curr Atheroscler. Rep.* 10: 117-120.

Drel, V.R. and Sybirna, N. (2010). Protective effects of polyphenolics in red wine on diabetes associated oxidative/nitrative stress in streptozotocin-diabetic rats. *Cell Biol. Int.*, 34: 1147-1153.

Dry, P.R. and Coombe, B.G. (2004). Viticulture-1 Resources, second edition. Winetitles, Adelaide, Australia.

Eglinton, J.M., Buckingham, L. and Henschke, P.A. (1993). Increased volatile acidity of white wines by chemical vitamin mixtures is grape juice dependent. pp. 197–198. *In*: Stockley, C.S. (Ed.). Proc. 8th Aust. Wine Ind. Tech. Conf., 25–29 Oct., 1992. Melbourne, Victoria Winetitles, Adelaide, Australia.

El Darra, N., Turk, M.F., Ducasse, M.A., Grimi, N., Maroun, R.G., Louka, N. and Vorobiev, E. (2016). Changes in polyphenol profiles and colour composition of freshly fermented model wine due to pulsed electric field, enzymes and thermovinification pretreatments. *Food Chem.* 194: 944-950.

Esteban-Fernández, A., Corona, G., Vauzour, D. and Spencer, J.P.E. (2016). Neuroprotective effects associated with wine and its phenolic constituents. pp. 279-292. *In*: "https://www.bookdepository.com/author/M-Victoria-Moreno-Arribas" M.V. Moreno-Arribas and "https://www.bookdepository.com/author/Begona-Bartolome-Sualdea" B.B. Sualdea (Eds.). Wine Safety, Consumer Preference and Human Health. Springer International Publishing, Switzerland.

Feher, J., Lengyel, G. and Lugasi, A. (2007). The cultural history of wine-theoretical background to wine therapy. *Cent. Eur. J. Med.* 2: 379-391.

Fiket, Z., Mikac, N. and Kniewald, G. (2011). Arsenic and other trace elements in wines of eastern Croatia. *Food Chem.* 126: 941-947.

Fogliano, V., Verde, V., Randazzo, G. and Ritieni, A. (1999). Method for measuring antioxidant activity and its application to monitoring the antioxidant capacity of wines. *J. Agri. Food Chem.* 47: 1035-1040.

Frankel, E.N., Waterhouse, A.L. and Kinsella, J.E. (1993). Inhibition of human LDL oxidation by resveratrol. *Lancet* 341: 1103-1104.

Frankel, E.N., Waterhouse, A.L. and Teissedre, P.L. (1995). Principal phenolic phytochemicals in selected California wines and their antioxidant activity in inhibiting oxidation of human low density lipoproteins. *J. Agric. Food Chem.* 43: 890-894.

Frias, S., Perez Trujillo, J.P., Pena, E.M. and Conde, J.E. (2001). Classification and differentiation of bottled sweet wines of Canary islands (Spain) by their metallic content. *Eur. Food Res. Technol.* 213: 145-149.

Fukui, M. and Yokotsuka, K. (2003). Content and origin of protein in white and red wines: Changes during fermentation and maturation. *Am. J. Enol. Vitic.* 54: 178-188.

Galani-Nikolakaki, S., Nallithrakas-Kontos, N. and Katsanos, A.A. (2002). Trace element analysis of cretan wines and wine products. *Sci. Total Environ.* 285: 155-163.

Ganan, M., Martinez-Rodriguez, A.J. and Carrascosa, A.V. (2009). Antimicrobial activity of phenolic compounds of wine against *Campylobacter jejuni*. *Food Control* 20: 739-742.

Geana, E.I., Popescu, R., Costinel, D., Dinca, O.R., Ionete, R.E., Stefanescu, I., Artem, V. and Bala, C. (2016). Classification of red wines using suitable markers coupled with multivariate statistic analysis. *Food Chem.* 192: 1015-1024.

Gebhardt, S. and Thomas, R.G. (2002). Nutritive Value of Foods. United States Department of Agriculture. Agricultural Research Service, Nutrient Data Laboratory.

German, J.B. and Walzem, R.L. (2000). The health benefits of wine. *Ann. Rev. Nutr.* 20: 561-593.

Ghiselli, A., Serafini, M., Natella, F. and Scaccini, C. (2000). Total antioxidant capacity as a tool to assess redox status: Critical view and experimental data. *Free Radic. Biol. Med.* 29: 1106-1114.

Gin, H., Morlat, P., Ragnaud, J.M. and Aubertin, J. (1992). Short-term effect of red wine (consumed during meals) on insulin requirement and glucose tolerance in diabetic patients. *Diabetes Care* 15: 546-548.

Gonzalez Hernandez, G., Hardisson de la Torre, A. and Arias Leon, J.J. (1996). Quantity of K, Ca, Na, Mg, Fe, Cu, Pb, Zn and ashes in DOC Tacoronte-Acentejo (Canary Islands, Spain) musts and wines. *Unters. Forsch.* 203: 517-521.

Grindlay, G., Mora, J., Gras, L. and de Loos-Vollebregt, M.T. (2011). Atomic spectrometry methods for wine analysis: A critical evaluation and discussion of recent applications. *Anal Chim Acta.* 691: 18-32.

Gronbaek, M., Becker, U., Johansen, D., Gottschau, A., Schnohr, P., Hein, H.O., Jensen, G. and Sorensen, T.I. (2000). Type of alcohol consumed and mortality from all causes, coronary heart disease and cancer. *Ann Intern Med.*, **133:** 411-419.

Guth, H. (1998). Comparison of different white wine varieties by instrumental and analyses and sensory studies. *In*: Waterhouse, L.A. and Ebeler, S.E. (Eds.). Chemistry of Wine Flavour. American Chemical Society, Washington, DC. ACS Symposium Series #714.

Harding, C. and Maidment, C. (1996). An investigation into the antibacterial effects of wine and other beverages. *J. Biol. Educ.* 30: 237-239.

Harding, G. (2005). A Wine Miscellany. Clarkson Potter Publishing, New York, USA.

Haseeb, S., Alexander, B. and Baranchuk, A. (2017). Wine and cardiovascular health: A comprehensive review. *Circulation* 136: 1434-1448.

He, S., Sun, C. and Pa, Y. (2008). Red wine polyphenols for cancer prevention. *Int. J. Mol. Sci.* 9: 842-853.

Hossain, M.K., Dayem, A.A., Han, J., Yin, Y., Kim, K., Saha, S.K., Yang, G., Choi, H.Y. and Cho, S. (2016). Molecular mechanisms of the anti-obesity and anti-diabetic properties of flavonoids. *Int. Mol. Sci.* 17. Doi: 10.3390/ijms17040569

Hou, C.Y., Lin, Y.S., Wang, Y.T., Jiang, C.M., Lin, K.T. and Wu, M.C. (2008). Addition of phenolic acids on the reduction of methanol content in wine. *J. Food Sci.* 73: C432-7.

Ibanez, J.G., Carreon-Alvarez, A., Barcena-Soto, M. and Casillas, N. (2008). Metals in alcoholic beverages: A review of sources, effects, concentrations, removal, speciation, and analysis. *J. Food Comp. Anal.* 21: 672-683.

Ifie, I., Marshall, L.J., Ho, P. and Williamson, G. (2016). *Hibiscus sabdariffa* (Roselle) extracts and wine: Phytochemical profile, physicochemical properties, and carbohydrase inhibition. *J. Agric. Food Chem.* 64: 4921-4931.

Izquierdo-Cañas, P.M., García-Romero, E., Mena-Morales, A. and Gómez-Alonso (2016). Effect of malo-lactic fermentation on colour stability and phenolic composition of petit verdot red wines. *Wines Studies* 5. Doi: 10.4081/ws.2016.5795.

Jackson, R. (2014). Wine Science: Principles and Applications - IV. Academic Press, New York, USA.

Jaganath, I.B. and Crozier, A. (2011). Flavonoid metabolism. pp. 285-312. *In*: Ashihara, H., Crozier, A. and Komamine, A. (Eds.). Plant Metabolism and Biotechnology. John Wiley and Sons. Chichester, UK.

Jagtap, U.B. and Bapat, V.A. (2015). Wines from fruits other than grapes: Current status and future prospectus. *Food Biosci.* 9: 80-96.

Jeandet, P., Breuil, A.C.D., Bessis, R., Debord, S., Sbaghi, M. and Adrian, M. (2002). Phytoalexins from the Vitaceae: Biosynthesis, phytoalexin gene expression in transgenic plants, antifungal activity and metabolism. *J. Agric. Food Chem.* 50: 2731-2741.

Ji, Z., Zhao, F., Meng, L., Zhou, C., Tang, W., Xu, F., Liu, L., Lv, H., Chi, J., Pen, F. and Guo, H. (2016). Chinese yellow wine inhibits production of matrix metalloproteinase-2 induced by homocysteine in rat vascular endothelial cells. *Int. J. Clin. Exp. Med.* 9: 838-852.

Jindal, P.C. (1990). Grape. pp. 85-92. *In*: Bose, T.K. and Mitra, S.K. (Eds.). Fruits, Tropical and Sub-tropical. Naya Prakashan. Calcutta.

Jones, P.R., Gawel, R., Francis, I.L. and Waters, E.J. (2008). The influence of interactions between major white wine components on the aroma, flavour and texture of model white wine. *Food Qual. Prefer.* 19: 596-607.

Jos, A., Moreno, I., Gonzalez, A.G., Lopez Artiguez, M. and Camean, A.M. (2004). Study of the mineral profile of Catalonian 'brut' cava using atomic spectrometric methods. *Eur. Food Res. Technol.* 218: 448-451.

Joshi, V.K., Attri, V., Singh, T.K. and Abrol, G.S. (2011). Fruit wines: Production technology. *In*: Joshi, V.K. (Ed.). Handbook of Enology: Principles, Practices and Recent Innovations - II. Asiatech Publishers, Inc. New Delhi, India.

Just, J.R. and Daeschel, M.A. (2003). Antimicrobial effects of wine on *Escherichia coli* O157:H7 and *Salmonella typhimurium* in a model stomach system. *J. Food Sci.* 68: 285-290.

Kanner, J., Frankel, E., Granit, R., German, B. and Kinsella, E. (1994). Natural antioxidants in grapes and wines. *J. Agric. Food Chem.* 42: 64-69.

Kartas, D.D., Aydin, F., Aydin, I. and Karatas, H. (2015). Elemental composition of red wines in southeast Turkey. *Czech, J. Food Sci.* 33: 228-236.

Katalinic, V., Milos, M. and Modun, D., Music, I. and Boban, M. (2004). Antioxidant effectiveness of selected wines in comparison with (+)-catechin. *Food Chem.* 86: 593-600.

Kennedy, J.A. (2008). Grape and wine phenolics: Observations and recent findings. *Cien. Inv. Agr.* 35: 107-120.

Kutleša, Z. and Mršić, D.B. (2016). Wine and bone health: A review. *J. Bone Min. Metab.* 34: 11-22.

Kwon, Y., Apostolidis, E. and Shetty, K. (2008). Inhibitory potential of wine and tea against α-amylase and α-glucosidase for management of hyperglycemia linked to type 2 diabetes. *J. Food Biochem.* 32: 15-31.

Lamuela-Raventós, R.M. and Estruch, R. (2016). Mechanism of the Protective Effects of Wine Intake on Cardiovascular Disease. Springer International Publishing, Switzerland.

Landrault, N., Poucheret, P., Azay, J., Krosniak, M., Gasc, F., Jenin, C., Cros, G. and Teissedre, P.L. (2003). Effect of a polyphenols-enriched Chardonnay white wine in diabetic rats. *J. Agric. Food Chem.* 51: 311-318.

Lara, R., Cerutti, S., Salonia, J.A., Olsina, R.A. and Martinez, L.A. (2005). Trace element determination of Argentine wines using ETAAS and USN-ICP-OES. *Food Chem. Toxicol.* 43: 293-297.

Lawrence, R., Tripathi, P. and Jeyakumar, E. (2009). Isolation, purification and evaluation of antibacterial agents from *Aloe vera*. *Braz. J. Microbiol.* 40: 906-915.

Lea, A.G.H. and Arnold, G.M. (1978). The phenolics of ciders, bitterness and astringency. *J. Sci. Food Agric.* 29: 478-483.

Lee, S.J., Rathbone, D., Asinont, S., Adden, R. and Ebeler, S.E. (2004). Dynamic changes in ester formation during Chardonnay juice fermentations with different yeast inoculation and initial Brix conditions. *Am. J. Enol. Vitic.* 55: 346-354.

Leighton, F., Cuevas, A., Guasch, V., Perez, D.D., Strobel, P., San Martin, A., Urzua, U., Diez, M.S., Foncea, R., Castillo, O., Mizon, C., Espinoza, M.A., Urquiaga, I., Rozowski, J., Maiz, A. and Germain, A. (1999). Plasma polyphenols and antioxidants, oxidative DNA damage, endothelial function in a diet and wine intervention study in humans. *Drugs Exp. Clin. Res.* 25: 133-141.

Lesschaeve, I. and Noble, A.C. (2005). Polyphenols: Factors influencing their sensory properties and their effects on food and beverage preferences. *Am. J. Clin. Nutr.* 81: 330S-335S.

Li, L., Adams, L.S., Chen, S., Killian, C., Ahmed, A. and Seeram, N.P. (2009). *Eugenia jambolana* Lam. berry extract inhibits growth and induces apoptosis of human breast cancer but not non-tumorigenic breast cells. *J. Agr. Food Chem.* 57: 826-831.

Li, H., Guo, A. and Wang, H. (2008). Mechanisms of oxidative browning of wine. *Food Chem.* 108: 1-13.

Lin, Y.T., Vattem, D., Labbe, R.G. and Shetty, K. (2005). Enhancement of antioxidant activity and inhibition of *Helicobacter pylori* by phenolic phytochemical-enriched alcoholic beverages. *Process Biochem.* 40: 2059-2065.

Lingua, M.S., Fabani, M,P., Wunderlin, D.A. and Baroni, M.V. (2016). From grape to wine: Changes in phenolic composition and its influence on antioxidant activity. *Food Chem.* 208: 228-238.

Loizzo, M.R., Bonesi, M., Di Lecce, G., Boselli, E., Tundis, R., Pugliese, A., Menichini, F. and Frega N.G. (2013). Phenolics, aroma profile, and *in vitro* antioxidant activity of Italian dessert passito wine from Saracena (Italy). *J. Food Sci.*, 78: C703-C708.

Lopez, F.F., Cabrera, C., Lorenzo, M.L. and Lopez, M.C. (1998). Aluminium levels in wine, beer and other alcoholic beverages consumed in Spain. *Sci. Total Environ.* 220: 1-9.

Lopez-Velez, M., Martinez-Martinez, F. and Del Valle-Ribes, C. (2003). The study of phenolic compounds as natural antioxidants in wine. *Crit. Rev. Food Sci. Nutr.* 43: 233-244.

Lugasi, A. and Hovari, J. (2003). Antioxidant properties of commercial alcoholic and non-alcoholic beverages. *Nahrung. Food* 47: 79-86.

Lutterodt, G.D., Ismail, A., Bashear, R.H. and Baharudin, H.M. (1999). Antimicrobial effects of *Psidium guajava* extracts as one mechanism of its anti-diarrhoeal action. *Malay. J. Med. Sci.* 6: 17-20.

Macheix, J.J. and Fleuriet, A. (1998). Phenolic acids in fruits. pp. 35-59. *In*: Rice-Evans, C.A. and Packer, L. (Eds.). Flavonoids in Health and Disease. Marcel Dekker, Inc. New York, USA.

Mandalari, G., Bennett, R.N., Bisignano, G., Trombetta, D., Saija, A., Faulds, C.B., Gasson, M.J. and Narbad, A. (2007). Antimicrobial activity of flavonoids extracted from bergamot (*Citrus bergamia* Risso) peel, a byproduct of the essential oil industry. *J. Appl. Microbiol.* 103: 2056-2064.

Mandalari, G., Bisignano, C., D'Arrigo, M., Ginestra, G., Arena, A., Tomaino, A. and Wickham, M.S.J. (2010). Antimicrobial potential of polyphenols extracted from almond skins. *Lett. Appl. Microbiol.* 51: 83-89.

Marais, J. and Pool, H.J. (1980). Effect of storage time and temperature on the volatile composition and quality of dry white table wines. *Vitis.* 19: 151-164.

Marais, J., van Rooyen, P.C. and du Plessis, C.S. (1979). Objective quality rating of pinotage wine. *Vitis.* 18: 31-39.

Marfella, R., Cacciapuoti, F., Siniscalchi, M., Sasso, F.C., Marchese, F., Cinone, F., Musacchio, E., Marfella, M.A., Ruggiero, L., Chiorazzo, G. and Liberti, D. (2006). Effect of moderate red wine intake on cardiac prognosis after recent acute myocardial infarction of subjects with Type 2 diabetes mellitus. *Diab. Med.* 23: 974-981.

Marimon, J.M., Bujanda, L., Gutierrez-Stampa, M.A., Cosme, A. and Arenas, J.I. (1998). Antibacterial activity of wine against *Salmonella enteritidis*: pH or alcohol? *J. Clin. Gastroenterol.* 27: 179-180.

Marjorie, M.C. (1999). Plant products as antimicrobial agents. *Clin. Microbiol. Rev.* 12: 564-582.

Markovic, J.M.D., Petranovic, N.A. and Baranac, J.M. (2000). A spectrophotometric study of the copigmentation of malvin with caffeic and ferulic acids. *J. Agric. Food Chem.* 48: 5530-5536.

Martinez-Rodriguez, A.J. and Pueyo, E. (2008). Influence of the elaboration process on the peptide fraction with angiotensin I - converting enzyme inhibitor activity in sparkling wines and red wines aged on lees. *Food Chem.* 111: 965-969.

Martinez-Rodriguez, A.J., González, R. and Carrascosa, A.V. (2004). Morphological changes in autolytic wine yeast during aging in two model systems. *J. Food Sci.* 69: 233-239.

Mazza, G. and Miniati, E. (1993). Grapes. pp. 149-199. *In*: Mazza, G. and Miniati, E. (Eds.). Anthocyanins in Fruits, Vegetables and Grains. CRC Press. Boca Raton, FL.

McDonald, J. (1986). A Symposium on Wine, Health and Society. *Nutrition*, Wine Institute, Washington DC, USA.

McRae, J.M., Ziora, Z.M., Kassara, S., Cooper, M.A. and Smith, P.A. (2015). Ethanol concentration influences the mechanisms of wine tannin interactions with poly (L-proline) in model wine. *J. Agric. Food Chem.* 63: 4345-4352.

Mezzano, D., Leighton, F., Martinez, C., Marshall, G., Cuevas, A., Castillo, O., Panes, O., Munoz, B., Perez, D.D., Mizon, C., Rozowski, J., San Martin, A. and Pereira, J. (2001). Complementary effects of Mediterranean diet and moderate red wine intake on haemostatic cardiovascular risk factors. *Eur. J. Clin. Nutr.* 55: 444-451.

Miller, M.B. and Bassler, B.L. (2001). Quorum sensing in bacteria. *Ann. Rev. Microbiol.* 55: 165-199.

Minussi, R.C., Rossi, M., Bologna, L., Cordi, L., Rotilio, D., Pastorea, G.M. and Duran, N. (2003). Phenolic compounds and total antioxidant potential of commercial wines. *Food Chem.* 82: 409-416.

Moretro, T. and Daeschel, M.A. (2004). Wine is bactericidal to food-borne pathogens. *J. Food Sci.* 69: M251-M257.

Mukamal, K.J., Jensen, M.K., Gronbaek, M., Stampfer, M.J., Manson, J.E., Pischon, T. and Rimm, E.B. (2005). Drinking frequency, mediating biomarkers and risk of myocardial infarction in women and men. *Circulation* 112: 1406-1413.

Nambiar, S.S., Venugopal, K.S., Shetty, N.P. and Appaiah, K.A.A. (2016). Fermentation induced changes in bioactive properties of wine from *Phyllanthus* with respect to atherosclerosis. *J. Food Sci. Technol.* 53: 2361-2371.

Napoli, R., Cozzolino, D., Guardasole, V., Angelini, V., Zarra, E., Matarazzo, M., Cittadini, A., Saccà, L. and Torella, R. (2005). Red wine consumption improves insulin resistance but not endothelial function in type 2 diabetic patients. *Metabolism* 54: 306-313.

Negi, B., Kaur, R. and Dey, G. (2013). Protective effects of a novel sea buckthorn wine on oxidative stress and hypercholesterolemia. *Food Funct.* 4: 240-248.

Nikicevíc, N. and Téševic, V. (2005). Possibilities for methanol content reduction in plum brandy. *J. Agric. Sci.* 50: 49-60.

Noble, A.C. and Bursick, G.F. (1984). The contribution of glycerol to perceived viscosity and sweetness in white wine. *Am. J. Enol. Vitic.* 35: 110-112.

Nohynek, L.J., Alakomi, H.L., Kähkönen, M.P., Heinonen, M., Helander, I.M., Oksman-Caldentey, K.M. and Puupponen-Pimiä, R.H. (2006). Berry phenolics: Antimicrobial properties and mechanisms of action against severe human pathogens. *Nutr. Cancer* 54: 18-32.

Nuengchamnong, N. and Ingkaninan, K. (2010). On-line HPLC–MS–DPPH assay for the analysis of phenolic antioxidant compounds in fruit wine: *Antidesma thwaitesianum* Muell. *Food Chem.* 118: 147-152.

Oszmianski, J., Romeyer, F.M., Sapis, J.C. and Macheix, J.J. (1986). Grape seed phenolics: Extraction as affected by some conditions occurring during wine processing. *Am. J. Enol. Vitic.* 37: 7-12.

Ough, C.S. (1991). Winemaking Basics. Food Products Press. New York, USA.

Ozeki, Y., Matsuba, Y., Abe, Y., Umemoto, N. and Sasaki, N. (2011). Pigment biosynthesis i. anthocyanins. pp. 155-181. *In*: Ashihara, H., Crozier, A. and Komamine, A. (Eds.). Plant Metabolism and Biotechnology. Wiley. Chichester, United Kingdom.

Paine, A. and Davan, A.D. (2001). Defining a tolerable concentration of methanol in alcoholic drinks. *Hum. Exp. Toxicol.* 20: 563-568.

Palmieri, D., Pane, B., Barisione, C., Spinella, G., Garibaldi, S., Ghigliotti, G., Brunelli, C., Fulcheri, E. and Palombo, D. (2011). Resveratrol counteracts systemic and local inflammation involved in early abdominal aortic aneurysm development. *J. Surg. Res.* 171: e237-e246.

Papadopoulou, C., Soulti, K. and Roussis, I.G. (2005). Potential antimicrobial activity of red and white wine phenolic extracts against strains of *Staphylococcus aureus*, *Escherichia coli* and *Candida albicans*. *Food Technol. Biotechnol.* 43: 41-46.

Pazzini, C.E.F., Colpo, A.C., Poetini, M.R., Pires, C.F., de Camargo, V.B., Mendez, A.S.L., Azevedo, M.L., Soares, J.C.M. and Folmer, V. (2015). Effects of red wine tannat on oxidative stress induced by glucose and fructose in erythrocytes *in vitro*. *Int. J. Med. Sci.*, 12: 478-486.

Porter, L.J., Hrstich, L.N. and Chan, B.G. (1986). The conversion of procyanidins and prodelphinidins to cyanidin and delphinidin. *Phytochem.* 25: 223-230.

Radovanovic, A., Radovanovic, B. and Jovancicevic, B. (2009). Free radical scavenging and antibacterial activities of southern Serbian red wines. *Food Chemistry* 17: 326-331.

Ramos, S. (2008). Cancer chemoprevention and chemotherapy: Dietary polyphenols and signaling pathways. *Mol. Nutr. Food Res.* 52: 507-526.

Rana, A. and Singh, H.P. (2013). Bio-utilisation of wild berries for preparation of high valued herbal wines. *Ind. J. Nat. Prod. Res.* 4: 165-169.

Rapp, A. and Güntert, M. (1986). Changes in aroma substances during the storage of white wines in bottles. pp. 141-167. *In*: Charalambous, G. (Ed.). The Shelf Life of Foods and Beverages. Elsevier. Amsterdam, Netherland.

Rapp, A. and Versini, G. (1996). Influence of nitrogen on compounds in grapes on aroma compounds in wines. *J. Int. Sci. Vigne. Vin.* 51: 193-203.

Reddy, L.V. and Reddy, O.V.S. (2005). Effect of enzymatic maceration on synthesis of higher alcohols during mango wine fermentation. *J. Food Qual.* 32: 34-47.

Reed, G. (1981). Use of microbial cultures: Yeast products. *Food Technol.* 35: 89-94.

Renaud, S.C. and De Lorgeril, M. (1992). Wine, alcohol, platelets and the French paradox for coronary heart disease. *Lancet.* 339: 1523-1526.

Renouf, V., Claisse, O. and Lonvaud-Funel, A. (2007). Inventory and monitoring of wine microbial consortia. *Appl. Microbiol. Biotechnol.* 75: 149-164.

Revilla, I. and González-San José, M.L. (1998). Methanol released during fermentation of red grapes treated with pectolytic enzymes. *Food Chem.* 80: 205-214.

Ribereau-Gayon, P., Glories, Y., Maujean, A. and Dubordieu, P. (2000). Glucids. pp. 65-90. *In*: Handbook of Enology - II. John Wiley and Sons Ltd. Chichester, England.

Ribereau-Gayon, P., Peynaud, E. and D'enologye, P.S.T. (1976). Sciences et technique du vin. analyse et controle des vins. Bordas, Paris 1: 471-513.

Rice-Evans, C., Miller, N.J. and Paganga, G. (1996). Structure – antioxidant activity relationships of flavonoids and phenolic acids. *Free Radical Biol. Med.* 20: 933-956.

Richter, C.L., Kennedy, A.D., Guo, L. and Dokoozlian, N. (2015). Metabolomic measurements at three time points of a Chardonnay wine fermentation with *Saccharomyces cerevisiae*. *Am. J. Enol. Vitic.* 66: 294-301.

Rimm, E.B., Giovannucci, E.L., Willett, W.C., Colditz, G.A., Ascherio, A., Rosner, B. and Stampfer, M.J. (1991). Prospective study of alcohol consumption and risk of coronary disease in men. *Lancet* 338: 464-468.

Rio, D.D., Rodriguez-Mateos, A., Spencer, J.P.E., Tognolini, M., Borges, G. and Crozier, A. (2013). Dietary (poly)phenolics in human health: Structures, bioavailability and evidence of protective effects against chronic diseases. *Antioxid Redox Signal.* 18: 1818-1892.

Rivero-Pérez, M.D., Pérez-Magariño, S. and González-San José, M.L. (2002). Role of melanoidins in sweet wines. *Anal Chim Acta.* 45: 169-175.

Robichand, J. and Noble, A.C. (1990). Astringency and bitterness of selected phenolics in wine. *J. Sci. Food Agric.* 55: 343-353.

Robinson, A.L., Ebeler, S.E., Heymann, H., Boss, P.K., Solomon, P.S. and Trengove, R.D. (2009). Interactions between wine volatile compounds and grape and wine matrix components influence aroma compound headspace partitioning. *J. Agric. Food Chem.* 57: 10313-10322.

Rodriguez Vaquero, M.J., Alberto, M.R. and de Nadra, M.C.M. (2007a). Antibacterial effect of phenolic compounds from different wines. *Food Control* 18: 93-101.

Rodriguez Vaquero, M.J., Alberto, M.R. and Mancadenadra, M. (2007b). Influence of phenolic compounds from wines on the growth of *Listeria monocytogenes*. *Food Control* 18: 587-593.

Ruf, J.C. (2003). Overview of epidemiological studies on wine, health and mortality. *Drugs Exp. Clin. Res*., 29: 173-179.

Sanchez-Moreno, C., Larrauri, J.A. and Saura-Calixto, F. (1999). Free radical scavenging capacity of selected red, rose and white wines. *J. Sci. Food Agr*. 79: 1301-1304.

Santos-Buelga, C. and Scalbert, A. (2000). Proanthocyanidins and tannin-like compounds nature, occurrence, dietary intake and effects on nutritional and health. *J. Sci. Food Agric*. 80: 1094-1117.

Sanz, M.L. and Castro, I.M. (2008). Carbohydrates. pp. 245- 735. *In*: Moreno-Arribas, M.V. and Carmen, P. (Eds.). Wine Chemistry and Biochemistry – IV. Springer International Publishing. New York, USA.

Sato, M., Ramarathnam, N., Suzuki, Y., Ohkubo, T., Takeuchi, M. and Ochi, H. (1996). Varietal differences in the phenolic content and superoxide radical scavenging potential of wines from different sources. *J. Agric. Food Chem*. 44: 37-41.

Satora, P., Tarko, T., Sroka, P. and Blaszczyk, U. (2014). The influence of *Wickerhamomyces anomalus* killer yeast on the fermentation and chemical composition of apple. *FEMS Yeast Res*. 14: 729-740.

Scalbert, A., Andres-Lacueva, C., Arita, M., Kroon, P., Manach, C., Urpi-Sarda, X.M. and Wishar, D. (2011). Databases on food phytochemicals and their health-promoting effects. *J. Agric. Food Chem*. 59: 4331-4348.

Shukla, A., Swati, S. and Srivastava, S. (2014). The immortals of Mluha and the science in their belief. *Int. J. Res. Humanities Arts Lit*. 2: 89-94.

Silva, P., Fernandes, E. and Carvalho, F. (2015). Dual effect of red wine on liver redox status: A concise and mechanistic review. *Arch Toxicol*. 89: 1681-1693.

Simonetti, P., Pietta, P. and Testolin, G. (1997). Polyphenol content and total antioxidant potential of selected Italian wines. *J. Agric. Food Chem*. 45: 1152-1155.

Singkong, W., Rattanapun, B. and Kaweewong, K. (2012). Promotion of safe winemaking practices using quantity comparison and methanol-reduction process for rice wine and whisky. *Asian J. Food Agro-Ind*. 5: 61-70.

Soleas, G.P., Diamendis, E.P. and Goldberg, D.M. (1997). Wine as a biological fluid: History, production and role in disease prevention. *J. Clin. Lab. Anal*. 11: 287-313.

Soni, S.K., Marwaha, S.S., Marwaha, U. and Soni, R. (2011). Composition and nutritional value of wine. pp. 89-145. *In*: V.K. Joshi (Ed.). Handbook of Enology: Principles, Practices and Recent Innovations. Vol. I. Asiatech Publisher. New Delhi, India.

Sorrentino, F., Voilley, A. and Richon, D. (1986). Activity coefficients of aroma compounds in model food systems. *Aiche J*. 32: 1988-1993.

Stringini, M., Comitini, F., Taccari, M. and Ciani, M. (2009).Yeast diversity during tapping and fermentation of palm wine from Cameroon. *Food Microbiol*. 26: 415-420.

Su, H.C., Hung, L.M. and Chen, J.K. (2006). Resveratrol, a red wine antioxidant, possesses an insulin-like effect in streptozotocin-induced diabetic rats. *Am. J. Physiol. Endocrinol. Metab*. 290: E1339-E1346.

Sugita-Konishi, Y., Hara-Kudo, Y., Iwamoto, T. and Kondo, K. (2001). Wine has activity against entero-pathogenic bacteria *in vitro* but not *in vivo*. *Biosci. Biotech. Bioch*. 65: 954-957.

Sumby, K.M., Grbin, P.R. and Jiranek, V. (2010). Microbial modulation of aromatic esters in wine: Current knowledge and future prospects. *Food Chem*. 1: 1-16.

Sumpio, B.J., Cordova, A.C. and Sumpio, B.E. (2016). Wine, polyphenols, and cardioprotection. pp. 97-108. *In*: D.F. Romagnolo and O. Selmin. (Eds.). Mediterranean Diet. Dietary Guidelines and Impact on Health and Disease. Springer International Publishing, Switzerland.

Swiegers, J.H., Bartowsky, E.J., Henschke, P.A. and Pretorius, I.S. (2005). Yeast and bacterial modulation of wine aroma and flavour. *Aust. J. Grape Wine Res*. 11: 139-173.

Teissedre, P.L., Vique, C.C., Cabanis, M.T. and Cabanis, J.C. (1998). Determination of nickel in French wines and grapes. *Am. J. Enol. Viticult*. 49: 205-210.

Teixeira, A., Eiras-Dias, J., Castellarin, S.D. and Gerós, H. (2013). Berry phenolics of grapevine under challenging environments. *Int. J. Mol. Sci*. 14: 18711-18739.

Trivedi, N. (2013). Process development for the production of herbal wines from *Aloe vera* and evaluation of their therapeutic potential. Thesis submitted to Panjab University, Chandigarh.

Trivedi, N., Rishi, P. and Soni, S.K. (2012). Production of a herbal wine from *Aloe vera gel* and evaluation of its effect against common food-borne pathogens and probiotics. *Int. J. Food Fer. Technol.* 2: 157-166.

Trivedi, N., Rishi, P. and Soni, S.K. (2015a). Protective role of *Aloe vera* wine against oxidative stress induced by salmonella infection in a murine model. *IJFANS.* 4: 64-76.

Trivedi, N., Rishi, P. and Soni, S.K. (2015b). Antibacterial activity of prepared Aloe vera-based herbal wines against common food-borne pathogens and probiotic strains. *Int. J. Home Sci.* 1: 91-99.

Ugliano, M. and Henschke, P.A. (2009). Yeasts and wine flavour. pp. 313-392. *In*: Moreno-Arribas, M.V. and Polo, M.C. (Ed.). Wine Chemistry and Biochemistry, Part II. Springer. New York, USA.

Van Rensburg, P. and Pretorius, I.S. (2000). Enzymes in winemaking: Harnessing natural catalysts for efficient biotransformations: A review. *S. Afr. J. Enol. Vitic.* 21: 52-73.

Vaquero, M.J.R., Alberto, M.R. and de Nadra, M.C.M. (2007). Antibacterial effect of phenolic compounds from different wines. *Food Control.* 18: 93-101.

Varela, C., Dry, P.R., Kutyna, D.R., Francis, L.L., Henschke, P.A., Curtin, C.D. and Chambers, P.J. (2015). Strategies for reducing alcohol concentration in wine. *Aust. J. Grape Wine Res.* 21: 670-679.

Vattem, D.A., Mihalik, K., Crixell, S.H. and McLean, R.J. (2007). Dietary phytochemicals as quorum sensing inhibitors. *Fitoterapia.* 78: 302-310.

Venturini, C.D., Merlo, S., Souto, A.A., Fernandes, M.D.C., Gomez, R. and Rhoden, C.R. (2010). Resveratrol and red wine function as antioxidants in the nervous system without cellular proliferative effects during experimental diabetes. *Oxid. Med. Cell Longev.* 3: 434-441.

Vichitphan, S. and Vichitphan, K. (2007). Flavonoid content and antioxidant activity of Krachai-Dum (*Kaempferia parviflora*) wine. *KMITL Sci. Tech. J.* 7: 97-104.

Villiers, A., Alberts, P., Tredoux, G.J.A. and Nieuwoudt, H.H. (2012). Analytical techniques for wine analysis. *Anal Chimic Acta.* 730: 2-23.

Vinson, J.A., Mandarano, M., Hirst, M., Trevithick, J.R. and Bose, P. (2003). Phenol antioxidant quantity and quality in foods: Beers and the effect of two types of beer on an animal model of atherosclerosis. *J. Agric. Food Chem.* 51: 5528-5533.

Visioli, F. and Galli, C. (1998). Olive oil polyphenols and their potential effects on human health. *J. Agric. Food Chem.* 46: 4292-4296.

Voguel, W. (2003). Que es el vino? *In*: S.A. Acribia (Ed.). Elaboración casera de vinos. Vinos de uvas, manzanas y bayas. Zaragoza, Spain.

Von Hellmuth, K.H., Fischer, E. and Rapp, A. (1985). Über das verhalten von spurenelementen und radionukliden in traubenmost bei der gärung und beim weinausbau, Dtsch. *Lebensm.-Rundsch* 81: 171-176.

Waterhouse, A.L. and Ebeler, S.E. (1998). Chemistry of wine flavour. *In*: Proceedings of a Symposium at the 213th National Meeting of the American Chemical Society. San Francisco, California, USA.

Waterhouse, A.L. (2002). Wine phenolics. *Ann. NY. Acad. Sci.* 957: 21-36.

Weisse, M.E., Eberly, B. and Person, D.A. (1995). Wine as a digestive aid: Comparative antimicrobial effects of bismuth salicylate and red and white wine. *Brit. Med. J.* 311: 1657-1660.

Williams, A.A. and Rosser, P.R. (1981). Aroma enhancing effects of ethanol. *Chem. Senses* 6: 149-153.

Yao, Q., He, G., Guo, X., Hu, Y., Shen, Y. and Gou, X. (2016). Antioxidant activity of olive wine, a byproduct of olive mill wastewater. *Pharmaceutical Biol.* 54: 2276-2281.

# Section 2
# Basics of Winemaking

# 4 Grape Varieties for Winemaking

**Ajay Kumar Sharma*, R.G. Somkuwar and Roshni R. Samarth**
ICAR-National Research Centre for Grapes, Pune – 412307, India

## 1. Introduction

Grape growing is adopted globally but location and climate affect the suitability of grape growing in a particular region. Grape originated from a temperate region and winemaking started traditionally. Wine production occurs over relatively narrow geographical and climatic ranges, most often in mid-latitude regions that are prone to high climatic variability (the vintage effect). Wine production worldwide has changed over the past decades and countries of the new wine world are becoming important players. Many factors, such as grape varieties, yeast strains, winemaking technologies and human practices contribute to the bio-chemical composition of wines and directly affect wine quality. Although these factors are essential to the wine quality, the grape variety is the most basic and important factor for making good quality wine (Son *et al.*, 2009).

## 2. Origin and Variability of Grapes

Consumption of fermented beverages started thousands of years back. According to various reports, fermented beverages were being produced and used in China as early as 7000 BC. However, the first evidence of winemaking appeared in the form of wine presses in the reign of Udimu (Egypt), which is about 5,000 years back (Petrie, 1923). Wine residues have also been found in wine amphora in many ancient Egyptian tombs (Guasch-Jané *et al.*, 2004). The centre of origin of grape is known to be the region covering northwestern Turkey, northern Iraq, Azerbaijan, and Georgia. It is believed that domestication of grapes began in a Neolithic village of Transcaucasian region in Georgia (Ramishvili, 1983). However, grapevine domestication occurred independently in Spain also (Núñez and Walker, 1989). The main change however occurred during the domestication process of grapes when domesticated forms were found to have hermaphrodite flowers, with domesticated forms of grapes self-pollinating. Grape domestication resulted in larger berries, larger clusters and complete flowers.

Desirable traits of grapevines have been identified to obtain quality wines. Growing of wine grape varieties is based upon the average climatic conditions of a particular region and determine the wine suitability. Production of quality grapes is influenced by the variability in weather conditions from year-to-year and is known as the vintage effect. In general, grape production and wine quality depend on the region, the climate variability like extreme cold temperature during winter, frost severity during spring and fall, high temperature events during summer, extreme rain or hail events and broad spatial and temporal drought conditions.

Grapevines belong to Vitaceae family and genus *Vitis*, that exhibit a climbing habit (Pongnicz, 1978). The genus *Vitis* is found mainly in temperate climatic conditions and comprises two sub-genera – Euvitis and Muscadinia. However, most of the species are associated with the genus *Vitis*. The number of different varieties under grapevine germplasm collection worldwide is estimated to be 10,000 (Alleweldt and Dettweiller, 1994). However, only a few hundred are grown in different regions to produce different types and styles of wines and adopt for commercial wine production (Truel *et al.*, 1980). Presently, a vast majority of the world producers use only about 20 cultivars out of thousands of available grape varieties.

The old wine world's regions mainly depend on traditionally-cultivated varieties due to the belief that newly-bred varieties cannot compete on wine quality while wild grapevines are on the verge of extinction as they are threatened in their natural habitats. Hence, some regions are giving high priority to collection and preservation of elite grape germ plasm (Forneck *et al.*, 2003). Indeed, the conservation and

*Corresponding author: ajaysharma.icar@gmail.com

preservation of wild populations of *V. vinifera* sp. *sylvestris* is getting high priority for the maintenance of genetic variability and resistance to genetic erosion (Cunha *et al.*, 2009). Presently, clonal selection is adopted for the improvement of a limited set of traditional varieties. Breeders/owners of nurseries or estate vineyards are observing natural mutations in vine offshoots (bud-sports). While these mutations are beneficial in terms of better colour, berry size, or ripening, the new forms considered as clones are propagated by vegetative means, specially cuttings. The clones of one variety differ from the larger population for their grape quality and are preferred for production of quality wines which are appreciated by consumers (Stefanini *et al.*, 2000). Thus, clones differ in some properties like yield, mass of the cluster, sugar content, acidity, anthocyanins content, etc. which is mostly the result of varietal specificity and experience less effect of growing conditions (ENTAV-INRA, 1995; Tebeica and Popa, 2005). Selected clones of Cabernet Sauvignon variety that are characterised by higher yield and clusters having more mass, produce low quality wine compared to the clones in lower yield of Cabernet Sauvignon (Fidelibus *et al.*, 2006). From a great number of Cabernet Sauvignon clones, wines with distinctive flavour of fruit aroma, higher content of tannins, anthocyanins, etc. are produced in France (Jones and Davis, 2000), Italy (Fidelibus *et al.*, 2006), Australia and other countries where variations in terroirs are clearly observed.

Wines became more and more popular in almost all countries across the continents. It is easily observed by looking at the trend of grape production for winemaking. Today winemaking is becoming a passion and is crossing the boundaries of traditional wine making countries. Interestingly, in the last 20 years, non-traditional grape and wine producing countries, such as China and New Zealand, have significantly increased their wine production due to an increase in their vineyard areas and availability of required natural resources. On the other hand, traditional wine producing countries (France, Spain and Italy) have decreased their area of harvested vineyards. However, the world's largest grape growing regions lie in old wine world only.

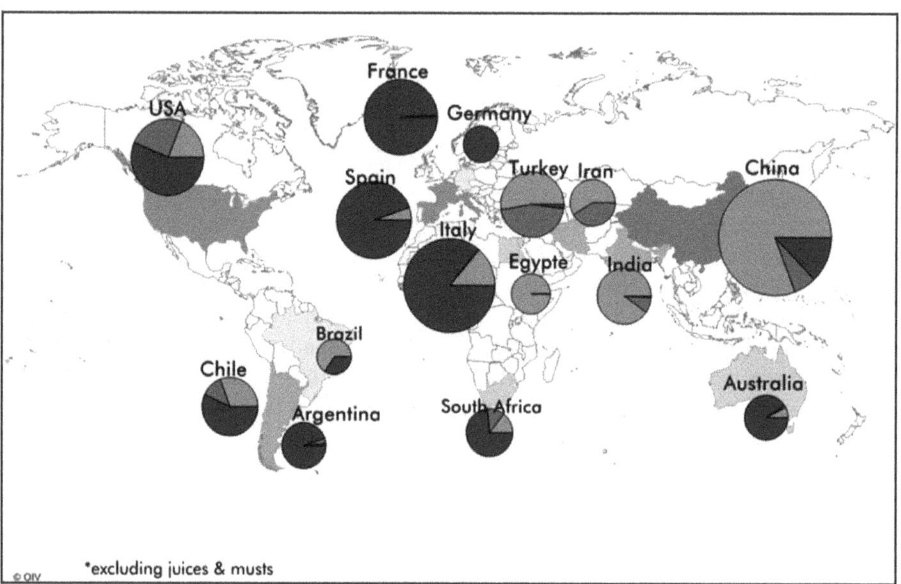


**Figure 1.** World's major grape producers
(*Source*: OIV statistical report on world vitiviniculture, 2016)

Color version at the end of the book

## 3. Wine Worlds

Vineyards are emerging in new and surprising places to meet the ever-growing demand. While up-and-coming wine regions are intriguing, around 80 per cent of wine comes from just 10 countries. Grape growing for winemaking is found suitable between the 30th and 50th degree latitude, across the regions in the Northern Hemisphere. The old wine world comprises countries or regions where winemaking first originated and are situated in countries of Europe, like France, Italy, Spain, Portugal, Greece and Germany. These countries have thousands years of history of wines. Wines made in these countries have uniqueness in quality. Old world wines are often described as tasting lighter, having less alcohol, higher acidity and less fruity aromas, while new world wines are from countries or regions where winemaking was imported during and after the age of exploration. These countries are United States, Australia, South Africa, Chile, Argentina and New Zealand. Wines from the new world are often found as tasting riper, possessing higher alcohol content, are less acidic in nature and have more fruity aromas. Varieties are, however, not limited to certain places and winemakers are improving their quality and applying their skill to create wines with differences. Grape growing and winemaking techniques, like adoption of suitable clone, harvesting of grapes at phenolic or aromatic maturity, unique yeast for fermentation of particular variety, modified maceration techniques, etc. are gaining importance in order to create wines with a difference.

Among the countries having tropical climatic conditions, grape growing is expanding at a faster rate. Countries in tropical belts of different continents, such as Bolivia, Colombia, Peru, Guatemala (in South America), Madagascar, Namibia, Tanzania (in Africa) and Vietnam, China, India (in Asia) have entered into winemaking (Jogaiah *et al*., 2013). Wine made in the tropical belt has higher sugar content and less acidity. In new areas, especially under tropical conditions, areas under grape production are increasing very fast and, winemaking is becoming a passion and occupying the market very fast.

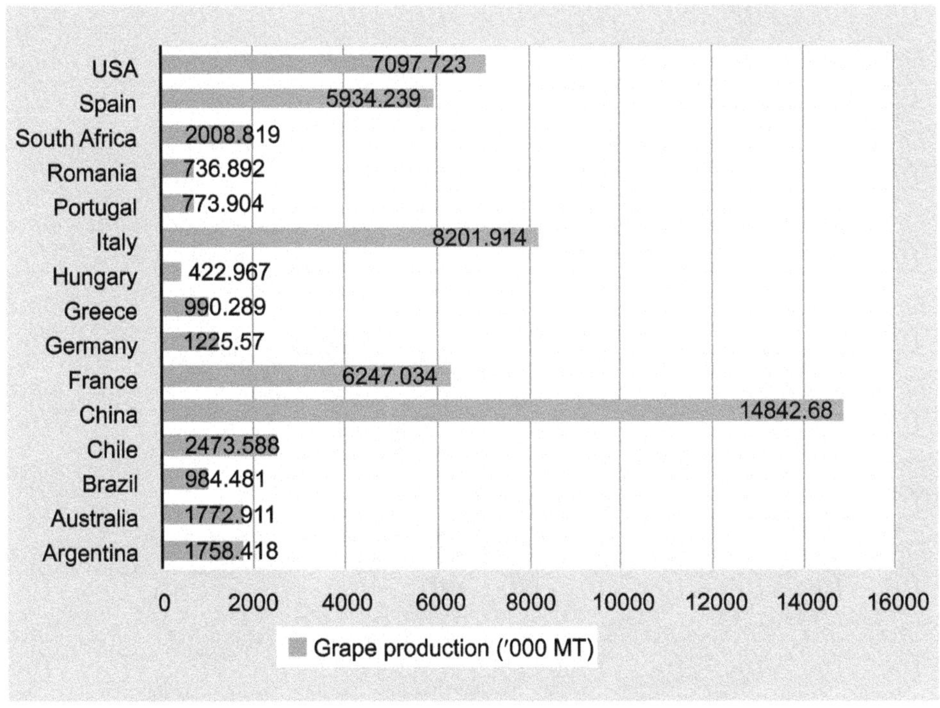


**Figure 2.** Worldwide grape production scenario during 2016 (Anonymous, 2016a)

## 4. Suitability of Wine Grape Varieties

Climate has long been a driving force behind the development of agriculture in any area. The shaping of current agro-ecological zones is determined by climatic conditions. Climatic conditions of a particular geographic location define the suitability based on yield and quality of grapes (Schultz and Jones, 2010). Wine-grape cultivation in particular is affected by conditions in a particular vintage and reflect the same tasting notes of wines. The grapes grown under climatic conditions of a particular region ultimately determine the quality of wine produced, but year-to-year variability can also have an obvious impact on wine quality. Warmer region produce wine with higher alcohol and lower acidity and tannins while a

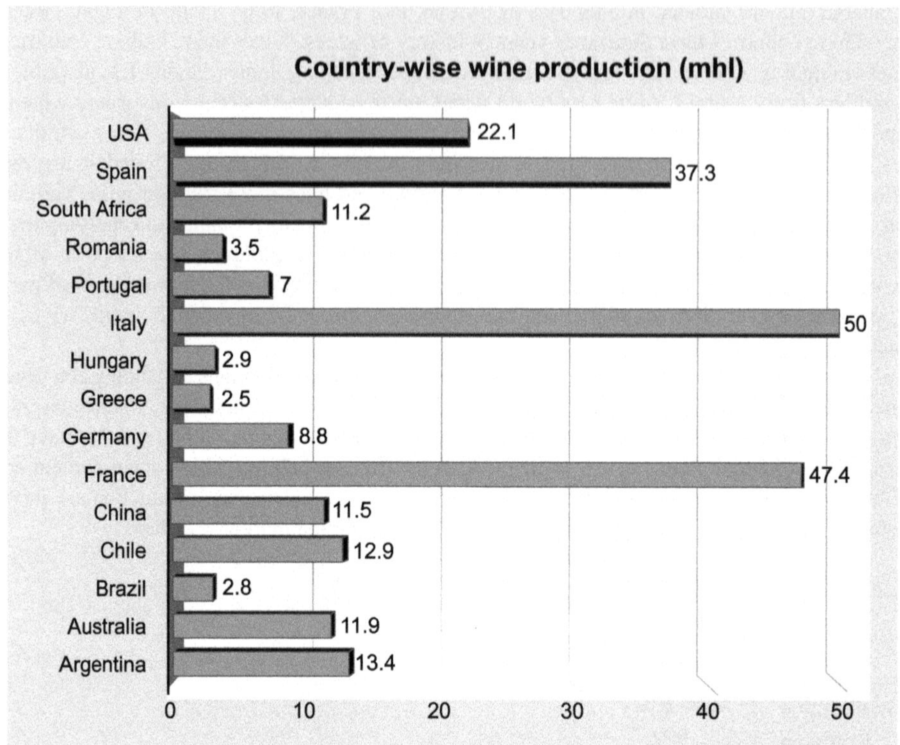


**Figure 3.** Wine production in major wine grape-producing countries (Anonymous, 2016b)

**Figure 4.** The wine world is accommodating new latitude regions
(*Source*: http://www.indianwineacademy.com/item_1716.aspx)

cooler region will tend to produce a wine which is more acidic in nature, but lacking in alcohol/sugars or balance (Jones, 2005; Jones *et al.*, 2005; Van Leeuwen and Seguin, 2006). A year with excess precipitation may produce diluted grapes, or one which is less robust, while an extremely dry year results in a low crop harvest.

In order to grapevine phenology to begin, the average temperature must steadily exceed 10°C, which is treated as the benchmark. The phenology of grape development can be categorised into several stages based on the occurrence of different stages of flowering and berry development. The amount of time taken between each phenological stage varies between the varieties and is dependent on the prevailing climatic conditions in the region and vineyard location. However, the timing of the stages is related to not only the yield, but also the quality of grapes. The earlier a stage is reached within the growing season would account for better and higher quality in that vintage (Jones, 2005).

Total rainfall in a particular area as well as time of rainfall in relation to growth stage is important for the suitability of the site to produce grapes with quality. Rainfall plays a vital role in vine-health management. Due to long or heavy rainfall, the soil gets waterlogged, which not only promotes root rot, but also causes physiological disturbances in berry development and ultimately in the quality of grapes. Ideally, the time frame after 'veraison' stage is to have limited amounts of precipitation, as an excess amount of precipitation can cause major quality issues, like fruit rot and excess vine vigour (Kubach, 2009). Increased precipitation causes grape berries to swell and split, which directly affects the quality. Precipitation that occurs just prior to harvest affects the sugar content in berries by diluting and causing the Brix to drop (Martinson, 2018). Such type of affected berries result in wines with lower alcohol.

## 5. Distribution of Wine Grape Varieties

In general, the grape varieties are grouped according to their specific or inter-specific origin. Most commercially adopted wine grape varieties are pure *V. vinifera* cultivars. American-French hybrids are the next largest group. These hybrids are derived from crosses between *V. vinifera* and one or more of the other species, like *V. riparia*, *V. rupestris* or *V. aestivalis*. Selection from indigenous grapevine is an early practice in US for identification of new varieties and hybridisation is initiated after considering the quality of French and adaptability of local types. Developed American-French hybrids have been found less affected by extreme weather events. Adaptation strategies include switching to warmer climate or more resilient grape varieties and re-locating to a higher latitude or altitude to retain the current mix of grape varieties, especially in the new wine world, where regions are still trying to identify their varietal comparative advantages and where no restrictions are prevalent on varietal choice. In these areas, wine grape growers are continually searching for attractive alternative varieties that are more adaptable to climatic conditions.

*Vitis vinifera* has several thousand different varieties, e.g. *Vitis vinifera* Chardonnay, *Vitis vinifera* Cabernet Sauvignon, etc. Each variety looks different and has a different taste in wine. Some varieties ripen early, others late; some are suitable for growing in warm climatic conditions, while others prefer cooler conditions; some like certain types of soil, others do not require specific soil requirement; some yield well, others are extremely shy in bearing, etc. In old wine world, each region makes wine in the way that they had been making for countless years. But now no barriers to growing areas for varieties are seen, for example, Riesling is a well-known variety grown in Germany and ice wine is made in Germany and Canada by using this variety. A few years back, Riesling was grown in tropical conditions of India and now wines are made from Riesling and are well appreciated by consumers. According to a report of OIV, maximum area in wine grapes is covered by Cabernet Sauvignon (red wine variety) followed by Merlot (Anonymous, 2017). In case of white wine grapes, Airen (originating in Spain) is the ruling variety followed by Chardonnay. The old world is producing well-established wine grape varieties according to wine types and styles, but the new world countries are establishing grape varieties as per their suitability and for producing specific wines. For example, New Zealand produces Sauvignon Blanc, Argentina: Malbec, Chile: Carmenere, South Africa: Pinotage and Chenin Blanc, Australia: Shiraz and Chardonnay; California goes in for Zinfandel and White Zinfandel.

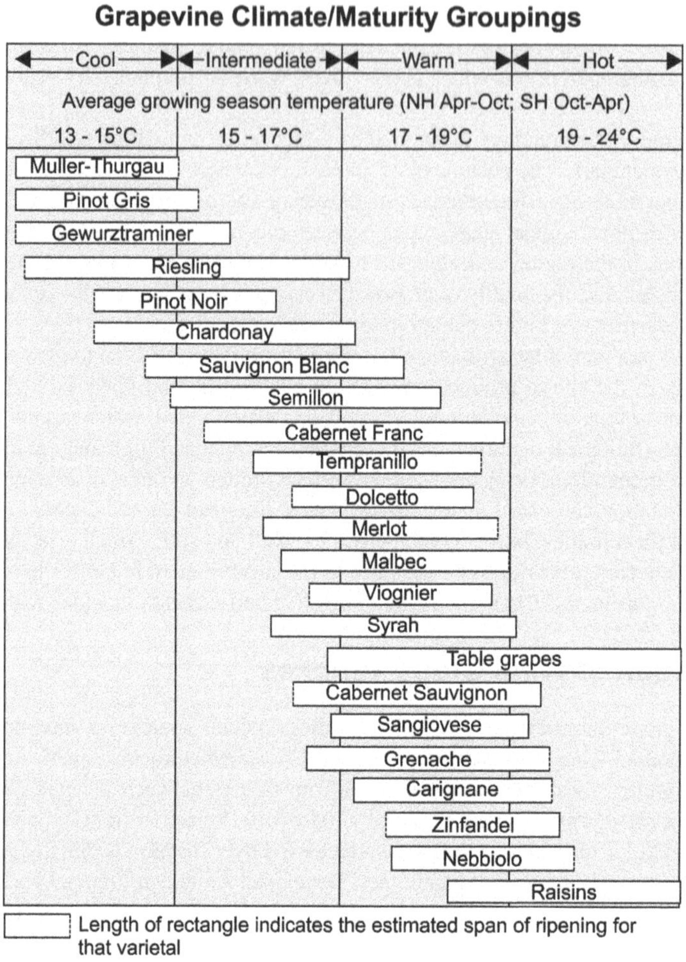

**Grapevine Climate/Maturity Groupings**

Cool | Intermediate | Warm | Hot

Average growing season temperature (NH Apr-Oct; SH Oct-Apr)

13 - 15°C | 15 - 17°C | 17 - 19°C | 19 - 24°C

Muller-Thurgau
Pinot Gris
Gewurztraminer
Riesling
Pinot Noir
Chardonay
Sauvignon Blanc
Semillon
Cabernet Franc
Tempranillo
Dolcetto
Merlot
Malbec
Viognier
Syrah
Table grapes
Cabernet Sauvignon
Sangiovese
Grenache
Carignane
Zinfandel
Nebbiolo
Raisins

Length of rectangle indicates the estimated span of ripening for that varietal

**Figure 5.** Suitability of varieties according to climatic conditions (*Source*: Jones, 2006)

**Table 1.** Worldwide Distribution of Wine Grape Varieties

| *Variety* | *Area ('000 ha)* |
|---|---|
| Cabernet Sauvignon | 340 |
| Merlot | 266 |
| Tempranillo | 231 |
| Airen | 218 |
| Chardonnay | 211 |
| Syrah | 190 |
| Grenache Noir | 163 |
| Sauvignon Blanc | 121 |
| Pinot Noir | 115 |
| Ugni Blanc | 111 |

*Source*: Anonymous, 2017

**Table 2.** Country-wise Distribution of Wine Varieties (Per cent Share of Total Grape Production of Respective Countries)

| *Countries* | *Wine varieties distribution* |
|---|---|
| Spain | Airen (22.3), Tempranillo (20.8), Bobal (6.4), Grenache Noir (6.4), Viura (4.7), Mourvèdre (4.4), Alicante H. Bouschet, (2.7), Pardina (2.6), Cabernet Sauvignon, (2.1), Syrah (2.1) |
| China | Cabernet Sauvignon (7.2), Carmenere (1.0), Merlot (0.8), Cabernet Franc (0.4), Chardonnay (0.3), Riesling (0.3), Syrah (0.2), Pinot Noir (0.1) |
| France | Merlot (13.9), Ugni Blanc (10.2), Grenache Noir (10.0), Syrah (7.9), Chardonnay (6.3), Cabernet Sauvignon (6.0), Cabernet Franc (4.1), Carignan Noir (4.1), Pinot Noir (4.0), Sauvignon Blanc (3.7) |
| Italy | Sangiovese (7.9), Montepulciano (4.0), Glera (4.0), Pinot Gris (3.7), Merlot (3.5), Catarratto Bianco Comune (3.1), Ugni Blanc (3.1), Chardonnay (2.9), Barbera (2.6) |
| United States | Chardonnay (9.7), Cabernet Sauvignon (9.3), Pinot Noir (5.6), Merlot (4.7), Zinfandel (4.3), Syrah (2.0), Pinot Gris (1.8), Colombard (1.8) |
| Argentina | Cot/Malbec (17.8), Cereza (12.9), Bonarda (8.4), Criolla Grande (7.1), Cabernet Sauvignon (6.7), Syrah (5.8), Pedro Ximenez (4.9), Torrontes Riojano (3.6), Moscatel Rosado (3.1), Chardonnay (2.7) |
| Chile | Cabernet Sauvignon (20.1), Sauvignon Blanc (7.0), Mission (6.1), Merlot (5.6), Chardonnay (5.6), Carmenere (5.1), Syrah (8.3) |
| Portugal | Tempranillo (9.0), Touriga Franca (7.5), Castelão (6.5), Fernão Pires (6.5), Touriga Nacional (6.0), Trincadeira (5.5), Baga (3.5), Síria (3.5), Arinto (3.0), Syrah (3.0) |
| Romania | Feteasca Regala (6.8), Feteasca Alba (6.3), Merlot (6.3), Riesling (3.1), Aligoté (3.1), Sauvignon Blanc (3.1), Cabernet Sauvignon (2.6), Muscat Ottonel (2.6), Feteasca Neagra (1.6), Rosioara (1.6) |
| Australia | Syrah (26.8), Cabernet Sauvignon (16.8), Chardonnay (14.1), Merlot (5.4), Sauvignon Blanc (4.0), Pinot Noir (3.4), Sémillon (3.4), Pinot Gris (2.7), Riesling (2.0), Muscat of Alexandria (1.3) |
| South Africa | Chenin Blanc (14.6), Colombard (10.0), Cabernet Sauvignon (9.2), Syrah (8.5), Sauvignon Blanc (7.7), Pinotage (6.2), Chardonnay (6.2), Merlot (4.6) |
| Greece | Savvatiano (10.3), Roditis (8.4), Muscat de Hambourg (2.8), Agiorgitiko (2.8), Liatiko (1.9), Xinomavro (1.9), Cabernet Sauvignon (1.9), Assyrtiko (1.9) |
| Germany | Riesling (23.3), Müller Thurgau (12.6), Pinot Noir (11.7), Dornfelder (7.8), Pinot Gris (5.8), Sylvaner (4.9), Pinot Blanc (4.9), Blauer Portugieser (2.9), Kerner (2.9), Blauer Trollinger (1.9) |
| Brazil | Bordo (11.5), Alphonse Lavallée (2.3), Couderc Noir (2.3), Cabernet Sauvignon (1.1) |
| Hungary | Blaufränkisch (10.3), Bianca (7.4), Cserszegi Fuszeres (5.9), Grasevina (5.9), Furmint (5.9), Cabernet Sauvignon (4.4), Chardonnay (4.4), Merlot (2.9), Zweigelt (2.9), Müller Thurgau (2.9) |

## 6. Varietal Characteristics of Wine Grapes

Varietal character of a particular type of grape defines the qualities and characteristics that a particular grape variety express when made into wine. Sometimes varietal characters of a particular grape variety appear in wine but other different qualities are also distinguished. These qualities include the taste, flavour, mouth feel and finish that are observed as and when the wines are tasted. The varietal characters vary not just with different type of grapes, but even the same type of grapes can give different characteristics to wine. These characteristics largely depend on the vintage, location of the vineyard, viticulture techniques and weather variations.

The grape variety has the maximum effect on the wine's flavour, but grapes can share the same name and express a different varietal character. This is due to some changes that occur, which may cause the wine to exhibit different characteristics despite being of the same variety. These chemico-sensory differences are most commonly related to environmental factors that influence the grapevine growth and development, involving interactions between environmental, temporal, geologic, plant-genetic, human and other factors (Van Leeuwen and Seguin, 2006). Presence and concentration of phenolic compounds in the fruits of wine grapes are one of the major quality factors to reflect the wine. The phenolic compounds have direct influence on important organoleptic properties of wines, such as colour, flavour, bitterness and astringency (Garrido and Borges, 2011). Textural properties of berries play an important role in releasing the phenolic compounds that are critical elements in deciding the wine quality. The good fragility of the skin cell wall facilitates extraction of anthocyanins, which synthesise in the vacuoles of the skin of red grapes into wine during the grape fermentation process (Río Segade *et al.*, 2011; Rolle *et al.*, 2011). Release of hydroxycinnamic acid, an important phenol of white wine, is affected by the textural profile of a whole berry (Urcan *et al.*, 2016). The anthocyanins are the principal source of colour in red grape skins and are released in wine during fermentation. Hence, the textural profile of berries has the impact on release of anthocyanins during fermentation and ultimately the colour level of wine. Highly coloured grapes do not necessarily produce highly coloured wines as it may be related to the ease with which anthocyanins are extracted from grape skins into musts. Monastrell grapes from the Jumilla area have the highest anthocyanin concentration (whether expressed as μg/g or mg/kg of berries), but the extractability assay indicated difficulty in their extraction. The extractability assay for Cabernet Sauvignon, Merlot and Syrah indicated that their anthocyanins can be extracted easily and this was confirmed through chromatic analysis of the produced wines (Romero-Cascales *et al.*, 2005). Hence, extractability of colour from skin during winemaking varies from variety to variety.

In comparison with other fruits, grape berries contain a higher concentration of phenolic compounds. Phenolic compounds of wine and wine by-products have attracted much interest due to their antioxidant and antimicrobial properties besides their potentially beneficial effects on human health (Sun *et al.*, 2002; Baydar *et al.*, 2006). Radical scavenging activities and reducing powers depend on the grape cultivars and different parts of grape and wine types (Baydar *et al.*, 2011).

Aroma is one of the important factors to decide on the acceptance of wine by consumers and has an impact on quality. Wine aromas consist of several hundred volatile compounds at concentrations ranging from several mg/L to a few ng/L or even less (Conde *et al.*, 2007). Some aromatic compounds are directly released by grape berries during fermentation while others are developed during the fermentation process. In case of neutral grape varieties, the most important flavour compounds in wine are those which arise from the fermentation process and include mainly ethyl esters, acetate esters, higher alcohol, fatty acids, and aldehydes. Voorpostel *et al.* (2014) concluded that grape flavour is also a major driver in preference for acceptance of the nectar. Concentration of aldehydes and terpenes differ in wines made by different varieties (Cheng *et al.*, 2015). Numerous studies have shown that the terpenoid compounds form the basis for the sensory expression of wine bouquet and can be used to differentiate grape varieties (Oliveira *et al.*, 2004). These compounds are responsible for the characteristic aroma in Muscat grapes and wines, although they are also present (at low concentrations) in simple flavoured varieties. Both free forms and odourless precursors, mainly glycosylated, have been identified in grapes and wines. Conde *et al.* (2007) classified grape varieties into three classes based on monoterpene concentration: (1) intensely-flavoured Muscats, in which total free monoterpene concentrations can be as high as 6 mg/L; (2) non-Muscat but aromatic varieties with total monoterpene concentration of 1-4 mg/L, including Traminer, Huxel and Riesling varieties; and (3) more neutral varieties not dependent upon monoterpenes for their flavour, including different wine grape cultivars, such as Cabernet Sauvignon, Sauvignon Blanc, Merlot, Shiraz and Chardonnay. In these cultivars, monoterpenes are available at a low concentration, generally below the perception threshold, such that these compounds can only play a minimal role in determining the varietal flavour.

Some grape varieties set themselves apart from others due to their region of origin, cost of production, quality of grapes and blending practices. An example of how these factors can affect a varietal's character is the flavour that transfers into a finished wine, which is derived from the terroir. Minerals in the soil and climate can influence the taste and character of a variety, which in turn, influences the finished wine. Some causes for differentiation in varietal characteristics are as under:

- Vineyard management (care and techniques)
- Clonal variations
- Individual histories
- Cultivation practices
- Blending procedures
- Topography and terroir

## 7. Important Wine Grape Varieties and Wine Quality

### 7.1. Red Wine Grape Varieties

#### *Cabernet Sauvignon*

It is believed that during the 17th century, Cabernet Franc and Sauvignon Blanc were crossed originally naturally and Cabernet Sauvignon appeared. After that, French wine-growers started crossing Cabernet Franc and Sauvignon Blanc. This new creation was named as Cabernet Sauvignon. Cabernet Sauvignon variety is used in making wines of different styles worldwide. Some wines made from this variety are riper, richer and more powerful, while other types offer a more elegant taste, texture and feel. In young Cabernet Sauvignon wines, the tannic structure makes it the perfect wine. Presently almost each wine grape-producing region across the continents is growing this variety, but France has the largest plantings of Cabernet Sauvignon. It is not because Cabernet Sauvignon has the ability to produce stand-alone wines, that is, wines are produced from only one grape variety, but it is also the key grape variety used widely in blending. This has been accomplished by different vineyard management techniques starting with canopy management to allow for more sunlight on the berries, green harvesting to reduce yields, picking fruit that is phenolically ripe, when possible, stem removal and extremely careful sorting. Variations in clonal selections of Cabernet Sauvignon from different growing regions of the world have been noted. Some known clones of Cabernet Sauvignon from different origins are Italy which has clone R5; Chile: clone ISV117; California: clones VCR8 and ISV2; France: clones 191, 341, 338, 169, 685; Argentina: clone ISV105, etc. Variations in wine quality of clonal selections are observed in different wine regions.

#### *Merlot*

Merlot is the most popular and widely grown wine grape variety in France. This started earning a reputation for producing quality wine in 1784, due to the growing fame of the wines produced on the right bank of Bordeaux. One states that the name is derived from the meaning 'little blackbird' in Bordeaux patois and the other is that the bird was in fact named after the grapes, since the bird's colour resembles that of the grapes on the Merlot vine and it is also the grape which this particular bird prefers to eat. This is a popular variety because of the early-mid season ripening. This variety is found suitable for climates that have high rainfall but which leads to the danger of rot. Wines made from Merlot are perfect for lunch or dinner and are found naturally soft textured and rich flavoured and work well with a diverse array of foods. Red berries of Merlot have easy tannins and wine contains a soft finish which is a specific characteristic. Merlot variety produces red wine with flavours and fruity aromas, such as black cherry, plum, chocolate and sometimes herbaceous aromas also. Merlot wine is popular with people who prefer drier, more complex red wines as it has other specific characteristics, like low tannins which make it easier to drink than other red wines. Cool climate Merlot is more structured with a higher presence of tannins and earthy flavours, like tobacco and tar while in warm climate Merlot wine has more fruit-flavour and less prevalent tannins. Flavour profile of Merlot is found similar to Cabernet Sauvignon as both are genetically closer (a descendent of Cabernet Franc). The main difference between these varieties is that Merlot has a thinner skin and tends to be less astringent due to fewer and softer tannins. Clonal variations are observed in Merlot also. Main clones from different regions are 181, 184, 346, 348, 481, 584, 595, 1056, 2096, VCR1, VCR 101, 022, 025, 029, RVC13, the Q45-15, the D3V14, the D3V5, the D3V7, the FPS06 (aka the '6R'), the FPS08 (aka the '8R'), the FPS18 and the SAVII02, etc. The performances of a clone varies with environmental conditions as environment and clone interaction have their own impact on the performance. Merlot clone wines are rich sources of bioactive phenolic compounds and have antioxidant potential.

### *Syrah*

Syrah is a red wine grape variety most popular and widely planted worldwide. In northern Rhone, it is only the red wine grape variety that is planted. This is an offspring of two ancient varieties, Dureza, which is a dark-skinned berry crossed with Mondeuse Blanche, a white-skinned grape. According to reports, this variety first occurred on the west bank of the Rhone. It is believed that both parental varieties were available in the Northern Rhone Valley, which is the home of Syrah. During the 1800s, Syrah started becoming more prevalent in the Rhone Valley and the Languedoc Roussillon area. But the true explosion in Syrah started slowly taking off in the 1960s. European countries started labelling their varietal wines by using the term Syrah. The term Shiraz is used almost exclusively in Australia and South Africa. However, the use of the word 'Syrah' for wines made by the old world style is trending now. These wines are less fruit forward and having floral aromas with medium tannin style, which is usually observed in the classic French and European wines. Syrah is a small, dark-skinned berry that grows in small bunches. It is felt that the grape takes on a shape that resembles an egg. Grapes of this variety have naturally high acidity and tannins. Wine made from this variety is deep coloured and concentrated, rich, and with the ability to age and evolve for years or multiple decades in the best case. Wines made from Syrah give a typical fruity aroma and flavour of blackberries, plums, black cherries; and other type of aromas which are mainly floral, spicy, earthy, chocolaty and peppery. Many of these secondary qualities develop when the wine ages. Under cold climates, whether in Northern Rhône or Victoria and parts of Western Australia, the presence of aromas of mint, pepper and spices is recorded. While in the warmer conditions, it changes from raspberry to blackberry, becomes chocolaty and, with age, tarries and gamey. The variety has higher shoot vigour with long shoots to protect its fruit and is sensitive to heat stress. After the vine matures and settles down, which seems to be after about ten years, it stabilises in performance. Clonal variations have been recorded in Syrah and are performing very well in different regions. The wines made from different clonal selections are appreciated by consumers. Main clones of Syrah are 100, 174, 300, 470, 525, 877 (Syrah 07); Shiraz 03; Durell (Syrah 08) R6WV28, BVRC12, 1654 (highest yielding), 1125, 1127, 2626, EVOVS3, BVOVS5, BVOVS10 and EVOVS12. The clone 877 has a larger berry and cluster than 174. It makes wines high in colour with lower in tannins and has earthy Syrah flavour. It has a small berries in cluster and makes dark, rich wine with higher tannins and darker berry flavour.

### *Tempranillo*

Tempranillo is a famous Spanish red wine grape variety used in many rich Spanish red wines. It is believed to have originated in Spain when the monks from Cluny distributed this variety along the Caminho de Santiago. Herrera (1513) mentions an Aragonez variety, whilst Valcárcel (1791) describes Tempranillo in Rioja and Navarre. Tempranillo wines have strong fruity flavours and aromas. Specific other characteristics of Tempranillo wines include deep red colour, low acidity, tannins having medium to strong nature and moderate levels of alcohol. Tempranillo wines are usually blended with other wines. Better flavour and colour are distinguished when Tempranillo is bottled as a varietal wine. This variety is especially prominent in wines from the Ribera del Duero and throughout the Rioja. Tempranillo is also a key blending varietal in Port, known by the name of *tintaroriz* in Portugal's Douro Valley wines. Tempranillo seems to prefer cooler climates, but the vines themselves tolerate heat well though undesirable flavours develop in warm climatic conditions. Sunny days allow development of thick-skinned berries to ripen fully and cooler nights help in retaining the natural acid balance in berries. The vines are very vigorous, having a tendency to over-crop and its clusters are usually large. Leaf thinning, suckering and crop thinning or 'green harvesting' are commonly required to insure a balanced crop that fully ripens. Tempranillo grapes tend to be low both in overall acidity and in sugar, but often are high in pH and nearly always high in tannin due to their thick skins, although this results in low colour intensity. Mindful of high tannins, many cool-climate producers advocate partial whole berry fermentation. Being an old variety, large clonal variations have been recorded in Tempranillo also. Main clones are 86, 1052, 336, 518, 501, 349, 280, 825, 807, 814, 318, 56, and 1084. According to a study, the clones 86 and 336 would be the most interesting accessions from an adaptation point of view.

### *Pinot Noir*

Pinot Noir is known as noble grape of Burgundy region of France. It is also called Red Burgundy in most parts of the world. Pinot Noir is among the most elegant wines coming out of France. This is thin-skinned, low-yielding grape variety which is sensitive to light exposure and susceptible to rot and various fungal diseases. It is pale coloured wine with flavour which is very subtle. The aromas related to minerality are easily detected in Pinot wines. Pinot Noir wines from Burgundy have fruity aromas like that of ripe red berries, sweet black cherries; however, flavour of mushrooms and what sommeliers call 'forest floor-like dump leaves' are also noted in these wines. Other than Burgundy, great and affordable Pinot Noir can be found in California, Oregon, Australia, Chile and New Zealand also. Pinot Noir from France, especially Burgundy, is found less rich in flavour and fruity taste than that grown in regions of the new wine world. The oldest Pinot clones date back to France of the 14th and 15th centuries, when Burgundian monks – practicing survival of the fittest – decided to multiply particular vines that had more positive characteristics than their siblings. Pinot Noir has recorded more clones than any other wine grape variety, and the maximum originate from France. About 100 Pinot Noir selections have been submitted for the registration programme of UC Davis, including French clones and heritage California selections. They are known as Dijon clones with designations like ENTAV-INRA® are 113, 114, 115, 165, 236, 375, 459, 667, 743, 777, and 943. Wines made from these clones are widely appreciated by consumers.

### *Grenache*

Grenache, or as it is known by its Spanish name, Garnacha, is obtained from one of the most widely planted red grapes in the world. It is also one of the key ingredients in some of the world's most famous wines. The Grenache grape was grown initially in the northern region of Spain, known as Aragon. Grenache has been popular in the Southern Rhone since the 17th century when it was used as a blending grape to give Burgundy wines the additional body and alcohol. Grenache comes in three versions – red Grenache or Grenache Noir, white Grenache, known better as Grenache Blanc which is used as a blending grape in the Rhone Valley and the semi-obscure version of white Grenache, known as Grenache Gris. Grenache, with its thin skin and tight clusters, is perfect for the hot, dry, sandy and stony soil of Chateauneuf du Pape. The grape enjoys a long growing season; the extra time on the vine is well spent. It allows the berry to develop high sugar levels. Grenache is one of the more alcoholic grapes. It often reaches 15 per cent alcohol by volume, but due to its level of sweet ripe fruits, when properly vinified, there is no sensation of heat. The grape was begun to be cultivated and was originally used for both single varietal wines as well as for blending. The unmistakable candied fruit roll-up and cinnamon flavour are what gives Grenache away to expert blind tasters. It has a medium-bodied taste due to its higher alcohol, but has a deceptively lighter colour and is semi-translucent. Depending on where it is grown, Grenache often has subtle aromas of orange rind and ruby-red grapefruit. Variations in Grenache Noir were identified and evaluated for wine quality. The identified main clonal selections are ENTAV-INRA® 70, 135, 136, 139, 362, 513, and 515. Other than these clones, some important selections are VCR 23, CAPVS1, CAPVS2, CAPVS5, CFC13, ISVICAPG, Rosso VCR3, ISVCVI17, etc. However, wine quality of these clones varies with the region.

### *Malbec*

Malbec is thin-skinned 'black grape' variety that originated in France but is well established and has become very famous in Argentina. Due to terroir effect, Malbec from Mendoza in Argentina is world famous. Malbec was also known as Cot in some regions. Initially it became very famous in Saint Emilion and in other right bank areas. In the 1700s, Sieur Malbek brought the grape from the right bank and planted in the Medoc. Due to its success and popularity in the left bank of Bordeaux, it was renamed as Malbec, in honour of Sieur Malbek. It is believed that the grape gained its popularity in Bordeaux by 1855. The Malbec grapes typically ripen midway through the growing season and look like small, intensely coloured berries. This variety is very sensitive to growing environment; so the level of ripeness has visible effects on the structure of the wine. Malbec wines have typically red, medium to full-body, dry with plenty of acidity and higher tannin and alcohol levels. Dark, inky purple colour and ripe fruit flavours of different fruits like plums, black cherry, and blackberry give decidedly a jammy character to wines. Different types of flavours and aromas, such as smoke, earthy, leathery, wild game, tobacco and white/black pepper give it a spicy type. Clonal selection is also proving to be vital to the success of Malbec in different regions. Most producers are using a selection of clones including 3, 4, 5, 9 and 595. Main clones of Malbec are FPS 4, FPS 6, and FPS 8; and 09 (Cot 180), 10 (Cot 46), and 12 Clone 10 (Cot 46), etc.

### *Zinfandel*

Zinfandel wine grapes originated in Europe and arrived in America in about 1820. By 1830, vines were being sold under the name of Zenfendal. In 1835, Zinfandel wine grapes gained popularity under the name of Zinfindal. Due to high productivity, by the middle of the 19th century, Zinfandal was established in northern California. The 'Zinfandel' grape is grown across multiple climatic regions and is able to produce many types of unique varietal wines. Usually uneven ripening is found in this variety, resulting in grape berries having different levels of ripening in a bunch at the time of harvesting. Due to this nature, the wine flavours are akin to cranberry and dried fig with a background of under-ripeness. If the bunches are picked too late, the wines swing between greenness and jam types, but in case of controlled and well managed yields, it reduces in balance. The Zinfandel variety can be a little confusing because there are distinctly red Zinfandels and white Zinfandels (which are actually pink). Both wines come from the same Zinfandel grapes and it is the process for making wine that differentiates the two. Zinfandel wine grapes perform best in hot, dry climates. Zinfandel grapes reach their best expression in sunny, dry weather during the day and cool temperatures at night. The vines are best suited to rocky, alluvial soil and hillside plantings. When

ripe, Zinfandel wine exudes ripe, red and black fruits including raspberries, jammy cherries, strawberries and red plums along with spice, pepper and jam characteristics. The wine prepared from Zinfandel has a great sweetness and alcohol. Most Zinfandel wine is best consumed within five years of the vintage, or less, to preserve its fruit and freshness. While some Zinfandel wine ages, most do not become more interesting with time. Popular clonal selections of Zinfandel are Zinfandel 1A, Zinfandel 2, Zinfandel 3, Primitivo 3, Primitivo 5, Primitivo 6, etc.

## 7.2. White Wine Grape Varieties

### *Chardonnay*

Among the white wine grape varieties, Chardonnay is most popular and widely planted across different climatic conditions. This variety was born in the Burgundy region of France, where it is known as White Burgundy. After recognition, Chardonnay became popular in the Champagne region also. After some years, it was found well adapted to making sparkling wine. However, Chardonnay is able to produce quality yield in wide types of terroirs, soils and climates, but finds best expression in soils having high concentrations of chalk, clay and limestone. Chardonnay grapes take on a very different characteristic in Champagne as compared to Chardonnay from Burgundy and this shows differences in wine quality. The berries are responsible for most fruity flavours found in Chardonnay wines. The flavours are observed as that from tropical fruits like banana, melon, pineapple and guava to stone fruits, such as peach, nectarine like citrus and apples. The type of fruity flavours in Chardonnay wines are governed by climatic conditions of the particular region. Warm regions of California, Chile and much of Australia tend to give more tropical styles while zones with temperate climate, such as southern Burgundy or northern New Zealand, create wines with stone-fruit notes. Chardonnay is grown in very cool areas like Chablis, Champagne and Germany, which produce wines with green-apple aromas. Chardonnay grapes produced from vineyards of very warm regions result in low acidity, flat nature and overblown wines. Most of the time, Chardonnay grapes are used to produce 100 per cent Chardonnay wines. However, the Chardonnay grapes are versatile in nature, so thay are used as blending grapes from time to time. The wide range of choices allows production of wines with different styles, ranging from flavours of green apples, pears, citrus, rocky and mineral-driven, steely or tropical honeyed fruit in nature. A lot of variability is recorded in Chardonnay in France, US and other countries. A few known clonal selections of Chardonnay are Clone-75, 76, 77, 78, 95, 96, 121, 124, 125, 277, 548, 809 (Dijon clones) FSP 15 (Prosser clone), I10V1, I10V5, G9V7, Mendosa, etc.

### *Sauvignon Blanc*

Sauvignon Blanc originated in the southwest of France as a native variety. The name of this variety originates from the French words 'sauvage' (wild) and 'blanc' (white). During the 18$^{th}$ century, Sauvignon Blanc gained recognition in the Loire Valley and Bordeaux vineyards of France. Vines develop buds late but interestingly the berries ripen early. This variety produces best grapes in diverse wine regions of temperate climate as Bordeaux, California, South Africa, Australia and New Zealand. However, the vines are relatively of robust type, vigorous and adapt readily to different types of growing environments. As the berries ripen early, this variety is grown in relatively cool climates, with its homeland 'Loire' being the

most obvious example. While even in warmer areas, its natural high acidic nature allows it to retain a level of freshness. The variety is well adapted in tropical regions of different countries. Sauvignon Blanc wines from warmer climate develop a rounder, riper, fig-scented fruitiness and denser texture. The wine tends to be earthier and more terroir-driven in France as compared to its new world counterparts. The characteristics of Sauvignon Blanc wines vary widely. In Bordeaux, the grape is blended with Sémillon to create some of the world's greatest dessert wines and crisp, refreshing, mineral-driven whites. Within two appellations (Bordeaux and Loire valley), chalky, gravel soils impart flavours of gunpowder and flint, with strong, smoky minerality. Prevailing temperature during grape fermentation also plays a vital role in deciding the wine's character. In high temperature conditions, more minerality is noted in wines, while low temperature conditions result in tropical fruit and citrus sensations. In general, wine is found with low alcohol and balanced structure. Inregions of the new world, it tends to be less flinty and earthy, with stronger emphasis on fruity. Major parts of Australia are too warm to preserve the characteristically 'green' (i.e. slightly under ripe) aroma of Sauvignon Blanc. Some fine examples of specific aroma have emerged in the Adelaide Hills of Tasmania and cooler spots in Victoria and New South Wales. A distinctively refreshing style of Semillon/Sauvignon Blanc blends have been developed in Western Australia. Known clonal selections from Sauvignon Blanc are FPS 03 (Jackson), 22 (Oakville), 23 (Kendall-Jackson Winery), 26 (Napa Valley), FPS 06 (ISV 5), 07 (ISV 2), 17 (ISV 1), 24 (ISV 3), FPS 14 (French 316), 18 (French 317), 20 (French 242), 21 (French 378), and 25 (French 378), etc.

### *Chenin Blanc*

The variety Chenin Blanc originated in the famous Loire valley of France. It is called Steen when grown in South Africa. Chenin Blanc grapes are used to produce wines of varying quality. According to available official French documents, Chenin Blanc was first mentioned as early as 845. The variety has appeared in various parts of the Loire valley under a multitude of synonyms. Usually grape growers obtain high yields from Chenin Blanc which allows flavours to concentrate and floral bouquets to develop. In this variety, sprouting of buds begins early and berries ripen late. This type of tendency of bud sprouting invites high risk from frost in cooler parts of the world. Bunches of Chenin Blanc ripen unevenly which leads to harvesting by hand in the cooler Loire valley of France. The presence of higher acidity level in less ripened grapes provides a strong base for making sparkling wines. Finally, at the end of the harvest season, the grapes picked in the last are overripen or affected with noble rot having higher sugar concentrations, which lead to rich flavours of orange marmalade, ginger and saffron. Chenin Blanc wine reached South Africa around middle of 17$^{th}$ century and became popular immediately due to productivity

and its ability to generate high acids, even in hot conditions. Wines from South African Chenin Blanc have tropical fruity flavours, such as that of melon, pineapple and banana and sometimes very distinguishing flavour of green apple. Hence, terroir effect reflects in wine. Even nominally similar wines made from Chenin Blanc grapes have different tastes if the grapes come from different regions or are produced by very different wineries. Hundreds of clones of Chenin Blanc have been identified though a few have importance in wine quality. Some known clones of Chenin Blanc are 220, 278, 416, 417, 624, 880, 982 and 1018; SN 1064, SN 1061, SN 64, 3/1061, 111/13, etc.

### *Riesling*

The variety Riesling is native to Germany's Rhein and Mosel river valleys. The Riesling wines are produced in a very wide range of styles in both dry to sweet variations as well as light to full-bodied. German Rieslings are categorised according to their style (levels of dryness) and the ripening levels at the time of harvest (i.e. *Kabinett, Spatlese, Auslese,* etc). For a dry style of Riesling, it is known as *Trocken* ('dry' in German) or *Halbtrocken* ('half-dry' in German translates to off-dry or semi-sweet on the palate). Riesling from the Mosel and its even cooler tributaries, the Saar and Ruwer, is one of the world's most distinctive wines with least imitable styles as light, crisp, racy, refreshing like mountain stream. Riesling is also a classic grape variety of the Rheingau region where it perhaps best reflects in a steely, lemony, sometimes mineral-scented way. The differences can be easily distinguished between Riesling wines made from even neighbouring vineyards. Riesling is recorded as an extremely fine candidate for producing botrytised sweet wine, though the noble rot tends to blur geographical differences and results in thick, almost raisiny deep golden wine usually labelled either Beerenauslese (BA) or Trockenbeerenauslese (TBA). Most of southern Europe is too warm for Riesling and in these warmer conditions, if grapes ripen fast and too early, wines are unable to produce any flavour. But it is more surprising to find an enclave of fine Riesling production in South Australia. Dry Australian Rieslings tend to display a pale colour, almost watery at times and occasionally with a greenish tinge. The young Riesling aromatics are often described as exotic perfumes, musky or oriental in character. Lemon, limes, green apple, blossoms, potpourri, musk sticks and minerality are general descriptions of Riesling wines. Late-harvest Riesling wines are rich and complex and have the ability to age in the bottle and develop nutty, honeyed qualities. German, Austrian and Canadian ice wines are made with grapes left to freeze on the vine so that sugars are concentrated and water crystalises as ice and separates out when the grapes are pressed. Riesling grapes also suffer from identity confusion in the United States, where unrelated cultivars and distant relatives have adopted different names, as Grey Riesling (Trousseau gris), Missouri Riesling, Hungarian Riesling (Italian Riesling progeny), Emerald Riesling (Muscadelle du Bordelais x Riesling). Often wines made in the 'German style' from high acid, light-coloured grapes, such as Sylvaner and Burger, were given the Riesling name even when Riesling grapes were not included in the blend, e.g. Hungarian Riesling, Grey Riesling, Kleinberger Riesling. Riesling clones, such as German: GM110, GM198 and GM239, 'Riesling Klon 90', Australia: SAVI 2, SAVI 3, SAVI 4, SAVI 7, SAVI 10, ENTAV-INRA: 1089, 1090, 1091, 1092, 1094, 1096, 1097, etc. are being adopted in different countries and are producing quality premium wines.

### *Airen*

Valdepeñas and La Mancha regions are known as origins of Airen grapes. A white grape wine variety is produced and is best suited to the southern plateau (Castillala Mancha) of Spain. In terms of total vine planted area in the world, Airen is placed among the top ten grape varieties. This variety is mostly planted in Spain's inland La Mancha and Valdepenas wine regions, where it shows tolerance to hot and dry conditions. Low-maintenance requirements and higher yield are advantages and people compromise on the quality attributes. Spain's brandy industry depends on Airen grapes traditionally as it produces oxidative and high-alcohol white wine. Loose and long bunches produce yellowish tinted grapes having soft pulp and with very slightly coloured juice. Traditionally, Airen grape variety is used to produce bulk wine as huge grapes are available in Spain. It has also been seen as an important component in blends, being very delicate and almost neutral in flavour. Airen produces a poor quality of wine, but the ease of the vine lends itself to producing base wines for brandy and high alcohol whites for Spain's local consumers. Airen wines blend with Cencibel to produce a light-bodied red wine. Most recently it is found that Airen grapes are giving rise to a crisp, dry white wine also.

### *Semillon*

The origin of Semillon grapes is unknown. This variety first arrived in Australia in the early $19^{th}$ century and by the 1820s, the grape covered over 90 per cent of South Africa's vineyards, where it was known as *Wyndruif*, meaning 'wine grape'. Semillon belongs to Bordeaux where it is used to make dry whites and sublime dessert wines. The wines of Semillon are commonly blended with Sauvignon. In California, Semillon grapes are often used to produce generic white wines; Semillon can also produce a good varietal wine which, however, is not popular in comparison to other white wines. It is widely planted in Australia, America and other regions also. Due to its thin skin berries, there is a risk of sunburn in hotter climates. So this variety is well suited to areas where sunny days and cool nights prevail. The grapes are rather heavy with low acidity and almost oily texture. Semillon is widely grown in Australia, particularly in the Hunter Valley in north of Sydney, where it was known as 'Hunter River Riesling'. Four styles of Semillon-based wines made there are very popular. This variety is unpopular and often criticised for lack of complexity and intensity. Semillon wines have aromas of green apple, blossoms, lemon and perhaps lanolin. Palates range from tart lime juice, green pear, crunchy green apples to tropical fruits, nuts, passion fruit and grass. Clonal variability has been recorded in Semillon and some known clones are ENTAV-INRA: 173, 299, 315, 380, 908, 909, 910; FPS: 02, 03, 04, 05, 06, etc.

## 8. Wine Grape Varieties: Attempts from India

In India, few attempts have been made to breed grapes for winemaking. Breeding work was initiated at Indian Agricultural Research Institute (New Delhi), Indian Institute of Horticultural Research (Bangalore), Agharkar Research Institute (Pune) and National Research Centre for Grapes (Pune). These institutes developed grape varieties mainly for table purpose. However, Arka Soma (Anab-e-Shahi x Queen of Vineyards) and Arka Trishna (Bangalore Blue and Convent Large Black) varieties were developed for white and red wines, respectively by IIHR. These varieties were not accepted by wineries. Chardonnay is well known white-wine-grape variety but couldn't perform well under Indian conditions. Considering the importance of Chardonnay, breeding work was initiated at NRCG, Pune and Chardonnay was crossed with Arkavati. After evaluation, four crosses of Chardonnay and Arkavati, named as Charark-1, Charark-2, Charark-3 and Charark-4, were found to be better. Juice and wines prepared from Charark series of grapes were found acceptable. Further studies are in progress. Teinturier varieties, namely Pusa Navrang and Manjari Medika have also been developed for juice purposes. Manjari Medika contains higher anthocyanins and phenols and can be promoted as a good variety for blending of wines with higher antioxidant activities.

## References

Alleweldt, G. and Dettweiller, E. (1994). The Genetic Resources of *Vitis*: World List of Grapevine Collections, Second ed. Geilweilerhof, Siebeldingen, Germany.

Anonymous (2017). OIV FOCUS 2017: Vine varieties distribution in the world. http://www.oiv.int/public/medias/5336/infographie-focus-oiv-2017-new.pdf

Anonymous (2016a). http://www.fao.org/faostat/en/#data/QC

Anonymous (2016b). World vitiviniculture situation, 2016. OIV Statistical Report on World Vitivin Culture. http://www.oiv.int/public/medias/5029/world-vitiviniculture-situation-2016.pdf

Baydar, N.G., Sağdıç O., Özkan, G. and Çetin, E.S. (2006). Determination of antibacterial effects and total phenolic contents of grape (*Vitis vinifera* L.) seed extracts. *International Journal of Food Science and Technology* 41: 799-804.

Baydar, N.G., Babalik, Z., Turk, F.H. and Çetin, E.S. (2011). Composition and antioxidant activities of wines and extracts of some grape varieties grown in Turkey. *Journal of Agricultural Sciences* 17: 67-76.

Cheng, G., Liu, Y., Yue, T. and Zhang, Z.W. (2015). Comparison between aroma compounds in wines from four Vitis vinifera grape varieties grown in different shoot positions. *Food Science and Technology* 35(2): 237-246. https://dx.doi.org/10.1590/1678-457X.6438

Conde, C., Silva, P., Fontes, N., Dias, A.C.P., Tavares, R.M., Sousa, M.J., Agasse, A., Delrot, S. and Gerós, H. (2007). Biochemical changes throughout grape berry development and fruit and wine quality. *Global Science Books* 1(1): 1-22.

Cunha, J., Santos, T., Carneiro, C., Fevereiro, P. and Eiras-Dias, J.E. (2009). Portuguese traditional grapevine cultivars and wild vines (*Vitis vinifera* L.) share morphological and genetic traits. *Genet. Resour. Crop Evol.* 56: 975-989. DOI 10.1007/s10722-009-9416-4

ENTAV-INRA (1995). Catalogue of Selected Wine Grape Varieties and Clones Cultivated in France. Ministry of Agriculture, Fisheries and Food. CTPS.

Fidelibus, M., Christenson, L., Katayama, D. and Verdenal, P. (2006). Yield components and fruit composition of six Cabernet Sauvignon grapevine select in the Central San Joaquin Valley, California. *Journal of the American Pomological Society* 60(1): 32-36.

Forneck, A., Walker, M., Schreiber, A. Blaich, R. and Schumann, F. (2003). Genetic diversity in *Vitis vinifera* ssp. *Sylvestrisgmelin* from Europe, the Middle East and North Africa. *Acta Hort.* (ISHS) 603: 549-552. doi: 10.17660/ActaHortic.2003.603.72

Garrido, J. and Borges, F. (2011). Wine and grape polyphenols – A chemical perspective. *Food Research International* 44(10): 3134-3148. http://dx.doi.org/10.1016/j.foodres.2011.11.001

Guasch-Jane, M.R., Ibern-Gomez, M., Andreslacueva, C., Jauregui, O. and Lamuela-Raventos, R.M. (2004). Liquid chromatography mass spectrometry in tandem mode applied for the identification of wine markers in residues from ancient Egyptian vessels. *Analytical Chemistry* 76: 1672-1677.

Jogaiah, S., Oulkar, D.P., Vijapure, A.N., Maske, S.R., Sharma, A.K. and Somkuwar, R.G. (2013). Influences of canopy management practices on fruit composition of wine grape cultivars grown in semi-arid tropical region of India. *African Journal of Agricultural Research* 8(26): 3462-3472. DOI: 10.5897/AJAR2013.7307

Jones, G., Michael, W., Owen, C. and Karl, S. (2005). Climate Change and Global Wine Quality. *Climatic Change* 73(3): 319-343.

Jones, G.V. (2005). Climate Change in the Western United States Grape Growing Regions. *Acta Horticulturae* 689: 41-60.

Jones, G.V. (2006). Climate and terroir: Impacts of climate variability and change on wine. pp. 247. *In*: Macqueen, R.W. and Meinert, L.D. (Eds.). *Fine Wine and Terroir – The Geoscience Perspective.* Geoscience Canada Reprint Series Number 9. Geological Association of Canada. St. John's, Newfoundland.

Jones, G.V. and Davis, R.E. (2000). Climate Influences on Grapevine Phenology, Grape Composition, and Wine Production and Quality for Bordeaux, France. *American Journal Enology and Viticulture* 51(3): 249-261.

Kubach, H.K. (2009). Wine grape suitability and quality in a changing climate: An assessment of Adams county, Pennsylvania (1950-2099). https://www.ship.edu/globalassets/geo-ess/kubach_answer_120502.pdf

Núñez, D.R. and Walker, M.J. (1989). A review of paleobotanical findings of early Vitis in the Mediterranean and of the origins of cultivated grape-vines, with special reference to prehistoric exploitation in the western Mediterranean. *Rev. Paleobot. Palynol.* 61: 205-237.

Martinson, T. (2018). Veraison to Harvest. Cornell Agri Tech. https://grapesandwine.cals.cornell.edu/sites/grapesandwine.cals.cornell.edu/files/shared/Veraison-To-Harvest-2018-Issue-3.pdf

Petrie, W.M.F. (1923). Social Life in Ancient Egypt. Methuen, London.

Pongracz, D.P. (1978). Practical Viticulture. David Philip Publisher Ltd., Claremont.

Ramishvili, R. (1983). New material on the history of viniculture in Georgia (in Georgian, with Russian summary) Matsne. *Hist. Archaeol. Ethnol. Art Hist. Ser.* 2: 125-140.

Río Segade, S., Vázquez, E.S., Orriols, I., Giacosa, S. and Rolle, L. (2011). Possible use of texture characteristics of winegrapes as markers for zoning and their relationship with anthocyanin extractability index. *Int. J. Food Sci. Technol.* 46: 386-394.

Rolle, L., Río Segade, S., Torchio, F., Giacosa, S., Cagnasso, E., Marengo, F. and Gerbi, V. (2011). Influence of grape density and harvest date on changes in phenolic composition, phenol extractability indices, and instrumental texture properties during ripening. *J. Agric. Food Chem.* 59: 8796-8805.

Romero-Cascales, I., Ortega-Regules, A., Lopezroca, J.M., Fernadez-Fernadez, J.I. and Gomez-Plaza, E. (2005). Differences in anthocyanin extractability from grapes to wines according to variety. *Am. J. Vitic. Enol.* 56: 212-219.

Schultz, H.R. and Jones, G.V. (2010). Climate-induced historic and future changes in viticulture. *Journal of Wine Research* 21(2/3): 137-145. doi: 10.1080/09571264 2010.530098

Son, H.S., Hwang, G.S., Ahn, H.J., Park, W.M., Lee, C.H. and Hong, Y.S. (2009). Characterization of wines from grape varieties through multivariate statistical analysis of $^{1}$H NMR spectroscopic data. *Food Research International* 42(10): 1483-1491.

Stefanini, M., Iacono, F., Colugnati, G., Bregnant, F. and Crespon, G. (2000). Adaption of same Cabernet Sauvignon clones to the environmental conditions of north-eastern Italian wine-growing areas. *Acta Horticulture* 528: 779-784.

Sun, J., Chu, Y.F., Wu, X. and Liu, R.H. (2002). Antioxidant and anti-proliferative activities of common fruits. *Journal of Agricultural Food Chemistry* 50: 7449-7454.

Tebeica, V. and Popa, C. (2005). Results Concerning Applying of Clonal Selection of Variety Sauvignon. The Romanian Society of Horticulturists, 'Ion Ionescu from Brad' Publishing House, Iasi, Year XLVIII Vol. I (48). I.S.S.N. 1454-7376.

Truel, P., Rennes, C. and Domergue, P. (1980). Identification in collection of grape vines. Proc III Int. Symp. on Grape Breed. Davis CA (USA), June 15-18, pp. 78-86.

Urcan, D.E., Lung, M.L., Giacosa, S., Torchio, F., Ferrandino, A., Vincenzi, S., Río Segade, S., Pop, N. and Rolle, L. (2016). Phenolic substances, flavour compounds and textural properties of three native Romanian wine grape varieties. *Int. J. Food Prop.* 19: 76-98.

Van Leeuwen, C. and Seguin, G. (2006). The concept of terroir in viticulture. *Journal of Wine Research* 17: 1-10, 10.1080/09571260600633135

# 5 Advances in Ripening, Maturation and Harvest of Fruits

**R. Gil-Muñoz[1] and E. Gómez-Plaza[2*]**

[1] Instituto de Investigación y Desarrollo Agroalimentario, Av. Ntra. Sra. de la Asunción, 24, 30520 Jumilla, Murcia, Spain

[2] Department of Food Technology, Universidad de Murcia, Campus de Espinardo, 30071 Murcia, Spain

## 1. Introduction

Grape ripening is one of the major events for both table and wine grapes. The quality of the harvested fruits and the obtained wines is strongly related with the characteristics achieved in the grapes during maturation.

Development of grape berries during ripening exhibits a double sigmoid growth pattern separated by a lag phase (Coombe and McCarthy, 2000). The first period of growth lasts from bloom to approximately 60 days afterwards and it is during the first period that the berry is formed and the seed embryos are produced. This is the time when rapid cell division also occurs.

The lag phase may vary in duration, depending on the variety, from non-existing to several days (Conde *et al.*, 2007). The beginning of the second phase of berry growth marks the beginning of fruit ripening that starts with the so-called 'veraison' (clearly seen in red grapes, due to the appearance of anthocyanins – the pigment responsible for red colour) and is the period when major changes occur, affecting physical and chemical characteristics, such as modifications in size, composition, colour, texture, and flavor, occur. Most of the important solutes in the grape berry accumulate during this second period of development.

## 2. Grape Ripening: Physical Changes in Grape Berries during the Ripening Period

### 2.1. Size and Volume

The most important physical changes that grapes undergo during development and maturation are in their size and texture. This property is important since the composition of wine can be manipulated by simply changing the berry size (Kennedy, 2002). Most of the important compounds that influence the wine quality are located in the skin of the grapes (phenolic and aroma compounds). So the largest the grapes, the higher is the dilution of these compounds in the must.

The potential size of the berry is controlled by three principal factors – number of cells, cell volume and sugar content. Studies demonstrate that the number of cells in a grape berry is established during the first three weeks after anthesis and no further cell division occurs after this period (Dokoozlian, 2000). Harris *et al.* (1968) found in Sultana grapes that berry pericarp growth is the product of both cell division and cell expansion up to approximately 25 days after anthesis. From then on, pericarp growth occurs due to cell expansion alone. So, during the first phase, besides an increase in the number of cells, cell volume increases significantly. However, significant differences can be found in size and volume of different grape varieties, indicating that genetics may also have an important role on this characteristic.

The increase in size also causes a change in berry conformation – mainly the expansion in the fruit equatorial plane which occurs during the second period of growth (Harris *et al.*, 1968).

Amongst all, the fruit growth is mainly a result of water accumulation. Water influx before veraison is due to water import from the xylem whereas most of the post-veraison gain is due to water import from

*Corresponding author: encarna.gomez@um.es

the phloem, since shortly after veraison the xylema vessels entering the berry are blocked (Keller, 2010). In fact, the increase in size and volume not always is maintained till harvest. Sometimes, this growth stops during phase II and shrinkage can be observed due to loss of water through transpiration, or due to blockage of the phloem elements into the berry (Kennedy, 2002).

Water deficit generally leads to formation of smaller berries since it inhibits both cell division and cell expansion (Conde *et al.*, 2007). If there is a water deficit, especially during the first phase of development, the final size of the berry will be smaller; more than if the deficit occurred during the second phase, since during the first phase the number of cells will be compromised whereas during the second phase, only the size of the cell will be affected (Smart and Coombe, 1983; Hardie and Considine, 1976; Van Zyl, 1984; Matthews *et al.*, 1987). This event is well known among viticulturist and is commonly used to manipulate the berry size.

## 2.2. Changes in Texture and Cell Wall Composition

The grape cells present a cell wall. These cell walls are highly complex and dynamic, composed of polysaccharides, phenolic compounds and proteins and stabilised by ionic and covalent linkages (Ortega-Regules *et al.*, 2006). Hemicellulose, pectins and structural proteins are inter-knotted with the network of cellulose microfibrils, the skeleton of cell walls.

The maturation process of berries is accompanied by a gradual 'weakening' of berry cell walls and a softening of the berry. The modification of the cell-wall polysaccharide matrix, including depolymerisation of pectins (Nunan, 1998; Nunan *et al.*, 2001) and xyloglucan molecules and a decrease of hemicellulose may be the cause of this softening. Such modifications are principally due to an increase in the enzymatic activity of pectin methylesterase, α-galactosidase and β-galactosidase, which were registered after *véraison* (Nunan *et al.*, 2001; Ortega-Regules *et al.*, 2008a).

It is clear that the quantity of isolated cell wall material decreases as ripening progresses (Fig. 1) as observed by Ortega-Regules *et al.* (2008a). Barnavon *et al.* (2000) also observed loss of cell-wall material during the last weeks of grape development and stated that the decrease seemed to be a consequence of an increase in cell volume, with cell walls becoming thinner at the end of maturation. This was confirmed by the results of transmission electron microscopy (Ortega-Regules *et al.*, 2008a), which clearly illustrated how the cell walls of Monastrell grapes became thinner as ripening progressed.

Looking at the changes in the polysaccharide composition of the cell walls, the levels of cellulosic glucose, high quantities of which have been correlated with firmness (Rosli *et al.*, 2004), did not change significantly during the maturation of grapes (Ortega Regules *et al.*, 2008a). However, Ishimaru and Kobayashi (2002) found that cellulose decreased gradually during berry softening in Kyoho grapes. More than glucose, galactose is the sugar detected in high concentrations in immature grapes and it substantially falls during ripening in all varieties. This decrease in galactose is one of the most common features of ripening fruits. In grapes, the decrease in $\beta$-(1$\rightarrow$ 4)-linked galactose residues during ripening has been attributed to a loss of galactans and has been reported as a crucial step associated with the initiation of softening.

All these changes affect the skin thickness (Coombe and McCarthy, 2000), decreasing as epidermis and hypodermis cells expand. Even the morphology changes, the cuboid parenchymatous pericarp cells of very young fruits become radially elongated (Fig. 2) as the berry matures (Harris *et al.*, 1968). Thus, the skin becomes thinner in the berry during ripening, between fruit set and maturity stages (Keller, 2010). The skin's thickness is one of the most important grape skin morphological characteristics affecting gas exchange regulation, the berry susceptibility to fungal diseases and the resistance to mechanical injuries (Rosenquist and Morrison, 1989; Kok and Celik, 2004).

It is clear that all these changes and softening of the berry, change the texture of the fruit (Le Moigne *et al.*, 2008). Preliminary research shows that the changes on the grape texture could recognise the veraison earlier than a visual identification performed in the field, which is of particular importance for white grapes where the colour change is slight (Rolle *et al.*, 2008, Robin *et al.*, 1997; Grotte *et al.*, 2001).

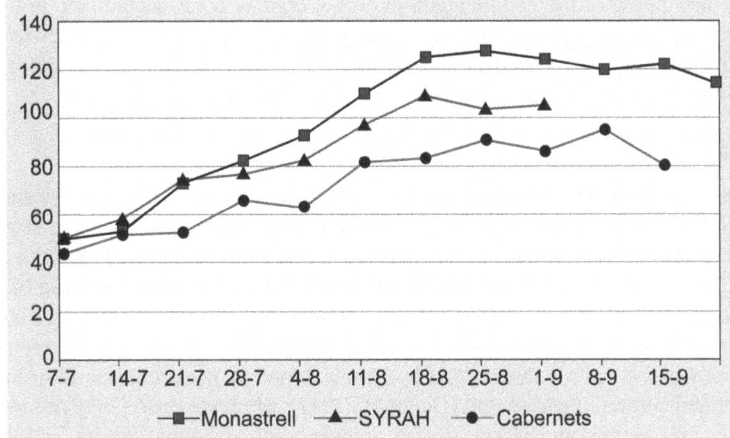


**Figure 1.** Evolution of berry weight in Monastrell, Syrah and Cabernet Sauvignon grapes from veraison to harvest in southeast of Spain

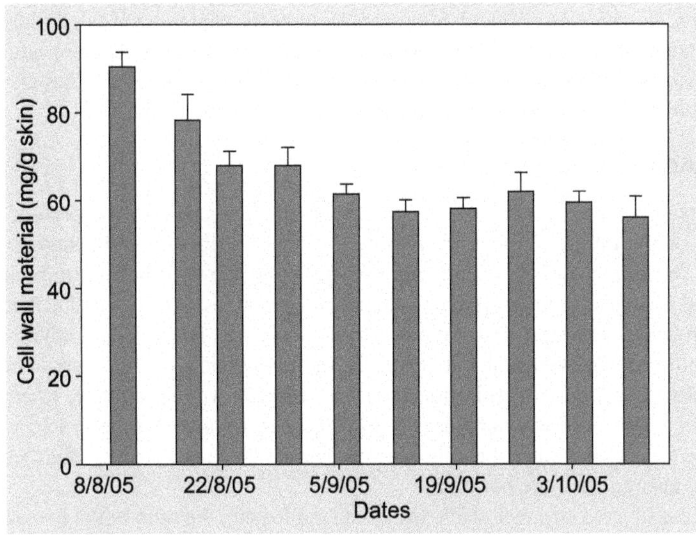


**Figure 2.** Evolution of cell-wall material in Monastrell grapes from veraison to harvest (*Source*: Adapted from Ortega-Regules, 2006)

## 3. Grape Ripening: Chemical Changes in Grape Berries during the Ripening Period

The chemical composition of grapes at harvest is a determinant of the final quality of wines, allowing grapes to develop their varietal characteristics (Gómez *et al.*, 1995). Grape-ripening period starts at veraison and lasts about approximately forty days, depending on the variety, environment and agricultural practices. The changes produced by the grapes during this period include physical changes (volume, weight, colour and rigidity) and chemical changes (pH, sugars, acidity, phenolics and volatile composition).

Over 700 compounds are known to be present in ripe grape berries. Many components increase in roughly the same manner as sugar, while for others, the increase is influenced by various factors. Yet others, such as flavour volatiles and some amino acids, show most of their accumulation later in ripening when sugar increase slows down.

During the first phase of the double sigmoid curve, organic acids accumulate in the vacuoles and tannins, hydroxycinnamates, and several phenolic compound precursors are synthesised. At the end of the lag phase, the period known as veraison, is characterised by the initiation of sugar accumulation and rapid pigmentation of berries by anthocyanins in red grape varieties.

## 3.1. Sugars in Grapes

Most soluble solids in grape juice are fermentable sugars (90-95 per cent) with glucose and fructose being the most abundant, and others, such as sucrose, raffinose and some minor sugars, appearing in small quantities. Glucose and fructose are in similar proportion at the moment of harvest, with their ratio being around 1. This ratio however, changes during the ripening period. So at the beginning of ripening period (veraison), glucose considerably exceeds fructose, but as sugar accumulation proceeds, the proportion of fructose increases. Sugar contents range from 15-25 per cent in grape juice at the moment of harvest.

Sugar composition is mainly determined by genotype and its concentration is strongly affected by the environment and cultural management (Dai *et al.*, 2011). Shiraishi *et al.* (2010) conducted a detailed analysis of table grape berry composition (sugars, organic acids, and amino acids) for 129 cultivars from Europe, North America, and Japan, that included 57 *Vitis vinifera*, and 72 *V. labruscana*×*V. vinifera hybrids*. They found separate groups of grapes based on their sugar composition; one group was classified by a glucose/(fructose+sucrose) ratio >0.8 and in the other, this ratio was <0.8. The differences were attributed to genetic or regional factors. *Vitis vinifera* cultivars only accumulate trace amounts of sucrose, while *V. labrusca, V. rotundifolia*, and interspecific hybrids contain non-negligible amounts of this compound. As stated before, most *Vitis vinifera* cultivars have a glucose/fructose ratio of 1 at maturity, while this ratio varies from 0.47 to 1.12 in wild species, with only a few species (*V. champinii* and *V. doaniana*) accumulating more glucose than fructose (Dai *et al.*, 2011).

## 3.2. Organic Acids in Grapes

Organic acids are present in lesser amounts as compared to sugars and are present in grapes in less than 1 per cent of the total juice weight. However, they contribute significantly to the overall taste (Nelson, 1985). The amounts of each acid vary in function of the variety, but can be influenced heavily by environmental changes, such as weather, viticultural practices and winemaker strategies. The most important acid in grape is tartaric acid, followed by malic, citric, succinic and other acids. Tartaric and malic acid accumulate before veraison, but during the ripening period, they suffer a decrease, especially in malic acid. The decrease is due to respiration and this is more in warmer climate, where more respiration occurs. Therefore, grapes from warmer places or warmer seasons have less malic acid and lower titratable acidity than those from cooler sites or cooler seasons, and it becomes a problem for these areas where the effects of climate change are more noticeable.

Since tartaric acid has a stronger acidic taste than malic acid, the ratio of the two acids will affect the flavour balance of the wine; however, the ratio of tartaric to malic acid is cultivar-specific and depends on the genetic background. It has been long known that species within the genus *Vitis* and individual varieties of the cultivated grapevine *V. vinifera* show wide variation in the natural acidity of berries. Kliewer *et al.* (1967) found differences in the levels of tartaric and malic acid in a study made on 26 species of *Vitis* and 78 varieties of *Vitis Vinifera.* Tartaric acid levels in wine grapes varied from a maximum of 1.08 g/100 ml (Emerald Riesling) to below 0.6 g/100 ml in cultivars,including Mataro, Pinot Gris and Semillon.

It is however, not only the acid concentration that influences the total acidity in grapes that ranges at harvest between 0.4-1.4 per cent at harvest, but strongly depends, not only on the acid concentration, but also on the ratio of concentrations between free organic acids and their potassium salt forms, which increase throughout ripening.

## 3.3. Phenolic Compounds in Grapes

Phenolic compounds are divided into two groups – flavonoid (anthocyanins, flavan-3-ols, condensed tannins and flavonols) and non-flavonoid compounds (phenolic acids and stilbenes). The composition of each family of polyphenols is directly responsible for the special characteristics in the different grape varieties and of the resulting wine. The concentration of polyphenols in grape berries depends on the grapevine variety and is influenced by viticultural and environmental factors (Guidoni *et al.*, 2008; Vacca *et al.*, 2009), as has occurred in most of the chemical compounds present in grapes.

Berry colouration is an important factor for market acceptance of grape cultivars, especially for red grapes and for the red wines produced (Fig. 3). Anthocyanins are responsible for the red colour. These compounds are located in the thick-walled hypodermal cells of the skin, although teinturier grape cultivars (also called dyers) synthesise anthocyanins, not only in the skin, but also in the pulp and other organs, such as pedicels, rachis, leaves, and stem epidermis (Jeong *et al.*, 2006; Guan *et al.*, 2012). Anthocyanins are mainly found in the skin of *Vitis vinifera* L. grapes as 3-O-monoglucosides, with malvidin-3-glucoside being the predominant anthocyanin in amounts varying according to variety, ecological conditions and viticultural practices (Kennedy *et al.*, 2006). Anthocyanin accumulation, which commences at veraison and continues throughout ripening, is frequently used as a fingerprint for cultivar recognition (Fig. 4). This accumulation during ripening usually achieves a maximum value followed by a decline before harvest.

Proanthocyanidins or condensed tannins accumulate mainly in berry skins and seeds before véraison, achieving their highest concentration at véraison. From this moment, seed and skins proanthocyanidins decline slowly until nearing grape ripeness, but thereafter they remain relatively constant (Kennedy *et al.*, 2000, Downey *et al.*, 2003).

**Figure 3.** Evolution of berry colour during ripening (green phase, veraison, maturity)

Color version at the end of the book

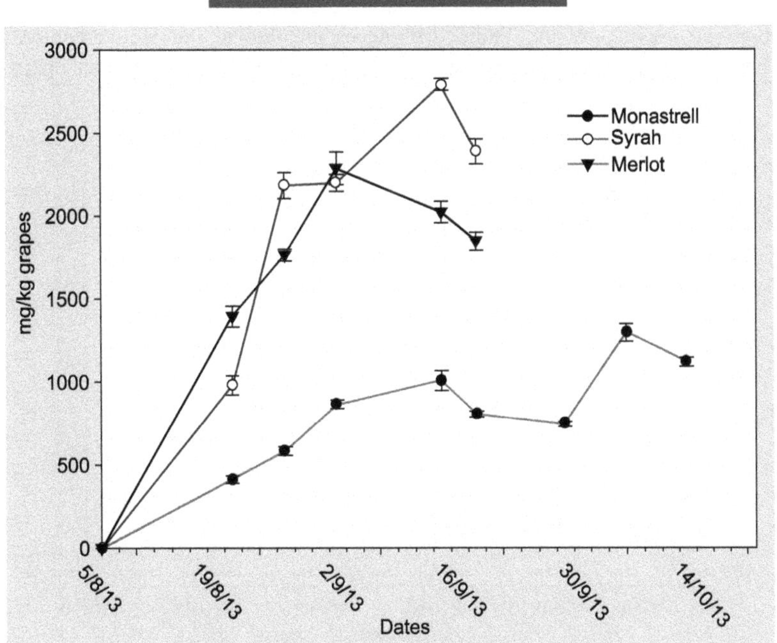


**Figure 4.** Evolution of anthocyanidin during ripening for three varieties of grapes (*Source*: Adapted from Gómez-Plaza *et al.*, 2017)

The proanthocyanidin content during ripening and other events occurring at the same time, such as histological and histochemical changes occurring in skin and seeds, will influence the extractability of these compounds into must-wine during winemaking and their final concentration in wines (Mattivi *et al.*, 2009).

Flavonols are the other important group of phenolic compounds, including kaempferol, quercetin, myricetin and others appearing in low concentration and they can be found in their aglicone form in wines, while in grapes, they are present in the form of glucosides, galactosides, glucuronides and rutinosides. In most cases however, flavonols are less abundant than the other phenolic compounds (Fig. 5) as discussed earlier (Revilla and Ryan, 2000; Kennedy *et al.*, 2006).

Among the non-flavonoid compounds, phenolic acids and stilbenes are present in grapes. Phenolic acids are located in the skin and pulp. In the skin, the proportion of these acids is similar to that of anthocyanins. The concentration of phenolic compounds is much lower in the pulp and the must. Numerous references to the presence of hydroxycinnamic acid derivatives, particularly tartaric acid esters, in grapes exist in the literature. These components, in particular caffeoyl tartaric acid (caftaric acid), play an important role in the browning reactions in grapes and wines. The tartaric acid ester contents decrease during growth and ripening in all the grape varieties; however, the decrease is much smaller from the end of growth/beginning of maturation (Fernández de Simón *et al.*, 1992).

The other important group of non-flavonoid compounds includes the stilbenes, their importance being closely related to their health-promoting properties. Their concentration in grapes depends on the phenological stage in the plant. Moreno *et al.* (2008) reported that an increase in total stilbenes is experimented at the end of the ripening phase. Similar results were documented earlier by Gatto *et al.* (2008) by studying 78 *Vitis vinifera* cultivars.

### 3.4. Volatile Compounds in Grapes

The volatile composition is one of the most important parameters responsible for wine quality and, hence, for consumer acceptance.

The aroma composition of grapes, which includes a wide range of components, is highly related with the technological potential of the wine-grape variety. These compounds will represent the wine varietal aroma and may allow identifying the variety used to elaborate a certain wine. Although the overall composition of most grape varieties is similar, there are clear and distinct aromas and flavour differences between most cultivars. These differences are due to the absence/presence of specific compounds but mostly due to the variations in the ratios of the compounds that make up the aroma

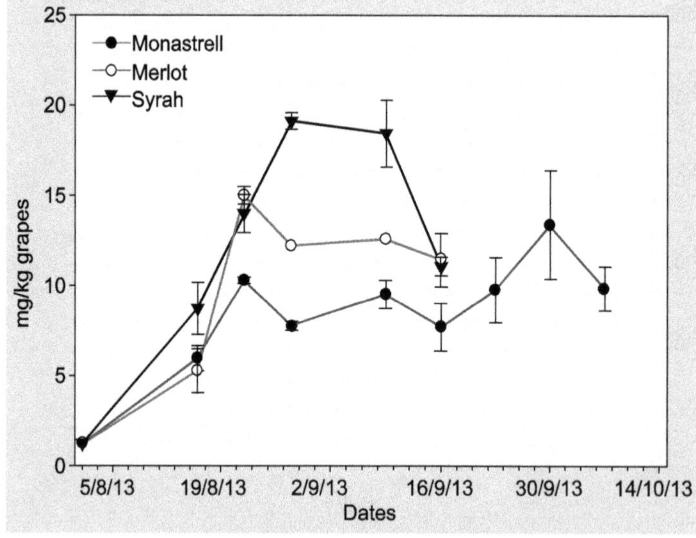


**Figure 5.** Evolution of flavonols during ripening for the three varieties (*Source*: Adapted from Gómez-Plaza *et al.*, 2017)

profile of grapes. Several families of compounds are responsible for the varietal aroma of grapes, such as monoterpenols, abundant in Muscat varieties, methoxypyrazines, which characterise the Cabernet family, C13-norisoprenoids, abundant in Chardonnay, volatile thiols, present in Sauvignon Blanc grapes, volatile phenols, common in Traminer aromatic grapes, and other minor compounds (Tominaga *et al.*, 2000; *Segurel et al.*, 2005).

The aroma components are found both in free and odourless glycosidically-bound forms in grape berries. Free forms are volatile compounds directly involved in the flavour, but in contrast, bound glycoside forms are non-volatile compounds that do not directly contribute to the aroma of the grapes. However, glycosides can be transformed into free volatile compounds through hydrolysis by specific enzymes, thus increasing the aromatic characteristics of grapes and influencing the flavour of the wine (Styger *et al.*, 2011). In general, bound glycoside forms are more abundant than free forms (Güñata *et al.*, 1985). The increase in concentration of free and glucosilated aroma compounds in the advanced stages of ripening occurs when sugar increased per berry has slowed down (González-Barrio *et al.*, 2013).

The concentration of varietal aroma compounds in grapes is influenced by several factors, such as grape variety and degree of maturity ,and vintage (Genisheva and Oliveira, 2009; Vilanova *et al.*, 2007), as well as environmental and agricultural factors (Kuhn *et al.*, 2014). The knowledge of grape varietal volatile composition can be used to estimate the wine aroma potential and therefore, some viticulturist determine the concentrations of varietal volatiles (terpenes, C13-norisoprenoids and C6-compounds) as a criterion to define the date of harvest (Salinas *et al.*, 2004).

Most of the volatile flavor components are produced after veraison until harvest. However, notable aroma compounds are produced during the first period of growth and then decline during fruit ripening (Belancic and Agosin, 2007). In red grapes, a maximum varietal volatile compound content is reached at maturity, as established by the sugar/acidity ratio and remains constant in the following weeks (Salinas *et al.*, 2004).

## 4. The Role of Cultural and Climatic Conditions on the Maturation of Grapes

Climate change is a major challenge in wine production. Temperatures are increasing worldwide and most regions are exposed to water deficits more frequently, with temperatures and microclimate changes largely affecting maturation as observed in Table 1.

**Table 1.** Composition of Monastrell Grapes from Different Zones in D.O. Jumilla

| *Area* | *Harvest date* | *ºBrix* | *Total acidity (mg/L)* | *pH* | *Tartaric acid (mg/L)* | *Malic acid (mg/L)* |
|---|---|---|---|---|---|---|
| Cañada del Judio | 10 Sep. 07 | 29.35 | 16.40 | 4.04 | 3.69 | 5.53 |
| El Carche | 21 Sep. 07 | 23.86 | 13.32 | 3.69 | 3.62 | 4.59 |
| Rubializas | 24 Sep. 07 | 22.34 | 12.47 | 4.86 | 3.47 | 5.95 |
| Cañada de Albatana | 19 Sep. 07 | 22.83 | 12.75 | 4.31 | 3.62 | 4.50 |
| Agüeros | 5 Oct. 07 | 24.92 | 13.96 | 4.16 | 3.49 | 4.82 |
| Varahonda | 22 Oct. 07 | 22.89 | 12.78 | 4.18 | 3.68 | 4.73 |

Higher temperatures trigger advanced phenology (Van Leeuwen and Darriet, 2016). Different strategies are being adopted in different parats of the world to aim at mitigating the effects of climate change. This effect is noticeable in crops, such as grapes – a very important crop in many places around the world with high social and economic impact. Viticulture is a vulnerable sector to changes in climate due to the sensitivity of grapevine phenology and fruit composition to temperature. In some areas, such as the Mediterranean region, the impact of new situations, like extreme summer temperatures and droughts (García-Ruiz *et al.*, 2011), could have irreversible consequences on the quality of wine.

Different strategies must be adopted in order to get an adaptation to this new situation. Some short-term strategies may include the use of different crop management practices, such as different irrigation systems, sunscreens for leaf protection, etc. On the other hand, some long-term measures should also be considered, such as varietal selection and land allocation changes (Fraga *et al.*, 2012). Finally, changes in enological practices may be the other tool to get positive effects on wine quality (Lobell *et al.*, 2006).

As commented before, the consequences of these new circumstances and scenario (warmer temperatures) will be reflected in the final quality of grapes, obtained with very high sugar content and lower acidity, leading to wines with high alcohol content. Different authors have shown the consequences of this situation, so Mira de Orduña (2010) argues that wine-making regions under extremely hot temperatures may have an increased risk of sensory degradation and wine spoilage.

High temperatures affect the composition of the grapes, and with especial importance to phenolic composition in grapes, inhibiting the biosynthesis of some of them, such as anthocyanin (Buttrose *et al.*, 1971), reducing grape colour (Downey *et al.*, 2006) and increasing the degradation of aroma compounds (Bureau *et al.*, 2000). It is clear that if grape quality decreases, the quality of wines will also be negatively affected. The most studied environmental conditions influencing the ripening period and the effect on grape quality are light conditions (Berli *et al.*, 2010), water status (Deis *et al.*, 2011), temperature (Pillet *et al.*, 2012) and pathogens (Vega *et al.*, 2011).

## 4.1. Variety

Varieties must adapt to the new scenario, meaning that, for example, some northern European regions may benefit with a wide range of varieties for winegrape growth in the future (Stock *et al.*, 2005), but, by contrast, areas in the southern of Europe will have to adapt to varieties more suitable to warmer and drier climates. Jones (2006) gives clues to varieties which are more adapted to warmer climates, such as Cabernet Sauvignon, Malbec, Merlot, Syrah and Tempranillo. This situation will also affect to white wine grape varieties, which will lose a part of their aromatic and fruity expression. In order to achieve these goals, new grapevine breeding programmes must be conducted to develop new varieties resistant to heat and droughts and with longer vegetative cycles.

## 4.2. Rootstocks

Rootstock is another factor that may affect yield, quality and other physiological parameters of the vineyard (Pavlousek, 2011). Its effect under warm and dry climates should be considered as rootstocks show complex interaction with soil water availability (Romero *et al.*, 2006). Rootstock choice should also be reconsidered when opting for those showing medium to high vigour and ablility to maintain the properly vine yield and a slower berry ripening; rootsocks such as: 110 Richter, 140 Ruggeri, 779 and 1103 Paulsen will be adequate for the purpose (Morando, 2001).

As an example, Koundouras *et al.* (2008), using Cabernet Sauvignon variety in Greece, compared the effect of two different rootstocks, concluding that 1103P is better for winegrape growth under semi-arid conditions, while S04 is preferable where no water limitation exists. The goal will be to assure proper choice of rootstock to combat new climate conditions as well as new pest and disease conditions (Phylloxera, nematodes, mildew, bacteria and viruses) (E-VitiClimate, 2012; Kirkpatrick, 2011).

## 4.3. Irrigation Management

In Europe, irrigated vineyards represent less than 10 per cent of the total area, but the tendency is towards an increase in irrigation. It is being increased to mitigate the effects of climate change and the more stressful environment. Irrigation has therefore expanded in dry regions of France, Spain, Portugal and Italy (De Leo *et al.*, 2015).

In arid and semi-arid environments, irrigation is a major tool to regulate soil water availability to vines. Water stress during the period from fruit set to veraison (i.e. the onset of ripening) heavily reduces the fruit size, as the limiting effect of soil water deficit on early fruit growth cannot be normally compensated even if water supply returns at full dosage later in the season (Ojeda *et al.*, 2001; Poni *et al.*, 1994). In this sense, a reduction in berry size might have a positive impact on wine phenolics content (Roby *et al.*, 2004) and sugar concentration (Trought and Tannock, 1996). Previous research (Intrigliolo and Castel, 2010) shows that in two out of four experimental seasons, a certain water restriction during the fruit set to veraison period leads to wines with higher colour and larger phenolic and anthocyanins concentrations than wines made from well-watered plots.

However, water is a scarce resource. Several deficit irrigation strategies (e.g. regulated deficit irrigation – RDI; partial root drying – PRD; sustained deficit irrigation – SDI) can be used to improve WUE (Water Use Efficiency), allowing optimal grape maturity and wine quality. So, regulated deficit irrigation (RDI) can be applied as a strategy to improve vine performance while minimising the possible negative impact of high water status on wine quality (Acevedo-Opazo *et al.*, 2010). Romero *et al.* (2010), using RDI by applying 60 per cent crop evapotranspiration water for irrigation during the growing season, found significant increases in WUE on Monastrell grapevines while Basile *et al.* (2011) reported that RDI may also affect sensory quality attributes of wine. In many viticultural areas around the world, the use of RDI practices has resulted in significant improvement in WUE and red wine quality (Kriedemann and Goodwin, 2003).

Partial root drying is another deficit-irrigation strategy that has been adapted for viticulture to improve WUE (Romero and Martinez-Cutillas, 2012). In this system, half of the plant root system is slowly dehydrated, whereas the other half is irrigated, decreasing water usage by 50 per cent. In theory, drying roots triggers hormonal signals that are transported to the shoots *via* the xylem, thereby reducing vegetative growth, stomatal conductance, transpiration and whole plant water use (Stoll *et al.*, 2000). Using this technique, Santos *et al.* (2005) reported no reduction in yield of Castelao grapevines. Recent studies in PRD also indicated some indirect benefits, such as a slight increase in water conservation associated with changes in soil evaporation losses and irrigation efficiency (Marsal *et al.*, 2008), direct benefits, such as improvement in water-use efficiency and fruit quality in table grapes (Du *et al.*, 2008) and changes in the composition and accumulation of anthocyanins (Bindon *et al.*, 2008). However, at present, there is still considerable controversy regarding the effects of PRD in wine grapes and other crops (Sadras, 2009) because some researchers have reported no improvement in vine performance, vine water use, crop yield or fruit quality compared to conventional drip irrigation at the same irrigation amounts in different soils and climatic conditions, and with different varieties (Pudney and McCarthy, 2004).

## 4.4. Tillage Treatments

Tillage systems are of key importance in mitigating the effects of climate change due to the importance of vineyard floor management. Nowadays it is common to suffer typical alternations of dry periods with some deficient precipitation or periods of excessive heat requiring a better management of ground water resources (Cass and Baumgartner, 2011). Therefore, different tillage treatments can affect yield and quality (Bahar and Yasasin, 2010) because floor management technique has an influence on the grapevine growth, weed management, soil conservation, disease attack and wine composition (Guerra and Steenwerth, 2012). These treatments may differ in both the start and duration of the tillage process.

In recent years, because of development of environment and consumption conscience, economic production demands, climatic changes and necessity of saving energy, using drastic changes in soil cultivation are being done. According to these changes, conservative soil tillage started to become widespread as an alternative to conventional soil tillage. According to the relationship to maturity grape, in a study made by Dobrei *et al.* (2015) using different tillage managements, the influence of treatment on sugar content from grape juice was low and insignificant. On the other hand, Palliotti *et al.* (2007) in their exploration concerning the effects of tillage and permanent cover crops on grape yield quality, found that cover crops influenced sugar content. Therefore, in this case it should be mentioned that the metabolism of the berry changed after veraison and the sugar accumulation was favoured (Lizana *et al.*, 2007).

## 4.5. Leaf and Canopy Management

Canopy microclimate is important in determining the fruit and wine quality (Sternad *et al.*, 2013). Different situations can be explained: a dense canopy with inadequate sunlight exposure will result in poor quality grapes (Morrison & Noble, 1990); a high-sunlight exposition that will result in higher total soluble solids, anthocyanins and phenolics and lower titratable acidity and malic acid content (Diago *et al.*, 2012); over-exposition to sunlight will lead to supra optimal berry temperature, leading to fruit sunburn and inhibition of colour development (Spayd *et al.*, 2002). Therefore, in order to achieve optimum sunlight expositon, the levels of leaf removal must be adapted to achieve high quality grapes.

On the other hand, basal leaf removal is a vineyard practice carried out in all the wine-growing regions and applied between the setting and veraison phonological stages of the grapevine. This practice may affect the ripening process in grapes, increasing sugar and polyphenolic compounds, improving its quality at harvest, but also causing a drop in titratable acidity (Poni *et al.*, 2006). Therefore, the selection of an adequate training system which is able to maintain bunches under a main regimen of diffused light by avoiding over-exposure and overheating of fruit during the hottest hours of the day, can be an agronomical winning choice.

## 5. Determining Harvest Date

The definition of optimal maturity is a crucial task for viticulturists and enologists and is not always easy to achieve. It may vary depending upon the desired style of wine to achieve and the content of some compounds, such that sugar, acids, the aromatic compounds and phenolic content will need to be followed so that the time of harvest can be optimised to meet the goals (Bisson, 2001, personal communication).

### 5.1. Technological Maturity

It is important to define optimal grape maturity for wine production and the chemical or biochemical compounds that can be used to define the optimum ripeness.

Sugar is the component most often used to determine ripeness. Moreover, sugar is very easy to assess, adding to its value as an index of ripeness. As already commented, its content increases during ripening. However, it is important to know that sometimes sugar accumulation may cease due to unfavourable environmental conditions, such as very high or low temperatures but most of the times, their accumulation continues once the conditions have changed, especially in the case of low temperatures. Elevated temperatures may irreversibly affect berry development, especially if the canopy is severely affected. It is important to be able to distinguish a transient effect from a permanent cessation of transport of material (Bisson, 2001).

Assessment of acidity is also used to define the optimal time of harvest. This can be evaluated as either pH or titratable acidity or both. The ratio of sugar: acidity has also been used. Besides these classic indices, there are other indices that are becoming more important when harvesting grapes for high quality wines, as discussed in the next section.

### 5.2. Phenolic Maturity

There are other metabolites more complicated to follow and which have been proposed for the assessment of optimum ripeness, for example, the phenolic content or, more exactly, the phenolic extractability. This issue has led to the definition of phenolic maturity. Studies have shown that grapes that are high in phenols do not necessarily produce wines that are also rich in phenolic compounds (Busse-Valverde *et al.*, 2012) and this is due, among other factors, to the phenolic compounds not being easily extracted (reasons being the morphology and composition of cell walls, the phenolic barriers). Moreover, sometimes it occurs that maximum phenolic extractability does not coincide with the maximum content of anthocyanins in the skins or with the optimum sugar ripeness (Bautista-Ortin *et al.*, 2006).

Glories and Augustine (1993) used the term 'grape phenolic maturity' for the first time to indicate the concentration of phenolic compounds in grapes and the ease with which they are released. This definition encompasses the anthocyanin concentration in the skin and its degree of extractability. The method proposed by Glories and Augustine consists of extracting the phenolic compounds from the crushed grapes under two different conditions – first at very low pH, favouring complete degradation of the cell walls and a second stage using a pH value of 3.2 which does not cause any further degradation of the cell membrane other than that normally reached during ripening and determining the concentration and subsequently, comparing the data. The smaller the difference in the parameters between pH 1 and pH 3.2, the greater is the level of phenolic maturation. This method and some variations have been widely

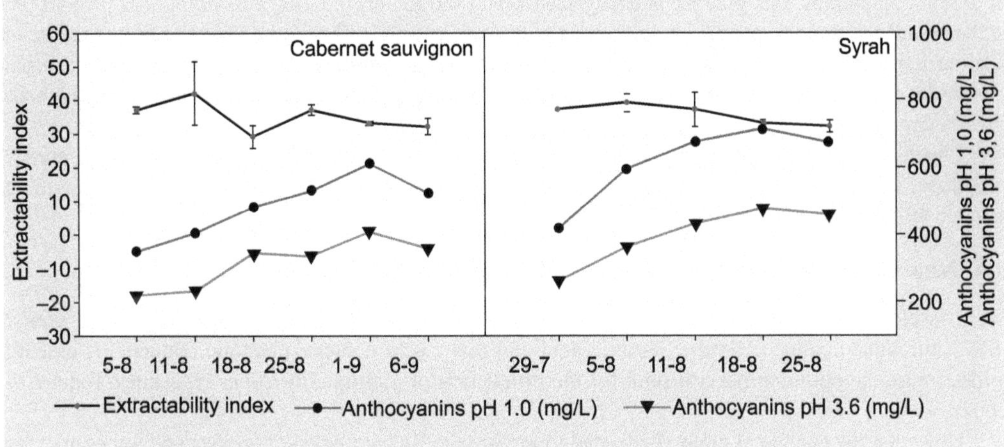


**Figure 6.** Evolution of phenolic maturity in Cabernet Sauvignon and Syrah grapes (*Source*: Adapted from Ortega-Regules *et al.*, 2008b)

used and now it is a grape characteristic that is usually followed and taken into account when deciding the harvest time for the elaboration of high-quality red wines.

Since the Glories method and similar methods used tedious laboratory methods, other approaches to the determination of phenolic maturity were attempted, such as the work of Rio-Segade *et al.* (2008), that tried to find a correlation between those parameters involved in the assessment of phenolic maturity, like total anthocyanin concentration, extractable anthocyanin concentration, cellular maturity index and seeds maturity index, and texture analysis variables to simplify the analytical methodology and to reduce the time involved in the analysis. Also, the evolution of the appearance of Carménère seeds was evaluated as a harvest criterion, comparing it with chemical (seed tannins percentage, the extractable anthocyanins, total anthocyanins and total polyphenols index, as well as the titratable acidity, soluble solids and pH) and phenolic ripening (Fredes *et al.*, 2010). They concluded that the observation of seed coat colour can be a reliable, simple and low-cost parameter to determine the correct ripeness of phenols in 'Carménère' grapevines.

For continuous monitoring of this maturation, other rapid and non-destructive methods are being developed. For example, an optical method was recently proposed, based on the screening of fruit chlorophyll fluorescence that allows both flavonol and anthocyanin contents of intact berry skin to be measured (Ben Ghozlen *et al.*, 2010).

## 5.3. Aromatic Ripeness

Aromatic ripeness is also very important, especially for white grapes. It has been stated that long after the increase of sugar content has stopped, the synthesis of aromatic compounds continues. For example, under dry ripening conditions, aroma synthesis is retarded in extreme cases, causing the wine from grapes of 25° Brix or more to remind one of an aroma profile of 17° Brix grapes (Schneider, personal communication). Deficiency of aromatic ripeness might lead to the absence of typical aroma, the appearance of an untypical flavour in wines and the appearance of veggie-green aromas in some varieties (due to high concentration of methoxypyrazines).

Free and bound terpene are the compounds more often used to assess berry aroma development and potential, but the problem is that this relies on assessment of a single family of components and the aroma of some varieties, the so-called neutral varieties, does not depend on terpene concentration. Since many types of aromatic compounds are present in the berry in the form of glycosidic precursors, the analysis of the total precursor level by assessment of the glycoside glucose (GG) content of the grapes may yield more information on the aromatic potential. This analysis is based on the fact that the sugars to which the grape-derived aroma compounds are bound can be of various types; nevertheless, one molecule of glucose is required per molecule of aroma precursor (Iland *et al.*, 1996). In contrast to the heterogeneous mixture of aroma compounds, this glucose is easily measured (Whiton *et al.*, 2002). Its amount is proportional to the bound aroma compounds, i.e. the potential primary aroma is liberated during the fermentation and post-fermentation processes. The technical procedure for measuring comprises the isolation of the aroma glycosides from a juice sample, its acidic hydrolysis by heat, and the measurement of the freed glucose (Iland *et al.*, 1996).

The disadvantage of this technique however, is that only the bound and therefore, sensorially inactive precursors of grape-derived aromas are measured, while the aromas already present in the free form and capable of being smelled are ignored.

### 5.4. New Rapid Methods for Optimum Maturity Determination

So, it is clear that measurement of certain grape quality parameters, e.g. total soluble solid content (SSC), titratable acidity, pH-value, tartaric acid and malic acid content, phenolic maturity, is essential to determine the optimum harvest time for the production of high-quality wines (Martínez-Toda *et al.*, 2007).

However, the analytical methods needed to measure these parameters in grapes and wines are slow, sometimes even tedious. Minimal sample preparation methods are becoming very popular as some of them give us real time information on the grape and wine quality parameters. Visible/near-infrared (Vis/NIR) spectroscopy is a rapid and non-destructive technique requiring minimal sample processing before analysis and coupled with chemometrics methods is a very good analytical tool for studying fruit quality and ripeness. Its main limitation is that accurate and good calibrations to predict ripeness parameters are needed. Using this technique, grape anthocyanins can be predicted accurately compared with the reference determination (Guideti *et al.*, 2010). Arana *et al.* (2005) also showed that NIR technology is suitable to determine the soluble solid content although it requires a calibration model for each variety. Also, NIR technique allows classification of grapes from different varieties, achieving up to 97 per cent of correct classification. Moreover, NIR spectroscopy can be used to measure quality properties of wine grapes (*Vitis vinifera* L.) directly on the field, during on-vine ripening, with similar results than when applied to the crushed berries in the laboratory (Gonzalez-Caballero *et al.*, 2010).

Another important modern development for rapid measurements and for determination of harvest time is the adoption of Precision Agriculture (PA) technologies to viticulture (PV). Researchers are developing sensing systems, such as biomass or leaf area index sensors, yield sensors and quality sensors, to provide information in a very fast way and at a high resolution (Tysseyre *et al.*, 2007). In this way, the use of remote sensing methods involving measuring properties by using remote satellites or aircraft-mounted sensors is becoming more and more interesting (Campbell and Whyme, 2011).

Precision viticulture has, potentially, the ability to provide a fast description of grapevine shape, size and vigour over an entire vineyard. This would allow, for example, to do differential harvest, that is, we can identify areas of different 'vigour' within a vineyard (using aerial images and sensors) and these different areas may need a differential harvest, so that we may either pick the block on the same day and segregate the different zones into different bins (and into different quality wines) or pick the different zones on different days when maturity and quality within each zone is considered optimum (Tysseyre *et al.*, 2007). The advantage of selective vintage lies, therefore, in greater economic benefit obtained when harvesting the two zones and processing the grapes separately. These tools will allow therefore better vineyard monitoring and crop segmentation according to quality.

## 6. Harvest Process and Its Influence on Grape Quality: Mechanical Harvest vs Manual Harvest

The choice of manual or mechanical harvesting of wine grapes depends on many factors including vineyard location, conducting system, value of the crop, the availability and cost of hand labour, the efficiency of fruit removal and transporting the harvested product (Clary *et al.*, 1990).

The first experimental mechanical harvesting of grapes began in California in the fifties. Improvements in harvesting equipment over the years have made possible extensive operations (Arfelli *et al.*, 2010). The benefits of the process can be economically important, but it is also essential to maintain the quality of the harvested grapes and to respect the vine integrity. Arfelli *et al.* (2010) stated that the most important parameters to be monitored during mechanical harvesting are the characteristics of harvested grapes (damaged bunch, amount of the dispersed must and presence of impurities such as leaves), the level of damage to plants and to the support system, and the losses during harvesting, from unpicked grapes and the ones fallen to the ground (Intrieri and Poni, 1990). These last two losses are generally grouped in the so-called hidden losses because they are less visible (Intrieri and Poni, 1990).

The most detrimental characteristic of mechanically harvested grapes is the berry damage that induces the release of free-run must (Intrieri and Poni, 1990) and decreases its quality since free-run must is exposed to biochemical reactions, such as enzymatic phenolic oxidation and uncontrolled microbial growth that may reduce the quality of the obtained must.

Several experiments have been carried out with different varieties. Clary *et al.* (1990) working with Chardonnay found no significant difference in weight of harvested fruit compared with manual harvest although the harvest contained higher stem content. The wines produced from the harvest treatments exhibited minor detectable differences but were all high-quality Chardonnay wines.

However, Noble *et al.* (1975) found that white wines made from damaged or machine-harvested grapes were significantly darker by visual assessment as well as by absorbance at 420 nm. The phenolic content of wines increased. Paetzold *et al.* (1990) observed that mechanically harvest grapes produced must with increased protein content compared to that from manual harvest grapes. A possible explanation is that physical damage caused by mechanical harvesting facilitates extraction of protein from the skins and other berry solids. It has been reported that grape skins contain 13-15 per cent protein by weight (Igartuburu *et al.*, 1991).

## 7. Future Research Needs

Future needs and research in viticulture must be routed in different directions and always looking for adaptation to climate change already present. For these objectives, different strategies can be used in the future:

- Models of simulation of integrated soil-plant-atmosphere analysis (SPA) being crucial to link together all different processes occurring in the SPA system and to correlate this information with to grape quality
- To deliver an innovative, science-driven and approachable precision viticulture platform to measure and manage sources of vineyard variation, using sensor technologies to get the fusion of the spatial data processing techniques and data to construct vineyard maps, showing relevant vineyard management zones
- The use of nano-particules to alleviate problems related to the agriculture sector as they can act efficiently as fungicides, insecticides, herbicides, pesticides and also as plant-growth-promoting factors
- Development of grape breeding programmes (hybrids), obtaining novel cultivars with greater fecundity in warm conditions and/or increased tolerance to the major grapevine pests and diseases

## References

Acevedo-Opazo, C., Tisseyre, B., Ojeda, H., Ortega-Farıas, S. and Guillaume, S. (2008). Is it possible to assess the spatial variability of vine water status? *Journal International des Sciences de la Vigne et du Vin* 42: 203-219.

Arana, I., Jaren, C. and Arazuri, S. (2005). Maturity, variety and origin determination in white grapes (*Vitis vinifera* L.) using near infrared reflectance technology. *Journal of Near Infrared Spectroscopy* 13: 349-357.

Arfelli, G., Sartini, E., Bordini, F., Caprara, C. and Pezzi, F. (2010). Mechanical harvesting optimisation and postharvest treatments to improve wine quality. *Journal International des Sciences de la Vigne et du Vin* 44: 101-115.

Bahar, E. and Yasasin, A.S. (2010). The yield and berry quality under different soil tillage and clusters thinning treatments in grape (*Vitis vinifera* L.) cv. Cabernet-Sauvignon. *African Journal of Agricultural Research* 5: 2986-2993.

Barnavon, L., Doco, T., Terrier, N., Ageorges, A., Romieu, C. and Pellerin, P. (2000). Analysis of cell wall neutral sugar composition, β-galactosidase activity and a related cDNA clone throughout the development of *Vitis vinifera* grape berries. *Plant Physiology and Biochemistry* 38: 289-300.

Basile, B., Marsal, J., Mata, M., Vallverdu, X., Bellvert, J. and Girona, J. (2011). Phenological sensitivity of Cabernet Sauvignon to water stress: Vine physiology and berry composition. *American Journal of Enology and Viticulture* 62: 452-461.

Bautista-Ortin, A.B., Fernandez-Fernandez, J.I., López.Roca, J.M. and Gómez-Plaza, E. (2006). The effect of grape ripening stage on red wine colour. *Journal International des Sciences de la Vigne et du Vin* 40: 15-24.

Belancic, A. and Agosin, E. (2007). Methoxypyrazines in grapes and wines of *Vitis vinifera* cv. Carmenere. *American Journal of Enology and Viticulture* 58: 462-469.

Ben Ghozlen, N., Cerovic, Z., Germain, C., Toutain, S. and Latouche, G. (2010). Non-destructive optical monitoring of grape maturation by proximal sensing. *Sensors* 10: 10040-10068.

Berli, F.J., Moreno, D., Piccoli, P., Hespanhol-Viana, L., Silva, M.F., Bressan-Smith, R., Cavagnaro, J.B. and Bottini, R. (2010). Abscisic acid is involved in the response of grape (*Vitis vinifera* L.) cv. Malbec leaf tissues to ultraviolet-B radiation by enhancing ultraviolet-absorbing compounds, antioxidant enzymes and membrane sterols. *Plant, Cell and Environment* 33: 1-10.

Bindon, K.A., Dry, P.R. and Loveys, B.R. (2008). Influence of partial rootzone drying (PRD) on the composition and accumulation of anthocyanins in grape berries (*Vitis vinifera* L. cv. Cabernet Sauvignon). *Australian Journal of Grape and Wine Research* 14: 91-103.

Bureau, S.M., Razungles, A.J. and Baumes, R.L. (2000). The environment of vine or bunch on volatiles and glycoconjugates. *Journal of Science Food and Agriculture* 80: 2012-2020.

Busse-Valverde, N., Bautista-Ortín, A.B., Gómez-Plaza, E., Fernández-Fernández, J.I. and Gil-Muñoz, R. (2012). Influence of skin maceration time on the proanthocyanidin content of red wines. *European Food Research and Technology* 235: 1117-1123.

Buttrose, M.S., Hale, C.R. and Kliewer, W.M. (1971). Effect of temperature on composition of Cabernet Sauvignon berries. *American Journal of Enology and Viticulture* 22: 71-75.

Busari, M.A., Kukal, S.S., Kaur, A., Bhatt, R. and Dulazzi, A.A. (2015). Conservation tillage impacts on soil crop and the environment. *International Soil and Water Conservation Research* 3: 119-129.

Campbell, J. and Wynne, R. (2011). Introduction to Remote Sensing, Fifth edition. Guilford Press, NY, USA.

Cass, A. and Baungartner, K. (2011). Soil health and wine quality: How vine health and fruit quality are influenced by soil properties. *In*: Blair, R.J., Lee, T.H. and Pretorious, I.S. (Eds.). Proceedings of the 14th Australian Wine Industry Technical Conference. Adelaide, 2010, pp. 226-235.

Clary, C., Steinhauer, R., Frisinger, J. and Peffer, T. (1990). Evaluation of machine vs hand-harvested Chardonnay. *American Journal of Enology and Viticulture* 41: 176-181.

Conde, C., Silva, P., Fontes, N., Dias, A., Tavares, R., Sousa, M.J., Agasse, A., Delrot, S. and Gerós, H. (2007). Biochemical changes throughout grape berry development and fruit and wine quality. *Food* 1: 1-22.

Coombe, B.G. and McCarthy, M. (2000). Dynamics of grape berry growth and physiology of ripening. *Australian Journal of Grape and Wine Research* 6: 131-135.

Dai, Z.W., Ollat, N., Gomès, E. *et al.* (2011). Ecophysiological, genetic and molecular causes of variation in grape berry weight and composition: A review. *American Journal of Enology and Viticulture* 62: 413-425.

De Leo, F., Minglietta, P.P. and Massari, S. (2015). Water sustainability assessment of Italian vineyards: Doc vs generic wines. Proceedings of Specialised Conference of the EuroMed Academy Business. Contemporary Trends and Perpectives in Wine and Agrifood Management, pp. 133-145.

Deis, L., Cavagnaro, B., Bottini, R., Wuilloud, R. and Fernanda Silva, M. (2011). Water deficit and exogenous ABA significantly affect grape and wine phenolic composition under in field and in-vitro conditions. *Plant Growth Regulation* 65: 11-21.

Diago, M.P., Ayestarán, B., Guadalupe, Z., Poni, S. and Tardaguila, J. (2012). Impact of pre-bloom and fruit-set basa leaf removal on the flavonol and anthocyanin composition of Tempranillo grapes. *American Journal of Enology and Viticulture* 63: 367-376.

Dobrei, A., Nistor, E., Sala, F. and Dobrei, A. (2015). Tillage practices in the context of climate change and a sustainable agriculture. *Notulae Scientia Biologicae* 7: 500-504.

Dokoozlian, N. (2000). Grape berry growth and development. pp. 30-37. *In*: Raisin Production Manual. University of California. Agricultural and Natural Resources Publication 3393, Oakland, CA.

Downey, M.O., Dokoozlian, N.K. and Krstic, M.P. (2006). Cultural practice and environmental impacts on the flavonoid composition of grapes and wine: A review of recent research. *American Journal of Enology and Viticulture* 57: 257-268.

Du, T.S., Kang, S.Z., Zhang, J.H., Li, F.S. and Yan, B.Y. (2008). Water use efficiency and fruit quality of table grape under alternate partial root-zone drip irrigation. *Agricultural Water Management* 95: 659-668.

E-VitiClimate, 2012. Lifelong Learning Project. E-VitiClimate, EuroProject Lifelong Learning Programme. Available at: http://www.eviticlimate.eu (accessed 28.11.13.)

Fernández de Simón, B., Perez-Ilzarbe, J., Hernandez, T., Gomez-Cordoves, C. and Estrella, I. (1992). Importance of phenolic compounds for the characterisation of fruit juices. *Journal of Agriculture and Food Chemistry* 40: 1531-1535.

Fraga, H., Santos, J.A., Malheiro, A.C. and Moutinho-Pereira, J. (2012). Climate change projections for the Portuguese viticulture using a multi-model ensemble. *Ciência e Técnica Vitivinicola* 27: 39-48.

Fredes, C., Von Bennewitz, E. and Saavedra, F. (2010). Relation between Seed Appearance and Phenolic Maturity: A Case Study Using Grapes cv. Carménère. *Chilean Journal of Agricultural Research* 70: 381-389.

García-Ruiz, J.M., López-Moreno, J.I., Vicente-Serrano, S.M., Lasanta, T. and Beguería, S. (2011). Mediterranean water resources in a global change scenario. *Earth Science Reviews* 105: 121-139.

Gatto, P., Vrhovsek, U., Muth, J., Segala, C., Romualdi, C., Fontana, P. *et al.* (2008). Ripening and genotype control stilbene accumulation in healthy grapes. *Journal of Agricultural and Food Chemistry* 56: 11773-11785.

Genisheva, Z. and Oliveira, J.M. (2009). Monoterpenic characterisation of white cultivars from *Vinhos verdes*: Appellation of Origin (North Portugal). *International Journal of the Institute of Brewing* 115: 308-315.

Glories, Y. and M. Augustin (1993). Maturité phénolique du raisin, consequences technologiques: Application aux millesimes 1991 et 1992. *Actes du colloque: Journée technique du CIVB*. Bordeaux, pp. 56.

Gómez, E., Martínez, A. and Laencina, J. (1995). Changes in volatile compounds during maturation of some grape varieties. *Journal of the Science of Food and Agriculture* 67: 229-233.

Gómez-Plaza, E., Bautista-Ortín, A.B., Ruiz-García, Y., Fernández-Fernández, J.I. and Gil-Muñoz, R. (2017). Effect of elicitors on the evolution of grape phenolic compounds during ripening period. *Journal of the Science of Food and Agriculture* 97: 977-983.

González-Barreiro, C., Rial-Otero, R., Cancho-Grande, B. and Simal-Gándara, J. (2013). Wine aroma compounds in grapes: A critical review. *Critical Reviews in Food Science and Nutrition* 55: 202-218.

González-Caballero, V., Sánchez, M.T., López, M.I. and Pérez-Marín, D. (2010). First steps towards the development of a non-destructive technique for the quality control of wine grapes during on-vine ripening and on arrival at the winery. *Journal of Food Engineering* 101: 158-165.

Grotte, M., Cadot, Y., Poussier, A., Loonis, D., Pietri, E., Dupart, F. and Barbeau G. (2001). Détermination du degré de maturité des baies de raisin par des mesures physiques: Aspects méthodologiques. *Journal Internationale des Sciences de la Vigne et du Vin* 35: 224-226.

Guan, L., Li, J-H., Fan, P-G., Chen, S., Fang, J-B., Li, S-H. and Wu, B-H. (2012). Anthocyanin accumulation in various organs of a teinturier grape cultivar (*V. vinifera* L.) during the growing season. *American Journal of Enology and Viticulture* 63: 177-184.

Guerra, B. and Steenwerth, K. (2012). Influence of floor management technique on grapevine growth disease, pressure and juice and wine composition: A review. *American Journal of Enology and Viticulure* 63: 149-164.

Guidetti, R., Beghi, R. and Bodria, L. (2010). Evaluation of grape quality parameters by a simple vis/NIR system. *Transactions ASABE* 53: 477-484.

Guidoni, S., Ferrandino, A. and Vittorino, N. (2008). Effects of seasonal and agronomical practices on skin anthocyanin profile of Nebbiolo grapes. *American Journal of Enology and Viticulure* 1: 22-29.

Günata, Y., Bayonove, C., Baumes, R. and Cordonnier, R. (1985). The aroma of grapes: Localisation and evolution of free and bound fractions of some grape aroma components cv. Muscat during first development and maturation. *Journal of Science Food and Agriculture* 36: 857-862.

Hardie, W.J. and Considine, J.A. (1976). Response of grapes to water-deficit stress in particular stages of development. *American Journal of Enology and Viticulture* 27: 55-61.

Harris, J.M., Kriedemann, P.E. and Possingham, J.V. (1968). Anatomical aspects of grape berry development. *Vitis* 7: 106-119.

Igartuburu, J.M., del Rio, R.M., Montiel, J.A. and Pando, E. (1991). Study of agricultural by-products: Extractability and amino acid composition of grape (*Vitis vinifera*) skin proteins from cv. Palomino. *Journal of the Science of Food and Agriculture* 57: 437-440.

Iland, P., Cynkar, W., Francis, I., Williams, P.J. and Coombe, B. (1996). Optimisation of methods for the determination of total and red-free glycosyl glucose in black grape berries of *Vitis vinifera. Australian Journal of Grape and Wine Research* 2: 171-178.

Intrieri, C. and Poni, S. (1990). A new integrated approach between training system and mechanical equipment for full mechanisation of quality vineyards. *In*: Williams, P.J., Davidson, D.M. and Lee, T.H. (Eds.). Proceedings of the 7th Australian Wine Industrial Technical Conference. Winetitles, Adelaide, pp. 35-50.

Intrigliolo, D.S. and Castel, J.R. (2010). Response of grapevine cv. 'Tempranillo' to timing and amount of irrigation: Water relations, vine growth, yield and berry and wine composition. *Irrigation Science* 28: 113-125.

Ishimaru, M. and Kobayashi, S. (2002). Expression of a xyloglucan *endo*-transglycosylase gene is closely related to grape berry softening. *Plant Science* 162: 621-628.

Jeong, S.T., Goto-Yamamoto, N., Hashizume, K. and Esaka, M. (2006). Expression of the flavonoid 3′-hydroxylase and flavonoid 3′,5′-hydroxylase genes and flavonoid composition in grape (*Vitis vinifera*). *Plant Science* 170: 61-69.

Jones, G.V. (2006). Climate and terroir: Impacts of climate variability and change on wine. Geoscience Canada Reprint Series 9: 1-14. Geological Association of Canada. St. John's Newfoundland.

Keller, M. (2010). The Science of Grapevines: Anatomy and Physiology. Academic Press, New York.

Kennedy, J. (2002). Understanding grape berry development. *Practical Winery and Vineyard*, 2002, July-August.

Kennedy, J.A., Saucier, C. and Glories, Y. (2006). Grape and wine phenolics: History and perspective. *American Journal of Enology and Viticulture* 57: 239-248.

Kirkpatrick, N. (2011). Will global warming affect the wine industry? Mmm. Com., Available at: http// www.mmm.com/food/beverages/stories/will-global-warming-affect-the-wine-industry

Kliewer, W. (1967). Concentration of tartrates, malates, glucose and fructose in the fruits of the genus *Vitis*. *American Journal of Enology and Viticulture* 18: 87-96.

Kök, D. and Çelik, S. (2004). Determination of characteristics of grape berry skin in some table grape cultivars (*V. vinifera* L.). *Journal of Agronomy* 3: 141-146.

Koundouras, S., Tsialtas, I.T., Zioziou, E. and Nikolaou, N. (2008). Rootstock effects on the adaptive strategies of grapevine (*Vitis vinifera* L. cv. Cabernet-Sauvignon) under contrasting water status: Leaf physiological and structural responses. *Agriculture, Ecosystems and Environment* 128: 86-96.

Kriedermann, P.E. and Goodwin, I. (2003). Regulated deficit irrigation and partial rootzone drying. National Programme for Sustainable Irrigation: Land and Water, Australia. Canberry, Australia.

Kuhn, N., Guan, L., Dai, Z.W., Wu, B.H., Lauvergeat, V., Gomès, E., Li, S.H., Godoy, F., Arce-Johnson, P. and Delrot, S. (2014). Berry ripening: Recently heard through the grapevine. *Journal of Experimental Botany* 65: 4543-4559.

Le Moigne, M., Maury, C., Bertrand, D. and Jourjon, F. (2008). Sensory and instrumental characterisation of Cabernet Franc grapes according to ripening stages and growing location. *Food Quality and Preference* 19: 220-231.

Lizana, L.A., Garcia de Cortazar, V., Pinto, M. and Miranda, J. (2007). Effect of night temperatures on Thompson seedless soluble solids and acidity evolution and berry growth. *Acta Horticulturae* 754: 191-196.

Lobell, D.B., Field, C.B., Cahill, K.N. and Bonfils, C. (2006). Impacts of future climate change on California perennial crop yields: Model projections with climate and crop uncertainties. *Agricultural and Forest Meteorology* 141: 208-218.

Marsal, J., Mata, M., del Campo, J. *et al.* (2008). Evaluation of partial rootzone drying for potential field use as a deficit irrigation technique in commercial vineyards according two different pipeline layouts. *Irrigation Science* 26: 347-356.

Martínez de Toda, F., Tardaguila, J. and Sancha, J. (2007). Estimation of grape quality in vineyards using a new viticultural index. *Vitis* 46: 168-173.

Matthews, M.A., Cheng, G. and Weinbaum, S.A. (1987). Changes in water potential and dermal extensibility during grape berry development. *Journal of American Society of Horticultural Science* 112: 314-319.

Mattivi, F., Vrhovsek, U., Masuero, D. and Trainotti, D. (2009). Differences in the amount and structure of extractable skin and seed tannins amongst red grape varieties. *Australian Journal of Wine and Grape Research* 15: 27-35.

Mira de Orduña, R.M. (2010). Climate change associated effects on grape and wine quality and production. *Food Research International* 43: 1844-1855.

Morrison, J.C. and Noble, A.C. (1990). The effects of leaf and cluster shading on the composition of Cabernet Sauvignon grapes and on fruit and wine sensory properties. *Am. J. Enol. Vitic* 41: 193-200.

Morando, A. (2001). Vigna Nuova: Materiali e Tecnique per L'impianto del vigneto, Edizioni Vit. En. Colosso, Asti, Italy, pp. 50-53.

Moreno, A., Castro, M. and Falque, E. (2008). Evolution of trans- and cis-resveratrol content in red grapes (*Vitis vinifera* L. cv. Mencia Abarello and Merenzao) during ripening. *European Food Research and Technology* 227: 667-674.

Morrison, J.C. and Noble, A.C. (1990). The effects of leaf and cluster shading on the composition of Cabernet Sauvignon grapes and on fruit and wine sensory properties. *American Journal of Enology and Viticulure* 41: 193-199.

Nelson, K.E. (1985). Harvesting and handling California table grapes for market. *Bulletin 1913*, 72 pp. University of California Press. DANR Publications. Oakland, California, USA.

Noble, A., Ough, C. and Kasimatis, A. (1975). Effect of leaf content and mechanical harvest on wine quality. *American Journal of Enology and Viticulture* 26: 158-163.

Nunan, K.J., Sims, I.M., Bacic, A. and Robinson, S.P. (1998). Changes in Cell Wall Composition during Ripening of Grape Berries. *Plant Physiology* 118: 783-792.

Nunan, K., Davies, C., Robinson, S. and Fincher, G. (2001). Expression patterns of cell wall-modifying enzymes during grape berry development. *Planta*, 214: 257-264.

Ojeda, H., Deloire, A. and Carbonneau, A. (2001). Influence of water deficits on grape berry growth. *Vitis* 40: 41-145.

Ortega-Regules, A., Romero-Cascales, I., Ros-García, J.M., López-Roca, J.M. and Gómez-Plaza, E. (2006). A First Approach Towards the Relationship Between Grape Skin Cell Wall Composition and Anthocyanin Extractability. *Analytica Chimica Acta* 563: 26-32.

Ortega-Regules, A., Ros-García, J.M., Bautista-Ortín, A.B., López-Roca, J.M. and Gómez-Plaza, E. (2008a). Changes in skin cell wall composition during the maturation of four premium wine grape varieties. *Journal of the Science of Food and Agriculture* 88: 420-428.

Ortega-Regules, A., Romero-Cascales, I., Ros-García, J.M., Bautista-Ortin, A.B., López-Roca, J.M., Fernández-Fernández, J.I. and Gómez-Plaza, E. (2008b). Anthocyanins and tannins in four grape varieties (*Vitis vinifera* L.): Evolution of their content and extractability. *Journal Internationale des Sciences de la Vigne et du Vin* 42: 1-10.

Paetzold, M., Dulau, L. and Dubourdieu, D. (1990). Fractionnement et caractérisation des glycoprotéines dans les moûts de raisins blancs. *Journal International des Sciences de la Vigne et du Vin* 24: 13-28.

Palliotti, A. (2011). A new closing Y-shaped training system for grapevines. *Australian Journal of Grape and Wine Research* 18: 57-63.

Pavlousek, P. (2011). Evaluation of drought tolerance of new grapevine rootstock hybrids. *Journal of Enviromental Biology* 32: 543-549.

Pillet, J., Egert, A., Pieri, P., Lecourieux, F., Kappel, C., Charon, J., Gomès, E., Keller, F., Delrot, S. and Lecourieux, D. (2012). VvGOLS1 and VvHsfA2 are involved in the heat stress responses in grapevine berries. *Plant and Cell Physiology* 53: 1776-1792.

Poni, S., Lakso, A.N., Turner, J.R. and Melious, R.E. (1994). The effects of pre- and post-veraison water stress on growth and physiology of potted Pinot Noir grapevines at varying crop levels. *Vitis* 32: 207-214.

Poni, L., Casalini, L., Berrnizzoni, F., Civardi, S. and Intrieri, C. (2006). Effect on early defoliation on shoot photosynthesis, yield components and grape composition. *American Journal of Enology and Viticulture* 57: 397-407.

Pudney, S. and McCarthy, M.G. (2004). Water use efficiency of field-grown Chardonnay grapevines subjected to partial root-zone drying and deficit irrigation. *Acta Horticulturae* 664: 567-573.

Revilla, E. and Ryan, J.M. (2000). Analysis of several phenolic compounds with potential antioxidant properties in grape extracts and wines by high performance liquid chromatography photodiode array detection without simple preparation. *Journal of Chromatography* A 881: 461-469.

Rio Segade, S., Rolle, L., Gerbi, V. and Orriols, I. (2008). Phenolic ripeness assessment of grape skin by texture analysis. *Subtropical Plant Science* 21: 644-649.

Robin, J.P., Abbal, P. and Salmon J.M. (1997). Fermeté et maturation du raisin: Définition et évolution de différents paramètres rhéologiques au cours de la maturation. *Journal Internationale des Sciences de la Vigne et du Vin* 31: 127-138.

Roby, G., Harbertson, J.F., Adams, D.A. and Matthews, M.A. (2004). Berry size and vine water deficits as factors in winegrape composition: Anthocyanins and tannins. *Australian Journal of Grape and Wine Research* 10: 100-107.

Rolle, L., Torchio, F., Zeppa, G. and Gerbi, V. (2008). Anthocyanin extractability assessment of grape skins by texture analysis. *Journal International des Sciences de la Vigne et du Vin* 42: 157-162.

Romero, P., Navarro, J.M., Perez-Perez, J., Garcia-Sanchez, F., Gomez-Gomez, A., Porras, *I. et al.* (2006). Deficit irrigation and rootstock: Their effects on water relations, vegetative development, yield, fruit quality and mineral nutrition of Clemenules mandarin. *Tree Physiology* 26: 1537-1548.

Romero, P., Fernandez-Fernandez, *J.I. and* Martinez-Cutillas, *A.* (2010). Physiological thresholds for efficient regulated deficit-irrigation management in winegrapes grown under semiarid conditions. *American Journal of Enology and Viticulture* 61: 300-312.

Romero, P. *and* Martinez-Cutillas, *A.* (2012). The effects of partial root-zone irrigation and regulated deficit irrigation on the vegetative and reproductive development of field-grown Monastrell grapevines. *Irrigation Sciences* 30: 377-396.

Rosenquist, J.K. and Morrison, J.C. (1989). Some factors affecting cuticle and wax accumulation on grape berries. *American Journal of Enology and Viticulture* 40: 241-244.

Rosli, H.M., Civello, P.M. and Martínez, G.A. (2004). Changes in cell wall composition of three *Fragaria x ananassa* cultivars with different softening rate during ripening. *Plant Physiology and Biochemistry* 42: 823-831.

Sadras, V.O. (2009). Does partial root-zone drying improve irrigation water productivity in the field? A meta-analysis. *Irrigation Science* 27: 183-190.

Salinas, M.R., Zalacain, A., Pardo, F. and Alonso, G.L. (2004). Stir bar sorptive extraction applied to volatile constituents evolution during *Vitis vinifera* ripening. *Journal of Agricultural and Food Chemistry* 52: 4821-4827.

Santos, T.P., Lopes, C.M., Rodriguez, M.L., Sourza, C.R., Silva, J.R., Maroco, J.S., Pereira, J.S. and Chaves, M.M. (2005). Effects of partial rootzone drying irrigation on cluster microclimate and fruit composition of Castelao field-grown grapevines. *Vitis* 44: 117-125.

Segurel, M.A., Razungles, A., Riou, C., Trigueiro, M.G. and Baumes, R.L. (2005). Ability of possible DMS precursors to release DMS during wine aging and in the conditions of heat-alkaline treatment. *Journal of Agriculture and Food Chemistry* 53: 2637-2645.

Shiraishi, M., Fujishima, H. and Chijiwa, H. (2010). Evaluation of table grape genetic resources for sugar, organic acid, and amino acid composition of berries. *Euphytica* 174: 1-13.

Smart, R.E. and Coombe, B.G. (1983). Water relations of grapevines. pp. 137-196. *In*: T.T. Kozlowsky (Ed.). Water Deficits and Plant Growth. Academic Press, London.

Spayd, S.E., Tarara, J.M., Dee, D.L. and Ferguson, J.C. (2002). Separation of sunlight and temperature effects on the composition of *Vitis Vinifera* cv. Merlot berries. *American Journal of Enology and Viticulture* 53: 171-182.

Sternad, M., Sivilotti, P., Franceschi, P., Wehrens, R. and Vrhovsek, U. (2013). The use of metabolite profiling to study grape skin polyphenolic behaviour as a result of canopy microclimate manipulation in 'Pinot noir' vineyard. *Journal of Agriculture and Food Chemistry* 61: 8976-8986.

Stock, M., Gerstengarbe, F.W., Kartschall, T. and Werner, P.C. (2005). Reliability of climate change impact assessments for viticulture. *Acta Horticulturae* 689: 29-39.

Stoll, M., Loveys, B. and Dry, P. (2000). Hormonal changes induced by partial root-zone drying of irrigated grapevine. *Journal of Experimental Botany* 51: 1627-1634.

Stoll, M., Scheidweiler, M., Lafontaine, M. and Schultz, H.R. (2009). Possibilities to reduce the velocity of berry maturation through various leaf area to fruit ratio modifications in *Vitis vinifera* L. Riesling. *Proceedings XVI GESCO Symposium*, Davis, USA, 12–15 July, pp. 93-96.

Styger, G., Prior, B. and Bauer, F.F. (2011). Wine flavor and aroma. *Journal of Industrial Microbiology and Biotechnology* 38: 1145-1159.

Tisseyre, B., Ojeda, H. and Taylor, J. (2007). New technologies and methodologies for site-specific viticulture. *Journal Internationale des Sciences de la Vigne et du Vin* 41: 63-76.

Trought, M.C.T. and Tannock, S. (1996). Berry size and soluble solids variation. *Proceedings of the Fourth International Cool Climate Symposium V*, pp. 70-73.

Tominaga, T., Baltenweck-Guyot, R., Peyrot des Gachons, C. and Dubourdieu, D. (2000). Contribution of volatile thiols to the aroma of white wines made from several *Vitis vinifera* grape varieties. *American Journal of Enology and Viticulture* 51: 178-181.

Vacca, V., Del Caro, A., Milella, G.G. and Nieddu, G. (2009). Preliminary characterisation of Sardinian red grape cultivars (*Vitis vinifera* L.) according to their phenolic potential. *South African Journal of Enology and Viticulture* 30: 93-100.

Van Leeuwen, C. and Darriet, P. (2016). The impact of climate change on viticulture and wine quality. *Journal of Wine Economics* 11: 150-167.

Van-Zyl, J.L. (1984). Response of *Colombard* grapevines to irrigation as regards quality aspects and growth. *South African Journal for Enology and Viticulture* 5: 19-28.

Vega, A., Gutiérrez, R.A., Peña-Neira, A., Cramer, G.R. and Arce-Johnson, P. (2011). Compatible GLRaV-3 viral infections affect berry ripening decreasing sugar accumulation and anthocyanin biosynthesis in *Vitis vinifera*. *Plant Molecular Biology* 77: 261-274.

Vilanova, M. and Martinez, C. (2007). First study of determination of aromatic compounds of red wine from *Vitis vinifera* cv. Castanal grown in Galicia (NW Spain). *European Food Research and Technology* 224: 431-436.

# 6 Varietal Impact on Wine Quality and Aroma

**Vishal S. Rana[1*], Neerja S. Rana[2] and Ravina Pawar[1]**

[1] Department of Fruit Science, Dr Y.S. Parmar University of Horticulture and Forestry, Nauni, Solan – 173230 (HP), India

[2] Department of Basic Sciences, Dr Y.S. Parmar University of Horticulture and Forestry, Nauni, Solan – 173230 (HP), India

## 1. Introduction

Worthy grapes is the first essential key to success in making suitable wine. Generally, it is said, 'great wines are made in the vineyards', which means that for commercial winemaking as a whole, cooperation is highly required between the grape growers and the wine-makers. The wine industry of the world is built upon one species, *Vitis vinifera* L., which is native to the area of Asia Minor near Black and Caspian seas and is generally termed as 'European grape'. Ice wines are a great example associated with the choice of vineyard and climatic changes which play a tempting role in the choice of suitable grape variety and wine type. They are produced when grapes freeze on the vine before harvest, producing a wine with intense flavour and fullness. Exceptional wine quality can be achieved only through an appropriate balance between fruit, fermentation and processing derived aromas and flavours. Conclusively, wine colour, aroma, flavour quality and stability of wines depend on the content and composition of a group of compounds which are available in the variety of grapes as well as the subsequent enological processes.

The in-mouth impression of wines having a contribution of different ethanol concentrations, sugar levels and spices extract exhibit significant differences for various sensory quality parameters. These sensory parameters may play a role in the acceptance and enjoyment of wine and are considered to be the most important factor. Alcohol is another important factor for wine sensory sensations but also by their interaction with other wine components, such as aromas and tannins, these influence wine viscosity and body and perceptions of astringency, sourness, sweetness, aroma, and flavour (Fontoin *et al.*, 2008; Meillon *et al.*, 2009). Wine aroma is determined by a complex balance of several volatiles. More than 800 volatile compounds have been identified in wine (Bayonove *et al.*, 1998; Noble, 1994). In grape, some potent aroma compounds are linked to sugars as 0-glycosides. This bound form can contribute to wine aroma upon hydrolysis, followed in some cases by chemical modifications (Cabaroglu *et al.*, 2002). Terpenes as well as other grape-derived flavour compounds are present as free volatiles and as sugar-bound precursors. These later glycosidic conjugates are non-volatile and tasteless at levels found in wine but represent a potential reservoir of flavour (Girard *et al.*, 2002).

Wine flavours arise as a result of compounds originating from the native fruit during the various processes used for its production, transformation by yeast during fermentation and subsequent aging process. A wide range of flavour compounds have been described in wines. Many of primary floral and fruit aromas exist in the form of higher terpene alcohols, such as citronellol, linolool and geraniols. These are mostly found in Muscat grape varieties. Beside this, esters also contribute to aroma by imparting fruity flavours to the wine and are tindispensable components of wine quality. Primary fruit aromas in red grapes are often identified as that of cherry, blank currant, strawberry, raspberry and plum, which are often used as flavour descriptors in the evaluation of wines made from several grape varieties.

Certain winemaking techniques allow a better extraction of volatile components, especially the phenols which result in overall improvement in the sensorial quality of wine. Fermentation in oak barrels leads to wines with much more complex sensory properties, largely attributed to the phenols extracted from oak wood. Wine-makers are aware of the importance of phenolic compounds, which give astringency and bitterness to wood-matured wines. Heating at the end of maceration and the use of

*Corresponding author: vishalranafrs@yspuniversity.ac.in

enzymes are also employed in attaining these objectives. Quantification of various aroma compounds in wine at very minute levels has been achieved through solvent extraction and concentration, followed by several sophisticated analytical techniques. Recently, Solid Phase Microextraction (SPME) has been investigated as a simple and rapid way to isolate and concentrate wine aroma components (Whiton and Zoecklein, 2000). This method does not involve the use and disposal of hazardous, toxic and expensive solvents. Head space (HS) – SPME has been applied more recently to analyse a variety of desirable and off odours (Chatonnet *et al.*, 1992; Butzke *et al.*, 1998).

From wine quality and aroma point of view, the volatile composition of fruit, must and wine along with their vinification process have been examined and discussed quite comprehensively in this chapter.

### 1.1. Varieties

Globally, wines are prepared from different grape varieties, each one imparting specific characteristics. The varieties noted for making wines with desirably distinctive flavours include Grenache, Cabernet Sauvignon, Muscat as a group, Merlot, Semillon and Riesling. In some regions, one variety is used to make a wine; in others, wine-makers blend several varieties into a single wine. From a wine-maker's point of view, distinctive and desirable flavoured varieties are sought for new flavoured wines and consideration is given to the existence, accessibility, size and price characteristics of the market. Grape varieties make up the largest and most simply recognised and controlled differences, affecting wine composition and quality. Clearly, the choice of variety is a crucial aspect. Other special characteristics important in the selection of varieties include duration of ripening and usual harvest date, retention of acidity at harvest and certain processing problems, though the processing problems can be overcome with modern enology. From the vintner's point of view, the flavour characteristics of the wines are probably the most crucial factors in varietal selection.

The wine industry of the world is based primarily on *Vitis vinifera,* i.e. European grape species. There may be a few who disagree, but all the world's greatest and most popular wines come from the fruits of *Vitis vinifera*. Other species, such as *V. labrusca*, *V. rotundifolia*, *V. rupestris*, *V. riparia*, are used to make wine or to produce hybrids by crossbreeding with *Vitis vinifera*. Muscat varieties are the most extensively planted grape types used for the production of Pisco, under Chilean legislation. Moscatel de Alejandria and Moscatel Rosada proved to be highly aromatic, while the terpene profiles of two little grown varieties, i.e. Early Muscat and Moscatel Amarilla, indicate that these could contribute much to the aroma of Pisco. Significant differences have also been detected in the concentrations of total free terpenols among different varieties.

The results of 20 years of wine variety evaluation and breeding at CSIRO, Division of Horticultural Research, Adelaide, South Australia indicate that there is considerable potential to improve wine flavour and quality in hot irrigated vineyards, which produce about 80 per cent of Australian wine. The combined analysis of sugar (°Brix), wine pH and titratable acid (g/l), the panel's comments regarding flavour and aroma, the panel's score and the vine yield contribute towards arbitrary interest ranking for promoting varieties.

Table 1 describes the rating system used for each evaluating parameter. The number of stars is totalled to determine the interest ranking for each variety.

**Table 1.** Rating Standards for Wine Grape Varieties

| *Rating* | *Brix* | *White wine pH* | *Red wine pH* | *Wine acid (g/L)* | *Wine flavour* | *Wine score* | *Yield (t/ha)* |
|---|---|---|---|---|---|---|---|
| * | <16 | >3.59 | >3.79 | <5.0 | Unpleasant | <13.0 | <10.0 |
| ** | 16-17.9 | 3.45-3.59 | 3.60-3.79 | 5.0-5.9 | Neutral | 13.0-13.9 | 10.0-19.9 |
| *** | 18-19.9 | 3.30-3.44 | 3.40-3.59 | 6.0-6.9 | Sour | 14.0-14.9 | 20.0-29.9 |
| **** | >19.9 | 3.15-3.29 | 3.20-3.39 | 7.0-7.9 | Good | 15.0-15.9 | 30.0-39.9 |
| ***** | - | <3.15 | <3.20 | >7.9 | Excellent | >15.9 | >39.9 |

*Source*: Clingeleffer, 1986

A valuation of performance characters of the most promising varieties, the country of their origin and ripening time are presented in Table 2.

**Table 2.** Detailed Ratings of the Assessment Parameters of Promising Grape Varieties

| *Variety* | *Country of origin* | *Harvest period* | *Brix* | *pH* | *Acid* | *Flavour* | *Score* | *Yield* |
|---|---|---|---|---|---|---|---|---|
| **A. Light red wine varieties** | | | | | | | | |
| Carignan | Spain | Mid March | **** | ** | ** | *** | *** | **** |
| Gamay | France | Mid February | **** | ***** | ***** | *** | ** | * |
| Grenache | Spain | Late February | **** | *** | ** | *** | ** | *** |
| Pinot noir | France | Mid February | **** | ** | ** | *** | ** | *** |
| Tarrango | Australia | Late March | **** | *** | *** | **** | *** | **** |
| Valdiguie | France | Early March | **** | ***** | **** | **** | **** | ** |
| Zinfandel | USA | Mid March | *** | *** | ***** | *** | * | * |
| **B. Full-bodied red wine varieties** | | | | | | | | |
| Cabernet Franc | France | Late February | **** | ** | *** | **** | *** | *** |
| Carbernet Sauvignon | France | Late February | **** | * | *** | **** | *** | ** |
| Chambourein | France | Mid February | **** | *** | ***** | **** | *** | *** |
| Malbec | France | Mid February | **** | ** | *** | **** | **** | ** |
| Roboso Piave | Italy | Late March | **** | **** | ***** | **** | *** | * |
| Ruby Cabernet | USA | Late February | **** | ** | *** | **** | *** | *** |
| Shiraz | France | Late February | **** | * | ** | *** | ** | **** |
| Tannat | France | Early March | **** | *** | **** | **** | *** | ** |
| **C. Delicate white wine varieties** | | | | | | | | |
| Emerald Riesling | USA | Late February | **** | ***** | **** | *** | **** | ** |
| GF 31-17-115 | Germany | Mid January | **** | ***** | **** | *** | ** | *** |
| Rieslina (CG38.049) | Argentina | Late February | **** | ***** | *** | *** | *** | *** |
| Riesling | Germany | Late February | **** | ***** | **** | **** | *** | ** |
| **D. Full-bodied white wine varieties** | | | | | | | | |
| Chenin Blanc | France | Late February | *** | **** | **** | *** | *** | *** |
| Colombard | France | Early March | **** | *** | *** | *** | *** | *** |
| Ehrenfelser | Germany | Early February | **** | **** | *** | *** | *** | ** |
| Goyura | Australia | Early March | **** | **** | **** | *** | ** | *** |
| Semillon | France | Late February | **** | ** | ** | *** | *** | *** |
| **E. Aromatic white wine varieties** | | | | | | | | |
| Bacchus | Germany | Late January | **** | *** | *** | **** | **** | *** |
| Sauvignon blanc | France | Early February | **** | *** | *** | *** | ** | *** |
| Schonburger | Germany | Mid January | **** | *** | **** | **** | *** | ** |
| Taminga | Australia | Early March | **** | **** | *** | **** | **** | ***** |
| Verdelet | France | Early February | ** | **** | **** | **** | *** | *** |

| **F. Muscat wine varieties** | | | | | | | | |
|---|---|---|---|---|---|---|---|---|
| Gordo Blanco | Egypt | Mid March | **** | * | * | **** | **** | **** |
| Irsay Oliver | Hungary | Mid January | *** | ***** | ***** | **** | ***** | ** |
| Muscat a' petits grains rouge | France | Mid February | **** | *** | * | **** | **** | *** |

# 2. Volatile Constituents Contributing Wine Aroma and Qualtiy

## 2.1. Glycosidic Precursors

Grapes, being a major horticultural crop of the world, are amongst the earliest of many fruits to have been studied for flavour precursors (Teranishi, 1989). The precursors in fruits that could act as a source of latent or potential flavour were recognised many years ago (Hewitt *et al.*, 1959). These precursors are non-volatile conjugates of mevalonic acid and shikimic acid derived secondary metabolites. The conjugation found most commonly in fruit flavour precursors is glycosidic (Williams, 1993).

Varieties of grapes, like various Muscats, Reisling and Gewurztraminer have highly distinctive flavours. Free monoterpene compounds are responsible for the sensory properties of these grapes. Monoterpene glycosides were first isolated and identified as flavour precursors in grape berries and wines made from these varieties (Strauss *et al.*, 1986). A number of varieties have been screened and classified into three divisions based on different monoterpene concentrations. These are i) intensely flavoured Muscats, in which total monoterpene concentrations can be as high as 6 mg/l, ii) non-Muscat but aromatic varieties with total monoterpene concentration of 1-4 mg/l and iii) neutral varieties which do not depend upon monoterpenes for their flavour.

**Table 3.** Classification of Some Grape Varieties based on Monoterpene Content

| *Muscat varieties* | *Non-Muscat aromatic varieties* | *Varieties independent of monoterpenes for flavour* | |
|---|---|---|---|
| Canada Muscat | Traminer | Bacchus | Merlot |
| Muscat of Alexandria | Huxel | Cabernet Sauvignon | Nobling |
| Muscat Petits | Kerner | Chardonnay | Rkaziteli |
| Grains Blanc | Morio Muskat | Carignan | Rulander |
| Moscato Bianco | Muller Thurgau | Chasselas | Sauvignon Blanc |
| Del Piemonte | Riesling | Chenin blanc | Semillon |
| Muscat Hamburg | Schemebe | Cinsault | Shiraz |
| Muscat Ottonel | Schonburger | Clairette | Sultana |
| | Siegerebe | Doradillo | Terret |
| | Sylvaner | Forta | Trebbiano |
| | | Grenache | Verdellho |
| | | | Viegnier |

*Source*: William *et al.*, 1980a,b, 1986

Chardonnay is a non-floral, white wine variety from which volatile aglycons are obtained after hydrolysis of the glycosides present in it. From a descriptive analysis, in relation to the volatiles and their possible biosynthetic precursors, it has been established that hydrolysed precursors from the juice of Chardonnay grapes have characteristic aroma properties. About 300 more volatile components were observed from the precursor fractions of Chardonnay grapes. The Chardonnay hydrolysate gave volatiles that were particularly rich in nor-isoprenoid compounds. As the floral grape varieties are monoterpene dependent for their flavour, Chardonnay can be analogously categorised as nor-isoprenoid dependent

(Sefton *et al.*, 1993). The volatile nor-isoprenoids are glycosidase-released aglycons. The great majority of nor-isoprenoid compounds found in Chardonnay grape precursor hydrolyzate could be derived from four major xanthophylls reported in grapes, i.e. lutein, antheraxanthin, violaxanthin and neoxanthin.

Initial research into the origins of these three grape nor-isoprenoids showed that these compounds were generated by acid hydrolysis from non-volatile precursor forms. The precursors could be isolated from juices and wines by a procedure used to obtain glycosidic conjugates of other grape volatiles (Williams *et al.*, 1982). Further, the identification and possible origins of 24 nor-isoprenoids in the juices of Chardonnay, Sauvignon Blanc and Semillon cultivars were reported (Sefton *et al.*, 1993). The disparate glycosidic nature of grape nor-isoprenoid precursors has been established in a recent study on Riesling wine (Winterhalter *et al.*, 1990).

## 2.2. Phenolic Substances

Sensory properties of wine, especially colour and taste, are largely related to the phenolic compounds extracted from grapes. These include flavonoids, including anthocyanins, flavonols and flavanols (Fig. 1). Presence of phenols in grapes is genetically controlled and grape varieties differ quantitatively and qualitatively over a wide range. Each of these phenolic classes comprises various structures, differing by the number and position of hydroxyl groups, which can also be diversely substituted.

Sensory analysis of six-month old red wines made from Cabernet Franc grapes grown in different locations pointed out the site-related characteristics, which were attributed to flavonoids (Asselin *et al.*, 1992). In a study, major differences in phenolic composition of Cabernet Franc grapes within Loir valley, were observed in respect of anthocyanin levels and anthocyanin to tannin ratios (Brossaud *et al.*, 1999). Higher anthocyanin levels were correlated with the earliness of vine. Qualitative anthocyanin compostion was same in all the samples and, therefore, was regarded as a varietal character. It must be kept in mind that the wine phenolic composition depends not only on that of grapes used to make the wine, but also on the kinetics of extraction of various compounds from grapes and of their subsequence reactions.

The tannins extracted from skins and seeds during alcoholic fermentation are the principal components of astringency in red wines by virtue of their ability to precipitate salivary proteins that provide lubrication in the mouth (Gawel, 1998). In grape tannin, which may have as many as 15 subunits, there could be more than $10^5$ unique chemical structures (Prieur *et al.*, 1994). Therefore, it is necessary to determine the total amount of tannin in berry skins and seeds. For wines, it is probably sufficient to estimate the total quantity of tannins that contribute to astringency.

Several methods for determination of total tannins have been described and the most useful are based on the ability to precipitate proteins from solution (Hagerman and Butler, 1978; Asquith and Butler, 1985; Makkar *et al.*, 1988; Makkar, 1989). One of the most widely used is protein precipitation method which has been regarded as reliable (Hagerman and Butler, 1978). Recently, a simple procedure for tannin analysis in grapes and wines based on co-precipitation of BSA and an alkaline phosphatase enzyme has been developed (Adams and Habertson, 1999). The protein precipitation assay was used to monitor tannin concentration separately in skins and seeds of three red *Vitis vinifera* wine grape varieties, viz. Cabernet Sauvignon, Syrah and Pinot noir (Habertson *et al.*, 2002).

Wines fermented in oak barrels havemore complex sensory properties and this is largely attributed to the phenols extracted from oak wood. Phenolic compounds give astringency and bitterness to wood-matured wines which is responsible for the so called 'woody' character in wine, indicating flavour intensity and complexity which are not acquired in wines fermented in stainless steel vats. Both qualitative and quantitative changes occur in wine during barrel aging (Aiken and Noble, 1984). The main phenolic species that give the typical character to the barrique wines are substituted benzoic, cinnamic acids and aldehydes. These are well-known products of lignin and tannin degradation (Puech, 1981). A characteristic oak-wood phenol, e.g. coniferaldhyde in white wines fermented in barrels was observed and the same was not detected in stainless steel vats (Gomez *et al.*, 2001).

## 2.3. Methoxypyrazines

Wines from Cabernet Sauvignon and Sauvignon Blanc grapes often have a characteristic aroma which is described as vegetative, herbaceous, grassy and green. This aroma has been attributed to methoxypyrazine

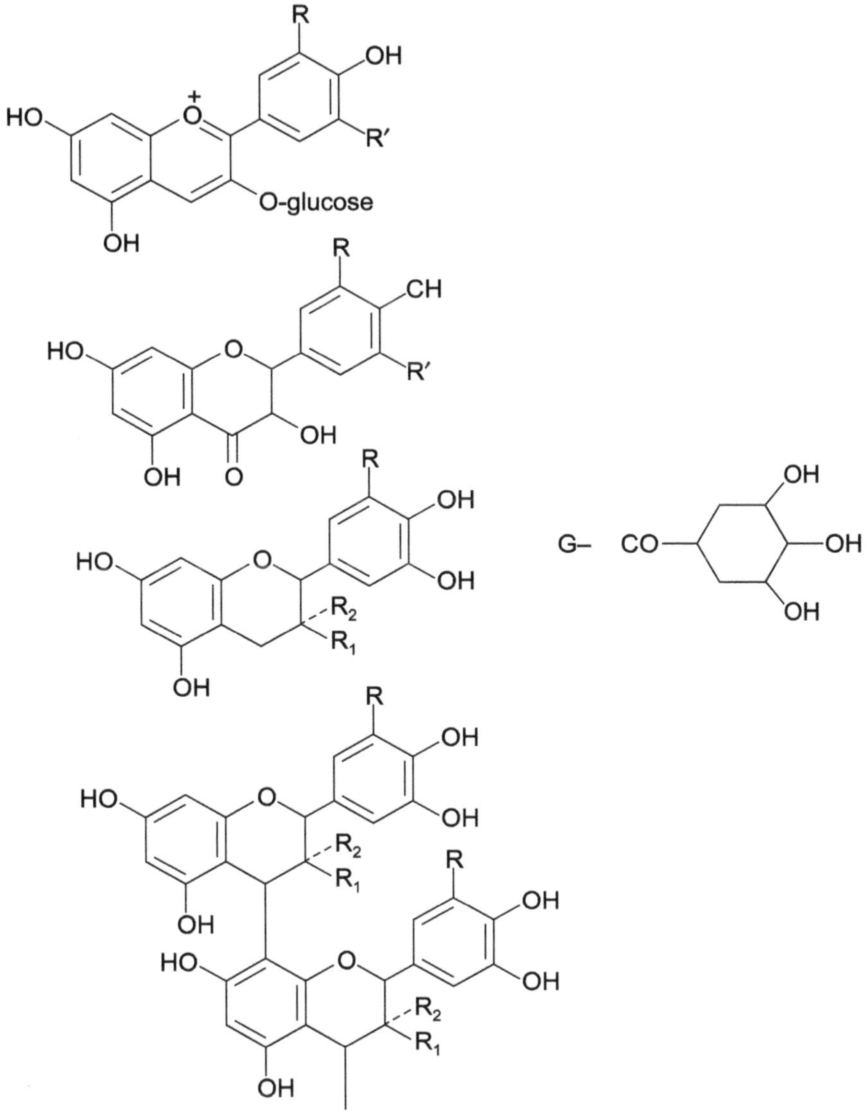

**Figure 1.** Structures of major grape flavonoids (*Source*: Brossaud *et al.*, 1999)

(MP) components (Fig. 2). The occurrence of 2-methoxy-3-(2-methyl propyl) pyrazine has been reported in Cabernet Sauvignon grapes (Bayonove *et al.*, 1975). The odour of isobutyl (2-methoxy-3-isobutyl pyrazine or 2-isobutyl-3-methoxy pyrazine) MP has been described as bell pepper (Buttery *et al.*, 1969). It contributes to the characteristic aroma of some vegetables (Maga and Sizer, 1973; Murray *et al.*, 1975) and has an extremely low sensory detection threshold in water of about 2 ng/l (2 parts per trillion) (Simpson, 1978). Rigorous identification and quantification were not possible in the early reports of their detections (Bayonove *et al.*, 1975; Augustyn *et al.*, 1982) and they were not detected in a thorough investigation of a Cabernet Sauvignon wine (Slingsby *et al.*, 1980). Furthermore, their odour detection threshold is more than two orders of magnitude below the detection limit of a quantitative method based on HPLC techniques (Heymann *et al.*, 1986). It can also be assayed in wine using stable isotope dilution gas chromatography – mass spectrometry (Allen *et al.*, 1994; De Boubee, 2000).

Methoxypyrazine levels in New Zealand wines were significantly higher than the Australian wines. Twenty-two wines of Australia, New Zealand and French origin were analysed together with 16 juice samples from four Australian regions (Table 4). Fruits grown under cool conditions gave higher grape

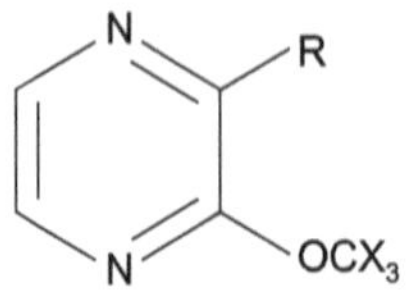


| R | Name of the methoxy pyrazine |
|---|---|
| a. $CH_2CH(CH_3)_2$ | 2-Methoxy-3-(methylethyl) pyrazine OR isobutyl MP |
| b. $CH(CH_3)_2$ | 2-Methoxy-3-isopropylpyrazine OR isopropyl MP |
| c. $CH\ (CH_3)\ CH_2CH_3$ | 2-Methoxy-3-Sec. Butylpyrazine OR Sec. butyl MP |

**Figure 2.** Methoxy pyrazine identified in Sauvignon blanc grapes and wines (*Source*: Kepner *et al.*, 1972)

**Table 4.** Methoxy Pyrazine (MP) Concentrations in Wines and Juices of Grapes of Four Australian Locations

| *Location* | *Mean Jan. temp.* | *Year* | *Sampling date* | *MP components (ng/l)* | | |
|---|---|---|---|---|---|---|
| | | | | *Isobutyl* | *Isopropyl* | *Sec. Butyl* |
| New South Wales (NSL) | 23.9 | 1987 | 30 Jan. | 35.4 | 1.1 | - |
| | | | 13 Feb. | 11.4 | - | - |
| | | | 27 Feb. | 2.3 | 0.3 | - |
| | | | 13 Mar. | 1.3 | 1.0 | - |
| | | 1988 | 8 Jan. | 30.7 | 1.6 | - |
| | | | 22 Jan. | 2.8 | 0.4 | - |
| | | | 1 Feb. | 0.7 | 0.2 | - |
| | | | 12 Feb. | 0.6 | 0.3 | - |
| Adelaide Hills, SA | 19.4 | 1987 | 16 Mar. | 78.5 | 6.8 | 0.6 |
| | | | 23 Mar. | 13.4 | 1.3 | - |
| Great Western, Vic | 20.2 | 1987 | 26 Mar. | 8.6 | 0.6 | - |
| | | | 2 Apr. | 11.5 | 0.5 | - |
| Coonawarra, SA | 19.6 | 1987 | 17 Mar. | 12.1 | 0.6 | - |
| | | | 23 Mar. | 15.9 | 0.7 | 0.1 |
| | | | 30 Mar, | 9.5 | 0.5 | 0.1 |

*Source*: Lacey *et al.*, 1991

methoxypyrazine levels than those in hot conditions. Grape methoxypyrazine levels were relatively high at veraison but decreased markedly with ripening.

Regardless of ripeness, these compounds were mainly located in the stems, skins and seeds, while the flesh contained very little. During ripening, the proportion of 1 BMP in stems and seeds decreased, while it increased in the skins. During the winemaking process, methoxy pyrazine was easily extracted from Sauvignon Blanc at the beginning of pressing and Cabernet Sauvignon grapes after 24 hours in the vat.

## 2.4. Lactones

Lactones, particularly gamma lactones, occupy a place of prominence not only in terms of their contribution to the total aroma and bouquet picture, but also because of their physiological properties. Table 5 represents the structures of lactones that were isolated and identified in wines. Among them, delta lactones are often associated with impact flavour compounds (Urbach *et al.*, 1972). The odour of delta lactone, i.e. 6-methyldihydro-2, 5 (3H)-pyrondione isolated from wine, is somewhat reminiscent of wine but it cannot be regarded as an 'impact' flavour compound.

**Table 5.** Lactones Isolated from Wines

| *Structure* | *Name* | *Occurrence* | *References* |
|---|---|---|---|
| 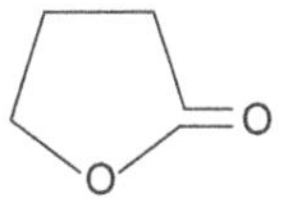 (five-membered lactone ring) | Gamma Butyrolactone | All wines | Web and Kepner, 1962, Web and Muller, 1972 |
| 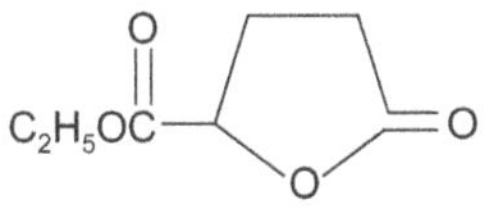 $C_2H_5OC(=O)$– (lactone ring) | 5-Carboethoxy-Dihydro-2(3H)-Furanone | Flor sherries, Cabernet Sauvignon, Ruby Cabernet | Web *et al.*, 1967, 1969 |
| 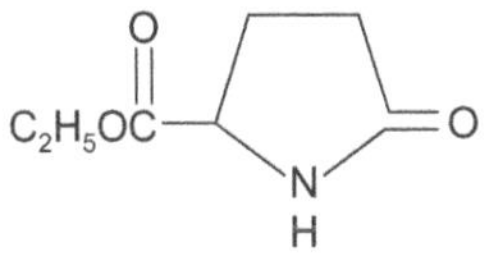 $C_2H_5OC(=O)$– (lactam ring, NH) | Ethyl Pyroglutamate | Flor sherries | Web *et al.*, 1967 |
| 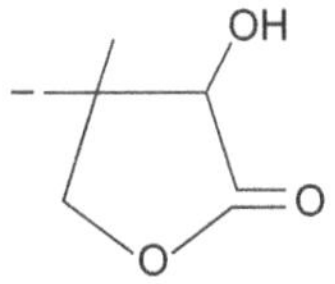 (dimethyl, OH lactone ring) | Pantolactone | Flor sherries | Web *et al.*, 1967 |
| 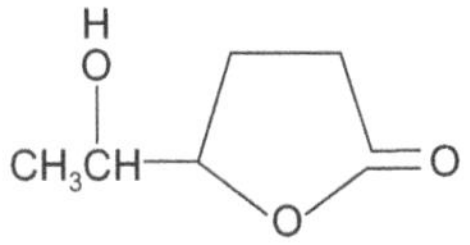 $CH_3CH(OH)$– (lactone ring) | (+) 4R:5S or 4S:5R & (-) 4R:5R or 4S:5S 4,5-Dihydroxyhexanoic Acid Gamma-Lactones | Flor sherries | Muller *et al.*, 1969 |
| 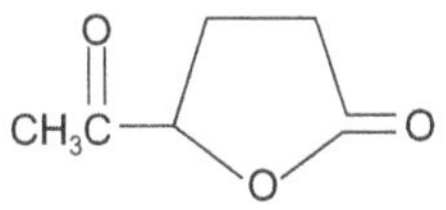 $CH_3C(=O)$– (lactone ring) | 5-Acetyldihydro-2(3H)-Furanone | All wines | Muller *et al.*, 1973 |
| 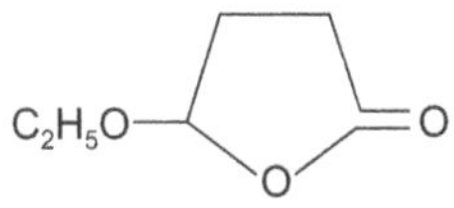 $C_2H_5O$– (lactone ring) | 5-Ethoxydihydro-2(3H)-Furanone | Ruby cabernet | Muller *et al.*, 1972 |
| 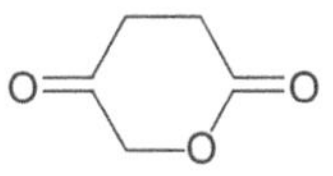 (six-membered ring, two C=O) | 6-Methyldihydro-2,5(3H)-Pyrandione | Ruby cabernet | Muller *et al.*, 1972 |
| 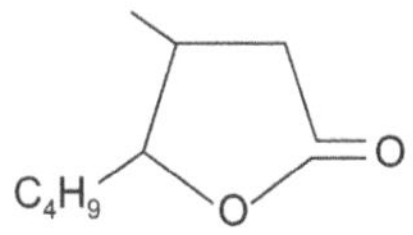 $C_4H_9$– (methyl lactone ring) | Trans-5-Butyl-4-Methyl-Dihydro-2(3H)-furanone | Oak-Wood-Aged cabernet | Kepner *et al.*, 1972 |

*Source*: Muller *et al.*, 1973

Lactonisation can occur by chemical means at pH of about 3.5 but enzyme-mediated lactonizations are more common in nature. In wines, succinic semiladehyde gives rise to two series of lactones, depending on the manner in which it reacts with other metabolites in the fermentation milieu. Direct reactions with alcohols produce alkoxy lactones, whereas reaction with 2-ketoacids, with concomitant decarboxylation of the ketoacid moiety, yields acyllactones. This latter head-to-head condensation of aldehydes and ketoacids is of common occurrence in nature.

## 2.5. Volatile Sulphur Compounds

Hydrogen sulphide, methionol, 3-methyl thiopropionic acid, ethyl 3-methylthiopropionate, 2-methyl thioethanol, 2-methyltetrahydro-thiophenone-3, cis- and trans-2-methylthioptrano-3-ol are some of the volatile sulphur compounds to have been investigated in white wine cultivars, Batiki and Muscat of Hamburg (Karagiannis and Panos, 1999). The formation of these substances during and after fermentation are directly influenced by the vinification parameters, like bisulphite addition to the must, must turbidity, fermentation temperature, inoculation with different yeast strains and period of contact of sulphited wines with their yeast sediment and pressings during must preparation. Furthermore, higher alcohols, fatty acids and their ethyl esters, which constitute the fermentation aroma, were also demonstrated to be negatively affected by the same treatments, which lead to the production of the off-flavour sulphur compounds.

The mechanisms of formation of volatile sulphur compounds (VSC) during wine fermentation are partially understood due to the complex nature of factors involved in winemaking. However, the production of VSCs during fermentation was studied in musts varying widely in composition as well as in model systems with nutrient deficiencies or with residual elemental sulphur. Volatile and non-volatile thiol-containing compounds which are potential precursors of VSCs during fermentation were studied in eight white grape musts, viz. one Thompson seedless, one Palomino, one Chenin Blanc, two Sauvignon Blanc and three Chardonnays (Park *et al.*, 2000). The prominent VSC, $H_2S$ was continuously produced throughout fermentation and was highest during rapid growth phase of yeast. Total $H_2S$ which ranged from 112-516 mg/l was inversely correlated with the concentration of assimilable amino acids and with total nitrogen content (Table 6).

**Table 6.** Days to Dryness, Concentration of Must Assimiliable Acids (EAA), Must Total Nitrogen and Wine Glutathione (mg/l) and Total Hydrogen Sulphide ($H_2S$) in Head Space (µg/l) at the End of Fermentation

| *Must* | *Days to dryness* | *EAA* | *Total nitrogen* | *Wine SH* | *Total $H_2S$* | *$H_2S$ at end* |
|---|---|---|---|---|---|---|
| Palomino | 7 | 2079 | 1196 | 5.10 | 112 | 9 |
| Thompson seedless | 8 | 1405 | 368 | 2.60 | 172 | 9 |
| Chenin blanc | 6 | 1290 | 284 | 2.00 | 177 | 9 |
| Sauvignon blanc-O | 9 | 838 | 189 | 0.64 | 330 | 20 |
| Chard.-L | 9 | 889 | 305 | 0.10 | 341 | 31 |
| S blanc-O | 9 | 1331 | 455 | 2.10 | 347 | 7 |
| 81 Chard.-H-O | 11 | 734 | 194 | 0.38 | 362 | 28 |
| Chard.-L-O | 14 | 480 | 84 | 0.40 | 516 | 41 |

Origin of wine O – Oak ville, L – Livermore, L- Low N, H – High N
*Source*: Park *et al.*, 2000

## 2.6. Volatile Thiols

The impact of these volatile thiols on the aromas of wines made from different grape varieties, namely, Gewurztraminer, Riesling, Colombard, Petit Manseng and Botrytized Semillon has been recorded (Tominaga *et al.*, 2000). It was found that the five volatile thiols identified in Sauvignon Blanc wines are also present in wines made from various other white grape varieties. 4-mercapto-4-methyl-pentane-2-one (4 MMP) and 3-mecaptohexyl acetate (A3MH), with their box tree odour, may have an undeniable impact on the aromas of Muscat d'Alsace and Colombard wines. 3-mercaptohexan-1-ol (3 MH) contributes to the grapefruit and passion fruit odours in wines made from Gewurztraminer, Riesling, Petit Manseng and Botrytized Semillon. These findings show that some olfactory analogies used empirically between the different wines and Sauvignon wines correspond to chemical similarities.

The olfactory impact of some of these compounds on the Muscat aroma of Gewurztraminer, Pinot Gris, Reisling, Muscat d'Alsace and Muscat d'Ottonel is undeniable. However, they only account for a part of the varietal aroma of wines made from these grape varieites. They are not responsible for all the nuances, particularly those revealed by alcoholic fermentation.

## 2.7. Neutral Volatile Components

The volatile constituents contributing to the aroma and bouquet of wines were investigated by several researchers. The development of gas chromatographic analytical techniques led to study a wide range of varietal wines. The volatile components in Sherry were investigated by several researchers (Diemar and Schams 1960; Webb *et al.*, 1964; Webb *et al.*, 1964; Galleto *et al.*, 1966). Some neutral components of a table wine made from 'Zinfandel' were isolated by methylene chloride extraction which was followed by basic extraction (Brander, 1974). A chromatogram of the neutral components separated by temperature programming on the STAP (Steroid Analysis Phase) gas-chromatographic column is represented in Fig. 3. The neutral volatiles of 'Zinfandel' wine, with their approximate concentration,are enlisted in Table 7. The numbers given correspond to peak numbers of the chromatogram.

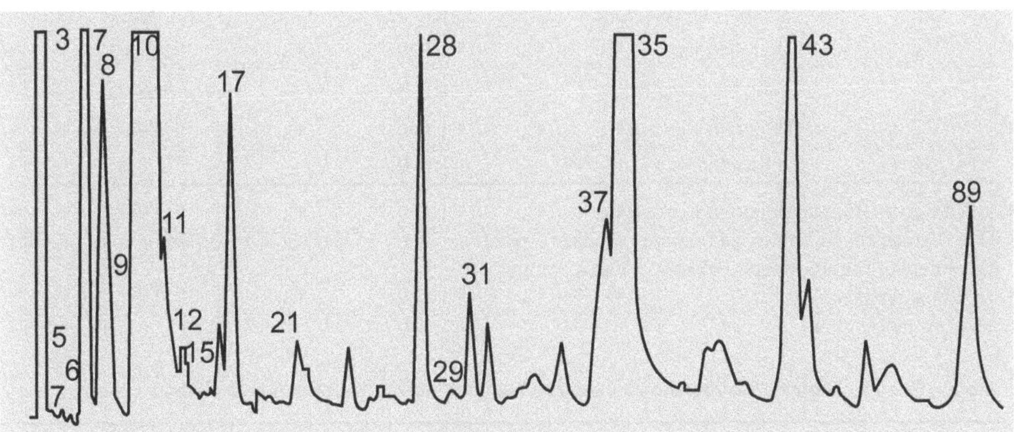


**Figure 3.** Chromatogram of neutral components extracted from 'Zinfandel' wine. Varian Aerograph series 1200 gas chromatograph, 5 ft × 1/8 in. STAP column. Program rate: 6°C/min from 60°C to 200°C, N, flow rate: 30 ml/min, sample; 1.0 μl (*Source*: Brander, 1974)

## 2.8 Additional Odour-active Compounds

The odour-active compounds also play an important role in relation to the study of wine flavour chemistry. However, it must be noted that most of the odour-active compounds in wine are present in a very low concentration. Several research groups have been trying to understand the biochemical mechanisms from which odour-active compounds are formed, either in grapes or during the winemaking process (Etievent, 1981; Nykanen, 1986; Rapp and Mandery, 1986; Tesniere *et al.*, 1989). The gas chromatography-mass spectrometry (GC-MS), gas chromatography-olfactometry (GCO) techniques are some of the methodologies which allow hundreds of volatile compounds to be identified and quantified in grapes (Mc Daniel *et al.*, 1989).

Several new odour-active compounds in Pinot noir wines have been reported (Lopez *et al.*, 1992). These include phenolic compounds, like acetovanillone; ethyl vanillate and methyl vanillate; sulphur compounds like methionol and fatty acids which include isovaleric, hexanoic, octanoic, decanoic and tridecanoic fatty acids (Table 8).

**Table 7.** Neutral Components Identified in Extract of 'Zinfandel' Wine

| *Peak No.*[a] | *Compound* | *Means of identification*[b] | *Approx. conc. (mg/l)* |
|---|---|---|---|
| 2 | Ethyl acetate | RT | NA[c] |
| 3 | Ethanol | RT | NA |
| 5 | Isobutyl acetate | RT | NA |
| 6 | n-Propanol | RT | 15-20 |
| 7 | Isobutanol | RT, IR | 35-40 |
| 8 | Isoamyl acetate | RT | 6-8 |
| 9 | n-Butanol | RT | 2-3 |
| 10a | 2-Methyl-1-butanol | RT, IR | 150-160 |
| 10b | 3-Methyl-1-butanol | RT, IR | |
| 11 | n-Pentanol | RT | 0.3-0.4 |
| 15 | Ethyl lactone | RT | NA |
| 17 | n-Hexanol | RT | NA |
| 21 | Ethyl n-caprylate | RT | 0.9-1.0 |
| 28 | Gamma-butyrolactone | RT, IR | NA |
| 29 | Ethyl n-caprate | RT | 0.2-0.3 |
| 31 | Diethyl succinate | RT, IR | 0.9-1.0 |
| 37 | Benzyl alcohol | RT | 0.1-0.2 |
| 38 | 2-Phenethanol | RT, IR | 15-18 |

[a] Peak numbers from chromatogram of Figure 1.
[b] RT = Agreement of known and unknown relative retention times on at least three different columns, IR = Agreement of infrared spectra of known and unknown
[c] NA = Not available
*Source*: Brander, 1974

**Table 8.** Additional Odour-active Compounds Found in Pinot Noir Wine

| *Chemical name* | *Concentration (mg/l)*[1] | *Odour* | *Reported in grape wine*[1] |
|---|---|---|---|
| Methylvanillate | 11-214 | Vanilla, herbal, spicy, caramel | Sinohara, 1985; Tesniere 1989 |
| Ethylvanillate | 31-13 | Sweet, vanilla, spicy | Strauss, 1987; McDaniel, 1989 |
| Acetovanillone | 63-1227 | Vanilla, spicy, molasses, clove-like | Tesniere, 1989; Etievant, 1981 |
| Methianol | 16-58 | Vegetable, cabbage, stinky | Nykanen, 1986; Strauss, 1987; Rapp and Mandery, 1986 |
| Hexanoic acid | 11-37 | Sour, vinegar, cheese, sweety, rancid, fatty, pungent | Schreier, 1976 |
| Octanoic acid | 11-41 | Oily, fatty, rancid, soapy, sweet, faint fruity | Rapp and Mander,y 1986; Schreier, 1976 |
| Decanoic acid | 0-54 | Fatty, unpleasant rancid, citrus | Rapp and Mandery, 1986 |

*Odour = Active compounds concentrations calculated from the six Pinot noir Freon extracts used in the present study.
[1] Previously reported in the literature cited in the present paper
*Source*: Lopez *et al*., 1992

# 3. Pre-harvest Factors Affecting Wine Quality and Aroma

## 3.1. Influence of Vineyard Site and Climate on Grapevine Phenology, Grape Composition, Wine Production and Quality

The topographical, agro-pedological and climatic environmental conditions that influence grape and wine composition and quality are referred to collectively by means of the French term 'terroir' (Jackson and Lombard, 1993). An understanding of the influence of the terroir allows grape growers not only to produce better-quality grapes in traditional wine-producing regions, but also to expand production into new viticulture areas in a rational fashion.

Vineyard site had the most significant influence on sensory scores and wine composition, followed by canopy management. Probably, the first feature recognised as favouring finer grapes was limited soil fertility (Robinson *et al*., 2011). Soils with just adequate nutrient levels restrict vegetative growth and permit a higher proportion of photosynthate to be directed towards fruit maturation; they thus favour flavour formation, which tends to develop near the end of ripening. Many low-nutrient soils are also highly porous, which results in periods of mild water deficit that further limit vegetative growth and also favours the rapid warming of the soil. This, in turn, can improve the microclimate around the vine and delay or minimize frost severity. Irrigation practices also have a variable effect on accumulation of sugars in grape berries. Irrigated vines have a greater concentration of glucose and fructose than non-irrigated vines (Intrigliolo *et al*., 2008). In recent years, there has been considerable emphasis on the importance of cool climate viticulture to achieve ultimate wine quality (Croser, 1983). In Australia, there has been a shift of premium quality table-wine vineyards from the fertile valleys to the mountain ranges. This altitude factor is considered to be an indication of site coolness.

Grape vines grown in distinct climatic regimes worldwide that provide ideal situations to produce high quality grapes (De Blij, 1983). Interactions between the local climate, soil and site locations play an important role in ontogeny and yield of the grape vines. Mild to cool and wet winters followed by warm springs, then warm to hot summers with little precipitation provide adequate growth potential and increase the likelihood of higher wine quality (Coombe, 1987; Gladstones, 1992; Jones, 1997). Over the last 30 years, the estimated alcohol levels of Riesling grapes increased due to warmer ripening periods and earlier phenology (Duchene and Schneider, 2005). A long term (1952-1997) climatology using reference vineyard observations was developed in Bordeaux, France (Jones *et al*., 2000). The procedure partitioned the season into growth intervals from one phenological event to the next, viz. bud burst, floraison, veraison and harvest. Over the last two decades, the phenology of grape vines in Bordeaux has tended towards earlier phenological events, a shortening of phenological intervals, and lengthening of a growing season. Merlot and Cabernet Sauvignon varieties have tended to produce higher sugar to total acid ratios, greater berry weights and greater potential wine quality.

## 3.2. Influence of Yield Manipulations on the Terpene Content of Juices and Wine

Since, the first report on the significance of aroma compounds 'monoterpenes' in Muscat of Alexandria grapes and wines in Australia (Williams *et al*., 1980a, b), substantial interest arose in these substances for their potential as a quality indicator of wines (Marais, 1983). The terpenes appeared important not only to Muscat varieties but also to other fruity varieties, such as Muller, Thurgau, Traminer and Riesling. Keeping this factor into consideration, it was realised that clear relationships between the concentration of monoterpenes of juices and the final wine quality wine might also be found in New Zealand, where wines typically show well-developed varietal characteristics (Ewart *et al*., 1984; Dimitridis and Strauss, 1985). More recently, there has been a strong trend towards determining juice quality by testing the flavour rather than the traditional parameters, such as sugars, titratable acidity and pH. This approach is used to identify the precise time when flavour development is most pronounced and therefore, likely to yield a better wine (Cootes, 1983; Kenworthy, 1984; De Boubee, 2000).

Terpene concentration was measured during ripening of Muller Thurgau grapes for providing objective information on grape and wine quality (Eschenbruch and Van Dam, 1985/1986). Reduction in yield and change of the canopy microclimate by shoot and cluster thinning resulted in a higher terpene concentration in 1986, which was unchanged during 1985 (Fig. 4). It was shown that the changes in

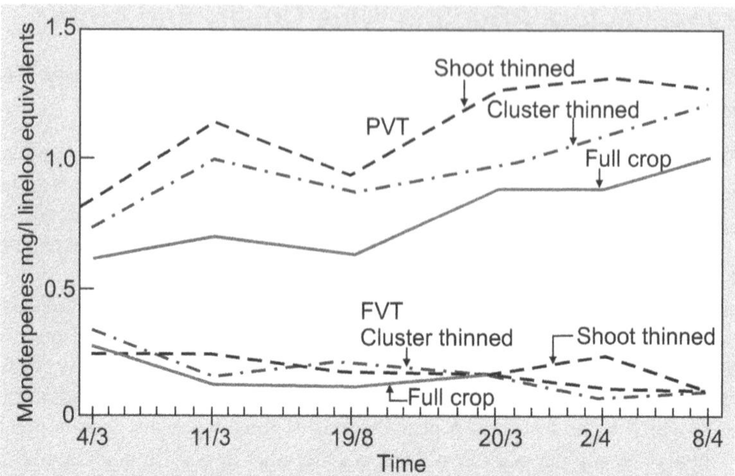


**Figure 4.** Free (FVT) and potential (PVT) volatile terpenes of Mul Thurgau grapes after canopy manipulation (*Source*: Eschenbruch *et al.*, 1985/86)

the canopy microclimate were inadequate to achieve a clear relationship between the concentration of terpenes in the grapes and the quality of wine.

California's north coast grape-growing regions, which include Napa, Sonowa and Mendocino counties, produce a substantial amount of Chardonnay (61 per cent), Pinot noir (68 per cent), Cabernet Sauvignon (57 per cent) and Sauvignon Blanc (37 per cent) grapes. Viticulturists and oenologists are now exploring different techniques that affect grape composition in the vineyard. The wine-growers believe that much can be done in the vineyard to affect grape flavour. However, many new experiments and trials on clones, rootstock, spacing, canopy management and water management are under progress.

## 4. Processing Factors Affecting Wine Flavour, Aroma and Quality

### 4.1. Influence of Vinification Treatments on Aroma Constituents and Sensory Descriptors

Wine colour, aroma and flavour depend primarily on the initial condition and composition of the grapes as well as subsequent enological processes. Fermentation temperature can impact significantly in red wines (Amerine and Ough, 1957; Francis *et al.*, 1994). Colour extraction in Pinot noir was a problem and raising temperature for 10°C to 21°C improved both colour intensity and flavour in the resultant wines (Amerine, 1955). When a range of fermentation temperatures, i.e. 12°C, 15.5°C, 21°C and 27°C were examined for Pinot noir wines, colour and flavour were adjudged the best in both (21°C and 27°C) the treatments (Amerine and Ough, 1957). Increasing fermentation temperature from 12°C to either 20°C or 30°C in Cabernet Sauvignon, Grenache, or Pinot noir wines increased both colour and tannin extraction (Ough and Amerine, 1960). Subsequent work showed a significant linear relationship between temperature and wine quality for these three cultivars (Ough and Amerine, 1961; Ough and Amerine, 1967). Most of the red wine fermentations in California are carried out at temperatures between 20°C and 35°C.

The aromatic properties of acid hydrolysates of glycosidic fractions from juices of Chardonnay, Semillon and Sauvignon Blanc grapes have recently been studied using sensory descriptive analysis. This work showed that aroma attributes, such as tea, lime and honey, which were characteristics of wines made from the same juices, were exhibited by the hydrolysates (Francis *et al.*, 1992). The quality of wine is also affected by yeast strain, grape cultivar, time of skin contact, oxygen level and type of suspended solids. Besides this, other factors like insoluble materials, i.e. grape solids, bentonite, diatomaceous earth, etc. can also influence the fermentation environment and rate of sugar conversion (Ough and Groat, 1978). However, studies on the effects of suspended solids on wine quality have been mainly restricted to white cultivars (Maga and Sizer, 1973; Singleton *et al.*, 1975, Buteau *et al.*, 1979).

The influence of vinification treatments on aroma constituents and sensory descriptiors of Pinot noir wines was studied (Girard, 1997). Three verfication methods were compared to assess their effects on chemical composition, sensory descriptors and head space volatile constituents. Two methods involved standard vinification at fermentation temperatures of 20°C ($VM_1$) and 30°C ($VM_2$), respectively, while the third method ($VM_3$) included a two-stage pre-fermentation treatment involving heat extraction, followed by fermentation at 15°C in contact with bentonite. $VM_2$ produced wines with highest colour intensity, current aroma and current flavour. However, $VM_3$ wines, however, contained twice the concentration of anthocyanins of $VM_1$ and $VM_2$ wines and possessed the most intense fruity aroma and flavour. Total ester concentration was fourfold higher in $VM_3$ compared to $VM_1$ and $VM_2$. Several esters were responsible for this difference but isoamyl acetate was pre-dominant (Figs 5 to 9).

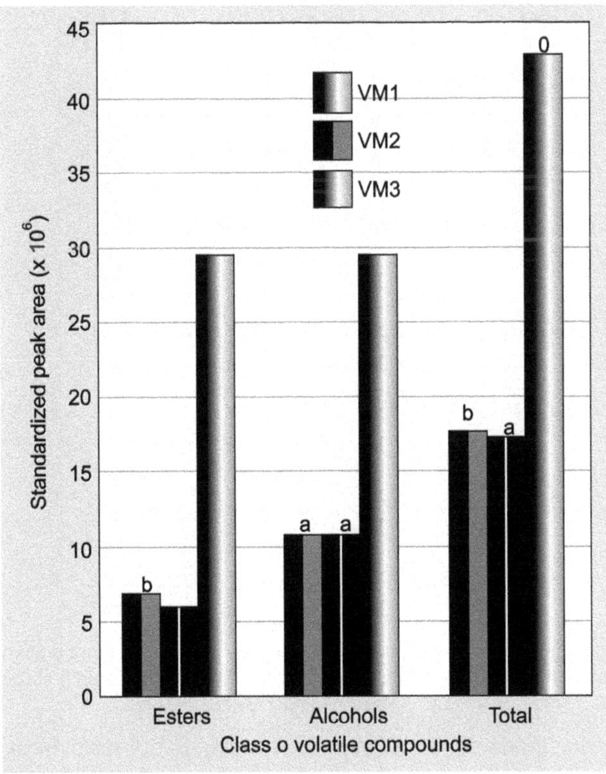


**Figure 5.** Effect of vinification methods on classes of volatile aroma compounds in Pinot noir wines. Vinification methods within volatile classes with different letters are significantly different at $p \leq 0.05$ (*Source*: Girard *et al.*, 1997)

## 4.2. Effect of Wine Yeast Strains on Wine Quality

Although the influence of yeast strains on wine fermentation aromas consisting of esters, higher alcohols, fatty acids has been long known (Suomalainen, 1971; Mateo *et al.*, 1992; Antonelli *et al.*, 1999), the effect of winemaking yeasts on grape aromas and their precursors has not been studied to any great extent. It is known that the monoterpenol composition and muscat aroma of wines vary little during fermentation (Gunata *et al.*, 1986; Dobourdieu *et al.*, 1988; Delcroix *et al.*, 1994) as the glycosidases of *Saccharomyces cerevisiae* have little effect on terpenic glycosides at normal pH of the must. It is not so in some of the varieties known as simple flavoured, which is not very intense in must but develops considerably during fermentation. The amplification of grape aroma by yeast has been clearly demonstrated for Sauvignon Blanc. The volatile thiols responsible for the box tree, grape fruit and passion fruit nuances of wines made from this varieties (Tominaga *et al.*, 1998; Tominaga *et al.*, 1998a) are principally formed by yeast from cysteinylated precursors in the must (Tominaga *et al.*, 1998a). The compounds involved are 4-mercapto-4-methyl pentan-2-one (4 MMP), 3-mercaptohexan-1-ol (3 MH), 4-mercapto-4-methylpentan-2-ol

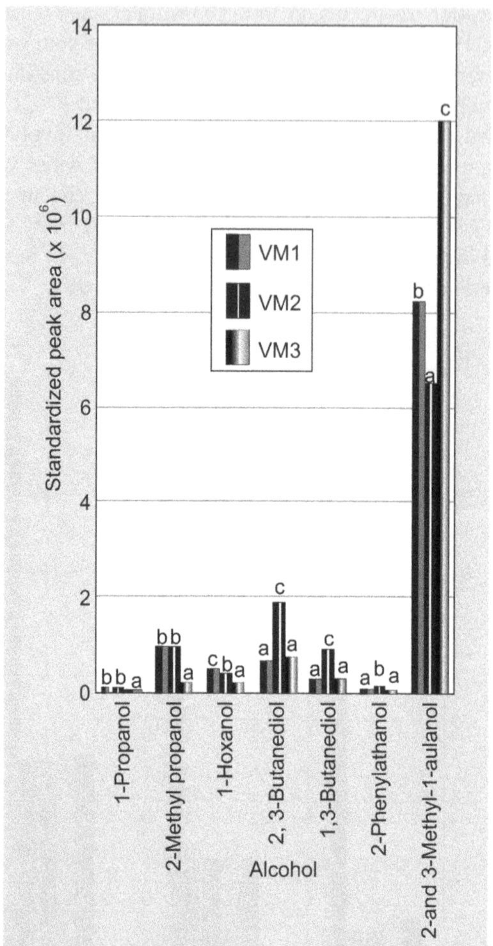


**Figure 6.** Influence of vinification methods on specific alcohols in Pinot noir wines. Vinification method within volatile compounds with different letters are significantly different at $p \leq 0.05$ (*Source*: Girard *et al.*, 1997)

(4 MMPOH), which are produced from grape precursors, S-4-(4-methylpentan-2-one)-L-cysteine, S-3-(hexan1-ol)-L-cystein and S-4-(4-methylpentan-20-ol)-L-cysteine, respectively. Some strains of *Metschnikowia pulcherrima* and *Kluyveromyces* yeasts are known to predominate during the intial stage of wine fermentation and are appropriate for lowering the alcohol yields by respiration (Quiros *et al.*, 2014).

The ability of four different industrial strains of *S. cerevisiae* to release certain Sauvignon Blanc aromas from their cysteinlated precursors were studied (Murat *et al.*, 2001). The comparison was made both in a model medium and in must under normal winemaking conditions in four Bordeaux wineries. The quantification of the wines showed statistically significant differences among the various yeast strains (Table 9).

It showed clearly that winemaking yeasts have a decisive influence on the concentration of major constituents, which vary according to the specific strains. The VL3C and EG8 strains were found to be quite effective in the enhancing the aroma components. During alcoholic fermentation, 4 MMP, 4 MMPOH and 3 MH are released from S-4-(4-methylpentan-2-one)-L-cysteine, S-4-(4-methylpentan-2-ol)-L-cysteine and S-3-(hexan-1-ol)-L-cysteine, respectively. The mechanisms by which these odourless precursors are converted into aromas by yeast during alcoholic fermentation have not been fully elucidated. Previous work has shown that β-lyase-type enzyme activity is found in several microorganisms, including Baker's and Brewer's yeasts which are capable of hydrolysing S-conjugate cysteins and releasing the corresponding volatile thiols (Tominaga *et al.*, 1998a).

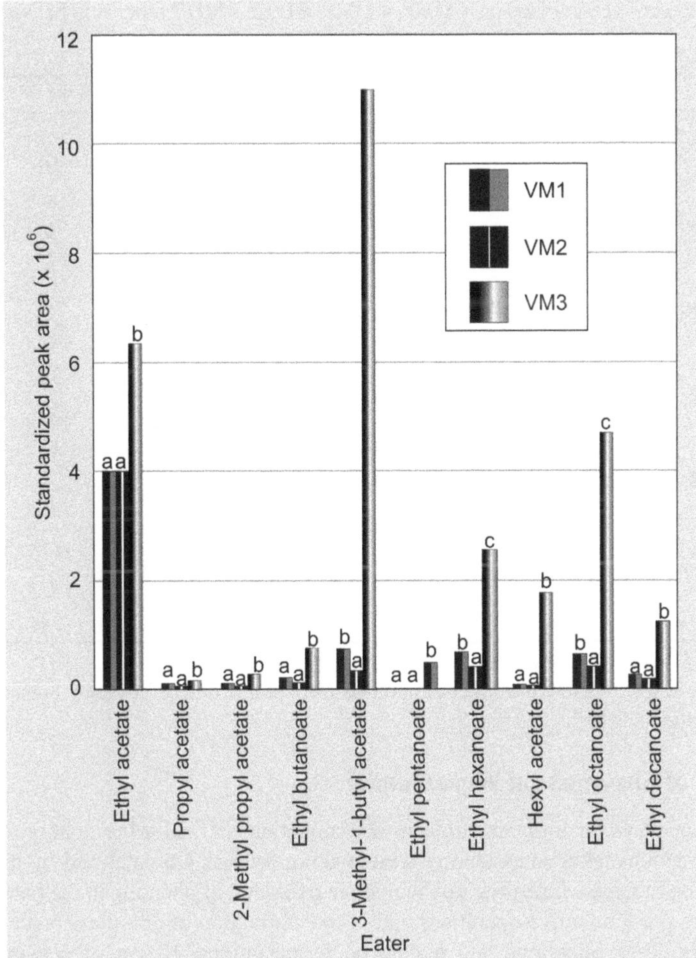


**Figure 7.** Influence of vinification methods on specific esters in Pinot noir wines. Vinification methods within volatile compounds with different letters are significantly different at $p \leq 0.05$ (*Source*: Girard *et al.*, 1997)

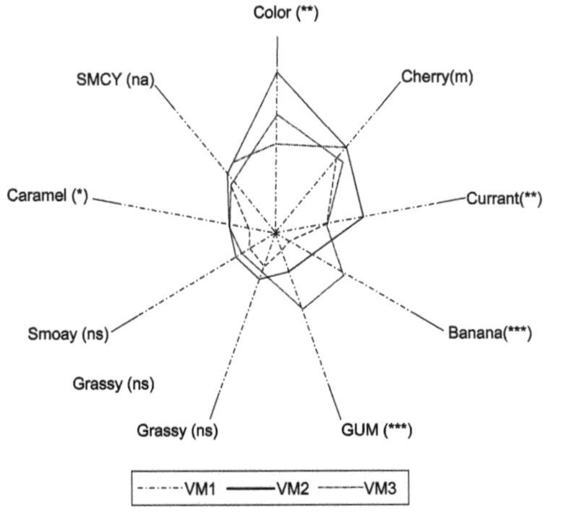


**Figure 8.** Sensory profile plot of intensity ratings and minimum significant differences (MSD, --) for color and aroma descriptors of Pinot noir wines (0 min., 10 max.) (*Source*: Girard *et al.*, 1997)

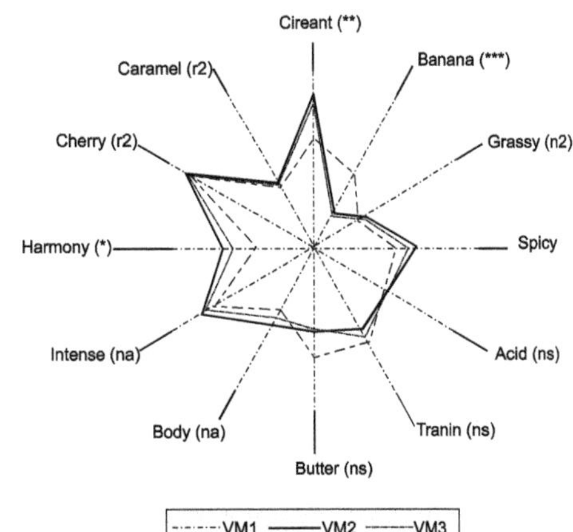


**Figure 9.** Sensory profile mean intensity ratings and minimum significant differences (MSD,____) for favour descriptors of Pinot noir wines (0 min., 10 max.) (*Source*: Girard *et al.*, 1997)

**Table 9.** Effect of yeast strains on 4 MMP, 4 MMPOH and 3 MH Amounts in Sauvignon Blanc Wines after Alcoholic Fermentation

| *Must* | *Wine a*** | *Wine b*** | *Wine c*** | *Wine d*** | *Average* |
|---|---|---|---|---|---|
| **4 MMP (ng/l): 4-Mercapto-4-methyl pentan-2-one** | | | | | |
| VL3C | 12 | 12 | 12 | 10 | 12a* |
| EG8 | 8 | 9 | 16 | 8 | 10a* |
| VL1 | 7 | 2 | 7 | 6 | 6b* |
| 522d | 0 | 0 | 0 | 0 | 0c* |
| **4 MMPOH (ng/l): 4-Mercapto-4-methyl pentan-2-ol** | | | | | |
| VL3C | 28 | 12 | 27 | 41 | 27a* |
| EG8 | 25 | 9 | 10.6 | 39 | 21ab* |
| VL1 | 25 | 7 | 9 | 38 | 20ab* |
| 522d | 25 | 6 | 2 | 32 | 16b* |
| **3 MH (ng/l): 3-Mercaptohexan-1-ol** | | | | | |
| VL3C | 2161 | 3261 | 413 | 991 | 1706* |
| EG8 | 2994 | 4581 | 460 | 1135 | 2267* |
| VL1 | 2077 | 2227 | 305 | 1457 | 1516* |
| 522d | 2128 | 2890 | 235 | 1184 | 160* |

* Values denoted by different letters are statistically different ($\alpha = 0.01$)
** Wine a, wine b, wine c and wine d represent musts from four different Bordeux, France vineyards
*Source*: Murat *et al*., 2001.

## 4.3. Influence of Enzymes on Wine Quality

Grape composition plays an important role in determination of final wine quality. A complex array of biochemical reactions takes place during wine making process are catalysed by different enzymes coming from various sources. Enzymes are involved in oxidation of phenolic in the formation of volatile compounds during pre-fermentive operations and in transformation of odourless precursors into odour-active compounds during alcoholic and malolactic fermentations. Extent of enzymatic reactions is considered to be the efficient technological step in the modern wine industry. Therefore, understanding the role played by enzymes during winemaking can help in development of rational and effective strategies for optimising wine processing.

The majority of wine sensorially-active compounds are formed or degraded during winemaking though pathways involving the intervention of enzymes. Addition of exogenous enzymes is currently the most commonly adopted practice in the wine industry to optimise the rate of enzymatic reactions during the winemaking process. In addition, promising results are being obtained with strain selection, particularly for applications related to wine aroma enhancement.

The use of enzymes in enology has been widely studied (Villetaz, 1995; Wightman *et al*., 1997; Ducruet *et al*., 1997; Ducruet *et al*., 2000) and today their use is more frequent. Commercial preparations are made from industrially grown fungi, *Aspergillus niger* and *Trichoderma harzianum*. The authorised enzymatic activities are limited to those of pectinase and β-1, 3-1, 6-glucanase but the process or the strain by which commercial preparations are made also result in secondary activities that may potentially have effects on the quality of the wine (Betrand, 1996; Guerrand, 2000). The activity of cinnamyl esterase releases phenol acids, particularly those seen in forms related to tartaric acid (Chatonnet *et al*., 1992; Dugelay *et al*., 1993). These free phenol acids can be transported into aromatic molecules, i.e. volatile phenols by the yeasts of *Brettanomyces* type (Larue *et al*., 1991).

Two techniques, i.e. heating at the end of macertion and the use of enological enzymes, can potentially cause an excess production of volatile phenols. Compared with the classic temperature of maceration without use of enzymes, one of these alternative techniques may triple the presence of volatile phenols even more so when both techniques were administered on the same wine (Gerbaux *et al*., 2002).

Cinnamyl esterase, a secondary activity seen in most enzymatic preparations produced by *Aspergillus niger*, is the cause of this problem. The use of a purified enzyme without cinnamyl esterase does not seem to induce an overproduction of volatile phenols. This property is already present in enzymes used in the vinification of white wines and it is important to use the same purified enzymes for red wine production.

## 4.4. Effect of Ultrafiltration and Fermentation on the Aroma and Flavour Characteristics of Wine

Ultrafiltration has been introduced for wine-juice processing in order to achieve protein stability in finished wines and it has been reported in some studies to cause no difference in flavour, whereas in other studies, beneficial and deleterious effects have been reported. However, no significant differences in the sensory assessment of the flavour of freshly processed kiwifruit juices by UF and conventional processing techniques were detected by other workers (Wilson *et al.*, 1984). For apple juice, the ultrafiltered apple juice was rated significantly lower in sensory flavour than the apple juice filtered by the conventional plate and frame method (Drake and Nelson, 1986).

Ultrafiltration had beneficial effect on the flavour of dry white wines (Balanutse *et al.*, 1985). In the same year, ultrafiltration did not affect wine taste, except for a slight reduction in the bitterness of an astringent red wine (Sachs *et al.*, 1983; Gaillard, 1985); however, in white wine, a sensory panel rated the controls as having a more pronounced odour and flavour initially than the UF wine, but the controls showed slight oxidation after four months and considerable oxidation after 12 months (Meglioli and Marchesini, 1985). In the sensory evaluation of ultrafiltered white wine and its bentonite-treated control, the ultrafiltered sample had a cleaner aroma (Testaniere, 1985).

Ultrafiltration did not affect on the aroma and flavour characteristics of white Riesling and Gewiirztraminer wines (Flores *et al.*, 1991). They did not observe any significant flavour changes which could be attributed to ultrafication. In the pilot scale wine trials, the UF-treated wines were consistently lower in intensity in most of the significant aroma descriptors. Intensity ratios were of the order: control unfiltered > bentonite fined > unfiltered.

Resveratrol is one of the phenolic compounds implicated in the health benefits associated with wine consumption. Resveratrol concentrations are higher in red wines as compared to white wines because it is found in the skin of grapes and red wines are fermented on the skins. These concentrations increased during fermentations on the skin. Cabernet Sauvignon, Cynthiana and Noble have high level of resveratrol amongst the red wine varieties.

## 4.5. Effect of Carbonic Maceration on Wine Aroma under Different Aging Conditions

Wines undergo a large number of alterations from their original compositions due to certain factors, like container type, length of maturation and environmental conditions. The changes in the volatile fractions may be due to the chemical reactions as in the case of ester formation and hydrolysis (Marais, 1978; Rapp *et al.*, 1985; de Mora *et al.*, 1986), oxidation of alcohols (Simpson, 1978), or aldehyde and acetal formation (Wildenradt and Singleton, 1974; Schreier, 1984; Somers and Wescombe, 1987), physical reactions, such as the evaporation of volatile substances or the solubilisation of extractive substances from wood (Pontallier *et al.*, 1982), or at a biological level, like malolactic fermentation and acetic acid formation (Chatonnet *et al.*, 1993).

Wines resulting from carbonic maceration (CM) have a distinct aroma in which vague fruity notes predominate as a consequence of the anaerobic metabolism of the grapes. Because they undergo carbonic anaerobiosis, the alcoholic fermentation of these wines is rapid, which means they are biologically stable before wines which undergo traditional vinification processes. Therefore, CM wines can be marketed sooner (Flanzy *et al.*, 1990). They are principally intended for immediate consumption and have certain characteristics, such as higher pH and a large proportion of higher molecular weight polyphenols which might suggest that CM wines are not suitable for aging in barrels. Several researchers have studied the aromatic composition of wines made by carbonic maceration and they found that carbonic macerated wines differ from traditionally-produced wines due to lower concentration of $C_6$ compounds, higher concentration of ethyl cinnamate and volatile phenols (Ducret *et al.*, 1983; Ducret, 1984; Dubois *et al.*,

1997) and significant differences in vinyl benzene, 2-phenylethyl acetate and benzaldehyde (Salinas *et al.*, 1992).

Length of anaerobiosis stage also influences the volatile composition of wines made from Monastrell grapes by carbonic maceration and the effect of aging in barrels of old and new wood (Salinas *et al.*, 1996). The optimal sensory qualities were shown by wines which had undergone a long period of maceration and then were aged in new woods, resulting in a higher concentration of ethylic esters and fatty acids and a lower alcoholic concentration.

## 4.6. Qualitative Changes in the Volatiles During Aging

Aging brings about perfect maturity in wines which bear their own distinctive sensory make-up. Wines made from Riesling and Vidal Blanc grapes are often consumed when they are one to two years old, since they quickly acquire their characteristic fruity aroma. Riesling is usually aged in glass and examination of changes in the chemical composition of the aroma compounds upon storage has led to the description of a varietal bouquet for Riesling wines (Simpson, 1979; Simpson and Miller, 1984; Rapp and Guntert, 1986).

Off odours do not contribute significantly to the aroma of aged Riesling and their role has been described by several workers. It has been found that the concentration of dimethyl sulphide in some white wines increases with time and temperature and makes a significant contribution to the bouquet of aged wines (Marais, 1979). 1, 1, 6-trimethyl-1, 2-dihydronapthalene (TDN) has been blamed for a characteristic kerosene aroma in Riesling wines, particularly those made from grapes grown in warmer climates such as South Africa and Australia (Simpson and Miller, 1983; Rapp *et al.*, 1985; Marais *et al.*, 1992). The level of TDN in wine can be used to estimate the aging potential of Riesling wine (Winterhalter *et al.*, 1990). The fate of monoterpenes during aging plays a critical role in the quality of aged wine. Riesling wines are not affected negatively by aging, whereas Muscat type wines, which have a higher level of monoterpenes, do not improve with bottle storage (Rapp and Madery, 1986).

Based on the aroma characteristics of aged Vidal Blanc wine, it was found that upon aging, Vidal Blanc lost much of its characteristic fruity aroma (Chisholm *et al.*, 1995). In some cases, the wine acquired a strong odour of asparagus after two to three years, while in other cases the dominant aroma of the three-year-old wine was found to resemble straw. The appearance of the asparagus aroma was not predictable and may depend upon the yeast used (de Mora *et al.*, 1986). These results suggest that the aging of Vidal Blanc wine is most affected by changes in the terpene composition and therefore, it does not maintain its varietal character with age, unlike Riesling wines. The amount of monoterpenes in young Vidal Blanc wine is higher than that of Riesling, but lower than that of Muscat wines.

Analysis of periodic changes in the volatile composition of Zinfandel wine during aging, starting with a newly fermented must and every three months thereafter year was carried out. The most significant changes occurred at six months, coinciding with the completion of malo-lactic fermentation. Approximately, 20 new compounds were identified after six months of storage. Figure 10 shows a portion of the chromatogram from 62 to 78 min. of the six months' batch of wine with the aroma descriptions. The peak identifications determined by GC-MS are 3, ethyl hexanoate; 3A, hexyl acetate; 4, β-phenyl ethenol; 5, diethyl succinate; 6, ethyl octanoate. According to the descriptions and retention time, the only possible assignment for the carnation aroma could be benzyl isoeugenol.

## 4.7. Effect of Thermal Processing on Wine Quality

A technique appropriate to the modern winemaking technology is the application of heat, either to a juice prior to fermentation or to a wine, in order to affect an *in situ* hydrolysis of the grape flavour precursors which are catalysed by the natural acids of the juice or wine. Heating wines for the purpose of accelerated aging has been practiced widely and for certain wine styles, thermal treatment can apparently result in an improvement of quality (Singleton, 1962). More recent work on the sensory evaluation of thermally processed young red wines showed that heating at 45°C under inert gas produces wine with chemical and sensory attributes of aged wines (Somers and Pocock, 1990).

Pre- and post-fermentation thermal processing decreased the floral character of the bottle-aged wines and enhanced characters, such as oak, honey and smoky in Chardonnay and Semillon wines (Francis *et al.*, 1994). Heat treatments of juices and wines at 90°C produced wines which were not significantly different from their respective controls. Thermal processing should be appraised as an additional procedure in enhancement of wine flavour complexity, viz. the malo-lactic fermentation and oak wood treatment.

The wine style and quality of red wine is having a convincing effect of fermentation temperature. This has frequently been used in varieties like Pinot noir for which colour extraction is a problem (Amerine, 1955; Amerine and Ough, 1957; Ough and Amerine, 1960; Gao *et al.*, 1997). Recently cold-soak treatments have been reported to increase glycosides in Cabernet Sauvignon (McMahon *et al.*, 1999). Influence of fermentation temperature on composition and sensory properties of Semillon and Shiraz wines was studied by combining pre-fermentation cold soak and post-fermentation maceration with increasing fermentation temperature (Reynolds *et al.*, 2001). They have concluded that increasing fermentation temperature in Semillon resulted in increased titratable acidity, ethanol content and fruitier aroma. The 20°C and 30°C temperatures resulted in the highest ethanol content. In contrast, reduced titratable acidity along with increase in pH and anthocyanin content were resulted with the increasing fermentation temperature, cold soak usage and extended maceration.

Use of cold soak increased colour (15°C) and black currant flavour, reduced herbaceous flavour and black currant aroma. Extended maceration enhanced wine colour and body, but reduced black currant aroma. Extended macertaion enhanced wine colour and body, but increased herbacious aroma. Increasing fermentation temperature from 15 to 30°C increased wine colour, reduced herbaceous flavour, increased black currant flavour, but increased perceived acidity. These results suggest that Semillon wines should be produced through the use of fermentation pomace contact and cool fermentation temperatures, whereas Shiraz wines would benefit from higher fermentation temperatures, use of cold soak and extended post fermentation maceration (Tables 10 and 11).

**Table 10.** Chemical Composition of Semillon Wines Fermented at Four Different Temperatures

| *Factors* | *Titratable acidity* | *pH* | *Ethanol (%)* |
|---|---|---|---|
| **Fermentation temperature** | | | |
| 15 | 9.7 | 3.17 | 11.7 |
| 20 | 9.7 | 3.15 | 12.4 |
| 25 | 9.8 | 3.15 | 11.6 |
| 30 | 10.4 | 3.16 | 12.7 |
| Standard error of mean | 0.11 | 0.04 | 0.13 |
| Significance | *** | * | *** |
| **Pre-fermentation treatment at 15°C** | | | |
| Crush and press | 10.00 | 3.02 | 11.4 |
| 25 hour pomace contact | 9.7 | 3.17 | 11.7 |
| Standard error of mean | 0.14 | 0.03 | 0.18 |
| Significance | * | *** | NS |

*, **, ***, NS significant at $p \leq 0.05$, 0.01, 0.001 or not significant respectively.
*Source*: Reynolds *et al.*, 2001

## 4.8. Effect of Sensory Evaluation on Appreciating Wine

Preference of a wine is influenced by many extrinsic factors in addition to flavour and appearance. Price, grape variety, vintage, origin, expert recommendations, winery reputation, bottle appearance and awards can influence quality perception of a wine. Furthermore, in-mouth impression of wine may influence quality perception of wine. In several food and beverage categories, a difference in preference between consumer panels and trained panels has been observed (Cardello *et al.*, 1982). The market in general appreciated full-body red wines with intense and complex flavour profiles produced from grapes with adequate phenolic ripeness, optimal flavour balance and lower acidity. In addition, consumers also perceive that high levels of alcohol affect sensory perceptions in wine, leading to unbalanced wines

**Table 11.** Effects of Fermentation Temperature, Cold Soak and Post Fermentation Extended Maceration on Composition of Shiraz Wines

| *Factors* | *Titratable acidity (g/l)* | *pH* | *Total anthocyanins (mg/l)* | *Ethanol (%)* |
|---|---|---|---|---|
| **Main effects: Initial fermentation (°C)** | | | | |
| 15 | 11.4 | 2.94 | 348.4 | 9.23 |
| 20 | 11.2 | 2.97 | 393.6 | 7.70 |
| 30 | 10.8 | 2.95 | 392.1 | 7.93 |
| Standard error of mean | 0.16 | 0.04 | 1.54 | 0.17 |
| Significance | ** | ** | *** | *** |
| **Main effects: Pre-fermentation treatment** | | | | |
| Crush only (C) | 11.4 | 2.94 | 360.4 | 8.03 |
| 10 day cold soak (CS) | 10.8 | 2.96 | 395.7 | 8.53 |
| Standard error of mean | 0.16 | 0.04 | 1.56 | 0.25 |
| Significance | *** | NS | *** | ** |
| **Pre- and post-fermentation treatment (20°C)** | | | | |
| Crush only (C) | 11.6a[b] | 2.96b | 370.4b | 7.50c |
| 10 day cold soak (CS) | 10.8ab | 2.99ab | 416.8b | 7.90b |
| Standard error of mean | 10.3b | 3.03a | 444.2a | 8.00a |
| Significance[a] | * | * | ** | ** |
| Interactions (Fermentation temperature (CS) | | | | |
| 15 C | 11.6 | 2.94 | 313.7 | 8.90 |
| 15 CS | 11.2 | 2.93 | 383.2 | 9.55 |
| 20 C | 11.6 | 2.96 | 370.4 | 7.50 |
| 20 CS | 10.8 | 2.99 | 416.8 | 7.90 |
| 30 C | 11.1 | 2.92 | 387.0 | 7.70 |
| 30 CS | 10.5 | 2.98 | 387.1 | 8.15 |
| Standard error of mean | 0.11 | 0.02 | 0.99 | 0.12 |
| Significance | NS | ** | ** | NS |

[a]*, **, **, NS significant at $p \leq 0.05$, 0.01, 0.001 or not significant, respectively.
[b] means followed by different letters are significant at $p \leq 0.05$, Duncan's Multiple Range test
*Source*: Reynolds *et al.*, 2001

(Kontoudakis *et al.*, 2011; Casassa *et al.*, 2013). Reduction in the alcohol content of oaked white wine results in a minimal impact on sensory composition and consumer preferences since no perceptible changes to the sensory profile were observed (King and Heymann, 2014). The reduction process increased the intensity of astringency, bitterness and acidity and has slightly affected the wine sensory profile (Lisanti *et al.*, 2011). These sensory parameters may play a role in the acceptance and enjoyment of wine and are considered to be the most important factor. Subjects, experienced with wine flavour description, have been shown to use more terms to describe the odour qualities of wines than in experienced tasters. Additionally, experienced judges or wine experts were better able to match previously written descriptions of a set of wines to the same wine than were less experienced tasters (Lawless, 1984).

Twelve red wines were evaluated by descriptive analysis in which 12 trained judges rated 14 attributes in duplicate. Fifty-seven subjects rated liking for these wines on a 9-point hedonic scale. Subjects were segmented based on their level of wine knowledge (written test) and sensory expertise (sensory test) and on overall expertise, as estimated by the combined scores. There was no correlation in performance between the two tests, indicating that sensory performance and wine knowledge are two distinctly different types of expertise and that one cannot be inferred from the other (Frost and Noble, 2002).

## 5. Future Outlook

Variety is considered to be one of the most important factors to influence grape and wine aromatic characteristics. Volatile organic compounds, carbohydrates, acids, phenolics, nitrogenous compounds, terpenoids, fats and lipids are vital for the quality of grape berries and wines. In each of these categories, usually there are many individual substances. New compounds in grapes continue to be reported and added to the list of several hundreds already identified. The ultimate quality and aroma of the wine depends upon several pre-harvest factors, affecting growth and development of grape varieties as well as the subsequent enological processes. The varietal aroma depends on the overall profile of odour-active compounds present in grapes and wines.

The volatile constituents contributing aroma and bouquet of wine have also been investigated by several researchers. An additional effort is required to characterise some of these volatile components, especially the odourants that appear during storage of wines. It is possible that these odourants arise from glycosidic aroma precursors. The many varieties that are world famous for distinctive noble wines are relatively few. Still fewer have been studied in a grape vineyard/winery situation. However, the development of several sophisticated analytical techniques has led to study a wide range of varietal wines. It can be said that the varietal differences contribute significantly in aroma and quality differences. Thus, understanding the timing of volatile compound production and cultivar differences will guide viticulture researchers and growers in the optimisation of vineyard strategies to enhance grape aroma attributes, contributing to wine aroma.

## References

Adams, D.O. and Harbertson, J.F. (1999). Use of alkaline phosphatase for the analysis of tannins in grapes and red wines. *Am. J. Enol. Vitic*. 50: 247.

Aiken, J.W. and Noble, A.C. (1984). Composition and sensory properties of Cabernet Sauvignon wine aged in French Versus American Oak barrels. *Vitis* 23: 27.

Allen, M.S., Lacey, M.J. and Boyd, S. (1994). Determination of methoxypyrazines in red wines by stable isotope dilution gas chromatography mass spectrometery. *J. Agri. Food Chem*. 42: 1734.

Amerine, M.A. (1955). Further studies on controlled fermentations, III. *Am. J. Enol*. 6: 1.

Amerine, M.A. and Ough, C.S. (1957). Studies on controlled fermentations, III. *Am. J. Enol. Vitic*. 8: 18.

Antonelli, A., Castellari, L., Zambonelli, C. and Carnacini, A. (1999). Yeast influence on volatile composition of wine. *J. Ag. Food. Chem*. 47: 1139.

Asquith, T.N. and Butler, L.G. (1985). Use of dye-labelled protein as spectrophotometric assay for protein precipitants such as tannin. *J. Chem. Ecol*. 11: 1535.

Asselin, C., Pages, J. and Morlat, R. (1992). Typologie sensorielles du Cabernet franc et influence du terroir utilisation de methodes satistiques multidimensionnelles. *J. Int. Sci. Vigne. Vin*. 26: 129.

Augustyn, O.P.H., Rapp, A. and Van Wyk, C.J. (1982). Some volatile aroma components of *V. vinifera* L. cv. Sauvignon blanc. *S. Afr. Enol. J. Vitic*. 3: 53.

Balanutse, A.O., Mustyatse, G.F., Kalyan, B.N. and Kitii, F.D. (1985). Effect of ultrafiltration on content and fractional composition of protein in wine sadovodstvo, vinogradorstvo. *Vinodelie Moldavii* 38(11): 35, Abstract. *In*: *FSTA* 8: Ho129.

Bayonove, C.L., Baumes, R.L., Crouzet, J. and Gunata, Z. (1998). Aromes. pp. 164. *In*: Oenologie C. Flanzy (Ed.). Technique and Documentation, Lavoisier, Paris.

Bayonove, C.R., Cordonnier, R. and Dubois, P. (1975). Etude d'une fraction caracteristique de l'arome du raisin de la variete Cabernet-Sauvignon, mise en evidence de la 2-methoxy-3-isobutyl pyrazine. *C.R. Acad. Sc.*, Paris, Ser. D. 281: 75.

Betrand, A. and Beloqui, A.A. (1996). Essais d utilisation d enzymes pectolytiques a activities complementaires de type glucosidase dans les vins: Influence sur l'arome. *Rev. Fr. Oenol.*, 157: 19.

Brander, C.F. (1974). Volatile composition of 'Zinfandel' table wine: Some neutral components. *Am. J. Enol. Vitic.* 25(1): 13.

Brossaud, F., Cheynier, V., Asselin, C. and Moutounet, M. (1999). Flavonoid compositional differences of grapes among site test plantings of Cabernet franc. *Am. J. Enol. Vitic.* 50(3): 277.

Buteau, C., Duitschaever, C.L. and Ashton, G.C. (1979). Vinification of three white grape varieties by three different methods: I. Fermentation process and wine composition. *Am. J. Enol. Vitic.* 30: 139.

Buttery, R.G., Teranishi, R. and Ling, L.C. (1988). Identification of damascenone in tomato volatiles. *Chem. Industr.* 238.

Butzke, C.E., Evans, T.J. and Ebeler, S.E. (1998). Detection of cork taint in wine using automated solid-phase microextraction in combination with GC/MS-SIM. pp. 208. *In*: A.L. Waterhouse and S.E. Ebeler (Eds.). Chemistry of Wine Flavour. ACS Symposium Series 714. American Chemical Society. Washington, DC.

Cabaroglu, T., Canbas, A., Lepoutre, J.P. and Gunata, Z. (2002). Free and Volatile composition of red wines of Vitis vinifera L cv. Okuzozii and Bogazkere grown in Turkey. *Am. J. Enol. VItic.* 53(1): 64.

Cardello, A.V., Maller, O., Kapsalis, J.G., Segars, R.A., Sawyer, F.M., Murphy, C. and Moskowitz, H.R. (1982). Preception of texture by trained and consumer panelists. *J. Food Sci.* 47: 1186.

Casassa, L.F., Beaver, C.W., Mireles, M., Larsen, R.C., Hopfer, H., Heymann, H. and Harbertson, J.F. (2013). Influence of fruit maturity, maceration length, and ethanol amount on chemical and sensory properties of merlot wines. *Am. J. Enol. Vitic.* 64: 437-449.

Chatonnet, P., Barbe, C., Canal-Llauberes, R.M., Dubourdieu, D. and Boldron, J.N. (1992). Incidence de certaines preparations pectolytiques sur la teneur en phenols volatils des vins blancs. *J. Int. Sci. Vigne Vin.* 26(4): 253.

Chatonnet, P., Boidron, J.N. and Dubourdieu, D. (1993). Influence of aging and sulphuring conditions of red wines stored in barrels on their content in acetic acid and ethylphenols. *J. Intern. Sci. Vigevin.* 27: 277.

Chisholm, M.G., Guiher, L.A. and Zaczkiewicz, S.M. (1995). Aroma characteristics of aged Vidal blanc wine. *Am. J. Enol. Vitic.* 56(1): 56.

Clingeleffer, P.R., Kerridge, G.H. and Possingham, J.V. (1986). Effect of variety on wine quality. *In*: Proceedings of the Sixth Australian Wine Industry Technical Conference, held at Adelaide, South Australia. 14-17 July 1986, p. 78.

Coombe, B.G. (1987). Influence of temperature composition and quality of grapes. *Acta Hortic.* 206.

Cootes, R.L. (1983). Grape juice aroma and grape quality assessment used in vineyard classification. *In*: Proceeding of the Fifth Australian Wine Industry Technical Conference. T.A. Lee, T.C. Somers and W.A. Perth (Eds.). Advances in Viticulture and Oenology for Economic Gain. 29 November-1 December 1983. Adelaide, SA, p. 275.

Croser, B.J. (1983). Flavour retention and enhancement in table wines: A winemaker's view point. *In*: Contenary Grape and Wine Symposium. 23 May 1983. Roseworthy S.A., p. 122.

De Blij, H.J. (1983). Geography of viticulture: Rationale and resource. *J. Geog.*, 112.

De Boubee, R.D., Leeuwen, C. and Dubourdieu, D. (2000). Organoleptic impact of 2-methoxy-3-isobutyl pyrazine on red Bordeaux and Loir wines: Effect of environmental conditions on concentration in grape during ripening. *J. Agri. Food Chem.* 48: 4830.

de Mora, S.J., Eschenbruch, R., Knowles, S.J. and Spedding, D.J. (1986). The formation of dimethyl-sulphide during fermentation using a wine yeast. *Food Microbiol.* 3: 27.

deBoubee, D.R., Cumsille, A.M., Pons, M. and Dubourdieu, D. (2002). Location of 2-methoxy-3-isobutylpyrazine in Cabernet Sauvignon grape bunches and its extractability during vinification. *Am. J. Enol. Vitic.* 53(1): 1.

Delcroix, A., Gunata, Y.Z., Sapis, J.C., Salwon, J.M. and Bayonove, C. (1994). Glycosidase activities of three ecological yeast strains during winemaking: Effect on the terpenol content of Muscat wine. *Am. J. Enol. Vitic.* 45: 291.

Diemar, W. and Schams, E. (1960). Gaschromatographie in der Lebensmittelanalktik. *Z. Lebensm-untersuch Forsch* 112: 457.

Dimitriads, E. and Strauss, C.R. (1985). The actinidois: Nor isoprenoid compounds in grapes, wines and spirits. *Phytochemistry*, 24: 767.

Dobourdieu, D.P., Darriet, C., Ollivier, J.N. Boidron and Ribereau Gayon (1988). Role de la levure Saccharomyces cerevisiae dans l' hydrolyse des heterosides lepeniques du jus de raisin. *C.R. Sc.* Paris, Series 111 306: 489.

Drake, S.R. and Nelson, S.W. (1986). Apple juice quality as influenced by ultrafiltration. *J. Food Qual.* 9: 399.

Dubois, P., Etievant, P., Dekimpe, J., Buret, M., Chambroy, Y. and Flanzy, C. (1977). Etude sur l'arome des vins de maceration carbonique. *C.R. Acad. Agric.* 63: 1183.

Duchêne, E. and Schneider, C. (2005). Grapevine and climatic changes: A glance at the situation in Alsace. *Agron. Sustain. Dev.* 24: 93-99.

Ducruet, J., An, D., Canal-Llauberes, R.M. and Glories, Y. (1997). Influence des enzymes pectolytiques selectionnees pour l, oenologie sur la qualite et la composition des vins rouges. *Rev. Fr. Oenol.* 166: 16.

Ducruet, J., Glories, Y. and Canal-Llauberes, R.M. (2000). Etude de 1' influence d'une preparation enzymatique de maceration sur le vin et le raisin rouge. *Rev. Oenologues* 97: 15.

Ducruet, V. (1984). Comparison of the headspace volatile of carbonic maceration and traditional wine. *Lebensm. Wiss. Technol.* 17: 217.

Ducruet, V., Flanzy, C., Bourzeix, M. and Chambroy, Y. (1983). Les constituants volatiles des vins jeunes de maceration carbonique. *Sci. Ailments.* 3: 413.

Dugelay, I., Gunata, Z., Sapis, J.C., Baumes, R. and Bayonove, C. (1993). Role of cinnamoyl esterase activities from enzymes preparations on the formation of volatile phenols during winemaking. *J. Agric. Food Chem.* 41: 2092.

Eschenbruch, R. and Van Dam, T.G.J. (1985/86). Small-scale Winemaking at Te Kauwhata. *South. Hortic. Grape Grower Wine Maker* 3: 79.

Etievant, P.X. (1981). Volatile phenol determination in wine. *J. Agric. Food Chem.* 29: 65.

Ewart, A.J.W., Eschenbruch, R. and Fisher, B.M. (1984). The determination of monoterpenes in grape juice. Oenological and Viticultural Bulletin No. 45. Te Kauwhata, New Zealand. Te Kauwhata Research Station.

Flanzy, C.M., Flanzy, M. and P. Benard (1990). La vinification par maceration carbonique (Ed.), Madrid Vicente, INRA, Paris, p. 143.

Flores, J.H., Heatherbell, D.A., Henderson, L.A. and McDaniel, M.R. (1991). Ultrafiltration of wine: Effect of ultrafiltration on the aroma and flavour characteristics of white Reisling and Gewurztraminer wines. *Am. J. Enol. Vitic.* 42(2): 91.

Fontoin, H., Saucier, C., Teissedre, P.L. and Glories, Y. (2008). Effect of pH, ethanol and acidity on astringency and bitterness of grape seed tannin oligomers in model wine solution. *Food Qual. Preference* 19: 286-291.

Francis, I.L., Sefton, M.A. and Williams, P.J. (1994). The sensory effects of pre- or post-fermentation thermal processing on Chardonnay and Semillon wines. *Am. J. Enol. Vitic.* 45: 243.

Francis, I.L., Sefton, M.A. and Williams, P.J. (1992). Sensory descriptive analysis of hydrolyzed precursor fractions from Semillon, Chardonnay and Sauvignon Blanc grape juices. *J. Sci. Food. Agric.* 59: 511.

Frost, M.B. and Noble, A.C. (2002). Preliminary study of the effect of knowledge and sensory experitise on liking for red wines. *Am. J. Enol. Vitic.* 53(4): 275.

Gaillard, M. (1985). Resultats de L' observation du fonctionnement d' un appareil d' ultrafiltration industriel wine processor Societe I.D.E. *In*: ITV Proceedings: Ultrafiltration et Microfiltration Tangentielle en Oenology. Institut Technique de La Vigne et du Vin. 21, rue Farncois 1er 75008 Paris, p. 109.

Galletto, W.G., Webb, A.D. and Kepner, R.E. (1966). Identification of some acetals in an extract of submerged culture flor sherry. *Am. J. Enol. Vitic.* 17: 11.

Gao, L., Girard, B., Mazza, G. and Reynolds, A.G. (1997). Simple and polymeric anthocyanins and colour characteristics of Pinot noir wines from different vinification processes. *J. Agric. Food Chem.* 45: 2003.

Gawel, R. (1998). Red wine astringency: A review. *Aust. J. Grape Wine Res.* 4: 74.

Gerbaux, V., Vincent, B. and Bertrand, A. (2002). Influence of maceration temperature and enzymes on the content of volatile phenols in Pinot noir wines. *Am. J. Enol. Vitic.* 53(2): 131.

Girard, B., Fukumoto, L., Mazza, G., Delaquis, P. and Ewert, B. (2002). Volatile terpene constituents in maturing Gewurztraminer grapes from British Columbia. *Am. J. Enol. Vitic.* 53(2): 99.

Girard, B., Kopp, T.G., Reynolds, A.G. and Cliff, M. (1997). Influence of vinification treatments on aroma constituents and sensory descriptors of Pinot noir wines. *Am. J. Enol. Vitic.* 48(2): 198.

Glaldstones, J. (1992). Viticulture and Environmental. Winetiltes Adelaide. 310.

Gomez, M.I., Lacueva, A., Raventos, R.M.L., Luque, C.L., Buxanderas, S. and de la Torre-Buxaderas, S. and de la Torre-Boronat, M.C. (2001). Differences in phenolic profile between oakwood and stainless steel fermentation in white wines. *Am. J. Enol. Vitic.* 52(2): 159.

Guerrand, D. (2000). Preparations enzymatiques: Profils d' activite et performances. *Rev. Fr. Oenol.* 183: 19.

Gunata, Y.Z., Bayonove, C.L., Baumes, R.L. and Cordonnier, R.E. (1986). Stability of free and bound fractions of some aroma components of grapes cv. Muscat during the wine processing. *Am. J. Enol. Vitic.* 37: 112.

Hagerman, A.E. and Butler, L.G. (1978). Protein precipitation method for the quantitative determination of tannins. *J. Agric. Food Chem.* 26: 809.

Harbertson, J.F., Kennedy, J.A. and Adams, D.O. (2002). Tannin in skins and seeds of Cabernet Sauvignon, Syrah and pinot noir berries during ripening. *Am. J. Enol. Vitic.* 53(1): 54.

Hewitt, E.J., Mackay, D.A.M., Konigsbacher, K. and Hasselstrom, T. (1959). Art. *Food Technol.* 10: 487.

Heymann, H., Noble, A.C. and Boulton, R.B. (1986). Analysis of methoxypyrazines in wines: I. Development of a quantitative procedure. *J. Agric. Food Chem.* 34: 268.

Intrigliolo, D.S. and Castel, J.R. (2008). Effects of Irrigation on the performance of grapevine cv. Tempranillo in Requena. *Am. J. Enol. Vitic.* 59: 30-38.

Jackson, D.I. and Lombard, P.B. (1993). Environmental and management practices affecting grape composition and wine quality – A review. *Am. J. Enol. Vitic.* 44: 409-430.

Jones, G.V. (1997). A synoptic climatological assessment of viticultural phenology. Dissertation. Dept. of Environ. Sci., University of Viginia, 394.

Jones, G.V. and Davis, R.E. (2000). Using a synoptic climatological approach to understand climatic viticulture relationships. *Inter. J. Clin.* 20: 813.

Kenworthy, D. (1984). Approaches to grape quality assessment at Simi Winery. Oenological and Viticultural Bulletin No. 43. Kauwhata Research Station, Te Kauwhata, New Zealand.

Kepner, R.E., Webb, A.D. and Muller, C.J. (1972). Identification of 4-hydroxy-3-methyloctanoic acid and gamma-lactone [5-butyl-4-methyldihydro-2(3H)-furanone] as a volatile component of oak-wood-aged wines of *V. vinifera* var. Cabernet Sauvignon. *Am. J. Enol. Vitic.* 23: 103.

King, E.S. and Heymann, H. (2014). The effect of reduced alcohol on the sensory profiles and consumer preferences of white wine. *J. Sens. Stud.* 29: 33-42.

Kontoudakis, N., Esteruelas, M., Fort, F., Canals, J.M. and Zamora, F. (2011). Use of unripe grapes harvested during cluster thinning as a method for reducing alcohol content and pH of wine. *Aust. J. Grape Wine Res.* 17: 230-238.

Lacey, M.J., Allen, M.S., Harris, R.L.N. and Brown, W.V. (1991). Methoxypyrazine in Sauvignon blanc grapes and wines. *Am. J. Enol. Vitic.* 42(2): 103.

Larue, F., Rozes, N., Froudiere, I., Couty, C. and Perreira, G.P. (1991). Incidence du developpment de Dekkera/Brettanomyces dans les mouts et les vins. *J. Int. Sci. Vigne Vin.* 25(3): 149.

Lawless, H.T. (1984). Flavour description and white wine by expert and non-expert wine consumers. *J. Food Sci.* 49: 120.

Lisanti, M.T., Gambuti, A., Piombino, P., Pessina, R. and Moio, L. (2011). Sensory study on partial dealcoholisation of wine by osmotic distillation process. *Bull. OIV* 84: 95-105.

Lopez, R.M., Libbey, L.M., Watson, B.T. and McDaniel, M.R. (1992). Identification of additional odor-active compounds in Pinot noir. *Am. J. Enol. Vitic.* 43(1): 90.

Maga, J.A. and Sizer, C.E. (1973). Pyrazines in foods: A review. *J. Agric. Food Chem.* 21: 22.

Makkar, H.P.S. (1989). Protein precipitation methods for quantitation of tannins: A review. *J. Agric. Food Chem.* 37: 1197.

Makkar, H.P.S., Dawra, P.K. and Singh, B. (1988). Determination of both tannin and protein in a tannin-protein complex. *J. Agric. Food Chem.* 36: 523.

Marais, J. (1978). The effect of pH on esters and quality of Colombard wine during maturation. *Vitis*. 17: 396.

Marais, J. (1979). Effect of storage time and temperature on the formation of dimethyl sulphide and on white wine quality. *Vitis.* 18: 254.

Marais, J. (1983). Terpenes in the aroma of grapes and wines: A review. *S. Afr. J. Enol. Vitic.* 4: 49.

Marais, J., Van Wyk, J. and Rapp, A. (1992). Effect of storage time, temperature and region on the levels of 1,1,6-trimethyl–1,2-dihydronapthanlene and other volatiles, and on quality of Weisser Riesling wines. *S. Afr. J. Enol. Vitic.* 13: 33.

Mateo, J.J., Jimenez, M., Huerta, T. and Pastor, A. (1992). Comparison of volatiles produced by four Saccharomyces cerevisiae strains isolated from Monostrell musts. *Am. J. Enol. Vitic.* 43: 206.

Mc Daniel, R.M., Miranda-Lopez, R., Watson, B.T., Micheals, N.J. and Libbey, L.M. (1989). Pinot noir aroma: A sensory/gas chromatographic. pp. 22. *In*: G. Charalambous (Ed.). 6th International Flavour Conference. Crete, Greece.

McMahon, H.M., Zoeklein, B.W. and Jasinski, Y.W. (1999). The effects of pre-fermentation maceration temperature and percent alcohol (v/v) at press on the concentration of Cabernet Sauvignon grape glycosides and glycoside fractions. *Am. J. Enol. Vitic.* 50: 385.

Meglioli, G. and Marchesini, C. (1985). Ultrafiltration: Its winemaking applications (trans.). *Enotecnico* 21(1): 39.

Muller, C.J., Kepner, R.E. and Webb, A.D. (1972). Identification of 4-ethoxy-4- hydroxybutyric acid α-lactone (5-ethoxylihydro-2(3H)-furanone) as an aroma component of wine from *Vitis vinifera* var. Ruby Cabernet. *J. Agric. Food Chem.* 20: 193.

Muller, C.J., Richard, K.E. and Webb, A.D. (1973). Lactones in wines: A review. *Am. J. Enol. Vitic.* 24(1): 4.

Murat, M.L., Masneuf, I., Darriet, P., Lavigne, V., Tominaga, T. and Dubourdieu, D. (2001). Effect of *Saccharomyces cerevisiae* yeast strains on the liberation of volatile thiols in Sauvignon blanc wine. *Am. J. Enol. Vitic.* 52(2): 136.

Murray, K.E., Shipton, J. and Whitfield, F.B. (1975). The occurrence of 3-alkyl-2-methoxypyrazines in raw vegetables. *J. Sci. Food Agric.* 26: 973.

Noble, A.C. (1994). Wine flavour. pp. 228. *In*: J.R. Piggott and A. Paterson (Eds.). Understanding Natural Flavours. Blackie Academic and Professioal, London.

Nykanen, L. (1986). Formation and occurrence of flavour compounds in wine and distilled alcoholic beverages. *Am. J. Enol. Vitic.* 37: 89.

Ough, C.S. and Amerine, M.A. (1960). Experiments with controlled fermentations IV. *Am. J. Enol. Vitic.* 11: 5.

Ough, C.S. and Amerine, M.A. (1961). Studies with controlled fermentation VI: Effects of temperature and handling on rates, composition and quality of wines. *Am. J. Enol. Vitic.* 12: 117.

Ough, C.S. and Amerine, M.A. (1967). Studies with controlled fermentation X: Effect of fermentation temp. on some volatile compounds in wine. *Am. J. Enol. Vitic.* 18: 157.

Park, S.K., Boulton, R.B. and Noble, A.C. (2000). Formation of hydrogen sulphide and glutathione during fermentation of white grape musts. *Am. J. Enol. Vitic.* 51(2): 91.

Pontallier, P., Salogoity-Auguste, M.H. and Ribereau-Gayon, P. (1982). Intervention du bois de chene dans l'elevage des vins runges eleves en barriques, Connaiss. *Vigne Vin.* 1: 45.

Prieur, C., Rigaud, J., Cheynier, V. and Moutounet, M. (1994). Oligomeric and polymeric procyanidins from grape seeds. *Phytochemistry* 36: 781.

Puech, J.L. (1981). Extraction and evolution of ligmin products in Armagnac matured in oak. *Am. J. Enol. Vitic.* 32: 111.

Quirós, M., Rojas, V., Gonzalez, R. and Morales, P. (2014). Selection of non-*Saccharomyces* yeast strains for reducing alcohol levels in wine by sugar respiration. *Int. J. Food Microbiol.* 181: 85-91.

Rapp, A. (1988). Wine analysis. pp. 29. *In*: H.F. Linskens and J.F. Jackson (Eds.). Modern Methods of Plant Analysis. New Series, vol. 6. Springer-Verlag, Berlin.
Rapp, A. and Mandery, H. (1986). Wine aroma. *Experientia* 42: 873.
Rapp, A. and Guntert, M. (1986). Changes in aroma substances during the storage of white wines in bottles. *Dev. Food Sci*. 12: 141.
Reynolds, A., Cliff, M., Girard, B. and Kopp, T.G. (2001). Influence of fermentation temperature on composition and sensory properties of Semillon and Shiraz wines. *Am. J. Enol. Vitic*. 52(3): 235.
Robinson, L., Anthony, Boss, K., Paul, Heymann, Hildegarde, Solomon, Peter, Trengove, Robert (2011). Influence of yeast strain, canopy management and site on the volatile composition and sensory attributes of Cabernet Sauvignon Wines from Western Australia. *Journal of Agricultural and Food Chemistry* 59: 3273-3284.
Sachs, S.B., Gaillard, M. and Cassignard, R. (1983). Application of tangential filtration to wine processing (trans.). *Vigne Vin*. 320: 23.
Salinas, M.R., Navarro, G., Alonso, G., Esteban-infantes, F.J. and Pardo, F. (1992). Constityentes volatiles de un vino de maceracion carbonica y de veniificacion tradicional. *Vitivinicultura* 22: 46.
Salinas, M.R., Alonso, G.L., Navarro, G., Pardo, F., Jimano, J. and Huerta, M.D. (1996). Evolution of the aromatic composition of wines undergoing carbonic maceration under different aging conditions. *Am. J. Enol. Vitic*. 47(2): 134.
Schreier, P. (1984). Formation of wine aroma. pp. 9. *In*: L. Nykanen and P. Lehtonen (Eds.). Flavour Research of Alcoholic Beverages – Instrumental and Sensory Analysis. Foundation for Biotechnical and Industrial Fermentation Research. Helsinki.
Schreier, P., Drawert, F. and Junker, A. (1976). Identification of volatile constituents from grapes. *J. Agric. Food Chem*. 24: 331.
Sefton, M.A., Francis, I.L. and Williams, P.J. (1993). The volatile composition of Chardonnay juices: A study by flavour precursor analysis. *Am. J. Enol. Vitic*. 44(4): 359.
Simpson, R.F. (1978). Aroma and compositional changes in wine with oxidation, storage and ageing. *Vitis* 17: 274.
Simpson, R.F. (1979). Aroma composition of bottle-aged white wine. *Vitis* 18: 148.
Simpson, R.F. and Miller, G.C. (1983). Aroma composition of aged Riesling wine. *Vitis* 22: 51.
Simpson, R.F. and Miller, G.C. (1984). Aroma composition of Chardonnay wine. *Vitis* 23: 143.
Simpson, R.F., Strauss, C.R. and Williams, P.J. (1977). Vitispirane: A $C_{13}$ spiro-ether in the aroma volatiles of grape juice, wines and distilled grape spirits. *Chem. Industr*. 663.
Singleton, V.L. (1962). Aging of wines and other spiritous products, accelaration by physical treatments. *Hilgardia* 32: 319.
Singleton, V.L., Sieberhagen, H.A., Dewet, P. and Van Wyk, C.J. (1975). Composition and sensory qualities of wines prepared from white grapes by fermentation with and without grape solids. *Am. J. Enol. Vitic*. 26: 62.
Sinohara, T.L. (1985). Importance des substances volatiles du vin.formation et. effects sur la qualite. *Bull. OIV* 641/642: 608.
Slingsby, R.W., Kepner, R.E., Muller, C.J. and Webb, A.D. (1980). Some volatile components of *Vitis vinifera* var. Cabernet Sauvignon wine. *Am. J. Enol. Vitic*. 31: 360.
Somers, T.C. and Pocock, K.F. (1990). Evolution of red wines, III: Promotion of the maturation phase. *Vitis* 29: 21.
Somers, T.C. and Wescombe, L.G. (1987). Evolution of red wines, II: An assessment of the role of acetaldehyde. *Vitis* 26: 27.
Strauss, C.R., Wilson, B., Gooley, P.R. and Williams, P.J. (1986). Biogeneration of aroma. *In*: T.H. Parliament and R.B. Croteau (Eds.). ACS Symposium Series 317. American Chemical Society, Washington, DC. pp. 222.
Strauss, C.R., Wilson, B., Anderson, R. and Williams, P.J. (1987). Development of precursors of $C_{13}$ norisoprenoid flavorants in Riesling grapes. *Am. J. Enol. Vitic*. 38: 23.
Suomalainen, H. (1971). Yeast and its effect on the flavour of alcoholic beverages. *J. Inst. Brew*. 77: 164.
Teranishi, R. (1989). Flavour chemistry trends and developments. *In:* R. Teranishi, R.G. Buttery, F. Shahidi (Eds). ACS Symposium Series No. 388. American Chemical Society, Washington, DC. p. 1.

Tesniere, G., Baume, R., Bayonove, C. and Flanzy, C. (1989). Effect of simulated alcoholic fermentation on aroma components of grape berries during anaerobic metabolism. *Am. J. Enol. Vitic.* 4: 183.

Testaniere, D. (1985). Resultants d'essai d'ultrafiltration de blanc appare industriel: Millipore. *In*: ITV Proceedings: Ultrafiltration et microfiltration tangentielle en œnologie. Institute Technique de la vigne et al., Vin., Rue Francois 1, 75008 Paris. p. 131.

Tominaga, T., Furrer, A., Henry, R. and Dubourdieu, D. (1998). Identification of new volatile thiols in the aroma of *Vitis vinifera* L. var. Sauvignon Blanc wines. *Flavour Fragrance J.* 13: 159.

Tominaga, T., Murat, M.L. and Dubourdieu, D. (1998a). Development of a method for analysing the volatile thiols involved in the characteristic aroma of wines made from *Vitis vinifera* L. cv. Sauvignon Blanc. *J. Agri. Food Chem.* 46: 1044.

Tominaga, T., Guyot, R.B., Gachons, C.P.D. and Dubourdieu, D. (2000). Contribution of volatile thiols to the aromas of white wines made from several *Vitis vinifera* grape varieties. *Am. J. Enol Vitic.* 51(2): 178.

Urbach, G., Stark, W. and Forss, D.A. (1972). Volatile compounds in butter oil, II: Flavour and flavour thresholds of lactones, fatty acids, phenols, indole and skatole in deodorised synthetic butter. *J. Dairy Res.* 39: 35.

Villetaz, J.C. (1995). Utilisation des enzymes oenologiques pour I' extraction de la couleur et pour I' extraction et la revelation des aromes. Presented at the 21 eme congres de la Vigne et du vin d'OIV, Puenta del Este., Uruguay.

Webb, A.D. and Muller, C.J. (1972). Volatile aroma components of wines and other fermented beverages. *In*: D. Perlman (Ed.). Advances in Applied Microbiology. Academic Press, New York, 15: 75.

Webb, A.D., Kepner, R.E. and Galetto, W.G. (1964). Comparison of the aromas of flor sherry, baked sherry and submerged culture sherry. *Am. J. Enol. Vitic.* 15: 1.

Webb, A.D., Kepner, R.E. and Maggiora, L. (1967). Sherry aroma, VI: Some volatile components of flor sherry of Spanish origin: Neutral substances. *Am. J. Enol. Vitic.* 18: 190.

Webb, A.D., Kepner, R.E. and Maggiora, L. (1969). Some volatile components of wines of *Vitis vinifera* varieties Cabernet Sauvignon and Ruby Cabernet – I: Neutral compounds. *Am. J. Enol. Vitic.* 20: 16.

Webb, A.D., Riberau-Gayon, P, and Boidron, J.N. (1964). Composition d'une essence extralite d'un Vin de *V. vinifera* (*variete* Cabernet Sauvignon). *Bull. Soc. Chem.*, France 1415.

Whiton, R.S. and Zoecklein, B.W. (2000). Optimisation of headspace solid-phase microextraction for analysis of wine aroma compounds. *Am. J. Enol. Vitic.* 51: 379.

Wightman, J.D., Price, S.F., Watson, B.T. and Wrolstad, R.E. (1997). Some effects of processing enzymes on anthocyanins and phenolics in Pinot noir and Cabernet Sauvignon wines. *Am. J. Enol. Vitic.* 48(1): 39.

Williams, P.J. (1993). Hydrolytic flavour release in fruit and wines through hydrolysis of non-volatile precursors. *In*: T.E. Acree and R. Teranishi (Eds.). Flavour Sciences: Sensible Principles and Techniques. ACS Professional Reference Book Series. American Chemical Society, Washington, DC.

Williams, P.J., Strauss, C.R. and Wilson, B. (1980a). New linalool derivatives in Muscat of Alexandria grapes and wines. *Phytochemistry*, 19: 1137.

Williams, P.J., Strauss, C.R. and Wilson, B. (1980b). Hydroxylated linalool derivatives as precursors of volatile monoterpenes of Muscat grapes. *J. Agric. Food Chem.* 28: 766.

William, P.J., Strauss, C.R., Aryan, A.P. and Wilson, B. (1986). Grape flavour: A review of some pre- and postharvest influences. pp. 111. *In*: Terry Lee (Ed.). Proceedings of Sixth Australian wine Industry Conference held at Adelaide, South Australia. 14-17 July, 1986.

Williams, P.J., Strauss, C.R., Wilson, B. and Massy-Westropp, R.A. (1982). Use of $C_{18}$ reversed phase liquid chromatography for the isolation of monoterpene glycosides and nor-isoprenoid precursors from grape juice and wines. *J. Chromatogr.* 235: 471.

Wilson, E.L., Hogg, M.C. and Burns, D.J.W. (1984). Sensory evaluation of clarified kiwifruit juices. *J. Food Qual.* 7(1): 43.

Winterhalter, P., Sefton, M.A. and Williams, P.J. (1990). Volatile $C_{13}$-norisoprenoid compounds in Riesling wine are generated from multiple precursors. *Am. J. Enol. Vitic.* 41: 277.

# 7 *Botrytis* and Wine Production

**Monika Cioch[1*], Dorota Semik-Szczurak[1] and Szymon Skoneczny[2]**

[1] Department of Fermentation Technology and Technical Microbiology, Faculty of Food Technology, Balicka 122, 30-149 Krakow, Poland

[2] Department of Chemical and Process Engineering, Cracow University of Technology, Warszawska 24, 31-155 Krakow, Poland

## 1. Introduction

During the ripening process of grapes, the fruits and the plant are exposed to many risk factors which can result in destruction of the bush or significant decreases in the quality and yield of the fruits. The noble mould *Botrytis* (grey mould) is a dream of connoisseurs. Thanks to it, the most noble sweet wines are created, characterised by a unique taste and aroma, but most often its presence means a lot of trouble for winemakers. Even a small percentage of grapes infected with grey mould has a profound effect on the wine aroma. In order to obtain a noble mould, strict conditions are required. The perfect conditions for its development are created by foggy mornings and sunny afternoons, during which water slowly evaporates and the fruit skin dries and wrinkles. As a result of this drying, raisin with a unique concentration of all flavours is obtained. The noble mould does not destroy varietal odours, but highlights them. As a result of this, there are rich, aromatic and long-lasting white wines are produced, like Hungarian Tokay, French Sauternes and German Trockenbeerenauslese.

Tokaj has a shorter history than many other wine regions in Hungary. The first vineyards of the Tokaj region were recorded in the twelfth century. Aszú wines have been made in the Tokaj Mountains since the late sixties of the 16th century. By the middle of the 16th century, Tokaj became the most prosperous wine region in Hungary. In the 18th century, regulations governing the production of wines were introduced, the boundaries of crops were established and the classification of vineyards was made. The Tokaj vineyards, through many centuries, were put to a severe test (including phylloxera) and survived. There are unique throughout the world conditions in this wine region. Above all, these include the variety of volcanic soils that allows producing extraordinary original wines and continental climate creating favourable conditions for a late harvest. The traditional way of Aszú wine making is worthy to cherish and to preserve for the future.

There are two faces of *Botrytis cinerea* in viticulture. The grey mould, i.e. the rot of the vine fruit caused by the *B. cinerea* fungus, is a disease that causes significant economic losses worldwide. In all areas of the cultivation of *Vitis vinifera* L., this fungus can cause significant damage manifested by a drastic decrease in the yield of berries and significantly affect the quality of wine produced from the infected fruits. The presence of *B. cinerea* on berries gives a characteristic aroma to grape musts. Aromatic components, including terpenes, are destroyed, but causes phenolic aromas that characterize some botrytised wines. The *Botrytis* fungus also synthesises esterase, which can decompose fruit esters that impart character to white wines. Pathogen *B. cinerea* produces complex polysaccharides. The fermentation of must from botrytised berries is difficult due to the high level of sugar in the grape juice. Due to the fact that the fruits of the grapevine lose water, the juice extraction is complicated. The contact of the peels with the must during maceration makes it easier to extract the juice, but also causes the glucan to enter the must, which causes difficulty in wine filtration, while heteropolysaccharides inhibit alcoholic fermentation. They can cause increase in the concentration of acetic acid and glycerol in wine. *B. cinerea* also uses a significant portion of available nitrogen in grapes and grape must and amino acids.

The other face of *Botrytis* infection in grape berries is related with fermentation of berries for wine production especially wine with characteristic aroma and flavour. The fungus *B. cinerea* can contribute to the creation of high-quality wines. Sweet wines made from grapes attacked by *B. cinerea* are called

*Corresponding author: monika.cioch@urk.edu.pl

'botrytised' wines. They are characterised by beneficial health properties. The most significant components present in them are polyphenols. Botrytised wines contain a much higher content of polyphenols than wines obtained from fruit not infected by *B. cinerea*. Resveratrol is thought to play a key role in wine. It occurs naturally in grapes and in most red wines but is not practically identified in dry white wines. It has a wide range of biological activity and has many pro-health properties. These include, above all, antioxidant and anti-inflammatory effects, which are important in the prevention of cardiovascular, neoplastic and nervous system diseases.

This chapter presents an overview of the most important information about the presence of *Botrytis cinerea* during vine cultivation and wine production. The life cycle of the fungus, changes in fruit, juice and wine composition caused by *B. cinerea*, pathogen detection, protection against grey mould, influence of fungus on grapes microflora and impact on human health were described. However, the main focus was on wines made from the noble rot grapes.

## 2. *Botrytis cinerea*: The Cause of Grey Mould Disease

Grey mould is a vine fruit rot caused by *Botrytis cinerea*, Pers. (Teleomorph *Botryotinia fuceliana*). This disease brings heavy, economically significant, losses across the world (Ellison *et al*., 1997; Viret *et al*., 2004). In 1950-2000, minimising the possible development of *B.cinerea* mould was largely based on the use of synthetic chemicals (Rosslenbroich and Stuebler, 2000). However, tight regulations limiting the use of chemicals have been imposed in view of the ease of chemical agents spreading and fungus resistance to them, as well as increased control over human health and environmental conditions (Latorre *et al*., 2002;Sergeeva *et al*., 2002; Leroux, 2004; Spadoro and Gullino, 2005; Elmer and Regliski, 2006). This fungus can cause significant damage in all the wine-growing zones as can be seen in a decrease in berry yield and which make a decisive contribution to the quality of wines produced from grapes infected with *B. cinerea* (Ellis, 2004). This pathogen disrupts mainly the ageing tissue of dicotyledonous plants and causes serious losses during the harvest of apparently healthy crops (Williamson *et al*., 2007). Although, in most of the traditional wine-producing countries like France, Italy, Chile and South Africa, moderate damage is usually reported, under favourable conditions for fungal growth very rapid outbreaks of disease may occur (Pearson *et al*., 1989; Ash *et al*., 1998).

*B. cinerea* requires high humidity (70-100 per cent and moderately high temperatures (15-25°C)) for its growth. The fungus does not develop well on the unripe vine fruit. However, it can penetrate the berries earlier, through the inflorescence remains in grapes and the scars left behind by dropped caps. Such infections might remain latent until the fruit ripens (Viret *et al*., 2004; Adamczewska-Sowińska *et al*., 2016). The latent form becomes active in cool climate where air humidity is higher which affects mycelium activation. Dampness during both phases, flowering and the pre-harvest period, promotes the growth of *B. cinerea*. (Viret *et al*., 2004; Jackson, 2008). A rainy weather favours development of spores on the berries and the infection appearance.

## 3. Infection of Leaves, Shoots, Flowers and Fruits

*B. cinerea* can infect all the plant organs. However, the greatest damage and economic loss that it can cause is related to the fruit infestation. During long rainy seasons, the fungus might infect parts of young stems, leaves and flowers that have been damaged by climatic factors such as hail, etc. The disease symptoms are relatively observed on the young leaves and shoots. They are more likely to appear on the unripe fruits (Evans and Emmett, 2013; Adamczewska-Sowińska *et al*., 2016) than the ripe fruits.

In early spring, the pathogen can grow on young leaves. Initially, it does not cause any darkening of the damaged tissue. The young shoots are likely to be infected as a result of which they become brown and wither away (Pearson, 1989). The fungal disease can cause a decrease in lignification of the vine shoots which, in turn, makes them more sensitive to freezing cold and frost during winter. Immediately prior to flowering, reddish, irregular spots (tissue necrosis), surrounded by yellow-green border may appear on young leaves. Then, this part of the plant dries out, becomes brown and is colonised by a delicate mycelium and conidia (Holz *et al*., 2004; Choquer *et al*., 2008).

Inflorescences infestation can occur at various stages of development, but it is most dangerous during the full bloom period. The spores penetrate into flowers through scars from the drop-out caps formed by five connate petals. When the first caps start to fall, this is the beginning of vine flowering. Some flowers infested by *B. cinerea* do not show any exogenous symptoms. It is due to the in-growth of mycelium to the receptacle and the transition to the latent phase (Elmer and Michailides, 2004; Viret *et al.*, 2004; Elmer and Reglinski, 2006; Kretschmer *et al.*, 2007).

The latent phase lasts a long time, predominantly from late April to the beginning of berry ripening. At this time, the vine berries are resistant to the infestation, which is associated with the production of immune components, such as resveratrol, glycolic acid, catechins, proanthocyanidins (Goetz *et al.*, 1999; Viret *et al.*, 2004; Pezet *et al.*, 2004; Elmer and Reglinski, 2006; Kretschmer *et al.*, 2007). The mycelium can activate its growth when berries start to ripe. In oenological literature, the time of immature grapes discolouration during initial period of ripening is termed veraison (Fillion *et al.*, 1999; Pratelli *et al.*, 2002; Cioch and Tuszyński, 2014). The beginning of this process is characterised by physical and biochemical changes in fruits. One of the most important factors among them is rapid reduction of acidity, which is linked directly to the higher content of sugars in berries. In addition, during the fruit development, the intensity of respiration decreases. Following ripening, the pathogen load in berries rises repeatedly under favourable conditions. With the infection development, the amount of fruit sugar that is the medium for *B. cinerea*, increases (Doehlemann *et al.*, 2005, 2006; Choquer *et al.*, 2008). The disease can intensify significantly if grape clusters are damaged by insects or because of weather. In years of favourable conditions for grey mould growth (high temperature, precipitation), precious fruit ripening or even drying of berries can occur (Ellis, 2004). *B. cinerea* is capable of producing laccase (benzenediol: oxygen oxidoreductase; EC 1.10.3.2) leading to the wine quality reduction. It is the enzyme that exhibits strong oxidising properties, causing browning of the red wines. This disadvantage can be eliminated but it requires expensive technologies. A secondary fungal invasion might be associated with spores produced by mycelium that resumed their growth after latent period, as well as germs from adjacent infectious grapes, and spreading from the other close located vineyards (Claire, 2000; Viret *et al.*, 2004). The fruit rotting occurs very often during the ripening stage. Grape varieties characterised by compact clusters tend to spoil easily. It is therefore, important to create a dry microclimate in the vineyard (higher shrubs, fewer shoots, scaffoldings for vine branches, careful cutting during vegetation, etc.).

## 4. Noble Rot Infection Process

*B. cinerea* can infect all vine organs, but its greatest harmfulness is associated with fruit infection. Although the pathogen does not grow well in immature berries, it can get into fruits earlier, for example, through the withering parts of the flowers or remnants of inflorescences remaining in the clusters. Such infections remain in latent form and the mycelium can activate when the berries ripen. This fungus thrives in conditions of high humidity and in the absence of airflow. Both humid period during flowering and pre-harvest period favours the growth of *Botrytis* and provides conditions for its further development.

### 4.1. Life Cycle of Mould

The *B. cinerea* life cycle begins with the infection of young, fast-growing tissues of plant and flower. This fungus is initially a pathogen thriving in living tissue and then subsequently becomes a saprophyte which feeds on dead plant material. The conidia, that initiate the primary infection, are produced by the mycelium which has overwintered on the vine shoots in the previous season (Pezet *et al.*, 2003). That spores can occur in the special sclerotia forms, that is a compact mass of fungal mycelium (Agrios, 2005). The conidia infect tissues covering shoots or leaves and then growing mycelium causes their slow dying by the production of enzymes and different metabolites. Tissue decay occurs in infected parts, while conidiophores with spores appear on the surface. In autumn, black sclerotia are created in these places. They form hard structures with diameter around 3 mm. In late winter and early spring, mainly in rainy seasons, sclerotia germinate and produce conidia, which are considered to be the main source of vine leaf infections (Gubler and Bettiga, 2012). Spores may also be derived from other infected plants or weeds present in the vineyard, which are hosts for this fungus (Williamson *et al.*, 2007; Mundy *et al.*, 2012). Studies show that insects and invertebrates can only be random vectors of the gray mould pathogen.

Although *B.cinerea* can be easily detected in the soil, it represents only a minor part of fungal biomass in that environment (Dorado *et al*., 2001).

The ability to infect many plant species and low genetic diversity make it difficult to reduce the source of infections caused by *B. cinerea*. Under these circumstances, the whole burden of preservation must be shifted to the prevention and control of infection, which unfortunately often involves the use of chemical fungicides (Ash *et al*., 1998; Coertze *et al*., 2001; Choquer *et al*., 2007; Jacometti *et al*., 2007; Kretschmer *et al*., 2007).

### 4.2. Germination

As the vast majority of fungi, *B.cinerea* spores also require some growth factors. In order to develop successfully, they must have a supply of water and nutrients, as well as the optimum temperature, so the grey mould can germinate and initiate infection. For example, at temperatures between 18 and 24°C, only two-hour free water supply is needed for germination and plant infestation, while at higher temperatures (above 27°C), spores need more time to develop. Nutrients may come from different sources. The main group of vital ingredients essential for fungal growth are monosaccharides, including glucose and fructose. Damage caused by insects or birds, as well as other mechanical fruit damage result in the release of juice, providing nutrients and water needed for germination. Certain ambient temperatures are also needed for this process to occur. Usually, higher temperatures might reduce spores development due to berries drying. Although, *B. cinerea* does not grow at temperatures exceeding 32°C, grey mould can develop in refrigerated conditions (even at 1°C) which contributes to infection during grape storage (Jackson, 2008; Gubler and Bettig, 2012). Vine has natural self-defence mechanisms against *B. cinerea* infection. The plant generates an outer protecting layer on the leafs and berries, called the cuticle and the second layer called epidermis which produce compounds with fungicidal properties, like resveratrol, as well as other phenolic compounds (Elad *et al*., 2007). Assuring appropriate microclimate conditions in vineyard is undoubtedly a significant factor in avoiding *Botrytis* infection. The land where vines are planted out should be well prepared to provide the best conditions. It requires vented and sunny location. Bushes must be properly exposed to light. Over-fertilisation of soil by nitrogen nutrients needs to be avoided. High nitrogen content in fruits makes plant more susceptible to the fungal infection (Claire, 2000). In culture aimed at obtaining *Botrytis* resistant varieties, the main factor to be taken into account during selection is the degree of berry density. Therefore, the selection of species with low berry density and the favourable structural factors of grapes are a great opportunity to overcome the problem of grey mould.

## 5. Changes in Fruit Composition Induced by *Botrytis cinerea*

Grey mould caused by *B. cinerea* is one of the most serious diseases of the vine. It lowers both the quantity and quality of fruits and leads to the modification of the chemical composition of grapes. This pathogen produces toxic compounds that cause inhibition of fermentation, which affects the quality of the final product – wine (Hong *et al*., 2011; Agudelo-Romero *et al*., 2015; Negri *et al*., 2017). From an economic point of view, *B. cinerea* infection might have very damaging consequences for growers and wine-makers. Its presence decreases the crop yield. Some reports show that in the region of Alsace (France), over the period 1976-1980, losses caused by grey mould reached 27 per cent. Taking into account average trends, it can be considered that 35 per cent of the infected berries after harvest affect the reduction of the red wine volume from 700 hl to 500 hl (Ky *et al*., 2012). The grapes attacked by *B. cinerea* show significant changes in chemical composition, depending largely on the nature and degree of fungus infection. During development, grey mould metabolizes 35 to 45 per cent of sugars present in the berries. The reduction of osmotic potential explains the higher sugar content in the infected fruits.

Additionally, the increasing concentration of glycerol is observed in wines obtained from berries damaged by the disease (1-10 g/l). It is a feature typical for the botrytised wines. Direct oxidation of glucose by glucose oxidase (β-D-glucose: oxygen 1-oxidoreductase, EC 1.1.3.4) leads to accumulation of gluconic acid in the fruit during the stationary phase of growth (1-5 g/l). Both glucose and glycerol are virtually undetectable in the healthy grapes. Their presence can therefore be assumed to be indicative of the *Botrytis* infestation. When acetic acid bacteria of the species *Gluconobacter* is present, the content of gluconic acid in berries increases. As acetic acid bacteria infect grapes attacked by *Botrytis*, they are probably responsible for most of the gluconic acid formation in the fruits (Sponholz and Dittrich, 1985). However, not only the fungus but also, the acetic acid bacteria can contribute to the appearance of these components in the vine berries. It has also been shown that the wild yeast activity may also be responsible for elevated glycerol and gluconic acid levels in grapes and, consequently, in wine (Sponholz *et al.*, 2004). These microorganisms can also produce acetic acid and ethyl acetate (Jackson, 2008).

Besides glycerol, other polyols, such as arabitol, erythritol, mannitol, sorbitol and xylitol might be synthesised. D-sorbitol and inositol accumulate in botrytis-infected grapes. Despite the metabolism of sugar, its content in berries increases due to the fruit dehydration. Depending upon climatic and geographic conditions, level of sugar may increase two to five times (Ribéreau-Gayon *et al.*, 2000). Its final concentration in aszú grapes may vary between 700 and 800 g/l. The sugar content is also modified by polysaccharide and pectin decomposition by the enzymes secreted by *B. cinerea*. They contribute to the accumulation of arabinose, rhamnose, galactose, mannose, xylose and galacturonic acid, which are partially oxidized to galactaric acid (mucic acid). Calcium salts of this acid precipitate in wine, especially in German botrytised wines and Tokaj (Magyar, 2011).

*B. cinerea* pathogen synthesises complex polysaccharides. The presence of glucans may make it difficult to filter wine, while heteropolysaccharides inhibit alcoholic fermentation and may give rise to higher acetic acid and glycerol content in the final product. Glucans are synthesized inside grapes and form a soft jelly between the flesh and the skin. Hence, gentle berry pressing prevents the release of large amounts of glucans into the grape must. The presence of glucans is not the only reason for the filtration problem. Relatively high alcohol content may also be associated with it. It can be assumed that alcohol increases the size of the colloidal polysaccharide aggregates. When the accumulated particles reach a sufficient size, they precipitate. This phenomenon occurs when an alcoholic strength is more than 17 per cent by vol. (Gayon *et al.*, 2006).

The processing of berries infected by *Botrytis* is the cause of the characteristic smell of the grape must. During mould development, aromatic ingredients, including terpenes, are destroyed. Linalool, geraniol and nerol are metabolized to less volatile compounds such as β-pinene, α-terpineol, as well as furan and pyran oxides (Bock *et al.*, 1988). These compounds may account for the phenolic aromas characterising some of the botrytised wines. In addition, fungus synthesizes also esterase, enzyme that can decompose fruit esters influencing the aroma of white wines. This feature is also dependent on the vine variety. For example, Muscat grapes often have less complex taste than Riesling and Semillon. *B. cinerea* may decompose 1, 1, 6-trimethyl-1, 2-dihydronaphthalenate (TDN) (Sponholz and Hühn, 1996). Besides, the presence of geosmin in grapes is also associated with gray mould occurrence (La Guerche, 2006). The increase in the content of iso-amyl alcohol and iso-butanol is attributed to the oxidative deamination of leucine and valine that are their free amino acid precursors (Cãmara *et al.*, 2006). Other volatile substances produced by the fungus are aldehydes (benzaldehyde, phenylacetaldehyde) and furfural (Sarrazin, 2007; Magyar, 2011; Pinar *et al.*, 2016). In addition, many lactones have been found in Tokaju Aszú (Miklósy and Kerényi, 2004; Miklósy *et al.*, 2004). These compounds are identified in old wines long aged with the access to oxygen. They are created in fruit due to the oxidative effect of *B.cinerea*, water loss or the Maillard's reaction (Miklósy *et al.*, 2004). It has also been shown that increased concentrations of homo-furonol, furanol, norfuranol and phenylacetaldehyde in wines are results of the fungal growth (Sarrazin *et al.*, 2007).

The laccase is one of the most significant enzymes produced by *B. cinerea*. It is stable in wine and can oxidise a wider range of polyphenolic compounds and is relatively resistant to elevated sulphur dioxide concentration. It can cause decomposition of anthocyanins and procyanidins which are important constituents of red wine (Goetz *et al.*, 1999; Van Baarlen *et al.*, 2007). Hence, even a low level of the infection can cause significant anthocyanin loss and the wine browning. In the presence of phenolic compounds, pectin can increase the synthesis of laccase. Its activity is however, inhibited at pH 3.4 and sulphur dioxide concentration above 50 mg/l.

Thiamin deficiency increases the synthesis of sulphur-binding compounds during the fermentation. Hence, its small addition to grape must before fermentation (0.5 mg/l) is recommended. Some *Botrytis* strains produce compounds that can stimulate or inhibit yeast metabolism. However, so far, a little is known about malo-lactic fermentation process in must obtained from the infected grapes (Jackson, 2008).

Changes in the acidity, depending on the grape variety and the wine region, have been observed in wines derived from botrytis-infected fruit. It has been associated with the metabolism of tartaric acid and to a lesser extent malic acid. Citric acid is also decomposed during winemaking. However, it can be produced in small quantities by *Botrytis*. The reduction of acids content in grapes is compensated by dehydration. Depending on the degree of dehydration, the acidity of the must may slightly decrease or increase. *B. cinerea* produces also small amounts of pyruvic and 2-ketoglutaric acids. All acids, including those produced by acetic acid bacteria, contribute to increased acidity and more complex chemical composition of botrytised grapes and wines. It is generally beneficial because it balances the high sugar content.

*B. cinerea* also uses a significant portion of assimilable nitrogen and amino acids in grapes (Doneche, 1993). Some studies have reported that levels of nitrogenous compounds may decrease by 30-80 per cent (Jackson, 2008).

The hydrolytic enzymes produced by fungi affect the integrity of the grape skins causing fruit cracking during harvest. Grey mould presence on grapes may also have an impact on the other microorganisms developed in berries, than in must during fermentation and consequently in finished wine. Some changes, i.e. cells number and species composition, in population of epiphytic microflora, mainly yeast, may occur during fungal infection. Studies have shown that yeasts of *Candida stellata*, *Torulaspora delbrueckii* and *Saccharomyces bayanus* can grow on grapes simultaneously with *B. cinerea*, (Jackson, 2008).

*B. cinerea* is a dangerous vine pathogen of *Vitis* species. It contributes to the instability of wine, the lack of varietal characteristic aromas, the susceptibility to oxidation, as well as acidity changes. However, noble mould is a connoisseurs dream. In suitable environmental conditions, *B. cinerea* interacts specifically with the infected berries, which ultimately results in the creation of unique, sweet wines called 'botrytised wines'. The noble mould does not destroy varietal odours, but even highlights them. Wonderful, rich, aromatic and long white liqueur wines, like Hungarian Tokay, French Sauternes, Coteaux de Layon and German Trockenbeerenauslese are produced, thanks to the *B. cinerea*.

## 6. Fermentation Process: Difficulties and Stopping Fermentation

Despite advances in wine technology and improvements in fermentation control, problems in the manufacturing process still affect the winemakers around the world. Over the years, many factors that could influence the winemaking process have been investigated and identified (Saigal and Ray, 2007). The fermentation invariably relates to the yeast life cycle. Although they adapt to adverse environmental conditions, sometimes situations when their life cycle is inhibited arise. If there are several slowing down factors, their synergistic effects usually lead to stopping the fermentation.

Problems with the fermentation of botrytis-infected grapes may be attributed to the presence of epiphytic yeasts and bacteria, low level of assimilable nitrogen, high sugar content and presence of the inhibitory substance called ''botryticin''. It stimulates the glycerol and acetic acid synthesis by yeast at the end of the fermentation process (Zoecklein *et al.*, 1999).

The must pressing is also made difficult by the water loss from the vine fruit. The grape skin contact during maceration greatly facilitates the extraction of juice. However, it also causes the glucan release into the must leading to the filtration difficulties.

## 7. Botrytised Wines

The late harvest wines are made from over-ripen grapes which, after reaching full physiological maturity, were still left on the vines, resulting in sugar concentration of more than 25°Brix. After the grapes are fully ripened, sugar deposition in berries is stopped and further growth of its content takes place only through the water loss by evaporation. In addition, concentration of other fruit components, including acids and polyphenols occurs, as well as a partial decomposition of malic acid and a formation of certain characteristic flavours (Jackson, 2008).

The botrytised wine is the specific variety of so-called late harvest wine. Berries affected by the noble mould are used for their production. This phenomenon is caused by *B. cinerea* fungus. The first condition for its appearance is the proper maturity of the fruit. At concentrations above 22° B, sugar starts to inhibit the mould growth. As a consequence, the fungus covers berry surface with a thin layer, but does not infiltrate its inside. It results in a delicate skin perforation, which accelerates the loss of water, but does not destroy the fruit. If *Botrytis* infection occurs at the lower stage of maturity, it usually takes the harmful form of grey mould. The vine variety itself is one of the major importance here as well. *Botrytis* tolerant *V. vinifera* varieties are the most suitable for the production of botrytised wines.

Other factors affecting *Botrytis* development are the suitable weather and environmental conditions. First of all, the right moisture content necessary for the fungus growth must periodically appears in the vineyard. However, when vines are exposed to dampness over a long period, the fungus might develop too intensively. After several hours of humid weather, dry and sunny days should come and stop the excessive growth of the mould. If the air becomes too dry the mould does not appear at all, but if too much moisture is present, *Botrytis* will develop into a grey mould, which would force vine growers to immediate fruit harvest. In regions known for the production of botrytised wines, such as Tokaj, Sauternes or the Moselle valley, the noble mould appears only in certain years. Appropriate habitat conditions for the development of the noble mould are usually found in river valleys and their neighbourhood as well as near water bodies situated on lower parts of slopes.

## 7.1. Tokaj Aszú Wine

Tokaj is a white dessert Hungarian wine made from dried grapes, cultivated in the wine region located on the confluence of two rivers - Bodrog and Tias, at the southern foot of the volcanic Tokaj mountains. Harvests in this region are usually late (early October to late November). The botrytised grapes (sometimes described as aszú berries) are brown with violet tones and completely dry. There are several types of Tokaj wine, i.e. Eszencia, Fordítás and Máslás. The most popular are Szamorodni, obtained from the must without the grapes pre-selection process, and aszú-sweetened by the addition of the selected dried grapes (Kirkland, 1996; Magyar, 2011). Tokaj is primarily produced from the Furmint grape variety, which accounts for 70 per cent of the cultivation area. Tokaj Furmint is from the Tokaj-Hegyalja region, the only one where wines may be labelled with this world famous name (Alkonyi, 2000). The variety is characterised by strong, intensively growing vinelet. Grape berries with a thick peel are gathered in not very large bunches. The Furmint is the late-maturing variety that easily dries on shoots. The grapes are characterised by high acidity and excellent aroma. In addition to the dominant Furmint, other grape varieties are approved for Tokaj wine production, for example, Hárslevelű, which gives a flower-honey aroma as well as Sárgamuskotály. The most precious form of sugar is the Aszú grapes. The specific geographic location of the Tokaj-Hegyalja region, as well as the vine varieties, has a great influence on the wine making process. Furmint and Hárslevelű varieties are characterised by the high amounts of aszú bunches mainly due to suitable climatic conditions. In Autumn, thanks to damp and misty mornings and the nocturnal formation of dew, *B. cinerea* grows rapidly, while during the day, warm sunshine and dry winds help to evaporate water and reduce noble mould development.

The skin of infected berry becomes thinner and pores start to appear over its surface as a result of which the majority of water contained in the fruit is evaporated in a few days. The grapes affected by the noble mould lose water which in turn leads to increase in the sugar concentration (Nyizsalovszki and Fórián, 2007; Macra *et al*., 2009). The existing soil conditions are also important for the grapes of Tokaj cultivation. The bedrock composition in the region is slightly variable according to the geographical location. It is mainly comprised of volcanic tuffs rich in minerals. In the vineyards located in the south of the region the top layer is lose soil whereas the northern part has primarily clay soil at the surface.

During harvesting, the aszú bunches are picked by hand and placed in a special containers. Must or wine from un-rotted berries of the same vintage is poured over botrytised fruits and this makes production technology unique (Alkonyi, 2000; Jonson and Robinson, 2001; Magyar, 2011). Before pressing, the must is subjected to a maceration process which lasts from a few to several hours. The sweetness of wine depends on the amount of aszú berries added to the barrels of non-botrytis grape must. In the past, puttons were used as its measure which they had about 25 litres capacity. The number of puttons given on the wine label informed how many such buckets filled with aszú berries were added to each barrel of must or young wine. The next stage of production is the must pressing, followed by the fermentation process. When the fermentation is over, young aszú wine is matured in oak barrels, at least for two years, and then poured into the bottles (Kirkland, 1996; Magyar, 2011). Previously, the wines were stored in barrels a few years longer and came into contact with the air because wooden containers were not completely filled. Today oxidation of ageing wine is considered to be rather an undesirable factor, depriving their intense and fresh fruit taste. The new oak barrel enriches the wine quality significantly. It improves its structure, helps to develop a range of interesting flavour notes – nuts, almonds, coffee, chocolate, caramel or tobacco.

During the aszú grapes storage, a small portion of the juice leaks out of the berries and is collected at the bottom of the container (Alkonyi, 2000; Jackson, 2008). This nectar, called Eszencia, is used for the production of one of the most exclusive type of wine. Tokaj Eszencia is characterized by a very high sugar content (500-800 g/l), low alcohol content (5-7 per cent) and intense flavour. It is rarely sold, but often used for blending with aszú wine (Jackson, 2008).

### 7.2. German Botrytised Wines

The most important wine regions in Germany are Baden, Franconia, Hessische Bergstraße, Mosel, Rheingau, Rheinhessen, Palatinate (Rheinpfalz) and Saxony. A lot of grape varieties are cultivated in Germany, i.e. Műller-Thurgau, Silvaner, Scheurebe and Pinot Gris, but Riesling is considered to be the most noble one (Jackson, 2008).

The label of 'quality' wine shall specify information, i.e. about the wine-growing region and level of fruit ripeness. The current German wine classification (*Qualitätswein mit Prädikat*) includes three types of wine, the production of which is related to the noble mould development on the grapes. These fruits must contain a certain sugar content which is measured in the Oechsle scale. Further:

- Auslese is the first type of the late harvest wines produced from specially selected ripe grapes. These wines are usually very sweet, with a low alcohol content of 6-8 per cent (Robinson, 2006; Jackson, 2008).
- The second one is Beerenauslese (BA), produced from individually selected, over-ripe fruits. It is an extremely rich, sweet dessert wine, well balanced by its acidity.
- Another type is Trockenbeerenauslese (TBA) wine, produced from specially selected, overripe shrivelled grapes characterised by rich, sweet honey flavour. These fruits are similar to the aszú berries harvested in the Tokaj region.

The above-mentioned wines contain relatively small amounts of alcohol. The high acidity is compensated by the increased concentration of sugar. They are rarely matured in the oak barrels. German botrytised wines are characterized by a rich apricot and honey flavours, and deep golden caramel colour (Robinson, 2006; Magyar, 2011).

The most famous Austrian botrytised wine included in Prädikatswein classification is Ruster Ausbruch. However, this category exists in Austria, not in Germany. It is made in the area of Rust, near Lake Neusiedl, mainly from the botrytszed berries of Furmint, Muscadel, Pinot Blanc, Pinot Gris, Chardonnay and Traminer varieties. During the production process, the grapes are crushed, poured with fresh must or young wine, belonging to the Auslese or Beerenauslese category, then macerated and pressed.

### 7.3. French Botrytised Wines

Sauternes is the most famous sweet wine of France, produced in the Bordeaux wine region. It is one of the most noble liqueur wines that is produced there and its name derives from the place of production. Harvesting and production processes cost a lot of money, making Sauternes very expensive. They are produced exclusively from three grape varieties: Semillon, Sauvignon Blanc and Muscadelle. Semillon gives sweetness that is well balanced by acidity of Sauvignon Blanc. Sauternes wines are characterised by fruit aromas, including apricot, peach and honey (Bailly *et al.*, 2006; Jackson, 2008).

## 8. Production Process of Botrytised Wines

The production of botrytised wines is a challenge for winemakers. In addition to the difficulty of the noble mould developing, which is mainly related to climatic conditions, low juice yield and a high risk of deterioration make these kind of wine one of the most expensive in the world.

## 8.1. Harvesting of Botrytised Grapes

Various techniques are used to harvest grapes infected by *B. cinerea*. One of them, the most labour-intensive, is going through the vineyard several times and picking by hand only the individual berries that have been sufficiently affected by the noble rot. The precision of these procedures contributes significantly to the quality of the final products – aszú and eszencia wines. The shrivelling of berries remains uneven. At the same time, in one bunch, a part of berries are often unripe but some of them are ready to be harvested. Botrytis infected, slightly dried, grapes are collected and stored separately, while firm, un-rotten fruits wait until November. The aszú berries are what makes the Tokaj wines sweet (Alkonyi, 2000).

The harvested berries are transported to the winery and stored in small containers for several weeks (Bene and Magyar, 2004; Magyar and Bene, 2006). These grapes contain specific microflora – *Metschnikowia pulcherrima* and *Hanseniaspora uvarum* are dominant microorganisms at the vineyard but, during storage, the reduction of their population is observed, because of *Candida zemplinina* that gains an advantage over them. The strains of *Zygosaccharomyces, Torulaspora* and *Kluyveromyces* are also present. Depending on their content, botrytised berries can be used in production of dry wines, late-harvest wines, or Szamorodni Tokaji (Kirkland, 1996).

Another method of selected harvesting is collecting of botrytised grape bunches, containing berries at different level of mould development, with the greatest proportion of fruit at the 'pourri roti' stage. This method, called Triage, is characteristic of the Sauternes wine production process. The selective picking of botrytis infected grapes is widely used in the traditional winemaking (like Ausbruch, BA (Beerenauslese), most TBAs (Trockenbeerenauslese). Definitely less popular and more expensive is the harvesting method, including gathering healthy and rotten bunches together. This technique is used in the production of new botrytised wines (Magyar, 2011).

## 8.2. Processing of Botrytised Grapes

The grapes infected by *B. cinerea* should be harvested very carefully to prevent damage of the berry skin and excessive glucan diffusion into the juice (Ribéreau-Gayon, 2000). Tokaj Aszú is defined as the wine produced by adding wine or fermenting must to the berries affected by noble mould. The botrytised fruits are stored in the winery, while the healthy ones are subjected to juice extraction using the press. The berry maceration is a common practice, during which the release of aromatic substances takes place. This process lasts a few, even several dozen hours. During the production of Tokaj Aszú, the juice obtained after maceration usually contains alcohol. Some wine-makers think it is good to use intensively fermented base wines, as the carbon dioxide bubbles released by them gently penetrate and float the previously crushed aszú grapes up. In this way, the fruit pulp is enriched with a new supply of sugar and the fermentation process starts smoothly. After its completion, there is a high content of unfermented residual-sugars in the wine.

The juice clarification, before the fermentation process is also a common method used in the production of botrytised wines. Depending on the local laws and regulations, in France, Germany and New World wine regions, the must composition can be adjusted by the sugar or acid addition. However, it is rarely used procedure. Sugar enrichment (chaptalisation) is not allowed for the production of Tokaj, French SGN and botrytised wines from Austria and South Africa (Madyar, 2011).

## 8.3. The Fermentation Process

Alcoholic fermentation is the substance of wine production. Its purpose is to convert sugar into ethyl alcohol in such a way that as low as possible unwanted by-products are created, and the natural aroma and taste of the fruits have been preserved in a possibly highest extent. During the fermentation, compounds are formed that are responsible for the wine's aroma and flavour profile.

### *8.3.1. Yeasts*

The grape must fermentation is a slow process. It can last from one to six months or even a year. It has been carefully examined in Bordeaux wines (Fleet *et al.*, 1984), Californian wines (Mills *et al.*, 2002) and Tokaj Aszú (Magyar *et al.*, 1999; Sipiczki *et al.*, 2001, Antunovics *et al.*, 2003). The results obtained clearly indicate a greater diversity of microorganisms during the process of the botrytised wines fermentation compared to wines produced in the traditional way without grapes infected with the noble mould (Magyar *et al.*, 1999; Mills *et al.*, 2002; Nisiotou *et al.*, 2007). The studies confirmed that *Candida zemplinina* may produce excess amounts of the volatile compounds but does not give any specific flavours characteristic for this type of wines (Tóth-Márkus *et al.*, 2002). However, these yeasts can contribute to the increase of glycerol content in wine. In addition to *Saccharomyes cerevisiae*, *Saccharomyces uvarum* species are also present (Sipiczki *et al.*, 2001; Antunovics *et al.*, 2003; Magyar *et al.*, 2008; Tosi *et al.*, 2009). This strain is very common in sweet cool climate wines, but not limited to. *S. uvarum* is cryotolerant and grows well at lower temperatures of 7-13° C (Kishimoto and Goto, 1995; Magyar, 2011). It ferments slowly, producing higher amounts of glycerol, succinic acid and 2-phenylethanol and lower amounts of ethanol and acetic acid than *S. cerevisiae* (Massoutier *et al.*, 1998; Magyar *et al.* 2008; Tosi *et al.*, 2009). Typically, Tokaj wines are fermented spontaneously in small wineries. However, some wine-makers use native yeast cultures selected from the local microbiota (Magyar, 2011).

### *8.3.2. Chemical Composition of Botrytised Grapes*

The high content of sugar in grape must significantly reduces the yeast growth as well as the rate of fermentation but on the other hand, initiates a secondary fermentation in wine. During this process, higher level of glycerol is observed in the botrytised wines (up to 30g/l) than in wines obtained from healthy fruits not affected by noble mould. Volatile acidity increases as a result of the acetic acid bacteria occurrence.

The volatile compounds of grapes are accumulated mainly in the berry skin. The aroma composition, which has a direct influence on the wine quality, depends on the grape variety, the terroir and the ripening stage of the fruit. It is extremely complex because volatile compounds can react with alcohol, glycerol, organic acids, sugars and polyphenols. Of the almost 900 volatile components found in wines, about 10 per cent play an important role in the creation of its aroma (Barócsi and Terjék, 2016). Flavours in wine come directly from the grapes as well as can be formed during their processing. Most of them, i.e. alcohols, acids and esters, are produced during the fermentation process (Pérez-Prieto *et al.*, 2003).

Some studies have shown higher concentration of carbonyl compounds (acetaldehyde, pyruvic acid, 2-ketoglutaric acid) in botrytised wines as compared to the wine obtained from healthy fruits. These compounds are accumulated at the time of thiamin deficiency in the affected by *B. cinerea* berries. Carbonyl compounds are responsible for the high $SO_2$ binding capacity of the botrytised wines. Addition of thiamin at the level of 0.5-0.6 mg/l improves decarboxylase activity and reduces the need to add $SO_2$ to the must (Jackson, 2008; Magyar, 2011). The nitrogen deficiency, in turn, can result slow the fermentation process and increase acetic acid production.

Botrytised wines are characterised by lower content of iso-amyl and 2-phenylethyl acetate. Also, the presence of ethyl esters is lower, probably due to the activity of esterase present in the must from infected fruits (Ribéreau-Gayon *et al.*, 2006). Differences in the content of components in wines may also be the result of different strains of yeast present on the infected grapes, which is related to their different ability to metabolize nitrogen (Carrau *et al.*, 2008).

Studies have also shown that the main compound responsible for the aroma of botrytised wines is sotolon (3-Hydroxy-4,5-dimethylfuran-2(5H)-one). This compound is also a dominant flavour in sherry wines (Martin and Etievant, 1991; Moreno *et al.*, 2005). Other studies have shown that there is no correlation between sotolone content and *Botrytis* infection in wines, and the presence of this ingredient is the result of Maillard's reaction proceeds during aging process (Sponholz and Hühn, 1993). The compounds typical of botrytised wines are δ-decalactone, γ-decalactone (Tokaj Aszú) and 2-phenylethanol, 3-mercaptohexan-1-ol, ethyl hexanoate, methional, furaneol (Sauternes) (Magyar, 2011).

### *8.3.3. Fermentation Technology*

Although in most wine regions stainless steel fermenters are used for fermentation, wooden barrels are still commonly used in the production of Sauternes and Tokaj wines. Sauternes can ferment at temperatures up to 28°C, whereas fermentation temperature for Tokaj wines ranges from 10 to 12° C. Due to the late harvesting of fruits, the low fermentation rate and the use of small barrels (200-230 l) the additional cooling is not necessary during Tokaj Aszú production. Wineries keep the fermentation temperature around 20°C for Tokaj wines (Magyar, 2011) and 20-24°C for Sauternes (Ribéreau-Gayon, 2000).

An especially important step in the production of botrytised wine is the completion of fermentation with the required residual sugar content. Traditional process stops spontaneously and obtained wine might consist different concentration of ethanol, which is usually higher than desirable, leading to insufficient residual sugar. During the production of Sauternes wines, $SO_2$ is added (at least 50 mg/l). Due to the fact that a large amount of $SO_2$ is bound to carbonyl compounds, especially keto acids, an addition of $SO_2$ at the level of 200-300 mg/l is required (Ribéreau-Gayon *et al.*, 2000). In Europe, the upper limit for the addition of $SO_2$ to botrytised wines is 400 mg/l. In addition, the high $SO_2$ content inhibits the production of laccase and oxidases by *Botrytis*, which limits the wine browning (Dittrich and Grossman, 2011). It also inhibits the growth of acetic acid bacteria. Tokaj Aszú wine usually contains less free $SO_2$ (20-30 mg/l) compared to German or French wines. However, moderate sulphitation, filtration and cooling might be necessary to stop the fermentation process during its production.

### *8.3.4. Aging and Stabilisation*

Botrytised wines are aged for several years in barrels (one to three years) and then in bottles. German BA and TBA wines rarely mature in barrels due to their low alcohol content, which increases the risk of secondary fermentation. Sauternes wines are kept in barrels for 12 to 18 months, or even two years and longer (Ribéreau-Gayon *et al.*, 2000). Similarly, during Tokaj Aszú production process aging time in the barrel is at least two years.

Chemical changes occurring during barrel aging have not been fully investigated. However, it is known that alcohols, aldehydes, esters, acetals and lactones oxidation occurs. The longer the aging, the more aromas of dried fruit, chocolate and coffee are created. The most basic aroma is the smell of peach, quince, honey. The complexity of the aroma increases as a result of the extraction of some components present in the wood that pass to wine during this process. Aging in oak barrel gives it a unique flavour and aroma. Most importantly, it provides tannins that protect the wine from oxidation, reducing the risk of unpleasant odours, and also highlights other aromas of wine and intensifies its colour (Pérez-Prieto *et al.*, 2003).

Before bottling, the wine is often stabilised. For this purpose, bentonite addition and cold stabilization are used to remove proteins and avoid the tartrate crystallisation. The unique feature of botrytised wines is the formation and precipitation of calcium mucate crystals, salt of galactaric acid or mucid acid (Dittrich and Grossmann, 2011). These compounds do not occur in traditional wines, while in bottles of old botrytised wines have a size from 1 to 3 mm. Technological problem during the production of botrytised wines is the presence of glucans, which even in low concentrations (2-3 mg/l), reduce the ability of filtration. The content of glucans at 50 mg/l makes filtration process impossible to conduct (Magyar, 2011).

## 9. Prevention, Treatment and Care

Despite the fact that moderate damage caused by *B. cinerea* has been observed over the years in most of the traditional wine producing countries, i.e. France, Italy and Chile, there can be an extremely sudden outbreak of the disease. It is a pathogenic fungus that is particularly common in Europe, especially in regions with cool and humid climate, such as Switzerland or eastern France. Gray mould is an extremely complex disease. It is the most striking example of the interactions between pathogen, host and environment described as so-called '"the plant disease triangle'. To prevent the attack and development of *Botrytis,* vines should be planted with distance between rows to allow for proper ventilation and enough sunlight exposure of canopy. It is also necessary to avoid excessive compaction of grape clusters on the plant. In order to limit the sources of infection, systematic weed, deep soil loosening, removal of fallen fruits and leaves as well as elimination of infected parts should be applied. In some countries, it is possible to use fungicides at the key stages of the cultivation (Evans, 2010).

The standard chemical protection strategy implies performing the treatments in four basic phases, as presented in Table 1. (Misc, 2002; Elad *et al.*, 2007; Moyer and Grove, 2011; Adamczewska-Sowińska *et al.*, 2016).

**Table 1.** Standard Periods of Application of Fungicides Protecting the Vine against the Grey Mould

| *Treatment period* | *Phenological stage of vine* |
|---|---|
| I | |
| I a | 10-20 cm shoots, 5 per cent of petals fallen |
| I b | 80 per cent of petals fallen |
| II | before bunch closure |
| III | beginning of berry ripening |
| IV | pre-harvest period |

Sometimes, the use of only two treatments may be sufficient and there is no need to do a full number of them. However, in most cases the standard performance of all four applications guarantees the highest efficiency. The benefits of earlier or later application vary from season to season and depend on the total precipitation and the susceptibility of the vine variety to *B. cinerea* infection.

As regards the grey mould life cycle, two types of fungicides can be used for treatments during flowering. These may be preventive or interventional. The strictly prophylactic fungicides should be used in the early flowering period before the sensitive floral tissue is exposed to mould spores. The use of such formulations, with a wide variety of activities, makes it difficult to develop resistance to fungicides in pathogen population. However, at the end of flowering it is necessary to use effective interventional preparations that can destroy the mycelium which has already entered the tissues of the various parts of the flowers. The use of the aniline pyrimidines at the end of flowering (Phase I) and in the period prior to the bunch closure (Phase II) is particularly important due to the specific type of their action. Such preparations operate on the principle of vapour pressure which means that they partly act in the gas phase, which makes it far easier to penetrate the flowers as well as the closing bunches. In addition, the aniline pyrimidines can be used in both, preventive and interventional treatments (Mazurek).

In cases where a small percentage of latent infections reactivates, the sudden outbreak of the disease might happen. The factors which stimulate grey mould development are the long-lasting humid weather and berry damages caused by insects and adverse weather conditions. In addition, healthy fruits can be infected by a direct contact with the infested berries. Fungicides that are used to remove gray mould from ripe berries should inhibit the growth of mycelium reactivated after latency (Phase III), reduce spore production and prevent its spreading (Phase IV) as well as protect healthy berries from infection. Cyprodynil (anilinopyrimidines) and fludioxonil formulations proactively inhibit the spore formation and are therefore widely used for the control of grey mould (Mazurek). They were applied for the first time in French vineyards in 1994 and in Switzerland in 1995. In Italy, cyprodynil was used in 1997. Anilinopyrimidines are highly effective against *B. cinerea*, but studies have shown an increase in pathogen resistance to this group of formulations, hence, they are used in mixture with phenylpyrroles (Fabreges and Birchmore, 1998). Fluoxonidyl belongs to the class of fungicides that influence osmoregulation. It is an inhibitor of spore germination (Topolovec-Pintarić, 2011).

Treatments in Phase III should be done at the beginning of berry ripening which requires a very precise capture of the '*veraison*'. Particularly useful are anilinepyrimidines fungicides that are capable of penetrating the fruit skin. Therefore, anilinopyrimidines, dicarboximides and hydroxyanilides can be successfully used in Phase IV. One of the hydroxyanilide fungicides is fenhexamide. Studies have shown a different mechanism of its action compared to other botryticides (Rosslenbroich and Stuebler, 2000). It is highly active against *B. cinerea* and easily to decompose. Lipophilic character of fenhexamide makes it rapidly penetrate into the grape skin, reducing the risk of flushing it by rain (Haenssler and Pontzen, 1999). Fenhexamid inhibits spore germination only in high concentrations; however, it is highly effective in prevention against the subsequent stages of the *Botrytis* infection (Topolovec-Pintarić, 2011). Another broad-spectrum botryticide is boxalid (carboxamide), highly effective against pathogenic fungi. It causes the inhibition of fungal respiration by blocking electron transport in Complex II of the respiratory chain, thereby depriving the fungus of a significant source of energy and building material for the synthesis important components for the pathogen cell growth. In laboratory studies, conducted on conidia and mycelium, the effects of this fungicide depended mainly on the type of the carbon source. In the presence of glucose, its activity was low. The highest efficacy of this agent was demonstrated when succinate or acetate were added to the nutrient medium. *B. cinerea* populations were tested in the Champagne vineyards and no moderately or highly resistant to boscalid strains were found. Lack of cross-immunity to benzimidazole, phenylcarbamates and anilinopyrimidine has also been demonstrated in the study (Leroux *et al.*, 2007).

In some situations, the control of grey mould in the vineyard is not satisfactory. This is due to the fact that there is the alternative oxidase (AOX) present in the *Botrytis* mitochondria. It allows electrons to pass by the blockage of the cytochrome bc1 pathway caused by strobilurins (Tamura *et al.*, 1999; Wood and Hollomon, 2003). *In vitro* study show that in the presence of salicylhydroxamic acid (SHAM) that is AOX inhibitor, or after inhibition of the AOX-encoding gene there was a greater susceptibility to strobilurin. On the other hand, the contribution of AOX during the interaction of *B. cinerea* with vines remains questionable. It has been found that the practical prevention against *B. cinerea* with strobilurin is dependent on the amount of natural inhibitors of AOX (e.g. flavonoids) in plant tissues (Tamura *et al.*, 1999). The plant pathogens have acquired natural fungicide resistance, which is due to the point mutation, which is the conversion of glycine to alanine at position 143 in the mitochondrial cytochrome b gene (Gisi *et al.*, 2002).

The chemical control of *Botrytis* infection is hampered by the development of fungus resistance to many botryticides, legal regulations and the negative public perception. Consequently, in many countries, the use of pesticides is limited (Gullino and Kuijpers, 1994). A case in point is France, where the number of possible fungicide applications is limited to one per season to avoid excess of the Maximum Residue Level (MRL). Some of them cannot be used near harvest (e.g. fludioxonil, dietofenkar) (Couteux and Lejeune, 2003). Available fungicides are classified in five categories due to their biochemical ways of action. These include substances affecting fungal respiration, anti-microtubule agents, fungicides affecting osmoregulatory function, inhibitors of sterol biosynthesis and toxicants, the effect of which are specifically reversed by the amino acids (Anonymous, 1988). Also in Poland, in the present legal status, there are no fungicides registered to protect the vine against grey mould. Certainly it is one of the factors limiting the ability to produce quality wine. The arguments for the possibility of selection of varieties moderately or completely resistant to grey mould for growing are not entirely convincing. The taste preferences vary across customers. Limiting consumers' choices to only two options, excellent wines produced from chemically-protected plantations and lower quality wines produced from organic plantations is not preferred.

## 10. Effect of Noble Rot on the Grape Microbiota

*Botrytis* is one of the most important pathogens attacking vine fruit, leading to damage (grey mould) or drying of berries (noble rot). The grey mould causes large losses during cultivation of vines, posing a serious risk to growers and wine producers. On the other hand, *B. cinerea* also known as noble rot allows to produce the best high-quality wines in the world (Fedrizzi *et al.*, 2016). There are only a few geographic areas in the world where sweet white wines could be produced – the world's most famous come from Sauternes-Barsac (France), Rheingau and Mosel-Saar-Ruwer (Germany), Tokaj (Hungary and Slovakia) and Rust (Austria). However, many other wine regions in France, Italy, Romania, Austria and Switzerland also have favourable conditions for their production. Outside of Europe, sweet white wines can be made in the USA (e.g. Napa Valley and Santa Barbara, California), Canada (Niagara Peninsula), Chile (Valle del Maule) or Australia (New South Wales and Victoria) (Sipiczki *et al.*, 2010).

The presence of *B. cinerea* in vineyards is related to the fungal penetration inside a fruit, which makes grapes susceptible to diseases caused by other microorganisms. Yeasts naturally living on berries or transmitted by insects can be secondary infectious agents. *Botrytis* infection and berry drying are associated with changes in the composition of the grape juice. Although the fungus can metabolise more than one-third of the sugar, its final concentration is about 60-70 per cent, depending on the degree of botrytisation. High concentration of sugar reduces yeast's growth and fermentation activity. Usually, glucose is metabolised more extensively than fructose, thus increasing the fructose/glucose ratio in the must. It may have an inhibitory effect on *Saccharomyces* yeast growth (Gafner and Schütz, 1996). In addition, compounds produced by *Botrytis*, such as botryticin, norboteryl acetate and botrylactone, exhibit fungicidal activity and may cause difficulties during fermentation process (Sipiczki *et al.*, 2010).

The main yeast species present in the botrytised fruit and grape must are *Candida zemplinina/stellata, Candida lactis-condensi, Saccharomyces cerevisiae, Zygosaccharomyces bailli, Zygosaccharomyces rouxii, Saccharomyces uvarum, Aureobasidium* (Tokaj) and *Torulopsis stellata, Hanseniaspora uvarum, Pulcherrima meschnichia, Candida krusi* (Sauternes). Although some of these species (e.g. *K. apiculata/ Hanseniaspora uvarum, Candida zemplinina/stellata*) naturally occur on healthy grapes as well as during must fermentation, many studies confirm that the development of the yeast population depends mainly on environmental conditions, grape varieties and terroir. The environment of the botrytised must is not conducive to the yeast growth. Some species rapidly die out, to be replaced by *Saccharomyces* strains (Sipiczki, 2010). Although these strains are usually not identified on the fruit, they dominate in the fermenting grape must. The presence of two species belonging to the genus *Saccharomyces, S. cerevisiae* and *S. uvarum,* is confirmed in the botrytised musts (Naumov *et al*., 2002; Antunoics *et al*., 2005; Masneuf-Pomerede *et al*., 2007). *Candida zemplinina* is the other strain capable of surviving to the end of the fermentation process (Cocolin *et al*., 2001; Mills *et al*., 2002). Interestingly, *S. uvarum* cultures are strongly glucophilic (Schütz and Gafner, 1995), as did *S. cerevisiae* (Berthels *et al*., 2004). In turn, *C. zemplinina* is highly fructophilic (Mills *et al*., 2002). It can therefore be considered that a balanced ratio between glucose and fructose is necessary for proper fermentation. In the case of Essencia wines produced from specially selected botrytised grapes, the high sugar content decreases the activity of *Saccharomyces* and *C. zemplinin* which in turn promotes the development of osmolytic species such as *Z. bailli, Z. rouxii* and *C. lactis-condenci* (Csoma and Sipiczki, 2007; Csoma and Sipiczki, 2008).

The presence of *Candida stellata* and *Kloecker apiculata* on grapes infected by *B. cinerea* is confirmed by a number of studies (Antunovics *et al*., 2003; Bene and Magyar, 2004; Magyar, 2011). It has been shown that *C. stellata* cultures isolated from Tokaj Aszu wine are significantly different. After determining the nucleotide sequences of the 26S and 5.8S-ITS rDNA regions, these yeasts were categorized as a novel species *C. zemplinina* (Mills *et al*., 2002; Sipiczki, 2003; Sipiczki, 2004; Barata *et al*., 2011). *C. stellata* strain was isolated from fermented Tokaj wine and botrytised grapes much less frequently (Magyar and Bene, 2006). Both species are phenotypically similar, resistant to high sugar levels and cryogenic (Sipiczki, 2004), which may explain their easy adaptation at the surface of grapes infected be *B. cinerea*. The study also showed the presence of *Metschnikowia pulcherrima* yeast (*C. aureus pulcherrima*) at the aszú grapes in the Tokaj wine region. In addition, *H. uvarum* (*K. apiculata*), *M. fructicola, C. zemplinina*, as well as other species belonging to the *Basidiomycota*, including *Rhodotorula, Cryptoccus, Sporidiobolus* also appear in this region (Barata *et al*., 2011). It has been observed that the population of *M. pulcherrima* is reduced after the grape harvesting and storage, while the amount of yeast *C. stellata/C. Zemplinina* rises (Bene and Magyar, 2004; Magyar and Bene, 2006). The presence of *M. pulcherrima* was confirmed by analysis of microflora of fresh grape must obtained from botrytised berry (Sauternes). The results show that *C. zemplinina* and *M. pulcherrima* are strong competitors for *B. cinerea*. *M. pulcherrima* strain inhibits the activity of some yeast strains, including *S. cerevisiae* (Nguyen and Panon, 1998). Although Magyar and Bene (2006) found that *S. cerevisiae* species do not occur in Tokaj Aszú wines, Naumov *et al*. (2002) showed the existence of *S. cerevisiae* and *S. uvarum* on grapes in the Tokaj wine region. The presence of species belonging to the genus *Aspergillus* and *Penicillium* as well as *Aureobasidium* were has also been shown to be significant (Magyar, 2011).

In addition, substantial amounts of acetic acid bacteria have also been observed on botrytised grapes who are responsible for the formation of acetic acid and other components in musts. Unlike the *Acetobacter* strains, *Gluconobacter oxydans* prefers a sugar-rich environment to produce gluconic, 2-ketogluconic, 5-ketogluconic and 2, 5-ketogluconic acids (Sponholz and Dittrich, 1985). These acids are partly responsible for the high $SO_2$ binding capacity in botrytised must and wine. Only small quantities of acetic acid are formed during the alcohol oxidation. The botrytised berry juice contains a significant amount of acetic acid, which is probably produced also by wild yeast species (Donèche, 1993). Little is known about the presence and significance of lactic acid bacteria on the fruits infected by *Botrytis*. A small number of these microorganisms (mainly *Pediococcus*) have been reported in fresh berries and remained stable until the end of the fermentation process. Although high sugar content could encourage their growth (Donèche, 1993), complex nutritional requirements and poor competitiveness of lactic acid bacteria decreased development (Magyar, 2011).

Many non-pathogenic microbes inactivate the growth of plant pathogens through competition for nutrients, as well as the metabolites production, which naturally reduces plant diseases in the environment. Numerous studies describe the isolation of antagonistic microbes in order to use their inhibiting effect against biological diseases, including grey mould infestations.

*Trichoderma* is one of the fungus species used in the fight against phytopathogenic *B. cinerea*. *Trichiderma harzianum*, originally isolated from cucumber, was the first species of the genus that was used to reduce this disease. It shows the ability to colonise senescent floral parts (stamens, calyx, and aborted fruitlets) and bunches green tissue, which are most likely to be exposed to *B. cinerea* (Holz *et al.*, 2003). It has been shown that the *T. harzianum* species is more effective in fight against pathogenic fungi in South African wines compared to *Gliocladium roseum* and *Trichosporon pullulans* (Holz and Volkmann, 2002). However, the effectiveness of this microorganism in relation to grey mould is not extremely high, contrary to *T. atroviride* growth inhibition. *Ulocladium atrum* shows similar properties. It has been shown that its use reduces *B. cinerea* infection by up to 67 per cent in various wine regions of Germany (Schoene *et al.*, 2000). *Cladosporium ssp.* and *Epicoccum nigrum* also exhibited antagonistic activity to the pathogenic fungus (Elmer and Reglinski, 2006).

Yeasts, which are epiphyte microflora of vines, may also have an inhibition effect on *B. cinerea*. The antagonistic species include *Rhodotorula glutinis* and *Cryptococcus laurentii*. Similar properties are characteristic for *Meshashkovia pulcherrima* (anamorf *Candida pulcherrima*) (Nigro *et al.*, 1999) and *M. fructicola* (Kurtzman and Droby, 2001). In turn, the *Pichia membranifaciens* strain was an effective *B. cinerea* antagonist on vine planting material grown in vitro (Masih *et al.*, 2001).

Antagonistic activity against *B. cinerea* has also been confirmed for some bacteria, i.e. *Bacillus* spp. (Paul *et al.*, 1998), *Brevibacillus brevis* (Seddon *et al.*, 2000), *Pseudomonas fluorescens* (Krol, 1998), *Baculus circulans* (Paul *et al.*, 1997) and *Serratia liquefaciens* (Whiteman and Stewart, 1998; Elmer and Reglinski, 2006).

## 11. Health-related Aspects of Botrytised Wines

Botrytised wines have been shown to offer some health benefits. For a long time it was thought that Tokaj Aszú has even the medicinal properties (Kállay *et al.*, 1999). The most significant components present in the wines are polyphenols. It has long been known that grapes and wines contain a significant amount of antioxidants, including resveratrol, catechins, epicatechin and proanthocyanidins. The influence of wine on health and its antioxidant activity have been extensively investigated (Bertelli, 2009; Dávalos and Lasunción, 2009).

The studies show that botrytised wines are richer in polyphenols than the wines obtained from healthy fruits (Pour Nickfardjam, 2002; Pour Nickfardjam *et al.*, 2006). It is believed that, among the wine polyphenols, resveratrol plays the most important role. It occurs naturally in grapes and most red wines, but is practically not identified in dry white wines (Pour Nikfajdjam, 2002). Resveratrol has a wide range of biological activity. Its antioxidant and anti-inflammatory properties are important in the prevention of cardiovascular diseases, cancer and diseases of the nervous system. The highest concentrations of resveratrol are found in wines from cool wine regions such as Canada and northern Europe. The climate, vine varieties susceptible to infection and higher levels of precipitation are conducive to *B. cinerea* growth. Because resveratrol is produced as a result of response to fungal infection, it may be expected that its concentration in botrytised wines will be higher than in wines from healthy grapes (Jeandet *et al.*, 1991; Ky *et al.*, 2012). Its level in the plant is the highest 24 hours after the occurrence of the harmful agent and starts to decrease after 42-72 hours. The resveratrol concentration also depends on the degree of damage, the grape variety, the maceration time, the filtration process as well as the storage of wine (AWRI, 2009). However, a study at Tokaj Aszú has revealed low levels of this compound (0.2-3.9 m/l). Pour Nikfajdam *et al.* (2006) showed, in turn, that resveratrol content in Tokaj Aszú was higher (2.5 mg/l) compared to German botrytised wines (0.9 mg/l). Probably, it was the result of a longer maceration process. The low concentration of resveratrol in botrytised wines can also be explained by the high activity of the stilbene oxidase (laccase) in *Botrytis*. This enzyme oxidizes resveratrol and converts it into inactive ingredients (Jeandet *et al.*, 1995; Van Baarlen *et al.*, 2007). Low levels of trans-astringin, trans-resveratrol and pallidol have been demonstrated during the development of the noble mould on grapes of the Sémillon and Sauvignon varieties (<0.5 mg/kg og grapes). Only viniferin was detected at relatively high concentrations (2 mg/kg) (Landrault *et al.*, 2002; Magyar, 2011).

Monoamines (also known as 'biogenic amines') are present in foods as products of protein metabolism due to the amino acid decarboxylation process. Histamine, tyramine, serotonin, phenylethylamine are contained in the wine. They are involved in inducing symptoms of intolerance (Maintz and Novak, 2007; Kaschak *et al.*, 2009). However, a number of studies on the biogenic amines in wine have been published, only few described botrytiwed wine (Hajós *et al.* 2000; Eder *et al.*, 2002; Kallay, 2003; Kiss *et al.*, 2006; Sass-Kiss and Hajós, 2005; Sass-Kiss *et al.*, 2008). Malo-lactic fermentation process is considered to be the one that makes the greatest contribution to their formation during winemaking (Marcobal *et al.*, 2006). Marques *et al.* (2008) investigated the carbendazim, iprodione and procymidone fungicides and their ability to reduce the levels of biogenic amines in grape must and wine. It was found that untreated wine is characterised by higher concentration of monoamines after the malo-lactic fermentation. This may indicate that metabolic activity of the fungus can directly contribute to its formation (especially iso-amylamine). Studies have shown that red wines contain much more biogenic amines, mainly histamine, tyramine and putrescine than white wines that usually do not undergo the malo-lactic fermentation (Romero *et al.*, 2002; Leitão *et al.*, 2005; Hernández-Orte *et al.*, 2006; Yildirim *et al.*, 2007). Studies by Edera *et al.* (2002) show that these compounds (mainly isopentylamine and phenylethylamine) are present in larger quantities in botrytised wines than in those obtained from healthy berries. In addition, higher polyamine content was also found in grape aszú from the Tokaj region (Kiss *et al.*, 2006). There, in the botrytised wines were also larger amounts of 3-methylbutylamine (isopentylamine), phenylethylamine, ibutylamine, agmantine and spermidine. In turn, lower histamine content was recorded (Marcobal *et al.*, 2006). Other studies by Sass-Kiss *et al.* (2008) demonstrate higher concentrations of 3-methylbutylamine and 2-methylbutylamine in Tokaj Aszú compared to other botrytised wines. In addition, serotonin quantities were substantial (> 100 mg/l) (Kállay, 2003, 2005). However, considering the moderate consumption of the botrytised wine, the excessive content of biogenic amines does not cause any health issues.

The presentation of ochratoxin A (OTA) in wines was first described by Zimmerli and Dick (1996). This mycotoxin is much more often identified in red wines than in white ones (Stratakou and van der Fels-Klerx, 2011). Its production, in warm climates, is mainly attributable to species belonging to the genus *Aspergillus* (*A. ochraceus, A. carbonarius*), while in the cool climate it is produced by *Penicillium* (Torelli *et al.*, 2005; Varga *et al.*, 2007). The maximum level of OTA in wines is 2 mg/kg. It is believed that botrytised wines may contain OTAs. Although *Botrytis* does not produce ochratoxin, the species of *Penicillium* and *Aspergillus* present in the vine fruit may be responsible for its production. Insects foraging in vineyards intensify the colonisation of grapes by mycotoxin-producing fungi. Only slight amounts of OTA have been detected in Tokaj aszú (Kállay and Bene, 2003; Valero *et al.*, 2008; Magyar, 2011).

## 12. Detection and Quantification Methods

There are several methods for the *B. cinerea* detection on grapes and in wine, including visual inspection by a moist incubation of samples, the enzyme-linked immunosorbent assay (ELISA), the real-time polymerase chain reaction (qPCR) and the spectroscopy method (near infrared – NRI and mid infrared – MIR) (Dewey and Yohalem, 2007; Scott *et al.*, 2010).

The moist incubation of samples is a technique used to assess the presence of pathogens in plant tissues. It can be used to evaluate the presence of *B. cinerea* latent infection. For this purpose, berries or bunches are harvested in the ripening stage (Holz *et al.*, 2003; Dewey and Yohalem, 2007). Then, they are sterilised, washed and frozen, or treated with herbicides (Holz *et al.*, 2003; Cadle-Davidson, 2008). Freezing or herbicide treatment destroys a natural protective mechanism of green berries against *B. cinerea*. After the moist incubation, vine tissue is evaluated for the fungal sporulation (Holz *et al.*, 2003; Cadle-Davidson, 2008). Moist incubation techniques provide a cheaper alternative to molecular methods. However, there are some limitations, including the time of analysis and the need to train the person identifying the pathogens that are assessed after incubation of the grape tissue (McCartney *et al.*, 2003). The use of molecular techniques in *B. cinerea* detection focuses on the analysis of the population diversity, isolation of the specific pathogenic genes and species identification (Zheng *et al.*, 2000; Moyano *et al.*, 2003; Schena *et al.*, 2004). It is also possible to quantify *B. cinerea* in grape tissue by using ELISA and quantitative PCR (qPCR or real-time PCR) (Rigotti *et al.*, 2002; Obanor *et al.*, 2004; Suarez *et al.*, 2005).

The ELISA tests are quick and easy methods for determining plant pathogens. Unlike the molecular PCR techniques, analyses do not require specialised equipment and hence are much cheaper. The advantages of the ELISA test are high sensitivity, the ability to evaluate multiple samples at one time, and quantitative results. Traditionally, these tests have been used to detect plant viruses, but they are also used to identify pathogenic fungi. These analyses are much more complex due to the complex structure of the fungal DNA as compared to the viruses; hence, their reliability in some cases is limited (Boonham *et al.*, 2008). ELISA tests are normally performed on polystyrene 96-well plates. Each of the analysed samples, diluted in appropriate proportion in a special buffer, is located in a separate well. The method consists in the antibodies attachment to the walls and the bottom of the well. This process is called 'solid phase' flattening. Unattached antibodies are removed with the buffer and the plate is washed with another buffer. Then the antigen is introduced into the same wells. It binds to the antibodies and, after that, its excess is removed. The wells are washed again with buffer. Next, a solution of antibodies is introduced into the well, which is connected to a suitable enzyme. Excess antibody is removed from the plate and rinsed again with buffer. The addition of an enzyme substrate is the last phase. The enzyme substrate is a colourless substance that the enzyme converts into a colour product. ELISA is therefore a colorimetric method. Colour change of the solution is measured spectrophotometrically. Numerous studies have demonstrated the use of ELISA to detect *B. cinerea* in grapes (Dewey *et al.*, 2005, 2008; Celik *et al.*, 2009). They were mainly used to estimate its concentration in juice during fruit crushing (Dewey *et al.*, 2005). Quick and easy analysis ensures that the necessary actions are taken (e.g. separation of infected berries) to limit the fungus entry into the grape must. Where more accurate analysis is required, ELISA tests are replaced by molecular techniques. Celika *et al.* (2009) show that the ELISA test works when an actively growing fungus is present on the vine fruit and symptoms of infestation are visible. Molecular methods allow for early identification of pathogen colonisation.

Among modern technologies, special attention is paid to immunochromatographic rapid tests. Nowadays, they gain an advantage over other diagnostic methods, such as traditional laboratory tests, due to the short analysis time, low cost, and the lack of necessity to use any special equipment. The principle of the immunochromatographic method is compared with the ELISA, with the difference that the immunological reaction is carried out on chromatographic paper (nitrocellulose membrane) using capillary action.

Increasingly, qPCR techniques are used to identify fungal plant pathogens. *B. cinerea* enters the latent phase during the vine flowering and the berries development, which makes the detection methods very limited. Unlike ELISA tests, molecular techniques can be used at every stage of infection to monitor latent infections and manage and control the use of the fungicides. The use of spectroscopy also allows the detection of *B. cinerea* in grapes and grape must. The advantage of this type of technique is that they require little sample preparation and are easy and quick to perform (Gishen *et al.*, 2005; Dewey and Yohalem, 2007; Gishen *et al.*, 2010).

## 13. The Positive and Negative Aspects of *B. cinerea* on Grapes and Wine

The positive and negative effects of the presence of *B. cinerea* on grapes and in wine have already been described. Nevertheless, it has long been the subject of research. The multidimensional interaction between the grape variety, the *Botrytis* fungus, the species of yeast that colonise the grapes and the climatic factors make the process of making botrytised wines very complex. The use of carefully selected strains of *Saccharomyces* and other cultures not belonging to the genus *Saccharomyces* could be a promising step in the fermentation process. Therefore, improvement of the alcoholic fermentation process as well as the microbiological and chemical stabilisation of wines are the main technological challenges in the production of butyritised wines.

The positive and negative effects of the presence of B. cinerea on grapes and in wine are summarised in Table 2.

Noble rot of the grape is a complex microbiological and biochemical process that has long been the subject of research. The multidimensional interaction between the grape variety, the *Botrytis* fungus, the species of yeast that colonise the grapes and the climatic factors make the process of making botrytised wines very complex. The use of carefully selected strains of *Saccharomyces* and other cultures not belonging to the genus *Saccharomyces* may be a promising direction in controlling the fermentation process, which is currently one of the most important problems in the production of botrytised wines Therefore, improvement of the alcoholic fermentation process as well as the microbiological and chemical stabilisation of wines are the main technological challenges in the production of wines with the participation of *B. cinerea*.

**Table 2.** Positive and Negative Effects of *B. cinerea* on Grapes and Wine

| *Positive effects of B. cinerea on grapes and wine* | *References* | *Negative effects of B. cinerea on grapes and wine* | *References* |
|---|---|---|---|
| Production of wines from overripe and botrytised grape (so-called wine from late harvest) | Ribéreau-Gayon *et al.*, 2000<br>Jackson, 2008 | Decrease in the yield of grapes, damage during the harvest of seemingly healthy crops | Ellis, 2004<br>Williamson *et al.*, 2007<br>Ky *et al.*, 2012 |
| Production of high quality Tokay wines | Kirkland, 1996<br>Magyar *et al.*, 1999<br>Alkonyi, 2000<br>Sipiczki *et al.*, 2001<br>Antunovics *et al.*, 2003<br>Bene i Magyar, 2004<br>Magyar i Bene, 2006<br>Magyar, 2011 | Physical and biochemical changes in grapes – reduction of berry acidity, increase in sugar content | Ribéreau-Gayon *et al.*, 2000<br>Doehlemann *et al.*, 2005<br>Doehlemann *et al.*, 2006<br>Choquer *et al.*, 2008 |
| Production of quality German wines with a controlled name of origin with a declaration (Qualitätswein mit Prädikat) | Robinson, 2006<br>Jackson, 2008<br>Magyar, 2011 | Production of laccase – lowering the quality of wine; laccase has strong oxidizing properties, causing the browning of red wines, decomposition of anthocyanins and procyanidins, which are important components of red wine | Pezet, 1998<br>Goetz *et al.*, 1999<br>Claire, 2000<br>Viret *et al.*, 2004<br>Van Baarlen *et al.*, 2007<br>Pinar *et al.*, 2017 |
| Production of French botrytised wines Sauternes | Bailly *et al.*, 2006<br>Jackson, 2008 | Occurrence of elevated glycerol content (1-10 g/l) and gluconic acid in grapes (1-5 g/l) in wines | Sponholz i Dittrich, 1985 Sponholz *et al.*, 2004<br>Dittrich i Grossmann, 2011 Magyar, 2011 |
| Favourable health properties of botrytised wines – polyphenols | Jeandet *et al.*, 1991<br>Jeandet *et al.*, 1995<br>Kállay *et al.*, 1999<br>Pour Nikfajdjam, 2002<br>Van Baarlen *et al.*, 2007<br>Bertelli, 2009<br>Dávalos i Lasunción, 2009<br>Ky *et al.*, 2012 | Production of complex polysaccharides; the presence of glucans may cause difficulties with wine filtration, while heteropolysaccharides inhibit alcoholic fermentation and may cause the increase of acetic acid and glycerol in wine | Hong *et al.*, 2011<br>Agudelo-Romero *et al.*, 2015 Negri *et al.*, 2017 |

(*Contd.*)

**Table 2.** *(Contd.)*

| *Positive effects of B. cinerea on grapes and wine* | *References* | *Negative effects of B. cinerea on grapes and wine* | *References* |
|---|---|---|---|
| | | Characteristic smell of grape musts | Rolle *et al.*, 2012 |
| | | Esterase synthesis can break down fruit esters that give character to white wines | Sponholz i Hühn, 1996 |
| | | Production of aldehydes (benzaldehyde, phenylacetaldehyde) and furfural | Sarrazin, 2007<br>Magyar, 2011<br>Pinar *et al.*, 2016 |
| | | The use of available nitrogen in grapes and grape must and amino acids | Doneche, 1993<br>Jackson, 2008 |
| | | Production of biogenic amines in botrytised wines | Hajós *et al.*, 2000<br>Eder *et al.*, 2002<br>Romero *et al.*, 2002<br>Kalláy, 2003<br>Kállay, 2005<br>Leitão *et al.*, 2005<br>Sass-Kiss i Hajós, 2005<br>Hernández-Orte *et al.*, 2006<br>Kiss *et al.*, 2006<br>Yildirim *et al.*, 2007<br>Sass-Kiss *et al.*, 2008 |
| | | The presence of OTA in botrytised wines | Kállay i Bene, 2003<br>Valero *et al.*, 2008<br>Magyar, 2011 |

## 14. Conclusions and Future Perspectives

The grey mould is one of the most common fungal diseases of *V. vinifera*. Symptoms of infection occur on both green and ligneous parts. This fungus causes the greatest damage on fruits. The growth of grey mould requires high humidity (70-100 per cent) and relatively high temperature (15-25°C). Symptoms on young branches and leaves are rather rare.

With favourable climatic and weather conditions, white mould, called noble, develops on the skins of grapes instead of grey mould. Botrytised wines produced from grapes affected by *B. cinerea* are perceived as the best and the most expensive in the world. This fungus attacks the grapes, removes a significant amount of water from the fruit, leaving in it a higher content of sugars, extract and a unique richness of aromas. The most famous botrytised wines are the French Sauternes and the Hungarian Tokaj Aszú. The production of botrytised wine is spreading all over the world. Currently, they are also produced in Italy, Australia, New Zealand and California. Wines from these regions vary in the degree of botrytisation of grapes, the ratio of healthy and infected berries, the amount of added sulphur dioxide, or the ageing procedure. So far, there have been many studies on the mechanism of infection of *B. cinerea*, as well as the effect of the fungus on changes in the physical and biochemical profile during the ripening of berries.

The most important include the increase of sugar and acidity in grapes, the accumulation of glycerol, gluconic acid, galacturonic acid, galactaric acid and polysaccharides (ß-glucans) and the emergence of a unique aroma of wines obtained with *Botrytis*. The aromatic compounds (esters of hydroxyl-, oxo- and dicarboxylic acid, acetals, lactones, thiols) occurring in botrytised wines have been studied and characterised extensively. However, additional research is necessary to identify active odour compounds specific for botrytised wines (Magyar, 2011). Wines produced with the participation of *B. cinerea* are rich in polyphenols. Their content was noted in Tokaj Aszú, despite the fact that the fungus produces resveratrol-oxidising enzymes (oxidases). Therefore, the concentration of this compound in botrytised wines is not significantly higher than in wines that were produced without the participation of *B. cinerea*.

## References

Adamczewska-Sowińska, K., Bąbelewski, P., Chohura, P., Czaplicka-Pędzich, M., Gudarowska, E., Krężel, J., Mazurek, J., Sosna, I. and Szewczuk, A. (2016). Agrotechniczne aspekty uprawy winorośli. Praca zbiorowa. Wrocław, 1-203.

Agudelo-Romero, P., Erban, A., Rego, C., Carbonell-Bejerano, P., Nascimento, T., Sousa, L., Martínez-Zapater, J.M., Kopka, J. and Fortes, A.M. (2015). Transcriptome and metabolome reprogramming in *Vitis vinifera* cv. Trincadeira berries upon infection with *Botrytis cinerea*. *Journal of Experimental Botany* 66(7): 1769-1785.

Ash, G., McDonald, C., Ellison, P., Mcdonaldb, C., Ellison, P. and McDonald, C. (1998). An Expert System for the Management of *Botrytis cinerea* in Australian Vineyards, I. *Development Agricultural Systems* 56(2): 185-207.

Bertelli, A. (2009). Grapes, wines, resveratrol and heart health. *Journal of Cardiovascular Pharmacology* 54: 468-476.

Bock, G., Benda, I. and Schreier, P. (1988). Microbial transformation of geraniol and nerol by *Botrytis cinerea*. *Applied Microbiology and Biotechnology* 27: 351-357.

Boonham, N., Glover, R., Tomlinson, J. and Mumford, R. (2008). Exploiting generic platform technologies for the detection and identification of plant pathogens. *European Journal of Plant Pathology* 121: 355-363.

Cadle-Davidson, L. (2008). Monitoring pathogenesis of natural *Botrytis cinerea* infections in developing grape berries. *American Journal of Enology and Viticulture* 59: 387-395.

Cãmara, J.S., Alves, M.A. and Marques, J.C. (2006). Changes in volatile composition of Madeira wines during their oxidative ageing. *Analytica Chimica Acta* 563: 188-197.

Celik, M., Kalpulov, T., Zutahy, Y., Ish-shalom, S., Lurie, S. and Lichter, A. (2009). Quantitative and qualitative analysis of *Botrytis* inoculated on table grapes by qPCR and antibodies. *Postharvest Biology and Technology* 52: 235-239.

Choquer, M., Fournier, E., Kunz, C., Levis, C., Pradier, J.M., Simon, A. and Viaud, M. (2008). *Botrytis cinerea* virulence factors: New insights into a necrotrophic and polyphageous pathogen. *FEMS Microbiology Letters* 277: 1-10.

Cioch, M. and Tuszyński, T. (2014). Biologiczne metody odkwaszania win gronowych. *Nauki Inżynierskie i Technologie* 1: 9-23.

Claire, D.J.R. (2000). Managing Bunch Rots. Proceedings ASVO Viticulture Seminar. Mildura Arts Centre, Mildura, Victoria.

Couteux, A. and Lejeune, V. (2003). Index phytosanitaire Acta 2003, 38è edition, ACTA, Paris, France.

Dávalos, A. and Lasunción, M.A. (2009). Health-promoting effects of wine phenolics. *In*: M.V. Moreno-Arribas and M.C. Polo (Eds.). Wine Chemistry and Biochemistry. Springer, New York.

Dewey, F.M., Hill, M. and DeScenzo, R. (2008). Quantification of *Botrytis* and laccase in wine grapes. *American Journal of Enology and Viticulture* 59: 47-54.

Dewey, F., Meyer, U. and Danks, C. (2005). Rapid immunoassays for stable *Botrytis* antigens in pre- and postsymptomatic grape berries, grape juice and wines. Proceedings of the American Society of Enology and Viticulture 56th Annual Meeting, 302A-303A.

Dewey, F.M. and Yohalem, D. (2007). Detection, quantification and immunolocalisation of *Botrytis* species. *In*: Y. Elad *et al.* (Eds.). Botrytis: Biology, Pathology and Control. Springer.

Dittrich, H.H. and Grossmann, H. (2011). Mikrobiologie des Weines, 4th ed. Verlag Ulmer, Stuttgart.

Doehlemann, G., Molitor, F. and Hahn, M. (2005). Molecular and functional characterisation of a fructose specific transporter from the grey mould fungus *Botrytis cinerea*. *Fungal Genetics and Biology* 42: 601-610.

Doehlemann, G., Berndt, P. and Hahn, M. (2006). Trehalose metabolism is important for heat stress tolerance and spore germination of *Botrytis cinerea*. Microbiology 152: 2625-2634.

Doneche, B.J. (1993). Botrytised Wines in Wine Microbiology and Biotechnology. G.H. Fleet (Ed.). Harwood Academic Publishers.

Dorado, N., Berneji, E., Gonzalez, J.L., Sanchez, A. and Luma, N. (2001). Development influence of *Botrytis cinerea* on grapes. *Advances in Food Science* 23: 153-159.

Eder, R., Brandes, W. and Paar, E. (2002). Influence of grape rot and fining agents on the contents of biogenic amines in musts and wines. *Mitteilungen Klosterneuburg* 52: 204-217.

Elad, Y., Williamson, B., Tudzynski, P. and Delen, N. (2007). Botrytis: Biology, Pathology and Control. Springer.

Ellis, M., Doohan, D., Bordelon, B., Welty, C., Williams, R., Funt, R. and Brown, M. (2004). Midwest Small Fruit Pest Management Handbook. The Ohio State University Extension.

Ellison, P., Ash, G. and McDonald, C. (1997). An Expert System for the Management of *Botrytis cinerea* in Australian Vineyards, I. *Development, Agricultural Systems* 56: 185-207.

Elmer, P.A.G. and Michailides, T.J. (2004). Epidemiology of *Botrytis cinerea* in orchard and vine crops. pp. 243-272. *In*: Y. Elad, B. Williamson, P. Tudzynski and N. Delen (Eds.). Botrytis: Biology, Pathology and Control. Dordrecht, Kluwer Academic Publishers.

Elmer, P.A. and Reglinski, T. (2006). Biosuppression of *Botrytis cinerea* in grapes. *Plant Pathology* 55: 155-177.

Evans, K.J. (2010). Botrytis Management. Integrated Management of Botrytis Bunch Rot. Fact Sheet. Tasmanian Institute of Agricultural Research, University of Tasmania.

Fabreges, C. and Birchmore, R. (1998). Pyrimethanil: Monitoring the sensitivity of *B. cinerea* in the vineyard. *Phytoma* 505: 38-41.

Fillion, L., Ageorges, A., Pacaud, S., Coutos-Thévenot, P., Lemoine, R., Romieu, C. and Derlot, S. (1999). American Society of Plant Physiologists. *Plant Physiology* 120: 1083-1093.

Fugelsang, K.C. (1997). Wine Microbiology. California State University at Fresno. Springer.

Gishen, M., Cozzolino, D. and Dambergs, R.G. (2010). The analysis of grapes, wine and other alcoholic beverages by infrared spectroscopy. pp. 539-556. *In*: E.C.Y. Li-Chan, P.R. Griffiths and J.M. Chalmers (Eds.). Applications of Vibrational Spectroscopy in Food Science, Analysis of Food, Drink and Related Materials. John Wiley & Sons Ltd. West Sussex, United Kingdom.

Gishen, M., Dambergs, R.G. and Cozzolino, D. (2005). Grape and wine analysis – Enhancing the power of spectroscopy with chemometrics: A review of some applications in the Australian wine industry. *Australian Journal of Grape and Wine Research* 11: 296-305.

Gisi, U., Sierotzki, H., Cook, A. and McCaffery, A. (2002). Mechanisms influencing the evolution of resistance to Qo inhibitor fungicides. *Pest Management Science* 58: 859-867.

Goetz, G., Fkyerat, A., Métais, N., Kunz, M., Tabacchi, R., Pezet, R. and Pont, V. (1999). Resistance factors to grey mould in grape berries: Identification of some phenolics inhibitors of *Botrytis cinerea* stilbene oxidase. *Phytochemistry* 52: 759-767.

Gullino, M.L. and Kuijpers, L.A.M. (1994). Social and political implications of managing plant diseases with restricted fungicides in Europe. *Annual Review of Phytopathology* 32: 559-579.

Haenssler, G. and Pontzen, R. (1999). Effect of fenhexamid on the development of *Botrytis cinerea*. *Pflanzenschutz-Nachrichten Bayer* 52: 158-176.

Hajós, G., Sass-Kiss, A., Szerdahelyi, E. and Bardocz, S. (2000). Changes in biogenic amine content of Tokaj grapes, wines and aszu wines. *Journal of Food Science* 65: 1142-1144.

Hernández-Orte, P., Peña-Gallego, A., Ibarz, M.J., Cacho, J. and Ferreira, V. (2006). Determination of the biogenic amines in musts and wines before and after malolactic fermentation using 6-aminoquinolyl-

N-hydroxysuccinimidyl carbamate as the derivatising agent. *Journal of Chromatography A* 1129: 160-164.

Holz, G., Coertze, S. and Williamson, B. (2004). The ecology of *Botrytis* on plant surfaces. *In*: Y. Elad, B. Williamson, P. Tudzynski and N. Delen (Eds.). Botrytis: Biology, Pathology and Control. Kluwer Academic Publishers. Dordrecht, The Netherlands.

Holz, G., Gütschow, M., Coertze, S. and Calitz, F. (2003). Occurrence of *Botrytis cinerea* and subsequent disease expression at different positions on leaves and bunches of grape. *Plant Disease* 87: 351-358.

Hong, Y.S., Cilindre, C., Liger-Belair, G., Jeandet, P., Hertkorn, N. and Schmitt- Kopplin, P. (2011). Metabolic influence of *Botrytis cinerea* infection in champagne base wine. *Journal of Agricultural and Food Chemistry* 59: 7237-7245.

Jackson, R.S. (2008). Wine Science: Principles and Applications. Academic Press, San Diego.

Jeandet, P., Bessis, R. and Gautheron, B. (1991). The production of resveratrol (3,5,4-trihydroxystilbene) by grape berries in different developmental stages. *American Journal of Enology and Viticulture* 42: 41-46.

Jeandet, P., Bessis, R., Sbaghi, M., Meunier, P. and Trollat, P. (1995). Resveratrol content of wines of different ages: Relationship with fungal disease pressure in the vineyard. *American Journal of Enology and Viticulture* 46: 1-4.

Kállay, M. (2003). Determination of biogenic amine-content of Tokaj wine specialities. *International Journal of Horticultural Science* 9(3-4): 87-90.

Kállay, M. (2005). Some Thoughts on Tokaji Wines. *Hung. Agricultural Research* 14(2): 4-8, Tokaj-special edition.

Kállay, M. and Bene, Z. (2003). Study on the penicillin-content of botrytised wines and noble rotted berries in Tokaj-region. *International Journal of Horticultural Science* 9: 91-94.

Kállay, M., Török, Z. and Korány, K. (1999). Investigation of the antioxidant effect of Hungarian white wines and Tokaj wine specialties. *International Journal of Horticultural Science* 5: 22-27.

Kaschak, E., Göhring, N., König, H. and Pfeiffer, P. (2009). Biogene Amine in deutsche Weinen: Analyse und Bewertung nach Anwendung verschiedener HPLC – Verfahren. *Deutsche Lebensmittel Rundschau* 105: 375-382.

Kiss, J., Korbasz, M. and Sass-Kiss, A. (2006). Study of amine composition of botrytised grape berries. *Journal of Agricultural and Food Chemistry* 54: 8909-8918.

Kretschmer, M., Kassemeyer, H.H. and Hahn, M. (2007). Age-dependent grey mould susceptibility and tissue-specific defence gene activation of grapevine berry skins after infection by *Botrytis cinerea. Journal of Phytopathology* 155: 258-263.

Ky, I., Lorrain, B., Jourdes, M., Pasquier, G., Fermaud, M., Gèny, L., Rey, P., Doneche, B. and Teissedre, P.L. (2012). Assessment of grey mould (*Botrytis cinerea*) impact on phenolic and sensory quality of Bordeaux grapes, musts and wines for two consecutive vintages. *Australian Journal of Grape and Wine Research* 18: 215-226.

La Guerche, S. (2006). Mouldy-earthy faults in wines: Geosmin identified as the main compound responsible. *International Journal of Viticulture and Enology* 5: 1-10.

Landrault, N., Larronde, F., Delaunay, J.C., Castagnino, C., Vercauteren, J., Merillon, J.M., Gasc, F., Cros, G. and Teissedre, P.L. (2002). Levels of stilbene oligomers and astilbin in French varietal wines and in grapes during noble rot development. *Journal of Agricultural and Food Chemistry* 50: 2046-2052.

Latorre, B.A., Spadaro, I. and Rioja, M.E. (2002). Occurrence of resistant strains of *Botrytis cinerea* to anilinopyrimidine fungicides in table grapes in Chile. *Crop Protection* 21: 957-961.

Leitão, M.C., Marques, A.P. and San Romão, M.V. (2005). A survey of biogenic amines in commercial Portuguese wines. *Food Control* 16: 199-204.

Leroux, P. (2004). Chemical control of *Botrytis* and its resistance to chemical fungicides. *In*: Y. Elad, B. Williamson, P. Tudzynski and N. Delen (Eds.). Botrytis: Biology, Pathology and Control. Kluwer Academic. Dordrecht, The Netherlands.

Leroux, P. (2007). Chemical control of Botrytis and its resistance to chemical fungicides. *In*: Y. Elad *et al*. Botrytis: Biology, Pathology and Conrol. Springer.

McCartney, H., Foster, S., Fraaije, B. and Ward, E. (2003). Molecular diagnostics for fungal plant pathogens. *Pest Management Science* 59: 129-142.

Magyar, I. (2011). Botrytised Wines. *Advances in Food and Nutrition Research* 63: 147-206.

Maintz, L. and Novak, N. (2007). Histamine and histamine intolerance. *American Journal of Clinical Nutrition* 85: 1185-1196.

Marcobal, A., Martin-Álvarez, P.J., Polo, M.C., Munoz, R. and Morno-Arribas, M.V. (2006). Formation of biogenic amines throughout the industrial manufacture of red wine. *Journal of Food Protection* 69: 391-396.

Marques, A.P., Leitão, M.C. and San Romão, M.V. (2008). Biogenic amines in wines: Influence of oenological factors. *Food Chemistry* 107: 853-860.

Mazurek, J. (2007). http://winnicejaworek.pl/pdf/Szkodliwosc%20i%20zwalczanie%20szarej%20 plesni.pdf.

Miklósy, E., Kalmár, Z. and Kerényi, Z. (2004). Identification of some characteristic aroma compounds in noble rotted grape berries and aszú wines from Tokaj by GC-MS. *Acta Alimentaria Hungary* 3: 215-226.

Miklósy, E. and Kerényi, Z. (2004). Comparison of volatile aroma components in noble rotted grape berries from two different locations of Tokaj wine district in Hungary. *Analytica Chimica Acta* 513: 177-181.

Misc. (2002). Good plant protection practice: Grapevine - PP 2/23(1). *EPPO Bulletin* 32(2): 371-392.

Moyano, C., Alfonso, C., Gallego, J., Raposo, R. and Melgarejo, P. (2003). Comparison of RAPD and AFLP marker analysis as a means to study the genetic structure of *Botrytis cinerea* populations. *European Journal of Plant Pathology* 109: 515-522.

Moyer, M. and Grove, G. (2011). Botrytis bunch rot in commercial Washington grape production. Washington State University Extension Fact Sheet FS046E, 1-5.

Mundy, D.C., Agnew, R.H. and Wood, P.N. (2012). Grape tendrils as an inoculum source of *Botrytis cinerea* in vineyards – A review. *New Zealand Plant Protection* 65: 218-227.

Negri, S., Lovato, A., Boscaini, F., Salvetti, E., Torriani, S., Commisso, M., Danzi, R., Ugliano, M., Polverari, A., Tornielli, G.B. and Guzzo, F. (2017). The Induction of noble rot (*Botrytis cinerea*) infection during postharvest withering changes the metabolome of grapevine berries (*Vitis vinifera* L., cv. Garganega). *Frontiers in Plant Science* 8: 1-12.

Obanor, F.O., Williamson, K., Mundy, D.C., Wood, P.N. and Walter, M. (2004). Optimisation of PTA-ELISA Detection and quantification of *Botrytis cinerea* infections in grapes. *New Zealand Plant Protection* 57: 130-137.

Pearson, G.D.A., Goheen, C.A. and Gadoury, D.M. (1989). Compendium of grape diseases. *Mycologia.* The American Phytopathological Society, 93.

Pezet, R. (1998). Purification and characterization of a 32-kDa laccase-like stilbene oxidase produced by *Botrytis cinerea* Pers., Fr. *FEMS Microbiology Letters* 167: 203-208.

Pezet, R., Viret, O. and Gindro, K. (2004). Plant-microbe interaction: The *Botrytis* grey mould of grapes – Biology, biochemistry, epidemiology and control management. pp. 71-116. *In*: A. Hemantaranjan (Ed.). Advances in Plant Physiology. Scientific Publishers. Jodhpur, India.

Pezet, R., Viret, O., Perret, C. and Tabacchi, R. (2003). Latency of *Botrytis cinerea* Pers.: Fr. and biochemical studies during growth and ripening of two grape berry cultivars, respectively susceptible and resistant to grey mould. *Journal of Phytopathology, Phytopathologische Zeitschrift* 151: 208-214.

Pinar, A.L., Rauhut, D., Ruehl, E. and Buettner, A. (2017). Effects of bunch rot (*Botrytis cinerea*) and powdery mildew (*Erysiphe necator*) fungal diseases on wine aroma. *Frontiers in Chemistry* 5: 1-12.

Pour Nikfardjam, M.S. (2002). Polyphenole in Weissweinen und Traubensäften und ihre Veränderung im Verlauf der Herstellung, Tectum. Verlag Marburg, Germany.

Pour Nikfardjam, M., Laszlo, G. and Dietrich, H. (2006). Resveratrol-derivatives and antioxidative capacity in wines made from botrytised grapes. *Food Chemistry* 96: 74-79.

Pratelli, R., Lacombe, B., Torregrosa, L., Gaymard, F., Romieu, C., Thibaud, J.-B. and Sentenac, H. (2002). A grapevine gene encoding a guard cell K+ channel displays developmental regulation in the grapevine berry. *Plant Physiology* 128: 564-577.

Ribereau-Gayon, P. (1988). *Botrytis*: Advantages and disadvantages for producing quality wines. Proceedings of the Second International Cool Climate Viticulture and Oenology Symposium. Auckland, New Zealand, pp. 319-323.

Ribéreau-Gayon, P., Dubourdieau, D., Donèche, B. and Lonvaud, A. (2000). Handbook of Oenology, Vol. 1: The Microbiology of Wine and Vinifications. John Wiley and Sons, Ltd. Chichester.

Rigotti, S., Gindro, K., Richter, H., and Viret, O. (2002). Characterisation of molecular markers for specific and sensitive detection of *Botrytis cinerea* Pers.: Fr. in strawberry (Fragaria × ananassa Duch.) using PCR. *Microbiology Letters* 209: 169-174.

Rolle, L., Giordano, M., Giacosa, S., Vincenzi, S., Río Segade, S., Torchio, F., Perrone, B. and Gerbi, V. (2012). CIEL*a*b* parameters of white dehydrated grapes as quality markers according to chemical composition, volatile profile and mechanical properties. *Analytica Chemica Acta* 732: 105-113.

Romero, R., Sánchez-Viñas, M., Gázquez, D. and Bagur, M.G. (2002). Characterisation of selected Spanish table wine samples according to their biogenic amine content from liquid chromatographic determination. *Journal of Agricultural and Food Chemistry* 50: 4713-4717.

Rosslenbroich, H.J. and Stuebler, D. (2000). *Botrytis cinerea* – History of chemical control and novel fungicides for its management. *Crop Protection* 19: 557-561.

Saigal, D. and Ray, R.C. (2007). Winemaking: Microbiology, biochemistry and biotechnology. pp. 1-33. *In*: Ramesh C. Ray and O.P. Ward (Eds.). Microbial Biotechnology in Horticulture, Vol. 3. Science Publishers. New Hampshire, USA.

Sarrazin, E., Dubourdieu, D. and Darriet, P. (2007). Characterisation of key-aroma compounds of botrytized wines, influence of grape botrytisation. *Food Chemistry* 103: 536-545.

Sass-Kiss, A. and Hajós, G. (2005). Characteristic biogenic amine composition of Tokaj Aszú wines. *Acta Alimentaria Hungary* 34: 227-235.

Sass-Kiss, A., Kiss, J., Havadi, B. and Adányi, N. (2008). Multivariate statistical analysis of botrytised wines of different origin. *Food Chemistry* 110: 742-750.

Schena, L., Nigro, F., Ippolito, A. and Gallitelli, D. (2004). Real-time quantitative PCR: A new technology to detect and study phytopathogenic and antagonistic fungi. *European Journal of Plant Pathology* 110: 893-908.

Scott, E.S., Dambergs, R.G. and Stummer, B. (2010). Fungal contaminants in the vineyard and wine quality. pp. 481-509. *In*: A.G. Reynolds (Ed.). Managing Wine Quality, Vol. 1. Viticulture and Wine Quality. Woodhead Publishing Ltd. CRC Press.

Sergeeva, V., Nair, N.G., Verdanega, J.R., Shen, C., Barchia, I. and Spooner-Hart, R. (2002). First report of anilinopyrimidine resistant phenotypes in *Botrytis cinerea* on grapevines in Australia. *Australasian Plant Pathology* 31: 299-300.

Spadoro, D. and Gullino, M.L. (2005). Improving the efficacy of biocontrol agents against soil-borne pathogens. *Crop Protection* 24: 601-613.

Sponholz, W.R., Brendel, M. and Periadnadi (2004). Bildung von Zuckersäuren durch Essigsäurebakterien auf Trauben und im Wein. *Mitteilungen Klosterneuburg* 54: 77-85.

Sponholz, W.R. and Dittrich, H.H. (1985). Über die Herkunft von Gluconsäure, 2-und 5-Oxo-Gluconsäure sowie Glucuron- und Galacturonsäure in Mosten und Weinen. *Vitis* 24: 51-58.

Sponholz, W.R. and Hühn, T. (1996). Aging of wine: 1,1,6-Trimethyl-1,2-dihycronaphthalene (TDN) and 2-aminoacetophenone. *In*: T. Henick-Kling et al. (Eds.). Proceedings of the 4th International Symposium on Cool Climate Viticulture and Enology. New York State Agricultural Experimental Station. Geneva, NY.

Stratakou, I. and van der Fels-Klerx, H.J. (2010). Mycotoxins in grapes and wine in Europe: Occurrence, factors affecting the occurrence and related toxicological effects. *World Mycotoxin Journal* 3: 283-300.

Suarez, M., Walsh, K., Boonham, N., O'Neill, T., Pearson, S. and Barker, I. (2005). Development of real-time PCR (Taqman®) assays for the detection and quantification of *Botrytis cinerea* in plants. *Plant Physiology and Biochemistry* 43: 890-899.

Tamura, H., Mizutani, A., Yukioka, H., Miki, N., Ohba, K. and Masuko, M. (1999). Effect of the methoxyiminoacetamide fungicide, SSF 129, on respiratory activity in *Botrytis cinerea*. *Pesticide Science* 55: 681-686.

The Australian Wine Research Institute (2009). Wine and Health Information. Australia.

Topolovec-Pintarić, S. (2011). Resistance to Botryticides. *In*: Thajuddin N. (Ed.). Fungicides – Beneficial and Harmful Aspects. InTech. Europe.

Torelli, E., Firrao, G., Locci, R. and Gobbi, E. (2005). Ochratoxin A producing strains of *Penicillium spp.* isolated from grapes used for the production of 'passito' wines. *International Journal of Food Microbiology* 106: 307-312.

Williamson, B., Tudzynski, B., Tudzynski, P. and Van Kan, J.A.L. (2007). *Botrytis cinerea*: The cause of grey mould disease. *Molecular Plant Pathology* 8(5): 561-580.

Wood, P.M. and Hollomon, D.H. (2003). A critical evaluation of the role of alternative oxidase in the performance of strobilurin and related fungicides acting at the Qo site of complex III. *Pest Management Science* 59: 499-511.

Valero, A., Marín, S., Ramos, A.J. and Sanchis, V. (2008). Survey: Ochratoxin A in European special wines. *Food Chemistry* 108: 593-599.

Van Baarlen, P., Legendre, L. and van Kan, J.A.L. (2007). Plant defence compounds against *Botrytis* infection. *In*: Y. Elad, B. Williamson, P. Tudzynski and N. Delen (Eds.). Botrytis: Biology, Pathology and Control. Springer.

Van Baarlen, P., Staats, M. and Van Kan, J. (2004). Induction of programmed cell death in lily by the fungal pathogen *Botrytis elliptica*. *Molecular Plant Pathology* 5: 559-574.

Varga, J., Koncz, Z., Kocsubé, S., Mátrai, T., Téren, J., Ostry, V., Skarkova, J., Ruprich, J., Kubatova, A. and Kozakiewitz, Z. (2007). Mycobiota of grapes collected in Hungarian and Czech vineyards in 2004. *Acta Aliment. Hungary* 36(3): 329-341.

Viret, O., Keller, M., Jaudzems, V.G. and Cole, F.M. (2004). *Botrytis cinerea* infection of grape flowers: Light and electron microscopical studies of infection sites. *Phytopathology* 94(8): 850-857.

Yildirim, H.K., Üren, A. and Yücel, U. (2007). Evaluation of biogenic amines in organic and non-organic wines by HPLC OPA derivatisation. *Food Technology and Biotechnology* 45: 62-68.

Zheng, L., Campbell, M., Murphy, J., Lam, S. and Xu, J.-R. (2000). The BMP1 gene is essential for pathogenicity in the gray mould fungus *Botrytis cinerea*. *Molecular Plant-Microbe Interactions* 13: 724-732.

Zimmerli, B. and Dick, R. (1996). Ochratoxin A in table wine and grape juice: Occurrence and risk assessment. *Food Additives & Contaminants* 13: 6656-6668.

Zoecklein, B.W., Fugelsang, K.C., Gump, B.H. and Nury, F.S. (1999). Wine Analysis and Production. Springer, New York.

# 8 Genetic Engineering in Fruit Crops

**K. Kumar[1*], R. Kaur[2], and Shilpa[2]**

[1] Faculty of Agriculture, Shoolini University, Bajhol-Solan (H.P.) 173229, India

[2] Department of Biotechnology, Dr Yashwant Singh Parmar University of Horticulture and Forestry, Nauni – Solan (H.P.) 173230, India

## 1. Introduction

Fruit crops are an important agricultural produce, adding billions of dollars annually to the overall economy and are major sources of income for developing countries. Approximately 100 million acres of land has been devoted to their production worldwide and the livelihood of millions of farming families depends on continued global trade. Fruits provide essential nutrients, vitamins, antioxidants and fibres in individual's diet (Rai and Shekhawat, 2014). However, the present world fruit industry is based on a few genotypes. Intense selection and fixation of genotypes by clonal propagation leads to a narrow germplasm base (Janick, 1992). Thus, it was needed to change the cultivars. This can be achieved by conventional breeding, but it is hampered by the long generation time and juvenile periods, complex reproductive biology, high levels of heterozygosity, limited genetic sources and linkage drag of undesirable traits from wild relatives. Chemical controls and different managements have also been practically used in commercial production of different fruits in order to overcome disease infection, to improve shelf-life and to generate dwarf trees. However, these practices are either inefficient or unfriendly to human health and environment. In contrast, genetic engineering offers a better possibility to improve plant traits (Zhu *et al.*, 2004). During the last about two decades, genetic engineering has been successfully used to improve tolerance to biotic and abiotic stresses, increased fruit yield, improved shelf-life of fruit, reduced generation time and production of fruit with superior nutritional value. Selectable marker genes are widely used for the efficient transformation of crop plants. Mostly antibiotic or herbicide resistance marker genes are preferred because they are more efficient. But due to consumer concerns, considerable effort is being put into developing strategies (homologous recombination, site-specific recombination, and co-transformation) to eliminate the marker gene from the nuclear or chloroplast genome after selection. The absence of selectable marker gene in the final product and the introduced gene(s) derived from the same plant will increase the consumer's acceptance (Krens *et al.*, 2004). This can be fulfilled by new genetic engineering approaches, like cisgenesis or intragenesis. Still, the development of transgenic fruit plants and their commercialisation are also hindered by many regulatory and social issues. Hence, the future use of transformed horticultural crops on a commercial basis depends upon thorough evaluation of the potential environmental and public health risks of the modified plants, stability of transgene over a long period of time and the effect of the gene on crop and fruit characteristics.

## 2. Methods of Genetic Transformation

Tissue culture plays a critical role in genetic transformation of a plant species because the initial requirement for most gene transfer systems is efficient regeneration of plantlets from cells carrying a foreign gene. Besides plant tissue culture, various methods of gene transfer are also necessary to obtain transgenic plants. Successful generation of transgenic plants requires a combination of the following:

- A cloned gene with elements, such as promoters, enhancing and targeting sequences for its regulatory expression
- A reliable method for delivery and stable integration of DNA into cells
- A 'tissue culture' method to recover and regenerate intact plant from transformed organs or tissues

*Corresponding author: drkrishankumar@gmail.com

Till date, a number of strategies have been developed for identification, isolation and cloning of desirable gene sequences from almost any organism. Next step is to deliver the desired gene sequence into the cell of target organism. This can be achieved by using one of the following methods:

- *Agrobacterium* – mediated gene transfer
- Particle bombardment or biolistic delivery
- Electroporation
- PEG-induced DNA uptake
- DNA delivery via silicon carbide fibres
- Laser-mediated DNA transfer
- Microinjection
- Liposome mediated gene transfer or lipofection
- DEAE-dextran (diethylaminoethyloethyl-dextran) mediated transfection
- Agroinfiltration
- Vacuum infiltration

Besides these methods, there are many other transformation methods, such as ultrasound, seed imbibitions and electrophoresis, but these are not as successful as the methods mentioned earlier. Since the methods listed above have proved successful and applicable to several plant species, their principles will be explained in brief.

## 2.1. Agrobacterium-*mediated Gene Transfer*

Most transgenic plants produced to date were created through the use of the *Agrobacterium* system. *A. tumefaciens* is the agent of crown gall disease and produces crown galls on infected species. The usefulness of this bacterium as a gene transfer system was first recognised when it was demonstrated that the crown galls were actually produced as a result of transfer and integration of genes from the bacterium into the genome of the plant cells (Hamilton and Fall, 1971). Virulent strains of *Agrobacterium* contain large Ti-(tumour inducing) plasmids, which are responsible for the DNA transfer and the following disease symptoms. Ti-plasmid contains two sets of sequences necessary for gene transfer to plants: one is T-DNA (transferred DNA) region that is transferred to the plant, and another is the *Vir* (virulence) genes which are triggered by phenolic compounds, especially acetosyringone and hydroxy acetosyringone secreted by wounded cells. The T-DNA regions is flanked by 25 base pairs on right and left borders and these are responsible for the T-DNA transfer to infected plant cells. Undesirable DNA sequences, e.g. genes for auxin, cytokinin or opine biosynthesis (sugar derivatives) present in the T-DNA of wild type strains have been deleted as phytohormones synthesis interferes with the plant regeneration and replaced with marker gene or desirable gene.

In addition to *A. tumefaciens, A. rhizogenes* is also capable of plant transformation by a similar process. Instead of tumours, *A. rhizogenes* causes 'hairy-roots' upon infection due to production of auxins in transformed cells. *A. rhizogenes* strains have also been engineered for transformation but are not widely used.

*Agrobacterium*-mediated transformation being simple and widely applicable, remains the method of choice to transform plants. Along with a suitable protocol for regeneration of complete plants from transformed cells, a large number of transformants can be generated. The whole procedure of DNA transfer and integration is very precise. Additionally, the resources required for transformation are not very difficult. The accurate site of integration of T-DNA into the host chromosome seems to be random. Independent transgenics for the same gene show wide variations in the levels of expression of the transgene. Hence, proper evaluation of transgenics for stable expression over generations is necessary.

## 2.2. Biolistic-mediated Transformation

Transformation of plant cells through microprojectile coated with DNA was developed by Sanford and Johnston (1985). In this method, the DNA to be used for transformation is coated with 1-2 μm tungsten or gold particles which are then mechanically 'shot' directly into the plant tissue by custom-made device, using either an electrical discharge, high gas pressure (such as compressed air, nitrogen or helium) or gun powder. The particles penetrate the cells and the adsorbed DNA is delivered into the cells. Use of biolistic gun is relatively simple and rapid. Its main advantage is its non-specificity resulting from the physical properties of the method. Moreover, any type of cells can be targetted, including embryos, pollen grains, microspores, leaf, stem, apical meristem, etc. Though particle bombardment has yielded success with a range of plant species, however, the high cost of equipments makes it expensive for individual laboratories with limited resources and also the recovery of transgenics is rather low.

## 2.3. Electroporation

Electroporation is perforation caused in the cell membranes by the use of electric current. Suitable ionic solution containing linearised recombinant plasmid DNA is used to suspend the protoplasts. This electroporation mixture is then exposed to the suitable voltage-pulse combination which depends upon the plant species and the source of protoplast. Protoplasts are then cultured in order to obtain cell colonies and plants. This method was initially developed for bacteria, yeast and animal cells and later adapted to plants. With the development of protoplast technology, electroporation was used to transform protoplasts (Fromm *et al.*, 1985). Later, the technique was found applicable to even intact tissues (D'Halluin *et al.*, 1992). This method has received less attention because of the wide success of particle bombardment technology. Yet, this method is widely valued for introduction of macromolecules into cells and to study transient expression of introduced DNA sequences.

## 2.4. Polyethylene Glycol (PEG)-mediated Uptake of DNA

PEG is widely used for fusion of animal cells and plant protoplasts (Kao and Michayluk, 1974). PEG precipitates the DNA on to the outer surface of plasma lemma and this precipitate is taken up by endocytosis. In this approach, protoplasts are suspended in a transformation medium followed by addition of linearised plasmid DNA containing the gene construct. After this, PEG is added to the mixture with pH adjustment of 8. The application of PEG to protoplast suspension leads to contraction of their volume. This results in endocytic vesiculation. DNA-cation complexes adsorbed to the membrane thus get introduced into the cells. PEG-mediated transformation is perhaps the least expensive of all the methods of transformation available at present.

## 2.5. Silicon Carbide Fibres

Silicon carbide fibre-mediated transformation involves vortexing of cells and fibres in a buffer that contains DNA. The fibres create holes in the cells and through the force of vortexing, DNA enters into the cell. This is a relatively recent method and hence has not been widely tested. The simplicity of the technique is its great desirability; however, the routine use of this method is hampered by hazardous nature of silicon fibres.

## 2.6 Microinjection

Microinjection is the direct mechanical introduction of DNA into cells under microscopical control. In this method, a glass needle or micropipette of fine tip (0.5-1.0 micrometre diameter) is directly used to inject the DNA into plant protoplasts or cells (specifically into the nucleus or cytoplasm). This method is used to introduce DNA into large cells, such as oocytes, eggs and the cells of early embryo. By examination through a microscope, a cell is held in place with gentle suction while being manipulated with the use of a blunt capillary. A fine pipette is then used to insert the DNA into the cytoplasm or nucleus. This technique is more effective and easier to transform protoplasts as compared to cells because of interference caused by the cell wall. To obtain high rates of transformation, the DNA should be introduced into the nucleus or cytoplasm. This method is more successful in non-vacuolated embryonic cells with dense cytoplasm than large vacuolated cells because in the presence of large vacuoles, the DNA is delivered into the vacuole which is then degraded. But this technique is not in routine use because it is more demanding and time consuming.

## 2.7. Liposome-mediated Gene Transfer or Lipofection

Liposomes are artificial circular lipid vesicles surrounded by a synthetic membrane of phospholipids. They have an aqueous interior which can carry nucleic acids. Liposomes encapsulate the DNA fragments and then adhere to the protoplast membranes to fuse with them and transfer the DNA fragments. Thus, the DNA enters the cell and then the nucleus. Lipofection is a very efficient technique used to transfer genes in bacterial, animal and plant cells, but is not in much use because of its low transformation frequencies. This process generally involves three steps:

- Adhesion of liposomes to the protoplast surface
- Fusion of liposomes at the site of adhesion
- Release of plasmid inside the protoplast or animal cell

## 2.8. DEAE-dextran (Diethylaminoethy Loethyl-dextran)-mediated Transfection

DEAE-dextran transfection is a method for introduction of DNA into the eukaryotic cells. DEAE-dextran facilitates DNA binding to cell membranes and entry of the DNA into the cell via endocytosis. As DEAE-dextran is toxic to cells, the transfection conditions for individual cell lines may require careful optimisation for both DEAE-dextran concentration and exposure times. In general, DEAE-dextran-mediated transfection is successful in transient, but not stable transfection of cells. At higher DEAE-dextran concentrations, the exposure time to cells can be shortened in order to minimise cell death.

## 2.9. Agroinfiltration

Agroinfiltration is a method in plant biology to induce transient expression of genes in a plant or to produce a desired protein. In this method, a suspension of *A. tumefaciens* is injected into a plant leaf, where it transfers the desired gene to plant cells.

First step of the protocol is to introduce a gene of interest to a strain of *Agrobacterium*. Subsequently the strain is grown in a liquid culture and the resulting bacteria are washed and suspended into a suitable buffer solution. This solution is then placed in a syringe (without a needle). The tip of the syringe is pressed against the underside of a leaf while simultaneously applying gentle counter pressure to the other side of the leaf. The *Agrobacterium* solution is then injected into the airspaces inside the leaf through stomata, or sometimes through a tiny incision made to the underside of the leaf. The benefits of agroinfiltration, when compared to traditional plant transformation, are its speed and convenience.

## 2.10. Vacuum Infiltration

Vacuum infiltration is another way to penetrate *Agrobacterium* deep into plant tissue. In this procedure leaf disks, leaves, or whole plants are submerged in a beaker containing the solution, and the beaker is placed in a vacuum chamber. The vacuum is then applied, forcing air out of the stomata. When the vacuum is released, the difference in pressure forces the solution through the stomata into the mesophyll.

Once inside the leaf, the *Agrobacterium* remains in the intercellular space and transfers the gene of interest in high copy numbers into the plant cells. The gene is then transiently expressed (no selection for stable integration is performed). The plant can be monitored for a possible effect in the phenotype, subjected to experimental conditions and then harvested and used for purification of the protein of interest.

## 2.11. Clustered Regularly Interspaced Short Palindromic Repeats (CRISPRs)

Recently, in 2012, a new technique of targetted genome editing through engineered nuclease emerged, known as clustered regularly interspaced short palindromic repeats (CRISPRs) (Sharma *et al.*, 2017). In general, CRISPR is a family of DNA sequences in bacteria containing snippets of DNA from virus that have attacked the bacteria which are used by the bacteria to detect and destroy DNA from further attack by similar virus. This basic technique now has been used in many fruit crops, like apple and grapes, for successful alteration of characters by delivering cas9 nuclease complexed with synthetic guide RNA (gRNA) into the cell. The cell's genome can be cut at a desired location, allowing the existing gene to be either silenced or removed and new ones added. Main steps followed in this technique are: 1) cas9 protein forms a complex with gRNA in the cell. 2) This complex matches to gDNA adjacent to a spacer. 3) cas9-RNA complex cuts the double strand of DNA. 4) Programmed DNA may be inserted at the cut.

The fruit crop-wise progress made through genetic transformation is presented as under:

## Apple

Apple has been successfully transformed using GUS (beta-glucuronidase), *npt*II (Neomycin phosphotransferase II) and the nopaline synthase gene from *A. tumefaciens.* In apple, genetic transformation has been carried out for various traits, like disease resistance, improved phenotypic characters, enhanced nutrition level and salt tolerance.

### *(i) Disease Resistance*

A lot of emphasis has been laid on transformation of apple for resistance to scab (*Venturia inequalis*). Attempts have been made to incorporate scab resistance by introducing various resistance genes, viz. a wheat puroindoline B (*pinB*) gene coding a protein with antifungal activity (Faize *et al.,* 2004), *Vfa1, Vfa2,* or *Vfa41* (Malnoy *et al*., 2008), barley hordothionin gene (*hth*) (Krens *et al*., 2011).

Another disease which causes great damage to apple crop is fire blight caused by *Erwinia amylovora*. Several studies were carried out to overcome this disease through genetic transformation in apple cultivars with resistance genes, namely, viral EPS-depolymerase gene (*dpo*) (Flachowsky *et al*., 2008), *fb_mr5* originating from crab apple accession *Malus robusta*5 (Mr5) (Giovanni *et al*., 2014). To increase resistance to this disease, a very recent technique, i.e. CRISPR has been used by Malnoy *et al*. (2016) in which they targeted *DIPM-1*, *DIPM- 2* and *DIPM-4* genes in the Golden Delicious variety.

Successful genetic transformation has also been done to provide powdery mildew (caused by *Podosphaera leucotricha*) resistance by Chen *et al*. (2012), who introduced *Malus hupehensis*-derived npr1 (*Mhnpr1*) gene into the 'Fuji' apple. Four transgenic apple lines were verified by PCR and RT-PCR. Furthermore, the transgenic apple plants resisted infection by *Podosphaera leucotricha* better than the wild-type plants. Rihani *et al*. (2017) modified flavonoid pathway in apple cvs i.e. 'Holsteiner Cox' and 'Gala' via *Agrobacterium*-mediated transformation by over-expressing the *MdMyb10* transcription factor to increased effect on plant disease resistance.

### *(ii) Improved Phenotypic Characters*

In apple, a breakthrough was made in the reduction of juvenile phase using transgenic approach. Kotoda *et al.* (2000) cloned the '*Mdtfl*' (*Malus* x *domesticatf*) gene homologous to Arabidopsisterminal flower 1 (*tfl*1) gene which suppresses floral meristem identity genes, leafy (*lfy*) and apetala1 (*ap1*). Transgenic apples, expressing antisense '*Mdtfl*' gene, flowered at about eight to 15 months after grafting on rootstocks; on the other hand, non-transformed control plants did not flower in five years. Expression of genes, i.e. '*Bpmads4*' gene of silver birch (Flachowsky *et al*., 2007), 'florigen-like' gene flowering locus T (*ft*) from poplar (Flachowsky and Hanke, 2012) also showed acceleration in the onset of flowering in apple plants.

The apple rootstock M26 (*Malus pumila*) is a very popular rootstock with semi-dwarf habits and gives high quality fruit at a young agem but it is prone to poor prop ability in soil. So young trees require staking in windy locations. To combat this problem, '*rolC*' gene was introduced into M26 by *A. tumefaciens* LBA4404 for dwarfism and enhanced rooting ability. The '*rolC*' transgenic lines showed reduced stem length and increased root number *in vitro*. Rooting ability was also examined in an isolated greenhouse after mound layering (Kim *et al*., 2009). When compared with non-transgenic M26, '*rolC*' transgenic line showed higher rooting ability.

A recent study showed knocking out of S-gene expression in the pistil and overcoming of self-incompatibility in apple cultivar (Broothaerts *et al.,* 2004). The apple cultivar 'Elstar' was transformed with T-DNA constructs containing the S3-RNase cDNA in anti-sense orientation. The resultant transgenic lines were screened for their ability to set fruits following self-pollination under controlled greenhouse conditions. In 12 lines, complete '*S3 S-RNase*' gene modified the self-incompatibility behaviour of the plant, leading to either intermediate or complete self-fertility (Dreesen *et al*., 2012).

Apple fruits turn brown quickly after slicing. Though a natural phenomenon, browning has been considered an undesirable trait that causes consumer inconvenience and unnecessary waste. To solve this issue of fruit browning in apple, Okanagan Specialty Fruits (OSF), a Canadian firm located in British Columbia, developed a series of new apple varieties from widely grown existing apple varieties, such as Golden Delicious, Granny Smith, Gala and Fuji, which do not turn brown for over two weeks under appropriate conditions. To achieve the non-browning trait in these apples, gene silencing was used to turn down the expression of polyphenol oxidase (PPO), an enzyme that causes browning, by transformation with '*pgas*' (a hybrid sequence carrying four gene groups, i.e. PPO2, GPO3, APO5 and pSR7) transgene, which was directly derived from the apple genome. These genetically-engineered non-browning apples have been named as Arctic Apples. Compared with their non-transgenic controls, Arctic Apples showed 76-82 per cent reduction in PPO activities in leaves and 90-91 per cent of reduction in the mature fruit. Because of high percentage reductions in PPO activities in mature fruit, Arctic Apples show little browning (Carter, 2012; Xu, 2013) (Fig. 1). OSF is looking for market access in Canada and the US for Arctic Golden Delicious and Arctic Granny Smith since the last few years (Lehnert, 2011). In 2012, a field test application was approved to conduct study of the Arctic Apple in the state of Washington followed by USDA approval in February, 2015 (Pollack, 2015) and by FDA in March, 2015 (FDA, 2015). It became the first genetically-modified apple approved for US sale (Tracy, 2015). In 2016, three varieties were approved by the USDA (Arctic Golden, Arctic Granny and Arctic Apple Fuji) and are expected for retail sale in 2017 with each apple bearing a 'snowflake' logo.

**Reduced PPO = Nonbrowning**

Arctic® apples are the same as their conventional counterparts except they won't brown when sliced, bitten or bruised!

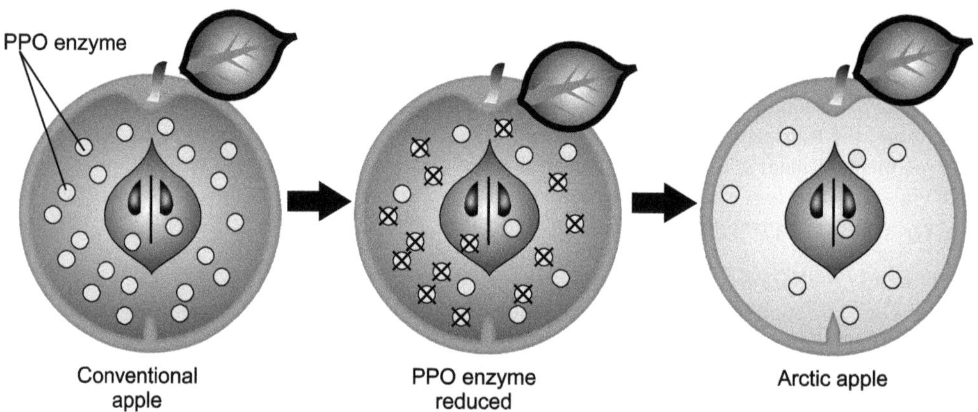


**Figure 1.** Production of Arctic Apple (*Source*: www.arcticapple.com)

### *(iii) Enhanced Nutrition Level and Salt Tolerance*

Nutrition is another important factor among the consumers. To fulfil this aspect, Cation Exchangers 1 (*cax1)* gene from *Arabidopsis thaliana* was introduced into Korean apple cv. 'Hongro' by *A. tumefaciens* to obtain transgenic apple with enhanced 'calcium' levels. Southern blot analysis of 'Hongro' transformants showed that two putative transgenic lines were integrated with '*cax1*' gene in genomic DNA. The *cax1* comparative expression levels of two transgenic lines were higher than that of non-transformants, when evaluated using a real-time PCR. These two lines were multiplied *in vitro* and micro-grafted on apple rootstocks M9 in the greenhouse. After two years of micro-grafting, the fruits came into bearing (Kim *et al*., 2010).

Soil salinity is a major factor limiting apple production in some areas. Tonoplast Na+/H+ antiporters play a critical role in salt tolerance. Li *et al.* (2010) isolated *Mdnhx1*, a vacuolar Na+/H+ antiporter from Luo-2, a salt-tolerant rootstock of apple and introduced it into apple rootstock M26 by *Agrobacterium*-mediated transformation. RT-PCR analysis indicated that the gene was highly expressed in transgenic plants, which conferred high tolerance to salt stress.

## *Apricot*

Genetic transformation studies in apricot (*Prunus armeniaca*) are relatively few and whatever attempts had been made were aimed at enhancing transformation protocols and incorporating resistance to Sharka (plum pox virus) disease. In an attempt, Petri *et al.* (2006) developed an efficient regeneration and transformation protocol in callus of mature tissues of apricot cultivars by integration of transgenes (Green

Fluorecsent Protein gene (*gfp*) and *nptII* marker genes). It was observed that a low selection pressure applied during the first days of co-culture, followed by a higher selection pressure afterwards, greatly improved transformation and selection efficiencies. Putative transformed shoots were *in vitro* multiplied and then rooted to confirm '*gfp*' expression in the roots.

It was observed that transformation protocols are genotype dependent. Wang *et al.* (2009, 2013) developed genotype independent protocols using meristematic cells and by direct shoot regeneration from the proximal zone of mature cotyledons.

*(i) Disease Resistance*

Sharka (plum pox virus) is a dreaded viral disease of genus *Prunus*. The integration of '*PPV-cp*' gene via *Agrobacterium*-mediated transformation into apricot genome has been confirmed using both GUS staining and PCR analysis (Machado *et al.,* 1992) in immature embryos. The resultant transformed plants were observed to be resistant to virus infection. These results raised the possibility of transforming commercial apricot cultivars for different traits (Sansavini, 2004).

## Avocado

Genetic transformation has played an important role in avocado to provide disease resistance, herbicide resistance and improved shelf life. Various strategies have been followed to improve the efficiency of genetic transformation in avocado (Cruz *et al.*, 1998; Ahmed *et al.*, 2012).

*(i) Disease Resistance*

In an experiment, antifungal protein (*afp*) gene was used by Litz *et al.* (2007) to transform cultivars 'Suardia' and 'Hass' of avocado, using embryogenic cultures to enhance resistance to root diseases. From transformed cultures, enlarged and opaque somatic embryos were obtained by plating on to semisolid MS medium with 20 per cent coconut water. But the rate of normal germination was low and most somatic embryos produced only small shoots. Therefore, micrografting on to *in vitro* seedlings was done to rescue shoots from somatic embryos. To obtain plants *ex vitro, in vitro*-derived shoots were also grafted on to three-week old seedlings. Transformation was confirmed by PCR and transgenic plants grew luxuriantly in the greenhouse.

*(ii) Herbicide Resistance*

Attempts were made to introduce resistance to herbicide basta in somatic hybrid of cultivar 'Fuerte' using genes *glucanase, chitinase, bar, uidA* (Raharjo *et al.*, 2008). In another study, the same team of workers transformed cultivar 'Hass' using *afp*, *npt II* and *uidA* for herbicide resistance.

*(iii) Improved Shelf-life*

Another problem with avocado fruits is that they ripen on the tree and have a poor shelf-life. To overcome this problem, avocado was transformed to extend the 'on-the-tree' stay of mature fruits and to increase the shelf-life of fruits. For this embryogenic cultures were genetically transformed using *A. tumefaciens* strain EHA101 harboring SAM (S-adenosyl-L-methionine) hydrolase, a gene that blocks ethylene biosynthesis (Efendi, 2003; Litz *et al.*, 2007). Transformation was confirmed using PCR. Transgenic plants were recovered successfully and were transferred to the greenhouse, where they showed good growth.

## Banana

Genetic transformation of banana focused on disease resistance, abiotic stress tolerance and control of fruit ripening. To achieve these goals, genetic transformation was achieved in several ways. It was observed that the most successful procedure is *Agrobacterium*-mediated transformation of embryogenic suspension cultures (Hernandez *et al.*, 1999). Embryogenic cultures are usually induced from immature male (Escalant *et al.,* 1994) and female flowers (Grapin *et al.,* 2000).

*(i) Disease Resistance*

Bananas have also been suggested as an appropriate vehicle for edible vaccines (Mor *et al.,* 1998) and bananas containing malaria epitopes have been generated (Hassler, 1995). The transgenic bananas will thus be first edible vaccines that hopefully protect millions of children in the developing world against bacterial and viral infections (Gleba *et al.*, 2004).

To induce resistance to BBTV, banana cv. 'Dwarf Brazilian' was transformed using *A. tumefaciens* containing replicase-associated protein (*rep*) gene of the Hawaiian isolate of BBTV. Twenty-one transformants were found resistant to BBTV challenge and showed no bunchy top symptoms, whereas all of the control plants were infected with BBTV after a period of six months monitored in greenhouse (Borth *et al.*, 2009; Krishna *et al.*, 2013).

Another most devastating disease of banana is wilt. To overcome this disease a number of genes viz., soybean endo- beta -1,3-glucanase gene (*pROKla-Eg*) (Maziah *et al.*, 2007), sweet pepper hypersensitivity response-assisting protein (*hrap*) gene (Tripathi *et al.*, 2010), rice thaumatin-like protein (*tlp*) gene (Hu *et al.*, 2013), were transferred to banana through *Agrobacterium*-mediated transformation to produce wilt resistant plants. Bioassay of transgenic banana plants with *Fusarium* wilt pathogen showed that expression of *tlp* enhanced resistance to *Fusarium* as compared to control plants.

Tripathi *et al.* (2017) demonstrated transgenic banana expressing sweet pepper genes, i.e. *Hrap* and *Pflp* providing resistance to Xanthomonas wilt. They also enhanced resistance to mixed population of nematodes by expressing cysteine proteinase inhibitor and synthetic peptide.

#### *(ii) Abiotic Stress Tolerance*

Several reports show rise in transgenic bananas being developed with salt tolerance. Musa Stress Associated Proteins1 (*sap1*) gene (Sreedharan *et al.*, 2012) and plasma membrane intrinsic protein gene (*Musapip1;2*) (Sreedharan *et al.*, 2013) were used to transform banana plants to provide abiotic stress tolerance. It was noted that greenhouse hardened transgenic plants were tolerant to drought, salt, cold, heat and oxidative stress and also showed faster recovery towards normal growth and development.

#### *(iii) Shortening of Life Cycle*

A long juvenile cycle of banana affects the crop productivity. To overcome this problem, Talengera *et al.* (2009) transformed embryogenic cells of banana cv. 'Sukalindiizi' with '*Arabidopsis thalianacyclin D2;1*' gene. Presence, integration and transcription of the transgene were confirmed by PCR, Southern blot and reverse transcriptase PCR analyses. Regenerants showed improved leaf elongation in two lines.

### *Cherry*

Genetic transformation is mainly aimed at development of enhanced phenotypic characters. Several transformants have been regenerated on the dwarfing cherry rootstock 'Inmil' (*P. incisa* x *serrula*) with *Agrobacterium rhizogenes*. Improvement in grafting and rooting rate was reported on further grafting on two wild cherry scions (Druart and Gruselle, 2007).

Investigations were also carried out to transform pear rootstock (*Cydonia oblonga*) and sweet cherry rootstock (*Prunus cerasus* x *P. canescens*) plants with rooting stimulating '*rolB*' gene. After PCR analysis it was confirmed that the gene integrated in one transformant of *Cydonia* and eight transformants of sweet cherry. All transformants rooted *in vitro*. It was established during the investigations that *rolB* transgene did not influence proliferation rate of genetically modified sweet cherry rootstocks *in vitro*. An increase in cold hardiness of sweet cherry rootstock was noted under *in vitro* conditions, though no significant difference was observed in transformed and control plants of *Cydonia* for cold hardiness (Rugienius *et al.*, 2009).

Prunus necrotic ringspot virus (PNRSV) is a major pollen-disseminated virus that adversely affects many *Prunus* species. In this study, Song *et* al. (2013) used RNA interference (RNAi)-mediated silencing with vector containing an inverted repeat (IR) region of PNRSV was transformed into two hybrid cherry rootstocks, 'Gisela 6' (GI 148-1) and 'Gisela 7'(GI 148-8)'. One year after inoculation with PNRSV and Prune Dwarf Virus, non-transgenic 'Gisela 6' exhibited no symptoms but a significant PNRSV titre, while the transgenic 'Gisela 6' had no symptoms and minimal PNRSV titre (Fig. 2).

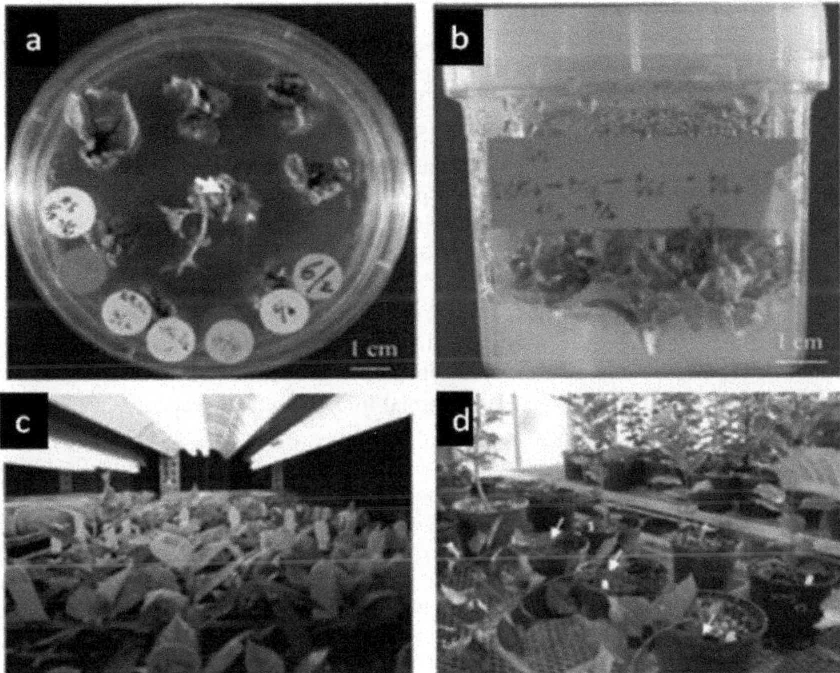


**Figure 2.** Transformation of cherry rootstocks (a) shoot regeneration from leaf explants; (b) proliferation of a putative transgenic; (c) rooting of transgenic plants; (d) growing of transgenic plants. The white arrows are showing three procumbent plants derived from one transgenic event of 'Gisela 6' (*Source*: Song *et al.*, 2013)

Color version at the end of the book

## Citrus

A lot of transformation work has been done on citrus and this helped to overcome several serious problems like disease resistance, abiotic stress tolerance, enhanced flower and fruit characters.

### *(i) Disease resistance*

Development of commercial cultivars with greater resistance to citrus canker is the best strategy for effective disease management. Several attempts were made to impart resistance against citrus canker by transforming citrus with different genes, viz. insect-derived attacin A gene (*attA*) (Boscariol *et al.*, 2006), Arabidopsis npr1 gene (*At npr1*), (Zhang *et al.*, 2010), dermaseptin coding sequence (Furman *et al.*, 2013) and *Fls2* from *Nicotiana benthamiana* (Hao *et al.*, 2016).

Citrus mosaic virus (CiMV) is one of the most harmful diseases of citrus. Rootstocks with coat protein gene were transformed which were tolerant to citrus mosaic virus (CiMV) (Iwanami, 2010). Azevedo *et al.* (2006) produced transgenic Rangpur lime plants with the '*bO*' gene, for fungus resistance via *A. tumefaciens*-mediated transformation and evaluated these plants for *Phytophthora nicotianae* resistance. One of the two transgenic lines showed greater tolerance to the fungal pathogen as compared to the control.

RNA-silencing is another genetic engineering technique to knock down the expression of gene. This technique has been used to knock down the '*acrts1*' transcripts encoding a hydroxylase involved in the biosynthesis of host-selective ACR-toxin in the rough lemon pathotype of *Alternaria alternata.* A genomic BAC clone, containing a portion of the ACRT cluster, was sequenced which led to identification of three open reading frames present only in the genomes of ACR-toxin producing isolates. The functional role of one of these open reading frames, '*acrts1*' encoding a putative hydroxylase, in ACR-toxin production was studied by homologous recombination-mediated gene disruption. The silenced transformants did not produce detectable ACR-toxin and were not pathogenic (Izumi *et al.*, 2012).

*(ii) Abiotic Stress Tolerance*

Genetic transformation for different abiotic stresses, viz. salt and drought tolerance was carried out. Cervera *et al.* (2000b) transformed cv. 'Carrizo' citrange with the yeast gene '*hal2*', which helps to tolerate salt stress. To enhance drought tolerance and antioxidant enzymatic activity, transgenic 'Swingle' citrumelo rootstocks were transformed with the P5CSF129A gene coding for key-enzyme for proline synthesis, under water deficit. It was observed that transgenic plants were able to survive with water deficit better than non-transformed controls. Also proline diminished the deleterious effects caused by oxidative stress due to its ability to increase the activity of antioxidant enzymes (Carvalho *et al.*, 2013).

*(iii) Enhanced Flower and Fruit Characters*

Enhanced flowering is an important aspect of fruit production. To achieve this aspect, flower initiation genes, i.e. Arabidopsis leafy (*lfy*) or the apetala1 (*ap1*) (Pena *et al.*, 2001) and Citrus flowering locus T (*Cift*) (Endo *et al.*, 2009) were genetically engineered in citrus. It was noted that transgenic plants flowered extremely early and started to produce normal fruit within two years of genetic transformation by *A. tumefaciens* infection.

Another factor contributing to fruit character is parthenocarpic or seedless fruits. To fulfil this goal, parthenocarpy gene (*DefH9-iaaM*) and chimeric ribonuclease gene (*cg1*-400) were introduced into citrus to obtain possible seedless transformants. After *A. tumefaciens*-mediated transformation, many transgenic citrus lines with seedless trait were obtained (Tan *et al.*, 2012).

## Grapes

In the 90s, development of transgenic grapevine rootstocks and scion cultivars (Nakano *et al.,* 1994; Franks *et al.*, 1998) took place. It was noted that efficiency of transformation in grapes was highly influenced by *Agrobacterium* strain, the genotype of grape and culture conditions (Torregrosa *et al.*, 2002).

*(i) Biotic Stress Tolerance*

Powdery mildew is a fungus that scars the mature fruit, infects buds and leaves a white powdery coat over grape leaves, leading to leaves falling off that the plant and can no longer produce enough sugar to create wine-quality grapes. To overcome this problem, Yamamoto *et al.* (2000) introduced rice chitinase gene (*RCC2*) into the somatic embryos of grapevine by *Agrobacterium*-mediated infection. Two transformants showed enhanced resistance against powdery mildew and slight resistance against Elisinoe ampelina inducing anthracnose, leading to reduction in disease lesions. Bornhoff *et al.* (2000) also attempted to improve fungal resistance by transforming grape cv. Seyval blanc with genes for chitinase and RIP (ribosome-inactivating protein). Transgenic plants were also generated by transferring rice *chitinase* gene to raise disease-resistant plants. The transgenic plants showed delayed onset of the disease and only small lesions formed after *in vitro* inoculation of powdery mildew. The transgenic plants grew well in the greenhouse without any phenotypic alterations (Nirala *et al.*, 2010). In another study, a susceptible gene '*MLO-7*' was targeted by Malnoy *et al.* (2016) in grape cultivar Chardonnay to increase powdery mildew resistance. Efficient protoplast transformation, the molar ratio of Cas9 and sgRNAs were optimised alongwith analysis of targeted mutagenesis insertion and deletion rate using targeted deep sequencing. This led to the conclusion that direct delivery of CRISPR/Cas9 RNPs to protoplast system enables targeted gene editing together with generation of DNA-free genome edited grapevine plants.

Another approach of transformation is cisgenesis, which involves isolation and modification of genetic elements from the host genome, which are reinserted to develop plant varieties with improved characteristics. As a first step towards production of fungal-disease-resistant cisgenic grapevines, the *Vitis vinifera* thaumatin-like protein (*vvtl*-1) gene was isolated from cv. 'Chardonnay' and re-engineered for constitutive expression in 'Thompson Seedless'. Among the engineered plant lines of 'Thompson Seedless', two exhibited seven to 10 days delay in powdery mildew disease development during greenhouse screening and decreased severity of black rot disease in field tests (Dhekney *et al.*, 2011).

Silencing of conserved root-knot nematodes (RKN) effector gene was achieved through RNA interference (RNAi) for inducing nematode resistance in grape. Two hairpin-based silencing constructs, containing a stem sequence of 42 bp (pART27-42) or 271 bp (pART27-271) of the *16D10* gene, a conserved RKN effector gene, were transformed into grape hairy roots and compared for their small interfering RNA (siRNA) production and efficacy on suppression of nematode infection. Transgenic hairy root lines, carrying either of the two RNAi constructs, showed less susceptibility to nematode infection as compared with control (Yang *et al.*, 2013).

*(ii) Abiotic Stress Tolerance*

Ferritin helps to protect plant cells from oxidative damage which is induced by abiotic stresses. Keeping in view this point, *Medicago sativa* (alfalfa) ferritin gene (*MsFer*) was used to transform *Vitis berlandieri* x *Vitis rupestris* cv. 'Richter 110' grapevine rootstock lines. The transformants exhibited increased production of ferritin, which led to improved abiotic stress tolerance (Zok *et al.*, 2010).

*(iii) Phenotypic Characters*

Key genes responsible for *flavonoid 3'-hydroxylase* (*F3'H*) and *flavonoid 3',5'-hydroxylase* (*F3'5'H*), for flavonoid hydroxylation (and also for their stability, colour and antioxidant capacity) have been cloned in red grapevine, cv. Shiraz. Also ectopic expression of their functionality was proven in *Petunia hybrida* (Bogs *et al.*, 2006). In grape, two kinds of anthocyanin active transporters localiszed to the tonoplast were discovered: two belonging to the Multidrug And Toxic Extrusion (MATE) family called *anthoMATE1-3* (*AM1 and AM3*), which can bind acylated anthocyanins and translocate them to the vacuole in the presence of MgATP (Gomez *et al.*, 2009) and an ABC-type transporter, *ABCC1*, shown to perform the transport of glucosylated anthocyanidins (Francisco *et al.*, 2013). More recently, three GSTs (*VviGST1*, *VviGST3*, *VviGST4*) have been tested for their ability to bind glutathione and monomers of different phenylpropanoids (anthocyanin, PAs, and flavonols). All the three genes displayed the binding activity with distinct specificity according the phenylpropanoid class (Perez-Diaz *et al.*, 2016).

CRISPR/Cas9 system has been used for efficient knockout of the *L-idonate dehydrogenase* gene (*IdnDH*), involved in the tartaric acid pathway (Ren *et al.*, 2016). In grape, a computational survey of all the CRISPR/Cas9 sites available in the genome revealed the presence of 35,767,960 potential CRISPR/Cas9 target sites, distributed across all chromosomes with a preferential localisation at the coding region level (Wang *et al.*, 2016). A Grape-CRISPR website of all possible protospacers and target sites was created and made available to the public (http://biodb.sdau.edu.cn/gc/index.html), so that future research can be facilitated (Gascuel *et al.*, 2017). Nakajima *et al.*, 2017 also reported successful targeted mutagenesis in grape cv. Neo Muscat using the CRISPR/Cas9 system. They targeted *Vitis vinifera* phytoene desaturase (VvPDS) gene. DNA sequencing confirmed that the VvPDS gene was mutated at the target site in regenerated grape plants (Fig. 3).

### *Kiwifruit*

Genetic transformation in kiwifruit has provided many improved varieties with disease resistance, enhanced salt tolerance, altered flower and fruit characters.

*(i) Disease Resistance*

Kiwifruit was transformed with a soybean β-1,3-endoglucanase cDNA (Nakamura *et al.*, 1999) and with a stilbene synthase (*sts*) gene, responsible for synthesis of resveratrol – an antifungal phytoalexin from *Vitis* (Kobayashi *et al.*, 2000). Although only an increased resistance to grey mould disease (*Botrytis cinerea*) was reported for plants transformed with the endoglucanase enzyme, the plants transformed with '*sts*' gene from *Vitis* produced piceid resveratrol – a glucoside, which may confer some beneficial effects on human health to the transformed kiwifruit plants.

Kiwifruit was also transformed with osmotin gene to attain resistant to *Botrytis cinerea* causing grey mould and *Cadophora luteo-olivacea,* causing post-harvest problem. Experiment was carried out to evaluate the resistance of stored fruits after separate artificial inoculation with *B. cinerea* and *C. luteo-olivacea*. A divergent degree of resistance to fungi and post-harvest damage was detected among the transgenic clones (Rugini *et al.*, 2011).

*(ii) Enhanced Salt Tolerance, Flower and Fruit Characters*

For maintaining high Na+/H+ ratio, '*At nhx1*' from Arabidopsis, was transferred into kiwifruit by

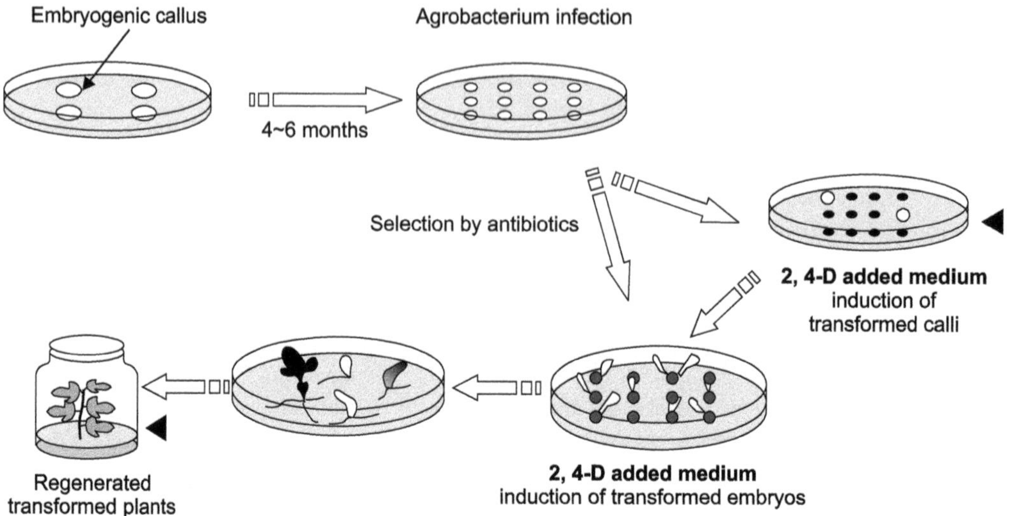


**Figure 3.** Schematic representation of transformation and regeneration process for gene editing using CRISPR/cas9. Red arrowheads indicate the point of mutation detection (*Source*: Nakajima *et al.*, 2017)

*Agrobacterium*-mediated protocol. Under salt stress, these transgenic lines accumulated more Na+ than control, due to an increased Na+/H+ antiporter activity (Tian *et al.*, 2011). Flowering and fruiting of transgenic plants was obtained within two years of transformation in greenhouse. GUS activity indicating stable expression of the *uidA* gene was observed in leaf, stem, root, petal and fruit tissues. Transgenic phenotypes were inherited in seedling progeny (Wang *et al.*, 2012).

## *Litchi*

In litchi, genetic transformation was aimed at imparting disease resistance and to produce parthenogenic fruits.

### *(i) Disease Resistance*

Rice *chitinase* gene was introduced for antifungal response. The transgenic plants showed delayed onset of the disease and smaller lesions following *in vitro* inoculation of die-back, leaf spots and blight pathogen (*Phomopsis* sp.). The transgenic plants were adapted to the greenhouse and no phenotypic variations were recorded (Das and Rahman, 2010, 2012).

### *(ii) Parthenocarpic Fruits*

One of the objectives of litchi transformation is recovery of parthenocarpic fruits (Yao *et al.*, 2001). For production of parthenocarpic fruits, embryogenic cultures of 'Brewster' ('Chen Tze') litchi derived from leaves of a mature tree were transformed with the pistillata (*pi*) cDNA in antisense orientation through *Agrobacterium*-mediated transformation. Transgene integration was confirmed by conventional and quantitative PCR (Padilla *et al.*, 2013) in four transformed lines.

### *(iii) Improved Shelf-life*

Das *et al.* (2016) isolated SAMDC cDNA from *Datura stramonium* and introduced into litchi genome by *Agrobacterium tumefaciens* through zygote disc transformation. Transgenic plants expressing Datura SAMDC showed 1.7- to 2.4-fold higher levels of spermidine and spermine than wild-type plants demonstrating that increasing polyamine biosynthesis in plants may be a means of creating improved fruit shelf-life.

## *Mango*

Genetic transformation of mango has been based on embryogenic cultures derived from the nucellus of young fruits (Litz, 1984).

The feasibility of gene transfer using the beta-glucuronidase (GUS) reporter gene showed that osmotic treatment and particle acceleration pressure had a major effect on GUS transitory expression (Cruz-Hernandez *et al.*, 2000; Samanta *et al.*, 2007).

*(i) Disease Resistance*

Rivera *et al.* (2011) performed the genetic transformation of somatic embryos of mango cv. 'Ataulfo' with the Bell pepper '*J1 defensin*' gene. *In vitro* tests showed that protein extracts from transformed embryos inhibited the growth of *Colletotrichum gloeosporioides*, *Aspergillus niger* and *Fusarium* sp.

## Papaya

Papaya is perhaps the only successful example among fruit crops where transgenic plants are being produced commercially and various transformation studies were conducted for biotic stress tolerance and to enhance fruit quality.

*(i) Biotic Stress Tolerance*

Papaya ring-spot virus is a devastating virus that causes severe damage to the papaya industry. Transgenic papaya cvs. 'Rainbow' and 'SunUp' resistant to papaya ringspot virus (PRSV) were released in Hawaii in 1998 (Manshardt, 1998). These transgenics were the joint venture of Cornell University, University of Hawaii, the USDA and Upjohn Company in USA to save papaya industry from destruction of PRSV. Some other attempts have also been made till date to provide resistance in papaya by various genes, viz. PRSV (Papaya leaf-distortion mosaic virus) coat protein (*cp*) (Wei *et al.*, 2008; Rola *et al.*, 2010). No changes to endogenous gene function and no allergenic reactions were predicted from analysis of the insertion site and flanking genomic DNA in transgenics, thus supporting a positive review of the appeal for importing and consumption of transgenic cv. 'Rainbow' and its derivatives (Fig. 4).

Several fungal diseases cause great damage to the papaya industry. So transgenic papaya plants were generated with stilbene synthase gene cloned from grapevine to impart fungal disease resistance. Greenhouse studies showed that the disease levels in transgenic plants were reduced to 35 per cent of the disease levels in non-transformed control plants (Zhu *et al.*, 2010).

Commercial papaya cv. 'Kapoho', which is highly susceptible to mites, was transformed with the snowdrop lectin (*Galanthus nivalis* agglutin [*gna*]) gene having insecticidal activity towards sap-sucking insects. A laboratory bioassay using carmine spider mites recorded improved pest resistance in the transgenic papaya plants (McCafferty *et al.*, 2008).

*(ii) Enhanced Fruit Quality*

Through anti-sense RNA technology, down regulation of the '*ACC (Aminocyclopropane-1-carboxylic acid) oxidase*' gene (responsible for the last step in ethylene formation) resulted in the suppression of ethylene production, thereby delaying fruit ripening thereby allowing more accumulation of amino acids and sugars, resulting in better quality papaya (Che *et al.*, 2011).

## Passionfruit

Transformation of passionfruit mediated by *A. tumefaciens* as well as direct methods has been reported for disease resistance and to enhance phenotypic characters.

*(i) Disease Resistance*

Trevisan *et al.* (2006) transformed passionfruit with a sequence derived from the replicase and coat protein (*cp*) genes from passionfruit woodiness virus (PWV) and the preliminary results suggested that this strategy could be used to control this virus disease. Particle bombardment method of gene transfer was used to transfer the bactericide '*attacin A*' gene driven by 35S promoter to yellow passionfruit (Vieira *et al.*, 2002), conferring resistance to *Xanthomonas campestris* pv. passiflorae.

*(ii) Enhanced Phenotypic Characters*

*A. rhizogenes*-mediated transformation of passionfruit species was checked using suspension culture. Hairy roots, differentiated at the inoculation sites to establish individual root clones were used to initiate long-term cultures on semi-solid medium. The clones retained their high growth rates and antibiotic

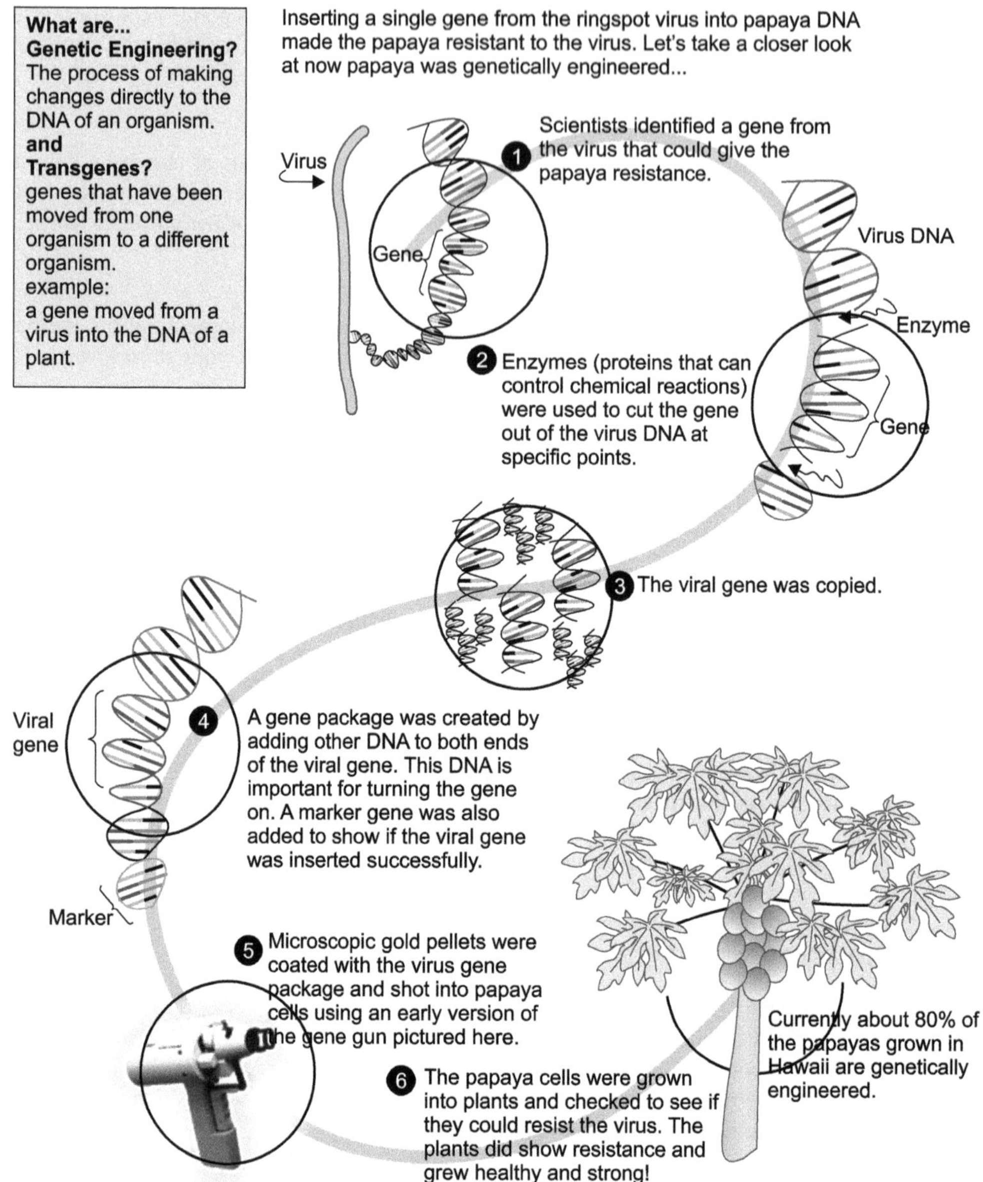


**Figure 4.** The process of genetic engineering in papaya (*Source*: Wieczorek and Wright, 2012)

resistance phenotypes. The regenerated roots displayed typical features of hairy roots, such as hairiness, branching and growth habit (Reis *et al.*, 2007).

## Peach

Though several reports of *Agrobacterium*-mediated genetic transformation of mature and immature peach tissues have been reported in the past, however, no transgenic plants were recovered and limited data were presented as evidence of transformation and stable integration of foreign DNA into the peach genome. Nevertheless, work is going on for development of efficient transformation protocols in peach.

Based upon green fluorescent protein (*gfp*) technology, Perez-Clemente and workers (2004) have developed *A. tumefaciens* based efficient, reliable transformation and regeneration system to produce transgenic peach plants using embryo sections of mature seeds as starting material. Survived shoots exhibited high-level of '*gfp*' expression mainly visible in the young leaves of the apex. *In vivo* monitoring of '*gfp*' expression reported an early, rapid and easy discrimination of both transgenic and escape buds. High levels of '*gfp*' expression were also maintained in the second generation of transgenic peach plants. Hence it is now possible to produce transgenic peach plants throughout the year without the need to isolate immature seeds from which the regeneration of transgenic plants and recovery of non-chimeric plants is difficult.

Padilla *et al.* (2006) performed a strategic study using *Agrobacterium* mediated transformation and *gfp* markers to assess transformation efficiency of various bacterial strains and explants. Highest rates of transformation were achieved with a combination of *A. tumefaciens* EHA105, plasmid pBIN19 and the CaMV35s promoter in peach epicotyl internodes (56.8 per cent), cotyledons (52.7 per cent), leaves (20 per cent), and embryonic axes (46.7 per cent). The study showed that internodes, cotyledons, and embryonic axes were superior to embryonic hypocotyls. Still, further optimisation will be required to make peach transformation and regeneration a routine.

## Pear

In pear, *Agrobacterium*-mediated genetic transformation was attempted to impart disease resistance, abiotic stress tolerance and to enhance phenotypic characters.

### *(i) Disease Resistance*

Transformation for disease resistance in pear against fire blight (caused by *Erwinia amylovora*) was done by Malnoy *et al.* (2005) and Djennane *et al.* (2009), who used an exopolysaccharide (*eps*) depolymerase and *ferritin* gene from pea.

### *(ii) Abiotic Stress Tolerance*

Over-expression of the apple spermidine synthase (*Mdspds1*) gene in pear conferred salt and multiple abiotic stress tolerance by altering polyamine titers in the plants. Selected lines were exposed to salt, osmosis and heavy metal stresses for evaluating their stress tolerances. One of the transgenic lines which was revealed to have the highest spermidine synthase accumulation and expression level of '*Mdspds1*', showed the strongest tolerance to these stresses (Wen *et al.*, 2008, 2009).

### *(iii) Enhanced phenotypic characters*

Rooting ability of dwarfing sweet cherry hybrid rootstock (*Prunus cerasus* x *P. canescens*) and pear rootstock (*Cydonia oblonga* P. Mill) was improved by transforming them with the '*rolB*' gene (cloned from *A. rhizogenes* plasmid pRiA4) using *A. tumefaciens* mediated gene transfer. As compared with the control, after co-cultivation with *A. tumefaciens*, the rooting rate of *Cydonia* regenerants *in vitro* increased by 6-44 per cent and of *Prunus* hybrid by 8-30 per cent. All transformants had well-formed roots (Staniene *et al.*, 2007).

Matsuda *et al.* (2009) transformed European pear 'La France' and 'Ballade' with the citrus flowering locus T (*Cift*) gene, which induces early flowering. Of the seven seedlings that expressed the '*Cift*' gene, five flowered within 10 months after their transfer to greenhouse, indicating that the '*Cift*' gene induced early flowering in the transformed pear plants. Usually pear plants flower after four to five years.

A transgenic line of 'Spadona', named Early Flowering-Spadona (*EF-Spa*), was produced using an '*Mdtfl1*' RNAi cassette targeting the native pear genes '*Pctfl1-1*' and '*Pctfl1-2*' to cut short the juvenile period. Pollination of '*EF-Spa*' trees generated normal-shaped fruits with viable $F_1$ seeds. The greenhouse-grown transgenic $F_1$ seedlings formed shoots and produced flowers one to 33 months after germination (Freiman *et al.*, 2012).

To inhibit the browning process in fruits of Yali pear, Antisense gene technique was used to reduce the expression of polyphenol oxidase (*ppo*) gene. Northern blot analysis and enzyme activity assay showed that the PPO activities in the transgenic Yali pear shoots were significantly decreased compared with the non-transformed shoots. This has paved the way for breeding new varieties of pears with browning resistance in future (Li *et al.*, 2011).

## Persimmon

There are very few studies on genetic transformation in persimmon. Some studies with respect to abiotic stress tolerance and enhanced phenotypic characters are as follows:

*(i) Abiotic Stress Tolerance*

Gao *et al.* (2000) were the first to successfully introduce the *cod* Agene for choline oxidase of *Arthrobacter globiformis* into persimmon cv. Jiro by *Agrobacterium*-mediated transformation using leaf discs as explants. Regenerants were recorded with the ability to synthesize glycinebetaine, hence, found salt stress tolerant.

*(ii) Enhanced Phenotypic Characters*

Studies conducted by Koshita *et al.* (2002) led to successful introduction of *rolC* gene from *A. rhizogenes* into cv. Saijo, indicating the possibility of producing dwarf plants. Recently, persimmon has been transformed with the gene encoding the pear fruit polygalacturonase inhibiting protein (PGIP) using *A. tumefaciens* EHA101 to enhance shelf-life of the fruit (Tamura *et al.*, 2004).

Gao *et al.* (2013) transformed the Japanese persimmon with Arabidopsis flowering locus T gene (*Atft*), and '*Pmtfl1*' gene- a *Prunus mume* ortholog of Arabidopsis terminal flower 1 (*tfl1*) gene. Ten lines of transgenic '*Pmtfl1*' shoot and two lines of transgenic '*Atft*' shoot were obtained. The '*Pmtfl1*' transgenic *in vitro* shoots did not show a different appearance compared with non transformed 'Jiro' shoots, however, the '*Atft*' transgenic shoots showed a bushy phenotype with the short internodes.

## Pineapple

Pineapple has been transformed by microprojectile bombardment and by co-cultivation with *A. tumefaciens* to impart disease resistance, herbicide tolerance and enhanced quality traits.

*(i) Disease Resistance*

Stewart *et al.* (2001) cloned a polyphenol oxidase (*ppo*) gene from pineapple fruits under conditions that produce blackheart, a fruit defect. The '*ppo*' gene has been silenced in transformed plants. A comparative analysis of transgenic pineapple lines [silenced for polyphenol oxidase (*ppo*)] gene and the untransformed control lines revealed that all of the control lines expressed blackheart and exhibited the greatest incidence and severity, while the transgenic lines were regarded as blackheart resistant, having no blackheart symptoms (Ko *et al.*, 2013).

*(ii) Herbicide Tolerance*

Pineapple plants transformed with the '*bar*' gene for bialaphos herbicide resistance were evaluated for transgene stability, gene expression and tolerance to gluphosinate ammonium, which is an active ingredient of herbicide BastaReg. X, under field conditions. Genetically-modified plants remained green and healthy following spraying with the herbicide. In contrast, non-transformed pineapple plants of the same cultivar became necrotic and died within 21 days of spraying of the herbicide (Sripaoraya *et al.*, 2006).

*(iii) Quality Traits*

Leaf bases of Pineapple cv. Queen were transformed with soybean ferritin cDNA. Few of the transgenic plants were hardened in the greenhouse and were grown to maturity to determine the enhanced iron and zinc accumulation in the fruits (Mhatre *et al.*, 2011).

In the USA, pineapples were genetically modified for viral and nematode resistance, delayed maturation, modified sugar composition and flowering time (Hanke and Flachowsky 2010). Recently, the Del Monte company obtained red-fleshed pineapple named 'Rose' by combining over-expression of a gene derived from tangerine and suppression of other genes in order to increase the accumulation of lycopene (Ogata *et al.*, 2016).

## Plum

Plum producers world-wide are facing multiple challenges including climate change, the need for reduced chemical inputs, the spread of native and exotic pests and pathogens, and consumer demands for improved fruit quality and health benefits. In an effort to develop new approaches genetic engineering was proved a successful technology.

### *(i) Disease Resistance*

The USDA-ARS Appalachian Fruit Research Station fruit-breeding programme in collaboration with partners in the USA and Europe have developed a genetic-engineering approach to target resistance to Plum pox virus (PPV) by transferring Plum pox virus-coat protein gene (*PPV-cp*). This program has resulted in the development of a transformed plum cultivar 'HoneySweet' which has been tested for 15 years in the European Union and USA and is highly resistant to PPV. 'HoneySweet' has received full regulatory approval in the USA and represents a new source of PPV resistance (Hily *et al.*, 2007; Mikhailov *et al.*, 2012; Scorza *et al.*, 2013).

Ravelonandro *et al.* (2011) studied the heritability of the virus transgenes engineered in 'HoneySweet' plum through different crosses with two commercial cultivars of *Prunus domestica* ('Prunier d'Ente 303' and 'Quetsche 2906') and one wild species, *P. spinosa* 2862 rootstock using 'HoneySweet' plum as the pollen donor. As much as 46 per cent of the $F_1$ progeny was transgenic. These results confirmed the high potential of 'HoneySweet' plum for PPV resistance breeding programmes.

Two transgenic lines of plum cv. 'Stanly' expressed Gastrodia antifungal gene (*gafp-1*) and recorded to exhibit resistance to the pathogen *Phytophthora cinnamomi* and the root-knot nematode *Meloidogyne incognita*. '*gafp-1*' lectin was identified within the roots, but not in the soft shoot or leaf tissues of the grafted, wild type scions. These results suggest that *gafp-1* mRNA and protein are not moving into the wild type scion tissues of chimeric-grafted plum trees (Nagel *et al.*, 2010). Rootstocks created from such transgenic lines will be more readily accepted by consumers, proving that foreign gene products are not migrating into a grafted, nontransgenic scion on which fruit is produced and they express only in the rootstock portion where they are required.

### *(ii) Enhanced fruit quality*

Plum hypocotyls were transformed with an antisense peach '*ACC oxidase*' gene (responsible for the last step in ethylene formation). Fruit quality data consisting of fruit firmness, colour, date of ripening, brix, and size as well as ethylene production rates were measured for two years on the fruiting lines which suggested that in some transgenic lines, ethylene production as well as softening were delayed relative to the 'Bluebyrd' parental line (Callahan and Scorza, 2007).

The flowering locus T1 (*ft1*) gene from *Populus trichocarpa* under the control of the 35S promoter was introduced into the European plum. Transgenic plants expressing higher levels of 'ft' flowered and produced fruits in the greenhouse within one to ten months. Plums with '*ft*' gene did not enter dormancy after cold or short day treatments, yet field planted '*ft*' plums remained winter hardy down to at least –10°C. The flowering and fruiting phenotype was found to be continuous in the greenhouse but limited to spring season and they tend to fall in the field (Srinivasan *et al.*, 2012).

## *Strawberry*

Efficient genetic transformation protocols have been used to impart biotic and abiotic stress tolerance and enhanced fruit quality in strawberry.

### *(i) Biotic Stress Tolerance*

The CP4EPSP synthase gene, which confers resistance to glyphosate, an active ingredient of herbicide Roundup, was introduced into cv. Camarosa through *Agrobacterium*-mediated transformation. The transformants showed a range of tolerance to the herbicides ranging from complete tolerance to death (Morgan *et al.*, 2002). 19 independent transgenic lines of cv. Firework and 15 lines of cv. Selekta were obtained via *Agrobacterium*-mediated transformation for taste improvement and enhanced disease resistance by introduction of thaumatin II (*thauII*) gene (Schestibratov and Dolgov, 2006).

Modification of strawberry plants by introducing the stilbene synthase (*sts*) gene resulted in production of the phytoalexin resveratrol and provided enhanced resistance against several pathogenic fungi (Hanhineva *et al.*, 2009). An effective system for chitinase (*chit42*) gene transformation mediated by *A. tumefaciens* was determined to obtain a disease-resistant strawberry plant (Xie *et al.*, 2008).

**Table 1.** Genetic Transformation Work done in Fruit Crops

| *S. No.* | *Crop* | *Trait* | *Remarks* | *References* |
|---|---|---|---|---|
| 1. | Apple | Scab | Wheat puroindoline B (*pinB*);<br>*Vfa1, Vfa2, or Vfa4* (identified within the Vf locus of *Malus floribunda* 821); Barley hordothionin (*hth*) genes conferred resistance to scab disease | Faize *et al.*, 2004;<br>Malnoy *et al.*, 2008;<br>Krens *et al.*, 2011 |
| | | Fire blight | Resistance to fire blight disease was conferred by Viral EPS-depolymerase gene (*dpo*);<br>*fb_mr5* originating from the crab apple accession *Malus robusta5* (Mr5) genes<br>*DIPM-1, DIPM- 2* and *DIPM-4* genes | Flachowsky *et al.*, 2008;<br>Giovanni *et al.*, 2014;<br>Malnoy *et al.*, 2016 |
| | | Powdery mildew | *Malus hupehensis*-derived NPR1 (*Mhnpr*1) gene conferred resistance against powdery mildew | Chen *et al.*, 2012 |
| | | Reduction of juvenile phase | *Mdtfl* (*Malus* x *domestica tf*); *Bpmads4* (*Betula pendula* mads-box); flowering locus T (*ft*) genes resulted in reduced juvenile phase | Kotoda *et al.*, 2000;<br>Flachowsky *et al.*, 2007;<br>Flachowsky and Hanke, 2012 |
| | | Dwarfism and enhanced rooting ability | Rooting locus C (*rolC*) provided dwarf phenotype along with improved rooting | Kim *et al.*, 2009 |
| | | Overcoming of self-incompatibility | *S3-RNase* gene was used to overcome self-incompatibility | Broothaerts *et al.*, 2004;<br>Dreesen *et al.*, 2012 |
| | | Nonbrowning | '*pgas*' (a hybrid sequence carrying four gene groups i.e. PPO2, GPO3, APO5 and pSR7) | Carter, 2012; Xu, 2013 |
| | | Enhanced calcium level | *cax1* (Cation Exchangers) gene provided increased calcium level | Kim *et al.*, 2010 |
| | | Salt stress tolerance | Tolerance to salt stress was achieved by transformation with gene *Mdnhx1* (*Malus domestica* Na+/H+) | Li *et al.*, 2010 |
| 2. | Apricot | Sharka (plum pox virus) | *PPV-cp* (Plum pox virus-coat protein) gene was used to provide disease resistance | Machado *et al.*, 1992;<br>Sansavini, 2004 |
| 3. | Avocado | Root diseases | Gene for antifungal protein (*afp*) provided resistance to root diseases | Litz *et al.*, 2007 |

(*Contd.*)

| | | | | |
|---|---|---|---|---|
| | | Resistance to herbicide Basta | Herbicide resistance was provided by *glucanase, chitinase, bar, uidA* and plant defensin gene (*pdf1.2*) genes | Raharjo *et al*., 2008 |
| | | Improved shelf life | Improvement in shelf life was achieved by transforming avocado with SAM (S-adenosyl –L-methionine) hydrolase | Efendi, 2003; Litz *et al*., 2007 |
| 4. | Banana | Banana bunchy top virus (BBTV) | Replicase-associated protein (*rep*) gene provided resistance to BBTV | Borth *et al*., 2009; Krishna *et al*., 2013; |
| | | Wilt | Soy bean endo- beta -1,3-glucanase gene (*pROKla-Eg*); Sweet pepper hypersensitivity response-assisting protein (*hrap*); Rice thaumatin-like protein (*tlp*) genes conferred resistance against wilt disease | Maziah *et al*., 2007; Tripathi *et al*., 2010; Hu *et al*., 2013 |
| | | Xanthomonas wilt | *Hrap* and *Pflp* providing resistance to Xanthomonas wilt | Tripathi *et al*., 2017 |
| | | Nematodes | Cysteine proteinase inhibitor and synthetic peptide | |
| | | Salt stress tolerance | Musa Stress Associated Proteins1 (*sap1*); Plasma membrane intrinsic protein gene (*Musapip1;2*) genes were used to provide salt stress tolerance | Sreedharan *et al*., 2012; Sreedharan *et al*., 2013 |
| | | Shortening of life cycle | *Arabidopsis thaliana* cyclin *D2;1* gene decreased life cycle of banana | Talengera *et al*. (2009) |
| 5. | Cherry | Stimulation of rooting | Improvement in rooting was conferred by rooting locus B (*rolB*) gene | Rugienius *et al*., 2009 |
| 6. | Citrus | Canker | attacin A (*attA*); *Arabidopsis thaliana* NPR1 (*At npr1*); Dermaseptin coding sequence; *Fls2* provided resistance to canker disease | Boscariol *et al*., 2006; Zhang *et al*., 2010; Furman *et al*., 2013; Hao *et al*., 2016 |
| | | Citrus mosaic virus (CiMV) | Coat protein (*cp*) gene was used to provide resistance to CiMV | Iwanami, 2010 |
| | | Fungal disease caused by *Phytophthora nicotianae* | Resistance to fungal disease was conferred by *bO* (bacterio-opsin) gene | Azevedo *et al*., 2006 |

(*Contd.*)

**Table 1.** *(Contd.)*

| *S. No.* | *Crop* | *Trait* | *Remarks* | *References* |
|---|---|---|---|---|
| | | Bacterial disease caused by *Alternaria alternata* | Acute cellular rejection TS1 (*acrts1*) gene provided bacterial disease resistance | Izumi *et al.*, 2012 |
| | | Salt stress tolerance | Tolerance to salt stress was conferred by histidine ammonia lyase 2 (*hal2*) gene | Cervera *et al.*, 2000b |
| | | Enhanced drought tolerance and antioxidant enzymatic activity | Transformation with pyrroline-5-carboxylate synthetase (P5CSF129A) gene resulted in better drought tolerance and antioxidant enzymatic activity | Carvalho *et al.*, 2013 |
| | | Enhanced flower initiation | Flower initiation was enhanced by transformation with Arabidopsis *leafy (lfy) or apetala1 (ap1)*; Citrus flowering locus T (*Cift*) genes | Pena *et al.*, 2001; Endo *et al.*, 2009 |
| | | Seedless fruits | Transformation with defensin H9-indole acetic acid M (*DefH9-iaaM*) and chimeric ribonuclease (*cg1-400*) genes was used to produce seedless fruits | Tan *et al.*, 2012 |
| 7. | Grapes | Powdery mildew | Rice chitinase gene (*RCC2*)<br>Rice *chitinase* gene conferred resistance to powdery mildew disease<br>*MLO-7* | Yamamoto *et al.*, (2000);<br>Nirala *et al.*, 2010;<br>Malnoy *et al.*, (2016) |
| | | Black rot | Resistance to black rot was conferred by *Vitis vinifera* thaumatin-like protein (*vvtl-1*) gene | Dhekney *et al.*, 2011 |
| | | Nematode resistance | Silencing of conserved Root Knot Nematode (RKN) effector gene, *16D10*, provided nematode resistance | Yang *et al.*, 2013 |
| | | Abiotic stress tolerance | *Medicago sativa* ferritin (*MsFer*) gene conferred tolerance to abiotic stresses | Zok *et al*, 2010 |

*(Contd.)*

| | | | | |
|---|---|---|---|---|
| | | Phenotypic characters | Flavonoid 3′-hydroxylase (*F3′H*) and flavonoid 3′,5′-hydroxylase (*F3′5′H*),<br>*anthoMATE13* (*AM1* and *AM3*)<br>ABC-type transporter, *ABCC1*,<br>*VviGST1, VviGST3, VviGST4*<br>*IdnDH* | Bogs *et al*., 2006<br>Gomez *et al*., 2009<br>Francisco *et al*., 2013<br>Perez-Diaz *et al*., 2016<br>Ren *et al*., 2016 |
| 8. | Kiwifruit | Grey mould disease | *b-1,3-endoglucanase; Stilbene synthase; Osmotin* genes provide resistance to grey mould disease | Nakamura *et al*., 1999; Kobayashi *et al*., 2000; Rugini *et al*., 2011 |
| | | Salt stress tolerance | *Arabidopsis thaliana* Na+/H+ antiporter (*At nhx1*) gene conferred salt tolerance | Tian *et al*., 2011 |
| | | Flower and fruit characters | *uidA* gene led to enhance flowering and fruiting | Wang *et al*., 2012 |
| 9. | Litchi | Die-back, leaf spots and blight pathogen | Rice *chitinase* gene provided disease resistance | Das and Rahman, 2010; 2012 |
| | | Parthenocarpic fruits | Parthenocarpic fruits were produced by transformation with pistillata (*pi*) gene | Padilla *et al*., 2013 |
| | | Improved shelf-life | SAMDC cDNA | Das *et al*., 2016 |
| 10. | Mango | Resistance against *Colletotrichum gloeosporioides*, *Aspergillus niger* and *Fusarium* sp. | Disease resistance was conferred by bell pepper *J1 defensin* gene | Rivera *et al*., 2011 |
| 11. | Papaya | Papaya ringspot virus (PRSV) | PRSV (Papaya ringspot virus) coat protein (*cp*) gene conferred resistance to PRSV | Wei *et al*., 2008; Rola *et al*., 2010 |
| | | Fungal disease | Stilbene synthase (*sts*) gene provided fungal disease resistance | Zhu *et al*., 2010 |

(*Contd.*)

**Table 1.** (*Contd.*)

| *S. No.* | *Crop* | *Trait* | *Remarks* | *References* |
|---|---|---|---|---|
| | | Mites | Resistance to mites was conferred by *Galanthus nivalis* agglutin [*gna*] gene | McCafferty *et al.*, 2008 |
| | | Enhanced fruit quality | *ACC oxidase* gene in antisense orientation was used to enhance fruit quality | Che *et al.*, 2011 |
| 12. | Passionfruit | Passionfruit woodiness virus (PWV) | Resistance to PWV was provided by replicase and capsid protein (*cp*) | Trevisan *et al.*, 2006 |
| | | Fungal disease caused by *Xanthomonas campestris* pv. Passiflorae | Transformation with *attacin A* gene resulted in fungal disease resistance | Vieira *et al.*, 2002 |
| | | Enhanced rooting | *A. rhizogenes* suspension culture provided enhancement in rooting | Reis *et al.*, 2007 |
| 13. | Peach | Efficient transformation protocols | Expression of Green fluorescent protein (*gfp*) resulted in efficient genetic transformation | Padilla *et al.*, 2006 |
| 14. | Pear | Fire blight | Exopolysaccharide (EPS) depolymerase and *ferritin* genes conferred resistance to fire blight disease | Malnoy *et al.*, 2005; Djennane *et al.*, 2009 |
| | | Abiotic stress tolerance | *Malus domestica* spermidine synthase (*Mdspds1*) gene conferred tolerance to salt, osmosis, and heavy metal stresses | Wen *et al.*, 2008 |
| | | Improve rooting | Improvement in rooting was conferred by rooting locus B (*rolB*) gene | Staniene *et al.*, 2007 |
| | | Early flowering | Transformation with Citrus flowering locus T (*Cift*) gene resulted in early flowering | Matsuda *et al.*, 2009 |
| | | Reduced juvenile period | *Malus domestica* terminal flower 1 (*Mdtfl1*) gene reduced the juvenile period | Freiman *et al.*, 2012 |
| | | Inhibition of browning | Altered expression of polyphenol oxidase (*ppo*) gene led to inhibition of browning | Li *et al.*, 2011 |

(*Contd.*)

| | | | | |
|---|---|---|---|---|
| 15. | Persimmon | Abiotic stress tolerance | Codeine A (*CodA*) gene provided salt stress tolerance | Gao *et al*., 2000 |
| | | Production of dwarf plants | Rooting locus C (*rolC*); Polygalacturonase inhibiting protein (PGIP); *Arabidopsis thaliana* flowering locus T (*Atft*) and *Pmtfl1*, a *Prunus mume* ortholog of Arabidopsis terminal flower 1 (*tfl1*) genes conferred production of dwarf plants | Koshita *et al.*, 2002; Tamura *et al*., 2004; Gao *et al*., 2013 |
| 16. | Pineapple | Blackheart resistance | Resistance to blackheart disease was conferred by silencing of polyphenol oxidase (*ppo*) gene | Stewart *et al*., 2001; Ko *et al*., 2013 |
| | | Bialaphos resistance | *bar* gene provided tolerance to herbicide | Sripaoraya *et al*., 2006 |
| | | Enhanced iron and zinc accumulation | *Ferritin* gene enhanced quality traits like accumulation of iron and zinc | Mhatre *et al*., 2011 |
| 17. | Plum | Plum pox virus (PPV) | Plum pox virus coat protein (*PPV-cp*) gen provided resistance to PPV | Hily *et al*., 2007; Mikhailov *et al*., 2012; Scorza *et al*., 2013 |
| | | Resistance to *Phytophthora cinnamomi* and the root-knot nematode *Meloidogyne incognita* | Gastrodia antifungal protein (*gafp-1*) conferred resistance to fungal disease and root-knot nematode | Nagel *et al*., 2010 |
| | | Delayed fruit softening | Silencingof *Aminocyclopropane-1-carboxylic acid (ACC) oxidase* led to enhanced fruit quality by delaying fruit softening | Callahan and Scorza, 2007 |
| | | Early flowering and fruiting | Transformation with flowering locus T1 (*ft1*) gene led to early flowering and fruiting | Srinivasan *et al*., 2012 |
| 18. | Strawberry | Glyphosate tolerance | CP4.Enolpyruvylshikimate-3-phosphate (CP4.EPSP) synthase conferred herbicide tolerance | Morgan *et al.,* 2002 |
| | | Fungal disease resistance | Stilbene synthase (*sts*); chitinase (*chit42*) genes led to fungal disease resistance | Hanhineva *et al*., 2009; Xie *et al*., 2008 |

(*Contd.*)

**Table 1.** (*Contd.*)

| *S. No.* | *Crop* | *Trait* | *Remarks* | *References* |
|---|---|---|---|---|
| | | Cold tolerance | CBF1 (C-repeat binding factor) gene provide cold tolerance | Jin *et al.*, 2007 |
| | | Salt stress tolerance | Tolerance to salt stress was conferred by tobacco *osmotin* gene | Husaini *et al.*, 2012 |
| | | Improved keeping quality | *Fragaria ananassa* ethylene receptor (*FaEtr1* and *FaEr2*); silencing of *FaplC* and *FaE31*; Rooting locus C (*rolC*) led to enhancement in fruit quality | Zhu *et al.*, 2009; Garcia *et al.*, 2012; Youssef *et al.*, 2013; Landi *et al.*, 2009 |
| | | Enhanced vitamin C level | Transformation with GDP-l-galactose phosphorylase (*ggp*) gene resulted in enhanced level of vitamin C | Bulley *et al.*, 2012 |

*(ii) Abiotic Stress Tolerance*

A foreign target-gene of cold-inducible transcription factor CBF1 (C-repeat binding factor) from *Arabidopsis thaliana* was introduced into strawberry by using the leaf-disc method via *A. tumefaciens*. All transgenic and control plants were subjected to a temperature of –2°C for seven days. It was noted that 60 per cent of transgenic plants and 85 per cent of control plants wilted at low temperature (Jin *et al.*, 2007). Over-expression of tobacco '*osmotin*' gene led to increased salt stress tolerance, total soluble protein and chlorophyll content as compared to the wild plants in strawberry (Husaini *et al.*, 2012).

*(iii) Enhanced Fruit Quality*

Improving the storing quality of strawberry fruit is a difficult problem. This is because of ethylene production which decreases shelf-life of fruit. Applying antisense gene strategy to control ethylene receptor is a new feasible method. Two antisense ethylene receptor genes *Fragaria ananassa* ethylene receptor '*FaEtr1*' and '*FaEr2*' were introduced into leaves of strawberry with *A. tumefaciens* EHA105. Northern blot analysis indicated that '*FaEtr1*' and '*FaEr2*' mRNA abundance was decreased in antisense strawberry plants (Zhu *et al.*, 2009).

'*rolC*' lines of strawberry cv. Calypso were produced by genetic transformation using *A. tumefaciens*. Yield and fruit quality of the control and transgenic lines were measured under open-field conditions, which showed 30 per cent greater yields in transformants than controls, due to 20 per cent more fruits per plant and an increased fruit weight along with better fruit quality. No significant increase towards tolerance to *Phytophthora cactorum* and symbiosis with root arbuscular mycorrhizal fungi (AMF) was observed (Landi *et al.*, 2009).

RNAi mediated transformation was used to down-regulate the genes '*FaplC*' and '*FaE3*' encoding a pectate lyase and a polygalacturonase enzyme, respectively. Notable reduction towards the loss of firmness at the red ripe stage was observed (Garcia *et al.*, 2012; Youssef *et al.*, 2013). Bulley *et al.* (2012) increased ascorbate (vitamin C) in strawberry, using a transgene, GDP-l-galactose phosphorylase (*ggp*).

## 3. Overview and Future

In fruit crops, widely used methods of transformation are *Agrobacterium*-mediated gene transfer and microprojectile bombardment. In nature, *Agrobacterium* infects host plants where there is an injury and liberation of phenolic compounds. The addition to the culture medium of phenolic compounds like acetosyringone stimulates transcription of virulence genes in *Agrobacterium*. Together with the gene of interest, other genes required for transformation are transferred including marker genes that allow selection of transformed cells. However, due to public concern with the introduction of antibiotic resistance genes into food, methods to eliminate them from the transformed plants and strategies that avoid selection of transformed cells with antibiotics are being developed. These new alternative methodologies have only begun to be applied to the production of transformed fruit trees.

One of such methodology is production of marker-free transgenics. Marker-free transgenic crops confer several advantages over transgenic crops containing selection genes coding, for example for antibiotic resistance. Efforts are on to avoid or minimise the inclusion of transgenes or sequences by promoting the use of clean vector systems. A positive selectable marker gene such as mannoseA (*manA*) that encodes for phosphomannose isomerase and provides transformed cells with a metabolic advantage over non-transformed cells was tested. The second strategy is the use of reporter genes, such as green fluorescent protein gene (*gfp*), b-glucuronidase (GUS), b-galactosidase (LacZ) and luciferase (LUC) for screening of transgenic plants.

In conclusion, transformation and regeneration of fruit crops especially tree crops is not routine, generally being limited to a few genotypes or to seedlings. The future of genetic transformation as a tool for breeding of fruit crops requires the development of genotype-independent procedures based on the transformation of meristematic cells with high regeneration potential and/or the use of regeneration-promoting genes. Yet another obstacle is that international laws will neither allow deliberate release of plants carrying antibiotic resistant genes nor their commercialiszation. Therefore, development of procedures to avoid the use of antibiotic-based selection or to allow elimination of marker genes from the transformed plant will be a research priority in the coming years.

## References


Ahmed, M.F., Kantharajah, A.S. and Holford, P. (2012). Genetic transformation studies on avocado cultivar 'hass' (*Persea americana*). *Am. J. Pl. Sci.* 3(9): 1225-1231.

Ainsley, P.J., Collins, G.G. and Sedgley, M. (2001). Factors affecting *Agrobacterium*-mediated gene transfer and the selection of transgenic calli in paper shell almond (*Prunus dulcis* Mill). *J. Hort. Sci. Biot.* 75: 522-528.

Azevedo, F.A., Mourao, F.F.A.A., Mendes, B.M.J., Almeida, W.A.B., Schinor, E.H., Pio, R., Barbosa, J.M., Guidetti, G.S., Carrer, H. and Lam, E. (2006). Genetic transformation of Rangpur lime (*Citrus limonia* Osbeck) with the bO (bacterio-opsin) gene and its initial evaluation for *Phytophthora nicotianae* resistance. *Pl. Mol. Biol. Rep.* 24(2): 185-196.

Bogs, J., Ebadi, A., McDavid, D. and Robinson, S.P. (2006). Identification of the flavonoid hydroxylases from grapevine and their regulation during fruit development. *Plant Physiol.* 140: 279-291.

Bornhoff, B.A., Harst, M., Zyprian, E. and Topfer, R. (2005). Transgenic plants of *Vitis vinifera* cv. Seyval Blanc. *Plant Cell Rep.* 24(7): 433-438.

Borth, W., Perez, E., Cheah, K., Chen, Y., Xie, W.S., Gaskill, D., Khalil, S., Sether, D., Melzer, M., Wang, M., Manshardt, R., Gonsalves, D. and Hu, J.S. (2009). Transgenic banana plants resistant to Banana bunchy top virus infection. *Acta Hort* 897: 449-457.

Boscariol, R.L., Monteiro, M., Takahashi, E.K., Chabregas, S.M., Vieira, M.L.C., Vieira, L.G.E., Pereira, L.F.P., Mourao, F.F.A.A., Cardoso, S.C., Christiano, R.S.C., Bergamin, F.A., Barbosa, J.M., Azevedo, F.A. and Mendes, B.M.J. (2006). Attacin A gene from Tricloplusia ni reduces susceptibility to *Xanthomonas axonopodis* pv. citri in transgenic *Citrus sinensis* 'Hamlin'. *J. Am. Soc. Hortic. Sci.* 131(4): 530-536.

Broothaerts, W., Keulemans, J. and Van Nerum, I. (2004). Selffertile apple resulting from S-RNase gene silencing. *Pl. Cell. Rpts.* 22: 497-501.

Bulley, S., Wright, M., Rommens, C., Yan, H., Rassam, M., Kui, L.W., Andre, C., Brewster, D., Karunairetnam, S., Allan, A.C. and Laing, W.A. (2012). Enhancing ascorbate in fruits and tubers through over-expression of the l-galactose pathway gene GDP-l-galactose phosphorylase. *Pl. Biotech. J.* 10(4): 390-397.

Callahan, A. and Scorza, R. (2007). Effects of a peach antisense ACC oxidase gene on plum fruit quality. *Acta Hort* 738: 567-573.

Carter, N. 2012. Petition for determination of non-regulated status: Arctic Apples (*Malus x domestica*) events GD743 (Arctic Golden Delicious) and GS784 (Arctic Granny Smith). *Okanagan Specialty Fruits Inc.*, Summerland, BC V0H 1Z0, Canada.

Carvalho, K., Campos, M.K.F., Domingues, D.S., Pereira, L.F.P. and Vieira, L.G.E. (2013). The accumulation of endogenous proline induces changes in gene expression of several antioxidant enzymes in leaves of transgenic *Swingle citrumelo*. *Mol. Biol. Rep.* 40(4): 3269-3279.

Cervera, M., Ortega, C., Navarro, L., Navarro, A. and Pena, L. (2000b). Generation of transgenic citrus plants with the tolerance-to-salinity gene HAL2 from yeast. *J. Hort. Sci. Biotech.* 75: 30.

Che, C.M.Z., Nurul, S.A.H., Naziratul, A.A.N. and Zainal, Z. (2011). Genetic transformation of antisense ACC oxidase in *Carica papaya* L. cv. sekaki via particle bombardment. *Malays. Appl. Biol.* 40(1): 39-45.

Chen, X.K., Zhang, J.Y., Zhang, Z., Du, X.L., Du, B.B. and Qu, S.C. (2012). Over-expressing MhNPR1 in transgenic Fuji apples enhances resistance to apple powdery mildew. *Mol. Biol. Rep.* 39(8): 8083-8089.

Cruz, H.A., Witjaksono, L.R.E. and Gomez, L.M.A. (1998). *Agrobacterium tumefaciens*-mediated transformation of embryogenic avocado cultures and regeneration of somatic embryos. *Pl. Cell Rpts.* 17: 497-503.

Cruz-Hernandez, A., Town, L., Cavallaro, A. and Botella, J.R. (2000). Transient and stable transformation in mango by particle bombardment. *Acta Hort.* 509: 237-242.

D'Halluin, E., Bonne, K.E., Bossut, M., Beuckeleer, M.D. and Leemans, J. (1992). Transgenic maize plants by tissue electroporation. *Pl. Cell* 4: 1495.

Dandekar, A.M., McGranahan, G.H., Vail, P.V., Uratsu, S.L., Leslie, S. and Tebbets, J.S. (1998). High levels of expression of full length cryIA(C) gene from *Bacillus thuringiensis* in transgenic somatic walnut embryos. *Pl. Sci.* 131: 181-193.

Das, D.K. and Rahman, A. (2010). Expression of a bacterial chitinase (ChiB) gene enhances antifungal potential in transgenic *Litchi chinensis* Sonn. (cv. Bedana). *Curr. Trends Biotechnol. Pharm.* 4(3): 820-833.

Das, D.K. and Rahman, A. (2012). Expression of a rice chitinase gene enhances antifungal response in transgenic litchi (cv. Bedana). *Plant Cell. Tiss. Org.* 109(2): 315-325.

Das, D.K., Prabhakar, M., Kumari, D. and Kumari, N. (2016). Expression of SAMDC gene for enhancing the shelf-life for improvement of fruit quality using biotechnological approaches into Litchi (*Litchi chinensis* Sonn.) cultivars. *Advances in Bioscience and Biotechnology* 7: 300-310.

DeWald, S.G., Litz, R.E. and Moore, G.A. (1989a). Optimising somatic embryo production in mango. *J. Am. Soc. Hort. Sci.* 114: 712-716.

Dhekney, S.A., Li, Z.T. and Gray, D.J. (2011). Grapevines engineered to express cisgenic *Vitis vinifera* thaumatin-like protein exhibit fungal disease resistance. *In Vitro Cell Dev. Biol. Plant* 47(4): 458-466.

Djennane, S., Cesbron, C., Sourice, S., Loridon, K. and Chevreau, E. (2009). Production of transgenic pear plants expressing ferritin gene from pea with the aim to reduce fire blight susceptibility. *Acta Hort* 814(2): 781-786.

Dreesen, R.S.G., Davey, M.W. and Keulemans, J. (2012). Genetic engineering of the self-incompatibility mechanism in 'Elstar' apple leads to distinct levels of self-fertility. *Acta Hort.* 967: 157-166.

Druart, P. and Gruselle, R. (2007). Behaviour of *Agrobacterium rhizogenes*-mediated transformants of 'Inmil' (*P. incisa* x *Serrula*) cherry rootstock. *Acta Hort.* 738: 589-592.

Efendi, D. (2003). Embryogenic avocado culture transformation with SAM-hydrolase and protocol development for cryopreservation. Ph.D. dissertation. University of Florida, Gainesville.

Endo, T., Shimada, T., Fujii, H., Nishikawa, F., Sugiyama, A., Nakano, M., Shimizu, T., Kobayashi, Y., Araki, T., Pena, L. and Omura, M. (2009). Development of a CiFT co-expression system for functional analysis of genes in citrus flowers and fruit. *J. Jpn. Soc. Hortic. Sci.* 78(1): 74-83.

Escalant, J.V., Teisson, C. and Cote, F. (1994). Amplified somatic embryogenesis from male flowers of triploid banana and plantain cultivars (*Musa* spp.). *In Vitro Cell Dev. Biol. Plant* 30: 181-186.

Faize, M., Sourice, S., Dupuis, F., Parisi, L., Gautier, M.F. and Chevreau, E. (2004). Expression of wheat puroindoline-b reduces scab susceptibility in transgenic apple (*Malus x domestica* Borkh.). *Plant Sci.* 167: 347-354.

FDA Arctic Apples and Innate Potatoes are safe for consumption (2015). United States Food and Drug Administration.

Flachowsky, H. and Hanke, M.V. (2012). Genetic control of flower development in apple and the utilisation of transgenic early flowering apple plants in breeding. *Acta Hort.* 967: 29-34.

Flachowsky, H., Peil, A., Sopanen, T., Elo, A. and Hanke, V. (2007). Over-expression of BpMADS4 from silver birch (*Betula pendula* Roth.) induces early-flowering in apple (*Malus* x *domestica* Borkh.). *Plant Breed.* 126(2): 137-145.

Flachowsky, H., Richter, K., Kim, W.S., Geider, K. and Hanke, M.V. (2008). Transgenic expression of a viral EPS-depolymerase (dpo) is potentially useful to induce fire blight resistance in apple. *Ann. Appl. Biol.* 153(3): 345-355.

Francisco, R.M., Regalado, A., Ageorges, A., Burla, B.J., Bassin, B., Eisenach, C., Zarrouk, O., Vialet, S., Marlin, T., Chaves, M.M., Martinoia, E. and Nagy, R. (2013). ABCC1, an ATP binding cassette protein from grape berry, transports anthocyanidin 3-O-glucosides. *Plant Cell* 25: 1840-1854.

Franks, T., He, D.G. and Thomas, M.R. (1998). Regeneration of transgenic *Vitis vinifera* L. Sultana plants: Genotypic and phenotypic analysis. *Mol. Breed.* 4: 321.

Freiman, A., Shlizerman, L., Golobovitch, S., Yablovitz, Z., Korchinsky, R., Cohen, Y., Samach, A., Chevreau, E., Roux, P.M., Patocchi, A. and Flaishman, M.A. (2012). Development of a transgenic early flowering pear (*Pyrus communis* L.) genotype by RNAi silencing of PcTFL1-1 and PcTFL1-2. *Planta* 235(6): 1239-1251.

Fromm, M., Taylor, L.P. and Walbot, V. (1985). Expression of genes transferred into monocot and dicot plant cells by electroporation. *Proc. Natl. Acad. Sci. USA* 82: 5824.

Furman, N., Kobayashi, K., Zanek, M.C., Calcagno, J., Garcia, M.L. and Mentaberry, A. (2013). Transgenic sweet orange plants expressing a dermaseptin coding sequence show reduced symptoms of citrus canker disease. *J. Biotechnol.* 167(4): 412-419.

Gao, M., Sakamoto, A., Miura, K., Murata, N., Sugiura, A. and Tao, R. (2000). Transformation of Japanese persimmon (*Diospyros kaki* Thunb.) with a bacterial gene for choline oxidase. *Mol. Breed.* 6: 501-510.

Gao, M., Takeishi, H., Katayama, A. and Tao, R. (2013). Genetic transformation of Japanese persimmon with flowering locus T (FT) gene and terminal flower 1 (tFl1) homologues gene. *Acta Hort.* 996: 159-164.

Garcia, G.J.A., Barcelo, M., Lopez, A.J.M., Munoz, B.J., Pose, S., Pliego, A.F., Mercado, J.A. and Quesada, M.A. (2012). Improvement of strawberry fruit softening through the silencing of cell wall genes. *Acta Hort.* 929: 107-110.

Gascuel, Q., Diretto, G., Monforte, A.J., Fortes, A.M. and Granell, A. (2017). Use of natural diversity and biotechnology to increase the quality and nutritional content of tomato and grape. *Front. Plant Sci.* 8: 652.

Giovanni, A.L.B., Thomas, W., Johannes, F., Thomas, D.K., Henryk, F., Andreas, P., Maria, V.H., Klaus, R., Andrea, P. and Cesare, G. (2014). Engineering fire blight resistance into the apple cultivar 'Gala' using the FB_MR5 CC-NBS-LRR resistance gene of *Malus robusta*. *Plant Biotech. J.* doi: 10.1111/pbi.12177. 1-6.

Gleba, Y., Klimyuk, V. and Marillonnet, S. 2004. Magnifection—A new platform for expressing recombinant vaccines in plants. *Methods Mol. Biol.* 267: 365-383.

Gomez, C., Terrier, N., Torregrosa, L., Vialet, S., Fournier-Level, A., Verries, C., Souquet, J.M., Mazauric, J.P., Klein, M., Cheynier, V. and Ageorges, A. (2009). Grapevine MATE-type proteins act as vacuolar $H^+$-dependent acylated anthocyanin transporters. *Plant Physiol.* 150: 402-415.

Grapin, A., Ortiz, J.L., Lescot, T., Ferriere, N. and Cote, F.X. (2000). Recovery and regeneration of embryogenic cultures from female flowers of False Horn Plantain. *Pl. Cell. Tiss. Org. Cult.* 61: 237-244.

Hamilton, R.H. and Fall, M.Z. (1971). The loss of tumor-initiating ability in *Agrobacterium tumefaciens* by incubation at high temperature. *Experientia* 27: 229.

Hanhineva, K., Anttonen, M.J., Kokko, H, Soininen, P., Laatikainen, R., Karenlampi, S., Rogachev, I. and Aharoni, A. (2009). Stilbene synthase gene transfer resulted in down regulation of endogenous chalcone synthase in strawberry (*Fragaria* x *ananassa*) and led to the identification of novel phenylpropanoid glucosides. *Acta Hort.* 839: 673-680.

Hanke, M.V. and Flachowsky, H. (2010). Fruit crops. pp. 307-348. *In*: Kempken, F. and Jung, C. (Eds.). Genetic Modification of Plants, Biotechnology in Agriculture and Forestry 64. Springer-Verlag, Berlin Heidelberg.

Hao, G., Pitino, M., Duan, Y. and Stover, E. (2016). Reduced susceptibility to *Xanthomonas citri* in transgenic citrus expressing the FLS2 receptor from *Nicotiana benthamiana*. *Mol. Plant Microbe. Interact.* 29(2): 132-142.

Hassler, S. 1995. Bananas and biotech consumers. *Biotechnology* 13: 417.

Hernandez, J.B.P., Remy, S., Sauco, V.G., Swennen, R. and Sagi, L. (1999). Chemotactic movement and attachment of *Agrobacterium tumefaciens* to banana cells and tissues. *J. Plant Physiol.* 155: 245-250.

Hily, J.M., Ravelonandro, M., Damsteegt, V., Bassett, C., Petri, C., Liu, Z.R. and Scorza, R. (2007). Plum pox virus coat protein gene intron-hairpin-RNA (ihpRNA) constructs provide resistance to plum pox virus in *Nicotiana benthamiana* and *Prunus domestica*. *J. Am. Soc. Hortic. Sci.* 132(6): 850-858.

Hu, C.H., Wei, Y.R., Huang, Y.H. and Yi, G.J. (2013). An efficient protocol for the production of chit42 transgenic Furenzhi banana (*Musa* spp. AA group) resistant to *Fusarium oxysporum*. *In Vitro Cell Dev. Biol. Plant* 49(5): 584-592.

Husaini, A.M., Abdin, M.Z., Salim, K., Xu, Y.W., Samina, A. and Mohammed, A. (2012). Modifying strawberry for better adaptability to adverse impact of climate change. *Curr. Sci.* 102(12): 1660-1673.

Iwanami, T. (2010). Properties and control of Satsuma dwarf virus. *Jpn. Agr. Res. Q* 44(1): 1-6.

Izumi, Y., Kamei, E., Miyamoto, Y., Ohtani, K., Masunaka, A., Fukumoto, T., Gomi, K., Tada, Y., Ichimura, K., Peever, T.L. and Akimitsu, K. (2012). Role of the pathotype-specific ACRTS1 gene encoding a hydroxylase involved in the biosynthesis of host-selective ACR-toxin in the rough lemon pathotype of *Alternaria alternate*. *Phytopathology* 102(8): 741-748.

Janick, J. (1992). Introduction. pp. xix-xxi. *In*: F.A. Hammerschlag and R.E. Litz (Eds.). Biotechnology of Perennial Fruit Crops. Wallingford, UK.

Jin, W.M., Dong, J., Yin, S.P., Yan, A.L. and Chen, M.X. (2007). CBF1 gene transgenic strawberry and increase freezing tolerance. *Acta. Bot. Boreal.-Occid. Sin.* 27(2): 223-227.

Kao, K.N. and Michayluk, C. (1974). A method for high frequency intergeneric fusion of plant protoplasts. *Planta* 115: 355.

Khaled, A.L., Hans-Jorg Jacobsen, R., Hofmann, T., Schwab, W. and Hassan, F. (2017). Metabolic engineering of apple by overexpression of the *MdMyb10* gene. *Journal of Genetic Engineering and Biotechnology* 15(1): 263-273.

Kim, J.H., Kwon, S., Shin, I.S., Cho, K.H., Heo, S.O. and Kim, H.R. (2009). The apple rootstock transgenic M.26 (*Malus pumila*) with enhanced rooting ability. *Korean J. Breed. Sci.* 41(4): 482-487.

Kim, J.H., Shin, I.S., Cho, K.H., Kim, S.H., Kim, D.H. and Hwang, J.H. (2010). Introduction of CAX1 into 'Hongro' apple via *Agrobacterium tumefaciens*. *Korean J. Breed. Sci.* 42(5): 534-539.

Ko, L., Eccleston, K., O'Hare, T., Wong, L., Giles, J. and Smith, M. (2013). Field evaluation of transgenic pineapple (*Ananas comosus* (L.) Merr.) cv. 'smooth Cayenne' for resistance to blackheart under subtropical conditions. *Scient. Hort.* 159: 103-108.

Kobayashi, S., Ding, C.K., Nakamura, Y., Nakajima, I. and Matsumoto, R. (2000). Kiwifruits (*Actinidia deliciosa*) transformed with a *Vitis* stilbene synthase gene produce piceid (resveratrol-glucoside). *Pl. Cell. Rpts.* 19: 904-910.

Koshita, Y., Nakamura, Y., Kobayashi, S. and Morinaga, K. (2002). Introduction of rolC gene into the genome of the Japanese persimmon causes dwarfism. *J. Jap. Soc. Hort. Sci.* 71: 529-531.

Kotoda, N., Wada, M., Komori, S., Kidou, S., Abe, S., Masuda, T. and Soejima, J. (2000). Expression pattern of homologues of floral meristem identity genes LFY and AP1 during flower development in apple. *J. Am. Soc. Hort. Sci.* 125: 398-403.

Krens, F.A., Pelgrom, K.T.B., Schaart, J.G., den Nij, A.P.M. and Rouwendal, G.J.A. (2004). Clean vector technology for marker-free transgenic fruit crops. *Acta Hort.* 663: 431-435.

Krens, F.A., Schaart, J.G., Groenwold, R., Walraven, A.E.J., Hesselink, T. and Thissen, J.T.N.M. (2011). Performance and long-term stability of the barley hordothionin gene in multiple transgenic apple lines. *Transgenic Res.* 20(5): 1113-1123.

Krishna, B., Kadu, A.A., Vyavhare, S.N., Chaudhary, R.S., Joshi, S.S., Patil, A.B., Subramaniam, V.R. and Sane, P.V. (2013). RNAi-mediated resistance against Banana bunchy top virus (BBTV) in 'Grand Nain' banana. *Acta Hort.* 974: 157-164.

Landi, L., Capocasa, F., Costantini, E. and Mezzetti, B. (2009). ROLC strawberry plant adaptability, productivity and tolerance to soil-borne disease and mycorrhizal interactions. *Transgenic Res.* 18(6): 933-942.

Lehnert, R. (2011). Firm seeks approval for transgenic apple (http://www.goodfruit.com/Good-Fruit-Grower/January-15th-2011/Firm-seeks-approval-for-transgenic-apple/), *Good Fruit Grower*.

Leslie, C.A., McGranahan, G.H., Vail, P.V. and Tebbets, J.S. (2001). Development and field testing of walnuts expressing the (RYIACC) gene for Lepidopteran insect resistance. *Acta Hort.* 544: 195-199.

Li, G.Q., Qi, J., Zhang, Y.X., Gao, Z.H., Xu, D.Q., Li, H.X. and Huo, C.M. (2011). Construction and transformation for the antisense expression vector of the polyphenol oxidase gene in Yali pear. *Front. Agric. China.* 5(1): 40-44.

Li, Y.H., Zhang, Y.Z., Feng, F.J., Liang, D., Cheng, L.L., Ma, F.W. and Shi, S.G. (2010). Overexpression of a *Malus* vacuolar $Na^+/H^+$ antiporter gene (MdNHX1) in apple rootstock M.26 and its influence on salt tolerance. *Plant Cell Tiss. Org.* 102(3): 337-345.

Litz, R.E. (1984). *In vitro* somatic embryogenesis from nucellar callus of monoembryonic mango. *Hort. Sci.* 19: 715-717.

Litz, R.E., Raharjo, S.H.T., Efendi, D., Witjaksono and Gomez, L.M.A. (2007). Plant recovery following transformation of avocado with anti-fungal protein and SAM hydrolase genes. *Acta Hort* 738: 447-450.

Machado, L.C.M., Machado, A.C., Hanzer, V., Weiss, H., Regner, F., Steinkeliner, H., Mattanovich, D., Plail, R., Knapp, E., Kaltho, B. and Katinger, H.W.D. (1992). Regeneration of transgenic plants of *Prunus armeniaca* containing the coat protein gene of Plum Pox Virus. *Pl. Cell Rpts.* 11: 25-29.

Malnoy, M., Faize, M., Venisse, J.S., Geider, K. and Chevreau, E. (2005). Expression of viral EPS depolymerase reduces fire blight susceptibility in transgenic pear. *Pl. Cell Rpts.* 23: 632-638.

Malnoy, M., Viola, R., Jung, M.H., Koo, O.J., Kim, S., Kim, J.S., Velasco, R. and Kanchiswamy, C.N. (2016). DNA-free genetically edited grapevine and apple protoplast using CRISPR/Cas9 ribonucleo proteins. *Front. Plant Sci.* 7: 1-9.

Malnoy, M., Xu, M.L., Borejsza-Wysocka, E., Korban, S.S. and Aldwinckle, H.S. (2008). Two receptor-like genes, Vfa1 and Vfa2, confer resistance to the fungal pathogen *Venturia inaequalis* inciting apple scab disease. *Mol. Plant Microbe In.* 21(4): 448-458.

Manshardt, R.M. (1998). 'UHF Rainbow' Papaya. Germplasm G.1., University of Hawaii, College of Tropical Agriculture & Human Resources, Honolulu.

Matsuda, N., Ikeda, K., Kurosaka, M., Takashina, T., Isuzugawa, K., Endo, T. and Omura, M. (2009). Early flowering phenotype in transgenic pears (*Pyrus communis* L.) expressing the CiFT gene. *J. Jpn. Soc. Hortic. Sci.* 78(4): 410-416.

Maziah, M., Sariah, M. and Sreeramanan, S. (2007). Transgenic banana Rastali (AAB) with beta-1,3-glucanase gene for tolerance to *Fusarium* wilt race 1 disease via *Agrobacterium*-mediated transformation system. *Plant Pathol. J.* 6(4): 271-282.

McCafferty, H.R.K, Moore, P.H. and Zhu, Y.J. (2008). Papaya transformed with the *Galanthus nivalis* GNA gene produces a biologically active lectin with spider mite control activity. *Plant Sci.* 175(3): 385-393.

McGranahan, G.H., Leslie, C.A., Uratso, S.L., Martin, L.A. and Dandekar, A.M. (1988). *Agrobacterium*-mediated transformation of walnut somatic embryos and regenration of plants. *Biotechnol.* 6: 800-804.

Mhatre, M., Lingam, S. and Ganapathi, T.R. (2011). Enhanced iron and zinc accumulation in genetically engineered pineapple plants using soybean ferritin gene. *Biol. Trace. Elem. Res.* 144(1/3): 1219-1228.

Mikhailov, R., Firsov, A., Shulga, O. and Dolgov, S. (2012). Transgenic plums (*Prunus domestica* L.) of 'Startovaya' express the Plum pox virus coat protein gene. *Acta Hort.* 929: 445-450.

Mor, T.S., Gomez-Lim, M.A. and Palmer, K.E. (1998). Edible vaccines: A concept coming of age. *Trends Microbiol.* 6: 449-453.

Morgan, A., Baker, C.M., Chu, J.S.F., Lee, K., Crandall, B.A. and Jose, L. (2002). Production of herbicide tolerant strawberry through genetic engineering. *Acta Hort.* 567(1): 113-115.

Nagel, A.K., Kalariya, H. and Schnabel, G. (2010). The Gastrodia antifungal protein (GAFP-1) and its transcript are absent from scions of chimeric-grafted plum. *Hort. Science* 45(2): 188-192.

Nakajima, I., Ban, Y., Azuma, A., Onoue, N., Moriguchi, T., Yamamoto, T., Toki, S. and Endo, M. (2017). CRISPR/Cas9-mediated targeted mutagenesis in grape. *PLoS ONE* 12(5): e0177966.

Nakamura, Y., Sawada, H., Kobayashi, S., Nakajima, I. and Yoshikawa, M. (1999). Expression of soybean β-1,3-endoglucanase cDNA and e.ect on disease tolerance in kiwifruit plants. *Pl. Cell. Rpts.* 18: 527-532.

Nakano, M., Hoshino, Y. and Mii, M. (1994). Regeneration of transgenic of grapvine (*Vitis vinifera* L.) via *A. rhizogenes*-mediated transformation. *J. Exptl. Bot.* 45: 649-656.

Nirala, N.K., Das, D.K., Srivastava, P.S., Sopory, S.K. and Upadhyaya, K.C. (2010). Expression of a rice chitinase gene enhances antifungal potential in transgenic grapevine (*Vitis vinifera* L.). *Vitis* 49(4): 181-187.

Ogata, T., Yamanaka, S., Shoda, M., Urasaki, N. and Yamamoto, T. (2016). Current status of tropical fruit breeding and genetics for three tropical fruit species cultivated in Japan: Pineapple, mango and papaya. *Breed Sci.* 66(1): 69-81.

Padilla, G., Perez, J.A., Perea, A.I., Moon, P.A., Gomez, L.M.A., Borges, A.A., Exposito, R.M. and Litz, R.E. (2013). *Agrobacterium tumefaciens*-mediated transformation of 'Brewster' ('Chen Tze') litchi (*Litchi chinensis* Sonn.) with the PISTILLATA cDNA in antisense. *In Vitro Cell Dev. Biol. Plant* 49(5): 510-519.

Padilla, I.M.G., Golis, A., Gentile, A., Damiano, C. and Scorza, R. (2006). Evaluation of transformation in peach (*Prunus persica*) explants using green fluorescent protein (GFP) and beta-glucuronidase (GUS) reporter genes. *Plant Cell Tiss. Organ. Cult.* 84: 309-314.

Pena, L., Martyn-Trillo, M., Juarez, J.A., Pina, J.A., Navarro, L. and Martynez-Zapater, J.M. (2001). Constitutive expression of Arabidopsis LEAFY or APETALA1 genes in citrus reduces their generation time. *Nat. Biotechnol.* 19: 263-267.

Perez-Clemente, R.M., Perez-Sanjuan, A., Garcia-Ferriz, L., Beltran, J.P. and Canas, L.A. (2004). Transgenic peach plants (*Prunus persica* L.) produced by genetic transformation of embryo sections using the green fluorescent protein (GFP) as an *in vivo* marker. *Mol. Breed.* 14: 419-427.

Perez-Diaz, R., Madrid-Espinoza, J., Salinas-Cornejo, J., Gonzalez-Villanueva, E. and Ruiz-Lara, S. (2016). Differential roles for VviGST1, VviGST3, and VviGST4 in proanthocyanidin and anthocyanin transport in *Vitis vinifera*. *Front. Plant Sci.* 7: 1166.

Petri, C., Lopez, N.S., Alburquerque, N. and Burgos, L. (2006). Regeneration of transformed apricot plants from leaves of a commercial cultivar. *Acta Hort.* 717: 233-235.

Pollack, A. (2012). That Fresh Look, Genetically Buffed (http://www.nytimes.com/2012/07/13/business/growers-fret-over-a-newapple-that-wont-turn-brown.html?_r=2&pagewanted=all) New York Times.

Raharjo, S.H.T., Witjaksono, N.F.N., Gomez, L.M.A., Padilla, G. and Litz, R.E. (2008). Recovery of avocado (*Persea americana* Mill.) plants transformed with the antifungal plant defensin gene PDF1.2. *In Vitro Cell Dev. Biol. Plant* 44(4): 254-262.

Rai, M.K. and Shekhawat, N.S. (2014). Recent advances in genetic engineering for improvement of fruit crops. *Plant Cell Tiss. Org.* 116(1): 1-15.

Ramesh, S.A., Kaiser, B.N., Franks, T., Collins, G. and Sedgley, M. (2006). Improved methods in *Agrobacterium*-mediated transformation of almond using positive (mannose/pmi) or negative (kanamycin resistance) selection-based protocols. *Pl. Cell. Rpts.* 25(8): 821-828.

Raquel, M.H., Miguel, C., Nolasco, G. and Oliveira, M.M. (2000). Construction and testing of a transformation vector aiming at introducing PDV resistance in almond. *Acta Hort.* 521: 73-81.

Ravelonandro, M., Scorza, R., Briard, P., Lafargue, B. and Renaud, R. (2011). Inheritance of silencing in transgenic plums. *Acta Hort.* 899: 139-144.

Reis, L.B., Lemes, S.M., Lima, A.B.P., Oliveira, M.L.P., Pinto, D.L.P., Lani, E.R.G. and Otoni, W.C. (2007). *Agrobacterium rhizogenes*-mediated transformation of passionfruit species: *Passiflora cincinnata* and *P. edulis* f. *flavicarpa*. *Acta Hort.* 738: 425-431.

Ren, C., Liu, X., Zhang, Z., Wang, Y., Duan, W., Li, S. and Liang, Z. (2016). CRISPR/Cas9-mediated efficient targeted mutagenesis in Chardonnay (*Vitis vinifera* L.). *Sci. Rep.* 6: 32289.

Rivera, D.M., Astorga, C.K.R., Vallejo, C.S., Vargas, A.I. and Sanchez, S.E. (2011). Transgenic mango embryos (*Mangifera indica*) cv. Ataulfo with the defens in J1 gen. *Revista Mexicana de Fitopatologia* 29(1): 78-80.

Rola, A.C., Elazegui, D.D., Nguyen, M.R., Magdalita, P.M., Dumayas, E.E. and Chupungco, A.R. (2010). Innovations in seed systems for potential biotechnology products: The case of genetically modified papaya in the Philippines. *Philipp. J. Crop. Sci.* 35(1): 80-91.

Rugienius, R., Stanys, V. and Staniene, G. (2009). Genetic transformation of pear and sweet cherry rootstocks by rooting stimulating rolB transgene and evaluation of its expression *in vitro*. *Sodininkyste ir Darzininkyste* 28(1): 43-54.

Rugini, E., Cristofori, V., Martignoni, D., Gutierrez, P.P., Orlandi, S., Brunori, E., Biasi, R., Muleo, R. and Magro, P. (2011). Kiwifruit transgenics for Osmotin gene and inoculation tests with *Botrytis cinerea* and *Cadophora luteo-olivacea*. *Acta Hort.* 913: 197-203.

Samanta, S., Ravindra, M.B., Dinesh, M.R., Anand, L. and Mythili, J.B. (2007). *In vitro* regeneration and transient gene expression in mango cv. 'Vellaikolumban'. *J. Hortic. Sci. Biotech.* 82(2): 275-282.

Sanford, J.C. and Johnston, S.A. (1985). The concept of parasite derived resistance deriving resistance genes from the parasite's own genome. *Theor. Biol.* 113: 395.

Sansavini, S. (2004). Where we are with the application of GMO biotechnology to fruit tree species? *Rivista-di-Frutticoltura-e-di-Ortofloricoltura* 66(1): 49-52.

Schestibratov, K.A. and Dolgov, S.V. (2006). Genetic engineering of strawberry cv. Firework and cv. Selekta for taste improvement and enhanced disease resistance by introduction of thauII gene. *Acta Hort.* 708: 475-481.

Scorza, R., Callahan, A.M., Dardick, C.D., Srinivasan, C., DeJong, T.M., Abbott, A.G. and Ravelonandro, M. (2013). Biotechnological advances in the genetic improvement of *Prunus domestica*. *Acta Hort.* 985: 111-117.

Sharma, S., Kaur, R. and Singh, A. (2017). Recent advances in CRISPR/Cas mediated genome editing for crop improvement. *Plant Biotechnol. Rep.* 11: 193-207.

Song, G., Sink, K.C., Walworth, A.E., Cook, M.A., Allison, R.F. and Lang, G.A. (2013). Engineering cherry rootstocks with resistance to Prunus necrotic ring spot virus through RNAi-mediated silencing. *Plant Biotechnol. J.* 11: 702-708.

Sreedharan, S., Shekhawat, U.K.S. and Ganapathi, T.R. (2012). MusaSAP1, a A20/AN1 zinc finger gene from banana functions as a positive regulator in different stress responses. *Plant Mol. Biol.* 80(4/5): 503-517.

Sreedharan, S., Shekhawat, U.K.S., Ganapathi, T.R. (2013). Transgenic banana plants over-expressing a native plasma membrane aquaporin MusaPIP1;2 display high tolerance levels to different abiotic stresses. *Plant Biotech. J.* 11(8): 942-952.

Srinivasan, C., Dardick, C., Callahan, A. and Scorza, R. (2012). Plum (*Prunus domestica*) trees transformed with poplar FT1 result in altered architecture, dormancy requirement, and continuous flowering. *PLoS-ONE* 7(7): e40715.

Sripaoraya, S., Keawsompong, S., Insupa, P., Power, J.B., Davey, M.R. and Srinives, P. (2006). Genetically manipulated pineapple: Transgene stability, gene expression and herbicide tolerance under field conditions. *Plant Breed.* 125(4): 411-413.

Staniene, G., Rugienius, R., Gelvonauskiene, D. and Stanys, V. (2007). Effect of rolB transgene on *Prunus cerasus* x *P. canescens* and *Cydonia oblonga* microshoot rhizogenesis. *Biologia* 53(1): 23-26.

Stewart, R.J., Sawyer, B.J.B., Bucheli, C.S. and Robinson, S.P. (2001). Polyphenol oxidase is induced by chilling and wounding in pineapple. *Aust. J. Pl. Physiol.* 28: 181-191.

Talengera, D., Beemster, G.T.S., Fabio, F., Inze, D., Kunert, K. and Tushemereirwe, W.K. (2009). Transformation of banana (*Musa* spp.) with a D-type cyclin gene from *Arabidopsis thaliana* (ArathCYCD2;1). *Asp. Appl. Biol.* 96: 45-53.

Tamura, M., Gao, M., Tao, R., Labavitch, J.M. and Dandekar, A.M. (2004). Transformation of persimmon with a pear fruit polygalacturonase inhibiting protein (PGIP) gene. *Scient. Hort.* 103: 19-30.

Tan, B., Li, D.L., Xu, S.X. and Guo, W.W. (2012). Creation of transgenic plants in Jincheng orange (*Citrus sinensis*) with CG1-400-RNase gene by *Agrobacterium*-mediated transformation. *Journal of Fruit Science* 29(4): 544-549.

Tian, N., Wang, J. and Xu, Z.Q. (2011). Over-expression of $Na^+/H^+$ antiporter gene AtNHX1 from *Arabidopsis thaliana* improves the salt tolerance of kiwifruit (*Actinidia deliciosa*). *S. Afr. J. Bot.* 77(1): 160-169.

Torregrosa, L., Iocco, P. and Thomas, M.R. (2002). Influence of *Agrobacterium* strain, culture medium and cultivar on the transformation efficiency of *Vitis vinifera* L. *Am. J. Enol. Vitic.* 53: 183.

Tracy, T. (2015). First Genetically Modified Apple Approved for Sale in U.S. *Wall Street Journal.* Article Retrieved February 13, 2015.

Trevisan, F., Mendes, B.M.J., Maciel, S.C., Vieira, M.L.C., Meletti, L.M.M. and Rezende, J.A.M. (2006). Resistance to passionfruit woodiness virus in transgenic passion flower expressing the virus coat protein gene. *Plant Disease* 90(8): 1026-1030.

Tripathi, L., Atkinson, H., Roderick, H., Kubiriba, J. and Tripathi, J.N. (2017). Genetically engineered bananas resistant to Xanthomonas wilt disease and nematodes. *Food and Energy Security* 10.1002/fes3.101.

Tripathi, L., Mwaka, H., Tripathi, J.N. and Tushemereirwe, W.K. (2010). Expression of sweet pepper Hrap gene in banana enhances resistance to *Xanthomonas campestris* pv. *Musacearum. Mol. Plant Pathol.* 11(6): 721-731.

Vieira, M.L.C., Takahashi, E.K., Falco, M.C., Vieira, L.G. and Pereira, L.F.P. (2002). Direct gene transfer to passion fruit (*Passiflora edulis*) with the attacin A gene. Proc. 10th Int. Assoc. for Plant Tissue Culture and Biotechnology Congress, Orlando, 103-A (abstract).

Walawage, S.L., Phu, M.L., Britton, M.T., Uratsu, S.L., Li, Y., Leslie, C.A. and Dandekar, A.M. (2013). Disease and pest resistant transgenic rootstocks: Analysis, validation, deregulation and stacking of RNAi-mediated resistance traits, *Walnut Research Reports,* doi: 10.13140/2.1.2041.2161. 59-73

Wang, T., Karunairetnam, S., Wu, R., Wang, Y.Y. and Gleave, A.P. (2012). High efficiency transformation platforms for kiwifruit (*Actinidia* spp.) functional genomics. *Acta Hort.* 929: 143-148.

Wang, Y., Liu, X., Ren, C., Zhong, G.Y., Yang, L., Li, S. and Liang, Z. (2016). Identification of genomic sites for CRISPR/Cas9-based genome editing in the Vitis vinifera genome. *BMC Plant Biol.* 16: 96.

Wei, J.Y., Liu, D.B., Chen, Y.Y., Cai, Q.F. and Zhou, P. (2008). Transformation of PRSV-CP dsRNA gene into Papaya by pollen-tube pathway technique. *Acta Bot. Boreal.-Occid. Sin.* 28(11): 2159-2163.

Wen, X.P., Ban, Y., Inoue, H., Matsuda, N. and Moriguchi, T. (2009). Aluminium tolerance in a spermidine synthase over-expressing transgenic European pear is correlated with the enhanced level of spermidine via alleviating oxidative status. *Environ. Exp. Bot.* 66: 471-478.

Wen, X.P., Pang, X.M., Matsuda, N., Kita, M., Inoue, H., Hao, Y.J., Honda, C. and Moriguchi, T. (2008). Over-expression of the apple spermidine synthase gene in pear confers multiple abiotic stress tolerance by altering polyamine titers. *Transgenic Res.* 17(2): 251-263.

Wieczorek, A.M. and Wright, M.G. (2012). History of agricultural biotechnology: How crop development has evolved. *Nature Education Knowledge* 3(10): 9.

Xie, Z.B., Zhong, X.H. and Deng, Z.N. (2008). Establishment of *Agrobacterium*-mediated transformation system of strawberry with chitinase gene. *Journal of Hunan Agricultural University* 34(1): 25-28.

Xu, K. (2013). An overview of arctic apples: Basic facts and characteristics. *New York Fruit Quarterly* 21(3): 8-10.

Yamamoto, T., Iketani, H., Ieki, H., Nishizawa, Y., Notsuka, K., Hibi, T., Hayashi, T. and Matsuta, N. (2000). Transgenic grapevine plants expressing a rice chitinase with enhanced resistance to fungal pathogens. *Plant Cell Rep.* 19(7): 639-646.

Yang, Y.Z., Jittayasothorn, Y., Chronis, D., Wang, X.H., Cousins, P. and Zhong, G.Y. (2013). Molecular characteristics and efficacy of 16D10 siRNAs in inhibiting root-knot nematode infection in transgenic grape hairy roots. *PLoS-ONE* 8(7): e69463.

Yao, J.L., Dong, Y.H. and Morris, B.A.M. (2001). Parthenocarpic apple fruit production conferred by transposon insertion mutations in a MADSbox transcription factor. *Proc. Natl. Acad. Sci., USA* 98: 1306-1311.

Youssef, S.M., Amaya, I., Lopez, A.J.M., Sesmero, R., Valpuesta, V., Casadoro, G., Blanco, P.R., Pliego, A.F., Quesada, M.A. and Mercado, J.A. (2013). Effect of simultaneous down-regulation of pectate lyase and endo-beta-1,4-glucanase genes on strawberry fruit softening. *Mol. Breed.* 31(2): 313-322.

Zhang, X.D., Francis, M.I., Dawson, W.O., Graham, J.H., Orbovic, V., Triplett, E.W. and Mou, Z.L. (2010). Over-expression of the Arabidopsis NPR1 gene in citrus increases resistance to citrus canker. *Eur. J. Plant. Pathol.* 128(1): 91-100.

Zhu, H.S., Wen, Q.F., Pan, D.M. and Lin, Y.Z. (2009). Transformation of antisense ethylene receptor FaEtr1 and FaErs1 genes in strawberry (*Fragaria ananassa*). *J. Agric. Biotech.* 17(5): 825-829.

Zhu, L.H., Li, X.Y., Ahlman, A., Xue, Z.T. and Welander, M. (2004). The use of mannose as a selection agent in transformation of apple rootstock M26 via *Agrobacterium tumefaciens*. *Acta Hort.* 663: 503-506.

Zhu, Y.J., Agbayani, R., Tang, C.S. and Moore, P.H. (2010). Developing transgenic papaya with improved fungal disease resistance. *Acta Hort.* 864: 39-44.

Zok, A., Olah, R., Hideg, E., Horvath, V.G., Kos, P.B., Majer, P., Varadi, G. and Szegedi, E. (2010). Effect of *Medicago sativa* ferritin gene on stress tolerance in transgenic grapevine. *Plant Cell Tiss. Org.* 100(3): 339-344.

# 9 Winemaking: Microbiology

**Albert Mas[1], M. Jesus Torija[1], Gemma Beltran[1] and Ilkin Yucel Sengun[2*]**

[1] Oenological Biotechnology Group, Department of Biochemistry and Biotechnology, Faculty of Oenology, University Rovira i Virgili, Marcel·li Domingo, 1. 43007, Tarragona, Spain

[2] Department of Food Engineering, Engineering Faculty, Ege University, 35100, Bornova, Izmir, Turkey

## 1. Introduction

The production of fermented products has accompanied humanity for the last 10,000 years or more. Wine, as fermented grape juice (must), has been one of the favourite products and one of the most prestigious one in different cultures. Since its appearance in Georgia around 7,000 years ago, it has been appreciated on the tables of kings, priests and almost all the humanity, especially in Europe and America. In the beginning of its history, it was considered a miraculous process by its producers, the priests. For old humans to understand how a sugary juice started boiling without the application of any heat and got transformed into a sugarless juice which when consumed led to euphoria was much beyond their comprehension. Nevertheless, the humanity had already a previous experience with the production of beer, knowing that these fermented products were healthy and the best way to drink water. At that time, water was a source of many diseases and drinking water in the form of wine or beer was recognised as safest.

The nature of the 'miracle' took very long to unveil. Although the chemical transformation was clearly set at the end of the 18$^{th}$ century, it was necessary to enter the 19$^{th}$ century to indicate clearly that living microorganisms were responsible for the process and until the 20$^{th}$ century to clearly describe it and even control it. Cagniard Latour, in 1836, was the first one to mention living beings responsible for alcoholic fermentation; however, the findings were overlooked until 30 years later. The father of modern biochemistry and microbiology, Louis Pasteur, devoted several years to understand the production of wine and beer, describing beyond any doubt its microbiological nature and already described a succession of microorganisms, later named as yeasts (he used the term mycoderma that meant surface fungi that were considered at that time a part of the plant kingdom). Furthermore, among these microorganisms, he also identified those responsible for spoilage and production of vinegar. Later on, with the development of microbiological methods, necessary for isolation and the study of isolated species and strains, researchers and winemakers reached an understanding of the process and started to control it. Hansen developed the use of selected starters initially for beer production way back in 1899, but it took a little bit longer to reach the wine cellars. The production of starters in liquid media was not the most appropriate for the wine sector due to its strong demand in a very short period (four-six weeks). The breakthrough of inoculation in winemaking came with the development of active dry yeast and the wide availability of the same commercially. Although many winemakers have used and still continue to use this commercial presentation, recent movements in winemaking challenge this practice with the return to old fashion uncontrolled process that led to production of 'natural wines'. In the process of wine making, besides yeasts, bacteria like lactic acid bacteria and acetic acid bacteria play a very significant role. The lactic acid bacteria are involved in the malo-lactic fermentation while the acetic acid bacteria are known to play a role mostly in wine spoilage. In this chapter,therefore the focus on these bacteria would also be made.

## 2. Methods of Analysis: Classical and Molecular

The traditional methods to detect and quantify different wine microorganisms are based on morphological tests supplemented with several physiological tests. Besides, an isolation of the microorganisms is needed before identifying or quantifying. The classification schemes described in Barnett *et al.* (2000) can be used to identify yeasts. It is necessary to conduct many tests for reliable identification of most yeasts at

*Corresponding author: ilkin.sengun@ege.edu.tr

species level. Thus, this work is time-consuming and an accurate interpretation requires considerable expertise.

At bacterial level, Gram stain and catalase test are routine analyses to distinguish between Lactic Acid Bacteria (LAB) and Acetic Acid Bacteria (AAB) in wine, although to distinguish at species level is quite difficult and sometimes physiological tests are not enough for such differentiation. One of the first microbiological tests performed was to examine the morphology of the microorganisms under a microscope, using phase-contrast microscopy. This examination yields information related to the shape, size and arrangements of the cells, but may lead to incorrect interpretations due to the appearance of microorganisms depending on age and culture conditions. Also, the morphology of the colonies grown in different specific media can be useful (Fugelsang and Edwards, 2007).

The estimation of microbiological population density and diversity plays an important role in the winemaking process. Population densities can be measured through many methods, but the two most commonly used by oenologists are counts under the microscope and direct plating. Microscope counting techniques are quickest but require a minimal population. An alternative for lower populations is concentration by filtration or direct plating methods. The combination of both can be applied to the wines suspected of having a low viable population. Microscope counting comprises quantification, using a microscope counting chamber (Neubauer or Thoma for instance). The main hurdle is the low limit of detection and the lack of discrimination between alive and dead cells.

Plate enumeration counts the colonies formed by different microorganisms on selected media. There are non-selective media that allow the growth of all microorganisms associated with fermentation. However, in samples with mixed species, the most prevalent species dominate on the media, preventing the detection of species found in low proportion. Several approaches are applied to circumvent this problem by using different selective media that will prevent the growth of the most dominant microorganisms. For instance, Lysine agar is an example of selective medium that prevents the detection of *Saccharomyces cerevisiae* because this yeast grows poorly in the presence of lysine as the sole nitrogen source. This medium is commonly used to isolate and enumerate non-*Saccharomyces* yeasts. Also, the use of antibiotics in the media that inhibit microorganisms is a good alternative. Finally, media enriched in different nutrients to facilitate the growth of different types of microorganisms are also used, for instance, MRS agar (De Man, Rogosa and Sharpe agar) for LAB or GYC agar (glucose yeast extract calcium carbonate agar) for AAB. GYC agar should also be considered as a differential medium because the acid produced by AAB dissolves the precipitates of calcium carbonate and forms a halo around the AAB colony. Other factors that can convert a general medium into a selective one are pH, temperature, an aerobiosis condition, etc. Often, a combination of the different conditions is used for a more efficient enumeration.

The identification of grape and/or wine microorganisms and the ecology was a very important step forward when the plating methods were accompanied by analysis of the colonies through molecular methods based on the analysis of DNA. The ribosomal RNA-coding regions became the most usual tool for identification of wine microorganisms.The ribosomal genes of any organism are grouped in tandem, forming transcription units that are repeated in the genome. In each transcription unit exist internal transcriber spacers (ITS), the external transcriber spacers (ETS) and the rRNA-codifying genes. The ribosomal genes are powerful tools to establish the phylogenetic relationship and to identify species (Kurtzman and Robnett, 1998). The ribosomal genes are highly conserved regions and their sequences are aligned with the sequences available in the databases, allowing the identification of microorganisms. Instead, the ITS regions are more variable, but can be also used to differentiate closely-related species that cannot be differentiated by ribosomal genes. Generally, a phylogenetic tree can be generated with the sequences deposited in the databases and used for the identification of microorganisms.

For sequencing ribosomal genes of yeast, the main regions are the domain D1 and D2 in the 26S gene (Kurtzman and Robnett, 1998). In the case of bacteria, the main gene used is 16S rRNA (Cole *et al.*, 2005). These regions are used to differentiate among yeast species (Montrocher *et al.*, 1998; Egli and Henick-Kling, 2001; Belloch *et al.*, 2002) and bacterial species (Le Jeune and Lonvaud-Funel, 1997) in wine. However, for routine analysis of a large number of samples as required in ecological studies, a cheaper alternative is the restriction analysis of ribosomal genes (Polymerase Chain Reaction-Restriction Fragment Length Polymorphism, PCR-RFLPs). This technique uses specific endonucleases to generate fragments that can be species-specific. The regions used for wine yeast identification lie between 18S and 26S rRNA genes for yeast, which includes the intergenic spacers ITS1, ITS2, and the 5,8S rRNA gene. The most RFLP used for bacteria is the 16S rRNA gene, which has been denominated Amplified Ribosomal DNA Restriction Analysis (ARDRA). The application to wine species was initiated by Guillamón *et al.* (1998) and Esteve-Zarzoso *et al.* (1999) and several studies used this technique later on (Torija *et al.*, 2001; Beltran *et al.*, 2002; Raspor *et al.*, 2006, etc.). ARDRA has been used to identify LAB (Rodas *et al.*, 2005) and AAB (Poblet *et al.*, 2000; Ruiz *et al.*, 2000; González *et al.*, 2006[a]; Gullo *et al.*, 2006; Vegas *et al.*, 2010). Additional species discrimination has been done with the 16S-23S intergenic spacer region (Ruiz *et al.*, 2000; Trček and Teuber, 2002).

The changes introduced in this century after the effort to fulfil the human genome made sequencing more accessible and affordable. Nowadays, only sequencing, alignment with sequences in databases and elaboration of genetic trees are accepted as criteria for the identification of microbial species. However, when a large number of samples is to be processed, grouping through RFLP of the appropriate ribosomal genes or ITS is treated as an initial step, assuming that all the isolates that present the same identification or banding pattern will belong to the same species. At least two or three representatives of each grouping should be sequenced.

The application of molecular-based methods on plate isolates allows the discrimination at strain level. The methods of strain genotyping use polymorphism and repeated sequences along the genome. The most basic technique is based on the random amplification of genomic DNA with a single primer sequence of 9 or 10 bases of length (RAPD). Each strain presents different amplification fragments in size and number. The amplification is followed by agarose gel electrophoresis, which yields a band pattern that should be characteristic of a given strain. This technique has been used to genotype wineyeasts (Cocolin *et al.*, 2004; Capece *et al.*, 2005; Martorell *et al.*, 2006) LAB strains of *Oenococcus oeni* (Cappello *et al.*, 2008) and AAB strains (Bartowsky, 2003). Other methods use the repetitive elements of the genome, all of them based on the design of oligonucleotides homologous to these repeated sequences that allow the amplification of these regions and obtaining a pattern of electrophoretic bands for each species or strain. There are different techniques for genotyping yeast and bacteria, which have been applied for the identification of different wine microorganisms. For instance, microsatellites are tandem repeat units of short DNA sequences, typically 1-10 nucleotide length in eukaryotic cells. The number of repeated sequences along the genome is very variable, making the distances between sequences highly polymorphic in size. Thus, the technique consists in amplification of the parts of the genome that are flanqued by these microsatellites, which yield an amplicon pattern that allows to differentiate the strains. The most common oligonucleotides used are $(GACA)_4$, $(GAG)_5$, $(GTG)_5$ and others. *S. cerevisiae* strains were differentiated by Lieckfeldt *et al.* (1993) and applied to wine strains by Maqueda *et al.* (2010). Gevers *et al.* (2001) used $(GTG)_5$-PCR (also named rep-PCR in bacteria) to differentiate a wide range of food-associated lactobacilli and other LAB species. Nowadays, $(GTG)_5$-PCR are extensively used to genotype AAB in wine vinegar production (Hidalgo *et al.*, 2010; Vegas *et al.*, 2010).

Different methods are used to genotype *S. cerevisiae* as the main microorganism in alcoholic fermentation. For instance, delta elements are conserved sequences that flank transposable Ty elements. The separation distance between these delta elements is variable and does not exceed 1-2kb, which determines that these delta elements are appropriate to amplify the region comprised between them. The separation by size of these bands can be used to differentiate *S. cerevisiae* strains. This method was developed for Ness *et al.* (1993) and Masneuf and Dubourdieu (1994) to genotype strains of *S. cerevisiae*. The facility to perform the PCR analysis without extraction of the DNA (using directly the colony) has made this technique the most widely used to differentiate *S. cerevisiae* strains. The other main technique to differentiate *S. cerevisiae* strains is restriction analysis of mitochondrial DNA (mtDNA-RFLP). The basis of this technique is to use specific restriction endonucleases to fragment the DNA into specific sites, generating fragments of variable size. These fragments are separated on agarose gel showing pattern strain-specific. This technique was first applied to Brewer's yeast and wine strains of *S. cerevisiae* by Aigle *et al.* (1984) and Dubourdieu *et al.* (1987), respectively. Querol *et al.* (1992) simplified the protocol by using a unique characteristic of the mtDNA with high proportion of AT. Then, the restriction pattern DNA with enzymes that target sequences, such as GCAT, will cut less frequently the mtDNA than the nuclear DNA. So far, this was the most used technique to genotype the strains of *S. cerevisiae* (Torija *et al.*, 2001; Beltran *et al.*, 2002; Nikolaou *et al.*, 2007; Maqueda *et al.*, 2010), although it still has the need to extract the DNA and is more time consuming than direct PCR that can be performed with delta elements.

Finally, the most traditionally used technique for typing is the Pulsed-Field Gel Electrophoresis (PFGE) based on the separation of the entire chromosomes electrophoresis with alternating electrical fields. The chromosomes should change their migration direction, thus enabling the separation of large fragments of DNA. This technique has been used to genotype wine strains of *S. cerevisiae* (Guillamón *et al.*, 1996), some non-*Saccharomyces* (Esteve-Zarzoso *et al.*, 2001, 2003) and *O. oeni* (Vigentini *et al.*, 2009). However, the main drawback of the methods based on plating is that they only quantify the microorganisms that are able to grow and thus, the cells that are able to form colonies (colony forming units, abbreviated as CFU or cfu). The population enumerated by this method is considered as the 'culturable' population. Although it is broadly used, this aspect together with the time required for some microorganisms to grow (two to five days in yeasts, two to 10 days in bacteria) is a main handicap for the wine industry. However, one of the main challenges of wine microbial ecology is that many microorganisms undergo states that are defined as Viable But Not Culturable (VBNC, Millet and Lonvaud-Funel, 2000). Microorganisms that are VBNC state are those that lose the ability to grow in a culture medium but still maintain some metabolic activity. This is one of the responses of many microorganisms when the environmental conditions are not optimal. The previous assumption was that these microorganisms were dead. However, these microorganisms are alive but they are not able to form colonies. The VBNC state involves the microorganism ability to recover and grow if they are allowed to recover in media without the stress that originates this status (Oliver, 2005). Although they have reduced their metabolism, they are able to spoil wines, for instance. In fact, during long periods when wines are settled (ageing, bottles, etc.), the chances of spoilage are greater even in the absence of culturable spoiling microorganisms. Even when the metabolism of the microorganisms is slow and the population is low, they have a lot of time to act and, thus, to alter the wine properties. These microorganisms maintain the basal metabolism to keep active the main cellular functions. For instance, they continue to express genes during VBNC states (Lleò *et al.*, 2000, 2001; Yaron and Mathews, 2002). Microorganisms entering VBNC state have different protein profiles from growing microorganisms (Heim *et al.*, 2002). Also, VBNC microorganisms modify the composition of fatty acids in the plasma membranes, which are essential in this state (Day and Oliver, 2004) to maintain the membrane potential (Porter *et al.*, 1995; Tholozan *et al.*, 1999).

Thus, as a consequence, in a microbial mixture (such as in the different stages of wine production), there are live cells, dead cells and several cells in transient states (Fig. 1). These transient cells could be old cells that still retain the ability to grow under optimal conditions, old cells that have impaired the ability to grow in regular plates but still fully viable with active metabolism and finally cells that enter the lytic process. The old cells that have lost the ability to grow on plates can often be recovered by providing a very rich medium, normally using a liquid medium with strong aeration (in case of yeast or aerobic bacteria, for instance) to promote their growth again (Wang *et al.*, 2016). Thus, culture-independent techniques use molecular techniques to identify and/or quantify wine microorganisms without previous cultivation of these microorganisms (Rantsiou *et al.*, 2005). These methods provide a better knowledge of the population, avoiding the biases that represent the microorganisms that are absent or do not grow well on a plate.

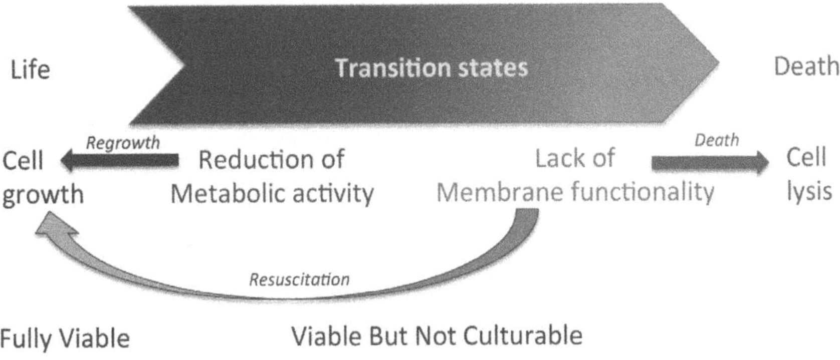


**Figure 1.** Cell statuses during alcoholic fermentation

Color version at the end of the book

As a result, most of the consolidated knowledge on wine microbiology has emerged from the use of plating and analysis of microorganisms that could be recovered on plates. However, enumeration and identification of the microorganisms recovered on plates underwent a strong change from the extension of the molecular biology techniques that targeted DNA as main element. This meant a quick and big step towards the determination of grape and wine ecology. The expansion of these molecular biology techniques for identification and typing allowed a step further – the use of these techniques directly from grapes or wines, without the steps of culturing the microorganisms on plates. These 'culture-independent' techniques have been used quite extensively since the beginning of the 21st century and still are very common. Many of these culture-independent techniques have some limitations, though. If the main target is DNA, this molecule is rather stable with time and does not allow the differentiation between live and dead cells. Several alternatives have been proposed to circumvent this limitation – targeting more labile molecules, such as RNA; quantification and identification through hybridisation of non-DNA molecules with short life, etc. For instance, a solution for appropriate differentiation between dead cells, VBNC cells and culturable alive cells has been the use of culture-independent techniques with some modifications to eliminate the DNA from dead cells or use RNA. Several studies used RNA instead of DNA to quantify or detect the viable population, since this molecule is rapidly degraded in the dead cells (Cocolin and Mills, 2003; Hierro *et al.*, 2006). However, it is very tedious to work with RNA because it is unstable and can be degraded during purification or analysis. Furthermore, rRNA might be more stable than required (Hierro *et al.*, 2006; Andorrà *et al.*, 2011). Successful alternatives to RNA have been developed with DNA binding dyes that only penetrate in the dead cells (damaged membranes) and block the amplification of this DNA (Rudi *et al.*, 2005; Nocker and Camper, 2006). Ethidium monoazide bromide (EMA) and propidium monoazide bromide (PMA) were proposed by Nogva *et al.* (2003) and Nocker *et al.* (2006), respectively, to detect the bacteria viable cells. Both chemicals penetrate only into dead cells, in fact, into cells with compromised membrane integrity but not into live cells with fully functional cell membrane. Upon binding to the DNA of dead cells, the photo-inducible azide group allows these dyes to be covalently cross-linked by exposure to bright light and precipitate the DNA (Nocker and Camper, 2006; Nocker *et al.*, 2006). Thus, only the DNA from live cells will be detected and quantified after the treatment with these dyes. This methodology has been applied successfully to wine microorganisms (Andorrà *et al.*, 2010a).

The control of winemaking process requires identification of the microorganisms present as well as the quantification of each species in different stages. The quantification is based on the correlation of the amount of the target molecules with the amount of biomass. This is true for DNA but is not completely valid for other molecules, such as RNA or proteins, as they are more related to the physiological statuses of the cells, which present strong changes during winemaking. In fact, almost all the relevant microorganisms in winemaking undergo complete life and growing cycles during the process.

Among the most usual techniques, the following have been discussed briefly:

- ***Direct Epifluorescence Technique (DEFT):*** This technique is based on the direct counting of viable cells with specific dyes that react with the cell's material or are incorporated in specific subcellular organelles. This technique is more applicable to routine wine control, especially since fluorescent dyes are now available, although it requires the fluorescence microscope. Despite the initial use of acridine orange (Froudière *et al.*, 1990), currently there are different kits available using mixtures of SYTO 9 green fluorescent nucleic acid stain and the red fluorescent nucleic acid stain propidium iodide. SYTO 9 stains all cells with both intact and damaged membranes, whereas propidium iodide only penetrates into cells with damaged membranes by competing with SYTO 9 for binding sites and reducing the activity of SYTO-9. Hence, cells with green fluorescence are considered as live, whereas cells in red are considered as dead. There are also other combinations, such as FUN 1 (red-fluorescent cylindrical intravacuolar structures in the metabolically active cells) with Calcofluor White M2R (stains the cell walls with a blue-fluorescent, regardless the metabolic state). However, the intravacuolar structures are sometimes difficult to detect and underestimate the viability (van Zandycke, 2003). This is a rapid and reliable technique that allows the quantification of viable and non-viable cells, although it cannot distinguish between the different genus or species. The use of this method has been the reference to demonstrate the presence of significant non-culturable populations of both bacteria and yeast in wine (Millet and Lonvaud-Funel, 2000; Du Toit *et al.*, 2005).

- ***Fluorescence In Situ Hybridisation (FISH):*** This technique involves direct hybridisation of some labelled probes to DNA or RNA. After a fixation step, cell membranes have to be permeable to allow the entrance of the probe into the cell. Generally these probes are 15 to 20 nucleotides in length and are labelled with a fluorescent dye. Targets are detected by the fluorescence microscope. This technique has been developed for wine microorganisms, such as yeasts (Xufré *et al.*, 2006; Wang *et al.*, 2014), LAB (Sohier and Lonvaud-Funel, 1998; Blasco *et al.*, 2003) and AAB (Blasco, 2003).
- ***Flow Cytometry:*** Flow cytometry measures and analyses simultaneously several physical characteristics of single particles, for instance, cells labelled by fluorescent dye. The particles flow through a beam of laser light that is scattered and emits fluorescence. It measures the particle size, granularity or internal complexity and fluorescence intensity. Obviously, this technique can be easily coupled with any of the previous ones in order to get appropriate identification or information about the viability statuses of the analysed microorganisms. In this way the metabolic state of yeast and bacteria in wine (Malacrino *et al.*, 2001; Boyd *et al.*, 2003; Chaney *et al.*, 2006; Herrero *et al.*, 2006) is analysed. Flow cytometry has been combined with FISH for a selective enumeration of mixed wine microbial populations with a high resolution and the advantage of an automated analysis (Andorrà *et al.*, 2011; Wang *et al.*, 2014).
- ***Real Time or Quantitative PCR (Q-PCR):*** Quantitative PCR is based on the continuous monitoring of PCR during the reaction. Q-PCR detects and quantifies a fluorescent donor, which increases simultaneously with the PCR product. The fluorescence can be obtained through binding agents or probes. SYBR Green is the common binding agent which intercalates into double stranded DNA. Also, probes are characterised by having combination of fluorochromes that, when both are bound in the probe, the fluorescence is quenched. When the *Taq* polymerase amplifies the DNA, the quenching is released and fluorescence is emitted. Real-time PCR provides sensitive detection of the DNA product, eliminates the need for post-PCR analysis, guarantees detection during the linear range of amplification and incorporates specialised software to simplify data analysis. Thus, this is used to evaluate the initial quantity of DNA or cells. These techniques have high specificity, sensibility and are quick. Nevertheless, all these parameters depend strictly on the primer design and only quantify the target microorganism. This technique has been used extensively to detect and quantify different wine microorganisms, like AAB (González *et al.*, 2006b; Torija *et al.*, 2010), LAB (Neeley *et al.*, 2005) and wine yeast (Martorell *et al.*, 2005; Hierro *et al.*, 2006).
- ***Massive Sequencing:*** Massive sequencing, metagenomic sequencing, high-throughput sequencing or pyrosequencing are different names for the same technique, based on sequencing of all the amplicons (in the form of libraries) that can be generated from a sample. If the primers are universal, all the organisms can be amplified. Normally, the libraries are generated for prokaryotes and eukaryotes. The high sensitivity is due to the vast amount of data that can be collected and the deepness of the analyses. However, it is often difficult to associate unambiguously a given sequence to a completely known taxon (genus or species). So the term Operational Taxonomic Unit (OTU) is used to refer to a series of sequences closely related. It has been used very recently to determine the relative abundance of microorganisms in vine, grape and wine fermentations (Setati *et al.*, 2012; Bokulich *et al.*, 2014; David *et al.*, 2014; Pinto *et al.*, 2014; Taylor *et al.*, 2014; Pinto *et al.*, 2015; Setati *et al.*, 2015). It highlighted the significant regional differences in vineyard biodiversity and thus a hypothesis of 'microbial terroir' was proposed for regional wine style (Bokulich *et al.*, 2014; Capozzi *et al.*, 2015; Setati *et al.*, 2015).

## 3. Grape Microbiome

Grapes, generally support a series of microorganisms that are mostly epiphytes (that grow on the grape surface). The substrates that allow the growth of microorganisms are normally the exudates from grapes, rich in saccharides.The yeast population on sound grapes can go from $10^2$ cfu/berry to $10^5$ cfu/berry, depending on the ripening state (Renouf *et al.*, 2005). Interestingly, the population quantity also changes during ripening of grapes, being the highest at the end of ripening (Renouf *et al.*, 2005; Clavijo *et al.*,

2010). The increased population at the harvest time is mostly due to increased nutrient availability because the berry cuticle softens and might have some microfissures not easily visible (Barata *et al.*, 2012). At full ripening, the grape musts obtained from healthy grapes contain yeast population, varying from $10^4$ to $10^6$ cfu/ml (Beltran *et al.*, 2002; Padilla *et al.*, 2016). Damaged grape berries can sustain the growth of many microorganisms, increasing considerably at least one log cycle of population (to $10^6$ or $10^8$ cfu/berry) due to nutrient availability (Barata *et al.*, 2012).

The yeasts present on the grape surface are mostly *Ascomycetous* moulds (yeast-like), *Basidiomycetous* and *Ascomycetous*. As the main species of the *Ascomycetous* moulds, *Aureobasidium pullulans* is the most common yeast-like mould occupying the grape surface. *Basidiomycetous* yeasts are also abundant on grape surface but the most frequent species belong to genera *Cryptococcus*, *Rhodotorula* and *Sporodiobolus*. Although *Ascomycetous* yeasts generally colonise intact grape berries, a great diversity is found in the worldwide surveys. Common *Ascomycetous* yeasts on grape surface include the genera *Hanseniaspora*, *Candida* (most of those found on grapes have been later reclassified within *Starmerella)*, *Issatchenkia*, *Debaryomyces*, *Metschnikowia* and *Pichia*. Species diversity of *Ascomycetous* yeasts is even higher, depending on a series of variations (climatic conditions, vineyard treatments, biotic factors, geographic location and vineyard factors including size, age, variety of grape and vintage year) (Barata *et al.*, 2012). However, some species of *Ascomycetous* have been found worldwide, such as *Hanseniaspora uvarum*, *Metschnikowia pulcherrima*, *Issatchenkia terricola* and *Issatchenkia orientalis*. *Saccharomyces cerevisiae* has hardly been found on sound grape berries, similar to some spoilage species, such as *Zygosaccharomyces bailii*. However, damaged or rotten berries can provide high nutrient to favour the growth of *Ascomycetous* yeast. When the whole bunch is harvested, some damaged berries may yield higher numbers of the *Ascomycetous* yeast. Therefore, the isolation of *S. cerevisiae* and other spoilage species from grape berry is suspected to be related with grape health and sampling approach (Barata *et al.*, 2012). *Ascomycetous* moulds and *Basidiomycetous* yeasts are considered residents of grape berries and phylloplanes. The population quantity and dynamics of *Ascomycetous* moulds and *Basidiomycetous* yeast keep a similar pace on both grape berries and phylloplanes. These oligotrophic residents are thought to be adapted to the environment with poor nutrient availability (Loureiro *et al.*, 2012). However, *Ascomycetous* yeasts are classified as copiotrophic opportunists because they are rarely detected on immature grape berries but are found on grape berries with high nutrient availability (veraison, harvest or damaged grape berries). This is supported by the uneven distribution of *Ascomycetous* yeasts: microcolonies gather around the sites with the most likely nutrient leaking from the berries (Loureiro *et al.*, 2012). Although all *Ascomycetous* yeasts are opportunist, it is difficult to isolate some species on sound berries even at harvest time and the classical representative is *S. cerevisiae* and its close relatives (other *Saccharomyces* yeast species), which reside primarily on tree barks and soils as spores, where they are detected all year long. Only in two months with grape growing from veraison to harvest or decay, the spores are dispersed on to the grape berries by some vectors, such as insects (Loureiro *et al.*, 2012).

## 4. The Succession of Microorganisms during Alcoholic Fermentation: Yeast Interactions

Yeasts on grape berries survive and grow in grape must during alcoholic fermentation. Yeasts metabolise the main nutrients (sugars) to ethanol, but also to other volatile compounds, imparting to the wine its particular character (Fig. 2). According to their fermentation capacity, competitiveness and contribution to wine, two main types of yeast can be considered in spontaneous wine fermentation: non-*Saccharomyces* yeasts and *Saccharomyces* yeasts. Non-*Saccharomyces* yeasts have lower fermentative capacity and are less competitive than *Saccharomyces* yeasts. However, today they are considered to increase wine complexity (Jolly *et al.*, 2014; Mas *et al.*, 2016).

The transformation of grape must into wine is a complex process that involves the sequential development of microbial species, mostly fungi, yeast, LAB and AAB. The microorganisms present on the berry surface are mainly yeasts. The microbiota associated with grapes varies constantly in response to grape variety, climatic conditions, viticultural practices, stage of ripening, physical damage (caused by moulds, insects and birds) and fungicides applied to vineyards (Pretorius *et al.*, 1999). Although

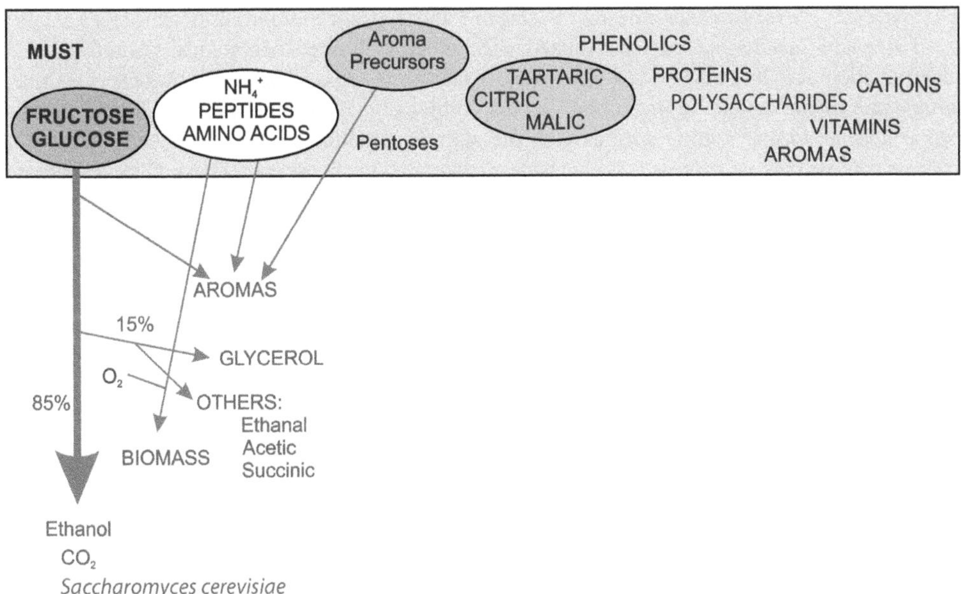


**Figure 2.** Alcoholic Fermentation

Color version at the end of the book

grape must is rather complete in nutrient content, its low pH and high sugar content yield a selective media where only a few bacteria and yeast species can grow. Furthermore, the oenological practice of adding sulphur dioxide as an antioxidant and antimicrobial preservative supposes an additional selection. This practice is meant to limit the growth of undesirable oxidative microbes and prevent oxidation of grape must. Another important factor derives from the anaerobic conditions created during fermentation, especially at the start due to massive production of carbon dioxide (Henschke, 1997). As a result, the alcoholic fermentation of grape juice into wine can be regarded as a heterogenous microbial process. The number of yeasts on the grape berry and grape must change, depending on the geographical situation of the vineyard, climatic conditions, sanitary state of the berries and pesticide treatments of the vineyard (Beltran *et al*., 2002; Romano *et al*., 2006; Padilla *et al*., 2016). At harvest time, the yeast population is quite complex and the major fermenting yeast, *S. cerevisiae*, is not very abundant (Beltran *et al*., 2002; Torija *et al*., 2001). Therefore, the non-*Saccharomyces* population is expected to be dominant in the early stages of grape must processing. Thus, non-*Saccharomyces* yeasts predominate during the early stages of wine fermentation (Fleet, 2003) and finally the *S. cerevisiae* yeast species – the most alcohol-tolerant yeast, dominates the fermentation. Besides, some species of non-*Saccharomyces* may be also present during fermentation and in the wine. Some of these yeast species are considered as spoilage microorganisms because they produce metabolites with an undesirable impact (Pretorius, 2000).

## 4.1 Non-*Saccharomyces* Yeasts

The term non-*Saccharomyces* has no taxonomical significance. According to Jolly *et al*. (2014), only yeast with a positive role in wine production is included in this description whereas spoilage yeasts, such as *Dekkera*/*Brettanomyces* should not be included in this denomination. However, this is not a widespread concept and many authors refer to all species regardless of their effect as non-*Saccharomyces*. In fact, many of those species considered as having a positive role in wine fermentation may have spoilage activity if their activity is prolonged during wine fermentation. Non-*Saccahromyces* yeasts are commonly known as wild yeasts because they are mostly present in grapes at the beginning of the fermentation (Fugelsang and Edwards, 2007).

There are around 15 non-*Saccharomyces* yeast genera involved in wine fermentation. These are: *Dekkera* (anamorph *Brettanomyces*), *Candida*/*Starmerella*, *Cryptococcus*, *Debaryomyces*, *Hanseniaspora* (anamorph *Kloeckera*), *Kluyveromyces*/*Lachancea*, *Metschnikowia*, *Pichia*, *Rhodotorula*,

*Saccharomycodes*, *Schizosaccharomyces*, *Torulaspora* and *Zygosaccharomyces* (Pretorius *et al.*, 1999). Most of the non-*Saccharomyces* wine-related species show limited oenological aptitudes, such as low fermentation activity and low $SO_2$ resistance (Ciani *et al.*, 2010). However, these species play an important role in the metabolic impact and aroma complexity of the final product. Furthermore, this species contribute to enzymatic reactions with the main enzymatic activities described for some non-*Saccharomyces* species as protease, β-glucosidase, esterase, pectinase and lipase (Esteve-Zarzoso *et al.*, 1998). Thus, the metabolic activities of various non-*Saccharomyces* yeast species during alcoholic fermentation have been a matter of interest. The potential applications of non-*Saccharomyces* yeasts in the wine industry are listed in Table 1. Some yeast species, such as *Torulaspora delbrueckii*, *Metschnikowia pulcherrima*, *Pichia kluyveri* and *Lachancea thermotolerans* are currently sold as commercial starters for wine production. The assessment of *Hanseniaspora uvarum*, *Starmerella bacillaris* (previously *Candida zemplinina*) and other species is still on the way to balance their positive contribution and negative impact on wine (Masneuf-Pomarede *et al.*, 2016). Another species, *Hanseniaspora vineae* has been successfully used in wines from Uruguay and Spain (Lleixà *et al.*, 2016; Martín *et al.*, 2016), although it is not present as a commercial product yet.

**Table 1.** Non-*Saccharomyces* Yeasts with Positive Impact Demonstrated

| *Features of interest in winemaking* | *Species/synonym* |
|---|---|
| Acetate ester production | *Hanseniaspora guillermondii*<br>*Hanseniaspora vineae*<br>*Lachancea thermotolerans* |
| Aroma and complexity | *Hanseniaspora uvarum*<br>*Metschnikowia pulcherrima*<br>*Pichia anomala*<br>*Pichia fermentans*<br>*Pichia kluyveri*<br>*Starmerella bacillaris*<br>*Torulaspora delbrueckii* |
| Enzymatic activities | *Debaryomyces hansenii* |
| Ester production | *Metschnikowia pulcherrima*<br>*Pichia membranifaciens* |
| Fructophily | *Starmerella bacillaris*<br>*Zygosaccharomyces bailii* |
| Glycerol production | *Lachancea thermotolerans*<br>*Starmerella bacillaris* |
| Killer against *Dekkera/Brettanomyces* | *Pichia anomala* |
| Reduced ethanol production | *Starmerella bacillaris* |
| Reduction of malic acid and total acidity | *Schizosaccharomyces pombe* |
| Volatile acidity reduction | *Torulaspora delbrueckii* |

*Source:* Adapted from Masneuf-Pomarede *et al.*, 2016; Jolly *et al.*, 2014

The negative impact of non-*Saccharomyces* is mainly the low fermentative activity and high level of undesirable flavours. The low fermentative activity can be overcome by mixed fermentation with *Saccharomyces* yeasts; the undesirable flavours are solved by olfactive perception experiments to screen acceptable or neutral strains (Bely *et al.*, 2013). The genetic and phenotypic performance of 115 *Hanseniaspora uvarum* strains were fully assessed by Albertin *et al.* (2016), as well as of 63 *Starmerella bacillaris* strains by Englezos *et al.* (2015), both being designed for exploitation of the two common non-*Saccharomyces* yeast species isolated in wine fermentation.

## 4.2 *Saccharomyces* Yeasts

*Saccharomyces* is the most useful and widely exploited yeast genus at the industrial level. The taxonomy of the genus *Saccharomyces* has undergone many revisions and reclassifications. In fact, many species considered as non-*Saccharomyces* were initially classified as *Saccharomyces*. According to Barnett *et al.* (2000) and Naumov *et al.* (2000), *Saccharomyces* yeasts were taxonomically separated into three groups: *Saccharomyces sensu stricto* group, containing *S. cerevisiae*, *S. bayanus*, *S. paradoxus*, *S. pastorianus*, *S. cariocanus*, *S. mikatae* and *S. kudriavzevii*, *Saccharomyces sensulato* group, including *S. dairensis*, *S. exiguus*, *S. unisporus*, *S. servazzi* and *S. castelli* in the second and the third group with only *S. kluyveri*. Later, *Saccharomyces* genus involved four species isolated from natural habitats, *S. cariocanus*, *S. kudriavzevii*, *S. mikatae* and *S. paradoxus* and three species associated with industrial fermentation processes – *S. bayanus*, *S. cerevisiae* and *S. pastorianus* (Barrio *et al.*, 2006). Nowadays, only *S. arboricolus* (not a wine species), *S. eubayanus* and *S. uvarum* are considered pure species, and the others are considered hybrids (Borneman and Pretorius, 2015). Physiological tests are not useful to differentiate the species of *Saccharomyces* and only their DNA sequences are reliable (Ribéreau-Gayon *et al.*, 2006). In fact, the *Saccharomyces* species of oenological interest are *S. cerevisiae* and *S. bayanus*. *S. cerevisiae* is the main species in alcoholic fermentation responsible for the metabolism of grape sugar to alcohol and carbon dioxide, but also important in the formation of secondary metabolites and conversion of grape aroma precursors to varietal wine aromas (Fig. 2). *S. bayanus* has been used for alcoholic fermentation at low temperature since it is cryotolerant (Tamai *et al.*, 1998); *S. bayanus* var. *uvarum* (synonym *S. uvarum*) is proved to be a good starter culture due to its reduced ethanol production, psychrophilism and acetate ester production (Masneuf-Pomarede *et al.*, 2010; Bely *et al.*, 2013; Csernus *et al.*, 2014). In addition to these species, it is important to remember that haploid cells or spores from the *Saccharomyces sensu stricto* species are able to mate with each other, resulting in viable hybrids (Querol *et al.*, 2003). Hybrid strains of *S. bayanus* and *S. cerevisiae* and of *S. cerevisiae* and *S. kudriavzevii* have been isolated in alcoholic fermentations (González *et al.*, 2006c). This phenomenon has a great possibility for the development of new species or strains. However, it is a source of taxonomic confusion due to the molecular and phenotypic classification analysis, for example, *S. cerevisiae* and *S. bayanus* are thought to be either two separate species, or the same species that differ slightly from physiological aspects (Fugelsang and Edwards, 2007). Also known is the physiological instability of strains belonging to *Saccharomyces sensu stricto* group (Ribéreau-Gayon *et al.*, 2006).

*Saccharomyces* genus possesses a series of unique characteristics that are not found in other genera. *Saccharomyces* yeasts have the ability to produce and accumulate ethanol even under aerobic conditions (crabtree effect) (Marsit and Dequin, 2015). Also, they have a high capacity to ferment sugars quickly and efficiently. This ability allows them to colonise sugar-rich media and efficiently overgrow other yeasts, which are not so tolerant to alcohol (Barrio *et al.*, 2006). However, the competition between *Saccharomyces* and non-*Saccharomyces* is more complex than the production of ethanol. In fact, there are many interactions among them, probably the most relevant being cell-to-cell contact, nutrient limitation or the secretion of antimicrobial peptides (Wang *et al.*, 2016). Although most of these mechanisms of interactions are shown by analysing the growth on plates, recent findings relate that they induce the VBNC states that can end with the cell death (Branco *et al.*, 2015; Wang *et al.*, 2016). Nissen and Arneborg (2003) describe cell-to-cell contact as a possible inducer of lack of cultivability, although the reported mechanism seems to be limited to S101 *S. cerevisiae* strain, as other strains do not show the same mechanism (Wang *et al.*, 2015).

## 4.3 Population Dynamics of Wine Yeast during Spontaneous Fermentation

The contribution of yeasts to wine is affected by their participation during alcoholic fermentation (Comitini *et al.*, 2011). Yeast species commonly found in spontaneous fermentation are divided into three groups: aerobic yeast (*Pichia*, *Debaryomyces*, *Rhodotorula*, *Candida/Starmerella*, *Cryptococcus*), apiculate yeast (*Hanseniaspora*) and fermentative yeast (*Kluyveromyces*, *Torulaspora*, *Metschnikowia*, *Zygosaccharomyces* and *Saccharomyces*). Generally, the succession of yeast involves initial domination of aerobic and apiculate yeasts which are present on the grape surface, these decrease, then the increase of fermentative yeasts during fermentation and finally, the domination of the *Saccharomyces* yeasts (Schütz and Gafner, 1993; Torija *et al.*, 2001; Beltran *et al.*, 2002; Combina *et al.*, 2005; Nurgel *et al.*, 2005; Di Maro *et al.*, 2007; Ocón *et al.*, 2010). The main yeast species isolated at the beginning of fermentation generally belong to *Hanseniaspora, Metschnikowia* and *Starmerella* genera.

The dominance of *S. cerevisiae* is needed to complete alcoholic fermentation (Jolly *et al.*, 2014). However, distinct fermentation dynamics are a result of the fermentation conditions and the relative levels of the main yeast species present. For instance, *Hanseniaspora* persists longer in fermentation at low temperature (Andorrà *et al.*, 2010b); *Zygosaccharomyces bailii* leads botrytis-affected spontaneous fermentation (Nisiotou *et al.*, 2007); *Pichia kudriavzevii* emerges along with *Saccharomyces* when relatively low ethanol (9 per cent) was obtained at the end of fermentation (Wang and Liu, 2013); *Starmerella (Candida)* was reported to co-dominate at late stages of fermentation (Llauradó *et al.*, 2002; David *et al.*, 2014) or to finish alcoholic fermentation (Clemente-Jimenez *et al.*, 2004).

Furthermore to the succession of different yeast species during wine fermentation, a dynamic change of strains within each species is also evident, based on molecular techniques for strain differentiation (Fleet, 2003). For *S. cerevisiae*, some dominant or co-dominant strains were found (Sabate *et al.*, 1998; Torija *et al.*, 2001) and in some cases, where a single strain dominates, the killer phenotype may be present (Schuller *et al.*, 2005). Strain diversity of non-*Saccharomyces* species has also been reported but focusses on their oenological interest rather than on the dynamic changes (Capece *et al.*, 2005; Masneuf-Pomarede *et al.*, 2015; Albertin *et al.*, 2016).

## 5. Control of Fermentation: From Spontaneous to Inoculated

Winemakers have traditionally seen non-*Saccharomyces* yeast as a source of wine spoilage. One of the means of microbiological control in fermentation is the use of starter cultures. In winemaking, the most common yeast used as starter culture is *S. cerevisiae.* The development of cellar-friendly Active Dry Wine Yeast (ADWY) has extended its use in wine production, helping the winemaker to control the fermentation. The selection of yeast to be used as starter culture has been developed using different tests and criteria. Nowadays, many different ADWY are commercially available and are meant to increase aromatic expression, resistance to ethanol, low or high temperature, etc. However, all of them have good fermentation potential and are generally sufficient to complete the alcoholic fermentation.

Furthermore, yeasts not only lead in alcoholic fermentation, but also have an important role to play in wine quality. The activity of different yeast species and strains has an important effect on the sensory profiles of wine, increasing its complexity and sensory richness (Ribereau Gayon *et al.*, 2006). Presently, wine producers use commercial starters of *S. cerevisiae* to ensure the control of fermentation and produce a predictable and reproducible wine. A side effect of the widespread practice is the elimination of the participation of native microbiota. This limited participation might result in wines with similar sensory and analytical properties, depriving them from the complexity, variability and personality, which define the typicality of a wine (Fleet, 1993). Thus, the use of indigenous or native yeasts can be a tool to protect the authenticity, since it has been documented that microbial diversity is distinctive for a given area (Bokulich *et al.*, 2014; Setati *et al.*, 2015). The microbial population characteristic of a given area can be defined as the '*microbial fingerprint*'. This microbial population will develop a distinctive character in the wine, measurable by the various components (molecules) that each microorganism leaves and which we can define as '*microbial footprint*'.

The different microbial footprints could be related to the presence of these microorganisms during the winemaking process. Information on evolution of yeast population during alcoholic fermentation is going on, since microbiology provides the appropriate methods. Obviously, as techniques have evolved, our knowledge has been expanded. Despite the fact that the population of *Saccharomyces* is very low in grapes (Beltran *et al.*, 2002), its development during alcoholic fermentation and extensive use of ADWY hase turned *S. cerevisiae* as the most common 'cellar-resident yeast' (Beltran *et al.*, 2002; Bokulich *et al.*, 2014). Thus, the population associated with grapes changes with the cellar environment (presses, pumps, tanks) contact, where they join the resident microbiota. This microbiota is not usually in new wineries with equipment without previous use (Constanti *et al.*, 1997).

In spontaneous fermentation, the native microbiota proliferate for several days and produce various compounds that improve the sensory quality of wine or at least give the wine a specific flavour. When the activities of these yeasts were analysed, the presence of enzymatic activities of great interest was detected – esterases, beta-glucosidase, pectinolytic, etc. (Jolly *et al*., 2014). Additionally, they may cause ethanol reduction (Gonzalez *et al*., 2013; Contreras *et al*., 2014), which has been proposed as a key objective in current winemaking practice due to the increased concentration of sugars, among other effects, derived from climate change (Mira de Orduna, 2010). Despite these favourable aspects, the traditional bias of winemakers against non-*Saccharomyces* yeast has limited their use. However, in recent years, there has been an increase in the interest in selecting non-*Saccharomyces* yeasts for use with *S. cerevisiae*. Thus, the key role of *S. cerevisiae* during alcoholic fermentation has been challenged (Fleet, 2003; Jolly *et al*., 2014).

The positive effects on wine quality form the main goal in the selection of non-*Saccharomyces* yeast. These include either the production of new aromas or the removal of detrimental compounds that decrease the wine quality. *Torulaspora delbrueckii* reduces the volatile acidity that is normally produced during winemaking (Renault *et al*., 2009) and has proved appropriate for the fermentation of botrytised grapes (Bely *et al*., 2008). Nowadays, it is possible to find various commercial preparations of this yeast. Another commercially available non-*Saccharomyces* yeast is *Metschnikowia pulcherrima*, which is recommended for the production of some aromas based on thiols and terpenes in white wines (González-Royo *et al*., 2015). Finally, another yeast is *Lachancea thermotolerans*, used for its production of lactic acid and glycerol (Gobbi *et al*., 2013). Although there are still a few commercial preparations of non-*Saccharomyces* yeasts, they will probably increase in the near future. These include *Starmerella bacillaris* that produces large amounts of glycerol (Ciani and Ferraro, 1996) and also because of its fructophilic character, which favours the end of fermentation (Soden *et al*., 2000). Other non-*Saccharomyces* species to be expected in commercial preparations are the typical apiculate yeasts from the *Hanseniaspora* genus, such as *H. uvarum* (Andorrà *et al*., 2010c), *H. vinae* (Medina *et al*., 2013) and *H. guilliermondii* (Moreira *et al*., 2008). Other species that can have some oenological interest are species of the genera *Hansenula, Pichia, Schizosaccharomyces, Zygosaccharomyces,* etc. though its possible commercial development seems unlikely (Jolly *et al*., 2014). Nevertheless, pure culture fermentations with non-*Saccharomyces* wine yeast generally increase metabolite contributions to noticeable negative levels and poor fermentation activities that generally exclude their use as single starter cultures. The most important spoilage metabolites produced by non-*Saccharomyces* yeast are acetic acid, acetoin, acetaldehyde and ethyl acetate (Ciani *et al*., 2010).

However, the use of non-*Saccharomyces* yeast in the production of wine has the goal to increase some characteristics of the final product, though it does not solve the main problem induced by the massive use of ADWY – the uniformity observed in inoculated wines. Some winemakers have eliminated or reduced the amount of starter cultures used in the production of 'natural' wine to increase the effect of the native microbiota. This practice increases the risks of uncontrolled fermentations, which may lead to economical losses as these wines may have much higher risks of presenting different levels of spoilage that will not be acceptable to the consumer.

The recommended solution to fight this uniformity is to exploit indigenous yeasts. Some years ago different yeast producers developed commercial 'local selection' yeasts in an attempt to protect the genuineness and authenticity of wines. However, in all the cases the focus was on strains of *S. cerevisiae*. This solution defends the policy of *terroir* and typicality by using the starter cultures from local selection. Therefore, the use of oenologically competent indigenous yeasts as suitable inocula for the production of conventional or organic wines can achieve this goal.

## 6. After Alcoholic Fermentation

### 6.1 Lactic Acid Bacteria (LAB)

LAB cell wall is essentially composed of a peptidoglycan, which implies that it is Gram-positive. It can have a round morphology (cocci) or elongated (bacilli). LAB are strict anaerobes and mesophilic with an optimal growth temperature between 15-30°C. Their names are derived from the main product from the metabolism of glucose, that is lactic acid. In the grape must and wine, the more common genera are *Lactobacillus*, *Leuconostoc*, *Oenococcus* and *Pediococcus* (Ribéreau-Gayon *et al*., 2006). LAB can be homofermentative (which produces exclusive lactic acid from glucose and/or fructose) and heterofermentative (which produces lactic acid and also carbon dioxide, ethanol, acetic acid, etc. from the same carbohydrates). *O. oeni* presents heterofermentative metabolism whereas *P. pentosaceus* and *P. damnosus* are homofermenative. *L. casei* and *L. plantarum* are described as facultative heterofermentative; *L. brevis* and *L. hilgardii* are considered as strict heterofermentative. *O. oeni* is the main LAB conducting the malo-lactic fermentation which occurs in several wines, mainly in red ones (Ribéreau-Gayon *et al.*, 2006). LAB species impact the quality of wine. Some species of *Lactobacillus* and *Pediococcus* are considered detrimental to the quality of wine. *Pediococcus* genus, which is represented by *P. damnosus, P. parvulus, P. pentosaceus* and *P. inopinatus*, is associated with grape and wine and produces excessive amounts of diacetyl, biogenic amines, degradation of glycerol, etc. (Lonvaud-Funel, 1999; Dicks and Endo, 2009). *Lactobacillus* may result in haze, sediment, excessive volatile acidity formation etc. in bottled wines (Fugelsang and Edwards, 2007).

During grape must fermentation, different types of LAB have been found. On grapes, the main Lactic Acid Bacteria (LAB) belong to the species *Lactobacillus plantarum, L. hilgardii* and *L. casei*. The species *O. oeni*, which became the predominant species conducting malo-lactic fermentation, is rarely detected in the beginning of fermentation. During the initial stages of alcoholic fermentation, the common LAB are *L. plantarum, L. casei, L. hilgardii, L. brevis, P. damnosus, P. pentosaceus, L. mesenteroides* and *O. oeni*. During alcoholic fermentation, their population remains between $10^2$ to $10^4$ cfu/ml. Their presence during alcoholic fermentation however, varies depending on the pH of the must and the concentration of the added $SO_2$. Although the levels of $SO_2$ used in winemaking limit the growth of LAB, they do not completely inhibit their growth. After alcoholic fermentation, LAB remain in a latent phase for a while.

Alcoholic fermentation is often followed by malo-lactic fermentation that can proceed spontaneously through the activity of native LAB (Fig. 3). Currently, small wineries proceed with spontaneous malo-lactic fermentation. However, as in alcoholic fermentation, malo-lactic spontaneous fermentation is very unpredictable either for its completion or for the lack of control over the possible negative aspects, such as undesired off-odours or the occurrence of biogenic amines. The beginning of malo-lactic fermentation depends on several factors, such as temperature, pH, ethanol, etc. It has an initial growth phase of LAB that takes several days and needs to raise the population to at least $10^7$ cfu/ml or even more. As soon as the malic acid is exhausted due to its transformation into lactic acid, the bacterial population declines. If the wine is not sulphited after malo-lactic fermentation, the bacteria can remain there for months together (Lonvaud-Funel, 1999).

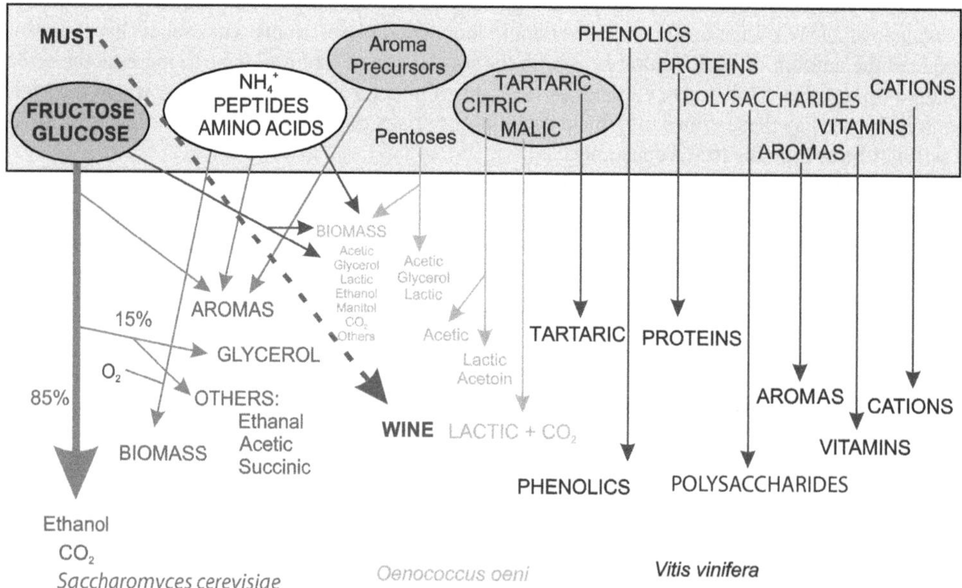


**Figure 3.** The conversion of must into wine

Color version at the end of the book

To control malo-lactic fermentation and reduce the possible deleterious effects (especially the production of biogenic amines), about 40 years ago, selected commercial starter cultures of LAB strains appeared. Most of these starter cultures were from species of *O. oeni*, due to its endurance to the wine pH and ethanol concentration. Despite the benefits in controlling malo-lactic fermentation, the number of strains of *O. oeni* used successfully is still very limited. Furthermore, the requirement for diverse strains to perform malo-lactic fermentation is very high due to the different wine conditions, some of which find it very difficult to support any microbial growth. Additionally, these strains should develop acceptable sensory properties and meet food security standards, especially lack of production of biogenic amines. Other LAB species can exhibit good malo-lactic fermentation performance. Some strains of *L. plantarum*, for instance, yielded promising results in wines with high pH, having low production of acetic acid and higher complex enzyme activities. LAB can be selected from the indigenous microbiota to serve the need to improve the typicality of the wines.

## 6.2 Acetic Acid Bacteria (AAB) and Production of Vinegar

Acetic Acid Bacteria (AAB) are Gram-negative or Gram-variable, obligately aerobic, non-spore forming, oxidase negative, catalase positive, ellipsoidal to rod-shaped, motile (with peritrichous or polar flagella) or non-motile microorganisms. They are mesophilic bacteria with optimum growth temperature at around 30°C. The optimum pH for the growth of AAB is 5-6.5, but most of them can grow at low pH values, between 3-4. Their names are derived from the main product due to metabolism of ethanol that is acetic acid (Asai, 1968; Sievers and Swings, 2005; Sengun, 2015).

AAB belong to the family *Acetobacteriaceae* and class *Alphaproteobacteria*. The first recognised genera of AAB were *Acetobacter* Beijerinck 1898 (the type genus of the family *Acetobacteriaceae*) and *Gluconobacter*Asai 1935. After their description, they have underwent many taxonomic changes and a great number of AAB species were identified. At the moment, 18 genera are recognised in the group of AAB as *Acetobacter*, *Gluconobacter*, *Acidomonas*, *Gluconacetobacter*, *Asaia*, *Kozakia*, *Swaminathania*, *Saccharibacter*, *Neoasaia*, *Granulibacter*, *Tanticharoenia*, *Ameyamaea*, *Neokomagataea*, *Komagataeibacter*, *Endobacter*, *Nguyenibacter*, *Swingsia* and *Bombella*, which are mostly monotypic, except for *Acetobacter*, *Asaia*, *Gluconobacte*r, *Gluconacetobacter* and *Komagataeibacter* (Malimas *et al*., 2017).

AAB are widespread microorganisms in nature, commonly found on different tropical flowers and fruits and food fermenters. *Gluconobacter* strains commonly occur in sugar-enriched environments while *Acetobacter* strains are more adapted to alcohol-enriched environments (Mercanoglu Taban and Saichana, 2017). Additionally, some members of AAB form symbiotic relationships with insects (Crotti *et al*., 2010). Although AAB are commonly known as non-pathogenic to humans and or animals, some species (e.g. *Acetobacter cibinongensis, Asaia bogorensis, Asaia lannaensis, Granulibacter bethesdensis, Gluconobacter* spp.) cause opportunistic infections in humans (Saichana *et al*., 2015).

AAB show poor growth on plates due to the aggregation of the cells, accumulation of their own oxidation products in the medium, lack of appropriate selective culture media or most probably due to the status of VBNC. AAB can be identified at genus level by phenotypic tests comprising acetate oxidation, lactate oxidation and acetic acid production from ethanol (Yamada and Yukphan, 2008). Since they have similar phenotypic characteristics, a polyphasic approach, which combines genotypic, phenotypic and chemotaxonomic data of strains, isrecommended to find phylogenetic relationships and perfect identifications of AAB (Cleenwerck and De Vos, 2008; Malimas *et al*., 2017).

AAB have a direct oxidation system for sugars, alcohols and sugar alcohols (traditionally known as oxidative fermentation) in which partially oxidised compounds are produced, such as acetic acid from ethanol, gluconic acid, 2-ketogluconic acid, 5-ketogluconic acid, and 2,5-diketogluconic acid from glucose, fructose from mannitol, L-sorbose from sorbitol and 5-ketofructose from fructose or sorbitol (Komagata *et al*., 2014). This special and unique metabolism brings them to an important place in the production of several industrial products, including several foods and beverages like vinegar, cocoa and Kombucha and other specific products, such as ascorbic acid and cellulose.

### *Production of Vinegar*

Vinegar is a special kind of condiment, which has long been used, not only as flavouring and preserving agent, but also in traditional and natural folk medicine for treating a variety of diseases (Karabiyikli and Sengun, 2017). It is produced from suitable sources containing fermentable carbohydrates through the activity of yeasts and AAB. The ethanol produced by yeast is oxidised to acetic acid by AAB in two serial reactions catalysed by the membrane-bound pyrroloquinoline quinine-dependent alcohol dehydrogenase (ADH) and aldehyde dehydrogenase (ALDH) (Matsushita et al., 1994).

There are many types of vinegar produced worldwide and which have different aroma and taste, depending on their raw material, production system and use (Solieri and Giudici, 2009). Vinegars can mainly be grouped as grain and fruit vinegars, based on raw materials used during production (Chen *et al.*, 2016). Vinegars produced in different countries include cereal vinegars, wine vinegars, vegetable vinegars, fruit vinegars and other vinegars produced from various raw materials, such as whey, honey, etc. (Sengun, 2015). Although numerous types of vinegars are described, the most popular vinegars on the European market include apple, wine and alcohol vinegars (Trček *et al.*, 2016). When distilled ethanol obtained from fermented raw materials is used, the product is called as 'spirit vinegar' or 'white vinegar' (Giudici *et al.*, 2017).

Vinegar production can be performed by two well-defined methods – the traditional (also called as 'surface culture method' or 'slow fermentation process' or 'Orleans method') and the submerged method (fast fermentation process). See the literature cited (Mas *et al.*, 2014; Giudici *et al.*, 2017). In the submerged method, AAB are submerged in the liquid and the oxidation of ethanol in liquid takes place between the air-liquid interfaces of air bubbles under controlled stirring conditions. This method provides high acidic vinegars, which are produced by fast fermentation processes, where acetification is completed within a few days (Tesfaye *et al.*, 2002; Bamforth, 2008; Mas *et al.*, 2014; Sengun, 2015; Rosma *et al.*, 2016).

Acetic acid is the main component in vinegar that is responsible for the basic sensorial characteristic of vinegar and is also recognised as an effective antimicrobial compound that prevents the growth of undesirable microorganisms. See the literature cited for the specifications of vinegar (FDA, 1995; CNS, 1999).

Many studies have examined the predominant AAB in vinegar fermentation (Table 2). Although 18 genera are currently recognised in this group, only *Acetobacter, Gluconacetobacter* and *Komagataeibacter* are employed in vinegar production. Moreover, each type of vinegar shows specific AAB profile, depending on the selective stress applied by the acetic acid concentration (Gullo *et al.*, 2014).

During alcoholic fermentation, the strains that can survive at high ethanol and low acetic acid concentrations are predominant, while acetic acid tolerant strains dominate the end of acetification. Thus, at the first stage of the acetification, *Acetobacter* species predominate and then *Komagataeibacter* take over the process. *K. europaeus*, *K. intermedius* and *K. oboediens* are the predominant species in wine vinegar since they are resistant against the highest concentration of acetic acid (>6 per cent) (Mas *et al.*, 2014; Trček *et al.*, 2016). Moreover, a mixed culture of quick start species, such as *A. pasteurianus* and high acetic acid tolerance species, such as *K. europaeus* could be promising for perfect vinegar production (Mas *et al.*, 2014). For obtaining high quality of persimmon vinegar, mixed culture containing *A. malorum* and *K. saccharivorans* would be a suitable choice (Hidalgo *et al.*, 2012). On the other hand, selected AAB are scarcely applied in the vinegar industry for technological and economical reasons. One of the main reasons restricting the use of selected AAB in vinegar production is the difficulty of cultivation of this group (Gullo *et al.*, 2014).

Although the final quality of vinegar depends on the AAB species taking part in the process, other factors, such as raw material, technological process, wood contact and aging process also affect the final composition (Hidalgo *et al.*, 2013a; Mas *et al.*, 2014). Aroma-active compounds have a significant effect on the quality of vinegar. Volatile compounds of vinegar are widely variable as also esters, higher alcohols and several aldehydes and ketones, such as acetoin, acetaldehyde, ethyl acetate, ethyl lactate, isoamyl alcohols, metanol, 2,3-butanediol and 2-phenylethanol (Roda *et al.*, 2017). Esters present in vinegar have important influence on aroma, while ethyl acetate and isoamyl acetate are among the compounds that have the highest odour activity in vinegar (Baena-Ruano *et al.*, 2010). A wide variety of phenolic compounds,

**Table 2.** Acetic Acid Bacteria (AAB) Isolated from Different Types of Vinegars

| *Type of vinegar* | *AAB species* | *References* |
|---|---|---|
| Fukuyama pot vinegar | *A. pasteurianus* | Okazaki *et al.*, 2010, Furukawa *et al.*, 2013 |
| Korean black raspberry vinegar | *A. aceti* | Song *et al.*, 2016 |
| Rice vinegar | *A. pasteurianus* | Nanda *et al.*, 2001, Haruta *et al.*, 2006 |
| Shanxi aged vinegar | *A. pasteurianus, A. indonesiensis, A. malorum, A. orientalis, A. senegalensis, G. oxydans* | Wu *et al.*, 2012 |
| Tianjin duliu mature vinegar | *A. pasteurianus* | Nie *et al.*, 2013 |
| Zhenjiang aromatic vinegar | *A. pasteurianus, A. pomorum, K. intermedius** | Xu *et al.*, 2011 |
| Red wine vinegar | *A. oboediens*<br>*K. europaeus*, K. xylinus**<br>*K. oboediens* | Sokollek *et al.*, 1998<br>Fernandez-Perez *et al.*, 2010<br>Trček *et al.*, 2016 |
| Traditional wine vinegar | *A. pasteurianus, K. europaeus**<br>*A. pasteurianus, K. europaeus*, K. hansenii*, K. intermedius, K. xylinus*, Gluconacetobacter* sp. | Ilabaca *et al.*, 2008<br>Vegas *et al.*, 2010<br>Vegas *et al.*, 2013 |
| White wine vinegar | *A. pasteurianus, K. europaeus*, K. xylinus** | Fernandez-Perez *et al.*, 2010 |
| Cider vinegar | *A. pomorum*<br>*A. pasteurianus, K. europaeus*, K. hansenii*, K. xylinus*,* | Sokollek *et al.*, 1998<br>Fernandez-Perez *et al.*, 2010 |
| High-acid industrial vinegar | *K. europaeus** | Sievers *et al.*, 1992 |
| High-acid sprit vinegar | *Ga. entanii* | Schüller *et al.*, 2000 |
| Apple-based kefir vinegar | *Acetobacter pasteurianus and Acetobacter syzygii* | Viana *et al.*, 2017 |
| Blueberry vinegar | *A. pasteurianus* | Hidalgo *et al.*, 2013b |
| Persimmon vinegar | *A. malorun, A. pasteurianus, A. syzygii, K. intermedius*, K. europaeus*, K. saccharivorans** | Hidalgo *et al.*, 2012 |
| Strawberry vinegar | *A. orleanensis*<br>*A. malorum,K. saccharivorans*, K. xylinus** | Mamlouk *et al.*, 2011<br>Hidalgo *et al.*, 2013a |

Abbreviations: *A.: Acetobacter, G.: Gluconobacter, Ga.: Gluconacetobacter, K.: Komagataeibacter*
* In the mentioned references, described as *Acetobacter* or *Gluconacetobacter*

including glycosylated and free compounds among different sub-classes of flavonoids, hydroxycinnamic acids, phenolic acids, benzendiols and coumarins have also been identified in vinegar, which significantly affect the quality (Roda *et al.*, 2017).

## 7. Spoilage Microorganisms in Winemaking

In the wine industry, where alcoholic fermentation is conducted by many microorganisms, it is difficult to distinguish between beneficial fermenting activity and spoilage activity. Microorganisms can spoil wines at several stages of production. Any inappropriate growth of microorganisms may produce undesirable flavours.

Wine that is exposed to air may develop fermentative or oxidative yeasts on its surface, usually species of *Candida* and *Pichia* (Fleet, 2003). These species oxidise ethanol, glycerol and acids, giving wines an unacceptably high level of acetaldehyde, esters and acetic acid. Other wines can also be spoiled by fermentative species of *Zygosaccharomcyes*, *Dekkera* (anamorph *Brettanomyces*), *Saccharomyces* and *Saccharomycodes*. In addition to causing excessive carbonation, sediments and haze, these species produce estery and acid off-flavours (Sponholz, 1993).

The winemaker's most feared spoilage yeast is *Dekkera/Brettanomyces*. This yeast produces off-flavour due to the synthesis of tetrahydropyridines and volatile phenols (4-ethylguaiacol and 4-ethylphenol). Generally, the production of these phenolic off-odours is noticed under a broad range of descriptors, such as 'barnyard-like, mousy, horsey, leather and pharmaceutical' (Grbin and Henschke, 2000; Du Toit and Pretorius, 2000). Among the species of this genus, *Dekkera bruxellensis* is the most representative in wines (Rodrigues *et al.*, 2001). Furthermore, it has been found that other species are able to produce volatile phenols, such as *Pichia guilliermondii*, which has the ability to produce 4-ethylphenol with efficiencies as high as those observed in *D. bruxellensis* (Dias *et al.*, 2003).

*Pichia anomala*, *Metschnikowia pulcherrima* and *H. uvarum* are known for producing high levels of ethyl acetate and acetic acid before and during initial fermentation steps, leading to serious wine deterioration (Romano *et al.*, 1992; Plata *et al.*, 2003). However, *H. uvarum* does not produce ethyl acetate when it is present during fermentation along with *S. cerevisiae* (Zohre and Erten, 2002).

Spoilage species of LAB and AAB may grow at different stages of winemaking – during storage in the cellar and after bottling (Sponholz, 1993; Fuselsang,1997; Fleet, 1998; Du Toit and Pretorius, 2000). LAB can spoil wine during winemaking or during maturation and bottle aging. In the first case, bacteria can start performing malo-lactic fermentation too early and before all the sugars have been consumed by yeasts. The fermentation of these carbohydrates by LAB leads to the production of lactic acid as a major metabolite, but acetic acid, ethanol and $CO_2$ are also produced. Ideally, during wine aging, no yeasts or bacteria should survive in wine. Not all the strains spoil wine; most depreciations and diseases are related to lactobacilli and pediococci, but they are normally destroyed during wine production. However, some strains demonstrate abnormal tolerance to the medium, especially to the ethanol concentration. Other undesirable compounds which are consequence of the LAB metabolism are the biogenic amines and ethylcarbamate (Lonvaud-Funel, 1999). These metabolites do not have an impact on the aroma of the wine but are considered as pernicious for the health of the wine consumer.

The AAB can also spoil wines at many stages of the winemaking process. AAB, that naturally occur in grapes, can survive in winemaking processes, depending on the environmental conditions and the technological practices carried out. Moreover, equipment and instruments used during winemaking could be a good vehicle of AAB to contaminate the product in which the hygienic conditions are disregarded. The AAB isolated from grapes of different origins include the species of *Acetobacter, Ameyamaea, Asaia, Gluconobacter* and *Komagataeibacter* genus (Joyeux *et al.*, 1984; Barbe *et al.*, 2001; González *et al.*, 2005; Bae *et al.*, 2006; Prieto *et al.*, 2007; Nisiotou *et al.*, 2011; Valera *et al.*, 2011; Barata *et al.*, 2012; Mateo *et al.*, 2014). On the other hand, the present view of microbial species associated with grapes, must and wines is much more complex than has been previously described in early studies based on culture-dependent methods (Portillo and Mas, 2016).

The AAB species found on grapes or in grape must show differences from those in wine, depending on the differences in environmental conditions. Recent studies based on next-generation sequencing technologies suggest that AAB are more abundant than previously thought during wine fermentation, independently of the grape variety (Portillo and Mas, 2016). AAB that are usually involved in the wine spoilage are strains belonging to the genera *Acetobacter, Gluconobacter, Gluconacetobacter, Komagataeibacter* and *Asaia* (Table 3).

**Table 3.** AAB Identified in Wine

| *Source* | *AAB species* | *References* |
|---|---|---|
| Austrian wine | *A. tropicalis* | Silhavy and Mandl, 2006 |
| Botrytized wine | *Acetobacter, Gluconobacter, Gluconacetobacter* | Bokulich *et al.*, 2012 |
| Bottled red wine | *A. pasteurianus* | Bartowsky and Henschke, 2008 |
| Grenache wine fermentation | *Gluconobacter, Gluconacetobacter, Acetobacter, Asaia* | Portillo and Mas, 2016 |
| Portugal red wine | *A. oeni* | Silva *et al.*, 2006 |
| Palm wine | *A. indonesiensis, A. tropicalis, G. oxydans, K. saccharivorans** | Ouoba *et al.*, 2012 |
| Red wine fermentation | *G. oxydans, K. hansenii*, A. aceti, A. nitrogenifigens* | González, 2005<br>Dutta and Gachhui, 2006 |
| South African red wine | *K. hansenii*, A. liquefaciens, A. pasteurianus, G.oxydans* | Du Toit and Lambrechts, 2002 |
| White wine | *A. aceti, A. pasteurianus,G.oxydans* | Joyeux *et al.*, 1984 |
| Wine | *A. aceti, G. oxydans K. hansenii** | Ruiz *et al.*, 2000 |

Abbreviations: *A.: Acetobacter, G.: Gluconobacter, Ga.: Gluconacetobacter, K.: Komagataeibacter*
* In the mentioned references described as *Acetobacter* or *Gluconacetobacter*

Factors affecting the growth of AAB during the winemaking process include availability of air, the ethanol concentration, acidity or pH, temperature and the amount of sulphur dioxide present. At the beginning of fermentation, ethyl alcohol and carbon dioxide are produced by indigenous or added starter yeasts, mainly *S. cerevisiae* strains under anaerobic conditions, which create unsuitable environment for the increase of AAB population (Sengun, 2015; Sun *et al.*, 2016). The levels of AAB are closely related to the dissolution of oxygen in wine. AAB species most commonly grow during transfer, agitation, fining and aeration processes of wine. Thus, aeration strongly affects fermentation processes and the quality of the final product. On the other hand, AAB can be isolated from the bottom of tanks and barrels, where the amount of oxygen is low and since they can survive in the presence of other electron acceptors, such as phenolic compounds and quinones (Drysdale and Fleet, 1989). A temperature higher than 10°C also encourages the development and the metabolism of AAB in wine and the spoilage occurs most rapidly at 20-35°C, which is the optimum growth temperature for the growth of AAB (Sengun, 2015; Valera *et al.*, 2017). Depending on the type of wine, the alcohol concentration typically ranges between 8-13 per cent by volume (Machve, 2009). Besides, ethanol is toxic for most of the AAB strains in concentrations of 5-10 per cent (v/v), while some resistant strains are able to survive in concentrations up to 15 per cent (v/v) (De Ley *et al.*, 1984). Normally, the number of *Gluconobacter* declines during alcoholic fermentation since they are sensitive to alcohol (Du Toit and Pretorius, 2002; González *et al.*, 2005). However, in recent studies using next-generation sequencing technologies, *Gluconobacter* has been described as the dominant genus during wine fermentation (Bokulich *et al.*, 2012; Portillo and Mas, 2016).

AAB are rather considered spoilage organisms because their major metabolites result in disagreeable wine sensory characteristics. AAB produce acetic acid as the main product of ethanol metabolism, but other products, as acetaldehyde and ethyl acetate, are also produced with similar negative influence in wine quality. Acetic acid is the main constituent of wine's volatile acidity and is considered to be undesirable in wine at concentrations exceeding 0.5-1.5 g/L, variability depending on the complexity of the wine. Sensorially, acetic acid is recognised in wine as producing a sour flavour with a vinegar-like aroma. Acetaldehyde can also contribute to the sensory spoilage of wine by giving distinctive aroma: sherry-like, bruised apple. The ethyl ester of acetic acid, ethyl acetate, has a pungent aroma, solvent-like and reminiscent of nailpolish remover or nutty (Bartowsky and Henschke, 2008). Hence, it is very important to control AAB presence and ulterior development to obtain high-quality wines.

Finally, filamentous fungi can also impact wine production at several stages – spoilage of the grapes in the vineyard, production of mycotoxins in grapes and their transfer to wines, production of metabolites that enhance or inhibit the growth of wine yeast and malo-lactic bacteria, and cause the earthy, corky taints in wines after growth in grapes, corks and wine barrels (Fleet, 2003).

In order to prevent wine spoilage, hygienic conditions should be maintained during wine production. Although high hygienic conditions limit the contaminant microorganisms, additional applications are mostly necessary to decrease the risk of spoilage. Sulphur dioxide ($SO_2$) is one of the most efficient additives used for prevention of wine spoilage. The effects of $SO_2$ depend on the kinds of organism suppressed and also pH value and sugar content of wine. A concentration ranging from 75 to 200 ppm sulphur dioxide is enough to inactivate spoilage microorganisms in must, while low concentrations of sulphur dioxide have a minimal effect on *A. pasteurianus* strain (Du Toit *et al.*, 2005). On the other hand, some metabolites synthesised by AAB, such as acetaldehyde from ethanol and dihydroxyacetone from glycerol, bind $SO_2$ and reduce the anti-microbial effect of this compound (Ribéreau-Gayon *et al.*, 2000; Valera *et al.*, 2017). In recent years, there has been a growing interest in developing emerging preservation technologies that can replace or complement the action of $SO_2$, since it might cause negative effects on health. These alternatives include the addition of anti-microbial agents (silver nanoparticles, bacteriocins, polyphenols, etc.) and the application of physical methods (high pressure, low electric current, pulsed electric field, pulsed light, ultrasound, UV and e-beam irradiation, etc.) (García-Ruiz *et al.*, 2015; Morata *et al.*, 2017).

## References

Aigle, M., Erbs, D. and Moll, M. (1984). Some molecular structures in the genome of larger brewing yeast. *American Society of Brewing Chemists* 42: 1-7.

Albertin, W., Miot-Sertier, C., Bely, M. and Mostert, T.T. (2016). *Hanseniaspora uvarum* from winemaking environments show spatial and temporal genetic clustering. *Frontiers in Microbiology* 6: 1569.

Andorrà, I., Esteve-Zarzoso, B., Guillamón, J.M. and Mas, A. (2010a). Determination of viable wine yeast using DNA binding dyes and quantitative PCR. *International Journal of Food Microbiology* 144: 257-262.

Andorrà, I., Landi, S., Mas, A., Esteve-Zarzoso, B. and Guillamón, J.M. (2010b). Effect of fermentation temperature on microbial population evolution using culture-independent and dependent techniques. *Food Research International* 43: 773-779.

Andorrà, I., Berradre, M., Rozés, N., Mas, A., Guillamón, J.M. and Esteve-Zarzoso, B. (2010c). Effect of pure and mixed cultures of the main yeast species on grape must fermentations. *European Food Research and Technology* 231: 215-224.

Andorrà, I., Monteiro, M., Esteve-Zarzoso, B., Albergaria, H. and Mas, A. (2011). Analysis and direct quantification of *Saccharomyces cerevisiae* and *Hanseniaspora guilliermondii* populations during alcoholic fermentation by fluorescence in situhybridisation, flow cytometry and quantitative PCR. *Food Microbiology* 28: 1483-1491.

Asai, T. (1968). Acetic Acid Bacteria: Classification and Biochemical Activities. University of Tokyo Press, Tokyo.

Bae, S., Fleet, G.H. and Heard, G.M. (2006). Lactic acid bacteria associated with wine grapes from several Australian vineyards. *Journal of Applied Microbiology* 100: 712-727.

Baena-Ruano, S., Santos-Duenas, I.M., Mauricio, J.C. and Garcia-Garcia, I. (2010). Relationship between changes in the total concentration of acetic acid bacteria and major volatile compounds during the acetic acid fermentation of white wine. *Journal of the Science of Food and Agriculture* 90: 2675-2681.

Bamforth, C.W. (2008). Food, Fermentation and Microorganisms. John Wiley & Sons. Chichester, GBR.

Barata, A., Malfeito-Ferreira, M. and Loureiro, V. (2012). Changes in sour rotten grape berry microbiota during ripening and wine fermentation. *International Journal of Food Microbiology* 154: 152-161.

Barbe, J.C., de Revel, G., Joyeux, A., Bertrand, A. and Lonvaud-Funel, A. (2001). Role of botrytized grape micro-organisms in $SO_2$ binding phenomena. *Journal of Applied Microbiology* 90: 34-42.

Barnett, J.A., Payne, R.W. and Yarrow, D. (2000). Yeasts: Characteristics and Identification. Third edition. Cambridge University Press. Cambridge, UK.

Barrio, E., González, S.S., Arias, A., Belloch, C. and Querol, A. (2006). Molecular mechanisms envolved in the adaptative evolution of industrial yeasts. pp. 153-174. *In*: A. Querol and G.H. Fleet (Eds.). Yeast in Food Beverages. Springer, Berlin.

Bartowsky, E.J., Xia, D., Gibson, R.L., Fleet, G.H. and Henschke, P.A. (2003). Spoilage of bottled red wine by acetic acid bacteria. *Letters in Applied Microbiology* 36: 307-314.

Bartowsky, E.J. and Henschke, P.A. (2008). Acetic acid bacteria spoilage of bottled red wine – A review. *International Journal of Food Microbiology* 125: 60-70.

Belloch, C., Fernández-Espinar, T., Querol, A., Dolores García, M. and Barrio, E. (2002). An analysis of inter- and intra-specific genetic variabilities in the *Kluyveromyces marxianus* group of yeast species for the reconsideration of the K. Lactis. *Taxon.* 19: 257-268.

Beltran, G., Torija, M.J., Novo, M., Ferrer, N., Poblet, M., Guillamon, J.M., Rozes, N. and Mas, A. (2002). Analysis of yeast populations during alcoholic fermentation: A six year follow-up study. *Systematic and Applied Microbiology* 25: 287-293.

Bely, M., Stoeckle, P., Masneuf-Pomarède, I. and Dubourdieu, D. (2008). Impact of mixed *Torulaspora delbrueckii-Saccharomyces cerevisiae* culture on high-sugar fermentation. *International Journal of Food Microbiology* 122: 312-320.

Bely, M., Renault, P., da Silva, T., Masneuf-Pomarede, I., Albertin, W., Moine, V., Coulon, J., Sicard, D., de Vienne, D. and Marullo, P. (2013). Non-conventional yeasts and alcohol level reduction. Conference on Alcohol Level Reduction in Wine, pp. 3-37.

Blasco, L., Ferrer, S. and Pardo, I. (2003). FISH application for the acetic acid bacteria present in wine. *OEnologie*, Editions TEC & DOC, Bordeaux, France, pp. 274-278.

Bokulich, N.A., Joseph, C.M.L., Allen, G., Benson, A.K. and Mills, D.A. (2012). Next generation sequencing reveals significant bacterial diversity of botrytised wine. *PLoS One* 7: e36357.

Bokulich, N.A., Thorngate, J.A., Richardson, P.M. and Mills, D.A. (2014). Microbial biogeography of wine grapes in conditioned by cultivar vintage, and climate. *Proceedings of the National Academy of Sciences* 111(1): 139-148.

Borneman, A.R. and Pretorius, I.S. (2015). Genomic insights into the *Saccharomyces sensu stricto* complex. *Genetics* 199: 281-291.

Boyd, A.R., Gunasekera, T.S., Attfield, P.V., Simic, K., Vincent, S.F. and Veal, D.A. (2003). A flow cytometric method for determination of yeast viability and cell number in a brewery. *FEMS Yeast Research* 3: 11-16.

Branco, P., Viana, T., Albergaria, H. and Arneborg, N. (2015). Anti-microbial peptides (AMPs) produced by *Saccharomyces cerevisiae* induce alterations in the intracellular pH, membrane permeability and culturability of *Hanseniaspora guilliermondii* cells. *International Journal of Food Microbiology* 205: 112-118.

Capece, A., Fiore, C., Maraz, A. and Romano, P. (2005). Molecular and technological approaches to evaluate strain biodiversity in *Hanseniaspora uvarum* of wine origin. *Journal of Applied Microbiology* 98: 136-144.

Capozzi, V., Garofalo, C., Chiriatti, M.A., Grieco, F. and Spano, G. (2015). Microbial terroir and food innovation: The case of yeast biodiversity in wine. *Microbiological Research* 181: 75-83.

Cappello, M.S., Stefani, D., Grieco, F., Logrieco, A. and Zapparoli, G. (2008). Genotyping by Amplified Fragment Length Polymorphism and malate metabolism performances of indigenous *Oenococcus oeni* strains isolated from Primitivo wine. *International Journal of Food Microbiology* 127: 241-245.

Chaney, D., Rodriguez, S., Fugelsanza, K. and Thornton, R. (2006). Managing high-density commercial scale wine fermentations. *Journal of Applied Microbiology* 100: 689-698.

Chen, Y., Bai, Y., Li, D., Wang, C., Xu, N. and Hu, Y. (2016). Screening and characterization of ethanol-tolerant and thermotolerant acetic acid bacteria from Chinese vinegar Pei. *World Journal of Microbiology and Biotechnology* 32: 14.

Ciani, M. and Ferraro, L. (1996). Enhanced glycerol content in wines made with immobilized Candida stellata cells. *Applied and Environmental Microbiology* 62: 128-132.

Ciani, M., Comitini, F., Mannazzu, I. and Domizio, P. (2010). Controlled mixed culture fermentation: A new perspective on the use of non-*Saccharomyces* yeasts in winemaking. *FEMS Yeast Research* 10: 123-133.

Clavijo, A., Caldertón, I.L. and Paneque, P. (2010). Diversity of *Saccharomyces* and non- *Saccharomyces* yeasts in three red grape varieties cultured in the Serranía de Ronda (Spain) vine-growing region. *International Journal of Food Microbiology* 143: 241-245.

Cleenwerck, I. and De Vos, P. (2008). Polyphasic taxonomy of acetic acid bacteria: An overview of the currently applied methodology. *International Journal of Food Microbiology* 125: 2-14.

Clemente-Jiménez, J., Mingorance-Cazorla, L., Martínez-Rodríguez, S., Heras-Vázquez, F.J.L. and Rodríguez-Vico, E. (2004). Molecular characterisation and oenological properties of wine yeasts isolated during spontaneous fermentation of six varieties of grape must. *Food Microbiology* 21: 149-155.

CNS (Chinese National Standard) (2005). Edible Vinegar. No. 14834, N5239, Ministry of Economic Affairs. Taiwan, Republic of China.

Cocolin, L. and Mills, D.A. (2003). Wine yeast inhibition by sulfur dioxide: A comparison of culture-dependent and independent methods. *American Journal of Enology and Viticulture* 54(2): 125-130.

Cocolin, L., Pepe, V., Comitini, F., Comi, G. and Ciani, M. (2004). Enological and genetic traits of *Saccharomyces cerevisiae* isolated from former and modern wineries. *FEMS Yeast Research* 5: 237-245.

Cole, J.R., Chai, B., Farris, R.J., Wang, Q., Kulam, S.A., McGarrell, D.M., Garrity, G.M. and Tiedje, J.M. (2005). The Ribosomal Database Project (RDP-II): Sequences and tools for high-throughput rRNA analysis. *Nucleic Acids Research* 33: 294-296.

Combina, M., Elía, A., Mercado, L., Catania, C., Ganga, A. and Martñinez, C. (2005). Dynamics of indigenous yeast populations during spontaneous fermentation of wines from Mendoza, Argentina. *International Journal of Food Microbiology* 99: 237-243.

Comitini, F., Gobbi, M., Domizio, P., Romani, C., Lencioni, L., Mannazzu, I. and Ciani, M. (2011). Selected non-*Saccharomyces* wine yeasts in controlled multistarter fermentations with *Saccharomyces cerevisiae*. *Food Microbiology* 28: 873-882.

Constantí, M., Poblet, M., Arola, L., Mas, A. and Guillamón, J.M. (1997). Analysis of yeast populations during alcoholic fermentation in a newly established winery. *American Journal of Enology and Viticulture* 48: 339-344.

Contreras, A., Hidalgo, C., Henschke, P.A., Chambers, P.J., Curtin, C. and Varela, C. (2014). Evaluation of non *Saccharomyces* yeasts for the reduction of alcohol content in wine. *Applied and Environmental Microbiology* 80(5): 1670-1678.

Crotti, E., Rizzi, A., Chouaia, B., Ricci, I., Favia, G., Alma, A., Sacchi, L., Bourtzis, K., Mandrioli, M., Cherif, A., Bandi, C. and Daffonchio, D. (2010). Acetic acid bacteria, newly emerging symbionts of insects. *Applied and Environmental Microbiology* 76: 6963-6970.

Csernus, O., Pomázi, A. and Magyar, I. (2014). Isolation, characterisation, and selection of wine yeast strains in etyek-buda wine district, Hungary. *Acta Alimentaria* 43(3): 489-500.

David, V., Terrat, S., Herzine, K., Claisse, O., Rousseaux, S, Tourdot-Marecha, R., Masneuf-Pomarede, I., Ranjard, L. and Alexandre, H. (2014). High throughput sequencing of amplicons for monitoring yeast biodiversity in must and during alcoholic fermentation. *Journal of Industrial Microbiology and Biotechnology* 41(5): 811-821.

Day, A.P. and Oliver, J.D. (2004). Changes in membrane fatty acid composition during entry of *Vibrio vulnificus* into the viable but nonculturable state. *The Journal of Microbiology* 42: 69-73.

De Ley, J., Gillis, M. and Swings, J. (1984). Family VI: Acetobacteraceae. *Bergey's Manual of Systematic Bacteriology* 1: 267-278.

Dicks, L.M.T. and Endo, A. (2009). Taxonomic status of lactic acid bacteria in wine and key characteristics to differentiate species. *South African Journal of Enology and Viticulture* 30: 72-90.

Di Maro, E., Ercolini, D. and Coppola, S. (2007). Yeast dynamics during spontaneous wine fermentation of the Catalanesca grape. *International Journal of Food Microbiology* 117: 201-210.

Dias, L., Dias, S., Sancho, T., Stender, H., Querol, A., Malfeito-Ferreira, M. and Loureiro, V. (2003). Identification of yeasts originated from wine related environments and capable of producing 4-ethylphenol. *Food Microbiology* 20: 567-574.

Drysdale, G.S. and Fleet, G.H. (1989). The growth and survival of acetic acid bacteria in wines at different concentrations of oxygen. *American Journal of Enology Viticulture* 40: 99-105.

Du Toit, M. and Pretorius, I.S. (2000). Microbial spoilage and preservation of wine: Using weapons from Nature's own arsenal – A review. *South African Society for Enology & Viticulture* 21: 74-96.

Du Toit, W.J. and Lambrechts, M.G. (2002). The enumeration and identification of acetic acid bacteria from South African red wine fermentations. *International Journal of Food Microbiology* 74: 57-64.

Du Toit, W.J. and Pretorius, I.S. (2002). The occurrence, control and esoteric effect of acetic acid bacteria in winemaking. *Annals of Microbiology* 52: 155-179.

Du Toit, W.J., Pretorius, I.S. and Lonvaud-Funel, A. (2005). The effect of sulphur dioxide and oxygen on the viability and culturability of a strain of *Acetobacter pasteurianus* and a strain of *Brettanomyces bruxellensis* isolated from wine. *Journal of Applied Microbiology* 98: 862-871.

Dubourdieu, D., Sokol, A., Zucca, J., Thalouarn, P., Datee, A. and Aigle, M. (1987). Identification *the souches de levures isolées de vins par l'analyse de leur* DNA mitocondrial. *Conn. Vigne Vin.* 4: 267-278.

Dutta, D. and Gachhui, R. (2006). Novel nitrogen-fixing Acetobacter nitrogenifigens sp. nov., isolated from Kombucha tea. *International Journal of Systematic and Evolutionary Microbiology* 56: 1899-1903.

EC (Council Regulation) (1999). No 1493/1999 of 17 May 9. 199 on the common organisation of the market in wine. *Official Journal of the European Communities* L179: 1-84.

Egli, C.M. and Henick-Kling, T. (2001). Identification of *Brettanomyces/Dekkera* species based on polymorphism in the rRNA Internal Transcribed Spacer Region. *American Journal of Enology and Viticulture* 52: 241-247.

Englezos, V., Rantsiou, K., Torchio, F., Rolle, L., Gerbi, V. and Cocolin, L. (2015). Exploitation of the non-*Saccharomyces* yeast *Starmerella bacillaris* (synonym *Candida zemplinina*) in wine fermentation: Physiological and molecular characterisations. *International Journal of Food Microbiology* 199: 33-40.

Esteve-Zarzoso, B., Manzanares, P., Ramón, D. and Querol, A. (1998). The role of non-Saccharomyces yeasts in industrial winemaking. *International Microbiology* 1: 143-148.

Esteve-Zarzoso, B., Belloch, C., Uruburu, F. and Querol, A. (1999). Identification of yeasts by RFLP analysis of the 5.8S rRNA gene and the two ribosomal internal transcribed spacers. *International Journal of Systematic Bacteriology* 49: 329-337.

Esteve-Zarzoso, B., Peris-Torán, M.J., Ramón, D. and Querol, A. (2001). Molecular characterisation of *Hanseniaspora* species. *Antonie Van Leeuwenhoek* 80: 85-92.

Esteve-Zarzoso, B., Zorman, T., Belloch, C. and Querol, A. (2003). Molecular characterization of the species of the genus *Zygosaccharomyces*. *Systematic and Applied Microbiology* 26: 404-411.

FDA (Food and Drug Administration) (1995). Compliance Policy Guides (CPG) Sec. 525.825, Vinegar, Definitions-Adulteration with Vinegar Eels. Available from: http://www.fda.gov/ICECI/ComplianceManuals/CompliancePolicyGuidance Manual/ucm074471.htm (Accessed: September 2017).

Fernandez-Perez, R., Torres, C., Sanz, S. and Ruiz-Larrea, F. (2010). Strain typing of acetic acid bacteria responsible for vinegar production by the submerged elaboration method. *Food Microbiology* 27: 973-978.

Fleet, G.H. (1993). Wine Microbiology and Biotechnology. Harwood, Chur, Switzerland.

Fleet, G.H. (1998). Microbiology of alcoholic beverages. pp. 217-262. *In*: B.J. Wood (Ed.). Microbiology and Fermented Foods. Blackie Academic & Professional. London, England.

Fleet, G.H. (2003). Yeast interactions and wine flavor. *International Journal of Food Microbiology* 86: 11-22.

Froudière, I., Larue, F. and Lonvaud-Funel, A. (1990). Utilisation de l'épifluorescence pour la détectiondes micro-organismes dans le vin. *Journal International des Sciences de la Vigne et du Vin*. 24: 43-46.

Fugelsang, K.C. (1997). Wine Microbiology. Chapman & Hall. New York, USA.

Fugelsang, K.C. and Edwards, C.G. (2007). Wine Microbiology: Practical Applications and Procedures. Springer Science Business Media, LLC. New York, USA.

Furukawa, S., Watanabe, T., Toyama, H. and Morinaga, Y. (2013). Significance of microbial symbiotic coexistence in traditional fermentation. *Journal of Bioscience and Bioengineering* 116: 533-539.

García-Ruiz, A., Crespo, J., Lopez-de-Luzuriaga, J.M., Olmosi, M.E., Monge, M., Rodríguez-Alfaro, M.P., Martín-Alvarez, P.J., Bartolome, B. and Moreno-Arribas, M.V. (2015). Novel biocompatible silver nano particles for controlling the growth of lactic acid bacteria and acetic acid bacteria in wines. *Food Control* 50: 613-619.

Gevers, D., Huys, G. and Swings, J. (2001). Applicability of rep-PCR fingerprinting for identification of Lactobacillus species. *FEMS Microbiology Letters* 205: 31-36.

Giudici, P., De Vero, L. and Gullo, M. (2017). Vinegars. Chap. 10. pp. 261-287. *In*: I.Y. Sengun (Ed.). Acetic Acid Bacteria: Fundamentals and Food Applications. CRC Press. Taylor & Francis Group, Boca Raton.

Gobbi, M., Comitini, F., Domizio, P., Romani, C., Lencioni, L., Mannazzu, I. and Ciani, M. (2013). *Lachancea thermotolerans* and *Saccharomyces cerevisiae* in simultaneous and sequential co-fermentation: A strategy to enhance acidity and improve the overall quality of wine. *Food Microbiology* 33: 271-281.

González, A. (2005). Application of molecular techniques for identification of acetic acid bacteria. PhD Thesis, Universitat Rovira I Virgili. Tarragona, Spain.

González, A., Hierro, N., Poblet, M., Mas, A. and Guillamón, J.M. (2005). Application of molecular methods to demonstrate species and strain evolution of acetic acid bacteria population during wine production. *International Journal of Food Microbiology* 102: 295-304.

González, A., Guillamón, J.M., Mas, A. and Poblet, M. (2006a). Application of molecular methods for routine identification of acetic acid bacteria. *International Journal of Food Microbiology* 108: 141-146.

González, A., Hierro, N., Poblet, M., Mas, A. and Guillamón, J.M. (2006b). Enumeration and detection of acetic acid bacteria by real-time PCR and nested-PCR. *FEMS Microbiology Letters* 254: 123-128.

González, S.S., Barrio, E., Gafner, J. and Querol, A. (2006c). Natural hybrids from *Saccharomyces cerevisiae, Saccharomyces bayanus,* and *Saccharomyces kudriavzevii* in wine fermentations. *FEMS Yeast Research* 6: 1221-1223.

Gonzalez, R., Quiros, M. and Morales, P. (2013). Yeast respiration of sugars by non-*Saccharomyces* yeast species: A promising and barely explored approach to lowering alcohol content of wines. *Trends in Food Science and Technology* 29: 55-61.

González-Royo, E., Pascual, O., Kontoudakis, N., Esteruelas, M., Esteve Zarzoso, B., Mas, A., Miquel Canals, J. and Zamora, F. (2015). Oenological consequences of sequential inoculation with non-*Saccharomyces* yeasts (*Torulaspora delbrueckii* or *Metschnikowia pulcherrima*) and *Saccharomyces cerevisiae* in base wine for sparkling wine production. *European Food Research and Technology* 240: 999-1012.

Grbin, P.R. and Henschke, P.A. (2000). Mousy off-flavour production in grape juice and wine by *Dekkera* and *Brettanomyces* yeasts. *Australian Journal of Grape and Wine Research* 6: 255-262.

Guillamón, J.M., Barrio, E. and Querol, A. (1996). Characterisation of wine yeast strains of the *Saccharomyces* genus on the basis of molecular markers: Relationships between genetic distance and geographic or ecological origin. *Systematic and Applied Microbiology* 19: 122-132.

Guillamón, J.M., Sabate, J., Barrio, E., Cano, J. and Querol, A. (1998). Rapid identification of wine yeast species based on RFLP analysis of the ribosomal internal transcribed spacer (ITS) region. *Archives of Microbiology* 169: 387-392.

Gullo, M., Caggia, C., De Vero, L. and Giudici, P. (2006). Characterisation of acetic acid bacteria in 'traditional balsamic vinegar'. *International Journal of Food Microbiology* 106: 209-212.

Gullo, M., Verzelloni, E. and Canonico, M. (2014). Aerobic submerged fermentation by acetic acid bacteria for vinegar production: Process and biotechnological aspects. *Process Biochemistry* 49: 1571-1579.

Haruta, S., Ueno, S., Egawa, I., Hashiguchi, K., Fujii, A., Nagano, M., Ishii, M. and Igarashi, Y. (2006). Succession of bacterial and fungal communities during a traditional pot fermentation of rice vinegar assessed by PCR-mediated denaturing gradient gel electrophoresis. *International Journal of Food Microbiology* 109: 79-87.

Heim, S., Lleo, M.D.M., Bonato, B., Guzman, C.A. and Canepari, P. (2002). The viable but nonculturable state and starvation are different stress responses of *Enterococcus faecalis*, as determined by proteome analysis. *Journal of Bacteriology* 184: 6739-6745.

Henschke, P.A. (1997). Yeast sugar metabolism. pp. 527-560. *In*: K.D. Entian and F.K. Zimmermann (Eds.). Wine Yeast. Technomic Publishing. Lancaster, Pennsylvania, USA.

Herrero, M., Quiros, C., Garcia, L.A. and Diaz, M. (2006). Use of flow cytometry to follow the physiological states of microorganisms in cider fermentation processes. *Applied and Environmental Microbiology* 72: 6725-6733.

Hidalgo, C., Vegas, C., Mateo, M., Tesfaye, W., Cerezo, A.B., Callejón, R.M., Poblet, M., Guillamon, J.M., Mas, A. and Torija, M.J. (2010). Effect of barrel design and the inoculation of *A. pasteurianus* in wine vinegar production. *International Journal of Food Microbiology* 141: 56-62.

Hidalgo, C., Mateo, E., Mas, A. and Torija, M.J. (2012). Identification of yeast and acetic acid bacteria isolated from the fermentation and acetification of persimmon (*Diospyros kaki*). *Food Microbiology* 30: 98-104.

Hidalgo, C., Mateo, E., Mas, A. and Torija, M.J. (2013a). Effect of inoculation on strawberry fermentation and acetification processes using native strains of yeast and acetic acid bacteria. *Food Microbiology* 34: 88-94.

Hidalgo, C., Garcia, D., Romero, J., Mas, A., Torija, M.J. and Mateo, E. (2013b). Acetobacter strains isolated during the acetification of blueberry (*Vaccinium corymbosum* L.) wine. *Letters in Applied Microbiology* 57: 227-232.

Hierro, N., Esteve-Zarzoso, B., González, A., Mas, A. and Guillamón, J.M. (2006). Real-time quantitative PCR (QPCR) and reverse transcription-QPCR (RT-QPCR) for the detection and enumeration of total yeasts in wine. *Applied and Environmental Microbiology* 72: 7148-7155.

Ilabaca, C., Navarrete, P., Mardones, P., Romero, J. and Mas, A. (2008). Application of culture-independent molecular biology-based methods to evaluate acetic acid bacteria diversity during vinegar processing. *International Journal of Food Microbiology* 126: 245-249.

Jolly, N.P., Varela, C. and Pretorius, I.S. (2014). Not your ordinary yeast: Non-*Saccharomyces* yeasts in wine production uncovered. *FEMS Yeast Research* 14: 215-237.

Joyeux, A., Lafon-Lafourcade, S. and Ribéreau-Gayon, P. (1984). Evolution of acetic acid bacteria during fermentation and storage of wine. *Applied and Environmental Microbiology* 48: 153-156.

Karabiyikli, S. and Sengun, I.Y. (2017). Beneficial effects of acetic acid bacteria and their food products. Chap. 13. pp. 321-342. *In*: I.Y. Sengun (Ed.). Acetic Acid Bacteria: Fundamentals and Food Applications. CRC Press. Taylor & Francis Group, Boca Raton.

Komagata, K., Iino, T. and Yamada, Y. (2014). The family Acetobacteraceae. pp. 3-78. *In*: E. Rosenberg, E.F. De Long, S. Lory, E. Stackebrandt and F. Thompson (Eds.). The Prokaryotes, Alphaproteobacteria and Betaproteobacteria. Springer-Verlag. Berlin Heidelberg.

Kurtzman, C.P. and Robnett, C.J. (1998). Identification and phylogeny of ascomycetous yeast from analysis of nuclear large subunit 26S ribosomal DNA partial sequences. *Antonie Van Leeuwenhoek* 73: 331-371.

Le Jeune, C. and Lonvaud-Funel, A. (1997). Sequence of DNA 16S/23S spacer region of *Leuconostoc oenos* (*Oenococcus oeni*): Application to strain differentiation. *Research in Microbiology* 148: 79-86.

Lieckfieldt, E., Meyer, W. and Börn, T. (1993). Rapid identification and differentiation of yeasts by DNA and PCR fingerprinting. *Journal of Basic Microbiology* 33: 413-426.

Llauradó, J., Rozés, N., Bobet, R., Mas, A. and Constantí, M. (2002). Low temperature alcoholic fermentations in high sugar concentration grape musts. *Journal of Food Science* 67: 268-273.

Lleixà, J., Martín, V., Portillo, M.C., Carrau, F., Beltran, G. and Mas, A. (2016). Comparison of fermentation and wines produced by inoculation of *Hanseniaspora vineae* and *Saccharomyces cerevisiae*. *Frontiers in Microbiology* 7: 338.

Lleò, M.M., Pierobon, S., Tafi, M.C., Signoretto, C. and Canepari, P. (2000). mRNA detection by reverse transcription-PCR for monitoring viability over time in an *Enterococcus faecalis* viable but nonculturable population maintained in a laboratory microcosm. *Applied and Environmental Microbiology* 66: 4564-4567.

Lleò, M.M., Bonato, B., Tafi, M.C., Signoretto, C., Boaretti, M. and Canepari, P. (2001). Resuscitation rate in different enterococcal species in the viable but non-culturable state. *Journal of Applied Microbiology* 91: 1095-1102.

Lonvaud-Funel, A. (1999). Lactic acid bacteria in the quality improvement and depreciation of wine. *Antonie Van Leeuwenhoek* 76: 317-331.

Loureiro, V., Ferreira, M.M., Monteiro, S. and Ferreira, R.B. (2012). The microbial community of grape berry. pp. 241-268. *In*: H. Gerós, M.M. Chanves and S. Delrot (Eds.). The Biochemistry of the Grape Berry. Bentham Science Publishers. Sharjah, United Arab Emirates.

Machve, K.K. (2009). Fermentation Technology. Mangalam Publishers. Delhi, India.

Malacrino, P., Zapparoli, G., Torriani, S. and Dellaglio, F. (2001). Rapid detection of viable yeasts and bacteria in wine by flow cytometry. *Journal of Microbiological Methods* 45: 127-134.

Malimas, T., Vu, H.T.L., Muramatsu, Y., Yukphan, P., Tanasupawat, S. and Yamada, Y. (2017). Systematics of acetic acid bacteria. Chap. 1. pp. 3-43. *In*: I.Y. Sengun (Ed.). Acetic Acid Bacteria: Fundamentals and Food Applications. CRC Press. Taylor & Francis Group, Boca Raton.

Mamlouk, D., Hidalgo, C., Torija, M.J. and Gullo, M. (2011). Evaluation and optimisation of bacterial genomic DNA extraction for no-culture techniques applied to vinegars. *Food Microbiology* 28: 1374-1379.

Maqueda, M., Zamora, E., Rodríguez-Cousiño, N. and Ramírez, M. (2010). Wine yeast molecular typing, using a simplified method for simultaneously extracting mtDNA, nuclear DNA and virus dsRNA. *Food Microbiology* 27: 205-209.

Marsit, S.M. and Dequin, S. (2015). Diversity and adaptive evolution of *Saccharomyces* wine yeast: A review. *FEMS Yeast Research* 15: fov067.

Martín, V., Mas, A., Carrau, F., Dellacasa, E. and Boido, E. (2016). Effect of yeast as similable nitrogen on the synthesis of phenolic aroma compounds by *Hanseniaspora vineae* strains. *Yeast* 33: 323-328.

Martorell, P., Querol, A. and Fernández-Espinar, M.T. (2005). Rapid identification and enumeration of Saccharomyces cerevisiae cells in wine by real-time PCR. *Applied and Environmental Microbiology* 71: 6823-6830.

Martorell, P., Barata, A., Malfeito-Ferreira, M., Fernández-Espinar, M.T., Loureiro, V. and Querol, A. (2006). Molecular typing of the yeast species *Dekkera bruxellensis* and *Pichia guilliermondii* recovered from wine-related sources. *International Journal of Food Microbiology* 106: 79-84.

Mas, A., Torija, M.J., García-Parrilla, M.C. and Troncoso, A.M. (2014). Acetic acid bacteria and the production and quality of wine vinegar. *The Scientific World Journal* 2014: 6, Article ID 394671.

Mas, A., Padilla, B., Esteve-Zarzoso, B., Beltran, G., Reguant, C. and Bordons, A. (2016). Taking advantage of natural biodiversity for winemaking: The WILDWINE Project. *Agriculture and Agricultural Science Procedia* 8: 4-9.

Masneuf, I. and Dubourdieu, D. (1994). Comparaison de deux techniques d'identification des souchesde levures de vinification basées sur le polymorphisme de l'ADN génomique: Réaction de polymérisationen chaine (PCR) et analyse des caryotypes (electrophorèse en champ pulsé). *Journal International des Sciences de la Vigne et du Vin* 28: 153-160.

Masneuf-Pomarede, I., Bely, M., Marullo, P., Lonvaud-Funel, A. and Dubourdieu, D. (2010). Reassessment of phenotypic traits for *Saccharomyces bayanus* var. *uvarum* wine yeast strains. *International Journal of Food Microbiology* 139: 79-86.

Masneuf-Pomarede, I., Juquin, E., Miot-Sertier, C., Renault, P., Laizet, Y.H., Salin, F., Alexandre, H., Capozzi, V., Cocolin, L., Colonna-Ceccaldi, B., Englezos, V., Girard, P., Gonzalez, B., Lucas, P., Mas, A., Nisiotou, A., Sipiczki, M., Spano, G., Tassou, C., Bely, M. and Albertin, W. (2015). The yeast *Starmerella bacillaris* (synonym *Candida zemplinina*) shows high genetic diversity in winemaking environments. *FEMS Yeast Research* 15(5): fov045.

Masneuf-Pomarede, I., Bely, M., Marullo, P. and Albertin, W. (2016). The genetics of non-conventional wine yeasts: Current knowledge and future challenges. *Frontiers in Microbiology* 6: 1563.

Mateo, E., Torija, M.J., Mas, A. and Bartowsky, E.J. (2014). Acetic acid bacteria isolated from grapes of South Australian vineyards. *International Journal of Food Microbiology* 178: 98-106.

Matsushita, K., Toyama, H. and Adachi, O. (1994). Respiratory chain and bioenergetics of acetic acid bacteria. pp. 247-301. *In*: A.H. Rose and D.W. Tempest (Eds.). Advances in Microbial Physiology 35. Academic Press. London.

Medina, K., Boido, E., Dellacassa, E. and Carrau, F. (2012). Growth of non-*Saccharomyces* yeasts affects nutrient availability for *Saccharomyces cerevisiae* during wine fermentation. *International Journal of Food Microbiology* 157: 245-250.

Medina, K., Boido, E., Fariña, L., Gioia, O., Gomez, M.E., Barquet, M., Gaggero, C., Dellacassa, E. and Carrau, F. (2013). Increased flavour diversity of Chardonnay wines by spontaneous fermentation and co-fermentation with *Hanseniaspora vineae*. *Food Chemistry* 141: 2513-2521.

Mercanoglu Taban, B. and Saichana, N. (2017). Physiology and biochemistry of acetic acid bacteria. Chap. 3. pp. 71-91. *In*: I.Y. Sengun (Ed.). Acetic Acid Bacteria: Fundamentals and Food Applications. CRC Press. Taylor & Francis Group, Boca Raton.

Millet, V. and Lonvaud-Funel, A. (2000). The viable but non-culturable state of wine micro-organisms during storage. *Letters in Applied Microbiology* 30: 136-141.

Mira de Orduna, R. (2010). Climate change associated effects on grape and wine quality and production. *Food Research International* 43: 1844-1855.

Montrocher, R., Verner, M.C., Briolay, J., Gautier, C. and Marmeisse, R. (1998). Phylogenetic analysis of the *Saccharomyces cerevisiae* group based on polymorphisms of rDNA spacer sequences. *International Journal of Systematic and Evolutionary Microbiology* 48: 295-303.

Morata, A., Loira, I., Vejarano, R., Gonzalez, C., Callejo, M.J. and Suarez-Lepe, J.A. (2017). Emerging preservation technologies in grapes for winemaking. *Trends in Food Science & Technology* 67: 36-43.

Moreira, N., Mendes, F., Guedes de Pinho, P., Hogg, T. and Vasconcelos, I. (2008). Heavy sulphur compounds, higher alcohols and esters production profile of *Hanseniaspora uvarum* and *Hanseniaspora guilliermondii* grown as pure and mixed cultures in grape must. *International Journal of Food Microbiology* 124: 231-238.

Nanda, N., Taniguchi, M., Ujike, S., Ishihara, N., Mori, H., Ono, H. and Murooka, Y. (2001). Characterisation of acetic acid bacteria in traditional acetic acid fermentation of rice vinegar (Komesu) and unpolished rice vinegar (Kurosu) produced in Japan. *Applied and Environmental Microbiology* 67: 986-990.

Naumov, G.I., James, S.A., Naumova, E.S., Louis, E.J. and Roberts, I.N. (2000). Three new species in the *Saccharomyces sensu* stricto complex: *Saccharomyces cariocanus, Saccharomyces kudriavzevii* and *Saccharomyces mikatae*. *International Journal of Systematic and Evolutionary Microbiology* 50: 1931-1942.

Neeley, E.T., Phister, T.G. and Mills, D.A. (2005). Differential real-time PCR assay for enumeration of lactic acid bacteria in wine. *Applied and Environmental Microbiology* 71: 8954-8957.

Ness, F., Lavalleé, F., Dubourdieu, D., Aigle, M. and Dulau, L. (1993). Identification of yeast strains using the polymerase chain reaction. *Journal of the Science of Food and Agriculture* 62: 89-94.

Nie, Z., Zheng, Y., Wang, M., Han, Y., Wang, Y., Luo, J. and Niu, D. (2013). Exploring microbial succession and diversity during solid-state fermentation of Tianjin duliu mature vinegar. *Bioresource Technology* 148: 325-333.

Nikolaou, E., Andrighetto, C., Lombardi, A., Litopoulou-Tzanetaki, E. and Tzanetakis, N. (2007). Heterogeneity in genetic and phenotypic characteristics of *Saccharomyces cerevisiae* strains isolated from red and white wine fermentations. *Food Control* 18: 1458-1465.

Nisiotou, A.A., Spiropoulos, A.E. and Nychas, G.E. (2007). Yeast community structures and dynamics in healthy and Botrytis-affected grape must fermentations. *Applied and Environmental Microbiology* 73(21): 6705-6713.

Nisiotou, A.A., Rantsiou, K., Iliopoulos, V., Cocolin, L. and Nychas, G.J. (2011). Bacterial species associated with sound and Botrytis-infected grapes from a Greek vineyard. *International Journal Food Microbiology* 145: 432-436.

Nissen, P. and Arneborg, N. (2003). Characterisation of early deaths of non-*Saccharomyces* yeasts in mixed cultures with *Saccharomyces cerevisiae*. *Archives of Microbiology* 180: 257-263.

Nocker, A. and Camper, A.K. (2006). Selective removal of DNA from dead cells of mixed bacterial communities by use of ethidium monoazide. *Applied and Environmental Microbiology* 72: 1997-2004.

Nocker, A., Cheung, C.Y. and Camper, A.K. (2006). Comparison of propidium monoazide with ethidium monoazide for differentiation of live *vs.* dead bacteria by selective removal of DNA from dead cells. *Journal of Microbiological Methods* 67: 310-320.

Nogva, H.K., Drømtorp, S.M., Nissen, H. and Rudi, K. (2003). Ethidium monoazide for DNA-based differentiation of viable and dead bacteria by 5'-nuclease PCR. *Bio Techniques* 34: 804-813.

Nurgel, C., Erten, H., Canbas, A., Cabaroglu, T. and Selli, S. (2005). Yeast flora during the fermentation of wines made from *Vitis vinifera* L. cv. Emir and Kalecik Karasi grown in Anatolia. *World Journal of Microbiology and Biotechnology* 21: 1187-1194.

Ocón, E., Gutiérrez, A.R., Garijo, P., Tenorio, C., López, I., López, R. and Santamaría, P. (2010). Quantitative and quanlitative analysis of non-*Saccharomyces* yeasts in spontaneous alcoholic fermentations. *European Food Research and Technology* 230(6): 885-891.

Okazaki, S., Furukawa, S., Ogihara, H., Kawarai, T., Kitada, C., Komenou, A. and Yamasaki, M. (2010). Microbiological and biochemical survey on the transition of fermentative processes in Fukuyama pot vinegar brewing. *Journal of General and Applied Microbiology* 56: 205-211.

Oliver, J.D. (2005). The viable but nonculturable state in bacteria. *The Journal of Microbiology* 43: 93-100.

Ouoba, L.I.I., Kando, C., Parkouda, C., Sawadogo-Lingani, H., Diawara, B. and Sutherland, J.P. (2012). The microbiology of Bandji, palm wine of Borassus akeassii from Burkina Faso: Identification and genotypic diversity of yeasts, lactic acid and acetic acid bacteria. *Journal of Applied Microbiology* 1364-5072.

Padilla, B., García-Fernández, D., González, B., Izidoro-Pacheco, I., Esteve-Zarzoso, B., Beltran, G. and Mas, A. (2016). Yeast biodiversity from DOQ priorat uninoculated fermentations. *Frontiers in Microbiology* 7: 930.

Pinto, C., Pinho, D., Sousa, S., Pinheiro, M., Egas, C.C. and Gomes, A. (2014). Unravelling the diversity of grapevine microbiome. *PLoS ONE* 9(1): e85622.

Pinto, C., Pinho, D., Cardoso, R., Custodio, V., Fernades, J., Sousa, S., Pinheiro, M., Egas, C. and Gomes, A.C. (2015). Wine fermentation microbiome: A landscape from different Portuguese wine appellations. *Frontiers in Microbiology* 6: 905.

Plata, C., Millán, C., Mauricio, J.C. and Ortega, J.M. (2003). Formation of ethyl acetate and isoamylacetate by various species of wine yeasts. *Food Microbiology* 20: 217-224.

Poblet, M., Rozès, N., Guillamón, J.M. and Mas, A. (2000). Identification of acetic acid bacteria by restriction fragment length polymorphism analysis of a PCR-amplified fragment of the gene coding for 16S rRNA. *Letters in Applied Microbiology* 31: 63-67.

Porter, J., Edwards, C. and Pickup, R.W. (1995). Rapid assessment of physiological status in Escherichia coli using fluorescent probes. *The Journal of Applied Bacteriology* 4: 399-408.

Portillo, M.C. and Mas, A. (2016). Analysis of microbial diversity and dynamics during wine fermentation of Grenache grape variety by high-throughput barcoding sequencing. *LWT – Food Science and Technology* 72: 317-321.

Pretorius, I.S., van der Westhuizen, T.J. and Augustyn, O.P.H. (1999). Yeast biodiversity in vineyards and wineries and its importance to the South African wine industry. *South African Journal of Enology and Viticulture* 20: 61-70.

Pretorius, I.S. (2000). Tailoring wine yeast for the new millennium: Novel approaches to the ancient art of winemaking. *Yeast* 16: 675-729.

Prieto, C., Jara, C., Mas, A. and Romero, J. (2007). Application of molecular methods for analysing the distribution and diversity of acetic acid bacteria in Chilean vineyards. *International Journal of Food Microbiology* 115: 348-355.

Querol, A., Barrio, E., Huerta, T. and Ramón, D. (1992). Molecular monitoring of wine fermentations conducted by active dry yeast strains. *Applied and Environmental Microbiology* 58: 2948-2953.

Querol, A., Fernández-Espinar, M.T., del Olmo, M. and Barrio, E. (2003). Adaptive evolution of wine yeast. *International Journal of Food Microbiology* 86: 3-10.

Rantsiou, K., Urso, R., Iacumin, L., Cantoni, C., Cattaneo, P. and Comi, G. (2005). Culture dependent and independent methods to investigate the microbial ecology of Italian fermented sausages. *Applied and Environmental Microbiology* 71: 1977-1986.

Raspor, P., Milek, D., Polanc, J., Mozina, S.S. and Cadez, N. (2006). Yeasts isolated from three varieties of grapes cultivated in different locations of the Dolenjska wine growing region, Slovenia. *International Journal of Food Microbiology* 109: 97-102.

Renault, P., Miot-Sertier, C., Marullo, P., Hernandez-Orte, P., Lagarrigue, L., Lonvaud-Funel, A. and Bely, M. (2009). Genetic characterisation and phenotypic variability in *Torulaspora delbrueckii* species: Potential applications in the wine industry. *International Journal of Food Microbiology* 134: 201-210.

Renouf, V., Claisse, O. and Lonvaud-Funel, A. (2005). Understanding the microbial ecosystem on the grape berry surface through numeration and identification of yeast and bacteria. *Australian Journal of Grape and Wine Research* 11: 316-327.

Ribéreau-Gayon, P., Dubourdieu, D., Doneche, B. and Lovaud, A. (2000). Handbook of Enology, vol. 1. The Microbiology of Wine and Vinifications. John Wiley & Sons Ltd. Chichester, England.

Ribéreau-Gayon, P., Dubourdieu, D., Doneche, B. and Lovaud, A. (2006). Handbook of Enology, vol. 1. The Microbology of Wine and Vinifications, second ed. John Wiley & Sons Ltd. West Sussex, England.

Roda, A., Lucini, L., Torchio, F., Dordoni, R., De Faveri, D.M. and Lambri, M. (2017). Metabolite profiling and volatiles of pineapple wine and vinegar obtained from pineapple waste. *Food Chemistry* 229: 734-742.

Rodas, A.M., Ferrer, S. and Pardo, I. (2005). Polyphasic study of wine *Lactobacillus* strains: Taxonomic implications. *International Journal of Systematic Evolutionary Microbiology* 55: 197-207.

Rodriguez, N., Gonçalves, G., Malfeito-Ferreira, M. and Loureiro, V. (2001). Development and use of a differential medium to detect yeasts of the genera *Dekkera/Brettanomyces*. *International Journal of Food Microbiology* 90: 588-599.

Romano, P., Suzzi, G., Comi, G. and Zironi, R. (1992). Higher alcohol and acetic acid production by apiculate wine yeasts. *Journal of Applied Bacteriology* 73: 126-130.

Romano, P., Capece, A. and Jespersen, L. (2006). Taxonomic and ecological diversity of food and beverage yeasts. pp. 55-82. *In*: A. Querol and G.H. Fleet (Eds.). Yeast in Food Beverages. Springer. Berlin, Germany.

Rosma, A., Nadiah, A.H.S., Raj, A., Supwat, T., Sharma, S. and Joshi, V.K. (2016). Acetic acid fermented product. pp. 598-635. *In*: V.K. Joshi (Ed.). Indigenous Fermented Foods of South Asia. CRC Press. Taylor & Francis Group, Florida.

Rudi, K., Naterstad, K., Dromtorp, S.M. and Holo, H. (2005). Detection of viable and dead *Listeria monocytogenes* on gouda-like cheeses by real-time PCR. *Letters in Applied Microbiology* 40: 301-306.

Ruiz, A., Poblet, M., Mas, A. and Guillamon, J.M. (2000). Identification of acetic acid bacteria by RFLP of PCR-amplified 16S rDNA and 16S–23S rDNA intergenic spacer. *International Journal of Systematic and Evolutionary Microbiology* 50: 1981-1987.

Sabate, J., Cano, J., Querol, A. and Guillamon, J.M. (1998). Diversity of *Saccharomyces* strains in wine fermentations: Analysis for two consecutive years. *Letters in Applied Microbiology* 26: 452-455.

Saichana, N., Matsushita, K., Adachi, O., Frébort, I. and Frebortova, J. (2015). Acetic acid bacteria: A group of bacteria with versatile biotechnological applications. *Biotechnology Advances* 33: 1260-1271.

Schuller, D., Alves, H., Dequin, S. and Casal, M. (2005). Ecological survey of *Saccharomyces cerevisiae* strains from vineyards in the Vinho Verde region of Portugal. *FEMS Microbiology Ecology* 51: 167-177.

Schüller, G., Hertel, C. and Hammes, W.P. (2000). Glucon acetobacter entanii sp. nov., isolated from submerged high-acid industrial vinegar fermentations. *International Journal of Systematic and Evolutionary Microbiology* 50: 2013-2020.

Schütz, M. and Gafner, J. (1993). Analysis of yeast diversity during spontaneous and induced alcoholic fermentations. *Journal of Applied Microbiology* 75: 551-558.

Sengun, I.Y. (2015). Acetic acid bacteria in food fermentations. pp. 91-111. *In*: D. Montet and R.C. Ray (Eds.). Fermented Foods: Part 1: Biochemistry and Biotechnology. CRC Press. Boca Raton, USA.

Setati, M.E., Jacobson, D., Andong, U. and Bauer, F. (2012). The vineyard of yeast microbiome, a mixed model microbial map. *PLoS One* 7(12): e52609.

Setati, M.E., Jacobson, D. and Bauer, F.F. (2015). Sequence-based analysis of the *Vitis vinifera* L. cv Cabernet Sauvignon grape must mycobiome in three south African vineyards employing distinct agronomic systems. *Frontiers in Microbiology* 6: 1358.

Sievers, M., Sellmer, S. and Teuber, M. (1992). Acetobacter europaeus sp. nov., a main component of industrial vinegar fermenters in central Europe. *Systematic and Applied Microbiology* 15: 386-392.

Sievers, M. and Swings, J. (2005). Family Acetobacteraceae. pp. 41-95. *In*: G.M. Garrity (Ed.). Bergey's Manual of Systematic Bacteriology, second ed. Springer, New York.

Silhavy, K. and Mandl, K. (2006). *Acetobacter tropicalis* in spontaneously fermented wines with vinegar fermentation in Austria. *Mitteilungen Klosterneuburg* 56: 102-107.

Silva, L.R., Cleenwerck, I., Rivas, R., Swings, J., Trujillo, M.E., Willems, A. and Velázquez E. (2006). *Acetobacter oeni* sp. nov., isolated from spoiled red wine. *International Journal of Systematic and Evolutionary Microbiology* 56: 21-24.

Soden, A., Francis, I.L., Oakey, H. and Henschke, P.A. (2000). Effects of co-fermentation with *Candida stellata* and *Saccharomyces cerevisiae* on the aroma and composition of Chardonnay wine. *Australian Journal of Grape and Wine Research* 6: 21-30.

Sohier, D. and Lonvaud-Funel, A. (1998). Rapid and sensitive *in situ* hybridisation method for detecting and identifying lactic acid bacteria in wine. *Food Microbiology* 15: 391-397.

Sokollek, S.J., Hertel, C. and Hammes, W.P. (1998). Description of *Acetobacter oboediens* sp. nov. and *Acetobacter pomorum* sp. nov., two new species isolated from industrial vinegar fermentations. *International Journal of Systematic Bacteriology* 48: 935-940.

Solieri, L. and Giudici, P. (2009). Vinegars of the World. Springer-Verlag.

Song, N.E., Cho, H.S. and Baik, S.H. (2016). Bacteria isolated from Korean black raspberry vinegar with low biogenic amine production in wine. *Brazilian Journal of Microbiology* 47: 452-460.

Sponholz, W. (1993). Wine spoilage by microorganisms. pp. 395-420. *In*: G.H. Fleet (Ed.). Wine Microbiolgy and Biotechnology. Harwood Academic Publishers. Chur, Switzerland.

Sun, S.Y., Gong, H.S., Liu, W.L. and Jin, C.W. (2016). Application and validation of autochthonous *Lactobacillus plantarum* starter cultures for controlled malo-lactic fermentation and its influence on the aromatic profile of cherry wines. *Food Microbiology* 55: 16-24.

Tamai, Y., Momma, T., Yoshimoto, H. and Kaneko, Y. (1998). Co-existence of two types of chromosome in the bottom fermenting yeast. *Saccharomyces pastorianus*, *Yeast* 14: 923-933.

Taylor, M.W., Tsai, P., Anfang, N., Ross, H.A. and Goddard, M.R. (2014). Pyrosequencing reveals regional differences in fruit-associated fungal communities. *Environmental Microbiology* 16(9): 2848-2858.

Tesfaye, W., Morales, M.L., García-Parrilla, M.C. and Troncoso, A.M. (2002). Wine vinegar: Technology, authenticity and quality evaluation. *Trends in Food Science and Technology* 13: 12-21.

Tholozan, J.L., Cappelier, J.M., Tissier, J.P., Delattre, G. and Federighi, M. (1999). Physiological characterisation of viable-but-non-culturable *Campylobacter jejuni* cells. *Applied and Environmental Microbiology* 65: 1110-1116.

Torija, M.J., Rozès, N., Poblet, M., Guillamón, J.M. and Mas, A. (2001). Yeast population dynamics in spontaneous fermentations: Comparison between two different wine-producing areas over a period of three years. *Antonie van Leeuwenhoek* 79: 345-352.

Torija, M.J., Mateo, E., Guillamón, J.M. and Mas, A. (2010). Identification and quantification of acetic acid bacteria in wine and vinegar by TaqMan-MGB probes. *Food Microbiology* 27: 257-265.

Trček, J., Mahnič, A. and Rupnik, M. (2016). Diversity of the microbiota involved in wine and organic apple cider submerged vinegar production as revealed by DHPLC analysis and next-generation sequencing. *International Journal of Food Microbiology* 223: 57-62.

Trček, J. and Teuber, M. (2002). Genetic restriction analysis of the 16S–23S rDNA internal transcribed spacer regions of the acetic acid bacteria. *FEMS Microbiology Letters* 19: 69-75.

Valera, M.J., Federico Laich, F., Sara, S., González, S.S., Torija, M.J., Mateo, E. and Mas, A. (2011). Diversity of acetic acid bacteria present in healthy grapes from the Canary Islands. *International Journal of Food Microbiology* 151: 105-112.

Valera, M.J., Torija, M.J. and Mas, A. (2017). Detrimental effects of acetic acid bacteria in foods. Chap. 2. pp. 299-320. *In*: I.Y. Sengun (Ed.). Acetic Acid Bacteria: Fundamentals and Food Applications. CRC Press. Taylor & Francis Group, Boca Raton.

Van Zandycke, S.M. (2003). Determination of yeast viability using fluorophores. *Journal of the American Society of Brewing Chemists* 61: 15-22.

Vegas, C., Mateo, E., González, A., Jara, C., Guillamon, J.M., Poblet, M., Torija, M.J. and Mas, A. (2010). Population dynamics of acetic acid bacteria during traditional wine vinegar production. *International Journal of Food Microbiology* 138: 130-136.

Vegas, C., González, A., Mateo, E., Mas, A., Poblet, M. and Torija, M.J. (2013). Evaluation of representativity of the acetic acid bacteria species identified by culture-dependent method during a traditional wine vinegar production. *Food Research International* 51: 404-411.

Viana, R.O., Magalhães-Guedes, K.T., Braga Jr., R.A., Dias, D.R. and Schwan, R.F. (2017). Fermentation process for production of apple-based kefir vinegar: Microbiological, chemical and sensory analysis. *Brazilian Journal of Microbiology* 48: 592-601.

Vigentini, I., Picozzi, C., Tirelli, A., Giugni, A. and Foschino, R. (2009). Survey on indigenous *Oenococcus oeni* strains isolated from red wines of Valtellina, a cold climate wine-growing Italian area. *International Journal of Food Microbiology* 136: 123-128.

Wang, C. and Liu, Y. (2013). Dynamic study of yeast species and *Saccharomyces cerevisiae* strains during the spontaneous fermentations of Muscat Blanc in Jingyang, China. *Food Microbiology* 33: 172-177.

Wang, C., Esteve-Zarzoso, B. and Mas, A. (2014). Monitoring of *Saccharomyces cerevisiae*, *Hanseniaspora uvarum*, and *Starmarella bacillaris* (synonim *Candida zemplinina*) populations during alcoholic fermentation by fluorescence *in situ* hybridisation. *International Journal of Food Microbiology* 191: 1-9.

Wang, C., Mas, A. and Esteve-Zarzoso, B. (2015). Interaction between *Saccharomyces cerevisiae* and *Hanseniaspora uvarum* during alcoholic fermentation. *International Journal of Food Microbiology* 206: 67-74.

Wang, C., Mas, A. and Esteve-Zarzoso, B. (2016). The Interaction between *Saccharomyces cerevisiae* and non-*Saccharomyces* yeast during alcoholic fermentation is species and strain specific. *Frontiers in Microbiology* 7: 502.

Wu, J.J., Mac, Y.K., Zhang, F.F. and Chen, F.S. (2012). Biodiversity of yeasts, lactic acid bacteria and acetic acid bacteria in the fermentation of 'Shanxi aged vinegar', a traditional Chinese vinegar. *Food Microbiology* 30: 289-297.

Xu, L., Huang, Z., Xiaojun, Z., Li, Q.Z., Lu, Z., Shi, J., Xu, Z.Z. and Ma, Y. (2011). Monitoring the microbial community during solid-state acetic acid fermentation of Zhenjiang aromatic vinegar. *Food Microbiology* 28: 1175-1181.

Xufre, A., Albergaria, H., Inacio, J., Spencer-Martins, I. and Girio, F. (2006). Application of fluorescence *in situ* hybridisation (FISH) to the analysis of yeast population dynamics in winery and laboratory grape must fermentations. *International Journal of Food Microbiology* 108: 376-384.

Yamada, Y. and Yukphan, P. (2008). Genera and species in acetic acid bacteria. *International Journal of Food Microbiology* 125: 15-24.

Yaron, S. and Matthews, K. (2002). A reverse transcriptase-polymerase chain reaction assay for detection of viable *Escherichia coli* O157:H7: investigation of specific target genes. *Journal of Applied Microbiology* 92: 633-640.

Zohre, D.E. and Erten, H. (2002). The influence of *Kloeckera apiculata* and *Candida pulcherrima* on wine fermentation. *Process Biochemistry* 38: 319-324.

# 10 Wine Fermentation Microbiome: An Overview of Yeasts' Ecology

**Cátia Carvalho Pinto[1*], Ana Catarina Batista Gomes[1,2] and João Salvador Simões[1]**

[1] Genomics Unit, Biocant - Biotechnology Innovation Center, Cantanhede, Portugal

[2] CNC-UC, Center for Neurosciences and Cell Biology – University of Coimbra, Portugal

## 1. Introduction

Viticulture and the winemaking process is an ancient art, remarkably related with cultural and social lifestyles. Nowadays, these activities are of great economic importance in different countries of the world. Globally, it was estimated that the area under viticulture was of 7.6 mha in 2017, in which Spain (14 per cent), China (11 per cent), France (10 per cent), Italy (9 per cent) and Turkey (7 per cent) represented 50 per cent of the world vineyard (OIV, 2017). The production of grapes reached 75.8 million tons and, among them, 47 per cent of this production is exclusive for wine production. Forecasts for 2017 indicated that the world wine production reached 246.7 mhl, excluding juice and must products, which represents a fall of 8.2 per cent compared with 2016. This decrease was due to unfavourable climatic conditions. Italy, France, Spain, USA and Australia are the four biggest wine producers (OIV, 2017).

Wine fermentation is a natural and spontaneously biological and dynamic process driven by a succession of indigenous microorganisms, such as different species of yeasts, filamentous fungi and bacteria, which interact between them and may have a direct or indirect role in the wine fermentation process (Fleet *et al.*, 1984; Pretorius, 2000; Fleet, 2003; Wang *et al.*, 2016). Among this microbial consortium, wine yeasts are of utmost interest as they are primarily responsible for initiating wine fermentation and, thereby the alcoholic fermentation (AF). Among them, *Saccharomyces cerevisiae* is responsible for fermenting the grape juice into wine, while lactic acid bacteria (LAB) are responsible for malo-lactic fermentation (MLF) (Fleet, 1993). However, several other microorganisms are present in grape berries until wine fermentation and, some of them, have an important oenological significance.

Given particularly the grape berries, these harbour a myriad of microbiomes. Indeed, it was reported in literature that there has been isolation and identification of 93 different yeast species belonging to 30 different genera, from 49 different grape varieties growing across 22 countries (Barata *et al.*, 2008; Barata *et al.*, 2012a; Kassemeyer and Berkelmann-Löhnertz, 2009). The predominant yeasts on the surface of grape berries belong to the genus *Hanseniaspora*, accounting for 50-75 per cent of total yeast population (Pretorius *et al.*, 1999; Romano *et al.*, 2006). Additionally, species of the genera *Candida – Hansenula, Kluyveromyces, Metschnikowia, Pichia, Rhodotorula* and *Torulaspora* are present in lower amounts (Fleet and Heard, 1993) and fermentative species of *Saccharomyces* are present in extremely low numbers (Martini *et al.*, 1996) (Table 1). Although great yeast species diversity is found in grape berries, the population densities are low on immature grapes ($10^1$-$10^3$ CFU/g), increasing at harvest time (to $10^3$-$10^6$ UFC/g) as a consequence of the increase in sugar content during the maturation process (Jolly *et al.*, 2003; Prakitchaiwattana *et al.*, 2004; Renouf *et al.*, 2005; Raspor *et al.*, 2006; Barata *et al.*, 2012a; Setati *et al.*, 2012). Moreover, this population dynamics may be influenced by several other factors, such as, grape berries condition and grape vine varieties, microclimatic and climatic conditions, animal transportation and/or agricultural practices (Bokulich *et al.*, 2014; Pinto *et al.*, 2014; Liu *et al.*, 2015).

Regarding the wine fermentative ecosystem, this is highly complex and is constituted by a vast diversity of microorganisms, whose origin is deeply related with grape berries' surface and winery environment (Sabate *et al.*, 2002; Bokulich *et al.*, 2014; Pinto *et al.*, 2015). Among them, microorganisms, such as *Aureobasidium. pullulans* or other yeasts as *Rhodotorula, Pichia, Candida, Metschnikowia, Hanseniaspora, Lachancea* and *Torulaspora* are present (Table 1). Although, these

*Corresponding author: catia.pinto@biocant.pt

**Table 1.** General Overview of the Yeast Communities Associated with Wine Fermentation

| *Type of approach* | *Wine appellation* | *Type of wine* | *Grape berries* | *Grape musts/Grape juice* | *Alcoholic fermentation* | | *Malolactic fermentation* | | *References* |
|---|---|---|---|---|---|---|---|---|---|
| | | | | | *Early stages* | *Final stages* | *Early stages* | *Final stages/Wine* | |
| Dependent - approach | Bourdeaux wine | Red wine | – | **Dominant:** *Kloeckera apiculata, Torulopsis stellata and Saccharomyces cerevisiae* | *Pichia, Rhodotorula, S. fermentati, K. apiculata and T. stellata* | *S. cerevisiae* | *S. cerevisiae and Pichia membranae faciens* | – | Fleet *et al.*, 1984 |
| | | | | **Lower levels:** *Pichia terricola, Pichia kudriavzevii, Rhodotorula, Pichia kluyveri, Saccharomyces fermentati and Rhodotorula glutinis* | | | | | |
| | | White wine | – | **Dominant:** *Hanseniaspora uvarum, Kloeckera apiculata, T. stellata and Saccharomyces cerevisiae* | *S. cerevisiae, T. stellata, H. uvarum, M. pulcherrima and C. krusei* | *S. cerevisiae and T. stellata* | – | | |
| | | | | **Lower levels:** *Metschnikowia pulcherrima, C. krusei, Rhodotorula graminis and Saccharomyces krusei* | | | | | |
| Dependent-approach | General overview | – | **Oxidative microorganisms:** *Filobasidium, Cryptococcus, Rhodotorula, Aureobasidium pullulans* **Oxidative and/or weakly** | **Occasional:** *Filobasidium, Cryptococcus, Rhodotorula* **Present:** *Aureobasidium pullulans, Hanseniaspora uvarum,* | **Oxidative or weakly fermentative:** *Hanseniaspora/ Kloeckera (H. uvarum/ K. apiculata), Candida* | **Oxidative or weakly fermentative:** *Candida (C. stellata or C. zemplinina), Pichia (P. membranifaciens, P. guilliermondii)* | *Oenococcus* sp., *Lactobacillus* spp., *Pediococcus* sp. | | Barata *et al.*, 2012a |

(Contd.)

**Table 1.** (*Contd.*)

| *Type of approach* | *Wine appellation* | *Type of wine* | *Grape berries* | *Grape musts/Grape juice* | *Alcoholic fermentation* | | *Malolactic fermentation* | | *References* |
|---|---|---|---|---|---|---|---|---|---|
| | | | | | *Early stages* | *Final stages* | *Early stages* | *Final stages/Wine* | |
| | | | **fermentative:** *Hanseniaspora uvarum, Kloeckera apiculata, Candida (C. stellata, C. zemplinina, Zygoascushellenicus, C. steatolytica), Metschnikowia (M. pulcherrima), Pichia (P. anomala, P.membranifaciens, P. guilliermondii), Debaromyces hansenii, Lachancea thermotolerans Fermentative: Torulaspora (T. delbrueckii), Zygosaccharomyces(Z. bailii, Z. bisporus), Dekkera/ Brettanomyces (D. bruxellensis), Saccharomyces (S. cerevisiae, S. bayanus, S. pastorianus), Schizosaccharomyces (Sc. pombe), Saccharomycodes (S. ludwigii)* LAB: *Oenoccocus* sp., *Lactobacillus* spp., *Pediococcus* spp., *Weisella* spp. | *Kloeckera apiculata, Candida (C. stellata, C. zemplinina, Zygoascushellenicus, C. steatolytica, T. delbrueckii, Z. rouxii* | *(C. stellata or C. zemplinina, Zygoascushellenicus/ C. steatolytica), Metschnikowia (M. pulcherrima), Pichia (P. anomala, P.membranifaciens, P. guilliermondii), Debaromyces (D. hansenii), Lachancea (L. thermotolerans, L. fermentati). Fermentative: Torulaspora (T. delbrueckii), Zygosaccharomyces (Z. bailii, Z. bisporus), Saccharomyces (S. cerevisiae, S. bayanus, S. paradoxus, S. pastorianus), Schizosaccharomyces (Sc. pombe), Saccharomycodes (S. ludwigii)* | | | | |
| Dependent-approach | General overview | – | **Dominant:** *Hanseniaspora uvarum, Kloeckera apiculata,* accounting 50-75% of the total yeast population **Others:** *Candida (C. stellata and C. pulcherrima), Brettanomyces, Cryptococcus, Kluyveromyces, Pichia, Rhodotorula* **Rare**: *S. cerevisiae* | – | **Dominant:** *Kloeckera, Hanseniaspora and Candida* **Followed by:** *Metschnikowia, Pichia* | **Dominant:** *Saccharomyces cerevisiae* **Others:** *Brettanomyces, Kluyveromyces, Schizosaccharomyces, Torulaspora, Zygosaccharomyces* | – | – | Pretorius *et al.*, 1999 |

*(Contd.)*

| Independent-approach | Portuguese wine appellations | – | – | **Dominant:** *Aureobasidium (A. pullulans), Rhodotorula (R. nothofagi), Hanseniaspora (H. uvarum), Lachancea (L. thermotolerans)* **Others:** *Saccharomyces, Botryotinia, Alternaria, Aspergillus, Metschnikowia, Filobasidiella, Candida* | **Dominant:** *Saccharomyces (S. cerevisiae), Hanseniaspora, Lachancea* **Others:** *Metschnikowia (M. pulcherrima and M. viticola), Torulaspora (T. delbrueckii), Schizosaccharomyces (S. japonicus), Candida (C. zemplinina), Issatchenkia (I. terricola), Pichia (P. kluyveri and P. kudriavzevii)* | *Saccharomyces, Hanseniaspora, Torulaspora (T. delbrueckii)* | – | – | Pinto *et al*., 2015 |
|---|---|---|---|---|---|---|---|---|---|
| Independent-approach | Growing regions of California | – | – | *Cladosporium sp., Botryotinia fuckeliana, Penicillium* spp., *Davidiella tassiana, Aureobasidium pullulans, S. cerevisiae, Hanseniaspora uvarum, Candida zemplinina* | – | – | – | – | Bokulich *et al*., 2013 |
| Independent-approach | Oakville, Napa Country, CA | Cabernet Sauvignon | – | **Dominant:** *Botryotinia fuckeliana, Cladosporium, S. cerevisiae* | **Dominant:** *Botryotinia fuckeliana, S. cerevisiae, Cladosporium* | **Dominant:** *S. cerevisiae, Botryotinia fuckeliana, Candida zemplinina* | **Dominant:** *Cladosporium, S. cerevisiae, Botryotinia fuckeliana, Candida zemplinina,* **Others:** *Cryptococcus* | **Dominant:** *Cladosporium, Botryotinia fuckeliana, S. cerevisiae, Candida zemplinina,* **Others:** *Cryptococcus, Sporobolomyces, Cryptococcus macerans, Davidiella* | Bokulich *et al*,. 2016 |

microorganisms can achieve a high population density at the start of fermentation, they decrease or even die off as soon as fermentation progresses (Fleet, 2003). Furthermore, some species, such as *Hanseniaspora uvarum* and *Starmerella bacillaris*, grown to a high density of $10^5$-$10^7$ cells/ml, dominate the other non-*Saccharomyces* species (Wang *et al.*, 2015). However, these species are soon replaced by *S. cerevisiae*, that become dominant or the only species occurring at the late stage of fermentation (Wang *et al.*, 2016). Overall, wine fermentation is started by yeasts of the genera *Kloeckera, Hanseniaspora* and *Candida*, and, to a lesser extent by *Metschnikowia* and *Pichia* (Pretorius *et al.*, 1999), which will then be replaced by *S. cerevisiae*. Though, other yeasts may be still present, such as, *Brettanomyces, Kluyveromyces, Schizosaccharomyces, Torulaspora* and *Zygosaccharomyces* (Pretorius *et al.*, 1999) (Table 1). It is assumed that non-*Saccharomyces* yeasts perish because these cells lose their ability to divide and grow. This loss of cultivability may arise from several factors, such as excreted compounds in the interaction between *Saccharomyces* and non-*Saccharomyces* yeasts (Ciani and Comitini, 2015; Liu *et al.*, 2015; Albergaria and Arneborg, 2016), or insufficient adaptability to environmental changes during the fermentations process. Indeed, alcoholic fermentation is characterised by a selective medium for microbial growth which is conducted by the low pH values (3.0-3.5), high initial sugar contents (140-260 g/l), low oxygen availability (Hansen *et al.*, 2001), low nitrogen concentrations (150-200 mg/l) (Monteiro and Bisson, 1991), high levels of ethanol (10-14 per cent v/v) (Fleet, 2003), organic acids, extrinsic factors such as $SO_2$ and other fermentative metabolites (Albergaria and Arneborg, 2016). Overall, during wine fermentation, yeasts consume the sugar from grape musts of 140-260 g/l to residual concentrations lower than 2 g/l, and produce fermentative metabolites and ethanol (10-14 per cent v/v) (Boulton *et al.*, 1996).

After alcoholic fermentation, the MLF can occur. The MLF is carried out by LAB and consists of decarboxylation of L-malic acid to L-lactic acid and $CO_2$ (Costello *et al.*, 2015). The LAB involved belong to the genera *Oenoccocus* sp., *Lactobacillus* sp., *Pediococcus* spp. and *Weissella* spp. (Barata *et al.*, 2012) (Table 1). The MLF process is associated with deacidification of wine and with enrichment of its sensory composition and is crucial as it confers a biological stability to the final product and may even improve the wine quality (Bartowsky *et al.*, 2009).

The knowledge of wine-associated microorganisms was initially acquired by performing a culture-dependent approach. Despite its importance, this technique is time-consuming and limits the analysis of all the microbial biodiversity, as it depends on the isolation of microorganisms, which may limit the growth of certain species due to microbial antibiosis and competition. Furthermore, microbial communities, that are naturally less abundant in Nature, will be difficult to identify and others that are not able to grow on standard microbiological media cannot be identified (Cocolin *et al.*, 2011). Advances of high-throughput techniques, which do not require the isolation of microorganisms, revolutionised our knowledge of this microbial ecology by allowing a more sensitive detection and identification of these microorganisms across wine fermentation (Pace, 1997; Cocolin *et al.*, 2011; Bokulich *et al.*, 2013; Pinto *et al.*, 2015; Bokulich *et al.*, 2016a).

The microbial biodiversity associated with wine fermentation is affected by several factors, namely vineyards, grapevine varieties, climatic conditions over different regions and countries, agricultural practices, regulatory constraints and even the methodology applied for microbiome analysis (dependent *vs* independent-approach) (Table 2). Furthermore, several of these variables are not independent and could be grouped by effects (e.g. climatic conditions and cultivar). The grapes-associated microbial biogeography is, however non-randomly associated with regional, varietal and climatic factors across multi-scale viticulture zones (Bokulich *et al.*, 2013; Bokulich *et al.*, 2014). Additionally, yeast biodiversity is affected by intra-vineyard spatial fluctuations, such as the observed heterogeneity of grape samples harvested from single vineyards at the same stage of ripeness (Setati *et al.*, 2012).

## 2. Yeast Ecology on Grape Berries

Grape berries harbour a complex microbial consortium, including bacteria, filamentous fungi and yeasts, with different characteristics and effects on wine production. Furthermore, grape berries are the primary source of microbial biodiversity at start of fermentationt (Barata *et al.*, 2012a), though the origin of this microbial community is still controversial (Belda *et al.*, 2017). As previously referred, several factors may have an impact on yeast population dynamics, such as condition of grape berries and grape vine varieties, microclimatic and climatic conditions, animal transportation and/or agricultural practices (Bokulich *et al*., 2014; Pinto *et al*., 2014; Liu *et al*., 2015) (Table 2).

**Table 2.** Summary of Abiotic and Biotic Factors that Influence Microbiome Diversity

| *Medium* | *Abiotic* | *Biotic* |
|---|---|---|
| Grapes berries | Agricultural practice | Grape vine variety |
| | Climatic condition | Grape berries conditions |
| | | Animal transportation |
| Grape Must Fermentation | Temperature | Microbiome competition for space |
| | Nitrogen content | Microbiome competition for nutrients |
| | Oxygen content | Production of killer factor |
| | Molecular sulfur dioxide | Cell-Cell contact |
| | pH | |
| | Ethanol content | |
| | Osmotic pressure | |
| | Molecular sulfur dioxide | |

Concerning the condition of grape berries, their characteristics change throughout the maturation, increase in surface area and size of each berry, along with and increase in nutrients availability, sugar concentration and decrease in acidity (Combina *et al*., 2005; Cadez *et al*., 2010). Furthermore, the health status of berries and grape skin integrity also influence the yeasts population. Indeed, an infection by the mold *Botrytis cinerea*, can penetrate the grape surface and release nutrients, changing grape surface conditions and influencing the microbial flora (Nisiotou *et al*., 2007; Barata *et al*., 2008). Additionally, yeasts of genus *Metschnikowia* can develop in berries affected by *B. cinerea*, inhibiting the development of other microorganisms (yeasts, filamentous fungi and bacteria), through a mechanism of iron sequestration (Sipiczki, 2006). Grape variety and berry colour also influence yeast population diversity (Sabate *et al*., 2002; Renouf *et al*., 2005; Nisiotou and Nychas, 2007). Briefly, a previous study showed that the genera most frequently isolated from Grenache grapes was *Cryptococcus,* whereas *Hanseniaspora* was frequently isolated from Carignan (Sabate *et al*., 2002). Regarding the influence of climatic and micro-climatic conditions on yeast population, results are contradictory: several authors point to a higher yeast count with high rainfall (Longo *et al*., 1991; Combina *et al*., 2005; Cadez *et al*., 2010), though the opposite is also reported (Rementeria *et al*., 2003). Furthermore, large-scale investigations did not find any relation between climatic conditions and yeast biodiversity (Barata *et al*., 2012a). The contribution of animals to yeast population variability on berries is based on yeast transport. Indeed, association was reported between yeasts and insects, such as, bees, social wasps and *Drosophila* (Stefanini *et al*., 2012; Liu *et al*., 2015). Additionality, it was suggested that migratory birds may serve as vectors of *S. cerevisiae* cell dispersion (Francesca *et al*., 2010). Moreover, agricultural practices may also influence yeast population diversity. Studies show that organic and conventional vineyards have different yeasts richness (Comitini *et al*., 2008; Cadez *et al*., 2010; Cordero-Bueso *et al*., 2011; Milanović *et al*., 2013; Pinto *et al*., 2014). Also, the application of chemical treatments in vineyards may have an effect on yeasts dynamics (Setati *et al*., 2012; Pinto *et al*., 2014).

## 3. Yeast Ecology on Wine Fermentation

Wine fermentation is a microbial dynamic process, where both environmental and other wine-associated microorganisms are present and actively interact. With the rapid evolution of fermentation, death and disappearance of environmental microorganisms occur, which is consistent with their weak fermentative metabolism and sensitivity to the increase in ethanol levels (Fleet *et al*., 1984). Indeed, these microbial interactions are influenced by several factors, namely abiotic (pH, ethanol, temperature, osmotic pressure, nitrogen, molecular sulphur dioxide, nutrient availability, oxygen) and biotic factors (microorganisms, killer factors or grape variety) (Ciani *et al.*, 2016) (Table 2).

Given the abiotic factors, the increase of ethanol in must is one of the major factors affecting yeast population diversity and the main producer of this compound is *S. cerevisiae*. During fermentation, ethanol concentration increases and yeasts with lower tolerance to ethanol perish, leading do a decline in yeast biodiversity (Beltran *et al.*, 2002; Constantí *et al.*, 1997; Combina *et al.*, 2005). Among them, non-*saccharomyces* yeasts are most susceptible to this phenomenon and although ethanol tolerance could vary greatly within a specific species, some indigenous yeasts from genera *Hanseniaspora*, *Candida*, *Pichia*, *Kluyveromyces*, *Metschnikowia* and *Issatchenkia*, usually do not survive above ethanol concentration ranging from 3-10 per cent (v/v) (Jolly *et al.*, 2014). Nonetheless, several non-*Saccharomyces* yeasts with higher ethanol tolerance can endure until the end of fermentation and these are *Torulaspora delbrueckii*, *Candida zemplinina*, *Zygosaccharomyces bailii*, *Schizosaccharomyces pombe* and *Pichia spp.* (Pina *et al.*, 2004 ; Combina *et al.*, 2005 ; Santos *et al.*, 2008 ; Jolly *et al.*, 2014).

Temperature is also an environmental factor that contributes to yeast population biodiversity during wine fermentation. Elevated temperatures in synergy with increasing ethanol concentration affect the cell membrane permeability and integrity. Additionally, when temperature is below 15°C, ethanol will decrease its selective pressure, which will not be advantageous to *S. cerevisiae* and, thus can result in persistence or dominance of non-*Saccharomyces* yeasts (Gao and Fleet, 1988; Ciani and Comitini, 2006). Gobbi showed that the antagonist effect between a co-culture of *S. cerevisiae* and *L. thermotolerans* was the temperature of the fermentation (Gobbi *et al.*, 2013).

Other factors that modulate yeast population biodiversity are the availability of resources during the wine-fermentation process. The anti-microbial activity of *M. pulcherrima* arises due to the production of pulcherriminic acid (precursor of pulcherrimin pigment) that depletes iron from the surrounding medium, making it unavailable to other yeasts (Sipiczki, 2006; Türkel and Ener, 2009; Oro *et al.*, 2014) such as, *Hanseniaspora, Brettanomyces/Dekkera* and *Pichia* genera (Oro *et al.*, 2014).

Oxygen is another compound that influences population dynamics through selective action among the various yeasts species (Ciani and Comitini, 2006; Brandam *et al.*, 2013; Jolly *et al.*, 2014; Taillandier *et al.*, 2014). During wine production, oxygen levels are low, which negatively impacts the aerobic yeasts, such as *Candida* spp., *Debaryomyces* spp., *Pichia* spp., *Rhodotorula* spp. and *Cryptococcus albidus* (Combina *et al.*, 2005; Jolly *et al.*, 2014). On the other hand, this low oxygen level will benefit the growth of anaerobic yeasts, such as *S. cerevisiae* (Hansen *et al.*, 2001). Remarkably, under aerobic conditions, *S. cerevisiae* can produce ethanol (crabtree effect) (Pronk *et al.*, 1996), with fast rates of sugar consumption and ethanol production (Visser *et al.*, 1990).

In wine fermentation, assimilable nitrogen and vitamins modulate yeast biodiversity. It was reported that the growth of *T. delbrueckii* during the first 48 hours can deplete nitrogen, blocking *S. cerevisiae* growth and leading to sluggish fermentation (Taillandier *et al.*, 2014). A similar effect was observed with a strain of *Kloeckeraapiculata,* that reduces thiamine availability, which then affects the *S. cerevisiae* growth (Mortimer, 2000). Furthermore, it was reported that different nitrogen sources have different impacts on the growth and fermentation behaviour of *S. cerevisiae, T. delbrueckii, Lb. thermotolerans, H. uvarum* and *M. pulcherrima* (Kemsawasd *et al.*, 2015).

Regarding the biotic factors, interactions between yeast strains also influence the yeast population diversity. Thus, one extreme negative interaction is the killer phenomenon, that consists in the production of extracellular proteins and glycoproteins by strains (killer yeast) inhibiting the development or even killing other strains (sensitive yeast) (Mehlomakulu *et al.*, 2015; Ciani *et al.*,2016) (Table 3). This phenomenon is extensively described in *S. cerevisiae* (Van Vuuren and Jacobs, 1992; Musmanno *et al.*, 1999; Gutiérrez *et al.*, 2001,), which influences succession of yeast strains during fermentation. It was reported that an initial proportion of 2-6 per cent of killer yeasts is enough to dominate and kill sensitive strains during wine fermentation (Pérez *et al.*, 2001). Although, killer toxins produced by *S. cerevisiae* are active only against strains of the same species, it was reported that the 2-10 KDa protein fraction of *S. cerevisiae* CCMI 885 supernatants expresses a fungicidal effect on *H. guilliermondii* and a fungistatic effect on *K. marxianus*, *K. thermotolerans* and *T. delbrueckii* (Albergaria *et al.*, 2010). The peptides that showed this activity were identified as derived from the glycolytic enzyme, glyceral-dehyde 3-phosphate dehydrogenase (GAPDH) (Branco *et al.*, 2014). Additionally, several non-*Saccharomyces*

**Table 3.** Summary of Killer Yeasts in Wine; Correspondent Killer Toxin Produced and Target-sensitive Yeast

| *Killer yeast* | *Killer toxin* | *Sensitive strain / Target Yeast* | *Application indications* | *References* |
|---|---|---|---|---|
| *Kluyveromyces wickerhamii* | Kwkt | *Dekkera/Brettanomyces* | Anti-Brett activity | Comitini *et al.*, 2004 |
| *Pichia membranifaciens* | PMKT2 | *Dekkera/Brettanomyces* | Anti-Brett activity | Santos *et al.*, 2009 |
| *Saccharomyces cerevisiae* | K2 type | *Saccharomyces cerevisiae* | Control of *S. cerevisiae* wild strains | Shimizu, 1993 |
| *Tetrapisi sporaphaffii* | Kpkt | *Hanseniaspora/Kloeckera* | Control of "apiculate" yeast | Comitini and Ciani, 2010 |
| *Torulaspora delbrueckii* | Kbarr-1 | *S. cerevisiae killer strains Broad anti-Wine* | Broad anti-wine yeast activity | Ramírez *et al.*, 2015 |
| *Torulaspora delbrueckii* | TdKT | *Pichia and Brettanomyces/Dekkera* | Spoilage wine yeasts | Villalba *et al.*, 2016 |
| *Ustilago maydis* | KP6 | *B. bruxellensis* | Anti-Brett activity | Santos *et al.*, 2011 |
| *Wickerhamomyces anomalus* | Pikt | *Dekkera/Brettanomyces* | Anti-Brett activity | Comitini *et al.*, 2004 |

Adapted from Ciani *et al.*, 2016 and Mehlomakulu *et al.*, 2015

yeasts were reported to produce killer toxins against yeasts, namely *K. phaffii* that produces a killer toxin (ZymocinKpKt) active against the genus *Hanseniaspora* (Ciani *et al.*, 2001). *Pichia anomala* and *K. wickerhamii* secrete two toxins, namely KwKt and PIKT that are active against yeasts of *Brettanomyces* genus (Comitini *et al.*, 2004). Also, *Pichiamembranifaciens* produces a toxin (PMKT2) which is active against *B. bruxellensis* (Santos *et al.*, 2009). Furthermore, other compounds formed during wine fermentation and produced by different yeasts could have negative effects against each other, such as short fatty acids, medium-chain fatty acids, acetic acid (including acetic, hexanoic, octanoic, and decanoic acids) and acetaldehyde (Bisson, 1999; Fleet, 2003; Giannattasio *et al.*, 2005; Ivey *et al.*, 2013).

Given the interactions between yeasts, these can also result in commensalism in relations. One of this example occurs when non-*Saccharomyces,* in particular yeasts belonging to *Hanseniaspora* and *Metschnikowia* genera, with high extracellular proteolytic activity, break the peptide bonds of proteins present in the medium, releasing amino acids that can be further used by *S. cerevisiae* (Dizy and Bisson, 2000; Fleet, 2003). Another event consists in the death of yeasts and autolysis process, which result in an increase of amino acids and nutrients. The death and autolysis of non-*Saccharomyces* yeasts after the initial stages of alcohol fermentation, will provide nutrients for *S. cerevisiae.* Subsequently, after alcohol fermentation, *S. cerevisiae* will also perish and release nutrients that will be used by spoilage yeast species, such as *Dekkera/Brettanomyces* (Guilloux-Benatier *et al.*, 2001), that have higher ethanol tolerance than other wild yeasts (Renouf *et al.*, 2007). Yeasts can also produce metabolites that benefit other yeast species. It was reported that mixed culture of *S. cerevisiae* and *S. cerevisiae -S. uvarum* hybrid strain exhibited a much larger population than the sum of the maximum population of the two yeasts growing in pure cultures (Cheraiti *et al.*, 2005). This difference arises from the production of enormous quantities of acetaldehyde by the hybrid strain that is used by *S. cerevisiae* strain, lowering its cellular NAD(P)H levels, changing its redox potential and increasing biomass and fermentation rate (Cheraiti *et al.*, 2005).

## 4. The Relevance of Yeast on Wine Sensory Properties

It is well recognised that the microbial consortium associated with both grapes and wine musts, and its metabolic and physiological characteristics, have a prominent role in the wine fermentation process as they impact the wine colour, aroma and flavour and therefore, wine quality and value. In the case of wine yeasts, this is no exception and is of utmost importance and, as pointed by Varela, non-*Saccharomyces* yeasts greatly influence wine composition and/or wine sensory properties (Varela, 2016) (Table 4). However, the organoleptic properties of wine are not only determined by its microbial ecology, but also by other factors, such as grape variety, geographic location, soil, viticulture management practices, winemaking practices or cellar technology (Fleet, 2003).

One class of molecules with major importance in wine are phenols, especially in red wines. The phenolic content is affected by grape variety, maceration temperature, length of grape pomace contact, yeast metabolism and other vinification conditions (Caridi *et al.*, 2004). Furthermore, the type, quality and quantity of phenolics, namely anthocyanins, flavonols, catechins and other flavonoids, greatly contribute to wine sensory characteristics, such as colour and astringency (Caridi *et al.*, 2004). Given anthocyanins, these are responsible for the red colour of wine, and are extracted from grapes during maceration (Suárez-Lepe and Morata, 2012). Furthermore, yeasts influence wine colour, by adsorption of anthocyanins through the yeast cell wall, or by the production of metabolites, namely acetaldehyde and pyruvic acid, which reacts with several phenolics, and stabiliszes colour pigments (Caridi *et al.*, 2004). Another colour stabilisation process carried out by yeasts, involves the hydroxycinnamate decarboxylase (HCDC) activity. Remarkably, one of the yeasts with higher HCDC activity is *P. guillermondii* (Benito *et al.*, 2011).

Aroma is also an important wine characteristic. During wine fermentation, yeasts influence wine aroma by producing higher alcohols and esters. Fusel alcohols and fusel acids are formed through the catabolism of sulphur containing amino acid (methionine), branched-chain amino acids (leucine, valine, and isoleucine) and aromatic amino acids (phenylalanine, tyrosine, and trytophan) (Hazelwood *et al.*, 2008). Additionally, these higher alcohols are involved in the 'winery' aroma of fermented musts

**Table 4.** Summary of Studies Evaluating the Influence of non-Saccharomyces Yeasts on Wine Composition and Sensory Properties (Adapted from Varela, 2016)

| *Non-Saccharomyces yeast* | *Impact on physical properties, chemical composition or sensory attributes* | *References* |
|---|---|---|
| *Candida sake* | Increased concentrations of terpenes and higher alcohols | Maturano *et al.*, 2015 |
| *Debaryomyces vanrijiae* | Increased concentrations of esters and fatty acids | Maturano *et al.*, 2015 |
| *Hanseniaspora uvarum* | Increased concentrations of several C13-norisoprenoids and some terpenes, Increased 'tropical fruit', 'berry', 'floral', and 'nut aroma' characters | Hu *et al.*, 2016 |
| | Increased concentration of acetate esters, esters of MCFAs, isoamyl alcohol, 2-phenylethanol and α-terpineol | Tristezza *et al.*, 2016 |
| *Hanseniaspora vineae* | Increased 'white prune', 'pear', 'citric fruits', and 'honey' attributes | Medina *et al.*, 2013 |
| | Increased concentration of phenyl-ethyl acetate, ethyl lactate and α-terpineol | Lleixa *et al.*, 2016 |
| | Increased 'floral' attribute | |
| *Kazachstania gamospora* | Increased 'flavour persistence', 'flavour intensity', 'floral' and several 'fruity' attributes | Dashko *et al.*, 2015 |
| *Lachancea thermotolerans* | Increased 'peach/apricot' characters | Benito *et al.*, 2015 |
| | Enhanced acidity and increased 'spicy' attributes | Gobbi *et al.*, 2013 |
| *Metschnikowia pulcherrima* | Increased concentration of mannoproteins | Domizio *et al.*, 2014 |
| | Increased 'citrus/grape fruit' and 'pear' attributes | Benito *et al.*, 2015 |
| | Increased foam persistence and 'smoky' and 'flowery' attributes | Gonzalez-Royo *et al.*, 2015 |
| *Pichia fermentans* | Increased concentration of mannoproteins | Domizio *et al.*, 2014 |
| *Pichia kluyveri* | Increased 'peach/apricot' characters | Benito *et al.*, 2015 |
| *Pichia kudriavzevii* | Decreased malic acid concentration, increased 'fruit' and 'cooked pears' characters | del Monaco *et al.*, 2014 |
| *Saccharomycodes ludwigii* | Increased concentration of mannoproteins | Domizio *et al.*, 2014 |
| *Schizosaccharomyces pombe* | Increased formation of vitisins and vinylphenolicpyranoanthocyanin | Loira *et al.*, 2015 |
| *Starmerella bacillaris* | Increased glycerol concentration, greater volatile complexity | Englezos *et al.*, 2016 |
| | Increased glycerol content, 'jam', 'softness', 'structure' and 'persistence' attributes, and decreased 'bitter' characters | Zara *et al.*, 2014 |
| *Torulaspora delbrueckii* | Increased acetate esters and medium-chain fatty acids (MCFAs) | Cordero-Bueso *et al.*, 2013 |
| | Increased concentration of thiols, 3-sulfanylhexan-1-ol (3SH) and 3-sulfanylhexyl acetate (3SHA) | Renault *et al.*, 2016 |
| | Increased concentration of terpenes α-terpineol and linalool | Cus and Jenko, 2013 |

(*Contd.*)

**Table 4.** (Contd.)

| *Non-Saccharomyces yeast* | *Impact on physical properties, chemical composition or sensory attributes* | *References* |
|---|---|---|
| | Increased 'aroma intensity', 'complexity' and 'persistence' | Azzolini *et al.*, 2015 |
| | Increased 'color intensity', 'overall impression' and 'aroma quality' | Belda *et al.*, 2015 |
| | Increased 'complexity' and 'fruity' attributes | Renault *et al.*, 2015 |
| | Increased foamability and foam persistence | Gonzalez-Royo *et al.*, 2015 |
| *Wickerhamomyces anomalus* | Increased concentration of acetate- and ethyl-esters and panel preferences | Izquierdo-Canas *et al.*, 2014 |
| *Zygosaccharomyces bailii* | Increased concentration of ethyl esters, ethyl acetate, ethyl octanoate and ethyl decanoate | Garavaglia *et al.*, 2015 |
| *Zygosaccharomyces kombuchaensis* | Increased 'flavour intensity' and several 'fruity' attributes | Dashko *et al.*, 2015 |
| *Zygotorulaspora florentina* | Increased 'fruity' and 'floral' notes and decreased values for 'astringency' | Lencioni *et al.*, 2016 |

(Lambrechts and Pretorius, 2000) and its production and composition are dependent upon the winemaking techniques, must composition and its initial oxygenation, fermentation conditions and the predominant yeasts strains (Suárez-Lepe and Morata, 2012).

The second most abundant volatile compounds in wine are esters, produced by esterases, alcohol acetyltransferases, and lipases (Sumby *et al.*, 2010), which are responsible for fruity aromas. Yeasts strains produce diverse amounts of esters, reflecting a variety of wine aromas profiles (Lambrechts and Pretorius, 2000; Cadière *et al.*, 2011). Additionally, the production and concentration of esters in wine depend on grape variety and amino acids profiles in grape, as well as on nitrogen content of the musts (Richter *et al.*, 2013). However, during wine production, spoilage microorganisms, such as yeasts, moulds, LAB and AAB, might negatively affect wine quality and even food safety (Pozo-Bayón *et al.*, 2012). Indeed, these microorganisms are of utmost concern as they can increase the acids and ester levels in wines and, thus, produce off flavours (Bartowsky, 2009). Among them, ochratoxin A, ethyl carbamate, and biogenic amines are the compounds that might endanger the consumer's health. Some yeast spoilage microorganisms are *Pichia membranaefaciens*, that has the ability to form pellicle on the surface of bulk wines and produce off-flavours; *Dekkera bruxellensis,* that produce off-flavour normally associated with aromas of barnyard, burnt plastic, wet animal and horse-sweat; and *Zygosaccharomyces bailii,* that induce sediment and cloudiness formation in wine (Barata *et al.*, 2012b). Among bacteria, some microorganisms, as *Oenococcus oeni*, present during the MLF and may have a positive contribution to wine sensory when present at concentrations below 4 $mg.L^{-1}$. However, higher concentrations can give rise to a butterscotch aroma (Bartowsky, 2009). Others, such as *Lactobacillus* sp., *Pediococcus*sp. and all AAB species, are considered as wine spoilage microorganisms (Bartowsky, 2009). The development of these spoilage microorganisms during wine fermentation has traditionally been controlled through the addition of sulphur dioxide; however, this procedure entails health problems for the final consumer and for this reason, other methods are being integrated (Bartowsky, 2009).

After wine fermentation, lees formed by dead microorganism's cells, inorganic compounds and tartaric acid can influence the wine-aging process (Rodrigues *et al.*, 2012). The aging of wines on lees leads to the reduction of bitterness and astringency, enhancement of structure, roundness and body of wine and increase in the complexity and persistence of wine aromatic notes (Del Barrio-Galán *et al.*, 2011). Furthermore, this may improve the colour stability of red wine during aging (Escot *et al.*, 2001). During this process, the dead microorganisms suffer a mechanism called autolysis, where the cell wall gets degraded and several compounds are released, namely amino acids, peptides, mannoproteins, polysaccharides, fatty acids, and lipids, whilch impact the wine characteristics (Tao *et al.*, 2014). Additionally, lees cell walls can also decrease the content of volatile compounds, namely 4-ethylphenol and 4-ethylguaiacol, throughout sorption (Pozo-Bayón *et al.*, 2009; Chassagne *et al.*, 2005). The molecules with amphilic character and polarity, and responsible for the adsorption phenomenon are 4-propylguaiacol, 4-methylguaiacol and 5-cell wall mannoproteins (Tao *et al.*, 2014). Indeed, mannoproteins are the mainly responsible for wine astringency reduction by retention of tannins (Rodrigues *et al.*, 2012). Considering that cell wall properties depend on yeasts species, wine aged over lees with *Schizosaccharomyces* have scored well in sensorial tests, when compared to wines aged over lees with *Saccharomyces* (Suárez-Lepe and Morata, 2012).

## 5. Spontaneous *vs.* Inoculated Wine Fermentations

Contrary to spontaneous wine fermentation, inoculated fermentation is induced through inoculation of selected strains of yeasts and even LAB at AF and MLF, respectively (Lonvaud-Funel, 1999). In these cases, these microorganisms interact with the indigenous microbial communities of grape musts, though it is expected that the natural flora still have an effect on wine fermentation. The application of starter cultures is a general practice in the fermentation industry, as it normally guarantees the consistency of the final product in response to large-scale wine production. Generally, the starter cultures are selected single-strains of *S. cerevisiae*, which are added to grapes after crushing. However, the inoculation of musts by using selected *Saccharomyces* strains does not ensure their dominance at the end of fermentation (Capece *et al.*, 2010). Because commercial strains will compete not only with non-*Saccharomyces* yeasts, but also with indigenous *S. cerevisiae* strains, which theoretically are well adapted to must conditions, this will take several days for commercial strains to take over fermentation (Barrajón *et al.*, 2011). Moreover, it is clear that both spontaneous and inoculated wine fermentations are affected by the microbial biodiversity associated with vineyard (grape berries) and winery (Pretorius *et al.*, 1999).

In the last few years, some *boutique* wineries have become more and more aware about the importance of the local microflora on winemaking, as these microorganisms can be regarded as a distinctive feature and are increasingly promoting spontaneous fermentation to produce new wine styles with more complexity, stylistic distinction and vintage variability (Pretorius, 2000; Steensels and Verstrepen, 2014). Thus, a better knowledge of the indigenous microbial communities across spontaneous fermentation will not only improve the understanding of their role in fermentation, but also their contribution in the production of premium wines with specific characteristics, aromatic and flavour profile.

Currently, there is a rising demand for autochthonous yeast starters, which are potentially adapted to a grape must and reflect the biodiversity of a particular area (Boundy-Mills, 2006; Bokulich *et al.*, 2014). In fact, in the wine industry, temporal and regional variations are accepted and recognised and are being defined in terms of vintage or *terroir* (Boundy-Mills, 2006). Indeed, the successive fermentation of a single food product in a place through long periods of time is a process of domestication (Scott and Sullivan, 2008). Registered appellations are the result of interaction between local yeasts, local soils, weather and local cave conditions, resulting in a distinct regional wine, even when the same grape cultivar is used in different regions (Scott and Sullivan, 2008).

## 6. Molecular Identification of Yeasts

An accurate identification of microorganisms is critical for the study of microbial communities. Classical culture-based methods were allowed to isolate and to identify more than 40 yeast species (Jolly *et al.*, 2014), 50 bacterial species (Barata *et al.*, 2012b) and 70 genera of filamentous fungi (Rousseaux *et al.*, 2014) associated with grapevine and wine fermentation processes. These methods are extremely laborious and dependent on morphological, physiological and biochemical state of cells. In the past, heterogeneous phenotypical results delivered incorrect results or did not permit species definition (Barata *et al.*, 2012b). Furthermore, only species that are able to grow on the culture media and under the cultivation conditions used can be isolated and identified. Additionally, species in low abundance or viable but non-culturable are generally overlooked (Abbasian *et al.*, 2015). Nonetheless, these methods are still significant because the microbial species and strains retrieved by these approaches can be exploited, depending on their biochemical or genetic profiles. Indeed, more than 100 commercial strains of *S. cerevisiae* (Fernández-

Espinar *et al.*, 2001; Guzzon *et al.*, 2014) and, more recently, strains such as *Torulasporadelbrueckii, Metschnikowia pulcherrima, Lachancea thermotolerans* and *Pichia kluyveri* are available for the wine industry (Lu *et al.*, 2016; Padilla *et al.*, 2016).

Currently, the identification and detection of wine-associated microorganisms is mainly based in DNA approaches. Contrary to classical identification, these have high specificity, are of easy application, allow delivery of information on the phylogeny of detected species and are independent of morphological, physiological and biochemical criteria (Boundy-Mills, 2006; Begerow *et al.*, 2010).

A simple and powerful method to identify several yeast species is based on PCR amplification of the ITS region, followed by restriction analysis of amplified products (Guillamón *et al.*, 1998). This approach allows identification of 191 yeasts associated with foodstuffs and beverages (Esteve-Zarzoso *et al.*, 1999; De Llanos Frutos *et al.*, 2004). In this method, the PCR product is digested with the restriction enzymes HaeIII, Hinfl, CfoI and DdeI and compared with known restriction profiles. In case the restriction profiles are absent or not conclusive, the best alternative is to sequence the 28S rDNA (Barata *et al.*, 2012b).

The bacterial small subunit ribosomal RNA gene 16S rRNA, as well as the fungal 5.8S, 26S (Kurtzman and Robnett, 1998) and 18S rRNA (James *et al.*, 1997) are the main target genes used for species identification and for estimating the phylogenetic diversity in microbial communities (Justé *et al.*, 2008; Cocolin *et al.*, 2013; Sun and Liu 2014). Although, 18S rRNA gene (SSU) is used for phylogenetic studies, it is less variable than is homolog 16S rRNA which when used in bacteria, is 26S rRNA gene (LSU) preferred for fungi species identification. Indeed, the DNA sequence availability for D1/D2 region of LSU in DNA databases permits the identification of yeast species when the sequence homology is greater than 99 per cent or when complemented with ITS (Schoch *et al.*, 2012; Kurtzman and Robnett, 1998; Scorzetti *et al.*, 2002).

DNA amplification detects all the present microbial communities from a sample, independent of the microorganisms being active or dormant. Additionally, complementary DNA synthesis by reverse transcription of messenger RNA (mRNA) detects only active microorganisms (Boundy-Mills, 2006).

The identification and quantification of microorganisms can be achieved by Real-Time qPCR. This technique is based on detection and quantification of a fluorescent donor dye emission and the emission intensity is directly proportional to the quantity of PCR product. This technique is highly specific and sensitive but is highly dependent on the correct design of primers and/or probes (Barata *et al.*, 2012b). The main genes or genomic regions selected for these approaches are the ITS and the D1/D2 rDNA regions (Kurtzman and Robnett, 1998), the nuclear gene actin (Daniel, 2003), and the mitochondrial gene COX2 (Belloch *et al.*, 2000; Kurtzman and Robnett, 1998). Previous studies have demonstrated the power of this technique by monitoring *Saccharomyces* and *Hanseniaspora* species during alcoholic ethanol fermentation (Hierro *et al.*, 2006; Hierro *et al.*, 2007). When the nucleotide sequence similarity between species undermines the design of primers, the construction of specific primers may be carried out by using the analysis of random amplified polymorphic DNA (RAPD), through the selection and sequencing of specific bands. The design of primers will then be made for this region (Barata *et al.*, 2012b; Martorell *et al.*, 2005; Salinas *et al.*, 2009).

## 7. Metagenomic Identification of Microbiomes

The deep study of microbial communities' structure and their functional role in ecosystems requires a metagenomics approach. Metagenomics is a culture-independent approach that allows a direct genetic analysis of the collective of genomes within an environmental sample (Thomas *et al.*, 2012). This approach does not require cloning or PCR amplification, permits the study of uncultured organisms (Huson *et al.*, 2007; Simon *et al.*, 2009) and is generally used to study the whole microbial ecosystem, since it permits the identification of the overall population, such as eukaryotic and prokaryotic microorganisms (Illeghems *et al.*, 2012). These technologies rely on a parallel process in which each single DNA fragment is sequenced independently and separated in clonal amplicons for subsequent analysis (Wooley *et al.*, 2010; Diaz-Sanchez *et al.*, 2013). In these methodologies, an uninterrupted operation of scanning process is used to read tens of thousands of matching strands (Schadt *et al.*, 2010). Furthermore, the next-generation sequencing technologies have steadily improved since its release, increasing the read length produced, improving assembly quality by reducing the number of gaps, becoming cost-effective and less labour-intensive (Forde *et al*., 2013). These improvements along with the increasing number of available genomes in public databases greatly contribute to the increase in robustness of this approach (Forde *et al*., 2013). Nonetheless, it should not be assumed that these genomes are representative of the microorganisms present in the ecosystem (Illeghems *et al*., 2012).

To unveil the yeast diversity in the microbial ecosystem, barcoding regions, that are normally present in all species, are used. These regions contain both highly conserved and variable regions fragments that facilitate the PCR primer's design, and allow for species discrimination, respectively (Justé *et al*., 2008; Cocolin *et al*., 2013; Sun and Liu, 2014). In bacteria, the 16S rRNA gene is typically targeted, while in fungi the ITS1, ITS2 and D2 rDNA genes are used (Illeghems *et al*., 2012; Pinto *et al*., 2014; Bokulich *et al*., 2014). Bokulich and Mills (2013) tested several ITS primers, revealing that ITS1 region shows higher levels of taxonomic classification (species and genus), maximised sequence coverage and smallest difference between amplicon lengths of Ascomycota and Basidiomycota phylum. Furthermore, Pinto and colleagues (Pinto *et al*., 2014; Pinto *et al*., 2015) tested the ITS2 region and D2 domain, showing that these two regions have similar taxonomic depth, but share only a portion of the observed OTUs with the ITS region provide a slightly higher coverage.

High throughput-sequencing techniques usually generate immense amounts of sequence data; subsequently, bioinformatics is the only viable option to handle such amount of information (Morgan *et al*., 2017). The analysis of amplicon sequencing data generally involves three basic steps, namely (i) quality trimming and de-noising, (ii) OTU picking/ clustering, and (iii) taxonomic assignment (Morgan *et al*., 2017). Quality trimming consists in the removal of erroneous reads obtained through PCR, sequencing instruments and sequencing reactions (Bokulich *et al*., 2013). Clustering and OTU-picking consist of pairwise comparison of sequencing with a set percentage identity threshold and is used to minimise the volume of data for annotation. Next, a single representative of sequences with high similarity is selected and annotated through BLAST or BLAT algorithms. The assignment of species or annotation of functional genes relies on correspondence and similarity to sequences in specific databases, such as NCBI, Green genes, SILVA, UNITE or SWISSPROT (Morgan *et al*., 2017).

The yeast communities associated with grapevine have mainly been investigated in must samples. Overall, the yeast population includes the genera *Hanseniaspora*, *Issatchenkia*, *Pichia*, *Candida*, *Rhodotorula*, *Lachancea*, *Metschnikowia*, *Cryptococcus*, *Filobasidiella*, *Sporobolomyces* and *Torulaspora* (Bokulich *et al*., 2014; David *et al*., 2014; Taylor *et al*., 2014; Pinto *et al*., 2015; Setati *et al*., 2015; Wang *et al*., 2015; Kecskeméti *et al*., 2016; De Filippis *et al*., 2017). This information suggests that most of these yeasts are cultivable; however due to their presence in minor concentrations they are often missed in culture-based studies (Morgan *et al*., 2017).

Several studies propose that microbial communities associated with grapevines are region-differentiated (Bokulich *et al*., 2014, 2016a,b; Taylor *et al*., 2014; Pinto *et al*., 2015; Wang *et al*., 2015). Pinto and colleagues (2015) showed that microbial dominance is different among several Portuguese wine appellation regions, namely *Lachancea* that prevailed in the Alentejo appellation, *Hanseniaspora* in Bairrada, *Lachancea* and *Rhodotorula* in Dão, and *Rhodotorula* was dominated in the Estremadura, Douro and Minho appellation. One of the main factors attributed to microorganism diversity is agronomic practices. Additionally, metagenomics approach has detected minor and rare species, such as yeast genera of *Kazachstania*, *Malassezia*, *Schizosaccharomyces*, and *Debaryomyces* (David *et al*., 2014; ,Pinto *et al*. 2015; Setati *et al*., 2015; Grangeteau *et al*., 2017), and was observed before with culture-dependent methods. Interestingly, *S. cerevisiae* is rarely encountered in grape must. In some cases, strong fermentative yeasts, such as *Lachancea*, *Starmerella* and *Schizosaccharomyces* were found to be present at high frequency in the initial population and persisted until the end of fermentation (Pinto *et al*., 2015; Wang *et al*., 2015; Bokulich *et al*., 2016a).

## 8. Conclusion

This chapter highlights the yeast ecology associated with the wine fermentation process. Overall, the winemaking process involves several steps, with wine must fermentation being the crucial one. Wine fermentation is an ecological complex process and involves the growth, development and a succession of microbial populations, including bacteria and/or yeasts, such as non-*Saccharomyces* and *Saccharomyces*.

These microorganisms coexist and interact between themselves, which may influence the dominance and persistence of wine-fermenting yeasts and, ultimately, may have a prominent role in the analytical profiles of wine. Furthermore, this microbial biodiversity is highly dynamic during the wine-fermentation process and is affected by both abiotic and biotic factors. However, knowledge of how these factors affect both bacteria and yeast species dynamics is still scarce. In this context, the importance and role of this wine microbiome in oenology is a main challenge that needs to be explored.

## References

Abbasian, F., Lockington, R., Megharaj, M. and Naidu, R. (2015). The integration of sequencing and bioinformatics in metagenomics. *Rev. Environ. Sci. Biotechnol.*, 14: 357-383.

Albergaria, H. and Arneborg, N. (2016). Dominance of *Saccharomyces cerevisiae* in alcoholic fermentation processes: Role of physiological fitness and microbial interactions. *Applied Microbiology and Biotechnology* 100: 2035-2046.

Albergaria, H., Francisco, D., Gori, K., Arneborg, N. and Gírio, F. (2010). *Saccharomyces cerevisiae* CCMI 885 secretes peptides that inhibit the growth of some non-Saccharomyces wine-related strains. *Applied Microbiology and Biotechnology* 86: 965-972.

Barata, A., Malfeito-ferreira, M. and Loureiro, V. (2012a). Changes in sour rotten grape berry microbiota during ripening and wine fermentation. *International Journal of Food Microbiology* 154: 152-161.

Barata, A., Seborro, F., Bellochand, C. and Loureiro, V. (2008). Ascomycetous yeast species recovered from grapes damaged by honeydew and sour rot. *Journal of Applied Microbiology* 104(4): 1182-1191.

Barata, A., Malfeito-Ferreira, M. and Loureiro, V. (2012b). The microbial ecology of wine grape berries. *International Journal of Food Microbiology* 153: 243-259.

Barrajón, N., Arévalo-Villena, M., Ubeda, J. and Briones, A. (2011). Enological properties in wild and commercial *Saccharomyces cerevisiae* yeasts: Relationship with competition during alcoholic fermentation. *World Journal of Biotechnology* 27: 2703-2710.

Bartowsky, E.J. (2009). Bacterial spoilage of wine and approaches to minimise it. *Letters in Applied Microbiology* 48: 149-156.

Bartowsky, E., Costello, P., Abrahamse, C., McCarthy, J., Chambers, P., Herderich, M. and Pretorius, I. (2009). Wine bacteria – friends and foes. *Wine Industry Journal* 24: 18-20.

Bauer, E.F. and Pretorius, L.S. (2000). Yeast stress response and fermentation efficiency: How to survive the making of wine – A review. *South African Journal for Enology and Viticulture* 27: 27-51.

Begerow, D., Nilsson, H., Unterseher, M. and Maier, W. (2010). Current state and perspectives of fungal DNA barcoding and rapid identification procedures. *Applied Microbiology and Biotechnology* 87: 99-108.

Belda, I., Navascués, E., Marquina, D., Santos, A., Calderon, F. and Benito, S. (2015). Dynamic analysis of physiological properties of *Torulaspora delbrueckii* in wine fermentations and its incidence on wine quality. *Applied Microbiology and Biotechnology* 99: 1911-1922.

Belda, I., Ruiz, J., Esteban-fernández, A., Navascués, E., Marquina, D., Santos, A. and Moreno-arribas, M.V. (2017). Microbial contribution to wine aroma and its intended use for wine quality improvement. *Molecules* 22: 1-29.

Belda, I., Zarraonaindia, I., Perisin, M., Palacios, A. and Acedo, A. (2017). From vineyard soil to wine fermentation: Microbiome approximations to explain the 'terroir' concept. *Frontiers in Microbiology* 8: 1-12.

Belloch, C., Querol, A., García, M.D. and Barrio, E. (2000). Phylogeny of the genus Kluyveromyces inferred from the Mitochondrial Cytochrome-c Oxidase II Gene. *International Journal of Systematic and Evolutionary Microbiology* 50: 405-416.

Beltran, G., Torija, M.J., Novo, M., Ferrer, N., Poblet, M., Guillamón, J.M. and Mas, A. (2002). Analysis of yeast populations during alcoholic fermentation: A six-year follow-up study. *Systematic and Applied Microbiology* 25: 287-293.

Benito, S., Hofmann, T., Laier, M., Lochbühler, B., Schüttler, A., Ebert, K. and Rauhut, D. (2015). Effect on quality and composition of Riesling wines fermented by sequential inoculation with non-*Saccharomyces* and *Saccharomyces cerevisiae*. *European Food Research and Technology* 241: 707-717.

Bisson, L.F. (1999). Stuck and sluggish fermentations. *American Journal of Enology and Viticulture* 50: 107-119.

Bokulich, N.A., Collins, T.S., Masarweh, C., Allen, G., Heymann, H., Ebeler, S.E. and Mills, D.A. (2016a). Associations among wine grape microbiome, metabolome, and fermentation behavior suggest microbial contribution to regional wine characteristics. *American Society for Microbiology* 7: 1-12.

Bokulich, N.A., Lewis, Z.T., Boundy-Mills, K. and Mills, D.A. (2016b). A new perspective on microbial landscapes within food production. *Curr. Opin. Biotechnol.* 37: 182-189.

Bokulich, N.A. and Mills, D.A. (2013). Improved selection of internal transcribed spacer-specific primers enables quantitative, ultra-high-throughput profiling of fungal communities. *Applied and Environmental Microbiology* 79: 2519-2526.

Bokulich, N.A., Thorngate, J.H., Richardson, P.M. and Mills, D.A. (2014). Microbial biogeography of wine grapes is conditioned by cultivar, vintage, and climate. *Proceedings of the National Academy of Sciences of the United States of America* 111: 39-48.

Boundy-Mills, K. (2006). Methods for investigating yeast biodiversity. pp. 67-100. *In*: C. Péter and G. Rosa (Eds.). Biodiversity and Ecophysiology of Yeasts. Springer, Berlin Heidelberg.

Branco, P., Francisco, D., Chambon, C., Hébraud, M., Arneborg, N., Almeida, M.G. and Albergaria, H. (2014). Identification of novel GAPDH-derived antimicrobial peptides secreted by *Saccharomyces cerevisiae* and involved in wine microbial interactions. *Applied Microbiology and Biotechnology* 98: 843-853.

Brandam, C., Lai, Q.P. and Julien-Ortiz, A. (2013). Influence of oxygen on alcoholic fermentation by a wine strain of *Torulaspora delbrueckii*: Kinetics and carbon mass balance influence of oxygen on alcoholic fermentation by a wine strain. *World Journal of Microbiology and Biotechnology* 77: 1848-1853.

Cadez, N., Zupan, J. and Raspor, P. (2010). The effect of fungicides on yeast communities associated with grape berries. *FEMS Yeast Research* 10: 619-630.

Capece, A., Romaniello, R., Siesto, G., Pietrafesa, R., Massari, C., Poeta, C. and Romano, P. (2010). Selection of indigenous *Saccharomyces cerevisiae* strains for Nero d'Avola wine and evaluation of selected starter implantation in pilot fermentation. *International Journal of Food Microbiology* 144: 187-192.

Cheraiti, N., Guezenec, S. and Salmon, J. (2005). Redox interactions between *Saccharomyces cerevisiae* and *Saccharomyces uvarum* in mixed culture under enological conditions. *Applied and Environmental Microbiology* 71: 255-260.

Ciani, M., Capece, A., Comitini, F., Canonico, L., Siesto, G. and Romano, P. (2016). Yeast interactions in inoculated wine fermentation. *Frontiers in Microbiology* 7: 1-7.

Ciani, M. and Comitini, F. (2006). Influence of temperature and oxygen concentration on the fermentation behaviour of *Candida stellata* in mixed fermentation with *Saccharomyces cerevisiae*. *World Journal of Microbiology and Biotechnology* 22: 619-623.

Ciani, M. and Comitini, F. (2015). Yeast interactions in multi-starter wine fermentation. *Current Opinion in Food Science* 1: 1-6.

Ciani, M., Fatichenti, F. and Bevan, S. (2001). Killer toxin of *Kluyveromyces phaffii* DBVPG 6076 as a biopreservative agent to control apiculate wine yeasts. *Applied and Environmental Microbiology* 67: 3058-3063.

Cocolin, L., Campolongo, S., Alessandria, V., Dolci, P. and Rantsiou, K. (2011). Culture independent analyses and wine fermentation: An overview of achievements 10 years after first application. *Annals of Microbiology* 61: 17-23.

Cocolin, L., Alessandria, V., Dolci, P., Gorra, R. and Rantsiou, K. (2013). Culture independent methods to assess the diversity and dynamics of microbiota during food fermentation. *International Journal of Food Microbiology* 167: 29-43.

Combina, M., Elía, A., Mercado, L., Catania, C., Ganga, A. and Martinez, C. (2005). Dynamics of indigenous yeast populations during spontaneous fermentation of wines from Mendoza, Argentina. *International Journal of Food Microbiology* 99: 237-243.

Comitini, F., Capece, A., Ciani, M. and Romano, P. (2017). Science direct new insights on the use of wine yeasts. *Current Opinion in Food Science* 13: 44-49.

Comitini, F. and Ciani, M. (2010). The zymocidial activity of *Tetrapisispora phaffii* in the control of *Hanseniaspora uvarum* during the early stages of winemaking. Letters in *Applied Microbiology* 50: 50-56.

Comitini, F., Ciani, M., Politecnica, U. and Bianche, V.B. (2008). Influence of fungicide treatments on the occurrence of yeast flora associated with wine grapes. *Annals of Microbiology* 58: 489-493.

Comitini, F., De Ingeniis, J., Pepe, L., Mannazzu, I. and Ciani, M. (2004). *Pichia anomala* and *Kluyveromyces wickerhamii* killer toxins as new tools against *Dekkera/Brettanomyces* spoilage yeasts. *FEMS Microbiology Letters* 238-240.

Constantí, M., Poblet, M., Arola, L., Mas, A. and Guillamón, J.M. (1997). Analysis of yeast populations during alcoholic fermentation in a newly established winery. *American Journal of Enology and Viticulture* 48: 339-344.

Cordero-Bueso, G., Arroyo, T., Serrano, A., Tello, J., Aporta, I., Vélez, M.D. and Valero, E. (2011). Influence of the farming system and vine variety on yeast communities associated with grape berries. *International Journal of Food Microbiology* 145: 132-139.

Costello, P., Déléris-Bou, M., Descenzo, R., Hall, N., Krieger, S., Lonvaud-Funel, A., Loubser, P., Heras, J.M., Molinari, S., Morenzoni, R., Silvano, A., Specht, G., Vidal, F. and Wild, C. (2015). Malo-lactic Fermentation – Importance of Wine Lactic Acid Bacteria in Winemaking. Lallemand Inc, Canada.

Daniel, H. (2003). Evaluation of ribosomal RNA and actin gene sequences for the identification of ascomycetous yeasts. *International Journal of Food Microbiology* 86: 61-78.

Dashko, S., Zhou, N., Tinta, T. and Sivilotti, P. (2015). Use of non-conventional yeast improves the wine aroma profile of Ribolla Gialla. *Journal of Industrial Microbiology and Biotechnology* 42: 997-1010.

David, V., Terrat, S., Herzine, K., Claisse, O., Rousseaux, S., Tourdot-Maréchal, R., Masneuf-Pomarede, I., Ranjard, L. and Alexandre, H. (2014). High-throughput sequencing of amplicons for monitoring yeast biodiversity in must and during alcoholic fermentation. *J. Ind. Microbiol. Biotechnol.* 41: 811-821.

De Filippis, F., La Storia, A. and Blaiotta, G. (2017). Monitoring the mycobiota during Greco di Tufo and Aglianico wine fermentation by 18S rRNA gene sequencing. *Food Microbiology* 63: 117-122.

De Llanos Frutos, R., Fernández-Espinar, M.T. and Querol, A. (2004). Identification of species of the genus candida by analysis of the 5.8S rRNA gene and the two ribosomal internal transcribed spacers. *Antonie Van Leeuwenhoek* 85: 175-185.

del Mónaco, S.M., OnacoBarda, N.B., Rubio, N.C. and Caballero, A.C. (2014). Selection and characterization of a Patagonian *Pichia kudriavzevii* for wine deacidification. *Journal of Applied Microbiology* 117: 451-464.

Diaz-Sanchez, S., Hanning, I., Pendleton, S. and D'Souza, D. (2013). Next-generation sequencing: The future of molecular genetics in poultry production and food safety. *Poultry Science* 92: 562-572.

Dizy, M. and Bisson, L.F. (2000). Proteolytic activity of yeast strains during grape juice fermentation. *American Journal of Enology and Viticulture* 51: 155-167.

Domizio, P., Liu, Y., Bisson, L.F. and Barile, D. (2014). Use of non-*Saccharomyces* wine yeasts as novel sources of mannoproteins in wine. *Food Microbiology* 43: 5-15.

Englezos, V., Torchio, F., Cravero, F., Marengo, F., Gerbi, V., Rantsiou, K. and Cocolin, L. (2016). Aroma profile and composition of Barbera wines obtained by mixed fermentations of *Starmerella bacillaris* (synonym *Candida zemplinina*) and *Saccharomyces cerevisiae*. *LWT – Food Science and Technology* 73: 567-575.

Esteve-Zarzoso, B., Belloch, C., Uruburu, F. and Querol, A. (1999). Identification of yeasts by RFLP analysis of the 5.8S rRNA gene and the two ribosomal internal transcribed spacers. *International Journal of Systematic Bacteriology* 49: 329-337.

Fernández-Espinar, M.T., López, V., Ramón, D., Bartra, E. and Querol, A. (2001). Study of the authenticity of commercial wine yeast strains by molecular techniques. *International Journal of Food Microbiology* 70: 1-10.

Fleet, G.H., Lafon-Lafourcade, S. and Ribereau-Gayon, P. (1984). Evolution of yeasts and lactic acid bacteria during fermentation and storage of Bordeaux wines. *Applied and Environmental Microbiology* 48: 1034-1038.

Fleet, G.H, (2003). Yeast interactions and wine flavour. *International Journal of Food Microbiology* 86: 11-22.

Forde, B.M. and O'Toole, P.W. (2013). Next-generation sequencing technologies and their impact on microbial genomics. *Briefings in Functional Genomics* 12: 440-453.

Francesca, N., Chiurazzi, M., Romano, R., Aponte, M., Settanni, L. and Moschetti, G. (2010). Indigenous yeast communities in the environment of "Rovellobianco" grape variety and their use in commercial white wine fermentation. *World Journal of Biotechnology* 26: 337-351.

Gao, C. and Fleet, G.H. (1988). The effects of temperature and pH on the ethanol tolerance of the wine yeasts, *Saccharomyces cerevisiae*, *Candida stellata* and *Kloeckera apiculata*. *Journal of Applied Bacteriology* 65: 405-409.

Garavaglia, J., De Cassia, R., Schneider, D.S., Denise, S., Mendes, C., Elisa, J. and Alcaraz, C. (2015). Evaluation of *Zygosaccharomyces bailii* BCV 08 as a co-starter in wine fermentation for the improvement of ethyl esters production. *Microbiological Research* 173: 59-65.

Giannattasio, S., Guaragnella, N. and Corte-real, M. (2005). Acid stress adaptation protects *Saccharomyces cerevisiae* from acetic acid-induced programmed cell death. *Gene* 354: 93-98.

Gobbi, M., Comitini, F., Domizio, P., Romani, C., Lencioni, L., Mannazzu, I. and Ciani, M. (2013. Lachanceathermotolerans and Saccharomyces cerevisiae in simultaneous and sequential co-fermentation: A strategy to enhance acidity and improve the overall quality of wine. *Food Microbiology* 33: 271-281.

González, E., Olga, R. and Nikolaos, P. (2015). Oenological consequences of sequential inoculation with non-Saccharomyces yeasts (*Torulasporadel brueckii* or *Metschnikowia pulcherrima*) and *Saccharomyces cerevisiae* in base wine for sparkling wine production. *European Food Research and Technology* 240: 999-1012.

Grangeteau, C., Roullier-Gall, C., Rousseaux, S., Gougeon, R.D., Schmitt-Kopplin, P., Alexandre, H. and Guilloux-Benatier, M. (2017). Wine microbiology is driven by vineyard and winery anthropogenic factors. *Microbial Biotechnology* 10: 354-370.

Guillamón, J.M., Sabaté, J., Barrio, E., Cano, J. and Querol, A. (1998). Rapid identification of wine yeast species based on RFLP analysis of the ribosomal internal transcribed spacer (ITS) region. *Archives of Microbiology* 169: 387-392.

Guilloux-Benatier, M., Chassagne, D., Alexandre, H., Charpentier, C. and Feuillat, M. (2001). Influence of yeast autolysis after alcoholic fermentation on the development of *Brettanomyces/Dekkera* in wine. *Journal International Des Sciences de La Vigne et Du vin*. 35: 157-164.

Gutiérrez, A.R., Epifanio, S., Garijo, P., López, R. and Santamaría, P. (2001). Killer yeasts: Incidence in the ecology of spontaneous fermentation. *American Journal of Enology and Viticulture* 4: 352-356.

Guzzon, R., Nicolini, G., Nardin, T., Malacarne, M. and Larcher, R. (2014). Survey about the microbiological features, the oenological performance and the influence on the character of wine of active dry yeast employed as starters of wine fermentation. *International Journal of Food Science & Technology* 49: 2142-2148.

Hansen, E.H., Nissen, P., Sommer, P., Nielsen, J.C. and Arneborg, N. (2001). The effect of oxygen on the survival of non-*Saccharomyces* yeasts during mixed culture fermentations of grape juice with *Saccharomyces cerevisiae*. *Journal of Applied Microbiology* 91: 541-547.

Hierro, N., Esteve-Zarzoso, B., González, A., Mas, A. and Guillamón, J.M. (2006). Real-time quantitative PCR (QPCR) and reverse transcription-QPCR for detection and enumeration of total yeasts in wine. *Applied and Environmental Microbiology* 72: 7148-7155.

Hierro, N., Esteve-Zarzoso, B., Mas, A. and Guillamón, J.M. (2007). Monitoring of *Saccharomyces* and *Hanseniaspora* populations during alcoholic fermentation by real-time quantitative PCR. *FEMS Yeast Research* 7: 1340-1349.

Hu, K., Qin, Y., Tao, Y., Zhu, X., Peng, C. and Ullah, N. (2016). Potential of glycosidase from non-*Saccharomyces* isolates for enhancement of wine aroma. *Journal of Food Science* 81: 934-943.

Huson, D.H., Auch, A.F., Qi, J. and Schuster, S.C. (2007). MEGAN analysis of metagenomic data. *Genome Research* 17: 377-386.

Illeghems, K., De Vuyst, L., Papalexandratou, Z. and Weckx, S. (2012). Phylogenetic analysis of a spontaneous cocoa bean fermentation metagenome reveals new insights into its bacterial and fungal community diversity. *PLoS One* 7: e38040.

Ivey, M., Massel, M. and Phister, T.G. (2013). Microbial interactions in food fermentations. *Annual Review of Food Science and Technology* 4: 141-162.

Izquierdo, P.M., Esteban, C., Romero, G., Heras, J.M., Mónica, M. and González, F. (2014). Influence of sequential inoculation of *Wickerhamomyces anomalus* and *Saccharomyces cerevisiae* in the quality of red wines. *European Food Research and Technology* 239: 279-286.

James, S.A., Cai, J., Roberts, I.N. and Collins, M.D. (1997). A phylogenetic analysis of the genus *Saccharomyces* based on 18S rRNA gene sequences: Description of *Saccharomyces Kunashirensis* Sp. Nov. and *Saccharomyces Martiniae* Sp. Nov. *International Journal of Systematic and Evolutionary Microbiology* 47: 453-460.

Jenko, M. (2013). The influence of yeast strains on the composition and sensory quality of Gewürztraminer Wine. *Food Technology and Biotechnology* 51: 547-553.

Jolly, N.P., Augustyn, O.P.H. and Pretorius, I.S. (2003). The occurrence of non-*Saccharomyces cerevisiae* yeast species over three vintages in four vineyards and grape musts from four production regions of the Western Cape, South Africa. *South African Journal of Enology and Viticulture* 24: 8-10.

Jolly, N.P., Varela, C. and Pretorius, I.S. (2014). Not your ordinary yeast: Non-*Saccharomyces* yeasts in wine production uncovered. *FEMS Yeast Research* 14: 215-237.

Justé, A., Thomma, B.P.H.J. and Lievens, B. (2008). Recent advances in molecular techniques to study microbial communities in food-associated matrices and processes. *Food Microbiology* 25: 745-761.

Kassemeyer, H.H. and Berkelmann-Löhnertz, B. (2009). Fungi of grapes. pp. 61-87. *In*: H. König, G. Unden and J. Fröhlich (Eds.). Biology of Microorganisms on Grapes, in Must and in Wine. Springer, Berlin Heidelberg.

Kecskeméti, E., Berkelmann-Löhnertzand, B. and Reineke, A. (2016). Are epiphytic microbial communities in the carposphere of ripening grape clusters (*Vitis vinifera* L.) different between conventional, organic, and biodynamic grapes? *PLoS One* 11: 1-23.

Kemsawasd, V., Viana, T., Ardö, Y. and Arneborg, N. (2015). Influence of nitrogen sources on growth and fermentation performance of different wine yeast species during alcoholic fermentation. *Applied Microbiology and Biotechnology* 99: 10191-10207.

Kurtzman, C.P. and Robnett, C.J. (1998). Identification and phylogeny of Ascomycetous yeasts from analysis of nuclear large subunit (26S) ribosomal DNA partial sequences. *Antonie Van Leeuwenhoek* 73: 331-371.

Leticia, M., Susana, J., Ariel, C. and Paula, M. (2016). TdKT, a new killer toxin produced by *Torulasporadel brueckii* effective against wine spoilage yeasts. *International Journal of Food Microbiology* 217: 94-100.

Liu, Y., Rousseaux, S., Tourdot-Maréchal, R., Sadoudi, M., Gougeon, R., Schmitt-Kopplin, P. and Alexandre, H. (2015). Wine microbiome, a dynamic world of microbial interactions. *Critical Reviews in Food Science and Nutrition* 57: 856-873.

Lleixà, J., Martín, V., Portillo, M.C., Carrau, F., Beltran, G. and Mas, A. (2016). Comparison of fermentation and wines produced by inoculation of *Hanseniaspora vineae* and *Saccharomyces cerevisiae*. *Frontiers in Microbiology* 7: 1-12.

Loira, I., Morata, A., Comuzzo, P., Callejo, M.J., González, C., Calderón, F. and Suárez-Lepe, J.A. (2015). Use of *Schizosaccharomyces pombe* and *Torulasporadel brueckii* strains in mixed and sequential fermentations to improve red wine sensory quality. *Food Research International* 76: 325-333.

Longo, E., Cansado, J., Agrelo, D. and Villa, T.G. (1991). Effect of climatic conditions on yeast diversity in grape musts from Northwest Spain. *American Journal of Enology and Viticulture* 42: 141-144.

Lonvaud-Funel, A. (1999). Lactic acid bacteria in the quality improvement and depreciation of wine. *Antonie Leeuwenhoek* 76: 317-331.

Lu, Y., Huang, D., Lee, P.R. and Liu, S.Q. (2016). Assessment of volatile and non-volatile compounds in durian wines fermented with four commercial non-*Saccharomyces* yeasts. *Journal of the Science of Food and Agriculture* 96: 1511-1521.

Martini, A., Ciani, M. and Scorzetti, G. (1996). Direct enumeration and isolation of wine yeasts from grape surfaces. *Journal of the Science of Food and Agriculture* 5: 2-7.

Martorell, P., Querol, A. and Fernández-Espinar, M.T. (2005). Rapid identification and enumeration of *Saccharomyces Cerevisiae* cells in wine by real-time PCR. *Applied and Environmental Microbiology* 71: 6823-6830.

Maturano, Y.P., Assof, M., Fabani, M.P., Nally, M.C., Jofré, V., Rodríguez Assaf, L.A. and Vazquez, F. (2015). Enzymatic activities produced by mixed *Saccharomyces* and non-*Saccharomyces* cultures: Relationship with wine volatile composition. *Antonie van Leeuwenhoek* 108: 1239-1256.

Medina, K., Boido, E., Fariña, L., Gioia, O., Gomez, M.E., Barquet, M. and Carrau, F. (2013). Increased flavour diversity of Chardonnay wines by spontaneous fermentation and co-fermentation with *Hanseniaspora vineae*. *Food Chemistry* 141: 2513-2521.

Mehlomakulu, N.N., Setati, M.E. and Divol, B. (2015). Non-*Saccharomyces* killer toxins: Possible biocontrol agents against *Brettanomyces* in wine. *South African Journal for Enology and Viticulture* 36: 94-104.

Milanović, V., Comitini, F. and Ciani, M. (2013). Grape berry yeast communities: Influence of fungicide treatments. *International Journal of Food Microbiology* 161: 240-246.

Morgan, H.H., du Toit, M. and Setati, M.E. (2017). The grapevine and wine microbiome: Insights from high-throughput amplicon sequencing. *Frontiers in Microbiology* 8: 820.

Mortimer, R.K. (2000). Kloeckeraapiculata concentrations control the rates of natural fermentations. *Rivista Di Viticoltura E Di Enologia* 53: 61-68.

Musmanno, R.A., Di Maggio, T., Coratza, G. and Molecolare, B. (1999). Studies on strong and weak killer phenotypes of wine yeasts: Production, activity of toxin in must, and its effect in mixed culture fermentation. *Journal of Applied Microbiology* 87: 932-938.

Nisiotou, A.A. and Nychas, G.E. (2007). Yeast populations residing on healthy or Botrytis-infected grapes from a vineyard in Attica, Greece. *Applied and Environmental Microbiology* 73: 2765-2768.

Nisiotou, A., Spiropoulos, A.E. and Nychas, G.J.E. (2007). Yeast community structures and dynamics in healthy and Botrytis-affected grape must fermentations. *Applied and Environmental Microbiology* 73: 6705-6713.

OIV (2017). World vitiviniculture situation. Statistical Report on World Vitiviniculture, OIV Publications.

Oro, L., Ciani, M. and Comitini, F. (2014). Antimicrobial activity of *Metschnikowia pulcherrima* on wine yeasts. *Journal of Applied Microbiology* 116: 1209-1217.

Pace, N.R. (1997). A molecular view of microbial diversity and the biosphere. *Science* 276: 734-740.

Padilla, B., Gil, J.V. and Manzanares, P. (2016). Past and future of non-*Saccharomyces* yeasts: From spoilage microorganisms to biotechnological tools for improving wine aroma complexity. *Frontiers in Microbiology* 7: 1-20.

Pérez, F., Ramírez, M. and Regodón, J.A. (2001). Influence of killer strains of *Saccharomyces cerevisiae* on wine fermentation. *Antonie van Leeuwenhoek* 79: 393-399.

Pina, C., Santos, C., Couto, J.A. and Hogg, T. (2004). Ethanol tolerance of five non-*Saccharomyces* wine yeasts in comparison with a strain of *Saccharomyces cerevisiae* – influence of different culture conditions. *Food Microbiology* 21: 439-447.

Pinto, C., Pinho, D., Sousa, S., Pinheiro, M., Egas, C. and Gomes, A.C. (2014). Unravelling the diversity of grapevine microbiome. *PloS One* 9: e85622.

Pinto, C., Pinho, D., Cardoso, R., Custódio, V., Fernandes, J., Sousa, S., Pinheiro, M., Egas, C. and Gomes, A.C. (2015). Wine fermentation microbiome: A landscape from different Portuguese wine appellations. *Frontiers in Microbiology* 6: 1-13.

Prakitchaiwattana, C.J., Fleet, G.H. and Heard, G.M. (2004). Application and evaluation of denaturing gradient gel electrophoresis to analyse the yeast ecology of wine grapes. *FEMS Yeast Research* 4: 865-877.

Pretorius, I.S., van der Westhuizen, T. and Augustyn, O.P. (1999). Yeast biodiversity in vineyards and wineries and its importance to the South African wine industry: A review. *South African Journal for Enology and Viticulture* 20: 61-70.

Pretorius, I.S. (2000). Tailoring wine yeast for the new millennium: Novel approaches to the ancient art of winemaking. *Yeast* 16: 675-729.

Pronk, J.T., Steensmays, Y.H. and Van Dijken, J.P. (1996). Pyruvate Metabolism in *Saccharomyces cerevisiae*. *Yeast* 12: 1607-1633.

Ramírez, M. (2015). Killer strain with broad antifungal activity and its toxin-encoding double-stranded RNA virus. *Frontiers in Microbiology* 6: 1-12.

Raspor, P., Mikli, D., Polanc, J. and Smole, S. (2006). Yeasts isolated from three varieties of grapes cultivated in different locations of the Dolenjska vine-growing region, Slovenia. *International Journal of Food Microbiology* 109: 97-102.

Rementeria, A., Rodriguez, J.A., Cadaval, A., Amenabar, R., Muguruza, J.R., Hernando, F.L. and Sevilla, M.J. (2003). Yeast associated with spontaneous fermentations of white wines from the "Txakoli de Bizkaia" region (Basque Country, North Spain). *International Journal of Food Microbiology* 86: 201-207.

Renault, P., Coulon, J., Moine, V., Thibon, C. and Bely, M. (2016). Enhanced 3-sulfanylhexan-1-ol production in sequential mixed fermentation with *Torulaspora delbrueckii/Saccharomyces cerevisiae* reveals a situation of synergistic interaction between two industrial strains. *Frontiers in Microbiology* 7: 1-10.

Renault, P., Coulon, J., De Revel, G., Barbe, J. and Bely, M. (2015). Increase of fruity aroma during mixed *T. delbrueckii/S. cerevisiae* wine fermentation is linked to specific esters enhancement. *International Journal of Food Microbiology* 70: 40-48.

Renouf, V., Claisse, O. and Lonvaud-funel, A. (2005). Understanding the microbial ecosystem on the grape berry surface through numeration and identification. *Australian Journal of Grape and Wine Research* 11: 316-327.

Renouf, V., Claisse, O. and Lonvaud-funel, A. (2007). Inventory and monitoring of wine microbial consortia. *Applied Microbiology and Biotechnology* 75: 149-164.

Romano, P., Capece, A. and Jespersen, L. (2006. Taxonomic and Ecological Diversity of Food and Beverage Yeasts BT – Yeasts in Food and Beverages. pp. 13-53. *In*: A. Querol and G. Fleet (Eds.). Springer. Berlin Heidelberg.

Rousseaux, S., Diguta, C.F., Radoï-Matei, F., Alexandre, H. and Guilloux-Bénatier, M. (2014). Non-Botrytis grape-rotting fungi responsible for earthy and moldy off-flavors and mycotoxins. *Food Microbiology* 38: 104-121.

Sabate, J., Cano, J., Esteve-zarzoso, B. and Guillamón, J.M. (2002). Isolation and identification of yeasts associated with vineyard and winery by RFLP analysis of ribosomal genes and mitochondrial DNA. *Microbiological Research* 157: 267-274.

Salinas, F., Garrido, D., Ganga, A., Veliz, G. and Martínez, C. (2009). Taqman real-time PCR for the detection and enumeration of *Saccharomyces cerevisiae* in wine. *Food Microbiology* 26: 328-332.

Santos, A., Mauro, M.S., Bravo, E. and Marquina, D. (2009). PMKT2, a new killer toxin from *Pichia membranifaciens*, and its promising biotechnological properties for control of the spoilage yeast Brettanomyces bruxellensis. *Microbiology* 155: 624-634.

Santos, J., Sousa, M., Cardoso, H., Inácio, J., Silva, S., Spencer-martins, I. and Leão, C. (2008). Ethanol tolerance of sugar transport, and the rectification of stuck wine fermentations. *Microbiology* 154: 422-430.

Schadt, E.E., Turner, S. and Kasarskis, A. (2010). A window into third-generation sequencing. *Human Molecular Genetics* 19: 227-240.

Schoch, C.L., Seifert, K., Huhndorf, S., Robert, V., Spouge, J.L., Levesque, C.A. and Chen, W. (2012). Nuclear ribosomal internal transcribed spacer (ITS) region as a universal DNA barcode marker for fungi. *Proceedings of the National Academy of Sciences of the United States of America* 109: 6241-6246.

Scorzetti, G., Fell, J.W., Fonseca, A. and Statzell-Tallman, A. (2002). Systematics of Basidiomycetous yeasts: A comparison of large subunit D1/D2 and internal transcribed spacer rDNA regions. *FEMS Yeast* 2: 495-517.

Scott, R. and Sullivan, W.C. (2008). Ecology of fermented foods. *Human Ecology Review* 15: 25-31.

Setati, M.E., Jacobson, D., Andong, U. and Bauer, F. (2012). The vineyard yeast microbiome, a mixed model microbial map. *PLoS ONE* 7.

Setati, M.E., Jacobson, D. and Bauer, F.F. (2015). Sequence-based analysis of the *Vitis vinifera* L. cv cabernet sauvignon grape must mycobiome in three South African vineyards employing distinct agronomic systems. *Front Microbiology* 6: 1-12.

Simon, C. and Daniel, R. (2009). Achievements and new knowledge unraveled by metagenomic approaches. *Applied Microbiology Biotechnology* 85: 265-276.

Sipiczki, M. (2006). *Metschnikowia* strains isolated from Botrytized grapes antagonize fungal and bacterial growth by iron depletion. *Applied and Environmental Microbiology* 72: 6716-6724.

Shimizu, K. (1993). Killer yeasts. pp. 243-264. *In*: Fleet, G.H. (Ed.). Wine Microbiology and Biotechnology. Harwood. Chur, Switzerland.

Steensels, J. and Verstrepen, K.J. (2014). Taming wild yeast: Potential of conventional and nonconventional yeasts in industrial fermentations. *Annual Review of Microbiology* 68: 61-80.

Stefanini, I., Dapporto, L., Legras, J., Calabretta, A. and Di, M. (2012). Role of social wasps in *Saccharomyces cerevisiae* ecology and evolution. *Proceedings of the National Academy of Sciences* 109: 13398-13403.

Sun, Y. and Liu, Y. (2014). Investigating of yeast species in wine fermentation using terminal restriction fragment length polymorphism method. *Food Microbiology* 38: 201-207.

Swiegers, J.H., Bartowsky, E.J., Henschke, P. and Pretorius, I.S. (2005). Yeast and bacterial modulation of wine aroma and flavour. *Australian Journal of Grape and Wine Research*, 11: 139-173.

Taillandier, P., Phong Lai, Q., Julien-ortiz, A. and Brandam, C. (2014). Interactions between *Torulasporadel brueckii* and *Saccharomyces cerevisiae* in wine fermentation: Influence of inoculation and nitrogen content. *World Journal of Biotechnol* 30: 1959-1967.

Taylor, M.W., Tsai, P., Anfang, N., Ross, H.A. and Goddard, M.R. (2014). Pyrosequencing reveals regional differences in fruit-associated fungal communities. *Environ. Microbiology* 16: 2848-2858.

Thomas, T., Gilbert, J. and Meyer, F. (2012). Metagenomics – A guide from sampling to data analysis. *Microbial Informatics and Experimentation* 2: 3.

Torija, M.J., Rozès, N., Poblet, M., Guillamón, J.M. and Mas, A. (2001). Yeast population dynamics in spontaneous fermentations: Comparison between two different wine-producing areas over a period of three years. *Antonie van Leeuwenhoek* 79: 345-352.

Tristezza, M., Tufariello, M., Capozzi, V., Spano, G., Mita, G. and Grieco, F. (2016). The oenological potential of Hanseniaspora uvarum in simultaneous and sequential co-fermentation with *Saccharomyces cerevisiae* for industrial wine production. *Frontiers in Microbiology* 7: 1-14.

Türkel, S. and Ener, B. (2009). Isolation and characterization of new *Metschnikowia pulcherrima* strains as producers of the antimicrobial pigment pulcherrimin. *Z Naturforsch* 64: 405-410.

Van Vuuren, H.J.J. and Jacobs, C.J. (1992). Killer yeasts in the wine industry: A review. *American Journal of Enology and Viticulture* 43: 116119-116128.

Varela, C. (2016). The impact of non-*Saccharomyces* yeasts in the production of alcoholic beverages. *Applied Microbiology and Biotechnology* 100: 9861-9874.

Varela, C. and Borneman, A.R. (2017). Yeasts found in vineyards and wineries. *Yeast* 34: 111-128.

Visser, W., Scheffers, W.A., Vegte, B. and Dijken, J.P. (1990). Oxygen requirements of yeasts. *Applied and Environmental Microbiology* 56: 3785-3792.

Wang, C., García-fernández, D., Mas, A. and Esteve-Zarzoso, B. (2015). Fungal diversity in grape must and wine fermentation assessed by massive sequencing, quantitative PCR and DGGE. *Frontiers in Microbiology* 6: 1-8.

Wang, C., Mas, A. and Esteve-Zarzoso, B. (2016). The interaction between *Saccharomyces cerevisiae* and non-*Saccharomyces* yeast during alcoholic fermentation is species and strain specific. *Frontiers in Microbiology* 7: 1-11.

Wooley, J.C., Godzik, A. and Friedberg, I. (2010). A primer on metagenomics. *PLoS Computational Biology* 6: e1000667.

Xufre, A., Albergaria, H., Inácio, J., Spencer-Martins, I. and Gírio, F. (2006). Application of fluorescence in situ hybridisation (FISH) to the analysis of yeast population dynamics in winery and laboratory grape must fermentations. *International Journal of Food Microbiology* 108: 376-384.

Zara, G., Mannazzu, I., Caro, A., Budroni, M., Pinna, M.B., Murru, M. and Zara, S. (2014). Wine quality improvement through the combined utilisation of yeast hulls and *Candida zemplinina/Saccharomyces cerevisiae* mixed starter cultures. *Australian Society of Viticulture and Oenology* 20: 199-207.

# 11 Wine Yeast: Physiology and Growth Factors

**María-Jesús Torija[1*], Albert Mas[1], Ilkin Yucel Sengun[2] and Gemma Beltran[1]**

[1] Oenological Biotechnology Group, Department of Biochemistry and Biotechnology, Faculty of Oenology, University Rovira i Virgili, Marcel·li Domingo, 1. 43007, Tarragona, Spain

[2] Department of Food Engineering, Engineering Faculty, Ege University, 35100, Bornova, Izmir, Turkey

## 1. Introduction: Yeast and Alcoholic Fermentation

The history of winemaking runs parallel to that of civilisation: there are many evidences to show that wine was made in Caucasus and Mesopotamia as early as 6000 BC. The expansion along the Mediterranean shores took around 4,000 years to become one of the main foods, along with bread and olive oil. The Romans spread it to the rest of Europe and set the agronomic and technological basis that lasted for almost 2,000 years. However, it was not until the 19th century that several scientists, among them Louis Pasteur, proved that yeasts, that naturally present on the surface of grape berries, are responsible for spontaneous fermentation. This was the beginning of scientifically-driven winemaking, beyond the trial-and-error practice.

The transformation of fermentable sugars (mostly glucose and fructose) into alcohols and $CO_2$ during fermentation is the basis for production of multiple alcoholic beverages, such as wine, beer, cider and sake, among others. However, this metabolic process encompasses many more transformations than ethanol production. The simple alcoholic fermentation or transformation of sugars into alcohol implies a more complex process, where many microorganisms, but mostly yeasts, develop and adapt to a changing medium. On the one side, the transformation is characterised by a series of yeasts, broadly named as non-*Saccharomyces*, that start transforming the medium, reducing the sugar concentration and producing ethanol but also many other compounds, both volatiles and non-volatiles. *Saccharomyces cerevisiae* normally takes over due to its adaptation to alcoholic fermentation as well as the capacity to produce new compounds, some of them with anti-microbial effects (Wang *et al.*, 2015). The increase of ethanol and continuous decrease of nutrients shape up the adaptation of microorganisms as well as the chemical composition of the final product, which is wine. Thus, the chemical and organoleptic characteristics of wine are the results of the physiological adaptation and the interactions of different microorganism that proliferate in the fermenting must. A typical development of an alcoholic fermentation can be observed in Fig. 1, where the main parameters (yeast population, density and ethanol) are represented. Density is the most common parameter to monitor the consumption of sugar that is used in the cellars.

Yeasts are strongly inclined to perform alcoholic fermentation under both aerobic and anaerobic conditions (van Dijken *et al.*, 1993). For high glucose concentrations (above 10-20 g/l), as it is the case of grape must, *Saccharomyces cerevisiae* only metabolises sugars through the fermentative pathway. In these conditions, even in the presence of oxygen, aerobic respiration is blocked in what is known as the Crabtree effect and catabolic repression by glucose or the Pasteur contrary effect (Ribereáu-Gayon *et al.*, 2006). In *S. cerevisiae,* glucose and fructose are metabolised to pyruvate through the glycolytic pathway and then, this pyruvate is decarboxylated to acetaldehyde, the latter being finally reduced to ethanol. Therefore, during alcoholic fermentation, a molecule of glucose or fructose yields two molecules of ethanol and $CO_2$. However, in a standard fermentation, only about 95 per cent of the sugars are transformed into ethanol and $CO_2$, being the remainder used for growth (1 per cent into cellular material) and production of other metabolites, such as glycerol (approximately 4 per cent).

*Corresponding author: mjesus.torija@urv.cat

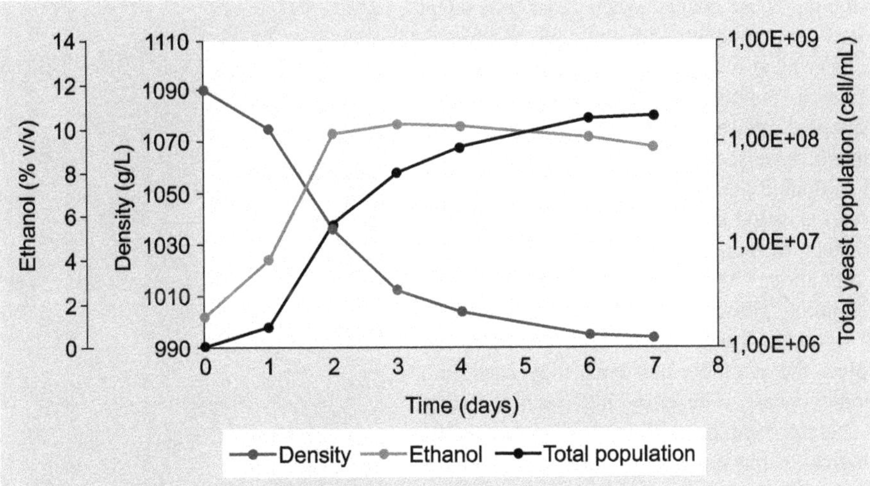


**Figure 1.** Evolution of alcoholic fermentation with the indication of the total yeast population, decrease of density (a typical measure of the fermentable sugars) and increase of ethanol

Color version at the end of the book

Although *S.cerevisiae* is considered the 'wine yeast' *par excellence*, the surfaces of grapes are primarily colonised by non-*Saccharomyces* yeasts, including *Hanseniaspora*, *Starmerella* (syn. *Candida*), *Hansenula*, or *Metschnikowia*, among others. Indeed, yeasts belonging to *Saccharomyces* genus are difficult to detect in must due to their low predominance on grapes. For this reason, non-*Saccharomyces* yeasts initiate spontaneous alcoholic fermentation and are then replaced by *Saccharomyces* throughout the process (Fleet, 2003; Heard and Fleet, 1988; Ribéreau-Gayon *et al*., 2006). *S. cerevisiae* presents the most efficient fermentative catabolism, ethanol tolerance and, therefore, this species is traditionally preferred for alcoholic fermentation (Pretorius, 2000).

## 2. Yeast Morphology and Physiology

Yeast is non-motile, non-photosynthetic and, like all fungi, may reproduce sexually and/or asexually. Perfect fungi undergo both reproductive cycles, while imperfect fungi only reproduce asexually (by mitosis). However, the most common cell division in yeasts occurs by budding, as in the case for *S. cerevisiae*, or by fission, as seen in the case of *Schizosaccharomyces pombe*. This vegetative cell cycle involves a succession of events whereby a cell grows and divides into two daughter cells. *S. cerevisiae* exists both as a haploid and a diploid cell and in the presence of adequate nutrients, both cell types undergo repetitive rounds of vegetative growth and mitosis. The different division systems derive in different yeast morphology that can be easily appreciated under the microscope. Due to subjectivity of the observation under the microscope, some selective media are used to differentiate the different microorganism present. One of the most usual one is the Wallenstein medium, which allows differentiation of the colonies by the shape and colour (Fig. 2). However, this is only indicative as some species present very similar morphologies. The most common practice is the use of molecular methods (*see* the chapter on yeast identification in this book).

Sexual reproduction in yeast takes place *via* conjugation of two haploid cells of different mating types and subsequent sporulation by meiosis, introducing genetic variation to a population, although this type of reproduction is very rare in some yeast species. *S. cerevisiae* presents two mating types in haploid cells – a and α, which mate to form a diploid cell. *S. cerevisiae* diploid cells can undergo vegetative multiplication by budding or under unfavourable conditions such as nutrient limitation, can induce sporulation, producing four haploid spores by meiosis, which will be released when environmental conditions improve.

Under optimal conditions, yeast cells demonstrate a typical microbial growth curve. This sigmoid curve is composed of four phases: lag phase, exponential phase, stationary phase and decline phase.

Initially the lag phase occurs, when yeast cells adapt to the new environment and the cell machinery prepares to grow at a higher rate. Population growth at this time is practically nill. Afterwards, during the exponential phase, optimal population growth occurs, in which yeast cells divide asexually. The time required to double the population, thus, to complete a cell cycle, is called the 'generation time', which is approximately 90 minutes long for *S. cerevisiae* under optimal conditions. At the end of this phase, when the deceleration phase occurs, yeast cells achieve maximal growth. Then, when proliferating yeast cells deplete the available nutrients, they enter into the stationary phase. The entry into such a resting state is characterised by cell cycle arrest, specific morphological, physiological and biochemical changes and an increased resistance to a variety of environmental stresses. Moreover, during this phase, cell numbers no longer increase and, after a certain time, dead cells begin to accumulate, leading to the death phase. The duration of different growth phases is different, with the death phase being three to four times longer than the growth phase (Ribéreau-Gayon *et al.*, 2006).

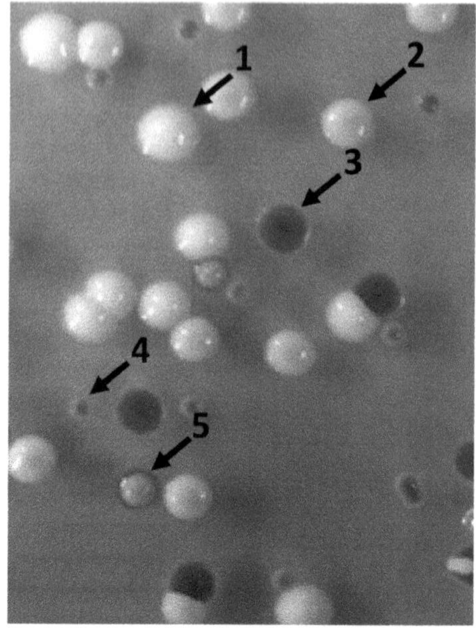


**Figure 2.** Different wine yeast morphologies in Wallenstein medium (WLN medium): 1. *S. cerevisiae*; 2. *T. delbrueckii*; 3. *H. uvarum*; 4. *St. bacillaris*; 5. *M. pulcherrima*

Color version at the end of the book

When yeast cells do not grow in an optimal medium or conditions, the growth curve is affected. During wine fermentation, many factors, such as non-optimal temperature or pH, nutrient (e.g. nitrogen or vitamins) or oxygen limitation, or the presence of toxic substances (such as ethanol, sulphites, killer toxins, etc), independently or jointly, may induce a decrease in the metabolism of the yeast cell, resulting in a decrease of cell viability, biomass production, and fermentation rate. Due to its economical importance, the factors responsible for the occurrence of stuck and sluggish fermentations have been widely studied (Fleet and Heard, 1993; Ribéreau-Gayon *et al.*, 2006).

On the other hand, these abiotic factors, together with the presence of other microorganisms and grape variety, influence the competition degree of each individual strain. The knowledge of the competitiveness of a strain in relation to all these factors is very important for the wine industry to select the most appropriate strain to be inoculated in each wine condition (García-Rios *et al.*, 2014).

## 2.1. Temperature

Temperature is a crucial parameter in wine fermentation, since it affects the yeast growth, the development of alcoholic fermentation and therefore, the sensory quality of the final product. In fact, the must and the fermentation temperature, condition the number of different species and their survival during alcoholic fermentation, affecting the chemical and sensory qualities of the wine (Fleet and Heard, 1993).

Moreover, it is important to highlight that during alcoholic fermentation, 40 kcal of free energy per molecule of glucose is liberated. Although this energy is partly used by yeast to form two molecules of ATP, the non-utilised energy is dissipated in the form of heat, resulting in an increase of the fermentation temperature (Ribéreau-Gayon *et al.*, 2006).

High temperatures in winemaking can accelerate fermentation, resulting invigorous alcoholic fermentation that causes a sharp temperature and ethanol increase. Moreover, this high temperature stress results in the disruption of hydrogen bonding and the denaturation of proteins and nucleic acids, which causes damage and, finally, the death of yeast cells (Walker and van Dijck, 2006). In fact, this can be another cause for stuck or sluggish fermentations and the temperature control of alcoholic fermentations is an absolute requirement for successful winemaking. For this reason, temperature monitoring and a system for cooling down the fermentation are common in the modern winemaking. It is relevant to consider that in most winemaking regions, harvest and alcoholic fermentation might overlap with hot Summer days and, thus the risks are very high.

Low temperatures (10-15°C) are routinely used in the production of white and rosé wines to improve its sensory quality. Low temperatures are believed to increase, not only the retention and the production of some desirable volatile compounds, such as fruity acetate esters, but also to decrease some undesirable compounds such as higher alcohols imparting solvent-like characters (Beltran *et al.*, 2006; Llauradó *et al.*, 2005; Molina *et al.*, 2007). However, these low temperatures are often suboptimal for yeast growth, which lead to the risk of stuck and sluggish fermentations. Previous studies have revealed that suboptimal temperatures can significantly affect, among others, cell membrane fluidity, enzymatic activity, protein translation and folding rates, heat shock protein regulation and RNA secondary structure stability (Jones and Inouye, 1994; Sahara *et al.*, 2002; Schade *et al.*, 2004). Some important effects of these low temperatures are: (i) modification of microorganism population, (ii) longer lag-phase with the risk of prevailing non-*Saccharomyces* strains, (iii) decrease in sugar consumption rate, (iv) longer fermentations, which increase the risk of stuck and sluggish fermentations, (v) different profile of secondary metabolites production, due to the modification of the metabolic activity of yeasts, (vi) modification of lipid membranes, resulting in a modification of the compounds transported through these membranes.

Low temperatures have an impact on yeast ecology and growth during alcoholic fermentation. Several authors have reported that some non-*Saccharomyces* species, such as *Kloeckera apiculata* and *Starmerella bacillaris* (previously named as *Candida stellata* or *Candida zemplinina*) have a better chance to grow at low temperatures than *Saccharomyces* strains (Sharf and Magalith, 1983; Heard and Fleet, 1988), due to their higher tolerance to ethanol in these conditions (Gao and Fleet, 1988). Fleet (2003) indicated that low temperatures decreased the sensitivity of non-*Saccharomyces* species to ethanol and thus, more non-*Saccharomyces* species could be found in fermentation at 15°C than at higher temperature. In fact, Andorrà *et al.* (2010) showed that although temperature has a limited influence on yeast diversity, at 25°C, *Hanseniaspora* disappeared quicker than at 13°C. Later, Salvadó *et al.* (2011) determined the growth rates of six yeast species in a range of temperatures (4-46°C), showing different optimal growth temperatures, depending on the yeast species. *Hanseniaspora uvarum*, *S. bacillaris*, *Torulaspora delbrueckii* and *Pichia fermentans* grow better around 25°C, *Kluyveromyces marxianus* at about 39°C and *S. cerevisiae* around 32°C. However, when temperature decreased to around 15°C, *S. cerevisiae* started to lose its growth superiority to other non-*Saccharomyces* species. Consistent with Salvadó *et al.* (2011), other studies on *S. cerevisiae* dominance also highlight the decisive role of temperature (Goddard *et al.*, 2008; Williams *et al.*, 2015).

These changes in wine ecology determine the changes in the chemical and sensory qualities of wines (Fleet and Heard, 1993). Torija *et al.* (2003a) studied the influence of the fermentation temperature in a mixed population of *S. cerevisiae*, demonstrating that different strains were better or less suited to carry out fermentations at different temperatures (Fig. 3A). Moreover, the number of strains that were able to endure and conduct the fermentation was higher at low temperatures. This ability to grow at different temperatures is often used as a criterion to select yeast starters (Torija *et al.*, 2003a).

Temperature also affects yeast viability. At low temperatures, maximal population remains throughout alcoholic fermentation, whereas at high temperatures, especially at 35°C, the viable cells decrease considerably (Lafon-Lafourcade, 1983; Torija *et al.*, 2003a) (Fig. 3B). This increase in yeast mortality may induce stuck fermentations, leaving a large sugar content not consumed.

Temperature has also a great impact on fermentation kinetics (Fleet and Heard, 1993; Ribéreau-Gayon *et al.*, 2006). Temperature affects the fermentation rate, resulting in slower and longer fermentations when low temperatures were used. The lag phase becomes shorter and the fermentation rate increases at high temperatures.

Besides, temperature also affects the yeast metabolism, and therefore the production of ethanol, secondary metabolites, such as glycerol, succinic acid, acetic acid and aromatic compounds, such as acetate esters, fatty acid ethyl esters and fusel alcohols. All these changes will modify the chemical composition and the sensory quality of wines (Fleet and Heard, 1993).

High temperatures are generally associated with lower alcohol yields, mainly related to a drop in the ethanol yield and a reduction of the substrate (Riberéau-Gayon *et al.*, 1975; Llauradó *et al.*, 2002; Torija *et al.*, 2003a). This reduction is also related to an increase in the products of other metabolic pathways,

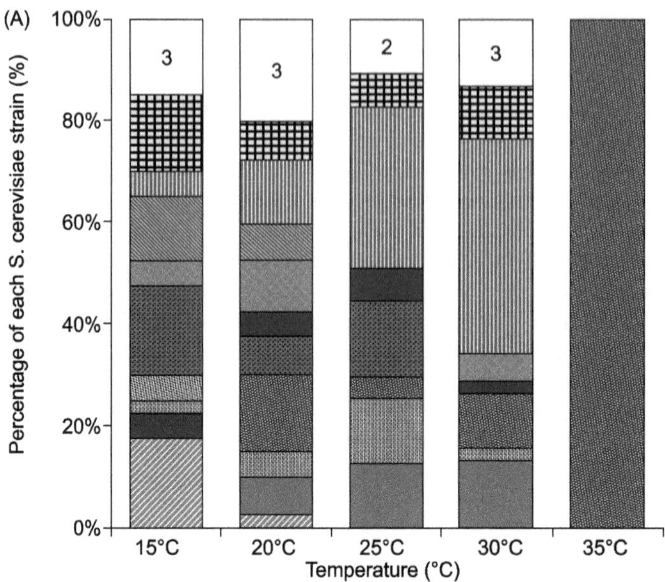


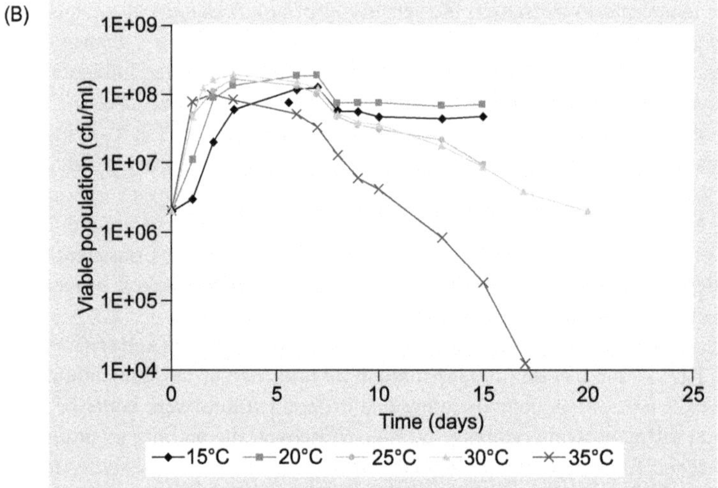


**Figure 3.** Effect of fermentation temperature on *S. cerevisiae* strain diversity at the end of mixed alcoholic fermentations (A) and on yeast viability during alcoholic fermentation (B). Each frame indicates a different *S. cerevisiae* strain. The numbers on the top of the bars indicate the number of different minority strains (<10 per cent) (Adapted from Torija *et al.*, 2003a)

Color version at the end of the book

such as acetic acid, acetaldehyde or glycerol. On the other hand, fermentations at low temperatures result in an increase of acetate esters, ethyl esters and overall medium-chain fatty acids (MCFA), although this increase seems to be strain dependent (Torija *et al.*, 2003a). Instead, fusel alcohols decrease with temperature (Castellari *et al.*, 1995; Llauradó *et al.*, 2002).

Finally, temperature also has impact on the membrane fluidity. Membrane fluidity is modulated in response to many environmental stresses by changing its dynamic and structural characteristics, by modifying its molecular composition or by adding foreign molecules that interact with membrane constituents (Beney and Gervais, 2001). Low temperatures cause a decrease in the membrane fluidity.

This is usually compensated by an increase in the production of unsaturated fatty acids and a decrease in the length of the fatty acid chains, what result in a less ordered structure and an increase in the membrane fluidity (Redón *et al.*, 2009; Torija *et al.*, 2003b). A decrease in the membrane fluidity also impacts the transport across, reducing the lateral diffusion of membrane proteins, decreasing membrane-associated enzyme activity and in consequence, lowering the membrane transport (Vigh *et al.*, 1998). The membrane permeases are highly dependent on the temperature, since temperature modification can produce changes in the conformation of these proteins (Entian and Barnett, 1992).

Nitrogen assimilation also depends on the fermentation temperature because low temperatures are related with a lower rate of amino acid assimilation (Lopez *et al.*, 1996), which results in lower rates of yeast growth and fermentation.

## 2.2. pH

The typical pH of grape must has no negative effect on the growth of *Saccharomyces* yeasts, which ranges between 2.8 and 4.2 (Heard and Fleet, 1988). In fact, the toxic effects of pH are due to the increased effects of ethanol (Pampulha and Loureiro-Dias, 1989) and sulfite (Farkas, 1988). Instead, potassium ions in the must can increase acidic pH tolerance, since potassium not only accelerates the rate of glucose consumption, but is also taken up by cells in exchange for hydrogen ions to counteract the acidification of the cytoplasm (Kudo *et al.*,1998). It is important to highlight that *S. cerevisiae* has to maintain their intracellular pH between 5.5-5.75 for a correct metabolism; therefore, when the cytoplasm is acidified, cells have to pump out protons by spending ATP to keep the intracellular pH within a physiological range optimum for the metabolism (Thomas *et al.*, 2002). In fact, at pH levels around 2.9-3.2, weak organic acids (acetic, malic, succinic and tartaric acids) present in the growth medium may enter into the cells in the undissociated form, because their corresponding pKa are higher than the external pH. Consequently, they increase the acidity of the cytosol and can lead to sluggish or stuck fermentation. For example, acetic acid and ethanol show synergistic effect and inhibit alcoholic fermentation by decreasing the internal pH, enolase activity and fermentation rate (Pampulha and Loureiro, 1989; Pampulha and Loureiro-Dias, 1989, 1990). Something similar happens with Medium Chain Fatty Acids (MCFA) which may also undergo dissociation because of the low pH, causing problems for yeasts (Borrull *et al.*, 2015). Therefore, in the case of wine, undissociated form of MCFAs and weak acids can pass across the plasma membrane by passive diffusion and dissociate in the neutral cytoplasm, causing a decrease in intracellular pH (Viegas and Sá-Correia, 1997) and possibly leading to the cell death. Instead, at physiological pH, although fatty acids also dissociate, they require a carrier to cross the yeast membrane.

On the other hand, low pH favours the hydrolysis of disaccharides, and therefore, fermentation, as well as, preventing spoilage microorganisms. Indeed, low pH of wine is one of the main preservatives for wine stability. Wines with higher pH (3.7-4.2) could promote the development of spoilage microorganisms, such as lactic acid bacteria and, thus are more problematic for aging or storage. Furthermore, high pH also reduces the antiseptic effectiveness of $SO_2$, as we will see later.

## 2.3. Sugar Concentration and Osmotic Pressure

In grape must, fermentable sugars are mainly glucose and fructose (relation 1:1). The total sugar concentration in must can vary between 150-250 g/l, being higher in musts used for the production of sweet wines (up to 350 g/l in Sauternes musts) (Ribéreau-Gayon *et al.*, 2006). The concentration of sugars can influence the selection of yeast strains to ensure the fermentation (Fleet, 1992). In fact, osmotic pressure is the first stress that yeast encounters after inoculation and is due to the presence of high concentration of sugars in the must (> 200 g/l). Previous reports described that fermentation with a few grams of sugar in the medium is slow and that the fermentation speed increases from 15-20 g/l and remains stable up to 200 g/l. However, above this concentration, the fermentation performance decreases (Ribéreau-Gayon *et al.*, 2006). Thus, a high amount of sugar hinders the yeast growth and decreases the maximum population. Additionally, the presence of these sugars involves a regulatory mechanism known as glucose or 'catabolite repression' that mainly acts at a transcriptional level (Gancedo, 2008). Due to this mechanism, the transcription of genes that are essential for the catabolism of slowly or non-fermentable carbon sources is repressed in the presence of glucose.

Due to osmotic stress, yeast cells first exhibit a lag-phase to adapt and then, start to reproduce rapidly. During the adaptation process, yeast cells accumulate metabolites that have osmo-protective effect, in particular glycerol. This molecule, which is produced throughout fermentation due to a need to balance the redox equilibrium *via* $NAD^+$ formation, has a fundamental action as osmo-protector to counteract the high sugar concentration in the early stages of the fermentation (Hohmann, 2002; Heinisch and Rodicio, 2009). Glycerol synthesis is accomplished through an increased activity of the enzymes, which divert glycolytic flux from triosephosphates. The signalling in this condition is mediated by the High Osmolarity Glycerol (HOG) pathway. This pathway stimulates the glycerol accumulation, but also induces the gene expression in response to heat shock (Varela *et al.*, 1992). During hyperosmotic stress, yeast cells usually balance the excess of $NAD^+$ formed during the production of glycerol by the formation of acetate using a $NAD^+$ dependent aldehyde dehydrogenase (Miralles and Serrano, 1995; Navarro-Aviño *et al.*, 1999). A recent study carried out with wine strains in natural grape must have concluded, after analysing the metabolic and transcriptomic data, that yeasts, under winemaking conditions, are able to respond and adapt themselves, after a few minutes, to the hyperosmotic environment by applying mechanisms that lead to the accumulation of intracellular glycerol until reaching an optimal level for osmoprotection. Once adapted, yeast cells can turn on their metabolism for biomass production, determining the exit from the lag phase (Noti *et al.*, 2015)

The ability of yeast to accurately detect and respond to adverse conditions is relevant during wine fermentation, especially in the first few hours. For this reason, several studies have revealed the transcription of genes just after the inoculation in the grape must. Rossignol *et al.* (2006) reported that surprisingly, inoculation into the grape must did not trigger a stress response despite the high concentrations of sugars, while Novo *et al.* (2007) linked an improved yeast vitality with the presence of fermentable carbon sources in the rehydration medium. Later, Jiménez-Martí *et al.* (2011), using the data obtained by transcriptomic approaches and the contents of intracellular glycerol, proposed that adaptation of yeast strains to the wine production conditions need a high expression of genes involved in both biosynthetic processes and glycerol biosynthesis, and the appropriate levels of intracellular glycerol. Recently, Ferreira *et al.* (2017) studied the effect of individual and combined stresses during the lag phase period in different wine yeast and found that low temperature and osmotic stress substantially affected all the strains, having as a result considerably extended lag phases.

## 2.4. Oxygen

Yeasts obtain energy from degradation of sugars, which can be carried out by either the respiratory or fermentative pathway. However, in grape must, due to catabolic repression of the respiration caused, as previously explained, by the high concentration of glucose, sugar degradation is performed almost exclusively by alcoholic fermentation in *S. cerevisiae* (Ribéreau-Gayon *et al.*, 2006). This is known as the Crabtree effect, a metabolic feature that strongly favours fermentative over respiratory metabolism despite oxygen availability (Pronk *et al.*, 1996), and must have been essential for the adaptation of this species to high sugar concentrations (Piskur *et al.*, 2006). Yeast species are classified according to how they regulate respiro-fermentative metabolism in Crabtree-positive or Crabtree-negative (Table 1).

Determination of this Crabtree status is generally performed using carbon limited chemostat conditions (Pronk *et al.*, 1996), but recent studies demonstrate that such assays are not best to predict the behaviour of yeast under the conditions found during wine fermentation (Quirós *et al.*, 2014; Morales *et al.* 2015). One of the problems for a good classification of yeast species is that until recently, very few yeast species have been systematically studied in relation to their carbon metabolism (Merico *et al.*, 2007). Recently, Hagman *et al.* (2014) studied the carbon metabolism of several yeast species in response to a glucose pulse, concluding that *Pichia*, *Debaryomyces* and *Kluyveromyces marxianus* do not form ethanol while *Lanchacea*, *Torulaspora* and *Saccharomyces* exhibits a rapid ethanol accumulation. This preferential respiration as a metabolic feature of some non-*Saccharomyces* species have been used as a tool for reducing the alcohol levels in wines (Quirós *et al.*, 2014; Morales *et al.*, 2015), selecting for this purpose some *Metschnikowia pulcherrima* strains and two species of *Kluyveromyces* (Quirós *et al.*, 2014).

Moreover, the oxygen concentration is also important to regulate the presence and dominance of yeast species during wine fermentation. Hansen *et al.* (2001) mentioned that *S. cerevisiae* is more tolerant

**Table 1.** Classification of Different Wine Yeast according to their Respiro-fermentative Metabolism

| *Respiro-fermentative metabolism* | |
|---|---|
| *Crabtree-positive yeasts* | *Crabtree-negative yeasts* |
| *Saccharomyces cerevisiae* | *Hanseniaspora uvarum* |
| *Zygosaccharomyces bailii* | *Pichia anomala* |
| *Brettanomyces intermedius* | *Candida utilis* |
| *Torulopsis glabrata* | *Hansenula neofermentans* |
| *Schizosaccharomyces pombe* | *Kluyveromyces marxianus* |
| *Hanseniaspora guilliermondii* | *Debaryomyces hansenii* |
| *Candida stellata* | |
| *Torulaspora delbrueckii** | |

*Despite being considered Crabtree-positive, several authors have reported a higher contribution of respiration in *T. delbrueckii* metabolism than in *S. cerevisiae* (Alves-Araújo *et al.*, 2007; González *et al.*, 2013; Merico *et al.*, 2007).

Adapted from Rodicio and Heinisch 2009

to low oxygen levels compared to other yeast. Thus, a reduced oxygen availability under grape juice fermentation might have an important role as a selective factor in mixed cultures. In fact, the low tolerance to low available oxygen of *Lachancea thermotolerans* (syn. *Kluyveromyces thermotolerans*) and *T. delbrueckii* could in part explain their relative competitiveness and rapid death when they are in the presence of *S. cerevisiae* (Hansen *et al.*, 2001). On the other hand, *M. pulcherrima*, that improves its growth and sugar consumption in aerobic conditions, survives for greater time in mixed cultures with *S. cerevisiae* when few quantities of oxygen (10-25 per cent air) were supplied during the fermentation (Morales *et al.*, 2015).

Despite all the findings explained so far, *S. cerevisiae* requires oxygen during the alcoholic fermentation for the synthesis of sterols and fatty acids and thus it can be stated that the amount of available oxygen is crucial for its growth and fermentation performance (Hanl *et al.*, 2005). Although the concentration of molecular oxygen is particularly low during wine fermentation, several practices that are employed during the first stages of winemaking, such as pumping-over or micro-oxygenation, can transiently but significantly increase the oxygen concentration. These techniques are mainly used in red wines; white wines are not usually oxygenated due to fear of oxidising the must and modifying the aromas. In general, aeration accelerates fermentation, resulting in an increased demand of nitrogen containing-nutrients. The addition of oxygen has a similar effect as the addition of sterols, which are considered to be oxygen substitutes (Ribéreau-Gayon *et al.*, 2006). Aeration can benefit fermentation when oxygen is added during the first stages of the process. It is important to highlight that it is not the must that needs to be aerated, but rather the yeasts that ferment the must. Sablayrolles and Barre (1986) defined a value around 10 mg/l as the oxygen needs of yeasts. In fact, aeration has been reported to improve cell viability during alcoholic fermentation (Alfenore *et al.*, 2004), especially in fermentations at low temperatures (Redón *et al.*, 2009) and with high ethanol content (Alexandre *et al.*,1994; Ding *et al.*, 2009). But oxygen is also dangerous for life as it can produce the accumulation of Reactive Oxygen Species (ROS) in cells. Oxygen is a highly reactive molecule that can be partially reduced to generate ROS, including superoxide anions ($O_2^{\cdot -}$), singlet oxygen ($^1O_2$), hydroxyl radicals ($OH^-$) or hydrogen peroxide ($H_2O_2$). In a biological context, ROS are natural byproducts of the normal metabolism of oxygen, which can cause yeast death and may lead to undesirable production of acetaldehyde and hydrogen sulphide as well as reduced production of aromatic esters (Nykänen, 1986). Moreover, ROS can also react with DNA, lipids and proteins, altering their functions. Cumulative ROS is also known as oxidative stress. Yeast cells have a range of responses to ROS that are dose-dependent. At very low doses, cells are able to adapt to become more resistant to the following lethal exposures (Borrull *et al.*, 2015). At higher doses, cells activate various antioxidant functions, including cell-division cycle delay. However, at even higher doses, a proportion of cells die by apoptosis (Perrone, 2008).

## 2.5. Ethanol

Ethanol is an inhibitor of yeast growth even at relatively low concentrations as it is able to inhibit cell division and to decrease both cell volume as well as specific growth rate. At high ethanol concentrations, the negative effects increase, reducing cell vitality and increasing cell death (Birch and Walker, 2000). Due to its small size and alcoholic hydroxyl group, ethanol is soluble both in aqueous and in lipid environments, allowing it to cross through the plasma membrane to enter into the cell and to increase the membrane fluidity and permeability (Navarro-Tapia *et al.*, 2016). Ethanol also affects cell metabolism and macromolecular biosynthesis. All these changes result in the induction of the production of heat shock-like proteins (HSP), the reduction of the rate of RNA and protein accumulation, the enhancement of the frequency of petite mutations and finally, in the denaturation of intracellular proteins and glycolytic enzymes (Hu *et al.*, 2007; Stanley *et al.*, 2010a).

Ethanol is one of the principal products of the metabolism of yeast and is the main stress factor during fermentation. It slows the assimilation of nitrogen and paralyses the yeast, by modifying cell active transport systems across the membrane (Henschke and Jiranek, 1993). The concentration of ethanol required to block the fermentation depends, among other factors, on yeast strain, temperature and aeration (Ribéreau-Gayon *et al.*, 2006). Rosa and Sa-Correia (1991) reported that concentrations of ethanol above 3 per cent (v/v) affect growth and fermentation rates, but also result in a potent activator of the plasma membrane $H^+$-ATPase, one of the mechanisms that yeasts have to regulate its cell internal pH.

Therefore, which are the mechanisms used by yeast species, such as *Saccharomyces* to cope with the stress generated by the ethanol? Several studies provide evidences to elucidate the molecular basis underlying the yeast response and the resistance to ethanol stress (Alexandre *et al.*, 2001; Chandler *et al.*, 2004; Li *et al.*, 2010; Stanley *et al.*, 2010b). Ethanol stress triggers multiple cell responses to antagonise fluidity caused by ethanol, by enhancing the stability of proteins and membranes and maintaining intracellular pH homeostasis. These mechanisms include (1) inhibition of cell cycle and propagation, (2) accumulation of trehalose and glycogen, (3) increase in heat shock proteins, (4) increase the activity of plasma membrane ATPase, the levels of oleic acid and ergosterol in membrane, (5) induction of genes encoding vacuolar proteases and their inhibitors and (6) increased activation of genes related with unfolded protein response and their transcription factors (Heinisch and Rodicio, 2009; Navarro-Tapia *et al.*, 2016). All these mechanisms that allow wine yeasts to endure ethanol stress have been basically studied in the model species *S. cerevisiae*. However, little is known about the mechanism related with ethanol tolerance of non-*Saccharomyces* yeasts.

*Saccharomyces* is a highly ethanol-tolerant species; however, the presence of relatively high ethanol concentrations also results in the inhibition of the cell growth and viability, and in a reduction of the fermentation productivity and ethanol yield (Galeote *et al.*, 2001; Aguilera *et al.*, 2006). In fact, this high tolerance to ethanol has been related to its ability to prevail during wine fermentations and similarly, the disappearance of non-*Saccharomyces* species has been attributed to their inability to survive the increasing concentrations of ethanol produced during this process. A concentration between 5-7 per cent (v/v) of ethanol has been often cited as the maximum tolerance for non-*Saccharomyces* species (Heard and Fleet, 1988; Gao and Fleet, 1988). However, it's noteworthy that some strains of non-*Saccharomyces* have been isolated at late stages and even at the end of spontaneous fermentations (Torija *et al.*, 2001; Llauradó *et al.*, 2002; Nurgel *et al.* 2005; Wang and Liu, 2013; Padilla *et al.*, 2016). Recent studies also found that some non-*Saccharomyces* strains of *H. uvarum*, *H. guilliermondii*, *T. delbrueckii* and *S. bacillaris* could tolerate ethanol of 10 per cent (Pina *et al.*, 2004; Pérez-Nevado *et al.*, 2006; Wang *et al.*, 2015), which means that these species may be more resistant to ethanol than it was considered in the past. Salvadó *et al.* (2011) studied the effect of increasing concentrations of ethanol (0-25 per cent) and temperatures (4-46°C) during wine fermentations performed in competition between *S. cerevisiae* and several non-*Saccharomyces* species (*H. uvarum, T. delbrueckii, S. bacillaris, P. fermentans, K. marxianus*). These authors suggested that high ethanol tolerance of *S. cerevisiae* only gives a clear advantage over the other species when the ethanol concentration exceeds 9-10 per cent (v/v). Therefore, their results cannot explain the imposition of *Saccharomyces* during the earlier stages of fermentation, or its dominance over *S. bacillaris*, which was the most resistant yeast in this study. Instead, their data show that a temperature increase of just a few degrees results in strong increase of *Saccharomyces* in the medium, as evident at temperatures above 15°C. Interestingly, *Saccharomyces* and also *Dekkera* have been reported to use the 'make-accumulate-consume'" strategy that consists in rapid consumption of sugars by these yeasts, transforming them into ethanol, which help them to establish a competitive dominance in the ecological niche, and finally catabolising this ethanol for energy (Piskur *et al.*, 2006; Dashko *et al.*, 2014; Marsit and Dequin, 2015).

## 2.6. Sulphites

The use of sulphur dioxide ($SO_2$) in winemaking is believed to be dated back to the end of 18th century. Its principal properties are: (i) as an antiseptic, it inhibits the development of most spoilage microorganisms, (ii) as an antioxidant, it binds dissolved oxygen; however, this reaction is slow, protecting wines only from chemical oxidation but not from enzymatic oxidations, (iii) as an antioxidasic, it instantaneously inhibits the activity of oxidation enzymes (tyrosinase, laccase) and can ensure their destruction over the time (Ribéreau-Gayon *et al.*, 2006). Sulphur dioxide exists in equilibrium among molecular $SO_2$, bisulphite and sulphite after dissolving in water and the dominant form depends on pH (Fig. 4). Molecular $SO_2$ is believed to be an antimicrobial form by various mechanisms, including breakage of protein disulphide bridges, reaction with $NAD^+$ and ATP, deamination of cytosine to uracil and reduction of crucial nutrients such as thiamin (Fugelsang and Edwards, 2007). Salma *et al.* (2013) reported that an increase in medium pH from 3.5-4.0 can reduce the toxicity of $SO_2$, allowing resuscitation of viable but non-culturable (VBNC) cells.

At the same time, high levels of sulphites can also have a negative impact on wine sensory properties, delay the onset of malo-lactic fermentation and cause some health concerns. For all these reasons, $SO_2$ levels in wines are regulated. On wine bottles, it is compulsory to include 'contains sulphites' on the label when the concentration exceeds 10 mg/l and the European legislation (Reg. ECNo 606/ 2009) has set

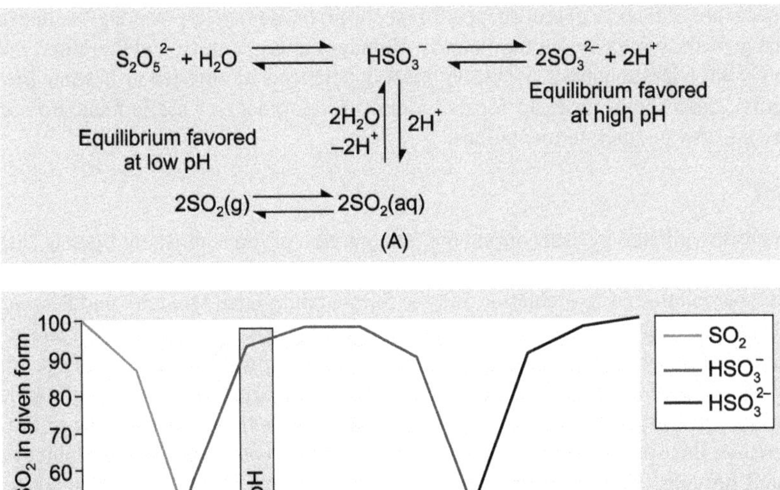


**Figure 4.** Equilibrium between the different sulphur dioxide forms (molecular, bisulphite and sulphite) in aqueous solution after supplementation in the form of metabisulphite (A) and percentage of each form in relation to the pH (B). The grey box indicates the range of pHs usually found in wines

Color version at the end of the book

a maximum limit to the sulphites in wines for the protection of human health. This limit is established as 150 mg/l for red wines and 200 mg/l for white and rosé wines, being higher for wines with a content of sugars above 5 g/l. In the case of organic wines, this limit is reduced to 100 m/l for red wines and 150 mg/l for white and rosé wines (Reg. EC No. 203/2012). It is important to highlight that yeasts also produce small quantities of $SO_2$ during fermentation, usually below 10 mg/l level but in some cases, it can exceed 30 mg/l (Ribéreau Gayon *et al.*, 2006). Consequently, in the winemaking process, it is important to control and manage the $SO_2$ content of wine for maintaining the lowest possible concentration while preserving its characteristic properties.

Non-*Saccharomyces* yeasts are known to be more sensitive to the combined toxicity of $SO_2$ and ethanol than *Saccharomyces* (Jolly *et al.*, 2014). The inhibitory effect of $SO_2$ on the growth of non-*Saccharomyces* has been investigated by culture-dependent and culture-independent techniques, including plating, quantitative Polymerase Chain Reaction (qPCR), PCR-Denaturing Gradient Gel Electrophoresis (PCR-DGGE) or Direct Epifluorescence Technique (DEFT) (Cocolin and Mills, 2003; Divol and Lonvaud-Funel, 2005; du Toit *et al.*, 2005; Andorrà *et al.*, 2008; Takahashi *et al.*, 2014). For example, Divol and Lonvaud-Funel (2005) used DEFT to observe metabolic activity of yeast cells under $SO_2$ stress and found that *S. bacillaris* is more sensitive to $SO_2$ than *Zygosaccharomyces bailii* and *S. cerevisiae*. However, $SO_2$ addition after crushing has small effect on some yeast species from genera of *Pichia*, *Saccharomycodes*, *Schizosaccharomyces* and *Zygosaccharomyces* (Fugelsang and Edwards, 2007). Moreover, Bokulich *et al.* (2015) investigated the impact of $SO_2$ treatments (from 0 to 150 mg/l) on microbial communities during grape must fermentation, using massive sequencing and finding that sulphite addition did not significantly affect the fungal population but did affect bacterial diversity.

## 3. Nutritional Requirements

Nutritional aspects are critical to obtain successful alcoholic fermentations. Among the nutrients needed to support yeast growth, sugars are by far the most abundant in grape must and therefore, they are not a limiting factor for wine fermentations. However, must deficiencies in nitrogen or in some growth factors (vitamins, sterols), could be a handicap for the correct development of the fermentation and must be controlled to avoid slow or stuck fermentations.

### 3.1. Nitrogen

Of all the nutrients assimilated by yeast during the fermentation of grape must, nitrogen is, quantitatively, the most important one after carbon compounds. In fact, the nitrogen composition of grapes affects the cell viability, the completion of fermentation and the fermentation rate (Monteiro and Bisson,1991). The nitrogen concentration of grape musts can vary between 0.1-1 g/l, including polypeptides (25-40 per cent), amino acids (25-30 per cent), proteins (5-10 per cent) and ammonium cation (3-10 per cemt of total nitrogen) (Riberéau-Gayon *et al.*, 2006). Yeasts utilise these nitrogen sources, mainly amino acids and ammonium cation, to incorporate them into the structural and functional components of the cell. In the case of *S. cerevisiae*, this species cannot fix atmospheric nitrogen (being non-diazotrophic) or assimilate polypeptides and proteins, since it is not able to hydrolyse them. Therefore, *S. cerevisiae* requires a supply of readily assimilable organic nitrogen (e.g. amino acids) or inorganic nitrogen (e.g. ammonium salts) for growth and fermentative metabolism. In fact, *S. cerevisiae* can survive with ammonia as sole nitrogen source because it is capable of synthesizing all amino acids from ammonium cation. Anyway, the addition of amino acids stimulates the yeast growth and the addition of mixtures of amino acids and ammonia are even more effective as growth stimulants.

Nitrogen deficiencies are one of the major causes of stuck or sluggish fermentation. To avoid these problems, oenologists add routinely nutritional supplements, usually inorganic sources of nitrogen (ammonium salts) to the grape must at the beginning of wine fermentation. In fact, these additions are usually made without previously determining the quantity of nitrogen present in the grape must or the nitrogen requirements of the yeast used in the cellar. Bely *et al.* (1990) proposed a concentration of 140 mg/l as the minimum amount of nitrogen necessary to carry out a normal fermentation without the need of an external nitrogen supplementation. In the case of nitrogen deficient musts, additions during the fermentation rapidly increase the fermentation rates and reduce the fermentation times. However, these additions could be even harmful when the initial content of nitrogen is around 200-350 mg/l, since yeasts have totally met their nitrogen needs and this excess of nitrogen can cause microbiological instability and the formation of toxic compounds, such as ethyl carbamate and biogenic amines. Moreover, the timing of the nitrogen addition also has a great impact on the fermentation, influencing the biomass and secondary metabolites production, the fermentation performance and the nitrogen pattern consumption (Beltran *et al.*, 2005). In the study performed by Beltran *et al.* (2005), a reduction of the total fermentation time was observed, regardless of the time of nitrogen addition. However, this reduction was higher when the nitrogen was added in the exponential phase. Furthermore, the addition of ammonium in the late stages of the fermentation seems to be counterproductive, since this compound is the preferred one for the biomass production at the beginning of the process, but it was hardly consumed when added in the final stages of the fermentation.

The effect of nitrogen concentration on fermentation kinetics is also dependent on sugar concentration, especially to complete fermentation. Yeast nitrogen requirements during grape must fermentations increase with increasing sugar concentrations (Martínez-Moreno *et al.*, 2012).

### *3.1.1. Nitrogen Metabolism*

Once inside the cell, nitrogen compounds can be used directly in biosynthetic processes, deaminated to ammonium or transaminated, transferring the amino group to α-ketoglutarate, to generate glutamate (Cooper, 1982; Magasanik, 1992). Therefore, ammonium, glutamate together with glutamine play an essential role in the nitrogen metabolism and the interconversion between them is called Central Nitrogen Metabolism (CNM). From glutamate and glutamine, all other nitrogen containing compounds in the cell can be produced (Fig. 3). Both glutamate and glutamine can be synthesised, using ammonia as the amino donor. Glutamate is synthesised from ammonium and α-ketoglutarate by the NADPH-dependent glutamate dehydrogenase (NADPH-GDH), whereas glutamine synthetase (GS) catalyse the synthesis of glutamine from ammonium and glutamate at the cost of one ATP (reviewed by Ljungdahl and Daignan-Fornier, 2012). The rest of nitrogen compounds are synthesised by using ammonia from either glutamate or glutamine as shown in Fig. 5. On the other hand, one molecule of glutamine and α-ketoglutarate are converted into two molecules of glutamate by the glutamate synthase (GOGAT) and this glutamate can be degraded into α-ketoglutarate and ammonia by the NAD dependent glutamate dehydrogenase (NAD-GDH). Finally, Cooper (1982) defined that around 85 per cent of the total cellular nitrogen is incorporated via glutamate and the remaining 15 per cent via glutamine.

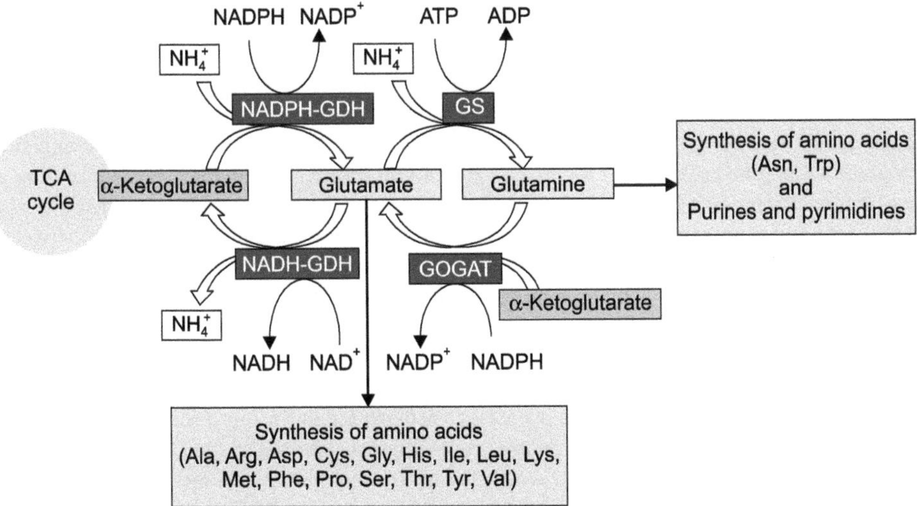


**Figure 5.** Central Nitrogen Metabolism (CNM). NADPH-GDH: NADPH-dependent glutamate dehydrogenase, GS: glutamine synthetase, NAD-GDH: NAD-dependent glutamate dehydrogenase, GOGAT: glutamate synthase (Adapted from Ljungdahl and Daignan-Fornier, 2012)

### *3.1.2 Nitrogen Source Classification*

The nitrogen compounds used by yeast are often classified as preferred (good) or non-preferred (poor) sources. This classification has been empirically based on two criteria – the growth rate supported by a particular nitrogen compound when it is present in the medium as sole nitrogen source and which systems for using alternative nitrogen sources are de-repressed during growth on a particular nitrogen source. The nitrogen sources that do not de-repress the pathways for the use of alternative nitrogen sources, such as those affected by nitrogen catabolite repression (NCR) are classified as preferred nitrogen sources. Instead, the nitrogen compounds that lead to the de-repression of alternative nitrogen pathways, what indicates an inactive NCR, are known as non-preferred nitrogen sources (Cooper, 1982; Magasanik, 1992; ter Schure *et al.*, 2000; Magasanik and Kaiser, 2002; Ljungdahl and Daignan-Fornier, 2012).

Although most nitrogen sources can be unambiguously classified as preferred and non-preferred, according to these two criteria, this classification is not absolute, since the repressive effects of some particular sources can vary significantly between different yeast strain backgrounds. For example, ammonium seems to be a preferred nitrogen source for the strains derived from *S. cerevisiae* ∑1278b, but not for S288c-derived strains, although for both strains, ammonia promotes high growth rates (Magasanik and Kaiser, 2002). Godard *et al.* (2007) compared the transcriptome of *S. cerevisiae* growing on 21 individual sources of nitrogen and found that a differential transcription response was triggered depending on the nitrogen source. Moreover, these authors classified the nitrogen sources in two groups according to their data – a first group was composed by alanine, ammonium, arginine, asparagine, aspartate, glutamate, glutamine and serine, all of them being compounds that support fast growth, have a highly active NCR and their catabolism yields carbon derivatives directly assimilable by the cell, and a second group that includes isoleucine, leucine, methionine, threonine, tryptophan and tyrosine, nitrogen compounds that support slower growth and promote the excretion of non-metabolised carbon compounds, such as fusel oils and the activation of the general control of amino acid biosynthesis.

### *3.1.3. Nitrogen Consumption During Wine Fermentation*

Spontaneous fermentation of grape must is driven by a complex community of microorganisms. Non-*Saccharomyces* species grow well during the early stages of the fermentation, when the ethanol concentration is still low, being later replaced by *Saccharomyces*, which are more tolerant to ethanol and more competitive to grow in media with high sugar concentration (Amerine *et al.*, 1982; Lafon-Lafourcade, 1983; Querol *et al.*, 1990). In general, this early growth of non-*Saccharomyces* species during wine fermentation reduces the presence of amino acids and vitamins in the medium, leading to limited growth of *S. cerevisiae* (Fleet, 2003; Ciani and Comentini, 2015). This fact was evidenced by Bisson (1999) and Taillandier *et al.* (2014), as these authors proved that the early growth of *H. uvarum* (*K. apiculata*) and *T. delbrueckii* leads to nitrogen exhaustion and therefore, to sluggish fermentation. In mixed fermentation, different results were obtained depending on the species used. For example, in sequential fermentations inoculated with *S. cerevisiae* and *M. pulcherrima* or *H. vinae,* an evident competition for nitrogen was highlighted (Medina *et al.*, 2012). Instead, in mixed fermentation of *S. cerevisiae* and *H. uvarum,* less consumption of nitrogen was observed in comparison to fermentation carried out with pure cultures (Ciani and Comentini, 2015), indicating no competition for assimilable nitrogen between these species. However, the delay in the inoculation of *S. cerevisiae* in mixed fermentation is affected by the availability of nitrogen and low concentrations of nitrogen together with late inoculation can result in sluggish or stuck fermentation (Lleixà *et al.*, 2016). Anyway, it is important to highlight that some non-*Saccharomyces* species can exhibit an important proteolytic activity that can contribute to the enrichment of the medium in nitrogen sources.

Nitrogen limitation probably impacts on yeast growth pattern during grape must fermentation, which depends on the different nitrogen requirements from different wine yeast species (Ciani and Comitini, 2015, Albergaria and Arneborg, 2016). Recently, the nitrogen requirements, metabolic mechanisms and genetic basis were studied for *S. cerevisiae* (Martínez-Moreno *et al.*, 2012; Gutiérrez *et al.*, 2013; Brice *et al.*, 2014, 2015), whereas, nitrogen requirements of non-*Saccharomyces* species have not yet been fully characterised. The addition to the medium of complex nutrients, such as yeast extract or peptone, has been reported to partially improve the fermentation capacity of *H. uvarum*, *H. guilliermondii* and *S. bacillaris*

(Albergaria, 2007). On the other hand, the analysis of nitrogen consumption by *S. cerevisiae*, *H. uvarum* and *S. bacillaris* in pure and mixed fermentation indicated that the two non-*Saccharomyces* yeasts were less effective converting amino acids into biomass than *S. cerevisiae* but contributed more to aromatic compounds formation (Andorrà *et al.*, 2012). Finally, Kemsawasd *et al.* (2015) investigated the effect of different nitrogen sources (single amino acids and ammonium sulphate and two multiple nitrogen sources) on growth, glucose consumption and ethanol production of *S. cerevisiae*, *L. thermotolerans*, *M. pulcherrima*, *H. uvarum* and *T. delbrueckii*. In general, the single amino acids with beneficial effects on fermentation performance were different for each wine species. Moreover, a positive effect of complex nitrogen mixtures was described on *S. cerevisiae* and *T. delbrueckii*, whereas on the other three non-*Saccharomyces* species, mixtures and single amino acids present a similar effect. However, not all single amino acids are able to support growth and alcoholic fermentation equally, with great differences that led to their classification in several levels of 'appropriateness' for alcoholic fermentation (Fig. 6).

Nitrogen source and concentration also influence the growth and fermentation profiles of different *S. cerevisiae* strains during wine fermentation and therefore, their imposition. Different strains of *S. cerevisiae* exhibit a particular nitrogen demand and ability to use a variety of nitrogen sources (Gutiérrez *et al.*, 2012, 2013). Brice *et al.* (2014) assessed the fermentation capacity of several yeast strains under nitrogen-deficient conditions and observed important differences between strains. Thus, it could be

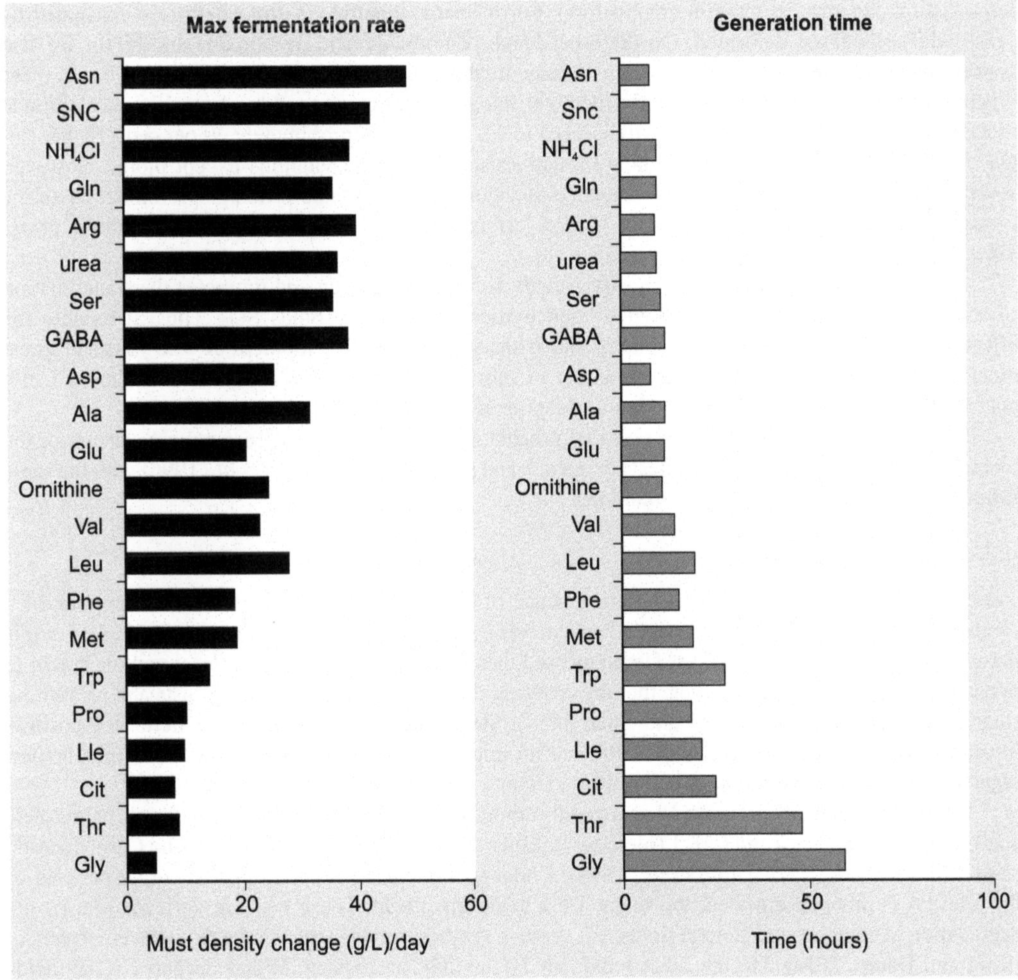


**Figure 6.** Effect of different nitrogen sources on fermentation kinetics and yeast growth on synthetic must (160 g/L sugar). The nitrogen concentration was 300 mg N/L in all cases. SNC: complete synthetic must (25 per cent $NH_4^+$ and 75 per cent organic nitrogen)

concluded that the nitrogen requirements of yeasts are strain-specific in terms of nitrogen concentration. These variances were not explained by differences in their ability to store nitrogen neither by their protein synthesis activity. However, the transcriptome analysis showed specific expression patterns for the strains with low or high nitrogen requirements.

## 3.2 Growth Factors

Growth factors affect cellular multiplication and activity, even at very low concentrations. They are indispensable to microorganisms and a deficiency in these substances disturbs their metabolism. Growth factors include vitamins, nucleotides and nucleosides, purines and pyrimidines, fatty acids and sterols. Grape musts should be able to provide these growth factors to yeast for alcohol fermentation. However, sometimes there are deficiencies in some growth factors that are corrected using commercially available products that supplement the medium.

### *3.2.1. Vitamins*

Vitamins essential for yeasts are biotin, thiamine, pyridoxine, pantothenic acid, mesoinositol, nicotinamide, riboflavin. They are essential components of coenzymes and are involved in metabolic reactions.

Although grape must has sufficient amounts of vitamins to ensure the yeast development, alcoholic fermentation can alter its vitamin composition. For instance, thiamine is almost entirely consumed by yeast, while riboflavin is formed. On the other hand, pantothenic acid, pyridoxine and biotin are also consumed by yeast but later, they are again released to the medium, therefore, the concentrations of these vitamins in the must and wine are nearly identical. Meosinositol is not used by yeast, like nicotinamide in red wines. However, the latter is partially used in white wines (approximately 40 per cent) (Ribéreau-Gayon *et al.*, 2006). *S. cerevisiae* is able to synthesise all essential vitamins except biotin. However, several studies have reported that the presence of vitamins in the medium is highly stimulatory to the growth and fermentation of this species (Lafon-Lafourcade and Ribéreau-Gayon, 1984; Ough *et al.*, 1989; Fleet and Heard, 1993).

A deficiency in thiamine reduces yeast growth, slows fermentation and promotes the accumulation of some metabolites such as acetaldehyde and pyruvic acid, which binds $SO_2$. Thus, it reduces the effectiveness of $SO_2$ as an anti-microbial or antioxidant agent. The development of wild yeasts or grape moulds, such as botrytis, diminishes thiamine in musts. The addition of thiamine is authorised in a concentration no more than 0.6 g/l, added as thiamine hydrochloride (Reg. EC No. 606/2009)

A deficiency in pantothenate, an essential precursor for the synthesis of coenzyme A, decreases the biomass yield and increases the synthesis of acetate and pyruvate (Taherzadeh *et al.*, 1996) with the same results as previously mentioned for thiamine.

### *3.2.2. Sterols and Fatty Acids*

Yeasts are unable to grow in the complete absence of oxygen because yeasts need this compound for synthesising unsaturated fatty acids (UFA) and sterols (Alexandre *et al.*,1994; Ding *et al.*, 2009). For this reason, for many years, winemakers supplemented their fermentations with ergosterol and oleic acid to prevent lack of these essential lipids, thereby ensuring the healthy growth of their yeasts and preventing sluggish fermentation (Andreasen and Stier, 1954). Sterol and fatty acid (FA) and sterol composition in yeast have been shown to be responsible for an important physical characterisation of membranes, regulating membrane permeability and fluidity (Daum *et al.*,1998).

The FA composition of *S. cerevisiae* is mainly composed by C16:0 (palmitic acid), C16:1 (palmitoleic acid), C18:0 (stearic acid) and C18:1 (oleic acid). Minor FAs are C14:0 (myristic acid) and C26:0 (cerotic acid) (Viljoen *et al.*, 1986; Tuller *et al.*, 1999). Under standard growth conditions, about 80 per cent of the total FA is monounsaturated, the major UFA being oleic acid. These FAs are derived from (i) the degradation of complex and storage lipids, (ii) *de novo* synthesis and (3) the uptake from the environment (Klug and Daum, 2014). On the other hand, MCFA, mainly octanoic (C8) and decanoic (C10) acids are produced by yeast during alcoholic fermentation. High concentrations of these MCFAs have been related with stuck and sluggish fermentations, because they are toxic for yeast (Taylor and Kirsop *et al.*, 1977; Lafon-Lafourcade *et al.*, 1984; Viegas *et al.*, 1989; Edwards *et al.*, 1990). Two hypotheses have been proposed for this synthesis (Bardi *et al.*, 1999): (i) MCFA are produced to counteract the lack of synthesis of UFAs in absence of oxygen, since the presence of a short chain increases membrane fluidity in a similar way to the presence of a double bond in a long chain FA, (ii) MCFAare produced due to the release of medium-chain acyl-CoA from the fatty acid synthetase complex and their further hydrolysis to recycle CoSH, then being released into the medium. Although Bardi *et al.* (1999) concluded that MCFAs are not immobilized in cell structures and are mainly released into the medium, Torija *et al.* (2003b) reported that *Saccharomyces* species clearly increase their percentages of MCFA in the membrane fatty acid profiles during fermentation, especially at 13°C, but they are also released into medium. Therefore, both hypotheses for MCFA synthesis seem to be correct.

On the other hand, sterols are essential to maintain the integrity of the yeast membranes. Yeast can synthesise and incorporate sterols into its membrane, ergosterol being the principal sterol in yeasts (Starr and Parks, 1962; Parks, 1978). Ergosterol consists of a four-ring structure with an acyl side chain and a hydrophilic hydroxyl group, which allow its insertion into the membranes (Klug and Daum, 2014).

As explained before, UFAs and sterols are poorly synthesised in hypoxic conditions, while the levels of MCFA, squalene and lanosterol increase (Ratledge and Evans, 1989). Oxygen has an influence on biosynthesis of cellular fatty acids, sterols and phospholipids during alcoholic fermentation by yeasts. In fact, several studies have established that ethanol tolerance of yeast is related to an increase in UFA and to the maintenance of ergosterol and phospholipid contents (Thomas *et al.*,1978; Mauricio *et al.*,1991; Alexandre *et al.*,1994, 1996).

It has been proven that yeasts can directly incorporate sterols and UFAs from grape must during alcoholic fermentations (Chen, 1980). In fact, Mauricio *et al.* (1998) described that *S. cerevisiae*, but not *T. delbrueckki,* was able to recover a normal cellular growth and fermentation activity by the addition of ergosterol (25 mg/l) and oleic acid (31 mg/l) to a culture in anaerobiosis. Instead, both species were able to recover their growth and fermentation activity when a short aeration (48 hours) was applied to these cultures in anaerobiosis.

Several studies have confirmed that lipid membranes of yeast are stress sensors and modulate their composition to reach the optimal proportion in each situation that improves yeast viability (Vigh *et al.*,1998; Shobayashi *et al.*, 2005). For instance, in Redón *et al.* (2009), cells incubated with C16:1 increased their viability but also significantly reduced the fermentation time at 13°C. These cells exhibited higher levels of C16:1 and ergosterol, a shorter chain length of FAs and higher sterol/phospholipid ratio.

## 4. Conclusions and Future Remarks

Wine fermentation is a complex microbial process based on the transformation of fermentable sugars into alcohols. It is performed by various yeast species occurring naturally on the grapes or from commercial yeast starter cultures. Non-*Saccharomyces* yeasts naturally associated with grapes initiate spontaneous alcoholic fermentation and are then, replaced by *Saccharomyces* throughout the process. However, this type of fermentation can lead to the spoilage of wine, because of the unpredictable nature of microbial interactions. Many factors, including unfavourable temperature and pH, nutrient deficiencies, high sugar concentrations, presence of some toxic residues, microbial incompatibility and enological practices may also lead to stuck and sluggish fermentations. Hence, the use of yeast starters provides control of fermentation process and improvement of the overall quality, complexity and analytical profile of the wine.

It is very important to know the factors that influence the yeast cell viability and success of the fermentation to select the most appropriate strains to be inoculated in each wine condition. *S. cerevisiae* is traditionally preferred for alcoholic fermentation and considered as the 'wine yeast', since it presents the most efficient fermentative catabolism and ethanol tolerance. Moreover, recent studies have shown that selected wild strains have great potential to be used as starter culture at industrial level by performing not only successful alcoholic fermentation, but also improving aromatic complexity.

Although *S. cerevisiae* is commonly used in wine industry as commercial starters to ensure a predictable and reproducible processes, non-*Saccharomyces* yeasts have also been highlighted to reduce the ethanol content and provide high quality wines by producing some beneficial metabolites.

In recent years, the use of mixed starter cultures, composed by *S. cerevisiae* and non-*Saccharomyces* species, has gained momentum in wine industry especially for improving sensorial profile of the product. Although the nutrient requirements, metabolic mechanisms and genetic basis have been well studied for *S. cerevisiae,* little is known about non-*Saccharomyces* yeasts. Hence, further efforts are desirable to understand the properties and requirements of non-*Saccharomyces* species and how each species contributes to fermentation. In this regard, investigations on yeast interactions and mixed fermentations that will provide proper management of wine fermentation, are required.

## References

Aguilera, F., Peinado, R.A., Millan, C., Ortega, J.M. and Mauricio, J.C. (2006). Relationship between ethanol tolerance, $H^+$-ATPase activity and the lipid composition of the plasma membrane in different wine yeast strains. *International Journal of Food Microbiology* 110: 34-42.

Alexandre, H., Ansanay-Galeote, V., Dequin, S. and Blondin, B. (2001). Global gene expression during short-term ethanol stress in *Saccharomyces cerevisiae. FEBS Letters* 498: 98-103.

Alexandre, H., Rousseaux, I. and Charpentier, C. (1994). Relationship between ethanol tolerance, lipid composition and plasma membrane fluidity in *Saccharomyces cerevisiae* and *Kloeckera apiculata. FEMS Microbiology Letters* 124: 17-22.

Alexandre, H., Mathieu, B. and Charpentier, C. (1996). Alteration in membrane fluidity and lipid composition and modulation of $H^+$-ATPase activity in *Saccharomyces cerevisiae* caused by decanoic acid. *Microbiology* 142: 469-475.

Albergaria, H. (2007). Physiological studies of non-*Saccharomyces* wine-related strains in single and mixed cultures with *Saccharomyces cerevisiae*. Ph.D. thesis, Catholic University of Portugal.

Albergaria, H. and Arneborg, N. (2016). Dominance of *Saccharomyces cerevisiae* in alcoholic fermentation processes: Role of physiological fitness and microbial interactions. *Applied Microbiology and Biotechnology* 100: 2035-2046.

Alfenore, S., Cameleyre, X., Benbadis, L., Bideaux, C., Uribelarrea, J.L., Goma, G., Molina-Jouve, C. and Guillouet, S.E. (2004). Aeration strategy: A need for very high ethanol performance in *Saccharomyces cerevisiae* fed-batch process. *Applied Microbiology and Biotechnology* 63: 537-542.

Alves-Araujo, C., Pacheco, A., Almeida, M.J., Spencer-Martins, I., Leao, C. and Sousa, M.J. (2007). Sugar utilisation patterns and respiro-fermentative metabolism in the baker's yeast *Torulaspora delbrueckii. Microbiology* 153: 898-904.

Amerine, M.A., Berg, H.W., Kunkee, R.E., Ough, C.S., Singleton, V.L. and Webb, A.D. (1982). The Technology of Winemaking. AVI Publishing Co., Westport.

Andorrà, I., Landi, S., Mas, A., Guillamón, J.M. and Esteve-Zarzoso, B. (2008). Effect of oenological practices on microbial populations using culture-independent techniques. *Food Microbiology* 25: 849-856.

Andorrà, I., Landi, S., Mas, A., Esteve-Zarzoso, B. and Guillamón, J.M. (2010). Effect of fermentation temperature on microbial population evolution using culture-independent and dependent techniques. *Food Research International* 43: 773-779.

Andorrà, I., Berradre, M., Mas, A., Esteve-Zarzoso, B. and Guillamón, J.M. (2012). Effect of mixed culture fermentations on yeast populations and aroma profile. *LWT – Food Science and Technology* 49: 8-13.

Andreasen, A.A. and Stier, T.J.B. (1954). Anaerobic nutrition of *Saccharomyces cerevisiae*, II: Unsaturated fatty acid requirement for growth in a defined medium,. *Journal of Cellular and Comparative Physiology* 43: 271-281.

Bardi, L., Cocito, C. and Marzon, M. (1999). *Saccharomyces cerevisiae* cell fatty acid composition and release during fermentation without aeration and in absence of exogenous lipids. *International Journal of Food Microbiology* 47: 133-140.

Beltran, G., Esteve Zarzoso, B., Rozès, N., Mas, A. and Guillamon, J.M. (2005). Influence of the timing of nitrogen additions during synthetic grape must fermentations on fermentations kinetics and nitrogen consumption. *Journal of Agricultural and Food Chemistry* 53: 996-1002.

Beltran, G., Novo, M., Leberre, V., Sokol, S., Labourdette, D., Guillamon, J.M., Mas, A., François, J. and Rozes, N. (2006). Integration of transcriptomic and metabolic analyses for understanding the global responses of low-temperature winemaking fermentations. *FEMS Yeast Research* 6: 1167-1183.

Bely, M., Sablayrolles, J.M. and Barre, P. (1990). Automatic detection of assimilable nitrogen deficiencies during alcoholic fermentation in oenological conditions. *Journal of Fermentation and Bioengineering* 70: 246-252.

Beney, L. and Gervais, P. (2001). Influence of the fluidity of the membrane on the response of microorganisms to environmental stresses. *Applied Microbiology and Biotechnology* 57: 34-42.

Birch, R.M. and Walker, G.M. (2000). Influence of magnesium ions on heat shock and ethanol stress responses of *Saccharomyces cerevisiae*. *Enzyme and Microbial Technology* 26: 678-687.

Bisson, L.F. (1999). Stuck and sluggish fermentations. *American Journal of Enology and Viticulture* 50: 107-119.

Bokulich, N.A., Swadener, M., Sakamoto, K., Mills, D.A. and Bisson, L.F. (2015). Sulphur dioxide treatment alters wine microbial diversity and fermentation progression in a dose-dependent fashion. *American Journal of Enology and Viticulture* 66: 73-79.

Borrull, A., López-Martínez, G., Poblet, M., Cordero-Otero, R. and Rozès, N. (2015). New insights into the toxicity mechanism of octanoic and decanoic acids on *Saccharomyces cerevisaie*. *Yeast* 32: 451-460.

Brice, C., Sanchez, I., Tesnière, C. and Blondin, B. (2014). Assessing the mechanism responsible for differences between nitrogen requirements of *Saccharomyces cerevisiae* wine yeasts in alcoholic fermentation. *Applied and Environmental Microbiology* 80: 1330-1339.

Brice, C., Sanchez, I., Bigey, F., Legras, J. and Blondin, B. (2015). A genetic approach of wine yeast fermentation capacity in nitrogen-starvation reveals the key role of nitrogen signalling. *BMC Genomics* 15: 495.

Castellari, L., Magrini, A., Passarelli, P. and Zambonelli, C. (1995). Effect of must fermentation temperature on minor products formed by cryo and non-cryotolerant *Saccharomyces cerevisiae* strains. *Italian Journal of Food Sciences* 7: 125-132.

Chandler, M., Stanley, G.A., Rogers, P. and Chambers, P. (2004). A genomic approach to defining the ethanol stress response in the yeast *Saccharomyces cerevisiae*. *Annals of Microbiology* 54: 427-454.

Chen, E.C.H. (1980). Utilisation of wort fatty acids by yeast during fermentations. *Journal of the American Society of Brewing Chemists* 38: 148-153.

Ciani, M. and Comitini, F. (2015). Yeast interactions in multi-starter wine fermentation. *Current Opinion in Food Science* 1: 1-6.

Cocolin, L. and Mills, D.A. (2003). Wine yeast inhibition by sulphur dioxide: A comparison of culture-dependent and independent methods. *American Journal of Enology and Viticulture* 54: 125-130.

Cooper, T.G. (1982). Nitrogen Metabolism in *Saccharomyces cerevisiae*. Cold Spring Harboour Monograph Archive 11B: 39-99.

Dashko, S., Zhou, N., Compagno, C. and Piskur, J. (2014). Why, when, and how did yeast evolve alcoholic fermentation? *FEMS Yeast Research* 14: 826-832.

Daum, G., Less, N.D., Brad, M. and Dickson, R. (1998). Biochemistry, cell biology and molecular biology of lipids of *Saccharomyces cerevisiae*. *Yeast* 14: 1471-1510.

Ding, J., Huang, X., Zhang, L., Zhao, L., Yang, D. and Zhang, K. (2009). Tolerance and stress response to ethanol in the yeast *Saccharomyces cerevisiae*. *Applied Microbiology and Biotechnology* 85: 253-263.

Divol, B. and Lonvaud-Funel, A. (2005). Evidence for viable but non-culturable yeasts in botrytis-affected wine. *Journal of Applied Microbiology* 99: 85-93.

Du Toit, W.J., Pretorius, I.S. and Lonvaud-Funel, A. (2005). The effect of sulphur dioxide and oxygen on the viability and culturability of a strain of *Acetobacter pasteurianus* and a strain of *Brettanomyces bruxellensis* isolated from wine. *Journal of Applied Microbiology* 98: 862-871.

EC (Council Regulation) (2009). No. 606/2009 of 10 July 2009 laying down certain detailed rules for implementing Council Regulation (EC) No. 479/2008 as regards the categories of grapevine products, oenological practices and the applicable restrictions. *Official Journal of the European Communities* L193: 1-59.

EC (Council Regulation) (2012). No 203/2012 of 8 March 2012 amending Regulation (EC) No. 889/2008 laying down detailed rules for the implementation of Council Regulation (EC) No. 834/2007, as regards detailed rules on organic wine. *Official Journal of the European Communities* L71: 42-47.

Edwards, C.G., Beehnan, R.B., Bartley, C.E. and McConnell, L.A. (1990). Production of decanoic acid and other volatile compounds and the growth of yeasts and malo-lactic bacteria during vinification. *American Journal of Enology and Viticulture* 41: 48-56.

Entian, K.D. and Barnett, J. (1992). Regulation of sugar utilisation by *Saccharomyces cerevisiae*. *Trends in Biochemical Sciences* 17: 506-510.

Farkas, J. (1988). Technology and Biochemistry of Wine, vols. 1 and 2. Gordon and Breach, New York.

Ferreira, D., Galeote, V., Sanchez, I., Legras, J.L., Ortiz-Julien, A. and Dequin, S. (2017). Yeast multistress resistance and lag-phase characterisation during wine fermentation. *FEMS Yeast Research* 17(6): fox051.

Fleet, G.H. (1992). Spoilage yeasts. *Critical Reviews in Biotechnology* 12: 1-44.

Fleet, G.H. (2003). Yeast interactions and wine flavor. *International Journal of Food Microbiology* 86: 11-22.

Fleet, G.H. and Heard, G.M. (1993). Yeast growth during fermentation in wine. pp. 27-54. *In*: G.H. Fleet (Ed.). Microbiology and Biotechnology. Harwood Academic Publishers. Switzerland.

Fugelsang, K.C. and Edwards, C.G. (2007). Wine microbiology. Practical Applications and Procedures, second ed., Springer Science+Business Media LLC, New York, USA.

Galeote, V.A., Blondin, B., Dequin, S. and Sablayrolles, J.M. (2001). Stress effects of ethanol on fermentation kinetics by stationary-phase cells of *Saccharomyces cerevisiae*. *Biotechnology Letters* 23: 677-681.

Gancedo, J.M. (2008). The early steps of glucose signalling in yeast. *FEMS Microbiology Reviews* 32: 673-704.

Gao, C. and Fleet, G.H. (1988). The effects of temperature and pH on the ethanol tolerance of the wine yeasts, *Saccharomyces cerevisiae*, *Candida stellata* and *Kloeckera apiculate*. *Journal of Applied Microbiology* 65: 405-409.

García-Ríos, E., López-Malo, M. and Guillamón, J.M. (2014). Global phenotypic and genomic comparison of two *Saccharomyces cerevisiae* wine strains reveals a novel role of the sulfur assimilation pathway in adaptation at low temperature fermentations. *BMC Genomics* 15: 1059.

Godard, P., Urrestarazu, A., Vissers, S., Kontos, K., Bontempi, G., van Helden, J. and André, B. (2007). Effect of 21 different nitrogen sources on global gene expression in the yeast *Saccharomyces cerevisiae*. *Molecular and Cellular Biology* 27: 3065-3086.

Goddard, M.R. (2008). Quantifying the complexities of *Saccharomyces cerevisiae*'s ecosystem engineering via fermentation. *Ecology* 89: 2077-2082.

Gonzalez, R., Quirós, M. and Morales, P. (2013). Yeast respiration of sugars by non-*Saccharomyces* yeast species: A promising and barely explored approach to lowering alcohol content of wines. *Trends in Food Science and Technology* 29: 55-61.

Gutiérrez, A., Chiva, R., Sancho, M., Beltran, G., Arroyo-López, F.N. and Guillamon, J.M. (2012). Nitrogen requirements of commercial wine yeast strains during fermentation of a synthetic grape must. *Food Microbiology* 31: 25-32.

Gutiérrez, A., Beltran, G., Warringer, J. and Guillamón, J.M. (2013). Genetic basis of variations in nitrogen source utilisation in four wine commercial yeast strains. *PLoS One* 8(6): e67166.

Hagman, A., Säll, T. and Piskur, J. (2014). Analysis of the yeast short-term crabtree effect and its origin. *The FEBS Journal* 281: 4805-4814.

Hanl, L., Sommer, P. and Arneborg, N. (2005). The effect of decreasing oxygen feed rates on growth and metabolism of *Torulaspora delbrueckii*. *Applied Microbiology and Biotechnology* 67: 113-118.

Hansen, H., Nissen, P., Sommer, P., Nielsen, J.C. and Arneborg, N. (2001). The effect of oxygen on the survival of non-*Saccharomyces* yeasts during mixed culture fermentations of grape juice with *Saccharomyces cerevisiae*. *Journal of Applied Microbiology* 91: 541-547.

Heard, G.M. and Fleet, G.H. (1988). The effects of temperature and pH on the growth of yeast species during the fermentation of grape juice. *Journal of Applied Microbiology* 65: 23-28.

Heinisch, J.J. and Rodicio, R. (2009). Physical and chemical stress factors in yeast. pp. 275-291. *In*: H. König, G. Unden and J. Fröhlich (Eds.). Biology of Microorganisms on Grapes, in Must and in Wine. Springer. Berlin, Heidelberg.

Henschke, P.A. and Jiranek, V. (1993). Yeasts – Metabolism of nitrogen compounds. pp. 77-164. *In*: G.H. Fleet (Ed.). Wine Microbiology and Biotechnology. Harwood Academic Publishers. Chur, Switzerland.

Hohmann, S. (2002). Osmotic stress signalling and osmo-adaptation in yeasts. *Microbiology and Molecular Biology Reviews* 66: 300-372.

Hu, X.H., Wang, M.H., Tan, T., Li, J.R., Yang, H., Leach, L., Zhang, R.M. and Luo, Z.W. (2007). Genetic dissection of ethanol tolerance in the budding yeast *Saccharomyces cerevisiae*. *Genetics* 175: 1479-1487.

Jiménez-Martí, E., Gomar-Alba, M., Palacios, A., Ortiz-Julien, A. and del Olmo, M.L. (2011). Towards an understanding of the adaptation of wine yeasts to must: Relevance of the osmotic stress response. *Applied Microbiology and Biotechnology* 89: 1551-1561.

Jolly, N.P., Varela, C. and Pretorius, I.S. (2014). Not your ordinary yeast: Non-*Saccharomyces* yeasts in wine production uncovered. *FEMS Yeast Research* 14: 215-237.

Jones, P.G. and Inouye, M. (1994). The cold-shock response – A hot topic. *Molecular Microbiology* 11: 811-818.

Kemsawasd, V., Viana, T., Ardö, Y. and Arneborg, N. (2015). Influence of nitrogen sources on growth and fermentation performance of different wine yeast species during alcoholic fermentation. *Applied Microbiology and Biotechnology* 99: 10191-10207.

Klug, L. and Daum, G. (2014). Yeast lipid metabolism at a glance. *FEMS Yeast Research* 14: 369-388.

Kudo, M., Vagnoll, P. and Bisson, L.F. (1998). Imbalance of potassium and hidrogen ion concentration as a cause of stuck enological fermentations. *American Journal of Enology and Viticulture* 49: 295-301.

Lafon-Lafourcade, S. (1983). Wine and brandy. pp. 81-163. *In*: H.J. Rehm and G. Reed (Eds.). Biotechnology, vol. 5: Food and Feed Production with Microorganisms. Verlag Chemie, Weinheim.

Lafon-Lafourcade, S. and Ribéreau-Gayon, P. (1984). Developments in the microbiology of wine production. pp. 1-45. *In*: M.E. Bushell (Ed.). Progress in Industrial Microbiology: Modern Applications of Transitional Biotechnologies, 1. Elsevier Biomedical Press. Amsterdam.

Lafon-Lafourcade, S., Geneix, C. and Ribéreau-Gayon, P. (1984). Inhibition of alcoholic fermentation of grape must by fatty acids produced by yeasts and their elimination by yeast ghosts. *Applied and Environmental Microbiology* 47: 1246-1249.

Li, B.Z., Cheng, J.S., Ding, M.Z. and Yuan, Y.J. (2010). Transcriptome analysis of differential responses of diploid and haploid yeast to ethanol stress. *Journal of Biotechnology* 148: 194-203.

Ljungdahl, P.O. and Daignan-Fornier, B. (2012). Regulation of amino acid, nucleotide, and phosphate metabolism in *Saccharomyces cerevisiae*. *Genetics* 190: 885-929.

Llauradó, J.M., Rozès, N., Bobet, R., Mas, A. and Constantí, M. (2002). Low temperature alcoholic fermentation in high sugar concentration grape must. *Journal of Food Science* 67: 268-273.

Llauradó, J.M., Rozès, N., Constantí, M. and Mas, A. (2005). Study of some *Saccharomyces cerevisiae* strains for winemaking after pre-adaptation at low temperatures. *Journal of Agricultural and Food Chemistry* 53: 1003-1011.

Lleixà, J., Martín, V., Portillo, M.C., Carrau, F., Beltran, G. and Mas, A. (2016). Comparison of fermentation and wines produced by inoculation of *Hanseniaspora vineae* and *Saccharomyces cerevisiae*. *Frontiers in Microbiology* 7: 338.

Lopez, R., Santamaria, P., Gutierrez, R.A. and Iñiguez, M. (1996). Changes in amino acids during the alcoholic fermentation of grape juice at different temperatures. *Sciences des Aliments* 16: 529-535.

Magasanik, B. (1992). Regulation of Nitrogen Utilisation. Cold Spring Harbour Monograph Archive 21B: 283-317.

Magasanik, B. and Kaiser, C.A. (2002). Nitrogen regulation in *Saccharomyces cerevisiae*. *Gene* 290: 1-18.

Marsit, S. and Dequin, S. (2015). Diversity and adaptive evolution of *Saccharomyces* wine yeast: A review. *FEMS Yeast Research* 15(7): fov067.

Martínez-Moreno, R., Morales, P., González, R., Mas, A. and Beltran, G. (2012). Biomass production and alcoholic fermentation performance of *Saccharomyces cerevisiae* as a function of nitrogen source. *FEMS Yeast Research* 12: 477-485.

Mauricio, J.C., Guijo, S. and Ortega, J.M. (1991). Relationship between phospholipid and sterol content in *Saccharomyces cerevisiae* and *Torulaspora delbrueckii* and their fermentation activity in grape musts,. *American Journal of Enology and Viticulture* 42: 301-308.

Mauricio, J.C., Millán, C. and Ortega, J.M. (1998). Influence of oxygen of the biosynthesis of cellular fatty acids, sterols and phospholipids during alcoholic fermentation by *Saccharomyces cerevisiae* and *Torulaspora delbrueckii*. *World Journal of Microbiology and Biotechnology* 14: 405-410.

Medina, K., Boido, E., Dellacassa, E. and Carrau, F. (2012). Growth of non-*Saccharomyces* yeasts affects nutrient availability for *Saccharomyces cerevisiae* during wine fermentation. *International Journal of Food Microbiology* 157: 245-250.

Merico, A., Sulo, P., Piskur, J. and Compagno, C. (2007). Fermentative lifestyle in yeasts belonging to the *Saccharomyces* complex. *FEBS Journal* 274: 976-989.

Miralles, V.J. and Serrano, R. (1995). A genomic locus in *Saccharomyces cerevisiae* with four genes up-regulated by osmotic stress. *Molecular Microbiology* 17: 653-662.

Molina, A.M., Swiegers, J.H., Varela, C., Pretorius, I.S. and Agosin, E. (2007). Influence of wine fermentation temperature on the synthesis of yeast-derived volatile aroma compounds. *Applied Microbiology and Biotechnology* 77: 675-687.

Monteiro, F.F. and Bisson, L.F. (1991). Amino acid utilisation and urea formation during vinification fermentations. *American Journal of Enology and Viticulture* 42: 199-208.

Morales, P., Rojas, V., Quirós, M. and Gonzalez, R. (2015). The impact of oxygen on the final alcohol content of wine fermented by a mixed starter culture. *Applied Microbiology and Biotechnology* 99: 3993-4003.

Navarro-Avino, J.P., Prasad, R., Miralles, V.J., Benito, R.M. and Serrano, R. (1999). A proposal for nomenclature of aldehyde dehydrogenases in *Saccharomyces cerevisiae* and characterization of the stress-inducible ALD2 and ALD3 genes. *Yeast* 15: 829-842.

Navarro-Tapia, E., Nana, R.K., Querol, A. and Pérez-Torrado, R. (2016). Ethanol cellular defense induce unfolded protein response in yeast. *Frontiers in Microbiology* 7: 189.

Noti, O., Vaudano, E., Pessione, E. and García-Moruno, E. (2015). Short-term response of different *Saccharomyces cerevisiae* strains to hyper-osmotic stress caused by inoculation in grape must: RT-qPCR study and metabolite analysis. *Food Microbiology* 52: 49-58.

Novo, M.T., Beltran, G., Rozès, N., Guillamon, J.M., Sokol, S., Leberre, V., François, J. and Mas, A. (2007). Early transcriptional response of wine yeast after rehydration: Osmotic shock and metabolic activation. *FEMS Yeast Research* 7: 304-316.

Nurgel, C., Erten, H., Canbas, A., Cabaroglu, T. and Selli, S. (2005). Yeast flora during the fermentation of wines made from *Vitis vinifera* L. cv. Emir and Kalecik Karasi grown in Anatolia. *World Journal of Microbiology and Biotechnology* 21: 1187-1194.

Nykänen, L. (1986). Formation and occurrence of flavour compounds in wine and distilled alcoholic beverages. *American Journal of Enology and Viticulture* 37: 84-96.

Ough, C.F., Davenport, M. and Joseph, K. (1989). Effects of certain vitamins on growth and fermentation rate of several commercial active dry wine yeasts. *American Journal of Enology and Viticulture* 40: 208-213.

Padilla, B., García-Fernández, D., González, B., Izidoro, I., Esteve-Zarzoso, B., Beltran, G. and Mas, A. (2016). Yeast diversity from DOQ Priorat uninoculated fermentations. *Frontiers in Microbiology* 7: 930.

Pampulha, M.E. and Loureiro, V. (1989). Interaction of the effects of acetic acid and ethanol on inhibition of fermentation in *Saccharomyces cerevisiae*. *Biotechnology Letters* 11: 269-274.

Pampulha, M.E. and Loureiro-Dias, M.C. (1989). Combined effect of acetic acid, pH and ethanol on intracellular pH of fermenting yeast. *Applied Microbiology and Biotechnology* 31: 547-550.

Pampulha, M.E. and Loureiro-Dias, M.C. (1990). Activity of glycolytic enzymes of *Saccharomyces cerevisiae* in the presence of acetic acid. *Applied Microbiology and Biotechnology* 34: 375-380.

Parks, L.W. (1978). Metabolism of sterols in yeast. *CRC Critical Reviews in Microbiology* 6: 301-340.

Pérez-Nevado, F., Albergaria, H., Hogg, T. and Girio, F. (2006). Cellular death of two non-*Saccharomyces* wine-related yeasts during mixed fermentations with *Saccharomyces cerevisiae*. *International Journal of Food Microbiology* 108: 336-345.

Perrone, G.G. (2008). Reactive oxygen species and yeast apoptosis. *Biochimica et Biophysica Acta* 1783: 1354-1368.

Pina, C., Santos, C., Couto, J.A. and Hogg, T. (2004). Ethanol tolerance of five non-*Saccharomyces* wine yeasts in comparison with a strain of *Saccharomyces cerevisiae* – Influence of different culture conditions. *Food Microbiology* 21: 439-447.

Piskur, J., Rozpedowska, E., Polakova, S., Merico, A. and Compagno, C. (2006). How did *Saccharomyces* evolve to become a good brewer? *Trends in Genetics* 22: 183-186.

Pretorius, I.S. (2000). Tailoring wine yeast for the new millennium: Novel approaches to the ancient art of winemaking. *Yeast* 16: 675-729.

Pronk, J.T., Steensma, H.Y. and van Dijken, J.P. (1996). Pyruvate metabolism in *Saccharomyces cerevisiae*. *Yeast* 12: 1607-1633.

Querol, A., Jimenez, M. and Huerta, T. (1990). A study on microbiological and enological parameters during fermentation of must from poor and normal grapes harvest in the region of Alicante (Spain). *Journal of Food Science* 55: 1603-1606.

Quirós, M., Rojas, V., Gonzalez, R. and Morales, P. (2014). Selection of non-Saccharomyces yeast strains for reducing alcohol levels in wine by sugar respiration. *International Journal of Food Microbiology* 181: 85-91.

Ratledge, C. and Evans, C.T. (1989). Lipids and their metabolism. pp. 367-455. *In*: A.H. Rose and J.S. Harrison (Eds.). The Yeasts: Metabolism and Physiology of Yeasts, vol, 3, second ed. Academic Press. San Diego, USA.

Redón, M., Guillamón, J.M., Mas, A. and Rozès, N. (2009). Effect of lipid supplementation upon *Saccharomyces cerevisiae* lipid composition and fermentation performance at low temperature. *European Food Research Technology* 228: 833-840.

Ribéreau-Gayon, J., Peynaud, E., Ribéreau-Gayon, P. and Sudraud, P. (1975). Sciences et Techniques du Vin, vol. 2. Dunod, Paris.

Ribéreau-Gayon, P., Dubourdieu, D., Donèche, B. and Lonvaud, A. (2006). Handbook of Enology: The Microbiology of Wine and Vinifications, second ed. John Wiley & Sons Ltd. The Atrium, Southern Gate, Chichester, West Sussex PO19 8SQ, England.

Rodicio, R. and Heinisch, J.J. (2009). Sugar metabolism by *Saccharomyces* and non-*Saccharomyces*. pp. 113-134. *In*: H. König, G. Unden and J. Fröhlich (Eds.). Biology of Microorganisms on Grapes, in Must and in Wine. Springer. Berlin, Heidelberg.

Rosa, M.F. and Sá-Correia, I. (1991). *In vivo* activation by ethanol of plasma membrane ATPase of *Saccharomyces cerevisiae*. *Applied and Environmental Microbiology* 57: 830-835.

Rossignol, T., Postaire, O., Storai, J. and Blondin, B. (2006). Analysis of the genomic response of a wine yeast to rehydration and inoculation. *Applied Microbiology and Biotechnology* 71: 699-712.

Sablayrolles, J.M. and Barre, P. (1986). Evaluation des besoins en oxygène de fermentations alcooliques en conditions oenologiques stimulées. *Science des aliments* 6: 373-383.

Sahara, T., Goda, T. and Ohgiya, S. (2002). Comprehensive expression analysis of time-dependent genetic responses in yeast cells to low temperature. *Journal of Biological Chemistry* 277: 50015-50021.

Salma, M., Rousseaux, S., Grand, A.S., Divol, B. and Alexandre, H. (2013). Characterisation of the viable but non-culturable (VBNC) state in *Saccharomyces cerevisiae*. *PLoS One* 8(10): e77600.

Salvadó, Z., Arroyo-López, F.N., Barrio, E., Querol, A. and Guillamón, J.M. (2011). Quantifying the individual effects of ethanol and temperature on the fitness advantage of *Saccharomyces cerevisiae*. *Food Microbiology* 28: 1155-1161.

Schade, B., Jansen, G., Whiteway, M., Entian, K.D. and Thomas, D.Y. (2004). Cold adaptation in budding yeast. *Molecular Biology of the Cell* 15: 5492-5502.

Sharf, R. and Magalith, P. (1983). The effect of temperature on spontaneous wine fermentation. *European Journal of Applied Microbiology and Biotechnology* 17: 311-313.

Shobayashi, M., Mitsueda, S., Ago, M., Fujii, T., Iwashita, K. and Iefuji, H. (2005). Effects of culture conditions on ergosterol biosynthesis by *Saccharomyces cerevisiae*. *Bioscience Biotechnology and Biochemistry* 69: 2381-2388.

Stanley, D., Bandara, A., Fraser, S., Chambers, P.J. and Stanley, G.A. (2010a). The ethanol stress response and ethanol tolerance of *Saccharomyces cerevisiae*. *Journal of Applied Microbiology* 109: 13-24.

Stanley, D., Chambers, P.J., Stanley, G.A., Borneman, A. and Fraser, S. (2010b). Transcriptional changes associated with ethanol tolerance in *Saccharomyces cerevisiae*. *Applied Microbiology and Biotechnology* 88: 231-239.

Starr, P.R. and Parks, L.W. (1962). Some factors affecting sterol formation in *Saccharomyces cerevisiae*. *Journal of Bacteriology* 82: 1042-1046.

Taherzadeh, M.J., Lidén, G., Gustafsson, L. and Niklasson, C. (1996). The effects of pantothenate deficiency and acetate addition on anaerobic batch fermentation of glucose by *Saccharomyces cerevisiae*. *Applied Microbiology and Biotechnology* 46: 176-182.

Taillandier, P., Lai, Q.P., Julien-Ortiz, A. and Brandam, C. (2014). Interactions between *Torulaspora delbrueckii* and *Saccharomyces cerevisiae* in wine fermentation: Influence of inoculation and nitrogen content. *World Journal of Microbiology and Biotechnology* 30: 1959-1967.

Takahashi, M., Ohta, T., Masaki, K., Mizuno, A. and Goto-Yamamoto, N. (2014). Evaluation of microbial diversity in sulphite-added and sulphite-free wine by culture-dependent and independent methods. *Journal of Bioscience and Bioengineering* 117: 569-575.

Taylor, G.T. and Kirsop, B.H. (1977). The origin of medium chain length fatty acids present in beer. *Journal of the Institute of Brewing* 83: 241-243.

Ter Schure, E.G., van Riel, N.A. and Verrips, C.T. (2000). The role of ammonia metabolism in nitrogen catabolite repression in *Saccharomyces cerevisiae*. *FEMS Microbiology Reviews* 24: 67-83.

Thomas, D.S., Hossack, J.A. and Rose, A.H. (1978). Plasma membrane lipid composition and ethanol tolerance in *Saccharomyces cerevisiae*. *Archives of Microbiology* 117: 239-245.

Thomas, K.C., Hynes, S.H. and Ingledew, W.M. (2002). Influence of medium buffer capacity on inhibition of *S. cerevisiae* growth by acetic acid and lactic acid. *Applied and Environmental Microbiology* 68: 1616-1623.

Torija, M.J., Rozès, N., Poblet, M., Guillamón, J.M. and Mas, A. (2001). Yeast population dynamics in spontaneous fermentations: Comparison between two different wine-producing areas over a period of three years. *Antonie van Leeuwenhoek* 79: 345-352.

Torija, M.J., Rozès, N., Poblet, M., Guillamón, J.M. and Mas, A. (2003a). Effects of fermentation temperature on the strain population of *Saccharomyces cerevisiae*. *International Journal of Food Microbiology* 80: 47-53.

Torija, M.J., Beltran, G., Novo, M., Poblet, M., Guillamón, J.M., Mas, A. and Rozès, N. (2003b). Effects of temperature and *Saccharomyces* species on the cell fatty acid composition and presence of volatile compounds in wine. *International Journal of Food Microbiology* 85: 127-136.

Tuller, G., Nemec, T., Hrastnik, C. and Daum, G. (1999). Lipid composition of subcellular membranes of an FY1679-derived haploid yeast wild-type strain grown on different carbon sources. *Yeast* 15: 1555-1564.

van Dijken, J.P., Weusthuis, R.A. and Pronk, J.T. (1993). Kinetics of growth and sugar consumption in yeasts. *Antonie van Leeuwenhoek* 63: 343-352.

Varela, J.C., van Beekvelt, C., Planta, R.J. and Mager, W.H. (1992). Osmostress-induced changes in yeast gene expression. *Molecular Microbiology* 6: 2183-2190.

Viegas, C.A., Rosa, F., Sà-Correia, I. and Novais, J.M. (1989). Inhibition of yeast growth by octanoic and decanoic acids produced during ethanol fermentation. *Applied and Environmental Microbiology* 55: 21-28.

Viegas, C.A. and Sá-Correia, I. (1997). Effects of low temperatures (9-33°C) and pH (3.3-5.7) in the loss of *Saccharomyces cerevisiae* viability by combining lethal concentrations of ethanol with octanoic and decanoic acids. *International Journal of Food Microbiology* 34: 264-277.

Vigh, L., Maresca, B. and Harvood, J.L. (1998). Does the membrane's physical state control the expression of heat shock and other genes? *Trends in Biochemical Sciences* 23: 369-374.

Viljoen, B.C., Kock, J.L. and Lategan, P.M. (1986). Long-chain fatty acid composition of selected genera of yeasts belonging to the *Endomycetale*. *Antonie Van Leeuwenhoek* 52: 45-51.

Walker, G.M. and van Dijck, P. (2006). Physiological and molecular responses of yeasts to the environment. pp. 111-152. *In*: A. Querol and G.H. Fleet (Eds.). Yeasts in Food and Beverages. Springer. Heidelberg.

Wang, C. and Liu, Y. (2013). Dynamic study of yeast species and *Saccharomyces cerevisiae* strains during the spontaneous fermentations of Muscat Blanc in Jingyang, China. *Food Microbiology* 33: 172-177.

Wang, C., Mas, A. and Esteve-Zarzoso, B. (2015). Interaction between *Hanseniaspora uvarum* and *Saccharomyces cerevisiae* during alcoholic fermentation. *International Journal of Food Microbiology* 206: 67-74.

Wang, C., Mas, A. and Esteve-Zarzoso, B. (2016). The Interaction between *Saccharomyces cerevisiae* and non-*Saccharomyces* yeast during alcoholic fermentation in species and strain specific. *Frontiers in Microbiology* 7: 502.

Williams, K.M., Liu, P. and Fay, J.C. (2015). Evolution of ecological dominance of yeast species in high-sugar environments. *Evolution* 69: 2079-2093.

# 12 Malo-lactic Bacteria in Winemaking

**Albert Bordons*, Isabel Araque, Mar Margalef-Català and Cristina Reguant**
Department of Biochemistry and Biotechnology, Facultat d'Enologia, Universitat Rovira i Virgili, Tarragona, Catalonia, Spain

## 1. Introduction

Malo-lactic fermentation (MLF), a so-called secondary fermentation, is a microbial process undertaken by some lactic acid bacteria (LAB) in several wines, mainly the red ones and which consists in the conversion of L-malic acid to L-lactic acid and $CO_2$. For this reason, LAB that carry out the MLF are often called as 'malolactic bacteria'. In winemaking, MLF is an important step, due to its several benefits – the main one being reducing the acidity of the L-malic acid, which can be relatively high in red wine. Nevertheless, LAB during MLF contribute also to other benefits in wine (Liu, 2002) such as sensory improvement and the microbial stabilisation. On the other hand, MLF is not favourable for all wines. For instance, in warmer areas, grapes tend to be less acid and a further decrease in acidity by MLF may be deleterious for the sensory properties and microbiological stability of wine (Versari *et al.*, 1999). In addition to its occurrence in wine, MLF occurs in other fermented beverages, such as cider (Sánchez *et al.*, 2014). In this chapter, all aspects connected with the malo-lactic acid fermentation have been described.

## 2. Malo-lactic Fermentation

Malo-lactic fermentation, also known as malo-lactic conversion, is a process in which tart-tasting malic acid, naturally present in grape must, is converted to softer-tasting lactic acid. MLF is most often performed shortly after the end of the primary alcoholic fermentation (AF), but can sometimes run concurrently with it.

### 2.1. Malic Acid in Grapes and Wine

The principal organic acids in grapes are L-tartaric and L-malic – L-tartaric acid is biosynthesised from ascorbic acid as the principal intermediary product, while L-malic acid is formed *via* glycolysis and the TCA cycle. During the green or herbaceous stage, immediately after flowering of grape, the berries undergo a rapid expansion with a significant increase in vacuolar size of berry cells due to the rapid storage of L-malic and L-tartaric acids. After that, berry growth ceases and its acidity reaches a maximum due to the continued accumulation of both the acids. At this moment, L-malic is the most abundant acid, up to 25 g/l, resulting in an internal pH of 2.5 in grape berries (Ribéreau-Gayon *et al.*, 2000).

Then, the berries begin the phase of ripening or *véraison*, which generally starts between six to eight weeks after flowering and lasts 35-55 days (Volschenk *et al.*, 2006). With the onset of *véraison*, the L-malic concentration rapidly decreases, with a concomitant increase in internal berry pH, around 3.5. This way, the juice of most grapes, except in very warm climates, contains a certain amount of L-malic acid (2-5 g/l), which has a strong bitter taste, not very nice.

Organic acids can contribute positively to the organoleptic character of wine if it is balanced with the other wine components. The sour-sweet balance is well known as a required sensory quality in wine. Acid-balanced wines are usually perceived as having refreshing sensory undertones, while excessive concentrations of organic acids are related with sour taste and specifically the L-malic acid, which resembles the taste of unripe apples. Moreover, wine acidity often accentuates the perception of other wine tastes. It masks excess sweetness but the perception of astringency is emphasised for low pH values

*Corresponding author: albert.bordons@urv.cat

(Fischer and Noble, 1994). On the other hand, the pH determines the degree of organic acid and amino acid ionisation in a wine solution. Therefore, a minor change in pH, coinciding with changes for instance of malic acid content, significantly influences the organoleptic perception of wine (Volschenk *et al.*, 2006).

L-malic and other organic acids, and pH, also play an important role in the development of specific flavour compounds during vinification. The release of stored L-malic and tartaric acids from the grapes, during crushing, is responsible for acid hydrolysis of non-volatile compounds, like monoterpene glycosides, norisoprenoids, benzyl alcohol and others. These flavour compounds are essential for the development of a complex flavour profile during vinification and subsequent ageing of wine (Winterhalter *et al.*, 1990).

## 2.2. Metabolism of Malo-lactic Fermentation

During alcoholic fermentation (AF), yeasts hardly degrade the malic acid and therefore, it is found in most wines. Usually, after the AF, though rarely simultaneously with this, LAB such as *Oenococcus oeni* can achieve this MLF. It is a decarboxylation reaction carried out by the malolactic enzyme (MLE), which decarboxylates the biacidic L-malate to the monoacidic L-lactate and releases $CO_2$, which appears as tiny bubbles in the wine. The MLE catalyses this one-step reaction in presence of $Mn^{2+}$ and $NAD^+$. Because malic acid is a dicarboxylic and lactic acid is monocarboxylic, it implies a reduction in the acidity of 0.1 to 0.5 pH units. Considering stoichimiometrics, 1 g L-malic acid produces 0.67 g L-lactic acid and 0.33 g $CO_2$.

The L-malic acid is taken into the cell by a mechanism widely studied (Cox and Henick-Kling, 1989; Loubiere *et al.*, 1992; Salema *et al.*, 1994). At the usual pH of wine (3.2-3.6), the L-malic acid enters the cell by passive transport, but at higher pH, L-malic is in its anionic form and it is uptaken by a malate permease, creating an electric potential gradient (Salema *et al.*, 1996). After decarboxylation of L-malate by the MLE, the L-lactate and $CO_2$ produced go out of the cell (Salema *et al.*, 1996), allowing the simultaneous exit of $H^+$ extrusion. The difference among internal and external pH, together with the electric potential gradient, favours a proton motive force, which facilitates the formation of ATP by the ATPase complex, thanks to the re-entry of $H^+$ (Fig. 1).

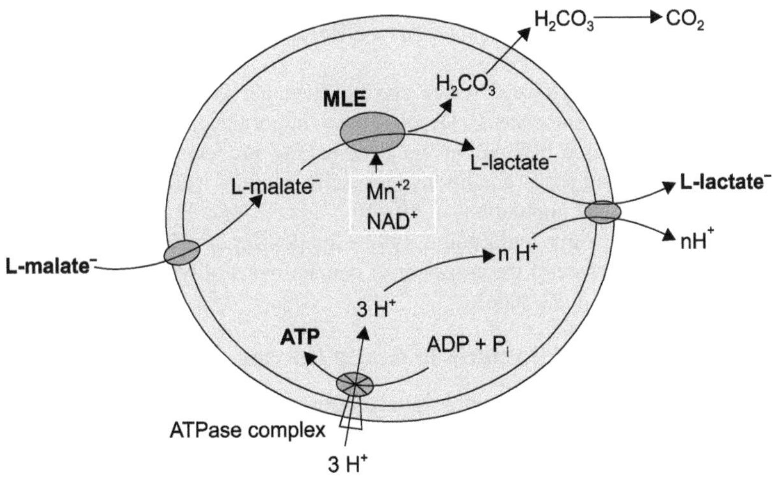


**Figure 1.** Malo-lactic reaction inside the LAB cells and its electrochemical gradient formation. MLE is the malo-lactic enzyme (Modified from Bartowsky, 2005)

This malo-lactic enzyme, key of the MLF, has around 540 amino acids and, according the resources of the NCBI (2017), it is found in several LAB (*Oenococcus, Lactobacillus, Fructobacillus, Leuconostoc, Pediococcus, Lactococcus* and *Streptococcus*) but it has also been described in some other firmicutes of the same bacilli class, such as some species of *Bacillus* and *Staphylococcus*.

From a metabolic point of view, MLF is not a classical fermentation. In these, for instance the lactic fermentation (homolactic or heterolactic fermentation), sugars are used as substrates and ATP is obtained by substrate-level phosphorylation reactions of glycolysis. On the other hand, MLF is only a decarboxylation and does not seem to obtain in principle energy benefits for the cells doing it. The only apparent benefit would be the rise of external pH. However, MLF is a peculiar fermentation with ATP synthase, where the output of L-lactic from cells (Fig. 1) is made by a simport with protons (Cox and Henick-Kling, 1989; Poolman *et al.*, 1991; Salema *et al.*, 1996). In parallel, protons can enter the cell by a favourable gradient, since external pH is 3-4 and the internal pH is higher than 6, and this gradient is attached to the ATP synthesis by an ATP synthase.

Thus, *O. oeni* can get some ATP by the decarboxylation of L-malic acid or MLF in the wine, a medium where there are almost no sugars and therefore, where bacteria cannot get ATP by lactic acid fermentation. Those few ATP, along with some nutrients coming from the yeast remains, may allow a slight growth of these bacteria in wine (Reguant *et al.*, 2005a, 2005b).

### 2.3. Effect of Malo-lactic Deacidification in Wine

From the oenological point of view, this deacidification involves also an improvement in wine quality, reducing the roughness and strong sense of acidity of malic acid, similar to the green apple tasting. In addition, L-lactic acid that appears has a nice, sweet and smooth tasting. The lactic acid impresses nicely the taste buds and assembly better with the astringency given by the phenolic compounds, unlike malic acid, that gives an immature acidity, an astringent taste and a certain green shade. So, the rough and bitter taste given by acidity and tannins is gradually erased to be replaced by a delicate and softly taste (Popescu-Mitroi *et al.*, 2009). Moreover, the small increase of pH gives the wine a softer character and decreases the bitterness sensation in mouth. In summary, these changes add to the wine – mostly the red one – a soft, fleshy and fatty character. Overall, the wines become more harmonious, more full and evolved.

## 3. Lactic Acid Bacteria in Winemaking: Presence, Benefits and Disadvantages

The first to recognise the presence of LAB in wine was Louis Pasteur in the 19th century (Pasteur, 1873), but he just correlated them with their negative effects, such as the production of lactic acid, which turns the wine sour. A century later, oenologists of Bordeaux identified LAB as positive in wine (Peynaud, 1956).

The LAB isolated from musts and wines can be from the genera *Lactobacillus*, *Pediococcus*, *Leuconostoc*, *Weissella* and *Oenococcus*. Despite their importance, few studies have specifically investigated the origin or source of LAB in the winemaking process. Freshly extracted grape juice or must, produced under commercial conditions, generally contains various species of *Lactobacillus*, *Pediococcus* and *Leuconostoc* at populations of $10^2$–$10^3$ CFU (Colony Forming Units) per ml (Costello *et al.*, 1983; Fleet *et al.*, 1984; Pardo and Zúñiga, 1992; Godálová *et al.*, 2016). It is considered that these bacteria originate from the surface of the grapes or as contaminants of winery equipment that is used to process the juice or must (Bae *et al.*, 2006).

### 3.1. Presence of Lactic Acid Bacteria in Grape Berries

The more frequently described LAB species isolated from several grape varieties are *Lactobacillus plantarum*, *Lb. casei*, *Lb. brevis*, *Lb. hilgardii*, *Lb. curvatus*, *Lb. buchneri*, *Leuconostoc dextranicum* and *Le. mesenteroides* (Lafon-Lafourcade *et al.*, 1983; Sieiro *et al.*, 1990; Suárez *et al.*, 1994; Godálová *et al.*, 2016).

The incidence and populations of LAB on wine grapes are low and the damaged grape berries have a greater presence of these bacteria than undamaged berries (Bae *et al.*, 2006). These authors did not find *O. oeni* on grapes, but they signalled that its recovery could be obscured by overgrowth from other species. On the other hand, Garijo *et al.* (2011) were able to isolate a colony of *O. oeni* from grapes, and the DNA of *O. oeni* was detected in grapes (Renouf *et al.*, 2005, 2007) by PCR-DGGE of the *rpoB* gene, although no *Oenococcus* was isolated in this work.

Nevertheless, in a recent work Franquès *et al.* (2017) gave the first evidence of the presence of viable *O. oeni* on grapes. Thereby, in order to clarify the presence of LAB in grapes, and especially *O. oeni*, a large survey on the autochthonous LAB from vineyards in the Catalan standout wine region of Priorat was carried out. A total of 254 LAB isolates from Grenache and Carignan grape berries were identified and typed (Table 1).

**Table 1.** Total Number of Identified, Isolated and Typed Strains for Different LAB Species, obtained from Berries of Grenache and Carignan Vineyards in Eight Properties along Two Vintages

| *LAB species* | *No. isolates* | *No. strains* |
|---|---|---|
| *Oenococcus oeni* | 53 | 16 |
| *Lactobacillus plantarum* | 123 | 43 |
| *L. sanfranciscensis* | 29 | 5 |
| *L. lindneri* | 14 | 6 |
| *L. mali* | 5 | 4 |
| *Fructobacillus tropaeoli* | 19 | 11 |
| *Pediococcus pentosaceus* | 11 | 4 |
| Total | 254 | 89 |

*Source*: Adapted from Franquès *et al*., 2017.

As seen, 53 (almost a 21 per cent) of them were *O. oeni*, which showed a considerable phylogenetic diversity. Some of these strains were also isolated from wine samples elaborated from the same grapes, which supports the fact that these isolated *O. oeni* on grapes were viable. As seen, non-*Oenococcus* species were also identified and typed, *Lactobacillus plantarum* being predominant in grapes.

Besides isolating and cultivating microorganisms, nowadays High Throughput Sequencing (HTS) makes it possible to analyse complex bacterial communities on grape berries. HTS technique, using short amplicons such as hyper variable domains of prokaryotic 16S rDNA (Caporaso *et al*., 2012), has been used in different grapes (Bokulich *et al*., 2014; Zarraonaindia *et al*., 2015). However, *Oenococcus* DNA has been detected by HTS (Portillo *et al*., 2016; Franquès *et al*., 2017) in the same samples where *O. oeni* has been isolated, which corroborates the culturing results. An average of almost 6 per cent of *Oenococcus* DNA was found, which means it has not a negligible presence in grapes. *Lactobacillus*, the most abundant genus isolated from grapes, was also detected by HTS (1 per cent), but to a lesser extent than *Oenococcus*.

## 3.2. Presence of Lactic Acid Bacteria and *Oenococcus* in Winemaking

*O. oeni*, known previously as *Leuconostoc oenos* (Dicks *et al*., 1995), is the species most commonly isolated from wines (Chalfan *et al*., 1977; Lafon-Lafourcade *et al*., 1983; Davis *et al*., 1985; Wibowo *et al*., 1985; Davis *et al*., 1986; Edwards *et al*., 1991; Lonvaud-Funel *et al*., 1991; Franquès *et al*., 2017), but its number during alcoholic fermentation is usually low. *O. oeni* and other LAB are inhibited by ethanol and other compounds produced by yeasts and by $SO_2$ added to the must to control bacterial population growth, especially of acetic acid bacteria. When AF is finished and the yeast die, some LAB can thrive and achieve some growth, up to $10^7$/ml in some cases, partly due to compounds released by yeasts. Then these LAB can produce some changes in the wine, of which the most interesting is the malo-lactic fermentation (MLF).

Several LAB species, including *Leuconostoc mesenteroides*, *Pediococcus damnosus* and different species of *Lactobacillus*, such as *Lb. hilgardii*, *Lb. plantarum* and *Lb. brevis*, are present in fermenting must during alcoholic fermentation, but there is an exclusive predominance of *O. oeni* during the MLF (Lonvaud-Funel *et al*., 1991). Numbers of total LAB during the AF are usually very low, around $10^3$ per ml.

On the other hand, it has been found that strains of *O. oeni* are predominant among the LAB found in the air of cellars (Garijo *et al*., 2009). Apparently, *O. oeni* are not detected until the first MLF in the cellar begins (Garijo *et al*., 2008), suggesting that they are scattered in the air by the $CO_2$ expelled from tanks during fermentation, as it happens with yeasts. This would be an explanation for the spontaneous MLF done by the same strains predominant in the cellar, *id est*, by the 'autochthonous cellar strains'.

### 3.3. Benefits of Lactic Acid Bacteria in Winemaking

As commented above, the first benefit of LAB in winemaking is malo-lactic fermentation and the main benefit is reduction in the acidity of L-malic acid, which can be relatively high in red wines. Nevertheless, bacteria doing the MLF contribute also to other benefits in wine, mainly the organoleptic improvement, and the microbial stabilisation after growth of bacteria doing the MLF. Another benefit due to the small rise in pH is that the colour of red wine, mainly due to anthocyanins as malvidine, evolves towards less intense colours and not so red, which makes them more interesting visually (Lonvaud-Funel, 1999).

LAB, and particularly *O. oeni*, exhibit many secondary metabolic activities during MLF, which can modify the sensory properties of wine. These secondary activities include the metabolism of organic acids, carbohydrates, polysaccharides and amino acids. Numerous enzymes such as glycosidases, esterases and proteases generate volatile compounds well above their odour detection threshold (Bartowsky and Borneman, 2011).

In this way, from the organoleptic point of view, MLF leads to an improvement in wine because the development of bacteria causes these changes in the components of the wine, in addition to deacidification and changing of L-malic acid for L-lactic acid. The main volatile compounds that appear are diacetyl, acetoin, ethyl lactate and a small quantity of acetic acid. Diacetyl is considered the most important due to its buttery or creamy aroma – dairy flavours – that characterises many wines that have made the MLF (Davis *et al.*, 1985). It appears in small concentrations (up to 2 mg/l) but has a very low threshold for sensory detection and is produced by degradation of citric acid, present in wine (Bartowsky and Henschke, 2004) (Fig. 2).

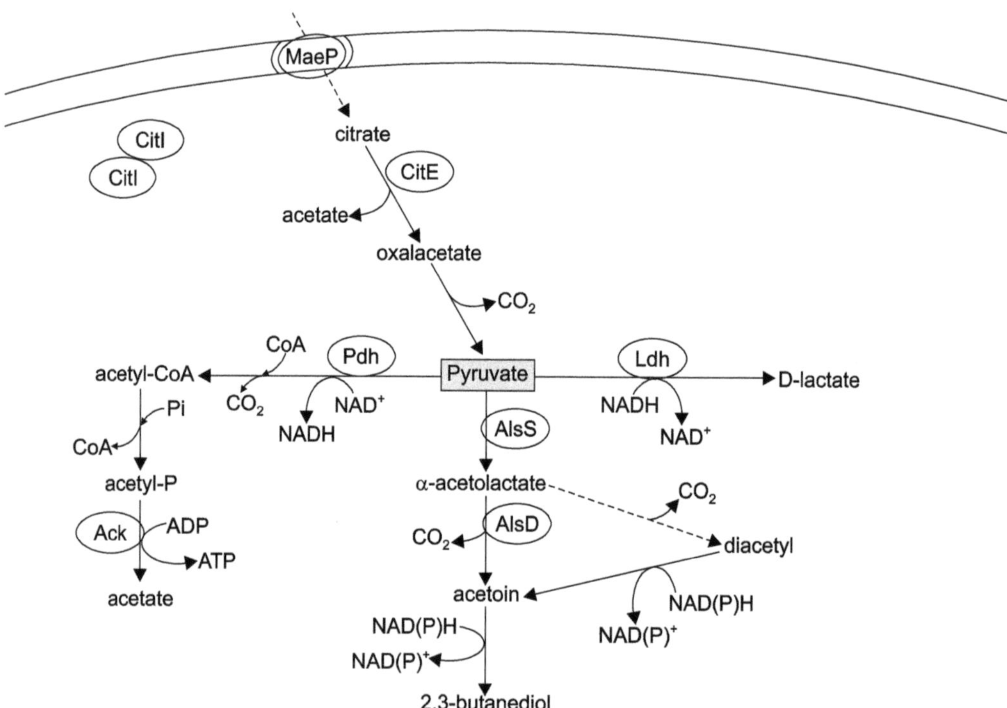


**Figure 2.** The citrate pathway in malolactic bacteria, of which the main enzymes are citrate lyase (CitE), pyruvate dehydrogenase (Pdh), lactate dehydrogenase (Ldh), acetolactate synthase (AlsS) and acetolactate decarboxylase (AlsD). Dashed line to diacetyl means a non-enzymatic reaction (more details in the original figure from Olguín *et al.*, 2009)

Moreover, the malo-lactic bacteria, such as *O. oeni*, can also produce small amounts of exopolysaccharide (Ciezack *et al.*, 2010) that bind with tannins, which is responsible for the astringency of young wines. In this way, LAB contribute to a lower astringency and wine becomes more tasteful. Other studies have reported the presence of tannase enzymes in some wine LAB such as *Lb. plantarum* (Vaquero *et al.*, 2004) and the importance of the non-volatile matrix in the aromatic perception of red wines (Matthews *et al.*, 2006).

Some strains of *O. oeni* can also contribute to the wine aroma, thanks to their glycosidase activities, being a useful alternative source of these enzymes in winemaking (Grimaldi *et al.*, 2005; Olguín *et al.*, 2011; Pérez-Martín *et al.*, 2015). The fruity aroma of red wines is widely affected by MLF (Antalick *et al.*, 2013). While these authors have shown that LAB β-glycosidases are weakly involved in the modulation of this fruity aroma, conversely they have found that esterase seems to play a key role and it has, with other enzymatic activities, an impact on sulphur-containing compounds, such as thiols.

Interestingly, harsh conditions of some wines, such as ethanol above 14 per cent, enhance the differences in the production of aromatic compounds between *O. oeni* strains (Costello *et al.*, 2012). These differences between strains are likely due to the variation in coding capacity found in the sequenced genomes of several strains. For instance, one *O. oeni* strain that enhances fruit-red berry aroma and attributed to red wines (Bartowsky *et al.*, 2008) was found to have annotated genes for additional glycosidases in its genome (Bartowsky and Borneman, 2011).

Besides deacidification and the improvement of organoleptic qualities, another very important benefit of MLF is the microbiological stability that is achieved (Davis *et al.*, 1985; Davis *et al.*, 1988). The wines where MLF has taken place can be bottled without the risk of a possible further bacterial development, which could lead to the formation of gas ($CO_2$) in the bottles. In addition, the bacterial depletion of malic acid and remains of other nutrients such as sugars during MLF makes much more difficult for the subsequent growth of any bacteria. In fact, some wineries promote MLF not for a desired decrease in acidity but because once malic acid is finished, the produced lactic acid is biologically very stable in the conditions of wine.

## 3.4. Negative Aspects of Lactic Acid Bacteria in Winemaking

Research has been also extensively done on negative aspects of some LAB of wine in order to control and avoid them. Strains of some *Lactobacillus*, *Pediococcus* and other species can cause wine spoilage, including undesirable substances for the health of the wine consumer, such as biogenic amines and ethyl carbamate.

Malo-lactic fermentation is not favourable for all wines. In fact, in warmer areas, grapes tend to be less acidic and a further decrease in acidity by MLF may be deleterious for the sensory properties and microbiological stability of wine (Versari *et al.*, 1999). Sometimes the development of LAB and the same MLF may have negative consequences for the quality of wines. Fortunately, most of these cases occur sporadically, especially when there have been other problems prior to fermentation, or little control of winemaking (Sponholz, 1993; Sumby *et al.*, 2014).

Of these possible damages, the most common is the *piqûre lactique*, the lactic taint, which is the production of a certain amount of lactic and acetic acids by LAB. When there is residual sugar that has not been consumed by yeast, the normal situation for bacteria is to consume it, making lactic acid fermentation. In this case it produces D-lactic acid whereas in MLF from L-malic acid appears as L-lactic acid. Therefore, the analysis of enzyme isomers of lactic acid allows distinguishing if there is a beginning of lactic taint. It can occur when there is a proliferation of LAB before yeasts have finished the alcoholic fermentation.

In some wines with a relatively high pH, the MLF carried out by LAB can lead to excessive deacidification, which is not desirable because the organoleptic quality of wine is lost if it is not acid, and the higher pH may favour the growth of other harmful bacteria. To avoid this unwanted deacidification, usually more sulphur dioxide is added at the end of alcoholic fermentation, in order to keep more than 25 mg/l free $SO_2$, or lysozyme can be added to inhibit LAB (Lonvaud-Funel, 1999).

Another occasional problem related to the quality of wine caused by LAB is the ropiness in wine. It involves the overproduction of glucans, especially by *Pediococcus* strains, resulting in a characteristic viscosity of wine (Manca de Nadra and Strasser de Saad, 1995).

### 3.5. Potential Negative Health Effects of Lactic Acid Bacteria in Wine

There are only two compounds occasionally present in wine which can affect health of consumers and that are attributed to LAB – biogenic amines and ethyl carbamate. They also can be a source of problems in commercial transactions since some countries have established maximum limits.

#### *3.5.1. Biogenic Amines*

Biogenic amines (BA) are organic bases, related with biological activity, that are commonly present in living organisms. They can be naturally present in many foods, such as fruits, vegetables, meats and juices (Silla-Santos, 1996) or can also be produced by amino acid decarboxylases of some microorganisms (Farías *et al.*, 1993; Suzzi and Gardini, 2003). Some symptoms that they may cause are headache, respiratory distress, heart palpitation, hyper- or hypotension, and several allergic reactions (Silla-Santos, 1996).

The BA most commonly found in wine are histamine, tyramine, putrescine, cadaverine, and 2-phenylethylamine (Lonvaud-Funel, 2001; Landete *et al.*, 2005; Sahoo *et al.*, 2015). Their total concentration in wine is quite variable, depending on amino acid concentration in must and on the winemaking conditions, i.e. pH, ethanol concentration, temperature, $SO_2$ concentration, turbidity and storage and handling conditions (Soufleros *et al.*, 1998; Marqués *et al.*, 2008; Moreno-Arribas and Polo 2008; Del Prete *et al.*, 2009). The LAB strains are the main organisms responsible for BA accumulation, especially for tyramine and histamine (Lonvaud-Funel, 2001; Landete *et al.*, 2005).

Some amines can be produced in the grape or the musts or can be formed by yeast during alcoholic fermentation, although quantitatively only very low concentrations are reached in these stages (less than 3 mg/l). MLF is the main mechanism of biogenic amine formation, especially of histamine, tyramine and putrescine (Marcobal *et al.*. 2006). During this stage, the increase in these amines is accompanied by a significant decline in their amino acid precursors in wines (Landete *et al.*. 2005; Landete *et al.*. 2007a).

Strategies based on PCR amplification of the corresponding genes have been designed to detect biogenic amine-producing LAB (De las Rivas *et al.*, 2006; Landete *et al.*, 2007b), in order to apply early control measures to avoid the development of these bacteria. Anyway, the use of a selected malo-lactic starter and non-producer should be used to avoid the presence of biogenic amines in wines.

Different works have shown the ability of different LAB (García-Ruiz, 2011; Capozzi *et al.*, 2012; Callejón *et al.*, 2014) or some vineyard fungi (Cueva *et al.*, 2012) for degrading biogenic amines. The highest potential for degradation seems to be related to *Lactobacillus* and *Pediococcus* species, while *O. oeni* demonstrated low and rare degradation characteristics. The enzymes responsible for amine degradation have been isolated and purified from *Lb. plantarum* and *P. acidilactici* strains. They have been identified as multicopper oxidases or laccases (Callejón *et al.*, 2014) and one of these from *Lb. plantarum* has been cloned and characterised (Callejón *et al.*, 2016). Briefly, these findings indicate a potential application of wine-associated LAB or their enzymes in order to reduce biogenic amines in wine.

#### *3.5.2. Ethyl Carbamate*

Wine, like most fermented foods and beverages, contains trace amounts, usually not higher than 4 mcg/l, of ethyl carbamate (EC), also known as urethane (Ough, 1976; Vahl, 1993). EC is a genotoxic compound *in vitro* and *in vivo*, since it binds covalently to DNA and it is an animal carcinogen (Schlatter and Lutz, 1990).

EC is formed by reaction between ethanol and N-carbamyl compounds, such as urea and citrulline, at acid pH levels, and its formation is dependent on reactant concentration (Ough *et al.*, 1988). This reaction is favoured by high temperature and acid pH. The EC content is therefore higher in wines that have been stored for a long time without temperature control, arriving at 20-30 mcg/l EC (Uthurry *et al.*, 2004).

Although the urea produced by yeast is the main potential EC precursor in wine, LAB can contribute to the formation of EC due to the production of citrulline from arginine (Liu *et al.*, 1994), which is one of the predominant amino acids in wine. The arginine catabolism by LAB strains involves the arginine deiminase (ADI) pathway, which is the most widespread anaerobic route for bacterial arginine degradation. This pathway includes three enzymes: ADI (EC 3.5.3.6), ornithine transcarbamylase (EC 2.1.3.3, OTC), and carbamate kinase (EC 2.7.2.2, CK), which catalyse the following three reactions:

$$\text{L-arginine} + H_2O \xleftrightarrow{\text{ADI}} \text{L-citrulline} + NH_3$$
$$\text{L-citrulline} + \text{Pi} \xleftrightarrow{\text{OTC}} \text{L-ornithine} + \text{Carbamyl-P}$$
$$\text{Carbamyl-P} + \text{ADP} \xleftrightarrow{\text{CK}} \text{ATP} + CO_2 + NH_3$$

This pathway contributes positively to growth and viability of LAB due to ATP formation and the acidity decrease caused by ammonium production (Tonon and Lonvaud-Funel, 2000). However, the degradation of arginine yields citrulline, which can react with ethanol to form EC.

Wine LAB vary in their ability to degrade arginine. All heterofermentative lactobacilli are found to be degradative (Liu and Pilone, 1998). Other LAB-degrading arginine include *O. oeni, Pediococcus pentosaceus* and some strains of *Leuconostoc mesenteroides* and *Lactobacillus plantarum,* being this ability strain depending (Araque *et al.*, 2009b), as we can see in Table 2.

The presence of *arc* genes for the arginine-deiminase pathway (ADI) has been studied in several strains of different species of LAB. Genes encoding the three enzymes of this pathway in LAB are clustered in an operon-like structure: *arcA* (ADI), *arcB* (OTC), and *arcC* (CK). These genes have been characterised in different wine LAB such as *Lactobacillus hilgardii* (Arena *et al.*, 2002), *L. plantarum* (Spano *et al.*, 2004) and *O. oeni* (Tonon *et al.*, 2001). However, Araque *et al.* (2009b) demonstrated that in some species, such as *Leuconostoc mesenteroides,* these genes are not always grouped together as an operon.

As seen in Table 2, a good correlation was found between the presence and absence of the *arc* genes, mainly with *arcA* gene, with the degradation or lack of degradation of arginine.

Citrulline has been described as the main EC precursor produced by several *Lactobacillus* species (Mira de Orduña *et al.*, 2000; Azevedo *et al.*, 2002). The relation between citrulline produced and arginine consumed has been shown to be clearly higher in the presence of ethanol (10-12 per cent) and at low pH (3.0) in *L. brevis* and *L. buchneri* strains (Araque *et al.*, 2011). Arginine consumption reduces the inhibitory effect of low pH because the ADI pathway produces ammonia.

**Table 2.** Degradation of Arginine (5 g/l) in MRS Medium at pH 4.5 and Presence of *Arc* Genes in Strains of Different LAB. Genes were Detected by Three Pairs of Degenerate Primers

| *Species* | *No. tested strains* | *% arginine consumption (%)* | *Presence of arc genes* | | |
|---|---|---|---|---|---|
| | | | *arcA* | *arcB* | *arcC* |
| *Oenococcus oeni* | 8 | 13-99 | + | + | + |
| *Lactobacillus brevis* | 8 | 33-100 | + | + | + |
| *Lb. hilgardii* | 3 | 88 | + | + | + |
| *Lb. buchneri* | 2 | 61-99 | + | + | + |
| *Lb. plantarum* | 2 | 0-93 | + (*) | + | + (*) |
| *Pediococcus pentosaceus* | 8 | 82-100 | + | + | + |
| *P. parvulus* | 4 | 0 | - | - | - |
| *Leuconostoc mesenteroides* | 4 | 0-99 | + (*) | + (*) | - |

(*): This gene was not found in some strains (arginine non-degrading) of the species (Adapted with permission from Araque *et al.*, 2009b; copyright 2009, American Chemical Society)

The production of EC precursors often is associated to spoilage LAB species in wine and in beer, such as *Lb. brevis* and *P. pentosaceus* (Bartowsky, 2009). A higher expression of the *arcC* gene in *P. pentosaceus* was found in the presence of ethanol (Araque *et al.*, 2013), which resulted in a lower excretion of citrulline and ornithine than in *Lb. brevis*. This suggests that *Lb. brevis* is the species more likely to produce these amino acids, which are precursors of EC and of putrescine. Regarding *O. oeni*, a recent study (Araque *et al.*, 2016) showed that all 44 studied strains contained the three *arc* genes, but considerable variability in the ability to degrade arginine among strains was found, due to different expression patterns. Anyway, the citrulline utilisation by *O. oeni* could decrease noticeably the possibility of EC formation (Arena *et al.*, 1999).

## 4. Malo-lactic Bacteria: *Oenococcus oeni*

*O. oeni* is the most known species of the genus *Oenococcus* and the main species responsible for MLF in wine due to its ability to adapt to wine conditions. There are other species found in wine which can perform the MLF, such as some *Lactobacillus* (*see* 6.3 and 6.4), but *O. oeni* remains the most adapted to wine and the most used as starter culture.

### 4.1. Phylogeny of *Oenococcus oeni*

The genus *Oenococcus* belongs to the class *Bacilli*, order *Lactobacillales* and family *Leuconostocaceae*. The species *O. oeni* was formerly known as *Leuconostoc oenos*. This species was considerated in the past as a *Leuconostoc* because they are cocci (usually forming small chains), are gram-positive and catalase-negative and they do the heterolactic fermentation, that is, the production – from sugars – of other compounds in addition to lactic acid (D-isomer), mainly $CO_2$ and ethanol and acetic acid in small quantities. The classification into a new species was proposed by Dicks *et al.* (1995) based on its unique features: i) their habitat is exclusively must and wine, ii) they can grow at the pH of the wine, between 3 and 4, iii) they can grow in the presence of ethanol, up to 10 per cent (v/v), and more. Above all, the most relevant data to suggest that *Oenococcus* is another different genus from *Leuconostoc* was the confirmation by comparison of ribosomal RNA sequences that they are phylogenetically well separated (Fig. 3). As can be seen, *Oenococcus* is also distant from other related LAB, as heterolactic *Lactobacillus*.

Initially *O. oeni* was the only species of the genus but later another two species have been proposed: *O. kitaharae* (Endo and Okada, 2006), isolated from a composting distilled shochu residue, and *O. alcoholitolerans* (Badotti *et al.*, 2014), isolated from cachaça fermentation and bioethanol plants. The three species are associated with different ethanol-containing environments, but they have different adaptive and metabolic capacities. *O. kitaharae* is more sensitive to ethanol than *O. oeni* (Endo and Okada, 2006) and has an optimal growth pH between 6 and 6.8, which is three orders of magnitude less acid than the conditions found normally in wine. In exchange, *O. kitaharae* possesses more genes of cell defence mechanisms (bacteriocins production, restriction-modification systems and a CRISPR locus) and also genes that code for amino acid biosynthesis pathways that are absent in *O. oeni* (Borneman *et al.*, 2012a). *O. alcoholitolerans*, despite its name, is less resistant to ethanol than *O. oeni*. The gene coding

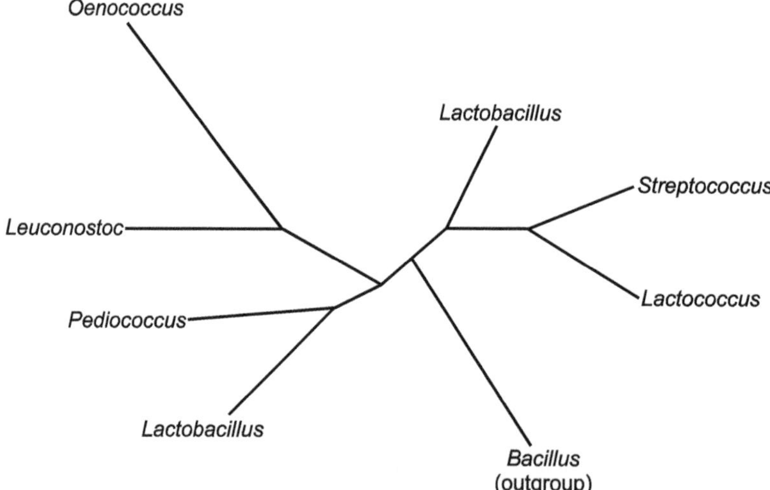


**Figure 3.** Unrooted phylogenetic tree of the main LAB including *Oenococcus*, based on 16S rRNA sequences (Modified from Dicks *et al.*, 1995)

for malo-lactic enzyme is intact in *O. alcoholitolerans*, so it is likely that this species is able to perform MLF, although there are no published data reporting it.

### 4.2. *Oenococcus oeni*: The Best Adapted to Wine Conditions

Such as other wine LAB, *O. oeni* has fastidious nutritional requirements. However, this species, due to selective pressure, has developed adaptive strategies that enable it to out-compete other potential MLF bacteria during the later stages of vinification and thus dominate in wine. It is, for example, well adapted to high ethanol concentrations (<15 per cent v/v), low pH (as low as 2.9) and limited nutrient availability (Bartowsky, 2005). Marcobal *et al.* (2008) suggested that the genus *Oenococcus* is hypermutable due to the loss of the mismatch repair pathway (genes *mutS* and *mutL*), which occurred with the divergence away from the *Leuconostoc* branch. This would explain the observed high level of allelic polymorphism (De Las Rivas *et al.*, 2004; Delaherche *et al.*, 2006; Zé-Zé *et al.*, 2008) among the known *O. oeni* isolates and had likely contributed to the high ecological competitiveness of this genus to acidic and alcoholic environments. In addition, the absence of genes for sporulation, catalase and other key enzymes of oxidative stress response make the study of this genus interesting because of its capacity to survive harsh environments using other mechanisms.

So, most LAB isolated in winemaking, especially after alcoholic fermentation, have the characteristics of *O. oeni*. It is the predominant species in MLF of wines from many different places, although in some cases it has been found that the fermentation is carried out by other species, mainly by *Lb. plantarum* (Du Plessis *et al.*, 2004; Lerm *et al.*, 2011; Bravo-Ferrada *et al.*, 2013).

### 4.3. Genetics of *O. oeni*

The first study about the genetic diversity of *O. oeni*, based on the diversity of 16S, 23S and 16-23S spacer sequences, suggested that the species was genetically homogeneous (Martínez-Murcia and Collins, 1990; Le Jeune and Lonvaud-Funel, 1997). The same conclusion was set from data of DNA-DNA homology and similarities between genetic maps (Dicks *et al.*, 1995; Zé-Zé *et al.*, 2008). However, this is not in accordance with other models that analysed the species' diversity at different levels (Tenreiro *et al.*, 1994). Further studies, which were based on a multi locus sequence typing (MLST) analysis of four housekeeping genes plus the *MleA* gene, indicated that the species was indeed heterogenic and composed of a panmitic population, with a structure shaped by recombination (De las Rivas *et al.*, 2004). Other studies brought evidence of the existence of at least two genetic groups of strains, namely A and B, that would have evolved independently (Bilhère *et al.*, 2009; Bridier *et al.*, 2010).

Nowadays, more than two hundred genome assemblies of *O. oeni* are available from the database. The first determined *O. oeni* genome, PSU-1 strain (Mills *et al.*, 2005), marked a significant new phase for wine-related research on LAB in which the physiology, genetic diversity and performance of starter cultures could be more rigorously examined. Before the availability of the *O. oeni* genomes, Zé-Zé *et al.* (1998, 2000) performed a detailed genome mapping on two strains (PSU-1 and GM), providing insight into the genetic organisation of the bacterium. Important phenotypes, such as malate degradation, citrate metabolism and diacetyl production, were mapped. This comparative analysis also revealed extensive conservation of loci order. However, further studies have indicated the genomic heterogeneity of *O. oeni* strains, also due to its loss of mismatch DNA repair (Marcobal *et al.*, 2008).

The comparative genome study of the MLF capacity of over 70 *O. oeni* strains (Bon *et al.*, 2009) in three wines indicated the presence of eight stress-responsive genes that could be associated with high MLF performance. Borneman *et al.* (2010) started using array-based comparative genome hybridisation and genome sequencing of three *O. oeni* strains (PSU-1, BAA-1163, and AWRIB429) revealing 10 per cent of genomic diversity. This variation in the *O. oeni* species protein-coding capacity would presumably be key to phenotypic differences between strains (Bartowsky and Borneman, 2011). Also Borneman *et al.* (2010), in order to ascertain the basis of these phenotypic differences, mapped the genomic content of ten wine strains of *O. oeni*. These strains comprised a genomically diverse group in which large sections of the reference genome were often absent from individual strains. The same authors (Borneman *et al.*, 2012b) provided 11 sequenced strains by whole-genome sequencing, which enhanced a broad insight into the genetic variation present within *O. oeni*.

Other works have been focused on the presence of specific traits related to wine conditions. *O. oeni* genome comprises different genes related to the synthesis of exopolysaccharides (EPS). Different comparative genome studies have been carried out using multilocus sequence typing (MLST), revealing the importance of EPS for the adaptation of the bacteria to wine and providing an inventory of the genes potentially involved in this biosynthesis (Dimopoulou *et al.*, 2012, 2014, 2016). Similar studies were conducted for the study of substrate transport and phosphorylation (Jamal *et al.*, 2013). These authors reported that as a part of the core genome, the phosphotransferase genes might contribute to the perfect adaptation of *O. oeni* to its singular ecological niche.

A wide study (Campbell-Sills *et al.*, 2015) involved 50 genomes from *Oenococcus* genus isolated from different wine products (red wine, white wine, champagne, and cider), different regions (France, Australia, Lebanon, United States, Italy and England) and from different years. Phylogenomic and population structure analyses revealed two major groups of strains, one related to wine and champagne and another related to wine and cider. The authors suggested that ancestral *O. oeni* strains were adapted to low ethanol-containing environment, such as overripe fruits, being domesticated to cider and wine, while another group suffered a process of further domestication to specific wines such as champagne.

Some of these mentioned studies have revealed that the major factor that affects genomic diversity is the presence of prophages, which generate large-size structural polymorphisms among *O. oeni* genomes (Zé-Zé *et al.*, 2008; Bon *et al.*, 2009; Borneman *et al.*, 2010). In addition, insertion sequences (IS) represent another class of variable genetic elements found in the pangenome of *O. oeni* (Zé-Zé *et al.*, 2008; El Gharniti *et al.*, 2012). IS are autonomous transposable elements (Mahillon and Chandler, 1998) and most of the IS-related sequences available in the species *O. oeni* have been characterised. They were linked to strain-specific genes involved in stress response and/or persistence in the niche, such as the ferritin-like protein *dpsB* (Bon *et al.*, 2009), the ornithine decarboxylase *odc* (Marcobal *et al.*, 2006) and the glucosyltransferase *gtf* genes (Dols-Lafargue *et al.*, 2008). IS elements may disrupt genes, but may also activate downstream genes (Treangen *et al.*, 2009) as well as increase the propensity of acquiring further adaptive mechanisms. Moreover, as observed in *Escherichia coli*, IS play a role in the inactivation and immobilisation of incoming phages and plasmids (Ooka *et al.*, 2009).

More extensive comparative genomics are needed in order to study the link between industrial characteristics and genomic features. Further, phylogenomics analysis involving a whole genome data could reconstruct the evolutionary history of microorganisms, such as *O. oeni*, and their acquired capacities related to its specific niche.

## 5. Adaptation Mechanisms of *Oenococcus oeni* to Wine Stress

Over centuries of selective pressure, *O. oeni* has adapted to the hostile conditions typical of wine: high ethanol concentrations (up to 15 per cent v/v), low pH (as low as 2.9) and limited nutrient availability. However, despite being *O. oeni*, the species is best adapted to wine but the induction of this fermentation remains sometimes uncertain. The introduction of commercial freeze-dried bacterial cultures of *O. oeni* for direct inoculation into wine has improved the control of the process ensuring better control of the time of onset and the rate of MLF (Nielsen *et al.*, 1996).

In wine, *O. oeni* deals with several stresses including low temperature, $SO_2$ concentration, short-chain fatty acid presence, phenolic compounds, low pH and ethanol content (Davis *et al.*, 1988; Lonvaud-Funel, 1999; Spano and Massa, 2006). The effects of all these factors on *O. oeni* have been studied in order to improve the knowledge of the cellular adaptation of this bacterium. Low temperatures affect growth rate and increase lag phase (Fugelsang, 1997). Sulphur dioxide reduces ATPase activity and decreases cell viability (Carreté *et al.*, 2002; Reguant *et al.*, 2005a), while some phenolic compounds produce breakdown of the LAB cell membrane (García-Ruiz *et al.*, 2011). Yeast fatty acids, such as decanoic and dodecanoic acid, are powerful inhibitors of LAB growth because, like ethanol, they alter the bacterial membrane (Lonvaud-Funel *et al.*, 1988). Finally, low pH reduces *O. oeni* growth and malo-lactic activity (Tourdot-Maréchal *et al.*, 1999). Other difficulties in MLF have been ascribed to phage attack (Poblet-Icart et al., 1998; Gindreau and Lonvaud-Funel, 1999). However, as the phages readily disappear through inactivation by wine components, it seems that they are not responsible for influencing MLF (Lonvaud-

Funel, 1999). Therefore, a better knowledge of stress physiology may be useful to optimise survival of starter cultures of *O. oeni* and improve the control of MLF in the wine industry (Beltramo *et al.*, 2006).

The multiple stresses present in wine provoke a complex response of *O. oeni*. Figure 4 reflects the most relevant stress response mechanisms of *O. oeni*. Changes in cell membrane composition and fluidity, activation of ATPase and stress proteins synthesis would be the main mechanisms against stress.

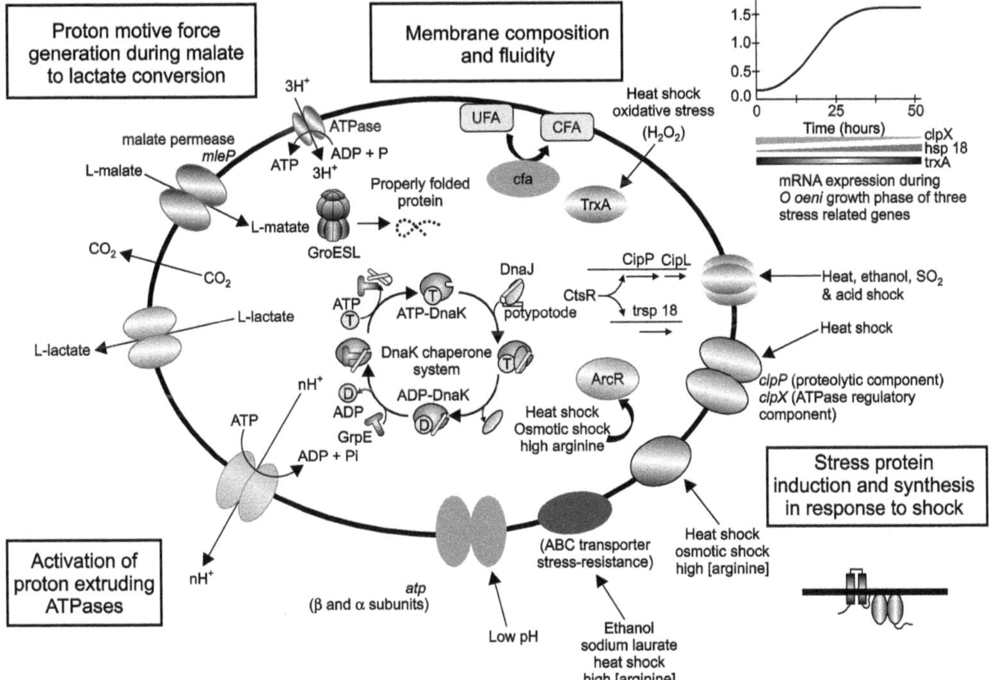


**Figure 4.** Mechanisms that are important in conferring, in *O. oeni*, the ability to survive in wine: the membrane composition, the proton motive force generated by malic acid metabolism, the activation of proton-extruding ATPase, and the stress protein induction and synthesis in response to shock (Adapted from Bartowsky, 2005)

Color version at the end of the book

## 5.1. Changes in Cell Membrane

Cell membrane is a selective barrier that has a specialised structure that facilitates its gatekeeping function, maintains homeostasis and consequently has an important role in adaptation to environmental changes (Šajbidor, 1997). Membrane proteins and lipids are targets for ethanol toxicity. The negative effect of ethanol is attributed to its interaction with the membrane at the aqueous interface. It results in a perturbed membrane structure and function (Weber and de Bont, 1996) which lead to a leakage of intracellular compounds, including enzymatic cofactors and ions essential for cell growth and fermentation as well as dissipation of the electrochemical gradient (Da Silveira *et al.*, 2002). As a result, membrane composition is adjusted to counteract the increase in membrane fluidity, like changing phospholipid content (Teixeira *et al.*, 2002; Grandvalet *et al.*, 2008).

The membrane phospholipids incorporate fatty acids (FA) of varying length and saturations. The lipid structure and the portion of saturated (SFA), unsaturated (UFA) and cyclopropane fatty acids (CFA) influence the fluidity of biological membranes. It has been described that low pH produced an increase of the ratio of UFA to SFA in membranes of *O. oeni* (Drici-Cachon *et al.*, 1996a). On the other hand, *O. oeni* cells respond to ethanol by increasing their CFA content, specifically lactobacillic acid (Teixeira *et al.*, 2002), by activating the gene *cfa* (encoding a CFA synthase) responsible for the conversion of UFA to CFA (Grandvalet *et al.*, 2008). The *cfa* gene has been related to the presence of ethanol and low pH (Grandvalet *et al.*, 2008). CFAs in the membrane reduced proton permeability (Da Silveira *et al.*, 2002) and increased membrane rigidity (Da Silveira *et al.*, 2003).

Changes in the fluidity of the cell membrane in response to stress have been measured for *O. oeni* using electro spin resonance (ESR) (Da Silveira *et al.*, 2003) and fluorescence anisotropy (Tourdot-Maréchal *et al.*, 2000; Chu-Ky *et al.*, 2005; Coucheney *et al.*, 2005; Weidmann *et al.*, 2010; Maitre *et al.*, 2012). The first work, focusing on the role of the membrane fluidity as a determinative factor in stress tolerance in *O. oeni*, presented the instantaneous variations of anisotropy values caused by heat (42°C), acid (pH 3.2) and ethanol (10 per cent v/v) shocks (Tourdot-Maréchal *et al.*, 2000). Heat or acid shocks decreased the anisotropy values (fluidising effects), whereas an ethanol shock increased the membrane rigidity. Chu-Ky *et al.* (2005) showed that cold shocks strongly rigidified plasma membrane. Ethanol shocks (10-14 per cent v/v) induced instantaneous membrane fluidisation followed by rigidification and acid shocks (pH 4.0 and pH 3.0) exerted a rigidifying effect on membrane without affecting cell viability. Other membrane fluidisation studies (Coucheney *et al.*, 2005; Weidmann *et al.*, 2010; Maitre *et al.*, 2012) have focused on the role of the small heat shock protein (sHsp) Lo18 in the membrane. These sHsps bind membranes to regulate bilayer fluidity (Nakamoto and Vígh, 2007; Horváth *et al.*, 2008) and Lo18, under ethanol exposure, was shown to interact with the cytoplasmic membrane and to stabilise liposome fluidity (Coucheney *et al.*, 2005).

The ethanol effect on membrane protein composition of *O. oeni* cells stressed with 12 per cent ethanol for one hour and cells grown in the presence of 8 per cent ethanol was reported by Silveira *et al.* (2004). The enzymes dTDT-glucose-4,6-dehydratase (encoded by *rmlB*) and D-alanine:D-alanine ligase (encoded by *ddl*), were increased in ethanol-adapted cells. These enzymes are known to be involved in lipopolysaccharide and peptidoglycan biosynthesis, respectively, suggesting that the cell wall is modulated during ethanol adaptation (Silveira *et al.*, 2004). Moreover, Elahwany (2012) reported an increase in the intracellular levels of amino acids under ethanol stress and aggregation of bacterial cells.

## 5.2. Activation of ATPase

ATP synthase or ATPase is a membrane-bound enzyme complex and ion transporter, which combines ATP synthesis and/or hydrolysis with the transport of protons across the membrane. ATPase is well known for its key role in the acid tolerance of bacteria. Induction of the activity and expression has been associated with increased resistance to low pH (Kobayashi, 1985; Kullen and Klaenhammer, 1999). Cox and Henick-Kling (1989) proposed that the malo-lactic conversion of L-malic acid to L-lactic acid, which is not energetic by itself, provides a proton motive force that is sufficient to drive ATP synthesis by membrane ATP synthase. Fortier *et al.* (2003) reported that in the absence of malic acid, ATPase of *O. oeni* is induced at low pH and its regulation seems to occur at transcription level. In *O. oeni*, no significant increase of ATPase activity was measured at low pH although mutants of the $F_1F_0$-ATPase isolated during a long-term survival screen at pH 2.6 were able to grow in acidic media and were characterised by a high $H^+$-ATPase activity at low pH (Drici-Cachon *et al.*, 1996b). In addition, preliminary results suggested the existence of several ATPases with different optimum pHs (Guzzo *et al.*, 2000) and a $K^+$-ATPase involved in pH homeostasis (Guchte *et al.*, 2002).

Tourdot-Maréchal *et al.* (1999) suggested that ATPase and malo-lactic activities of *O. oeni* are linked to each other and play a crucial role in the mechanism of resistance to an acid stress. This link was demonstrated later by Galland *et al.* (2003). On the other hand, ATPase activity can be inhibited by $SO_2$, pesticides as copper and fatty acids; moreover, this effect increases with ethanol presence (Carreté *et al.*, 2002).

Transcriptional studies report the increased expression of the *atpB* gene encoding the β-subunit of the $F_1F_0$ ATPase due to low pH and ethanol adaptation (Beltramo *et al.*, 2006; Olguín *et al.*, 2010). Although some transcriptomic studies have reported a non-differential expression due to ethanol presence in growth media (Olguín *et al.*, 2015), proteomic studies describe an increased concentration of ATPase protein in response to pH and ethanol (Costantini *et al.*, 2015; Margalef-Català *et al.*, 2017a). Recently, a genome-scale reconstruction of the metabolic network reported that *O. oeni* grown in the presence of 12 per cent ethanol (v/v) spent 30 times more ATP to stay alive than in the absence of ethanol. Most of this ATP would be employed for extruding protons outside of the cell (Mendoza *et al.*, 2017).

## 5.3. Synthesis of Stress Proteins

Microorganisms undergo complex programmes of differential gene expression, involving a rapid increase in the concentrations of specific sets of proteins to counteract the harsh environmental conditions. In *O. oeni*, some heat shock proteins (Hsps) from operons *GroESL* (*groES* - *groEL*) and DnaK (*grpE* - *dnaK* - *dnaJ*) and small Hsps (sHsps) perform chaperone functions, by stabilising new proteins to ensure correct folding or by helping to refold proteins that were damaged by the cell stress. Other stress proteins have protease activity, such as the Clp ATP-dependent protease, HtrA and FtsH, which degrade incorrectly folded proteins (Grandvalet *et al.*, 2005).

The most known sHsp in *O. oeni* is Lo18 protein. This protein is coded by the *hsp18* gene and its expression seems to be controlled at the transcriptional level (Coucheney *et al.*, 2005). This protein synthesis was determined after heat (42°C), acid (pH 3) and ethanolic shock (12 per cent v/v) (Guzzo *et al.*, 1997; Jobin *et al.*, 1997) and in the stationary growth phase (Guzzo *et al.*, 1997). The over-expression of the gene *hsp18* was also confirmed under heat shock and during the stationary phase in strain ATCC BAA-1163 (Desroche *et al.*, 2005). The sHps can bind membranes to regulate bilayer fluidity (Nakamoto and Vígh, 2007; Horváth *et al.*, 2008). Since Lo18 protein was found to be peripherally associated with the membrane of *O. oeni* by Jobin *et al.* (1997), several studies have focused on its role in this cellular barrier. Delmas *et al.* (2001) showed that the prokaryotic Lo18 can function as a molecular chaperone *in vitro* and its membrane association depends on the temperature upshift. In addition, expression of this sHsp was induced by administration of a membrane fluidiser, the benzyl alcohol, suggesting that Lo18 expression could be regulated by the level of membrane fluidity. Coucheney *et al.* (2005) suggested that Lo18 could be involved in an adaptive response allowing the maintenance of membrane integrity during temperature ramping in presence of Lo18 protein in *O. oeni* cells. More recently, Weidmann *et al.* (2010) suggested using transformed *E. coli* cells under heat shock, that a major part of Lo18 is associated with the cytoplasmic membrane. The same group has reported Lo18 preventing the thermal aggregation of proteins and playing a crucial role in membrane quality control (Maitre *et al.*, 2014).

Other stress proteins play a relevant role in *O. oeni* response to stress. Some of them are Clp proteins, responsible for abnormal protein degradation. This type of stress proteins consist of two subunits – the ATP-dependent proteolytic component (ClpP) and the ATPase regulatory component. Clp ATPases are classified in two categories, first class ATPases (ClpA, ClpB, ClpC, ClpD, ClpE and ClpL), also known as Hsp100, and smaller proteins called second class ATPases, such as ClpX (Beltramo *et al.*, 2004a). Several studies in *O. oeni* revealed the overexpression of ClpL, ClpX and ClpP in response to stress conditions (Jobin *et al.*, 1999a; Guzzo *et al.* 2000; Beltramo *et al.*, 2006). In *O. oeni*, CtsR regulates most of the major molecular chaperone genes, including the *clp* genes and the *groES* and *dnaK* operons (Grandvalet *et al.*, 2005). Thioredoxin gene (*trxA*), involved in the cell redox balance maintenance, has been also described as stress responsive gene (Jobin *et al.*, 1999b; Beltramo *et al.*, 2006; Olguín *et al.*, 2010; Margalef-Català *et al.*, 2017a; Margalef-Català *et al.*, 2017b). Also, the metabolism of the antioxidant tripeptide glutathione has been associated to stress response in *O. oeni*. Silveira *et al.* (2004) and Cecconi *et al.* (2009) reported that glutathione reductase was induced in ethanol-adapted cells, suggesting that maintenance of the redox balance plays an important role in alcohol adaptation. Although *O. oeni* is not able to synthesise glutathione, the uptake of this compound from the growth medium improves *O. oeni* cell survival and growth (Margalef-Català *et al.*, 2016b, Margalef-Català *et al.*, 2017c).

The mechanisms of stress-response regulation among the different Gram-positive bacteria are different. For *O. oeni,* the master regulator for many molecular chaperone genes is *CtsR* (Grandvalet *et al.*, 2005). Indeed, in the genomes of strains PSU-1 and ATCC BAA-1163, there are no genes encoding for other main transcriptional regulators, such as *hrcA* or alternative sigma factors, and no CIRCE elements were found (Beltramo *et al.*, 2004a; Grandvalet *et al.*, 2005; Mills *et al.*, 2005).

Gaining a better knowledge of the genes involved in the response to harsh wine conditions is the key to establish basic stress characters in order to select better starters able can overcome cell mortality after inoculation into wine.

## 6. Technological Aspects of Malo-lactic Fermentation and Its Control

From the viewpoint of performing the MLF in the cellar, there are some technological aspects that must be taken in account, in order to assure its development or to control it and the main bacteria responsible, that is, *O. oeni*.

## 6.1. Identification of *Oenococcus oeni*

In order to monitor MLF in the cellar and testing if the bacterial population evolves correctly, it is necessary to have efficient identification techniques to distinguish different species and strains of LAB and from other microorganisms. Several molecular techniques have been developed in the last few years, enabling the study of population dynamics and the confirmation of which strain has been the responsible of the MLF. These techniques are also required for any experimental study to improve the process or to do research on MLF, for example, to see the influence of various wine conditions on cells growth and the performance of MLF.

The classical phenotypic identification methods, based on biochemical tests, have many disadvantages, such as high variability or the number of tests to be carried out, and there is often uncertainty, since the descriptions of manual identification are based on data from a limited number of strains, while the variability of isolated samples from wine is higher. Furthermore, the identification of *O. oeni* is especially difficult because it has a slow growth and there is a large variability among strains in some tests and this species has remarkable poverty fermentation. Therefore, it is advisable to use molecular methods, which are reliable, accurate and fast.

One of the most commonly used method to identify whether an isolate is *O. oeni*, is a species-specific PCR, performed with specific primers (On1 and On2) for a gene fragment of the malo-lactic enzyme of *O. oeni* (Zapparoli *et al.*, 1998). It gives a 1025 bp amplified DNA, easily detectable by electrophoresis.

These usual identifications are done with the DNA extracted from isolated colonies, but some methods have been developed for detecting and quantifying *O. oeni* directly in wine. For instance, a real-time quantitative PCR (RT-qPCR) method has been developed with the specific primers for the malo-lactic enzyme and fluorogenic probes in order to quantify genomic DNA from wine samples without sample plating (Pinzani *et al.*, 2004).

Another possibility is the direct detection of cells of *O. oeni* and other species present in wine samples, using fluorescent *in situ* hybridisation (FISH) with 16S rDNA specific probes labelled with fluorophores (Blasco *et al.*, 2003). It allows the direct identification and quantification of bacterial species at microscopic level without previous cultivation. Similar FISH protocols for fast identification of *O. oeni*, based on the 5S rDNA and the ITS-2 region (23S-5S internal transcribed spacer), have been also developed (Hirschhäuser *et al.*, 2005).

## 6.2. Typing Strains of *Oenococcus oeni*

Besides knowing that an isolate belongs to *O. oeni* or other species, often it is necessary to recognise the strain, that is, to type it. For example, to check whether a particular strain used as starter is really the predominant strain at the end of MLF. Therefore, methods are needed that allow differentiating strains of *O. oeni* in order to study their population dynamics under different conditions (Reguant *et al.*, 2005a; Carreté *et al.*, 2006; Ruiz *et al.*, 2010; Franquès *et al.*, 2017).

One of the several methods developed to type strains of *O. oeni* is the total DNA macro restriction and separation of large fragments obtained by pulsed field gel electrophoresis (PFGE) (Zapparoli *et al.*, 2000; Ruiz *et al.*, 2008). It provides for each strain-specific profiles of bands very reproducible and easy to analyse, but it is quite a laborious technique. Another typing technique is the RAPD-PCR (random amplified polymorphic DNA PCR), which includes an arbitrary single short oligonucleotide (10-15 bp) as primer. It is simple, quick and discriminatory between strains. The only drawback is its low reproducibility. For increasing the reproducibility, this method was optimized with a RAPD multiplex technique (Reguant and Bordons, 2003), where two primers are used simultaneously: one of the most discriminant primers already used in RAPD, called Coc (Cocconcelli *et al.*, 1995) along with one of the two primers of the aforementioned species-specific PCR, called On2 (Zapparoli *et al.*, 1998). This multiplex method yields a profile of discriminating bands for each strain that is reproducible in different trials. A multiplex variation of this method allows the simultaneous species identification and strain typing of *O. oeni* (Araque *et al.*, 2009a).

A good and accurate method for typing *O. oeni* strains is using multilocus variable number of tandem repeats (VNTR). It is based on the presence of these VNTR at a specific locus due to DNA polymerase slippage during replication. The complete tandem repeat is amplified and sized, using a capillary electrophoresis system. It has been developed typing an extensive collection of *O. oeni* strains (Claisse and Lonvaud-Funel, 2012), and it has been improved by multiplexing amplifications in two separate PCR mixtures for five loci, taking advantage of the high performance of capillary electrophoresis (Claisse and Lonvaud-Funel, 2014). This VNTR method has been successfully applied for typing many isolates of *O. oeni* from French wines (Garofalo *et al.*, 2015; El Khoury *et al.*, 2017) and from the Catalan Priorat region (Franquès *et al.*, 2017).

### 6.3. Identifying other Wine LAB Species and Typing Their Strains

One of the most practical methods for identifying different species of LAB is 16S-ARDRA, used also for a variety of microorganisms. In this method, 16S rDNA is amplified by PCR with specific primers and then digested by restriction enzymes *Mse*I, *Bfa*I and *Alu*I (Rodas *et al.*, 2003). In this way, these authors were able to discriminate 32 LAB species by their band profiles, including several *Lactobacillus* species – *Leuconostoc mesenteroides*, *O. oeni*, *Pediococcus parvulus* and *P. pentosaceus*. In case of doubtful profiles, it is useful to sequence the fragment 16S rDNA to confirm the species.

Typing strains of different wine *Lactobacillus* strains can also be carried out with restriction fragment length polymorphism, followed by pulsed-field gel electrophoresis (RFLP-PFGE), as shown by Rodas *et al.* (2005). Finally, another good technique used both for identifying species and typing strains of *Lactobacillus* is the rep-PCR fingerprinting, using the $GTG_5$ primer, which targets these bacterial DNA repetitive elements, and that is suitable for a high-throughput of strains (Gevers *et al.*, 2001).

### 6.4. Using *O. oeni* Starters for MLF and Criteria of Their Selection

Even though MLF occurs spontaneously in wines, it starts randomly and any delay in the starting of MLF can lead to an alteration of wine quality (Henick-Kling, 1995). Therefore, most middle-size and great wineries use malo-lactic starters for direct inoculation in wines (Nielsen *et al.*, 1996; Maicas, 2001; Gindreau *et al.*, 2003). Most of the starters for MLF contain strains of *O. oeni*, distributed by several companies: Lallemand and its related Danstar, Chr. Hansen, Laffort, Agrovin, Enartis, Oenobrands, Scott Lab., Bioprox, Wyeast, 2B-FermControl, and others. Nevertheless, some of the malo-lactic starters contain strains belonging to other LAB species, mainly *Lactobacillus plantarum*, such as Viniflora® plantarum from Chr. Hansen, or ML Prime™ from Lallemand.

In the past, starter culture preparation required rehydration of the dried culture followed by activation and acclimatisation to the wine environment and then, growth of the culture to an appropriate volume (Henick-Kling, 1993; Krieger *et al.*, 1993). Since the 1990s, several commercial starter cultures have been developed that require minimal preparation, such as rehydration, or can be added directly to wine without preparation, usually with freeze-dried cultures (Nielsen *et al.*, 1996; Gockowiak and Henschke, 2003; Lallemand, 2005).

When the starter is added in sufficiently large amounts, with a minimum bacterial population of $10^6$ CFU/ml, it is not forced to grow and multiply under wine conditions, which means that the MLF will begin without waiting for the development of a sufficiently large number of cells. Commercial cultures are available in the frozen liquid or freeze-dried states. A number of the freeze-dried cultures have been biophysically adapted to withstand the rigours of the wine media and can be added directly to wine without an activation step. When properly executed, it will normally ensure a complete MLF, even in the most difficult of wines, because malo-lactic starter cultures have been selected for their resistance to adverse wine conditions (Lallemand, 2005).

The inoculation of malo-lactic starters is usually carried out after alcoholic fermentation (AF), but sometimes it is done during the same AF (Zapparoli *et al.*, 2008; Bartowsky *et al.*, 2015). In this case, the benefits include, among others, the higher content of potential nutrients for bacteria, lower sulphur dioxide and ethanol levels and an early completion of MLF, which means that the winemaker can make an earlier treatment with $SO_2$ post fermentation, in order to protect the wine from oxidation and spoilage microbes. Obviously, the disadvantages for early inoculation include the competence of yeast and LAB for resources and the potential production of the undesirable acetic acid by *O. oeni* metabolising the glucose still present in the must. Moreover, the MLF after the AF has the advantage of the lees being a nutrient source for bacteria, through the autolysis of the dead yeast cells (Fugelsang and Edwards, 2007).

However, induction of MLF by inoculation with commercially available strains of *O. oeni* is not always successful. The difficulty in inducing MLF in wine remains problematic because wine is a very harsh environment for bacterial growth. For this reason, a good selection of strains for getting a starter is needed.

### 6.4.1. Selection of Strains for Wine Inoculation

This is usually performed by classical tests based essentially on the survival in wine and monitoring the consumption of L-malic acid (Henick-Kling *et al.*, 1989). Some of the criteria to be taken into account when developing bacterial cultures (Torriani *et al.*, 2011) include resistance to ethanol and $SO_2$, resistance to bacteriophages (Poblet-Icart *et al.*, 1998), ability to grow at low pH levels, no health hazard for the end consumer, and resistance to technological stress – freezing, freeze-drying, hydration and inoculation into wine, among others (Henick-Kling, 1995). Additional criteria are that strains have a high malo-lactic performance in different types of wine, the production of desirable flavours, a low acetic acid production, no production of ropy polysaccharides neither off-flavours, and a good compatibility with the yeast strains used for alcoholic fermentation (Torriani *et al.*, 2011).

DNA-based molecular methods, besides being fundamental for species identification, can prove very useful for screening both positive and negative traits under the control of one or a few more genes. For instance, selection of strains with higher β-glycosidase activity can be done with both phenotypic and genotypic assays (Barbagallo *et al.*, 2004; Grimaldi *et al.*, 2005). Also, the selection of strains with increased production of diacetyl from citric acid can be improved by measuring the enzymatic activity and at the same time the expression of the genes involved in citric acid metabolism (Olguín *et al.*, 2009). As commented in earlier Section 5, stress genes of *O. oeni* related with wine adaptation, including those of redox balance (Beltramo *et al.*, 2006; Margalef-Català *et al.*, 2017a; Margalef-Català *et al.*, 2017c), are also good markers in order to select strains, since they play a key role in adaptation (Margalef-Català *et al.*, 2016a; Margalef-Català *et al.*, 2017b).

Another trend in selecting *O. oeni* strains is the practice of isolating indigenous malo-lactic bacteria from the same cellar or its environment to develop starter cultures that can be used to enhance the regional identity of wines. There is evidence that regional branding is an effective means for marketing purposes and especially because is mounting evidence that the local microbiota contributes to a wine's terroir (Bartowsky *et al.*, 2015; Franquès *et al.*, 2017).

Once selected, even the best strains require great care. Adequate production, storage and inoculation protocols are essential to guarantee the viability of the selected malo-lactic starters. It is also important to monitor the evolution of malo-lactic starters after inoculation (Torriani *et al.*, 2011).

### 6.4.2. Strain Improvement

In terms of strain improvement, *O. oeni* has the limitation that this species seems to be refractory for genetic transformation. Several attempts have been made to incorporate new genes in *O. oeni* by electroporation (Dicks, 1994; Zúniga *et al.*, 2003; Beltramo *et al.*, 2004b; Assad-García *et al.*, 2008); however, no further data has been published about stably transformed *O. oeni* strains. Other non-recombinant strategies for improving performance of *O. oeni* and its tolerance to harsh properties of wine have been assayed, such as random UV-mutagenesis for increasing the malo-lactic conversion rate (Li *et al.*, 2015), or the directed evolution for obtaining isolates withstanding higher concentrations of ethanol (Betteridge *et al.*, 2015).

### 6.4.3. Lactobacillus as Starters

Strains of *Lactobacillus* have several relevant characteristics that allow considering them as starter cultures of MLF, especially in low acidity wines, with pH higher than 3.5 (Lucio *et al.*, 2017). The best potential has been shown in *Lb. hilgardii* and especially in *Lb. plantarum* (Du Toit *et al.*, 2011). For inoculating them, the best results are achieved when *Lactobacillus* is inoculated in grape must prior to yeasts. After their inoculation, *Lactobacillus* can carry out the MLF usually during the first 48 hours of culture before yeasts are added (Lucio *et al.*, 2017).

These good characteristics are the ability to function well at low pH conditions, the tolerance of ethanol up to 14 per cent, the similar $SO_2$ tolerance than *O. oeni*, and that *Lb. plantarum* can synthesise a large amount of beneficial volatile compounds that improve wine flavour (Du Toit *et al.*, 2011). *Lb. plantarum* has a more diverse array of enzymes than *O. oeni*, which could lead to more aroma compounds being produced. The wines inoculated with *Lb. plantarum* strains have a softer palate and more fruity aromas (Matthews *et al.*, 2004; Spano *et al.*, 2005; Mtshali *et al.*, 2010). Several studies have shown the genetic potential of Lactobacilli for fermenting carbohydrates and for the production of metabolites (Spano *et al.*, 2002; Du Toit *et al.*, 2011; Lerm *et al.* 2011; Iorizzo *et al.*, 2016). Besides, *O. oeni* is heterofermentative for hexoses and can synthesise acetic acid, whereas *Lb. plantarum* is homofermentative and is, therefore, unable to synthesise acetic acid (Ribéreau-Gayon *et al.*, 2000).

Moreover, several *Lactobacillus* have been shown to produce bacteriocins that could be important in the survival and dominance of the strain in a competitive environment (Saranaraj *et al.*, 2013; Iorizzo *et al.*, 2016). Another advantage of *Lactobacillus* as MLF starter would be that most species of this genus have more possibilities for genetic manipulation than *O. oeni* (Du Toit *et al.*, 2011).

### *6.4.4. Controlling and Inhibiting the LAB in Wine*

Sometimes the LAB population in wine need to be controlled, for instance, when some of the problems above commented have been detected. In addition, there is the need to inhibit LAB growth when the MLF is not desired because the wine has already a high pH, or for organoleptic reasons. The last is the case of many sparkling wines, such as the Spanish 'cava' of Catalan Penedès region, where the final wine must keep its relatively acid pH, where lactic or buttery features are not desired and foaming properties must be kept without changes due to MLF (Andrés-Lacueva *et al.*, 1996).

The usual way to control LAB is the addition of sulphite, the same chemical preservative generally added to wine for inhibiting other microorganisms that endanger the quality of the end product, mainly acetic acid bacteria. Another alternative chemicals is sorbic acid, which is not effective against LAB, or the effective dimethyldicarbonate, known as Velcorin (Du Toit and Pretorius, 2000; Divol *et al.*, 2005). Nevertheless, the increasingly consumer bias against chemical preservatives favours the searching for the use of alternative bio-preservatives and reducing sulphite in wine. Among these antimicrobials of biological origin, the most known are lysozyme, bacteriocins and phenolic compounds.

- *Lysozyme* is an animal enzyme usually obtained from egg white, with glycoside hydrolase activity on the components of the peptidoglycan in the bacterial cell wall, which can lyse Gram-positive bacteria, including wine LAB. It provides a practical method for delaying or preventing the growth of *O. oeni* and consequently the onset of MLF. Lysozyme is not affected by alcohol and is active in the pH range of the winemaking process. However, its activity is affected by its reaction with tannins, pigments and bentonite, and some decrease in red wine colour has been observed after its addition (Bartowsky and Henschke, 2004). As lysozyme does not have antioxidative properties, it cannot replace $SO_2$ but enables the use of reduced levels. The International Organisation of Vine and Wine (OIV) has approved the addition of lysozyme to the winemaking process, but the economic implications of using it are still a limiting factor (Du Toit and Pretorius, 2000, Lerm *et al.*, 2010).
- *Bacteriocins* are proteinaceous toxins produced by bacteria to inhibit the growth of similar or closely related bacterial strains by destabilising the function of the cytoplasmic membrane. Nisin is the only bacteriocin produced by a LAB, *Lactococcus lactis*, with the GRAS (Generally Regarded As Safe) status and it is approved for usage in many countries. Most of wine-associated LAB are inhibited for low concentration of nisin in wine (Radler, 1990), and the wine yeasts are not affected and neither is the sensorial quality of the wine. Pediocin PD-1, another bacteriocin from *Pediococcus damnosus*, has also been shown to be effective for controlling wine LAB, including *O. oeni* (Bauer *et al.*, 2003). Although bacteriocins are a safe alternative to chemical preservation, their low efficiency and high cost are limiting factors (Du Toit and Pretorius, 2000), and their addition into wine is not yet authorised (Bauer and Dicks, 2004).
- The potential use of *phenolic* compounds as antimicrobial agents to control the growth of LAB in wine has also been proposed (Garcia-Ruiz *et al.*, 2008). The effect of the different phenolic compounds on LAB metabolic activity and growth has been extensively studied. For instance, *p*-coumaric has been shown to inhibit MLF by *O. oeni* (Reguant *et al.*, 2000). Due to complexity and variability of the effect of the different phenolic compounds, a good alternative is to use phenolic

extracts from different origins. In this way, a eucalyptus extract added to a red wine has shown to delay the progress of both inoculated and spontaneous MLF (García-Ruiz *et al*., 2012).

- Recently, *chitosan* has also received some attention due to its antimicrobial activity (Aider, 2010). It is a cationic polysaccharide produced by the deacetylation of chitin, which is a natural polymer found in the exoskeletons of crustaceans and insects and in the cell walls of fungi. It is recognised as safe and is economically advantageous over other named compounds as lysozyme or bacteriocins. It has been shown that 0.2 g/l chitosan is effective inhibiting *O. oeni*, other LAB and also *Brettanomyces*, while *S. cerevisiae* exhibited strongest resistance, with a MIC greater than 2 g/l. Nevertheless, its efficacy is variable due to the complex interactions with wine ingredients and it needs further investigation as an alternative or complementary preservative to $SO_2$ in wine industry (Elmacı *et al*., 2015).
- Besides these compounds, *physical* methods have also been proposed to control microbial growth in oenology, such as the application of low electric current or pulsed electric fields technology (Lustrato *et al*., 2006; Puértolas *et al*., 2009). However, no new compound or strategy has been identified to be a real substitute for $SO_2$ with all its oenological properties, including the antioxidative and antiseptic action in winemaking (Suzzi *et al*., 1985; García-Ruiz *et al*., 2008).

## 7. Conclusion and Future Perspectives

In spite of the increasing knowledge of the mechanisms of MLF and the malo-lactic bacteria *O. oeni*, as described in this chapter, this process remains difficult to control for many winemakers, basically because wine is a hostile medium for bacteria and is very variable, depending on vintage, varietal and technological characteristics of vinification. For this reason, more in depth studies on this fermentation and on this bacterium must be carried out. Among the main perspectives, there are the studies of systems biology, such as proteomics and metabolomics of *O. oeni*, in order to unveil the complex interactions of its molecular mechanisms related to survival, adaptation and performance of MLF. Other future perspectives are the studies related to new tendencies in winemaking, such as the use of autochthonous malo-lactic strains in order to maintain the *terroir* characteristics and the study of the influence of using non-*Saccharomyces* together with *S. cerevisiae*-, on *O. eni* and MLF.

## References

Aider, M. (2010). Chitosan application for active bio-based films production and potential in the food industry: Review. *LWT-Food Science and Technology* 43: 837-842.

Andrés-Lacueva, C., López-Tamames, E., Lamuela-Raventós, R.M., Buxaderas, S. and De la Torre-Boronat, M.C. (1996). Characteristics of sparkling base wines affecting foam behaviour. *Journal of Agricultural and Food Chemistry* 44: 989-995.

Antalick, G., Perello, M.C. and De Revel, G. (2013). Characterisation of fruity aroma modifications in red wines during malolactic fermentation. *Journal of Agricultural and Food Chemistry* 60: 12371-12383.

Araque, I., Bordons, A. and Reguant, C. (2009a). A multiplex PCR method for simultaneous species identification and strain typification of *Oenococcus oeni*. *World Journal of Microbiology and Biotechnology* 25: 15-18.

Araque, I., Gil, J., Carreté, R., Bordons, A. and Reguant, C. (2009b). Detection of *arc* genes related with the ethyl carbamate precursors in wine lactic acid bacteria. *Journal of Agricultural and Food Chemistry* 57: 1841-1847.

Araque, I., Reguant, C., Rozès, N. and Bordons, A. (2011). Influence of wine-like conditions on arginine utilisation by lactic acid bacteria. *International Microbiology* 14: 225-233.

Araque, I., Bordons, A. and Reguant, C. (2013). Effect of ethanol and low pH on citrulline and ornithine excretion and *arc* gene expression by strains of *Lactobacillus brevis* and *Pediococcus pentosaceus*. *Food Microbiology* 33: 107-113.

Araque, I., Gil, J., Carreté, R., Constantí, M., Bordons, A. and Reguant, C. (2016). Arginine deiminase pathway gene and arginine degradation variability in *Oenococcus oeni* strains. *Folia Microbiologica* 61: 109-118.

Arena, M.E., Saguir, F.M. and Manca de Nadra, M.C. (1999). Arginine, citrulline and ornithine metabolism by lactic acid bacteria from wine. *International Journal of Food Microbiology* 52: 155-161.

Arena, M.E., Manca de Nadra, M.C. and Muñoz, R. (2002). The arginine deiminase pathway in the wine lactic acid bacterium *Lactobacillus hilgardii* $X_1B$: Structural and functional study of the *arcABC* genes. *Gene* 301: 61-66.

Assad-García, J.S., Bonnin-Jusserand, M., Garmyn, D., Guzzo, J., Alexandre, H. and Grandvalet, C. (2008). An improved protocol for electroporation of *Oenococcus oeni* ATCC BAA-1163 using ethanol as immediate membrane fluidising agent. *Letters in Applied Microbiology* 47: 333-338.

Azevedo, Z., Couto, J.A. and Hogg, T. (2002). Citrulline as the main precursor of ethyl carbamate in model fortified wines inoculated with *Lactobacillus hilgardii*: A marker of the levels in a spoiled fortified wine. *Letters in Applied Microbiology* 34: 32-36.

Badotti, F., Moreira, A.P.B., Tonon, L.A.C., de Lucena, B.T.L., Gomes, F. de C.O., Kruger, R., Thompson, C.C., de Morais, M.A., Rosa, C.A. and Thompson, F.L. (2014). *Oenococcus alcoholitolerans* sp. nov., a lactic acid bacteria isolated from cachaça and ethanol fermentation processes. *Antonie van Leeuwenhoek* 106: 1259-1267.

Bae, S., Fleet, G.H. and Heard, G.M. (2006). Lactic acid bacteria associated with wine grapes from several Australian vineyards. *Journal of Applied Microbiology* 100: 712-727.

Barbagallo, R.N., Spagna, G., Palmeria, R. and Torriani S. (2004). Assessment of β-glucosidase activity in selected wild strains of *Oenococcus oeni* for malo-lactic fermentation. *Enzyme and Microbial Technology* 34: 292-296.

Bartowsky, E.J. (2005). *Oenococcus oeni* and malo-lactic fermentation moving into the molecular arena. *Australian Journal of Grape and Wine Research* 11: 174-187.

Bartowsky, E.J. (2009). Bacterial spoilage of wine and approaches to minimise it. *Letters in Applied Microbiology* 48: 149-156.

Bartowsky, E.J. and Borneman, A.R. (2011). Genomic variations of *Oenococcus oeni* strains and the potential to impact on malo-lactic fermentation and aroma compounds in wine. *Applied Microbiology and Biotechnology* 92: 441-447.

Bartowsky, E.J. and Henschke, P.A. (2004). The 'buttery' attribute of wine – diacetyl-desirability, spoilage and beyond. *International Journal of Food Microbiology* 96: 235-252.

Bartowsky, E.J., Costello, P. and McCarthy, J. (2008). MLF – Adding an 'extra dimension' to wine flavour and quality. *Australian & New Zealand Grapegrower & Winemaker* 533: 60-65.

Bartowsky, E.J., Costello, P.J. and Chambers, P.J. (2015). Emerging trends in the application of malolactic fermentation. *Australian Journal of Grape and Wine Research* 21: 663-669.

Bauer, R., Hannes, A.N. and Dicks, L.M.T. (2003). Pediocin PD-1 as a method to control growth of *Oenococcus oeni* in wine. *American Journal of Enology and Viticulture* 54: 86-91.

Bauer, R. and Dicks, L.M.T. (2004). Control of malolactic fermentation in wine: A review. *South African Journal of Enology and Viticulture* 25: 74-88.

Beltramo, C., Grandvalet, C., Pierre, F. and Guzzo, J. (2004a). Evidence for multiple levels of regulation of *Oenococcus oeni* clpP-clpL locus expression in response to stress. *Journal of Bacteriology* 186: 2200-2205.

Beltramo, C., Oraby, M., Bourel, G., Garmyn, D. and Guzzo, J. (2004b). A new vector, pGID052, for genetic transfer in *Oenococcus oeni*. *FEMS Microbiology Letters* 236: 53-60.

Beltramo, C., Desroche, N., Tourdot-Maréchal, R., Grandvalet, C. and Guzzo, J. (2006). Realtime PCR for characterising the stress response of *Oenococcus oeni* in a wine-like medium. *Research in Microbiology* 157: 267-274.

Betteridge, A., Grbin, P. and Jiranek, V. (2015). Improving *Oenococcus oeni* to overcome challenges of wine malolactic fermentation. *Trends in Biotechnology* 33: 547-553.

Bilhère, E., Lucas, P.M., Claisse, O. and Lonvaud-Funel, A. (2009). Multilocus sequence typing of *Oenococcus oeni*: Detection of two subpopulations shaped by intergenic recombination. *Applied and Environmental Microbiology* 75: 1291-1300.

Blasco, L., Ferrer, S. and Pardo, I. (2003). Development of specific fluorescent oligonucleotide probes for in situ identification of wine lactic acid bacteria. *FEMS Microbiology Letters* 225: 115-123.

Bokulich, N.A., Thorngate, J.H., Richardson, P.M. and Mills, D.A. (2014). Microbial biogeography of wine grapes is conditioned by cultivar, vintage and climate. *Proceedings of the National Academy of Sciences*, USA 111: E139-E148.

Bon, E., Delaherche, A., Bihère, E., De Daruvar, A., Lonvaud-Funel, A. and Le Marrec, C. (2009). *Oenococcus oeni* genome plasticity is associated with fitness. *Applied and Environmental Microbiology* 75: 2079-2090.

Borneman, A.R., Bartowsky, E.J., McCarthy, J. and Chambers, P.J. (2010). Genotypic diversity in *Oenococcus oeni* by high-density microarray comparative genome hybridisation and whole genome sequencing. *Applied Microbiology and Biotechnology* 86: 681-691.

Borneman, A.R., McCarthy, J.M., Chambers, P.J. and Bartowsky, E.J. (2012a). Functional divergence in the genus *Oenococcus* as predicted by genome sequencing of the newly-described species, *Oenococcus kitaharae*. *PLoS ONE* 7: e29626.

Borneman, A.R., McCarthy, J.M., Chambers, P.J. and Bartowsky, E.J. (2012b). Comparative analysis of the *Oenococcus oeni* pan genome reveals genetic diversity in industrially relevant pathways. *BMC Genomics* 13: 373.

Bravo-Ferrada, B.M., Hollmann, A., Delfederico, L., Valdés La Hens, D., Caballero, A. and Semorile, L. (2013). Patagonian red wines: Selection of *Lactobacillus plantarum* isolates as potential starter cultures for malo-lactic fermentation. *World Journal of Microbiology and Biotechnology* 29: 1537-1549.

Bridier, J., Claisse, O., Coton, M., Coton, E. and Lonvaud-Funel, A. (2010). Evidence of distinct populations and specific subpopulations within the species *Oenococcus oeni*. *Applied and Environmental Microbiology* 76: 7754-7764.

Callejón, S., Sendra, R., Ferrer, S. and Pardo, I. (2014). Identification of a novel enzymatic activity from lactic acid bacteria able to degrade biogenic amines in wine. *Applied Microbiology and Biotechnology* 98: 185-198.

Callejón, S., Sendra, R., Ferrer, S. and Pardo, I. (2016). Cloning and characterisation of a new laccase from *Lactobacillus plantarum* J16 CECT8944 catalysing biogenic amines degradation. *Applied Microbiology and Biotechnology* 100: 3113.

Campbell-Sills, H., El Khoury, M., Favier, M., Romano, A., Biasioli, F., Spano, G., Sherman, D.J., Bouchez, O., Coton, E., Coton, M., Okada, S., Tanaka, N., Dols-Lafargue, M. and Lucas, P.M. (2015). Phylogenomic analysis of *Oenococcus oeni* reveals specific domestication of strains to cider and wines. *Genome Biology and Evolution* 7: 1506-1518.

Caporaso, J.G., Lauber, C.L., Walters, W.A., Berg-Lyons, D., Huntley, J., Fierer, N., Owens, S.M., Betley, J., Fraser, L., Bauer, M., Gormley, N., Gilbert, J.A., Smith, G. and Knight, R. (2012). Ultra-high-throughput microbial community analysis on the Illumina HiSeq and MiSeq platforms. *The ISME Journal* 6: 1621-1624.

Capozzi, V., Russo, P., Ladero, V., Fernández, M., Fiocco, D., Alvarez, M.A., Grieco, F. and Spano, G. (2012). Biogenic amines degradation by *Lactobacillus plantarum*: Toward a potential application in wine. *Frontiers in Microbiology* 3: 122.

Carreté, R., Vidal, M.T., Bordons, A. and Constantí, M. (2002). Inhibitory effect of sulfur dioxide and other stress compounds in wine on the ATPase activity of *Oenococcus oeni*. *FEMS Microbiology Letters* 211: 155-159.

Carreté, R., Reguant, C., Rozès, N., Constantí, M. and Bordons, A. (2006). Analysis of *Oenococcus oeni* strains in simulated microvinifications with some stress compounds. *American Journal of Enology and Viticulture* 57: 356-362.

Cecconi, D., Milli, A., Rinalducci, S., Zolla, L. and Zapparoli, G. (2009). Proteomic analysis of *Oenococcus oeni* freeze-dried culture to assess the importance of cell acclimation to conduct malolactic fermentation in wine. *Electrophoresis* 30: 2988-2995.

Chalfan, Y., Goldberg, I. and Mateles, R.I. (1977). Isolation and characterisation of malo-lactic bacteria from Israeli red wines. *Journal of Food Science* 42: 939-943.

Chu-Ky, S., Tourdot-Maréchal, R., Maréchal, P.A. and Guzzo, J. (2005). Combined cold, acid, ethanol shocks in *Oenococcus oeni*: Effects on membrane fluidity and cell viability. *Biochimica Biophysica Acta – Biomembranes* 1717: 118-124.

Ciezack, G., Hazo, L., Chambat, G., Heyraud, A., Lonvaud-Funel, A. and Dols-Lafargue, M. (2010). Evidence for exopolysaccharide production by *Oenococcus oeni* strains isolated from non-ropy wines. *Journal of Applied Microbiology* 108: 499-509.

Claisse, O. and Lonvaud-Funel, A. (2012). Development of a multilocus variable number of tandem repeat typing method for *Oenococcus oeni*. *Food Microbiology* 30: 340-347.

Claisse, O. and Lonvaud-Funel, A. (2014). Multiplex variable number of tandem repeats for *Oenococcus oeni* and applications. *Food Microbiology* 38: 80-86.

Cocconcelli, P.S., Porro, D., Galandini, S. and Senini, L. (1995). Development of RAPD protocol for typing of strains of lactic acid bacteria and enterococci. *Letters in Applied Microbiology* 21: 376-379.

Costantini, A., Ratsiou, K., Majumder, A., Jacobsen, S., Pessione, E., Svensson, B., García-Moruno, E. and Cocolin, L. (2015). Complementing DIGE proteomics and DNA subarray analyses to shed light on *Oenococcus oeni* adaptation to ethanol in wine-simulated conditions. *Journal of Proteomics* 123: 114-127.

Costello, P.J., Morrison, G.J., Lee, T.H. and Fleet, G.H. (1983). Numbers and species of lactic acid bacteria in wine during vinification. *Food Technology in Australia* 35: 14-18.

Costello, P.J., Francis, I.L. and Bartowsky, E.J. (2012). Variations in the effect of malolactic fermentation on the chemical and sensory properties of Cabernet Sauvignon wine: Interactive influences of *Oenococcus oeni* strain and wine matrix composition. *Australian Journal of Grape and Wine Research* 18: 287-301.

Coucheney, F., Gal, L., Beney, L., Lherminier, J., Gervais, P. and Guzzo, J. (2005). A small HSP, Lo18, interacts with the cell membrane and modulates lipid physical state under heat shock conditions in a lactic acid bacterium. *Biochimica Biophysica Acta* 1720: 92-98.

Cox, D.J. and Henick-Kling, T. (1989). Chemiosmotic energy from malolactic fermentation. *Journal of Bacteriology* 171: 5750-5752.

Cueva, C., García-Ruiz, A., González-Rompinelli, E., Bartolomé, B., Martín-Álvarez, P.J., Salazar, O., Vicente, M.F., Bills, G.F. and Moreno-Arribas, M.V. (2012). Degradation of biogenic amines by vineyard ecosystem fungi: Potential use in winemaking. *Journal of Applied Microbiology* 112: 672-682.

Da Silveira, M.G., San Romão, M.V., Loureiro-Dias, M.C., Rombouts, F.M. and Abee, T. (2002). Flow cytometric assessment of membrane integrity of ethanol-stressed *Oenococcus oeni* cells. *Applied and Environmental Microbiology* 68: 6087-6093.

Da Silveira, M.G., Golovina, E.A, Hoekstra, F.A., Rombouts, F.M. and Abee, T. (2003). Membrane fluidity adjustments in ethanol-stressed *Oenococcus oeni* cells. *Applied and Environmental Microbiology* 69: 1-8.

Davis, C.R., Wibowo, D., Eschenbruch, R., Lee, T.H. and Fleet G.H. (1985). Practical implications of malolactic fermentation: A review. *American Journal of Enology and Viticulture* 36: 290-301.

Davis, C.R., Wibowo, D., Lee, T.H. and Fleet, G.H. (1986). Growth and metabolism of lactic acid bacteria during and after malolactic fermentation of wines at different pH. *Applied and Environmental Microbiology* 51: 539-545.

Davis, C.R., Wibowo, D., Lee, T.H. and Fleet, G.H. (1988). Properties of wine lactic acid bacteria: Their potential enological significance. *American Journal of Enology and Viticulture* 39: 137-142.

De Las Rivas, B., Marcobal, A. and Munoz, R. (2004). Allelic diversity and population structure in *Oenococcus oeni* as determined from sequence analysis of housekeeping genes. *Applied and Environmental Microbiology* 70: 7210-7219.

De Las Rivas, B., Marcobal, A., Carrascosa, A.V. and Muñoz, R. (2006). PCR detection of food-borne bacteria producing the biogenic amines histamine, tyramine, putrescine, and cadaverine. *Journal of Food Protection* 69: 2509-2514.

Del Prete, V., Costantini, A., Cecchini, F., Morassut, M. and Garcia-Marino, E. (2009). Occurrence of biogenic amines in wine: The role of grapes. *Food Chemistry* 112: 474-481.

Delaherche, A., Bon, E., Dupé, A., Lucas, M., Arveiler, B., De Daruvar, A. and Lonvaud-Funel, A. (2006). Intraspecific diversity of *Oenococcus oeni* strains determined by sequence analysis of target genes. *Applied Microbiology and Biotechnology* 73: 394-403.

Delmas, F., Pierre, F., Coucheney, F., Divies, C. and Guzzo, J. (2001). Biochemical and physiological studies of the small heat shock protein Lo18 from the lactic acid bacterium *Oenococcus oeni*. *Journal of Molecular Microbiology and Biotechnology* 3: 601-610.

Desroche, N., Beltramo, C. and Guzzo, J. (2005). Determination of an internal control to apply reverse transcription quantitative PCR to study stress response in the lactic acid bacterium *Oenococcus oeni*. *Journal of Microbiological Methods* 60: 325-333.

Dicks, L.M.T. (1994). Transformation of *Leuconostoc oenos* by electroporation. *Biotechnology Techniques* 8: 901-904.

Dicks, L.M.T., Dellaglio, F. and Collins, M.D. (1995). Proposal to reclassify *Leuconostoc oenos* as *Oenococcus oeni* [corrig.] gen. nov., comb. nov. *International Journal of Systematic Bacteriology* 45: 395-397.

Divol, B., Strehaiano, P. and Lonvaud-Funel, A. (2005). Effectiveness of dimethyldicarbonate to stop alcoholic fermentation in wine. *Food Microbiology* 22: 169-178.

Dimopoulou, M., Hazo, L. and Dols-Lafargue, M. (2012). Exploration of phenomena contributing to the diversity of *Oenococcus oeni* exopolysaccharides. *International Journal of Food Microbiology* 153: 114-122.

Dimopoulou, M., Vuillemin, M., Campbell-Sills, H., Lucas, P.M., Ballestra, P., Miot-Sertier, C., Favier, M., Coulon, J., Moine, V., Doco, T., Roques, M., Williams, P., Petrel, M., Gontier, E., Moulis, C., Remaud-Simeon, M. and Dols-Lafargue, M. (2014). Exopolysaccharide (EPS) synthesis by *Oenococcus oeni*: From genes to phenotypes. *PLoS ONE* 9: e98898, pages 1-15.

Dimopoulou, M., Bardeau, T., Ramonet, P.Y., Miot-Certier, C., Claisse, O., Doco, T., Petrel, M., Lucas, P. and Dols-Lafargue, M. (2016). Exopolysaccharides produced by *Oenococcus oeni*: From genomic and phenotypic analysis to technological valorisation. *Food Microbiology* 53: 10-17.

Dols-Lafargue, M., Lee, H.Y., Le Marrec, C., Heyraud, A., Chambat, G. and Lonvaud-Funel, A. (2008). Characterisation of *gtf*, a glucosyltransferase gene in the genomes of *Pediococcus parvulus* and *Oenococcus oeni*, two bacterial species commonly found in wine. *Applied and Environmental Microbiology* 74: 4079-4090.

Drici-Cachon, Z., Cavin, J.F. and Diviès, C. (1996a). Effect of pH and age of culture on cellular fatty acid composition of *Leuconostoc oenos*. *Letters in Applied Microbiology* 22: 331-334.

Drici-Cachon, Z., Guzzo, J., Cavin, J.-F. and Diviès, C. (1996b). Acid tolerance in *Leuconostoc oenos*: Isolation and characterisation of an acid-resistant mutant. *Applied Microbiology and Biotechnology* 44: 785-789.

Du Plessis, H.W., Dicks, L.M.T., Pretorius, I.S., Lambrechts, M.G. and du Toit, M. (2004). Identification of lactic acid bacteria isolated from South African brandy base wines. *International Journal of Food Microbiology* 91: 19-29.

Du Toit, M., Engelbrecht, L., Lerm, E. and Krieger-Weber, S. (2011). *Lactobacillus*: The next generation of malolactic fermentation starter cultures – An overview. *Food Bioprocess Technology* 4: 876-906.

Du Toit, M. and Pretorius, I.S. (2000). Microbial spoilage and preservation of wine: Using weapons from nature's own arsenal – A review. *South African Journal of Enology and Viticulture* 21: 74-96.

Edwards, C.G., Jensen, K.A., Spayd, S.E. and Seymour, B.J. (1991). Isolation and characterization of native strains of *Leuconostoc oenos* from Washington State wines. *American Journal of Enology and Viticulture* 42: 219-226.

El Gharniti, F., Dols-Lafargue, M., Bon, E., Claisse, O., Miot-Sertier, C., Lonvaud, A. and Le Marrec, C. (2012). IS30 elements are mediators of genetic diversity in *Oenococcus oeni*. *International Journal of Food Microbiology* 158: 14-22.

Elahwany, A.M.D. (2012). Genetic and physiological studies on *Oenococcus oeni* PSU-I in response to ethanol stress. *Acta Biologica Hungarica* 63: 128-137.

El Khoury, M., Campbell-Sills, H., Salin, F., Guichoux, E., Claisse, O. and Lucas, P.M. (2017). Biogeography of *Oenococcus oeni* reveals distinctive but non-specific populations in wine-producing regions. *Applied and Environmental Microbiology* 83: e02322-16.

Elmacı, B.S., Gülgör, G., Tokatlı, M., Erten, H., İşci, A. and Özçelik, F. (2015). Effectiveness of chitosan against wine-related microorganisms. *Antonie van Leeuwenhoek* 107: 675-686.

Endo, A. and Okada, S. (2006). *Oenococcus kitaharae* sp. nov., a non-acidophilic and nonmalolactic-fermenting *Oenococcus* isolated from a composting distilled *shochu* residue. *International Journal of Systematic and Evolutionary Microbiology* 56: 2345-2348.

Farias, M.E., Manca de Nadra, M.C., Rollan, G.C. and Strasser de Saad, M. (1993). Histidine decarboxylase activity in lactic acid bacteria from wine. *Journal International des Sciences de la Vigne et du Vin* 27: 191-199.

Fischer, U. and Noble, A.C. (1994). The effect of ethanol, catechin concentration and pH on sourness and bitterness of wine. *American Journal of Enology and Viticulture* 45: 6-10.

Fleet, G.H., Lafon-Lafourcade, S. and Ribéreau-Gayon, P. (1984). Evolution of yeasts and lactic acid bacteria during fermentation and storage of Bordeaux wines. *Applied and Environmental Microbiology* 48: 1034-1038.

Franquès, J., Araque, I., Palahí, E., Portillo, M.C., Reguant, C. and Bordons, A. (2017). Presence of *Oenococcus oeni* and other lactic acid bacteria in grapes and wines from Priorat (Catalonia, Spain). *LWT – Food Science and Technology* 81: 326-334.

Fortier, L.C., Tourdot-Maréchal, R., Diviès, C., Lee, B.H. and Guzzo, J. (2003). Induction of *Oenococcus oeni* $H^+$-ATPase activity and mRNA transcription under acid conditions. *FEMS Microbiology Letters* 222: 165-169.

Fugelsang, K.C. and Edwards, C.G. (2007). Wine Microbiology: Practical Applications and Procedures, second edn. Springer, New York.

Galland, D., Tourdot-Maréchal, R., Abraham, M., Chu, K.S. and Guzzo, J. (2003). Absence of malo-lactic activity is a characteristic of $H^+$-ATPase-deficient mutants of the lactic acid bacterium *Oenococcus oeni*. *Applied and Environmental Microbiology* 69: 1973-1979.

García-Ruiz, A., Bartolomé, B., Martinez-Rodriguez, A.J., Pueyo, E., Martin-Alvarez, P.J. and Moreno-Arribas, M.V. (2008). Potential of phenolic compounds for controlling lactic acid bacteria growth in wine. *Food Control* 19: 835-841.

García-Ruiz, A., González-Rompinelli, E.M., Bartolomé, B. and Moreno-Arribas, M.V. (2011). Potential of wine-associated lactic acid bacteria to degrade biogenic amines. *International Journal of Food Microbiology* 148: 115-120.

García-Ruiz, A., Cueva, C., González-Rompinelli, E.M., Yuste, M., Torres, M., Martín-Álvarez, P.J., Bartolomé, B. and Moreno-Arribas, M.V. (2012). Antimicrobial phenolic extracts able to inhibit lactic acid bacteria growth and wine malo-lactic fermentation. *Food Control* 28: 212-219.

Garijo, P., Santamaría, P., López, R., Sanz, S., Olarte, C. and Gutiérrez, A.R. (2008). The occurrence of fungi, yeasts and bacteria in the air of a Spanish winery during vintage. *International Journal of Food Microbiology* 125: 141-145.

Garijo, P., López, R., Santamaría, P., Ocón, E., Olarte, C., Sanz, S. and Gutiérrez, A.R. (2009). Presence of lactic bacteria in the air of a winery during the vinification period. *International Journal of Food Microbiology* 136: 142-146.

Garijo, P., López, R., Santamaría, P., Ocón, E., Olarte, C., Sanz, S. *et al.* (2011). Presence of enological microorganisms in the grapes and the air of a vineyard during the ripening period. *European Food Research and Technology* 233: 359-365.

Garofalo, C., El Khoury, M., Lucas, P., Bely, M., Russo, P., Spano, G. and Capozzi, V. (2015). Autochthonous starter cultures and indigenous grape variety for regional wine production. *Journal of Applied Microbiology* 118: 1395-1408.

Gevers, D., Huys, G. and Swings, J. (2001). Applicability of rep-PCR fingerprinting for identification of *Lactobacillus* species. *FEMS Microbiology Letters* 205: 31-36.

Gindreau, E. and Lonvaud-Funel, A. (1999). Molecular analysis of the region encoding the lytic system from *Oenococcus oeni* temperate bacteriophage phi 10MC. *FEMS Microbiology Letters* 171: 231-238.

Gindreau, E., Keim, H., de Revel, G., Bertrand, A. and Lonvaud-Funel, A. (2003). Use of direct inoculation malolactic starters: Settling, efficiency and sensorial impact. *Oeno One* 37: 1.

Grandvalet, C., Coucheney, F., Beltramo, C. and Guzzo, J. (2005). CtsR is the master regulator of stress response gene expression in *Oenococcus oeni*. *Journal of Bacteriology* 187: 5614-5623.

Grandvalet, C., Assad-García, J.S., Chu-Ky, S., Tollot, M., Guzzo, J., Gresti, J. and Tourdot-Maréchal, R. (2008). Changes in membrane lipid composition in ethanol- and acid-adapted *Oenococcus oeni* cells: Characterisation of the *cfa* gene by heterologous complementation. *Microbiology* 154: 2611-2619.

Grimaldi, A., Bartowsky, E. and Jiranek, V. (2005). A survey of glycosidase activities of commercial wine strains of *Oenococcus oeni*. *International Journal of Food Microbiology* 105: 233-244.

Gockowiak, H. and Henschke, P.A. (2003). Interaction of pH, ethanol concentration and wine matrix on induction of malolactic fermentation with commercial 'direct inoculation' starter cultures. *Australian Journal of Grape and Wine Research* 9: 200-209.

Godálová, Z., Kraková, L., Puskárová, A., Bucková, M., Kuchta, T., Piknová, L. and Pangallo, D. (2016). Bacterial consortia at different wine fermentation phases of two typical Central European grape varieties: *Blaufränkisch* (*Frankovka modrá*) and *Grüner Veltline* (*Veltlínske zelené*). *International Journal of Food Microbiology* 217: 110-116.

Guchte, M. van de, Serror, P., Chervaux, C., Smokvina, T., Ehrlich, S.D. and Maguin, E. (2002). Stress responses in lactic acid bacteria. *Antonie Van Leeuwenhoek* 82: 187-216.

Guzzo, J., Delmas, F., Pierre, F., Jobin, M.-P., Samyn, B., Van Beeumen, J., Cavin, J.-F. and Divies, C. (1997). A small heat shock protein from *Leuconostoc oenos* induced by multiple stresses and during stationary growth phase. *Letters in Applied Microbiology* 24: 393-396.

Guzzo, J., Jobin, M.P., Delmas, F., Fortier, L.C., Garmyn, D., Tourdot-Maréchal, R., Lee, B. and Diviès, C. (2000). Regulation of stress response in *Oenococcus oeni* as a function of environmental changes and growth phase. *International Journal of Food Microbiology* 55: 27-31.

Henick-Kling, T. (1993). Malolactic fermentation. pp. 289-326. *In*: G.H. Fleet (Ed.). Wine Microbiology and Biotechnology. Harwood Academic Publishers. Chur, Switzerland.

Henick-Kling, T. (1995). Control of malolactic fermentation in wine: Energetics, flavour modification and methods of starter culture preparation. *Journal of Applied Bacteriology Symposium Supplement* 79: 29S-37S.

Henick-Kling, T., Sandine, W.E. and Heatherbell, D.A. (1989). Evaluation of malolactic bacteria isolated from Oregon wines. *Applied and Environmental Microbiology* 55: 2010-2016.

Hirschhäuser, S., Fröhlich, J., Gneipel, A., Schönig, I. and König, H. (2005). Fast protocols for the 5S rDNA and ITS-2 based identification of *Oenococcus oeni*. *FEMS Microbiology Letters* 244: 165-171.

Horváth, I., Multhoff, G., Sonnleitner, A. and Vígh, L. (2008). Membrane-associated stress proteins: More than simply chaperones. *Biochimica Biophysica Acta* 1778: 1653-1664.

Iorizzo, M., Testa, B., Lombardi, S.J., García-Ruiz, A., Muñoz-González, C., Bartolomé, B. and Moreno-Arribas, M.V. (2016). Selection and technological potential of *Lactobacillus plantarum* bacteria suitable for wine malo-lactic fermentation and grape aroma release. *LWT – Food Science and Technology* 73: 557-566.

Jamal, Z., Miot-Sertier, C., Thibau, F., Dutilh, L., Lonvaud-Funel, A., Ballestra, P., Le Marrec, C. and Dols-Lafargue, M. (2013). Distribution and functions of phosphotransferase system genes in the genome of the lactic acid bacterium *Oenococcus oeni*. *Applied and Environmental Microbiology* 79: 3371-3379.

Jobin, M.P., Delmas, F., Garmyn, D., Diviès, C. and Guzzo, J. (1997). Molecular characterization of the gene encoding an 18-kilodalton small heat shock protein associated with the membrane of *Leuconostoc oenos*. *Applied and Environmental Microbiology* 3: 609-614.

Jobin, M.P., Garmyn, D., Diviès, C. and Guzzo, J. (1999a). The *Oenococcus oeni clpX* homologue is a heat shock gene preferentially expressed in exponential growth phase. *Journal of Bacteriology* 181: 6634-6641.

Jobin, M.P., Garmyn, D., Diviès, C. and Guzzo, J. (1999b). Expression of the *Oenococcus oeni trxA* gene is induced by hydrogen peroxide and heat shock. *Microbiology* 145: 1245-1251.

Klaenhammer, T.R., Altermann, E., Pfeiler, E., Buck, B.L., Goh, Y.-J., O'Flaherty, S., Barrangou, R. and Duong, T. (2008). Functional genomics of probiotic *Lactobacilli*. *Journal of Clinical Gastroenterology* 42: S160- S162.

Kobayashi, H. (1985). A proton-translocating ATPase regulates pH of the bacterial cytoplasm. *Journal of Biological Chemistry* 260: 72-76.

Krieger, S.A., Hammes, W.P. and Henick-Kling, T. (1993). How to use malo-lactic starter cultures in the winery. *Australian and New Zealand Wine Industry Journal* 8: 153-160.

Kullen, M.J. and Klaenhammer, T.R. (1999). Identification of the pH-inducible, proton-translocating F1F0-ATPase (atpBEFHAGDC) operon of *Lactobacillus acidophilus* by differential display: Gene structure, cloning and characterisation. *Molecular Microbiology* 33: 1152-1161.

Lafon-Lafourcade, S., Carré, E. and Ribéreau-Gayon, P. (1983). Occurrence of lactic acid bacteria during the different stages of vinification and conservation of wines. *Applied and Environmental Microbiology* 46: 874-880.

Lallemand (2005). Malo-lactic Fermentation in Wine: Understanding the Science and the Practice (Ed.). Lallemand Inc.

Landete, J.M., Ferrer, S., Polo, L. and Pardo, I. (2005). Biogenic amines in wines from three Spanish regions. *Journal of Agricultural and Food Chemistry* 53: 1119-1124.

Landete, J.M., Pardo, I. and Ferrer, S. (2007a). Biogenic amine production by lactic acid bacteria, acetic bacteria and yeast isolated from wine. *Food Control* 18: 1569-1574.

Landete, J.M., de las Rivas, B., Marcobal, A. and Muñoz, R. (2007b). Review: Molecular methods for the detection of biogenic amine-producing bacteria on foods. *International Journal of Food Microbiology* 117: 258-269.

Le Jeune, C. and Lonvaud-Funel, A. (1997). Sequence of DNA 16S/23S spacer region of *Leuconostoc oenos* (*Oenococcus oeni*): Application to strain differentiation. *Research in Microbiology* 148: 79-86.

Lerm, E., Engelbrecht, L. and du Toit, M. (2010). Malolactic fermentation: The ABC's of MLF. *South African Journal of Enology and Viticulture* 31: 186-212.

Lerm, E., Engelbrecht, L. and du Toit, M. (2011). Selection and characterisation of *Oenococcus oeni* and *Lactobacillus plantarum* South African wine isolates for use as malolactic fermentation starter cultures. *South African Journal of Enology and Viticulture*. 32: 280-295.

Li, N., Duan, J., Gao, D., Luo, J., Zheng, R., Bian, Y., Zhang, X. and Ji, B. (2015). Mutation and selection of *Oenococcus oeni* for controlling wine malo-lactic fermentation. *European Food Research and Technology* 240: 93-100.

Liu, S.Q. (2002). Malo-lactic fermentation in wine beyond deacidification. *Journal of Applied Microbiology* 92: 589-601.

Liu, S.Q., Pritchard, G.G., Hardman, M.J. and Pilone, G.J. (1994). Citrulline production and ethyl carbamate (urethane) precursor formation from arginine degradation by wine lactic acid bacteria *Leuconostoc oenos* and *Lactobacillus buchneri*. *American Journal of Enology and Viticulture* 45: 235-242.

Liu, S.Q. and Pilone, G.J. (1998). A review: Arginine metabolism in wine lactic acid bacteria and its practical significance. *Journal of Applied Microbiology* 84: 315-327.

Lonvaud-Funel, A. (1999). Lactic acid bacteria in the quality improvement and depreciation of wine. *Antonie van Leeuwenhoek* 76: 317-331.

Lonvaud-Funel, A. (2001). Biogenic amines in wines: Role of lactic acid bacteria. *FEMS Microbiology Letters* 199: 9-13.

Lonvaud-Funel, A., Joyeux, A. and Desens, C. (1988). Inhibition of malo-lactic fermentation of wines by products of yeast metabolism. *Journal of Agricultural and Food Chemistry* 44: 183-191.

Lonvaud-Funel, A., Joyeux, A. and Ledoux, O. (1991). Specific enumeration of lactic acid bacteria in fermenting grape must and wine by colony hybridisation with non-isotopic DNA probes. *Journal of Applied Bacteriology* 71: 501-508.

Loubiere, P., Salou, P., Leroy, M.J., Lindley, N.D. and Pareilleux, A. (1992). Electrogenic malate uptake and improved growth energetics of the malo-lactic bacterium *Leuconostoc oenos* grown on glucose-malate mixtures. *Journal of Bacteriology* 174: 5302-5308.

Lucio, O., Pardo, I., Heras, J.M., Krieger-Weber, S. and Ferrer, S. (2017). Use of starter cultures of *Lactobacillus* to induce malolactic fermentation in wine. *Australian Journal of Grape and Wine Research* 23: 15-21.

Lustrato, G., Alfano, G., Belli, C., Grazia, L., Iorizzo, M. and Ranalli, G. (2006). Scaling-up in industrial winemaking using low electric current as an alternative to sulphur dioxide addition. *Journal of Applied Microbiology* 101: 682-690.

Maicas, S. (2001). The use of alternative technologies to develop malolactic fermentation in wine. *Applied Microbiology and Biotechnology* 56: 35-39.

Maitre, M., Weidmann, S., Rieu, A., Fenel, D., Schoehn, G., Ebel, C., Coves, J. and Guzzo, J. (2012). The oligomer plasticity of the small heat-shock protein Lo18 from *Oenococcus oeni* influences its role in both membrane stabilization and protein protection. *Biochemical Journal* 444: 97-104.

Maitre, M., Weidmann, S., Dubois-Brissonnet, F., David, V., Covés, J. and Guzzo, J. (2014). Adaptation of the wine bacterium *Oenococcus oeni* to ethanol stress: Role of the small heat shock protein Lo18 in membrane integrity. *Applied and Environmental Microbiology* 80: 2973-2980.

Manca de Nadra, M.C. and Strasser de Saad, A.M. (1995). Polysaccharide production by *Pediococcus pentosaceus* from wine. *International Journal of Food Microbiology* 27: 101-106.

Marcobal, A., Martin-Alvarez, P.J., Polo, M.C., Muñoz, R. and Moreno-Arribas, M.V. (2006). Formation of biogenic amines throughout the industrial manufacture of red wine. *Journal of Food Protection* 69: 397-404.

Marcobal, A.M., Sela, D.A., Wolf, Y.I., Makarova, K.S. and Mills, D.A. (2008). Role of Hypermutability in the Evolution of the Genus *Oenococcus*. *Journal of Bacteriology* 190: 564-570.

Margalef-Català, M., Araque, I., Bordons, A., Reguant, C. and Bautista-Gallego, J. (2016a). Transcriptomic and proteomic analysis of *Oenococcus oeni* adaptation to wine stress conditions. *Frontiers in Microbiology* 7: 1554.

Margalef-Català, M., Araque, I., Weidmann, S., Guzzo, J., Bordons, A. and Reguant, C. (2016b). Protective roles of glutathione addition against wine-related stress in *Oenococcus oeni*. *Food Research International* 90: 8-15.

Margalef-Català, M., Araque, I., Bordons, A. and Reguant, C. (2017a). Genetic and transcriptional study of glutathione metabolism in *Oenococcus oeni*. *International Journal of Food Microbiology* 242: 61-69.

Margalef-Català, M., Felis, G.E., Reguant, C., Stefanelli, E., Torriani, S. and Bordons, A. (2017b). Identification of variable genomic regions related to stress response in *Oenococcus oeni*. *Food Research International*, online 19/09/2017.

Margalef-Català, M., Stefanelli, E., Araque, I., Wagner, K., Felis, G.E., Bordons, A., Torriani, S. and Reguant, C. (2017c). Variability in gene content and expression of the thioredoxin system in *Oenococcus oeni*. *Food Microbiology* 61: 23-32.

Marqués, A.P., Leitao, M.C. and San Romao, M.V. (2008). Biogenic amines in wines: Influence of oenological factors. *Food Chemistry* 107: 853-860.

Martinez-Murcia, A.J. and Collins, M.D. (1990). A phylogenetic analysis of the genus *Leuconostoc* based on reverse transcriptase sequencing of 16S rRNA. *FEMS Microbiology Letters* 70: 73-83.

Matthews, A., Grimaldi, A., Walker, M., Bartowsky, E., Grbin, P. and Jiranek, V. (2004). Lactic acid bacteria as a potential source of enzymes for use in vinification. *Applied and Environmental Microbiology* 70: 5715-5731.

Matthews, A., Grbin, P.R. and Jiranek, V. (2006). A survey of lactic acid bacteria for enzymes of interest to oenology. *Australian Journal of Grape and Wine Research* 12: 235-244.

Mendoza, S.N., Cañón, P.M., Contreras, A. and Agosín, E. (2017). Genome-scale reconstruction of the metabolic network in *Oenococcus oeni* to assess wine malo-lactic fermentation. *Frontiers in Microbiology* 8: 534.

Mills, D., Rawsthorne, H., Parker, C., Tamir, D. and Makarova, K. (2005). Genomic analysis of PSU-1 and its relevance to winemaking. *FEMS Microbiology Reviews* 29: 465-475.

Mira de Orduña, R., Liu, S.Q., Patchett, M.L. and Pilone, G.J. (2000). Kinetics of the arginine metabolism of malo-lactic wine lactic acid bacteria *Lactobacillus buchneri* CUC-3 and *Oenococcus oeni* Lo111. *Journal of Applied Microbiology* 89: 547-552.

Moreno-Arribas, M.V. and Polo, M.C. (2008). Occurrence of lactic acid bacteria and biogenic amines in biologically aged wines. *Food Microbiology* 25: 875-881.

Mtshali, P.S., Divol, B.T., Van Rensburg, P. and Du Toit, M. (2010). Genetic screening of wine-related enzymes in *Lactobacillus* species isolated from South African wines. *Journal of Applied Microbiology* 108: 1389-1397.

Nakamoto, H. and Vígh, L. (2007). The small heat shock proteins and their clients. *Cellular and Molecular Life Sciences* 64: 294-306.

NCBI (2017). National Centre for Biotechnology Information: www.ncbi.nlm.nih.gov.

Nielsen, J.C., Prahl, C. and Lonvaud-Funel, A. (1996). Malolactic fermentation in wine by direct inoculation with freeze-dried *Leuconostoc oenos* cultures. *American Journal of Enology and Viticulture* 47: 42-48.

Olguín, N., Bordons, A. and Reguant, C. (2009). Influence of ethanol and pH on the gene expression of the citrate pathway in *Oenococcus oeni*. *Food Microbiology* 26: 197-203.

Olguín, N., Bordons, A. and Reguant, C. (2010). Multigenic expression analysis as an approach to understanding the behaviour of *Oenococcus oeni* in wine-like conditions. *International Journal of Food Microbiology* 144: 88-95.

Olguín, N., Alegret, J.O., Bordons, A. and Reguant, C. (2011). Beta-glucosidase activity and *bgl* gene expression of *Oenococcus oeni* strains in model media and Cabernet Sauvignon wine. *American Journal of Enology and Viticulture* 62: 99-105.

Olguín, N., Champomier-Vergès, M., Anglade, P., Baraige, F., Cordero-Otero, R., Bordons, A., Zagorec, M. and Reguant, C. (2015). Transcriptomic and proteomic analysis of *Oenococcus oeni* PSU-1 response to ethanol shock. *Food Microbiology* 51: 87-95.

Ooka, T., Ogura, Y., Asadulghani, M., Ohnishi, M., Nakayama, K., Terajima, J., Watanabe, H. and Hayashi, T. (2009). Inference of the impact of insertion sequence (IS) elements on bacterial genome diversification through analysis of small-size structural polymorphisms in *Escherichia coli* O157 genomes. *Genome Research* 19: 1809-1816.

Ough, C.S. (1976). Ethyl carbamate in fermented beverages and foods, I: Naturally occurring ethyl carbamate. *Journal of Agricultural and Food Chemistry* 24: 323-328.

Ough, C.S., Crowell, E.A. and Gutlove, B.R. (1988). Carbamyl compound reactions with ethanol. *American Journal of Enology and Viticulture* 39: 239-242.

Pardo, I. and Zúñiga, M. (1992). Lactic acid bacteria in Spanish red rosé and white musts and wines under cellar conditions. *Journal of Food Science* 57: 392-395.

Pasteur, L. (1873). *Études sur le vin*. Librairie F. Savy, Paris.

Pérez-Martín, F., Izquierdo-Cañas, P.M., Seseña, S., García-Romero, E. and Palop, M.L. (2015). Aromatic compounds released from natural precursors by selected *Oenococcus oeni* strains during malo-lactic fermentation. *European Food Research and Technology* 240: 609-618.

Peynaud, E. (1956). New information concerning biological degradation of acids. *American Journal of Enology and Viticulture* 7: 150-156.

Pinzani, P., Bonciani, L., Pazzagli, M., Orlando, C., Guerrini, S. and Granchi, L. (2004). Rapid detection of *Oenococcus oeni* in wine by real-time quantitative PCR. *Letters in Applied Microbiology* 38: 118-124.

Poblet-Icart, M., Bordons, A. and Lonvaud-Funel, A. (1998). Lysogeny of *Oenococcus oeni* and study of their induced bacteriophages. *Current Microbiology* 36: 365-369.

Poolman, B., Molenaar, D., Smid, E.J., Ubbink, T., Abee, T., Renault, P.P. and Konings, W.N. (1991). Malo-lactic fermentation: Electrogenic malate uptake and malate/lactate antiport generate metabolic energy. *Journal of Bacteriology* 173: 6030-6037.

Popescu-Mitroi, I., Radu, D. and Stoica, F. (2009). Researches concerning the malic acid content of wines during storage period. *Journal of Agroalimentary Processes and Technologies* 15: 414-420.

Portillo, M.C., Franquès, J., Araque, I., Reguant, C. and Bordons, A. (2016). Bacterial diversity of Grenache and Carignan grape surface from different vineyards at Priorat wine region (Catalonia, Spain). *International Journal of Food Microbiology* 219: 56-63.

Puértolas, E., López, N., Condón, S., Raso, J. and Álvarez, I. (2009). Pulsed electric fields inactivation of wine spoilage yeast and bacteria. *International Journal of Food Microbiology* 130: 49-55.

Radler, F. (1990). Possible use of nisin in winemaking, I. Action of nisin against lactic acid bacteria and wine yeasts in solid and liquid media. *American Journal of Enology and Viticulture* 41: 1-6.

Reguant, C., Bordons, A., Arola, L. and Rozès, N. (2000). Influence of phenolic compounds on the physiology of *Oenococcus oeni* from wine. *Journal of Applied Microbiology* 88: 1065-1071.

Reguant, C. and Bordons, A. (2003). Typification of *Oenococcus oeni* strains by multiplex RAPD-PCR and study of population dynamics during malo-lactic fermentation. *Journal of Applied Microbiology* 95: 344-353.

Reguant, C., Carreté, R., Constantí, M. and Bordons, A. (2005a). Population dynamics of *Oenococcus oeni* strains in a new winery and the effect of $SO_2$ and yeast strain. *FEMS Microbiology Letters* 246: 111-117.

Reguant, C., Carreté, R., Ferrer, N. and Bordons, A. (2005b). Molecular analysis of *Oenococcus oeni* population dynamics and the effect of aeration and temperature during alcoholic fermentation and malo-lactic fermentation. *International Journal of Food Science and Technology* 40: 451-459.

Renouf, V., Claisse, O. and Lonvaud-Funel, A. (2005). Understanding the microbial ecosystem on the grape berry surface through numeration and identification of yeast and bacteria. *Australian Journal of Grape and Wine Research* 11: 316-327.

Renouf, V., Claisse, O. and Lonvaud-Funel, A. (2007). Inventory and monitoring of wine microbial consortia. *Applied Microbiology and Biotechnology* 75: 149-164.

Ribéreau-Gayon, P., Dubordieu, D., Donéche, B. and Lonvaud-Funel, A. (2000). Handbook of Enology. Wiley & Sons Ltd. Chichester, UK.

Rodas, A.M., Ferrer, S. and Pardo, I. (2003). 16S-ARDRA, a tool for identification of lactic acid bacteria isolated from grape must and wine. *Systematic and Applied Microbiology* 26: 412-422.

Rodas, A.M., Ferrer, S. and Pardo, I. (2005). Polyphasic study of wine *Lactobacillus* strains: Taxonomic implications. *International Journal of Systematic and Evolutionary Microbiology* 55: 197-207.

Ruiz, P., Izquierdo, P.M., Seseña, S. and Palop, M.L. (2008). Intraspecific genetic diversity of lactic acid bacteria from malolactic fermentation of Cencibel wines as derived from combined analysis of RAPD-PCR and PFGE patterns. *Food Microbiology* 25: 942-948.

Ruiz, P., Izquierdo, P.M., Seseña, S. and Palop. M.L. (2010). Analysis of lactic acid bacteria populations during spontaneous malolactic fermentation of Tempranillo wines at five wineries during two consecutive vintages. *Food Control* 21: 70-75.

Šajbidor, J. (1997). Effect of some environmental factors on the content and composition of microbial membrane lipids. *Critical Reviews in Biotechnology* 17: 87-103.

Salema, M., Poolman, B., Lolkema, J.S., Loureiro-Dias, M.C. and Konings, W.N. (1994). Uniport of monoanionic L-malate in membrane vesicles from *Leuconostoc oenos*. *European Journal of Biochemistry* 225: 289-295.

Salema, M., Lolkema, J.S., San Romao, M.V. and Loureiro Dias, M.C. (1996). The proton motive force generated in *Leuconostoc oenos* by L-malate fermentation. *Journal of Bacteriology* 178: 3127-3132.

Sánchez, A., de Revel, G., Antalick, G., Herrero, M., García, L.A. and Díaz, M. (2014). Influence of controlled inoculation of malo-lactic fermentation on the sensory properties of industrial cider. *Journal of Industrial Microbiology and Biotechnology* 41: 853-867.

Sahoo, L., Panda, S.K., Paramethiotes, S., Zodlac, N. and Ray, R.C. (2015). Biogenic amines in fermented foods. pp. 303- 317. *In*: Didier Montet and R.C.Ray (Eds.). Fermented Foods, Part 1: Biochemistry and Biotechnology. CRC Press, Florida.

Saranraj, P., Naidu, M.A. and Sivasakthivelan, P. (2013). Lactic acid bacteria and its antimicrobial properties: A review. *International Journal of Pharmaceutical and Biological Archive* 4: 1124-1133.

Schlatter, J. and Lutz, W.K. (1990). The carcinogenic potential of ethyl carbamate (urethane): Risk assessment at human dietary exposure levels. *Food and Chemical Toxicology* 28: 205-211.

Sieiro, C., Cansado, J., Agrelo, D., Velázquez, J.B. and Villa, T.G. (1990). Isolation and enological characterisation of malo-lactic bacteria from the vineyards of Northwestern Spain. *Applied and Environmental Microbiology* 56: 2936-2938.

Silla-Santos, M.H. (1996). Biogenic amines: Their importance in foods. *International Journal of Food Microbiology* 29: 213-231.

Silveira, M.G., Baumgärtner, M., Rombouts, F.M. and Abee, T. (2004). Effect of adaptation to ethanol on cytoplasmic and membrane protein profiles of *Oenococcus oeni*. *Applied and Environmental Microbiology* 70: 2748-2755.

Soufleros, E., Barrios, M.L. and Bertrand, A. (1998). Correlation between the content of biogenic amines and other wine compounds. *American Journal of Enology and Viticulture* 49: 266-277.

Spano, G., Beneduce, L., Tarantino, D., Zapparoli, G. and Massa, S. (2002). Characterization of *Lactobacillus plantarum* from wine must by PCR species-specific and RAPD-PCR. *Letters in Applied Microbiology* 35: 370-374.

Spano, G., Chieppa, G. Beneduce, L. and Massa, S. (2004). Expression analysis of putative *arcA, arcB* and *arcC* genes partially cloned from *Lactobacillus plantarum* isolated from wine. *Journal of Applied Microbiology* 96: 185-193.

Spano, G., Rinaldi, A., Ugliano, M., Moio, L., Beneduce, L. and Massa, S. (2005). A β-glucosidase gene isolated from wine *Lactobacillus plantarum* is regulated by abiotic stresses. *Journal of Applied Microbiology* 98: 855-861.

Spano, G. and Massa, S. (2006). Environmental stress response in wine lactic acid bacteria: Beyond *Bacillus subtilis*. *Critical Reviews in Microbiology* 32: 77-86.

Sponholz, W. (1993). Wine spoilage by microorganisms. pp. 395-429. *In*: Fleet, G.H. (Ed.). Wine Microbiology and Biotechnology. Switzerland: Harwood Academic Publishers.

Suárez, J.A., González, M.C., Callejo, J.J., Colomo, B. and González, A. (1994). Contribution to the study of varietal wines from Rioja and Navarra, I: Microbial growth trends during grape maturation. *Bulletin de l'O.I.V.* 759: 389-407.

Sumby, K.M., Grbin, P.R. and Jiranek, V. (2014). Implications of new research and technologies for malolactic fermentation in wine. *Applied Microbiology and Biotechnology* 98: 8111-8132.

Suzzi, G., Romano, P. and Zambonelli, C. (1985). *Saccharomyces* strain selection in minimizing $SO_2$ requirement during vinification. *American Journal of Enology and Viticulture* 36: 199-202.

Suzzi, G. and Gardini, F. (2003). Biogenic amines in dry fermented sausages: A review. *International Journal of Food Microbiology* 88: 41-54.

Teixeira, H., Gonçalves, M.G., Rozès, N., Ramos, A. and San Romão, M.V. (2002). Lactobacillic acid accumulation in the plasma membrane of *Oenococcus oeni*: A response to ethanol stress. *Microbial Ecology* 43: 146-153.

Tenreiro, R., Santos, M.A., Paveia, H. and Vieira, G. (1994). Inter-strain relationships among wine leuconostocs and their divergence from other *Leuconostoc* species, as revealed by low frequency restriction fragment analysis of genomic DNA. *Journal of Applied Bacteriology* 77: 271-280.

Tonon, T. and Lonvaud-Funel, A. (2000). Metabolism of arginine and its positive effect on growth and revival of *Oenococcus oeni*. *Journal of Applied Microbiology* 89: 526-531.

Tonon, T., Bourdineaud, J.P. and Lonvaud-Funel, A. (2001). The *arcABC* cluster encoding the arginine deiminase pathway of *Oenococcus oeni*, and arginine induction of a CRP-like gene. *Research in Microbiology* 152: 653-661.

Torriani, S., Felis, G.E. and Fracchetti, F. (2011). Selection criteria and tools for malolactic starters development: An update. *Annals of Microbiology* 61: 33-39.

Tourdot-Maréchal, R., Fortier, L.C., Guzzo, J., Lee, B. and Diviès, C. (1999). Acid sensitivity of neomycin-resistant mutants of *Oenococcus oeni*: A relationship between reduction of ATPase activity and lack of malolactic activity. *FEMS Microbiology Letters* 178: 319-326.

Tourdot-Maréchal, R., Gaboriau, D., Beney, L. and Divies, C. (2000). Membrane fluidity of stressed cells of *Oenococcus oeni*. *International Journal of Food Microbiology* 55: 269-273.

Treangen, T.J., Abraham, A.L., Touchon, M. and Rocha, E.P.C. (2009). Genesis, effects and fates of repeats in prokaryotic genomes. *FEMS Microbiology Reviews* 33: 539-571.

Uthurry, C.A., Varela, F., Colomo, B., Suárez Lepe, J.A., Lombardero, J. and García del Hierro, J.R. (2004). Ethyl carbamate concentrations of typical Spanish red wines. *Food Chemistry* 88: 329-336.

Vahl, M. (1993). A survey of ethyl carbamate in beverages, bread and acidified milks sold in Denmark. *Food Additives and Contaminants* 10: 585-592.

Vaquero, I., Marcobal, A. and Muñoz, R. (2004). Tannase activity by lactic acid bacteria isolated from grape must and wine. *International Journal of Food Microbiology* 96: 199-204.

Versari, A., Parpinello, G.P. and Cattaneo, M. (1999). *Leuconostoc oenos* and malolactic fermentation in wine: A review. *Journal of Industrial Microbiology and Biotechnology* 23: 447-455.

Volschenk, H., van Vuuren, H.J.J. and Viljoen-Bloom, M. (2006). Malic acid in wine: Origin, function and metabolism during vinification. *South African Journal of Enology and Viticulture* 27: 123-136.

Weber, F.J. and De Bont, J.M. (1996). Adaptation mechanisms of microorganisms to the toxic effects of organic solvents on membranes. *Biochimica Biophysica Acta – Review Biomembranes* 1286: 225-245.

Weidmann, S., Rieu, A., Rega, M., Coucheney, F. and Guzzo, J. (2010). Distinct amino acids of the *Oenococcus oeni* small heat shock protein Lo18 are essential for damaged protein protection and membrane stabilisation. *FEMS Microbiology Letters* 309: 8-15.

Wibowo, D., Eschenbruch, R., Davis, C.R., Fleet, G.H. and Lee, T.H. (1985). Occurrence and growth of lactic acid bacteria in wine: A review. *American Journal of Enology and Viticulture* 36: 302-312.

Winterhalter, P., Sefton, M.A. and Williams, P.J. (1990). Volatile C13 norisoprenoid compounds in Riesling wine are generated from multiple precursors. *Journal of Agricultural and Food Chemistry* 41: 277-283.

Zapparoli, G., Torriani, S., Pesente, P. and Dellaglio, F. (1998). Design and evaluation of malolactic enzyme gene targeted primers for rapid identification and detection of *Oenococcus oeni* in wine. *Letters in Applied Microbiology* 27: 243-246.

Zapparoli, G., Reguant, C., Bordons, A., Torriani, S. and Dellaglio, F. (2000). Genomic DNA fingerprinting of *Oenococcus oeni* strains by pulsed-field electrophoresis and randomly amplified polymorphic DNA-PCR. *Current Microbiology* 40: 351-355.

Zapparoli, G., Tosi, E., Azzolini, M., Vagnoli, P. and Krieger, S. (2008). Bacterial inoculation strategies for the achievement of malo-lactic fermentation in high-alcohol wines. *South African Journal of Enology and Viticulture* 30: 49-55.

Zarraonaindia, I., Owens, S.M.,Weisenhorn, P.,West, K., Hampton-Marcell, J., Lax, S., Bokulich, N.A., Mills, D.A., Martin, G., Taghavi, S., van der Lelie, D. and Gilbert, J.A. (2015). The soil microbiome influences grapevine-associated microbiota. *mBio* 6: e02527e14.

Zé-Zé, L., Tenreiro, R., Brito, L., Santos, M.A. and Paveia, H. (1998). Physical map of the genome of *Oenococcus oeni* PSU-1 and localisation of genetic markers. *Microbiology* 144: 1145-1156.

Zé-Zé, L., Tenreiro, R. and Paveia, H. (2000). The *Oenococcus oeni* genome: Physical and genetic mapping of strain GM and comparison with the genome of a 'divergent' strain, PSU-1. *Microbiology* 14: 195-204.

Zé-Zé, L., Chelo, I.M. and Tenreiro, R. (2008). Genome organisation in *Oenococcus oeni* strains studied by comparison of physical and genetic maps. *International Microbiology* 11: 237-244.

Zúñiga, M., Pardo, I. and Ferrer, S. (2003). Conjugative plasmid pIP501 undergoes specific deletions after transfer from *Lactococcus lactis* to *Oenococcus oeni*. *Archives of Microbiology* 180: 367-373.

# 13 Biochemistry of Winemaking

**V.K. Joshi[1*], H.P. Vasantha Rupasinghe[2], Ashwani Kumar[3], and Pooja Kumari[4]**

[1] Department of Food Science and Technology, Dr. Y.S. Parmar University of Horticulture and Forestry, Nauni, Solan- HP, India

[2] Department of Plant, Food, and Environmental Sciences, Faculty of Agriculture, Dalhousie University, Truro, Nova Scotia, Canada

[3] Department of Food Technology and Nutrition, Lovely Professional University, Phagwara, Punjab, India

[4] Department of Pharmaceutical Sciences, Lovely Professional University, Phagwara, Punjab, India

## 1. Introduction

The history of winemaking traces back to 7000 to 6600 BC and is among one of the oldest techniques known to mankind. Wines are produced from a variety of fruits but still grape wine and apple cider are the most popular fruit-fermented alcoholic beverages at a global scale. The process of winemaking is one of the most commercially prosperous biotechnological processes and is considered as one of the greatest arts even today (Moreno-Arribas and Polo 2005; Pinu *et al.*, 2014). The quality of the wine is affected by a range of factors including the variety of grapes, location of the vineyard, agricultural engineering, harvesting stage and the production technology (Kučerová and Široký, 2014). The conversion of a juice into wine is a complex biochemical process involving the enzymes originating from the fruit, yeast, lactic acid bacteria or those from the contaminating microorganisms. In brief, winemaking is the result of a series of biochemical transformations brought about by the action of several enzymes from different microorganisms. The primary fermentation, i.e. alcoholic fermentation (AF) is carried out by the yeast predominantly *Saccharomyces cerevisiae* (Amerine *et al.*, 1980; Moreno-Arribas and Polo, 2005). During winemaking, yeast use sugars and other grape juice components to produce alcohols and different odour active compounds. However, lactic acid bacteria play an important role in secondary malolactic fermentation which is responsible for reducing the acidity of the high acid wines (Bartowsky, 2011). Commercial enzyme preparations such as pectin esterase or amylolytic enzymes are widely used as supplements in wine making.

The changes involved in AF have been studied since the pioneering research of Louis Pasteur about 140 years ago and focus on various biochemical reactions involved and the end products formed (Amerine, 1985; Rana and Rana, 2011). AF involves the utilization of sugars and other constituents of grape juice or any other fruit juice for the growth of yeast and conversion of sugars into ethanol, carbon dioxide and other metabolites (Goyal, 1999; Rebordinos *et al.*, 2011). All these metabolites along with the products of secondary fermentation, contribute to the chemical composition and sensory characteristics of the wine. Hence, a thorough knowledge of the yeast growth and the biochemistry of fermentation can help to understand the fundamentals of winemaking. Nowadays, wines with different flavour bouquet are available in the market, and the production of quality wines is not as simple as the process discovered by Pasteur (Moreno-Arribas and Polo, 2005). Ample research work has been carried out in the last decade to understand the chemistry and interactions of yeast, lactic acid bacteria, and other microorganisms during the winemaking process. This work has greatly benefitted the winemakers to produce high-quality wines with different flavour bouquets. Fruit wines have long been served in moderate quantities as dessert wines and recognized as a natural source of essential minerals and many bioactive phytochemicals (Joshi, 1997; Joshi *et al.*, 1999b).

This chapter discusses the various metabolic events, specifically those leading to the translocation and metabolism of major ingredients like sugars, nitrogen, organic acids, and sulfur compounds. The biochemistry of the fruit maturity, composition and nutrients in relation to the changes in fermentation aspects is also illustrated, but no attempts have been made to discuss the very basics of biochemical

*Corresponding author: vkjoshipht@rediffmail.com

aspects for which the readers can refer to the literature cited (Amerine *et al.*, 1980; Goyal, 1999; Rana and Rana, 2011).

## 2. Alcoholic Fermentation

Alcoholic fermentation (AF) is one of the oldest and important processes carried out by the yeast especially *Saccharomyces cerevisiae* to produce a variety of beverages i.e. wines, beers and distilled liquors (Amerine *et al.*,1980; Joshi *et al.*, 2011; Rebordinos *et al.*, 2011). The primary end products of alcoholic fermentation in all the wines are ethanol and carbon dioxide; however, there are some critical differences between the winemaking of grapes and other fruits. Sugar and acid content, the percentage of juice released as well as the unique chemical fingerprint of each fruit that contributes to its flavour are some of the significant factors responsible for differences in the process of winemaking (Yang, 1953; Amerine *et al.*, 1980; Joshi *et al.*, 2011). For winemaking, the grapes are bred to near perfection and are allowed to fully ripe so that a desired sugar -to- acid ratio, low protein and pectin content are obtained, which results in increased yield and clarified juice (Buglass, 2011).

## 3. Biochemistry of Winemaking

During winemaking, the most important changes take place during vinification. This involve the conversion of sugars, especially hexoses (glucose and fructose) to ethanol and carbon dioxide along with the generation of a large number of minor by-products (Amerine *et al.*, 1980; Moreno-Arribas and Polo, 2005). It is noteworthy that the process of vinification depends largely on the growth and metabolic ability of yeast (Alexandre *et al.*, 1992; Angulo *et al.*, 1993; Santamariia, *et al.*, 1995; Dubois *et al.*, 1996; Charoenchai, *et al.*, 1998; Calderoin *et al.*, 2001; Rebordinos *et al.*, 2011) which in turn is affected by many factors, such as: temperature of fermentation, pH, sugar concentration of the juice and the inhibitory effect of ethanol on the specific growth rates and viability of *S. cerevisiae.* In addition to these, other stress factors like high levels of sulphur dioxide and carbon dioxide, the presence of competing microorganisms (for example lactic acid bacteria identified as *Lactobacillus kunkeei* spp. nov.) and the killer toxins produced by the killer yeasts also affect the fermentation of wine (Van Vuuren and Jacobs, 1992; Hidalgo and Flores, 1994; Boulton *et al.*, 1996;Cocolin and Comi, 2011). A pictorial representation of the metabolic events in yeast is shown in Fig. 1.

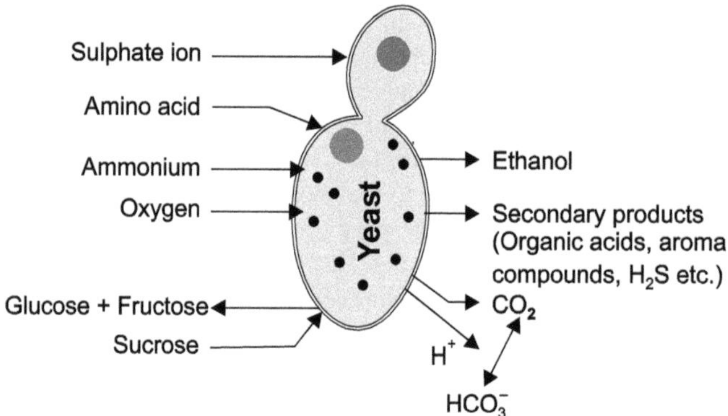


**Figure 1.** Diagram showing the utilization and formation of various compounds by yeast during alcoholic beverage (such as wine) production

Metabolism of carbon, nitrogen and sulfur compounds is the major biochemical event that occurs during AF. Among these, the carbon metabolism is the prime pathway of winemaking, and it is the initial concentration of sugars in the must/juice and prevailing conditions during fermentation which determines the ultimate concentration of ethanol in wine (Amerine *et al.*, 1980; Goyal, 1999; Rana and Rana, 2011). In case of fruits with low sugar content, the sugar balance can be corrected by amelioration with sugar or sugar syrup (Rupasinghe *et al.*, 2017). Nitrogen is the second important element that influences both fermentation kinetics and wine quality. It is an important nutritional factor for yeasts growth and metabolism due to its function in protein synthesis and sugar transport. Nitrogen is also essential for the biosynthesis of wine quality markers like higher alcohols, thiols, and esters by the wine yeast. Although, the nitrogen plays an important role in yeast growth and the flavour of wine but the quantity and the type of nitrogen content of the must can sometimes lead to problems like sluggish and stuck fermentation (Navarro and Navarro, 2011), the formation of reduced sulfur compounds, and certain other end products like ethyl carbamate (Ough *et al.*, 1988a).

During fermentation, many organic acids are either synthesized or degraded by yeast or lactic acid bacteria, which considerably affects the stability and sensory acceptability of the wine (Radler, 1993). A better understanding of the pathways of the synthesis and degradation of desirable and undesirable acids can help to control the quality of the wine (Radler, 1993). The major odour-active components (OAC) in wines are esters, aldehydes, alcohols, and terpenes. The final concentrations of various OAC are highly dependent on the yeast. The use of mixed strains of *Torulaspora delbrueckii* and *S. cerevisiae* during alcoholic fermentation of Sauvignon Blanc has shown a synergistic interaction between the two species, resulting in higher levels of 3SH (3-sulfanylhexan-1-ol) and its acetates. There was also an increased degradation of Glut-3SH (glutathionylated conjugate precursor) and production of Cys-3SH (cysteinylated conjugate precursor). The over production of Cys-3SH was dependent on the *T. delbrueckii* biomass (Renault *et al.*,2016). The Sulphur is also utilized by the yeast for biosynthesis of several sulfur compounds including hydrogen sulfide that is known to impart an unpleasant aroma to the wine.

## 3.1. Metabolism of Carbon

The degradation of sugars by yeast depends upon the aerobic or anaerobic conditions and can be carried out *via* two metabolic pathways, i.e. alcoholic fermentation and respiration. The operation of these pathways is also determined by the substrate availability and other conditions prevailing during the fermentation. When the concentration of sugars is high in the must, *S. cerevisiae* can only metabolize sugars by the fermentative route. The *Saccharomyces* species can metabolize only a limited number of carbon compounds i.e. monosaccharides (glucose and fructose) and disaccharides (sucrose, maltose, and melibiose). The trisaccharide, i.e., raffinose can also serve as the substrate in a few yeast strains, but pentoses are not utilized by the *Saccharomyces* wine strains (Goyal, 1999). However, a non-conventional yeast *Kluyveromyces marxianus* can utilize a wide range of sugars including pentose sugars viz., xylose. It has the ability to convert lactose, fructose, glucose, and xylose into ethanol (Pentjuss *et al.*, 2017).

Glycolysis is the major pathway employed for glucose and fructose catabolism (Amerine *et al.*, 1980; Rana and Rana, 2011) and is common in yeast and lactic acid bacteria with the production of pyruvate, which is further metabolized in different microorganisms to yield different compounds. Glucose is the most preferred sugar source to any other sugar by *S. cerevisiae.* It quickly takes up and utilizes glucose available in the grape juice at the beginning of the fermentation (Meijer *et al.*, 1998; Weinhandl *et al.*, 2014). Fructose is however, consumed in lesser amounts as compared to glucose (Broach, 2012; Pinu *et al.*, 2014) and once glucose and fructose are nearly depleted, *S. cerevisiae* starts to use other preferred carbon sources (Galeote *et al.*, 2010; Gutteridge *et al.*, 2010). In a study conducted by Pinu *et al.*, (2014) on nitrogen and carbon assimilation by *S. cerevisiae* during Sauvignon blanc juice fermentation the consumption of other hexoses such as mannose and sorbose was also observed. Mannose is a preferred carbon source to *S. cerevisiae,* but this result was unexpected for sorbose as it has been reported to be a non-metabolisable sugar by *S. cerevisiae* (Galeote *et al.*, 2010). Therefore, *S. cerevisiae* is capable of consuming sorbose, most likely under non-glucose repression conditions. The choice for the preferred source of sugar also varies with the type of the yeast. The yeast, *Kluyveromyces marxianus*, is capable of alcoholic fermentation and has a maximum preference for theoretical ethanol-to-substrate ratios for lactose (4), followed by inulin/glucose (2), and xylose (1.6) (Pentjuss *et al.*, 2017).

### *3.1.1. Glycolysis*

Glycolysis is the central pathway of carbohydrate metabolism, occurring in almost every living cell. The pathway involves the conversion of each glucose molecule into two molecules of pyruvate by a series of biochemical reactions involving a number of enzymes (Fig. 2). Glycolysis is carried out entirely in the cytosol of the cell and is operational under both aerobic and anaerobic conditions (Rana and Rana, 2011).

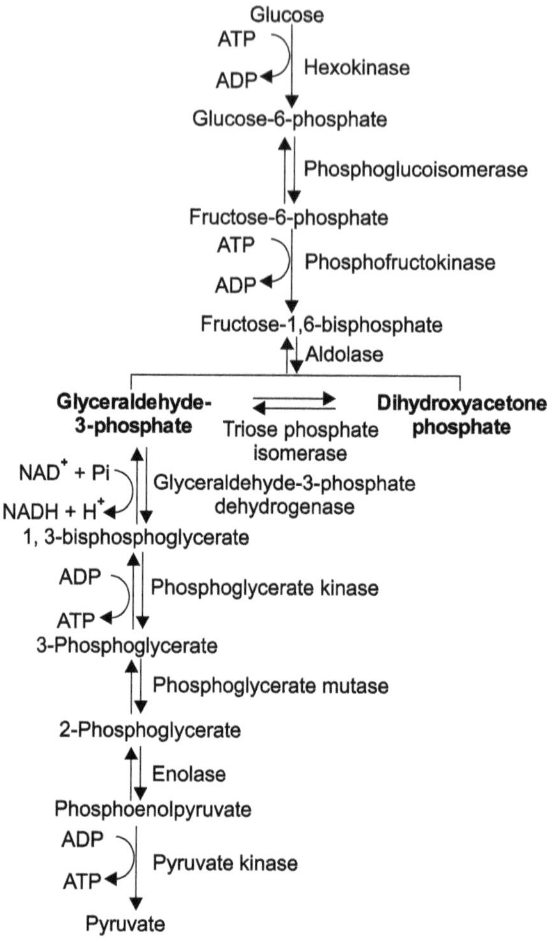


**Figure 2.** Flow diagram representation of the glycolytic pathway

Glucose metabolism for wine production includes three basic steps, viz. transport of sugars into the cell, their phosphorylation and conversion of glucose into glucose-6-phosphate, and finally, into pyruvate. The transport of sugars into the cell occurs *via* two types of mechanisms viz., passive transport and active transport. The former does not require an input of metabolic energy while the latter does need the same. For more details see the literature cited (Amerine *et al.*, 1980; Goyal, 1999; Rana and Rana, 2011). Under anaerobic conditions, the transport of sugar is a key step for the control of glycolytic flux. Generally, micro-organisms growing in sugar-containing media do not accumulate intermediates involved in sugar metabolism. The sugar transport in *Saccharomyces* is quite complex and is known to involve multiple carriers (Kruckeberg and Bisson, 1990). Mostly, glucose is utilized at a faster rate than fructose in fermentation. Glucose transport has also been studied in an industrial strain of *S. cerevisiae* (UCD522) during grape juice fermentation (McClellan *et al.*, 1989). At high concentration of glucose, more than one molecule tends to bind simultaneously to the carrier that inhibits its activity. In the next step glucose is phosphorylated to glucose-6-phosphate with the help of enzyme hexokinase. *S. cerevisiae* has two hexokinases (PI and PII) which are capable of phosphorylating glucose as well as fructose. Hexokinase PI functions mainly in catabolic role, whereas, hexokinase PII is essential and is mainly active during the yeast log phase in a medium having high sugar concentration. Hexokinase PI is partially repressed by glucose and is not active until the stationary phase is reached (Bisson, 1991). Glucose-6-phosphate is then, converted into its isomeric form, fructose-6-phosphate by the enzyme phosphofructokinase (PFK). Then another molecule of ATP is utilized for phosphorylation of fructose-6-phosphate to yield fructose-1,6-

bisphosphate. This enzyme also requires $Mg^{2+}$ ions for the reaction to proceed. Fructose-1,6-bisphosphate is then cleaved into two triose phosphate sugars viz., glyceraldehyde-3-phosphate and dihydroxyacetone phosphate (Fig. 3) by the enzyme aldolase (Amerine, 1985). Of the two products of aldolase reaction, only glyceraldehyde-3-phosphate is required for the next reaction in glycolysis. To prevent the loss of other triose phosphate sugar from the glycolytic pathway, dihydroxyacetone phosphate is isomerized to glyceraldehyde-3-phosphate in a reversible reaction catalyzed by triose phosphate isomerase. After this reaction, the original molecule of glucose has been converted to two molecules of glyceraldehyde-3-phosphate.

Triosephosphate isomerase

Glyceraldehyde-3-phosphate

Dihydroxyacetone phosphate

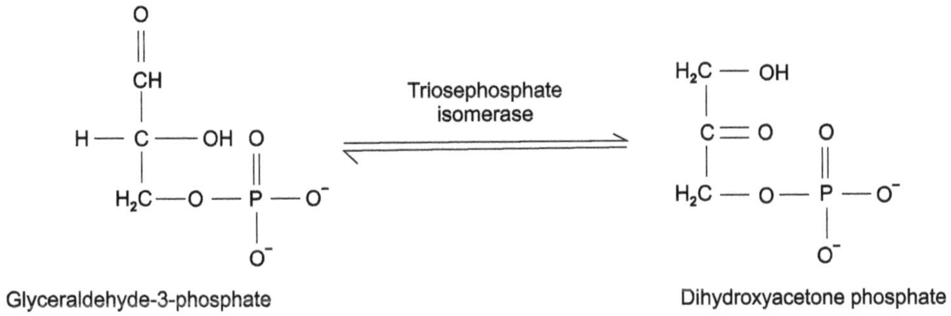

**Figure 3.** Interconversion of triose sugars

Glyceraldehyde-3-phosphate undergoes oxidation and phosphorylation to yield 1,3-bisphosphoglycerate in the presence of enzyme glyceraldehyde-3-phosphate dehydrogenase. The energy released in this oxidation process is conserved through phosphorylation of carboxylic group with inorganic phosphate and with the formation of NADH (Goyal, 1999). The high energy substrate formed in the previous reaction undergoes a transfer of phosphoryl group to form ATP from ADP and a low energy substrate, i.e., 3-phosphoglycerate in a reaction catalyzed by phosphoglycerate kinase. This phenomenon of generation of ATP from high energy substrate is called as substrate-level phosphorylation (Rana and Rana, 2011).

The next step of glycolysis, i.e., conversion of 3-phosphoglycerate to 2-phosphoglycerate is brought about by phosphoglycerate mutase which catalyzes the transfer of phosphoryl group from 3-hydroxyl position to 2-hydroxyl position to form 2-phosphoglycerate. In the next step, the enzyme enolase requiring $Mg^{2+}$ions, catalyzes the dehydration of 2-phosphoglycerate to produce phosphoenolpyruvate (PEP) in a reversible reaction. The last step of the glycolytic pathway yields pyruvate and is catalyzed by the enzyme pyruvate kinase that requires $Mg^{2+}$ and $K^+$ ions for its activity. This reaction also generates ATP by phosphorylating ADP in substrate level phosphorylation. Besides, the enzyme is also under the control of allosteric effectors (Rhodes *et al.*, 1986). The activity of the enzyme is inhibited by ATP, citrate, acetyl-CoA, long chain fatty acids and modulated by ADP and Pi. In glycolysis, thus, there is a net production of two molecules of ATP and NADH.

### *3.1.2. Metabolism of Pyruvate*

The microorganisms like *S. cerevisiae* metabolize carbohydrates through glycolysis with simultaneous utilization of $NAD^+$ and ADP to produce NADH and ATP, respectively (Amerine *et al.*, 1980). Both $NAD^+$ and ADP are present in small quantities in cells. Thus, the basic necessity of the cell is to generate $NAD^+$ and ADP if metabolism is to be continued. Therefore, each living organism has been provided with a mechanism to oxidize NADH for the continuity of substrate oxidation. For metabolism of pyruvate and regeneration of $NAD^+$, mainly three pathways i.e., fermentation, aerobic respiration and anaerobic respiration are utilized. Fermentation, especially in relation to wine has been described in this section, but only a brief account of other processes is given. A generalized view of pyruvate metabolism in *S. cerevisiae* is depicted in Fig. 4.

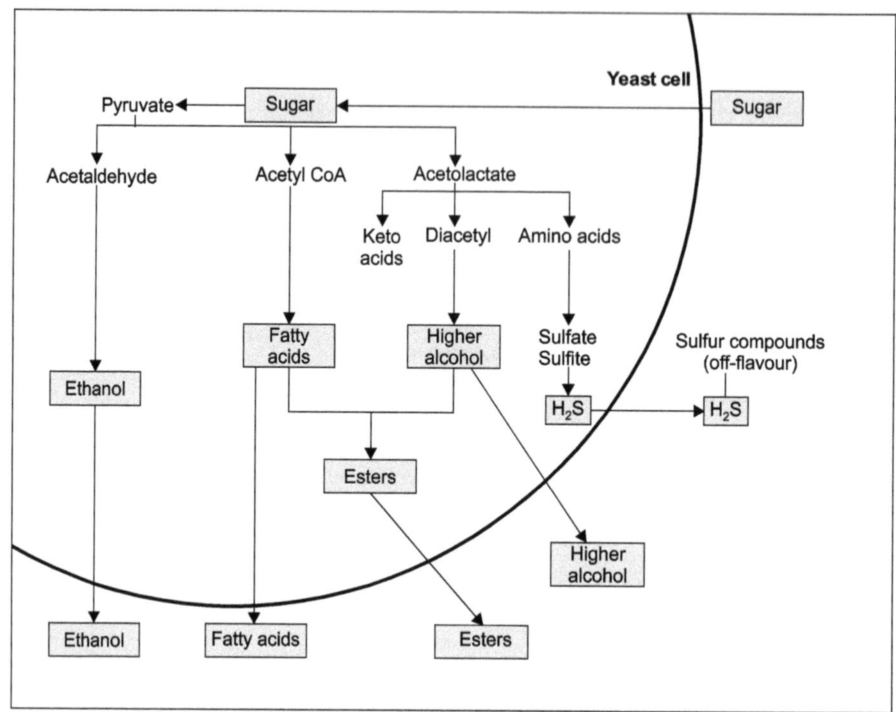


**Figure 4.** A generalized overview of metabolism in the yeast and the products formed including the conversion of pyruvate into various products

### *3.1.3. Citric Acid Cycle*

The amount of energy released during glycolysis is very little in comparison to the energy released *via* the citric acid cycle (TCA) and oxidative phosphorylation (Rana and Rana, 2011). After glycolysis, the final product pyruvate is formed which acts as a central metabolite, from which a number of products are formed (Fig. 4) by various enzymes found in different microorganisms (Goyal, 1999). Under aerobic conditions, pyruvate generates acetyl CoA catalyzed by the enzyme pyruvate dehydrogenase by oxidative decarboxylation in the presence of coenzyme A (CoA) and $NAD^+$. Pyruvate dehydrogenase is present in the interior of the mitochondria. The acetyl CoA produced earlier then enters the citric acid cycle, also called the tricarboxylic acid cycle (TCA cycle) and Krebs cycle (Fig. 5). In Krebs cycle, acetyl CoA is oxidized into carbon dioxide in the matrix of mitochondria. The cycle starts with the condensation of acetyl CoA with oxaloacetate to produce a tricarboxylic acid; citric acid. Further, a series of four oxidation-reduction reactions regenerate the oxaloacetate. Each complete cycle produces two $CO_2$ molecules, three NADH molecules and $FADH_2$ molecule (Ribereau-Gayon *et al.*, 2006). Various reactions involved in the TCA cycle are depicted in Fig. 5.

When the cellular energy level is high, i.e., ATP is in excess, the rate of citric acid cycle decreases and acetyl-CoA begins to accumulate. Under these conditions, acetyl-CoA can be employed for the synthesis of fatty acids or ketone bodies. However, $NAD^+$ used during glycolysis must be regenerated if glycolysis is to be continued. Under aerobic conditions, $NAD^+$ is regenerated by the re-oxidation of NADH *via* the electron transport chain. However, when oxygen is limiting as in muscular contractions, the re-oxidation of NADH to $NAD^+$ by the electron transport chain becomes insufficient to maintain glycolysis. Under these conditions, $NAD^+$ is regenerated by conversion of pyruvate to lactate by the enzyme, lactate dehydrogenase. The formation of lactic acid from pyruvate by lactic acid bacteria can significantly increase the acidity of the wine. The formation of lactic acid has been discussed in details later in this chapter.

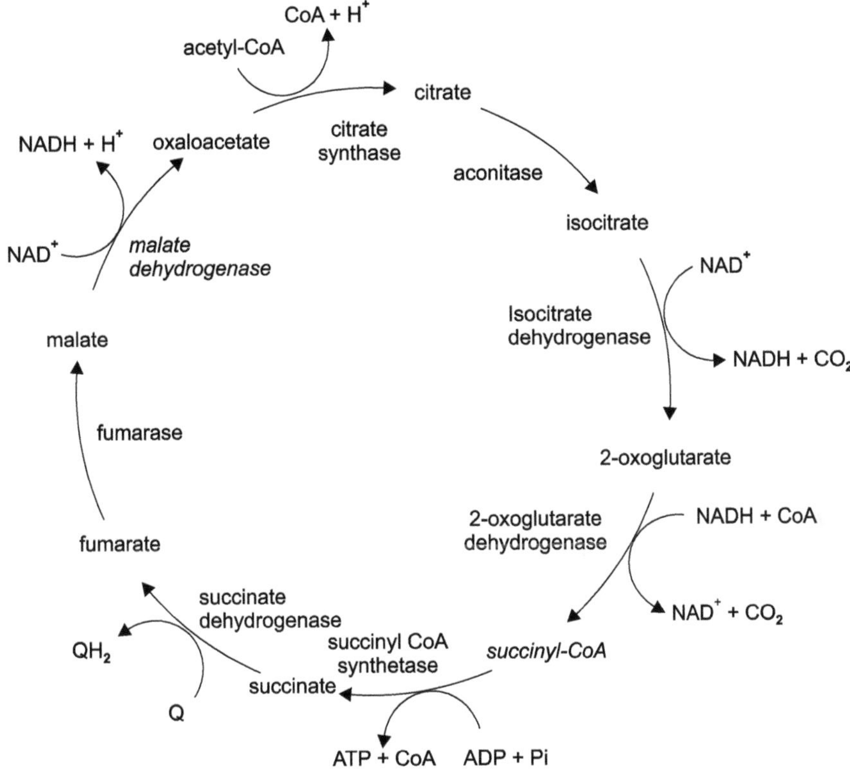


**Figure 5.** Various reactions involved in the tricarboxylic acid cycle (TCA) (*Source*: Cannon, 2014)

In *S. cerevisiae,* pyruvate is converted into ethanol (the main component of alcoholic beverages) and carbon dioxide (Fig. 6). This pathway in yeast is employed for the production of various ethanolic beverages, as discussed in other parts of the text. Pyruvate is converted into ethanol *via* acetaldehyde as the main intermediate. Besides ethanol, many other products of the different metabolic pathways of pyruvate are formed in different microorganisms (also depicted in Fig. 4). In a typical hetero-lactic acid fermentation, products like diacetyl or acetoin are produced from pyruvate which plays an important role in flavour development. Separate products can be formed from pyruvate by *Acetobacter* or *Gluconobacter* as is the case with the formation of entirely different products from pyruvate by *E. coli.*

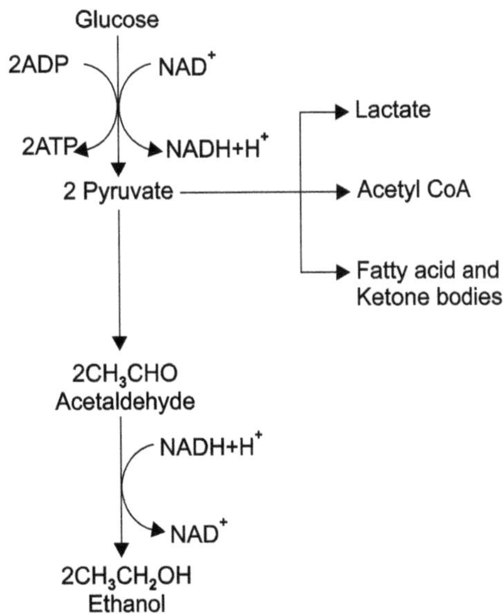


**Figure 6.** Production of ethanol from pyruvate in *Saccharomyces cerevisiae*, $NAD^+$ molecule regeneration and conversion to pyruvate to ethanol during alcoholic fermentation (*Source*: Goyal, 1999)

### *3.1.4. Fermentation and its Types*

Fermentation is an anaerobic breakdown of pyruvate (Goyal, 1999; Rana and Rana, 2011). Depending upon the end product, fermentation has been classified into various types as discussed here.

### Neuberg's First Fermentation (Alcoholic Fermentation)

The breakdown of pyruvate to produce ethanol is also called Neuberg's first fermentation (Fig. 6) as named after the German scientist Carl Neuberg. The yeast strains of *Saccharomyces* produce alcohol under anaerobic conditions by converting pyruvate to ethanol and carbon dioxide. This metabolism of pyruvate is also useful to regenerate $NAD^+$ molecule which was reduced earlier during glycolysis to form NADH. Alcoholic fermentation takes place in two steps.

In the first step, pyruvate undergoes irreversible decarboxylation to produce acetaldehyde, catalyzed by enzyme pyruvate decarboxylase. The enzyme requires thiamin pyrophosphate (TPP) as a coenzyme and $Mg^{2+}$ ions for its activity. The second step reduces acetaldehyde into ethanol at the expense of NADH. This reaction is catalysed by alcohol dehydrogenase whose active site contains a $Zn^{2+}$ ion. The reaction is favoured at pH 7.0. At higher pH, i.e., 9.5 and in the presence of an excess of $NAD^+$, the reaction tends to proceed in the backward direction.

### Neuberg's Second Form of Fermentation

When the glucose is fermented by yeast at pH 6 or below then very small amounts of glycerol and other products are formed. In the presence of sulphites the fermentation of glucose by yeast produces equivalent quantities of glycerol, carbon dioxide and acetaldehyde in its bisulfitic form (Ribereau-Gayon *et al.*, 2006). Acetaldehyde produced by the action of carboxylase during ethanol production combines with sulfite to form an additional compound. As a result, acetaldehyde is no longer able to act as a hydrogen acceptor for reduced $NAD^+$, which accumulates and is exhausted soon unless some other hydrogen acceptor is provided.

### Neuberg'sThird Form of Fermentation

Under the alkaline condition, dihydroxyacetone phosphate, an intermediate in glycolytic pathway replaces acetaldehyde as a hydrogen acceptor and is reduced to glycerol-3-phosphate which is dephosphorylated to glycerol with the regeneration of $NAD^+$(Goyal, 1999) as shown in Fig. 7. Many osmophilic yeasts carry out this fermentation, corresponding to Neuberg's third form in the absence of alkaline steering agents. The inhibition of alcohol dehydrogenase or pyruvate decarboxylase activities can produce glycerol in proportionally higher yields. The different species of *S.cerevisiae* vary in their ability to produce glycerol which ranges from 4.2 to 10.4 g/L (Radler and Schutz, 1982).

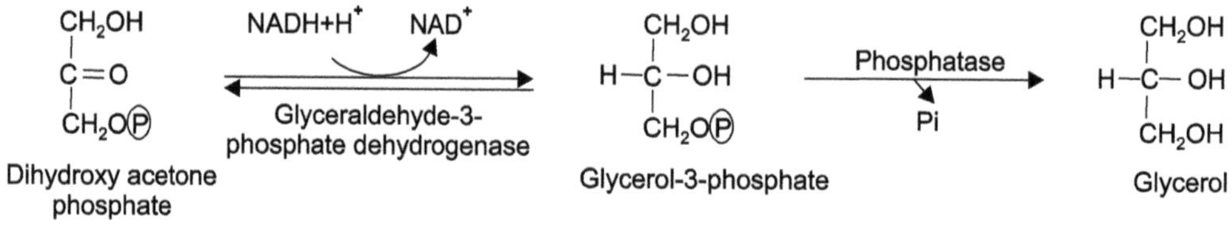

Dihydroxy acetone phosphate ($CH_2OH$–C=O–$CH_2O$Ⓟ) ⇌ [NADH+H⁺ → NAD⁺; Glyceraldehyde-3-phosphate dehydrogenase] Glycerol-3-phosphate ($CH_2OH$–H–C–OH–$CH_2O$Ⓟ) → [Phosphatase; Pi] Glycerol ($CH_2OH$–H–C–OH–$CH_2OH$)

**Figure 7.** Neubergs third form of fermentation (*Source*: Rana and Rana, 2011)

### *3.1.5. The Yield of Ethanol and By-products*

The yield of ethanol in alcoholic fermentation with a surplus amount of fermenting sugars depends on ethanol tolerance by the yeasts. A higher level of ethanol is inhibitory to the microbial enzymes and disintegrates the cell membrane through unfavourable interaction with cell wall components. Ethanol tolerance of yeast is also linked with membrane fluidity. A comprehensive review on this aspect in yeast had been compiled earlier (Casey and Ingledew, 1986). During alcoholic fermentation, one molecule of glucose produces two molecules, each of ethanol and $CO_2$ under anaerobic conditions. In other words, 180 g of glucose (1 mole) should yield 92 g of ethanol (2 moles) and 88 g $CO_2$ (2 moles). The theoretical yield of ethanol production, therefore, comes out to be 51 percent. Under practical conditions, a yield of 47% ethanol can be achieved. The metabolism though yields an equimolar quantity of $CO_2$ and ethanol, the actual amount of $CO_2$ liberated is less than theoretical because of the utilization of $CO_2$ in anaerobic carboxylation reactions (Ough *et al.*, 1988). In a fermentation process, it has been estimated that starting with 22 to 24 per cent sugars, 95 per cent of sugars are converted into ethanol and carbon dioxide, one per cent is converted to cellular biomass, and the remaining 4 per cent is converted to other end products. The amount of ethanol produced per unit of sugar during wine fermentation is of considerable commercial importance.

### *3.1.6. Regulation of Carbon Metabolism and Glycolysis in Yeast*

Various industrial processes, including food and beverage production like wine, bread, beer cheese, etc. utilizes yeast and widely depends on central carbon metabolism (CCM) by the yeast. For instance, the carbon metabolism determines the rate of fermentation, amount of ethanol production, acetate production and aroma production (Nidelet *et al.*, 2016). Various studies has been carried out on glycolytic pathway in *S. cerevisiae* both under aerobic and anaerobic conditions using $^{13}C$ and $^{31}P$ NMR spectroscopy, combined with the use of $^{14}C$ labeled glucose and determination of the ratio of different end products (Den Hollander *et al.*, 1986; Reibstein *et al.*, 1986; Campbell *et al.*, 1987) to understand the control of carbon flux through glycolysis. The study revealed that the control during sugar utilization is either at the step of transportation or phosphorylation or both and the loss of enzyme activity (hexokinase, phosphofructokinase, and pyruvate kinase) at any of the three irreversible steps of glycolysis immediately effects the glucose transportation. In eukaryotic cells, the glycolytic enzymes exist as a multi-enzyme complex (Green *et al.*, 1965; Walsh et al. 1989) and hence, certain kinds of regulatory interactions may be possible which cannot be duplicated with purified enzymes in an *in vitro* system.

The availability of other nutrients like nitrogen, also modulates the control of glycolytic flux. The nitrogen limitation causes a loss of glucose transporter activity (Busturia and Lagunas, 1986) leading to a decrease in overall glycolytic rate (Salmon, 1989). Ammonium nitrogen ($NH^{4+}$) is an allosteric effector of both the phosphofructokinase (PFK) and pyruvate kinase (PK), so it indirectly affects the transporter activity (Sols, 1981; Rhodes *et al.*, 1986). The mechanism of glucose inactivation i.e., inhibition of activity and subsequent proteolytic destruction of many proteins is unknown. The fermentation parameters vary for different compounds to obtain the maximum production. For example, the availability of fatty acid is the main factor which affects the synthesis of ethyl esters while the production of acetate depends on the activity of alcohol acetyltransferases (Rollero *et al.*, 2015).

The availability of molecular oxygen also plays a role in the control of sugar metabolism *via* the substrates level inhibition and the Pasteur effect. The phenomenon of substrate-level inhibition is observed in winemaking when at high sugar concentration, both uptake and consumption of sugar are inhibited, due to the inhibition of sugar transport as explained earlier. In the presence of oxygen, respiration is favoured in comparison to fermentation known as the Pasteur effect. This is operative in *Saccharomyces* only under very special circumstances of nitrogen limitation (Lagunas, 1986). Aceituno *et al.*,(2012), studied the effect of oxygen on wine production using *S. cerevisiae* EC1118, when it was grown under altered ecological conditions of sufficient carbon and limited nitrogen. In this study, when the concentration of dissolved oxygen was increased from 1.2 to 2.7 μM, a few metabolic changes were observed in yeast. Yeast cells became mixed respirofermentative from fully fermentative. This transitional change resulted in a switch in the operation of the TCA and activation of NADH shuttling from the cytosol to mitochondria. This further showed the Crabtree effect and fermentative ethanol production remained the major cytosolic NADH sink under all oxygen conditions. The study suggested that the limitation of mitochondrial NADH reoxidation was the major cause of the Crabtree effect. The Crabtree effect is defined as the occurrence of alcoholic fermentation under aerobic conditions. This is reinforced by the induction of several key respiratory genes by oxygen, despite the high sugar concentration. Oxygen significantly affects other genes associated with processes, such as proline uptake, cell wall remodeling, and oxidative stress. The results of this study indicated that respiration is responsible for a substantial part of the oxygen response in yeast cells during alcoholic fermentation.

### *3.1.7. End Products (Minor) of Sugar Metabolism*

As stated earlier, ethanol and carbon dioxide are the primary products of alcoholic fermentation in winemaking. However, a much lower concentration of secondary metabolites such as higher alcohols, polyols, esters, organic acids, vicinal diketones, aldehydes, lactones and terpenes (Fig. 4) are also produced by *S. cerevisiae*. Their concentration in wine ranges from 0.8-1.0 g/L. These compounds, synthesized mainly during the exponential phase of yeast growth, are very important flavour congeners in fermented beverages and determine the characteristic aroma and acceptance of any alcoholic beverage.

An important by-product, glycerol, is produced (Fig. 8) by the yeast *S. cerevisiae* in alcoholic fermentation called as Neuberg's fermentation (Goyal, 1999). In the normal process of the fermentation, the glyceraldehyde-3-phosphate is converted to dihydroxyacetone phosphate, where, it is reduced to glycerol phosphate by the enzyme dihydroxy actone phosphate reductase. However, the Neuberg's third fermentation is accomplished by addition of sulfite to the fermentation medium, where, glyceraldehyde-3-phosphate is converted to glycerol. Most of the glycerol is produced during the early stages of fermentation. Increase in sugar content, decreases the glycerol production relative to ethanol while the factors like low temperature, high tartaric acid and sulfur dioxide favour the production of glycerol in alcoholic fermentation. Although glycerol is produced only in trace amounts, it is the main by-product of alcoholic fermentation and contributes to the mouthfeel and consistency of wine. The sensory importance of glycerol is due to its sweet taste and oiliness (Goyal, 1999).

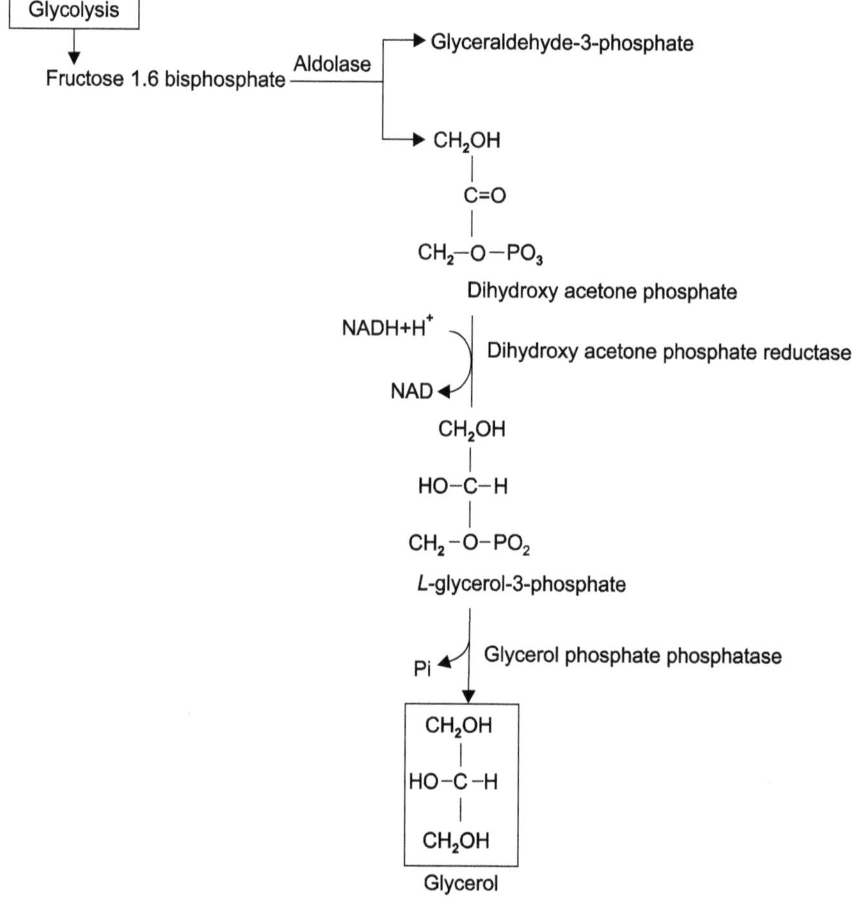


**Figure 8.** Pathway showing glycerol production (Goyal, 1999)

Methanol or methyl alcohol ($CH_3OH$) in wines and spirits is produced by the hydrolysis of a specific methyl-ester bond at the C-6 of galacturonic acid in the homogalacturonan linear chains of pectins by enzyme pectin methyl esterase (Fig. 9). This reaction leads to the release of a methanol molecule and a proton. The methanol content in wine therefore, depends on the content of pectin in the must, but it is also affected by a number of factors such as the ripening stage of the fruit, pomace maceration conditions, concentration and activity of the fruits pectolytic enzymes, the addition of commercial enzyme preparations, the pressure applied during squeezing of the pomace and fermentation conditions

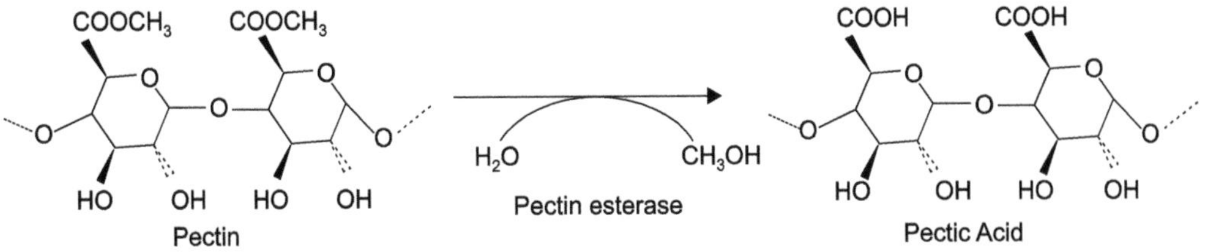

COOCH$_3$ COOCH$_3$ → COOH COOH; $H_2O$ → $CH_3OH$; Pectin esterase; HO OH HO OH; Pectin; Pectic Acid

**Figure 9.** Formation of methanol in alcoholic fermentation by action of pectin esterase enzyme on pectin (Goyal, 1999)

(pH, temperature, mixing intensity and frequency). Sanitary conditions of the grapes (fruit) can greatly affect the amount of methanol in the wine. For example, grey mold (*Botrytis cinerea*) synthesizes pectin methyl esterase of very high activity. Methanol is firstly oxidized to formaldehyde and then to formic acid, which accumulates in the blood. Methanol exerts relatively low toxicity; however, products obtained from its metabolic transformations, i.e. formaldehyde and formic acid are more toxic to humans through ingestion and inhalation. Eyes have very high sensitivity towards formaldehyde, the immediate product of methanol, hence, these are affected first. The pectinases that are employed for clarification and enhancing the yield of juice results in the formation of methanol sometimes beyond the acceptable range.

Absorption of methanol by the human body takes place very rapidly, either through the gut wall or of the respiratory system, and it usually takes 30 to 60 minutes to attain the maximum concentration in the blood. The oral lethal dose of methanol is in the range of 300 to 1000 µg/kg of body weight. An amount of methanol >200 mg/L in blood is considered to be a cause of severe poisoning, while death occurs when the concentration in the blood is above 1.5 to 2.0 g/L. Symptoms of methanol poisoning are manifested through headaches, severe abdominal pain, choking, a weakening pulse, a drop in body temperature, blindness, and even death (Miljić *et al.*, 2016)

The pectin rich fruits like plum and apricot give wine with a higher amount of methanol (Woidich and Pfannhauser, 1974). Methanol has a lower boiling point than ethanol, hence, its amount in the distillates can be reduced substantially by discarding the first fraction of distillate (Amerine *et al.*, 1980; Joshi, 1997).

### *3.1.8. End Products of Amino Acid Metabolism*

Amino acids that are deaminated metabolically in order to release their nitrogen components leave behind carbon skeleton which is regarded as a waste product. Higher alcohols are produced by the deamination or decarboxylation of amino acids (Fig. 10) in a pathway, known as Ehrlich Pathway, a catabolic route which comes into play when there are excess of amino acids in the medium.

Higher alcohols are collectively referred to as fuel alcohols or fusel oils and constitute a major portion of secondary products of yeast metabolism (Amerine *et al.*, 1980). When amino acids are deficient, an alternative anabolic route, called the Biosynthesis Pathway, becomes operational deriving higher alcohols from α-keto acid intermediates (Nykanen, 1986). These alcohols contain more than two carbons and the examples are n-propanol, isobutyl alcohol (2- methyl-1-propanol), 2- methyl butanol (optically active amyl alcohol), isoamyl alcohol and 2 phenyl ethanol. The corresponding amino acid in the fermentation medium determines the levels of individual fusel oil in beverages. For example, phenylalanine stimulates phenylethanol production leading to a rose-like aroma (Walker and Stewart, 2016). Fruit maturation also affect their concentration due to the qualitative and quantitative differences in amino acid composition. Among the assorted higher alcohols, isoamyl alcohol (3-methyl butanol) accounts for more than 50 per cent of the total (Muller *et al.*, 1993). These higher alcohols have higher boiling points, and molecular weights than ethanol. The production of these alcohols have been found to be linearly linked with the growth of the yeast (Pierce, 1982). Some other factors like fruit maturation, composition of amino acids, strain of yeast and fermentation conditions viz., temperature, pH, aeration (Webb and Ingraham, 1963; Rankine 1967; Sinton *et al.*, 1978; Ough and Bell, 1980; Vos, 1981; Rapp and Versini, 1991; Henschke and Jiranek, 1993) also affects the production of higher alcohols.

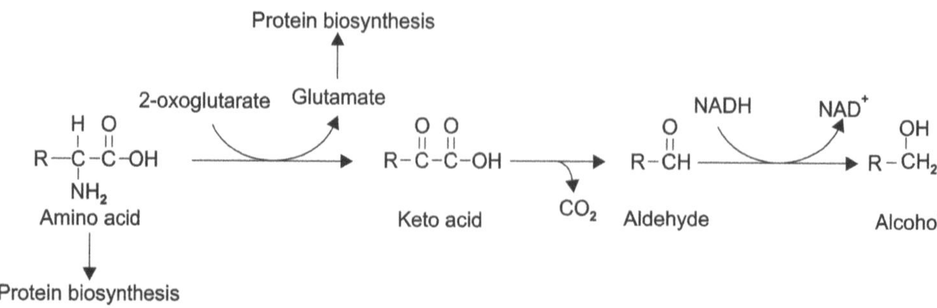


**Figure 10.** Pathway of higher alcohol formation from amino acid (*Source*: Jansen *et al.*, 2003)

These alcohols however, have a very little impact on the sensory properties of wine, but in distilled beverages these are more concentrated and contribute to the characteristic aroma of the distillates. A concentration below 300-400 mg/L of higher alcohols reinforces the flavour of the alcoholic beverages, while, concentration higher than this is often thought-about as undesirable as it imparts a fuesl (foul) smell and being a reason behind hangover in the consumers of such beverages (Amerine *et al.*, 1980; Webb and Ingraham, 1963). The concentrations of higher alcohols in white wine has been reported to be in the range from 162 to 266 mg/L while, 140 to 417 mg/L in red wines (Gumyon and Heintz, 1952). However, there is scattered documented information on the quantities of higher alcohols in non-grapes wines (Joshi *et al.*, 1999b).

### *3.1.9. Volatile and Non-volatile Organic Acids*

Organic acids play an important role in the sensory perception of wine and directly influence the overall sensory characteristics of the wine. Some of these acids are present naturally in the fruits or can be synthesized or degraded from existing compounds during alcoholic fermentation. These acids are responsible for the sharpness of fermented beverages and thus, improve the sensory quality of wine. Different organic acids have different organoleptic properties, and hence, the specific concentration of each acid impacts the sensory perception of wine.

The concentrations of organic acids can vary greatly and depends upon the condition and maturity of fruits. The organic acids derived from the grapes are tartaric acid, malic acid and citric acid. The concentration of tartaric acid in grapes ranges from 4.5 to 10 g/L and depends upon the factors like grape variety and soil type of vineyard (Ribereau-Gayon *et al.*, 2006). The concentration of tartaric acid is also affected by the temperature and a higher concentration, i.e. 6 g/L or above is found in cold climates while lower concentrations of 2 - 4 g/L are more commonly observed in warm climates. The concentration of malic acid ranges from 2 to 6.5 g/L of L-malic acid in mature grapes (Ribéreau-Gayon *et al.*, 2000). The grapes grown in warmer climates tend to have a faster rate of L-malic acid respiration compared to those of the cooler climates, and hence an excessively high amount of malic acid, i.e. 15 to 16 g/L may be present in grapes harvested from exceptionally cool-climatic regions (Gallander, 1977). Citric acid is an intermediate of the TCA cycle and the concentrations of citric acid in must and wine prior to malolactic fermentation are usually relatively low, between 0.5 and 1 g/L (Kalathenos *et al.*, 1995).

In alcoholic fermentation, the limiting activities of pyruvate decarboxylase and alcohol dehydrogenase coupled with high sugar concentration lead to the formation of many organic acids. The activities of pyruvate decarboxylase and alcohol dehydrogenase are not highly expressed in respiring cells. These enzymes are inducible by glucose (Sharma and Tauro, 1986) and as a consequence, compounds other than ethanol are produced at the beginning of fermentation. Succinic acid, malic acid, and acetic acid account for the major organic acids produced by the yeasts, however, ratio of these acids vary with yeast strains (Table 1).

Succinic acid is a dicarboxylic acid produced by the yeasts during wine fermentation (Fig. 11), and its concentration goes up to 2.0 g/L (Radler, 1993). It is produced mainly as an intermediate of TCA cycle during aerobic respiration and the enzyme 2-oxoglutarate dehydrogenase present in fermenting yeast is responsible for the production of succinate. The addition of sugars to the resting cells of yeasts results in rapid formation of succinic acid. The amino acid, glutamate also favours the production of succinic acid.

**Table 1.** Organic Acids Present in Grapes and Wine

| *Fixed acids* | | *Volatile acids* | |
|---|---|---|---|
| *Major acids* | *Minor acids* | *Major acids* | *Minor acids* |
| L-tartaric acid (5-10 g/L) | Pyruvic acid | Acetic acid | Formic acid |
| L-malic acid* (2-6.5 g/L) | α-ketoglutaric acid | | Propionic acid |
| L-lactic acid (1-3 g/L) | Isocitric acid | | 2-methylpropionic acid |
| Citric acid** (0.5-1 g/L) | 2-oxoglutaric acid | | Butyric acid |
| Succinic acid (0.5-1.5 g/L) | Dimethylglyceric acid | | 2-methylbutyric acid |
| Amino acids | Citramalic acid | | 3-methylbutyric acid |
| | Gluconic acid*** | | Hexanoic acid |
| | Glacturonic acid | | Octanoic acid |
| | Glucuronic acid | | Decanoic acid |
| | Mucic acid | | |
| | Coumaric acid | | |
| | Ascorbic acid | | |

*15-16 g/L L-malic acid has been reported in cool climate regions
** >0.3 g/L when wines are stabilised for metal precipitation
*** Present in wine with *Botrytis cinerae* infection
*Source*: Volschenk *et al.* (2006)

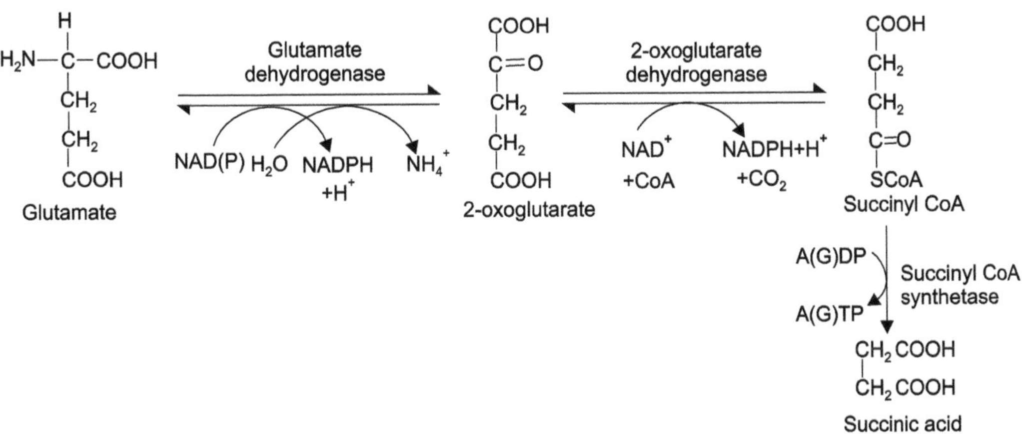


**Figure 11.** Oxidative pathway of synthesis for the succinic acid by yeast (*Source*: Rana and Rana, 2011)

An oxidative pathway involving succinyl-CoA becomes operative when the yeast is grown aerobically on glutamate leading to synthesis of succinic acid (Goyal, 1999). Small amounts of succinate can also be produced from oxaloacetate, malate or aspartate by the enzymes pyruvate decarboxylase of the reductive pathway (Schwartz and Radler, 1988). It is also produced as the fermentation end-product of anaerobic metabolism.

Succinic acid is sour with a salty, bitter taste and has a threshold concentration of approximately 35 mg/L (Chidi, 2016). Owing to bitter taste, winemakers need to pay much attention to succinic acid levels in wine.

The production of malic acid is favoured by the concentration of sugars (20-30%), nitrogen compounds (100-200 mg N/L), presence of $CO_2$ and pH (nearly 5)(Radler and Lang, 1982). *S. cerevisiae* leads to the formation of roughly equal quantities of succinate and malate, whereas, strain *S. uvarum* produces about 10 times more malate than succinate. The main volatiles of wine have been listed in Table 2.

**Table 2.** Sensory Properties of the Main Wine Volatiles

| | *Regular concentration in wine (mg/L)* | *S. cerevisiae production (mg/L)* | *Sensory threshold (mg/L)* | *Description of smell* |
|---|---|---|---|---|
| Acetic acid | >150 | 150-61400 | 600 | Vinegar |
| Acetaldehyde | 30-60 | 5-6405 | 1-62* | Nut/Over-ripe apple |
| Ethyl acetate | 5-60 | 5-687 | 12-630 | Pineapple/Nail polish |
| 2-butanol | 20-660 | 9-670 | 50 | Sweet apricot |
| Isobutanol | 9-6174 | 15-6237 | 75-6100 | Whiskey tusel |
| Isoamyl alcohol | 6-6490 | 6-6385 | 50-660 | Pear/Pungent |
| 2,3-butanediol | 80-6170 | 6-6186 | 600 | Fruity, creamy, buttery |

* Free acetaldehyde
*Source*: Malik, 2014

Acetic acid is produced by many strains of the yeast and is produced only under conditions of strict anaerobiosis (0.3 C-mmol g $DW^{-1}$ $h^{-1}$). It is the main volatile acid in the fermented beverages and constitutes more than 90 percent of the volatile acidity in wine (Amerine *et al.*, 1980). Its presence beyond a certain limit; however, spoils the alcoholic beverages.

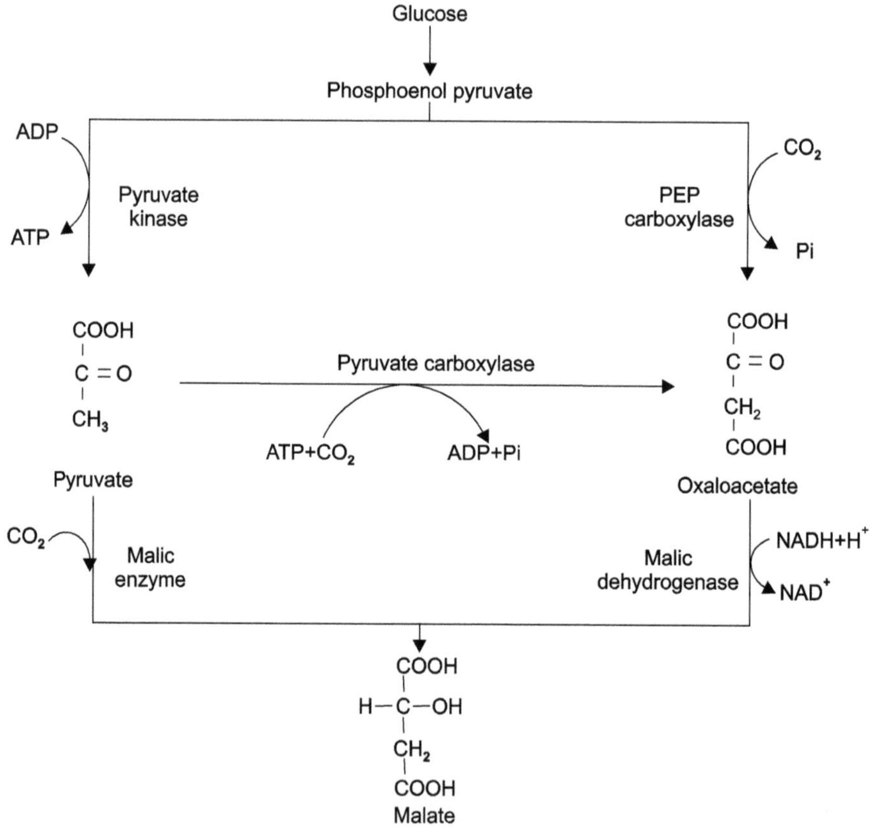


**Figure 12.** Synthesis of malate by yeast during fermentation

In addition to oxygen, the production of acetic acid depends on many factors like sugar content, nitrogen content, phytosterols, yeast strain, and temperature. Under anaerobic conditions, glucose is converted to pyruvate, which is further converted to acetic acid instead of ethanol by the enzyme aldehyde dehydrogenase. The specific activity of pyruvate decarboxylase is increased by the aeration resulting in higher acetic acid production than anaerobic pitching yeast. In a study conducted by Caridi (2003), a sugar content of above 400 g/L in Riesling Ice wine juices resulted in very low ethanol yields and high acetate production. The production of acetic acid was also increased when glutamate was used as a nitrogen source in comparison to ammonium and other amino acids. The production of acetic acid was low at high concentrations of phytosterols which might be due to the low requirement for acetyl-CoA, a precursor for lipid biosynthesis, in the presence of exogenous lipids (phytosterols) (Rollero *et al.*, 2015). Citrate present in the grape must is also converted to acetic acid by wine lactic acid bacteria which have an enzyme, citrate lyase, that can split citrate into oxaloacetate and acetate. The oxaloacetate is again decarboxylated to pyruvate and further to acetate and other metabolites (Whiting, 1976).

Yeast strains also play an important role in the acetate production. High amount of acetic acid is produced by some yeast strains like *Hansenula* and hence are not employed for brewing (Amerine *et al.*, 1980; Joshi, 1997; Joshi et al.,1999). This acid seems to be produced in the initial stages of the fermentation (Whiting, 1976) and usually varies from 100 to 200 mg/L in case of genus *Saccharomyces*. The high concentration of acetic acid degrades the sensory attributes of wine and hence, its production during grape juice or any other juice fermentation is highly undesirable. When the acetate content in wine exceeds 20 mg/L, it is legally regarded as spoiled. High acetic acid concentration is also associated with *Acetobacter* spoilage.

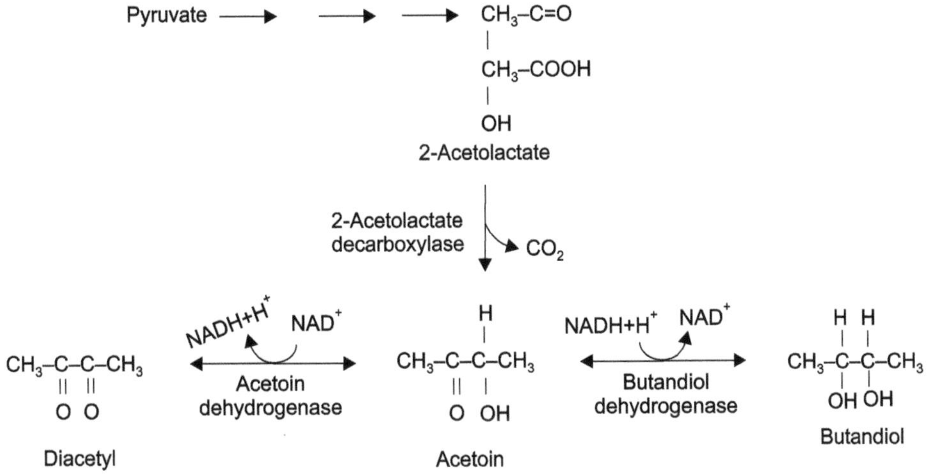

Pyruvate → → → $CH_3$–C=O | $CH_3$–COOH | OH

2-Acetolactate

2-Acetolactate decarboxylase → $CO_2$

Diacetyl ($CH_3$–CO–CO–$CH_3$) ⇄ Acetoin dehydrogenase (NADH+H$^+$, NAD$^+$) ⇄ Acetoin ($CH_3$–CO–CHOH–$CH_3$) ⇄ Butandiol dehydrogenase (NADH+H$^+$, NAD$^+$) ⇄ Butandiol ($CH_3$–CHOH–CHOH–$CH_3$)

**Figure 13.** Metabolism of acetolactate to diacetyl and butandiol by lactic acid bacteria

Lactic acid is a weak acid with a slight odour. It is synthesized from sugars by reduction of pyruvate by a lactate dehydrogenase and is a continuous byproduct of alcoholic fermentation (Genitsariotis, 1979). Only a small amount of carboxylic acid (0.04 to 0.75 g/L) is produced by yeast, and lactic acid constitutes a small fraction of these acids. *Torulopsis pretorians* (Synonym *S. pretoriensis*) which resemble *S. cerevisiae* is the largest producer of lactic acid. The formation of lactic acid is shown in Fig. 14.

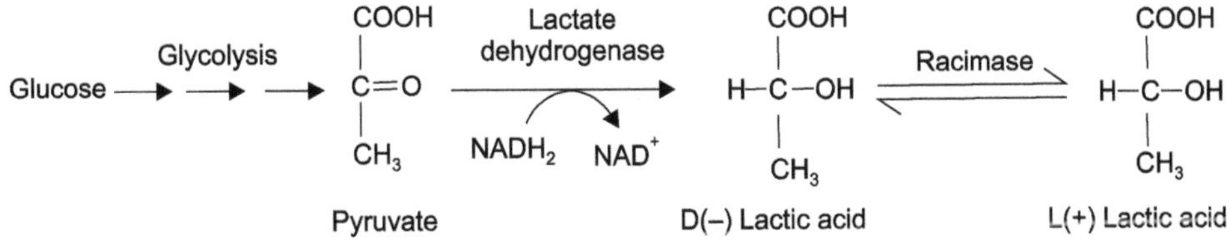

Glucose → Glycolysis → → Pyruvate (COOH–C=O–$CH_3$) → Lactate dehydrogenase ($NADH_2$ → NAD$^+$) → D(–) Lactic acid (COOH–H–C–OH–$CH_3$) ⇄ Racimase ⇄ L(+) Lactic acid (COOH–H–C–OH–$CH_3$)

**Figure 14.** Mechanism of lactic acid formation (Goyal, 1999)

Malolactic fermentation is also carried in wines to reduce the acidity and to produce smoother tasting wines. Malolactic fermentation involves the decarboxylation of L-malic acid to L-lactic acid and carbon dioxide (Walker and Stewart, 2016). Grapes grown in the cool regions naturally contain the high levels of organic acids and thus, the reduction of acidity is beneficial to improve the quality of wines prepared from such grapes.

Several keto/oxo acids like 2-ketoglutarate, 2-ketobutyrate, 2-keto-isovalerate, 2-keto-3-methylvalerate, and 2-keto-isocaproate are produced by yeast as products of amino acid metabolism in relatively small amounts. These keto acids play an important role in the chemical and microbiological stability of wines along with the formation of stable wine pigments. Therefore, grape must, or any fruit must for wine production should also contain low concentrations of pyruvate and 2-oxoglutarate. During fermentation, however, their amount increases and pyruvate is later partially metabolized by the yeast. Both the acids may be formed from the corresponding amino acids, alanine, and glutamate but they are also excreted by yeast cells growing in the presence of low concentrations of nitrogen compounds. Keto acids react with anthocyanins present in fruit must and produce more stable pyranoanthocyanins. They also bind to $SO_2$ and lower the content of free $SO_2$ in wine needed for the safe preservation of wine.

Citric acid is present in trace amounts in must but plays an important role in winemaking. It chelates the metals like iron and copper, reduces haziness and imparts stability to wines. It is usually added to wines which are nearing the end of processing just before bottling. The metabolic pathway of citric acid leads to the production of acetic acid and increases the volatile acidity of the wine. The fermentation of citrate also results in the production of diacetyl and other acetonic compounds, which affect the aroma of the wine. Although, acetic acid is the main volatile acid comprising of about 90 per cent, a variety of other fatty acids like hexanoic acid, octanoic acid, decanoic acid, propionic acid and butyric acid are also present in trace amounts in the wine (Sponholz and Dittrich, 1979). The pathway of the conversion of citrate into acetic acid and other metabolites is shown in Fig. 15.

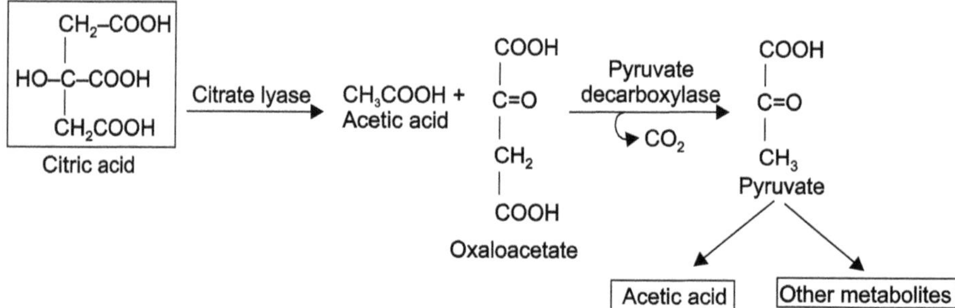


**Figure 15.** Pathway involved in the conversion of citrate to acetate

### *3.1.10. Decomposition of Organic Acids*

Malate metabolism by wine yeasts has been studied and well documented (Peynaud *et al.*, 1964; Rankine, 1966; Shimazu and Watanabe, 1981; Wenzel et al. 1982; Rodriquez and Thornton, 1990; Saayman and Viljoen-Bloom, 2006). The yeast species that have the ability to metabolize extracellular L-malate fall into either the K(+) or K(-) yeast groups. This grouping is based on their ability to utilize L-malate and other TCA cycle intermediates as a sole carbon of energy source (Saayman and Viljoen-Bloom, 2006). The K(+) group yeasts have the ability to utilize TCA cycle intermediates as sole carbon sources and include *Candida sphaerica*, *C. utilis*, *H. anomala*, *P. stipitis*, and *K. marxianus*. The K(-) group includes *S. cerevisiae*, *Schizosaccharomyces pombe*, *S. pombe* var. *malidevorans* and *Zygosaccharomyces bailii*. This group of yeasts is capable of utilizing TCA cycle intermediates only in the presence of glucose or other assimilable carbon sources. The K(-) grouped yeasts also display significant differences in their ability to degrade L-malate. Depending on the yeast strain i.e. *Kloeckera*, *Candida*, *Pichia*, and *Hansenula* only a small amount of malate (3-45%) is metabolized during fermentation. *S. cerevisiae* is regarded as a poor metabolizer of extracellular malate due to the lack of a mediated transport system for the acid (Salmon, 1989) and cause only minor changes to the total acidity of the wine. However, in the presence of glucose or other assimilable carbon sources the strains of *S. pombe, S. malidevorans* and *Z. bailii* can degrade high concentrations of L-malate to ethanol (Rankine, 1966; Coloretti *et al.*, 2002) as shown in Fig. 16.

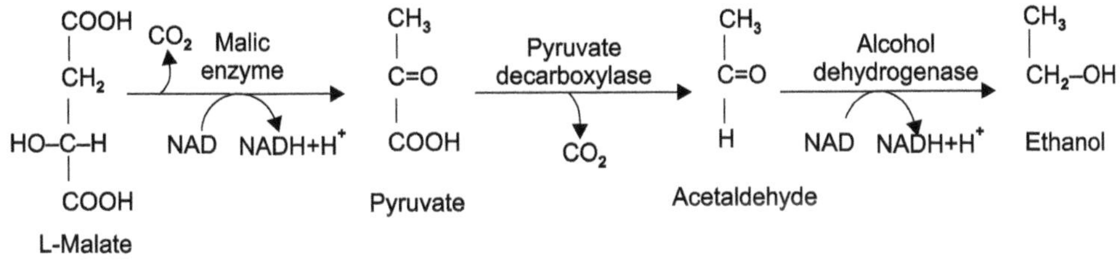

COOH | $CH_2$ | HO–C–H | COOH
L-Malate
$CO_2$ Malic enzyme
NAD NADH+H$^+$
$CH_3$ | C=O | COOH
Pyruvate
Pyruvate decarboxylase
$CO_2$
$CH_3$ | C=O | H
Acetaldehyde
Alcohol dehydrogenase
NAD NADH+H$^+$
$CH_3$ | $CH_2$–OH
Ethanol

**Figure 16.** Pathway showing metabolism of malate to ethanol in *Schizosaccharomyces pombe*

The high malic acid in fruits like plums, apples, and grapes produces highly acidic unpalatable wines (Joshi *et al.*, 1991b). Commercial wine yeast strains of *Saccharomyces* are unable to degrade L-malic acid effectively in must during alcoholic fermentation. Therefore, fermentation of juices from such fruits is conducted by making use of the yeast, *Schizosaccharomyces pombe*, *S. malidevorans* and *Z. bailii* which can degrade the malic acid into ethanol, thus reducing the acidity of the final product. Joshi *et al.*, 1991b studied the effect of varying levels of pH, ethanol, $SO_2$, and nitrogen sources on the deacidification activity of *Schizosaccharomyces pombe* during plum must fermentation. It was concluded that the deacidification activity of the yeast was found to be rapid at pH 3.0-4.5 but was adversely affected at pH 2.5 in the initial stages of fermentation. 150 ppm $SO_2$ was effective in enhancing the activity but the deacidification activity of the yeast was quite susceptible to a higher concentration of ethanol (5-15%). More efficient degradation of malate to lactate has been reported by using a genetically engineered strain of *Saccharomyces cerevisiae* made by using *Schizosaccharomyces pombe* and *Lactococcus lactis* in three days of fermentation in Cabernet Sauvignon and Shiraz grape must at 20° C. *Zygosaccharomyces baili* is yeast that gives strong (40-100%) degradation of L-malic acid (Volschenk *et al.*, 1997).

One of the other methods to reduce the acidity of wine naturally is malolactic fermentation induced by the addition of malolactic starter cultures. As the name suggests, in malolactic fermentation, malate is metabolized to lactic acid and $CO_2$. The biological deacidification of wine results in the direct transformation of malic acid (dicarboxylic acid) into lactic acid (monocarboxylic acid) and $CO_2$ that is catalyzed by the malolactic enzyme, degrading the malic acid in wine (Bartowsky, 2011).This fermentation efficiently decreases the acidic taste of wine, improves the microbial stability and modifies to some extent the organoleptic character of the wine. The phenomenon is of immense significance to the enologists as the malate is a dicarboxylic acid while the lactate has only one carboxylic acid, so this conversion reduces the acidity of wine (Henick-Kling, 1993; Ribereau-Gayon *et al.*, 1998; Versari*et al.*, 1999; Moreno-Arribas and Lonvaud-Funel, 2001).The detailed information on malo-lactic fermentation has been provided in Chapter 12.

### 3.1.11. Metabolic Pathways of LAB

Grape must or wines contain two major families representing three genera of lactic acid bacteria. The *Lactobacillaceae* are represented by the genus *Lactobacillus*, and the *Streptococcaceae* are represented by the genera *Oenococcus* and *Pediococcus*. Lactic acid bacteria utilize simple carbohydrates (hexoses, pentoses) as the main carbon source to form lactic acid either by the homo- or hetero-fermentative pathway. In addition to monosaccharides, these can also metabolize disaccharides such as lactose, maltose, or sucrose, but only amylolytic LAB can degrade starch. The different LAB groups differ in their mechanism of carbohydrate metabolism and the type of final fermentation products produced. The carbohydrate metabolism of different species of the LAB in various types is different and the biochemistry of lactic acid producing fermentation products also vary (Fig. 17 and Fig. 18). The obligate homofermenters, i.e. *L. delbrueckii* and *L. jensenii*, results in the transformation of glucose to pyruvate through EMP or glycolysis. Glycolysis (the EMP pathway) results in almost exclusively lactic acid as the end product under normal conditions, and this type of metabolism is referred to as homolactic fermentation (Neti *et al.*, 2011). The reduction of the pyruvic acid to lactate takes place by the action of lactate dehydrogenases, which needs NAD+/NADP+ as coenzymes. The lactate dehydrogenase enzyme varies in their stereo-specificity and can yield d- or l-lactic acid in the homolactic fermentation. The presence of active form of *lactate racemase* in the microbial cells results in production of racemic mixtures (Burulearu *et al.*, 2010).

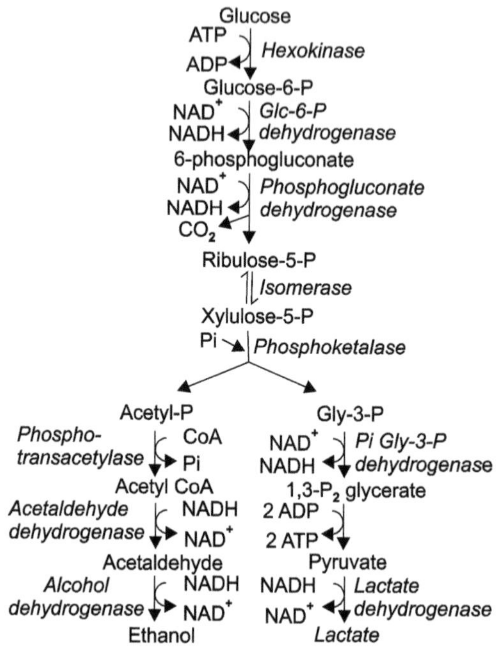


**Figure 17.** Formation of ethanol and lactic acid by hetero-fermentative lactic acid

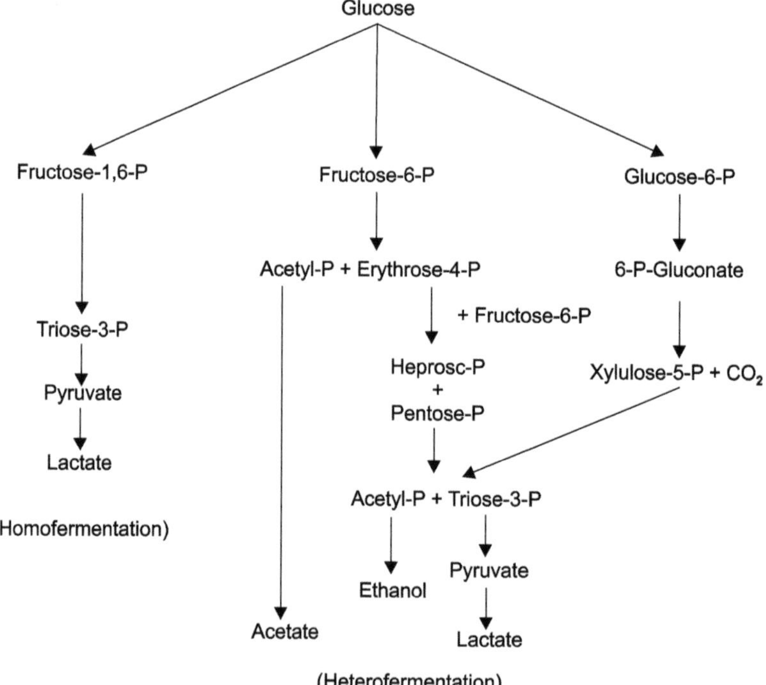


**Figure 18.** Formation of different products in the homolactic and heterolactic pathway (*Source*: Faria-Oliveira *et al.*, 2015)

Obligate hetero-fermentative LAB i.e. *O. oeni*, *L. brevis*, *L. hilgardii*, *L. fructivorans*, and *L. kunkeei* lack aldolase and divert the flow of carbon through pentose phosphate or phosphoketolase pathway and from 1 mole of glucose. Heterofermentative bacteria produce 1 mole each of lactate, $CO_2$, and either acetic acid or ethanol (Axelsson, 2004; Mayo *et al.*, 2010). Through heterolactic fermentation, *Lactobacillus brevis* produces lactic acid, acetic acid, and carbon dioxide while, *Leuconostoc mesenteroides* produces lactic acid, ethanol, and carbon dioxide. The utilization of co-substrates such as oxygen or fructose as electron acceptors by obligate hetero-fermentative *Lactobacilli* is coupled with increased production of acetate. The addition of hydrogen phosphate as a nitrogen source in facultative and obligately hexose fermentation spares the nitrogenous compounds of the must, restricts the production of higher alcohols, and thus, the quality of alcoholic beverage remains desirable. Formation of different products from pyruvate in lactic acid bacteria is depicted in Fig. 19.

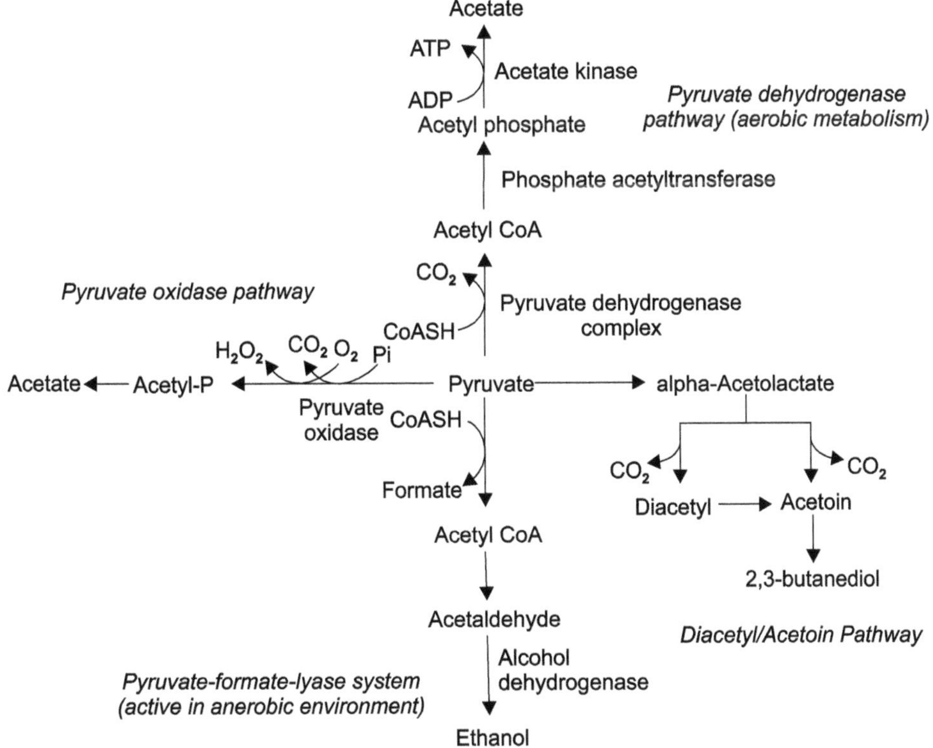


**Figure 19.** Formation of different products in lactic acid fermentation
(*Source*: Adopted and modified from Caplice and Fitzgerald, 1999)

## 3.2. Nitrogen Metabolism

An exogenous source of nitrogen is an important requirement for the growth and metabolism of yeast; mainly required for the synthesis of proteins and nucleic acid. Hence, nitrogenous compounds found in grape must are important nutrient substances and their availability play a key role in winemaking. Wine production has both positive and negative implications of nitrogen. Nitrogen deficiency or low nitrogen concentration in grape must/juice has been shown to result in slow and incomplete fermentation leading to problems like fermentation arrests and hydrogen sulfide ($H_2S$) production (Mendes-Ferreira *et al.*, 2011). However, nitrogen is also responsible for the production of reduced sulfur compounds (Henschke and Jiranek, 1991; Jiranek and Henschke, 1991) and the formation of ethyl carbamate. Because of the main role of nitrogen in the biosynthesis of volatile and non-volatile compounds, it plays an important role in enhancing wine aroma and eliminates sluggish fermentation (Navarro and Navarro, 2011). An overview of the complete metabolism of nitrogen is shown in Fig. 20.

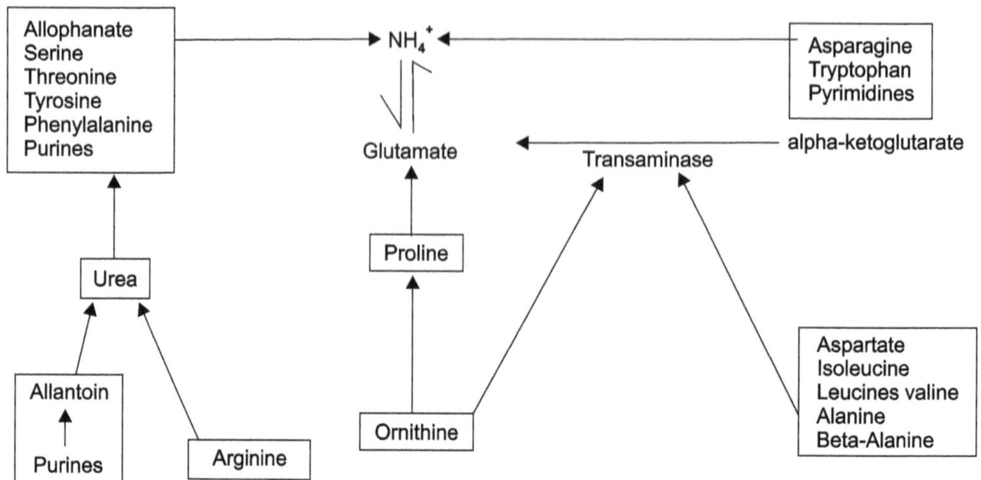


**Figure 20.** Major products of degradation of nitrogenous compound

*Saccharomyces cerevisiae* can grow on a diverse range of nitrogenous compounds including ammonium, urea, amino acids, small peptides, purines and pyrimidine-based compounds (Cooper, 1982; Large, 1986; Albers *et al.*, 1996). Generally, the nitrogenous content of grape juice ranges form 60 to 2400 mg/L and is mainly composed of amino acids, ammonium, peptides and proteins (Mendes-Ferreira *et al.*, 2011). All these nitrogeneos compounds are usually present in must in varing proportions and need to be transported into the cell before utilization. The effectiveness of the utilization depends on the expression, regulation, and efficiency of the transport system as well as regulation and energetics of subsequent catabolic and anabolic processes. Therefore, quantity and nature of the nitrogen sources available determines the growth, fermentation rate, and yield of yeast biomass.

### *3.2.1. Nitrogen Sources and Nitrogen Supplements*

Fruit must contain full complements of nutrients in the form of carbon and nitrogen. The nitrogenous compounds in the grapes are present in the form of amino acids, ammonium, peptides, nucleotides, proteins, and vitamins. The concentration of these compounds in fruits ranges from 60 to 2400 mg/L as stated earlier while the amino acids concentration in grape-musts is highly variable and found to range from 300 to 1600 mg/L. The major amino acids present in the grape juice are proline, arginine, glutamate, glutamine, serine, and threonine with proline and arginine, accounting for 30-65% of the total amino acid content (Mendes-Ferreira *et al.*, 2011). The sum of the amino acid and ammonium concentrations available in the grape juice at the start of fermentation, excluding proline, is commonly referred to as yeast assimilable nitrogen (YAN). Generally, a preferred yeast nitrogen source is one which is most readily converted biosynthetically into useful nitrogen compounds such as ammonia, glutamate or one that requires least energy input or cofactors for mobilization of nitrogen moiety and more importantly should not have toxic effect and be economically feasible (Amerine *et al.*,1980; Joshi,1997). *S. cerevisiae* can grow on ammonium, urea and most of the amino acids and does not have any absolute requirement of amino acids. The YAN content has been reported to exert a remarkable impact on yeast biomass yield and fermentation rate. A YAN content of 400 mg/L or more is reported to yield maximum yeast biomass and fermentation rate, while, a YAN content of below 150 mg/L is reported to cause stuck fermentation (Ugliano *et al.*, 2007). Therefore, it is recommended to supplement low YAN must with assimilable nitrogen to avoid the problems associated with nitrogen deficiency during fermentation. Diammonium phosphate (DAP) is mostly used as a YAN supplement and added to the must at 200 mg/L prior to inoculation with yeast, and further addition of 100 mg/L is done in response to the production of hydrogen sulfide (Henschke and Ough, 1991). DAP contains 21% N, therefore, for convenience, we can consider 100 mg DAP provides 20 mg of YAN. Nowadays, commercial preparations of organic nitrogen sources that contain inactivated yeast or yeast products, which also contain lipids and sometimes other nutrients, have become commercially available. Urea and other commercial yeast foods are other alternate nitrogen supplements (Ingledew and Kunkee, 1985). However, the addition of urea is prohibited in most wine producing countries due to its involvement in the production of ethyl carbamate (Ingledew *et al.*, 1987; Ough *et al.*, 1988; Monteiro *et al.*, 1989), as discussed in the latter section of the chapter.

### 3.2.2. Amino Acid Utilization Profile

Amino acids act as a source of nitrogen for yeast during fermentation and have a direct influence on the aromatic composition of wines (Drdak *et al.*, 1993). Amino acids can be present naturally in the fruits, may be excreted by some of the yeasts at the end of fermentation, or can be released by the autolysis of dead yeasts or produced by the enzymatic degradation of the proteins. The amino acids indigenous to fruits can be partially or totally metabolized by the yeasts during the growth phase. The initial distribution of nitrogeneos compounds in the medium but the preferences of the yeast strain determine the pattern of amino acid utilization during fermentation. In a medium with mixture of amino acids, the most important source of nitrogen for yeasts was arginine, providing 30 to 50% of the total nitrogen followed by lysine, serine, threonine, leucine, aspartate, and glutamate, while glycine, tyrosine, tryptophan, and alanine were the least utilized. This pattern of amino acid utilization was found to be similar irrespective of yeast strain, the degree of aeration and sugar and ammonium concentration of the medium (Jiranek *et al.*, 1990). Thomas and Ingledew (1990) conducted a study on the effect of amino acids on the fermentation and growth of yeast cells in very-high-gravity or very high sugar concentration conditions. It was found that the mixtures of amino acids (as in yeast extract or casamino acids) or single amino acid (such as glutamic acid) stimulated growth and decreased the fermentation time. In an another study, Radler and Schutz (1982), studied the effect of nature and the amount of the nitrogen source on anaerobic glycerol formation and found that some single amino acids (alanine, asparagine, serine, and valine) caused a decreased level of glycerol formation compared with that when a mixture of amino acids was used. Albers *et al.*, 1996 conducted a study on anaerobic growth of *S. cerevisiae* with three different nitrogen sources, i.e. ammonium salt, glutamic acid, and a mixture of amino acids, with 20 g of glucose per liter as the carbon and energy source. It was found that a higher ethanol yield for growth could be obtained on both glutamic acid (by 9%) and the mixture of amino acids (by 14%). Glycerol yields were also low with glutamic acid (0.17 mol/mol of glucose) or with the mixture of amino acids (0.10 mol/mol) as compared to ammonium-grown cultures (0.21 mol/mol). Glutamic acid increased the production of α-ketoglutaric acid, succinic acid, and acetic acid as compared to other nitrogen sources. Cultures grown on amino acids had a higher specific growth rate (0.52 $h^{-1}$) than the cultures of both ammonium-grown (0.45 $h^{-1}$) and glutamic acid-grown (0.33 $h^{-1}$) cell (Table 3).

### 3.2.3. Uptake and Transport of Nitrogen Compounds

In the fermentation of wine, nitrogenous compounds having low concentration in fruit juices are taken up very quickly. However, before degradation of compounds for a nitrogen source, the biosynthetic pool of amino acid is filled before yeast growth. Once this is achieved and the growth has started, various nitrogen compounds are taken up and degraded in a specific order of preference by a mechanism called as nitrogen catabolite repression (NCR) (Beltran *et al.*, 2004). Most preferred nitrogen sources which are utilized directly in biosynthesis are ammonium ion, glutamate, and glutamines. These nitrogen sources are depleted first from the medium before utilization of other nitrogen sources. The transport of nitrogen compound into the cell is the major factor controlling their utilization. There are three basic means of transport viz., simple diffusion, facilitated diffusion, and active transport. Comprehensive reviews on the amino acid transport mechanisms of *S. cerevisiae* have been published (Horak and Kotyk, 1977; Eddy, 1982; Wiame *et al.*, 1985; Casey and Ingledew, 1986; Henschke and Rose, 1991; Mendes-Ferreira *et al.*, 2011). The ammonium ion is transported inside the cell by facilitated diffusion. The cellular transport of ammonium in *S. cerevisiae* involves three permeases; Mep1p, Mep2p, and Mep3p, which are expressed under low ammonium concentrations and regulated by nitrogen catabolite repression. Mep2p displays the highest affinity for $NH^+_4$ ($K_m$ 1.4–2.1 μM), followed closely by Mep1p ($K_m$ 5–10 μM) and finally by Mep3p, whose affinity is much lower ($K_m$ 1.4–2.1 mM) (Beltran *et al.*, 2004).

**Table 3.** Growth Rates, Nitrogen Uptake, Product Yields, $Y_{ATP}$, and Balances for Anaerobic Growth of *S. cerevisiae* on Glucose with Three Different Nitrogen Sources

| *Parameter and unit* | *Value with the following nitrogen source* | | |
|---|---|---|---|
| | *Ammonium* | *Glutamic acid* | *Mixture of amino acids* |
| **Specific growth rate** ($h^{-1}$) | | | |
| From $OD_{610}$ | 0.45±0.04[b] | 0.33±0.01[b] | 0.52±0.04[b] |
| From heat[c] | 0.41±0.05 | 0.35±0.10 | 0.51±0.05 |
| **Nitrogen uptake** | | | |
| N-mol/mol of glucose | 0.12±0.04 | 0.13±0.04 | 0.18±0.02 |
| N-mmol/g (dry wt) of biomass | 7.3±2.8 | 7.7±2.2 | 7.6±0.9 |
| **Biomass** (C-mol/mol of glucose) | 0.61±0.05 | 0.62±0.03[d] | 0.92±0.06[d] |
| **Carbon dioxide** | | | |
| mol/mol of glucose | 1.48±0.05 | 1.53±0.06 | 1.75±0.09 |
| mmol/g (dry wt) of biomass | 89±2 | 94±3 | 70±8 |
| **Heat** | | | |
| kJ/mol of glucose | 108±4 | 98±7 | 112±14 |
| kJ/g (dry wt) | 6.5±0.5 | 6.0±0.4 | 4.5±0.5 |
| **Ethanol** | | | |
| mol/mol of glucose | 1.43±0.03 | 1.54±0.11 | 1.63±0.02 |
| mmol/g (dry wt) of biomass | 88±5 | 94±6 | 66±5 |
| **Glycerol** | | | |
| mol/mol of glucose | 0.21±0.02 | 0.17±0.02 | 0.10±0.01 |
| mmol/g (dry wt) of biomass | 13±2 | 10±2 | 4±1 |
| **Acetic acid** (mmol/mol of glucose) | 9±4 | 25±6 | 14±2 |
| **Pyruvic acid** (mmol/mol of glucose) | 5±1 | 5±3 | 3±1 |
| **Succinic acid** (mmol/mol of glucose) | 3±2 | 22±7 | 5±1 |
| **a-Ketoglutaric acid** (mmol/mol of glucose) | 0.5±0.0 | 64±12 | 1±0 |
| **Fumaric acid** (mmol/mol of glucose) | <0.08 | 0.3±0.1 | <0.08 |
| **Carbon recovery** (%) | 93.5±3.1 | 94.1±4.1 | 95.2±0.8 |
| **Energy recovery** (%) | 96.0±2.9 | 88.9±6.3 | 95.3±1.6 |
| $Y_{ATP}$ (g [dry wt] of biomass/mol of ATP) | 13.4±0.7 | 11.9±0.9 | 16.3±1.3 |

[a] Except as noted, all values are expressed as means dried from three independent growth experiments and maximal deviations. One C-mol is the amount of an (organic) compound containing 1 mol (i.e., 12 g) of carbon. $OD_{610}$ optical density at 610 nm.

[b] Mean and deviation from two experiments.

[c] Heat production rate.

[d] Biomass yields corrected for the carbon taken up directly as amino acids became 0.51 C-mol/mol of glucose when either glutamic acid or the mixture of amino acids was used as a nitrogen source. The corrections were made as follows: For glutamic acid all amino acids belonging to the glutamate family were considered to be derived directly from glutamic acid, and the weight of these amino acids (excluding the weight of water released in the polymerization reactions and the weight of nitrogen) was subtracted from the weight of the biomass. For the cultivations with amino acids, the weight of the amino acids taken up (excluding the weight of water released in the polymerization reactions and the weight of nitrogen) was subtracted from the weight of the biomass.

*Source*: Albers *et al.* (1996)

The next group of nitrogen compounds regarding preference includes alanine, serine, threonine, aspartate, asparagine, urea, arginine, proline, glycine, lysine, histidine, and pyrimidine. This order of preference however, may change depending upon environmental, physiological and strain-specific factors. In *S. cerevisiae*, the transport of amino acids across the plasma membrane occurs *via* a proton symport mechanism through a number of more-or-less specific amino acid permeases. The ability of the cell to excrete proton by ATPase hydrogen ion pump, which uses energy from the hydrolysis of one ATP molecule for each hydrogen ion pumped out of the cell, is an important regulatory factor for amino acid uptake (Roon *et al.*, 1975; Roon *et al.*, 1977). In addition to specific amino acid permeases, both D and L isomers of neutral and basic amino acid and proline to a limited extent are transported by group-specific transport system called as general amino acids permease (GAP). This is responsible for the uptake of all naturally occurring L-amino acids and related compounds, such as ornithine and citrulline, some D-amino acids, toxic amino acid analogs, and azetidine-2-carboxylate, as well as for the polyamines putrescine and spermidine. The GAP appears to operate as a nitrogen scavenger system. During the initial stage of fermentation, permeases are specific for a small number of amino acids, but GAP is the principal transport system for amino acids during the later stages of fermentation (Rose, 1987). Gap1permease functions as a transporter as well as amino acid sensor and hence, play an important role in wine fermentation. Other amino acid permeases with low specificity are the low-affinity amino acid permease Agp1p that accept asparagine, glutamine and other amino acids, and the high-affinity glutamine permease Gnp1p, which accepts leucine, serine, threonine, cysteine, methionine, glutamine and asparagine (Mendes-Ferreira *et al.*, 2011). Most of the strains of *Saccharomyces* utilizes thiamine directly as a biosynthetic precursor and not as a source of nitrogen. MUP genes aid in the uptake of methionine, Tat2p transports tryptophan and tyrosine, and Put4p is required for the transport of proline. Proline can be assimilated by *S. cerevisiae* but only under aerobic conditions. Can1p, Lyplp, and Alp1p are all specific for the cationic amino acids, lysine, and arginine.

The transportation of urea in the yeast is carried out by two mechanisms. Firstly, an active transport with an apparent Km of 14 mM having sensitivity to nitrogen repression (Colowick, 1973). The second mechanism is a passive or facilitated diffusion operating at external urea concentrations of greater than 0.5 mM (Cooper and Sumrada, 1975; Cooper, 1982). The protein and peptide utilization in yeast is dependent upon its ability to either transport these compounds or degrade them extracellularly. The transportation of di- and tri-peptidesis carried out using a general peptide transport system, but the utilization of larger peptides or protein depends on the capability of yeast. However, neither of these mechanisms of protein utilization has been reported in *S. cerevisiae* (Rosi *et al.*, 1987; Sturley and Young, 1988; Lagace *et al.*, 1990). Moreover, the importance of peptides to yeast metabolism is yet to be clarified.

### *3.2.4. Regulation of Nitrogen Transport*

The regulation of transport systems is done by several mechanisms varing in their specificity, response time and level of influence. The main processes involved in regulation of transport mechanisms are feedback inhibition, trans-inhibition, inactivation/reactivation and repression. Feedback inhibition is the phenomenon where the accumulated substrate inhibits further accumulation by the same transport carrier. The process is evident by the progressive decrease of substrate accumulation rate (Rana and Rana, 2011). This mechanism for regulation of transportation is of least significance when the internal substrate concentrations are kept low by catabolism or compartmentation into subcellular organelles.

Trans-inhibition refers to the competition for uptake exhibited between the substrates which do not share a common transport mechanism (Goyal, 1999). One such example is the inhibition of accumulation of allantoin by asparagine, structurally and metabolically unrelated nitrogen compounds (Cooper, 1982). Several ammonium sensitive permeases of *S. cerevisiae*, including the general amino acid permease (GAP), utilizes inactivation-reactivation processes (Grenson and Acheroy, 1982; Grenson, 1983a, b). When the yeast is grown on a poor nitrogen source such as proline, the permease is synthesized *via* the expression of the structural gene, GAP. Both ammonium-sensitive permeases and degradative enzymes can also be regulated by nitrogen catabolite repression (NCR). The regulation by repression may be acting at the level of transcription, translation or processing of RNA or precursor protein.

### *3.2.5. Factors Affecting Nitrogen Accumulation*

The composition of nitrogen-containing compounds in grape juice depends on various factors like variety of grape, time of harvest and vineyard management factors like nitrogen fertilization, berry maturation, vine water status, soil type, and fungal infection. Further, the accumulation of nitrogen compounds in grape must depend on the cultural conditions, medium composition, and yeast strain factors. Different nitrogen-containing compounds, i.e. amino acids, urea, ammonium, nitrogen bases, and purine derivatives can be used by yeast as a source of nitrogen. The general aspect of nitrogen metabolism in yeasts have also been reviewed earlier (Jones and Fink, 1982; Cooper, 1982; Cartwright *et al.*, 1986; Davis, 1986; Large, 1986; Nykanen, 1986; Pierce, 1987; Hinnesbusch, 1988; Cartwright *et al.*, 1989; Bisson, 1991; Schwenke, 1991). These nitrogenous compounds enter cells *via* permeases and can be immediately used in the biosynthesis of new compounds or converted to a related compound or catabolized to release nitrogen in the form of ammonium (*via* deamination), glutamate (*via* transamination), or both (Magasanik and Kaiser, 2002). The amino acid accumulated may be utilized as one of the followings:

- An amino acid is directly incorporated into the protein.
- An amino acid can be degraded to liberate nitrogen and can be used as a source of nitrogen for the biosynthesis of other nitrogen cell constituents.
- The carbon component of an amino acid is released and used for the biosynthesis of other cell carbon constituents.

Fundamentally, all the nitrogen compounds accumulated are degraded to either of the products, ammonium, glutamate or glutamine. The nitrogen catabolic pathways of yeasts have been summarized (Large, 1986). The reactions which convert ammonium and glutamine into glutamate are referred to as the central nitrogen metabolism (CNM). Ammonium and glutamate are the two important nitrogen compounds required by the cell as from these two compounds all other nitrogen-containing compounds in the cell are produced. These end products of nitrogen metabolism, however, are interconvertible. Generally, depending upon the available nitrogen source, two types of glutamate dehydrogenases (GDH) are expressed and these two types are rarely expressed equally (Cooper, 1982). When cells have ammonium ions as the sole source of nitrogen, the $NADP^+$ glutamate dehydrogenase, encoded by GDH1, is responsible for the synthesis of glutamate by combining ammonium with α-ketoglutarate. The glutamate thus, synthesized can combine with ammonium in a reaction catalyzed by glutamine synthetase (GS). When glutamate is the sole nitrogen source, the release of ammonia for the synthesis of glutamine from glutamate is carried out by $NAD^+$ glutamate dehydrogenase, encoded by GDH2. Glutamate synthetase (GOGAT), the product of GLT1, catalyzes the synthesis of glutamate from glutamine. NAD-GDH is expressed when glutamate, aspartate or alanine are the sole sources of nitrogen. Lysine, histidine, cysteine and glycine are considered as a good source of nitrogen for many type of yeast, but none of these amino acids is utilized efficiently by *Saccharomyces* as a nitrogen source. Different amino acids or nitrogen compounds catabolize to yield glutamate or $NH^{4+}$ used for various biosynthetic processes. Glutamine is a preferred nitrogen source as it generates both glutamate and $NH^{4+}$ hence, utilized first followed by alanine, serine, threonine, aspartate, asparagine, urea, and arginine. Under aerobic conditions, proline is the nitrogen source of choice. Most of the yeast strains do not utilize glycine, lysine, histidine, and pyrimidine as a source of nitrogen, however, these can be readily and directly utilized as biosynthetic precursors. Aromatic amino acids are not metabolized during fermentation as the process is complex and require oxygen and other cofactors. Arginine has the ability to support high growth rates as it is rich in nitrogen and for transportation requires only one proton making it economical to transport. Furthermore, urea excreted during arginine metabolism is reabsorbed and degraded further when the concentration of assimilable nitrogen becomes low (Henschke and Ough, 1991; Monteiro and Bisson, 1991). Sulfur-containing amino acids, like cysteine and methionine are synthesized by incorporating reduced sulfur into carbon compounds.

### *3.2.6. Enological Significance of Nitrogen Metabolism*

Nitrogen metabolism promotes yeast growth essentially, by supplying the precursors of protein and nucleic acid synthesis. It is a rich supply of nitrogen which allows high rates of growth, biomass yield and stimulate fermentation activity. The completion of wine fermentation and development of flavours is also dependent upon the availability and type of nitrogenous compounds present in the must. The deficiency of nitrogenous compounds leads to slow or incomplete fermentation. The presence of sulfur-containing amino acids like cysteine and methionine leads to the formation of sulfur compounds like hydrogen sulfide and methanethiol that results in a serious fermentation problem encountered in winemaking. Considerable research work has been done to tackle this problem (Agenbach, 1977; Eschenbruch *et al.*, 1978). Fermentation rate is directly affected by availability of nitrogen as it is utilized for synthesizing proteins for the glycolytic pathway and $NH^{4+}$ ions serve as an allosteric effector for phosphofructokinase activity, the main regulatory enzyme of the glycolytic pathway. Hence, nitrogen limitation causes decrease in fermentation rate, resulting in a sluggish or incomplete fermentation (Salmon, 1989; Navarro and Navarro, 2011). Elemental sulfur used as vineyard agrochemical or inorganic sulfur used as an antioxidant and antimicrobial compound acts as the main source for the formation of hydrogen sulfide in wine. Hydrogen sulfide ($H_2S$), has a low sensory threshold of (10-100 mg/L) and gives a positive yeast-like smell at low concentrations of 20-30 $mg^{-1}$ but at high concentrations, it imparts an unpleasant aroma of rotten eggs. The amount of sulfur containing amino acids and vitamins and fermentation conditions along with the genetic background of the yeast determines the amount of hydrogen sulfide released during winemaking (Spiropoulos *et al.*, 2000; Mendes-Ferreira *et al.*, 2011). Nitrogen metabolism is also one of the several mechanisms accounting for its formation (Monk, 1986; Henschke and Jiranek, 1991). In the process of winemaking, there are two broad phases of $H_2S$ production. The first phase occurs during active fermentation and is responsive to supplementation with assimilable nitrogen. On the other hand, the second phase is most frequently observed near the depletion of sugars from the must and proceeds in the presence of assimilable nitrogen. The total amount of $H_2S$ produced in the fermentation of grape juice is inversely related to the levels of assimilable amino acid (AAAs) present. In the limited availability of nitrogen, depending upon the amino acid, composition of the medium, cysteine is accumulated by various specific and general permeases (Yct1p, Gap1p, Mup1p). The enhanced wine flavour and urea formation is linked to higher concentrations of assimilable nitrogen and particular amino acid, but the extent to which the nitrogen availability affects the composition and sensory aspects of wine, is not clear.

## 3.3. Metabolism of Sulphur: Chemistry of Production of Off-Flavour

Yeasts require sulfur for growth due to its role in sulfur-containing amino acids, peptides, and proteins. The assimilable sulfur can be obtained from both the organic and inorganic sources like amino acids, peptides, elemental sulfur, sulfate and sulfite. A low level of organic sulfur compounds like cysteine and methionine are present in grapes and sulfate act as the major source of sulfur in winemaking (Waterhouse *et al.*, 2016). In the metabolism of sulfur, the formation of volatile sulfur compounds (VSC) takes place which results in the production of off-flavours. The production of off-flavours is the main problem in the production of quality wine and the main components responsible for off-flavours are acetic acids, sulfur-containing volatiles, free amino nitrogen, vitamins, and other factors. Volatile sulfur compounds in wine are categorized on the basis of their boiling point as highly volatile sulfur compounds (BP 90 °C) and less volatile sulfur compounds (BP more than 90 °C). The major high volatile sulfur compounds are hydrogen sulfide, ethanethiol and methanethiol, and their higher concentration above the threshold limits results in off-smell like rotten eggs, garlic, onion, and cabbage. Other highly volatile sulfur compounds of less significance are carbonyl sulfide and dimethyl sulfide. The less volatile sulfur compounds, 3-(methylthio)-1-propanol (methionol) are present in wines at concentrations up to 5 mg/L. The concentrations above the threshold value (1.2 mg/L), contributes a cauliflower aroma to the wine. Other low volatile sulfur compounds in wine are 2-mercaptoethanol (poultry-like aroma), 2-methyl tetrahydrothiophene-3-one (metallic, natural gas odour), 2-methylthioethanol (French bean), ethyl-3-methylthiopropionate (metallic, sulphur aroma), acetic acid-3-(methylthio) propyl ester (cooked potatoes) and 4-methylthiobutanol (chive-garlic aroma); and these are usually found in concentrations lower than threshold limit (Moreira *et al.*, 2002). The mechanism of the formation of volatile sulfur compounds (VSC) during wine fermentation is only partially understood due to the complex nature of the factors involved in winemaking. The composition of VSCs produced varies widely in model systems with nutrient deficiencies or with residual elemental sulfur. The VSCs are spoilage causing compounds which are very volatile and have unpleasant odours, generally described as odour of rotten egg, skunk aroma, garlic or onion etc. These compounds are present in a very minute concentration (10-100 mg/L) but pose a significant problem in winemaking due to their huge sensory impact.

Among the volatile sulfur-containing compounds, hydrogen sulfide ($H_2S$) creates the major problem with rotten egg-like flavour. During fermentation, $H_2S$ production may be due to several factors like the presence of elemental sulfur in grape skin (Rankine, 1963; Acree *et al.*, 1972; Schutz and Kunkee, 1977; Wenzel and Dittrich, 1978; Thomas *et al.*, 1993), inadequate levels of free α-amino nitrogen (FAN) in the must (Vos and Gray, 1979; Monk, 1986), a deficiency of pantothenic acid (Tanner, 1969) or pyridoxine (Vidal-Carou *et al.*, 1991) or high levels of cysteine in the juice and yeast strains (Acree *et al.*, 1972; Suzzi *et al.*, 1985; Radler, 1993). The biosynthesis of S-containing amino acids requires reduction of sulfur to sulfide ($S^{2-}$). Yeast employs the sulfate reduction sequence (SRS) pathway to form sulfide from sulfate, which uses several enzymes for the uptake and activation of sulfate, followed by its reduction to sulfite and then sulfide. After the reduction sequence, the incorporation of sulfide into cysteine and methionine *via* coupling with O-acetylhomoserine (O-AHS) takes place through additional enzymatic steps. While sulfite ($SO_3^{2-}$) is produced as an intermediate in the SRS pathway, extracellular sulfite (i.e. from the addition of bisulfite during winemaking) can also be utilized after diffusing into the cell. The odour threshold value of $H_2S$ in wine varies from 50 to 80 mg/L, however, at a concentration of 240mg/L and above it causes off-flavours. Sulfate uptake and reduction are regulated by sulfur-containing amino acid, i.e., methionine. A study was conducted on eight grape juices having different levels of assimilable amino acids and the total production of $H_2S$ was noted. In general, the highest concentration of $H_2S$ was produced during the rapid phase of fermentation. The $H_2S$ production at the later stages of fermentation may be due to deficiency of assimilable amino acids (Park *et al.*, 2000).

The unavailability or deficiency of free amino nitrogen in the must also cause $H_2S$ formation by yeast (Monk, 1986; Henschke and Jiranek, 1991; Boudreau *et al.*, 2018). Reduction in $H_2S$ formation has been reported in juices from 100 mg/L at 100 mg/L free amino nitrogen (FAN) to nearly zero at 300-400 mg/L FAN (Vos and Gray, 1979). Low concentration of FAN might also stimulate proteases, which cause hydrolysis of juice proteins. However, in a study conducted by Bohlscheid *et al.* (2011), a non-significant ($P>0.05$) effect of yeast assimilable nitrogen (YAN) on $H_2S$ production was found in synthetic apple juice using two strains of *Saccharomyces cerevisiae* (UCD 522 and EC1118). The deficiency of vitamins was found to increase the $H_2S$ production significantly, with maximal cumulative $H_2S$ production in media with 60 mg/L YAN, 10 µg/L biotin, and 50 µg/L pantothenic acid, while minimum production was observed with 250 mg/L YAN and 250 µg/L pantothenate for strain UCD 522 .The strain EC1118, produced the most $H_2S$ with 250 mg/L YAN, 0.5 µg/L biotin, and 50 µg/L pantothenic acid and the least in media that contained 250 mg/L YAN and 250 µg/L pantothenic acid. A deficiency of pantothenate leads to the deficiency of CoA, which is essential for methionine biosynthesis. Pantothenate deficiencies may cause low levels of methionine production in the cells, which results in $H_2S$ formation. Pyridoxin is necessary for several reactions in the pathway of methionine synthesis and a large amount of $H_2S$ is produced if the concentration is below 2 mg/L. Thiamine also serves as a coenzyme in sulfite and sulfate reductions, which affects hydrogen sulfide formation. In addition to these, the combined treatment of Cu(2+) and $SO_2$ significantly increased $H_2S$ formation in Verdelho wines samples that were not previously treated with either Cu(2+) or $SO_2$ (Bekker *et al.*, 2016). Normally, most of the $H_2S$ formed during fermentation is carried away with carbon dioxide. In most of the cases, $H_2S$ disappears after fermentation, therefore, is not detectable despite a detection limit of 1.5 mg/L. Increased amounts of $H_2S$ can be purged out or oxidized by aeration or oxidized by treatment with sulfite (Tanner, 1969).

Along with $H_2S$ several other sulfur compounds such as dimethyl sulphide, diethyl sulphide, mercaptans and thioester like S-methyl thioacetate are also detected which are responsible for the off-flavour. Dimethylsulfide (found in concentration ranging from 0 to 47 mg/L) is synthesized from S-methyl methionine (Loubser and Du Plessis, 1976) and diethylsulfide has a flavour of cooked vegetables, onion and garlic (Fuck and Radler, 1972; Januik, 1984; Goniak and Noble, 1987). Mercaptans having rotten egg-like flavour (Fuck and Radler, 1972) are formed during fermentation in synthetic media containing cysteine, methionine or sulfate. The factors leading to the production of these sulphur compounds have not been thoroughly elucidated. As discussed earlier, the amino acid composition of the juices results in the production of sulphur compounds that give off-character to the wine, hence, metabolism of sulfur-containing amino acids may be responsible for the synthesis of compounds giving off flavours. The effect of added threonine and methionine on $H_2S$ production has also been documented (Wainwright, 1971).

A considerable emphasis has been placed on reducing the undesirable hydrogen sulfide ($H_2S$) and other volatile sulphur compounds associated with off-flavours of wine (Rauhut et al, 1996; Park *et al.*, 2000; Spiropoulos *et al.*, 2000) and to a lesser extent on diacetyl and other related carbonyl compounds

(Martineauand Henick-Kling, 1995; Romanoand Suzzi, 1996). The elemental sulfur can be controlled by monitoring the vineyard sprays. In addition to this, settling of must with bentonite and yeast hulls can improve the elimination of pesticides during settling. The level of total solids should also be controlled, as a low level of solids results in nitrogen deficiency, while high levels can create a reductive environment, increasing the risk of sulfur off-aroma production. The nitrogen content also affects the off-flavour production, and it is important to measure YAN in must, as both high and low level of nitrogen can cause yeast to produce $H_2S$ (Mendes-Ferreira *et al.*, 2011). The selection of low sulfur producing yeast strains can further help to control $H_2S$ production. A supply of 8-10 mg/L of oxygen during the growth phase of yeast aids in the creation of sterols, strengthen cell wall and provide the ability to tolerate harsh environmental conditions and prevent the production of off-flavours (Ugliano *et al.*, 2009).

High molecular weight sulfur compounds such as N-(3-methylthiopropyl) acetamide and 3-methylthiopropionic acid, which are formed as by-products of metabolism of the amino acids: cysteine, methionine and homomethionine by yeast (Anocibar-Beloqui *et al.*, 1995; Lavigne and Dubourdieu, 1996) also needs attention of the researchers.

## 3.4. Acetic Acid Fermentation

Acetic acid bacteria are considered of little importance in winemaking due to anaerobic conditions, but its ability to survive and multiply in anaerobic or semi-anaerobic conditions has been evidenced in many studies (Drysdale and Fleet, 1988; Du Toit and Pretorious, 2002). The acetic acid bacteria are divided into the genera *Acetobacter*, *Gluconobacter*, *Acidomonas,* and *Gluconacetobacter*. Of these, *Gluconobacter oxygen*, *Acetobacter aceti*, *Acetobacter pasteurianus*, *Gluconacetobacter liquefaciens,* and *Gluconacetobacter hansenii* are normally associated with grapes and wine (Drysdale and Fleet, 1988). Formation of acetic acid during alcoholic fermentation may be in small amounts, takes place by the pathway shown in Fig. 21.

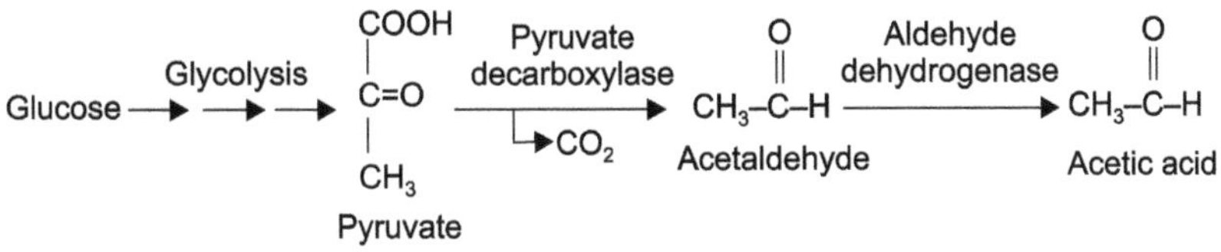

$$\text{Glucose} \xrightarrow{\text{Glycolysis}} \underset{\text{Pyruvate}}{\text{COOH–C(=O)–CH}_3} \xrightarrow[\rightarrow CO_2]{\text{Pyruvate decarboxylase}} \underset{\text{Acetaldehyde}}{CH_3\text{–C(=O)–H}} \xrightarrow{\text{Aldehyde dehydrogenase}} \underset{\text{Acetic acid}}{CH_3\text{–C(=O)–H}}$$

**Figure 21.** Mechanism of acetic acid synthesis

However, in *Acetobactor* and *Gluconobactor*, acetic acid is the major product and the process is called acetic acid fermentation. The first step in the conversion of ethanol to acetic acid is the formation of acetaldehyde according to the equation:

$$\underset{\text{Ethanol}}{CH_3CH_2OH} + O_2 \xrightarrow{\text{Alcohol dehydrogenase}} \underset{\text{Acetaldehyde}}{CH_2CHO} + H_2O$$

The second step is the formation of acetic acid from acetaldehyde. The latter first reacts with water to yield hydrated acetaldehyde, which in turn is oxidized or dehydrogenated to yield acetic acid.

$$\underset{\text{Acetaldehyde}}{CH_3CHO} + H_2O \longrightarrow H_3C\text{–C(H)(H)–OH} + [O] \xrightarrow{\text{Aldehyde dehydrogenase}} \underset{\text{Acetic acid}}{CH_3COOH} + H_2O$$

These bacteria can tolerate 14 to 15% of alcohol concentration (v/v). An alcohol concentration of 15.5% (v/v) is recommended for sherry stock to inhibit the growth of acetic acid bacteria (Drysdale and Fleet, 1988). A temperature of 25-30° C and pH of 5.5 to 6.3 is optimum for the growth of acetic acid bacteria, but it can survive at low pH values of 3.0 to 4.0 (du Toit and Pretorious, 2002). Acetic acid bacteria use oxygen as the terminal electron acceptor during respiration and the presence of oxygen, pH values and temperature can effect the ethanol sensitivity of these bacteria.

## 4. Role of Enzymes in Winemaking

Enzymes originate from the skin of grapes, yeasts and other microbes associated with vineyards. These are involved in many biotransformation reactions from pre-fermentation to fermentation, post-fermentation and aging. In addition to these commercial enzymatic preparations are also used due to their benefits like shorter maceration, increased juice yield (Joshi *et al.*, 1991a), clarification of wine (Largac and Bission,1990; Joshi and Bhutani, 1991) improved color and enhanced flavour (Claus and Mojosov, 2018). Current commercial enzymatic preparations are usually cocktails of different activities, such as glucosidases, glucanases, pectinases, and proteases. These commercial enzymatic preparations favour the natural process by reinforcing the enzymatic activities of fruits and yeast, giving the winemakers more control over the process. In vinification, the addition of these commercial enzymes is a common practice to resolve clarification and filtration problems (pectinases, xylanases, glucanases, proteases) to release varietal aromas (glycosidases) or to reduce ethyl carbamate (urease) production (Claus and Mojosov, 2018). These products are not new since they were first used in the 1970s.

## 5. Use of Antimicrobials

Since ancient times, sulfur dioxide ($SO_2$) is used as an antimicrobial and antioxidant agent in the wine industry but the antimicrobial activity of $SO_2$ is reduced in high pH wines. The antioxidant property of $SO_2$ is due to its reaction with oxygen which prevents oxidation of sensitive wine components (Richard *et al.*, 1997; Clarke and Bakker, 2004) and the antioxidant capacity is dependent on catechol concentration (Danilewiex *et al.*, 2008). $SO_2$ distributes itself into multiple molecular species (ionized and non-ionized) each having distinct properties (Rose, 1993). In solutions, the antioxidant activity of $SO_2$ is carried out by a number of mechanisms like oxidase inhibition and oxygen scavenging. All these properties of $SO_2$ are actually responsible for the antioxidant mechanisms, antimicrobial activity and colour stabilization (Guerrero and Cantos-Villar, 2015) in wines.

However, with increasing health concerns and consumer preferences for more natural foods, there is a worldwide trend to reduce $SO_2$ (du Toit and Pretorius, 2000). Several studies have described the efficacy of lysozyme to control or inhibit malolactic fermentation (Gerbaux *et al.*, 1997; Pilatte *et al.*, 2000; Bartowsky, 2003). The antimicrobial activity of phenolic acids and their derivatives, flavonoids, and tannins has been studied by many researchers (Campos *et al.*, 2003; Cueva *et al.*, 2012). Eugenol, cinnamaldehyde, thymol, and carvacrol are some examples of phytochemicals which can be used as natural preservatives. Phenolic acids and their derivatives, being lipophilic in nature, can cross the bacterial cell membrane in their undissociated form, acidify the cytosol, can cause protein denaturation and interfere with cellular activity. The flavonols quercetin, kaempferol and myricetin were found to be inhibitory against several pathogenic bacteria including *E. coli, S. aureus* and *Pseudomonas aeruginosa* (Xu and Lee, 2001; Vaquero *et al.*, 2007). Chitosan, a linear heteropolysaccharide of N-acetyl-2-amino-2-deoxy-d-glucopyranose and 2-amino-2-deoxy-d-glucopyranose derived from chitin by deacetylation was also found to be effective as an antimicrobial agent in wines. Chitosan at a concentration of 0.5 mg/mL and above 1 mg/mL resulted in longer lag phases for the *Bacillus intermedius, B. bruxellensis*, respectively. Ferreira *et al.* (2013) exhibited that chitosan inhibits the growth of *Brettanomyces/Dekkera* at concentrations ranging from 0.2 to 0.5 mg/mL. Bacteriocins like nisin have also been found effective to control different taxa (*Oenococcus, Lactobacillus, Leuconostoc* and *Pediococcus*, acetic acid bacteria and yeast). The combined use of bacteriocins and $SO_2$ on wine isolates was investigated by several researchers. Interestingly, synergistic effects on bacterial growth inhibition were observed suggesting that appropriate combinations of bacteriocins and metabisulphite could allow a decrease in the levels of $SO_2$ currently used in the winemaking process (Rojo-Bezares *et al.*, 2007; Díez *et al.*, 2012, García-Ruiz *et al.*, 2013).

## 6. Malolactic Fermentation

The grapes cultivated in cold climate have more malic acid which when made into wine has more acidity than palatable. Winemaking involves two consecutive fermentation stages, namely alcoholic fermentation

(AF) and malolactic fermentation (MLF). In AF, yeast metabolizes sugars and converts these into alcohol and carbon dioxide. MLF involves the action of lactic acid bacteria (LAB) which converts L-malate into L-lactate and carbon dioxide. This conversion may take place *via* one of the three different enzymatic pathways (Radler, 1986; Lerm *et al.*, 2010). AF is usually preceded by MLF. MLF can be a spontaneous process (initiated by LAB naturally present on the fruit) or a voluntary action which can be executed at the discretion of the winemaker. This step fulfills three major objectives: (1) deacidification of the wine and concomitant increase in pH, (2) removal of malic acid and provide microbial stability, (3) the modification of the wine aroma profile (Ugliano *et al.*, 2003; Bauer and Dicks 2004; Lerm *et al.*, 2010). Also, fermentation of additional sugars and production of desired aromatic compounds occurs during MLF. These compounds, so produced act as important determinants of quality of a wide variety of wines viz., red wines, white wines, classic sparkling wines and fruit wines. Only strains of *Leuconostoc*, *Lactobacillus*, *Pediococcus*, and *Oenococcus* are able to survive the low pH (<3.5), high $SO_2$ (50 ppm), and ethanol levels of 10% (v/v) (Van Vuuren and Dicks 1993; Lonvaud-Funel, 1999; Gao and Rupasinghe, 2013). The literature pertaining to the wine from grapes is well studied and well documented but that on fruit wines are not researched systematic and available literature is scanty in this field. There are many gaps in the literature on the specific effects of MLF on the all varieties of fruit wines, though these fruits contain malic acid besides citric acid (Table 4). For the detailed information on MLF readers can refer to Chapter 12 of this text.

**Table 4.** Amounts of Malic and Citric Acid Present in Selected Fruit Varieties

| *Fruit source* | *Malic acid (g L-1)* | *Citric acid (g L-1)* |
|---|---|---|
| Apple | 10.12 ± 0.23 | 0.36 ± 0.02 |
| Apricot | 4.59 ± 0.04 | 4.13 ± 0.05 |
| Pear | 2.49 ± 0.09 | 1.64 ± 0.07 |
| Kiwi | 2.66 ± 0.13 | 11.00 ± 0.14 |
| Orange | 2.13 ± 0.01 | 11.71 ± 0.17 |
| Strawberry | 1.74 ± 0.10 | 7.13 ± 0.34 |
| Pineapple | 1.43 ± 0.09 | 6.52 ± 0.18 |

Adapted from Rupasinghe *et al.* (2017).

## 7. Fermentation Bouquet and Yeast Flavour Compounds – Chemical Changes

The flavour bouquet of wine is the result of a number of variations in production including grapes variety, microbes used, and type of wood used for storage. A varietal distinction exists in the flavour compounds of grapes. Some of the sensorially important volatile compounds are produced by yeast and bacteria during AF and MLF. Important volatile metabolites, i.e. esters, higher alcohols, carbonyls, volatile fatty acids, and sulfur compounds, are derived from sugar and amino acid metabolism. Yeast fermentation that produces a special fruity aroma without any negative characteristic in the wine is called fermentation bouquet (Singleton *et al.*, 1975). Yeasts, during fermentation of grape sugars, produces a range of volatile compounds in low concentration from pyruvate. These compounds in togetherness form the so-called 'fermentation bouquet' (Fig. 22). The concentration of aroma compounds produced by yeast depends upon fermentation conditions like temperature, grape variety, micronutrients, vitamins and nitrogen composition of the must. The main group of compounds is higher alcohols, fatty acids (Valero *et al.*, 1998; Guitart *et al.*, 1999) aldehydes, and esters (Herraiz and Ough, 1992; Zea *et al.*, 1994; Ferreira *et al.*, 1995; Lema *et al.*, 1996; Vianna and Ebeler, 2001).

The compounds responsible for the special fruity odour are hexyl acetate, ethyl caproate, and isoamyl acetate in the ratio of about 3:2:1.51. Hexyl acetate plays a major role while isoamyl acetate appears to play the least important role in the special fermentation bouquet in determining the odour of

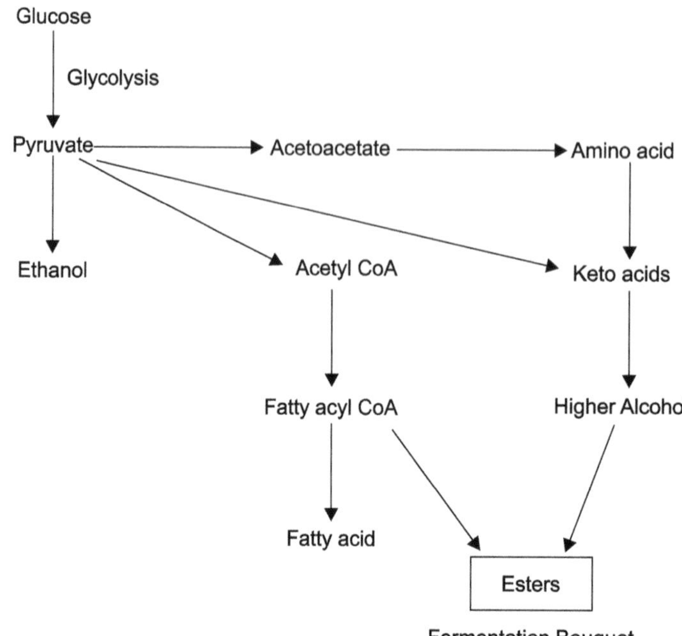


**Figure 22.** Pathway of synthesis of fermentation Bouquet

wine. Low temperature is the only method available yet for stabilization of these compounds. Apart from these esters, other compounds present in the fermentation bouquet are organic acids, higher alcohols and some aldehydes. Production of these compounds is influenced by the content and type of nitrogen source (Rapp and Versini, 1991). Fermentation at low temperatures i.e. about 15 °C encourages the production of volatile esters by yeasts (Killian and Ough, 1979). Fermentation at low temperatures encourages the production of the ethyl esters of straight chain fatty acid and acetate of higher alcohol. Some of the fatty acids CoA take part in ester formation by reacting with alcohol. Formation of fatty acid CoA and important esters of fermentation bouquet is positively correlated with the must total nitrogen (Bell *et al.*, 1979). In a study conducted by Carrau *et al.* (2008) a low initial nitrogen concentration, resulted in higher quantities of esters and fatty acids in strain KU1, whereas, higher concentrations of isoacids, gamma-butyrolactone, higher alcohols, and 3-methylthio-1-propanol was produced in M522 strain of *Saccharomyces cerevisiae*. The acetate esters of ethanol are majorly responsible for the aroma of freshly prepared wines. Their concentration is more dependent on yeast strains and sugar concentration than a fatty acid ester.

Esters can also be produced from the alcoholysis of acyl-CoA compounds which occur when fatty acid biosynthesis or degradation is interrupted in the cell and generate free coenzymes. Another source of ester formation is the carbon skeleton of amino acid. The formation of fermentation bouquet in total is little influenced by the cultivars involved or yeast strain conducting fermentation (Houtman *et al.*, 1980; Houtman and du Plessis, 1981). However, the concentration of individual esters can increase by certain yeast strains (Soles *et al.*, 1982). The fermentation bouquet compounds such as acetaldehyde, acetic acid, ethyl acetate, higher alcohols, and diacetyl if present in excess are regarded as undesirable.

Acetyl-CoA produced from sugar, amino acid, and sulfate during yeast metabolism react with higher alcohols. Small acyl chain esters are typically fruity or floral while the longer acyl chain esters are sweeter or soap like. If the total number of carbon atoms is 12, volatility is reduced and therefore, has less odour effect. The formation of esters is influenced primarily by temperature, the amino nitrogen content (Bell *et al.*, 1979; Ough and Lee, 1981) of the juice and the yeast strain employed (Soles *et al.*, 1982; Vos *et al.*, 1978).

Over one thousand chemical compounds and eight hundred aromatic compounds produced by yeast makeup the flavour of wine (Barbe, 2008; Styger *et al.*, 2011). Out of these, more than 400 compounds are produced during fermentation (Romano *et al.*, 2003). As discussed earlier, yeast metabolism during alcoholic fermentation is mainly responsible for the production of flavouring compounds. The contribution of several volatiles thiols to the aromas made from different grape varieties have recently been involved (Tominaga *et al.*, 1998; Tominaga *et al.*, 2000). Among them, mercapto-4 methyl-2-One (4MMP), 3- mercaptohexan-1-ol (3MH) and 3-mercaptohexyl acetate (A3MH) were found predominant in certain grape varieties.

Latest researches are focused on the role of yeast in the transformation of odourless molecules. These odourless molecules are believed to be the precursors of the aromatic substances. Currently, several strains of *S. cerevisiae*are marketed as "enhancers of varietal expression", claimed to have an ability to hydrolyze conjugated aroma precursors in juice, thereby, playing an important role in the enhancement of wine aroma (Zoecklein *et al.*, 1997; Mahon *et al.*, 1999). Aroma is an important characteristic and plays an important role in determining the quality of the wines. Hence, future research into the development of methods and strategies for the accurate quantification of aroma compounds is expected.

Various strains of *S. cerevisiae,* winemaking yeasts have been extensively investigated. The differences in the composition of wines made from different yeast species appear to be quantitative rather than qualitative (Romano *et al.*, 2003). A few studies have been conducted to analyze the contribution that 'non-*Saccharomyces*' wine yeast can make to the quality of the wine. These have been shown to contribute to the production of esters and other pleasant volatile compounds (Lema *et al.*, 1996; Romano *et al.*, 1997; Ciani and Maccarelli, 1998; Eglinton *et al.*, 2000; Ferraro *et al.*, 2000; Soden *et al.*, 2000). The non-*Saccharomyces* strains of the genera *Kloeckera* and *Hanseniaspora* have been reported to be having protein protease activity, which has, in turn, profound impact on the protein profile of finished wines (Charoenchai *et al.*, 1997; Dizy and Bisson, 2000). Non-*Saccharomyces* yeast have been studied in the context of their ability to produce β-glycosidase enzymes, which are involved in the flavour-releasing processes (McMahon *et al.*, 1999; Yanai and Sato, 1999; Manzanares *et al.*, 2000; Mendes-Ferreira *et al.*, 2001; Cordero-Otero *et al.*, 2003). Recently, the acetate esters formed by enzymatic activities of yeast strains belonging to the genera *Hanseniaspora* and *Pichia* have been studied in detail (Rojas *et al.*, 2001; Rojas *et al.*, 2003). In a study conducted by Romano *et al.* (2003), high production of acetoin and ethyl acetate and low production of higher alcohols were found in *C. stellate* as compared to *H. uvaraum. S. cerevisiae* was reported to produce a high concentration of isoamyl alcohol and 2,3-butanediol and low production of acetoin. The wine spoilage yeast, i.e. *S. ludwigii*, can produce a high level of acetoin, ethyl acetate, isoamyl alcohol, and isobutanol. More research on the non-*Saccharomyces* wine yeasts can show the potential of these yeasts in improving the flavour profile of the wine.

## 8. Chemical Changes Occurring During Sparkling Wine and Fortified Wines

### 8.1. Sparkling Wines

The traditional method used for the preparation of sparkling wines is called *method champenoise*. This method is predominantly used in the Champagne region of France and the Cava region in Spain (Amerine *et al.*, 1980). It is characterized by two successive fermentation processes and aging of the wine with the yeasts responsible for the second fermentation in grape wines (Jendeat *et al.*, 2011). However, when it comes to preparing sparkling wines from non-grapes fruits, information is limited. Although, scattered literature is available on the preparation process of apple wine or cider, plum wine or the sparkling wine from orange (Joshi *et al.*, 1995; Joshi *et al.*, 1999b; Joshi *et al.*, 2011b). Sparkling wines from apples are made by either carbonation in tank or bottle and then by secondary fermentation of apple base wine. Since the process is similar to grape wines, the relevant chemical aspects of the sparkling wines have been discussed taking grape wine as an example.

Aging on lees is the most characteristics aspect of the sparkling wines. The period of aging for different type of sparkling wines depend on the type of wine and kind of legislations prevalent in different countries. However, the aging period of wine cannot be less than nine months. During the process of aging, yeasts undergo autolysis, resulting in enzymatic auto-degradation of the cellular constituents that begins immediately after the death of the yeast cells. Consequently, different compounds and enzymes, which bring about important changes in the properties of the wine, are released into the medium that influence the sensorial quality of the resultant wine. For more information on yeast autolysis and its effects on wine quality, structural changes in the yeast cell walls and the biochemical composition of the wine, see the literature cited (Char-pentier and Feuillat, 1993; Martinez-Rodriguez and Polo, 2001).

During the process of autolysis, hydrolysis of the cell wall takes place which is mostly formed of polysaccharides. This break-down mainly depends on a strong endo- and exo-β-(1,3)-glucanase enzymatic activity in the yeast cell wall and on the α-mannosidases and end cellular proteases. These enzymes are responsible for the release of several compounds at different stages, beginning with polysaccharides and short-chain oligosaccharides in the early stages of autolysis, followed by the release of mannoproteins as the cell wall is broken down further (Charpentier and Feuillat, 1993). As revealed by electron microscopy, changes in the ultrastructure of the yeast also take place in the model solution (Kollar *et al*., 1993; Hernawan and Fleet, 1995). Autolysis of yeast results in the release of different products into the wine with special emphasis on proteolysis and the production of nitrogenous compounds (amino acids, peptides and proteins) (Moreno-Arribas *et al*., 1996; Moreno-Arribas *et al*., 1998; Martinez-Rodriguez and Polo 2000; Martinez-Rodriguez, *et al*., 2001), the release of polysaccharides, nucleic acids and lipids (Moreno-Arribas *et al*., 2000; Pueyo *et al*., 2000; Aussenac *et al*., 2001; de la Presa-Owens *et al*., 1998) and on the formation and degradation of volatile compounds (Pozo-Bayon *et al*., 2003; Postel and Ziegler, 1991). The major compounds released during autolysis are peptides and amino acids. These compounds are responsible for foam formation in the sparkling wine. Amino acids are the most studied compounds amongst all nitrogenous compounds released during the second fermentation and autolysis. Other important nitrogenous compounds released during autolysis of wine are proteins. Characteristic roles played by proteins in sparkling winemaking and the changes that occur in the process of secondary fermentation and aging have been widely studied and well documented (Luguera *et al*., 1997; Luguera *et al*., 1998).

Lipid metabolism is an important process during autolysis as lipids play an important role in membrane structure and composition. A large body of research literature is available on the characterization and release of fatty acids and other lipids by the yeasts (Hernawan and Fleet, 1995; Gallart *et al*., 1997; Le Fur *et al*., 1997; Pueyo *et al*., 2000). Polysaccharides, nucleic acids, fatty acids, and volatile compounds are also released during the process. A rise in polysaccharide concentration during the ageing of sparkling wines in the presence of yeasts have been documented (Llauberes *et al*., 1987; Feuillat *et al*., 1988; Cavazza *et al*., 1990; Pueyo *et al*., 1995; Andreis-Lacueva *et al*., 1997; Le Fur *et al*., 1997; Gallart *et al*., 1997; Luguera *et al*., 1997; Luguera *et al*., 1998; Moreno-Arribas *et al*., 2000; Loipez-Barajas *et al*., 2001).

Rise in volatile compounds, especially esters have been recorded during secondary fermentation and aging of the yeast (Cavazza *et al*., 1990; Postel and Ziegler, 1991; Zoecklein *et al*., 1997; Francioli *et al*., 2003; Pozo-Bayon *et al*., 2003). Due to the low storage temperature of wine, the enzymatic activity gets reduced and results in extremely slow autolysis of yeast. The addition of yeast autolysate or yeast extracts preparations have been attempted in order to accelerate the *in vitro* natural process of aging of sparkling wines (Feuillat, 1987). An another strategy to enhance autolysis during the manufacture of sparkling wines is to use mixed cultures of killer yeasts and the yeasts sensitive to their toxins (Todd *et al*., 2000; Cocolin and Comi, 2011). During aging, slight oxidation of Champagne wines can also happen as the gas exchange may occur and oxygen can penetrate due to the loss of carbon dioxide. Light exposure of the wine may result in the vitamin B2-related photodegradation of methionine, leading to the formation of volatile sulfur compounds such as methanethiol and dimethyldisulfide (DMDS), which give the wine cooked cauliflower or wet wool smells. Studies have shown that hydrolyzed autolysate had a favourable effect on secondary fermentation of apple wine (with alcohol content 5-6%) and was found suitable as a nitrogen source.

The bitterness, astringency and haze in the final product may be related to the presence of phenolic acids like chlorogenic acid and *p*-coumorylquinic acid, along with dihydrochalcones such as phloridzin and phloretin 2-xylo glucosides and flavonols (Picinelli *et al*., 2000). Apple wine produced with a high initial polyphenol level tasted very rough, which could be a consequence of the high content of flavon-3-ols, chlorogenic acid and the residual amount of malic acid, since incomplete malolactic fermentation was detected during studies. Changes in the content of polyphenols during the sparkling winemaking process have been documented in Table 5. Joshi *et al.* (1999b) has reported the effect of different concentrations of sugar (1, 1.5, 2%) and diammonium hydrogen phosphate (0.1 to 0.3%) on two strains of *Saccharomyces viz.* UCD 595 and UCD 505 in the preparation of sparkling wine from plum. A more elaborate research in these areas is expected in the future.

**Table 5.** Polyphenolic Contents of Sparkling Apple Wine during Different Days of Fermentation

| *Polyphenols* | *Time (days)* | | | |
|---|---|---|---|---|
| | *0* | *4* | *40* | *113* |
| (+)-Catechin | 12.7 | 12.8 ± 1.2 | 11.5±0.9 | 11.1±1.2 |
| B1 | 17.1 | 20.1 ±0.7 | 18.7±2.1 | 16.8±0.7 |
| (-)-Epicatechin | 92.2 | 102.3±3.3 | 78.5±11.4 | 66.9±9.5 |
| B2 | 100.5 | 113.9±1.2 | 80.9±13.8 | 75.6±4.5 |
| C1 | 15.2 | 16.9±0.5 | 11.2±2.4 | 9.5±1.0 |
| Tetramer Phloretin | 17.8 | 22.9±0.6 | 15.2±1.4 | 10.6±4.9 |
| 2-xyloglucoside | 62.3 | 72.4±1.0 | 54.4±7.7 | 58.0±5.1 |
| Phloridzin | 255.9 | 286.8±7.0 | 262.4±36.4 | 260.5±22.1 |

*Source*: Picinelli *et al.* (2000).

### 8.2. Biologically Aged Wines

Biologically aged wines such as sherry are produced using a several decades old method (Amerine *et al.*, 1980; Jendeat *et al.*, 2011); where soon after the completion of fermentation young wines are selected and fortified by adding vinous alcohol until the alcohol contents of wines reaches 15-15.5%. It is then, followed by the transfer of these wines to the oak barrels for aging. Wine aging occurs using 'Solera and Criaderas' system under the flor film of yeasts that grow on the surface of the wine, with an ethanol content of 15.5% or higher and low fermentable sugar content. The most significant metabolic change that occurs during biological aging of wines is the production of large quantities of acetaldehyde, which contributes to the sensory attributes of wine and is also considered as the best marker of the biological aging. Glycerol and acetic acid are used as carbon source in the absence of glucose. The moderate ethanol metabolism (Suarez-Lepe *et al.*, 1990; Suarez-Lepe, 1997; Mauricio *et al.*, 1997) with simultaneous consumption of all the amino acids especially proline has also been reported (Mauricio and Ortega, 1997; Berlanga *et al.*, 2001).

Acetaldehyde has been found to be an intermediate in ethanol metabolism by flor yeast and is the precursor for several other compounds, such as 2,3-butanodiol, acetic acid and acetoin which are found in sherry wines (Martinez *et al.*, 1995; Martinez *et al.*, 1997). These changes during aging of wines have been the objective of several recent articles. Many researches have also been devoted to understand the metabolic pathways for the production of these aroma compounds in relation to several factors, such as period of aging, the flor yeast strain used and the effect of acceleration conditions utilized to shorten the duration of this process (Plata *et al.*, 1998; Cortes *et al.*, 1998; Cortes *et al.*, 1999; Moyano *et al.*, 2002; Benitez *et al.*, 2003; Peinado *et al.*, 2003). However, there is no information on the preparation of such wines from the non-grape fruits that can be cited, but the future can foresee such changes.

## 9. Toxic Metabolites of Nitrogen Metabolism

The study of the metabolic role of lactic acid bacteria in the formation of undesirable compounds with implicit health effects such as ethyl carbamate, nitrosamines and biogenic amines is one of the most active research areas (Moreno-Arribas *et al.*, 2005). These compounds are generated due to the incomplete metabolism of nitrogen-containing compounds during the fermentation process.

## 9.1. Amines

Biogenic amines (BAs) are the other class of toxic compounds having important repercussions on health. These are low molecular weight organic bases with a high biological activity with undesirable effects in foods and beverages (Moreno-Arribas *et al.*, 2005). The total concentration of biogenic amines in wine has been reported to range from a few mg per liter to about 50 mg/L, depending upon the quality of the wine. BAs can be generated at the different stages of wine production due to decarboxylation of the corresponding amino acid by the activity of exogenous enzymes released by various microorganisms. Some of BAs are also formed in the vineyard, but their amount amounts vary with variety, soil type, fertilization, and climatic conditions during growth and degree of maturation. The high amounts of amines may also be due to unsanitary conditions during the wine making process (Zee *et al.*, 1981; Vidal-Carou *et al.*, 1991). Oenological parameters like maceration, must treatment, fermentation time, alcohol content, sulphur dioxide concentration, must composition, added nutrients, pH, temperature, clarification agents, enzymes used have been also reported to influence BAs in wine. These amines can be produced by the action of both the yeast and bacteria during AF and MLF. Most of the BAs are mainly derived from the activity of LAB that produce enzymes capable of decarboxylating the corresponding precursor amino acids. Yeast strain directly affects the content of amines and amino acids has no relationship with them (Goni and Azpiliceuta, 2001). Some of these amines are listed in Table 6.

**Table 6.** Some of the Amines Found in Wines

| | | |
|---|---|---|
| Ammonia | Hexylamine | Phenylamine |
| Butylamine | Histamine | Phenethylamine |
| Cadaverine | Indole | Piperidine |
| 1,5-Diaminopentane | Isopentylamine | Propylamine |
| Diethylamine | Isopropylamine | Putrescine |
| Dimethylamine | Methylamine | Pyrrolidine |
| Ethanolamine | 2-Methylbutylamine | 2-Pyrrolidine |
| Ethylamine | Morpholine | Serotonin |
| | | Tyramine |

Although 25 different biogenic amines have been described in wine, the most common ones are histamine, tyramine, putrescine, and cadaverine (Lehtonen, 1996; Gloria *et al.*, 1998; Vazquez-Lasa *et al.*, 1998; Mafra *et al.*, 1999) which are produced by the decarboxylation of histidine, tyrosine, ornithine/arginine, and lysine, respectively. Among these histamine and tyramine are the most studied BAs. Histamine has been reported to cause headaches, hyotension and digestive system problems whereas tyramine is associated with migranes and hypertension (Silla-Santos, 1996). Histamine plays a special role as indicator amine to assess the freshness and quality of the wine. When there is no histamine, there are no other biogenic amines (Askar and Treptow, 1986). The recommended upper limit for histamine in wine is 2 mg/L in Germany, 5-6 mg/L in Belgium, 8 mg/L in France and 10 mg/L in Switzerland. Detailed information on the factors influencing amine formation by LAB in the wines has been provided by Lonvaud-Funel (2001). Carcinogenic nitrosamines can be produced by the reaction of secondary amines like spermine, spermidine with nitrite.

## 9.2. Ethyl Carbamate

Ethyl carbamate (also referred to as urethane) is identified as an animal carcinogen and is a natural component of fermented foods and beverages (Ough, 1993). The formation of ethyl carbamate is related to concentrations of urea and ethanol (Fig. 23), and the rate of formation increases exponentially with increased wine storage temperature.

The urea produced by the yeast during alcoholic fermentation and citruline and carbamyl phosphate produced by lactic acid bacteria during malolactic fermentation, are the major carbamylic compounds in the wine. Urea may combine with ethanol to form ethyl carbamate or urethane (Monteiro *et al.*, 1989; Ough *et al.*, 1988a; Ough *et al.*, 1988b). Approximately 0.8 mg ethyl carbamate/L was formed after 48 hours in a model solution of urea and ethanol (12%) under acidic conditions of 3.2

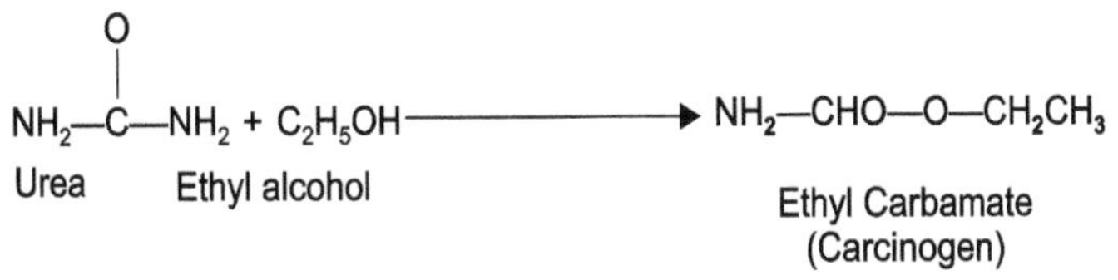

$$NH_2{-}\overset{\overset{\displaystyle O}{\|}}{C}{-}NH_2 + C_2H_5OH \longrightarrow NH_2{-}CHO{-}O{-}CH_2CH_3$$

Urea Ethyl alcohol Ethyl Carbamate (Carcinogen)

**Figure 23.** Formation of ethyl carbamate from urea

pH and 71 °C temperature. Freshly prepared wines do not contain significant concentrations of ethyl carbamate, though it is formed at the rate of about 4.8 mg/year at 13 °C from an initial concentration of 10 mg urea (Ough *et al.*, 1990). Urea is the principal precursor of ethyl carbamate, therefore, to control the concentration of ethyl carbamate much attention has been given to the control of urea and its precursor arginine. During storage, urea can react with ethanol to produce ethyl carbamate and therefore, to prevent its production the addition of the commercial preparations of the enzyme and urease has been admitted by the International Organisation of Vine and Wine and European Economic Community legislation. Solid phase microextraction and gas chromatography has been used for the determination of ethyl carbamate in wines (Whiton and Zoecklein, 2002).

## 10. Biochemical Basis of Wine Spoilage

### 10.1. Spoilage by Acetic Acid Bacteria

Acetic acid bacteria are a group of aerobic microorganisms which requires oxygen for their growth. It consists of 10 genera in the family *Acetobacteraceae*. The acetic acid bacteria generally, do not grow during alcoholic fermentation due to the anaerobic conditions but do exhibit exceptionally good survival properties during storage. The exposure of wine to air during pumping and transfer operations can quickly stimulate their growth. The chemistry involved in the metabolism of ethanol into acetic acid has been described in the earlier section on acetic acid fermentation. The high concentration of acetic acid i.e. above 1.2-1.3 g/L is objectionable in wines as it gives vinegar taints to wine (Margalith, 1981). In fully aerated wines, 50-60 percent of ethanol could be oxidized by acetic acid bacteria to produce 1.5-3.75 g/L acetic acid (Drdak *et al.*, 1993).

### 10.2. Spoilage by Lactic Acid Bacteria

Lactic acid bacteria play a key role in the malolactic fermentation of wines but it may also cause the spoilage of wines (Lonvaud, 1999). The heterofermentative species of the genera *Lactobacillus* and *Leuconostoc* produce acetic acid and D-lactic acid on the fermentation of sugars. Homofermentative species of *Lactobacillus* and *Pediococcus* also produce D-lactic acid through the glycolytic metabolism of sugars. However, both acetic acid and lactic acid is produced by these species on fermentation of pentose sugars.

Some LAB strains belonging to genera *Lactobacillus* convert glycerol to 3-hydroxypropionaldehyde (3-HPA) by glycerol dehydratase (Boulton *et al.*, 1996). 3-HPA has a potential for spontaneous conversion into acrolein under wine making conditions. Acrolein is highly toxic but its concentration in wines is very low. Therefore, acrolein as a single component is not problematic; however, when it reacts with anthocyanins and tannins, it can produce an unpleasant bitterness (Moreno-Arribas *et al.*, 2005). Acrolein concentration of 10 ppm can cause acrolein taint (Margalith, 1981). Complete degradation of glycerol can take place in apple and pear wine, where, the growth of LAB is not totally repressed.

Another type of LAB spoilage is mannitol taint. This problem occurs only in those wines which have high pH and contain a significantly high amount of residual sugar, especially fructose. Some hetero-fermentative LAB of *Lactobacillus brevis* can produce mannitol by the enzymatic reduction of fructose as reviewed earlier (Moreno-Arribas *et al.*, 2005). Manitol taint is a form of spoilage with high complexity because it is also accompanied by high concentrations of acetic acid, D-lactic acid, n-propanol, 2 butanol and diacetyl taint. A spoiled wine has a vinegary-estery taste and contains approximately 9 g/L mannitol, 3 g/L acetic acid, 3 g/L D-lactic acid (Sponholz, 1988; Sponholz, 1989).

Lactic acid bacteria are also responsible for the production of diacetyl. A high concentration of diacetyl gives a buttery or whey like flavour which is undesirable in wines. Wines produced by the yeast activity contain 0.2-0.3 mg/L of diacetyl. When the concentration of diacetyl in wines is more than 1 mg/L they are considered as faulty. The metabolic pathway for the production of diacetyl by LAB is unclear. However, the possible mechanism might be the conversion of sugars to pyruvate and pyruvate to acetolactate which is then converted to diacetyl (El-Gendy *et al.*, 1983; Kandler, 1983). Additionally, it may come from the metabolism of citric acid by citrate lyase which cleaves this acid into acetic acid and oxaloacetate. The oxaloacetate is then, metabolized into diacetyl through pyruvate. Sorbic acid may be added to wines as an antimicrobial agent to control the growth of yeast certain strains of lactic acid bacteria metabolize sorbic acid and impart geranium off odour to the wines. The substance responsible for the geranium odour has been identified as 2-ethoxyhexa-3,5-diene. If wines or juices are to be stored in the presence of sorbic acid, they should contain sufficient amount of free $SO_2$ to prevent the growth of LAB and development of geranium taint.

Lactic acid bacteria species are also associated with the spoilage of the taste of wine known as *'piqure lactique'* (Moreno-Arribas *et al.*, 2005). This kind of spoilage can occur at any stage of winemaking. The spoilage occurs if the LAB grows before the transformation of all the must sugars to ethanol. LAB ferment hexoses to ethanol, $CO_2$ and acetic acid (Strasser de Saad and Manca de Nadra, 1992). If excess amounts of lactic acid and acetic acid are formed under these conditions, the volatile acidity of the wine increases considerably. The D-isomer of lactic acid is associated with spoilage, whereas L-lactic acid is produced during malolactic fermentation (Fugelsang, 1997, Sponholz 1993).

Ropiness is another type of wine quality deterioration which is caused by genera *Leuconostoc* and *Pediococcus* (van Vuuren and Dicks, 1993; Manca de Nadra and Strasser de Saad, 1995). The LAB utilizes the residual sugars and synthesizes the extra cellular polysaccharides which results in the increased viscosity and slimmy appearance of wines. Such wines are known as 'ropy'. Some other strains are able to synthesize an exopolysaccharide characterized as a β-D-glucan (Llauberes et al., 1990). This type of wine spoilage can occur during vinification and, in most cases, after bottling. An odour similar to 'mouse urine' or acetamide is also produced in wines by some heterofermentative *Lactobacilli* (Costello and Henschke, 2002).

Moreover, the tendency of some *Dekkera* and *Brettanomyces* yeast to produce a mousy off-flavour has also been described (Grbin and Henschke, 2000). These olfactory defects of the wine are attributed to the formation of 3-heterocyclic volatile bases: 2-acetyltetrahydropyridine, 2-ethyltetrahydropyridine and 2-acetyl-1-pyrroline.

## 11. Summary and Future Outlook

Winemaking is a process which involves a series of biochemical transformations brought about by the microorganisms and their enzymes. The term wine is mainly used to an alcoholic product produced from fermentation of grapes; however, the fruits other than grapes are also employed to produce wines. The primary fermentation (alcoholic fermentation) in wines is carried out by yeast especially *Saccharomyces cerevisiae*. However, a secondary fermentation known as malolactic fermentation can also be carried out by LAB to naturally reduce the acidity of wines. Beside yeast and bacteria, fruit must is also exposed to numerous enzymes originating from other sources. Extensive research has been carried on the role of the organic acids on the quality of wines. Studies have been also carried on the compounds like urea, ethyl carbonate, methanol, hydrogen sulfide and amines but more studies should be carried to understand the production, interaction and toxicity of these compounds to ensure the safety of the wine consumers. Also, the studies on the wines produced from the fruits other than the grapes and yeasts other than the *Saccharomyces cerevisiae* is very limited. Focus also needs to be made on the metabolites produced by microorganisms other than the *Saccahromyces cerevisiae* and LAB. The production of healthier wines could be pursued by increasing the knowledge of the biochemical mechanisms of yeasts and lactic acid bacteria that give rise to the different compounds of wine responsible for its sensory and biological characteristics and health-related quality to enhance the components with positive effects and suppress those with negative ones.

## References


Aceituno, F.F., Orellana, M., Torres, J., Mendoza, S., Slater, A.W., Melo, F. and Agosin, E. (2012). Oxygen response of the wine yeast *Saccharomyces cerevisiae* EC1118 grown under carbon-sufficient, nitrogen-limited enological conditions. *Appl. Environ. Microbiol.* 78(23): 8340-8352.

Acree, T.E., Sonoff, E.P. and Splittstoesser, D.F. (1972). Effect of yeast strain and type of sulfur compound on hydrogen sulfide production. *Amer. J. Enol. Vitic.* 23: 6-9.

Agenbach, W.A. (1977). A study of must nitrogen content about incomplete fermentations, yeast production, and fermentation. *Proc. S. Afric. Soc. Enol. Vitic.* 66-87.

Albers, E., Larsson, C., Lidén, G., Niklasson, C. and Gustafsson, L. (1996). Influence of the nitrogen source on *Saccharomyces cerevisiae* anaerobic growth and product formation. *Appl. Environ. Microbiol.* 62(9): 3187-3195.

Alexandre, H., Berlot, J.P. and Charpentier, C. (1992). Effect of ethanol on membrane fluidity of protoplasts from *Saccharomyces cerevisiae* and *Kloeckera apiculata* grown with or without ethanol, measured by fluorescence anisotropy. *Biotechnol. Tech.* 5: 295-300.

Amerine, M.A., Kunkee, R.E., Ough, C.S., Singleton, V.L. and Webb, A.D. (1980). Technology of Wine Making. AVI Publ. Co. Westport, Connecticut.

Amerine, M.A. (1985). Winemaking. pp. 67. *In*: H. Koprowski and S.A. Plotin (Eds.). Worlds Debt to Pasteur. New York, Alan R. Liss in Incorporated.

Andres-Lacueva, C., Lamuela-Raventós, R.M., Buxaderas, S. and Torre-Boronat, M.C. (1997). Influence of variety and aging on foaming properties of cava (sparkling wine). *J. Agric. Food Chem.* 45(7): 2520-2525.

Angulo, L., Lema, C. and Lopez, J.E. (1993). Influence of viticultural and enological practices on the development of yeast population during winemaking. *Am. J. Enol. Vitic.* 44: 405-408.

Anocibar-Beloqui, A., Guedes de Pinho, P. and Bertrand, A. (1995).Importance of N-3-(methylthiopropyl) acetamide and 3-methylthiopropionic acid in wines. *J. Int. Sci. Vigne Vin.* 29: 17-26.

Askar, A. and Treptow, H. (1986). Biogene Amine in Lebensmitteln: Vorkommen, Bedeutung und Bestimmung. Ulmer, Stuttgart. pp. 197.

Aussenac, J., Chassagne, D., Claparols, M., Charpentier, C., Duteurtre, B., Feuillat, M. and Charpentier, C. (2001). Purification method for the isolation of monophosphate nucleotides from champagne wine and their identification by mass spectrometry. *J. Chromatogr. A.* 907: 155-164.

Axelsson, L. (2004). Lactic acid bacteria: Classification and physiology. pp. 1-66. *In*: A.V. Salminen and A.O. Wright (Eds.). Lactic Acid Bacteria, Microbiological and Functional Aspects. Ouwehand. Marcell Dekker, New York.

Barbe, J.C., Pineau, B. and Ferreira, A.C. (2008). Instrumental and sensory approaches for the characterization of compounds responsible for wine aroma. *Chem. Biodiv.* 5(6): 1170-1183.

Bartowsky, E. (2003). Lysozyme and winemaking. *Aus. J. Grape Wine Res.* 473a: 101-104.

Bartowsky, E.(2011). Malolactic fermentation. pp. 526-563. *In*: Joshi.V.K. (Ed.). Handbook of Enology Vol. 3. Asia Tech Publication, New Delhi.

Bauer, R. and Dicks, L.M.T. (2004). Control of malolactic fermentation in wine. A review. *S. Afr. J. Enol. Vitic.* 25(2): 74-88.

Bell, A.A., Ough, C.S. and Kliewer, W.H. (1979). Effects on must and wine composition, rates of fermentation and wine quality of nitrogen fertilization of *Vitis vinifera* var. Thompson Seedless grapevines. *Amer. J. Enol. Vitic.* 30: 124-129.

Beltran, G., Novo, M., Rozes, N., Mas, A. and Guillamón, J.M. (2004). Nitrogen catabolite repression in *Saccharomyces cerevisiae* during wine fermentations. *FEMS Yeast Res.* 4(6): 625-632.

Benitez, P., Castro, R. and Garcia-Barroso, C. (2003). Changes in the polyphenolic and volatile contents of 'Fino' sherry wine exposed to ultraviolet and radiation during storage. *J. Agric. Food Chem.* 51: 6482-6487.

Berlanga, T.M., Atanasio, C., Mauricio, J.C. and Ortega, J.M. (2001). Influence of aeration on the physiological activity of flour yeast. *J. Agric. Food Chem.* 49: 3378-3384.

Bisson, L.F. (1991). Influence of nitrogen on yeast and fermentation of grapes. pp. 172-184. *In*: J. Rantz and Davis, C.A. (Eds.). Proceedings of International Symposium on Nitrogen in Grapes and Wine. ASEV, Annual Meeting, Seattle USA. 136.

Boudreau IV, T.F., Peck, G.M., O'Keefe, S.F. and Stewart, A.C. (2018). Free amino nitrogen concentration correlates to total yeast assimilable nitrogen concentration in apple juice. *Foodsci. Nutr.* 6(1): 119-123.

Boulton, R.B., Singleton, V.L., Bisson, L.F. and Kunkee, R.E. (1996). Principles and Practices of winemaking. New York: Chapman Hall.

Broach, J.R. (2012). Nutritional control of growth and development in yeast. *Genetics* 192(1): 73-105.

Buglass, A. (Ed.) (2011). The Handbook of Alcoholic Beverages: Technical, Analytical, and Nutritional Aspects. Volume 1. John Wiley & Sons, United Kingdom.

Burulearu, L., Nicolescu, C.L., Bratu, M.G., Manea, J. and Avram, D. (2010). Study regarding some metabolic features during lactic and fermentation at vegetable juices. *Rom. Biotechnol. Lett.* 15: 5177–5188.

Busturia, A. and Lagunas, R. (1986). Catabolite inactivation of the glucose transport system in *Saccharomyces cerevisiae*. *J. Gen. Micro*. 132: 379-385.

Calderoin, F., Varela, F., Navascueis, E., Colomo, B., Gonzailez, M.C. and Suarez, J.A. (2001). Influence of pH and temperature in the biosynthesis of malic acid in wines by *Saccharomyces cerevisiae*. *Bull. OIV* 74(845-46): 474-486.

Campbell, B., den Hollander, J.A., Alger, J.R. and Shulman, R.G. (1987). $^{31}$PNMR saturation-transfer and $^{13}$C NMR kinetic studies of glycolytic regulation during anaerobic and aerobic glycolysis. *Biochem*. 26: 7493-7500.

Campos, F.M., Couto, J.A. and Hogg, T.A. (2003). Influence of phenolic acids on growth and inactivation of *Oenococcus oeni* and *Lactobacillus hilgardii*. *J. App. Microbiol*. 94(2): 167-174.

Cannon, W.R. (2014). Concepts, challenges and successes in modeling thermodynamics of metabolism. *Front. Bioeng. Biotechnol.* 2: 1-10.

Caplice, E. and Fitzgerald, G.F. (1999). Food fermentations: Role of microorganisms in food production and preservation. *Int. J. Food Microbiol*. 50: 131-149.

Caridi, A. (2003). Identification and first characterization of lactic acid bacteria isolated from the artisanal ovine cheese Pecorino del Poro. *Int. J. Dairy Technol*. 56(2): 105-110.

Carrau, F.M., Medina, K., Farina, L., Boido, E., Henschke, P.A. and Dellacassa, E. (2008). Production of fermentation aroma compounds by *Saccharomyces cerevisiae* wine yeasts: Effects of yeast assimilable nitrogen on two model strains. *FEMS Yeast Res.* 8(7): 1196-1207.

Cartwright, C.P., Juroszek, J.R., Beavan, M.J., Ruby, M.S., DeMorias, M.F. and Rose A.H. (1986). Ethanol dissipates the proton-motive force across the plasma membrane of *Saccharomyces cerevisiae*. *J. Gen. Micro*. 132: 369-377.

Cartwright, C.P., Rose, A.H., Calderbank, J. and Keenan, M.H.J. (1989). Solute transport. pp. 5. *In*: A.H Rose and J.S Harrison (Eds). The Yeasts. Vol. 3, 2nd edn. Academic Press, London.

Casey, G.P. and Ingledew, W.M. (1986). Ethanol tolerance in yeasts. *Crit. Rev. Microbiol*. 13: 219-280.

Cavazza, A.,Versini, G., Grando, H.W.S. and Price, K.R. (1990). Variabilita indotta dai ceppi di lievito nella rifermentazione dei vini spumanti. *Ind. Bevande.* 19: 225-228.

Charoenchai, C., Fleet, G. and Henscke, P.A. (1998). Effects of temperature, pH and sugar concentration on the growth rates and cell biomass of wine yeasts. *Am. J. Enol. Vitic.* 49: 283-288.

Charoenchai, C., Fleet, G.H., Henschke, P.A. and Todd, B.E.N. (1997). Screening of non-Saccharomyces wine yeast for the presence of extracel-lularhydrolyticenzymes. *Aust. J. Grape Wine Res.* 3: 2-8.

Charpentier, C. and Feuillat, M. (1993). Yeast autolysis. pp. 225-242. *In:* G.H. Fleet (Ed.). Wine Microbiology and Biotechnology. Switzerland: Harwood Academic Publishers.

Chidi, B.S. (2016). Organic acid metabolism in *Saccharomyces cerevisiae*: Genetic and metabolic regulation. PhD thesis: Institute of Wine Biotechnology, Faculty of AgriSciences, Stellenbosch University, South Africa.

Ciani, M. and Maccarelli, F. (1998). Oenological properties of non-*Saccharomyces* yeast associated with wine-making. *World J. Microbiol. Biot*. 14: 199-203.

Clarke, R.J. and Bakker, J. (2004). Wine Flavour Chemistry. Blackwell, Oxford.

Claus, H. and Mojosov, K. (2018). Enzymes for wine fermentation: Current and perspective applications. *Fermentation* 4(3): 52.

Cocolin, L. and Comi, G. (2011). Killer Yeasts in Winemaking. pp. 565-590. *In*: V.K. Joshi (Ed.). Handbook of Enology. Vol. III. Asia Tech Publishers, New Delhi.

Coloretti, F., Zambonelli, C., Castellari, L., Tini, V. and Rainieri, S. (2002). The effect of DL-malic acid on the metabolism of L-malic acid during wine alcoholic fermentation. *Food Tech. Biotechnol.* 40(4): 317-320.

Colowick, S.P. (1973). The Hexokinases. *In*: P.D. Boyer (Ed.). The Enzymes. New York Academic Press. 9: 1.

Cooper, T.C. and Sumrada, R. (1975). Urea transport in *Saccharomyces cerevisiae*. *J. Bacteriol.* 121: 571-576.

Cooper, T.G. (1982). Nitrogen metabolism in *Saccharomyces cerevisiae*. pp. 39-99. *In*: J.N. Strathern, E.W. Jones and J.R. Broach (Eds.). The Molecular Biology of the Yeast *Saccharomyces cerevisiae*: Metabolism and Gene expression. Cold Spring Harbor, New York.

Cordero-Otero, R.R., Ubeda-Iranzo, J.F., Briones-Perez, A.I., Potgieter, N., Villena, M.A., Pretorius, I.S. and Rensburg, P.V. (2003). Characterization of the beta-glucosidase activity produced by enological strains of non-*Saccharomyces* yeasts. *J. Food. Sci.* 68: 2564-2569.

Cortes, M.B., Moreno, J., Zea, L., Moyano, L. and Medina, M. (1998).Changes in aroma compounds of sherry wines during their biological aging carried out by *Saccharomyces cerevisiae* races bayanus and capensis. *J. Agric. Food Chem.* 46: 2389-2394.

Costello, P.J. and Henschke, P. (2002). Mousy off-flavor of wine: Precursors and biosynthesis of the causative N-heterocycles 2-ethyltetrahydropyridine, 2-acetyltetrahydropyridine, and 2-acetyl-1-pyrroline by Lactobacillus hildargiiDSM 20176. *J. Agric. Food Chem.* 50: 7079-7087.

Cueva, C., Mingo, S., Muñoz-González, I., Bustos, I., Requena, T., Del Campo, R. and Moreno-Arribas, M.V. (2012). Antibacterial activity of wine phenolic compounds and oenological extracts against potential respiratory pathogens. *Lett. App. Microbiol.* 54(6): 557-563.

Danilewiex, J.C., Seccombe, J.T. and Whelan, J. (2008). Mechanism of interaction of polyphenols, oxygen and sulphur dioxide in Model wine and wine. *Am. J. Enol. Vitic.* 59: 2128-2136.

Davis, R.H. (1986). Compartmental and regulatory mechanism in the arginine pathways of *Neurospora crassa* and *Saccharomyces cerevisiae*. *Microbiol. Rev*. 50: 280-313.

De la Presa-Owens, C., Lamuela-Raventoos, R.M., Buxaderas, S. and de la Torre-Boronat, M.C. (1998). Characterization of Macabeo, Xarel.lo, and Parellada white wines from the Penedes region II. *Am. J. Enol. Vitic.* 46: 539-541.

Den Hollander, J.A., Ugurbil, K., Brown, T.R., Bednar, M., Redfield, C. and Shulman, R.G. (1986). Studies of anaerobic and aerobic glycolysis in *Saccharomyces cerevisiae*. *Biochem*. 25: 203-211.

Díez, L., Rojo-Bezares, B., Zarazaga, M., Rodríguez, J.M., Torres, C. and Ruiz-Larrea, F. (2012). Antimicrobial activity of pediocin PA-1 against *Oenococcus oeni* and other wine bacteria. *Food Microbial*. 31(2): 167-172.

Dizy, M. and Bisson, L.F. (2000). Proteolytic activity of yeast strains during grape juice fermentation. *Am. J. Enol. Vitic.* 51: 155-167.

Drdak, M., Rajniakova, A., Buchtova, V. and Simko, P. (1993). Free amino acid content of various red wines. *Food/Nahrung*. 37(1): 77-78.

Drysdale, G.S. and Fleet, G.H. (1988). Acetic acid bacteria in winemaking: A review. *Am. J. Enol. Vitic.* 39(2): 143-154.

Du Toit, M. and Pretorius, I.S. (2000). Microbial spoilage and preservation of wine: Using weapons from nature's own arsenal—A review. *S. Afr. J.Enol. Vitic.* 21: 74-96.

Du Toit, W. J. and Pretorius, I.S. (2002). The occurrence, control and esoteric effect of acetic acid bacteria in winemaking. *Annal. Microbiol.* 52(2): 155-179.

Dubois, C., Manginot, C., Roustan, J.L., Sablayrolles, J.M. and Barre, P. (1996). Effect of variety, year, and grape maturity on the kinetics of alcoholic fermentation. *Am. J. Enol. Vitic.* 47: 363-368.

Eddy, A.A. (1982). Mechanism of solute transport in selected eukaryotic microorganisms. *Advances in Microbiol. Physiol.* 24: 1-78.

Eglinton, J.M., McWilliam, S.J., Fogarty, M.W., Francis, I.L., Kwiatkowski, M.J., Hoj, P.B. and Henschke, P.A. (2000). The effect of Saccharomyces bayanus-mediated fermentation on the chemical composition and aroma profile of Chardonnay wine. *Aust. J. Grape Wine Res.* 6: 190-196.

El-Gendy, S.M., Abdel-Galil, H., Shahin, Y. and Hegezi, F.Z. (1983). Acetoin and diacetyl production by homo and hetero-fermentative lactic acid bacteria. *J. Food Prot.* 46(5): 420-425.

Eschenbruch, R., Bonish, P. and Fisher, B.M. (1978). The production of $H_2S$ by pure culture yeasts. *Vitis.* 17: 67.

Faria-Oliveira, F., Diniz, R.H.S., Godoy-Santos, R., Piló, F.B., Mezadri, H., Castro, I.M. and Brandão, R.L. (2015). The Role of Yeast and Lactic Acid Bacteria in the Production of Fermented Beverages. *In*: Prof. Ayman Amer Eissa (Ed.). South America, Food Production and Industry. InTech, DOI: 10.5772/60877.

Ferraro, L., Fatichenti, F. and Ciani, M. (2000). Pilot scale vinification process using immobilized Candida stellata and *Saccharomyces cerevisiae*. *Process Biochem.* 35: 1125-1129.

Ferreira, V., Fernandez, P., Pena, C., Escudero, A. and Cacho, J.F. (1995). Investigation on the role played by fermentation esters in the aroma of young Spanish wines by multivariate analysis. *J. Sci. Food Agric.* 67: 381-392.

Ferreira, D., Moreira, D., Costa, E.M., Silva, S., Pintado, M.M. and Couto, J.A. (2013). The antimicrobial action of chitosan against the wine spoilage yeast Brettanomyces/Dekkera. *Journal of Chitin and Chitosan Science* 1(3): 240-245.

Feuillat, M. (1987). Preparation d'autolysats de levures et addition dans les vins effervescents elabores selon la methode champenoise. *Rev. Fr. Oenol.* 28: 36-45.

Feuillat, M., Freyssenet, M. and Charpentier, C. (1988). Production de colloides par les levures dans les vins mousseux elabores selon la methode champenoise. *Rev. Fr. Oenol.* 28: 36-45.

Francioli, S., Torrens, J., Riu-Aumatell, M., Loipez-Tamames, E. and Buxaderas, S. (2003). Volatile compounds by SPME-GC as age markers of sparkling wines. *Am. J. Enol. Vitic.* 54: 158-162.

Fuck, E. and Radler, F. (1972). Apfelsaurestoffwechsel bei Saccharomyces I. Der abnaerobe Apfelsaureabbau bei *Saccharomyces cerevisiae*. *Archiv. Mibrobiol.* 87(2): 149-164.

Fugelsang, K.C. and Edwards, C.G. (1997). Wine Microbiology: Practical Applications and Procedures. (2nd ed.). Springer, New York.

Fugelsang, K.C. (1997). *In*: K.C. Fugelsang (Ed.). Wine Microbiology. New York: Chapman & Hall.

Galeote, V., Novo, M., Salema-Oom, M., Brion, C., Valerio, E., Gonçalves, P. and Dequin, S. (2010). FSY1, a horizontally transferred gene in the *Saccharomyces cerevisiae* EC1118 wine yeast strain, encodes a high-affinity fructose/$H^+$ symporter. *Microbiol.* 156(12): 3754-3761.

Gallander, J.F. (1977). Deacidification of Eastern table wines with *Schizosaccharomyces pombe*. *Am. J. Enol. Vitic.* 28: 65-68.

Gallart, M., Francioli, S., Viu-Marco, A., Loipez Tamames, E. and Buxaderas, S. (1997). Determination of free fatty acids and their esters in musts and wines. *J. Cromatogr. A.* 776: 283-291.

Gao, J. and Rupasinghe, H.P.V. (2013). Characterization of malolactic conversion by *Oenococcus oeni* to reduce the acidity of apple juice. *Int. J. Food Sci. Technol.* 48: 1018-1027.

García-Ruiz, A., Tabasco, R., Requena, T., Claisse, O., Lonvaud-Funel, A., Bartolomé, B. and Moreno-Arribas, M.V. (2013). Genetic diversity of *Oenoccoccus oeni* isolated from wines treated with phenolic extracts as antimicrobial agents. *Food Microbiol.* 36(2): 267-274.

Genitsariotis, R. (1979). Uber die Bildung von L(+) Lactat bei Sachharomyces pretoriensis und die NAD-abhiangige Lactat-dehydrogenase. Dissertation Mainz.

Gerbaux, V., Villa, A., Monamy, C. and Bertrand, A. (1997). Use of lysozyme to inhibit malolactic fermentation and to stabilize wine after malolactic fermentation. *Am. J. Enol. Vitic.* 48: 49-54.

Gloria, M.B., Watson, B.T., Simon-Sarkadi, L. and Daeschel, M.A. (1998). A survey of biogenic amines in Oregon Pinot noir and Cabernet Sauvignon wines. *Am. J. Enol. Vitic.* 49: 279-282.

Goni, D.T. and Azpilicueta, C.A. (2001). Influence of yeast strain on biogenic amines content in wines: Relationship with utilization of amino acid during fermentation. *Am. J. Enol. Vitic.* 52(3): 185-190.

Goniak, O.J. and Noble, A.C. (1987). Sonsory study of selected volatile sulphur compounds in white wines. *Am. J. Enol. Vitic.* 38: 223-227.

Goyal, R.K. (1999). Biochemistry of fermentation. pp. 87-172. *In*: V.K. Joshi and A.Pandey (Eds.). Biotechnology: Food Fermentation. Vol. I. Asiatech Publishers Inc., New Delhi.

Grbin, P.R. and Henschke, P.A. (2000). Mousy off-flavour production in grape juice and wine. *Aus. J. Grape Wine Res.* 6: 255-262.

Green, D.E., Murer, E., Hultin, H.O., Richardson, S.H., Salmon, B., Brierly, G.P. and Baum, H. (1965). Association of integrated metabolic pathways with membranes I. Glycolytic enzymes of the red corpuscle and yeast. *Arch. Biochem. Biophys.* 112: 635-647.

Grenson, M. (1983a). Inactivation-Reactivation process and repression of permease formation regulate several ammonia-sensitive permeases in the yeasts *Saccharomyces cerevisiae*. *J. Biochem.* 133: 135-139.

Grenson, M. (1983b). Study of the positive control of the general amino-acid permease and other ammonia-sensitive uptake systems by the product of the NPR1 gene in the yeast *Saccharomyces cerevisiae*. *J. Biotech.* 133: 141-144.

Grenson, M. and Acheroy, B. (1982). Mutations affecting the activity and the regulation of the general amino-acid permease of *Saccharomyces cerevisiae*. Localisation of the cis acting dominant PGR regulatory mutation in the structural gene of this permease. *Mole. Gen. Genet.* 188: 261-265.

Guerrero, R.F. and Cantos-Villar, E. (2015). Demonstrating the efficiency of sulphur dioxide replacements in wine: A parameter review. *Trends Food Sci. Technol.* 42(1): 27-43.

Guitart, A., HernandezOrte, P., Ferreira, V., Pena, C. and Cacho, J. (1999). Some observations about the correlation between the amino acid content of musts and wines of the Chardonnay variety and their fermentation aromas. *Am. J. Enol. Vitic.* 50: 253-258.

Gumyon, J.F. and Heintz, J.E. (1952). The fusel oil content of Californian wine. *Food Technol.* 6: 359.

Gutteridge, A., Pir, P., Castrillo, J.I., Charles, P.D., Lilley, K.S. and Oliver, S.G. (2010). Nutrient control of eukaryote cell growth: A systems biology study in yeast. *BMC Biol.* 8(1): 68.

Henick-Kling, T. (1993). Malolactic fermentation. pp. 289-326. *In*: G.H. Flect (Ed.). Wine Microbiology and Biotechnology. Switzerland: Harwood Academic Publishers.

Henschke, P.A. and Jiranek, V. (1993). Yeast-metabolism of nitrogen compounds. pp. 77-164. *In*: G.H. Fleet (Ed.). Wine Microbiology and Biotechnology. Harwood Academic Publishers, Australia.

Henschke, P.A. and Ough, C.S. (1991). Urea accumulation in fermenting grape juice. *Am. J. Enol. Vitic.* 42: 317-321.

Henschke, P.A. and Rose, A.H. (1991). Plasma Memberanes. pp. 297-345. *In*: A.H. Rose. and J.S. Harrison (Eds). The Yeasts. Vol. 4. Academic Press, London.

Henschke, P.A. and Jiranek, V. (1991). Hydrogen sulfide formation during fermentation: Effects of nitrogen composition in model grape musts. pp. 172-175. *In*: J.M. Rantz and Davis, C.A. (Eds.). Proceedings of the International Nitrogen Symposium on Grapes and Wine. Seattle, Washington, USA.

Hernawan, T. and Fleet, G. (1995). Chemical and cytological changes during the autolysis of yeast. *J. Ind. Microbiol.* 14: 440-450.

Herraiz, T. and Ough, C.S. (1992). Identification and determination of amino acid ethyl esters in wines by capillary gas chromatography and mass spectrometry. *J. Agric. Food Chem.* 40: 1015-1021.

Hidalgo, P. and Flores, M. (1994). Occurrence of the killer character in yeasts associated with Spanish wine production. *Food Microbiol.* 11: 161-167.

Hinnesbusch, A.G. (1988). Mechanisms of gene regulation in the general control of amino acid biosynthesis in *Saccharomyces cerevisiae*. *Microbiol. Rev.* 52: 248-273.

Horak, J. and Kotyk, A. (1977). Temperature effects in amino acid transport by *Saccharomyces cerevisiae*. *Exp. Mycol.* 1: 63-68.

Houtman, A.C. and du Plessis, C.S. (1981). The effect of juice clarity and several conditions promoting yeast growth or fermentation rate, the production of aroma components and wine quality. *S. Afric. J. Enol. Vitic.* 2: 71-81.

Houtman, A.C., Marais, J. and Du Plessis, C.S. (1980). The possibilities of applying present-day knowledge of wine aroma components: Influence of several juice factors on fermentation rate and ester production during fermentation. *S. Afric. J. Enol. Vitic.* 1: 27-33.

Ingledew, W.M. and Kunkee, R.E. (1985). Factors influencing sluggish fermentations of grape juice. *Am. J. Enol. Vitic.* 36: 65-76.

Ingledew, W.M., Magnus, C.A. and Patterson, J.R. (1987). Yeast foods and ethyl carbamate formation in wine. *Am. J. Enol. Vitic.* 38: 332-325.

Jansen, M., Veurink, J.H., Euverink, G.J.W. and Dijkhuizen, L. (2003). Growth of the salt-tolerant yeast *Zygosaccharomyces rouxii* in microtiter plates: Effects of NaCl, pH and temperature on growth and fusel alcohol production from branched-chain amino acids. *FEMS Yeast Res.* 3(3): 313-318.

Januik, M.T. (1984). The development of a technique to quantify and identify wine sulfur volatiles by gas chromatography. Thesis, University of California, Davis.

Jeandet, P., Vasserot, Y., Liger-Belair, G. and Marchal, R. (2011). Sparkling wine production. pp: 1064-1115. *In*: V.K. Joshi (Ed.). Handbook of Enology. Vol. III. Asia Tech Publishers, INC. New Delhi.

Jiranek, V. and Henschke, P.A. (1991). Assimilable nitrogen: Regulator of hydrogen sulphide production during fermentation. *Aust. Grape Wine Res.* 325: 27-32.

Jiranek, V., Langridge, P. and Henschke, P.A. (1990). Nitrogen requirement of yeast during wine fermentation. pp. 166-171. *In*: Proceedings of the Seventh Australian Wine Industry Technical Conference.Australian Industrial Publishers Adelaide, Australia.

Jones, E.W. and Fink, G.R. (1982). Regulation of amino acid and nucleotide biosynthesis in yeast. pp. 181. *In*: J.N. Strathern, E.W. Jones and J.B. Broach (Eds.). Molecular Biology of the Yeast Saccharomyces. Metabolism and Gene Expression. Cold Spring Harbor Laboratory, New York.

Joshi, V.K. (1997). Fruit Wines. (2nd ed.). DTP System, Direcorate of Extension Education. Dr. Y.S. Parmar University of Horticulture and Forestry, Nauni, Solan (HP), 226 pp.

Joshi, V.K. (2016) (Ed). Indigenous Fermented Foods of South Asia. Rob Nout and Prabir Sarkar, Series Editors. The Fermented Foods and Beverages Series, CRC Press, Roca. FL, 849 pp.

Joshi, V.K. and Bhutani, V.P. (1991). The influence of enzymatic clarification on the fermentation behaviour, composition and sensory qualities of apple wine. *Sci. Des Aliments* 11(3): 491-496.

Joshi, V.K. and Pandey, A. (1999). Biotechnology: Food Fermentation. pp. 1-45. *In*: V.K. Joshi and A. Pandey (Eds.). Biotechnology: Food Fermentation. Vol. I. Asiatech Publishers Inc., New Delhi.

Joshi, V.K., Bhutani, V.P. and Thakur, N.K. (1999a). Composition and Nutrition of Fermented Products. pp. 259-320. *In*: V.K. Joshi and Ashok Pandey (Eds.). Biotechnology: Food Fermentation. Vol. I. Educational Publishers and Distributors, New Delhi.

Joshi, V.K., Bhutani, V.P. and Sharma, R.C. (1990). Effect of dilution and addition of nitrogen source on chemical mineral and sensory qualities of wild apricot wine. *Am. J. Enol. Vitic.* 41: 229-232.

Joshi, V.K., Chauhan, S.K. and Lal, B.B. (1991). Extraction of juices from plum, peach and apricot by the pectolytic enzyme treatment. *J. Food Sci. Technol.* 28(1): 64-65.

Joshi, V.K., Sharma, S.K., Goyal, R.K. and Thakur, N.S. (1999b). Sparkling plum wine: Effect of method of carbonation and the type of base wine on physico-chemical and sensory qualities. *Braz. Arch. Bio. Technol.* 42: 315-321.

Joshi, V.K., Sharma, S.K. and Thakur, N.S. (1995). Technology and quality of sparkling wine with special reference to plum – an overview. *Indian Food Packer.* 49: 49-63.

Joshi, V.K., Thakur, N.S., Bhatt, A. and Garg, C. (2011). Wine and Brandy: A perspective. pp. 3-45. *In*: V.K. Joshi (Ed.). Handbook of Enology. Vol. 1. Asia Tech Publishers, Inc. New Delhi.

Joshi, V.K., Sandhu, D.K. and Thakur, N.S. (1999). Fruit based alcoholic beverages. pp. 647-744. *In*: V.K. Joshi and Ashok Pandey (Eds.). Biotechnology: Food Fermentation. Vol. II. Educational Publishers and Distributors, New Delhi.

Kalathenos, P., Sutherland, J.P. and Roberts, T.A. (1995). Resistance of some wine spoilage yeasts to combinations of ethanol and acids present in wine. *J. Appl. Bacteriol.* 78: 245-250.

Kandler, O. (1983). Carbohydrate metabolism in lactic acid bacteria. *Antonie vanm Leeuwenhoek.* 49: 209-224.

Killian, E. and Ough, C.S. (1979). Fermentation esters – Formation and retention as affected by fermentation temperature. *Amer. J. Enol. Vitic.* 30(4): 301-305.

Kollar, R., Vorisek, J. and Sturdik, E. (1993). Biochemical, morphological, and cytochemical studies of enhanced autolysis of *Saccharomyces cerevisiae*. *Folia Microbiol.* 38: 479-485.

Kruckeberg, A.L. and Bisson, L.F. (1990). The HXT2 gene of *Saccharomyces cerevisiae* is required for high affinity glucose transport. *Mol. Cell. Biol.* 10(11): 5903-5913.

Kučerová, J. and Široký, J. (2014). Study of changes organic acids in red wines during malolactic fermentation. *Acta Universitatis Agriculturae et Silviculturae Mendelianae Brunensis* 59(5): 145-150.

Lagace, L.S. and Bisson, L.F. (1990). Survey of yeast acid proteases for effectiveness of wine haze reduction. *Am. J. Enol. Vitic.* 41(2): 147-155.
Lagunas, R. (1986). Misconceptions about the energy metabolism of *Saccharomyces cerevisiae*. *Yeast.* 2(4): 221-228.
Large, P.J. (1986). Degradation of organic nitrogen compounds by yeasts. *Yeast* 2: 1-34.
Lavigne, V. and Bourdieu, D. (1996). Demonstration and interpretation of the yeast lees ability to adsorb certain volatile thiols contained in wine. *J. Int. Sci. Vigne. Vin.* 30: 201-206.
Le Fur, Y., Maume, G., Feuillat, M. and Maume, B.F. (1997). Characterization by gas chromatography/ mass spectrometry of sterols in *Saccharomyces cerevisiae* during autolysis. *J. Agric. Food Chem.* 47: 2860-2864.
Lehtonen, P. (1996). Determination of amines and amino acids in wine: A review. *Am. J. Enol. Vitic.* 47: 127-133.
Lema, C., Garcia-Jares, C., Orriols, I. and Angulo, L. (1996). Contribution of Saccharomyces and Non-Saccharomyces populations to the production of some components of Albarino wine aroma. *Am. J. Enol. Vitic.* 47: 206-216.
Lerm, E., Engelbrecht, L. and Du Toit, M. (2010). Malolactic fermentation: the ABC's of MLF. *S. Afr. J. Enol. Vitic.* 31(2): 186-212.
Llauberes, R.M., Dubourdieu, D. and Villetaz, J.C. (1987). Exocellular polysaccharides from Saccharomyces in wine production. *Ind. Bevande.* 41: 277-286.
Llauberes, R.M., Richard, B., Lonvaud-Funel, A. and Dubourdieu, D. (1990). Structure of an exocellular B-D-glucan from *Pediococcus* sp., a wine lactic acid bacterium. *Carbohyd. Res.* 203: 103-107.
Loipez-Barajas, M., Loipez-Tamames, E., Buxaderas, S., Suberbiola, G. and de la Torre-Boronat, M.C. (2001). Influence of wine polysaccharides of different molecular mass on wine foaming. *Am. J. Enol. Vitic.* 52: 146-150.
Lonvaud-Funel, A. (1999). Lactic acid bacteria in the quality improvement and depreciation of wine. *Anton. Leeuw.* 76: 317-331.
Lonvaud-Funel, A. (2001). Biogenic amines in wines: Role of lactic acid bacteria. *FEMS Microbiol. Lett.* 199: 9-13.
Loubser, G.J. and Du Plessis, C.S. (1976). The quantitative determination and some values of dimethyl sulfide in white table wines. *Vitis.* 15: 248-252.
Luguera, C., Moreno-Arribas, V., Pueyo, E. and Polo, M.C. (1997). Capillary electrophoretic analysis of wine proteins. Modifications during the manufacture of sparkling wines. *J. Agric. Food Chem.* 45: 3766-3770.
Luguera, C., Moreno-Arribas, V., Pueyo, E., Bartolomei, B. and Polo, M.C. (1998). Fractionation and partial characterization of protein fractions present at different stages of the production of sparkling wines. *Food Chem.* 63: 465-471.
Mafra, I., Herbert, P., Santos, L., Barros, P. and Alves, A. (1999). Evaluation of biogenic amines in some Portuguese quality wines by HPLC fluorescence detection of OPA derivatives. *Am. J. Enol. Vitic.* 50: 128-132.
Magasanik, B. and Kaiser, C.A. (2002). Nitrogen regulation in *Saccharomyces cerevisiae*. *Gene.* 290(1-2): 1-18.
Mahon, H.M.M., Zoecklein, B.W. and Jasinski, Y.W. (1999). The effects of pre-fermentation maceration temperature and percent alcohol (v/v) at Press on the concentration of Cabernet Sauvignon grape glycosides and glycoside fractions. *Am. J. Enol. Vitic.* 50: 385-390.
Malik, K.D.H.H.F. (2014). Influence of different nutrition conditions on main volatiles of wine yeasts. *J. Food Nutr. Res.* 53(4): 304-312.
Manca de Nadra, M.C. and Strasser de Saad, A.M. (1995). Polysaccharide production by Pediococcus pentosaceous from wine. *Int. J. Food Microbiol.* 27: 101-106.
Manzanares, P., Rojas, V., Genoves, S. and Valles, S. (2000). A preliminary search for anthocyanin-B-D-glucosidase activity in non-Saccharomyces wine yeasts. *Int. J. Food Sci. Technol.* 35: 95-103.
Margalith, P.Z. (1981). Flavour Microbiology. Charles C Thomas Publisher.
Martinez, P., Codon, A.C., Perez, I. and Benitez, T. (1995). Physiological and molecular characterization of flor yeasts: Polymorphism of flor yeast populations. *Yeast* 11: 1399-1411.

Martinez, P., Perez, Rodriguez, L. and Benitez, T. (1997). Evolution of flor yeast population during the biological aging of fino sherry wine. *Am. J. Enol. Vitic.* 48: 160-168.

Martinez-Rodriguez, A.J. and Polo, M.C. (2000). Characterization of the nitrogen compounds released during yeast autolysis in a model wine system. *J. Agric. Food Chem.* 48: 1081-1085.

Martinez-Rodriguez, A.J. and Polo, M.C. (2001). Enological aspects of yeast autolysis. pp. 285-301. *In*: S.G. Pandalai (Ed.). Recent Research and Development in Microbiology. Trivandum, India: Research Signpost.

Martinez-Rodriguez, A.J., Carrascosa, A.V. and Polo, M.C. (2001). Release of nitrogen compounds to the extracellular medium by three strains of *Saccharomyces cerevisiae* during induced autolysis in a model wine system. *Int. J. Food Microbiol.* 68: 155-160.

Martinez-Rodriguez, A.J., Polo, M.C. and Carrascosa, A.V. (2001). Structural and ultrastructural changes in yeast cells during autolysis in a model wine system and in sparkling wine. *Int. J. Food Microbiol.* 71: 45-51.

Mauricio, J.C. and Ortega, J.M. (1997). Nitrogen compounds in wine during its biological aging by two flor film yeasts: An approach to accelerated biological aging of dry sherry-type wines. *Biotechnol. Bioeng.* 53: 159-167.

Mauricio, J.C., Moreno, J.J. and Ortega, J.M. (1997). In Vitro specific activities of alcohol and aldehyde dehydrogenases from two flor yeasts during controlled wine aging. *J. Agric. Food. Chem.* 45: 1967-1971.

Mayo, B., Aleksandrzak-Piekarczyk, T., Fernández, M., Kowalczyk, M., Álvarez-Martín, P. and Bardowski, J. (2010). Updates in the metabolism of lactic acid bacteria. pp. 3-33. *In*: Mozzi, F., Raya, R.R. and Vignolo, G.M. (Eds.). Biotechnology of Lactic Acid Bacteria: Novel Applications. Wiley-Blackwell, Lowa, USA.

Mc Clellan, C.J., Does, A.L. and Bisson, L.F. (1989). Characterization of hexose uptake in wine strains of *Saccharomyces cerevisiae* and *Saccharomyces bayanus*. *Am. J. Enol. Vitic.* 40: 9-15.

McMahon, H., Zoecklein, B.W., Fugelsang, K. and Jasinki, Y. (1999). Quantification of glycosidase activities in selected yeasts and lactic acid bacteria. *J. Ind. Microbiol. Biotechnol.* 23: 198-203.

Meijer, M.M., Boonstra, J., Verkleij, A.J. and Verrips, C.T. (1998). Glucose repression in *Saccharomyces cerevisiae* is related to the glucose concentration rather than the glucose flux. *J. Biol. Chem.* 273(37): 24102-24107.

Mendes-Ferreira, A., Barbosa, C., Lage, P. and Mendes-Faia, A. (2011). The impact of nitrogen on yeast fermentation and wine quality. *Ciência Téc. Vitiv.* 26(1): 17-32.

Mendes-Ferreira, A., Climaco, M.C. and Mendes-Faia, A. (2001). The role of non-Saccharomyces species in releasing glycosidic bound fraction of grape aroma components—a preliminary study. *J. Appl. Microbiol.* 91: 67-71.

Miljić, U., Puškaš, V., Velićanski, A., Mašković, P., Cvetković, D. and Vujić, J. (2016). Chemical composition and in vitro antimicrobial and cytotoxic activities of plum (*Prunus domestica* L.) wine. *J. Inst. Brew.* 122(2): 342-349.

Monk, P.R. (1986). Formation, utilization and excretion of hydrogen sulphide by wine yeast. *Aust. N. Z. Wine Ind. J.* 1: 10-16.

Monteiro, F.F. and Bisson, L.F. (1991). Amino acid utilization and urea formation during vinification. *Am. J. Enol. Vitic.* 42: 199-208.

Monteiro, F.F., Trousdale, E.K. and Bisson, L.F. (1989). Ethyl carbamate formation in wine: Use of radioactively labeled precursors to demonstrate the involvement of urea. *Am. J. Enol. Vitic.* 40(1): 1-8.

Moreira, N., Mendes, F., Pereira, O., de Pinho, P.G., Hogg, T. and Vasconcelos, I. (2002). Volatile sulphur compounds in wines related to yeast metabolism and nitrogen composition of grape musts. *Anal. Chim. Acta.* 458(1): 157-167.

Moreno-Arribas, M.V. and Polo, M.C. (2005). Winemaking biochemistry and microbiology: Current knowledge and future trends. *Crit. Rev. Food Sci. Nutr.* 45: 265–286.

Moreno-Arribas, V. and Lonvaud-Funel, A. (2001). Lactic acid bacteria. Involvement in wine quality. pp. 481-504. *In*: S.G. Pandalai (Ed.). Recent Research and Development in Microbiology. Trivandum, India: Research Signpost.

Moreno-Arribas, V., Bartolome, B., Pueyo, E. and Polo, M.C. (1998). Isolation and characterization of individual peptides from wine. *J. Agric. Food Chem.* 46: 3422-3425.

Moreno-Arribas, V., Pueyo, E. and Polo, M.C. (1996). Peptides in musts and wines. Changes during the manufacture of cavas (Sparkling wines). *J. Agric. Food Chem*. 44: 3783-3788.

Moreno-Arribas, V., Pueyo, E., Nieto, J., Martin-Alvarez, P.J. and Polo, M.C. (2000). Influence of the polysaccharides and the nitrogen compounds on foaming properties of sparkling wines. *Food Chem.* 70: 309-317.

Moreno-Arribas, V., Pueyo, E., Polo, M.C. and Martin-Alvarez, P.J. (1998). Changes in the amino acid composition of the different nitrogenous fractions during the aging of wine with yeast. *J. Agric. Food Chem.* 46: 4042-4051.

Moyano, L., Zea, L., Moreno, J. and Medina, M. (2002). Analytical study of aromatic series in sherry wines subjected to biological aging. *J. Agric. Food Chem.* 50: 7356-7361.

Muller, C.J., Fugelsang, K.C. and Wahlstorm, V.L. (1993). Capture and use of volatile flavour constituents emitted during wine fermentations. pp. 219. *In*: B.H. Gump (Ed.). Beer and Wine Production: Analysis, Characterization and Technological Advances. Am. Chem. Society Series 536, Washington. D.C.

Navarro, G. and Navarro, S. (2011). Stuck and sluggish fermentation. pp. 591-617. *In*: V.K. Joshi, (Ed.). Handbook of Enology: Principles, Practices, and Recent Innovations. Vol 2. Asiatech Publishers Inc., New Delhi.

Neti, Y., Erlinda, I.D. and Virgilio, V.G. (2011). The effect of spontaneous fermentation on the volatile flavor constituents of durian. *Int. Food Res. J.* 18(2): 635-641.

Nidelet, T., Brial, P., Camarasa, C. and Dequin, S. (2016). Diversity of flux distribution in central carbon metabolism of *S. cerevisiae* strains from diverse environments. *Microb. Cell Fact.* 15(1): 58.

Nykanen, L. (1986). Formation and occurrence of flavour compounds in wine and distilled alcoholic beverages. *Am. J. Enol. Vitic*. 37: 84-86.

Ough, C.S. (1993). Lead in wines—A review of recent reports. *Am. J. Enol. Vitic.* 44: 464–467.

Ough, C.S. and Bell, A.A. (1980). Effects of nitrogen fertilization of grapevines on amino acid metabolism and higher-alcohol formation during grape juice fermentation. *Am. J. Enol. Vitic.* 31: 122-123.

Ough, C.S. and Lee, T.H. (1981). Effect of vineyard nitrogen fertilization level on the formation of some fermentation esters. *Am. J. Enol. Vitic*. 32: 125-127.

Ough, C.S., Crowell, E.A. and Gutlove, B.R. (1988a). Carbamyl compound reactions with ethanol. *Am. J. Enol. Vitic.* 39: 239-242.

Ough, C.S., Crowell, E.A. and Monney, L.A. (1988b). Formation of ethyl carbamate precursers during grape juice (Chardonnary) fermentation. Addition of amino acid, urea and ammonia effects of fortification on nitrocellular and extracellular precursors. *Am. J. Enol. Vitic.* 39: 243-249.

Ough, C.S., Stevens, D., Sendovski, T., Huang, Z. and An, D. (1990). Factors contributing to urea formation in commercial fermented wines. *Am. J. Enol. Vitic*. 41(1): 68-73.

Park, S.K., Boulton, R.B. and Noble, A.C. (2000). Formation of hydrogen sulfite and glutathione during fermentation of white grape musts. *Am. J. Enol. Vitic.* 51: 91-97.

Peinado, R.A., Moreno, J.J., Ortega, J.M. and Mauricio, J.C. (2003). Effect of gluconic acid consumption during stimulation of biological aging of sherry wines by a flor yeast strain on the final volatile compounds. *J. Agric. Food Chem.* 51: 6198-6203.

Pentjuss, A., Stalidzans, E., Liepins, J., Kokina, A., Martynova, J., Zikmanis, P. and Fell, D.A. (2017). Model-based biotechnological potential analysis of *Kluyveromyces marxianus* central metabolism. *J. Ind. Microbiol Biotechnol*. 44(8): 1177-1190.

Peynaud, E., Domercq, S., Boidron, A., Lafon-Lafourcade, S. and Guimberteau, G. (1964). Etude der les levures Schizosachharomyces metabolisant I'acide L-malique. *Arch. Microbiol.* 48: 150.

Picinelli, A., Suarez, B., Garcia L. and Mangas, J.J. (2000).Changes in phenolic contents during sparkling wine making. *Am. J. Enol. Vitic.* 51(2): 144-149.

Pierce, J.S. (1982). The Margaret Jones Memorial Lecture: Amino acids in malting and brewing. *J. Inst. Brew.* 88(4): 228-233.

Pierce, J.S. (1987). Horace Brown memorial lecture; the role of nitrogen in brewing. *J. Inst. Brew.* 93(5): 378-381.

Pilatte, E., Nygaard, M., Cai Gao, Y., Krentz, S., Power, J. and Lagarde, G. (2000). Etude de l'effect du lisozyme sur differentes souches d' Oenococcus oeni. Applications dans la gestion de la fermentation malolactique. *Rev. Fr. Oenol.*, Novembre/Decembre N°185.

Pinu, F.R., Edwards, P.J., Gardner, R.C. and Villas-Boas, S.G. (2014). Nitrogen and carbon assimilation by *Saccharomyces cerevisiae* during Sauvignon blanc juice fermentation. *FEMS Yeast Res.* 14(8): 1206-1222.

Plata, M.C., Mauricio, J.C., Millan, C. and Ortega, J.M. (1998). *In vitro* activity of alcohol acetyltransferase and esterase in two flor yeast strains during biological aging of sherry wines. *Fement. Bioeng.* 85: 369-374.

Postel, W. and Ziegler, L. (1991). Influence of the duration of yeast contact and of the manufacturing process on the composition and quality of sparkling wines II. Free amino acids and volatile compounds. *Wein-Wissenschaft Viticultural and Enological Sciences* 46: 26-32.

Pozo-Bayon, M.A., Pueyo, E., Martin-Alvarez, P.J., Martinez-Rodriguez, A.J. and Polo, M.C. (2003). Influence of yeast strain, bentonite addition, and aging time on volatile compounds of sparkling wines. *J. Enol. Vitic.* 54: 273-278.

Pueyo, E., Martinez-Rodriguez, A.J., Polo, M.C., Santa-Maria, G. and Bartolome, B. (2000). Release of lipids during yeast autolysis in a model wine system. *J. Agric. Food Chem.* 48: 116-122.

Pueyo, E., Olano, A. and Polo, M.C. (1995). Neutral monosacharides composition of the polysaccharides from must, wines, and cava wines. *Rev. Esp. Cienc. Technol. Aliment* 35: 191-201.

Radler, F. and Schutz, H. (1982). Glycerol production form various strains of *Saccharomyces. Am. J. Enol. Vitic.* 33(1): 36-40.

Radler, F. (1993). Yeasts Metabolism of organic acids. pp. 165-182. *In*: G.H. Fleet (Ed.). Wine Microbiology and Biotechnology. Harwood Academic Publishers, Chur Switzerland.

Radler, F. (1986). Microbial biochemistry. *Experientia.* 42(8): 884-893.

Rana, N.S. and Rana, V.S. (2011). Biochemistry of wine preparation. pp. 618-678. *In*: V.K. Joshi (Ed.). Handbook of Enology: Principles, Practices, and Recent Innovations. Vol 2. Asiatech Publishers Inc., New Delhi.

Rankine, B.C. (1963). Nature, origin and prevention of hydrogen sulphide aroma in wines. *J. Sci. Food Agri.* 14(2): 79-91.

Rankine, B.C. (1966). Decomposition of L-malic acid by wine yeasts. *J. Sci. Food Agric.* 17(7): 312-316.

Rankine, B.C. (1967). Formation of higher alcohols by wine yeasts and relationship to taste thresholds. *J. Sci. Food. Agric.* 18(12): 583-589.

Rapp, A. and Versini, G. (1991). Influence of nitrogen compounds in grapes on aroma compounds of wine. pp. 156. *In*: J. Rantz and Davis (Eds.). Proceedings of the International Symposium on Nitrogen in Grapes and Wines. CA.

Rauhut, D., Kiirbel, H., Dittrich, H.H. and Grossmann, M. (1996). Properties and differences of commercial yeast strains with respect to their formation of sulfur compounds. *Vitic. Enol. Sci.* 51: 187-192.

Rebordinos, L., Infante, J.J., Rodriguez, M.E., Vallejo, I. and Cantroal. J.M. (2011). Wine Yeast Growth and Factor Affecting. pp. 406-434. *In*: Joshi, V.K. (Ed.). Handbook of Enology. Vol. 2. Asia Tech Publication, New Delhi.

Reibstein, D., den Hollander, J.A., Pilkis, S.J. and Shulman, R.G. (1986). Studies on the regulation of yeast phosphofructo-I-kinase: Its role in aerobic and anaerobic glycolysis. *Biochem.* 25(1): 219-227.

Renault, P., Coulon, J., Moine, V., Thibon, C. and Bely, M. (2016). Enhanced 3-sulfanylhexan-1-ol production in sequential mixed fermentation with *Torulaspora delbrueckii/Saccharomyces cerevisiae* reveals a situation of synergistic interaction between two industrial strains. *Front. Microbial.* 7: 293.

Rhodes, N., Morris, C.N., Ainsworth, S. and Kinderlerer, J. (1986). The regulatory properties of yeast pyruvate kinase. *Biochem. J.* 234: 705.

Ribéreau-Gayon, P., Dubourdieu, D., Donéche, B. and Lonvaud, A. (2000). Handbook of Enology. Wiley & Sons, Ltd., Chichester, England.

Ribereau-Gayon, P., Dubourdieu, D., Doneche, B. and Lonvaud, A. (2006). Biochemistry of alcoholic fermentation and metabolic pathways of wine yeasts. pp. 53-77. *In*: Handbook of Enology. Vol 1. The Microbiology of Wine and Vinification (2 edn.). John Wiley and Sons.

Ribereau-Gayon, P., Dubourdieu, D., Doneche, B. and Lonvaud, A. (1998). Le metabolism des bacteries lactiques. pp. 171-193. *In*: Traite d'Oenologie, Microbiologie du vin, Vinifications. Paris: Dunod.

Richard, P.V., Ellen, M.H., Theresa, B. and Cheri, W. (1997). Enology (Winemaking). pp. 95-145. *In*: Winemaking, from Grape Growing to Market Place. The Chapman and Hall Enology Library.

Rodriquez, S. and Thornton, R. (1990). Factors influencing the utilization of L-malate by yeasts. *FEMS Microbiol. Lett.* 72: 17-22.

Rojas, V., Gil, J.V., Pinaga, F. and Manzanares, P. (2001). Studies on acetate ester production by non-Saccharomyces wine yeast. *Int. J. Food Microbiol.* 70: 283-289.

Rojas, V., Gil, J.V., Pinaga, F. and Manzanares, P. (2003). Acetate ester formation in wine by mixed cultures in laboratory fermentations. *Int. J. Food Microbiol.* 86: 181-188.

Rojo-Bezares, B., Sáenz, Y., Zarazaga, M., Torres, C. and Ruiz-Larrea, F. (2007). Antimicrobial activity of nisin against *Oenococcus oeni* and other wine bacteria. *Int. J. Food Microbiol.* 116(1): 32-36.

Rollero, S., Bloem, A., Camarasa, C., Sanchez, I., Ortiz-Julien, A., Sablayrolles, J.M. and Mouret, J.R. (2015). Combined effects of nutrients and temperature on the production of fermentative aromas by *Saccharomyces cerevisiae* during wine fermentation. *App. Microbiol. Biotechnol.* 99(5): 2291-2304.

Romano, P. and Suzzi, G. (1996). Origin and production of acetoin during wine yeast fermentation. *Appl. Environ. Microbiol.* 62: 309-312.

Romano, P., Fiore, C., Paraggio, M., Caruso, M. and Capece, A. (2003). Function of yeast species and strains in wine flavour. *Int. J. Food Microbial.* 86(1): 169-180.

Romano, P., Suzzi, G., Domizio, P. and Fatichenti, F. (1997). Secondary products formation as a tool for discriminating non-*Saccharomyces* wine strains. *Anton. Leeuw.* 71: 239-242.

Rose, A.H. (1987). Responses to the chemical environment. pp. 5. *In*: A.H. Rose and J.S. Harrison (Eds.). The Yeasts. Vol. 2, 2nd ed., Academic Press, London.

Rose, A.H. (1993). Sulphur dioxide and other preservatives. *J. Wine Res.* 4(1): 43-47.

Rosi, I., Costamagna, L. and Bertuccioli, M. (1987). Screening for extracellular acid protease(s) production by wine yeasts. *J. Inst. Brew.* 93(4): 322-324.

Rupasinghe, H.V., Joshi, V.K., Smith, A. and Parmar, I. (2017). Chemistry of fruit wines. pp. 105-176. *In*: Kosseva, M.R., Joshi, V.K., Panesar, P. (Eds.). Science and Technology of Fruit Wine Production. Elsevier's Academic Press.

Saayman, M. and Viljoen-Bloom, M. (2006). The biochemistry of malic acid metabolism by wine yeasts – A review. *S Afr. J. Enol. Vitic.* 27(2): 113-122.

Salmon, J.M. (1989). Effect of sugar transport inactivation in *Saccharomyces cerevisiae* on sluggish and stuck enological fermentations. *Appl. Environ. Microbiol.* 55(4): 953-958.

Santamariia, P., Loi Pez, R., Gutierrez, R. and Garciia-Escudero, E. (1995). Evolution des acides gras totaux pendant la fermentation a differents temperatures. *J. Int. Sci. Vigne Vin.* 29: 101-104.

Schutz, M. and Kunkee, R.E. (1977). Formation of hydrogen sulfide from elemental sulfur during fermentation by wine yeast. *Am. J. Enol. Vitic.* 28(3): 137-144.

Schwartz, H. and Radler, F. (1988). Formation of L (-) malate by *Saccharomyces cerevisiae* during fermentation. *Appl. Microbiol. Biotech.* 27: 553-560.

Schwenke, J. (1991). Vacuoles, internal membranous systems and vesicles. pp. 347. *In*: A.H. Rose and J.S. Harrison (Eds.). The Yeasts. Vol. 4. 2nd ed. Academic Press, London.

Sharma, S. and Tauro, P. (1986). Control of ethanol production by yeast: Role of pyruvate decarboxylase and alcoholic dehydrogenase. *Biotech. Lett.* 8(10): 735-738.

Shimazu, Y. and Watanabe, M. (1981). Effects of yeast strains and environmental conditions on forming of organic acids in must during fermentation (Japanese). *J. Ferment. Tech.* 59(1): 27-32.

Siila Santos, M.H. (1996). Biogenic amines: Their importance in foods. *Int. J. Food Microbiol.* 29: 213-231.

Singleton, V.L., Sieberhagen, H.A., De Wet, P. and Van Wyk, C.J. (1975). Composition and sensory qualities of wines prepared from white grapes by fermentation with and without grape solids. *Am. J. Enol. Vitic.* 26(2): 62-69.

Sinton, T.H., Ough, C.S., Kissler, J.J. and Kasimatis, A.N. (1978). Grape juice indicators for prediction of potential wine quality. I. Relationship between crop level, juice and wine composition, and wine sensory ratings and scores. *Am. J. Enol. Vitic.* 29(4): 267-271.

Soden, A., Francis, I.L., Oakey, H. and Henschke, P.A. (2000). Effects of co-fermentation with *Candida stellata* and *Saccharomyces cerevisiae* on the aroma and composition of Chardonnay wine. *Aus. J. Grape Wine Res.* 6: 21-30.

Soles, R.M., Ough, C.S. and Kunkee, R.E. (1982). Ester concentration differences in wine fermented by various species and strains of yeasts. *Am. J. Enol. Vitic.* 33(2): 94-98.

Sols, A. (1981). Multimodulation of enzyme activity. *In*: Current Topics in Cellular Regulation, some metabolic features during lactic and fermentation at vegetable juices. *Rom. Biotechnol.* 19: 77.

Spiropoulos, A., Tanaka, J., Flerianos, I. and Bisson, L.F. (2000). Characterization of hydrogen sulfide formation in commercial and natural wine isolates of *Saccharomyces*. *Am. J. Enol. Vitic.* 51(3): 233-248.

Sponholz, W.R. (1988). Alcohols derived from sugars and other sources and full bodiedness of wines. p. 147. *In*: H.F. Linskens and J.F. Jackson (Eds.). Modern Methods of Plant Analysis. New series Vol. 6. Wine Analysis. Springer, Berlin, Heidelberg.

Sponholz, W.R. (1989). Fehlerhafter und Unerwunschte Erscheinungen in wein. pp. 385. *In*: G. Wurdig and R. Woller (Eds.).Chemic des Weines. Ulmer, Stuttgart, Germany.

Sponholz, W.R. (1993). Wine spoilage by microorganisms. pp. 395-420. *In*: G.H. Fleet (Eds.). Wine Microbiology and Biotechnology. Switzerland: Harwood Academic Publishers.

Sponholz, W.R. and Dittrich, H.H. (1979). Analytiche vergleiche von Mosten und Weinen aus gesunden and essigstichigen traubenbeeren. *Wein Wissenschaft.*

Strasser de Saad, A.M. and Manca de Nadra, M.C. (1992). Sugar and malic acid utilization and acetic acid formation by Leuconostoc oenos. *World J. Microbiol. Biot.* 8: 280-283.

Sturley, S.L. and Young, T.W. (1988). Extracellular protease activity in a strain of *Saccharomyces cerevisiae*. *J. Inst. Brew.* 94(1): 23-27.

Styger, G., Prior, B. and Bauer, F.F. (2011). Wine flavor and aroma. *J. Ind. Microbiol Biotechnol.* 38(9): 1145.

Suarez-Lepe, J.A. (1997). E1 caracter filmoogeno de las levaduras y otras propiedades de interes para vinificaciones especiales. pp. 171-196. *In*: Mundi-Prensa (Ed.). Levaduras vinicas, Funcionalidad y uso en bodega. Mundrid: Mundi-Prensa.

Suarez-Lepe, J.A., Hugo and Leal, B. (1990). Vinificaciones especiales desde el punto de vista microbioloogico: Los vinos con crianza biolgica. Microbiologia Enologica, Fundamentos de vinific. Salmon, J.M., Sablayrolles, J.M. and Rosenfeld, E. acwn. Mundi-Prensa ed. Madrid: Mundi-Prensa. pp. 501-538.

Suzzi, G., Romano, P. and Zambonelli, C. (1985). *Saccharomyces* strain selection in minimizing $SO_2$ requirement during vinification. *Am. J. Enol. Vitic.* 36(3): 199-202.

Tanner, H. (1969). Der weinböckser, entstehung und beseitigung. Zeitschrift für Obst-und Weinbau. 78: 105.

Thomas, C.S., Gubler, W.D., Silacci, M.W. and Miller, R. (1993). Changes in elemental sulfur residues on Pinot noir and Cabernet sauvignon berries during the growing season. *Am. J. Enol. Vitic.* 44(2): 205-210.

Thomas, K.C. and Ingledew, W.M. (1990). Fusel alcohol production: Effects of free amino acid nitrogen on fermentation of very-high-gravity wheat mashes. *Appl. Environ. Microbiol.* 56: 2046-2050.

Todd, B.E.N., Fleet, G.H. and Henschke, P.A. (2000). Promotion of autolysis through the interaction of killer and sensitive yeasts: Potential application in sparkling wine production. *Am. J. Enol. Vitic.* 51: 65-72.

Tominaga, T., Gachon, C.P. and Dubourdieu, D.(1998). A new type of flavour precursors in *Vitis vinifera* L. cv. Sauvignon Blanc, S-cysteine conjugates. *J. Agri. Food Chem.* 46(12): 5215-5219.

Tominaga, T., Guyot, R.B., Gachons, C.P. and Dubourdieu, D. (2000). Contribution of volatile thiols to the aromas of white wines made from several *Vitis vinifera* grape varieties. *Am. J. Enol. Vitic.* 51(2): 178-181.

Ugliano, M., Henschke, P.A., Herderich, M.J. and Pretorius, I.S. (2007). Nitrogen management is critical for wine flavour and style. *Wine Ind. J.* 22(6): 24-30.

Ugliano, M., Kwiatkowski, M.J., Travis, B., Francis, I.L., Waters, E.J., Herderich, M.J. and Pretorius, I.S. (2009). Post-bottling management of oxygen to reduce off-flavour formation and optimize wine style. *Aust. NZ Wine Ind. J.* 24(5): 24-28.

Valero, E., Millan, M.C., Mauricio, J.C., and Ortega, J.M. (1998).Effect of grape skin maceration on sterol, phospholipid, and fatty acid contents of *Saccharomyces cerevisiae* during alcoholic fermentation. *Am. J. Enol. Vitic.* 49: 119-124.

Van Vuuren, H.J.J. and Dicks, L.M.T. (1993). Leuconostoc oenos: A review. *Am. J. Enol. Vitic.* 44: 99-112.

Van Vuuren, H.J.J. and Jacobs, C.J. (1992). Killer yeasts in the wine industry: A review. *Am. J. Enol. Vitic.* 43: 119-128.

Vaquero, M.R., Alberto, M.R. and de Nadra, M.M. (2007). Antibacterial effect of phenolic compounds from different wines. *Food Control* 18(2): 93-101.

Vazquez-Lasa, M.B., Iniguez-Crespo, M., Gonzalez-Larraina, M. and Gonzalez-Guerrero, A. (1998). Biogenic amines in Rioja wines. *Am. J. Enol. Vitic.* 49(3): 229.

Versari, A., Parpinello, G.P. and Cattaneo, M. (1999). *Leuconostoc oenos* and malolactic fermentation in wine: A review. *J. Ind. Microbiol. Biot.* 23: 447-455.

Vianna, E. and Ebeler, S.E. (2001). Monitoring ester formation in grape juice fermentation using solid phase microextraction coupled with gas chromatography-mass spectrometry. *J. Agric. Food Chem.* 49: 589-595.

Vidal-Carou, M.C., Codony-Salcedo, R. and Marine-Font, A. (1991). Change in the concentration of histamine and tyramine during wine spoilage at various temperatures. *Am. J. Enol. Vitic.* 42(2): 145-149.

Volschenk, H., Van Vuuren, H.J.J. and Viljoen-Bloom, M. (2006). Malic acid in wine: Origin, function and metabolism during vinification. *S. Afr. J. Enol. Vitic.* 27(2): 123-136.

Volschenk, H., Viljoen, M., Grobler, J., Baur, F., Lonvaud-Funel, A., Denayrolles, M., Subden, R.E. and Vanvuuren, H.J.J. (1997). Malolactic fermentation in grape must by a genetically engineered strain of *Saccharomyces cerevisiae. Am. J. Ecol. Vitic.* 48(2): 193-197.

Vos, P.J.A. and Gray, R.S. (1979). The origin and control of hydrogen sulfide during fermentation of grape must. *Am. J. Enol. Vitic.* 30(3): 187-197.

Vos, P.J.A. (1981). Assimilable nitrogen – A factor influencing the quality of wines. pp. 163. *In*: International Association for Modern Winery Technology and Management. 6th International Oenological Symposium (28-30 April 1981). Mainz, Germany.

Vos, P.J.A., Zeeman, W. and Heymann, H. (1978). The effect on wine quality of diammonium phosphate additions to musts. In: *Proc. S. Afric. Soc. Enol. Vitic.* Stellenbosch, Cap Town, South Africa: 87.

Wainwright, T. (1971). Production of $H_2S$ by wine yeast: Role of nutrients. *J. Appl. Bacteriol.* 34(1): 161-171.

Walker, G. and Stewart, G. (2016). *Saccharomyces cerevisiae* in the production of fermented beverages. *Beverages* 2(4): 30.

Walsh, J.L., Keith, T.J. and Knull, H.R. (1989). Glycolytic enzyme interactions with tubulin and microtubules. *Biochimica et Biophysica Acta*, 999(1): 64-70.

Waterhouse, A.L., Sacks, G.L. and Jeffery, D.W. (2016). Understanding Wine Chemistry. John Wiley & Sons. pp. 420.

Webb, A.D. and Ingraham, J.L. (1963). Fusel oil. *Adv. Appl. Micro.* 5: 317-353.

Weinhandl, K., Winkler, M., Glieder, A. and Camattari, A. (2014). Carbon source-dependent promoters in yeasts. *Microb. Cell Fact.* 13(1): 5.

Wenzel, K. and Dittrich, H.H. (1978). Zur beeinflussung der schwefelwasserstoff-buildung der hefedurch trub, stickstoffgehalt, molecularen schwefel und kupfer bei der vergarung von traubenmost. *Wein Wissen.* 33: 200.

Wenzel, K., Dittrich, H.H. and Pletzonka, B. (1982). Untersuchungen zur Beteiligung von Hefen am Apfelsaureabbau bei der Weinbereitung. *Wein-Wissenschaft.* 37: 133.

Whiting, G.C. (1976). Organic acid metabolism of yeasts during fermentation of alcoholic beverages: A review. *J. Inst. Brew.* 82(2): 84-92.

Whiton, R.S. and Zoecklein, B.W. (2002). Determination of ethyl carbamate in wine by solid phase microextraction and gas chromatography/mass spectrometry. *Am. J. Enol. Vitic.* 53(1): 60-63.

Wiame, J.M., Grenson, M. and Arst, H.N., Jr. (1985). Nitrogen catabolite repression in yeast and filamentous fungi. *Adv. Micro. Physiol.* 26: 1-88.

Woidich, H. and Pfannhauser, W. (1974). Zur Gaschromatographichen Analyse Von Branntweinen: Quantitative best mming Von Acetaldenyd WEssigsaureanethylester, Essigsaureanthylester, methanol, Butanol-1, Butanol-2, Propanol-1, 2-methuylpropanol-1, Amylal Koholen und Hexanol-1. Mitt. Hoeschesen Bundesleher-Versuchsanst. *Wein Obstbau. Klosterneuburg.* 24: 155.

Xu, H.X. and Lee, S.F. (2001). Activity of plant flavonoids against antibiotic-resistant bacteria. *Phytotherapy Research: An International Journal Devoted to Pharmacological and Toxicological Evaluation of Natural Product Derivatives* 15(1): 39-43.

Yanai, T. and Sato, M. (1999). Isolation and properties of B-glucosidase produced by *Debaromyces hansenii* and its application in winemaking. *Am. J. Enol. Vitic.* 50: 231-235.

Yang, H.Y. (1953). Fruit wines: Requisites for successful fermentation. *J. Agri. Food Chem.* 1(4): 331-333.

Zea, L., Moreno, J., Medina, M. and Ortega, M.J. (1994). Evolution of C6, C8 and C10 acids and their ethyl esters in cells and musts during the aging with three *Saccharomyces cerevisiae* races. *J. Ind. Microbiol.* 13: 269-272.

Zee, J.A., Simard, R.E. and Roy, A. (1981). A modified automated ion exchange method for the separation and quantitation of biogenic amines. *Can. Inst. Food. Sci.* 14(1): 71-75.

Zoecklein, B.W., Hackney, C.H., Duncan, S.F. and Marcy, J.E. (1997). Effect of fermentation, aging and thermal storage on total glycosides, phenol-free glycosides and volatile compounds of white Riesling (*Vitis vinifera* L.) wines. *J. Ind. Microbiol. Biot.* 22: 100-107.

# 14 Genetic Engineering of Microorganisms in Winemaking

**Gargi Dey**

School of Biotechnology, Kalinga Institute of Industrial Technology (KIIT), Deemed to be University, Bhubaneswar – 751024, India

## 1. Introduction

Winemaking has come a long way from being an ancient art. The wines that are now available globally no longer resemble the wines that ancient Egyptians or Romans made. The wine fermentation and post-fermentation steps have become more sophisticated and automated. Earlier, wine fermentation was thought to be simplistic with assumed dominance of *Saccharomyces cerevisiae*. Over the years, investigations made by major research groups on the microbiology, biochemistry and molecular biology of the microorganisms involved in winemaking have established that yeast plays a more complex role in development of wine character. Consumer demand for newer styles of wines has been providing the oenologists with newer challenges and opportunities for innovation in wine fermentation. The concept of boutique-style products and products geared toward specific desires of the consumers is making the large-scale producers now look at both *Saccharomyces* and non-*Saccharomyces* strains to employ in controlled fermentation. With the growing understanding and knowledge in this field, designing transgenic yeasts for creative tailoring of wine seems to be a possibility in near future. This chapter discusses various aspects of transgenic wine yeasts and wine-associated lactic acid bacteria.

## 2. Genetics in Wine Microorganisms

While yeasts have been used for production of wines for several centuries now, the genetics of wine yeast has evolved more recently. *S. cerevisiae* are diploid/aneuploid, occasionally polyploidy, mainly homothallic, exhibiting chromosomal length polymorphism, possessing multiple translocations (Bidenne *et al*., 1992; Rachidi *et al*., 2000), while the lab-bred strains are haploid or diploid. The polyploidy state may be favouring the industrial yeast in terms of providing higher dosage of genes, which may be required for fermentation productivity (Salmon, 1997). The *S. cerevisiae* strains have 30-55 copies of retro-transposons, or Ty elements and 50-100 copies of 2μm plasmid DNA (6.3 kb extrachromosomal element) (Pretorius, 2000).

In the past, several research groups were involved in improvement of wine yeast recombinant DNA (Barre *et al*., 1993; Blondin and Dequin, 1998; Butzke and Bisson, 1996; Pretorius, 2000). Ten years back, the genome of five wine yeasts (AWRI 696, QA23, VIN7, VIN13, VL3 have been sequenced (Borneman *et al*., 2011; Borneman *et al*., 2012). Currently a database called *Saccharomyces* Genome Database (SGD; www.yeastgenome.org) is available, which contains not only the genomic, but transcriptomic, proteomic and metabolomics data sets.

## 3. Genetic Engineering of Yeast

In order to design new wine yeasts for commercial applications, several classical approaches were successfully applied. However, in several cases the classical techniques failed, which paved way for the development of more modern approaches. The following section discusses both the approaches.

### 3.1. Methods

The selection of a suitable method of genetic improvement depends on whether the desirable trait

E-mail: drgargi.dey@gmail.com; gargi.dey@gmail.com

is monogenic or polygenic. Desirable traits, like fermentative vigour, ethanol yield are multiple loci dependent (QTLs). For instance, the trait of ethanol tolerance is coded by 250 genes (Pretorius, 2000). In spite of the low phenotype-genotype correlation, especially in polygenic traits, several DNA technologies have evolved in the last few decades which may be used for designing strains with desirable recombinant traits.

### 3.1.1. Classical Approaches

#### *3.1.1.1. Sexual Hybridisation*

Many traits, like temperature profile, ethanol tolerance, etc. are under the control of multiple polymorphic loci and are broadly distributed throughout the QTL (Marullo *et al.*, 2004). For these traits, genome-wide approaches are more useful. Sexual hybridisation is a classical method used for improvement of such traits. One of the biggest advantages of this approach is probably that it allows exploitation of the enormous genetic diversity that is found naturally. This can be performed between gametes of single spore culture, or it can be performed directly between wild types. The former is more suited for monogenic traits and the latter is more appropriate for traits controlled by QTLs (Giucidi *et al.*, 2005). Several researchers in the past have applied this technique; this approach, for instance, to fuse non-$H_2S$ producing strain with flocculating strain (Romano *et al.*, 1985) or for increasing the fermentation rate and enhancing the aroma profile (Shinohara *et al.*, 1994). The main roadblock to this technique was the time intensive screening process, decreased sporulation efficiency and decreased spore viability (Gimeno-Alcaniz and Matallano, 2001).

#### *3.1.1.2. Rare Mating*

Some of the problems faced in sexual hybridisation can be overcome by rare-mating or alternate hybridisation. The biggest advantage of rare mating is that it does not require sporulating strains (Spencer and Spencer, 1996). In 2002, De Barros Lopes *et al.* showed that rare mating was possible between *S. cereviasiae* and *S. paradoxus* and *S. bayanus*, establishing interspecific hybridisation (De Barros Lopes *et al.*, 2002). Rare mating of cryotolernt *S. uvarum* and *S. cervisiae* was established (Giudici *et al.*, 2005). More recently in 2013, Bellon *et al.* applied rare mating and designed a new yeast, using inter-specific hybridisation between *S. cerevisiae* and *S. mikatae* (closely related to but ecologically distant member of *Saccharomyces sensu* strict *clade*) (Bellon *et al.*, 2013). The diploid *S. cerevisiae* cells undergo mating switch, giving rise to diploid homozygous a/a and α/α (Gunge and Nakatomi, 1972). Using this information Bellon and co-workers (2013) performed rare mating between opposite mate types resulting in triploid genome which was confirmed using comparative genome hybridisation and fluorescence flow cytometry. Evaluation of their fermentative ability showed that the hybrid strains were able to grow at temperature ranges 22°C and 4°C and at high ethanol concentration (14 per cent). The major achievement of this cross may have been with respect to the volatile metabolites that were generated, which are generally associated with non-*Saccahromyces* species. The only disadvantage of the rare mating approach is that the F1 generation is sterile.

#### *3.1.1.3. Random Mutagenesis*

This refers to the use of physical and chemical agents for genetic improvement. Industrial wine yeasts are usually diploid or polyploid because of the presence of two copies or more; they lack auxotrophic mutations, restricting the selection process of mutants. The common mutagens that have been used so far are UV rays, chemical mutagens, ethyl methane sulphonate (EMS), methyl methane sulphonate, N-methyl-N′-nitro-N-nitrosoguanidine (MNNG) and nitrous acid. The selection of mutants is done by positive selection, 5-fluoroorotic acid for URA3 mutants, α-aminoadipic acid for Lys 2 mutant or 5-fluoroanthranilic acid for the Trp 1 mutant. Using random mutagenesis, the mutants that were developed showed decreased levels of fusel alcohol (Giudici and Azinnato, 1983), better autolytic behaviour of second-fermentation for sparkling wine yeasts (Gonzalez *et al.*, 2003), improved nitrogen assimilation and fermentation kinetics (Salmon and Barr, 1998), increased ester formation (Wohrmann and Lange, 1980). Induced mutations using chemical mutagens followed by selection through replica plating (Pretorius, 2000) has been mainly used for removal of unwanted monogenic trait, for e.g. strains with specific amino acid auxotrophy stopped producing the corresponding alcohol, which led to improvement in flavour profile of wine (Rous *et al.*, 1983).

### 3.1.2. Modern Approach

#### 3.1.2.1. Adaptive Evolution/Evolutionary Engineering

This is a versatile technique for generation of industrially robust strain. It is based on the rationale that deliberate selective pressure forces yeast to acclimatize and makes it more efficient under oenological conditions. McBryde and co-workers (2006) used this evolutionary engineering on the commercial strain L-2056 and its haploid C-9 derivative resulting in more rapid catabolism of available sugars (McBryde *et al.*, 2006). This technique can be used to 'train' the yeasts and tailor them for specific wine fermentation conditions. However, one of significant aspect of adaptive evolution is accurate selection of the strain (Mangado *et al.*, 2015). Over the years this approach has been featuring as one of the available non-GMO techniques.

In 2012, wine yeasts were designed, using adaptive evolution which demonstrated higher rate of fermentation, lower acetate formation, higher amounts of phenyl ethanol, isobutanol isoamyl alcohol, ethyl acetate (Cadiere *et al.*, 2012, Mouret *et al.*, 2014). Likewise, this technique has also been used to develop a strain which has produced 41% more glycerol compared to parent strain (Kutyna *et al.*, 2012). In this case the 'training' of the yeast was done by using sulfite as selective pressure.

Two years later, in 2014, Tilloy and coworkers stimulated the HOG (high osmolarity glycerol) MAP kinase pathway by culturing diploid heterozygous commercial wine yeasts in presence of KCl (Tilloy *et al.*, 2014). The adaptive evolution technique resulted in higher glycerol content, higher succininc acid and 2,3-butanediol and lowered ethanol production by 9% compared to the control strain.

#### 3.1.2.2. Recombinant DNA Technology and Metabolic Engineering

This refers to direct manipulation of pathways by either deleting genes, adding new genes or modifying gene functions by modifying the strength of the promoter. It is no doubt scientifically more sound than random mutagenesis. However, this approach has been efficiently applied in case of traits where the gene encoded is well characterised, for instance, *S. cerevisiae* having marked pectinolytic, xylinolytic and glucanolytic activities show the desirable endpoint of efficient clarification (Laing and Pretorius, 1993).

The approach of metabolic engineering has been the backbone of almost all the genetic modification strategies that have been developed especially in 1990 and 2000 to address the industrial 'pain points' of wineries. Mainly heterogonous expression has been used with strong promotes and terminators. Multicopy shuttle vectors have facilitated the yeast engineering process. Some of the popular vectors that have been used for yeast transformation are yeast episomal plasmid (YEP), yeast replicating plasmid (YRP), yeast centromere (YCP) and yeast integrating plasmids (YIP) (Fig. 1a-c).

## 3.2. Selection Markers

Selectable transformation markers are essential for strain improvement by genetic methods. An important reason why classical genetic approach of spontaneous/induced mutation and sexual hybridisation has been less frequently used is lack of adequate selection procedures. In comparison, strain improvement strategies, based on genetic or metabolic engineering approaches, has resulted in several successful

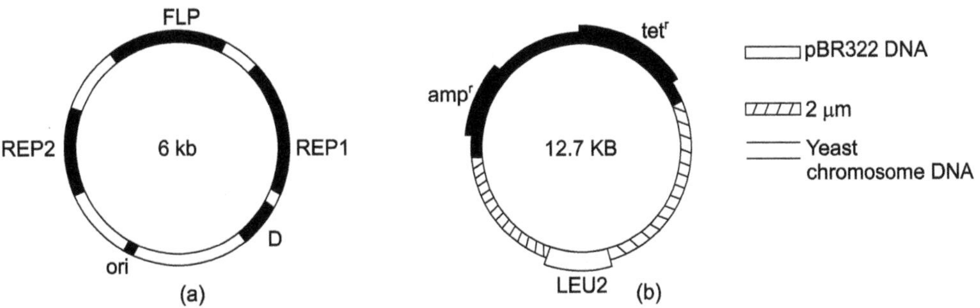


**Figure 1a.** Yeast episomal plasmids (a) showing 2 μm circle. REPI and REP2 are involved in replication of the plasmid and FLP codes for protein that can convert a form of the plasmid to B form, (b) YEP-pJDB219 (Adapted from Brown, 1990)

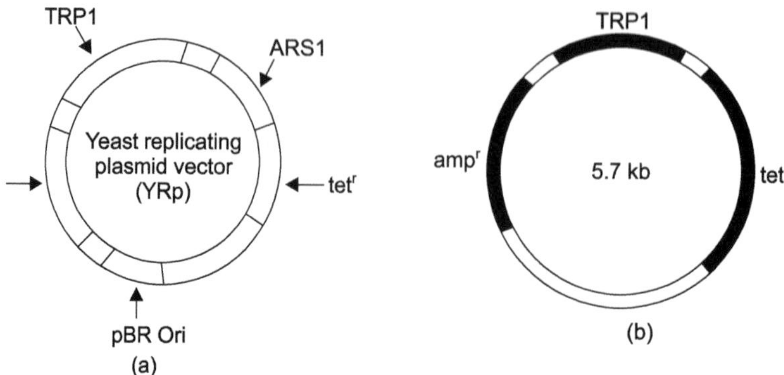


**Figure 1b.** Yeast replicating plasmid (a) (YRP) and (b) Yeast YRP7 (Adapted from Srivastava and Raverkar, 1999)

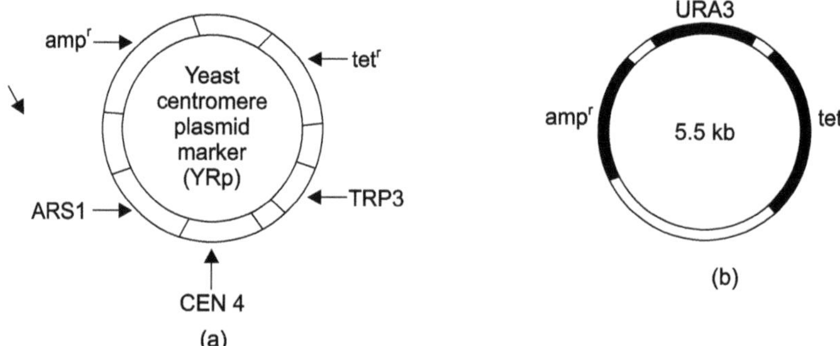


**Figure 1c.** (a) Yeast centromere plasmid vector and (b) Yeast integrating plasmid vector YIP (Adapted from Skipper and Bussey, 1977 and Brown, 1990)

transformants with better flavour profile, higher glycerol production, better flocculation, less ethyl carbamate production and other desirable traits. The success of this approach has been largely because of the use of suitable selection markers. The markers that have been used widely in past are mainly two types:

- dominant ones, like antibiotic resistance, like hygromycin, kanamycin, chloramphenicol, or *CUP1, SFA1* and *SMRI*
- recessive types, like auxotrophic markers *URA3, HIS3, LEU2, LYS2, TRP1, ADE2.*

In 2004, Cebollero and colleagues compared two selection markers *ARO4-OFP* which makes the transformant resistant to p-fluoro-DL-phenylalanine and *FZF1-4* which confers sulphite resistance on to the transformants. The markers were tried in both episomic and centromeric plasmids. Among the two markers, *ARO4-OFP* was found to be more efficient since it gave higher transformation frequencies. Also there was appearance of spontaneous resistant colonies or false positives in case of *FZF1-4* transformation (Cebollero *et al.*, 2004). However, since industrial strains lack auxotrophic markers (*URA2, LEU 2*), drug resistance markers have been used by yeast-derived cycloheximide resistance gene *CYH2*, or geneticine resistance *G418* markers or phleomycine resistance, *ble.*

## 4. Transgenic Wine Yeasts

Winemaking is not merely alcoholic fermentation; modern wineries rely on carefully selected starter cultures for obtaining consistent and reliable products. In spite of several centuries of fine tuning of the process, there still exist several bottlenecks or 'pain points' (Fig. 2) as we may call them; several lose ends that need to be tied up. In today's competitive world, there is very little scope of errors and large scope of improvements and value-addition. In the following section, seven most vital 'pain points' of global wineries have been discussed along with the genetic engineering strategies that have been developed over the years to counter them.

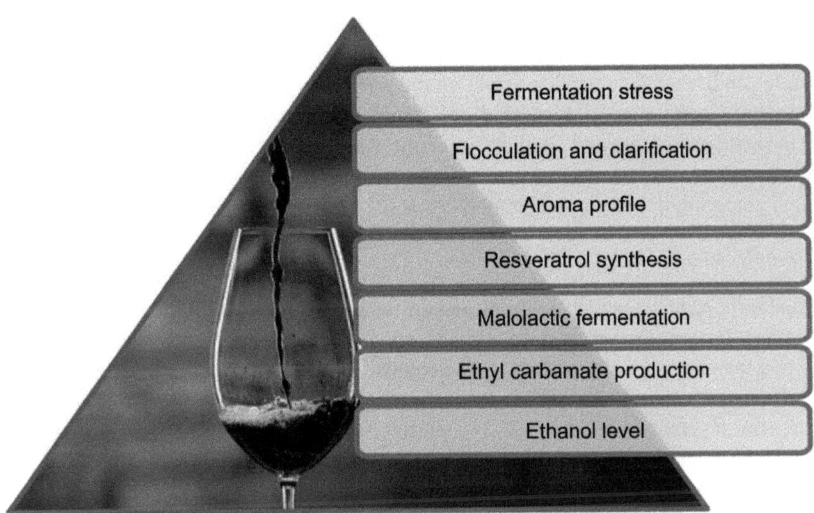


**Figure 2.** "Pain points" of the wineries

### 4.1. 'Pain Points' of Wine Industry

The concept of winemaking is going through a period of change, experimentation and adaptation. The bottlenecks or 'pain points' of this industry, like the improvement of fermentation process, simplification of downstreaming, improvement of the product quality, are all now being revisited. The powerful genetic engineering tools that have been developed for *Saccahromyces* are being put to use to address the 'pain points'. The subsequent sections discuss elaborately the reason of the 'pain points' and the strategies that are evolving to overcome these bottlenecks.

#### *4.1.1. Stress Resistance*

Individual yeast strains are lyophilised for long-term maintenance, which causes membrane lipid transition. The cycle of freeze-thawing also leads to reduction in cell viability or loss in cell vitality because of the induction of respiratory deficient variants. During large-scale culturing of yeasts, an increase in protein phosphate results in excess hydration of protein, which in turn negatively affects the drying procedure and viability (Degre, 1993). Trehalose functions as a cryoprotectant by forming a hydration shell around protein and glycerol functions as an osmoprotectant by controlling the intracellular solute potential, counteracting the dehydration effects (Walker, 1998). Ideally, therefore, it would be beneficial to have industrial yeast cells which retain higher amounts intracellular trehalose and glycerol content.

After the reconstitution of lyophilised culture and once the vinification process has been initiated, the yeast cells face a series of stresses like osmotic shock from high glucose concentration, low must pH (below 4.0), ethanol accumulation, high/low temperature, high $SO_2$, concentration, oxidative shock.

Ethanol production in wine is a double-edged sword. On the one hand, it is the major desirable product of wine fermentation while on the other it is a chemical stress factor which may result in sluggish and stuck fermentation (Boulton *et al.*, 1999). The loss of viability of yeast cell is mainly due to disintegration of plasma membrane, hydrophobic proteins, nuclear, mitochondrial endoplasmic and vacuolar membranes. Therefore, an industrially robust yeast strain with higher freeze-thaw and ethanol tolerance is favoured.

#### *4.1.2. Flocculation and Clarification*

After fermentation, a major downstream processing step is the removal of yeast cells, processing the removal of yeast cells from wine. Self-clearing wine yeasts at the end of fermentation are a highly desirable trait. But at the same time flocculation must not occur prematurely before the completion of alcohol fermentation.

Apart from yeast cells, it is vital to remove precipitates and partially soluble compounds to ensure physicochemical stability during storage, hence winemaking involves an extensive fining and clarification process. Finning can have a far-reaching effect on the sensory attributes of wine. Effective fining should remove haze-forming proteins, polysaccharides, tannic acid polyphenols, etc. Bentonite treatment effectively removes the proteins and smaller molecular weight compounds. The polysaccharides that are found in wine are originally from grapes, mainly pectins and glucans. Unfortunately, the endogenous pectinolytic and gluconolytic enzymes of *S. cerevisiae* are inadequate for effective fining and clarification.

### *4.1.3: Flavour Profile of Wine*

One of the most important organoleptic properties of wine is its 'bouquet' which comprises of desirable flavour compounds present in correct ratio and absence of off flavours. The variety of grapes, the strain of yeast used and the oenological practices – all strongly influence the flavour profile of a wine product.

The metabolites which are responsible for the varietal flavour of grapes are mainly present as non-volatile O-glucosides. Canal-Llauberes (1993) reported that hydrolytic glycosidase could liberate aroma precursor. For instance, geraniol and nerol can be converted into their aromatic metabolite after hydrolysis with terpenyl-glycosidaes, but the major 'pain point' is the non-efficiency of the grape glycosidases in the presence of glucose, low pH, high ethanol levels (Henschke, 1997). Exogenous addition of fungal β-glycosidases was suggested as an alternative; however, this was considered as an unnatural intervention and hence discouraged.

Similarly another important contribution to wine bouquet comes from alcohol acetates and fatty acid ethyl esters that are produced by yeasts during primary and secondary fermentations. The level of ester formation is influenced by the yeast strain, maceration, fermentation temperature and malo-lactic fermentation (Cole and Noble, 1995; Houtman *et al.*, 1980).

Another point to be addressed regarding the flavour profile is the thiol aromas. While wine yeasts may contain enzymes to release thiol aromas, the capacity to release monoterpene aroma compound is deficient because they lack the monoterpene synthases enzyme (Herrero *et al.*, 2008). Thus there seem to be several 'pain points' and scope for improvement that may be addressed through genetic modification of yeasts.

### *4.1.4. Health Benefits of Wine*

In the olden days wine played an important role in inhibiting pathogens, especially because of the ethanol and acidity. In the later years, however, it became established that prudent wine consumption in moderation can also be effective in stress management and reduction of risk of coronary heart disease (Seiman and Creasy, 1992). One of the reasons for this beneficial action of wine was attributed to the stilbene, resveratrol. In subsequent years it was further established that pure resveratrol had antioxidant and anti-inflammatory activities, inhibiting platelet aggregation (Frenkel *et al.*, 1993; Vinson *et al.*, 1995; Kallithraka *et al.*, 2001). The only source of resveratrol in wine is the skin cell of grape and the amount found is too low to show any physiological responses. This paved the way for yet another target in wine improvement.

### *4.1.5. Malo-lactic Fermentation*

Wine acidity has a major impact on the sensory quality of wine as well as the physic-chemical and microbial stability of wine during the long shelf-life. Apart from this, wine acidity can also influence the efficiency of post-fermentation practices, like bentonite clarification, colour formation, solublisation of proteins and potency of antimicrobial agents (Caputi and Ryan, 1996). The major fraction of titratable acidity of wine comes from tartaric acid and malic acid. The source of these acids is the grape during its maturation and subsequent conversion during fermentation. Imbalances in acids occur during temperature fluctuations while grapes mature or during the vinification process. Adjustment of pH and acidity of wine is an important part of wine downstream processing. Addition of calcium carbonate or tartarate, malate or citrate was practiced earlier but this compromised the wine quality and required extensive labour and capital input. The other alternative available was the bacterial malo-lactic fermentation (MLF) that resulted in shelf-stable wine. The MLF refers to conversion of L-malic acid to L-lactic acid in the presence of enzyme L-malate-NAD$^+$ carboxylase) (Naouri *et al.*, 1990). The predominating bacteria known for efficiently conducting MLF after the completion of alcoholic fermentation is *Oenococcus oeni* (Dicks *et al.*, 1995). However, nutrient limitation, low temperature, high alcohol, acidic pH conditions limit the growth and function of *O. oeni* and result in spoilage of wine due to stuck or sluggish malo-lactic fermentation. Interestingly *Schizosaccharomyces pombe* can also convert malate, but is accompanied by production of off-flavour.

Lactic acid bacteria (LAB) are also metabolically equipped to produce lactic acid; however, a common complication associated with LAB is the production of biogenic amines, like histamine, cadaverine, phenylethylamine, putrescine, tyramine (LonVaud-Funel, 2001). The biogenic amines are known to cause migraine, hypertension, diarrhoea (Soufleros *et al.*, 1998). Therefore, alternate technologies for effective acidity management are the need of the hour.

#### *4.1.6. Ethyl Carbamate Production*

Ageing wine, brandy and fortified wines support a chemical reaction between urea and ethyl alcohol, which leads to a high level of ethyl carbamate, which is a suspected carcinogen (Schlatter, 1986; Hubner *et al.*, 1997). The primary route through which ethyl carbamate gets into the food chain is via consumption of alcoholic beverage and fermented foods. In wine, specifically there are two routes of entry – one is through application of urea containing fertilisers to the grapevines, the other through supply of urea containing nutrients during yeast fermentation. Further the arginine catabolism via the *CAR1* encoded arginase also results in urea formation. Interestingly all *S. cerevisiae* strain secrete and reabsorb urea; however, the extent of re-absorption is strain specific (Ough *et al.*, 1988; An and Ough, 1993). The higher the vinification temperature, the higher is the urea secretion.

Thus from the health point of view ethyl carbamate seems to be a big challenge in alcohol beverage industry. Physical and chemical technologies focused on removal of ethyl carbamate include application of filtration, controlling vinification temperature, addition of cyanide catalyst precursor, addition of exogenous urease to degrade urea. Most of these methods have had limited success or complicated the situation further by generation of ammonia (Zhao *et al.*, 2013).

#### *4.1.7. Maintenance of Alcohol Level*

Elsewhere in this chapter, the production of ethanol during fermentation has been discussed with reference to ethanol stress on yeasts. In this section, the final ethanol concentration is being discussed in the purview of wine quality, financial consideration and consumer health considerations. In the present time, it is not uncommon to find table wines within the range of 15-16 per cent v/v (Varela *et al.*, 2015). Firstly, contrary to belief, a high ethanol concentration does not automatically mean a good quality wine. Desirable sensory attributes demand that there be a balance between acidity and alcohol content of the wine. In fact high ethanol content negatively impacts mouth-eel, aroma, flavour intensity and textural properties (Gawel *et al.*, 2007a, b); secondly, many countries levy higher tax on high alcohol products which imposes a financial burden on the consumers. Thirdly, and probably the most significant 'pain point' is the health care burden imposed by high alcohol product. While it has been established that moderate wine consumption is beneficial for reducing risk of coronary heart disease, it is equally well known that higher alcohol consumption leads to health perils. In order to address these issues, wine industries globally are now focusing on low ethanol wines.

### 4.2. Strategies to Address the 'Pain Points'

#### *4.2.1. Stress Tolerance*

In *S. cerevisiae* the key enzymes responsible for trehalose are the *TPS1*-encoded trehalose-6-phospate synthetase and *TPS2*-encoded trehalose-6-phosphatase (Francois *et al.*, 1997). Trehalose not only plays an important role in freeze-thaw and osmo-stress, but also in nutrient-starvation growth resumption and growth rate and hence is considered to be a worthy target to make yeast cell-resistant to stress. Likewise, glycogen is another carbohydrate which influences the viability and reactivation of yeast cells. The key enzymes in glycogen synthesis exist as two isozymes – gsy1p and gsy2p. The encoding genes *GSY1* is constitutively expressed and *GSY2* functions mainly at the end of lag phase and results in glycogen accumulation. Work by Perez-Torrado *et al.* has established that overproduction of *GSY2* gene results in enhanced viability of yeast strains even under glucose deprivation (Perez-Torrado *et al.*, 2002). Figure 3 depicts the trehalose and glycogen synthetic pathway.

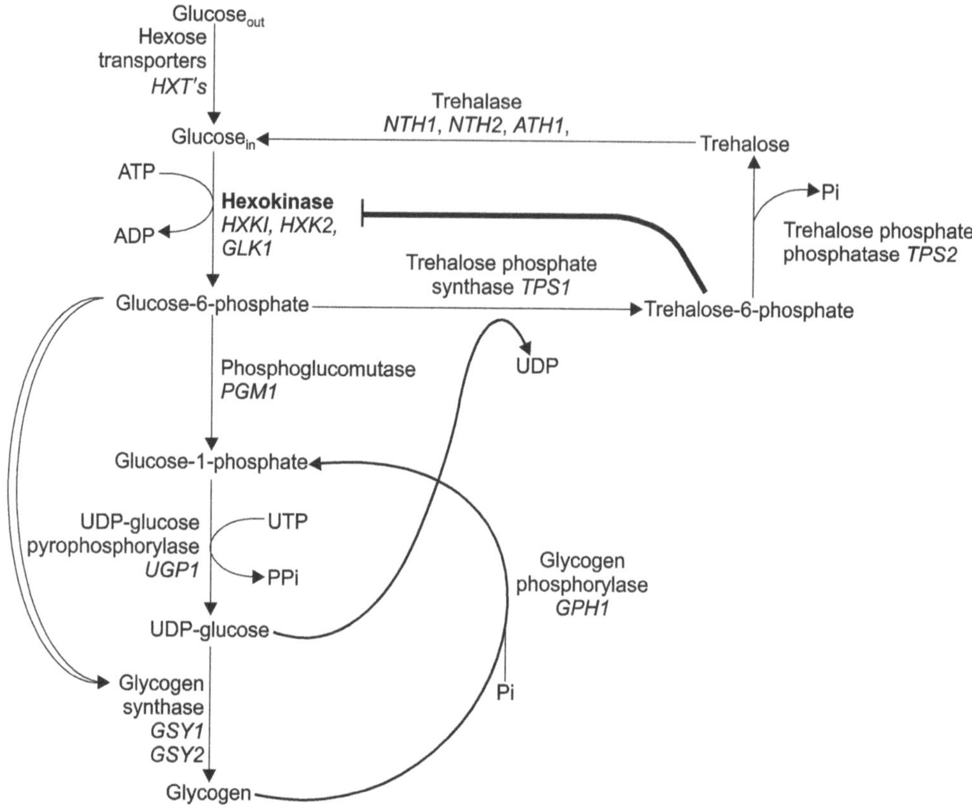


**Figure 3.** Trehalose and glycogen metabolism; TPS-trehalose phosphate synthetase and GSY (Glycogen synthase) (Adapted from De Silva-Udawatta and Cannon, 2001)

On the other hand, glycogen breakdown during nutrient depletion leads to sterol formation. Sterol is an essential lipid, which increases the robustness of the yeast strain. Hence overexpression of genes (*SUT1* and *SUT 2*) coding for sterol uptake has also to be explored as a strategy for making yeast stress resistance (Pretorius, 2000).

In 2007, Cardona and co-workers used a different approach to construct stress-resistant yeast strain. They targeted the transcription factor *MSn2*, replaced its promoter by *SPI1* gene promoter and in effect were able to induce *MSn2* protein in the late log phase rather than normal phase induction. This gene manipulation resulted in self-regulated expression *MSn2* under stress conditions. When assessed, the transformants were found to be more resistant to temperature stress, glucose depletion stress. Better fermentation behaviour was exhibited at temperatures 15°C-30°C, with higher glucose consumption during the initial days of vinification in spite of the osmotic stress (Cardona *et al.*, 2007).

Ethanol tolerance is a polygenic trait involving many genes; so a specific strategy cannot be chalked out. However, overexpressing trehalose synthesis genes, enhancing SOD activity and increased alcohol dehrogenase activities seem to be the practical targets for making ethanol tolerant yeast strains (Walker, 1998).

In 2010, some scientists focused their attention onto oxidative stress faced by the active dry yeast due to the ROS accumulation. Overexpression of thioredoxin gene (*TRX2*) was shown to enhance the oxidative stress resistance of the transformant since thioredoxin is an integral part of glutathione/ glutaredoxin defence system (Grant, 2001). The *Trx 2p* regulates the transcriptional response against ROS. The overexpression of *TRX2* gene showed higher antioxidant enzyme (Sod1p, Sod2p, catalase) activities, reduced lipid peroxidation and increased biomass production (Gomez-Pastor *et al.*, 2010).

In order to handle the complexities of metabolic landscapes, 'global transcription machinery engineering' (gTME) was introduced by Alper and colleagues (Alper *et al.*, 2006). The gTME involves global perturbations of transcriptome to optimise phenotypes. Very recently, Zhao and co-workers applied the gTME to create plasmid-based mutagenesis libraries of transcription factors *SPT15* and *TAF25* and used for transformation of yeast strains. The transformants showed tolerance to oxidative stress when subjected to $H_2O_2$ shock and also showed shorter lag phase (Zhao *et al.*, 2014).

### 4.2.2. Improved Flocculation and Clarification

The phenotype of Ca-dependent aggregation of yeast cells to form flocs is under the control of *FLO* gene-encoded flocculins. In the past, several authors have identified *FLO1, FLO2, FLO3, FLO4, FLO5, FLO6, FLO7, FLO11* genes (Lambrechts *et al.*, 1996; Lo and Dranginis, 1996; Teunissen and Steensma, 1995). Several workers in part have shown that of all the FLO genes, *FLO1* was capable of transforming non-flocculent yeasts into flocculent yeasts (Hammond, 1996: Ishida-Fuji 1998; Watari *et al.*, 1993).

In more recent years, Govander and colleagues (Govander *et al.*, 2011) replaced the promoters of *FLO1, FLO5, FLO11* gene of two non-flocculent strains (BM 45 and VIN13) and added promoters of *HSP30* and *ADH2* genes. The resulting transformant showed stable expression and corresponding flocculent phenotype. At the same time, several other researches had been focusing on strategies for enhancing the polysaccharide clarification. The strategy of heterologous co-expression of pectate lyase gene (*PEL E*) from *Erwinia chrysantheni* and polygalacturonase gene (*PEH I*) in *S. cerevisiae* resulted in the transformant to degrade pectin efficiently (Laing and Pretorius, 1993b). Similarly heterologous expression of endoglucanase genes from several other sources, like *Bacillus subtilis* (*BEG1*), *Phanerochaete chrysosporium* (*CBH1*), has also been successfully used to enhance the glucan clarification in *S. cerevisiae* (Van Rensburg *et al.*, 1996, 1997). The heterologous expression of xylanase gene from *Aspergillus nidulans* has also been established as a useful strategy for improving polysaccharide clarification in *S. cerevisiae* (Perez-Gonzalez *et al.*, 1996).

### 4.2.3. Improved Aroma Profile

The limitation of non-functioning grape glycosidases was overcome by expression of β-glucosidase gene (*BGL 1*) from *Saccharomycopsis fibuligera* into *S. cerevisiae*. Unlike grape glucosidases, the yeast enzymes are fully functional under vinification conditions and were found to be efficient in releasing the varietal aroma (Van Rensburg *et al.*, 1998). Earlier, in 1993, Perez-Gonzalez and co-workers constructed a transgenic yeast having the β-1,4-glucanase gene from *Trichoderma longibrachitum*, which resulted in a desirable end point of intensified aroma (Perez-Gonzalez *et al.*, 1993). More recently, in 2003, Manzanares and colleagues succeeded in expressing the α-L-rhamnosidase (*rhaA*) gene in an industrial wine yeast strain (Manzanares *et al.*, 2003). The trial fermentation with the transgenic yeast expressing *rhaA* gene along with another strain expressing a β-glucosidase gene resulted in higher accumulation of linalool.

Similarly, the desirable wine alcohol acetates and ethyl esters are hexyl acetate, ethyl caproate and caprylate (apple aroma), isoamylacetate (banana aroma), 2-phenylethyl acetate (fruity aroma). Three important enzymes have been identified which are majorly responsible for the synthesis of these flavour compounds in *S. cerevisiae;* they are alcoholacyltransferase (AAT), ethanol acetyltransferase (EAT) and isoamyl alcohol acetyltranferase (IAT) (Fujii *et al.*, 1996; Fujii *et al.*, 1997). The alcohol acetyltransferase encoded (*ATF-1* gene) was successfully overexpressed in the industrial yeast strain VIN13 under the PGK1 promoter and terminator sequences (Lilly *et al.*, 2000). The genetic modification resulted in enhancement of ethyl acetate, isoamyl acetate and 2-phenylethyl acetate.

Much later in 2007, Swiegers *et al.* (2007) developed a designer yeast using two strains of Σ 1278b, lab strain and the industrial strain VIN13. The tryphtophannase gene (*TnaA* gene) codes for cysteine-β-lyase enzyme in *S. cerevisiae* strains. A multi-copy episomal plasmid was used to introduce the *TnaA* gene construct. The resulting yeast transformants expressing the lyase activity was able to release 25 times more 4-mercapto-4-mthylpentan-2-on2 (4MMP) and 3-mercaptohexan-1-ol (3MH), which gave the final wine product an intense passion-fruit aroma.

Very recently, in 2016, pioneering work was reported on pathway engineering for synthesis of raspberry ketone in *S. cerevisiae* strain (Lee *et al.*, 2016). In this work, four gene coding for phenyl ammonia lyase (PAL/TAL), cinnamate-4-hydroxylase (C4H), coumarate-coA-ligase (4CL) and benzalacetone synthase (BAS) were heterologously expressed. Exogenous feeding with *p-coumaric* acid, the pathway precursor, showed 7.5 mg/l of raspberry ketone [4-(4-hydroxyphenyl)butan-2-one] and 0.68 mg/l of the ketone when the transgenic yeast was grown in minimal media and anaerobically fermented with Chardonnay juice under winemaking conditions.

### 4.2.4. Improved Health Benefit through Resveratrol Production

One of the obvious approaches to increase resveratrol in wine was to construct designer yeast which was equipped to synthesise this compound. By 2000, when the review by Pretorius was published, the key information that was available was that phenylproponoid pathway needed to be tweaked for production of p-coumaryl-coenzyme A, the precursor for resveratrol synthesis (Fig. 4). By then it was also known that *S. cerevisiae* strain had to be transformed with phenylammonia lyase (*PAL*) gene, cinnamate-4-hydroxylase gene (*C4H*) and coenzyme ligase (*4CL216*) genes. By 2003, Becker and co-workers (Becker *et al.*, 2013) reported a metabolic strategy for resveratrol synthesis, albeit in a lab strain of *S. cerevisiae.* The strain selected could already metabolise *p-coumaric acid.* To this strain, they co-expressed *4CL216*

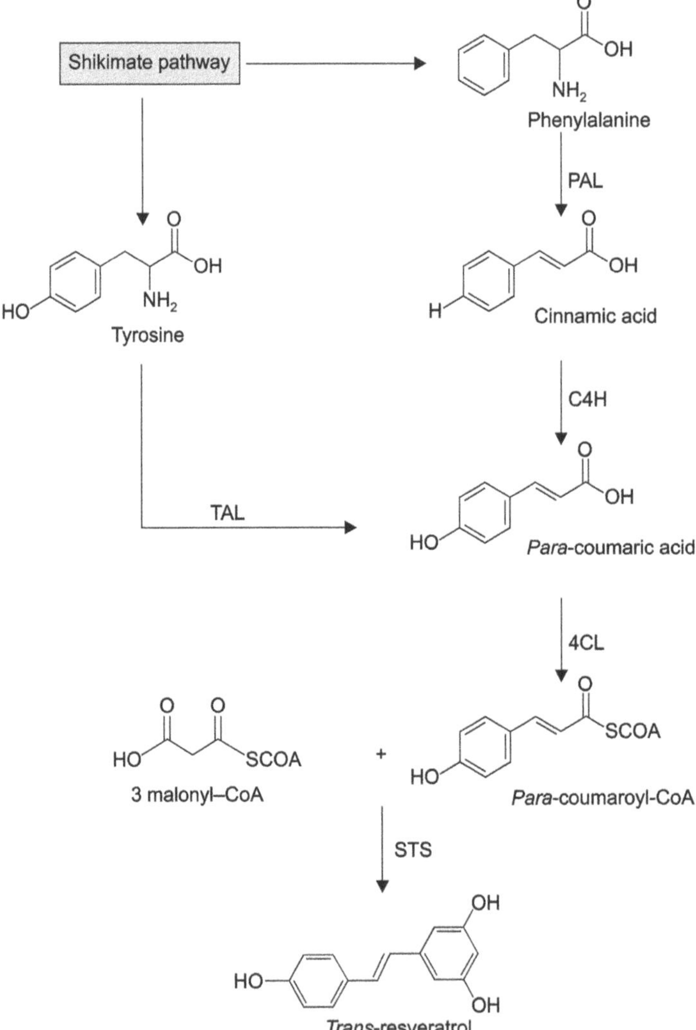


**Figure 4.** Steps in resveratrol synthesis via the phenylalanine/polymalonate pathway. PAL, phenylalanine ammonia lyase; TAL, tyrosine ammonia lyase; C4H, cinnamate-4-hydroxylase; 4CL, para-coumaric acid; coenzyme A ligase; STS, stilbene synthase (Adapted from Pretorius, 2000)

from a hybrid poplar and resveratrol synthesis gene (*vst1*) from grapevine. The resulting transformant produced the resveratrol-β-glucoside, piceid.

An altogether different approach was used by Gonzalez-Candelas and co-workers (Gonzalez-Candelas *et al.*, 2000). The α-L-arabinofuranosidase (*abfB*) gene from *A. niger* or the β-glucosidase bg1N gene from *Candida molischiana* was expressed in *S cerevisiae*. The transformant was able to produce *trans* and *cis*-piceid. Comparatively, *BgiN* gene showed a higher resveratrol content. The yeast strain used was a Spanish strain T73 (CECT1894) isolated from Monastrell must. Since then, several other research groups have reported on resveratrol production in different lab and industrial yeasts strains (Halls and Yu, 2008; Moglia *et al.*, 2010; Sydor *et al.*, 2010; Trantas *et al.*, 2009).

In 2011, Wang and colleagues published a work on accumulation of resveratrol in white wines comparative to that of red wines (Wang *et al.*, 2011). They used multiple approaches for this. The tyrosine ammonia lyase (*TAL*) gene from bacterium *Rhodobacter sphaeroides* was site directed mutaganised to replace the bacteria codons with yeast-preferred codons. The transformants produced 2.44 fold higher resveratrol than control strains.

### 4.2.5. Malolactic Fermentation

One of the earlier approaches which was tested for increasing lactate concentration in wines involved heterologous expression of lactate dehydrogenase gene (LDH gene) from *Lactobacillus casei* into eight wine yeasts. (Dequin *et al.*, 1994;, 1999; Porro *et al.*, 1995). However, the major drawback of this method was that it lowered the fermentation rate (Dequin *et al.*, 1999).

By now scientists realised that the solution may be to engineer a single wine yeast that could efficiently perform both alcoholic and malo-lactic fermentation. The initial efforts to engineer the malo-lactic fermentatation into *S. cerevisiae* involved introduction of NAD-dependent malo-lactic enzyme (*mleS*) from *Lactococcus lactis* (Ansanay *et al.*, 1993; Bony *et al.*, 1997; Denayvolles *et al.*, 1995) and *mleA* from *O.oeni* (Labarre *et al.*, 1996). Cloning of these genes did not result in efficient MLF in the transformed yeast strains. Later in 1997, Volschenk *et al.* explained that *S. cereviae* lacked the malate transport system due to which the earlier experiment of cloning malo-lactic enzyme had failed. The same group (Volschenk *et al.*, 1997) successfully constructed a yeast strain which demonstrated efficient MLF when it was transformed with mleS gene from *Lb. lactis* and *mae1* gene coding for malate permease from *Schizocaccharomyces pombe*. They have also succeeded in constructing designer yeast exhibiting efficient MLF by using the strategy of co-expression of mae1 permease gene and *mae2* malic enzyme gene, both from *S. pombe*. However, these successes were demonstrated only in lab strains.

A few years later, in 2006, Husnik and co-workers (Husnik *et al.*, 2006) laid the foundation of the historic work where they were able to apply metabolic engineering on an industrial yeast strain (ML01), which decarboxylated 5.5 g/l malic acid during alcoholic fermentation of Chardonnay grape must. This deserves a special mention since this is one of the only two transgenic yeasts which have been conferred the GRAS status by FDA and is being used for commercial applications. In the linear malo-lactic cassette; *mae1* (malate permease gene *S. pombe*) and *mleA* (malol-actic gene from *O. oeni*, were out under *S. cerevisiae* PGK1 promoter and terminator sequence (Fig. 5). The cassette was integrated into the URA3 locus of *S. cerevisiae* S92 and co-transformed with the expression plasmid pUT332, carrying phleomycin resistance (Tn5ble). This was inserted in only chromosome V. Though ML01 is efficient in MLF, the biggest advantage of this strain is probably in the fact that wines produced are totally free of health hazards of biogenic amine.

**Figure 5.** Linear malolactic cassette of ML01 strain (Adapted from Husnik *et al.*, 2006)

### 4.2.6. Elimination of Ethyl Carbamate

At one time it appeared that selection for low-urea-accumulating strain may be the potential solution to ethyl carbamate accumulation. However, with the growing success of genetic engineering, scientists were now keen on addressing this issue through genetic or metabolic engineering approaches. Deletion of *CAR-1* encoded arginase gene in *S. cerevisiae* seemed to be a likely target to reach the endpoint. Suizu and co-workers, in 1990 (Suizu *et al.*, 1990) constructed a lab strain knockout of *CAR-1* gene. The deletion mutant lost the ability to produce urea without losing any of its fermentative ability. Encouraged by this result, Kitamoto and colleagues constructed arginase-deficient sake yeast by double disruption of sake yeast by double disruption of *CAR-1* gene (Kitamoto *et al.*, 1991). However, these lab-based successful stories could not be reproduced in real life scenario since the sake mash was often contaminated with wild yeast.

Meanwhile, researchers found another route to reducing urea accumulation. They found that urea amidolayse (coded by *DUR1* and *DUR 2* gene) and urea permease (*DUR 3* gene) were the key enzymes for the urea transport and metabolism. Finally in 2006, Coulon and co-workers successfully integrated the *DUR1, DUR 2* gene into the industrial wine yeast genome (Coulon *et al.*, 2006). The metabolically-engineered wine yeast ECMo01 could reduce almost 90 per cent of ethyl carbamate in Chardonnay wine. In the history of genetic engineering of wine yeasts, this work is another landmark, since this was cleared by American and Canadian regulatory bodies and has become the second transgenic yeast to be made commercially available. The ECMo01 has an extra copy of *S. cerevisiae DUR1, DUR* 2 gene which is regulated by the PGK1 regulatory sequences, converting urea into ammonia, which is used as the preferred nitrogen source (Pretorius *et al.*, 2012). Figure 6 depicts the gene targets that have been used for reduction of ethyl carbamate. It needs to be stated here that EcMo01 is 'self cloned' and does not possess any foreign DNA but because the extra copy of gene addition was done through *in vitro* manipulation, it has been classified as GMO by the regulatory bodies of several countries (Pretorius *et al.*, 2012).

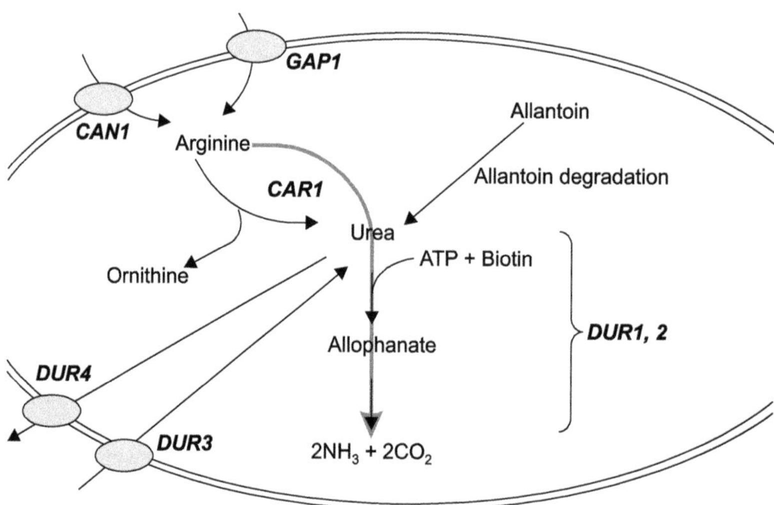


**Figure 6.** The gene targets reduction of ethyl carbamate CAR (arginase gene) and DUR (urea amidolyase) (Adapted from Coulon *et al.*, 2006)

### *4.2.7. Low Ethanol Wines*

Techniques like membrane filtration, vacuum distillation, blending of wines, limited dilution of must, addition of fungal glucose oxidase are routinely used by vintners to reduce alcohol in their wines. However, in this chapter, the focus will be on the genetic engineering approaches that have been developed and are currently available.

For reduction of ethanol an obvious strategy has been the redirection of carbon flux to other endpoints, like glycerol. The key enzymes of glycerol biosynthetic pathway (Fig. 7) are glycerol-3-phosphate dehydrogenase (encoded by *GPD1* and *GPD2* genes) (Scanes *et al.*, 1998; Remize *et al.*, 1999). Also the key enzyme for glycerol utilisation pathway is glycerol kinase (encoded by *GUT-1*). The glycerol transport facilitator protein that controls the influx and efflux is encoded by *FPS1*. Overexpression of *GPD1* and *GPD2* in engineered yeast strains has not only resulted in excess glycerol production but

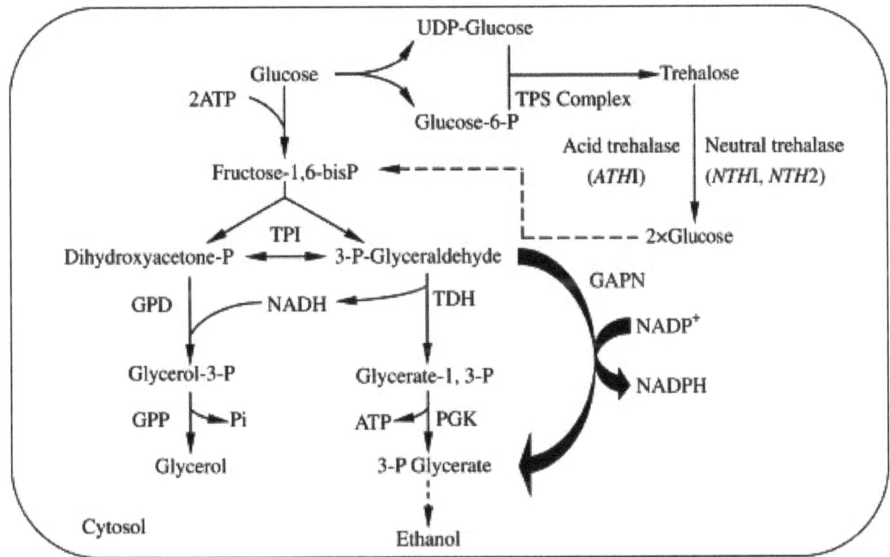


**Figure 7.** Glycerol biosynthetic pathway (Adapted from Scanes *et al.*, 1998).

more significantly decreased the ethanol levels in the wines (Overkamp *et al.*, 2002). But this strategy was accompanied by an added burden of acetate production mainly in order to protect the cell from redox imbalance. It has now been established that overexpression of *GPD1* and *GPD2* along with deletion of *ALD6* isomer of aldehyde dehrogenase gene, efficiently reduces the alcohol content from 15-12 per cent (Eglington *et al.*, 2002; Cambon *et al.*, 2006).

Since the targeted end point was reduction in ethanol level, direct reduction of ethanol via deletion of alcohol dehydrogenase enzyme was also explored as a strategy. Alcohol dehyrogenase enzyme is coded by *ADH 1, ADH 3, ADH 4, ADH* 5. Deletion mutants of *ADH 1* and *ADH 3* and *ADH 4* were assessed in trial fermentation. Though the genetic modifications resulted in decreasing the ethanol yield, the deletion mutants suffered from impaired growth (Cordier *et al.*, 2007; de Smidt *et al.*, 2011). Hence this strategy was not explored further.

Another approach has been the glycolytic pathway branchpoint, catalysed by the triose phosphate isomerase (encoded by the *TPI1*). The deletion of this gene resulted in NADH imbalance and drastically compromised the vitality of the deletion mutants.

Though these attempts failed, there were some other attempts which succeeded. For instance, heterologous expression of *A. niger* glucose oxidase (encoded by *GOX* gene) converted glucose to gluconate and resulted in decreased ethanol content (Malherbe *et al.*, 2003).

Recently, Varela and co-workers (Varela *et al.*, 2012) carried out an interesting work and compared all the genetic engineering approaches for low ethanol wine on to industrial yeast strain AWRI2531. Among the modifications assessed, they confirmed that the most successful strategy that has been so far developed for low alcohol wines, was the glycerol overexpression with *ALD 6* deletion.

A tabulation of all the strategies and gene modification targets that have been used so far to address the bottlenecks of wineries has been given in Table 1.

## 5. Synthetic Yeast Using Synthetic Genome Engineering

Currently approaches of genetic engineering and metabolic engineering are now slowly being replaced by genome engineering techniques. System biology and functional genomics assist in selection of special functionalities and criteria as per the consumer, production and environmental demands, through the use of genome-wide bioinformatic tools (Borneman *et al.*, 2007; Pizarro *et al.*, 2007). High precision CRISPR editing technologies further assist in genome shuffling.

**Table 1.** Gene Manipulation Targets to Address the "Pain Points" of Wineries

| *"Pain points"* | *Strategies adopted to address the "pain points"* |
|---|---|
| Stress Tolerance | TPS1 and TPS 2 overexpression for trehalose-6-phosphatase |
| | GSY1 and GSY2 for overexpression of glycogen synthetase |
| | SPI1- Overexpression if sterol uptake gene |
| | MSN2- Overexpression of transcription factor |
| | TRX-2 Overexpression of thioredoxin gene |
| | gTME-mutagenesis and expression f SPT15 and TAF25 transcription factor |
| Flocculation & Clarification | FLO1, FLO5 and FLO 11-Overexpression of flocculin gene |
| | PEL E-Expression of pectate lyase |
| | BEG-1 Expression of bacillus endonuclease |
| | CBH1-Expression of fungal endoglucanase |
| Improvement of Aroma profile | BGL 1-Expression of β-endoglucosidase |
| | rhaA-Expression α-L-rhamnosidase |
| | ATF-1-Expression of alcohol acyl transferase gene |
| | TNaA-Tryptophanase gene for cystein-β-thiolyase |
| | PAL/TAL, C4H. 4CL and BAS-Expression for raspberry ketone synthesis |
| Resveratrol accumulation | PAL, C4H, 4CL216-Expression for resveratrol synthesis |
| | ABFB and Bg1N- Expression of α-L-arabinofuranosidase and β-glucosidase |
| MLF | Mae 1-Expression of malate permease |
| | Mle A-Expression of malolactic enzyme |
| Ethyl carbamate reduction | CAR-1 Expression of arginase gene |
| | DUR 1, 2, 4-Expression of urea amidolyase gene |
| Ethanol reduction | GPD-1, GPD2-Expression of glycerol -3-phophate dehydrogenase |
| | ALD 6-Deletion of aldehyde dehydrogenase |
| | ADH1 and ADH 3-Deletion of alcohol dhydrogenase |
| | TPI1-Deletion of triose phosphate isomerase |
| | GOX- Expression of glucose oxidase |

In 2014, *S. cerevisiae* was made famous as it became the first eukaryotic organism to acquire a fully functional synthetic chromosome (Annaluru *et al.*, 2014). The synthetic yeast genome project (Sc 2.0) has been commissioned to synthesise all the 16 chromosomes of the lab strain S288c by 2018 (www.syntheticyeast.org). When completed, it would be a historical milestone of crossing over of the *in silico* design to *de novo* synthesis. So far pioneering work has been done on synthesis of chromosome 3 of the S288c strain; the synthetic chromosome (syn3) is 14 per cent shorter than the native chromosome, the natural telomere has been replaced, 21 retrotransposons have been removed, 98 LoxP site has been added for enabling SCRaMBLE genome shuffling and TAG stop codon has been replaced with TAA stop codon (Table 2) (Annaluru *et al.*, 2014). Table 2 summarises the differences in the chr 3 and syn 3. In spite of the differences, the yeast is viable and has not shown any phenotypic changes so far. The first flavour of what may be considered as the 'gain points' from this project and as a result of genome editing approaches is the de novo synthesis of raspberry ketone compound, 4-[4-hydroxyphenyl] butan-2-one, in a wine yeast strain (AWRI1631). The other potential 'gain points' that are envisioned from this synthetically engineered wine yeast is the de novo synthesis 3-mthoxy-4-hydroxybenzaldehyde, better known as vanillin and trans-3,5,4'-trihydroxystilbene, better known as resveratrol (Pretorius, 2017).

**Table 2.** Differences in the Natural (chr3) and Synthetic (syn 3) Chromosome 3 of *S. cerevisiae* 288c (Annaluru *et al.*, 2014, Pretorius, 2016)

| *S. cerevisiae S288c (chr3)* | *S. cerevisiae S288c (syn3)* |
|---|---|
| 316,617 bp | 272,871 bp |
| Natural telomere | Synthetic telemere |
| tRNA genes intact | 10 tRNA genes transformed from chr3 |
| Retrotransposons intact | 21 retrotransposons removed |
| Introns intact | Introns removed |
| TAG stop codon | TAA stop codon |

## 6. Genetic Engineering in Winery-associated Lactic Acid Bacterial LAB

The predominant genera of lactic acid bacteria (LAB) that are associated with wine making are *Lactobacillus, Pediococcus, Leuconostoc and Oenococcus*. These LABS are known to cause several biochemical transformations, like malo-lactic fermentation, conversion of citric acid into diacetyl and production of exopolysaccharide. These transformations form an essential part in the sensory attributes of wine (Lonvaud-Funel 2001). Unlike *S. cerevisiae*, a unique feature of LAB is that carbon and nitrogen metabolism has almost no overlap. Hence, genetic intervention in either of the metabolism leaves the other practically undisturbed. Their relatively simple molecular genetics and easier molecular control have resulted in quite a few successful metabolically-engineered LABs. For instance, end points, like higher alcohol production by disruption of lactate dehydrogenase (*LDH*), formation of diacetyl by overexpression of NADH oxidase (*NOX*) with simultaneous deletion of α-acetolactate decarboxylase (*ALDB*), biosynthesis of folic acid by engineering the pathway genes in *L. lactis* have been thoroughly reviewed by Hugenholtz and Kleerebezen (1999), though these end points may not be relevant directly to winemaking.

Evaluating from enological point of view, de Revel and co-workers (de Revel *et al.*, 2005) reported that wine LAB could enhance vanillin accumulation when in contact with wood, suggesting that the precursor of vanillin was present in wood. Taking cue from this, Bloem *et al.* (2006) evaluated *O. oeni, L. brevis, L. plantarum, P. damnosus and L. hilgardii* for their ability to synthesise vanillin from phenols, like eugenol, isoeugenol and vanillic acid. This work laid the ground for an important aspect. That is, MLF and the associate LABs could be used for unique aroma production during vinification. Reiteration of MLF significance brought back the research focus on the process and the organism responsible for it, *O. oeni*. Molecular biologists have directed their efforts into eliminating the 'pain points' associated with growth and functioning of *O. oeni*. The main hurdles that this bacteria faces is the low nutritional supplements, low pH, low processing temperature, high ethanol content and sulphur dioxide levels. Unlike other LABs, the exploration on genetic manipulation of *O. oeni*, has just started. The classical approach of random mutagenesis has been applied successfully by Li *et al.* (2015). The *O. oeni* strain was UV irradiated and selected that the mutagenised strain showed more efficient MLF than the parent strain. Meanwhile the *O. oeni* AWRIB429 strain has revealed a novel gene with potential glycosidase activity (Bartowsky and Borneman, 2011).

It will take a few more years for genetic manipulation techniques of *O.oeni* and it is envisaged that this strain will be mainly be explored for MLF and the engineering of novel aroma profile that may be possible through it.

## 7. Legislation

As per Regulation 1830/2003 and Regulation 65/2004 on traceability and labelling of GMOs and traceability of food and feed products, transgenic yeasts have to be labelled with unique identifier for GMO (EC 2003, EC 2004). In EU, wine made from transgenic yeast has to be considered transgenic and thus appropriately labelled. In the USA, as per the FDA guidelines, detailed prior analysis of genetic modifications is performed but specific labelling is not required when food is sufficiently similar in its nutritional composition (Federal Register of May 29, 1992 57 FR 22984). Latin America, Australia, Japan follow the model similar to that of FDA (Schuller and Casal, 2005).

For those arguing against transgenic yeast, a frequently cited reason is the risk of dissemination of GM yeasts and possible reduction in microbial diversity, due to the competitive advantage of recombinant strains. A study was performed throughout France and Portugal (Valero *et al.*, 2005). Interestingly, 94 per cent of the strains were recovered from the vicinity of winery (100-200 m); also, they were found in post harvest samples. Annual analysis of yeast population revealed that there was no occurrence of permanent implantation. It was also reported water runoff may felicitate yeast dissemination. It must be stated that an efficient system for assessment of potential and theoretical risks associated with GM yeasts will decide the scale and extent of their future use. With reference to assumed selective advantage of GM yeast, another study was performed by Bauer *et al.* (2003), where they are compared for fitness advantage of yeast (with overexpressed glucanase and pectinase) with that of commercial strains. The GM yeast did not affect the ecological balance of the vineyards microbiology. The list of concerns regarding GM yeast includes potential toxicity, possible antibiotic resistance, potential allergenicity and carcinogenicity. As per Annexure III of Directive 20017187EC, the GMOs need to be evaluated for their health and environmental impact (EC, 2001). This includes comparative nutritional composition analysis of transgenic product and conventional product, allergenicity study and toxicity studies on animals (Ramon *et al.*, 2005). Transgenic yeast-expressing xylanase from *Aspergillus nidulans* was evaluated for the above tests and was found to contain no risk as compared to conventional wine yeast (Pico *et al.*, 1999).

Along with strikingly different regulatory systems operative in different parts of the world, another complication that arises especially with wine is the international (OIV), national and local (appellation d'origine) regulations. Many producers oppose the GMO technology as they believe that tradition is the main strength for wine brands and production regions (Cebollero *et al.*, 2007). However, it is time to take a more pragmatic approach towards GM or transgenic yeasts.

## 8. Consumer Opinion

The question to whether there is a market for transgenic wine may not yield a straightforward and simple answer; it will be decided by the properties and attributes of wine, the country it is being produced and more importantly, the country where it is being sold. Several ground-breaking efforts on development of designer or transgenic yeasts has been made by research groups in USA, Latin America, Canada, France, Spain and South Africa. However, these technologies have not yet been translated by wineries into commercial products. Some authors claim that consumer acceptance of transgenic wine may be ruled by their feeling towards GM food in general. Going by this notion one may expect a more liberal consumer acceptance from USA, Latin America, Australia, South Africa, Spain and Portugal in the coming years. One may also surmise that transgenic wines produced from designer microorganisms, which target a better aroma or colour or increase in a compound with health benefits have a better chance consumer acceptance. The regulatory authorities of USA and Canada have approved two transgenic wine yeasts, one for reduction of ethyl carbamate (ECMo01) which has an extra copy of the *DUR1,2* gene (Coulon *et al.*, 2006) and one for improvement of malo-lactic fermentation ML01 which carries the *Schizosaccharomyces pombe* malate transporter gene (mae1) and the *Oenococcus oeni* malolactic enzyme gene (mleA) (Husnik *et al.*, 2006). This strain is able to perform both alcoholic and malo-lactic fermentations at once to simplify the winemaking process. In order to harness the commercial potential of transgenic wine, an appropriate legal framework is required to convince the consumers of the benefits being offered. Winemakers and yeast researchers face challenges of public opinion. The scientific interventions in winemaking have led to unfair criticism that the wine is no longer 'natural'. In view of the fact that many scientific concepts are unknown to public, appropriate education regarding attributes of transgenic wine may convert the sceptical public. A clear regulatory framework for adequate evaluation of safety of transgenic yeast and wine along with necessary labeling may assist consumers to make an informed choice.

## 9. Conclusion and Future Perspectives

While winemaking has been steeped in tradition, time has come to tilt the scale towards inventive science by proactive but prudent application of metabolic engineering and synthetic genome engineering. With the still prevailing anti-GM sentiments among the producers and consumers it is not possible to predict the extent of future applications of these synthetically engineered wine yeast. An educative and communicative dialogue has to be pursued between the policy makers, industrialists, researchers, regulators and consumers in order to harness the obvious benefits that these modern techniques can bring forth.

## References

Alper, H., Moxley, J., Nevoigt, E., Fink, G.R. and Stephanopoulos, G. (2006). Engineering yeast transcription machinery for improved ethanol tolerance and production. *Science* 314: 1565-1568.

An, D. and Ough, C.S. (1993). Urea excretion and uptake by wine yeasts as affected by various factors. *American Journal of Enology and Viticulture* 44: 35-40.

Annaluru, N., Muller, H., Mitchell, L.A., Ramalingam, S., Stracquadanio, G., Richardson, S.M., Dymond, J.S., Kuang, Z., Scheifele, L.Z., Cooper, E.M., Cai, Y., Zeller, K., Agmon, N., Han, J.S., Hadjithomas, M., Tullman, J., Caravelli, K., Cirelli, K., Guo, Z., London, V., Yeluru, A., Murugan, S., Kandavelou, K., Agier, N., Fischer, G., Yang, K., Martin, J.A., Bilgel, M., Bohutski, P., Boulier, K.M., Capaldo, B.J., Chang, J., Charoen, K., Choi, W.J., Deng, P., DiCarlo, J.E., Doong, J., Dunn, J., Feinberg, J.I., Fernandez, C., Floria, C.E., Gladowski, D., Hadidi, P., Ishizuka, I., Jabbari, J., Lau, C.Y., Lee, P.A., Li, S., Lin, D., Linder, M.E., Ling, J., Liu, J., London, M., Ma, H., Mao, J., McDade, J.E., McMillan, A., Moore, A.M., Oh, W.C., Ouyang, Y., Patel, R., Paul, M., Paulsen, L.C., Qiu, J., Rhee, A., Rubashkin, M.G., Soh, I Y., Sotuyo, N.E., Srinivas, V., Suarez, A., Wong, A., Wong, R., Xie, W.R., Xu, Y., Yu, A.T., Koszul, R., Bader, J.S., Boeke, J.D. and Chandrasegaran, S. (2014). Total synthesis of a functional designer eukaryotic chromosome. *Science* 344: 55-58.

Ansanay, V., Dequin, S., Blondin, B. and Barre, P. (1993). Cloning, sequence and expression of the gene encoding the malo-lactic enzyme from *Lactococcus lactis*. *FEBS Letters* 332: 74-80.

Barre, P., VeÂzinhet, F., Dequin, S. and Blondin, B. (1993). Genetic improvement of wine yeast. pp. 421-447. *In*: Fleet, G.H. (Ed.). Wine Microbiology and Biotechnology. Harwood Academic, Reading.

Bartowsky, E. and Borneman, A. (2011). Genomic variations of *Oenococcus oeni* strains and the potential to impact on malo-lactic fermentation and aroma compounds in wine. *Applied Microbiology and Biotechnology* 92: 441-447.

Bauer, F.F., Dequin, S., Pretorius, I.S., Shoeman, H., Wolfaardt, G., Schroeder, M.B. and Grossmann, M.K. (2004). The assessment of the environmental impact of genetically modified wine yeast strains. *Bulletin de l'OIV-Office International de la Vigne et du Vin* 77: 515-528.

Becker, J.V., Armstrong, G.O., van der Merwe, M.J., Lambrechts, M.G., Vivier, M.A. and Pretorius, I.S. (2003). Metabolic engineering of *Saccharomyces cerevisiae* for the synthesis of the wine-related antioxidant resveratrol. *FEMS Yeast Research* 4: 79-85.

Bellon, J.R., Schmid, F., Capone, D.L., Dunn, B.L. and Chambers, P.J. (2013). Introducing a new breed of wine yeast: Interspecific hybridisation between a commercial *Saccharomyces cerevisiae* wine yeast and *Saccharomyces mikatae*. *PLoS One* 8: e62053, 1-14.

Bidenne, C., Blondin, B., Dequin, S. and Vezinhet, F. (1992). Analysis of the chromosomal DNA polymorphism of wine strains of *Saccharomyces cerevisiae*. *Current Genetics* 22: 1-7.

Bloem, A., Bertrand, A., Lonvaud-Funel, A. and De Revel, G. (2007). Vanillin production from simple phenols by wine-associated lactic acid bacteria. *Letters in Applied Microbiology* 44: 62-67.

Blondin, B. and Dequin, S. (1998). Perspectives dans l'amélioration des levures en oenologie. pp. 415-444. *In*: Flanzy, C. (Ed.). Oenologie–fondements scientifiques et technologiques. Lavoisier, Paris.

Bony, M., Bidart, F., Camarasa, C., Ansanay, V., Dulau, L., Barre, P. and Dequin, S. (1997). Metabolic analysis of *S. cerevisiae* strains engineered for malo-lactic fermentation. *FEBS Letters* 410: 452-456.

Borneman, A.R., Chambers, P.J. and Pretorius, I.S. (2007). Yeast systems biology: Modelling the winemaker's art. *Trends in Biotechnology* 25: 349-355.

Borneman, A.R., Desany, B.A., Riches, D., Affourtit, J.P., Forgan, A.H., Pretorius, I.S., Egholm, M. and Chambers, P.J. (2011). Whole-genome comparison reveals novel genetic elements that characterise the genome of industrial strains of *Saccharomyces cerevisiae*. *PLoS Genetics* 7: 1-10.

Borneman, A.R., Desany, B.A., Riches, D., Affourtit, J.P., Forgan, A.H., Pretorius, I.S., Egholm, M. and Chambers, P.J. (2012). The genome sequence of the wine yeast VIN7 reveals an allotriploid hybrid genome with *Saccharomyces cerevisiae* and *Saccharomyces kudriavzevii* origins. *FEMS Yeast Research* 12: 88-96.

Boulton, R.B., Singleton, V.L., Bisson, L.F. and Kunkee, R.E. (1999). Yeast and biochemistry of ethanol fermentation. pp. 102-192. *In*: Principles and Practices of Winemaking. Springer, Boston, Massachusetts.

Brown, T.A. (1990). Cloning vectors for organisms other than *E. coli*. p. 128. *In*: Gene Cloning, second ed. Chapman and Hall, New York.

Butzke, C.E. and Bisson, L.F. (1996). Genetic engineering of yeast for wine production. *Agro-Food Ind Hi-Tech.* (Jul./Aug.): 26-30.

Cadière, A., Aguera, E., Caillé, S., Ortiz-Julien, A. and Dequin, S. (2012). Pilot-scale evaluation the enological traits of a novel, aromatic wine yeast strain obtained by adaptive evolution. *Food Microbiology* 32: 332-337.

Cambon, B., Monteil, V., Remize, F., Camarasa, C. and Dequin, S. (2006). Effects of GPD1 overexpression in *Saccharomyces cerevisiae* commercial wine yeast strains lacking ALD6 genes. *Applied and Environmental Microbiology* 72(7): 4688-4694.

Canal-Llauberes, R.M. (1993). Enzymes in winemaking. pp. 477-506. *In*: Fleet, G.H. (Ed.). Wine Microbiology and Biotechnology. Harwood Academic, Reading.

Caputi, A.J. and Ryan, T. (1996). Must and Wine Acidification. Presentation at a meeting of the OIV Expert Group Technologie du Vin, Paris.

Cardona, F., Carrasco, P., Perez-Ortin, J.E., Olmo, M.I.D. and Aranda, A. (2007). A novel approach for improvement of stress resistance in wine yeasts. *International Journal of Food Microbiology* 114: 83-91.

Cebollero, E. and Gonzalez, R. (2004). Comparison of two alternative dominant selectable markers for wine yeast transformation. *Applied and Environmental Microbiology* 70: 7018-7023.

Cebollero, E., González-Ramos, D., Tabera, L. and González, R. (2007). Transgenic wine yeast technology comes of age: Is it time for transgenic wine? *Biotechnology Letters* 29: 191-200.

Cole, V.C. and Noble, A.C. (1995). Flavour chemistry and assessment. pp. 361-385. *In*: Lea, A.G.H. and Piggott, J.R. (Eds.). Fermented Beverage Production. Blackie Academic and Professional, London.

Cordier, H., Mendes, F., Vasconcelos, I. and François, J.M. (2007). A metabolic and genomic study of engineered *Saccharomyces cerevisiae* strains for high glycerol production. *Metabolic Engineering* 9: 364-378.

Coulon, J., Husnik, J.I., Inglis, D.L., van der Merwe, G.K., Lonvaud, A., Erasmus, D.J. and van Vuuren, H.J. (2006). Metabolic engineering of *Saccharomyces cerevisiae* to minimise the production of ethyl carbamate in wine. *American Journal of Enology and Viticulture* 57: 113-124.

De Barros Lopes, M., Bellon, J.R., Shirley, N.J. and Ganter, P.F. (2002). Evidence for multiple interspecific hybridisation in *Saccharomyces* sensu stricto species. *FEMS Yeast Research* 1: 323-331.

de Revel, G., Bloem, A., Augustin, M., Lonvaud-Funel, A. and Bertrand, A. (2005). Interaction of *Oenococcus oeni* and oak wood compounds. *Food Microbiology* 22: 569-575.

de Smidt, O., du Preez, J.C. and Albertyn, J. (2012). Molecular and physiological aspects of alcohol dehydrogenases in the ethanol metabolism of *Saccharomyces cerevisiae*. *FEMS Yeast Research* 12: 33-47.

Degre, R. (1993). Selection and commercial cultivation of wine yeast and bacteria. *Wine Microbiology and Biotechnology* 421-448.

Denayrolles, M., Aigle, M. and Lonvaud-Funel, A. (1995). Functional expression in *Saccharomyces cerevisiae* of the *Lactococcus lactis* mleS gene encoding the malo-lactic enzyme. *FEMS Microbiology Letters* 125: 37-43.

Dequin, S. and Barre, P. (1994). Mixed lactic acid–alcoholic fermentation by *Saccharomyes cerevisiae* expressing the *Lactobacillus casei* L (+)–LDH. *Nature Biotechnology* 12: 173-177.

Dequin, S., Baptista, E. and Barre, P. (1999). Acidification of grape musts by *Saccharomyces cerevisiae* wine yeast strains genetically engineered to produce lactic acid. *American Journal of Enology and Viticulture* 50: 45-50.

Dicks, L.M.T., Dellaglio, F. and Collins, M.D. (1995). Proposal to reclassify *Leuconostoc oenos* as Oenococcus oeni [corrig.] gen. nov., comb. nov. *International Journal of Systematic and Evolutionary Microbiology* 45: 395-397.

EC (2001). Directive 2001/18/EC of the European Parliament and of the Council of 12 March 2001 on the deliberate release into the environment of genetically modified organisms and repealing Council Directive 90/220/EC. *Official Journal of the European Communities* L106, 12.3.2001: 1-38.

EC (2003). Regulation (EC) No 1830/2003 of the European Parliament and of the Council of 22 September 2003 concerning the traceability and labelling of genetically modified organisms and the traceability of food and feed products produced from genetically-modified organisms and amending Directive 2001/ 18/EC. *Official Journal of the European Communities* L268, 18.10.2003: 24-28

EC (2004). Regulation (EC) No 65/2004 of 14 January 2004 establishing a system for the development and assignment of unique identifiers for genetically modified organisms. *Official Journal of the European Communities* L10, 16.1.2004: 5-10.

Eglinton, J.M., Heinrich, A.J., Pollnitz, A.P., Langridge, P., Henschke, P.A. and de Barros Lopes, M. (2002). Decreasing acetic acid accumulation by a glycerol overproducing strain of *Saccharomyces cerevisiae* by deleting the ALD6 aldehyde dehydrogenase gene. *Yeast* 19: 295-301.

Francois, J., BlaÂzquez, M.A., Arino, J. and Gancedo, C. (1997). Storage carbohydrates in the yeast *Saccharomyces cerevisiae*. pp. 285-311. *In*: Zimmermann, F.K. and Entian, K.D. (Eds.). Yeast Sugar Metabolism. Technomic Publication, Lancaster, Pennsylvania.

Frankel, E.N., Waterhouse, A.L. and Teissedre, P.L. (1995). Principal phenolic phytochemicals in selected California wines and their antioxidant activity in inhibiting oxidation of human low-density lipoproteins. *Journal of Agricultural and Food Chemistry* 43: 890-894.

Fujii, T., Nagasawa, N., Iwamatsu, A., Bogaki, T., Tamai, Y. and Hamachi, M. (1994). Molecular cloning, sequence analysis and expression of the yeast alcohol acetyltransferase gene. *Applied and Environmental Microbiology* 60: 2786-2792.

Fujii, T., Yoshimoto, H. and Tamai, Y. (1996). Acetate ester production by *Saccharomyces cerevisiae* lacking the ATF1 gene encoding the alcohol acetyl transferase. *Journal of Fermentation and Bioengineering* 8: 538-542.

Fujii, T., Kobayashi, O. Yoshimoto, H., Furukawa, S. and Tamai, Y. (1997). Effect of aeration and unsaturated fatty acids on expression of the *Saccharomyces cerevisiae* alcohol acetyl transferase gene. *Applied and Environmental Microbiology* 63: 910-915.

Gawel, R., Francis, L. and Waters, E.J. (2007a). Statistical correlations between the in-mouth textural characteristics and the chemical composition of Shiraz wines. *Journal of Agricultural and Food Chemistry* 55: 2683-2687.

Gawel, R., Sluyter, S.V. and Waters, E.J. (2007b). The effects of ethanol and glycerol on the body and other sensory characteristics of Riesling wines. *Australian Journal of Grape and Wine Research* 13: 38-45.

Gimeno-Alcaniz, J.V. and Matallana, E. (2001). Performance of industrial strains of *Saccharomyces cerevisae* during wine fermentation is affected by manipulation strategies based on sporulation. *Systematic and Applied Microbiology* 24: 639-644.

Giudici, P. and Azinnato, A.Z. (1983). Influenza dell'uso di mutanti nutrizionali sulla produzione di alcoli superiori. *Vignevini* 10: 63-65.

Giudici, P., Solieri, L., Pulvirenti, A.M. and Cassanelli, S. (2005). Strategies and perspectives for genetic improvement of wine yeasts. *Applied Microbiology and Biotechnology* 66: 622-628.

Gonzâlez, R., Martínez-Rodríguez, A.J. and Carrascosa, A.V. (2003). Yeast autolytic mutants potentially useful for sparkling wine production. *International Journal of Food Microbiology* 84: 21-26.

Gomez-Pastor, R., Perez-Torrado, R., Cabiscol, E., Ros, J. and Matallana, E. (2010). Reduction of oxidative cellular damage by overproduction of the thioredoxin gene improves yield and quality of wine yeast dry active biomass. *Microbial Cell Factories* 9: 1-14.

Gonzalez-Candelas, L., Gil, J.V., Lamuela-Raventos, R.M. and Ramon, D. (2000). The use of transgenic yeasts expressing a gene encoding a glucosyl hydroxylase as a tool to increase resveratrol content in wine. *International Journal of Food Microbiology* 59: 179-183.

Govander, P., Kroppenstedt, S. and Bauer, F.F. (2011). Novel wine-mediated FLO11 floculation phenotype of commercial *S. cerevisiae* wine yeast strains with modified FLO gene expression. *FEMS Microbiology Letters* 317: 117-126.

Grant, C.M. (2001). Role of the glutathione/glutaredoxin and thioredoxin systems in yeast growth and response to stress conditions. *Molecular Microbiology* 39: 533-541.

Gunge, N. and Y. Nakatomi (1972). Genetic mechanisms of rare matings of the yeast *Saccharomyces cerevisiae* heterozygous for mating type. *Genetics* 70: 41-58.

Halls, C. and Yu, O. (2008). Potential for metabolic engineering of resveratrol biosynthesis. *Trends in Biotechnology* 26: 77-81.

Hammond, J.R.M. (1996). Yeast genetics. pp. 45-82. *In*: Priest, F.G. and Campbell, I. (Eds.). Brewing Microbiology. Chapman and Hall, London.

Henschke, P.A. (1997). Wine yeast. pp. 527-560. *In*: Zimmermann, F.K. and Entian, K.D. (Eds.). Yeast Sugar Metabolism. Technomic Publishing, Lancaster, Pennsylvania.

Herrero, O., Ramon, D. and Orejas, M. (2008). Engineering the *S. cerevisiae* isoprenoid pathway for de novo production of aromatic mono terpenes in wine. *Metabolic Engineering* 10: 78-86.

Houtman, A.C., Manns, J. and Du Plhssis, C. (1980). Factors affecting the reproducibility of fermentation of grape juice and of the aroma composition of wines l: Grape maturity, sugar, inoculum concentration, aeration. *Vitis* 19: 37-54.

Hubner, P., Groux, P.M., Weibel, B., Sengstag, C., Horlbeck, J., Leong-Morgenthaler, P.M. and Luthy, J. (1997). Genotoxicity of ethyl carbamate (urethane) in *Salmonella*, yeast and human lymphoblastoid cells. *Mutation Research* 390: 11-19.

Hugenholtz, J. and Kleerebezem, M. (1999). Metabolic engineering of lactic acid bacteria: Overview of the approaches and results of pathway rerouting involved in food fermentations. *Current Opinion in Biotechnology* 10: 492-497.

Husnik, J.I., Volschenk, H., Bauer, J., Colavizza, D., Luo, Z. and van Vuuren, H.J. (2006). Metabolic engineering of malo-lactic wine yeast. *Metabolic Engineering* 8: 315-323.

Husnik, J.I., Delaquis, P.J., Cliff, M.A. and van Vuuren, H.J. (2007). Functional analyses of the malo-lactic wine yeast ML01. *American Journal of Enology and Viticulture* 58: 42-52.

Ishida-Fujii, K., Goto, S., Sugiyama, H., Takagi, Y., Saiki, T. and Takagi, M. (1998). Breeding of flocculent industrial alcohol yeast strains by self-cloning of the flocculation gene FLO1 and repeated-batch fermentation by transformants. *The Journal of General and Applied Microbiology* 44: 347-353.

Kallithraka, S., Arvanitoyannis, I., El-Zajouli, A. and Kefalas, P. (2001). The application of an improved method for trans-resveratrol to determine the origin of Greek red wines. *Food Chemistry* 75: 355-363.

Kitamoto, K.A.T.S.U.H.I.K.O., Oda, K., Gomi, K. and Takahashi, K.O.J.I.R.O. (1991). Genetic engineering of a sake yeast producing no urea by successive disruption of arginase gene. *Applied and Environmental Microbiology* 57: 301-306.

Kutyna, D.R., Varela, C., Stanley, G.A., Borneman, A.R., Henschke, P.A. and Chambers, P.J. (2012). Adaptive evolution of *Saccharomyces cerevisiae* to generate strains with enhanced glycerol production. *Applied Microbiology and Biotechnology* 93: 1175-1184.

Labarre, C., Guzzo, J., Cavin, J.F. and Divies, C. (1996). Cloning and characterisation of the genes encoding the malolactic enzyme and the malate permease of *Leuconostoc oenos*. *Applied and Environmental Microbiology* 62: 1274-1282.

Laing, E. and Pretorius, I.S. (1999). Co-expression of an *Erwinia chrysanthemi* pectate lyase-encoding gene (pelE) and an *E. carotovora* polygalacturonase-encoding gene (peh1) in *Saccharomyces cerevisiae*. *Applied Microbiology and Biotechnology* 39: 181-188.

Lambrechts, M.G., Bauer, F.F., Marmur, J. and Pretorius, I.S. (1996). Muc1, a mucin-like protein that is regulated by Mss10, is critical for pseudohyphal differentiation in yeast. *Proceedings of the National Academy of Sciences* 93: 8419-8424.

Lee, D., Lloyd, D.R., Pretorius, I.S. and Borneman, A.R. (2016). Hetelogous production of raspberry ketone in wine yeast *S. cerevisiae* via pathway engineering and synthetic enzyme fusion. *Microbial Cell Factories* 15: 49-55.

Li, N., Duan, J., Gao, D., Luo, J., Zheng, R., Bian, Y., Zhang, Z. and Ji, B. (2015). Mutation and selection of *Oenococcus oeni* for controlling wine malo-lactic fermentation. *European Food Research and Technology* 240: 93-100.

Lilly, M., Lambrechts, M.G. and Pretorius, I.S. (2000). Effect of increased yeast alcohol acetyl transferase activity on flavour profiles of wine and distillates. *Applied and Environmental Microbiology* 66: 744-753.

Lo, W.S. and Dranginis, A.M. (1996). FLO11, a yeast gene related to the STA genes, encodes a novel cell surface flocculin. *Journal of Bacteriology* 178: 7144-7151.

Lonvard-Funel, A. (1999). Lactic acid bacteria in the quality improvement and depreciation of wine. *Antonie van Leeuwenhoek* 76: 317-331.

Lonvaud-Funel, A. (2001). Biogenic amines in wines: Role of lactic acid bacteria. *FEMS Microbiology Letters* 199: 9-13.

Malherbe, D.F., Du Toit, M., Otero, R.C., Van Rensburg, P. and Pretorius, I.S. (2003). Expression of the *Aspergillus niger* glucose oxidase gene in *Saccharomyces cerevisiae* and its potential applications in wine production. *Applied Microbiology and Biotechnology* 61: 502-511.

Mangado, A., Tronchoni, J., Morales, P., Novo, M., Quirós, M. and Gonzalez, R. (2015). An impaired ubiquitin ligase complex favours initial growth of auxotrophic yeast strains in synthetic grape must. *Applied Microbiology and Biotechnology* 99: 1273-1286.

Manzanares, P., Orejav, M., Gil, J.V., deGraff, L.H., Visser, J. and Ramon, D. (2003). Construction of genetically modified wine yeast strain expressing the *Aspergillus aculeatus* rhaA gene, encoding an α-L-rhamnosidase of enological interest. *Applied and Environmental Microbiology* 69: 7558-7562.

Marullo, P., Bely, M., Masneuf-Pomarede, I., Aigle, M. and Dubourdieu, D. (2004). Inheritable nature of enological quantitative traits is demonstrated by meiotic segregation of industrial wine yeast strains. *FEMS Yeast Research* 4: 711-719.

McBryde, C., Gardner, J.M., de Barros Lopes, M. and Jiranek, V. (2006). Generation of novel wine yeast strains by adaptive evolution. *American Journal of Enology and Viticulture* 57: 423-430.

Moglia, A., Comino, C., Lanteri, S., de Vos, R., de Waard, P., van Beek, T.A., Goitre, L., Retta, S.F. and Beekwilder, J. (2010). Production of novel antioxidative phenolic amides through heterologous expression of the plant's chlorogenic acid biosynthesis genes in yeast. *Metabolic Engineering* 12: 223-232.

Mouret, J.R., Cadiere, A., Aguera, E., Rollero, S., Ortiz-Julien, A., Sablayrolles, J.M. and Dequin, S. (2014). Dynamics and quantitative analysis of the synthesis of fermentative aromas by an evolved wine strain of *Saccharomyces cerevisiae*. *Yeast* 32: 257-269.

Naouri, P., Chagnaud, P., Arnaud, A. and Galzy, P. (1990). Purification and properties of a malo-lactic enzyme from *Leuconostoc oenos* ATCC 23278. *Journal of Basic Microbiology* 30: 577-585.

Ough, C.S., Crowell, E.A. and Gutlove, B.R. (1988). Carbamyl compound reactions with ethanol. *American Journal of Enology and Viticulture* 39: 239-242.

Overkamp, K.M., Bakker, B.M., Kötter, P., Luttik, M.A., van Dijken, J.P. and Pronk, J.T. (2002). Metabolic engineering of glycerol production in *Saccharomyces cerevisiae*. *Applied and Environmental Microbiology* 68: 2814-2821.

Pérez-González, J.A., Gonzalez, R., Querol, A., Sendra, J. and Ramon, D. (1993). Construction of a recombinant wine yeast strain expressing beta-(1,4)-endoglucanase and its use in microvinification processes. *Applied and Environmental Microbiology* 59: 2801-2806.

Pérez-Gonzalez, J.A., De Graaff, L.H., Visser, J. and Ramon, D. (1996). Molecular cloning and expression in *Saccharomyces cerevisiae* of two *Aspergillus nidulans* xylanase genes. *Applied and Environmental Microbiology* 62: 2179-2182.

Pérez-Torrado, R., Gimeno-Alcañiz, J.V. and Matallana, E. (2002). Wine yeast strains engineered for glycogen overproduction display an enhanced viability under glucose deprivation conditions. *Applied and Environmental Microbiology* 68: 3339-3344.

Picó, Y., Fernández, M., Rodríguez, R., Almudéver, J., Mañes, J., Font, G., Marín, R., Carda, C., Manzanares, P. and Ramón, D. (1999). Toxicological assessment of recombinant xylanase X(22) in wine. *Journal of Agriculture and Food Chemistry* 47: 1597-1602.

Pizarro, F., Vargas, F.A. and Agosin, E. (2007). A systems biology perspective of wine fermentations. *Yeast* 24: 977-991.

Pretorius, I.S. (2000). Tailoring wine yeast for the new millennium: Novel approaches to the ancient art of winemaking. *Yeast* 16: 675-729.

Pretorius, I.S., Curtin, C.D. and Chambers, P.J. (2012). The winemaker's bug. *Bioengineered Bugs* 33: 147-156.

Pretorius, I.S. (2017). Synthetic genome engineering forging new frontiers for wine yeast. *Critical Reviews in Biotechnology* 37: 112-136.

Porro, D., Brambilla, L., Ranzi, B.M., Martegani, E. and Alberghina, L. (1995). Development of metabolically engineered *Saccharomyces cerevisiae* cells for the production of lactic acid. *Biotechnology Progress* 11: 294-298.

Rachidi, N., Barre, P. and Blondin, B. (1999). Multiple Ty-mediated chromosomal translocations lead to karyotype changes in a wine strain of *Saccharomyces cerevisiae*. *Molecular and General Genetics* 261: 841-850.

Ramon, D., Genoves, S., Gil, J.V., Herrero, O., MacCabe, A., Manzanares, P., Matallana, E., Orejas, M., Uber, G. and Valles, S. (2005). Milestones in wine biotechnology. *Minerva Biotechnology* 17: 33-45.

Remize, F., Roustan, J.L., Sablayrolles, J.M., Barre, P. and Dequin, S. (1999). Glycerol overproduction by engineered *Saccharomyces cerevisiae* wine yeast strains leads to substantial changes in by-product formation and to a stimulation of fermentation rate in stationary phase. *Applied and Environmental Microbiology* 65: 143-149.

Romano, P., Soli, M.G., Suzzi, G., Grazia, L. and Zambonelli, C. (1985). Improvement of a wine *Saccharomyces cerevisiae* strain by a breeding programme. *Applied and Environmental Microbiology* 50: 1064-1067.

Rous, C.V., Snow, R. and Kunkee, R.E. (1983). Reduction of higher alcohols by fermentation with a leucine-auxotrophic mutant of wine yeast. *Journal of the Institute of Brewing* 89: 274-278.

Salmon, J.M. (1997). Enological fermentation kinetics of an isogenic ploidy series derived from an industrial *Saccharomyces cerevisiae* strain. *Journal of Fermentation and Bioengineering* 83: 253-260.

Salmon, J.M. and Barre, P. (1998). Improvement of nitrogen assimilation and fermentation kinetics under enological conditions by derepression of alternative nitrogen-assimilatory pathways in an industrial *Saccharomyces cerevisiae* strain. *Applied and Environmental Microbiology* 64: 3831-3837.

Scanes, K.T., Hohrnann, S. and Prior, B.A. (1998). Glycerol production by the yeast *Saccharomyces cerevisiae* and its relevance to wine: A review. *South African Journal of Enology and Viticulture* 19: 17-24.

Schlatter, J. (1986). The toxicity of urethane (ethylcarbamate). *Proceedings of Euro Food Tox II, Interdisciplinary Conference on Natural Toxicants in Food* Zurich, pp. 249-254.

Schuller, D. and Casal, M. (2005). The use of genetically modified *Saccharomyces cerevisiae* strains in the wine industry. *Applied Microbiology and Biotechnology* 68: 292-304.

Shinohara, T., Saito, K., Yanagida, F. and Goto, S. (1994). Selection and hybridisation of wine yeasts for improved winemaking properties: Fermentation rate and aroma productivity. *Journal of Fermentation and Bioengineering* 77: 428-431.

Siemann, E.H. and Creasy, L.L. (1992). Concentration of the phytoalexin resveratrol in wine. *American Journal of Enology and Viticulture* 43: 49-52.

Silva-Udawatta, D., Mihiri, N. and Cannon, J.F. (2001). Roles of trehalose phosphate synthase in yeast glycogen metabolism and sporulation. *Molecular Microbiology* 40: 1345-1356.

Skipper, N. and Bussey, H. (1977). Mode of action of yeast toxins: Eenergy requirement for *Saccharomyces cerevisiae* killer toxin. *J. Bacteriol.* 129: 668.

Soufleros, E., Barrios, M.L. and Bertrand, A. (1998). Correlation between the content of biogenic amines and other wine compounds. *American Journal of Enology and Viticulture* 49: 266-278.

Spencer, J.F. and Spencer, D.M. (1996). Rare-mating and cytoduction in *Saccharomyces cerevisiae*. pp. 39-44. *In*: Yeast Protocols, Humana Press.

Srivastava, D.K. and Raverkar, K.P. (1999). Genetic manipulation of industrially important microorganisms. p. 173. *In*: V.K. Joshi and A. Pandey (Eds.). Biotechnology: Food Fermentation Microbiology, Biochemistry and Technology, vol. I. Educational Publishers and Distributors, New Delhi.

Suizu, T., Iimura, Y., Gomi, K., Takahashi, K., Hara, S. and Yoshizawa, K. (1990). Construction of urea non-producing yeast *Saccharomyces cerevisiae* by disruption of the CAR1 gene (Microbiology & Fermentation Industry). *Agricultural and Biological Chemistry* 54: 537-539.

Swiegers, J.H., Capone, D.L., Pardon, K.H., Elsey, G.M., Sefton, M.A., Francis, I.L. and Pretorius, I.S. (2007). Engineering volatile thiol release in *Saccharomyces cerevisiae* for improved wine aroma. *Yeast* 24: 561-574.

Sydor, T., Schaffer, S. and Boles, E. (2010). Considerable increase in resveratrol production by recombinant industrial yeast strains with use of rich medium. *Applied and Environmental Microbiology* 76: 3361-3363.

Trantas, E., Panopoulos, N. and Ververidis, F. (2009). Metabolic engineering of the complete pathway leading to heterologous biosynthesis of various flavonoids and stilbenoids in *Saccharomyces cerevisiae*. *Metabolic Engineering* 11: 355-366.

Teunissen, A.W.R.H. and Steensma, H.Y. (1995). The dominant flocculation genes of *Saccharomyces cerevisiae* constitute a new subtelomeric gene family. *Yeast* 11: 1001-1013.

Tilloy, V., Ortiz-Julien, A. and Dequin, S. (2014). Reduction of ethanol yield and improvement of glycerol formation by adaptive evolution of the wine yeast *Saccharomyces cerevisiae* under hyperosmotic conditions. *Applied and Environmental Microbiology* 80: 2623-2632.

Valero, E., Schuller, D., Cambon, B., Casal, M. and Dequin, S. (2005). Dissemination and survival of commercial wine yeast in the vineyard: A large-scale, three-year study. *FEMS Yeast Research* 5: 959-969.

Van Rensburg, P., van Zyl, W.H. and Pretorius, I.S. (1996). Co-expression of a Phanerochaete chrysosporium cellobiohydrolase gene and a Butyrivibrio fibrisolvens endo-β-1,4-glucanase gene in *Saccharomyces cerevisiae*. *Current Genetics* 30: 246-250.

Van Rensburg, P., van Zyl, W.H. and Pretorius, I.S. (1997). Overexpression of the *Saccharomyces cerevisiae* exo-β-1,3-glucanase gene together with the *Bacillus subtilis* endo-β-1, 3-1,4-glucanase gene and the *Butyrivibrio fibrisolvens* endo-β-1, 4-glucanase gene in yeast. *Journal of Biotechnology* 55: 43-53.

Van Rensburg, P., van Zyl, W.H. and Pretorius, I.S. (1998). Engineering yeast for efficient cellulose degradation. *Yeast* 14: 67-76.

Varela, C., Kutyna, D.R., Solomon, M.R., Black, C.A., Borneman, A., Henschke, P.A., Pretorious, I.S. and Chambers, P.J. (2012). Evaluation of gene modification strategies for the development of low-alcohol-wine yeasts. *Applied and Environmental Microbiology* 78: 6068-6077.

Varela, C., Dry, P.R., Kutyna, D.R., Francis, I.L., Henschke, P.A., Curtin, C.D. and Chambers, P.J. (2015). Strategies for reducing alcohol concentration in wine. *Australian Journal of Grape and Wine Research* 21: 670-679.

Vinson, J.A., Jang, J., Dabbagh, Y.A., Serry, M.M. and Cai, S. (1995). Plant polyphenols exhibit lipoprotein-bound antioxidant activity using an in vitro oxidation model for heart disease. *Journal of Agricultural and Food Chemistry* 43: 2798-2799.

Volschenk, H., Viljoen, M., Grobler, J., Petzold, B., Bauer, F., Subden, R.E., Young, R.A., Lonvaud, A., Denayrolles, M. and van Vuuren, H.J. (1997). Engineering pathways for malate degradation in *Saccharomyces cerevisiae*. *Nature Biotechnology* 15: 253-257.

Wang, Y., Halls, C., Zhang, J., Matsuno, M., Zhang, Y. and Yu, O. (2011). Stepwise increase of resveratrol biosynthesis in yeast *Saccharomyces cerevisiae* by metabolic engineering. *Metabolic Engineering* 13: 455-463.

Watari, J., Nomura, M., Sahara, H., Koshino, S. and Keränen, S. (1994). Construction of flocculent brewer's yeast by chromosomal integration of the yeast flocculation gene FLO1. *Journal of the Institute of Brewing* 100: 73-77.

Wöhrmann, K. and Lange, P. (1980). The polymorphism of esterases in yeast (*Saccharomyces cerevisiae*). *Journal of the Institute of Brewing* 86: 174-177.

Zhao, J., Du Guocheng Zou, H., Fu, J., Zhao, J. and Chen, J. (2013). Progress in preventing the accumulation of ethyl carbamate in alcoholic beverages. *Trends in Food Science and Technology* 32: 97-107.

Zhao, H., Li, J., Han, B., Li, X. and Chen, J. (2014). Improvement of oxidative stress tolerance in *S. cerevisiae* thorugh global transcription machinery engineering. *Journal of Industrial Microbiology and Biotechnology* 41: 869-878.

# 15 Oenological Enzymes

**Harald Claus**

Institute of Molecular Physiology, Microbiology & Wine Research, Johannes Gutenberg-University, 55099 Mainz, Germany

## 1. Introduction

Oenological enzymes are used for various transformation reactions in wine technology. Industrial enzymes provide quantitative benefits (increased juice yield), qualitative advantages (enhancement of aroma and colour) and processing benefits (shorter maceration, settlement and filtration time) (van Rensburg and Pretorius, 2000; Ugliano, 2009; Mojsov, 2013; Mojsov *et al.*, 2015; Hüfner and Hasselbeck, 2017). This study provides an overview on important enzymes for vinification and the effects of commercial enzyme preparations on process technology and the quality of the final product. In addition, it highlights on the perspective use of beneficial enzymes from yeasts and lactic acid bacteria.

Commercial enzymes for winemaking are currently prepared from filamentous fungi, such as *Aspergillus* or *Trichoderma* species and are often mixtures of different activities (Hüfner and Haßelbeck, 2017). Although applications of enzyme cocktails bear some technical advantages, there is an increasing demand for enzymes with improved and more specific characteristics. In this context, the high endogenous enzymatic potential of wine and grape-associated microorganisms may support the wine industry to meet upcoming technical and consumer challenges. In contrast to filamentous fungi, selected wine yeasts and lactic acid bacteria with beneficial enzymatic equipment can be directly inoculated as starter cultures for must fermentations.

Enzymes used in the cellar industry should increase the yield during pressing, eliminate clarification and filtration problems, positively influence the taste of the wine, preserve the wine, prevent the formation of sensory deficiencies and the development of harmful substances (Table 1).

**Table 1.** Oenological Enzymes and Their Applications

| *Enzymatic activity* | *Aim of application* |
|---|---|
| Pectinases | Degradation of must viscosity (pectin).<br>Improvement of clarification |
| Pectinolytic enzymes with side activities (cellulases, hemicellulases) | Hydrolysis of plant cell wall polysaccharides. Improvement of skin maceration, extraction efficiency, colour |
| Glycosidases (occur as side activities of bulk enzyme preparations) | Improvement of aroma by splitting sugar residues from odourless precursors |
| Glycosidases | Improvement of bioavailability of beneficial polyphenols |
| Glucanases | Lysis of yeast cell walls, release of mannoproteins |
| Glucanases | Lysis of bacterial and fungal exopolysaccharides to improve clarification |
| Urease | Hydrolysis of urea, preventing formation of harmful ethyl carbamate |
| Lysozyme | Control of bacterial growth |
| Proteases | Prevention of protein haze.<br>Reduction of bentonite dosages |

E-mail: hhclaus@arcor.de

Special enzymes for winemaking are produced and marketed by companies, like *Oenobrands* (France), *AEB* (Italy), *Erbslöh Geisenheim AG* (Germany), *Darleon* (South Africa), *DSM* (Switzerland), *Novo Nordisk* (Denmark). They are usually prepared from fungi as less well-defined enzyme cocktails (van Rensburg and Pretorius, 2000; Ugliano, 2009; Mojsov, 2013; Mojsov *et al.*, 2015; Hüfner and Haßelbeck, 2017). The use of bulk enzyme preparations is advantageous because it fulfils several functions. Examples are liquefaction enzymes which contain cellulases and hemicellulases in addition to pectinases.

## 2. Enzymes Useful to Improve Extraction Efficiency and Juice Yield

In white wine production, the grapes are usually directly milled after harvest, i.e. the berry skins are crushed so far that the berry juice can easily escape. The resulting mash is then pressed immediately; the expired juice is fermented directly. The juice drain is greatly facilitated when the mash is left for a few hours, e.g. overnight. The reason for the time expenditure lies in the high pectin content of the grapes, which is only slightly counteracted by intrinsic pectolytic enzymes.

The basic building block of all pectin substances is polygalacturonic acid partially esterified with methanol. Pectin consists either only of galacturonic acid or includes additional sugars, such as rhamnose. Side chains of mono-, di- and polysaccharides are attached to these long-chain molecules.

Pectins are colloidally dissolved in water and can be precipitated by alcohol. Nevertheless, pectin clouding in wine does not occur because the pectins are rapidly demethylated and depolymerised by pectolytic enzymes already present in the grape.

There are hardly any technological problems with regard to pressing and yield, as long as grapes are healthy and enough time to leave the mash. However, if the grapes are partially rotted (mainly by *Botrytis cinerea*), there are also increased levels of pectin in the mash. The natural pectinase capacity of the grape is then no longer sufficient to reduce the increased pectin content, which makes juice extraction from the highly viscous mashes inefficient. The filtration of the wines obtained from such mash is also much more difficult. But even with healthy grapes there are problems when pressing, if large quantities of grapes are to be processed in a short time.

### 2.1. Technical Pectinases

The complete degradation of pectin requires the cooperation of several enzymes to break down the complex molecule into small fragments (Mojsov, 2013; Mojsov *et al.*, 2015). They include various enzymatic activities:

- Polygalacturonase (homogalacturonan hydrolase) (PG): Hydrolytic depolymerisation of the polygalacturonic acid chain. One can distinguish enzymes which either cleave single galacturonic acid units from the end of the chain (exo-activity, exoPG, EC 3.2.1.67) or in the middle of the chain (endo-activity, endoPG, EC 3.2.1.15)
- Pectinlyase/pectate lyase (EC 4.2.2.2 and 4.2.2.9): Non-hydrolytic cleavage of the polygalacturonic acid chain
- Pectinesterase (EC 3.1.1.11): Hydrolytic cleavage of methanol from the D-galacturonic acid chain
- Acetylesterase (EC 3.1.1.6): Cleaves acetyl residues from D-galacturonic acid with release of acetic acid

As a prerequisite, the enzymes must be effective under acidic wine conditions, moderate temperatures and in the presence of ethanol (Venturi *et al.*, 2013). The optimum temperature of the pectolytic enzymes is between 40-55°C. In most cases, the mash must therefore be warmed up. If this is technically not possible or not desirable for qualitative reasons, the enzyme dosage must be increased. There is also a strong pH dependence – from pH 2.7 to pH 4.0, the efficacy increases threefold.

Industrial pectinase preparations contain the active enzymes (2-5 per cent) and additives (sugars, inorganic salts, preservatives) that stabilise and standardise their specificities (Mojsov, 2013). Factors that generally inhibit proteins reduce the effectiveness of the enzymes. This includes juice clarification with bentonite, alcohol levels above 17 per cent (v/v) and $SO_2$ concentrations above 500 mg/l (van Rensburg and Pretorius, 2000). Tannin-rich wines show reduced enzyme activity because phenolic polymers react with the proteins and render them useless.

Under optimal conditions, the lifetime of normal mash is reduced by the addition of pectolytic enzymes in concentrations of 1-5 g/l from eight hours to one to two hours. Mashes of rotted grapes can be pressed without any problems; the self-clarification of the wines and the filtration capacity are significantly improved (Mojsov, 2013; Mojsov *et al.*, 2015).

Immobilisation on solid carriers is a common strategy to maintain the desirable properties of enzymes for biotechnological applications. Various methods have been described for pectinases, such as inclusion in alginate (Reyes *et al.*, 2006), physical adsorption to anionic resins (Sario *et al.*, 2001) and covalent bonding to supports such as porous glass (Romero *et al.*, 1989) and nylon (Lozano *et al.*, 1990). An *Aspergillus niger* pectinase immobilised on chitosan-coated slides retained 100 per cent of its original activity after several cycles of reuse (Ramirez *et al.*, 2016).

Under natural conditions, the release of aromatic compounds and colour from grapes is facilitated by increased ethanol concentrations in the course of alcoholic fermentation. However, the extraction is incomplete because the grape skin provides a physical barrier to the diffusion of anthocyanins, tannins and aromas from the cells. Therefore, various oenological techniques have been developed that result in wines with good visual properties that are as stable as possible (Sommer and Cohen, 2018). In particular, wines produced by pectinase treatment showed higher anthocyanin and total phenol concentrations as well as greater color intensity and optical clarity compared to untreated control wines (Mojsov *et al.*, 2013, 2015).

### 2.2. Yeast Pectinases

*Saccharomyces cerevisiae* strains usually show no or only weak pectinase activity. In contrast, many so-called wild yeasts have been identified as pectinase producers (Charoenchai *et al.*, 1997; Fernández *et al.*, 2000; Strauss *et al.*, 2001; Mateo and Maicas, 2016).

It is believed that grape fermentations at low temperatures (15-20°C) protect the volatile compounds and thereby improve the aromatic profile of wines. Therefore, cold-active enzymes are required for both extraction and clarification (Merín *et al.*, 2011). Psychrophilic yeasts are natural sources of such biocatalysts (Sahay *et al.*, 2013). The pectinolytic enzymes of *Cystofilobasidium capitatum* and *Rhodotorula mucilaginosa* are effective at oenological pH values and temperature conditions. Pectinases from several *Aurobasidium pullulans* strains also remain active at wine-relevant concentrations of glucose, ethanol or $SO_2$ and have the potential as processing aids for low temperature wine fermentations (Merín and de Ambrosini, 2015).

### 2.3. Yeast Lipases

Lipids in wine are derived directly from the grape berry (Gallander and Peng, 1980) and by autolysis of wine yeast (Pueyo *et al.*, 2000). It has been reported that the lipid composition undergoes significant changes during wine fermentation (Fragopoulou *et al.*, 2002).

Authentic lipases (E.C. 3.1.1.3) are mainly active at the oil-water interface of emulsified substrates with long fatty acid chains. Triglycerides are cleaved, yielding glycerol and fatty acids. In contrast, carboxylic ester hydrolases (*see below*) hydrolyse soluble esters with relatively short fatty acid chains (Vakhlu and Kour, 2006). The transition between the two activities, however, seems a bit flowing. Lipolytic activities were only detected in a few wine-relevant strains of *Lactobacillus* (Matthews *et al.*, 2006), but in different genera of yeast isolated from natural environments (Madrigal *et al.*, 2013; Molnárova *et al.*, 2014). Theoretically, lipases for winemaking could be used to break down lipoid cell membranes, thereby improving the colour yield from red grape berries.

## 3. Enzymes Useful for Remedy of Clarification and Filtration Problems

Polysaccharides in must and wine originate from the grape berries (cellulose, hemicellulose and pectins) and cell walls of yeasts during growth and autolysis (β-glucans, chitin). Multiple strains of lactic acid bacteria (especially *Pediococcus spp.*) and the grape fungus *Botrytis cinerea* produce capsular or extracellular polysaccharides that form viscous substances that interfere with wine filtration (Dimopoulou *et al.*, 2017). The colloidal polysaccharides cannot be removed from wine by flocculants, adsorbents or filtration. Thus, commercial products with glucanase activities, e.g. those of *Trichoderma* sp. and *Taleromyces versatilis* are useful for reducing the viscosity of must and wine caused by microbial contamination (Hüfner and Haßelbeck, 2017).

Some yeast cell constituents, especially the wall, can have a significant impact on the technological and sensory properties of wine. The cell wall consists of ß-glucans (~ 60 per cent), mannoproteins (~ 40 per cent) and chitin (~ 2 per cent). In particular, the mannoprotein fraction has aroused increasing interest in wine fermentation to stabilise tartaric acid and protein, improve mouthfeel and reduce astringency (Marchal and Jeandet, 2010; Pozo-Bayón *et al.*, 2012; Campos *et al.*, 2016). The use of mannoproteins of the cell wall and chitin as binding elements for the removal of undesirable compounds, such as ochratoxin A and toxic heavy metals, has been proposed. For these reasons, glucanases are also useful to break off and deliberate mannoproteins from yeast cells.

### 3.1. Technical Glucanases

Two types of glucanases are relevant to the fermentation:

- exo-β-1,3-glucanases cleave β-glucan chains by sequentially cleaving glucose residues from the non-reducing end and releasing glucose as the sole hydrolysis product
- endo-β-1,3-glucanases catalyze the intramolecular hydrolysis of ß-glucans to release oligosaccharides. As prerequisite, the enzymes must be effective under acidic wine conditions, moderate temperatures and in the presence of ethanol (Venturi *et al.*, 2013).

On an average they need 6-10 days to degrade the mucilage, the time varies depending on the pH (optimum: between 3-4) and temperature (optimal: between 12-50°C). If wine is treated and not must, twice the amount of enzyme is needed because of the inhibition by ethanol.

### 3.2. Microbial Glucanases

An important microbial source of polysaccharide-degrading exoenzymes are yeasts belonging to the genera *Kloeckera, Candida, Debaryomyces, Rhodotorula, Pichia, Zygosaccharomyces, Hanseniapora, Kluyveromyces* and *Wickerhamomyces* (van Rensburg and Pretorius, 2000; Strauss *et al.*, 2001). Glucanolytic enzyme activities have also been demonstrated in wine-relevant lactic acid bacteria (Matthews *et al.*, 2006).

## 4. Enzymes Useful for Enhancement of Aroma and Nutritional Value

The organoleptic properties of wine are determined by a variety of different compounds that are already present in the grape (aroma) or arise during fermentation or storage (bouquet). Acids such as tartaric acid or citric acid affect the taste, but the characteristic odour and taste is mainly due to volatile organic substances, such as esters, alcohols, thiols or terpenes (Ugliano and Henschke, 2009; Styger *et al.*, 2011; Hjelmeland and Ebeler, 2015).

### 4.1. Technical Glycosidases

Due to their low odour threshold, terpenes dominate the wine taste. Like other flavouring agents (C13 norisoprenoids, benzene derivatives, aliphatic alcohols, phenols), they are secondary metabolites mainly derived from the grape skin. About 90 per cent of these compounds do not exist in a free form, but are conjugated with mono- or disaccharides, forming water-soluble and odourless complexes. The most commonly occurring flavour precursors in grape varieties are the glycosidically bound terpenes linalool, nerol and geraniol. The sugar residues consist of rutinoside (rhamnose-glucose), arabinoside (arabinose-glucose) or apioside (apiose-glucose) (Hjelmeland and Ebeler, 2015).

The enzymatic hydrolysis of sugar-conjugated precursors releases very aromatic, volatile terpenes (aglycones). Usually, the terminal sugars are first cleaved off by a rhamnosidase, an arabinosidase or an apiosidase. In a second step, the terpenes are released by a β-D-glucopyranosidase. This means that the latter activity alone can release only terpene compounds bound to a single glucose residue. In addition to a stepwise reaction, some glucosidases are capable of hydrolysing the glycosidic linkage to the aglycone, regardless of the number of sugar residues (Mateo and DiStefano, 1997; Maicas and Mateo, 2016). Commercial enzyme preparations are now available that can hydrolyse the disaccharide directly from the terpene in a single step (Hüfner and Haßelbeck, 2017). In many bulk enzyme preparations, glycosidase activities occur as side activities along with pectinase and glucanase activities.

## 4.2. Microbial Glycosidases

An important microbial source of wine-related enzymatic activities are lactic acid bacteria (Matthews *et al.*, 2006). Perez-Martin *et al.* (2012) investigated > 1000 isolates for glycosidases. The β-glucosidase activities were found only in cells, but not in the supernatants of the cultures. Four *Oenococcus oeni* isolates retained their enzymatic activity under the conditions of winemaking. In a similar study, cell-bound glucosidase and arabinosidase activities of *O. oeni* strains released high levels of monoterpenes from natural substrates under optimal conditions (Michlmayer *et al.*, 2012). The enzymes showed broad substrate specificities (release of both primary and tertiary terpene alcohols) and remained active in grape juice.

Glycosidase activities have also been demonstrated in various non-*Saccharomyces* yeasts (*Candida, Hanseniaspora, Pichia, Metschnikowia, Rhodotorula, Trichosporon, Wickerhamomyces*) (Delcroix *et al.*, 1994; Zoecklein *et al.*, 1997; Iranzo *et al.*, 1998; Rodriguez *et al.*, 2004; Gonzales-Pombo *et al.*, 2011; Sabel *et al.*, 2014; Schwentke *et al.*, 2014; López *et al.*, 2015; Wang *et al.*, 2015; Hu *et al.*, 2016; Padilla *et al.*, 2017). Several experiments on the technical application of yeast glycosidases to improve the organoleptic quality of wines have yielded positive results (Delcroix *et al.*, 1994; Zoecklein *et al.*, 1997; Gonzales-Pombo *et al.*, 2011; Hu *et al.*, 2016; Maicas and Mateo, 2016).

Polyphenols have gained increasing public and scientific interest due to their supposedly beneficial effects on human health (Cosme and Jordão, 2014). Much of the polyphenols in nature are conjugated to sugars or organic acids, making them more hydrophilic and less bioavailable to humans. The amount of glycosylated forms of resveratrol, known as piceid or polydatin, is up to 10 times higher in red wines. Since these modified forms are less bioactive, experiments with β-glucosidases from various fungal sources have been undertaken to increase the trans-resveratrol content in wines by hydrolysis of glycosylated precursors. The multifunctional glucanase WaExg2 of *W. anomalus* AS1 released the aglycones from the model compounds arbutin, salicin, esculin and polydatin (Schwentke *et al.*, 2014). WaExg2 was active under typical wine conditions such as low pH (3.5-4.0), high sugar concentrations (up to 20 per cent w/v), high ethanol concentrations (10-15 per cent v/v), presence of sulfites and different cations. Therefore, this yeast strain could be useful in wine production for various purposes: to increase the levels of sensory and useful compounds by cleavage of glycosylated precursors or reduction of viscosity by hydrolysis of polysaccharide slimes. In accordance, Madrigal *et al.* (2013) underlined that glucose and ethanol tolerant enzymes from *Wickerhamomyces* are of great interest to the wine industry.

## 4.3. Esterases

Esters (e.g. ethyl acetate, isoamyl acetate, ethylhexanoate, ethyloctanoate and ethyl decanoate) contribute to the most desirable fruity wine aromas (Ugliano and Henschke, 2010; Styger *et al.*, 2011). They are synthesised by the grapes, but also produced by yeasts in the course of alcoholic fermentation (Saerens *et al.*, 2008). Significant changes in the concentration of individual esters were observed during malolactic fermentation (Maicas *et al.*, 1999). Presence of alcohol acyltransferases (ester synthesis) and esterases (ester hydrolysis) in wine yeast (Ugliano and Henschke, 2009) and lactic acid bacteria (Matthews *et al.*, 2007) is well documented.

Depsides are esters of aromatic hydroxycarboxylic acids or phenolic acids with each other or with other carboxylic acids of the grape, such as tartaric acid. These compounds can be hydrolysed by cinnamoyl esterases ('depsidases'), which often appear as side activities in enzyme preparations from *A. niger* (Hüfner and Haßelbeck, 2017). The fission products can negatively affect the wine quality. Phenolcarboxylic acids such as caffeic acid or coumaric acid can be converted by the yeast metabolism into the volatile phenol derivatives 4-vinylguajacol and 4-vinylphenol, which are unpleasant off-flavours in wine (Hüfner and Haßelbeck, 2017). Therefore, commercial pectinase preparations should be free of depsidase side activities. In a recent study with 15 commercial enzyme preparations, approximately half of the samples yielded significant cinnamoyl esterase activities (Fia *et al.*, 2016).

## 5. Enzymes Useful to Minimise Protein Haze and Risky Compounds

Proteins in wine originate from the grapes and from microbial cells and their activities (van Sluyter *et al.*, 2015). Another important sources are protein-based wine supplements (lysozyme, ovalbumin, gelatin, casein), which can exert allergy-like reactions to consumers (Weber *et al.*, 2009; Peñas *et al.*, 2015; Rizzi *et al.*, 2016). Most of these proteins disappear after the completion of wine fermentation and subsequent fining procedures. However, so-called pathogen-related (PR) proteins (β-glucanases, chitinases, thaumatin-related proteins) may still be present. They are synthesised by the plants to ward off bacterial or fungal infections and in response to abiotic stress (Selitrennikoff, 2001). Due to their compact structure, they are resistant to acidic wine conditions, heat and proteolysis (van Sluyter *et al.*, 2015).

In combination with other wine ingredients, PR proteins can cause undesirable clouding, especially during storage of white wines with negative economic consequences (van Sluyter *et al.*, 2015). Currently, protein removal is achieved mainly by the addition of bentonite (Jaeckels *et al.*, 2015), a process that can be associated with decreased wine quantity and quality. Bentonite acts essentially as a cation exchanger and individual wine proteins adsorb to varying degrees on the clay (Jaeckels *et al.*, 2015). Proteins that are negatively charged at wine pH (about 3.5) and/or are highly glycosylated as *Botrytis cinerea* laccases are less bound by bentonite. Thus, new fining agents are desired to remove proteins from wine.

### 5.1. Proteases from Fungal and Plant Sources

The enzymatic degradation of wine proteins appears to be an attractive alternative to bentonite treatment, as volume and aroma losses are minimised. As a prerequisite, suitable proteases must be active under specific wine conditions (acidic pH, presence of ethanol, sulphites, phenols) and preferably act at low temperatures. Another challenge is the resistance of PR proteins to proteolysis due to their specific molecular properties such as several disulphide bonds and glycosylations. However, other grape proteins may be more susceptible and thus proteases can help to reduce effective dosages of bentonite. At present proteases from plants (papain, bromelain) have been tested with some promising results (Esti *et al.*, 2013; Benucci *et al.*, 2014). A fungal protease of *Aspergillus sp.* (Aspergilloglutamic peptidase) has already been approved for Australian winemaking (Manrangon *et al.*, 2012). The enzymatic process involves the flash pasteurisation of grape must and is therefore restricted to specialised wineries. In this context, it has been reported that a protease from *Botrytis cinerea* BcAp8 hydrolyses the grape chitinase at moderate temperatures (van Sluyter *et al.*, 2013).

### 5.2. Microbial Proteases

Microbial proteases may be an alternative or supplement to bentonite treatment for the removal of unwanted wine proteins. Most *Saccharomyces cerevisiae* strains show no extracellular protease activity on diagnostic agar media (Charoenchai *et al.*, 1997; Dizzy and Bisson, 2000; Fernández *et al.*, 2000; Strauss *et al.*, 2001; Mateo and Maicas, 2016). However, a 72 kDa extracellular pepsin-like aspartic protease was obtained from a PIR1 strain (Younes *et al.*, 2011, 2013). The enzyme was active during grape juice fermentations, although it did not affect turbidity-inducing proteins unless the wine was incubated at 38°C for extended time.

Proteinase A (PrA, saccharomycin, EC 3.4.23.25) is the major vacuolar protease of *S. cerevisiae* encoded by the PEP4 gene. As a result of yeast autolysis, PrA enters the wine during alcoholic fermentation. Far more it was found that under stress conditions (e.g. nutrient limitations) PrA does not target the vacuole but is misdirected to the cell membrane and secreted in the medium (Song *et al.*, 2017). This would be advantageous for winemaking in terms of protein turbidity reduction. Interestingly, the same situation is undesirable for brewery because PrA degrades proteins (e.g. lipid transfer protein 1) necessary for beer foam formation. Apart from PrA, *S. cerevisiae* expresses various cell-bound proteases, some of which are not fully characterised (Kang *et al.*, 2000).

Non-*Saccharomyces* yeasts are important sources of extracellular enzymes including proteases (Lagace and Bisson, 1990; Charoenchai *et al.*, 1997; Strauss *et al.*, 2001; Reid *et al.*, 2012; Mateo *et al.*, 2015). Strains of *Metschnikowia pulcherrima* and *Wickerhamomyces anomalus* secreted aspartate proteases and degraded a model protein (bovine serum albumin) during growth in grape juice (Schlander *et al.*, 2017). In a recent study, the heterologous expressed aspartate protease MpAPr1 of *M. pulcherrima* (Theron *et al.*, 2017) was added to a Sauvignon Blanc must. It was shown that the enzyme was active during fermentation and degraded wine proteins to some extent (Theron *et al.*, 2018). An alternative strategy would be to carry out wine fermentations with suitable protease-positive starter cultures. In addition to cost reductions, there are no administrative restrictions on yeast applications in must and wine, which must be taken into account for enzyme preparations.

The occurrence of proteolytic activities in lactic acid bacteria is also well documented (Matthews *et al.*, 2006). The growth of *Oenococcus oeni* depends on the presence of amino acids in the culture medium due to a lack of appropriate synthetic routes. This bacterium secretes several proteases that can facilitate access to rare nitrogen sources during malo-lactic fermentation (Folio *et al.*, 2008).

### 5.3. Urease

Increased amounts of urea in wine can be formed by yeast activities and then be converted by a chemical reaction into the carcinogenic substance urethane (ethyl carbamate). During malolactic fermentation, lactic acid bacteria can produce other precursors of ethyl carbamate, such as arginine-derived citrulline and carbamyl phosphate. Especially at higher temperatures, fermented wines may contain excessive amounts of urethane (Lonvaud-Funel, 2016). Therefore, appropriate precautions should be taken to prevent the production of urethane. These include, for example, the selection of suitable starter cultures for malo-lactic fermentations and the reduction of arginine concentrations in the grape. Urease was introduced in 1997 by the EU as a new enzymatic wine treatment agent and can be used in exceptional cases. The enzyme splits urea into ammonia and carbon dioxide, preventing urethane formation. The commercial urease from *Lactobacillus fermentum* is effective with urea in doses of 50 mg/l in red wines and 25 mg/l in white wines (Pozo-Bayón *et al.*, 2012).

### 5.4. Laccases

Spontaneous and enzymatic oxidations exert dramatic effects on the final phenolic composition from the grape berry to bottled wine. Once the integrity of the berries is destroyed, oxidative enzymes (phenoloxidases) and their phenolic substrates are exposed to the air, resulting in enzymatic browning. There are two classes of copper enzymes responsible for these reactions (Claus and Decker, 2006; Claus *et al.*, 2014; Fronk *et al.*, 2015): Tyrosinases (E.C. 1.14.18.1) which originate from grape berries and laccases (E.C. 1.10.3.2) from epiphytic fungi, especially *Botrytis cinerea.* Both enzymes oxidize phenolic compounds and thus alter their molecular, antioxidant and antimicrobial properties (Riebel *et al.*, 2015; 2017; Sabel *et al.*, 2017). Especially the fungal laccase is the responsible catalyst of browning reactions in must and wine, due its broad substrate spectrum and stability.

Although laccase activities are not very welcome by winemakers, several studies have concluded that controlled laccase treatments could even promote stability and improve sensory properties of wines and juices (Maier *et al.*, 1990; Brenna and Bianchi, 1994; Servilli *et al.*, 2000; Minussi *et al.*, 2007; Lettera *et al.*, 2016). Volatile phenols particular produced by *Brettanomyces/Dekkera* sp. yeasts are associated with an unpleasant "horse taint" flavour in wine. Lustrata *et al.* (2015) used a laccase from *T. versicolor* to reduce concentrations of 4-ethyl-guaiacol and 4-ethylphenol in a synthetic model wine.

Biogenic amines are another class of undesirable compounds in wine (Smit *et al.*, 2008; Guo *et al.*, 2015; Preti *et al.*, 2016). They arise from the grape berries or are formed during the fermentation by activities of decarboxylase-positive microorganisms (Morena-Arribas and Polo, 2010; Sebastian *et al.*, 2011; Christ *et al.*, 2012: Kushnereva *et al.*, 2015; Henríquez-Aedo *et al.*, 2016). Although more common in foods such as cheese, biogenic amines have received much attention in wine, as ethanol can exacerbate the adverse effects on human health by inhibiting the enzymes responsible for detoxifying these compounds (Moreno-Arribas and Polo, 2010).

Biogenic amines such as tyramine, phenylethylamine, tryptamine or serotonin are also substrates for laccases (Claus, 2017). Callejón *et al.* (2014) found enzymatic activities responsible for biogenic amine degradation in lactic acid bacterial strains isolated from wine. Responsible enzymes were isolated and purified from *Lactobacillus plantarum* J16 and *Pediococcus acidilactici* CECT 5930 strains and identified as intracellular laccase-like multicopper oxidases. When the *L. plantarum* J16 laccase was overexpressed in *Escherichia coli*, it oxidized some BA, mainly tyramine (Callejón *et al.*, 2016).

The enzymatic degradation of biogenic amines is usually catalyzed by various classes of oxidases (Yagodina *et al.*, 2002; Klinman, 2003). Depending on the type of prosthetic group, they can be classified into FAD-dependent (E.C. 1.4.3.4) and copper-containing amine oxidases (CAOs, E.C. 1.4.3.6). The latter have been found in various yeasts such as *Kluyeromyces marxianus* or *Debaryomyces hansenii* (Corpillo *et al.*, 2003; Bäumlisberger *et al.*, 2015). These enzymes belong to the class of copper proteins of type 2 or "non-blue" oxidases, which, with equimolar consumption of molecular oxygen and formation of hydrogen peroxide and ammonia, convert primary amines into the corresponding aldehydes.

## 6. Enzymes Useful for Control of Microbial Growth

Yeast, lactic acid and acetic acid bacteria have a significant influence on wine quality (König *et al.*, 2017). The microbial growth of musts and wines is usually controlled by addition of sulfur dioxide. However, the presence of sulphites in alcoholic beverages, especially in wines, can cause pseudoallergic reactions with symptoms ranging from gastrointestinal problems to anaphylactic shock (Pozo-Bayón *et al.*, 2012). Other antimicrobial agents, such as sorbic acid and dimethyl carbonate, are primarily active against yeasts, but have limited activity against bacteria (Marchal and Jeandet, 2010; Pozo-Bayón *et al.*, 2012).

### 6.1. Lysozyme from Hen Egg White

Lysozyme (EC 3.2.1.17) is a muramidase commonly used to control microbial growth in foods such as cheese and wines (Liburdi *et al.*, 2014). Extensive enzymatic hydrolysis of the bacterial cell wall peptidoglycan, a polymer of N-acetyl-D-glucosamine units that are β-1,4 linked to N-acetylmuramic acid, results in cell lysis and death in hypoosmotic environments. Some lysozymes can kill bacteria by stimulating autolysin activity. In addition, bactericidal mechanisms involving membrane damage without enzymatic hydrolysis of peptidoglycan were reported for c-type lysozymes such as chicken egg white lysozyme (Blättel *et al.*, 2009). Gram-negative bacteria (i.e. acetic acid bacteria) are quite resistant to lysozyme because the outer membrane acts as a barrier.

Lysozyme, commercially produced from egg white, was approved by the International Organisation of Vine and Wine for wine production in 2001 (Sommer *et al.*, 2018). The added amount is usually between 250-500 mg/L. Four major applications and dosages are: (a) prevention of the onset of malo-lactic fermentation (early addition of 100-150 mg/L); (b) complete inhibition of bacterial activity and malo-lactic fermentation (500 mg/L); (c) protection of the wine with suboptimal alcoholic fermentation (250-300 mg/L); (d) stabilisation of the wine after malo-lactic fermentation (250-300 mg/L). Lysozyme can be eliminated by adding fining agents, among which bentonite and metatartaric acid are the most efficient.

It should be noted that chicken egg lysozyme can deliver pH dependent chitinase side activities. Under unfavourable conditions, yeast cell walls (which contain 2-4 per cent chitin mainly in the bud scar regions) can be attacked by lysozyme with significant effects on the vitality and stress response of *Saccharomyces cerevisiae* during wine fermentation (Sommer *et al.*, 2018).

### 6.2. Microbial Muramidases

It has been reported that various Gram-positive strains of *Pediococcus sp.*, *Lactobacillus sp.* and *Oenococcus oeni* were not efficiently hydrolysed by chicken egg lysozyme (Blättel *et al.*, 2009). Reasonable explanations are structural modifications of the peptidoglycan, such as the N-deacetylation and O-acetylation of the glycan chains or the amidation of free carboxyl groups of amino acids in the peptide chains (Blättel *et al.*, 2009). In addition, the presence of peripheral S-layer proteins may decrease the susceptibility of lactobacilli to cell wall hydrolases (Dohm *et al.*, 2011). As a possible alternative to chicken egg white lysozyme, exoenzymes (protease and muramidase) from *Streptomyces* species showed a broad bacteriolytic spectrum under wine conditions (Sebastian *et al.*, 2014).

## 7. Regulations for Enzyme Applications

The use of enzymes in wine production in the European Union is regulated by the International Organisation of Vine and Wine (OIV). Specific resolutions define general aspects of enzymes in winemaking, the permitted enzyme activities, the mode of application and enzyme activity measurements. The USA, Canada and China have national regulations for winemaking (Hüfner and Haßelbeck, 2017).

The OIV has decided that *Aspergillus niger* and *Trichoderma* sp. can be used as source organisms (i.e. have GRAS status, 'generally considered safe' (Mojsov, 2013; Mojsov *et al.*, 2015)). Selected *A. niger* strains are fermented under aerobic conditions in optimised growth media for the production of pectinases, hemicellulases and glycosidases. *Trichoderma* species are used for the production of glucanases and *Lactobacillus fermentum* for urease (Pozo-Bayón *et al.*, 2012).

Today's enzyme production is based either on specially selected wild-type strains or on genetically modified organisms (GMOs). The use of GMO production strains has considerable advantages - the product yields with GMOs are much higher than those of wild strains and undesirable side activities are minimised. This makes it more efficient to produce and guarantee the purity of the enzyme products. The designation is regulated in the resolution OIV-OENO 485-2012.

Although the number of *Saccharomyces* strains for winemaking has been considerably extended by genetic modifications (Pretorius, 2000), only two GMOs were authorised for wine production in three countries. The first recombinant strain officially approved by the Food Safety Authorities (US and Canada) was the malo-lactic wine yeast ML01. The GMO carries the *Schizosaccharomyces pombe* malate permease gene (*mae1*) and the *Oenococcus oeni* malolactic gene (*mleA*). The second strain, ECMo01, constitutively expresses the urease gene to prevent the formation of urethane (Gonzales *et al.*, 2016). Whether yeast strains obtained by protoplast fusion should be considered GMOs is in legal limbo.

## 8. Future Prospects

Use of technical enzymes is nowadays a well-established strategy to improve wine quantity and quality. Currently they are mainly produced by *Aspergillus* species and applied as bulk preparations with several side activities. In view of consumer safety, more defined activities and alternative biological producers seem to be preferable. Yeasts, naturally occurring on grapes, have been found to be a rich source of oenological interesting enzymes. Their activities can be exploited in form of new enzyme products or directly as starter cultures for wine fermentation. This would satisfy the increasing trend to produce more individual wines with the aid of non-*Saccharomyces* yeasts (Pretorius, 2016).

## References

Bäumlisberger, M., Moellecken, U., König, H. and H. Claus (2015). The potential of the yeast *Debaryomyces hansenii* H525 to degrade biogenic amines in food. *Microorganisms* 3: 839-850.

Benucci, I., Esti, M. and Liburdi, K. (2014). Effect of free and immobilised stem bromelain on protein haze in white wine. *Australian Journal of Grape and Wine Research* 20: 347-352.

Blättel, V., Wirth, K., Claus, H., Schlott, P., Pfeiffer, P. and König, H. (2009). A lytic enzyme cocktail from *Streptomyces sp.* B578 for the control of lactic and acid bacteria in wine. *Applied Microbiology and Biotechnology* 83: 839-848.

Brenna, O. and Bianchi, E. (1994). Immobilised laccase for phenolic removal in must and wine. *Biotechnology Letters* 16: 35-40.

Callejón, S., Sendra, R., Ferrer, S. and Pardo, I. (2014). Identification of a novel enzymatic activity from lactic acid bacteria able to degrade biogenic amines in wine. *Applied Microbiology and Biotechnology* 98: 185-198.

Callejón, S., Sendra, R., Ferrer, S. and Pardo, I. (2016). Cloning and characterisation of a new laccase from *Lactobacillus plantarum* J16 CECT 8944 catalysing biogenic amine degradation. *Applied Microbiology and Biotechnology* 100: 3113-3124.

Campos, F.M., Couto, J.A. and Hogg, T. (2016). Utilisation of natural and by-products to improve wine safety. pp. 27-49. *In*: Morena-Arribas, M.V., B. Bartolomé Sualdea (Eds.). Wine Safety, Consumer Preference, and Human Health. Springer International Publishing, Switzerland.

Charoenchai, C., Fleet, G.H., Henschke, P.A. and Todd, B.E.N. (1997). Screening of non-*Saccharomyces* wine yeasts for the presence of extracellular hydrolytic enzymes. *Australian Journal of Grape and Wine Research* 3: 2-8.

Claus, H. (2017). Laccases of *Botrytis cinerea*. pp. 339-356. *In*: König, H., Unden, G. and Fröhlich, J. (Eds.). Biology of Microorganisms on Grapes, in Must and in Wine, second edn. Springer International Publishing, Cham, Switzerland.

Christ, E., König, H. and Pfeiffer, P. (2012). Bacterial formation of biogenic amines in grape juice: Influence of culture conditions. *Deutsche Lebensmittel-Rundschau* 108: 73-78.

Claus, H., Sabel, A. and König, H. (2014). Wine phenols and laccase: An ambivalent relationship. pp. 155-185. *In*: Rayess, Y.E. (Ed.). Wine, Phenolic Composition, Classification and Health Benefits. Nova Publishers, New York.

Claus, H. and Decker, H. (2006). Bacterial tyrosinases. *Systematic and Applied Microbiology* 29: 3-14.

Corpillo, D., Valetti, F., Giuffrida, M.G., Conti, A., Rossi, A., Finazzi-Agrò, A. and Giunta, C. (2003). Induction and characterisation of a novel amine oxidase from the yeast *Kluyveromyces marxianus*. *Yeast* 20: 369-379.

Cosme, F. and Jordão, A.M. (2014). Grape phenolic composition and antioxidant capacity. pp. 1-40. *In*: Rayess, Y.E. (Ed.). Wine, Phenolic Composition, Classification and Health Benefits. Nova Publishers, New York.

Delcroix, A., Gunata, Z., Sapis, L.C., Salmon, J.M. and Baynone, C. (1994). Glycosidase activities of three enological yeast strains during winemaking: Effect on the terpenol content of Muscat wine. *American Journal of Enology and Viticulture* 45: 291-296.

Dimopoulou, M., Lonvauf-Funel, A. and Dols-Lafargue, M. (2017). Polysaccharide production by grapes must and wine microorganisms. pp. 635-658. *In*: König, H., Unden, G. and Fröhlich, J. (Eds.). Biology of Microorganisms on Grapes, in Must and in Wine, second ed. Springer International Publishing, Cham, Switzerland.

Dizzy, M. and Bisson, L.F. (2000). Proteolytic activity of yeast strains during grape juice fermentation. *American Journal of Enology and Viticulture*. 51: 155-167.

Dohm, N., Petri, A., Schlander, M., Schlott, B., König, H. and Claus, H. (2011). Molecular and biochemical properties of the S-layer protein from the wine bacterium *Lactobacillus hilgardii* B706. *Archives of Microbiology* 193: 251-261.

El Rayess, Y. (2014). Wine: Phenolic Composition, Classification and Health Benefits. Nova Publishers, New York. ISBN 978-1-63321-048-6

Esti, M., Benucci, I., Lombardelli, C., Liburdi, K. and Garzillo, A.M.V. (2013). Papain from papaya (*Carica papaya* L.) fruit and latex: Preliminary characterisation in alcoholic-acidic buffer for wine application. *Food and Bioproducts Processing* 91: 595-598.

Fernández, M., Úbeda, J.F. and Briones, A.I. (2000). Typing of non-*Saccharomyces* yeasts with enzymatic activities of interest in winemaking. *International Journal of Food Microbiology* 59: 29-36.

Fia, G., Oliver, V., Cavaglioni, A., Canuti, V. and Zanoni, B. (2016). Side activities of commercial enzyme preparations and their influence on hydroxycinnamic acids, volatile compounds and nitrogenous components of white wine. *Australian Journal of Grape and Wine Research* 22: 366-375.

Folio, P., Ritt, J.F., Alexandre, H. and Remize, F. (2008). Characterisation of EprA, a major extracellular protein of *Oenococcus oeni* with protease activity. *International Journal of Food Microbiology* 127: 26-31.

Fragopoulou, E., Antonopoulou, S. and Demopoulos, C.A. (2002). Biologically active lipids with antiatherogenic properties from white wine and must. *Journal of Agriculture and Food Chemistry* 50: 2684-2694.

Fronk, P., Hartmann, H., Bauer, M., Solem, E., Jaenicke, E., Tenzer, S. and Decker, H. (2015). Polyphenoloxidase from Riesling and Dornfelder wine grapes (*Vitis vinfera*) is a tyrosinase. *Food Chemistry* 183: 49-57.

Gallander, J.F. and Peng, A.C. (1980). Lipid and fatty acid composition of different wine grapes. *American Journal of Enology and Viticulture* 31: 24-27.

Gonzales, R., Tronchoni, J., Quirós, M. and Morales, P. (2016). Genetic improvement and genetically modified microorganisms. pp. 71-96. *In*: Morena-Arribas, M.V. and Bartolomé Sualdea, B. (Eds.). Wine Safety, Consumer Preference and Human Health. Springer International Publishing, Switzerland.

Gonzales-Pombo, O., Farina, L., Carreau, F., Batista-Viera, F. and Brena, B.M. (2011). A novel extracellular beta-glucosidase from *Issatschenkia terricola*: Immobilisation and application for aroma enhancement of white Muscat wine. *Process Biochemistry* 46: 385-389.

Guo, Y.Y., Yang, Y.P., Peng, Q. and Han, Y. (2015). Biogenic amines in wine: A review. *International Journal of Food Science and Technology* 50: 1523-1532.

Henríquez-Aedo, K., Durán, D., Garcia, A. Hengst, M.B. and Aranda, M. (2016). Identification of biogenic amines-producing lactic acid bacteria isolated from spontaneous malo-lactic fermentation of Chilean red wines. *LWT – Food Science and Technology* 68: 183-189.

Hjelmeland, A.K. and Ebeler, S.E. (2015). Glycosidically bound volatile aroma compounds in grapes and wine: A review. *American Journal of Enology and Viticulture* 66: 1-10.

Hu, K., Zhu, X.L., Mu, H., Ma, Y., Ullah, N. and Tao, Y.S. (2016). A novel extracellular glycosidase activity from *Rhodotorula mucilaginosa*: Its application potential in wine aroma enhancement. *Letters in Applied Microbiology* 62: 169-176.

Hüfner, E. and Haßelbeck, G. (2017). Application of microbial enzymes during winemaking. pp. 635-658. *In*: König, H., Unden, F. and Fröhlich, J. (Eds.). Biology of Microorganisms on Grapes, in Must and in Wine, second edn. Springer International Publishing, Cham, Switzerland.

Iranzo, J.F.U., Pérez, A.I.B. and Cañas, P.M.I. (1998). Study of oenological characteristics and enzymatic activities of wine yeasts. *Food Microbiology* 15: 399-406.

Jaeckels, N., Tenzer, S., Rosch, A., Scholten, G., Decker, H. and Fronk, P. (2015). ß-Glucosidase removal due to bentonite fining during winemaking. *European Food Research and Technology* 241: 253-262.

Kang, H.A., Choi, E.S., Hong, W.K., Kim, J.Y., Ko, S.M., Sohn, J.H. and Rhee, S.K. (2000). Proteolytic stability of recombinant human serum albumin secreted in the yeast *Saccharomyces cerevisiae*. *Applied Microbiology and Biotechnology* 53: 575-582.

Klinman, J.P. (2003). The multi-functional topa-quinone copper amine oxidases. *Biochemistry Biophysica Acta* 1647: 131-137.

König, H., Unden, F. and Fröhlich, J. (2017). Biology of Microorganisms on Grapes, in Must and in Wine, second edn. Springer International Publishing, Cham, Switzerland.

Kushnereva, E.V. (2015). Formation of biogenic amines in wine production. *Applied Biochemisty and Microbiology* 51: 108-112.

Lagace, L.S. and Bisson, L.F. (1990). Survey of yeast acid proteases for effectiveness of wine haze reduction. *American Journal of Enology and Viticulture* 41: 147-155.

Lettera, V., Pezzella, C., Cicatiello, P., Piscitelli, A., Giacobelli, V.G., Galano, E., Amoresano, A. and Sannia, G. (2016). Efficient immobilisation of a fungal laccase and its exploitation in fruit juice clarification. *Food Chemistry* 196: 1272-1278.

Liburdi, K., Benucci, I. and Esti, M. (2014). Lysozyme in wine: An overview of current and future applications. *Comprehensive Reviews in Food Science and Food Safety* 13: 1062-1073.

Lonvaud-Funel, A. (2016). Undesirable compounds and spoilage microorganisms in wine. pp. 3-26. *In*: Morena-Arribas, M.V. and Bartolomé Sualdea, B. (Eds.). Wine Safety, Consumer Preference and Human Health. Springer International Publishing, Switzerland.

López, M.C., Mateo, J.J. and Maicas, S. (2015). Screening of β-glucosidase and β-xylosidase activities in four non-*Saccharomyces* yeast isolates. *Journal of Food Science* 80: C1696-C1704.

Lozano, P., Manjon, A., Iborra, J.L., Canovas, M. and Romojaro, F. (1990). Kinetic and operational study of a cross-flow reactor with immobilised pectolytic enzymes. *Enzyme and Microbial Technology* 12: 499-505.

Lustrato, G., De Leonardis, A., Macciola, V. and Ranalli, G. (2015). Preliminary lab scale of advanced techniques as new tools to reduce ethylphenols content in synthetic wine. *Agro FOOD Industry Hi-tech* 26: 51-54.

Madrigal, T., Maicas, S. and Tolosa, J.J.M. (2013). Glucose and ethanol tolerant enzymes produced by *Pichia (Wickerhamomyces)* isolates from enological ecosystems. *American Journal of Enology and Viticulture* 64: 126-133.

Maicas, S., Gil, J.V., Pardo, I. and Ferrer, S. (1999). Improvement of volatile composition of wines by controlled addition of malo-lactic bacteria. *Food Research International* 32: 491-496.

Maicas, S. and Mateo, J.J. (2016). Microbial glycosidases for wine production. *Beverages* 2: 20, doi:10.3390/beverages2030020

Maier, G., Dietrich, H. and Wucherpfennig, K. (1990). Winemaking without $SO_2$ – with the aid of enzymes? *Weinwirtschaft-Technik* 126: 18-22.

Manrangon, M., van Sluyter, S.C., Robinson, E.M.C., Muhlack, R.A., Holt, H.E., Haynes, P.A., Godden, P.W., Smith, P.A. and Waters, E.J. (2012). Degradation of white wine haze proteins by *Aspergillopepsin* I and II during flash pasteurisation. *Food Chemistry* 135: 1157-1165.

Marchal, R. and Jeandet, P. (2010). Use of enological additives for colloids and tartrate salt stabilisation in white wines and for improvement of sparkling wine foaming properties. pp. 127-158. *In*: Morena-Arribas, M.V. and Polo, M.C. (Eds.). Wine Chemistry and Biochemistry. Springer Science & Business Media, New York.

Mateo, J.J. and DiStefano, R. (1997). Description of the beta-glucosidase activity of wine yeasts. *Food Microbiology* 14: 583-591.

Mateo, J.J. and Maicas, S. (2016). Application of non-*Saccharomyces* yeasts to winemaking process. *Fermentation* 2: 14, doi: 10.3390/fermentation2030014

Mateo, J.J., Maicas, S. and Thießen, C. (2015). Biotechnological characterisation of extracellular proteases produced by enological *Hanseniaspora* isolates. *International Journal of Food Science and Technology* 50: 218-225.

Matthews, A., Grbin, P.R. and Jiranek, V. (2006). A survey of lactic acid bacteria for enzymes of interest in oenology. *Australian Journal of Grape and Wine Research* 12: 235-244.

Matthews, A., Grbin, P.R. and Jiranek, V. (2007). Biochemical characterisation of the esterase activities of wine lactic acid bacteria. *Applied Microbiology and Biotechnology* 77: 329-337.

Merín, M.G. and de Ambrosini, V.I.M. (2015). Highly cold-active pectinases under wine-like conditions from non-*Saccharomyces* yeasts for enzymatic production during winemaking. *Letters in Applied Microbiology* 60: 467-474.

Merín, M.G., Mendoza, L.M., Farías, M.E. and de Ambrosini, V.I.M. (2011). Isolation and selection of yeast from wine grape ecosystem secreting cold-active pectinolytic activity. *International Journal of Food Microbiology* 147: 144-148.

Michlmayer, H., Nauer, S., Brandes, W., Schumann, C., Kulbe, K.D., del Hierro, A.M. and Eder, R. (2012). Release of wine monoterpenes from natural precursors by glycosidases from *Oenococcus oeni*. *Food Chemistry* 135: 80-87.

Minussi, R.C., Rossi, M., Bolgna, L., Rotilio, D., Pastore, G.M. and Durán, N. (2007). Phenols removal in musts; strategy for wine stabilisation by laccase. *Journal of Molecular Catalyis B: Enzymatic,* 45: 102-107.

Mojsov, K. (2013). Use of enzymes in wine making: A review. *International Journal of Technology Marketing* 3: 112-127.

Mojsov, K., Andronikov, D., Janevski, A., Jordeva, S. and Zezova, S. (2015). Enzymes and wine – The enhanced quality and yield. *Advanced Technologies* 4: 94-100.

Molnárova, J., Vadkertiová, R. and Stratilová, E. (2014). Extracellular enzymatic activities and physiological profiles of yeasts colonizing fruit trees. *Journal of Basic Microbiology* 51: S74-S84.

Moreno-Arribas, M. and Polo, M.C. (2010). Amino acids and biogenic amines. pp. 163-189. *In*: Morena-Arribas, M.V. and Polo, M.C. (Eds.). Wine Chemistry and Biochemistry. Springer Science & Business Media, New York.

Padilla, B., Gil, J.V. and Manzanares, P. (2016). Past and future of non-*Saccharomyces* yeasts: From spoilage microorganisms to biotechnological tools for improving wine aroma complexity. *Frontiers in Microbiology* 7: 111, doi: 10.3389/fmicb.2016.004111

Peñas, E., di Lorenzo, C. and Uberti, F. (2015). Allergenic proteins in enology: A review on technological applications and safety aspects. *Molecules* 20: 13144-13174.

Perez-Martin, F., Sesena, S., Miguel Izquierdo, P., Martin, R. and Llanos Palop, P. (2012). Screening for glycosidase activities of lactic acid bacteria as a biotechnological tool in oenology. *World Journal of Microbiology and Biotechnology* 28: 1423-1432.

Pozo-Bayón, M.A., Monagas, M., Bartolomé, B. and Moreno-Arribas, M.V. (2012). Wine features related to safety and consumer health: An integrated perspective. *Critical Reviews in Food Science and Nutrition* 52: 31-54.

Preti, R., Vieri, S. and Vinci, G. (2016). Biogenic amine profiles and antioxidant properties of Italian red wines from different price categories. *Journal of Food Composition and Analysis* 46: 7-14.

Pretorius, I.S. (2016). Conducting wine symphonics with the aid of yeast genomics. *Beverages* 2: 36, doi:10.3390/beverages2040036

Pretorius, I.S. (2000). Tailoring wine yeast for the new millennium: Novel approaches to the ancient art of winemaking. *Yeast* 16: 675-729.

Pueyo, E., Martinez-Rodriquez, A., Polo, M.C., Santa-Maria, G. and Bartomé, B. (2000). Release of lipids during yeast autolysis in model wine. *Journal of Agricultural and Food Chemistry* 48: 116-122.

Ramirez, H.L., Gómez Brizuela, L., Úbeda Iranzo, J., Areval-Villena, M. and Briones Pérez, A.I. (2016). Pectinase immobilization on a chitosan-coated chitin support. *Journal of Food Processing and Engineering* 39: 97-104.

Reid, V.J., Theron, L.W., du Toit, M. and Divol, B. (2012). Identification and partial characterisation of extracellular aspartic protease genes from *Metschnikowia pulcherrima* IWBT Y1123 and *Candida apicola* IWBT Y1384. *Applied and Environmental Microbiology* 78: 6838-6849.

Reyes, N., Rivas Ruiz, I., Dominguez-Espinosa, R. and Solis, S. (2006). Influence of immobilisation parameters on endopolygalacturonase productivity by hybrid *Aspergillus sp.* HL entrapped in calcium alginate. *Biochemical Engineering Journal* 32: 43-48.

Riebel, M., Sabel, A., Claus, H., Fronk, P., Xia, N., Li, H., König, H. and Decker, H. (2015). Influence of laccase and tyrosinase on the antioxidant capacity of selected phenolic compounds on human cell lines. *Molecules* 20: 17194-17207.

Riebel, M., Sabel, A., Claus, H., Xia, N., Li, H., König, H., Decker, H. and Fronk. P. (2017). Antioxidant capacity of phenolic compounds on human cell lines as affected by grape-tyrosinase and *Botrytis*-laccase oxidation. *Food Chemistry* 229: 779-789.

Rizzi, C., Mainente, F., Pasini, G. and Simonato, B. (2016). Hidden exogenous proteins in wine: Problems, methods of detection and related legislation – A review. *Czechian Journal of Food Science* 34: 93-104.

Rodríguez, M.E., Lopes, C.A., van Broock, M., Valles, S., Ramón, D. and Caballero, A.C. (2004). Screening and typing of Patagonian wine yeasts for glycosidase activities. *Journal of Applied Microbiology* 96: 84-95.

Romero, C., Sanchez, S., Manjon, S. and Iborra, J.L. (1989). Optimisation of the pectinesterase/endo-D-polygalacturonase immobilisation process. *Enzyme and Microbial Technology* 11: 837-843.

Sabel, A., Bredefeld, S., Schlander, M. and Claus, H. (2017). Wine phenolic compounds: Antimicrobial properties against yeasts, lactic acid and acetic acid bacteria. *Beverages* 3: 29, doi: 10.3390/beverages3030029

Sabel, A., Martens, S., Petri, A., König, H. and Claus, H. (2014). *Wickerhamomyces anomalus* AS1: A new strain with potential to improve wine aroma. *Annals of Microbiology* 64: 483-491.

Saerens, S.M.G., Delvaux, F., Verstrepen, K.J., van Dijck, P., Thevelein, J.M. and Delvaux, F.R. (2008). Parameters affecting ethyl ester production by *Saccharomyces cerevisiae* during fermentation. *Applied and Environmental Microbiology* 74: 454-461.

Sahay, S., Hanid, B., Singh, P., Ranjan, K., Chauhan, D., Rana, R.S. and Chaurse, V.K. (2013). Evaluation of pectinolytic activities for oenological uses from psychrotrophic yeasts. *Letters in Applied Microbiology* 57: 115-121.

Sario, K., Demir, N., Acar, J. and Mutlu, M. (2001). The use of commercial pectinase in the fruit industry, part 2: Determination of the kinetic behaviour of immobilised commercial pectinase. *Journal of Food Engineering* 47: 271-274.

Schlander, M., Distler, U., Tenzer, S., Thines, E. and Claus, H. (2017). Purification and properties of yeast proteases secreted by *Wickerhamomyces anomalus* 227 and *Metschnikowia pulcherrima* 446 during growth in a white grape juice. *Fermentation* 3: 2, doi:10.3390/fermentation3010002

Schwentke, J., Sabel, A., Petri, A., König. H. and Claus, H. (2014). The wine yeast *Wickerhamomyces anomalus* AS1 secretes a multifunctional exo-β-1,3 glucanase with implications for winemaking. *Yeast* 31: 349-359.

Sebastian, P., Claus, H. and König, H. (2014). Studies on two exoenzymes which lyse wine-spoiling bacteria. *Advances in Microbiology* 4: 527-538.

Sebastian, P., Herr, P., Fischer, U. and König, H. (2011). Molecular identification of lactic acid bacteria occurring in must and wine. *South African Journal of Enology and Viticulture* 32: 300-309.

Selitrennikoff, C.P. (2001). Antifungal proteins. *Applied and Environmental Microbiology* 67: 2883-2894.

Servili, M., de Stefano, G., Piacquadio, P. and Sciancalepore, V. (2000). A novel method for removing phenols from grape must. *American Journal of Enology and Viticulture* 51: 357-361.

Smit, A.A., du Toit, W.J. and du Toit, M. (2008). Biogenic amines in wine: Understanding the headache. *South African Journal of Enology and Viticulture* 29: 109-127.

Sommer, S. and Cohen, S.D. (2018). Comparison of different extraction methods to predict anthocyanin concentration and color characteristics of red wines. *Fermentation* 4: 39, doi:10.3390/fermentation4020039

Sommer, S., Wegmann-Herr, P., Wacker, M. and Fischer, U. (2018). Influence of lysozyme addition on hydroxycinnamic acids and volatile phenols during wine fermentation. *Fermentation* 4: 5, doi:10.3390/fermentation4010005

Song, L., Chen, Y., Du, Y., Wang, X., Guo, X., Dong, J. and Xiao, D. (2017). *Saccharomyces cerevisiae* proteinase A excretion and wine making. *World Journal of Microbiology and Biotechnology* 33: 210, doi 10.1007/s11274-017-2361-z

Strauss, M.L.A., Jolly, N.P., Lambrechts, M.G. and van Rensburg, P. (2001). Screening for the production of extracellular hydrolytic enzymes by non-*Saccharomyces* wine yeasts. *Journal of Applied Microbiology* 91: 182-190.

Styger, G., Prior, B. and F.F. Bauer (2011). Wine flavour and aroma. *Journal of Industrial Microbiology* 38: 1145-1159.

Theron, L.W., Bely, M. and Divol, B. (2017). Characterisation of the enzymatic properties of MpAPr1, an aspartic protease secreted by the wine yeast *Metschikowia pulcherrima*. *Journal of the Science of Food and Agriculture* 97: 3584-3593.

Theron, L.W., Bely, M. and Divol, B. (2018). Monitoring the impact of an aspartic protease (MpApr1) on grape proteins and wine properties. *Applied Microbiology and Biotechnology* 102: 5173-5183.

Ugliano, M. (2009). Enzymes in winemaking. pp. 103-126. *In*: Morena-Arribas, M.V. and Polo, M.C. (Eds.). Wine Chemistry and Biochemistry. Springer Science & Business Media, New York.

Ugliano, M. and Henschke, P.A. (2009). Yeasts and wine flavour. pp. 312-392. *In*: Morena-Arribas, M.V. and M.C. Polo (Eds.). Wine Chemistry and Biochemistry. Springer Science & Business Media, New York.

Vakhlu, J. and Kour, A. (2006). Yeast lipases: Enzyme purification, biochemical properties and gene cloning. *Electronic Journal of Biotechnology* 9, doi: 10.2225/vol9-issue1-fulltext-9

van Rensburg, P. and Pretorius, I.S. (2000). Enzymes in winemaking: Harnessing natural catalysts for efficient biotransformations: A review. *South African Journal of Enology and Viticulture* 21: 52-73.

van Sluyter, S.C., McRae, J.M., Falconer, R.J., Smith, P.A., Bacic, A., Waters, E.J. and Marangon, M. (2015). Wine protein haze: Mechanisms of formation and advances in prevention. *Journal of Agricultural and Food Chemistry* 63: 4020-4030.

van Sluyter, S.C., Warnock, N.I., Schmidt, S., Anderson, P., van Kan, J.A.L., Bacic, A. and Waters, E.J. (2013). Aspartic acid protease from *Botrytis cinerea* removes haze-forming proteins during white winemaking. *Journal of Agricultural and Food Chemistry* 61: 9705-9711.

Venturi, F., Andrich, G., Quartacci, M.F., Sanmartin, C., Andrich, L. and Zinnai, A. (2013). A kinetic method to identify the optimum temperature for glucanase activity. *South African Journal of Enology and Viticulture* 34: 281-286.

Wang, Y., Zhang, C., Xu, Y. and Li, J. (2015). Evaluating potential applications of indigenous yeasts and their ß-glucosidases. *Journal of the Institute for Brewery* 121: 642-648.

Weber, P., Kratzin, H., Brockow, K., Ring, J., Steinhart, H. and Paschke, A. (2009). Lysozyme in wine: A risk evaluation for consumers allergic to hen's egg. *Molecular Nutrition and Food Research* 53: 1469-1477.

Yagodina, O.V., Nikol′skaya, E.B., Khovanskikh, A.E. and Kormilitsyn, B.N. (2002). Amine oxidases of microorganisms. *Journal of Evolutionary Biochemistry and Physiology* 38: 251-258.

Younes, B., Cilindre, C., Jeandet, P. and Vasserot, Y. (2013). Enzymatic hydrolysis of thermo-sensitive grape proteins by a yeast protease as revealed by a proteomic approach. *Food Research International* 54: 1298-1301.

Younes, B., Cilindre, C., Villaume, S., Parmentier, M., Jeandet, P. and Vasserot, Y. (2011). Evidence for an extracellular and proteolytic activity secreted by living cells of *Saccharomyces cerevisiae* PIR1: Impact on grape proteins. *Journal of Agricultural and Food Chemistry* 59: 6239-6246.

Zoecklein, B., Marcy, J., Williams, S. and Jasinski, Y. (1997). Effect of native yeasts and selected strains of *Saccharomyces cerevisiae* on glycosyl glucose, potential volatile terpenes and selected aglycons of white Riesling (*Vitis vinfera* L.) wines. *Journal of Food Composition and Analysis* 10: 55-65.

# 16 Additives, Adjuvants, Packages, Closures and Labels in Oenology

**L. Veeranjaneya Reddy[1*] and V.K. Joshi[2]**

[1] Department of Microbiology, Yogi Vemana University, Kadapa, Andhra Pradesh – 516003, India
[2] Department of Food Science and Technology, Dr. Y.S. Parmar University of Horticulture and Forestry, Nauni, Solan, HP, India

## 1. Introduction

Wine production is the process by which yeast converts grape sugar into alcohol and carbon dioxide. But because of wine – the word sounds simple enough – today's winemakers are facing a number of problems beginning from the vineyard to the bottle. Since the past few decades, scientific knowledge of the wine process has become so extensive as to provide many choices to wine makers (Gardner, 2008). To process the products, including wine, several methods are employed to preserve through use of different additives, adjuvants and preservatives. The main reason wine adjuvants and additives are popular is because of their harmlessness (when used correctly) and their ability to improve the sensory qualities (taste, smell, etc.), stability, colour, clarity and age-worthiness of the wine. Common additives are added generally before fermentation. A good rule of thumb to follow is that if the wine needs corrective additives of some kind, then, something has to be wrong with the quality of the grapes, the region (climate), or the process of winemaking. Of course, finding out what additives a winemaker has used is a bit challenging because there is a shroud of secrecy and consumer fear around the topic (Marchal and Jeandet, 2009).

In the wine world, there are many different wine additives, some of which have been used for hundreds of years with no ill effects. The intention of these additives is not to adulterate the wine, but to stabilise it. When it is stable, it is possible for the wine to have a longer shelf life. Many of these are not really additives; instead they glom (with molecular attraction) on to unwanted particles and are removed from the finished wine. Many adjuvants/additives that help to restart a stuck fermentation have been proposed, for example, addition of ammonium does not raise any counter indications with little improvement of the second fermentation. The addition of ammonium sulphate should be limited to 5 g/hl due to the limited use of nitrogen by yeast. The winemaker must react accordingly and, if need be, use additives such as nitrogen, vitamins and yeast hulls whose effectiveness has been clearly established. In addition to the above elemental operations, notably aeration and temperature control must also be standardised. Fresh wine is turbid with very high particle content, consisting of yeast lees and other grape debris. Clarity and stability of wine are the essential qualities required to convince the consumers. Particles in suspension (that form a haze or disperse through the liquid) not only spoil the presentation, but also affect the flavour. Clarity is achieved by gradual settling, followed by racking to eliminate the solids. Wine treatments are determined by its intended purposes. For instance, filtration clarifies but does not stabilise; fining does both and the treatment with gum arabic stabilises wine, but does not clarify it. Efficient packaging is necessary for both kinds of fresh or processed food and it is an essential link between the food producer and the consumer. The basic function of packaging is to identify the product and safe transportation of the product through the distribution system to the consumer. Including the above preserving the farm or processor freshness or preventing physical damage, cost effectiveness is also very important in the designing and pacaking process. Packages must be easy and safe to handle, simple to open and use, and pose no problems in their disposal (Paine and Paine, 1992). Traditionally, wines have been packaged in glass bottles, but new developments include packaging in plastic bottles, laminate lined bag-in-box systems, laminated paperboard cartons and metal cans (Markowski, 1989; Buchner et al., 1988; Anelli, 1988). The basic purpose of a closure is mainly to seal the container and

*Corresponding author: lvereddy@yahoo.com

protect the contents from contamination, provide a means of decoration and act as a dispensing measure/measure cup. Cork is traditionally used as a protective seal of the wine bottle. Cork stoppers are used for wine bottles because of their impermeability to liquids and air (prevent wine oxidation), compressibility, resilience and chemical inertness. It is believed that this sets it apart from others and if done correctly, it proves an advantage in any retail environment (Muellera and Szolnoki, 2010). Labels give details of the product, like its origin, contents and quality. It is vital that the labels are not only different against the competitors, should also express the quality of the wine itself (Main and Morris, 1994). The chapter focuses on all of these aspects.

## 2. Additives

Wine is generally assumed to be a natural product – it is often seen as fermented juice of grapes. However, due to the possibility of spoilage from chemical reactions or microorganisms and the desire for an aesthetically appealing product, a few substances are added to wine in the production process. The substances that are added during winemaking together the various processes that wine undergoes, some of which remove substances from wine or must, are referred to as oenological additives (Robinson, 1999). Additives can be used in the process of winemaking for various reasons, and they can also be removed for various reasons. The Codex Alimentarius Commission (which is the international food safety standards body associated with the World Trade Organisation) gives the following definition:

"*Food additive means any substance not normally consumed as a food by itself and not normally used as a typical ingredient of the food, whether or not it has nutritive value, the intentional addition of which to food for a technological (including organoleptic) purpose in the manufacture, processing, preparation, treatment, packing, packaging, transport or holding of such food results, or may be reasonably expected to result, (directly or indirectly) in it or its by-products becoming a component of or otherwise affecting the characteristics of such foods. The term does not include 'contaminants' or substances added to food for maintaining or improving nutritional qualities*" (FAO/WHO, 2001).

Additives have been used in wine since at least the Roman times (Juban, 2000). The purpose of addition of these additives is not to adulterate the wine, but to stabilise it and impart it a longer shelf-life. The winemaker must react accordingly and, if need be, use additives, such as sugars, nitrogen, vitamins, sulphur and yeast hulls whose effectiveness is clearly established. Wine additives are broadly classified into two categories – safe (GARS) and unsafe substances. The two categories further divided into two sub-categories – accepted and unaccepted. A detailed wine additives classification is as given in Fig. 1.

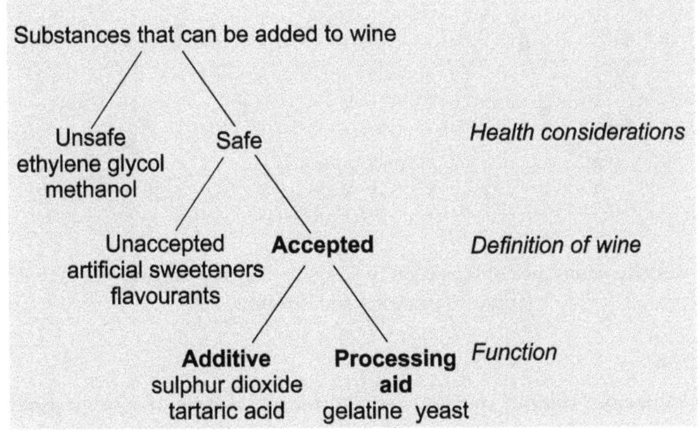


**Figure 1.** Classification of substances that can be added to the wine

### 2.1 Sulphur

Sulphur dioxide ($SO_2$) is commonly added as an antioxidant and antimicrobial compound, which reduces the population of indigenous yeasts and bacteria. Sulphur dioxide utilisation in winemaking dates back to Egyptian and Roman times (Bioleui, 1912; Danilewicz *et al.*, 2008). It is known to be used to kill unwanted bacteria and yeasts in the winemaking process. However, its role in wine is often not clearly understood because of multiple activities and reactions involved therein. Since 1987, American producers are made to mention the presence of sulphur dioxide if it exceeds 10 parts per million (ppm) in the finished wine. The EU recently passed a similar labelling law in 2005. These laws are designed to help and protect the small percentage of people who are sensitive to sulphur to not get confused with the myth that sulphites in wine can give a 'wine headache' (Rauhut *et al.*, 1993).

Sulphfur dioxide can be added to musts or wines in the form of compressed gas, potassium metabisulphite ($K_2S_2O_5$), or by burning candles containing sulphur in an enclosed container, such as a barrel. Most commonly, sulphur dioxide is incorporated into grape must or wine as potassium metabisulphite. Theoretically, two moles of $SO_2$ can be derived from each mole of $K_2S_2O_5$. Once dissolved in water, sulphur dioxide can exist in equilibrium between molecular $SO_2$ ($SO_2{\bullet}H_2O$), bisulphite ($HSO_3^{\ -}$) and sulphite ($SO_3^{\ 2-}$) species as illustrated below:

$$SO_2 + H_2O = SO_2{\bullet}H_2O$$

$$SO_2{\bullet}H_2O = HSO_3\ H^+$$

$$HSO_3 = SO_3^{2-}\ H^+$$

This equilibrium is however, dependent on pH, with the dominant bisulphite anion at pH 3 to 4 (Fig. 2). Besides being in equilibrium with the molecular and sulphite species, bisulphite also exists in 'free' and 'bound' forms. Here, the molecule will react with carbonyl compounds (e.g. acetaldehyde) to form addition products or adducts, such as hydroxysulphonic acids.

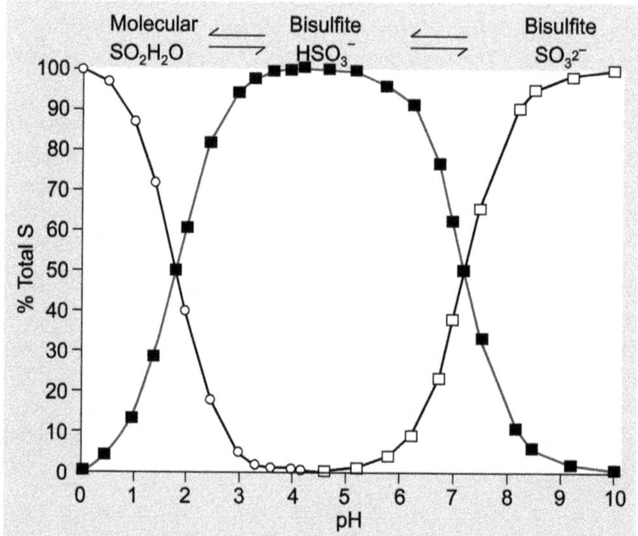


**Figure 2.** Relative abundance of molecular $SO_2$, bisulphite, and sulphite at different pH values (*Source*: Fugelsang and Edwards, 2007)

## *Mechanism of Action*

It is known that the molecular sulphur species is an antimicrobial form of sulphur dioxide. When added to the must, it forms $SO_2{\bullet}H_2O$, which does not have a charge and yields bisulphite and sulphite at a pH near 6.5. It inhibits the growth of microbes by different mechanisms, including rupture of disulphide bridges in proteins and reaction with co-factors, including $NAD^+$ and FAD. It also reacts with ATP and brings about deamination of cytosine to uracil, increasing the likelihood of lethal mutations. It also reduces the availability of crucial nutrients (Ough, 1993b; Romano and Suzzi, 1993). Specifically, $SO_2$ can cleave the vitamin thiamin into components not metabolically useable (Fig. 3). In the United States, the maximum allowable addition of thiamin hydrochloride is 0.6 mg/L.

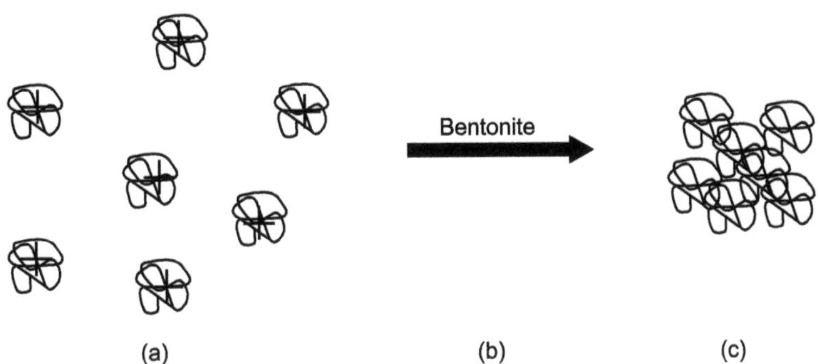


**Figure 3.** Electrostatic charges between a) positively-charged proteins and b) negatively charged bentonite, result in c) a neutralised complex, which will eventually clump together and precipitate from the solution

The concentration of molecular $SO_2$ needed to prevent growth of microorganisms varies with wine/juice pH, temperature, population density and diversity, stage of growth, alcohol level and other factors. Beech *et al.* (1979) suggested that addition of 0.8 mg/L molecular $SO_2$ is the amount required to bring about a $10^4$ CFU/mL reduction in 24 hours in white wines with several spoilage microorganisms. However, differences in sensitivity to $SO_2$ between various genera of yeasts and bacteria found in wines are known to exist and suggest that a concentration of 0.8 mg/L molecular $SO_2$ does not completely eliminate the bacteria (Du Toit *et al.*, 2005).

Sulphur metabolism is important in winemaking since, depending on the yeast strain and nutrient composition, it is the source of most unpleasant compounds, including $H_2S$ and mercaptans (Rauhut, 1993). Along with elemental sulphur, grape musts also contain other forms of inorganic sulphur. Sulphate is indeed usually present in excess amounts (up to 700 mg/L) and sulphite (up to 100 mg/L) is often added as an antioxidant and antimicrobial compound (Henschke and Jiranek, 1991). Inorganic sulphur forms are the main source of $H_2S$ formation during fermentation (Martineau and Henick-Kling, 1995; Nielsen and Richelieu, 1999).

## 2.2. Dimethyl Dicarbonate

Dimethyl dicarbonate (DMDC) is an approved additive used instead of sulphur dioxide. It sells under the trade name Velcorin™. The antimicrobial effect of DMDC results from inactivation of microbial enzymes (Ough, 1993a). Porter and Ough (1982) reported its mechanism of action as denaturation of the fermentative-pathway enzymes glyceraldehyde-3-phosphate dehydrogenase and alcohol dehydrogenase. This chemical is toxic by ingestion and inhalation, is a skin and eye irritant and is combustible when exposed to an open flame. As such, safety precautions must be taken when using Velcorin™ in the wine-production process. The equipment for dose reduces safety risks given the toxicity of the chemical. The additive does not possess any residual activity because DMDC undergoes hydrolysis to yield carbon dioxide and methanol. This is approved in the United States for table as well as low-alcohol and dealcoholised wines at a maximum concentration of 200 mg/L over the life of the wine (Ough, 1993a). The United States FDA also permits its use in juice and juice beverages.

## 2.3. Lysozyme

Lysozyme is a low molecular weight protein (14,500 Da) present in egg white. It is well known that lysozyme participates in lysis of the cell wall of Gram-positive bacteria (*Oenococcus*, *Lactobacillus*, and *Pediococcus*). Activity towards Gram-negative bacteria (*Acetobacter* and *Gluconobacter*) is limited because of the protective outer layers in this group and at the same time, lysozyme has no effect on yeasts and moulds (Conner, 1993).

Gerbaux *et al.* (1997) reported that addition of 500 mg/L is enough to inhibit MLF before the initiation of fermentation. Because of its specific inhibition of Gram positive bacteria, lysozyme finds application in white, rose and blush wine producers wanting to prevent malo-lactic fermentation as well as wineries wanting to reduce initial population of lactic acid bacteria before fermentation (Nygaard *et al*., 2002; Delfini *et al*., 2004). Lysozyme is more active in white wines than reds, most likely due to the differences in polyphenolic content (Daeschel *et al*., 2002; Delfini *et al*., 2004). Lysozyme may also influence protein stability in white wines; thus, fining trials need to be conducted prior to bottling wines treated with an enzyme. Lysozyme is now approved for use in the United States at concentrations up to 500 mg/L.

## 2.4. Sorbic Acid

Sorbic acid (2,4-hexandienoic acid) is a short-chain fatty acid that is used in grape juices and in sweetened, bottled wines to prevent re-fermentation by *Saccharomyces* (De Rosa *et al*., 1983; Renee Terrell *et al*., 1993). Sorbic acid is generally effective in controlling *Saccharomyces*, but other yeasts exhibit differential resistance (Warth, 1977, 1985). Mechanisms of inhibition are not fully understood but probably due to morphological differences in cell structure, changes ae seen in genetic material, alteration in cell membranes, as well as inhibition of enzymes or transport functions (Sofos and Busta, 1993). Bacteria are not affected by sorbic acid and in fact, several species can metabolise the acid to eventually yield 2-ethoxyhexa-3,5-diene – a compound that imparts a distinctive 'geranium' odour/tone to wines. Sorbates are recommended to use in sparkling wine production. The maximum concentration allowed in the United States is 300 mg/L, whereas the Office International de la Vigne et du Vin (OIV) places the limit at 200 mg/L. In practice, concentrations of 100 to 200 mg/L are typically used, but to inhibit *Schizosacccharomyces pombe* and *Zygosaccharomyces bailii* at least 672 mg/L are required (Warth, 1985).

## 2.5. Sugar

Addition of sugar to grape juice is a process known as chaptalisation which will increase the final alcohol level in the finished wine. Addition of sugar does not make a wine sweeter because the sugar is consumed by the yeast when it is fermented into alcohol. Chaptalisation can add up to 3 per cent ABV to a wine. It is legal in areas where grapes struggle with ripeness, such as Bordeaux, France and Oregon. Generally, winemakers add sugarcane juice to the grape must and this is called as sweet reserve (*süssreserve*). But adding cane sugar is not legal in California, Argentina, Australia, southern France and South Africa. Producers can add sugar-rich grape concentrate to simulate the same results, as the use of grape concentrate is not considered chaptalisation (Jackson, 2014).

## 2.6. Nutrients

The lack of some nutrients in grape juice can cause serious problems during fermentation. Nitrogen, vitamins, minerals, etc. may be deficient in grape juice. For this reason, yeast activators are usually added in wineries.

### *2.6.1. Nitrogen*

The composition of nitrogenous compounds and concentration in grape juice play an important role in satisfying the nutritional requirements of wine-fermenting microorganisms. *Saccharomyces* will utilise most of the amino acids in must, except proline, which is not metabolised under anaerobic conditions (Bisson, 1999). These compounds are present in juice or must and are available to yeast as ammonium salts (NH4 +) and amino acids primarily and generally, known as 'yeast assimilable nitrogen' (YAN). In addition to YAN, grape juice contains a variety of peptides and proteins as well. The composition of YAN depends mainly on farm practices along with the stage of harvesting and post-harvesting methods. In general, many grape musts are considered to be deficient in nitrogen, based on the estimated minimal requirement of 140-150 mg N/L. The lack of some nutrients in grape juice can also cause serious problems during fermentation.

In view of the above, generally nitrogen deficiency is corrected through supplementation of urea or diammonium phosphate (27 per cent $NH_4$ + 73 per cent $PO_4^{3-}$). Urea as a nitrogen source is no longer approved in many countries owing to its demonstrated involvement in the formation of ethyl carbamate. In case of $(NH_4)_2HPO_4$ the maximum level legally permitted to correct nutritional deficiencies is 960 mg/L, 300 mg/L and 400 mg/L in the United States, Europe and Australia respectively (Henschke and

Jiranek, 1993). In addition to $(NH_4)_2HPO_4$, some winemakers advocate the use of balanced nutritional formulations that also contain amino acids, minerals, vitamins, and/or other ingredients important for yeast growth. Addition of nitrogen at yeast inoculation is convenient. Addition of nitrogen after 48-72 hours of post-inoculation suggest avoidance of the problem of non-*Saccharomyces* yeasts (Beltran *et al.*, 2005).

Yeast hulls (ghosts) are also utilised by some winemakers to satisfy the nutritional demands of yeast in wine fermentation. These are available as byproducts in brewing industries; hulls represent the remnants of yeast cell walls. Yeast hulls provide some assimilable nitrogen and other nutrients, such as medium-chain fatty acid. The development of a 'rancid' character upon extended storage of wines with yeast hulls is common. This may be due to the fact that lipids (fats) oxidise upon exposure to oxygen; hence, usage of hulls is not advisable due to the potential for imparting these off-odours to the wine. At present, a new generation of yeast activators are present in the market and these new activators are prepared from yeast and provide several other interesting but substances, such as sterols, unsaturated fatty acids (UFA), minerals, panthothenic acid, etc. These activators are very useful during the later stages of fermentation.

### 2.6.2. Vitamins

Like other microorganisms, yeasts require a number of growth and survival factors, such as vitamins and minerals. Yeast benefits from vitamins, minerals or any chemical compound that helps to keep the yeast alive in grape juice during fermentation. The winemaker may elect to use one of the several commercial yeast formulations that contain vitamins. Yeasts require a variety of vitamins, such as riboflavin, thiamin, pantothenic acid, pyridoxine, nicotinamide, biotin and inositol for its good growth and can sustain in harsh conditions. However, the requirement depends on yeast species and growth conditions (Monk, 1994); for instance, *thiamine hydrochloride* is a B vitamin which helps keep yeast happy in high alcohol wines (14 per cent v/v). In general, all *Saccharomyces* strains need biotin and pantothenic acid while some also need inositol and/or thiamine (Walker, 1998). Pantothenic acid is an essential part of coenzyme A, a molecule required for sugar and lipid metabolism. Deficiency of pantothenic acid can also lead to $H_2S$ formation. Biotin is involved in carboxylation of pyruvic acid and the synthesis of nucleic acids, proteins and fatty acids. Thiamin may or may not be required because some strains of yeast can also synthesise the vitamin. Sulphur dioxide can cleave thiamine, making the vitamin unavailable. Finally, nicotinic acid is used in the synthesis of NAD+ and NADP+, and inositol is required for cell division. Nutrient requirements for yeasts, other than *Saccharomyces*, are widely different.

### 2.6.3. Minerals

Yeasts require a number of growth and survival factors, such as vitamins and minerals. Minerals, such as phosphate, potassium, magnesium, sulphur and trace minerals, such as calcium, chlorine, copper, iron and manganese are essential and are also present in yeast cell biomass (Monk, 1994; Walker, 1998). Yeast must be supplied with a source of phosphate, which is incorporated into nucleic acids, phospholipids, adenosine-5′-triphosphate (ATP), and other compounds. Potassium is necessary for uptake of phosphate and its deficiency may be linked to sluggish alcoholic fermentations (Kudo *et al.*, 1988). Other minerals required also *Saccharomyces* during fermentation and which is used primarily as enzyme activators, but also have a variety of functions.

### 2.6.4. Tannin and Oak Chips

Use of oak chips and tannin powder in wine is widely accepted in Europe. Tannin is one substance that makes wines age-worthy. Wine grapes are full of seeds which are rich in tannins. The seeds are crushed with the grapes to add structure to wine. Oak aging also adds small amounts of tannin as the wine is exposed to oakwood. While oak chips are not as *romantic as a room full of oak barrels,* they are better for forests and are cheaper to transport (Garc′ıa-Ruiz *et al.*, 2008b).

### 2.6.5. Acidifiers and De-acidifiers

The pH of wine is crucial to check how it tastes and how long a wine will last. On a perfect vintage, the wines will be more naturally in balance. So what to do when it is not perfect? Adding calcium carbonate (chalk) as a de-acidifier to wine will reduce the high acid levels and increase the pH. This practice is common in areas that have cooler weather and ripening is challenging. Winemakers add acidifiers when there is not enough acidity in wine. Tartaric acid, malic acid and citric acid or any blend thereof, can help balance the wine. Many people claim they can taste simulated acids in wine. Adding acid is a common practice with lower acidity grapes in warmer regions (Maujean, 2000).

# 3. Adjuvants

In general, different types of compounds are added to the wine to obain better taste and quality. The compounds/agents that are removed after their function from wine and do not remain in wine but are used as technological 'assistants', called adjuvants (Galpin, 2006). The Codex Alimentarius Commission (which is the international food safety standards body associated with the World Trade Organisation) makes the following definition:

"*Processing aid means any substance or material, not including apparatus or utensils, and not consumed as a food ingredient by itself, intentionally used in the processing of raw materials, foods or its ingredients, to fulfil a certain technological purpose during treatment or processing and which may result in the non-intentional but unavoidable presence of residues or derivatives in the final product*" (FAO/WHO, 2001, p. 43) (Fig. 1).

The process of addition of some agents of a reactive or an adsorptive substance (in some cases, two and even three fining agents are used simultaneously) to wine to remove or reduce the concentration of one or more undesirable constituents is called 'fining'. Fining agents are added for the purpose of achieving clarity, colour, flavour and/or stability modification in juices and wines. Most fining agents react within seconds and the contact time between the fining agent and the wine should be as short as possible. Carbon and PVPP can be filtered out immediately or a few hours after fining. At the opposite extreme, formation of flocculates requires a few days when proteins are used (depending on wine temperature) and they require a week or two to settle (Zoecklein, 1988a; Brissonnet and Maujean, 1993). The products are considered essential but are always used in specific cases and with a good understanding of the mode of action. The use of different clarification substances and oenological adjuvants, such as bentonite or PVPP, in anormal dose used in winemaking can affect the concentration of biogenic amines in wines, since these compounds have the ability for adsorption of some amines (Alcaide-Hidalgo *et al.*, 2008; Guilloux-Benatier and Feuillat, 1991). List and functions of adjuvants are given in Table 1.

Adjuvants used as fining agents are grouped according to their general nature (Zoecklein, 1988b):

- Earths: montmorillonite, bentonite, kaolinite
- Animal proteins: gelatin, isinglass, caseins
- Plant proteins: wheat gluten, soya, lupin, garden pea
- Carbon: wood charcoal
- Synthetic polymers: PVPP
- Silicon dioxide
- Metal chelators and enzymes (pectinases)

## 3.1. Earths

### *3.1.1. Kieselguhr or Diatomaceous Earth*

Diatomite is a siliceous sedimentary rock formed from the accumulation of microscopic fossil algae shells, or diatomaceous earth with 10 μm to several hundred micrometre dimensions. These rocks have different compositions, depending on their origins (marine or lacustrine). These are known to be from 60-100 million-years old. There are many widespread deposits in Massif Central, France, California, United States and North Africa. Grinding of these fossil earths up to siliceous powder produce the diatomaceous earth, also known as infusorial earth or kieselguhr ('small silica particle' in German). Diatomaceous earth has been used as a filtration adjuvant since the late 19th century, due to the extreme porosity of the powder obtained by processing the rock. The filter layer represents 80 per cent of the total mass, with a surface of 20–25 $m^2/g$. These characteristics are highly favourable for filtration. Around 1920, a

**Table 1.** List of Adjuvants Used as Fining/Clarification Agents in Wine Production

| *S. No.* | *Nature of adjuvant* | *Usage* | *Dosage* |
|---|---|---|---|
| 1. | **Earth** | | |
| | (a) Diatomaceous Earth | Filtration aid | |
| | (b) Bentonite | Polypeptides and proteins adsorption | 80 g/hl |
| | (c) Perlite | Filtration aid | |
| 2. | **Animal proteins** | | |
| | (a) Gelatin | Clarify and stabilise | |
| | (b) Isinglass | Reduce phenolic compounds; add fruitiness to wine | 1.25 to 2.5 g/hl |
| | (c) Albumin and Egg White | Soften wines with a high tannin content and excess astringency | 5-15 g/hl |
| | (d) Blood by-products | Soften wines | 10 to 20 g/hl |
| | (e) Casein | Precipitate and clarify | 10–30 g/hL |
| 3. | **Plant derived adjuvants** | | |
| | (a) Wheat gluten | Reduce haze and tannin concentrations | 10–20 g/hl |
| | (b) Alginate | Clarify wine by flocculation | 4 to 8 g/hl |
| | (c) Tannins | Remove excess protein in white wines | 5 to 10 g/hl |
| | (d) Wood charcoal | Decolourise and deodorise | 10–50 g/hL |
| 4. | **Synthetic polymers** | | |
| | (a) Polyvinyl Polypyrrolidone (PVPP) | Stabilise wine and reduce concentrations of tannoid substances | 20–30 g/hl |
| | (b) Kieselsol | Reduce phenolic compounds | |
| | (c) Nylon | Reduce polyphenols | |

new treatment process was developed for making high-permeability diatomaceous earths. Three types of diatomaceous earth are currently used as given below:

1. Natural diatomaceous earth, grey in colour, is crushed and dried to form fine particles. Filtration is very fine with good clarification, but throughput is very low and this medium is hardly ever used today. It may also contain residues of organic matter.
2. Diatomaceous earth calcinated at 1000°C, pink or red in colour, is crushed and sorted to produce powders free of organic matter, with coarse particles that are capable of fine filtration at satisfactory flow rates.
3. Fritted diatomaceous earth, i.e. activated by calcination at 1100/1200°C in the presence of a flux (calcium chloride or carbonate), is sorted to produce a white powder with even larger particles and looser structure. Filtration is less fine, but faster. There are different qualities of kieselguhr, differentiated by particle size, which control permeability, i.e. the rate at which a liquid passes through the material. It must also be kept away from odorous products, as it may easily fix volatile substances that could later be released into wine.

### *3.1.2. Bentonite*

Bentonite is the most commonly used fining adjuvant in the wine industry. It is a volcanic material which was deposited millions of years ago in broad layers, which weathered and changed from a fragile glassy state into a mineral one. This mineral is generally called montmorillonite, which refers to a small French town (Montmorillon), where it was first discovered. In the USA, bentonite is principally mined at Wyoming, hence the term 'Wyoming clay'. Bentonite is a complex, hydrated aluminum silicate with exchangeable cationic components: (Al, Fe, Mg) $Si_4O_{10}$ $(OH)_2$ ($Na^+$, $Ca^{++}$). The most commonly used bentonite form in enology is sodium bentonite. Sodium bentonite has enhanced protein-binding capability over calcium bentonite (Marchal *et al.*, 1995). The main uses of bentonite are clarification and protein stability. A major problem encountered in juice and wine production (white and rose wines) is protein stability that is the removal of heat-unstable proteins (Hsu and Heatherbell, 1987; Waters *et al.*, 1992). This form of instability, together with potassium bitartrate precipitation, is the most common non-microbiological defects in commercial wines. Bentonite is used to remove both stable and unstable proteins. Usage of small amounts of bentonite corrects the protein levels at which precipitation in the bottle will not occur.

*Mechanism of action*

Bentonite exists as small plates which, when hydrated, separate to form a colloidal suspension with enormous surface area (300–900$m^2/g$). Good-quality bentonite has practically no flavour or odour of its own. Its subsequent activity in solution is like that of a multiplated negatively charged structure which is able to exchange its cations with positively-charged components of the juice or the wine (not only proteins). Bentonite absorption of uncharged molecules also occurs, if they are polar. Additionally, due to the fact that the platelet edges are positively charged, some limited binding of negatively charged proteins may occur (Fig. 3).

The mechanism of polypeptides and proteins adsorption on clay minerals has been studied by Gougeon *et al.* (2002, 2003) with the help of homopolyamino acids (polylysine and polyglutamic acid) and a synthetic montmorillonite. The data acquired from $^{13}C$ NMR, BET measurements, X-ray diffraction and adsorption/desorption isotherms have led to understand the interactions between 'positively' charged proteins and 'negatively' charged plate surfaces in synthetic wine. From the above observations it can be proved that polypeptides are adsorbed at the periphery of the montmorillonite particles through specific interactions between the protein side chains and the silicate sheets, whereas the polypeptide backbones do not enter the interlayer spaces (Gougeon *et al.*, 2002, 2003).

Including the proteins, bentonite can also adsorb some phenolic compounds *via* binding with proteins that have complexed with phenolics indirectly. However, the amount of phenols removed is usually not significant. Bentonite is also known to enhance membrane filterability. Presumably, this is due to a reduction in the colloidal particle number in suspension. Vast literature on protein instability is available, but at which level/concentration actually wine proteins will remain stable is unknown. Proteins present in wine are not of single type; rather, are a mixture of probably more than 100 proteins derived from the grape, wine microorganisms (yeast, autolysed yeast and at times, *Botrytis cinerea* (Dambrouck et *al.*, 2003; Charpentier *et al.*, 1986; Cilindre *et al.*, 2007). Yeast proteins, however, have not been shown to play a role in white-wine protein clouding. Many wine proteins are not free but bounded to a minor quantity of grape phenolics (flavonoids). Bentonite removes different amounts of grape protein fractions (Moine-Ledoux and Dubourdieu, 2007).

Excessive lees production is the commonly expressed problem with sodium bentonite. Its volumes often range from 5-10 per cent. Centrifugation/filtration with a rotary vacuum filter is best among the several methods employed to minimise this problem. Proteins react with bentonite within the first minute of contact (Blade and Boulton, 1988); hence it needs only minutes to react with proteins and precipitate them. It needs longer periods when the capacity of the tank is very high. The possibility of leaching or 'sloughing off' of proteins from the bentonite platelets perhaps may occur. Gelatin is used to help bentonite flocculation and possibly aid in lees compaction. Multiple finings with bentonite small amounts rather than a single large addition is also a successful approach in reducing the overall bentonite requirement and lees too.

The quantity of bentonite to be added to the juice to reach a protein-stable wine is generally determined using a heat test (Dubourdieu *et al.*, 1988; Marchal *et al.*, 2002). In addition to protein stability, bentonite fining helps to enhance the wine filterability *via* general removal of suspended solids. Protein removal is not proportional to the amount of bentonite added, but follows a power law (Marchal *et al.*, 2002). Although complete removal of wine proteins can generally be achieved by the use of bentonite (except for yeast proteins, but the latter do not really participate to protein haze), it has been recognised that this may not be necessary to obtain protein stability and may have detrimental effects on the sensory qualities of wine (wine body, colour and possibly imparting an earthy, freshly 'laundered' smell). However, when bentonite is used in white wines in high doses (above 80 g/hl), it may attenuate their organoleptic characteristics. Nevertheless, there are reports that say that the use of bentonite alters the organoleptic characteristics of wines (Main and Morris, 1994; Muhlack *et al*., 2006) (Tables 2a and 2b).

**Table 2a.** Bentonite Effects on Sensory Quality of Wine

| *Attributes* | *Wine without bentonite addition* | *Wine with bentonite addition* |
|---|---|---|
| Visual aspect (0-9) | 2.6 | 3.9 |
| Aroma intensity (0-18) | 5.2 | 6.5 |
| Aroma quality (0-18) | 4.8 | 6.8 |
| Taste intensity (0-18) | 6.4 | 6.3 |
| Taste quality (0-27) | 7.5 | 9.8 |
| Harmony (0-27) | 8.4 | 9.8 |
| Total[a] | 34.9 | 45.7 |

[a] 0-7 = Excellent; 8-23 = Very good; 24-44 = Good; 45-62 = Correct; 63-78 = Regular; 79-90 = Inadequate; 90 = Eliminated (*Source*: Martínez-Rodriguez and Polo, 2003)

**Table 2b.** Impact of Bentonite Treatments on the Concentration of Must and Wine Aroma Ccompounds

| *Compounds* | *Control* | *Must fining* | *Wine fining* |
|---|---|---|---|
| Carbonyl compounds | 2.29 | 7.68 | 7.31 |
| Acetates | 54.1 | 53.86 | 61.2 |
| Linear ethyl esters | 1.76 | 1.89 | 2.22 |
| Branched ethyl esters (ug/l) | 15.25 | 17.26 | 18.0 |
| Cinnamate esters (ug/l) | 0.15 | 0.15 | 0.15 |
| Alcohols | 196 | 176.2 | 195.55 |
| Acids | 438.09 | 483.2 | 436.44 |
| Monoterpenes | 11.06 | 18.95 | 18.30 |
| Norisoprenoids | 3.35 | 5.93 | 5.29 |
| Phenols | 400.3 | 348.08 | 392.5 |
| Lactones | 2388.4 | 2478.1 | 2678.6 |
| Vanilline derivatives | 23.6 | 28.7 | 30.1 |

*Source*: Vela *et al*., 2017

### *3.1.3. Perlite*

This consists of spherical, pearl-shaped aluminium silicate particles made by processing volcanic rock. This rock contains 2-5 per cent of interstitial water and occluded gases, giving it the property of expanding 10-20 times on heating to 1000°C. This treatment reduces the density of the powder and increases its porosity. After grinding and sorting, a range of light, white powders of varying particle sizes is obtained by adjusting the processing conditions. Perlite makes it possible to run longer filtration cycles as it is much more porous than diatomaceous earth and its low density (20-30 per cent lower) reduces the weight of the adjuvant required. However, perlite has a lower adsorbent capacity and is most efficient in a fine precoat perlite which is used to filter must and liquids with a high solid content. It is abrasive and may cause rapid wear of injection pumps (Jackson, 2014).

## 3.2. Animal Proteins

### *3.2.1. Gelatine*

Gelatine is a protein prepared from collagen and has long molecules with many different side chains, which are responsible for their varying properties. The protein chain is amphoteric and can carry either a positive or a negative charge, depending on the pH of the medium. Gelatines used in enology are usually prepared from pig skin (Type A) by acid hydrolysis and from limed hide or limed ossein (Type B gelatines). Type A has isoelectric points between 6 and 9. Gelatines with a high gel strength (Bloom strength) have the higher pI and gelatins with a low Bloom strength have a pI closer to 6. Type B has a pI close to 5 (Calderon *et al.*, 1968).

The use of gelatine has also been challenged, even though it is mainly a pork by-product. It is, however, still widely used for its excellent clarification and stabilisation capacities, particularly in red wines. In Europe, Type A gelatine usage is higher and from the literature available, it can be suggested that low Bloom strength gelatine is quite suitable for fining (Maury *et al.*, 2001). Hence it is important to ensure that when a warm gelatine solution is added, it is added at a point of very intense agitation such that the small amounts of gelatine are intimately mixed into a large bulk of beverages before any gelling can occur. Gelatine is primarily used to soften red wines, but is also used to reduce the phenol level and brown colour in white juices, before fermentation. The potential for overfining with gelatine is great. Kieselsol, a negatively-charged silica compound, is recommended for white wines. Kieselsol helps to moderate the effect of gelatine on wine flavour and reduces the amount of gelatine needed and the volume of lees produced (Marchal *et al.*, 1993).

*Mechanism of Action*

The primary reaction occurring with gelatine is a complex formation between polyphenols in the wine and the protein of gelatine to give the desired floccular precipitate. Major polyphenols that are present in wine are tannins and anthocyanins. Both contain benzene rings with adjacent hydroxyl groups which are a major source for hydrogen bonds and can easily form a complex between gelatine and tannins or anthocyanins in wines. At the same time, gelatine is significantly suitable for hydrogen bonding because of its amino acid composition (Fig. 4) (one-third of the amino acids are glycine, where R=H, and hence, steric hindrance to hydrogen bonding would be far less than with proteins containing less glycine). However, the tannin/gelatine complex is pH dependent and dissociates at close to pH 8 (Table 3), due to both molecules becoming negatively charged and hence, mutually repulsive (Marchal *et al.*, 1993).

**Table 3.** Influence of the Quantity and Type of Gelatine on the Elimination of Tannins from Wine

| *Gelatine added (mg/l)* | *Tannins eliminated (mg/l)* | |
|---|---|---|
| | *Heat-soluble* | *Cold-soluble* |
| 50 | 50 | 50 |
| 100 | 160 | 90 |
| 200 | 310 | 230 |

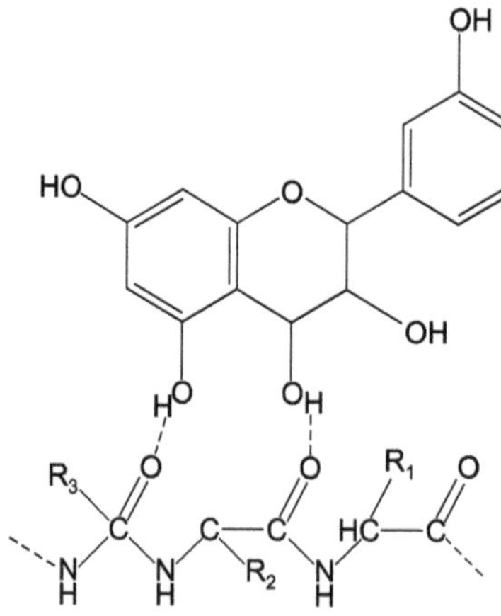

**Figure 4.** Hydrogen bonding between gelatine and phenol compound (Zoecklein *et al.*, 1995)

The second reaction is the complex formation between the natural wine proteins and the gelatine. Proline, which is a very important amino acid of gelatine, imparts a twist to the chain and affects the shape of the protein molecule and its rigidity. Generally, wine has pH of 2.9-3.6; at this pH most of the amino groups are positively charged and most of the acidic groups are uncharged. Proteins present in wine are essentially derived from grapes (protein released during must preparation and wine yeast (protein secreted by yeasts). During aging, some of these proteins can also associate to form insoluble precipitates and participate to 'protein instability'. For protein-protein interactions (polar associations), it is necessary that the two proteins should have opposite charges at the beverage pH. These reactions lead to a reduction of hydrophilic sites and hence, precipitation forms. Also, further hydrophobic bonding due to association of hydrophobic sites in an aqueous media can lead to an increase in effective molecular weight and precipitation (Marchal *et al*., 1993).

Gelatine has an isoelectric point anywhere between pH 9 and pH 5, depending upon the source and method of production. Type A has isoelectric points between 6 and 9. Some enological gelatines, sometimes derived from limed hide or limed ossein (Type B gelatines) and all of them have a pI close to 5. The higher the pI, the greater the cationic charge on the molecule at the wine pH. In other words, at pH 2.9-3.6, all gelatines would be positively charged, but the charge density would be far higher for high pI gelatins (Fig. 5) (Marchal and Waters, 2010).

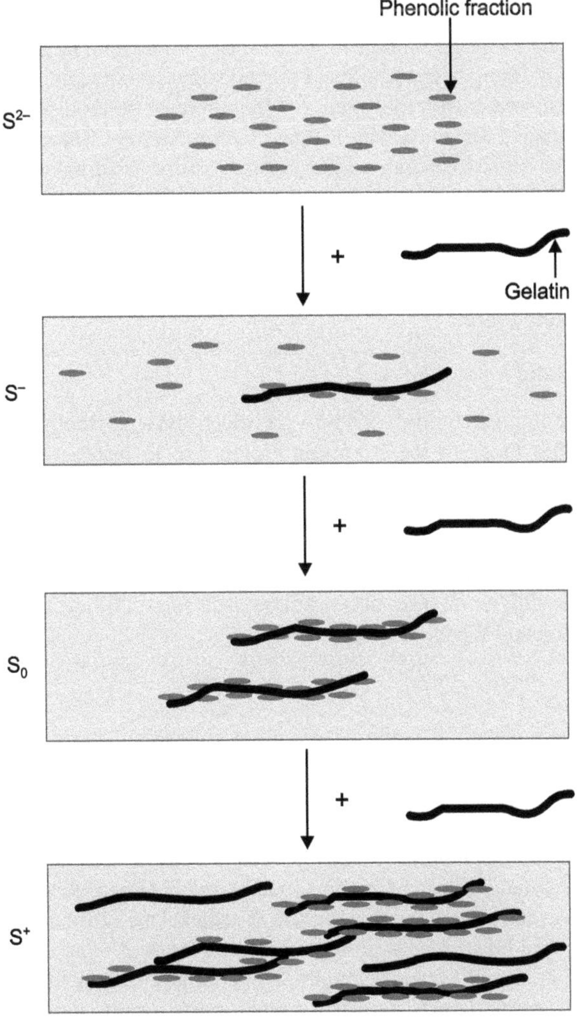


**Figure 5.** Schematic representation of the fining mechanism of gelatine

### 3.2.2. Isinglass

Isinglass has been used as a fining agent since the 18th century. It is a raw, unprocessed product from the swim bladder of certain fish, such as sturgeon. It consists mainly of collagen fibres and is available in sheets, strips, whitish chips or coarse vermiculated powders. The preparation process is as follows: first the dry isinglass is soaked in acidulated, sulphured water (0.5 ml HCl per litre + 200 mg/l $SO_2$) for about 10 days at a cool temperature and then sieved to obtain a homogeneous jelly. The vermiculated form of isinglass swells easily, without lumps. It must, however, be used immediately after preparation or hydrolysis converts it into gelatine. At present this fining agent is available in the form of ready-to-use jelly, prepared from fish cannery waste (skin, cartilage, etc.). Normally, isinglass is used in the dose of 1.25-2.5 g/hl in white wines. This concentration enhances their brilliance and reinforces the yellow colour. As bentonite lees, it also forms bulky lees and these lees make racking more difficult by clogging the filter surface. (Marchal *et al*., 2003).

### 3.2.3. Albumin

It is being used since a long period as a fining agent and is universally recognised for its qualities as a fining agent for red wines. The white portion of a fresh egg is used as egg albumin and it consists of several proteins (12.5 per cent of fresh egg white). Ovalbumin is the major component of egg albumin and it may be used in flake form. The colour of flakes varies from white to golden yellow. It has always been presented as the only fining agent employed for red wines. It is recommended for softening wines with a high tannin content and excess astringency. Albumin must be used with care on light wines and is therefore not recommended for white wines. Egg albumin does not flocculate but it precipitates as a compact deposit. Three to eight fresh egg whites or 5-15 g/hl of dried flakes are used. Egg albumin is heat sensitive; so it is advised to not warm the preparation to facilitate dissolving of the powder. One of the disadvantages of albumin is its relative solubility. If used in excess, it may donate a detectable meringue-like attribute to the treated wine. In addition, it may cause problems for those sensitive to egg products (Blade and Boulton, 1988; Jackson, 2014). Currently egg albumin is available commercially in ready-to-use, sterilised liquid form.

### 3.2.4. Blood By-products

Blood by-products were popular fining agents in the olden days; currently its usage is prohibited in many countries, including those in the European Union due to health reasons. But these products give good results in fining young red and white wines. They ares highly effective and attenuate any herbaceous character. Risk of overfining is low because of its low sensitivity to protective colloids and less tannin requirement for flocculation. Bitter, stalky, young red wines, with a robust tannic structure, are nicely softened. The dose must be adjusted between 10-20 g/hl according to the wine's tannin content. Herbaceous white wines with an intense, heavy aroma, lose some of their coarseness after fining with doses of 5-10 g/hl (Allison and Wilson, 2015).

### 3.2.5. Casein

Casein is an excellent fining agent for white wines with a 'refreshing' effect on their colour and flavour. It is a phosphorus containing heteroprotein and is produced through coagulation of skimmed milk. It also has a curative effect on yellowing and maderiszation. It works based on flocculation, which occurs exclusively due to the acidity of the medium, but the presence of tannins is necessary for precipitation and clarification. Hence, some of the tannins are also removed at the time of treatment. Casein is a positively-charged protein that flocculates in acidic media, such as wine. When added to wine, casein adsorbs and mechanically removes the suspended materials as it settles. One of the most desirable properties of the flocculating agent is rapid distribution through the entire mass of wine before it flocculates. Casein powder is not very soluble in pure water, but addition of sodium bicarbonate or potassium can enhance the solubility. In general, casein is used to remove undesirable odours, to bleach colour and to clarify white wines. It is sometimes used in place of charcoal for colour modification of juice and white wine.

It also has its effect on phenols, either by eliminating them or, most probably, by protecting them from oxidation. Normal dosage levels are 10-30 g/hL but 50 g/hl or more may be used in curative treatment (Marchal and Jeandet, 2006; Jackson, 2014).

### 3.3. Plant-derived Adjuvants

The occurrence of the 'mad cow disease' (bovine spongiform encephalopathy) caused grave concern on the use of proteins derived from animal sources in the food supply and winemakers are encouraged to discontinue use of gelatines (especially bovine gelatine) and more generally, animal proteins. Due to the above problems, research interest in replacing animal-origin adjuvants with plant-based products has considerably increased. Martin Vialatte Research Company (BP 1031, 51319 Epernay, France) first started studying the properties and assessing the possibilities of using plant proteins as fining agents for wine (Lefebvre *et al.*, 2000). Initial results have been found to be promising with several powdered products. Many investigations were carried out with wheat prolamins, commonly called gluten, selected as white musts and wine-clarifying agents (Marchal *et al.*, 2002b, 2002c).

#### *3.3.1. Wheat Gluten*

Maury *et al.* (2003) tested white lupine protein, two wheat gluten-based preparations and two chemical hydrolysates of gluten, using two unfiltered wines and a model solution prepared with phenolic compounds extracted from Syrah wine. To conclude, it should be possible to use plant proteins as fining agents in wine, but each preparation behaves in a specific way (Table 4). It will, therefore, be necessary to test a large number of products to determine which ones give the best results with different types of wine and define the most effective doses, likely to be around 10-20 g/hl. An application has been submitted to the appropriate authorities (*Office International de la vigne et du Vin*, OIV) for approval of these products and authorisation to use them in fining wine (Lefebvre *et al.*, 2003).

**Table 4.** Gluten and Plant Protein Effects on Turbidity and Removal of Phenols and Aroma Compounds

| *Name of the plant protein* | *Turbidity change (NTU after 60 h)* | *Removal % of phenols* | *% loss of aroma compounds* |
|---|---|---|---|
| Soy protein isolate | 13 | 77 | 20 |
| Pea protein isolate | 9 | 70 | 60 |
| Lentil flour | 8 | 65 | 40 |
| Gluten | 13 | 56 | 40 |
| Reference wine | 30 | - | - |

However, as wheat gluten is capable of producing allergic phenomena, it is important to ensure that no residues are left in wine. Lefebvre *et al.* (2003) showed that treatment with high concentrations, say up to 50 g/hl, did not contain any wheat protein. The treated wine was also tested for immunoreactivity and presented no risk of triggering allergic phenomena. Further researchers also evaluated residues in wine treated with pea and lupine extracts. The results indicate that from a health standpoint, there are no problems to use plant proteins for fining wine.

#### *3.3.2. Alginate*

Sodium alginate is a salt of alginic acid and is extracted from various *phaeophyceae* algae, especially kelp, by alkaline digestion and purification. Sodium alginate is a polymer of mannuronic acid, consisting of chains with a basic motif consisting of two mannuronic cycles. It may be effective in clarifying wine, although it is not a protein fining agent. It is available as a practically odourless, flavourless, white or yellowish powder, consisting of fibre fragments that are visible under a microscope. Sodium alginate in water at pH between 6-8 produces a viscous solution. It is insoluble in alcohol and in most organic solvents. It can be precipitated by the addition of 20 per cent calcium chloride solution ($10^{-1}$). If the calcium chloride is replaced by 10 per cent dilute sulphuric acid, gelatinous matter also precipitates due to the formation of alginic acid. Alginic acid has a p*K* of 3.7 and is displaced from its salts by relatively strong acids and then precipitates at acid values ≤3.5 as it is insoluble in water. Tannins are not involved in the process. Generally, in treatment of wine, alginate with molecular weights between 80,000 and 190,000 is used. Flocculation is very fast if the wine has a high acidity (pH ≤3.5), but the deposit settles slowly as the particles are very light. The main advantage with alginates is that they make it possible to filter wines just a few hours after fining. Normal usage dose is in the range of 4-8 g/hl. (Cabello-Pasini *et al.*, 2005).

### *3.3.3. Tannins*

According to Enological Codex, tannin is whitish yellow or brown coloured with an astringent taste. It is soluble in water, but rarely soluble in ethanol. It is commercially produced from gall nut, chestnut, oak and grape pomace which contain a high amount of tannin. These preparations are mixtures of two mail gropus: procyanidin-based condensed tannins from grapes and ellagitannin and gallotannin-based hydrolysable tannins from oak and chestnut wood, or gall nuts. Gallotannins are used to prevent oxidation in must made from botrytised grapes. Seed tannins stabilise anthocyanins and wine colour during fermentation, deepen the colour of new wine by co-pigmentation, and facilitate ageing. Tannins also cause partial precipitation of excess protein matter and may be used to facilitate clarification of a new wine and in fining white wines. In reality, it is difficult to achieve total elimination, even with doses of tannins as high as 100-150 mg/l that noticeably harden the wine. However, adding tannin to white wines is controversial. Normal used dosage varies between 5-10 g/hl for red wines and approximately 5 g/hl for white wines (Zoecklein *et al.*, 1995).

### *3.3.4. Yeast Mannoproteins*

Yeast mannoproteins extracted by heating yeast or by enzymatic activity, such as α-1,3 glucanase and β-1,6 glucanase and by ultrafiltration, have been the objective of many studies (Charpentier, 2000; Caridi, 2006). The following functions of mannoproteins have been described: aroma enhancers (Dupin *et al.*, 2000; Chalier *et al.*, 2007); inhibitors of tartrate salt crystallisation (Moine-Ledoux and Dubourdieu, 1999); preventers of protein haze (Waters *et al.*, 2005). Moine- Ledoux and Dubourdieu (2007) showed that an N-glycosylated 31.8 kDa that corresponds to a partial invertase fragment of *S. cerevisiae* improved protein stability of wines. The use of mannoproteins as oenological adjuvants has been approved by the *Organisation Internationale de la Vigne et du Vin* in 2001 (resolution OENO 4/2001) but, under EU law, their use is currently permitted only for experiment trials (EU Regulation 1622/2000, Art. 41).

## 3.4. Enzymes

Enzymes in winemaking are considered as processing aids and, as such, are no longer active in the finished wines due to either inhibition by the wine components (tannins, for example) or removal during the process (for example, by bentonite fining). Enzymes in winemaking,used mainly in clarification through their use in flavour enhancement, is a recent phenomenon. Skin and pulp of grapes contain significant amounts of pectic compounds along with other constituents (Vidal *et al.*, 2003). During the production of white wines, a part of pectic compounds from the berry is released into the juice with grape crushing and pressing and forms a colloid that slows or prevents sedimentation of solid particles, particularly skin fragments. Therefore, elimination of solid particles is a key step in the production of good quality white wines. To improve the efficiency of the clarification process, commercial preparations of pectinases can be added. These preparations possess a range of pectinase activities that can efficiently hydrolyse the pectic substances present in the juice, allowing sedimentation of the suspended solid particles (Usseglio-Tomasset, 1978). Clarification coadjutants as pectinases can be used at various stages of the winemaking process. During the degradation of cell walls, pectinases improve the release of grape skin constituents. Particularly, during the production of white wine, pectinase can be added when pre-fermentative cold maceration is applied, in order to improve the extraction of aroma compounds and precursors located in the skins. In red winemaking, pectinase preparations, often in combination with cellulase and hemicellulase, are often used to increase the degradation of skin cell walls and obtain increased pressing yields and improved extraction of colour and aroma precursors during maceration.

Pectinases can also be used to prevent filter-clogging prior to bottling (Ducruet *et al.*, 2000; Gil and Valles, 2001).

### 3.5. Activated Carbon

Activated charcoal/carbons are potential adsorptive agents made from wood but they are non-specific. The sponge like carbon binds with weakly polar molecules, especially those containing benzene rings. Carbon effectively removes phenolic compounds, especially small phenolic compounds. Compounds larger than dimers are too large to be adsorbed. Stripping of wine is often a problem with carbon because of its low selectivity; so great care has to be taken with its use. Carbon also contains a large quantity of air and oxidation sometimes follows carbon addition if the carbon is not quickly and thoroughly removed. The addition of carbon to juice rather than wine helps to diminish carbon-induced oxidation. Normal dosage levels are in the range of 10-50 g/hL (Bowyer, 2008a).

### 3.6. Synthetic Polymers

#### *3.6.1. Polyvinyl Polypyrrolidone (PVPP)*

PVPP is a high-molecular weight fining agent made of cross-linked monomers of polyvinlypyrrolidone. It has been used since 1961 to stabilise beer and reduce concentrations of tannoid substances. Polymerisation of vinylpyrrolidone produces water-soluble polyvinylpyrrolidone (PVP), but in the presence of alkali the insoluble polyvinylpolypyrrolidone (PVPP) is produced by braking the pyrrolidone cycle that forms complexes with phenolic and polyphenolic components in wine by adsorption and attracts low-molecular-weight catechins. It removes bitter compounds and browning precursors in both red and white wines. PVPP is a quick-acting compound with no preparation required. Wines must however be filtered to remove the PVPP, but wines may appear more astringent when the bitter compounds are removed. PVPP is also utilised to correct discolouration in white wines to eliminate unwanted pigments either on its own or, more effectively, in combination with decolourising vegetable carbon. PVPP reduces the astringency and softens excessively tannic red wines and it also fixes the most reactive tannins (200-300 mg/l of tannin for 250 mg/l of PVPP), although it has less effect on anthocyanins. The normal doses are within the range of 20-30 g/hl. At this level, it prevents browning without producing any negative changes in the sensory quality of the wine (Laborde *et al.*, 2006; Bowyer, 2008b).

### 3.7. Kieselsol

Kieselsol is a common name for aqueous silicon dioxide, a by-product in the glass industry. Kieselsols are produced mainly in Germany. The primary use of kieselsol is for clarification and as a replacement for tannins during gelatine fining of white wines. Kieselsols are negatively charged, so they bind electrostatically and adsorb positively-charged proteins and further initiate flocculation and settling. Several different kieselsol formulations are available at a variety of pH levels. It is however necessary to use a kieselsol that is recommended for wine. In general, kieselsol is used at a rate that is seven times the amount of gelatine. Gelatine should be added first and further fining trials must be done to insure proper settling (Bowyer, 2008b).

## 4. Packages

Appropriate and adequate packaging is necessary for finished wine to reach the end consumers. Although packaging does not improve the quality, it preserves the quality of product and also protects against contamination. Glass bottles, plastic and bag inbox containers with appropriate closures are used for packaging alcoholic beverages (Eilert, 2005).

### 4.1. Glass Bottles

Generally glass containers of different shapes and sizes have been used for packaging wine and brandy since ages. Glass is an inert and transparent material; it seems to be the most favoured agent to be utilised as a packaging media, particularly for beverages. Thus bottles have been employed for this purpose.

Different shapes and sizes of glass containers are used for packaging in the form of ampoules, vials and containers. Though glass containers are considered to be very fragile, they are one of the toughest containers used for packaging of food and wine. Another most important characteristic of the glass containers is its top-to-bottom compression resistance (Marwaha *et al.*, 2011).

The process of preparation of glass into bottles of different shapes through glass blowing was invented by the Romans. In the early stages, bottled wine was not found attractive by the consumers. This may be because of unavailability of correct volume and labelling facilities. Then colours given to bottles attracted the consumers. The traditional Burgundy shape in 'dead-leaf' green continued to be very popular for US-grown Chardonnay and Pinot Noir during the 1980s. Most folks believe that the wine they drink should be in a glass with a stem and come from a bottle with a cork. And, to be fair, glass bottles do a great job of protecting wine, regardless of whether they are sealed with a cork or a screw cap or a synthetic stopper. For sparkling wines or wines that will be stashed away for a good while, glass is generally considered the best option, even though barrier properties of plastics have been improved (Marwaha *et al.*, 2011).

While glass bottles are the leader in most countries, there is a good bit of regional variability in terms of packaging preferences. In China, consumers are pretty stubbornly committed to the 'classic wine experience' that includes wine (often French, usually red) in a glass bottle and sealed with a cork. In Australia, much of the wine for the domestic market is sold in bags-in-boxes (or 'casks') and the wine that is bottled is usually under a screw cap. Meanwhile, 'Old World' countries like France and Italy take a more pragmatic approach, favouring traditional packaging for finer vintages and wines heading for export markets, but using plastic bottles and bags-in-boxes for everyday reds and whites at home. The US has been historically slow to accept new packaging formats, but is coming around to the idea of seeing wine in a box, or under a screw cap.

Shapes and sizes of wine bottles, varying from one area to another, mainly evolve from area tradition. We have 'classic' shapes (Fig. 6a) in general use by the majority of producers from any given area and 'modern' shapes or 'arty' variations of the classics. Every bottle has a neck where the bottle narrows and allows the cork sealing and flat or 'punted bottom' that allows standing the bottle on the floor (Fig. 6b) (Marwaha *et al.*, 2011). Different volumetric sizes of bottles are also utilised according to the requirements of oenological practices (Fig. 6c). The advantages and disadvantages of glass bottles are tabulated in Table 5.

## 4.2. Plastic Bottles

In recent years, an alternative wine-preservation technology has been developed, using plastic packaging materials and at present these are posing a tight competition for conventional glass bottles/containers.

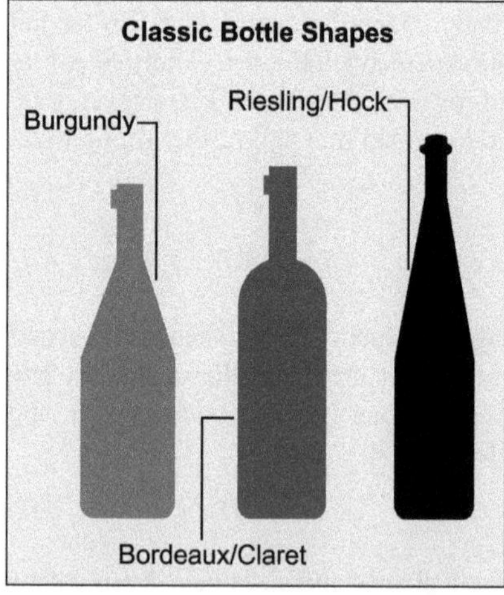


**Figure 6a.** Classic shapes of wine bottles
(*Source*: Marwaha *et al.*, 2011)

**Figure 6b.** Modern shapes of wine bottles
(*Source*: https://vinepair.com/wine-blog/why-wine-bottles-come-in-different-shapes)

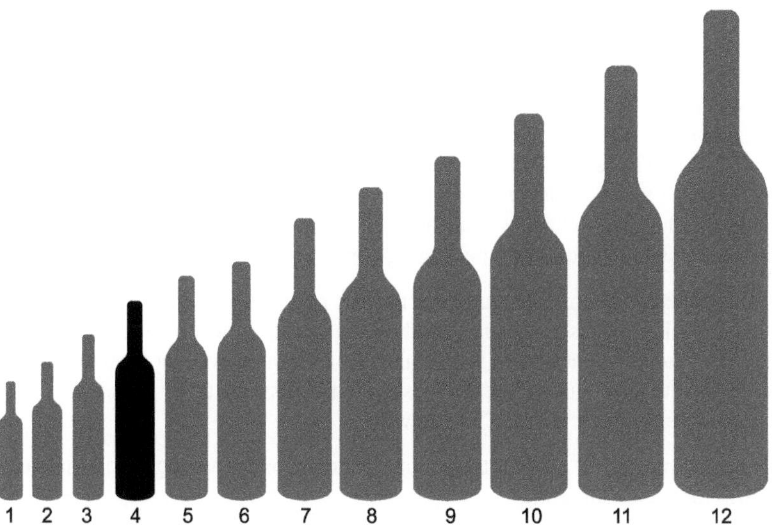


**Figure 6c.** Bottle sizes used in oenology
1. Piccolo – 187 ml (or 1/4 bottle), 2. Chopine – 250 ml (or 1/3 bottle), 3. Fillette (or Half Bottle, Demi, Split) - 375 ml (or 1/2 bottle), 4. Bouteille (or standard bottle) – 750 ml (or 1 bottle), 5. Magnum – 1.5 litres (or 2 bottles), 6. Jeroboam (or Double Magnum) – 3 litres (or 4 bottles), 7. Rehoboam – 4.5 litres (or 6 bottles), 8. Methuselah (or Imperial) – 6 litres (or 8 bottles), 9. Salmanazar – 9 litres (or 12 bottles), 10. Balthazar (or Belshazzar) – 12 литра (or 16 bottles), 11. Nebuchadnezzar – 15 litres (or 20 bottles), 12. Solomon (or Melchior) – 18 litres (or 24 bottles)
(*Source*: kata.bg/index.php/en/articles/6903-wine-bottle-shapes-sizes)

This is due to their cheaper prices, light weight, transparency, good strength, convenience in usage and rise of flexible packaging and finding more and more use in food packaging (Ghidossi *et al*., 2012). Various types of food grade plastics, such as Low Density Polyethylene (LDPE), Linear Low Density Polyethylene (LLDPE), High Density Polyethylene (HDPE), Polypropylene (PP), Polypropylene (PP), Biaxially Oriented Polypropylene (BOPP), Polypropylene (PP), Polyvinylidine Chloride (PVDC) and Polyester (PET) are available for wine packaging also (Osborn and Jenkins, 1992). PET has excellent gloss and clarity with high mechanical properties and good resistance to gases, water vapour and chemicals. PET has dimensional stability over a wide range of temperature. If metallised, it gives additional advantages (Jackson, 2014).

**Table 5.** Advantages and Disadvantages of Glass Bottles

| *S. No.* | *Advantages* | *Disadvantages* |
|---|---|---|
| 1. | Glass bottles are recyclable and it is actually more economical for wineries to use virgin glass because of the cost required to transport used or recycled glass over long distances. | Cost of production and emission of green house gasses are very high. |
| 2. | Glass bottles protect the quality of the wine by reducing oxygen permutation through the container. | Heavy in weight, high breakability and less flexibility for design. |
| 3. | Wine that has the potential to be aged longer than a year should be bottled in glass because of its superior ability to prevent deterioration due to oxygen. | High transport cost and a glass wine bottle will produce 1.8 times more $CO_2$ gas then a PET bottle travelling the same distance. |

### 4.3. Bag-in-Box

Furthermore, since bag-in-box has emerged, the wine industry has shown considerable interest in and found numerous applications for this technology. BIB consists of a resistant bladder (or plastic bag), usually comprising several layers in the same way as PET multi-layered bottles. For protection, the bag is housed in a sturdy cardboard box and a valve fitment is attached to the bag through which the product is filled and dispensed (Rapp, 2005). BIB packaging improves distribution efficiency, enhances end-use convenience and increases cost-effectiveness (Fu, Lim and Mc Nicholas, 2009). According to Revi *et al.* (2014), this type of packaging is widely used for medium-quality table wines. Due its large volume capacity, wines packaged in BIB are usually consumed over an extended time after the package is opened, during which its secondary shelf-life can be affected by oxygen ingress through the dispensing fitment (Fu *et al.*, 2009; Lee *et al.*, 2011).

Ghidossi *et al.* (2012) have studied the quality decay of white and red wines as affected by packaging material (glass bottles, PET multi-layer bottles, PET mono-layer bottles, and bag-in-box) and further examined chemical and sensory changes related to the phenomenon of oxidation. Below are given the significant differences due to the different packaging configurations:

- White wine was noticeably affected after six months of conservation in polyethylene terephthalate bottles. So, packaging has a significant impact on the quality of white wine during storage.
- In polyethylene terephthalate bottles, containing an oxygen scavenger as well as in a bag-in-box container, the results were much closer to those noted in glass bottles.
- No significant differences were noted for red wine.

The above results need to be taken into account when choosing packaging of wine as a function of the mode of distribution and marketing.

## 5. Closures

Wine closure is a protective seal of wine bottle and is used to inhibit the extensive contact with oxygen, which causes oxidation of the wine (Robinson, 2006). Since wine has been manufactured, various types of closure and bottle styles have been utilised. Amphoras were used by ancient Romans and Greeks to inhibit the oxidation of wine completely (Robinson, 2006). However, with the introduction of glass bottles in the 17th century, a sealing system was required for wine storage. Cork became the primary closure type due to the fact that other materials were not capable of making an airtight seal to maintain the wine from turning into vinegar (Phillips, 2000; Robinson, 2006). Other methods used during that time period included glass stoppers and decanters, both of which are still applied in the industry today (Robinson, 2006). Development of wine bottles and corks for packaging of wine led to the development of sparkling wines and ports, which until the 17$^{th}$ century had never been seen before (Phillips, 2000).

### 5.1. Cork

Cork stoppers are used for wine bottles because of their impermeability to liquids and air (preventing oxidation of wine), compressibility, resilience and chemical inertness (Snakkers *et al.*, 2000; David *et* al., 2003). In many countries, for several centuries, cork had proved to be the most effective closure for wine, protecting its qualities and allowing it to develop and improve over time. The technology of stopping wine bottles with clean, unsealed cork was perfected by Benedictine monks in the seventh century (Gil and Cortic, 1998). Then, in 1680, the first use of cork to seal champagne by Don Pierre Pérignon started a revolution in wine bottling (Borges and Cunha, 1985; Gil and Cortic, 1998; Rosa *et al.*, 2002). Wine consumers used to enjoy from the step of cork removal during drinking. Different types of cork stoppers (natural, colmated, agglomerated) can be distinguished. Cork is the bark of the oak (*Quercus suber* L.) which is harvested usually every 9-12 years from the tree (Fig. 7). Natural cork stoppers are punched direct from the best quality cork bark. Portugal is the major producer of cork throughout the world and the manufacturing of cork stoppers is the major economy resource for the Portuguese (Goode, 2010). The impermeability of cork to liquids and gases, derived from the fact that its closed cell walls are made up mainly of suberin, and its high compressibility and flexibility make it ideal for sealing bottles. Cork is recommended for bottles of reserve wines and wines that need to age in the bottle.

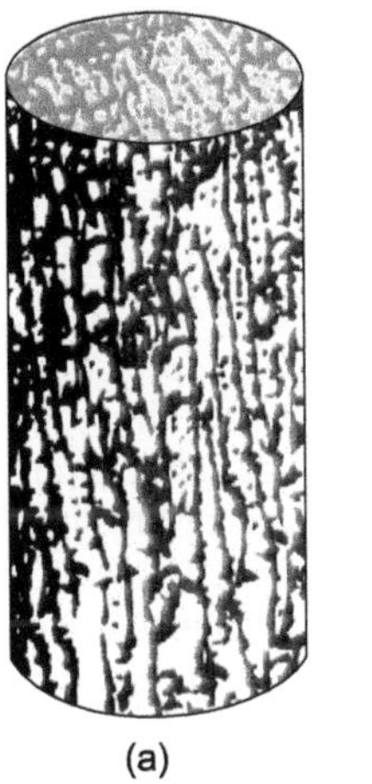
(a)

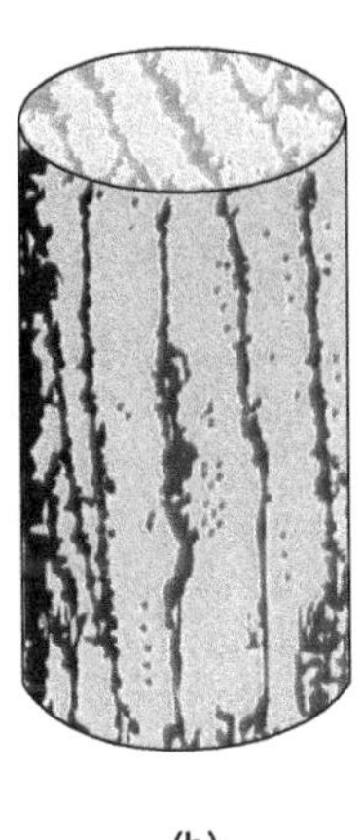
(b)

**Figure 7.** Representative diagrams of (a) high-quality corks, derived from slow-growing cork oak and (b) lower-quality cork, derived from rapid-growth cork oak. Note the large number of growth rings ($\geq 9$) in the cork on the left compared with the one on the right ($\leq 7$)

Cork is not an inert biological material; it contains some 'free' components, not chemically linked to the main structure and thus easily extractable with solvents. Some of these components are responsible for the organoleptic properties of wine (Insa *et al.*, 1992). The possible contribution to the aromatic character of wine was analysed and thereby extracted more than 70 volatile compounds from cork by Boidron *et al.* (1994). They concluded that a healthy cork will transfer favourable complex aromatic compounds to the wine aroma. On the other hand, musty flavours caused by the cork are called cork taint. The occurrence of taint in bottled wine is responsible for economical losses in the wine industry because of unpleasant alterations in wine flavour or aroma. The true cork-taint (a musty/mouldy aroma) is rare and easily recognised by wine specialists and reported to be affecting 0.5-6 per cent of total bottled wines (Lee and Simpson, 1993; Rebereau-Gayon *et al.*, 1998). The most common causes of musty taint in wine are generally by cork closure or contaminated oak products; 2,4,6-trichloroanisole (TCA) or 2,4,6-tribromoanisole (TBA) are the major chemical compounds that are present in cork taint. TCA is a compound which does not occur naturally. It is created when some fungi are treated with chlorinated phenolic compounds, which are a type of antimicrobial agent used in the processing of wood (Silva *et al.*, 2000). Very low quantities of this compound (nano grams) can cause off-flavour in the wine.

## 5.2. Special and Alternative Closures

Fifty years ago, all wine bottles were sealed with natural cork. There was no other closure discussion; nobody even knew about the musty taint ('corked' wine) issue, but seemed to tolerate it simply because there was no other practical way to seal wine bottles. There was no pressure to do a better job in the cork production industry because they had no competitors. In the 1980s, the problem of cork taint became serious due to the rising levels of cork taint. The cork taint increased in the 1980s and 1990s due to two key reasons: first, the Portuguese revolution in 1974 led to common ownership instead of traditional management for maximum gain. Various pesticides were introduced into the forests and cork was harvested before it should have been. The damage led to a drop in the quality of the raw materials used for cork production and the chemicals employed in the forests may have encouraged the production of taint compounds in the bark. Second, the growth of demand for cork stoppers in the 1980s led to a large increase in the popularity of bottled wines (Silva *et al.*, 2005; Reynolds, 2010; Goode, 2010). This increased demand led to a drop in quality. Another factor is advancement in wine sensory analysis for better spotting of cork taint. The consumer awareness on cork taint led to the search for alternatives to cork and to look for ways of eradicating 2,4,6-trichloroanisole (TCA) (Silva *et al.*, 2000).

### *5.2.1. Synthetic Closures*

Synthetic corks were the most commonly used alternative to natural cork. Metal Box, UK developed synthetic closure for wine in the early 1970s, but it was not promoted and did not popularise. The first synthetic cork that made a big impact was SupremeCorq. The first commercial packaging was started in 1995 with a 1994 Safeway Semillon Chardonnay, using a yellow Supreme Corq later on. Synthetic closures became popular in UK and Australian domestic wine (Goode, 2010).

Synthetic corks have some advantages when compared with other alternative closures, like corks – these are in-neck closure and robust synthetics have consistent performance unlike natural cork and do not need new equipment to use them, nor do they need different bottles. These can be made in various colours and are cost-effective – one of the cheapest ways of sealing a bottle of wine (Fig. 8a). There are two different types of synthetic cork – the first type is injection moulded, as the name suggests and are produced using moulds which are then injected with a plastic copolymer; the second type is extruded synthetics and these closures come in two types, with one made by extruding a fine honeycomb plastic core, which is then coated by a smooth cylindrical layer and chopped up into cork-sized pieces (Nomacorc and Neocork) and second one is produced in a different single-extrusion process, where heat is used to produce a smooth outside to the core of the closure (NuKorc) (Robinson, 2006). Extruded closures thus have a rather spongy looking core and a smooth outside. Both types have a silicone- or paraffin-based material coating applied to them and it is crucial for closure performance. Even though synthetics have some of the above-said advantages, they fail on all the four objectives (a) to seal bottles tightly enough to prevent oxidation of the wine while still being (b) easy enough to extract, (c) easy enough to take off the corkscrew and (d) easy enough to reinsert into the bottle, should the contents not be drunk in a single session.

### *5.2.2. Screw Caps*

Screw caps, also known as roll-on-tamper-evident (ROTE), were developed by a French company, La Bouchage Mecanique in 1959, as an alternative to cork for bottling wine and introduced to the market under the trade name of Stelcap-vin as an alternative by Australian Consolidated Industries Ltd (ACI) in 1970 (Goode, 2010). Screw caps with different wadding materials and a cork were tested on six wines (three white and three red). The results suggested that screw caps with right wadding material were a satisfactory seal for wine bottles (Eric *et al.*, 1976; Madigan, 2004). The major problem with screw caps was that these were seen by consumers as fit for cheap wines only. Figure 8b shows a screw-cap image. To seal, screw caps contain liners opposite to the rim or lip of the bottle. There are two different types of liners in usage. The first one is composed of polyethylene wadding with an inert Saranex layer, while the second one has a thin tin layer (sandwiched between polyethylene and a thin PVDC outer skin) that acts as an oxygen barrier. In Europe, and particularly Switzerland, mainly the Saranex-only liner is used, which has a higher rate of oxygen transmission while in Australia and New Zealand, the second type is used, which has extremely low oxygen transmission. It is very important to remember that not all screw caps are equal (Goode, 2007, 2010).

Screw caps seem to have good longevity and are 'guaranteed' to last at least 10 years, if the bottle and application machinery are correct. In practice screw caps last 25-30 years without a problem. Recent estimates show that screw caps now have over 10 epr cent of the bottled wine market. This total market is estimated to be some 18.3 billion annually worldwide. Of this, screw caps are now around 2.5 billion with synthetic corks around 3.7 billion and natural cork accounting for most of the remainder. This is a dramatic rise from the situation five years ago, when screw caps were estimated to be around 200 million worldwide (Sue Courtney, 2004).

### *5.2.3. Diam Closures*

Diam closures have both natural cork and alternative closure characters with an in-neck closure with both cork and synthetic components. DIAM is a technological wine-bottle closure that addresses two major threats to the wine industry: closure consistency and cork taint. This was first made by Oeneo Bouchage with small granules of cork that have been cleaned with supercritical $CO_2$, combined with synthetic microspheres (Penn, 2007a) (Fig. 8c). The manufacturing process of Diam closures involves four different steps (Fig. 9). The prepared closures are consistent and in theory should have lower risk of TCA taint than normal corks because the lignin-rich material that surrounds the lenticels is considered to harbour the major portion of TCA contamination. Initially, these closures were tremendously successful because they were initially marketed as being taint-free. Buta study conducted by the AWRI reported that each of the bottles sealed with Diam closures contained a TCA-like aroma with detectable concentrations of TCA. Diam is still a relatively new closure; so it is hard to give solid assurances about its stability and performance over time. However, according to the research reports, it will be good for 60 months but will not be suitable for wines destined for longer ageing.

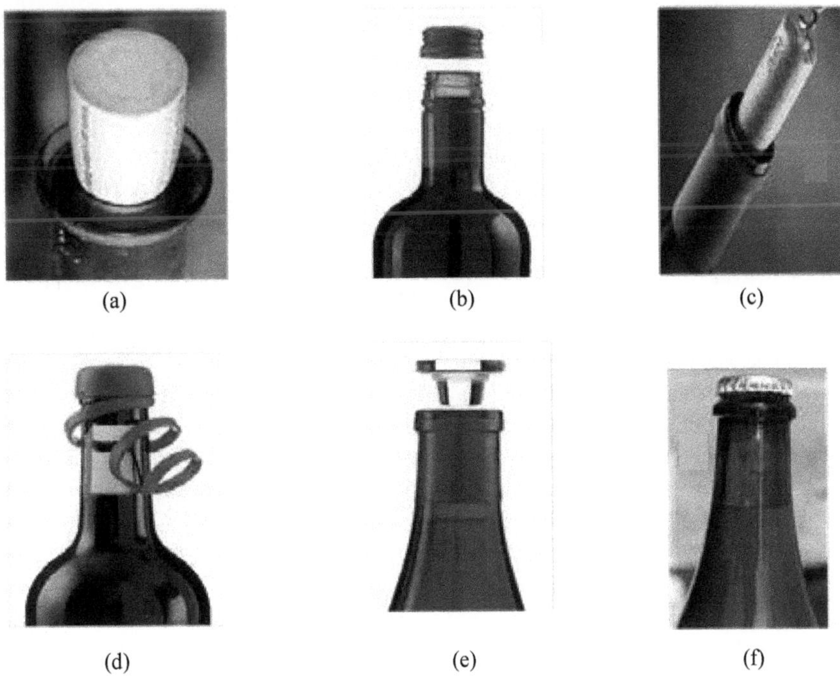


**Figure 8.** Alternative corks: a. synthetic corks, b. screw caps, c. diam closures, d. zork closures, e. vino-Lok closures, f. crown cap closure
(*Source*: https://en.wikipedia.org/wiki/Alternative_wine_closure)

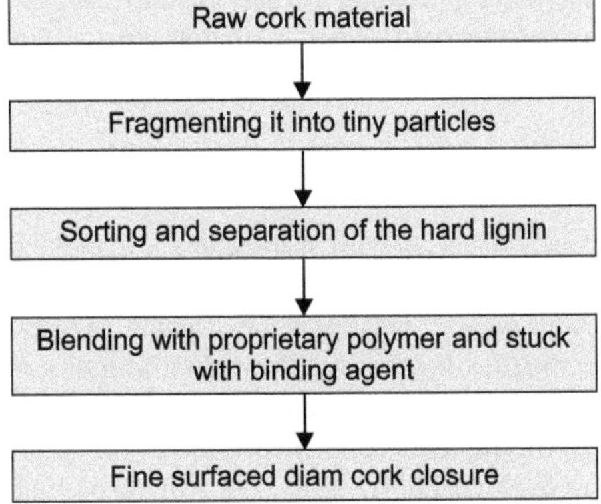


**Figure 9.** Schematic flow diagram of steps involves in Diam cork closure

### *5.2.4. Zork Closures*

Zork is perhaps the best choice among alternative closures and consists of three main bits – the first is the visible bit, the cap, which provides a tamper-evident clamp that locks on to the band that surrounds the outside rim of most standard wine bottles; the second is the foil, which provides an oxygen barrier similar to a screw cap, and is found inside the cap made of aluminium; the third component is the internal plunger, which creates a 'pop' on extraction and reseals after use and is made of polyethylene (Robinson, 2006; Penn, 2007b) (Fig. 8d). In studies, initially, the concentrations of free and total $SO_2$ decreased more rapidly under Zork than under screw cap, but this was thought to be because of the oxygen present in the plunger (part of the Zork closure that goes into the neck of the bottle). The shelf-life of wines sealed with Zork was calculated to be 5.1 years. Thus Zork is fine for most commercial applications, but is not really recommended for wines that will be aged for more than five years. Zork is also quite expensive as a closure, but its striking appearance could be incorporated into the brand design (Robinson, 2006; Penn, 2007).

### *5.2.5. Vino-Lok Closures*

A German company, Alcoa came up with the new idea (Vino-Lok®) to prepare wine closures in the early 20th century. Alcoa's Vino-Lok® is modern enough but looks fair as chemistry-lab bottles sealed with glass stoppers. However, the difference with Vino-Lok is that the seal is not one of ground glass, butis actually formed by a plastic 'O' ring (made of ethylene vinyl acetate) that acts as an interface between the bottle rim and stopper. It is quite a complicated closure solution because the glass stopper is held in place by an aluminium cap, which has to be removed before the stopper can be extracted and a special bottle neck is needed to accommodate this closure. It is also expensive and could prove the biggest barrier to its widespread acceptance (Gardner, 2008). Figure 8e illustrates the appearance of Vino-Loks. AERI research revealed OTR data of 0.0026-0.0031ml/o2/bottle/day, which means that wines with Vino-Lok never show reductive aromas even after several years of aging. The slow impact of oxygen perfectly holds the aroma and terrior expression. Vino-Loks are already popular in the German and Austrian markets. In the USA, there is a variant of Vino-Lok in the market and is called VinTegra™. Vino-Lok is widely popular among wine lovers because of the easy opening and closing of the bottle; no corkscrew is needed. Also, the bottle can be easily re-closed (Glories, 2006).

### *5.2.6. Crown Cap Closures*

The crown cap is widely used for beer and is also used by sparkling wine producers to store wines before the dosage stage when the bottles are finally sealed with corks (Fig. 8f), but it failed to win many converts as an alternative closure for wine. The seal of a crown cap is dependent on the lining material. Crown caps are currently being used to seal the high-end sparkling wines from Domaine Chandon in Australia's Yarra Valley. Crown caps could have quite a future as alternative closures for sparkling wines, given that it has not been easy to adapt screw caps for fizz, but consumer acceptance is currently a problem. The crown cap is affordable and highly functional, providing the best handling, highest speed of application and true tamper resistance. Caps can be decorated with logos or other designs, offering another tool for advertising the distinct identity of a brand (Capone *et al.*, 2003).

## 6. Cartons for Packing

A cartoning system combines a special carton with the machinery to make it erect from a flat condition, fill it with a product and close it. The machinery varies from simple hand-fed to automatic stations coupled with means for packing the cartons directly into cases for dispatch. Whatever the system employed, three main operations are performed: (1) forming the container, (2) filling the container and (3) closing/sealing (Paine and Paine, 1992). When designing boxes, there are several important factors to consider – such as customer-related factors, design of graphics, convenience for carrying and storing, dimensions to allow easy handling and to efficiently cover a pallet when stacked with no overlap at the edge, board quality and direction of flute to give maximum stacking strength, type of finish on the board if the surface is sealed efficiently as glue will not adhere correctly and bag-to-box ratio, e.g. for four-litre casks, a head space of about 10 per cent is recommended (Rocchi and Stefani, 2006). The cask has to be free of bulge and warp, yet resist the bulge on the filling line, fold crisply along score lines, fit within the outer box on high speed palletising lines and the performance around the spigot hole must open cleanly but burst open when travelling the filling line (Orth and Malkewitz, 2008).

It's the packaging that does this and while packaging does not stir the kind of emotion that the wine inside may, many people are very opinionated about how the wines they drink are packaged. In most countries, the packaging (whether it's a $5 or $5,000 wine) is going to be a glass bottle, even though any number of containers can be used. Cans, pouches, aseptic cartons (tetrapaks), plastic bottles and bag-in-box containers can all keep most wines safely (Fig. 10). Most wine is made to be drunk young, within two years of when it's produced; and those younger wines can be packaged perfectly well in a can or a bag. They tend to cost more than other packaging options and the costs tend to matter to just about everyone. Also, glass bottles are heavy and expensive to ship, plus they can break if you're not nice to them. In contrast, empty boxes and cartons are lighter, do not break and can be shipped when flat. This is not to say that glass bottles are without green credentials, though they're recyclable; and producers have also introduced lighter weight bottles (Paine and Paine, 1992; Buiatti *et al.*, 1997; Capitello *et al.*, 2012).

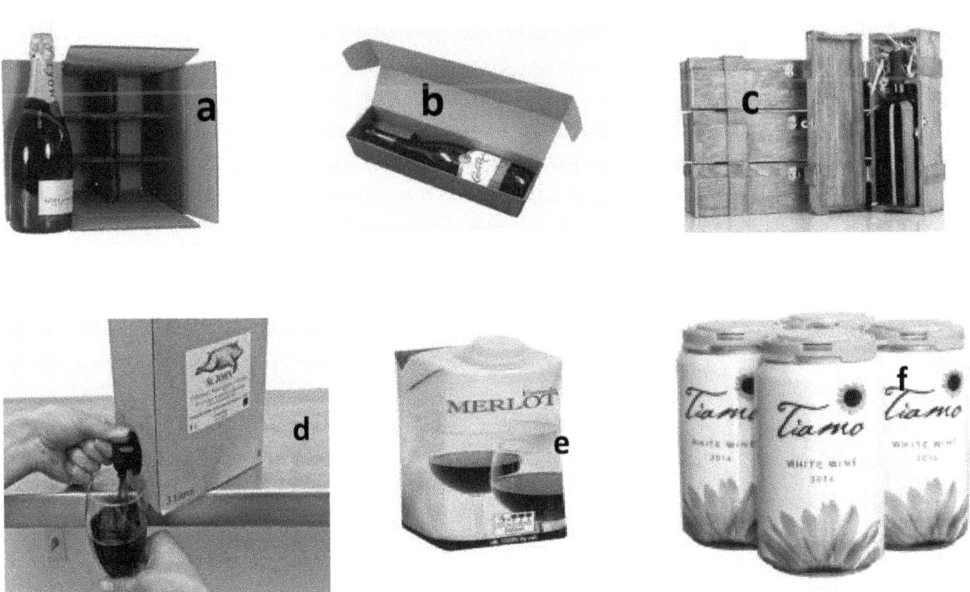


**Figure 10.** Cartons shapes and sizes used in oenology: a. standard carton for multiple bottles, b. single bottle carton, c. wooden carton, d. box-in-bag, e. tetrapak, f. tin bottles (*Source*: www.vinkempackaging.com.au/index.php/cartons)

## 7. Wine Labels

Label is something which can convey information and self-importance in the consumer's cellar. It also gives details of products, like origin, contents and quality. It is vital that the labels are not only different from what the competition is doing, but also express the quality of the wine itself. Label designs can be categorised into three types: traditional, contemporary and novelty. Traditional designs typically include images of coats-of-arms, chateaux and vineyards (Caldeway and House, 2000; Batt and Dean, 2000). Contemporary designs provide less information about the wine's origin and more of a sense of style and no longer show images of the vineyard or stately visuals that represent the winemaker (i.e. a family crest). With the contemporary design, labels freely explore and espouse art (for example, the brands Yellow Tail and Luna di Luna). The third type of novelty labels tend to feature images of caricaturised animals (e.g. frogs, roosters) and the novelty labelling promotes the label as fun. These labels use humour to capture the heart of the consumer. The popular novelty label design brands include 3 Blind Moose, Old Fart, and Arrogant Frog (Finkelstein and Quiazon, 2007). The descriptions of wine on labels influence the consumers far more than anybody thought. Emotive descriptions will elicit more positive emotions than those with simple descriptors and help in convincing consumers to pay more for a bottle of wine, as well as increase their appreciation of it. Barber *et al.* (2006) conducted research on the importance of well-written and accurate wine descriptions and suggested that that the information on the label is likely to influence the consumer's wine-drinking experience and behaviour. Consumers seek out wine labels when choosing wine by going through the retail aisles of wine and reading labels rather than seeking guidance from other information sources (Olsen *et al.*, 2003; Barber and Almanza, 2006).

Functionality of wine-labelling is equally important (Boulton *et al.*, 1996; Mueller and Lockshin, 2008). Wine labels cannot slip off from wet bottles; white labels must stay white even after a bath in an ice bucket. We have the right label material to express the unique brand story – from textured, natural wine label paper that evokes an earth-friendly approach to sleek, foil custom liquor labels that speak to style-minded consumers. Another important part of the label rationale is accoutrements, such as tags, stickers and other supplemental attention-getters (Sherman and Tuten, 2011). It is common to find vintners taking advantage of medals by significantly increasing the price for medal-winning wines. Most importantly, the price increase generates additional profits, which might be needed to lower the prices to make non-medal winners more attractive. Figure 11 illustrates some well-designed labels which have a successful market.

## 7.1. Label Shapes

At present labels are in many shapes, ranging from the conventional square or rectangle to odd shapes. But the traditional shapes (square or rectangle) are in common use and also economic when compared with modern labels (Fig. 11a). The labelling problems and increased labelling cost due to design and printing of odd shapes generally require customised dies for cutting. Non-traditional shapes cause greater paper waste. Labelling machines usually require expensive custom parts, so fabricated as to handle custom shapes (Boudreaux and Palmer, 2007; Marwaha *et al.*, 2011; Sherman and Tuten; 2011).

**Figure 11.** Wine label images, a. different colours and shapes, b. intelligent labels (RFID) (*Source*: https://www.google.co.in/search?q=wine+bottle+labels&dcr=0&tbm=isch&tbo=u&source =univ&sa=X&ved=0ahUKEwjZwY3co_nWAhUNSI8KHQvDDowQsAQITg

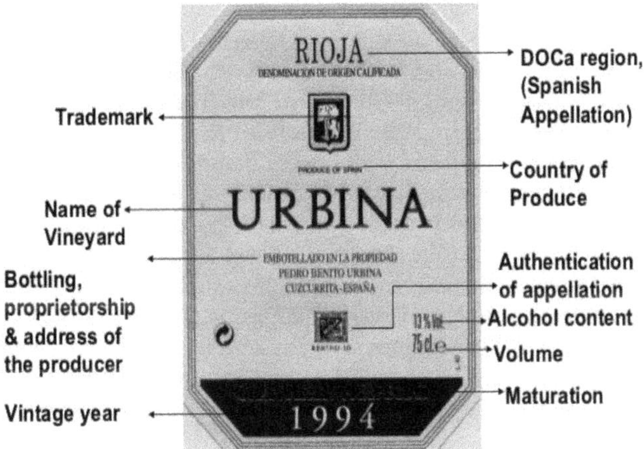


**Figure 12.** The minimum information required on label
(*Source*: https://www.google.co.in/search?q=wine+bottle+labels&dcr=0&tbm=isch&tbo=u&source =univ&sa=X&ved=0ahUKEwjZwY3co_nWAhUNSI8KHQvDDowQsAQITg&biw=1280&bih=651)

### 7.2. Label Colours

Attractive label colours are very important to attract the consumers. This is a bit expensive but absolutely essential in order to properly reach each wine to its target market. Colour psychology seems to differ with each consumer – some colours self-explain the quality of product (Marwaha *et al.*, 2011). Obviously, wine labels should be totally representative of the product and the producer, but there have been many wine failures because of overdone labels. Vintners must understand that no matter what the quality of their wine is, it is the package which creates the desire for a prospective consumer to try it. There are gold-medal-winning wines which go abegging because of inept packaging; conversely, there are mediocre wines which sell actively because of clever packaging (Vine, 1999). There can be no point in arguing that colour and its integration in wine packaging should be placed in the hands of professionals (Sherman and Tuten, 2011).

#### *7.2.1. Intelligent Labels or 'Smart Labels'*

Intelligent labels or 'smart labels' are inlaid with Radio Frequency Identification (RFID) technology and are rapidly improving many aspects of how we live – right from the clothes we wear and the food we eat to our cars, medicine and air travel (Fig. 11b). Intelligent label solutions include UHF, HF and NFC solutions to boost efficiency, enhance visibility and increase margins, while providing customers with greater convenience, reliability and safety. The minimum information required for bottles to be sold in the US (whether foreign or domestic) includes five categories (Fig. 12).

## 8. Future Perspectives

Once the fermentation is completed, wines, before bottling, undergo a process called maturation for some period which is also termed as 'adjustment'. Maturation involves the deposition and/or removal of particulate and colloidal material. During this period, the wine undergoes a variety of physical, chemical and biological changes that usually sustain or develop its sensory virtues. The number of adjustments is carried out through addition/deletion of some compounds for correcting the deficiencies present in the grapes and/or sensory imbalances that developed during fermentation. The adjustment of wine was done with the help of additives and adjuvants. Over-usage/intervention of these additives and adjuvants can disrupt the wine's inherent attributes, rendering any intervention deleterious. Hence, judicious adjustment after vinification should be made. Present-day wine consumers are switching to organic wine, without any adjustments. Winemakers have to think of producing virgin wines with better quality.

Cork still remains the bottle closure of choice for the majority of producers. From the current literature, the picture that emerges is that there is no such thing as a perfect closure. Winemakers are faced with many options, but still too little is known about the precise details of post-bottling wine chemistry to give definitive advice on closure choice based on firm data. The screw-cap tin layer of the liner will provide specified levels of oxygen transmission by means of micro perforations. In addition, glass-on-glass closures for high-end red wine and specified oxygen transmission levels are engineered by means of sintering the glass. Such means of closure would be expensive, but for wine retailing at $300, the cost of closure is not a huge issue. But what is needed are more data on exactly what oxygen transmission levels we would like from the ideal closure. The hope is that future studies may assist winemakers in obtaining more detailed information on the role of closure in post-bottling wine development. The choice of suitable packaging involves a number of considerations. A proper design is the prime means of providing 'a package which protects what it sells and sells what it protects'. Moreover, some compounds associated with freshness and fruitiness of wines are present in higher levels in wines sealed with natural cork stoppers, meaning that the sensory quality of these wines is much more preserved than in wines packed in Bag-in-Box.

## References

Allison, C. and Wilson, L. (2015). Post-Fermentation Clarification: Wine Fining Process. Project report submitted to Worcester Polytechnic Institute.

Anonymous (2004). Taint risk reduced, workshop told. *Wine Industry Journal* 19(5): 80-82.

Arriagada-Carrazana, J.P., Sáez-Navarrete, C. and Bordeu, E. (2005). Membrane filtration effects on aromatic and phenolic quality of Cabernet Sauvignon wines. *Journal of Food Engineering* 68: 363-368.

Barber, N., Almanza, B.A. and Donovan, J. (2006). Motivational factors of gender, income and age on selecting a bottle of wine. *International Journal of Wine Business Research* 18(3): 218-232.

Barber, N. and Almanza, B. (2006). Influence of wine packaging on consumers' decision to purchase. *Journal of Foodservice Business Research* 9(4): 83-98.

Batt, P.J. and Dean, A. (2000). Factors influencing the consumers' decision to purchase wine from a retail liquor store in Perth, Western Australia. *Australian and New Zealand Wine Industry Journal* 15(4): 34-41.

Beech, F.W., Burroughs, L.F., Timberlak, C.F. and Whiting, G.C. (1979). Progrès récents sur l'aspect chimique et l'action anti-microbienne de l'anhydride sulfureux ($SO_2$). *Bull. OIV* 52: 1001-1022.

Beltrán, G., Esteve-Zarzoso, B., Rozés, N., Mas, A. and Guillamón, J.M. (2005). Influence of timing of nitrogen additions during synthetic grape must fermentations on fermentation kinetics and nitrogen consumption. *Journal of Agriculture Food Chemistry* 53: 996-1002.

Bisson, L. (1999). Stuck and sluggish fermentations. *American Journal of Enology and Viticulture* 50: 107-119.

Blade, W.H. and Boulton, R.B. (1988). Adsorption of protein by bentonite in a model wine solution. *American Journal of Enology and Viticulture* 39: 193-199.

Boidron, J.N. (1994). Preparation and maintenance of barrels. pp. 71-82. *In*: The Barrel and the Wine: Scientific Advances of a Traditional Art. Sequin Moreau, USA, Napa, CA.

Boudreaux, C.A. and Palmer, S. (2007). A charming little Cabernet: Effects of wine label design on purchase intent and brand personality. *International Journal of Wine Business Research* 19(3): 170-186.

Boulton, R.B. (2005). The physics of wine bottle closures. *In*: The Science of Closures Seminar. American Society for Enology and Viticulture. 56th Annual Meeting, Seattle, WA.

Boulton, R.B., Singleton, V.L., Bisson, L.F. and Kunkee, R.E. (Eds.) (1996). Principles and Practices of Winemaking. NY: Chapman, New York,.

Bowyer, P. (2008). Part 1: Proteinaceous fining agents. *The Australian and New Zealand Grape Grower and Winemaker* June: 55-63.

Bowyer, P. (2008). Part 2: Non-proteinaceous fining agents. *The Australian and New Zealand Grapegrower and Winemaker* July: 65-71.

Brajkovich, M., Tibbits, N., Peron, G., Lund, C.M., Dykes, S.I. and Kilmaretin, P.A. (2005). Effect of screw cap and cork closures on $SO_2$ levels and aromas in a Sauvignon Blanc wine. *Journal of Agriculture Food Chemistry* 53: 10006-10011.

Brissonnet, F. and Maujean, A. (1993). Characterisation of foaming proteins in a Champagne-based wine. *American Journal of Enology and Viticulture* 44: 297-307.

Buiatti, S., Celotti, E., Ferrarini, R. and Zironi, R. (1997). Wine packaging for market in containers other than glass. *Journal of Agricultural and Food Chemistry* 45(6): 2081-2084.

Silva Pereire, C., Figueiredo Marques, J.J. and San Romao, M.V. (2000). Cork taint in wine: Scientific knowledge and public perception – A critical review. *Critical Reviews in Microbiology* 26(3): 147-162.

Calderon, P., Van Buren, J.P. and Robinson, W.B. (1968). Factors influencing the formation of precipitates and hazes by gelatine and condensed hydrolysable tannins. *Journal of Agricultural and Food Chemistry* 16: 479-482.

Caldewey, Jeffrey (2003). Icon: Art of the Wine Label. Wine Appreciation Guild Limited.

Casey, J. (2002). A commentary on the AWRI closure report. *Australian and New Zealand Grape Grower and Winemaker* 457: 65-69.

Capitello, R., Begalli, D. and Agnoli L. (2012). Is cellar door an opportunity for bag-in-box? A consumer preferences analysis in the Italian wine market. 5th Annual EuroMed Conference of the EuroMed Academy of Business, October 4th-5th, 2012. Glion-Montreux, Switzerland.

Capone, D., Sefton, M., Pretorius, I., Høj, P. (2003). Flavour 'scalping' by wine bottle closures – The 'winemaking' continues post-vineyard and winery. *Australian and New Zealand Wine Industry Journal* 18(5): 18-20.

Caridi, A. (2006). Enological functions of parietal yeast mannoproteins. *Antonie van Leeuwenhoek* 89: 417-422.

Charpentier, C., Nguyen Van Long, T., Bonaly, R. and Feuillat, M. (1986). Alteration of cell wall structure in *Saccharomyces cerevisiae* and *Saccharomyces bayanus* during autolysis. *Applied Microbiology and Biotechnology* 24: 405-413.

Cilindre, C., Jégou, S., Hovasse, A., Castro, A.J., Schaeffer, C., Clément, C., Van Dorsselaer, A., Jeandet, P. and Marchal, R. (2008). Proteomic approach to identify Champagne wine proteins as modified by *Botrytis cinerea* infection. *Journal of Proteome Research* 7: 1199-1208.

Daeschel, M.A., Musafi ja-Jeknic, T., Wu, Y., Bizzarri, D. and Villa, A. (2002). High-performance liquid chromatography analysis of lysozyme in wine. *American Journal of Enology and Viticulture* 53: 154-157.

Dambrouck, T., Marchal, R., Cilindre, C., Parmentier, M. and Jeandet, P. (2005). Determination of the grape invertase content (using PTA-ELISA) following various fining treatments *vs* changes in the total protein content of wine: Relationships with wine foamability. *Journal of Agricultural and Food Chemistry* 53: 8782-8789.

Danilewicz, J.C., Seccombe, J.T. and Whelan, J. (2008). Mechanism of interaction of polyphenols, oxygen and sulphur dioxide in model wine and wine. *Journal of Agricultural and Food Chemistry* 59: 128-136.

David, J.M., Santos, F.A., Guedes, M.L.S., David, J.P. (2003). Flavonóides e triterpenos de Stigmaphyllom paralias. *Quim Nova* 26: 484-487.

De Rosa, T., Margheri, G., Moret, I., Scarponi, G. and Versini, G. (1983). Sorbic acid as a preservative in sparkling wine: Its efficacy and adverse flavour effect associated with ethyl sorbate formation. *American Journal of Enology and Viticulture* 34: 98-102.

Delfini, C.M., Cersosima, V., Del Prete, M., Strano, G., Gaetano, A., Pagliara and Ambrò, S. (2004). Resistance screening essay of wine lactic acid bacteria on lysozyme: Efficacy of lysozyme in unclarified grape musts. *Journal of Agricultural and Food Chemistry* 52: 1861-1866.

Denise Gardner (2008). Innovative Packaging for the Wine Industry: A Look at Wine Closures. Virginia Tech Food Science and Technology. Duck Pond, Dr Blacksburg, VA 24061, 2008.

Du Toit, W.J., Pretorius, I.S. and Lonvaud-Funel, A. (2005). The effect of sulphur dioxide and oxygen on the viability and culturability of a strain of *Acetobacter pasteurianus* and a strain of *Brettanomyces bruxellensis* isolated from wine. *Journal of Applied Microbiology* 98: 862-871.

Dubourdieu, D., Serrano, M., Vannier, A.C. and Ribereau-Gayon, P. (1988). Etude compaŕee destests de stabilité protéique. *Conn. Vigne Vin* 22: 261-273.

Ducruet, J., Dong, A.N., Rose Marie Canal, L. and Glories, Y. (1997). Influence des enzymes pectolytiques séléctionnés pour l'œnologie sur la qualité et la composition des vins rouges. *Revue Francaise d' Onologie* 5: 1-16.

Dupin, I.V.S., Stockdale, V.J., Williams, P.J., Jones, G.P., Markides, A.J. and Waters, E.J. (2000). *Saccharomyces cerevisiae* mannoproteins that protect wine from protein haze: Evaluation of extraction methods and immunolocalisation. *Journal of Agricultural and Food Chemistry* 48: 1086-1095.

Eilert, S.J. (2005). New packaging technologies for the 21st century. *Meat Science* 71: 122-127.

FAO/WHO (2001). Codex Alimentarius Commission: Procedural Manual (12th ed.).

FAO / World Health Organisation, Retrieved 18 July 2005, from ftp://ftp.fao.org/docrep/fao/005/Y2200E/Y2200E00.pdf.

Fleet, G.H. and Heard, G. (1993). Yeast growth during fermentation. pp. 27-57. *In*: G. Fleet (Ed.). Wine Microbiology and Biotechnology. Chur, Switzerland: Harwood Academic Publishers.

Finkelstein, J. and Quiazon, R. (2007). Liquid images: Viewing the wine label. *Journal of Hospitality and Tourism Management* 14(1): 17-23.

Francis, L., Lattey, K. and Smyth, H. (2003). 'Reduced' aroma in screw-cap bottled white wines. *AWRI Technical Review* 142(10): 51-53.

Fu, Y., Lim, L.T. and Mc Nicholas, P.D. (2009). Changes on enological parameters of white wine packaged in bag-in-box during secondary shelf-life. *Journal of Food Science* 74: 608-618.

Fugelsang, K.C. and Edwards, C.G. (2007). Wine Microbiology: Practical Applications and Procedures. NY: Springer, New York.

Gerbaux, V., Villa, A., Monamy, C. and Bertrand, A. (1997). Use of lysozyme to inhibit malo-lactic fermentation and to stabilise wine after malo-lactic fermentation. *American Journal of Enology and Viticulture* 48: 49-54.

Galpin, V.C. (2006). A comparison of legislation about winemaking additives and processes, assignment submitted in partial requirement for the Cape Wine Master Diploma.

Garcıa-Ruiz, A., Bartolome, B., Martnez-Rodrıguez, A., Pueyo, E., Martın-Alvarez, P.J. and Moreno-Arribas, M.V. (2008a). Potential of phenolic compounds for controlling lactic acid bacteria growth in wine. *Food Control* 19: 835-841.

Gergely, S., Bekassy-Molnar, E. and Vatai, Gy. (2003). The use of multi-objective optimisation to improve wine filtration. *Journal of Food Engineering* 58: 311-316.

Ghidossi, R., Poupot, C., Thibon, C., Pons, A., Darriet, P., Riquier, L., De Revel, G. and Mietton Peuchot, M. (2012). The influence of packaging on wine conservation. *Food Control* 23: 302-311.

Gil, J.V. and Vallés, S. (2001). Effect of macerating enzymes on red wine aroma at laboratory scale: Exogenous addition or expression by transgenic wine yeasts. *Journal of Agricultural and Food Chemistry* 49: 5515-5523.

Gil, L. (1998). Cortic¸a: produc¸ão, tecnologia e aplicac¸ão, Lisbon, INETI.

Godden., P., Lattey, K., Francis, L., Gishen, M., Cowey, G., Holdstock, M., Robinson, E., Waters, E., Skouroumounis, G., Sefton, M., Capone, D., Kwiatkowski, M., Field, J., Coulter, A., D'Costa, N. and Bramley, B. (2005). Towards offering wine to the consumer in optimal condition – The wine, the closures and other packaging variables: A review of AWRI research examining the changes that occur in wine after bottling. *Wine Industry Journal* 20: 20-30.

Godden, P., Francis, L., Field, J., Gishen, M., Coulter, A., Valente, P., Hoj, P. and Robinson, E. (2001). Wine bottle closures: Physical characteristics and effect on composition and sensory properties of a Semillon wine 1: Performance up to 20 months post-bottling. *Australian Journal of Grape and Wine Research* 7: 62-108.

Goode, Jamie (2007). In Defence of Screw Caps. wineanorak.com

Goode, Jamie (2008). Wines & Vines 'Finding Closure'. Archived from the original on 2009-04-10, wineanorak.com

Goode, J. (2010). Alternatives to cork in wine bottle closures. pp. 255-269. *In*: Reynolds, A.G. (Ed.). Managing Wine Quality, vol. 2, Enology and Wine Quality. Woodhead Publishing Ltd., Great Abington, Cambridge CB21 6AH, UK.

Gougeon, R.D., Reinholdt, M., Delmotte, L., Miéhe-Brendle, J., Chezeau, J.M., Le Dred, R., Marchal, R. and Jeandet, P. (2002). Direct observation of polylysine side-chain interaction with smectites interlayer surfaces through 1H-27Al heteronuclear correlation NMR spectroscopy. *Langmuir* 18: 396-398.

Gougeon, R.D., Soulard, M., Reinholdt, M., Miéhé-Brendlé, J., Chézeau, J.M., Le Dred, R., Marchal, R. and Jeandet, P. (2003). Polypeptide adsorption on to a synthetic montmorillonite: A combined solid-state NMR, X-ray diffraction, thermal analysis and N2 adsorption study. *European Journal of Inorganic Chemistry* 7: 1366-1372.

Guilloux-Benatier, M. and Feuillat, M. (1991). Utilisation d'adjuvants d'origine levurienne pour améliorer l'ensemencement des vins en bactéries sélectionnées. *Rev. Fr .Oenol.* 132: 51-55.

Henschke, P.A. and Jiranek, V. (1991). Hydrogen sulphide formation during fermentation: Effect of nitrogen composition in model grape must. pp. 172-184. *In*: J. Rantz (Ed.). Proceedings of the International Symposium on Nitrogen in Grapes and Wine. Seattle, USA. Davis CA: American Society for Enology and Viticulture.

Hsu, J.C. and Heatherbell, D.A. (1987). Heat-unstable protein in wine, I: Characterisation and removal by bentonite fining and heat-treatment. *American Journal of Enology and Viticulture* 38: 11-16.

Insa, S., Salvadó, V. and Anticó, E. (2004). Development of solid-phase extraction and solid-phase microextraction methods for the determination of chlorophenols in cork macerate and wine samples. *J. Chromatography* A1047: 15-20.

Kudo, M., Vagnoli, P. and Bisson, L.F. (1988). Imbalance of pH and potassium concentration as a cause of stuck fermentation. *American Journal of Enology and Viticulture* 49: 295-301.

Laborde, B., Moine-Ledoux, V., Richard, T., Saucier, C., Dubourdieu, D. and Monti, J.-P. (2006). PVPP-polyphenol complexes: A molecular approach. *Journal of Agricultural and Food Chemistry* 54: 4383-4389.

Lee, D.H., Kaang, B.S. and Park, H.J. (2011). Effect of oxygen on volatile and sensory characteristics of Cabernet Sauvignon during secondary shelf-life. *Journal of Agricultural and Food Chemistry* 59: 11657-11666.

Lee, T.H. and Simpson, R.F. (1993). Microbiology and chemistry of cork taints in wine. *In*: Graham H. Fleet (Ed.). Wine Microbiology and Biotechnology. Harwood Academic Publishers. Chur, Switzerland, 353, London.

Lopes, P., Saucier, C. and Glories, Y. (2005). Nondestructive colourimetric method to determine the oxygen diffusion rate through closures used in winemaking. *Journal of Agricultural and Food Chemistry* 53: 6967-6973.

Lopes, P., Saucier, C., Teissedre, P.L. and Glories, Y. (2007). Main routes of oxygen ingress through different closures into wine bottles. *Journal of Agricultural and Food Chemistry* 55: 5167-5170.

Madigan, A. (2004). The screw cap revolution rolls on. *Wine Industry Journal* 19(5): 59-65.

Main, G.L. and Morris, J.R. (1994). Colour of Seyval Blanc juice and wine as affected by juice fining and bentonite fining during fermentation. *American Journal of Enology and Viticulture* 45: 417-422.

Marchal, R., Barret, J. and Maujean, A. (1995). Relation entre les caract´eristiques physicochimiques d'une bentonite et son pouvoir d'adsorption. *J. Inter. Sci. Vigne Vin.* 29: 27-42.

Marchal, R., Weingartner, S., Voisin, C., Jeandet, P. and Chatelain, F. (2002d). Utilisation de modèles mathématiques pour optimiser les doses de bentonite gonflée et non gonflée lors du collage des vins blancs. Partie I: Clarification et stabilisation collöıdale. *J. Inter. Sci. Vigne Vin.* 36: 169-176.

Marchal, R. and Waters, E.J. (2010). New directions in stabilisation, clarification and fining of white wines: Managing wine quality, vol. 2. *In*: Andrew G. Reynolds (Ed.). Oenology and Wine Quality. Woodhead Publishing Ltd.

Marchal, R. and P. Jeandet (2009). Use of enological additives for colloid and tartrate salt stabilisation in white wines and for improvement of sparkling wine foaming properties. pp. 127-155. *In*: M.V. Moreno-Arribas and M.C. Polo (Eds.). Wine Chemistry and Biochemistry. Springer Science+Business Media, LLC.

Martineau, B. and Henick-Kling, T. (1995). Formation and degradation of diacetyl in wine during alcoholic fermentation with *Saccharomyces cerevisiae* strain EC1118 and malo-lactic fermentation with *Leuconostoc oenos* strain MCW. *American Journal of Enology and Viticulture* 46: 442-448.

Martınez-Rodriguez, A.J. and Polo, M.C. (2003). Effect of the addition of bentonite to the tirage solution on the nitrogen composition and sensory quality of sparkling wines. *Food Chemistry* 81: 383-388.

Marwaha, S.S., Soni, S.K. and Marwaha, U. (2011). Packaging of wines. pp. 927-949. *In*: V.K. Joshi (Ed.). Handbook of Enology: Principles and Practices, vol. II. Asiatech Publishers, Inc. San Francisco, CA.

Maujean, A. (2000). Organic acids in wine. pp. 3-39. *In*: Handbook of Enology, vol. 2. The Chemistry of Wine, Stabilization and Treatments. New York: John Wiley & Sons, Ltd.

Maury, C., Sarni-Manchado, P., Lefebvre, S., Cheynier, V. and Moutounet, M. (2001). Influence of fining with different molecular weight gelatines on proanthocyanidin composition and perception of wines. *American Journal of Enology and Viticulture* 52: 140-145.

Moine-Ledoux, V. and Dubourdieu, D. (2007). Role of mannoproteins in wine stabilisation. pp. 393-401. *In*: P. Jeandet, C. Clément and A. Conreux (Eds.). *M*acromolecules and Secondary Metabolites of Grapevine and Wine. Paris, London. New York: Intercept/Lavoisier.

Monk, P.R. (1994). Nutrient requirements of wine yeast. pp. 58-64. *In*: T. Henick-Kling (Ed.). Proceedings of the New York Wine Industry Workshop. Geneva, NY.

Mueller, S. and Lockshin, L. (2008). How important is wine packaging for consumers? On the reliability of measuring attribute importance with direct verbal versus indirect visual methods. 4th AWBR Conference, Siena.

Muellera, S. and Szolnoki, G. (2010). Wine packaging and labelling – Do they impact market price? A hedonic price analysis of US scanner data, referred paper. 5th International Academy of Wine Business Research Conference, 8-10 Feb. 2010. Auckland, NZ.

Muhlack, R., Nordestgaard, S., Waters, E.J., O'Neill, B.K., Lim, A. and Colby, C.B. (2006). In-line dosing for bentonite fining of wine or juice: Contact time, clarification, product recovery and sensory effects. *Australian Journal of Grape and Wine Research* 12: 221-234.

Nielsen, J.C. and Richelieu, M. (1999). Control of flavour development in wine during and after malo-lactic fermentation by *Oenococcus oeni*. *Applied Environmental Microbiology* 65: 740-745.

Nygaard, M., Petersen, L., Pilatte, E. and Lagarde, G. (2002). Prophylactic use of lysozyme to control indigenous lactic acid bacteria during alcoholic fermentation. Abstr., 53rd American Society of Enology and Viticulture Annual Meeting, Portland. *American Journal of Enology and Viticulture* 53: 240A.

Orth, U.R and Malkewitz, K. (2008). Holistic package design and consumer brand impressions. *Journal of Marketing* 72(3): 64-81.

Osborn, K.D. and Jenkins, W.A. (1992). Plastic Films: Packaging Applications. Technomic Publishing, Lancaster.

Olsen, J., Thompson, K. and Clarke, T. (2003). Consumers' self-confidence in wine purchases. *International Journal of Wine Marketing* 15(3): 40-52.

Ough, C.S. (1993a). Dimethyl dicarbonate and diethyl dicarbonate. Chapter 9, pp. 343-368. *In*: P.M. Davidson and A.L. Branen (Eds.). Antimicrobials in Foods, second ed. Marcel Dekker, Inc., New York, NY.

Ough, C.S. (1993b). Sulphur dioxide and sulfites. Chapter 5, pp. 137-190. *In*: P.M. Davidson and A.L. Branen (Eds.). Antimicrobials in Foods, second ed. Marcel Dekker, Inc., New York, NY.

Paine, F.A. and Paine, H.Y. (1992). Handbook of Food Packaging, second ed. Blackie Academic & Professional Publishers.

Penn, C. (2004). Newest cork-cleaning treatment comes to market. *Wine Business Monthly* September 2004: 25-26.

Penn, C. (2007a). Cork and closure research update. *Wine Business Monthly* February 2003: 116-117, Wine Communications Group, Inc., California.

Penn, C. (2007b). Here comes Zork. *Wine Business Monthly* September 2007: 30-31, Wine Communications, Inc., California.

Porter, L.J. and Ough, C.S. (1982). The Effects of ethanol, temperature and dimethyl dicarbonate on viability of *Saccharomyces cerevisiae* montrachet No. 522 in Wine. *American Journal of Enology and Viticulture* 33: 222-225.

Rapp, A., Pretorus, P. and Kugler, D. (1992). Foreign and undesirable flavours in wine. pp. 485. *In*: Charalambous, G. (Ed.). Off-flavours in Foods and Beverages. Elsevier Science, Amsterdam.

Rauhut, D., Kurbel, H. and Dittrich, H. (1993). Sulphur compounds and their influence on wine quality. *Die Wein-Wiss* 48: 214-218.

Renee Terrell, F., Morris, J.R., Johnson, M.G., Gbur, E.E. and Makus, D.J. (1993). Yeast inhibition in grape juice containing sulphur dioxide, sorbic acid and dimethyldicarbonate. *Journal of Food Science* 58: 1132-1134.

Revi, M., Badeka, A., Kontakos, S. and Kontominas, M.G. (2014). Effect of packaging material on enological parameters and volatile compounds of dry white wine. *Food Chemistry* 152: 331-339.

Ribereau-Gayon, P., Glories, Y., Maujean, A. and Dubourdieu, D. (1998). Traite d'Oneologie. 2-Chimie du vin. *Stabilisation et traitements* Dunod, Paris, France, 519.

Ribéreau-Gayon, P., Dubourdieu, D., B. Donéche and Lonvaud, A. (2006). Handbook of Enology, vol. 1. The Microbiology of Wine and Vinifications, second ed. Chichester, UK: John Wiley & Sons.

Robinson, J. (2006). Down with Synthetic Corks. www.jancisrobinson.com

Robinson, J. (2006). Zork Talk. www.jancisrobinson.com

Rocchi, B. and Stefani, G. (2006). Consumers' perception of wine packaging: A case study. *International Journal of Wine Marketing* 18(1): 33-44.

Romano, P. and Suzzi, G. (1993). Sulphur dioxide and wine microorganisms. pp. 373-393. *In*: G.H. Fleet (Ed.). Wine Microbiology and Biotechnology. Harwood Academic Publishers. Chur, Switzerland.

Rosa, M.E., Fortes, M.A. and Nunez, R.V. (2002). Proc. 1st International Materials Symposium, Advanced Materials Forum I, April 2001, Coimbra, Portugal. *Key Eng. Mater*., 230-232, 295-299.

Saxby, M.J. (1996). A survey of chemicals causing taints and off-flavours in food. pp. 41. *In*: Saxby, M.J. (Eds.). Food Taints and Off-Flavours. Blackie Academic & Professional, Glasgow, UK.

Sherman, S. and Tuten, T. (2011). Message on a bottle: The wine label's influence. *International Journal of Wine Business Research* 23(3): 221-234, Doi: 10.1108/17511061111163050

Silva, S.P., Sabino, M.A., Fernandes, E.M., Correlo, V.M., Boesel, L.F. and Reis, R.L. (2005). Cork: Properties, capabilities and applications. *International Materials Reviews* 50(6): 345-365.

Snakkers, G., Nepveu, G., Guilley, E. and Cantagrel, R. (2000). Variabilités géographique, sylvicole et individuelle de la teneur en extractibles de chênes sessiles français (Quercus petraea Liebl): polyphénols, octalactones et phénols volatils. *Annals of Forest Science* 57: 251-260.

Springett, M.B. (1996). Formation of off-flavours due to microbiological and enzymatic action. pp. 274. *In*: Saxby, M.J. (Eds.). Food Taints and Off-Flavours. Blackie Academic & Professional, Glasgow, UK.

Sue Courtney (2004). The History and Revival of Screw Caps, http://www.wineoftheweek.com/screwcaps/history.html

Vine, R.P. (1999). Enology (Winemaking). *In*: Winemaking. Aspen Publishers Inc., Gailthersburg, Maryland.

Walker, G.M. (1998). Yeast: Physiology and Biotechnology. John Wiley & Sons, New York, NY.

Warth, A.D. (1977). Mechanism of resistance of *Saccharomyces bailii* to benzoic, sorbic and other weak acids used as food preservatives. *Journal of Applied Bacteriology* 43: 215-230.

Warth, A.D. (1985). Resistance of yeast species to benzoic and sorbic acids and to sulphur dioxide. *Journal of Food Protection* 48: 564-569.

Waters, E.J., Alexander, G., Muhlack, R., Pocock, K.F., Colby, C., O'Neill, B.N., Høj, P.B. and Jones, P.R. (2005). Preventing protein haze in bottled wine. *Australian Journal of Grape and Wine Research* 11: 215-225.

Waters, E.J., Wallace, W. and Williams, P.J. (1992). Identification of heat-unstable wine proteins and their resistance to peptidases. *Journal of Agricultural and Food Chemistry* 40: 1514-1519.

Zoecklein, B.W. (1988a). Protein Stability Determination in Juice and Wine. Virginia Co-operative Extension, Publication No. 463-015.

Zoecklein, B.W. (1988b). Protein Fining Agents for Wines and Juices. Virginia Co-operative Extension, Publication No. 463-012.

Zurn, F. (1982). Flaschensterilisation mit Ozon. *Weinwirt* 118: 793-796, 800.

# 17 Biogenic Amines in Wine

**Spiros Paramithiotis**[1*] **and Ramesh C. Ray**[2]

[1] Laboratory of Food Quality Control and Hygiene, Department of Food Science and Human Nutrition, Agricultural University of Athens, Athens, Greece

[2] ICAR-Central Tuber Crops Research Institute, Bhubaneswar, Odisha, India

## 1. Introduction

Biogenic amines are low-molecular-weight nitrogen compounds that are implicated in many physiological processes in animals, plants and microorganisms. Their formation may occur through decarboxylation of amino acid or amination and transamination of aldehydes and ketones. Their presence has been already documented in nearly all the food commodities. In the case of non-fermented foods, they may serve as an indicator of undesired microbial development and concomitantly, improper hygienic conditions during preparation. Regarding fermented foods, the use of microorganisms capable of producing biogenic amines inevitably increases their amount (Halasz *et al.*, 1994; Silla Santos, 1996).

Intake of excessive amounts of biogenic amines may result in a series of adverse effects, including headache, skin irritation, hypo- or hypertention, tachycardia, etc. (Ladero *et al.*, 2010). Histamine is the most toxic one, followed by tyramine and phenylethylamine. Putrescine, cadaverine and volatile amines (e.g. isobutylamine, ethylamine, isopropylamine) are not toxic to humans; however, the former two may impair the effect of the toxic ones whereas volatile amines may affect negatively the wine aroma. Finally, secondary or tertiary amines (e.g. spermine, spermidine, dimethylamine) may react with nitrous anhydride to nitrosamines (Scanlan, 1983), which have been associated with gastric cancer (Song *et al.*, 2015).

The biogenic amine content of wines depends upon their occurrence in grapes, the quantity of the precursor amino acids in the must, the presence of microorganisms that possess decarboxylase activity as well as aging and storage conditions. More accurately, factors, such as grape variety and ripening stage, soil type and fertilisation, irrigation, agricultural practices and climatic conditions determine the amount of biogenic amines that is formed in the grape. Then, technological parameters, such as the amount of precursor amino acids, the ability of the microorganisms to form biogenic amines, the pH value and the use of sulphur dioxide determine the level of their increase after alcoholic and malo-lactic fermentations. Finally, parameters associated with aging and storage of wine, including type of barrels, contact with the lees and storage temperature may play a decisive role on their fate during wine aging and storage.

The level of biogenic amines in wines from all over the world, the factors that affect their accumulation as well as the available strategies for their removal have been extensively studied and reviewed. The aim of the present chapter is to offer an update regarding ongoing research on the aforementioned topics, integrate them with issues that are equally important in understanding occurrence of biogenic amines generally in food, i.e. the methodological approaches used for their qualitative and quantitative determination as well as their physiological role in plants and microorganisms and present them in a concise and comprehensive way.

## 2. Methodological Approaches in Biogenic Amine Determination

In general, the selected analytical determination dictates the necessary preparation steps. High performance liquid chromatography employing C18 reverse-phase column is the most commonly applied approach for the determination of biogenic amines in wine. This method requires chemical derivatisation that in most cases is performed before the determination (pre-column). Several chemical agents have been used for this purpose, including dansyl chloride, benzoyl chloride, o-phthalaldehyde, 6-aminoquinolyl-n-

*Corresponding author: sdp@aua.gr

hydroxysuccinimidyl carbamate (AQC), etc. Specificity and stability of the derivatives are the properties that an agent should offer. However, most of the agents currently employed that offer the latter lack the former. Thus, a pretreatment able to remove interfering substances is necessary and is mostly performed through extraction of the target solutes by techniques such as solid-phase (micro)extraction, liquid-liquid extraction and dispersive liquid-liquid microextraction (Montes *et al.*, 2009; Basozabal *et al.*, 2013; Ramos *et al.*, 2014).

Other chromatographic methods that may be used include ultra-performance liquid chromatography (UPLC), ion chromatography (IC), thin-layer chromatography (TLC) and gas chromatography (GC). An optimisation of the first using post-column derivatisation with o-phthaldialdehyde (OPA) was reported by Lijima *et al.* (2013). Before analysis, samples were diluted 50 times and filtrated through a 0.2 μm filter; the analysis time was significantly reduced compared to conventional High-Performance Liquid Chromatography (HPLC) without significant differences in the limit of detection (LOD). UPLC coupled to triple quadruple mass spectrometer was recently applied by Tasev *et al.* (in press) for the assessment of the biogenic amine content of white, rose and red wines. Prior to detection, dilution of the samples with 0.1 per cent (v/v) formic acid in 1:3 ratio and filtration through 0.45-μm filter took place. As in the previous case, time of analysis was significantly reduced with LOD remaining at comparable levels. Ion chromatography was reported by Palermo *et al.* (2013) as an alternative for biogenic amine detection in various food matrices, including wine. Sample preparation included dilution with water and purification through solid-phase extraction. This approach was characterised by the authors as simple, fast, reproducible, sensitive and selective. Cunha *et al.* (2011) developed a gas chromatography – a mass spectrometry method for biogenic amine quantitative assessment in Port wines. Sample preparation included a two-phase derivatisation procedure with isobutyl chloroformate in a toluene medium followed by treatment with alkaline methanol. The method performance was satisfactory in terms of recovery, limit of detection and quantification, linearity and repeatability.

Capillary electrophoresis may also be regarded as an alternative to chromatographic methods. Indeed, Jastrzedska *et al.* (2014) performed a comparative study between isotachophoresis (ITP) and Reversed Phase (RP)-HPLC for simultaneous detection of eight biogenic amines; the advantages of ITP, namely speed, accuracy and low detection and quantification limits, were adequately highlighted.

## 2.1. Physiological Role of Biogenic Amines

Biogenic amines are produced by plants and bacteria because they have specific biological roles, which have been extensively studied and reviewed. The aim of this paragraph is to provide with merely an outline of these roles so that their importance is understood.

Spermidine and putrescine are the most abundant polyamines in bacteria and are considered as normal grape constituents (Gloria and Vieira, 2007; Kusano *et al.*, 2008). Their polycationic nature dictates their implication in many biological processes through their interactions with DNA, RNA and proteins. Thus, an effect on gene transcription and translation, free radical scavenging, interaction with hormones and ethylene biosynthesis, stabilisation of cell membranes, apoptosis and modulation of cell signalling has been claimed (Tabor and Tabor, 1985; Igarashi and Kashiwagi, 2000; Seiler and Raul, 2005).

In plants, cell division is generally accompanied by increased levels of polyamines (Walker *et al.*, 1985). Putrescine and spermidine are essential during embryogenic growth (Montague *et al.*, 1979; Garrido *et al.*, 1995; Yadav and Rajam, 1997; Pedroso *et al.*, 1997) as well as floral initiation and development (Torrigiani *et al.*, 1987a; Gerats *et al.*, 1988; Bais *et al.*, 2000b). Putrescine accumulation was reported during root growth and differentiation (Baraldi *et al.*, 1995; Bais *et al.*, 1999a, b, 2000a), fruit development (Teitel *et al.*, 1985; Biasi *et al.*, 1991; Alabadi *et al.*, 1996, Alabadi and Carbonell, 1998) and vegetative growth (Jirage *et al.*, 1994; Lee and Lin, 1996). Finally, increase of putrescine, spermine and spermidine levels has been associated with senescence (Cheng *et al.*, 1984; Downs and Lovell, 1986). Their role in abiotic stress, such as drought (Kusano *et al.*, 2007a, b; Capell *et al.*, 2004), salinity (Urano *et al.*, 2004), chilling (Kasukabe *et al.*, 2004), ozone (Navakoudis *et al.*, 2003) as well as multiple stresses resistance (Kasukabe *et al.*, 2004, 2006; Wen *et al.*, 2007; Yamagushi *et al.*, 2006, 2007) has been described. Moreover, a role in biotic stress resistance for spermine mainly through signal transduction (Takahashi *et al.*, 2003) and polyamine-catabolised $H_2O_2$ (Cona *et al.*, 2006) as well as spermine and spermidine as nitric acid inducers (Tun *et al.*, 2006; Yamasaki and Cohen, 2006) have also been exhibited. In most of the aforementioned cases, direct effects have been described (Bais and Ravishankar, 2002; Kaur-Sawhney *et al.*, 2003; Liu *et al.*, 2007; Kusano *et al.*, 2008); however, further interactions yet to be characterised may not be excluded.

Regarding microbial physiology, several roles have also been assigned to biogenic amines. From a winemaking perspective, the most important is maybe the response mechanism against acidic stress. This mechanism is depicted in Fig. 1. The membrane antiport (MA) possesses a central role as it couples amino acid uptake and biogenic amine excretion. The amino acid is decarboxylated in the cell by amino acid decarboxylases (AAD) with simultaneous consumption of protons. The biogenic amine is then excreted and drives ATP synthesis through proton motive force (Molenaar *et al.*, 1993; EFSA, 2011). Such mechanisms have been reported for histidine/histamine, tyrosine/tyramine, ornithine/putrescine and lysine/cadaverine (Molenaar *et al.*, 1993; Soksawatmaekhin *et al.*, 2004; Wolken *et al*,. 2006).

In addition, a potential role in gene expression (Pastre *et al.*, 2005; Lindemose *et al.*, 2005), protection against oxidative stress both as free radical scavengers and by upregulating *oxyR*, *katG* genes and the SOS regulon in *Escherichia coli* (Ha *et al.*, 1998; Kim and Oh, 2000; Tkachenko *et al.*, 2001; Chattopadhyay *et al.*, 2003), as well as possible roles in virulence (Polissi *et al.*, 1998; Ware *et al.*, 2006) signalling (Sturgill and Rather, 2004; Stevenson and Rather, 2006) and biofilm formation (Karatan *et al.*, 2005; Patel *et al.*, 2006) have been described.

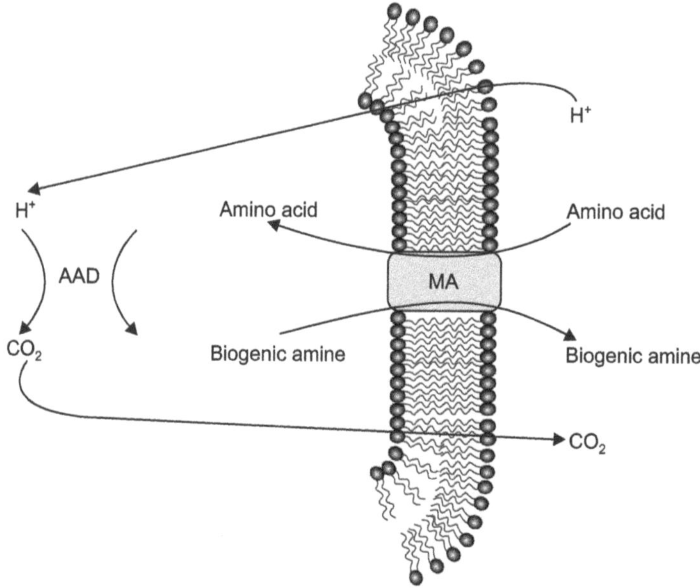


**Figure 1.** Response mechanism against acidic stress present in bacteria that involves production of biogenic amines

## 3. Factors Affecting Biogenic Amine Production

Biogenic amines may originate from the grapes or formed by microorganisms of vinification during fermentation, maturation or storage of wine. The qualitative and quantitative effect of agricultural practices, oenological parameters as well as maturation and storage conditions has been studied to some extent. Contemporary studies confirm the basic principle underlying biogenic amine production by the microorganisms, i.e. the strain-dependent character of this property. At the same time the importance of technological parameters has been adequately highlighted. Especially regarding the latter, each technological parameter may affect each biogenic amine in a different way, suggesting that generalisations may create confusion and thus great caution should be exercised.

Traditionally, the lactic acid bacteria of malo-lactic fermentation along with spoilage microorganisms are held responsible for biogenic amine production in wine (Russo *et al.*, 2016) despite the fact that several yeast strains have also been found to possess this property (Caruso *et al.*, 2002; Granchi *et al*., 2005; Chang *et al*., 2009; Tristezza *et al*., 2013). Thus, several factors that may affect microbiota development and concomitantly biogenic amine accumulation have been studied. In general, the higher the pH value the greater the risk of biogenic amine accumulation due to an increase of the biodiversity of the microecosystem (Lopez *et al*., 2012; Wang *et al*., 2014). Marcobal *et al*. (2006) reported that addition of sulphur dioxide prevents biogenic amine production during wine aging but not during alcoholic fermentation. On the other hand, addition of yeast autolysate results in an increase in biogenic amine content during malo-lactic fermentation, but not during alcoholic fermentation (Marco *et al*., 2006). In general, the selection of strains, especially lactic acid bacteria that are not capable of producing biogenic amines, is advised. To this end, Benito *et al*. (2016) proposed the use of a combination of *Lachancea thermotolerans* and *Schizosaccharomyces pombe* to carry out malo-lactic fermentation – the latter consumes malic acid while the former produces lactic acid. The final product obtained exhibited improved organoleptic properties and ethanol index compared to wines produced conventionally. Regarding biogenic amines, the amount of histamine formed was significantly lower; on the contrary, similar amounts of tyramine, phenylethylamine, putrescine and cadaverine were produced.

Amines, such as ethanolamine, ethylamine, putrescine, tyramine, agmatine, spermidine and spermine may be found in musts before fermentation since they possess a physiological role in plants (Bouchereau *et al*., 1999). Indeed, Del Prete *et al*. (2009) reported the presence of ethanolamine, ethylamine and putrescine but not agmatine and tyramine in Merlot, Syrah, Cabernet Franc., Carmenerc, and Cesanese d'Affile musts in both 2004 and 2005 vintages. All the aforementioned biogenic amines but putrescine were also detected in Montepulciano and Sangiovese musts. Similarly, Smit *et al*. (2014) stated that histamine, isopentylamine, ethylamine, phenylethylamine, putrescine, spermine, spermidine and cadaverine were present in Riesling musts in 2005 and 2007 vintages. Interestingly, their fate was rather different during fermentation; in many cases, such as in ethylamine and putrescine in 2005 vintage, an increase was noted, while in other, such as isopentylamine and putrescine in 2007 vintage, a decrease was evident. In the same study, enhanced nitrogen fertilisation was reported to affect significantly the biogenic amine content in both musts and wine. Moreover, the effect of vintage was also underlined. Especially regarding the latter, a number of studies are in agreement (Marques *et al*., 2008; Martinez-Pinilla *et al*., 2013; Ortega-Heras *et al*., 2014); however, the use of the term 'vintage' may create misunderstandings since it encompasses the majority of technological parameters related to vinification, such as grape variety, ripening stage, infections as well as cultivation practices and climatic conditions that have their own separate importance (Marques *et al*., 2008; Guo *et al*., 2007; Kiss *et al*., 2006; Sass-Kiss *et al*., 2000) and therefore should be carefully used. The diverse fate of the biogenic amines formed was also highlighted by Garcia-Marino *et al*. (2010). In that study, the effect of conventional and organic winemaking in the amount of biogenic amines after each step of the winemaking procedure was assessed. Histamine, ethylamine, putrescine and cadaverine were detected already in the must. On the contrary, isoamylamine, agmatine, serotonin and isobutylamine were not detected throughout the experiment while small amounts of tyramine, tryptamine and phenylethylamine were only detected in the final product of organic winemaking and not the conventional one. The amount of histamine increased after alcoholic and malolactic fermentation but the increase was more pronounced during maturation and much more intense in organic rather than conventional winemaking. A significant increase in the amount of ethylamine was observed after alcoholic fermentation; then a decrease was noted during maturation in organic winemaking. The amount of putrescine increased during maturation in organic winemaking, whereas the amount of cadaverine was stable in all cases. These results underline the existence of additional factors affecting the amount of biogenic amines that are yet to be studied.

The concentration of the precursor amino acids has been positively correlated to the amount of biogenic amines formed – either this takes place in the field or during fermentation. Regarding the former, the positive effect of irrigation and nitrogen supply on the amount of free amino acids has been exhibited (Lee and Schreiner, 2010; Ortega-Heras *et al*., 2014). The latter was verified by Lorenzo *et al*. (in press) upon supplementation with amino acids but only in the case of histamine formation; tyramine synthesis was mostly affected by temperature and alcoholic degree, tryptamine by temperature and phenylethylamine by alcoholic degree.

The effect of maceration in the presence of stems, a technique applied to improve organoleptic quality and colour stability in red wines, on biogenic amine content, was studied by Basile *et al*. (2017).

It was concluded that maceration in the presence of stems had no effect on the biogenic amine content of the wine. The effect of sulphur dioxide reduction with simultaneous addition of lysozyme or dimethyl carbonate on the biogenic amine content of Garnacha wine was assessed by Ancin-Azpilicueta *et al.* (2016). The results obtained were biogenic amine dependent. More accurately, the amount of tyramine, dimethylamine, pyrrolidine, diethylamine and amylamine was higher and the amount of hexylamine was lower when fermentation took place only with sulphur dioxide. The amount of putrescine was higher and the amount of phenylethylamine + spermidine and spermine was lower when fermentation took place only with sulphur dioxide or with partial replacement of sulphur dioxide with lysozyme compared to the respective when sulphur dioxide was partially replaced by dimethyl carbone. Histamine was only detected when sulphur dioxide was partially replaced with lysozyme. Finally, no effect was observed on cadaverine, ethylamine and isobutylamine concentrations.

The evolution of biogenic amines during storage has been addressed by Moreno *et al.* (2003) and Marco *et al.* (2006). In the first study the evolution of biogenic amines in red wine (Merlot) stored in oak barrels was studied for 243 days. It was reported that histamine and tyramine increased during storage, reached a peak after ca. 90 days and then decreased to their original level by the end of the experiment. Putrescine, cadaverine, phenylethylamine and spermidine increased during storage while spermine, dimethylamine, isobutylamine, pyrrolidine, ethylamine, isopropylamine, diethylamine, amylamine and hexylamine remained stable. Marco *et al.* (2006) examined the evolution of biogenic amines during storage of bottled wine (Chardonnay) at 4, 20 and 35°C for 105 days. Histamine, pyrrolidine and hexylamine presented an increase in the concentration. In the fir st case it was more pronounced at 20°C, in the second at 35°C and in the latter at all temperatures examined. The amount of putrescine, tyramine, cadaverine, ethylamine, isobutylamine and isopropylamine had no statistically significant change; the amount of diethylamine was reduced while spermine depleted.

## 4. Biogenic Amines in Wines

The biogenic amine content of wines has been exhaustively studied and adequately documented. In Table 1 an example of such studies is summarised. Most of the studies focus on detection and quantification of the toxic ones, viz. histamine, tyramine and phenylethylamine; volatile amines are also frequently assessed. Histamine, cadaverine, tyramine and putrescine seem to be the most commonly detected biogenic amines in wines. A great variability is observed in the biogenic amine content regarding both the type and amount present due to the effect of the raw materials, the agricultural practices as well as technological parameters, as already discussed in paragraph 4. Based on that, Galgano *et al.* (2011) used biogenic amine content as a quality parameter that along with total polyphenol and cis- and trans-resveratrol content helped discriminate 73 wines from southern Italy according to their region. More recently, Preti and Vinci (2016) used only biogenic amines as a marker for effective characterisation and classification of 56 wines according to their protected designations of origin.

## 5. Biogenic Amine Removal

The use of biogenic amine degrading microorganisms and enzymes has been extensively considered and studied with very promising results.

### 5.1. Biogenic Amine Degrading Nicrobiota

A wide variety of bacteria, including *Lactobacillus plantarum*, *Lb. sakei*, *Lb. casei*, *Pediococcus pentosaceus*, *Oenococcus oeni*, *Bacillus amyloliquefaciens*, *B. subtilis*, *Staphylococcus carnosus*, *St. intermedius*, *St. xylosus* and molds including *Penicillium citrinum*, *Alternaria* sp., *Phoma* sp., *Ulocladium chartarum* and *Epicoccum nigrum* have been reported to degrade biogenic amines either *in vitro* or in a food matrix in a strain-specific manner (Leuschner *et al.*, 1998; Dapkevicius *et al.*, 2000; Martuscelli *et al.*, 2000; Fadda *et al.*, 2001; Zaman *et al.*, 2010, 2011; Garcia-Ruiz *et al.*, 2011; Herrero-Fresco *et al.*, 2012; Capozzi *et al.*, 2012; Cueva *et al.*, 2012; Toy *et al.*, 2015; Xie *et al.* 2016). Garcia-Ruiz *et al.* (2011) proved the potential of wine-associated LAB as a promising strategy to reduce the biogenic amine content in wine. Strains belonging to the *Lactobacillus* spp. and *Pediococcus* spp. found to exhibit the

**Table 1a.** Biogenic Amine Content of Various Wines

| *Origin* | *Type* | *AGM* | *CAD* | *ETA* | *ETM* | *HIM* | *ISM* | *MEM* |
|---|---|---|---|---|---|---|---|---|
| Greece[1] | White (n=47) | - | 0.81 [nd$^{(11)}$-4.44] | - | 0.39 [nd$^{(4)}$-1.96] | 0.41 [nd$^{(26)}$-5.95] | 0.77 [nd$^{(4)}$-3.22] | 0.13 [nd$^{(12)}$-2.0] |
| | Red (n=45) | - | 0.52 [nd$^{(9)}$-3.21] | - | 0.40 [nd$^{(2)}$-1.61] | 0.31 [nd$^{(19)}$-2.11] | 0.75 [nd$^{(15)}$-8.17] | 0.10 [nd$^{(13)}$-0.66] |
| | Rose (n=8) | - | 1.20 [nd$^{(1)}$-4.49] | - | 0.44 [nd$^{(1)}$-1.32] | 1.03 [nd$^{(5)}$-4.43] | 0.25 [nd$^{(3)}$-1.44] | 0.14 [nd$^{(1)}$-0.5] |
| Greece[2] | White (n=17) | nd | nd$^{(6)}$-0.13 (0.02) | nd$^{(6)}$-0.17(0.02) | - | 0.18 (0.02)-1.13 (0.03) | 0.42 (0.02)-2.41 (0.03) | nd$^{(8)}$-0.49 (0.05) |
| | Red (n=15) | nd$^{(7)}$-0.63 (0.04) | nd$^{(3)}$-0.21 (0.02) | nd$^{(6)}$-0.71(0.04) | - | 0.98 (0.02)-1.65 (0.03) | nd$^{(6)}$-0.95 (0.02) | nd$^{(9)}$-0.59 (0.03) |
| Italy[3] | Red (n=60) | nd-5.18 | 0.33-9.90 | - | 1.01-16.40 | nd-3.16 | - | 0.15-1.63 |
| Italy[4] | | | | | | | | |
| Basilicata | Red (n=27) | 7.89 (4.16) | 1.74 (1.04) | 24.51 (7.22) | - | 1.95 (1.13) | - | 0.76 (0.41) |
| Calabria | Red (n=19) | 8.84 (3.45) | 1.94 (0.51) | 20.60 (8.01) | - | 3.01 (1.06) | - | 0.35 (0.18) |
| Campania | Red (n=11) | 14.31 (5.73) | 3.21 (1.24) | 19.89 (5.25) | - | 0.22 (0.09) | - | 0.27 (0.07) |
| Puglia | Red (n=16) | 4.75 (3.19) | 4.62 (2.18) | 38.07 (18.10) | - | 7.62 (4.07) | - | 0.61 (0.33) |
| Italy[5] | | | | | | | | |
| Aglianico del Vulture | Red (n=13) | 9.20 (1.90) [7.2-13.23] | 0.08 (0.05) [nd -0.21] | - | 0.17 (0.12) [nd -0.31] | 0.09(0.07) [nd -0.18] | - | 0.19 (0.05) [0.11-0.28] |
| Etna Rosso | Red (n=13) | 2.56 (0.74) [0.47-3.36] | 0.45 (0.14) [0.13-0.60] | - | 0.20 (0.10) [nd -0.97] | 0.12 (0.09) [nd -0.26] | - | 0.15 (0.10) [nd -0.31] |
| Primitivo di Manduria | Red (n=15) | 7.88 (1.65) [4.8-10.43] | 0.06 (0.04) [0.01-0.14] | - | 0.08 (0.10) [nd -0.38] | 0.22 (0.05) [0.13-0.29] | - | nd |
| Syrah Sicilia | Red (n=15) | 0.31 (0.12) [nd -0.47] | 0.49 (0.15) [0.24-0.78] | - | 2.35 (0.60) [1.53-3.28] | 0.22 (0.04) [0.12-0.26] | - | 0.22 (0.07) [0.13-0.39] |
| Austria[6] | Red (n=100) | - | 0.58 [nd-3.27] | - | - | 7.20 [0.52-27.0] | 0.25 [0.02-4.34] | - |

(*Contd.*)

**Table 1a.** *(Contd.)*

| *Origin* | *Type* | *AGM* | *CAD* | *ETA* | *ETM* | *HIM* | *ISM* | *MEM* |
|---|---|---|---|---|---|---|---|---|
| Spain[7] | White (n=6) | - | nd$^{(1)}$-1.16 (3.8) | - | - | nd$^{(1)}$-2.85 (4.4) | - | - |
| | Red (n=6) | - | nd$^{(2)}$-3.15(9.0) | - | - | 0.39 (10.3)-4.74 (8.4) | - | - |
| | Rose (n=7) | - | nd$^{(3)}$-2.87 (1.3) | - | - | nd$^{(1)}$-3.24 (3.1) | - | - |
| Spain[8] | | | | | | | | |
| La Rioja | Red (n=45) | - | - | - | - | 8.2 (5.5) | - | - |
| Utiel-Requena | Red (n=46) | - | - | - | - | 2.4 (1.1) | - | - |
| Tarragona | Red (n=12) | - | - | - | - | 4.5 (3.1) | - | - |

AGM: Agmatine; CAD: Cadaverine; ETA: Ethanolamine; ETM: Ethylamine; HIM: Histamine; ISM: Isoamylamine; MEM: Methylamine
nd: not detected; -: not determined.

[1]Soufleros *et al.* (2007). The mean value of each biogenic amine is presented. The range of values determined is given in square brackets. The superscript number in parenthesis indicates the number of samples in which the biogenic amine was not detected.

[2]Proestos *et al.* (2008). The range of values obtained is presented. Standard deviation is given in parenthesis. The superscript number in parenthesis indicates the number of samples in which the biogenic amine was not detected.

[3]Preti *et al.* (2016). The range of values obtained is presented.

[4]Galgano *et al.* (2011). The mean value is presented. Standard deviation is given in parenthesis.

[5]Preti and Vinci (2016). The mean value is presented. Standard deviation is given in parenthesis. The range of values determined is given in square brackets.

[6]Konakovsky *et al.* (2011). The mean value is presented. The range of values determined is given in square brackets.

[7]Romero *et al.* (2002). The range of values obtained is presented. Standard deviation is given in parenthesis. The superscript number in parenthesis indicates the number of samples in which the biogenic amine was not detected.

[8]Landete *et al.* (2005). The mean value is presented. Standard deviation is given in parenthesis.

**Table 1b.** Biogenic Amine Content of Various Wines

| *Origin* | *Type* | *PHM* | *PUT* | *SPD* | *SPM* | *TRYP* | *TYM* |
|---|---|---|---|---|---|---|---|
| Greece[1] | White (n=47) | 0.50 [nd$^{(28)}$-7.82] | 0.98 [nd$^{(4)}$-3.22] | 0.17 [nd$^{(29)}$-1.26] | 0.27 [nd$^{(33)}$-4.85] | -- | 0.42 [nd$^{(16)}$-2.54] |
| | Red (n=45) | 0.59 [nd$^{(25)}$-8.15] | 1.17 [nd$^{(3)}$-5.23] | 0.27 [nd$^{(23)}$-1.56] | 0.12 [nd$^{(31)}$-1.62] | - | 0.43 [nd$^{(13)}$-3.65] |
| | Rose (n=8) | 0.32 [nd$^{(5)}$-1.68] | 0.74 [nd$^{(1)}$-1.85] | 0.06 [nd$^{(6)}$-0.29] | 0.20 [nd$^{(5)}$-0.91] | - | 0.36 [nd$^{(2)}$-1.64] |
| Greece[2] | White (n=17) | - | nd$^{(2)}$-9.07 (0.02) | - | - | nd$^{(15)}$-0.51 (0.02) | nd$^{(7)}$-1.16 (0.03) |
| | Red (n=15) | - | nd$^{(1)}$-2.70 (0.03) | - | - | nd$^{(11)}$-1.32 (0.04) | nd$^{(6)}$-0.46 (0.02) |
| Italy[3] | Red (n=60) | nd-1.75 | 0.84-25.40 | nd-1.03 | nd-1.37 | | 0.22-34.99 |
| Italy[4] | | | | | | | |
| Basilicata | Red (n=27) | 4.20 (1.65) | 3.24 (1.18) | 0.86 (0.42) | 0.49 (0.15) | 11.34 (5.62) | 2.61 (1.41) |
| Calabria | Red (n=19) | 5.85 (3.22) | 5.99 (3.31) | 0.23 (0.02) | 0.09 (0.01) | 2.91 (0.84) | 1.22 (0.16) |
| Campania | Red (n=11) | 3.18 (1.42) | 2.23 (0.73) | 0.72 (0.11) | 1.27 (0.42) | 5.63 (2.38) | 1.14 (0.04) |
| Puglia | Red (n=16) | 3.71 (0.78) | 13.50 (7.55) | 0.80 (0.14) | 0.17 (0.09) | 14.65 (8.35) | 6.37 (2.99) |
| Italy[5] | | | | | | | |
| Aglianico del Vulture | Red (n=13) | 0.06 (0.06) [nd -0.19] | 0.55 (0.12) [0.4-0.77] | 0.01 (0.01) [nd -0.04] | nd | - | 0.08 (0.09) [nd -0.28] |
| Etna Rosso | Red (n=13) | 0.09 (0.08) [0.01-0.27] | 1.71 (0.54) [0.84-2.61] | 0.01 (0.01) [nd -0.04] | nd | - | 0.09 (0.17) [nd -0.62] |
| Primitivo di Manduria | Red (n=15) | 0.27 (0.11) [0.04-0.42] | 0.34 (0.13) [0.16-0.69] | 0.02 (0.01) [nd -0.05] | nd | - | 0.01 (0.03) [nd -0.1] |
| Syrah Sicilia | Red (n=15) | 0.06 (0.03) [nd -0.10] | 2.09 (0.21) [1.73-2.48] | 0.06 (0.07) [nd -0.21] | nd | - | 1.97 (0.67) [0.78-2.9] |
| Austria[6] | Red (n=100) | 0.16 [nd-1.74] | 19.4 [2.93-122] | 1.79 [0.03-4.96] | - | 0.06 [nd-1.59] | 3.52 [1.07-10.7] |
| Spain[7] | White (n=6) | nd | 1.48 (7.9)-12.61 (3.7) | nd | nd | nd | nd$^{(3)}$-1.44 (5.5) |
| | Red (n=6) | nd | 4.67 (3.2)-19.10 (5.3) | nd | nd$^{(4)}$-0.57 (8.6) | nd | nd$^{(1)}$-5.91 (4.3) |

**Table 1b.** *(Contd.)*

| *Origin* | *Type* | *PHM* | *PUT* | *SPD* | *SPM* | *TRYP* | *TYM* |
|---|---|---|---|---|---|---|---|
| | Rose (n=7) | nd | 3.25 (1.5)-9.60 (2.2) | nd$^{(6)}$-0.64 (5.1) | nd | nd | 0.33 (13.5)-2.59 (6.8) |
| Spain[8] | | | | | | | |
| La Rioja | Red (n=45) | 0.9 (0.4) | 47.3 (12.1) | - | - | - | 1.9 (1.1) |
| Utiel-Requena | Red (n=46) | 1.0 (0.4) | 7.5 (1.3) | - | - | - | 2.3 (0.8) |
| Tarragona | Red (n=12) | 0.8 (0.3) | 34.1 (10.5) | - | - | - | 1.8 (0.9) |

PHM: 2-phenylethylamine; PUT: Putrescine; SPD: Spermidine; SPM: Spermine; TRYP: Tryptamine; TYM: Tyramine
nd: not detected; -: not determined

[1]Soufleros *et al.* (2007). The mean value of each biogenic amine is presented. The range of values determined is given in square brackets. The superscript number in parenthesis indicates the number of samples in which the biogenic amine was not detected.

[2]Proestos *et al.* (2008). The range of values obtained is presented. Standard deviation is given in parenthesis. The superscript number in parenthesis indicates the number of samples in which the biogenic amine was not detected.

[3]Preti *et al.* (2016). The range of values obtained is presented.

[4]Galgano *et al.* (2011). The mean value is presented. Standard deviation is given in parenthesis.

[5]Preti and Vinci (2016). The mean value is presented. Standard deviation is given in parenthesis. The range of values determined is given in square brackets.

[6]Konakovsky *et al.* (2011). The mean value is presented. The range of values determined is given in square brackets.

[7]Romero *et al.* (2002). The range of values obtained is presented. Standard deviation is given in parenthesis. The superscript number in parenthesis indicates the number of samples in which the biogenic amine was not detected.

[8]Landete *et al.* (2005). The mean value is presented. Standard deviation is given in parenthesis.

greatest amine degradation ability. Capozzi *et al.* (2012) suggested that two *Lb. plantarum* strains (named NDT 09 and NDT 16) isolated from red wine undergoing malo-lactic acid fermentation were able to degrade biogenic amines, such as putrescine and tyramine.

### 5.2. Enzyme Activities Involved in Biogenic Amines Degradation

Monoamine-oxidases (MAO, E.C. 1.4.3.4.) are flavoproteins that catalyse the oxidase deamination of a number of biogenic amines in wine (Alvarez and Moreno-Arribas, 2014). On the contrary, information on the enzymes responsible for biogenic amine degradation is generally lacking (Alvarez and Moreno-Arribas, 2014). Only recently did the amine-degrading enzymes from *Pd. acidilactici* CECT 5930 and *Lb. plantarum* J16 studied in some detail (Callejon *et al.*, 2014). Both enzymes were identified as multicopper oxidases; in the case of the latter strain, the respective locus was identified and the protein was purified and biochemically characterised (Callejon *et al.*, 2016). In the case of wine only a couple of studies have been performed. Cueva *et al.* (2012) reported the isolation of 44 fungi strains from grapevine and vineyard soil with amine-degrading capacity. Furthermore, the cell-free supernatant of *Penicillium citrinum* CECT 20782 grown in yeast carbon base supplemented with histamine, tyramine or putrescine was used to degrade these amines in red, white and synthetic wines with very promising results. More recently, the enzyme histamine oxidase identified as a soluble flavoprotein containing non-covalently-attached flavin adenine dinucleotide as a co-factor and suggested that $Cu^{2+}$ is essential for the expression of enzymatic acivity. The multicopper complex of histamine oxidase found to be responsible for putrescine degradation in wine were isolated and purified from *Lb. plantarum* and *Pd. acidilactici* Callejón *et al.*, 2014).

### 5.3. Advanced Strategies in Biogenic Amines Degradation

Recently an alternative approach was provided by Vlasova *et al.* (2006, 2011). The adsorption of histamine, tryptamine and tyramine from aqueous solutions on the surface of highly dispersed silica was assessed (Vlasova *et al.*, 2006). Among them, the most stable complex was formed with histamine and the least with tyramine. Adsorption of tyramine and tryptamine improved when the surface was modified by the adsorption of bovine serum albumin. In that case, tyramine adsorption was higher than the respective of tryptamine (Vlasova *et al.*, 2011). More recently, Amghouz *et al.* (2014) examined the potential of zirconium phosphate subjected to annealing in NaCl-NaOH aqueous solution to adsorb histamine, putrescine, cadaverine and tyramine in synthetic wine. The fastest-adsorbed amine was histamine followed by cadaverine, putrescine and tyramine.

## 6. Conclusions and Future Perspectives

Accumulation of biogenic amines in wine depends upon a series of factors related to agricultural practices, oenological parameters, aging and storage conditions. Some of these factors have been assessed to some extent. The only effective intervention that has been applied is the selection of appropriate starter cultures, i.e. microorganisms that lack decarboxylase activities. However, a lot of research is still necessary in order to identify suitable field procedures that may prevent biogenic amine built-up in the grape.

## References

Alabadi, D. and Carbonell, J. (1998). Expression of ornithine decarboxylase is transiently increased by pollination, 2,4- dichlorophenoxyacetic acid, and gibberellic acid in tomato ovaries. *Plant Physiology* 118: 323-328.

Alabadi, D., Aguero, M.S., Perez-Amador, M.A. and Carbonell, J. (1996). Arginase, arginine decarboxylase, ornithine decarboxylase and polyamines in tomato ovaries: Changes in unpollinated ovaries and parthenocarpic fruits induced by auxin or gibberellin. *Plant Physiology* 112: 1237-1244.

Alvarez, M.A. and Moreno-Arribas, M.V. (2014). The problem of biogenic amines in fermented foods and the use of potential biogenic amine-degrading microorganisms as a solution. *Trends in Food Science & Technology* 39: 146-155.

Amghouz, Z., Ancin-Azpilicueta, C., Burusco, K.K., Garcia, J.R., Khainakov, S.A., Luquin, A., Nieto, R. and Garrido, J.J. (2014). Biogenic amines in wine: Individual and competitive adsorption on a modified zirconium phosphate. *Microporous and Mesoporous Materials* 197: 130-139.

Ancin-Azpilicueta, C., Jimenez-Moreno, N., Moler, J.A., Nieto-Rojo, R. and Urmeneta, H. (2016). Effects of reduced levels of sulphite in wine production using mixtures with lysozyme and dimethyl dicarbonate on levels of volatile and biogenic amines. *Food Additives & Contaminants: Part A* 33: 1518-1526.

Bais, H.P. and Ravishankar, G.A. (2002). Role of polyamines in the ontogeny of plants and their biotechnological applications. *Plant Cell, Tissue and Organ Culture* 69: 1-34.

Bais, H.P., George, J. and Ravishankar, G.A. (1999a). Influence of polyamines on growth of hairy root cultures of witloof chicory (*Cichorium intybus* L. cv. Lucknow local) and formation of coumarins. *Journal of Plant Growth Regulation* 18: 33-37.

Bais, H.P., Sudha, G. and Ravishankar, G.A. (1999b). Putrescine influences growth and production of coumarins in hairy root cultures of *Cichorium intybus* L. cv. Lucknow local (witloof chicory). *Journal of Plant Growth Regulation* 18: 159-165.

Bais, H.P., Bhagyalakshmi, N., Rajasekaran, T. and Ravishankar, G.A. (2000a). Influence of polyamines on growth and production of secondary metabolites in hairy root cultures of *Beta vulgaris* and *Tagetes patula*. *Acta Physiologicae Plantarum* 22: 151-158.

Bais, H.P., Sudha, G. and Ravishankar, G.A. (2000b). Putrescine and silver nitrate influences shoot multiplication, *in vitro* flowering and endogenous titres of polyamines in *Cichorium intybus* L. cv. Lucknow local. *Journal of Plant Growth Regulation* 19: 238-248.

Baraldi, R., Bertazza, G., Bregoli, A.M., Fasolo, F., Rotondi, A., Predieri, S., Serafini-Fracassini, D., Slovin, J.P.and Cohen, J.D. (1995). Auxins and polyamines in relation to differential in vitro root induction on microcuttings of two pear cultivars. *Journal of Plant Growth Regulation* 11: 21-31.

Bardocz, S. (1995). Polyamines in food and their consequences for food quality and human health. *Trends in Food Science and Technology* 6: 341-346.

Basile, T., Alba, V., Suriano, S., Savino, M. and Tarricone, L. (2017). Effects of ageing on stilbenes and biogenic amines in red grape winemaking with stem contact maceration. *Journal of Food Processing and Preservation* 2017: e13378.

Basozabal, I., Gomez-Caballero, A., Diaz-Diaz, G., Guerreiro, A., Gilby, S., Goicolea, M.A. and Barrio, R.J. (2013). Rational design and chromatographic evaluation of histamine imprinted polymers optimised for solid-phase extraction of wine samples. *Journal of Chromatography A* 1308: 45-51.

Benito, A., Calderon, F. and Benito, S. (2016). Combined use of *S. pombe* and *L. thermotolerans* in winemaking: Beneficial effects determined through the study of wines' analytical characteristics *Molecules* 21: 1744.

Biasi, R., Costa, G. and Bagni, N. (1991). Polyamine metabolism is related to fruit set and growth. *Plant Physiology and Biochemistry* 29: 497-506.

Bouchereau, A., Aziz, A., Larher, F. and Martin-Tanguy, J. (1999). Polyamines and environmental challenges: Recent developments. *Plant Science* 140: 103-125.

Callejon, S., Sendra, R., Ferrer, S. and Pardo, I. (2014). Identification of a novel enzymatic activity from lactic acid bacteria able to degrade biogenic amines in wine. *Applied Microbiology and Biotechnology* 98: 185-198.

Callejon, S., Sendra, R., Ferrer, S. and Pardo, I. (2016). Cloning and characterisation of a new laccase from *Lactobacillus plantarum* J16 CECT 8944 catalysing biogenic amines degradation. *Applied Microbiology and Biotechnology* 100: 3113-3124.

Capell, T., Bassie, L. and Christou, P. (2004). Modulation of the polyamine biosynthetic pathway in transgenic rice confers tolerance to drought stress. *Proceedings of the National Academy of Sciences, USA* 101: 9909-9914.

Capozzi, V., Russo, P., Ladero, V., Fernandez, M., Fiocco, D., Alvarez, M.A., Grieco, F. and Spano, G. (2012). Biogenic amines degradation by *Lactobacillus plantarum*: Toward a potential application in wine. *Frontiers in Microbiology* 3: 122.

Caruso, M., Fiore, C., Contursi, M., Salzano, G., Paparella, A. and Romano, P. (2002). Formation of biogenic amines as criteria for the selection of wine yeasts. *World Journal of Microbiology and Biotechnology* 18: 159-163.

Chang, S.-C., Lin, C.-W., Jiang, C.-M., Chen, H.-C., Shih, M.-K., Chen, Y.-Y.and Tsai, Y.-H. (2009). Histamine production by bacilli bacteria, acetic bacteria and yeast isolated from fruit wines. *LWT – Food Science and Technology* 42: 280-285

Chattopadhyay, M.K., Tabor, C.W. and Tabor, H. (2003). Polyamines protect *Escherichia coli* cells from the toxic effect of oxygen. *Proceedings of the National Academy of Sciences*, USA 100: 2261-2265.

Cheng, S.H., Shyr, Y.Y. and Kao, C.H. (1984). Senescence in rice leaves, XII: Effect of 1,3-diaminopropane, spermidine and spermine. *Botanical Bulletin of Academia Sinica* 25: 191-196.

Cona, A., Rea, G., Angelini, R., Federico, R. and Tavladoraki, P. (2006). Functions of amine oxidases in plant development and defence. *Trends in Plant Science* 11: 80-88.

Cueva, C., Garcıa-Ruiz, A., Gonzalez-Rompinelli, E., Bartolome, B., Martın-Alvarez, P.J., Salazar, O., Vicente, M.F., Bills, G.F. and Moreno-Arribas, M.V. (2012). Degradation of biogenic amines by vineyard ecosystem fungi: Potential use in winemaking. *Journal of Applied Microbiology* 112: 672-682.

Cunha, S.C., Faria, M.A. and Fernandes, J.O. (2011). Gas chromatography-mass spectrometry assessment of amines in port wine and grape juice after fast chloroformate extraction/derivatisation. *Journal of Agricultural and Food Chemistry* 59: 8742-8753.

Dapkevicius, M.L.N.E., Nout, M.J.R., Rombouts, F.M., Houben, J.H. and Wymenga, W. (2000). Biogenic amine formation and degradation by potential fish silage starter microorganisms. *International Journal of Food Microbiology* 57: 107-114.

Del Prete, V., Costantini, A., Cecchini, F., Morassut, M. and Garcia-Moruno, E. (2009). Occurrence of biogenic amines in wine: The role of grapes. *Food Chemistry* 112: 474-481.

Downs, C.G. and Lovell, P.H. (1986). The effect of spermidine and putrescine on the senescence of cut carnations. *Physiologia Plantarum* 66: 679-684.

EFSA (2011). Scientific opinion on risk-based control of biogenic amine formation in fermented foods. *European Food Safety Authority Journal* 9: 2393-2486.

Fadda, S., Vignolo, G. and Oliver, G. (2001). Tyramine degradation and tyramine/histamine production by lactic acid bacteria and Kocuria strains. *Biotechnology Letters* 23: 2015-2019.

Galgano, F., Caruso, M., Perretti, G. and Favati, F. (2011). Authentication of Italian red wines on the basis of the polyphenols and biogenic amines. *European Food Research and Technology* 232: 889-897.

Galston, A.W. and Kaur-Sawhney, R. (1987). Polyamines as endogenous growth regulators. pp. 280-295. *In*: P.J. Davies (Ed.). Plant Hormones and Their Role in Plant Growth and Development. Martinus Nijhoff, Dordrecht.

Garcia-Marino, M., Trigueros, A. and Escribano-Bailon, T. (2010). Influence of oenological practices on the formation of biogenic amines in quality red wines. *Journal of Food Composition and Analysis* 23: 455-462.

Garcia-Ruiz, A., Gonzalez-Rompinelli, E.M., Bartolome, B. and Moreno-Arribas, M.V. (2011). Potential of wine-associated lactic acid bacteria to degrade biogenic amines. *International Journal of Food Microbiology* 148: 115-120.

Garrido, R.L., Chibi, F. and Matila, A. (1995). Polyamines in the induction of *Nicotiana tabacum* pollen embryogenesis by starvation. *Journal of Plant Physiology* 145: 731-735.

Gerats, A.G.M., Kaye, C., Collins, C. and Malmberg, R.L. (1988). Polyamine levels in *Petunia* genotypes with normal and abnormal floral morphologies. *Plant Physiology* 86: 390-393.

Gloria, M.B.A. and Vieira, S.M. (2007). Technological and toxicological significance of bioactive amines in grapes and wines. *Food* 1: 258-270.

Gonzalez Marco, A. and Ancin Azpilicueta, C. (2006). Amine concentrations in wine stored in bottles at different temperatures. *Food Chemistry* 99: 680-685.

Granchi, L., Romano, P., Mangani, S., Guerrini, S. and Vincenzini, M. (2005). Production of biogenic amines by wine microorganisms. *Bullettin OIV* 78: 595-609.

Guo, Y.S., Guo, X.W. and Zhang, H.E. (2007). Changes in polyamine contents of ovules during grape (*Vitis vinifera* L.) embryo development and abortion. *Plant Physiology* 43: 53-56.

Ha, H.C., Sirisoma, N.S., Kuppusamy, P., Zweier, J.L., Woster, P.M. and Casero, R.A. Jr. (1998). The natural polyamine spermine functions directly as a free radical scavenger. *Proceedings of the National Academy of Sciences*, USA 95: 11140-11145.

Halasz, A., Barath, A., Simon-Sarkadi, L. and Holzapfel, W. (1994). Biogenic amines and their production by microorganisms in food. *Trends in Food Science and Technology* 5: 42-49.

Herrero-Fresno, A., Martinez, N., Sanchez-Llana, E., Diaz, M., Fernandez, M., Martin, M.C., Ladero, V. and Alvarez, M.A. (2012). *Lactobacillus casei* strains isolated from cheese reduce biogenic amine accumulation in an experimental model. *International Journal of Food Microbiology* 157: 297-304.

Igarashi, K. and Kashiwagi, K. (2000). Polyamines: Mysterious modulators of cellular functions. *Biochemical and Biophysical Research Communications* 271: 559-564.

Iijima, S., Sato,Y., Bounoshita, M., Miyaji, T., Tognarelli, D.J. and Saito, M. (2013). Optimisation of an online post-column derivatization system for Ultra High-Performance Liquid Chromatography (UHPLC) and its applications to analysis of biogenic amines. *Analytical Sciences* 29: 539-545.

Jastrzebska, A., Piasta, A. and Szlyk, E. (2014). Simultaneous determination of selected biogenic amines in alcoholic beverage samples by isotachophoretic and chromatographic methods. *Food Additives and Contaminants A* 31: 83-92.

Jimenez Moreno, N., Torrea Goni, D. and Ancin Azpilicueta, C. (2003). Changes in amine concentrations during aging of red wine in oak barrels. *Journal of Agricultural and Food Chemistry* 51: 5732-5737.

Jirage, D.B., Ravishankar, G.A., Suvarnalatha, G. and Venkataraman, L.V. (1994). Profile of polyamines during sprouting and growth of saffron (*Crocus sativus* L.) corms. *Journal of Plant Growth Regulation* 13: 69-72.

Karatan, E., Duncan, T.R. and Watnick, P.I. (2005). NspS, a predicted polyamine sensor, mediates activation of *Vibrio cholerae* biofilm formation by norspermidine. *Journal of Bacteriology* 187: 7434-7443.

Kasukabe, Y., He, L., Nada, K., Misawa, S., Ihara, I. and Tachibana, S. (2004). Over-expression of spermidine synthase enhances tolerance to multiple environmental stresses and up-regulates the expression of various stress-regulated genes in transgenic *Arabidopsis thaliana. Plant and Cell Physiology* 45: 712-722.

Kasukabe, Y., He, L., Watakabe, Y., Otani, M., Shimada, T. and Tachibana, S. (2006). Improvement of environmental stress tolerance of sweet potato by introduction of genes for spermidine synthase. *Plant Biotechnology* 23: 75-83.

Kaur-Sawhney, R., Tiburcio, A.F., Altabella, T. and Galston, A.W. (2003). Polyamines in plants: An overview. *Journal of Cell and Molecular Biology* 2: 1-12.

Kim, I.G. and Oh, T.J. (2000). SOS induction of the *recA* gene by UV-, γ-irradiation and mitomycin C is mediated by polyamines in *Escherichia coli* K-12. *Toxicology Letters* 116: 143-149.

Kiss, J., Korbaz, M. and Sass-Kiss, A. (2006). Study of amine composition of botrytised grape berries. *Journal of Agricultural and Food Chemistry* 54: 8909-8918.

Kusano, T., Berberich, T., Tateda, C. and Takahashi, Y. (2008). Polyamines: Essential factors for growth and survival. *Planta* 228: 367-381.

Kusano, T., Yamaguchi, K., Berberich, T. and Takahashi Y. (2007a). Advances in polyamine research in 2007. *Journal of Plant Research* 120: 345-350.

Kusano, T., Yamaguchi, K., Berberich, T. and Takahashi, Y. (2007b). The polyamine spermine rescues Arabidopsis from salinity and drought stresses. *Plant Signalling and Behaviour* 2: 250-251.

Ladero, V., Calles-Enriquez, M., Fernandez, M. and Alvarez, M. (2010). Toxicological effects of dietary biogenic amines. *Current Nutrition and Food Science* 6: 145-156.

Landete, J. Ma, Pardo, I. and Ferrer, S. (2006). Histamine, histidine and growth-phasemediated regulation of the histidine decarboxylase gene in lactic acid bacteria isolated from wine. *FEMS Microbiology Letters* 260: 84-90.

Lee, J. and Schreiner, R.P. (2010). Free amino acid profiles from "Pinot Noir" grapes are influenced by vine N-status and sample preparation method. *Food Chemistry* 119: 484-489.

Lee, T.M. and Lin, Y.H. (1996). Opposite effects of Fusicoccin and IAA on putrescine synthesis of rice coleoptiles. *Physiologia Plantarum* 97: 63-68.

Leuschner, R.G., Heidel, M. and Hammes, W.P. (1998). Histamine and tyramine degradation by food fermenting microorganisms. *International Journal of Food Microbiology* 39: 1-10.

Lindemose, S., Nielsen, P.E. and Mollegaard, N.E. (2005). Polyamines preferentially interact with bent adenine tracts in double-stranded DNA. *Nucleic Acids Research* 33: 1790-1803.

Liu, J.-H., Kitashiba, H., Wang, J., Ban, Y. and Moriguchi, T. (2007). Polyamines and their ability to provide environmental stress tolerance to plants. *Plant Biotechnology* 24: 117-126.

Lopez, R., Tenorio, C., Gutierrez, A.R., Garde-Cerdan, T., Garijo, P., Gonzalez-Arenzana, L., Lopez-Alfaro, I. and Santamaria, P. (2012). Elaboration of Tempranillo wines at two different pHs. Influence on biogenic amine contents. *Food Control* 25: 583-590.

Lorenzo, C., Bordiga, M., Perez-Alvarez, E.P., Travaglia, F., Arlorio, M., Salinas, M.R., Coisson, J.D. and Garde-Cerdan, T. (2017). The impacts of temperature, alcoholic degree and amino acids content on biogenic amines and their precursor amino acids content in red wine. *Food Research International*, http://dx.doi.org/10.1016/j.foodres.2017.05.016

Marco, A.G., Moreno, N.J. and Azpilicueta, C.A. (2006). Influence of addition of yeast autolysate on the formation of amines in wine. *Journal of the Science of Food and Agriculture* 86: 2221-2227.

Marcobal, A., Martin-Alvarez, P.J., Polo, M.C., Munoz, R. and Moreno-Arribas, M.V. (2006). Formation of biogenic amines throughout the industrial manufacture of red wine. *Journal of Food Protection* 69: 397-404.

Marques, A.P., Leitao, M.C. and San Romao, M.V. (2008). Biogenic amines in wines: Influence of oenological factors. *Food Chemistry* 107: 853-860.

Martınez-Pinilla, O., Guadalupe, Z., Hernandez, Z. and Ayestaran, B. (2013). Amino acids and biogenic amines in red varietal wines: The role of grape variety, malo-lactic fermentation and vintage. *European Food Research and Technology* 237: 887-895.

Martuscelli, M., Crudele, M.A., Gardini, F. and Suzzi, G. (2000). Biogenic amine formation and oxidation by *Staphylococcus xylosus* strains from artisanal fermented sausages. *Letters in Applied Microbiology* 31: 228-232.

Molenaar, D., Bosscher, J.S., Ten Brink, B., Driessen, A.J.M. and Konings, W.N. (1993). Generation of a proton motive force by histidine decarboxylation and electrogenic histidine/histamine antiport in *Lactobacillus buchneri*. *Journal of Bacteriology* 175: 2864-2870.

Montague, M.J., Armstrong, T.A. and Jarworski, E.G. (1979). Polyamine metabolism in embryogenic cells of *Daucus carota*. II: Changes in arginine decarboxylase activity. *Plant Physiology* 63: 341-345.

Montes, R., Rodriguez, I., Ramil, M., Rubi, E. and Cela, R. (2009). Solid-phase extraction followed by dispersive liquid-liquid microextraction for the sensitive determination of selected fungicides in wine. *Journal of Chromatography A* 1216: 5459-5466.

Navakoudis, E., Lutz, C., Langebartels, C., Lutz-Meindl, U. and Kotzabasis, K. (2003). Ozone impact on the photosynthetic apparatus and the protective role of polyamines. *Biochimica Biophysica Acta* 1621: 160-169.

Ortega-Heras, M., Perez-Magarino, S., Del-Villar-Garrachon, V., Gonzalez-Huerta, C., Gonzalez, L.C.M., Rodriguez, A.G., Sanchez, S.V., Gonzalez, R.G. and Helguera, S.M. (2014). Study of the effect of vintage, maturity degree and irrigation on the amino acid and biogenic amine content of a white wine from the Verdejo variety. *Journal of the Science of Food and Agriculture* 94: 2073-2082.

Palermo, C., Muscarella, M., Nardiello, D., Iammarino, M. and Centonze, D. (2013). A multiresidual method based on ion-exchange chromatography with conductivity detection for the determination of biogenic amines in food and beverages. *Analytical and Bioanalytical Chemistry* 405: 1015-1023.

Pastre, D., Pietrement, O., Landousy, F., Hamon, L., Sorel, I., David, M.O., Delain, E., Zozime, A. and Le Cam, E. (2006). A new approach to DNA bending by polyamines andits implication in DNA condensation. *European Biophysics Journal* 35: 214-223.

Patel, C.N., Wortham, B.W., Lines, J.L., Fetherston, J.D., Perry, R.D. and Oliveira, M.A. (2006). Polyamines are essential for the formation of plague biofilm. *Journal of Bacteriology* 188: 2355-2363.

Pedroso, M.C., Primikiros, N., Roubelakis, A. and Pais, M.A. (1997). Free and conjugated polyamines in embryogenic and nonembryogenic leaf regions of Camellia leaves before and during direct somatic embryogenesis. *Plant Physiology* 101: 213-217.

Polissi, A., Pontiggia, A., Feger, G., Altieri, M., Mottl, H., Ferrari, L. and Simon, D. (1998). Large-scale identification of virulence genes from *Streptococcus pneumoniae*. *Infection and Immunity* 66: 5620-5629.

Preti, R. and Vinci, G. (2016). Biogenic amine content in red wines from different protected designations of origin of Southern Italy: Chemometric characterisation and classification. *Food Analytical Methods* 9: 2280-2287.

Ramos, R.M., Valente, I.M. and Rodrigues, J.A. (2014). Analysis of biogenic amines in wines by salting-out assisted liquid-liquid extraction and high-performance liquid chromatography with fluorimetric detection. *Talanta* 124: 146-151.

Rollan, G.C., Coton, E. and Lonvaud-Funel, A. (1995). Histidine decarboxylase activity of *Leuconostoc oenos* 9204. *Food Microbiology* 12: 455-461.

Russo, P., Capozzi, V., Spano, G., Corbo, M.R., Sinigaglia, M. and Bevilacqua, A. (2016). Metabolites of microbial origin with an impact on health: Ochratoxin A and biogenic amines. *Frontiers in Microbiology* 7: 482.

Sass-Kiss, A., Szerdahelyi, E. and Hajos, G. (2000). Study of biologically active amines in grape and wines by HPLC. *Chromatographia Supplement* 51: S316-S320.

Scanlan, R.A. (1983). Formation and occurrence of nitrosamines in food. *Cancer Research* 43: 2435s-2440s.

Sebastian, P., Herr, P. and Fischer, U. (2011). Molecular identification of lactic acid bacteria occurring in must and wine. *South African Journal for Enology and Viticulture* 32: 300-309.

Seiler, N. and Raul, F. (2005). Polyamines and apoptosis. *Journal of Cellular and Molecular Medicine* 9: 623-642.

Silla-Santos, M.H. (1996). Biogenic amines: Their importance in foods. *International Journal of Food Microbiology* 29: 213-231.

Smit, I., Pfliehinger, M., Binner, A., Großmann, M., Horst, W.J. and Lohnertz, O. (2014). Nitrogen fertilisation increases biogenic amines and amino acid concentrations in *Vitis vinifera* var. Riesling musts and wines. *Journal of the Science of Food and Agriculture* 94: 2064-2072.

Soksawatmaekhin, W., Kuraishi, A., Sakata, K., Kashiwagi, K. and Igarashi, K.(2004). Excretion and uptake of cadaverine by CadB and its physiological functions in *Escherichia coli*. *Molecular Microbiology* 51: 1401-1412.

Song, P., Wu, L. and Guan, W. (2015). Dietary nitrates, nitrites and nitrosamines intake and the risk of gastric cancer: A meta-analysis. *Nutrients* 7: 9872-9895.

Stevenson, L.G. and Rather, P.N. (2006). A novel gene involved in regulating the flagellar gene cascade in *Proteus mirabilis*. *Journal of Bacteriology* 188: 7830-7839.

Sturgill, G. and Rather, P.N. (2004). Evidence that putrescine acts as an extracellular signal required for swarming in *Proteus mirabilis*. *Molecular Microbiology* 51: 437-446.

Tabor, C.W. and Tabor, H. (1985). Polyamines in microorganisms. *Microbiological Reviews* 49: 81-99.

Takahashi, Y., Berberich, T., Miyazaki, A., Seo, S., Ohashi, Y. and Kusano, T. (2003). Spermine signalling in tobacco: Activation of mitogen-activated proteinkinases by spermine is mediated through mitochondrial dysfunction. *The Plant Journal* 36: 820-829.

Tasev, K., Ivanova-Petropulos, V. and Stefova, M. (2017). Ultra-Performance Liquid Chromatography-Triple Quadruple Mass Spectrometry (UPLC-TQ/MS) for evaluation of biogenic amines in wine food. *Analytical Methods*, DOI 10.1007/s12161-017-0936-9

Teitel, D.C., Cohen, E., Arad, S., Birnbaum, E. and Mizrahi, Y. (1985). The possible involvement of polyamines in the development of tomato fruits *in vitro*. *Plant Growth Regulation* 3: 309-317.

Tkachenko, A., Nesterova, L. and Pshenichnov, M. (2001). The role of the natural polyamine putrescine in defence against oxidative stress in *Escherichia coli*. *Archives of Microbiology* 176: 155-157.

Torrigiani, P., Altamura, M.M., Pasqua, G., Monacelli, B., Serafini-Fracassini, D. and Bagni, N. (1987b). Free and conjugated polyamines during *de novo* floral and vegetative bud formation in thin cell-layers of tobacco. *Physiologia Plantarum* 70: 453-460.

Toy, N., Ozogul, F. and Ozogul, Y. (2015). The influence of the cell free solution of lactic acid bacteria on tyramine production by food borne-pathogens in tyrosine decarboxylase broth. *Food Chemistry* 173: 45-53.

Tristezza, M., Vetrano, C., Bleve, G., Spano, G., Capozzi, V., Logrieco, A., Mita, G. and Grieco, F. (2013). Biodiversity and safety aspects of yeast strains characterized from vineyards and spontaneous fermentations in the Apulia Region, Italy. *Food Microbiology* 36: 335-342.

Tun, N.N., Santa-Catarina, C., Begum, T., Silveira, V., Handro, W., Floh, E.I. and Scherer, G.F. (2006). Polyamines induce rapid biosynthesis of nitric oxide (NO) in *Arabidopsis thaliana* seedlings. *Plant and Cell Physiology* 47: 346-354.

Urano, K., Yoshiba, Y., Nanjo, T., Ito, T., Yamaguchi-Shinozaki, K. and Shinozaki, K. (2004). Arabidopsis stress-inducible gene for arginine decarboxylase AtADC2 is required for accumulation of putrescine in salt tolerance. *Biochemical and Biophysical Research Communications* 313: 369-375.

Vlasova, N.N., Markitan, O.V. and Golovkova, L.P. (2011). Adsorption of biogenic amines on albumin modified silica surface. *Colloid Journal* 73: 24-27.

Vlasova, N.N., Markitan, O.V. and Stukalina, N.G. (2006). The adsorption of biogenic amines on the surface of highly dispersed silica from aqueous solutions. *Colloid Journal* 68: 384-386.

Walker, M.A., Roberts, D.R., Shih, C.Y. and Dumbroff, E.B. (1985). A requirement for polyamines during the cell division phase of radicle emergence in seeds of *Acer saccharum*. *Plant and Cell Physiology* 26: 967-972.

Wang, Y.Q., Ye, D.Q., Zhu, B.Q., Wu, G.F. and Duan, C.Q. (2014). Rapid HPLC analysis of amino acids and biogenic amines in wines during fermentation and evaluation of matrix effect. *Food Chemistry* 163: 6-15.

Ware, D., Jiang, Y., Lin, W. and Swiatlo, E. (2006). Involvement of *potD* in *Streptococcus pneumoniae* polyamine transport and pathogenesis. *Infection and Immunity* 74: 352-361.

Wen, X.P., Pang, X.M., Matsuda, N., Kita, M., Inoue, H., Hao, Y.-J., Honda, C. and Moriguchi, T. (2007). Over-expression of the apple spermidine synthase gene in pear confers multiple abiotic stress tolerance by altering polyamine titers. *Transgenic Research* 17: 251-263.

Wolken, W.A., Lucas, P.M., Lonvaud-Funel, A. and Lolkema, J.S. (2006). The mechanism of the tyrosine transporter TyrP supports a proton motive tyrosine decarboxylation pathway in *Lactobacillus brevis*. *Journal of Bacteriology* 188: 2198-2206.

Xie, C., Wang, H., Deng, S. and Xu, X.L. (2016). The inhibition of cell-free supernatant of *Lactobacillus plantarum* on production of putrescine and cadaverine by four amine-positive bacteria *in vitro*. *LWT – Food Science and Technology* 67: 106-111.

Yadav, J.S. and Rajam, M.V. (1997). Spatial distribution of free and conjugated polyamines in leaves of *Solanum melongena* associated with differential morphogenetic capacity efficient somatic embryogenesis with putrescine. *Journal of Experimental Botany* 48: 1537-1545.

Yamaguchi, K., Takahashi, Y., Berberich, T., Imai, A., Miyazaki, A., Takahashi, T., Michael, A. and Kusano, T. (2006). The polyamine spermine protects against high salt stress in *Arabidopsis thaliana*. *FEBS Letters* 580: 783-788.

Yamaguchi, K., Takahashi, Y., Berberich, T., Imai, A., Takahashi, T., Michael, A. and Kusano, T. (2007). A protective role for the polyamine spermine against drought stress in Arabidopsis. *Biochemical and Biophysical Research Communications* 352: 86-90.

Yamasaki, H. and Cohen, M.F. (2006). No signal at the crossroads: Polyamine-induced nitric oxide synthesis in plants? *Trends in Plant Science* 11: 522-524.

Zaman, M.Z., Bakar, F.A., Jinap, S. and Bakar, J. (2011). Novel starter cultures to inhibit biogenic amines accumulation during fish sauce fermentation. *International Journal of Food Microbiology* 145: 84-91.

Zaman, M.Z., Bakar, F.A., Selamat, J. and Bakar, J. (2010). Occurrence of biogenic amines and amines degrading bacteria in fish sauce. *Czech Journal of Food Sciences* 28: 440-449.

# 18 Immobilised Yeast in Winemaking

**Steva Lević[1*], Verica Đorđević[2], Ana Kalušević[1], Radovan Đorđević[1], Branko Bugarski[2] and Viktor Nedović[1]**

[1] Department of Food Technology and Biochemistry, Faculty of Agriculture, University of Belgrade, Nemanjina 6, 11 080, Belgrade-Zemun, Serbia

[2] Department of Chemical Engineering, Faculty of Technology and Metallurgy, University of Belgrade, Karnegijeva 4, 11 120 Belgrade, Serbia

## 1. Introduction

Numerous food products have been produced using living cells, such as bacteria, yeasts or fungus. In the traditional food processes, these products are usually obtained by metabolic activity of live and free microbial cells. Sometimes, the cells remain in the product, while some processes require removal of microbial biomass in order to preserve the product and fulfil sensorial standards. However, in recent years, the growing demand for process intensifications, rationalisation and creation of products with new sensorial properties have opened the door for new types of biocatalysts. New enzymatic systems, specially selected cells stains and new reactor design, provide numerous possibilities for products development. One of the proposed approaches for process intensification is based on cells' additional protection by cells immobilisation, using adequate protective material.

Cells immobilisation may be defined as a process of cell protection and localisation, using protecting materials (i.e. carriers) (Margaritis and Kilonzo, 2005). The main goals of cells immobilisation are preserving the cell's catalytic activity, and at the same time, achievement of some other benefits from cell immobilisation, such as:

- Simplified cells manipulation and separation
- High cells concentration inside reactor
- Cells protection against unfavourable conditions
- Intensification of continuous fermentation processes

The immobilised cells are linked to some sort of supporting materials, mainly of sizes much bigger than the size of the individual cell. Thus, the separation of immobilised cells from fermentation medium and further manipulation is facilitated as compared to free cells. Another advantage of immobilised cells is the possibility of their reuse in the consecutive fermentation processes.

High cells concentration obtained by immobilisation may significantly reduce the fermentation time. Using of the immobilised cells could increase productivity and overall improve economical aspects of the fermentation processes. The increased reaction rate and better end products, especially regarding the clarity of beverages, are just some of the additional advantages of immobilised cells as compared to fermentation processes with free suspended cells (Neves *et al.*, 2014). Also, high cell concentration in the immobilised forms may also reduce the size of bioreactors needed for successful beverages fermentation (Kourkoutas *et al.*, 2004).

Protection role of immobilisation is also important, especially when harmful compounds are present in the fermentation media. These compounds could decrease productivity or completely block cells metabolism. For example, Lalou *et al.* (2013) used immobilised yeast cells for production of bioflavours from hydrolysed orange peel. Prior to fermentation, yeast cells were immobilised into calcium alginate matrix in the form of spherical beads. Alginate showed a protective role and successfully prevented the cell inhibition by harmful compounds present in the fermentation medium. Also, the system is suitable for repeated batch fermentations in multiple production cycles. In such cases, the application of free (i.e. non-protected) cells is almost impossible or economically unjustified due to cells inhibition.

*Corresponding author: slevic@agrif.bg.ac.rs

The protective role of immobilisation is also important when the cells are exposed to high ethanol concentrations. It was noticed that immobilisation reduced ethanol's negative effect on cells viability and acted as stress reduction factor (Kourkoutas *et al.*, 2004).

Although, not directly related to fermentation processes, the other benefits of immobilisation technology should be also mentioned. Immobilisation may be used for controlled delivery of active compounds and taste masking (e.g. the unpleasant taste of some food additives could be masked by formation of sensorial neutral protective layer around them) (Zuidam and Shimoni, 2010a; Nedović *et al.*, 2013). In general, such protection of food active compounds was found to be promising and today is the standard procedure in production of food additives (e.g. aromas) (Zuidam and Heinrich, 2010b). Also, this trend of introduction of immobilisation in the food sector has become more popular in the recent years for immobilisation of enzymes (Lević *et al.*, 2016) and cells (e.g. probiotics or for fermentations) (Nedović, 1999; Đorđević *et al.*, 2015; Kourkoutas *et al.*, 2004).

Besides its numerous advantages, especially compared to processes with free cells, the immobilisation also has some disadvantages. The primarily obstacle in the wider applications of immobilised cells in the fermentation processes are the operational costs, viz. the introduction of immobilisation into the production process means additional investments and adoption of new technology. The necessary investments are needed due to the production of cell biomass and new equipment for immobilisation procedures. Also, if the immobilisation procedure requires further cells stabilisation after immobilisation (e.g. drying), then the costs of these operations must be included too. According to Zuidam and Shimoni (2010a), the introduction of such new technological processes in the food industry is only justified if standard procedures do not fulfil the requirements.

The second very important negative aspect of cells immobilisation is the possible influence on cell metabolic activity and physiological functions. The influences of immobilisation procedure on cells viability, growth and physiology have been studied intensively in last several decades (Kourkoutas *et al.*, 2004), viz. the limited diffusion of nutrients and cell's metabolic products caused by immobilisation has significant influence on the cell's growth kinetic and reactions stoichiometry (Mensour *et al.*, 1996).

Another important aspect of cell immobilisation is the influence of immobilisation process on formation of volatile and generally aroma compounds that are essential for consumer's perception of final product. Some of the main issues related to aroma formation using immobilised yeast cells, especially during wine fermentation, are discussed below.

The choice of immobilisation method has a strong influence on metabolism and viability of immobilised cells. For example, cells entrapped in gel structure are usually affected by limited nutrients diffusion and by the accumulation of biomass inside gel. As a consequence, the cell growth inside the gel structure is not uniform, leading to cell starvation, especially in the case of cells that are located deeper inside gel matrix (Freeman and Lilly, 1998). Pajic-Lijakovic *et al.* (2015) showed that the structure of Ca-alginate gel beads changed during cell growth, but also it was affected by composition of fermentation medium. However, as fermentation progressed, formed cells clusters acted as a reinforcement. The same author also noticed the problem of cell's intensive growth close to the bead's surface rather than inside the beads (i.e. inside the core region). The problem related to non-uniform cells growth within gel matrix may be partially overcome by regulation of beads diameter. Nedović *et al.* (2001) noticed that smaller dimension of gel carriers is generally less affected on the cell's physiological activity. The mass transfer resistances are even more pronounced when additional coatings are added to protect immobilised cells. For example, alginate coated with chitosan to prevent the cell's leakage during fermentation could reduce production rates up to 40 per cent. The solution for this problem could be to increase the initial cells' concentration (Zhou *et al.*, 1998).

Also, some components that may be present in the fermentation medium (e.g. phosphates) can weaken the structure of some carriers, such as Ca-alginate (Lević *et al.*, 2015). The solution for this problem may be the synthetic materials that are relatively inactive and not affected by compounds present in the fermentation medium (Bezbradica *et al.*, 2007; Wandrey *et al.*, 2010), or by using of composites of natural and synthetic polymers as cells carriers (Bezbradica *et al.*, 2004).

Besides these limitations, cell immobilisation is now considered as a solution of many technological problems in food production. Currently, some of the immobilisation yeast cell system are commercialised and adopted as suitable biocatalysts for wine fermentation (*see* below).

## 2. Cell Immobilisation Methods

Generally, the methods for cell immobilisation can be classified into four groups (Verbelen *et al.*, 2010):

- Cell immobilisation by adsorption (i.e. surface attachment)
- Entrapment into gel matrix
- Cell immobilisation behind a barrier
- Cell aggregation

### 2.1. Cell Immobilisation by Adsorption

Adsorption of the cells (Verbelen *et al.*, 2010) as well as the enzymes (Lević *et al.*, 2016) on to organic or inorganic carrier materials is relatively an easy method for production of immobilised biocatalyst. Generally, the immobilisation procedure consists of mixing cells with carrier material (i.e. by making the contact between carrier and cells). Further, the cells are spontaneous binding on the carrier's surface by the electrostatic interactions or via covalent bonds. The cellulose derivates show good properties as carriers for cell immobilisation by adsorption (Verbelen *et al.*, 2010), while the other supporting materials, such as wood, sawdust, porous porcelain and porous glass may also be used. Even more, the basic carrier could be modified by addition of other materials in order to improve cell adsorption on to the carrier's surface (Kourkoutas *et al.*, 2004). More recently, a new class of materials with magnetic properties has attracted attention as potential carriers for immobilisation of biocatalysts by adsorption. The easy manipulation with biocatalysts immobilised on to such carriers under magnetic field offers numerous possibilities for better process control and biocatalyst reuse (Alftrén and Hobley, 2014).

### 2.2. Cell Entrapment into Gel Matrix

Cell immobilisation into gel matrix is one of the most frequently used methods for cell protection in biotechnology. Entrapment of cells inside the gel matrix is a relatively easy method and can be combined with natural and synthetic carrier materials (Nedović *et al.*, 2013). The main advantage of the gel carriers is the fact that they can be processed into a form of spherical beads with broad diameter sizes (Nedović *et al.*, 2001) as well as in the form of nano-sized fibres (Lević *et al.*, 2014). Natural polysaccharides were found to be very suitable for cell immobilisation using this method. According to Prüsse *et al.* (2008), uniform gel beads based on gelling properties of alginate can be produced using various extrusion techniques. This offers numerous possibilities for control of carrier size and intensification of immobilisation process.

Among available extrusion techniques, electrostatic extrusion has numerous advantages, especially when small and spherical carriers are required. Electrostatic extrusion is a very easy method for the production of a large quantity of gel beads with immobilised yeast cells. Alginate was found to be especially suitable for extrusion by this method. The electrostatic force between needle (i.e. alginate solution) and gelling bath (e.g. water solution of $Ca^{2+}$ salt) is sufficient to cut the liquid flow and ensure the production of Ca-alginate beads with diameter in the range 0.25-2.00 mm. The beads with small diameters showed as better carriers for yeast cell growth due to reduction in diffusion barrier that is more pronounced in the case of larger beads (Nedović *et al.*, 2001). Electrostatic extrusion is also suitable when the immobilisation procedure requires application of very viscous polymer solutions (Lević *et al.*, 2015) or combination of synthetic and natural carrier materials (Levic *et al.*, 2013).

### 2.3. Cell Immobilisation Behind a Barrier

For cell immobilisation behind porous barrier various carrier designs can be used. The barrier could be formed as porous fibres, membrane modules or as microcapsules (Verbelen *et al.*, 2010). Generally, membranes are suitable systems for cell containment inside bioreactors, especially in the processes that do not require high mass transfer of cell product and nutrients. During the previous several decades, the membrane modules were developed for cell immobilisation for wine fermentation too. However, the problems related to mass transfer obstruction are major disadvantages of this type of cell immobilisation (Kourkoutas *et al.*, 2004).

### 2.4. Cell Aggregation

Immobilisation by formation of cell aggregates (i.e. cell flocculation) is based on interaction between cells in the suspension, causing formation of larger particles (i.e. cell floccules). As a consequence, the cells can be more rapidly removed from fermentation broth, which is especially important in beer production, where cell separation is critical for the product's sensorial properties (Kourkoutas *et al.*, 2004). The ability of cells to form aggregates (i.e. to flocculate) is based on interaction between compounds (specific protein/carbohydrate residues interaction) located on the cell's surface. Formation of the aggregates depends on the cell's genetic properties, which further defines the chemical composition of the cell's wall. Also, the external influences, such as the presence of oxygen, specific cations and bioreactor conditions may contribute to efficiency of flocculation process (Soares, 2010). Moreover, the immobilisation by cell aggregation may be additionally improved, using the compounds that support the cell linkage and further stabilise aggregates (Kourkoutas *et al.*, 2004).

## 3. Immobilised versus Free Cell Technology in Winemaking

Immobilised cell technology has been intensively developed in food fermentation processes, particularly due to many technical and economical advantages of immobilised cells as compared to free cell systems. As pointed out above, the immobilisation provides high biomass concentration inside the reactor and consequently leads to rapid fermentations at ambient or extremely low temperatures, with higher productivities. Additionally, the cell immobilisation technology can be also implemented in the other fermentation process related to wine production, such as malo-lactic fermentation, which is important for sensorial properties of wines.

One of the main advantages of immobilised cell technology is better cell resistance against inhibitory substances present in the fermentation medium. Inhibitor effects of the ethanol, sulphur dioxide, heavy metals, phenolic compounds and essential oils as well as negative influences of external factors (e.g. extreme temperatures) may be reduced by cell immobilisation (Diviés and Cachon, 2005; Lalou *et al.*, 2013; Nedović *et al.*, 2015). Also, immobilised cells can provide easier separation of the yeast biomass at the end of fermentation. This is especially important in sparkling wine production where the separation of yeast cells at the end of bottle fermentation is a very expensive and inefficient process. Despite the modernisation of traditional sparkling wine bottle fermentation, this still remains a relatively slow operation. To overcome this particular problem, immobilisation of yeast cells into suitable forms could enable easy cell separation. One of the interesting concepts for this purpose could be the use of carrier materials with magnetic properties for cell immobilisation and consequently easy cell separation under magnetic field. For example, Berovic *et al.* (2014) succeeded in completing the separation of the magnetised biomass (magnetic nanoparticles absorbed on yeast) in the bottle using relatively weak magnetic-field gradient for approximately 15 min.

Other advantages of immobilised cell technology versus that with free cells is the possibility of storage of immobilised microorganisms for further use and reuse in repeated cycles of production. In this respect, it is preferable to formulate immobilised biocatalysts as products capable for storage, distribution and application in the agricultural and food marketplace. The process of freeze-drying provides protection to immobilised biocatalyst from contamination or infestation during storage, long viability and ease of biocatalyst distribution. However, this process has some undesirable effects and may cause a decrease in the viability of cells and some changes in their metabolic performances. For example, Sipsas *et al.* (2009) have found that after storage of freeze-dried immobilised *Saccharomyces cerevisiae* for six months, the fermentation process resulted in change of the aromatic profile of the final product. Thus, scientists have been searching for a proper biocatalyst which would give wine a satisfactory quality after re-activation of the freeze-dried immobilised cells (Iconomopoulou *et al.*, 2002; Kandylis *et al.*, 2010a; Sipsas *et al.*, 2009).

Immobilised cell technology offers the possibility for continuous industrial production, by which very high productivities are possible to achieve. Bakoyianis *et al.* (1997) have reported higher ethanol productivities at room and at low temperatures by continuous winemaking, in comparison to batch fermentations performed with free cells. However, the majority of fermenters usually used in wine industry are of batch type, especially in the production of high quality wines. Sipsas *et al.* (2009) showed that immobilised yeast cells applied in the special designed bioreactor could provide wine with satisfactory sensorial properties. Even more, the proposed immobilisation procedure is suitable for both batch and continuous fermentations.

Despite many benefits of immobilised cells, in the case of winemaking, full industrial use is still limited to the production of sparkling wines. There are several reasons for this situation. Firstly, there is a considerable impact on wine sensorial and oenological properties. Then, wine consumers usually have more trust in conventional than in modern winemaking practices. However, recently developed and commercially available immobilised yeast cells could be the basis for a new revolution in winemaking, overcoming some limitations of traditional fermentation processes.

## 4. Materials Used as Cell Supports in Winemaking

The carrier materials that can be used for cell immobilisation in winemaking should be primarily natural compounds of food grade quality (Table 1). Further, the carrier should be of low cost from sustainable sources and suitable for specific fermentation temperature and medium. Also, carrier for cell immobilisation should improve the wine quality, or at least not degrade wine's sensorial properties. The cells are either attached/adsorbed to a support or entrapped in a porous matrix. Adhesion of yeast cells is essentially dependent upon electrostatic interactions between the support and the cell's surface. Immobilisation of the yeast cells on the solid support may occur as a result of hydrogen bonding, entrapment of the cells inside the pores of the carrier, and due to the Van der Waals forces (Reddy *et al.*, 2011). The cells grow inside the porous structure of the carrier, resulting in increase of cell population. Consequently, during fermentation, cells number increase more near the surface of the support due to the greater availability of nutrients and some cells are released in the fermenting medium (Pajic-Lijakovic *et al.*, 2015).

All of these specific properties of carrier materials must be taken into consideration prior to the decision regarding the type of support and immobilisation procedure to be applied. Besides natural carrier materials, here should be mentioned that synthetic polymers also can be used for yeast cell immobilisation. The examples of synthetic materials for yeast immobilisation in fermentation processes are provided below.

### 4.1. Cell Immobilization in the Gel Matrix

The main focus of the first research trials in the 1980s and 1990s was on entrapment of bacteria and yeast cells into various gel systems, such as calcium alginate (Naouri *et al.*, 1991; Spettoli *et al.*, 1987), κ-carrageenan (Crapisi *et al.*, 1987a,b), polyacrylamide (Rossi and Clementi, 1984) and polyvinyl alcohol (Martynenko *et al.*, 2004) to perform alcoholic or malo-lactic fermentation of wine. Among these gel matrixes, calcium alginate gels are one of the most frequently used materials for immobilisation of yeast and other microorganism, at laboratory and industrial scale. Gel entrapment is mostly done by mixing cells with a polymer solution under sterile conditions followed by gelation, resulting in beads of a predetermined diameter range. The entrapment of yeast cells within calcium alginate hydrogel matrixes is performed by dropwise addition of sodium alginate with cell suspension to a solution containing divalent ions (usually $Ca^{2+}$), which causes spherical gel formations (i.e. alginate beads) with entrapped cells (Pajic-Lijakovic *et al.*, 2015). In general, alginate entrapped cells compared to free cells have improved stress tolerance and a better fermentation performance (Djordjević *et al.*, 2015), which depends on alginate concentration and beads diameter (Sevda and Rodrigues, 2011). However, alginate-based carriers have some limitations, such as gel degradation and poor mechanical stability during long-lasting fermentations and, in particular, mass transfer restrictions (Pajic-Lijakovic *et al.*, 2015). Kregiel *et al.*, (2013) observed reductions in enzyme activity (succinate dehydrogenase and pyruvate decarboxylase) and ATP content in yeast (*S. cerevisiae*) immobilised in foamed alginate. Furthermore, with this technique it was found that the critical factor is the leakage of the cells from the gel beads in the bottle during sparkling wine production (Fumi *et al.*, 1988; Gòdia *et al.* 1991). Oliveira *et al.* (2011) observed the breakage of the calcium alginate spheres during the fermentation process of fruit wine produced from cagaita (fruit grown in the central savannah region of Brazil), due to the increased cell density within the immobilised matrix.

**Table 1.** Immobilised Yeast Cell Cultures and Carrier Materials Proposed for Wine Fermentation and Stabilisation

| *Carrier material* | *Cell culture* | *Type of wine produced or treated by the immobilised yeast cells* | *References* |
|---|---|---|---|
| Ca-alginate beads | *S. cerevisiae* | Cabernet Sauvignon and Pinot Noir fermentation | Neves *et al.*, 2014 |
| Sugarcane pieces | *S. cerevisiae* | Bangalore blue fermentation | Reddy *et al.*, 2011 |
| Ca-alginate beads | Commercial available yeast strain | Sparkling wine production from several grape varieties | Miličević *et al.*, 2017 |
| κ-carragenate in the form of cylindrical beads | *S. cerevisiae* | Prevention of the browning in white wines | Merida *et al.*, 2007 |
| κ-carragenate in the form of cylindrical beads and alginate in the form of spherical beads | *S. cerevisiae* | Correction of colour of the white wines | Lopez-Toledano *et al.*, 2007 |
| Ca-alginate beads | *Candida intermedia* | Adsorption of ochratoxin A from grape juice | Farbo *et al.*, 2016 |
| Gluten pellets | *S. cerevisiae* | Batch and continuous fermentation of grape must | Sipsas *et al.*, 2009 |
| Double-layer alginate-chitosan | *S. cerevisiae* | Reduction of volatile acidity in wines | Vilela *et al.*, 2013 |
| Spheres of *Penicillium chrysogenum* | *S. cerevisiae* | Sweet wine production | García-Martínez *et al.*, 2015 |
| Co-immobilisation *S. cerevisiae*/*Penicillium chrysogenum* | *S. cerevisiae* | Sparkling wine production | Puig-Pujol *et al.*, 2013 |

Coating the beads with a gel layer without microorganisms is a way to avoid the problem of cell leakage. Double-layer alginate beads immobilisation has been tested in sparkling wine production (Gòdia *et al.*, 1991; Yokotsuka *et al.*, 1997) and the technology was also introduced in commercial production (*see* below). Namely, with the yeast immobilised in double-layer beads, the wine of fine clarity was obtained due to less leakage of viable cells from the gel matrices into the wine at the end of the fermentation process, compared with that of single-layer beads. Furthermore, double-layer bead-immobilised yeast was also more tolerant than free yeast to inhibitory effect of high $SO_2$ concentration (Yajima and Yokotsuka, 2001). Due to these advantages, the fermentation can be succeeded with a lower dosage of the external-layer alginate biocatalyst than with the simple alginate beads (Gòdia *et al.*, 1991).

One of the main disadvantages of gel matrix-type immobilisation is sensitivity of carrier material towards external stress conditions, viz. when gel beads are exposed to high pressure (e.g. in the bioreactor conditions), this may cause their weakening and cell release. In order to prevent negative effects of pressure on gel's mechanical properties, a multi-stage bioreactor was proposed for continuous wine fermentation (Ogbonna *et al.*, 1989).

Although immobilisation of yeast cells could reduce the fermentation time or provide the product with acceptable sensorial properties, the whole winemaking process depends on many other production parameters. Neves *et al.* (2014) showed that wines produced from Cabernet Sauvignon and Pinot Noir grape varieties, using thermovinification and immobilised yeast cells exhibited lower concentrations of anthocyanins, flavonoids and total phenolic compounds. The reasons for these losses could be the aging of wine in the bottle as well as thermovinification that could not achieve satisfactory extraction of tannins. This points out that immobilisation of cells could be beneficial, but the quality of final product depends on many other factors, especially in winemaking where grape characteristics (i.e. variety), ecological factors and other technological process steps (e.g. must extraction, wine clarification and aging) could limit the benefits that are derived from immobilised cells.

## 4.2. Cellulose-based Materials

Cellulose is the most abundant natural polymer and is widely used in immobilised cell technology. The main advantages of cellulosic carriers are their low cost, non-toxicity and high mechanical strength. However, due to chemical properties of these materials, they usually require chemical modification prior to use as cells carriers. Some early examples of modified cellulose-based carriers for cell immobilisation in fermentation processes, include the delignified cellulose (Bardi and Koutinas, 1994; Iconomou *et al.*, 1996; Kourkoutas *et al.*, 2002a). Other sources of cellulose are bacterial cells. Bacterial cellulose (mostly synthesised by *Acetobacter* strains) has attracted a lot of attention. The main differences between this and plant cellulose is that bacterial cellulose is free of lignin, pectin and hemicelluloses (Iguchi *et al.*, 2000; Qiu and Netravali, 2014). Also, bacterial cellulose has better mechanical properties, higher water adsorption capacity and biocompatibility (Astley *et al.*, 2001; Aber *et al.*, 2014). Immobilisation of wine yeast in bacterial cellulose was reported by Nguyen *et al.* (2009) and Ton *et al.* (2010, 2011). Generally, cellulose provide higher metabolic activity of yeast cells as well as better protection against stress conditions during fermentation. Ton *et al.* (2011) showed high potential of yeast cell immobilisation in bacterial cellulose for winemaking. The immobilised cells had good fermentation potential even after 10 consecutive cycles of the repeated batch fermentations.

## 4.3. Other Organic Materials

Various organic materials, usually by-products of food production processes, have been investigated as cell supports for wine fermentations. The materials, such as gluten pellets (Iconomopoulou *et al.*, 2002; Mallouchos *et al.*, 2003; Sipsas *et al.*, 2009), Brewer's spent grains (Mallouchos *et al.*, 2007; Kopsahelis *et al.*, 2012) and sugarcane (Reddy *et al.*, 2011) were used for cell immobilisation. The mechanism of cell immobilisation using these supporting materials is mainly based on cell entrapment into its porous structure and due to cell's physical adsorption (Fig. 1). The main disadvantage of this type of carriers is the cells release into fermentation medium as a consequence of biomass expansion during fermentation. Nevertheless, it has been showed that immobilisation of microorganisms into such natural porous structures can provide stability to cell biocatalysts during several repeated fermentation cycles. Some of the organic materials require chemical modifications (e.g. delignification) in order to increase the efficiency of the immobilisation process (Kandylis *et al.*, 2010a, b). On the other hand, carrier materials,

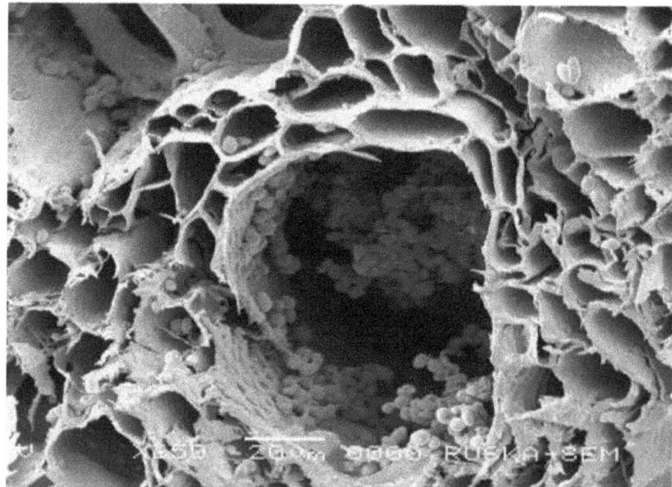

**Figure 1.** Scanning electron microscopic image of yeast cells inside the parenchyma of sugarcane pieces (Adapted from Reddy *et al.*, 2011, with permission of Springer)

such as barley, do not need any chemical preparation or modification prior to cell immobilisation process (Kandylis *et al.*, 2012b).

Natural materials, such as starch gel, corn cobs, corn grains and potatoes have been investigated in winemaking by several authors (Kandylis and Koutinas, 2008; Kandylis *et al.*, 2008 2012a; Genisheva *et al.*, 2013). The results of fermentation studies using these carrier materials for cell immobilisation indicated that wines with improved aroma have been produced. Also, one of the advantages of these carriers are good storage properties, allowing the use of such immobilised cells for a long period. For example, Genisheva *et al.* (2013) stored at 5°C for 31 days corn cobs with immobilised *Oenococcus oeni* after four consecutive malo-lactic fermentations. The system showed a good performance, with satisfactory malic acid conversion. It seems that the key factor for good protection of cells is the porous surface of the corn cobs, in which immobilised cells are protected.

Besides these single-material supports, composite materials were investigated as carriers for cell immobilisation in winemaking too. One such example of composite carrier is a biocatalyst consisting of two-layer composite of delignified cellulosic material with entrapped *Oenococcus oeni* cells, covered with starch gel containing the alcohol resistant and cryotolerant strain, *S. cerevisiae*. The biocatalyst was suggested for simultaneous alcoholic and malo-lactic fermentation of wine in the same bioreactor (Servetas *et al.*, 2013).

Another example of natural organic carrier materials for cell immobilisation are fruit species, such as quince (Kourkoutas *et al.*, 2003), apple (Kourkoutas *et al.*, 2002b), pear (Mallios *et al.*, 2004), pineapple (Diep *et al.*, 2009), dried raisin berries (Tsakiris *et al.*, 2004), guava (Reddy *et al.*, 2006), watermelon (Reddy *et al.*, 2008) and papaya (Maragatham and Panneerselvam, 2011). These materials have been proposed as ideal for yeast immobilisation for winemaking, since fruit-based supports can be easily accepted by the consumer. Also, the prices of such supporting materials are relatively low, making them attractive for cell immobilisation (Mallouchos *et al.*, 2002; Genisheva *et al.*, 2012,;Genisheva *et al.*, 2013; Genisheva *et al.*, 2014a). Overall procedures for cell immobilisation using fruits consist of washing the material with water, drying, cleaned from undesired parts and sterilisation. Further, the cells are immobilised spontaneously by adsorption. Fruit supports have long operation stability, even longer when compared to some other organic materials. Genisheva *et al.* (2013) have shown that *O. oeni* cells immobilised on grape skins were able to conduct consecutive fermentations for a total period of 192 days (seven batches) compared to 174 days (four batches) achieved with grape stems and only 150 days (three batches) with corn cobs. However, fruit supports are not inert materials since yeasts can metabolise some fruity compounds, which would affect sensorial properties of the final product. For example, Genisheva *et al.* (2012) reported more intensive colour of the white wine produced by immobilised yeast on grape pomace in comparison to the colour of the control wine since some colour compounds of grape pomace were released into the wine.

## 4.4. Inorganic Materials

Inorganic carrier materials have been far less used in winemaking with very few examples and include kissiris (porous volcanic mineral) (Bakoyianis *et al.*, 1993; Kana *et al.*, 1992), alumina (Loukatos *et al.*, 2000; Kana *et al.*, 1992) and white foam glass (Bonin and Skwira, 2008). They are inert and cheap and can preserve cell activity for a long time. One of the negative aspects of these materials is the release of mineral residues into the final product (Loukatos *et al.*, 2000).

The new concept of magnetised yeast cells has attracted greater attention in recent years due to the possibility of rapid separation of yeast biomass. After immobilisation using materials with magnetic properties, the cells can be easily separated under the action of a magnetic field (Dauer and Dunlop, 1991). Recently, Berovic *et al.* (2014) developed novel super-paramagnetic nano particles of iron oxide coated with a thin layer of silica and grafted with (aminoethylamino) propylmethyldimethoxysilane for absorption of yeast cells. Using this process for modification of yeast's surface, it is possible to fix the magnetic nano particles and secure a stable system even after fermentation (Fig. 2). In contrast to immobilisation in gel beads, the magnetisation of yeast cells does not hinder the natural cell mobility and promotes the production of aromatic compounds (Tataridis *et al.*, 2005). Berlot *et al.* (2013) showed that exposure of yeast cells to magnetic field for periods of one to three days caused better physiological cell response, rapid fermentation and acceptable wine sensorial properties. One of the explanations of these results may be the presence of free iron ions in the fermentation medium and their positive effect on cell mitochondria (Lill *et al.*, 2012).

### 4.5. Co-immobilisation of Yeast Cells with Other Microorganisms

An alternative to solid support for cell immobilisation could be the biomass of other microorganisms. Puig-Pujol *et al.* (2013) showed an innovative way for yeast cell immobilisation, where yeast cells of a *S. cerevisiae* strain were spontaneously co-immobilised with a filamentous fungus (*Penicillium chrysogenum*). These biocapsules, according to the same authors, could be a cost-effective and efficient method for yeast immobilisation and further application in sparkling wine production. Similar concept was also reported by García-Martínez *et al.* (2015) and was applied for yeast immobilisation into the form of capsules of filamentous fungus and further used for sweet wine production. A more detailed explanation of this concept is provided below.

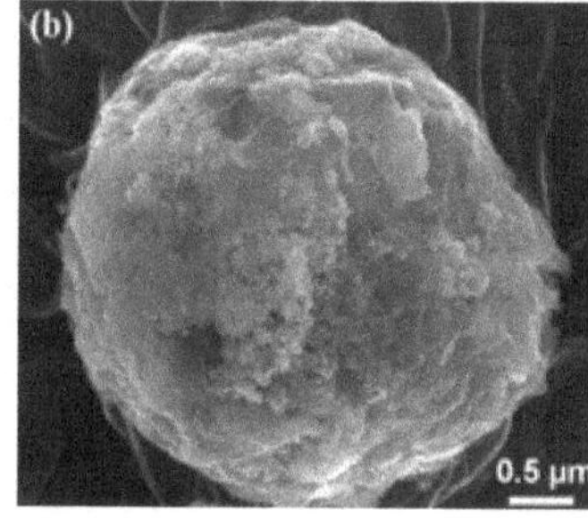

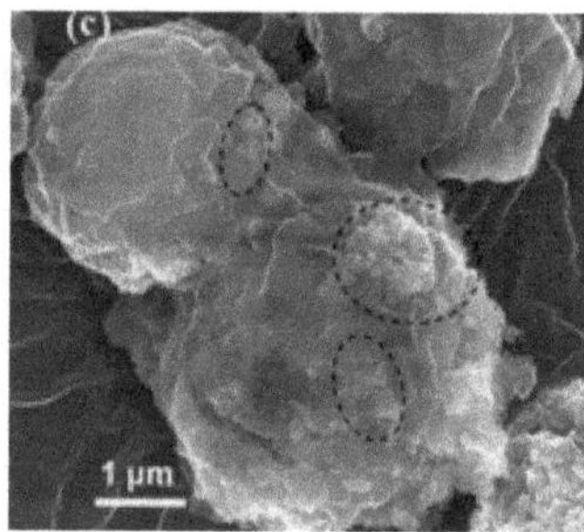


**Figure 2.** Scanning electron microscopic images of the yeast cells before the magnetisation (a), after magnetisation with amino-functionalised magnemite nano particles before (b) and after the fermentation (c); circles mark the nano particles still present on yeast cells after fermentation (the nano particles-to-yeast ratio of 1:50) (Adapted from Berovic *et al.*, 2014 with permission of Elsevier)

## 5. Impact of Immobilised Cells on Wine Properties

The immobilisation of yeast cells has two main consequences: first, the immobilisation changes the metabolism of immobilised cells and as consequence, the changes in metabolism lead to changes in the chemical and sensorial profile of the final product. Hence, immobilisation may cause positive or negative influence on wine properties and consequently on the consumer's perception of product.

Some examples of impact of immobilisation technology on wine's chemical properties are listed in the Table 2.

During fermentation, yeast cells produce various organic compounds that are usually considered as constituents of the product's flavour profile. This includes alcohols, esters, organic acids, sulphur and carbonyl compounds. The concentration of flavour compounds depends on selected yeast strain, chemical composition of fermentation medium and external physical factors (e.g. fermentation temperature). Generally, the yeast cell immobilisation is considered as a process that improves the wine's aromatic profile and contributes to its stability. Immobilisation of yeast cells also contributes to stability of fermentation process at different temperatures, enabling wine production that has a similar aromatic profile regardless of applied temperatures. Even more, for example, the concentration of esters can be increased using immobilised cells, providing wines with a more pleasant aroma. At the same time, the concentration of less desirable products (e.g. higher alcohols) can be reduced to a more acceptable level (Nedović *et al.*, 2015).

As can be seen in Table 2, the fermentation processes with immobilised cells result in products that exhibit differences in the aromatic profile compared to products obtained using conventional fermentation procedures. In some of the presented processes, the concentrations of pleasant flavours increased, while in the some other fermentations, the unfavourable compounds were detected. These findings indicate that introduction of immobilisation into commercial fermentation processes requires optimisation in order to preserve the sensorial properties of wine that are recognisable by consumers.

**Table 2.** Impact of Immobilisation Technology on Wine's Chemical Properties

| *Carrier* | *Cell cultures* | *Fermentation medium* | *Impact on flavour* | *Other characteristics* | *References* |
|---|---|---|---|---|---|
| Ca-alginate | *S. bayanus* and *Leuconostoc oenos* (new name *Oenococcus oeni* ) | Apple juice | Lower concentrations of isobutanol, isoamyl alcohol, propanol and isoamyl acetate, higher concentration of diacetyl | Time savings Higher productivity | Nedovic *et al.*, 2000 |
| Ca-alginate | *Leuconostoc oenos* (new name *Oenococcus oeni* ) | Apple juice | Higher concentration of ethyl acetate and methanol | Acetaldehyde the most discriminating variable in the ciders obtained at different temperatures. | Cahranks *et al.*, 1998 |
| Ion exchange sponge | *S. cerevisiae* | Apple juice | Higher concentration of ethanol | Decreased batch fermentation time and increased final ethanol concentration | O'Reilly and Scott, 1993 |
| Ca-alginate | *Oenococcus oeni* | Apple juice | Reduced ethyl acetate and acetic acid concentrations, higher concentrations of alcohols | Acetic acid content detected in the fermentation process with immobilized cells | Herrero *et al.*, 2001 |
| Tubular delignified cellulosic material and starch gel | *S. cerevisiae* AXAZ-1 and *Oenococcus oeni* | Grape must | Reduced amyl alcohols, isobutanol and total alcohol | Fermentations at 10,15 and 25°C | Servetas *et al.*, 2013 |
| Ca-alginate | selected yeast (Fermol® Bouquet) | Welschriesling, Pinot blanc, Pinot gris, Pinot noir, Chardonnay, Rheinriesling, Traminer | Reduced ethyl acetate and 1-hexanol | slightly lower values of total and volatile acidity | Miličević *et al.*, 2017 |
| White grape skin | *S. cerevisiae* | White grape must | Higher concentration of methanol and higher alcohols, and lower concentration of acetaldehyde | | Genisheva *et al.*, 2004 |

Table 2. (*Contd.*)

| *Carrier* | *Cell cultures* | *Fermentation medium* | *Impact on flavour* | *Other characteristics* | *References* |
|---|---|---|---|---|---|
| *P. chrysogenum strain* H3 | *S. cerevisiae, strain G1* (ATCC: MYA-2451) | The must obtained from sun-dried grapes of the Pedro Ximenez variety | Higher concentration of methanol and propanol and 2,3-Butanediol; reduced 2-Phenylethanol and ethyl acetate | Production of sweet wine | García-Martínez *et al.*, 2015 |
| Wheat grains | *S. cerevisiae* AXAZ-1 | Glucose and grape must at various temperatures | Higher concentration of esters, reduction of total organic acids and carbonyl compounds | | Kandylis *et al.*, 2010a |
| Amino-functionalised magnetic nanoparticles absorbed on cells surface | *S. cerevisiae* for PI *Saccharomyces bayanus* for SC | Grape juice Sauvignonasse PF in STR (15L) SC in bottle (0.75L) | Lower concentrations* of lactic, succinic and acetic acids | More intensive fermentation process* | Berovic *et al.*, 2014 |

PI – primary fermentation; SC – secondary fermentation; STR – stirred tank reactor
* Compared to the control (free cell system)

## 5.1. Other Applications of Immobilised Yeast Cells in Winemaking

Besides fermentation, immobilised yeast cells are used for stabilisation of wines. Merida *et al.* (2007) studied application of immobilised *S. cerevisiae* for prevention of the browning that occurs in white wines. As model systems, they used commercial sherry type white wines, while the yeast cells were immobilised using κ-carragenate cylindrical beads. The experiment was designed to facilitate wine browning so tht the effect of added yeast could be detected. According to the same authors, the addition of immobilised yeast cells led to reduction and delays the browning. In some cases, the immobilised cells significantly stabilised wine, even after one year of storage period. Interestingly, such a procedure for wine stabilisation did not change significantly the sensorial properties of treated wines that remain sensorially acceptable. Also, according to Lopez-Toledano *et al.* (2007), the κ-carragenate beads loading yeast cells could be used for wine colour correction and as a potential alternative for conventional clarification processes.

One promising application of immobilised yeast cells could be the adsorption of harmful compounds from wine. Farbo *et al.* (2016) proposed use of live or autoclaved immobilised *Candida intermedia* for adsorption of ochratoxin A from grape juice. The yeast cells were immobilised in calcium alginate beads. Additionally, the Ca-alginate beads were formulated with iron oxide magnetic nano powder in order to facilitate cell removal after juice treatments. The reduction of ochratoxin A in the grape juice treated by yeast cells was >80 per cent, depending on cell pre-treatments (i.e. autoclaving). Immobilisation had no effect on yeast adsorption performance, while thermal treated cells exhibited better toxin adsorption.

Control of wine's acidity is very important, especially the volatile acidity that could be critical for sensorial properties. Besides conventional deacidification methods, yeast could also be used for reduction of volatile acidity, viz. yeast cells, under specific conditions could metabolise acetic acid and consequently reduce wine acidity. Vilela *et al.* (2013) studied reduction of volatile acidify using *S. cerevisiae* immobilised into double-layer alginate-chitosan beads. The bead production was relatively simple and includes primarily yeast immobilisation into Ca-alginate beads, followed by additional beads coating with chitosan. According to the same authors, the immobilised yeast cells reduce the volatile acidity of wines up to 60 per cent. This is a promising way to control such a huge problem of winemaking, especially regarding current standards and regulations that strictly limit the level of volatile acidity.

### 5.2. Immobilised Yeast Cells for Sparkling Wines Production

Technology of sparkling wines production has changed significantly in the last decades. The new producers offer interesting tastes and wines suitable for all customers. Also, biotechnological companies have isolated new yeast strains that are optimised for specific fermentations and aroma profile formation.

One of the main problems in sparkling wine production is secondary fermentation and formation of adequate $CO_2$ concentration, parallel to preservation of desired sugar content and flavour formation. For these purposes, two main processes have been established:

- Secondary fermentation by yeast cells
- Carbonation of wine up to required $CO_2$ content

Carbonation method is usually considered as suitable for production of less prestigious sparkling wines and does not require secondary fermentation. However, wine used for carbonation must have adequate quality (Jackson, 2008). All the sparkling-wine-production processes exhibit some advantages and disadvantages. However, regarding the topic of this chapter, the main focus is on the improvement of secondary fermentation, especially by applying immobilised yeast cells.

The major problem of sparkling wine production is control of the fermentation process and removing of yeast cells from the system (e.g. bottles) at the end of fermentation. This has been challenging, since yeast cells are relatively small and their removal from the liquid is a complex task, linked with loss of product and possible contamination. Immobilisation of yeast cells has been proposed as a solution for these problems and already there are commercially available immobilised yeast cells for sparkling wine production (*see* below).

Entrapment of cells into calcium alginate matrix is one of the most efficient methods for yest cell immobilisation for the sparkling wine production. This procedure is relatively simple and requires only mixing of yeast cells into previously prepared alginate solution, followed by extrusion of such prepared suspension into gelling solution (e.g. $CaCl_2$ water solution). The beads could be easily recovered from the gelling solution and applied for bottle secondary fermentation during production of sparkling wines. It was observed that there are no significant variations of oenological parameters compared to sparkling wines produced by using traditional (free cells) method. The properties of sparkling wines, besides the type of fermentation (immobilised or free cells) are also affected by grape variety, where some varieties were found to be more suitable (Miličević *et al.*, 2017).

However, immobilised cells at some point could be released from capsules and continue to reproduce out of alginate beads. This portion of free cells could be defined as uncontrollable population of cell biocatalyst and could undermine benefits achieved by immobilisation. The logical solution for this problem is to create another protective layer (or more layers) of carrier material around the beads and limit the cell release. The one approach for such additional protection by layer formation could be the use of conventional and well-established techniques for coating of active ingredients, based on fluid bed technology. The protecting material is added to particles that contain the active compound and the whole material is kept in the fluid state by gas flow. Final formation of protective layer is achieved by evaporation of the solvent from coating material or by cooling of the particles (Agnihotri *et al.*, 2012). Such an approach has found application in the protection of probiotic bacterial strains, usually by lipid coatings or by proteins or carbohydrates as well. This is a proven technology for probiotic protection, like a Probiocap™ from Lallemand, a freeze-dried probiotic bacteria coated by low melting lipids using fluid bed coating (Chávarri *et al.*, 2012). However, rigid lipid barrier is not suitable for protection of immobilised yeast cells intended for fermentation since such a layer would be a great barrier to diffusion of nutrients and cell products. As alternative, additional coating of immobilised yeast cells could be performed by other polymer materials. One of the very frequently studied approaches for additional cell protection is formation of chitosan coatings around the alginate beads. Zhou *et al.* (1998) showed that the coating of alginate beads with low molecular weight chitosan reduced cells released during milk fermentation with *Lactococcus lactis.* The proposed procedure for bead coating is simple and is composed of immersion of alginate beads (with immobilised cells) into chitosan solution for a short period of time. After immersion, beads were filtered and washed. According to the same authors, coating of beads with chitosan significantly reduced the cell release, but also reduced the lactic acid production rate due to mass transfer resistance. As already mentioned above, a similar procedure was also employed in immobilisation of yeast cells for reduction of wine volatile acidity (Vilela *et al.*, 2013). Pereira *et al.* (2014) tested the coating procedure where alginate beads were additionally coated with chitosan. However, according to the same authors, the chitosan coating did not result in a significant reduction of free yeast cells.

Even more, the alginate beads could be protected by alginate too. This means that there is no need for a different polymer to be used as a coating agent. For example, Ehrhart *et al.* (2013) used alginate beads and additional alginate layers for protection of immobilised cells against the negative influence of the environment. They tested several approaches for multiphase beads coating in the alginate solutions with different polymer concentrations. Basically, the procedure could be simple described as a step-by-step layer formation by absorbing of alginate from the diluted solution. Additionally, it is possible to create even mineral deposits within the alginate layer, securing better layers connections. Using such an approach, it is possible to create numerous alginate layers around alginate beads, avoiding the use of the other polymers.

However, such additional coatings could create new obstacles for nutrient diffusion and limit productivity of immobilised cells. As shown above, the yeast cells immobilised in the matrix type systems suffer from lack of nutrients and thus tend to grow on the beads surfaces. This could be even more pronounced in the case of multi-layer coated gel beads. Besides these limitations, complex particles consisting of double layers of carrier materials have been commercialised in winemaking, and according to available data, could significantly improve wines production (*see* below).

### 5.3. Application of Immobilized Yeast Cells in the Continuous Wine Fermentation

The production of high quality products using biocatalysts, such as yeast and bacterial cells, requires precise process control. Specially designed bioreactors have been used in modern biotechnology to maintain the physical parameters during the fermentation process (e.g. temperature, pH, etc.) and consequently provide high quality products. Hence, the decision regarding which type of bioreactor should be used in specific production is one of the key questions.

Primarily, the processes with biocatalysts may be roughly divided into batch and continuous operations. The main problem during continuous fermentation is loss of cells from bioreactor by liquid flow. This could be overcome by cell linkage to solid supporting materials or by forming a permeable membrane between the cells and fermentation medium. For such purposes, both inorganic and organic materials are applied.

As pointed out before, the continuous operation is one of the main advantages of the immobilised yeast cells. However, the design of the reactor, regardless of the type of operation (i.e. batch or continuous) must provide the stability of immobilised cells. This is especially important due to the mechanical stability of carrier materials that limit the usage of some type of bioreactor. For example, some fragile carrier materials cannot be used in the bioreactor that operates under intensive mixing.

To maximise the efficiency of continuous operations, the multiphase system can be used. In this type of bioreactor, the three phases-liquid (fermentation medium), solid (carrier) and gas (in winemaking inert gases, such as $N_2$ and $CO_2$) can be present. The multiphase system may be used in several types of bioreactors: packed bed reactor, fluidised bed reactor, bubble column and air-lift reactor. The type of applied bioreactor depends on the method for cell immobilisation (Genisheva *et al.*, 2014b). Packed bed reactors were studied in the earlier attempts at application of immobilised yeast cells in the fermentation processes. However, due to poor performances of carrier materials, these systems had limited results. With development of new carrier materials, the reactor configurations have been evolved further, enabling production of beverages with more uniform properties. The new continuous bioreactor systems provide good stability of immobilised cells, with high cell concentration and excellent mass transfer (Kourkoutas *et al.*, 2004).

One of the first applications of continuous fermentation in winemaking is the production of sparkling wines in the former Soviet Union, from where the process has been expanded to other countries (Jackson, 2008). Since then, numerous publications have been published about continuous winemaking and especially about application of immobilised yeast cells in such processes.

The primarily studied materials for yeast immobilisation in continuous winemaking are natural polymers or organic materials (e.g. alginate, cellulose and fruit pieces), while the reactor configuration required for full achieving of sensorial properties of wine are usually multistage reactors (Genisheva *et al.*, 2014b).

Among natural carrier materials, fruit cuts were found to be a stable immobilisation support for yeast cell immobilisation in continuous wine fermentation. Kourkoutas *et al.* (2002b) studied apple cuts for immobilisation of yeast cells. The immobilised cells were tested under low temperatures for wine fermentation in the continuous bioreactor. The results showed high efficiency of fermentation process, while the final product exhibited good sensorial properties.

Also, continuous fermentation processes may be based on cells immobilised in the form of cells aggregates. Viegas *et al.* (2002) used flocculent yeast strains in the tower reactor system for continuous ethanol fermentation. The system provided a good environment for yeast cells, leading to high ethanol yield. This could be promising also for immobilisation of yeast cells for continuous wine fermentation. Moreover, using flocculent yeast and a carefully controlled process, it could be possible to achieve stability of immobilised cells in the continuous fermentation mode. Also, flocculated yeast cells may be a good alternative, for example, to gel beads (especially Ca-alginate beads) that according to literature data, have low mechanical and chemical stability (Kourkoutas *et al.*, 2004; Pajic-Lijakovic *et al.*, 2015).

Also, here should be mentioned the membrane systems as a promising solution for continuous winemaking. One of the earliest attempts at dry wine fermentation using membrane module in continuous regime was made by Takaya *et al.* (2002). According to the same authors, the two systems were tested: single-vessel membrane bioreactor (with cross-flow membrane) and double-vessel membrane bioreactor (combined continuous stirrer and membrane bioreactor). It was noticed that the double-vessel system was more suitable for dry wine production, providing good residual sugar concentration and high productivity.

However, one of the main disadvantages of continuous processes with immobilised yeast cells is limited life of such a biocatalyst, viz. yeast cells usually have up to 30 division cycles, after which they are become inactive and cannot replicate further. The ageing of yeast cells is also expressed by physiological and morphological changes of cells that also affect the production properties of biocatalysts. Also, the selection of appropriate yeast strains for immobilisation and continuous processes is important, where strains with better physiological properties and longevity are generally preferable (Brányik *et al.*, 2005).

## 5.4. Fruit and Special Wines Production Using Immobilised Cells

The production of wine from berry fruits has increased over the last two decades, especially in the last several years in contrast to cider and perry production from apples and pears which has been carried out since the 18th century.

Generally, there are the two approaches to fruit wine production. The first approach, and more common, is based on maintaining the wine fruit character and preserving the characteristic fruit aroma of the final product. The second approach tends to create a fruit wine that is more similar to the conventional grape wines. Production of fruit wines is generally very similar to traditional winemaking. The main advantage of fruit wine production is the fact that many fruits give significant juice yield, with a good sugars/acids ratio (McKay *et al.*, 2011).

Berry fruits represent a potentially valuable source of polyphenolic compounds, particularly flavonoids, such as anthocyanin pigments. Besides polyphenolic compounds, berry juice contains specific acid and sugar contents (pH 3.2-3.6 and about 14.5 °Brix) that make them suitable for fruit wine production. Berries are also attractive due to high fruit yields, a long harvest season and resistance to plant diseases. Also, the berries are suitable for the machine harvest. Berry fruits which cannot be used for consumption may be used in the production of fruit wines (Duarte *et al.*, 2010).

In recent years there has been an upsurge of interest in the use of immobilised cells for wine production due to the attractive technical and economic advantages of immobilised cell technology compared to the conventional free cell systems (Margaritis and Merchant, 1984; Nedovic *et al.*, 2000). Many immobilisation carriers have been proposed for use in winemaking, such as Ca-alginate, Kissiris, gluten pellets, delignified cellulosic materials, DEAE-cellulose and fruit pieces (Kourkoutas *et al.*, 2004). Immobilised cell systems, not only increase productivity and improve the economic efficiency of bioprocesses, but also influence yeast metabolism and, consequently, wine aroma, taste and product quality (Dervakos and Webb, 1991; Bakoyianis *et al.*, 1993; Bardi and Koutinas, 1994; Melzoch *et al.*, 1994; Sipsas *et al.*, 2009).

Djordjević *et al.* (2015) investigated the differences in composition and sensory properties of raspberry wines fermented with immobilised (in Ca-alginate beads) and suspended yeast cells of two strains of *S. cerevisiae.* Analysis of aroma compounds, glycerol, acetic acid and ethanol as well as kinetics of fermentation and sensory evaluation of wines was performed. Samples fermented with immobilised yeast cells had a shorter lag phase and faster utilisation of sugars and ethanol production than those fermented with suspended cells. Further, samples fermented with immobilised cells of both strains showed better sensorial properties than those fermented with suspended cells, although there were no significant differences in aroma composition.

Oliveira *et al.* (2011) elaborated a fruit wine from cagaita (*Eugenia dysenterica* DC). The aims of this study were to adapt a methodology for producing fruit wine from cagaita pulp and to compare fermentations conducted with free cells and with Ca-alginate immobilised cells, using two selected *S. cerevisiae* strains. The two yeast strains used in this study were able to ferment the cagaita must and a better performance was obtained with immobilised cells in calcium alginate than with free cells. Considering the time of fermentation, the fermentation system with immobilised yeast cells was 60 per cent faster than free cells of both strains. The cagaita wines produced by immobilised cells were characterised by the attributes of taste, aroma and overall character and were well accepted (at least 70 per cent for overall aspects), especially bearing in mind the fact that the tasters were not familiar with this beverage. This technology may be an alternative way of using the pulp of this tropical fruit and may provide a new industrial outlet for it.

Ca-alginate beads were also used for immobilisation of yeast cells and production of pomegranate wine. The results showed that fermentation with immobilised cells was more efficient than in the case of free cells. Also, the efficiency of immobilised yeast depended on alginate concentration, beads diameter and initial cells concentration (Sevda and Rodrigues, 2011).

Mango wine can be also efficiently produced using immobilised yeast cells. For this purpose, mango peel can be used as a carrier for yeast cell immobilisation. The system provided the product with satisfactory sensorial properties, while the immobilised yeast cells can be used in repeated batch fermentations of mango juice (Varakumar *et al.*, 2012).

From the commercial point of view, cider is one of the most important fruit wine produced worldwide. Cider is traditionally produced by fermentation of apple juice, using freely suspended yeast cells. In order to produce cider, two successive fermentation processes of apple juice (alcoholic and malo-lactic) have to be performed. More precisely, the first process implies alcoholic fermentation of sugars into ethanol proceeded by yeasts and the second one represents the malo-lactic fermentation of *L*-malic acid into *L*-lactic acid by lactic acid bacteria. Various disadvantages of traditional cider production led to adoption of the use of selected pure culture starters. Some of the main issues of traditional production were: very slow process, low control of the flavour formation and risks of spontaneous fermentation by indigenous microbial flora (Durieux *et al.*, 2005). The most dominant selected yeast type used for alcoholic fermentation in cider production is *Saccharomyces* sp. since it provides neutral sensorial notes (Nedović *et al.*, 2015). On the other hand, *Hanseniaspora* sp. gives 'fruity' sensory characteristics to cider, due to the presence of particular esters (De Arruda Moura Pietrowski *et al.*, 2012). Variations of yeast microflora and inconsistent composition of apple (juice/must) are the main reasons for such diversity in sensorial profiles of cider, including off-flavours. Apart from ethanol (1.2-8.5 per cent v/v) cider contains many side products of yeast and bacterial metabolism. Alcoholic and malo-lactic fermentations are sequential and simultaneous processes, and that is why significantly affect the flavour formation. Immobilisation of microorganism's cells as biocatalysts in those fermentations additionally influences the flavour formation and may contribute to control of these processes. Opposite to free cell fermentation, as pointed out above, cell immobilisation offers various advantages, like enhanced fermentation productivity, cell stability and viability, ability for cell recycling, application of continuous configurations and quality improvement (Kourkoutas *et al.*, 2010).

For example, co-immobilisation of selected yeast and bacteria cells as an integrated system is one of the solutions for advanced cider production. This technology gives the opportunity to co-immobilise different kinds of microorganisms within the same matrix, allowing the effectuation of these two fermentation processes simultaneously into one biocatalytic system. Some of implemented cultures were *Saccharomyces bayanus* and *Leuconostoc oenos* (new name *Oenococcus oeni*) co-immobilised into alginate matrix (Nedovic *et al.*, 2000). The suspension of cells was transformed into alginate spherical beads by the electrostatic extrusion technique. Further, continuous fermentation with reconstituted apple juice was performed in the packed bed bioreactor at 30°C. The continuous process contributed to a more rapid alcoholic and malo-lactic fermentation, high volumetric productivity and stability of the process as well as control of flavour development (Nedovic et al., 2000). Same effects were demonstrated with different cultures and matrixes, such as *S. cerevisiae* and *Lactobacillus plantarum* co-immobilised in sponge (Scott and O'Reilly, 1996).

Besides, fruit wines, immobilised yeast cells could be applied for production of special wines that are specific due to their technological process or fermentation medium. Such wines are more related to a broader category of fermented beverages and are clearly distinguished from traditional winemaking.

A good example of such a fermented beverage is mead. Mead is a fermented beverage produced from diluted honey, with alcohol content varying between 8-18 per cent v/v. In order to produce an adequate amount of mead's flavour compounds, the yeast cells could be protected in the form of calcium alginate beads or double-layer calcium alginate beads, where the second is made of chitosan. The immobilisation does not affect the fermentation time, but however improves the production of volatile compounds. It was also noticed that double-layer beads show a similar release of cells during fermentation as compared to uncoated beads (Pereira *et al.*, 2014).

## 6. Commercial Applications of Cell Immobilisation Technology in Winemaking

Although immobilisation of yeast cells has attracted attention of the scientific community, till recently the number of companies that have offered such products were limited. However, modern winemaking starts to realise the potential of immobilised cell technology and currently several immobilised yeast brands are available. Also, there are more new companies that offer the immobilisation technologies suitable for various applications (Table 3).

Regarding winemaking, there are two options when it comes to application of immobilised biocatalysts – first, the wineries can purchase already immobilised yeast cells optimised for specific processes in winemaking or for specific types of wines; second option is immobilisation of own yeast strains in order to preserve the authentic oenological properties of wine.

Immobilisation of own yeast stains could be realised using various immobilisation apparatus and techniques available on the market. In recent years, numerous companies have offered their immobilisation systems and carriers optimised for specific biocatalyst immobilisation. As food industry demands strict regulation of materials used in production processes, thus the main focus is primarily on use of the natural carrier materials, but also some synthetic and nontoxic polymers have been applied too.

Spherical shaped carriers are the most popular forms of immobilised biocatalysts. To produce the spherical beads, several companies offer the dispersion techniques, usually based on polysaccharide gels as matrix for cells immobilisation. For example, Nisco Engineering AG (Switzerland) is a company that produces modular units for biocatalyst immobilisation. Nisco's immobilisation systems are based on various dispersion techniques, such as electrostatic force, vibration dispersion, coaxial air flow dispersion, gravitational dripping as well as a combination of these techniques. For production of larger quantities of immobilised biocatalysts, multi-nozzle head can be used in the gravitational dripping process and electrostatic extrusion (Fig. 3a, b.) (www.nisco.ch).

Also, a Swiss-based company, BÜCHI Labortechnik AG, is a well-known producer of equipment for immobilisation of various active ingredients. The BÜCHI immobilisation units offer numerous possibilities for control of bead diameter and more importantly, the units are suitable for immobilisation of cells under sterile conditions, which is very important for the production of starter cultures (www.buchi.com). BRACE GmbH (Germany) offers various units optimised for production of different types of capsules suitable for food applications. This company produces lab-scale and pilot-scale units, with possibilities for efficient control of the immobilisation process (www.brace.de). For production of large quantities of immobilised biocatalysts, the geniaLab® company offers JetCutter® technology (Fig. 3c).

**Figure 3.** Commercial systems suitable for yeast cells immobilization: a) Head for 30 nozzles and b) encapsulation Unit VAR W10 (with permission of Nisco Engineering AG) (www.nisco.ch); c) Immobilisation process using JetCutter® technology and d) LentiKats® carriers (with permission of Institute of Agricultural Technology, Johann Heinrich von Thünen Institute, Germany) (www.genialab.com, www.thuenen.de)

The JetCutter® technology is based on liquid 'cutting' by a mechanical rotating wheel into the liquid droplets that further undergo the solidification (gelling) process (www.genialab.com).

However, for some applications, the natural polymers may not be suitable since the degradation of the carrier could lead to cell release and loss of performance of the biocatalyst. As alternative for natural polymers, the geniaLab® company and partner company LentiKat's a.s (www.lentikats.eu) offer the Lentikats® immobilisation technology based on polyvinyl alcohol (PVA) as carrier material (www.genialab.com). Besides possible application in food fermentation processes (Mathew *et al.*, 2013), the Lentikats® technology is also suitable for immobilisation of other types of biocatalysts, such as enzymes (Bajić *et al.*, 2017) or microbial cells for wastewater treatments (Boušková *et al.*, 2017) and production of chemical compounds (Dolejš *et al.*, 2014). According to available data, Lentikats® carrier material (PVA) is crosslinked by drying of PVA liquid (with added biocatalyst) at room temperature, followed by rehydration in the stabilisation solution. Due to specific extrusion of polymer solution (i.e. on the solid surface), the final form of PVA carrier has a lens-like shape (Fig. 3d). Immobilisation of *Oenococcus oeni* using a Lentikats® immobilisation technology was found to be a promising solution for malo-lactic fermentation in wines. It showed that immobilised cells can be reused in several fermentation cycles, maintaining high fermentation activity of up to 75 per cent. The viability of cells improved, most probably as a result of immobilisation and protective role of PVA against inhibitory effects of ethanol (Rodríguez-Nogales *et al.*, 2013). The immobilisation procedure using Lentikats® can be intensified by using Lentikat® Printer, a device optimised for production of uniform carrier particles (Parascandola *et al.*, 2006). The combination of Lentikats® carrier and Lentikat® Printer was also used in the beer fermentation, where PVA carriers showed good mechanical stability, preserving yeast cells activity for about six months (Bezbradica *et al.*, 2007). One of the main advantages of cell immobilisation in Lentikats® carriers is the possibility for applying of immobilised cells in the continuous fermentation processes (Mathew *et al.*, 2014). Also, Lentikats® carriers have been produced on a pilot scale (Rebroš *et al.*, 2016; www.lentikats.eu), which is an important advantage of this technology, allowing production of bigger quantities of immobilised biocatalyst suitable for large-scale operations.

As pointed out above, the second important approach in applying the immobilisation technology in industrial-scale fermentation processes, and especially in winemaking, is use of already immobilised cell strains selected by specialised companies. These strains are selected on the basis of their activity and the fact that they can provide aroma composition for specific types of wine. Also, such strains have certain fermentation activity and in some cases can be adapted to high sugar or ethanol content.

Company PROENOL (Portugal) is one of the leading producers of immobilised yeast cells for various fermentation steps during winemaking. Also, the company offers immobilised yeast stains, specially selected for specific types of wines (www.proenol.com):

- ProMalic®, *Schizosaccharomyces pombe* immobilised in double-layered calcium alginate for fermentation of malic acid into ethanol;
- ProDessert BA11®, *S. cerevisiae* immobilised in double-layered calcium alginate for fermentation of residual sugars and control of final sugars concentration in wine;
- ProElif®, *S. ceverisiae* yeasts immobilised in double-layered calcium alginate for sparkling wine production;
- ProRestart®, immobilised yeast cells in alginate for restart of stuck or sluggish fermentation.

As can be seen, PROENOL's immobilisation technology (for some products in collaboration with company Lallemand) is based on application of double-layered calcium alginate beads as carriers for yeast cells immobilisation. Calcium alginate beads are enveloped by a second layer of alginate, which is a good strategy for prevention of cell leakage into the fermentation medium. As pointed out above, the cells in the structure of calcium alginate beads intensively grow close to the bead surface, from where they can be easily released into the medium. The second alginate layer prevents or limits such release. Beads are usually around 2 mm in diameter and can be applied as free beads or packed in the special bags.

**Table 3.** Some Commercially Available Immobilisation Technologies and Immobilised Yeast Cells for Winemaking

| *Product, Company* | *Carrier material* | *Comments* | *References* |
|---|---|---|---|
| Lentikats®, (geniaLab®) (special carrier materials and immobilization equipment) | Primarily polyvinyl alcohol (PVA), other materials | Lens-like gel carriers (PVA) made using laboratory or pilot scale units. Other systems also available (*e.g.* JetCutter® technology) | www.genialab.com www.lentikats.eu |
| Nisco Engineering AG (immobilization equipment) | Natural and synthetic materials | Carriers of various dimensions | www.nisco.ch |
| BÜCHI Labortechnik AG (immobilization equipment) | Natural and synthetic materials | Carriers of various dimensions | www.buchi.com |
| BRACE GmbH (immobilization equipment) | Natural and synthetic materials | Carriers of various dimensions | www.brace.de |
| PROENOL (immobilized yeast strains) | Ca-alginate (coated) | Spherical beads (diameter around 2 mm) | www.proenol.com |

The calcium alginate beads with immobilised yeast cells are very stable since they are produced in the dried form, enabling easy handling and storage. According to the PROENOL's protocols, prior to use in the winemaking process, alginate beads with immobilised yeast cells should be rehydrated under appropriate conditions in order to secure proper cell activation and high metabolic activity. This procedure usually takes several hours, during which alginate beads are rehydrated in the sugar solution. After rehydration procedure, the immobilised cells are ready for fermentation.

As mentioned above, some winemaking processes require specific contact time between substrate (e.g. production of dessert wines) and yeast cells. In order to control contact time and proper removal of the immobilised yeast cells from tanks or barrels, the company PROENOL offers the already prepared bags (ProMesh®) with adequate immobilised yeast strain that can be easily inserted or removed from the fermentation tank. The mass of bags with immobilised yeast cells is optimised in accordance with the fermentation volume and there are two basic configurations – for application in tanks and for fermentation in barrels. The mass of applied beads with immobilised microbial culture also depends on the process type (i.e. alcohol or malic acid fermentation). Bags are fixed inside the tank and also the appropriate weight is added in order to prevent the bags floating on the must surface. Also, it is recommended that several bags should be fixed at different heights to ensure better contact with must. Since $CO_2$ accumulates in the bags, they should be shaken several times a day to remove gas. The bags are easily removed from the tank at desired moments, enabling full process control. This is especially important during the production of dessert wines, where stopping alcoholic fermentation is crucial in preservation of optimal sugar content in the final product (www.proenol.com).

## 7. Future Trends in Application of Immobilised Yeast Cells for Wine Production

Future developments in yeast immobilisation technology for winemaking could be in the direction of the study of the new materials for yeast immobilisation, new immobilisation techniques and new bioreactor designs.

Looking for the new carrier materials for yeast cells immobilisation is challenging since such materials must fulfil strict requirements, like safety and preservation of wine's sensorial properties. One interesting approach for yeast cell immobilisation was presented by Peinado *et al.* (2006). They proposed the concept of 'yeast biocapsules' based on the immobilisation of yeast cells (*S. cerevisiae*) into the matrix of *Penicillium* filamentous fungus (i.e. the co-immobilisation of biocatalysts). Here the fungus biomass acts as a carrier for yeast cells and the immobilisation procedure does not require additional carrier materials for immobilisation. The immobilisation procedure is relatively simple and includes incubation of yeast cells with fungus in the growth media composed of gluconic acid as the carbon source and mineral salts. The flasks containing biomass were mixed for seven days, during which the spontaneous immobilisation of yeast cells occurred. The obtained capsules were hollow and about a few millimetres to several centimetres in diameter. The 'yeast biocapsules' were tested for fermentation of several media, including the grape must. The system provided satisfactory ethanol production during wine fermentation, good composition of major flavour compounds and reduction in the fermentation time. The same 'yeast biocapsules' concept was proposed by García-Martínez *et al.* (2015) for production of sweet wine. According to the same authors, such a system could reduce the operational costs by reusing of immobilised yeast cells instead of the production of new inoculums for each batch. Also, the cells in the immobilised form were far more suitable for removal from the fermentation broth at the end of the process.

Further, co-immobilisation of various enzymes in the form of enzyme cascades has been intensively studied as a potential alternative for conventional enzymatic and chemical processes. In such a system, enzymes from various sources are combined, where, for example, products of one enzyme reaction can be the substrate for the next enzyme reaction (Kazenwadel *et al.*, 2015). Also, co-immobilisation of enzymes and cells offers even more opportunities for new process development and improvement of existing processes. For example, Bandaru *et al.* (2006) used co-immobilised amyloglucosidase and *Zymomonas mobilis* for simultaneous sago starch hydrolysis and production of ethanol. They used combined chitin particles and alginate as carriers for enzyme/cells co-immobilisation. In the first step, the enzyme was immobilised on powdered chitin, followed by mixing with cells and finally with sodium alginate. Cells/enzyme/sodium alginate mixture was then extruded into $CaCl_2$ to form calcium-alginate beads. Such a prepared co-immobilised biocatalyst was applied successfully for ethanol production under optimised conditions. As can be seen, the co-immobilised biocatalysts can reduce the process time and save energy. It can be expected that such a complex immobilised system will be applied in winemaking too.

In recent years, the new concept of cell modification by addition of various materials and coatings on to cell's surface has attracted attention, primarily as a solution for some analytical procedures. This new concept of cell modification by non-cellular materials is popularly called 'cyborg cells'. Addition of magnetic particles, polymer layers or formation of mineral shell can significantly improve the cell's metabolic property (Fakhrullin *et al.*, 2012), especially magnetically modified yeast cells have been studied as a relatively suitable tool for various applications in analytical procedures, biotechnology and bioremediation (Safarik *et al.*, 2015). Also, the new methods for production of 'cyborg cells' have been proposed and optimised for uniformly coating of microbial cells. For example, specially designed microfluidic device can be used for rapid and efficient individual cell coating by polymers, even in the form of multilayered protected cells (i.e. cells protected by multiple polymer layers) (Tarn *et al.*, 2013).

Such modified cells could be useful in winemaking too. Dušak *et al.* (2016) studied the possibilities for production of magnetised *Oenococcus oeni* for malo-lactic fermentation in wine. The procedure for production of such modified bacterial cells is based on bonding of functionalised magnetic nano particles to the bacterial surface. The magnetic nano particles were obtained by chemically modified nano particles coated with silica. Further, the cells were mixed with nano particles in the suspension, where positively charged nano particles were coated on to the negatively charged cell surface. The magnetised *O. oeni* cells were used for malo-lactic fermentation, where they showed satisfactory results. The main advantage of modified cells is the possibility for cell removal from the fermentation medium at a predefined moment using the magnetic force. Malo-lactic fermentation is a process that requires strict control, especially the control of fermentation duration. Thus, immobilised or, as in this case, specially functionalised cells could be used for easy process control and at the same time provide the product with desirable chemical composition.

Carrier materials with magnetic properties offer possibilities for development of new bioreactor designs that may improve characteristics of conventional fermentation processes. Liu *et al.* (2009) showed that *S. cerevisiae* immobilised in alginate magnetic particles could be used for ethanol production in the specially-constructed magnetically-stabilised fluidised bed reactor (MSFBR). They reported high performance of such a reactor system, especially regarding ethanol production, which was greater in MSFBR than in the fluidised bed reactor.

Immobilised cells provide many benefits, such as increased product yield, greater stability, easy manipulation, repeated use and possibility for continuous operations. On the other hand, high operational costs and changes in cell metabolism are the main problems related to immobilisation. However, with the growing number of small and medium sized wineries around the world, it can be expected that new immobilised forms of yeast cells will be released in the market. The opportunities for development of new immobilised yeast for winemaking are great and the examples of industrial applications of immobilised yeast cells in winemaking can be interesting for other researchers to start collaborations with the industry and implement own knowledge and capacities for developing of new immobilised biocatalysts and immobilisation procedures at industrial scale.

## Acknowledgments

This work was supported by the Ministry of Education, Science and Technological Development, Republic of Serbia (Project nos. III46010 and III46001) and FP7 Project AREA 316004.

## References

Aber, M.M., Amin, M.C.M. and Martin, C. (2014). A review of bacterial cellulose-based drug delivery systems: Their biochemistry, current approaches and future prospects. *Journal of Pharmacy and Pharmacology* 66: 1047-1061.

Agnihotri, N., Mishra, R., Goda, C. and Arora, M. (2012). Microencapsulation – A novel approach in drug delivery: A review. *Indo Global Journal of Pharmaceutical Sciences* 2: 1-20.

Alftrén, J. and Hobley, J.T. (2014). Immobilisation of cellulase mixtures on magnetic particles for hydrolysis of lignocellulose and ease of recycling. *Biomass and Bioenergy* 65: 72-78.

Astley, O.M., Chanliaud, E., Donald, A.M. and Gidley M.J. (2001). Structure of acetobacter cellulose composites in the hydrated state. *International Journal of Biological Macromolecules* 29: 193-202.

Bajić, M., Plazl, I., Stloukal, R. and Žnidaršič-Plazl, P. (2017). Development of a miniaturised packed bed reactor with ω-transaminase immobilized in LentiKats®. *Process Biochemistry* 52: 63-72.

Bakoyianis, V., Kana, K., Kaliafas, A. and Koutinas, A.A. (1993). Low-temperature continuous wine making by kissiris-supported biocatalyst: Volatile by products. *Journal of Agricultural and Food Chemistry* 41: 465-468.

Bakoyianis, V., Koutinas, A.A., Agelopoulos, K. and Kanellaki, M. (1997). Comparative study of kissiris, γ-alumina, and calcium alginate as supports of cells for batch and continuous wine-making at low temperatures. *Journal of Agriculture and Food Chemistry* 45: 4884-4888.

Bandaru, V.V.R., Somalanka, S.R., Mendu, D.R., Madicherla, N.R. and Chityala, A. (2006). Optimization of fermentation conditions for the production of ethanol from sago starch by co-immobilized amyloglucosidase and cells of *Zymomonas mobilis* using response surface methodology. *Enzyme and Microbial Technology* 38: 209-214.

Bardi, E.P. and Koutinas, A.A. (1994). Immobilisation of yeast on delignified cellulosic material for room temperature and low-temperature winemaking. *Journal of Agriculture and Food Chemistry* 42: 221-226.

Berlot, M., Rehar T., Fefer, D. and Berovic, M. (2013). The influence of treatment of *Saccharomyces cerevisiae* inoculum with a magnetic field on subsequent grape must fermentation. *Chemical and Biochemical Engineering Quarterly* 2: 423-429.

Berovic, M., Berlot, M., Kralj, S. and Makovec, D. (2014). A new method for the rapid separation of magnetised yeast in sparkling wine. *Biochemical Engineering Journal* 8: 77-84.

Bezbradica, D., Matić, G., Nedović, V., Čukalović-Leskošek, I. and Bugarski, B. (2004). Immobilisation of brewing yeast in PVA/alginate microbeads using electrostatic droplet generation. *Chemical Industry* 58: 118-120.

Bezbradica, D., Obradovic, B., Ida Leskosek-Cukalovic, I., Bugarski, B. and Nedovic, V. (2007). Immobilisation of yeast cells in PVA particles for beer fermentation. *Process Biochemistry* 42: 1348-1351.

Bonin, S. and Skwira, J. (2008). Effect of fermentation of fruit must on yeast cells. *Food Technology and Biotechnology* 46: 164-170.

Boušková, A., Mrákota, J., Stloukal, R., Trögl, J., Pilařová, V., Křiklavová, L. and Lederer, T. (2017). Three examples of nitrogen removal from industrial wastewater using Lentikats Biotechnology. *Desalination* 280: 191-196.

BRACE GmbH, Germany, www.brace.de

Brányik, T., Vicente, A.A., Dostálek, P. and Teixeira, J.A. (2005). Continuous beer fermentation using immobilised yeast cell bioreactor systems. *Biotechnology Progress* 21: 653-663.

BÜCHI Labortechnik AG, Switzerland, www.buchi.com

Cahranks, C., Moreno, J. and Mangas, J.J. (1998). Cider production with immobilised *Leuconostoc oenos*. *Journal of the Institute of Brewing* 104: 127-130.

Chávarri, M., Marañón, I. and Villarán, M.C. (2012). Encapsulation technology to protect probiotic bacteria, probiotics. *In*: Rigobelo, E. (Ed.). Tech., DOI: 10.5772/50046; available from: https://www.intechopen.com/books/probiotics/encapsulation-technology-to-protect-probiotic-bacteria

Crapisi, A., Nuti, M.P., Zamorani, A. and Spettoli, P. (1987a). Improved stability of immobilised *Lactobacillus* sp. cells for the control of malo-lactic fermentation in wine. *American Journal of Enology and Viticulture* 38: 310-312.

Crapisi, A., Spettoli, P., Nuti, M.P. and Zamorani, A. (1987b). Comparative traits of *Lactobacillus brevis*, *Lactobacillus fructivorans* and *Leuconostoc oenos* immobilised cells for the control of malo-lactic fermentation in wine. *Journal of Applied Bacteriology* 63: 513-521.

Dauer, R.R. and Dunlop, E.H. (1991). High gradient magnetic separation of yeast. *Biotechnology and Bioengineering* 37: 1021-1028.

De Arruda Moura Pietrowski, G., dos Santos, C.M.E., Sauer, E., Wosiacki, G. and Nogueira, A. (2012). Influence of fermentation with *Hanseniaspora* sp. yeast on the volatile profile of fermented apple. *Journal of Agricultural and Food Chemistry* 60: 9815-9821.

Dervakos, A.G. and Webb, C. (1991). On the merits of viable cell immobilisation. *Biotechnology Advances* 9: 559-612.

Diep, T.X. and Le, V.V.M. (2009). Immobilisation of yeast on pineapple pieces (*Ananas comosus*) for use in pineapple winemarking. pp. 743-748. *In*: Q.H. Tu (Ed.). Proceedings of the National Conference on Biotechnology. Thai Nguyen University and Institute of Biotechnology, Thai Nguyen (Thái Nguyên), Vietnam.

Diviés, C. and Cachon, R. (2005). Wine production by immobilised cell systems. pp. 285-293. *In*: V. Nedovic,and R. Willaert (Eds.). Applications of Cell Immobilisation Biotechnology. Springer, New York.

Djordjević, R., Gibson, B., Sandell, M., Billerbeck, G.M., Bugarski, B., Leskosek-Cukalovic, I., Vunduk, J., Nikicevic, N. and Nedović, V. (2015). Raspberry wine fermentation with suspended and immobilised yeast cells of two strains of *Saccharomyces cerevisiae*. *Yeast* 32: 271-279.

Dolejš, I., Krasňan, V., Stloukal, R., Rosenberg, M. and Rebroš. M., (2014). Butanol production by immobilised *Clostridium acetobutylicum* in repeated batch, fed-batch, and continuous modes of fermentation. *Bioresource Technology* 169: 723-730.

Đorđević, V., Balanč, B., Belščak-Cvitanović, A., Lević, S., Trifković, K., Kalušević, A., Kostić, I., Komes, D., Bugarski, B. and Nedović, V. (2015). Trends in encapsulation technologies for delivery of food bioactive compounds. *Food Engineering Reviews* 7: 1-39.

Duarte, F.W., Dias, R.D., Oliveira, M.J., Vilanova, M., Teixeira, A.J., Almeida e Silva, B.J. and Schwan, F.R. (2010). Raspberry (*Rubus idaeus* L.) wine: Yeast selection, sensory evaluation and instrumental analysis of volatile and other compounds. *Food Research International* 43: 2303-2314.

Durieux, A., Nicolay, X. and Simon, J.P. (2005). Application of immobilisation technology to cider production: A review. pp. 275-284. *In*: V. Nedovic and R. Willaert (Eds.). Applications of Cell Immobilisation Biotechnology. Springer, New York.

Dušak, P., Benčina, M., Turk, M., Bavčar, D., Košmerl, T., Berovič, M. and Makovec, D. (2016). Application of magneto-responsive *Oenococcus oeni* for the malo-lactic fermentation in wine. *Biochemical Engineering Journal* 110: 134-142.

Ehrhart, F., Mettler, E., Böse, T., Weber, M.M., Vásquez, J.A. and Zimmermann, H. (2013). Biocompatible coating of encapsulated cells using ionotropic gelation. *PLOS ONE* 8: e73498.

Fakhrullin, R.F., Zamaleeva, A.I., Minullina, R.T., Konnova, S.A. and Paunov, V.N. (2012). Cyborg cells: Functionalisation of living cells with polymers and nano materials. *Chemical Society Reviews* 41: 4189-4206.

Farbo, M.G., Urgeghe, P.P., Fiori, S., Marceddu, S., Jaoua, S. and Migheli, Q. (2016). Adsorption of ochratoxin A from grape juice by yeast cells immobilised in calcium alginate beads. *International Journal of Food Microbiology* 217: 29-34.

Freeman, A. and Lilly, M.D. (1998). Effect of processing parameters on the feasibility and operational stability of immobilised viable microbial cells. *Enzyme and Microbial Technology* 23: 335-345.

Fumi, M.D., Trioli, G., Colombi, M.G. and Colagrande, O. (1988). Immobilisation of *Saccharomyces cerevisiae* in calcium alginate gel and its application to bottle-fermented sparkling wine production. *American Journal of Enology and Viticulture* 39: 267-272.

García-Martínez, T., Moreno, J., Mauricio, J.C. and Peinado, R. (2015). Natural sweet wine production by repeated use of yeast cells immobilised on *Penicillium chrysogenum*. *LWT – Food Science and Technology*. 61: 503-509.

genialab GmbH, Germany, www.genialab.com

Genisheva, Z., Macedo, S., Mussatto, S.I., Teixeira, J.A. and Oliveira, J.M. (2012). Production of white wine by *Saccharomyces cerevisiae* immobilised on grape pomace. *Journal of the Institute of Brewing* 118: 163-173.

Genisheva, Z., Mussatto, S.I., Oliveira, J.M. and Teixeira, J.A. (2013). Malo-lactic fermentation of wines with immobilised lactic acid bacteria – Influence of concentration, type of support material and storage conditions. *Food Chemistry* 138: 1510-1514.

Genisheva, Z., Vilanova, M., Mussatto, S.I., Teixeira, J.A. and Oliveira, J.M. (2014a). Consecutive alcoholic fermentations of white grape musts with yeasts immobilised on grape skins – Effect of biocatalyst storage and $SO_2$ concentration on wine characteristics, *LWT – Food Science and Technology* 59: 1114-1122.

Genisheva, Z., Teixeira, J.A. and Oliveira, J.M. (2014b). Immobilised cell systems for batch and continuous winemaking. *Trends in Food Science & Technology* 40: 33-47.

Gòdia, F., Casas, C. and Solà, C. (1991). Application of immobilised yeast cells to sparkling wine fermentation. *Biotechnology Progress* 7: 468-470.

Herrero, M., Laca, A., García, L.A. and Díaz, M. (2001). Controlled malo-lactic fermentation in cider using *Oenococcus oeni* immobilised in alginate beads and comparison with free cell fermentation. *Enzyme and Microbial Technology* 28: 35-41.

Iconomopoulou, M., Psarianos, K., Kanellaki, M. and Koutinas, A.A. (2002). Low temperature and ambient temperature winemaking using freeze dried immobilised cells on gluten pellets. *Process Biochemistry* 37: 707-717.

Iconomou, L., Kanellaki, M., Voliotis, S., Agelopoulos, K. and Koutinas, A.A. (1996). Continuous wine making by delignified cellulosic materials supported biocatalyst. *Applied Biochemistry and Biotechnology* 60: 303-313.

Iguchi, M., Yamanaka, S. and Budhiono, A. (2000). Bacterial cellulose – a masterpiece of nature's arts. *Journal of Materials Science* 35: 261-270.

Institute of Agricultural Technology, Johann Heinrich von Thünen Institute, Germany, www.thuenen.de

Jackson, R.S. (2008). Specific and Distinctive Wine Styles. Wine Science: Principles and Applications (3rd Edition). Academic Press, San Diego, California, USA, pp. 520-576.

Kana, K., Kanellaki, M. and Koutinas, A.A. (1992). Volatile by-products formed in batch alcoholic fermentations: Effect of γ-alumina and kissiris supported biocatalysts. *Food Biotechnology* 6: 65-74.

Kandylis, P. and Koutinas, A.A. (2008). Extremely low temperature fermentations of grape must by potatoes supported yeast-strain AXAZ-1: A contribution is performed to catalysis of alcoholic fermentation. *Journal of Agricultural Food Chemistry* 56: 3317-3327.

Kandylis, P., Goula, A. and Koutinas, A.A. (2008). Corn starch gel for yeast cell entrapment: A view for catalysis of wine fermentation. *Journal of Agricultural Food Chemistry* 56: 12037-12045.

Kandylis, P., Manousi, M.E., Bekatorou, A. and Koutinas, A.A. (2010a). Freeze-dried wheat supported biocatalyst for low temperature wine making. *LWT – Food Science and Technology* 43: 1485-1493.

Kandylis, P., Drouza, C., Bekatorou, A. and Koutinas, A.A. (2010b). Scale-up of extremely low temperature fermentations of grape must by wheat supported yeast cells. *Bioresource Technology* 101: 7484-7491.

Kandylis, P., Mantzari, A., Koutinas, A.A., Ioannis, A. and Kookos, K. (2012a). Modelling of low temperature wine-making, using immobilised cells. *Food Chemistry* 133: 1341-1348.

Kandylis, P., Dimitrellou, D. and Koutinas, A.A. (2012b). Winemaking by barley-supported yeast cells. *Food Chemistry* 13: 425-431.

Kazenwadel, F., Franzreb, M. and Rapp, B.E. (2015). Synthetic enzyme supercomplexes: Co-immobilisation of enzyme cascades. *Analytical Methods* 7: 4030-4037.

Kopsahelis, N., Bosnea, L., Kanellaki, M. and Koutinas, A.A. (2012). Volatiles formation from grape must fermentation using a cryophilic and thermo-tolerant yeast. *Applied Biochemistry and Biotechnology* 167: 1183-1198.

Kourkoutas, Y., Dimitropoulos, S., Kanellaki, M., Marchant, R., Nigam, P., Banat, I.M. and Koutinas, A.A. (2002a). High temperature alcoholic fermentation of whey using *Kluyveromyces marxianus* IMB3 yeast immobilised on delignified cellulosic material. *Bioresource Technology* 82: 177-181.

Kourkoutas, Y., Koutinas, A.A., Kanellaki, M., Banat, I.M. and Marchant, R. (2002b). Continuous wine fermentation using a psychrophilic yeast immobilised on apple cuts at different temperatures. *Food Microbiology* 19: 127-134.

Kourkoutas, Y., Komaitis, M., Koutinas, A.A., Kaliafas, A., Kanellaki, M., Marchant, R. and Banat I.M. (2003). Wine production using yeast immobilized on quince biocatalyst at temperatures between 30 and 0°C. *Food Chemistry* 82: 353-360.

Kourkoutas, Y., Bekatorou, A., Banat, I.M., Marchant, R. and Koutinas, A.A. (2004). Immobilisation technologies and support materials suitable in alcohol beverages production: A review. *Food Microbiology* 21: 377-397.

Kourkoutas, Y., Manojlović, V. and Nedović, V.A. (2010). Immobilisation of microbial cells for alcoholic and malo-lactic fermentation of wine and cider. pp. 327-343. *In*: N.J. Zuidam and V.A Nedovic (Eds.). Encapsulation Technologies for Food Active Ingredients and Food Processing. Springer, Dordrecht.

Kregiel, D., Berlowska, J. and Ambroziak, W. (2013). Growth and metabolic activity of conventional and non-conventional yeasts immobilised in foamed alginate. *Enzyme and Microbial Technology* 53: 229- 234.

Lalou, S., Mantzouridou, F., Paraskevopoulou, A., Bugarski, B., Levic, S. and Nedovic, V. (2013). Bioflavour production from orange peel hydrolysate using immobilized *Saccharomyces cerevisiae*. *Applied Microbiology and Biotechnolgy* 97: 9397-9407.

LentiKat's Biotechnologies, Czech Republic, www.lentikats.eu

Levic, S., Djordjevic, V., Rajic, N., Milivojevic, M., Bugarski, B. and Nedovic, V. (2013). Entrapment of ethyl vanillin in calcium alginate and calcium alginate/poly(vinyl alcohol) beads. *Chemical Papers* 67: 221-228.

Lević, S., Đorđević, V., Knežević-Jugović, Z., Kalušević, A., Milašinović, N., Branko Bugarski, B. and Nedović, V. (2016). Encapsulation technology of enzymes and applications in food processing. pp. 469-502. *In*: R.C. Ray and C.M. Rosell (Eds.). Microbial Enzyme Technology in Food Applications. CRC Press, Taylor & Francis Group, Boca Raton.

Lević, S., Lijaković, I.P., Đorđević, V., Rac, V., Rakić, V., Knudsen, T.Š., Pavlović, V, Bugarski, B. and Nedović, V. (2015). Characterization of sodium alginate/D-limonene emulsions and respective calcium alginate/D-limonene beads produced by electrostatic extrusion. *Food Hydrocolloids* 45: 111-123.

Lević, S., Obradović, N., Pavlović, V., Isailović, B., Kostić, I., Mitrić, M., Bugarski, B. and Nedović, V. (2014). Thermal, morphological, and mechanical properties of ethyl vanillin immobilised in polyvinyl alcohol by electrospinning process. *Journal of Thermal Analysis and Calorimetry* 118: 661-668.

Lill, R., Hoffmann, B., Molik, S., Pierik, A.J., Rietzschel, N., Stehling, O., Uzarska, M.A., Webert, H., Wilbrecht, C. and Mühlenhoff, U. (2012). The role of mitochondria in cellular iron-sulfur protein biogenesis and iron metabolism. *Biochimica et Biophysica Acta* 182: 1491-1508.

Liu, C.Z., Wang, F. and Ou-Yang, F. (2009). Ethanol fermentation in a magnetically fluidised bed reactor with immobilised *Saccharomyces cerevisiae* in magnetic particles. *Bioresource Technology* 100: 878-882.

Lopez-Toledano, A., Merida, J. and Medina, M. (2007). Colour correction in white wines by use of immobilised yeasts on κ-carragenate and alginate gels. *European Food Research and Technology* 225: 879-885.

Loukatos, P., Kiaris, M., Ligas, I., Bourgos, G., Kanellaki, M., Komaitis, M. and Koutinas, A.A. (2000). Continuous wine making by γ-alumina supported biocatalyst. *Applied Biochemistry and Biotechnology* 89: 1-13.

Mallios, P., Kourkoutas, Y., Iconomopoulou, M., Koutinas, A.A., Psarianos, C., Marchant, R. *et al.* (2004). Low-temperature winemaking using yeast immobilised on pear pieces. *Journal of the Science of Food and Agriculture* 84: 1615-1623.

Mallouchos, A., Komaitis, M., Koutinas, A.A. and Kanellaki, M. (2003). Wine fermentations by immobilised and free cells at different temperatures: Effect of immobilisation and temperature on volatile by-products. *Food Chemistry* 80: 109-113.

Mallouchos, A., Loukatos, P., Bekatorou, A., Koutinas, A.A. and Komaitis, M. (2007). Ambient and low temperature winemaking by immobilised cells on brewer's spent grains: Effect on volatile composition. *Food Chemistry* 104: 918-927.

Mallouchos, A., Reppa, P., Aggelis, G., Kanellaki, M., Koutinas, A.A. and Komaitis, M. (2002). Grape skins as a natural support for yeast immobilisation. *Biotechnology Letters* 24: 1331-1335.

Maragatham, C. and Panneerselvam, A. (2011). Wine production from papaya piece using immobilised yeast (*Saccharomyces cerevisiae*) and its physicochemical analysis. *Research Journal of Pharmacy and Technology* 4: 798-800.

Margaritis, A. and Merchant, F.J.A. (1984). Advances in ethanol production using immobilised cell systems. *Critical Reviews in Biotechnology* 2: 339-393.

Margaritis, A., and Kilonzo, P.M. (2005). Production of ethanol using immobilised cell bioreactor systems. pp. 375-406. *In*: V. Nedović, and R. Willaert (Eds.). Applications of Cell Immobilisation Biotechnology. Springer Dordrecht.

Martynenko, N.N., Gracheva, I.M., Sarishvili, N.G., Zubov, A.L., El'-Registan, G.I. and Lozinsky, V.I. (2004). Immobilisation of champagne yeasts by inclusion into cryogels of polyvinyl alcohol: Means of preventing cell release from the carrier matrix. *Applied Biochemistry and Microbiology* 40(2): 158-164.

Mathew, A.K., Crook, M., Chaney, K. and Humphries, A.C. (2014). Continuous bioethanol production from oilseed rape straw hydrosylate using immobilised *Saccharomyces cerevisiae* cells. *Bioresource Technology* 154: 248-253.

Mathew, A.K., Crook, M., Chaney, K. and Humphries, A.C. (2013). Comparison of entrapment and biofilm mode of immobilisation for bioethanol production from oilseed rape straw using *Saccharomyces cerevisiae* cells. *Biomass and Bioenergy* 52: 1-7.

McKay, M., Buglass, A.J. and Gook Lee, C. (2011). Fermented beverages: Beers, ciders, wines and related drinks. pp. 419-434. *In*: A.J. Buglass (Ed.). Handbook of Alcoholic Beverages: Technical, Analytical and Nutritional Aspects. John Wiley & Sons, Chichester.

Melzoch, K., Rychtera, M. and Habova, V. (1994). Effects of immobilization upon the properties and behavior of *Saccharomyces cerevisiae* cells. *Journal of Biotechnology* 32: 59-65.

Mensour, N.A., Margaritis, A., Briens, C.L., Pilkington, H. and Russell, I. (1996). Application of immobilised yeast cells in the brewing industry. pp. 661-671. *In*: R.M. Buitelaar, C. Bucke, J. Tramper and R.H. Wijffels (Eds.). Immobilised Cells: Basics and Applications, vol. 11, first ed. Elsevier, Amsterdam.

Merida, J., Lopez-Toledano, A. and Medina, M. (2007). Immobilised yeasts in κ-carragenate to prevent browning in white wines. *European Food Research and Technology* 225: 279-286.

Miličević, B., Babić, J., Ačkar, Đ., Miličević, R., Jozinović, A., Jukić, H., Babić, V. and Šubarić, D. (2017). Sparkling wine production by immobilised yeast fermentation. *Czech Journal of Food Science* 35: 171-179.

Naouri, P., Bernet, N., Chagnaud, P., Arnaud, A. and Galzy, P. (1991). Bioconversion of L-malic acid into L-lactic acid using a high compacting multiphase reactor (HCMR). *Journal of Chemical Technology and Biotechnology* 51: 81-95.

Nedović, V. (1999). Imobilisani ćelijski sistemi u fermentaciji piva. *Zadužbina Andrejević*, Beograd.

Nedovic, V.A., Durieux, A., Van Nedervelde, L., Rosseels, P., Vandegans, J., Plaisant, A.M. and Simon, J.P. (2000). Continuous cider fermentation with co-immobilised yeast and *Leuconostoc oenos* cells. *Enzyme and Microbial Technology* 26: 834-839.

Nedović, V., Kalušević, A., Manojlović, V., Petrović, T. and Bugarski, B. (2013). Encapsulation systems in the food industry. pp. 229-254. *In*: S. Yanniotis, P. Taoukis, N.G. Stoforos and V.T. Karathanos (Eds.). Advances in Food Process Engineering Research and Applications. Springer, New York, USA.

Nedović, V., Gibson, B., Mantzouridou, T.F., Bugarski, B., Djordjević, V., Kalušević, A., Paraskevopoulou, A., Sandell, M., Šmogrovičová, D. and Yilmaztekin, M. (2015). Aroma formation by immobilised yeast cells in fermentation processes. *Yeast* 32: 173-216.

Nedović, V.A., Obradović, B., Leskošek-Čukalović, I., Trifunović, O., Pešić, R. and Bugarski, B. (2001). Electrostatic generation of alginate microbeads loaded with brewing yeast. *Process Biochemistry* 37: 17-22.

Neves, N.A., Pantoja, L.A. and dos Santos, A.S. (2014). Thermovinification of grapes from the Cabernet Sauvignon and Pinot Noir varieties using immobilised yeasts. *European Food Research and Technology* 238: 79-84.

Nguyen, D.N., Ton, N.M.N. and Le, V.V.M. (2009). Optimisation of *Saccharomyces cerevisiae* immobilization in bacterial cellulose by 'adsorption-incubation' method. *International Food Research Journal* 16: 59-64.

Nisco Engineering AG, Switzerland, www.nisco.ch

Ogbonna, C.J., Amano, Y., Nakamura, K., Yokotsuka, K., Shimazu, Y., Watanabe, M. and Hara, S. (1989). A multistage bioreactor with replaceable bioplates for continuous wine fermentation. *American Journal of Enology and Viticulture* 40: 292-298.

Oliveira, M.E.S., Pantoja, L., Duarte, W.F., Collela, C.F., Valarelli, L.T., Schwan, R.F. and Dias, D.R. (2011). Fruit wine produced from cagaita (*Eugenia dysenterica* DC) by both free and immobilised yeast cell fermentation. *Food Research International* 44: 2391-2400.

O'Reilly, A. and Scott, J.A. (1993). Use of an ion-exchange sponge to immobilise yeast in high gravity apple based (cider) alcoholic fermentations. *Biotechnology Letters* 15: 1061-1066.

Pajic-Lijakovic, I., Levic, S., Hadnađev, M., Stevanovic-Dajic, Z., Radosevic, R., Nedovic, V. and Bugarski, B. (2015). Structural changes of Ca-alginate beads caused by immobilised yeast cell growth. *Biochemical Engineering Journal* 103: 32-38.

Parascandola, P., Branduard, P. and de Alteriis, E. (2006). PVA-gel (Lentikats®) as an effective matrix for yeast strain immobilization aimed at heterologous protein production. *Enzyme and Microbial Technology* 38: 184-189.

Peinado, R.A., Moreno, J.J. and Villalba, J.M., González-Reyes, J.A., Ortega, J.M. and Mauricio, J.C. (2006). Yeast biocapsules: A new immobilisation method and their applications. *Enzyme and Microbial Technology* 40: 79-84.

Pereira, A.P., Mendes-Ferreira, A., Oliveira, J.M., Estevinho, L.M. and Mendes-Faia, A. (2014). Effect of *Saccharomyces cerevisiae* cells immobilisation on mead production. *LWT – Food Science and Technology* 56: 21-30.

Proenol-Indústria Biotecnológica SA, Portugal, www.proenol.com

Prüsse, U., Bilancetti, L., Bučko, M., Bugarski, B., Bukowski, J., Gemeiner, P., Lewinska, D., Manojlovic, V., Massart, B., Nastruzzi, C., Nedovic, V., Poncelet, D., Siebenhaar, S., Tobler, L., Tosi, A., Vikartovská, A. and Vorlop, K.D. (2008). Comparison of different technologies for alginate beads production. *Chemical Papers* 62: 364-374.

Puig-Pujol, A., Bertran, E., García-Martínez, T., Capdevila, F., Mínguez, S. and Mauricio, J.C. (2013). Application of a new organic yeast immobilisation method for sparkling wine production. *American Journal of Enology and Viticulture* 64: 386-394.

Qiu, K. and Netravali, A.N. (2014). A review of fabrication and applications of bacterial cellulose-based nanocomposites. *Polymer Reviews* 54: 598-626.

Rebroš, M., Dolejš, I., Stloukal, R. and Rosenberg, M. (2016). Butyric acid production with *Clostridium tyrobutyricum* immobilised to PVA gel. *Process Biochemistry* 51: 704-708.

Reddy, L., Reddy, Y., Reddy, L. and Reddy, O. (2008). Wine production by novel yeast biocatalyst prepared by immobilisation on watermelon (*Citrullus vulgaris*) rind pieces and characterisation of volatile compounds. *Process Biochemistry* 43: 748-752.

Reddy, L.V., Reddy, L.P., Wee, Y.J. and Reddy, O.V.S. (2011). Production and characterisation of wine with sugarcane piece immobilised yeast biocatalyst. *Food Bioprocess Technology* 4: 142-148.

Reddy, L.V.A., Reddy, Y.H.K. and Reddy, O.V.S. (2006). Wine production by guava piece immobilised yeast from Indian cultivar grapes and its volatile composition. *Biotechnology* 5: 449-454.

Rodríguez-Nogales, J.M., Vila-Crespo, J. and Fernández-Fernández, E. (2013). Immobilisation of *Oenococcus oeni* in Lentikats® to develop malo-lactic fermentation in wines. *Biotechnology Progress* 29: 60-65.

Rossi, J. and Clementi, F. (1984). L-malic acid catabolism by polyacrylamide gel entrapped *Leuconostoc oenos*. *American Journal of Enology and Viticulture* 36: 100-102.

Safarik, I., Maderova, Z., Pospiskova, K., Baldikova, E., Horska, K. and Safarikova, M. (2015). Magnetically responsive yeast cells: Methods of preparation and applications. *Yeast* 32: 227-237.

Scott, J.A. and O'Reilly, A.M. (1996). Co-immobilisation of selected yeast and bacteria for controlled flavour development in an alcoholic cider beverage. *Process Biochemistry* 31: 111-117.

Servetas, I., Berbegal, C., Camacho, N., Bekatorou, A., Ferrer, S., Nigam, P., Drouza, C. and Koutinas, A.A. (2013). *Saccharomyces cerevisiae* and *Oenococcus oeni* immobilised in different layers of a cellulose/starch gel composite for simultaneous alcoholic and malo-lactic wine fermentations. *Process Biochemistry* 48: 1279-1284.

Sevda, S.B. and Rodrigues, L. (2011). The making of pomegranate wine using yeast immobilised on sodium alginate. *African Journal of Food Science* 5: 299-304.

Sipsas, V., Kolokythas, G., Kourkoutas, Y., Plessas, S., Nedovic, V.A. and Kanellaki, M. (2009). Comparative study of batch and continuous multi-stage fixed-bed tower (MFBT) bioreactor during wine-making using freeze-dried immobilised cells. *Journal of Food Engineering* 90: 495-503.

Soares, E.V. (2010). Flocculation in *Saccharomyces cerevisiae*: A review. *Journal of Applied Microbiology* 110: 1-18.

Spettoli, P., Nuti, M.P., Crapisi, A. and Zamorani, A. (1987). Technological improvement of malo-lactic fermentation in wine by immobilised microbial cells in a continuous flow reactor. *Annals of the New York Academy of Sciences* 501: 386-389.

Takaya, M., Matsumoto, N. and Yanase, H. (2002). Characterisation of membrane bioreactor for dry wine production. *Journal of Bioscience and Bioengineering* 93: 240-244.

Tarn, M.D., Fakhrullin, R.F., Paunov, V.N. and Pamme, N. (2013). Microfluidic device for the rapid coating of magnetic cells with polyelectrolytes. *Materials Letters* 95: 182-185.

Tataridis, P., Ntagas, P. Voulgaris, I. and Nerantzis, E.T. (2005). Production of sparkling wine with immobilized yeast fermentation: Production and characterisation of wine with sugarcane piece immobilised yeast biocatalyst electron E. *Journal of Science and Technology* 1: 1-21.

Ton, N.M.N. and Le, V.V.M. (2011). Application of immobilised yeast in bacterial cellulose to the repeated batch fermentation in winemaking. *International Food Research Journal* 18: 983-987.

Ton, N.M.N., Nguyen, M.D., Pham, T.T.H. and Le, V.V.M. (2010). Influence of initial pH and sulphur dioxide content in must on wine fermentation by immobilised yeast in bacterial cellulose. *International Food Research Journal* 17: 743-749.

Tsakiris, A., Bekatorou, A., Psarianos, C., Koutinas, A.A., Marchant, R. and Banat, I.M. (2004). Immobilization of yeast on dried raisin berries for use in dry white winemaking. *Food Chemistry* 87: 11-15.

Varakumar, S., Naresh, K. and Reddy, O.V.S. (2012). Preparation of mango (*Mangifera indica* L.) wine using a new yeast-mango-peel immobilised biocatalyst system. *Czech Journal of Food Science* 30: 557-566.

Verbelen, P.J., Nedović, V.A., Manojlović, V., Delvaux, F.R., Laskošek-Čukalović, I., Bugarski, B. and Willaert, R. (2010). Bioprocess intensification of beer fermentation using immobilised cells. pp. 303-326. *In*: N.J. Zuidam and V.A. Nedovic (Eds.). Encapsulation Technologies for Food Active Ingredients and Food Processing. Springer, Dordrecht.

Viegas, M.C., Andrietta, S.R. and Andrietta, M.G.S. (2002). Use of tower reactors for continuous ethanol production. *Brazilian Journal of Chemical Engineering* 19: 167-173.

Vilela, A., Schuller, D., Mendes-Faia, A. and Côrte-Real, M. (2013). Reduction of volatile acidity of acidic wines by immobilised *Saccharomyces cerevisiae* cells. *Applied Microbiology and Biotechnology* 97: 4991-5000.

Wandrey, C., Bartkowiak, A. and Harding. E.S. (2010). Materials for encapsulation. pp. 31-100. *In*: N.J. Zuidam and V.A. Nedovic (Eds.). Encapsulation Technologies for Food Active Ingredients and Food Processing. Springer, Dordrecht.

Yajima, M. and Yokotsuka, K. (2001). Volatile compound formation in white wines fermented using immobilized and free yeast. *American Journal of Enology and Viticulture* 52: 210-218.

Yokotsuka, K., Yajima, M. and Matsudo, T. (1997). Production of bottle-fermented sparkling wine using yeast immobilised in double-layer gel beads or strands. *American Journal of Enology and Viticulture* 48: 471-481.

Zhou, Y., Martins, E., Groboillot, A., Champagne, C.P. and Neufeld, R.J. (1998). Spectrophotometric quantification of lactic bacteria in alginate and control of cell release with chitosan coating. *Journal of Applied Microbiology* 84: 342-348.

Zuidam, N.J. and Shimoni, E. (2010a). Overview of microencapsulates for use in food products or processes and methods to make them. pp. 3-29. *In*: N.J. Zuidam and V.A. Nedovic (Eds.). Encapsulation Technologies for Food Active Ingredients and Food Processing. Springer, Dordrecht.

Zuidam, J.N. and Heinrich, E. (2010b). Encapsulation of aroma. pp. 127-160. *In*: N.J. Zuidam and V.A. Nedovic (Eds.). Encapsulation Technologies for Food Active Ingredients and Food Processing. Springer, Dordrecht.

# 19 Winemaking: Control, Bioreactor and Modelling of Process

**Steve C.Z. Desobgo[1*] and Emmanuel J. Nso[2]**

[1] Department of Food Process and Quality Control, University Institute of Technology of the University of Ngaoundere, P.O. Box 455, Ngaoundere, Cameroon

[2] Department of Process Engineering, National School of Agro-Industrial Sciences (ENSAI), University of Ngaoundere, P.O. Box 455 ENSAI, Ngaoundere, Cameroon

## 1. Introduction

Agribusiness is embedded in production systems; hence the many current procedures subject to in-depth studies on methods to develop better systems for safety and quality. It is the way to ensure the quality and respect for the requirements of security, the system costs and natural effect. Among the classification (specifically structuring, preservation, separation and bioconversion) of nourishment forms, bioconversion incorporates most likely the biggest class of procedures. It is focal in the generation of maturing foods with some outstanding cases from the winemaking factory. Winemaking is customarily considered to be beginning with the blending of the berries and the introduction of yeast to realise fermentation. Innovation, smashing and regular squeezing could be viewed as the final stages of the vineyard operation. The control of quality of wine is basically essential. Institution study has accentuated the necessity to examine vulnerability to oxygen, extraction management of the substances from the skin, temperature monitoring amid fermentation, observing of sugar depletion and monitoring of microbes and malo-lactic fermentation. At present, the search is on for better quality items, keeping in mind that the general utilisation of wine is reducing, interest in astounding wines is expanding. Buyers are drinking less though 'better'. Wine producers throughout the world are consolidating winemaking techniques of centuries with new methodologies and thoughts, to satisfy purchaser's interest for better item quality and a maintainable and healthy lifestyle.

This chapter presents an overview of winemaking, monitoring, safety and quality control, which display the activities concerning each unit operation, the bioreactor characteristics and uses and finally, innovative approaches aimed at optimising the process efficiency.

## 2. Overview of Winemaking

### 2.1 Juice Extraction

As soon as grapes are received in the winery, they ought to be destemmed as well as squeezed with the particular ultimate objective to extract the juice (Soufleros, 1997a; 1997b). Likewise, care should be taken to keep the seeds in place. At the point when the outer securing seed shell is cracked, the huge measures of phenolic substances that the seeds hold, will concede to the wine and impart it an astringent taste (Ough, 1992). In the wake of stemming and pulverising, the juice moves into either a device used for draining, or a significant holding vessel or the concerned red grapes inside a fermenter. Within white cultivars, the prompt expulsion of juice in the peels and seeds is basic, as there are important measures of tannin-like substances in the peels. The touch between peel and juice (in the wake of pulverising) outperforms at 12 hours as the usual basement temperature might be destructive to the consequent wine and the degree of sensory characteristics that are involved (Ough, 1992). The must, to be transformed for white wine, is expelled from the skins, as they remain an important wellspring of regular microbial action and the level of phenol removal from the skins is restricted (Boulton *et al.*, 1996). Mash should be right away removed

*Corresponding author: desobgo.zangue@gmail.com

from the factory as it rapidly attracts dreadful little animals and diverse aggravations (Vine *et al.*, 1997). Of course, the juice from red grapes is less fragile than the white must and is not trailed by the systems, previously fermentation, since they are performed prior to it. Besides, the contact between peel and juice is appealing in red wines since the phenolic substances should be expelled from the seeds and should persist in the completed wine. The slurry is sent for pressing and the squeezed juice is incorporated in the key squeeze must and the crushed slurry ought to be separated and sent to the winery.

White juice is cleaned in the wake of pressing with a particular but ultimate objective to diminish the suspended grape compounds. Universally, a depletion of particles to less than 0.5 per cent is appropriate. This methodology can be reached by crisp falling or using mechanical methods. Sometimes, the extension or addition of pectolytic enzymes can help in clarification of juice. Starting late, there has been a more noteworthy use of them with a particular objective to quicken the wine clarification, subsequent to fermentation. It is perfect to incorporate the catalysts at this advanced step because the high ethanol contents achieved following fermentation tend to repress the activity of the enzyme (Tucker and Woods, 1996). Particles in juice provide a site to mature yeasts for $CO_2$ and ethanol release. Extraordinary refining of juice reduces the number of cells in the normal yeast concentration, reducing or annihilating their duty regarding ethanol (Wood, 1998).

Juices are much of the time put away to be utilised as a refreshing part or to raise the time of fermentation. Capacity states must be checked ($PCO_2 < 3.5$ atm, pH: 3.0 to 3.5, $T < 2$ °C) to hinder the improved decay of microbial cells (Fugelsang, 1997). Juices should be set up in one of the following ways: sulphating, chilling, juice concentration and cross-stream microfiltration (Boulton *et al.*, 1996).

## 2.2. Juice Preparation

Modification of the must before fermentation engages the winemaker to begin fermenting with every juice part. This process much of the time requires no less than one of the backing tasks: nutrient, $SO_2$ and catalyst incorporation, acidity, oxidation of juice (Boulton *et al.*, 1996). The fitness of the additional substances and their estimations need to be mastered (Tartaric acid: 0.5 to 1 g/L, Tannin: 5 g/L, $CaCO_3$: 0.5 to 1 g/L,) (Soufleros, 1997a).

At the point when the juice stabilises, the fermenters are loaded with juice having enough $SO_2$ and the yeast inoculum is incorporated. Care must however, be taken for the estimation of sulphur dioxide (<200 μg/mL juice), as over-the-top measure of it can realise yeast restraint and give a sulphur dioxide odour to the completed wine (Fugelsang, 1997). It is a direct but essential operation to incorporate an active yeast strain of *Saccharomyces* in the form of inoculum, to finish the alcoholic fermentation instead of relying upon the local microbial population(Lea and Piggott, 1995). The yeast strains might be observed imperceptibly to ensure that the ferments are of the vital sort and not that oxidative ferments and organisms are present (Fugelsang, 1997).

## 2.3. Fermentation

Fermentation ought to be the 'centre' of winemaking as the sugars of grape are transformed into alcohol by *Saccharomyces cerevisiae*. The nearness of increased heat (10-30°C) in the midst of fermentation can provoke destruction of the yeast inoculum and the impact of the all the more thermo-resistant microorganisms to finish the fermentation and plan unwanted side effects (Fugelsang, 1997). Berries, not strongly treated with pesticides in the farm can in addition be a source of the problem (Sala *et al.*, 1996). The extraordinary contamination with moulds, lactic and acetic acid organisms on grapes previously accumulated can convey some components that may ruin or obstruct yeast development in the midst of ethanolic fermentation (Wood, 1998). Also, an extreme extension of $SO_2$ can eradicate the predominant piece of appealing and unwanted cell and present a damaging impact on the wine flavour (Fugelsang, 1997). The nearness of ethyl carbamate concentration (<30 ppb) is a creation threat that might be viewed as critical as it is accepted to be cancer-causing (Fleet, 1994). That chemical is conveyed in the midst of the fermentation when having high heat in the direction of the completion of fermentation. Ferment strains that make a little estimation of urea are introduced and the farm is arranged vivaciously with characteristic compounds (Boulton *et al.*, 1996).

## 2.4. Malo-lactic Fermentation (MLF)

After ethanolic fermentation, wine regularly encounters the malo-lactic fermentation (MLF), that continues around 14 days to a month. Lactic acid organisms, occupant in the wine, are responsible for the MLF; however, various winemakers enable this deacidification by incorporation with strains of *Leuconostoc oenos*. The MLF realises a reduction in the taste of wine and an increase in its pH by around 0.3-0.5 units. That fermentation, however, is not helpful to every wine. Wines made of grapes developed in more hotter environments showed the tendency of being not so much acidic (pH > 3.5) and additionally fall in acidity should be malevolent to sensory characteristics. Also, the growth of MLF increases their pH to levels where crumbling microorganisms are more prone to develop (Wood, 1998). In cold climates, deacidification by MLF is desired so as to make the wine drinkable (Lea and Piggott, 1995).

At the point, when aging bubbles have crossed out of a direct recurrence, the novel wine might be racked off the gross stay in clean storage tanks (Vine *et al.*, 1997). Beginning hereon, for the term of the wine life, it should be essential to ensure that it is secured in compartments which are finished fully. Preventing air contact with the wine diminishes the first experience with oxygen which is a section for oxidation and the advanced deterioration of living cells (Soufleros, 1997b).

The common tried guidelines in mixing of wine are tantamount to imparting the assorted characteristics of distinctive wines in the wine getting a more extensive and fulfilling solicitation. Wines might be stirred prior to stabilisation in view of the fact that the various components required for steadiness, once in a while result in stable wines combined to shape a balanced wine (Vine *et al.*, 1997).

Just prior to fermenting and packing of wines, they are cleared up by utilising more than one method, which consolidates extension of fining materials (bentonite, egg whites, isinglass, gelatine) layer filtration and centrifugation (Ough, 1992). Frost stabilisation could then begin. It is finished by putting away the wine in a chiller at 3-2°C for not less than 21 days. This operation can generally be reduced by the incorporation of potassium bitartrate powder, that should be separated by using filtration into residue (Vine *et al.*, 1997). Wine precariousness can be expedited by numerous substances (synthetic, microbiological) and negatively affects wine quality. Moreover, Cu and the high iron compound can raise medical issues and need to be watched and cut down (Cu < 3 μg/mL wine, Fe < 12 μg/mL wine) by mixing. The genuine issue is metal in wine after the fermentation (Soufleros, 1997b). It can happen from components settling in the wine or vessel comprising these metals. Thus, it is essential to screen the metal substance (Pb <0.3 μg/mL wine, As <0.01 μg/mL wine) (Soufleros, 1997a).

## 2.5. Maturation

White wines start to mature when the yeast ends fermentation. When these yeasts are eliminated, wines are close to being drinkable. Certain white wines are matured in barrels. This generally is on the side of wines which will be traded at high expenses. Red wine maturation is more expanded than white (Ough, 1992). During maturation, each and every wooden vessel loses a little wine content because of absorption and leakage. It is basic to avoid air entrance to the wine in containers, especially to maintain a strategic distance from oxygen-devouring microbial growth. Each barrel might be precisely analysed at the minimum of four weeks in order to make up to the full capacity of wine. A couple of winemakers prefer ''wet-bung'' barrels prior to filling as a means to remember the ultimate objective of diminishing the risk of bacterial contamination in barrels via the bunghole (Vine *et al.*, 1997). Exactly as soon as red wines begin developing in barrels (or something else), the mind should be applied to inspecting for unwanted microbes and changes in shade, smell or flavour. Mix-ups in too little sulphur dioxide can generate wine spoilage by acetic acid bacteria and yeast (Ough, 1992).

Barrels are hard to clean and routinely hard to sanitise in case they appear to be corrupted with unwanted microbes and could be ousted from the winery. However, it is inconceivable since it could be damaging due to ethyl carbamate formation that is tumour-causing. Plus, ethyl carbamate could be formed in the midst of maturation as soon as there are urea build-ups which elevate the temperature. In this way, it must be measured prior to wine bundling (Gump and Pruett, 1993).

## 2.6. Packaging

The packing procedure is usually a wonder among the most joyful processes. During winemaking, genuine consideration must be given to the observance of hygienic practices (Vine *et al.*, 1997). The bundled wine is relied upon to be free of residual microbes. Wines are clarified through depth filters prior to their entry into the bundling vessel. Almost all issues arose due to the wrong utilisation of hygienic filtration procedure. The filter and notwithstanding the bundling line might be cleaned before the wine output. Air expulsion or counteractive action of the packing line is basic. The essential site where the wine is in connection with air is at the level of the filler bowl. The volume above the liquid in the bottle, in order to cover, could be blown with carbon dioxide or $N_2$ ahead of plugging (Ough, 1992). New bottles should be flushed frequently with extremely hot water and used jugs ought to have been sanitised when they were depleted, and held topsy turvy. Apparatus could be steam-sanitised. The filling equipment was seen to be the best explanation behind tainting, alongside the corking (Vine *et al*., 1997).

The stopper ought to be accurately treated and the neck of the jug should be of the most ideal size. Attention should be paid that the plugs don't have a bit of fragment or striations which could create spillage. The moistness of the plug is pivotal and might be 5-7 per cent (Ough, 1992). If it is bigger, it will be airtight too quickly. Stopper pollution is regarded as a noteworthy imperfection in packed wine. Stopper disease lead to a foul and off-flavour. It is seen that the most ideal approach to reduce the event of plug imperfection is to restrain the extreme conditions for microbial growth on the stopper by regulating the water activity (Fleet, 1994). Up-to-date stopper suppliers furnish sterile stoppers. In case of vulnerability, plugging might be done prior to being used in a solution of sulphur dioxide concentration of 10 g/hL (Marriott, 1994). Labels are a straightforward piece of trading wine. The paper should be rub-confirm, water-safe and preserved through fast bundling lines. Beyond this, it is basic to be coded in the event that there is an issue in the packed wine (Ough, 1992). Transport and storage of wines are normally not beneath the winemaker's mastery. Noteworthy harm is expected in packed wine as soon as it is predisposed to over abundance of heat or chill (Ough, 1992).

Just before the red wine is involved, de-stemming is vital to be executed cautiously in the production of red wine, as the stalk stores are not cancelled prior to the completion of fermentation and thus have a detrimental impact on the sensory attributes. At whatever point, the stalks are withdrawn from the contact with the pounded grapes by using a broadened time of less than two hours, a 'stemmy' off-nature might occur in wine. Also, the methodologies of skin dissociation and wet mash crushing are realised by following the fermentation method utilised for white wine production (Ough, 1992).

Wine is unreasonably acidic due to the development of pathogens. Various life forms don't survive in lesser pH medium (Speck, 1984). Microbes giving rise to wine deterioration are predominantly primitive yeasts and other bacteria. Basic deterioration yeasts include *Candida*, *Pichia* and different *Saccharomyces* spp. that cause development of films on the top of the wine. Wine-crumbling microorganisms are fundamentally lactic acid and acetobacters bacteria (Forsythe and Hayes, 1998). The best control method is keeping the perishing grapes away.

## 3. Monitoring Safety and Quality Control

### 3.1. Gathering/Harvesting

Grape gathering is a Critical Control Point (CCP1) accommodating chemical and physical risks (Table 1). Materially, grapes might be gathered in the absence of ruined parts, principally oxidation and pollution from microorganisms which can quickly grow.

In this way, gathering might be dovetailed with the best feasible precautions and a systematic contamination control method must be executed (Ellison *et al*., 1998; Dibble and Steinke, 1992). Pesticides play an unequivocal part in disturbance administration; in any case they should be taken care of deliberately in the light of the chemical dangers they present (Maner and Stimmann, 1992). At the season of gathering, the grapes should have accomplished proper development when acidity levels and Brix display matureness of fruit. Because chemical sediments on top of the berries represent chemical risks, studies suggest a quick and fundamental gas chromatographic technique for their estimation (Oliva *et al*., 1999). The best progress borders for insecticides in wines and grapes are granted by the Codex Alimentarius (Codex, 1998) and OIV (Organisation International du Vin) (OIV, 1994). In the end, mass receptacles utilised for grapes transferral must be successfully sanitised to stay away from any microbial contamination.

**Table 1.** Activities Concerning Security and Quality Control for Harvesting

| | | *Quality* | *Safety* |
|---|---|---|---|
| Harvesting/ Gathering (CCP 1) | Risks/Cause | • Untimely grapes gathering<br>• Overripe grapes gathering<br>• The hurt of grapes due to lack of precautions<br>• Piteous gathering techniques<br>• Mould contagion from affected grapes<br>• *Penicillium* and *Aspergillus* infection of grapes<br>• Growing of *Acetobacter* on grapes<br>• Traces of iprodione, vinclozolin, procymidone in the grapes | • Pesticide trace<br>• Unwanted substance from the soil<br>• Infection of gathering equipment |
| | Precaution measures | • Quotidian precaution amid gathering<br>• Mature grapes gathering<br>• Gathering workers with experience<br>• Conscientious compliance of MRLs<br>• Control of sugars and acid concentration in grapes<br>• Acceptable cleaning of the gathering machines<br>• Use $SO_2$ for mould contamination of grape | • Apply attention during gathering<br>• Harvest uncontaminated grapes<br>• Scrupulous conformity of MRLs<br>• Cleanliness rehearses application to stay away from the pollution of grapes |
| | Severe factors/ limits/ Controls | • Determination of grapes density<br>• Determination of grapes acidity<br>• Checking on grapes integrity<br>• Determination of insecticide residue content of grapes<br>• Check-up of grapes amid harvesting<br>• Investigation of cleanliness application amid collecting | • Determination of insecticide residue concentration<br>• Auditing of hygienic methods amid gathering<br>• Inspection of harvesting equipment hygiene |

*Source*: Adapted from Kourtis and Arvanitoyannis, 2001

### 3.2. Stemming

Stemming (CCP 2, Table 2) considers the elimination of leaves, grape stalks and stems prior to beating. This system has a few purposes of interest since the entire volume of the disposed items falls by 30 per cent, as needs demand bringing about littler tanks and subsequently increasing the ethanol concentration. Regardless of this, the completion of fermentation and the ethanol concentration of the completed wine rely generally on the sugar concentration of grapes. Stemmers mostly comprise a pierced cylinder for berries to experience yet preserve the area of stems and stalks.

### 3.3. Blending

Blending/crushing (CCP 3, Table 3) ordinarily instantly takes place in the wake of stemming. Released juice is especially prone to browning due to oxidation and microorganism pollution. The immensely acknowledged squeezing forms incorporate pressing the fruits in contact with a punctured equipment or directing the berries via rollers. It is fundamental to go without pulverising the seeds to protect against polluting the must with oils from the seed. Its oxidation could produce unwanted smells and represent an unpleasant fountainhead of acrid tannins. Essentially, it is fundamental to have the most ideal treatment of the product, since wrong arranging may incite an unexpected start of ethanolic fermentation and therefore lead to highr temperature of fermentation. though a recess may generate microbe pollution and browning (Zoecklein *et al.*, 1994).

**Table 2.** Activities Concerning Security and Quality Control for Stemming

| | | *Quality* | *Safety* |
|---|---|---|---|
| Stemming (CCP 2) | Risks/Cause | • Stem residue in grapes (red winemaking)<br>• *Botrytis cinera* infection of grapes<br>• Grapes contamination by foreign matter coming from equipment | • Infection of grapes coming from bad cleaning<br>• Foreign matter of grapes coming from equipment |
| | Precaution measures | • Destemming maintenance of equipment<br>• Manual elimination of external material from grapes<br>• Prevention of grapes degradation using $SO_2$<br>• The utilisation of cold water for adequate cleaning of destemmer | • GMP (Good Manufactured Practice) and sanitation amid destemming<br>• Equipment and environment sanitation in the winery |
| | Severe factors/ limits/controls | • GMP control (red winemaking)<br>• Convenient expulsion of mold tainted grapes<br>• Sulphur dioxide (40 mg/L) estimation of the grapes<br>• Monitoring of destemmer refreshing | • Mastery of GMP and hygiene amid destemming<br>• Mastery of hygienic techniques for traces of microorganisms and traces of clearing in the destemmer |

*Source*: Adapted from (Kourtis and Arvanitoyannis, 2001

**Table 3.** Activities Concerning Security and Quality Control for Crushing/Blending

| | | *Quality* | *Safety* |
|---|---|---|---|
| Crushing/ blending (CCP 3) | Risks/Cause | • Oxidation of must<br>• The increment in the measure of mass in the must<br>• Contamination of must with metal compound coming from apparatus | • Infection of must via insufficient clearing (deposits of microbes, traces of chemicals)<br>• Must contamination by external matter coming from apparatus |
| | Precaution measures | • Application of GMP during crushing<br>• Sufficient space between crushing cylinders<br>• The absence of air amid squashing | • GMP and sanitation amid crushing<br>• Utilisation of authorised cleaning agents |
| | Severe factors/ limits/controls | • Monitoring and application of GMP amid crushing<br>• The environment of crushing air control<br>• Crushing equipment cleaning and control | • Control of GMP and hygiene amid destemming<br>• Must supply time in the blenders <2 h<br>• Cleaning of blending equipment at the end of two days<br>• Dissociation of blending equipment: maximum possible<br>• Sanitation mastery and GMP request amid blending |

*Source*: Adapted from Kourtis and Arvanitoyannis, 2001

### 3.4. Maceration/Squeezing/Pressing

Maceration is the dislocation of grapea by mashing them. As long as maceration is continually required in the underlying period of red wine fermentation, the long-time practice has led to less soaking in the manufacture of white wine. Span and temperature of mashing depend on wine and grape cultivar.

Conventionally for rose wines and white wines, the time of maceration is below 24 hours – red scheduled for early use, is macerated during three to five days and red for fermentation, is soaked between 120 hours to 21 days. Fermentation all the more regularly occurs in the midst of this or in the direction of the termination of maceration. The quantity of the antimicrobes to be utilised, generally in addition to the musts of white wine which is most sensible to oxidation, relies on the gathering prosperity and maceration heat. $SO_2$ has an extraordinary favoured viewpoint above alternative antimicrobial substances, as a consequence of the comparative passiveness of the wine ferments to its activity. Notwithstanding this, it is moreover deadly, or hindering, to nearly all yeasts and microorganisms (*Hansenula, Pichia* and *Candida*) in little amounts (Farkas, 1984) and has a fairly reduced retentiveness limit following the clarification stage (Gnaegi *et al.*, 1983). The juice is permitted to stay in the press for a time, in the midst of which juice flows out under its own gravity. Being dependent on the press, the obtained must and wine portions differ in regards to their physico-chemical characteristics. Joining various wine parts, the winemaker influences the wine character. In any case, a potential peril might exist in the reaction of oxidation if there is an interference in the procedure (Lichine, 1985).

### 3.5. Ethanolic Fermentation

Ethanolic fermentation is ordinarily completed by *Saccharomyces cerevisiae* strains since this type is particularly resistant to the large amounts of sugar, ethanol and $SO_2$ and besides,to lesser pH (3.2-4) for grape juice. The strains of *Saccharomyces cerevisiae* are one constituent of the endogenous microbial population or might be to some degree included in attaining a density of approximately $10^5$-$10^6$ cells/mL in the juice (CCP 4, Table 4) (Constanti *et al.*, 1997).

Feasible pollution of juice with 'killer' yeasts (quality generally displayed in undomesticated *Saccharomyces* strains, furthermore in the alternative genus of yeast, for instance, *Cryptococcus, Torulopsis, Pichia, Kluyveromyces, Hansenula, Debaryomyces* and *Candida,*) could lead to bad fermentation (Van-Vuuren and Jacobs, 1992). Thought should be paid to the extra measure of $SO_2$ (175-225 µg/mL for white and red wine, independently) remembering the true objective is to prevent, if not to eradicate, the majority of wild yeast masses of grapes (Sudraud and Chauvet, 1985) and furthermore acidity management, and to Brix and tannin amount of the must. In fermentation, the accomplished chemical risks contain toxic metals (As <0.2 µg/mL, Cd <0.01 µg/mL, Cu <1 µg/mL, Pb <0.3 µg/mL), methanol amounts (300 µg/mL and 150 µg/mL for red wine and white wine, exclusively), EC amounts, insecticide traces and detergents (non-attendance) and ethylene glycol (non-appearance).

Attention should be paid with respect to the EC amount, in light of the fact that there is no enactment opposed to it in Europe, but it is so in USA (<60 ppb and <15 ppb for dessert and table wines, exclusively). The latter is confirmed from the chemical reaction of ethanol with materials rich in amino acids, basically, amino acids and urea like citrulline and arginine. Its management including gas chromatography (GC) measurement and evasion can be done by preventing concentrated fertiliser treatment of vines, elevated temperatures for conclusion or after ethanolic fermentation, utilising yeast strains for smaller ethyl carbamate and urea creation, using an enzyme and testing urea when prolonged storage is required.

The temperature of fermentation is noteworthy among the most basic factors affecting the metabolism of yeast, both clearly and in a roundabout way. For red and white wines, the appealing temperature vacillates to the extent of 8-15°C and 25-28°C, separately. Any existence of leftover sugars (fructose, glucose, sucrose) before the completion of fermentation is a risk that may create microbial destabilisation of wine.

The system of fermentation needs no oxygen. Nonetheless, residual oxygen towards the beginning of the exponential stage of yeast development quickens the fermentation, considering that the yeast cell number rises and the ordinary cell gets viability augmented. The pH (<3.0) may impact the procedure exactly at extraordinary levels where the advancement of fermentation yeasts is quelled (Zoecklein *et al.*, 1994).

At long last, the fungicide in the must may accept improvement of yeast inhibition and hinder the sensory characteristics of wine by affecting biosynthetic metabolisms (Pilone, 1986; Cabras *et al.*, 1988; Fatichenti *et al.*, 1984).

**Table 4.** Activities Concerning Security and Quality Control for Fermentation

| | | *Quality* | *Safety* |
|---|---|---|---|
| Fermentation (CCP 4) | Risks/Cause | • Development of troublesome bacteria in the fermenters<br>• Stuck fermentation<br>• Loss of wine aroma profile<br>• Acetic acid and $H_2S$ production<br>• Oxidation because of air entrance in the fermenters<br>• Sugar fermentation into lactic acid<br>• Glycerol lactic fermentation<br>• Tartaric acid lactic fermentation<br>• Augmentation of wine viscosity<br>• Abnormal proceeding amid fermentation<br>• Fermentors breaking because of high temperature or $CO_2$ | • EC production<br>• Cleaning chemicals residues in fermentors<br>• Other residues from pre-fermentation phases (yeasts, bentonite)<br>• Extreme injection of $SO_2$ in the fermented must |
| | Precaution measures | • Use of $SO_2$ to prevent wine spoilage<br>• Application of authorised $SO_2$ limitations<br>• Injection of favoured ferment strains into preceding inoculation<br>• Introduction of yeasts nutrients<br>• Maintain fermentation range temperature by utilising the automatic cooling system<br>• Clean fermentors<br>• Stabilise fermentation temperature<br>• Pump-over process (for the manufacture of red wine) | • GMP and sanitation amid destemming<br>• Insertion of sulphur dioxide < 200 μg/mL fermented juice<br>• Sanitation of fermenters<br>• Setup of small temperature in bioreactors<br>• injection of special yeasts inside the preceding inoculation |
| | Severe factors/ limits/ controls | • White wine fermentation temperature: 10-21°C<br>• Red wine fermentation temperature: 20-30°C<br>• Must aeration amid the first two days of fermentation<br>• Monitor yeast injection<br>• Authorised $SO_2$< 200 mg/L must<br>• Juice gravity control amid fermentation<br>• Monitoring of pump-over process | • Mastery of GMP and cleaning amid destemming<br>• EC content: <30 ppb in the fermented must<br>• Authorised sulphur dioxide < 200 μg/mL fermented juice<br>• Fermentor cleaning management<br>• Yeast purity control and safety<br>• Temperature control amid fermentation |

*Source*: Adapted from Kourtis and Arvanitoyannis, 2001

## 3.6. Malo-lactic Fermentation (MLF)

Early beginning and achievement of MLF incite development of $SO_2$ and stockpiling at chill temperatures and clearing. It is driven, using lactic acid (LA) organisms (*Oennococcus oenos*) that clearly decarboxylate the L-malic acid to L-lactic acid. This change achieves acidity reduction and pH increase, that are linked to the extended drinkability and creaminess of red wines (Davis *et al*., 1985; Guzzo *et al*., 1998). The underlying pH, the sulphite capture (Vaillant *et al*., 1995), the anthocyanin and the phenolic amount

(Vivas *et al.*, 1997) of must/wine unambiguously impact in the case, giving rise of MLF. Phages could truly interrupt MLF by affecting the *Oennococcus oenos* along these lines, creating destabilisation of wine microflora (Gnaegi and Sozzi, 1983). In this way, to ensure the progression of MLF, winemakers inject the fermented juice with no less than one *Oennococcus oenos* strains (CCP3, Table 5) (Nielsen *et al.*, 1996; Nault *et al.*, 1995). After fermenting, the wine's accepted total acidity is believed to differ inside to the extent of 0.55-0.85 per cent. At any point, total acidity beats the breaking points and fermentation and deacidification techniques are set up (Jackson, 1998).

**Table 5.** Activities Concerning Security and Quality Control for MLF

| | | *Quality* | *Safety* |
|---|---|---|---|
| Malolactic fermentation (CCP 5) | Risks/Cause | • Augmentation of wine pH<br>• Reduction of wine acidity<br>• Degradation of wine taste | • Microbiological contamination |
| | Precaution measures | • Injection (inoculation) of malolactic yeasts | • Certified suppliers, strictly following instructions |
| | Severe factors/limits/controls | • Controlling the pH of the wine<br>• Controlling the acidity of the wine<br>• Wine pH <3.5 | • Microbial analysis |

*Source*: Adapted from Kourtis and Arvanitoyannis, 2001

### 3.7. Maturation/Aging

The maturation step regularly keeps going from six months to one year in oak barrels. Amid maturation, a score of chemical and physical interactions occur inside the barrel, the encompassing environment and the wine in maturation, prompting a change of savour and characteristics of wine (Martinez *et al.*, 1996). At this level, we have a CCP (CCP 6, Table 6) regarding the oak vessel, which is expected to be flaw-free and ought to have been subjected to disinfecting processing.

The wood likewise should be exempted from noticeable or unpleasant smells, which pollute the fermented must (Mosedale and Puech, 1998). During the aging period, a few compounds of the wood are deleted to tannin of wine (Viriot *et al.*, 1993; Towey and Waterhouse, 1996). Since oak tannins could essentially increase wine savour, white wines are generally aged in oak for a smaller time than red wines and in prepared oak containers to discharge a smaller amount of extractable tannin (Popock *et al.*, 1984; Quinn and Singleton, 1985).

One more CCP is marked with the restraint of air infiltration along wood or amid racking and inspection of wine. In spite of the fact that a little oxidation is alluring, a more substantial one could generate different sensory modifications, for example, oxidised smell, browning, colour loss in red wines, yeast activation and bacteria spoilage, ferric casse development and tannin precipitation (Ranken *et al.*, 1997). Restrains on free and total sulphur dioxide amounts in completed wine vary from nation to nation.

### 3.8. Clarification

Clarification includes physical methods for evacuating the floating particles. Must clarification by filtration, centrifugation, or racking frequently enhances the savour improvement in white wine and supports the avoidance of spoilage by microorganisms. Assuming that an adequate period is given, fining and racking could create stable, completely clear wines; however, now that premature packaging in months or 14 days following fermentation is utilised, filtration and centrifugation are done to facilitate the required clearance amount (Ribereau-Cayon *et al.*, 1998). Pollution of wine by microorganisms amidst the previously stated techniques causes a possible issue for its steadiness (Ubeda and Briones, 1999). Racking is likewise powerful on insecticide traces and lessening of wine (Gennari *et al.*, 1992).

**Table 6.** Activities Concerning Security and Quality Control for Maturation/Aging

| | | *Quality* | *Safety* |
|---|---|---|---|
| Aging/ Maturation (CCP 6) | Risks/ Cause | • Wine sensory characteristics modification<br>• Barrel flavour in the wine<br>• Oxidation of wine<br>• *Dekkera, Brettanomyces, Pechia, Candida,* development in the fermented must<br>• *Acetobacter* development in the fermented must | • DMDG wine residue<br>• EC in wine<br>• Wine contamination by the development of microorganisms in barrels<br>• Wine contamination from the dirty winery |
| | Precaution measures | • Prevention of wine spoilage by adding $SO_2$<br>• The use of $N_2$ to remove $O_2$ from wine<br>• The barrels must be kept totally full<br>• Barrels must be carefully cleaned<br>• Tight-bunged barrel<br>• Maturation of wine using always wetted bung<br>• Maturing wine temperature (<12°C) | • Attentive barrel clearing<br>• Utilisation of new oak vessels<br>• Attentive winery clearing<br>• Keeping small temperature amid maturation |
| | Severe factors/ limits/ controls | • Monitoring of $SO_2$ concentration (>3 μg/ mL wine)<br>• Monitoring of keeping temperature (<12°C)<br>• Monitor odour of empty barrels<br>• Monitor oxygen absence amid wine maturation<br>• Monitoring spoilage bacteria in wine<br>• Monitoring barrel cleaning methods | • EC in wine measurement<br>• Monitoring of barrel clearing methods<br>• Mastery of the suitability of the barrels<br>• Monitoring of winery cleaning methods<br>• $SO_2$ measurement in wine (>3 mg/L) |

*Source*: Adapted from Kourtis and Arvanitoyannis, 2001

### 3.9. Fining/Stabilisation

The purpose behind fining is the creation of a lastingly clear and flavoured flaw-free wine. Nearly all essential strategies incorporate a) fining using tartrate by cooling the matured wine to close to its freezing temperature and afterward filtration or centrifugation is executed to evacuate the solids, b) protein fining with fixing, neutralisation, or degradation by bentonite is carried out (Blade and Boulton, 1988), c) polysaccharide expulsion is done with enzymes which hydrolyse the macromolecule, perturbing its defensive colloidal activity and membrane stopping characteristics (Ribereau-Cayon *et al.*, 1998), and d) stabilisation of metal casse (Fe, Cu) (CCP 7, Table 7) is initiated.

Ferric casse is monitored using the expansion of bentonites and proteins by adjusting the aggregation of ferric complexes which are insoluble, though wines with Cu content more noteworthy than 0.5 μg/mL are especially vulnerable to Cu casse development (Langhans and Schlotter, 1985). Legitimate remaining Cu levels in completed wines fluctuate and not all strategies for Cu evacuation are authorised in all the countries.

### 3.10. Bottling

Wine is packed in glass containers covered with stopper. The container might pass a sanitising stage and an examination to ensure the non-appearance of any inadequacy (CCP 8, Table 8) and the steadiness of the wine until its gobbling (Cooke and Berg, 1984).

The stopper must be well sized, 6-7 mm higher than the inside neck diameter of the bottle, to abstain from any feasible leaks. In packaging, all three risks might be found. In special, stopper microorganism, heavy metals traces, $SO_2$, insecticides and detergents, and non-appearance of cracks, scrapes and fissures in the lute speak for physical, chemical and microbiological risks.

**Table 7.** Activities Concerning Security and Quality Control for Stabilisation/Fining

| | | *Quality* | *Safety* |
|---|---|---|---|
| Fining/ Stabilization (CCP 7) | Risks/Cause | • Fining chemical residue in the wine<br>• The residue of lees in the wine<br>• Wine over fining<br>• Agents of adsorption in the wine<br>• Brown cloudiness of wine<br>• Cloudiness of wine due to microbes<br>• Cloudiness of wine due to $Fe^{2+}$<br>• The turbidity of wine due to $Cu^{2+}$<br>• The turbidity of wine due to colloidal substances | • Impure addition compounds in the wine<br>• Traces of stabilisation chemicals in the wine<br>• Traces of poisonous metals in wine<br>• Residues of chemical substances in wine |
| | Precaution measures | • Use of dosage pump to add a fining agent<br>• Dissolution of fining chemical in water<br>• Quick withdrawal of lees residues from wine<br>• Introduction in chilly weather environment of the fining agent<br>• Prevention of deterioration by adding $SO_2$<br>• Storage of wine far from sun and air<br>• Addition of bentonite | • Authorized substances addition according to legislation<br>• Addition of authorised substances for wine stabilisation<br>• Authorised substances for addition |
| | Severe factors/limits/ controls | • Monitoring of the addition of agent solution in the wine<br>• Monitoring of lees traces in the wine<br>• Control of over treatment of trub in the wine<br>• Control of weather environments amid fining<br>• Monitoring of fining chemical traces in the wine<br>• Oxidase measurement in wine<br>• Microscopic check-up of wine for microbes<br>• $Fe^{2+}$ <12 μg/mL wine<br>• $Cu^{2+}$ <3 μg/mL wine | • Monitor the purity of fining agents<br>• Monitor the authorised substances<br>• Monitor the residues of fining chemicals in the wine<br>• Monitoring of authorised additives<br>• Metal limits estimation (As <0.01 μg/mL, Cu <0.1 μg/ mL, Pb <0.3 μg/mL wine) |

*Source*: Adapted from Kourtis and Arvanitoyannis 2001

Although cork is important for its non-reactive property when touching the wine, it could generate unwanted flavours when polluted (Simpson *et al.*, 1986; Simpson, 1990) or when winemakers are not executing functional quality control (Neel, 1993). The control for the stopper is non-appearance of yeast and LAB and which could be verified by microbial test. When a long maintenance period of wine is predicted, higher and denser stoppers are favoured since a long exposure bit by bit influences the stopper integrity. When forcing the cork into the bottle neck, attention should be paid to stop the development of microbes inside the equipment (Malfeito-Ferreira *et al.*, 1997; Ubeda and Briones, 1999) and the lead transfer to wine through the wine-stopper-capsule method (Eschnauer 1986), and the oxidation during packing by washing out the glass containers with $CO_2$. Stopper placing might also happen under vacuum. The empty space occupied by oxygen could impact the item quality by giving rise to the disease of the 'bottle'. The containment limit for sulphur dioxide is 175-225 μg/mL respectively for red wine and white wine. Nearly <0.2 μg/mL, Cd <0.01 μg/mL, Cu <1 μg/mL, Pb <0.3 μg/mL traces of insecticides and pesticides in the completed item, are given by Office International de la Vigne et du Vin (OIV, 1994).

**Table 8.** Activities Concerning Security and Quality Control for Bottling

| | | Quality | Safety |
|---|---|---|---|
| Bottling (CCP 8) | Risks/Cause | • Deterioration microbes in wine bottles<br>• The growth of moulds in wine bottles<br>• Leakage of wine from bottles<br>• Oxidation of wine enhancing loss of sensory characteristics<br>• Foreign substances in wine coming from bottles | • Undesirable substances in the wine and coming from glass containers and filling equipment<br>• Residues of clearing agents in the fermented must<br>• Pollution of wine from the winery<br>• Development of microbes in glass containers polluting the wine<br>• Development of microbes in filling equipment pollute the wine |
| | Precaution measures | • Mechanical and chemical cleaning of bottles<br>• Bottling as stated to legislation<br>• $SO_2$ addition in wine before bottling<br>• Removal of air in wine using $N_2$ | • Cleaning of bottles<br>• Sanitation of bottles line<br>• Sanitation of winery |
| | Severe factors/ limits/controls | • Monitoring of wine cleaning methods<br>• Control of bottles visually and microbiologically<br>• GMP monitoring application during the bottling of wine | • Cleaning of bottles methods<br>• Monitoring of GMP during the bottling of wine<br>• Hygiene control measurement for bottles line, bottles and environment<br>• Microbial control measurement for bottles line, bottles and environment |

*Source*: Adapted from Kourtis and Arvanitoyannis, 2001

### 3.11. Storage

Storage and shipping of wine at high temperatures could induct fast modifications in wine flavour and colour. Straight subjection to sunlight reflects the influence of hot storage temperature. It impacts the reaction speeds implicated in maturation, for instance, the speeding up of the terpene fragrance loss and aromatic ester hydrolysis (De-la-Presa-Owens and Noble, 1997). Temperature can influence the volume of wine, alleviating the stopper seal, generating oxidation, leakage and eventually microbial growth due to bottled wine spoilage (CCP 9, Table 9).

**Table 9.** Activities Concerning Security and Quality Control for Storage

| | | Quality | Safety |
|---|---|---|---|
| Storage (CCP 9) | Risks/Cause | • Alteration of cartons and labels of wine bottles due to a humid area<br>• High temperature provoking wine leakage | |
| | Precaution measures | • Storage in low humidity environment<br>• Storage at environment temperature between 12-15°C | |
| | Severe factors/ limits/controls | • Control of storage condition | |

*Source*: Adapted from Kourtis and Arvanitoyannis, 2001

# 4. Bioreactors Typology, Technology and Uses

Winemaking innovation has seen amazing progressions all through the most recent 20 years, upgrading the nature of wines and furthermore formulating it to deliver wines with an extensive extent of traits. On this ground, some innovative progressions like enzymatic actions, use of picked yeasts, modification of microbe starters and immobilisation are of key importance (Fig. 1). These headways have influenced all aspects of winemaking, with wine remaining the last consequence of a mechanical chain that joins the handling and must treatment, fermentation, aging and packing. This inventive advance has upgraded the nature of the wines made. Quality wine is evaluated by intensity, fineness, advancement in smell and taste, and physic-chemical and microbiological stability (Dubourdieu, 1986; Noble, 1988; Rapp and Mandey, 1986; Schreier, 1979).

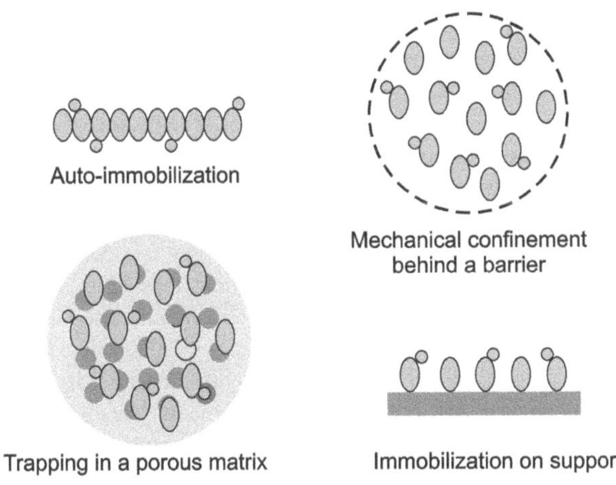


**Figure 1.** Yeast immobilisation system

## 4.1. Bioreactors Shape and Size

Fermenters of a broad collection of forms are straight-tube, barrel, V, square vessels, external forms etc. are utilised for wine manufacture (Boulton *et al.*, 1996). Forms of different fermenters used in fermented beverages (Maule, 1986; Moresi, 1989) are shown in Fig. 2. In most of the fermenters, floor inclining is done in the direction of the front. Bioreactors with domed or hemispherical bases are used in winemaking. Despite positive conditions in mash discharge in red winemaking, the usage of the funnel-shaped-based bioreactors has not sharpened. One of the most exorbitant and freshest advancements in the fermentation of red wine are the turning stainless-steel bioreactors that are uncommonly gainful with respect to the degree of energy and time anticipated that would build perfect skin introduction in the aging wine and inconsequential oxygen open to deterioration microbes. Despite having focal points, the usage of rolling and barrel-shaped vessels as a differentiating alternative to standard fermenters has found affirmation in red wines (Peyron and Feuillat, 1985).

There is a noteworthy distinction in measure, shape, outline and advancement materials used as a piece of fermentation tank in the manufacture of wine, inciting a fluctuated grouping of matured wines (Moresi, 1989). Any non-porous and non-perilous tank could be used as a bioreactor. Each tank could be ordered in two fundamental classes (tanks and vats). Vats are open at the top, although tanks are closed at the top. Earlier vats were used for red wine manufacture in view of the fact that a prompt access to the highest point of skins and seeds is desired in the midst of aging. White wines could be manufactured in tanks and are in a position to disallow air from the maturing juice. Most of the bioreactors outlined are rudimentary and a problem in planning occurs since their volume is extended. The development in volume moreover decreases the surface area for heat transfer. In red wine manufacturing, the unpredictability of the tank technology depends on the method used for the top submersion. Usually, the batch fermentation technique is used in wineries. Continuous bioreactors are moreover open but rarely used. As of now, fermentations used to be finished in 2.25-2.28 hL tanks or 6-12 hL vats (Diviès, 1988). The wooden or concrete vessels used earlier have now been replaced with especially planned stainless-steel bioreactors. A diagram demonstrating particular fragments of a stainless-steel bioreactor is shown in Fig. 2. Small fermenters are often used in red wine aging as a result of the problem in achieving an agreeable top submersion (Jackson, 1999). Business wineries are using bioreactors of 20 $m^3$ or more prominent limits. They are sensible to the extent of capital cost, computerisation and automation.

## 4.2. Types of Bioreactors

Efficiency in the fermentation of wine can be increased by utilising elevated yeast cell density by expanding the operative cell density or size by accumulation or cell immobilisation on a certain support. These methods are called high-cell-density reaction techniques. In addition, these methods are insensitive to unforeseen changes in working conditions or other characteristics of the must. Thus the total amount of organisms is maintained, the fermentation activity being restored once the problem is solved. The flow propels the procedures of immobilisation and have led to the improvement of proficient immobilised fermenters to completely make use of the benefits of biocatalysts and cell immobilisation (Fig. 2). The utilisation of the procedure of immobilisation for fermentation of wine accordingly, needs the improvement of a deliberately planned and reasonably constructed fermenter. Non-stop alcohol fermentation methods utilising immobilised cells have been widely reviewed (Gôdia *et al.*, 1987) and inferred that immobilised systems have many preferences over the customary suspended cell systems. In the bioreactors, the collection of ethanol inhibits the productivity (Goma, 1978). Thus, it is useful to complete in consecutive bioreactors the fermentation or in a gradient of concentration in reactors. A continuous procedure for the must fermentation to utilise serially associated fermenters was licensed in USA (Epchtein, 1984). A bioreactor with multistage systems, utilising expendable fixed bioplates, has likewise been produced for fermentation of wine in a continuous way (Ogbonna *et al.*, 1989). An overview of bioreactor technologies created demonstrated that developments in the last vious couple of years has occurred primarily in three zones: outlines, double phasic responses and environmental fermenters (Deshusses *et al.*, 1997). There are two noteworthy methods (heterogeneous and homogeneous) for immobilisation cell or limiting biomass (Diviès *et al.*, 1994). The harmonised method comprises identical dispensation of biomass as

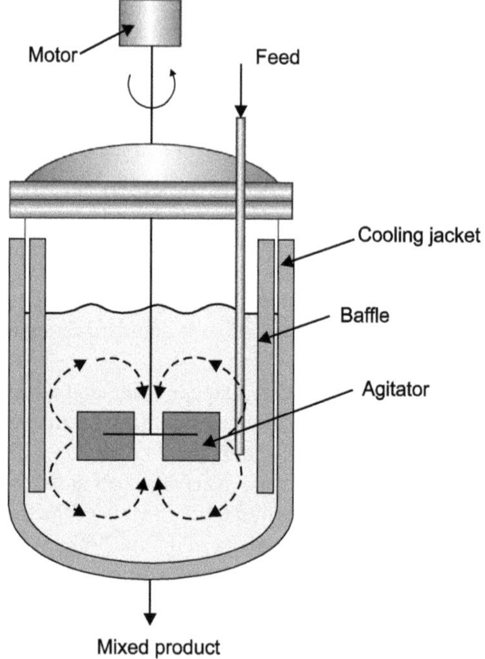


**Figure 2.** Classical agitated bioreactor

free organisms in the milieu. Rehashed utilisation of weight of organisms could be done by flocculation, centrifugation of yeast with outside or inside decanter or membrane bioreactor where the cells are introduced. Then again, the heterogeneous technique has two different stages, like fluid milieu that is supposed to be changed and a particulate phase having the cells. In this technology, biomass is restricted by way of support, auto flocculation and entrapment in gels.

### *4.2.1. Heterogeneous Bioreactors*

In heterogeneous bioreactors, microbes are immobilised by using bonding. The essential and vital thing to do is to augment the concentration of cells and keep their life in recycled or in a continuous method.

#### *4.2.1.1. Continuous Stirred Tank Bioreactor (CSTB)*

Stirred tank bioreactors (Fig. 2) comprise of a stirred tank where crisp milieu is continuously introduced and compared, the volume of the fluid substance is evacuated. They are well blended by the utilisation of impellors. The fluid component of the reactor is equivalent in the constitution, like the convergence of the surge. With the immobilisation of cells, high liquid speeds are expected to accomplish a steady provision of product and substrate expulsion. CSTBs or back-mix bioreactors, as they are occasionally named, are inexpensive, adaptable and particularly satisfactory when fluid phase reaction is required. The gas provision, temperature and pH control are simple. New agents can be effectively introduced to the tank and particulate substrate materials can be endured without a problem. In any case, the moderately strong-power input needed to give effective stirring in CSTB is obviously a weakness and it might cause erosion or destruction of the immobilised cell in view of the high cutting forces at the impellor surface.

In any case, the continuous stirred tank bioreactor provides the finest blending qualities and air exchange. The medium density in a continuous stirred tank bioreactor is normally smaller than the fluidised bed and packed bed bioreactors used in smaller average speeds. But bringing down the concentration of substrate might be favourable for hindered organism culture. Quick mixing speeds in the continuous stirred tank bioreactor in elevated shear stresses increase organism spillage from alignate (Margaritis and Wallace, 1982), or carrageenan beads (Jain *et al.*, 1985), cell separation from ion exchange resins (Bar *et al.*, 1987) and floc disruption (Fein *et al.*, 1983).

#### *4.2.1.2. Packed Fixed Bed Bioreactor (PBB)*

The packed (fixed) bed bioreactor is oftentimes utilised in immobilisation of the cell reactor for alcohol production (Gôdia *et al.*, 1987). The immobilised cells are loaded in a column at its most extreme density between which the milieu solution moves and the level of conversion of the substrate increments with the length of the column occurs (Fig. 3).

**Figure 3.** Packed bed bioreactor with the counter current flow

If the fluid momentum profile is completely flat, the packed (fixed) bed bioreactor works as a seal-flow bioreactor, which has a perfect behaviour. The efficiency of the packed (fixed) bed bioreactor for a specific biocatalyst relies upon the kind of fixation. High cell loadings are frequently accomplished by entrapment, bringing about enhanced productivity. The particle size for cell attachment likewise impacts efficiency. On a basic level, it is conceivable to accomplish full transformation into an item so that these bioreactors are perfect where full expulsion of a medium is required (detoxification).

The packed (fixed) bed bioreactor has the benefit of effortlessness operation and low cost-effective flow through the bed. It additionally can be flimsy amid long-term procedures in the light of non-stop biomass amassing, mass exchange restrictions and $CO_2$ holdup resulting in channelling and formation of dead spaces (Ghose and Bandyopadhyay, 1980) and even matrix disruption (Webb *et al.*, 1990). Gas developed may likewise lead to back blending, bringing a deviation from perfect seal-flow trend. Horizontal packed (fixed) bed bioreactor was utilised to help gas evacuation. Consequently, decreased channelling and gas fixation occur in the non-stop system (Shiotani and Yamane, 1981). The horizontal packed (fixed) bed bioreactor has been 1.5 times more profitable than the vertical packed (fixed) bed bioreactor. A lessening in channelling by $CO_2$ can likewise be acquired by partitioning the column into isolated stages with perforated plates (Grote *et al.*, 1980).

#### *4.2.1.3. Fluidised Bed Bioreactor (FBB)*

Fluidised bed bioreactor gives conditions which are middle to the one of the CSTB and packed (fixed) bed bioreactor. Blending is of higher quality in the packed (fixed) bed bioreactor yet it brings down amounts of shear when contrasted with CSTB. FBB comprises a column in which the cell particles are kept suspended in respect to every other by a non-stop flow of the medium or gas at the highest flow speeds (Fig. 4). The benefits of FBB can be seen in numerous studies. The high concentration of yeasts that aggregate in the reactor makes the system suitable for working at high productivities. As indicated by the literature, the fluid flow (must) at the inlet of the FBB might be near that of the outlet and is regulated by the working conditions utilised. This can be viewed as an extraordinary favourable advantage, particularly for reactions repressed by the product, as for alcoholic fermentation (Gôdia *et al.*, 1987; Gôdia and Solâ, 1995; Viegas *et al.*, 2002).

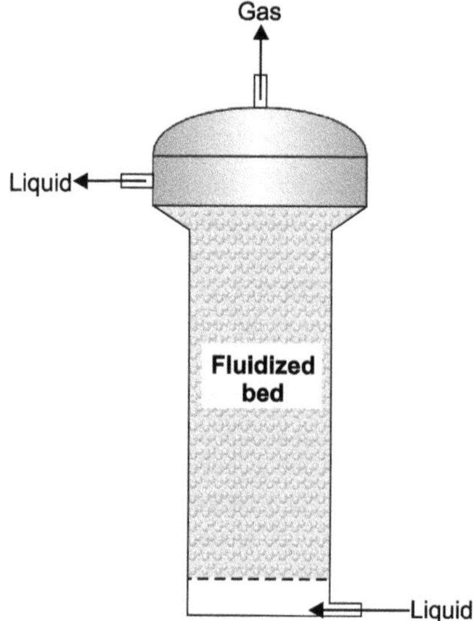


**Figure 4.** Fluidised bed bioreactor (tower fermenter)

The reduced pressure of the liquid flow underpins the mass of the bed. The FBB provides higher efficiency than CSTB in light of the fact that fluid estimates plug flow-like the packed (fixed) bed bioreactor. But, the FBB is more profitable for fermentation with medium hindrance than the packed (fixed) bed bioreactor as a result of the blending created by liquid flow. These reactors advance great mass exchange. The dead organisms are expelled in the process (Andrews, 1988) and expansive volumes of $CO_2$ are discharged without channelling (Keay *et al.*, 1990) and limits pressure decrease. Fluidisation abstains from such issues as pollution, damage of shear and constraints to scaling up, related to impellor shafts and sharp edges in mixed tanks (Dempsey, 1990). FBB can extend to suit developing organism mass so that they are less sensible to plugging and more helpful for cultures where oxygenation is required. FBB needs less energy (four times) contribution than a mechanically-agitated bioreactor. However, it is more energy consuming than the packed (fixed) bed bioreactor (Brodelius and Vandamme, 1987).

#### *4.2.1.4. Rotating Disc Bioreactor (RDB)*

It is made up of fixed cell units, for example, polyurethane froth sheets (Amin and Doelle, 1990) or fibre discs (Parekh *et al.*, 1989) appended to a pivoting shaft (Fig. 5). It is gradually blended, thus permitting complete blending and expulsion of dead organisms, residues and the developed $CO_2$. The energy needed for RDB is not as much as that for STB as a result of its moderate blending speed. This bioreactor can take industrial media-holding particle suspensions to attain high efficiency. No problem occurs with the elevated solid milieu in this sort of reactor (Parekh *et al.*, 1989).

#### *4.2.1.5. Air (gas) Lift Bioreactor (ALB)*

In ALB (Fig. 6), the liquid volume of the tank is separated into two joined areas by means of a bewilder – one area is sparged with air and another area that gets no gas is called down-comer. Bubbles convey the fluid, causing a lessening in fluid specific gravity. Gas runs away from the summit and the fluid fails in the down-comer. An outer loop method might substitute the internal bewilder for the distribution of fluid in some reactors. Stirring in the ALB because of gas move derives little shear with effective blending and mass transfer. The dimension of the bewilder impacts the hydrodynamics of the bioreactor. ALB are exceedingly energy productive in respect to mixed bioreactors.

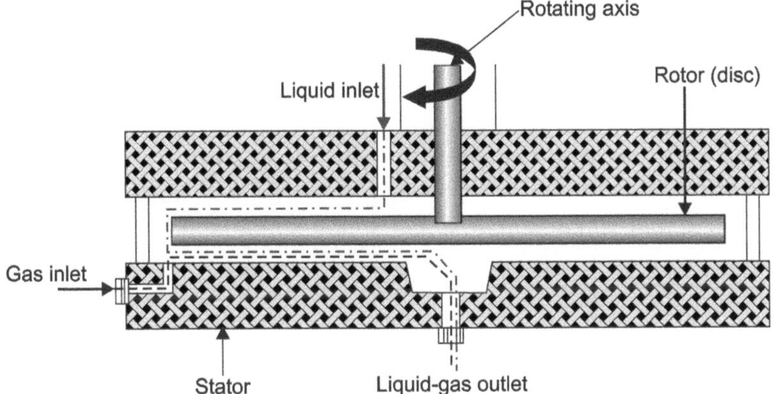


**Figure 5.** Classical rotating disc bioreactor

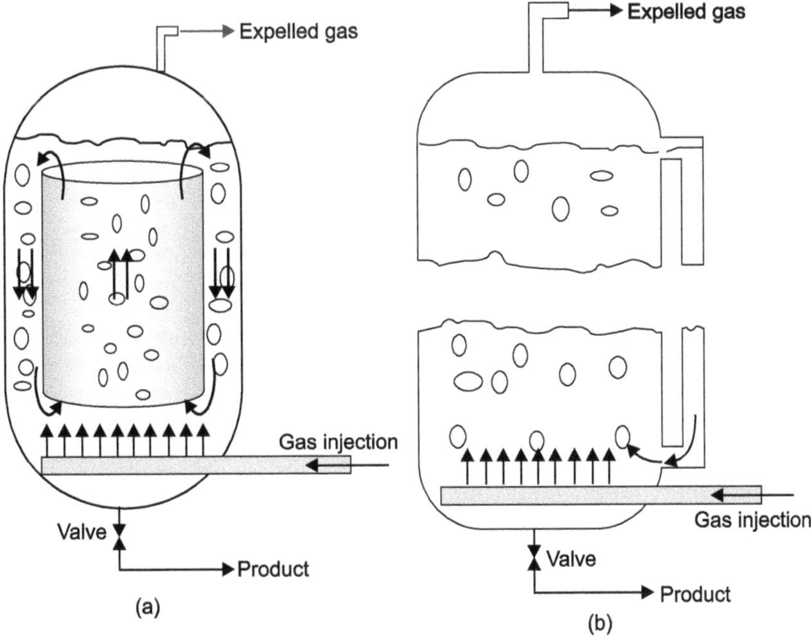


**Figure 6.** Airlift bioreactor: (a) Classical; (b) External-loop

### *4.2.2. Cell Recycle Batch Bioreactors (CRBB)*

The CRBB (Fig. 7), whose main development is the numerous progressive utilisation of the identical microbe starter in various batch fermentations remains the roughly adequate non-ordinary method in wine production. Not like continuous, the CRBB doesn't need the entire modifications in the winery methods nor undertakes it in the new machine. Indeed, yeast to get reused can be recuperated via natural settling or by membrane separation or centrifugation with equipment effectively existing in nearly all wineries. Five distinct procedures were studied (Guidoboni, 1984); the majority of them utilise a centrifugation stage which accompanies a unique bioreactor for reuse of yeast.

Increase in cell weight and also in efficiency was attained by utilising a fractional vacuum technique (Cysenski and Wilke, 1978). The principal disservices of centrifugation method are the reduction in life of the microbial biomass because of the tension as they are exposed to centrifugation equipment. The utilisation of membrane reactor is an optional technique to realise centrifugation in CRBB, where the yeast cells are held in the fermenter possessing a membrane with pore size of under 0.45 mm.

The medium is directed by the membrane reactor and the changed item moves downstream from the membrane. The productivity of the change can be expanded by reusing the item by the reactor (Diviès

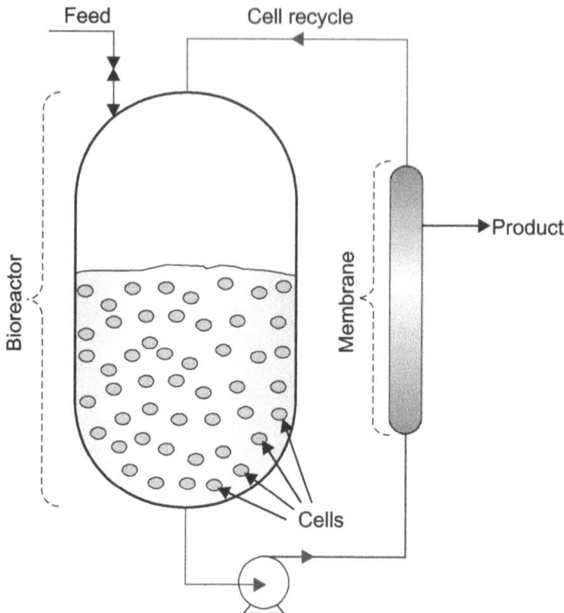


**Figure 7.** Cell recycle batch bioreactors

*et al.*, 1994). The principal restriction of this is the stopping and the membrane unclogging (Mehaia and Cheryan, 1990).

In order to examine the general legitimacy of nearly all proficient methods of fermentation to diminish the wine production costs, an off-skin fermentation of clear *Trebbianotoscano* juice was done by utilisation of a non-regular cell reuse batch fermentation system (Rosini, 1986). The procedure decreased the fermentation length and also changed the ethanol efficiency and yield. It can be advantageously used in the production of conventional table wines.

Numerous investigations have exhibited the likelihood of controlling MLF by utilising a bioreactor with cell immobilisation or enzymes. Utilisation of bioreactor introduces various benefits in the ordinary wine deacidification. Starter microbes could be reutilised. The diminished development of auxiliary fermentation and products could be ended and initiated at the right moment by the wine producer. But, contamination by phage, a transient decline in activity and a little change in the sensory characteristics of the treated wine cannot be precluded always (Maicas, 2001).

The utilisation of cell immobilisation in fermentation procedures over the utilisation of free organisms offers a few benefits, like increment in efficiency or giving a more protective condition and enhancing the resulting separation of the cell. The concentration of catalytic activity in a decreased volume enables the winemakers to lessen the dimension of the fermenters and recoup the final items more effectively in batch or continuous production methods. In spite of the fact that the characteristics of cell immobilisation were similar to the one of free cells, however, the immobilised cells are interestingly simple to recoup and reuse. The fixation and attachment/adsorption are the two principal immobilisation systems used to prompt MLF in wine. However, entrapment is a well-known strategy because of utilisation of non-dangerous chemicals in agreement with food manufacture (Cassidy *et al.*, 1996). The change of immobilisation strategies for deacidification of wine was long examined by utilising alginates (Shieh and Tsay, 1990), polyacrylamide (Clementi, 1990), ê-carrageenan (Crapisi *et al.*, 1987; Crapisi, *et al.*, 1990) and κ-carrageenan, etc. For example, an increase was noted in the operational stability of immobilised cells of *Lactobacillus* sp. in a κ-carrageenan matrix (Crapisi, *et al.*, 1987). The combined use of bentonite silica and this polymer has produced an effective bioreactor to develop the MLF of wine. The immobilised cells have shown great efficacy in decreasing L-malic acid, the conversion rate and reduction of titratable acidity being about 60 per cent. These studies have been extended to several species of lactic acid bacteria, including *O. oeni* and *Lactobacillus* (Crapisi, *et al.*, 1987).

The decision of the immobilised matrix must be done as per the long-term protection of cell life and for beverage manufacture because of its acknowledgment as GRAS (Generally Recognized As Safe). Alginate is just an appropriate matrix in both the considerations, has the size to reduce the diffusion limitation for the media and the items, and to increase the biomass dissemination. However, it is reversible and the existence of chelating compounds in the milieu could prompt leakage and incomplete matrix dissolution of the packed biomass. Bioreactor technologies comprising the high density of MLF microorganisms immobilised in alginate supports or carrageenan or packed between membranes was developed (Colagrande *et al.*, 1994). In MLF microscopic organisms react as biocatalyst and without development, quickly convert malic acid to lactic acid in wine that has gone through the bioreactor on a non-stop basis (Gao and Fleet, 1994).

#### *4.2.3. Continuous Bioreactors for Winemaking*

The continuous reactor is 'open'. There is a constant flow consisting of entry of the substrate on one hand and output of the product on the other. The main specificity of the continuous reactor is the opportunity to achieve dynamic equilibrium, that is to say that the system operates on the basis of the equilibrium state. Continuous reactors are widely used in chemical and food industries among others. Most operating reactors are multiphasic, including fixed bed, fluidised bed, bubble column and to lift air (Verbelen *et al.*, 2006). Multiphase reactors are structured in three phases: gas (air or other), solid (support) and liquid (the medium). In terms of the production of wine, inert gas ($CO_2$ or $N_2$) may replace the air to avoid oxidation of the wine. The continuous fermentation technique appears as an option that would reduce manufacturing costs and increase the ethanol yield (Ribéreau-Gayon *et al.*, 2006). According to the literature, it is proposed to use higher levels of $SO_2$ in continuous fermenters to stop contamination. The advantages of continuous fermentation are higher and faster substrate conversion rate; increased homogeneity of the wine; lower losses; best environmental management practices; better control of fermentation; and consistency in the quality of finished wines (Clement *et al.*, 2011; Genisheva *et al.*, 2014; Ribéreau-Gayon *et al.*, 2006).

## 5. Optimising Winery Unit Operation

Wineries nowadays are confronted with the increasing expenses of trading. In the course of recent years, the cost of gas and electricity for manufacturing has expanded and this expanding pattern is probably going to proceed. These expanded utility costs put additional pressure on business. The outcome is an industry confronting more tightly net revenues and an increased significance on the selection of procedures and technologies that empower quality wine to be manufactured at lower cost. Optimisation should be therefore a business mentality concentrating on executing procedures and innovations to lessen costs, increase speed and improve asset utilisation. Numerous enterprises have effectively embraced optimisation as a foundation for staying focused in nearby and worldwide markets.

### 5.1. Computerisation

Computerisation can take different structures inside a winery. It can be as essential as computerising areas of a refrigeration framework, or as elaborate as a completely mechaniszed winery. Computerisation is useful as it permits the change of each of the procedure productivity measures, i.e. production, work, materials, water and energy. It is accomplished by advancing procedure gear, permitting round-the-clock operations, enhancing quality and decreasing human mistakes. The computerisation can optimise the process in the winery by the means shown in Table 10.

### 5.2. Cross Flow Filtration

Cross stream (flow) filtration has developed as a productive filtration method, with differing application possibilities for both the quick moving customer merchandise enterprises and the wine factory. Several wineries have actualised this innovation; however many others have not executed this innovation yet. Cross-stream filtration is customised and can include various applications inside a winery. It can be basically more energy proficient than conventional winery filtration while permitting fast-filtration

**Table 10.** Optimisation Aspects of Computerisation

| *Process optimisation* | *Effect of chance on process effectiveness measure* |
|---|---|
| Production rate | Critical increments to fabrication speed and improvement of process apparatus |
| Work | Reductions difficult work enabling staff to be used all the more fittingly |
| Materials | Gives more noteworthy process control and accordingly decreases material waste |
| Energy | Gives large amounts of control and can fundamentally lessen energy– particularly in the systems of refrigeration for wineries |
| Water | Computerisation can give extra water productivity relying upon the particular use of the robotisation system |

speeds. Cross-stream filtration can enhance material effectiveness by removing the requirement for added substances (e.g. filter aid) and lessening wine misfortune caused by development through numerous filtration exercises. The way cross-stream (flow) filtration could optimise the filtration efficiency as summarised in Table 11.

**Table 11.** Optimisation Aspects of Cross-flow Filtration

| *Process optimisation* | *Effect of chance on process effectiveness measure* |
|---|---|
| Production rate | Augment filtration speeds by enabling numerous filtration operations to be embraced in one wine development.<br>Can finished filtration in one stage rather than numerous means. |
| Materials | Can diminish wine development and decline wine waste. Can recuperate wine from dregs expanding production.<br>Can diminish the utilisation of bentonite |
| Energy | Cross stream filtration is to a greater extent energy-saving per litre of wine obtained in contrasted to other filtration procedures |
| Water | Cross-stream filtration can be utilised to purify and recover process water |

## 5.3. Frosty Adjustment Methods

A typical issue in the wine factory amid vintage is the absence of enough fermentation vessels. A quicker essential fermentation rate enables more tanks to be reutilised amid vintage. Moderate or blocked fermentation not just decreasess the manufacture speed, it can expand material use by expecting added substances to 'restart' aging and require extra energy for the control of temperature. Expanding fermentation productivity includes both grape juice attributes, remedying for any basic differences (i.e. potassium accessibility, pH, etc.) and choosing yeast strains that are most appropriate to the juice characteristics. This guaranteed fermentation is attempted in a controlled and optimised manner, increasing the manufacture efficiency.

## 5.4. Continuous Processing

Continuous manufacture systems have more prominent efficiencies in contrast to batch procedure systems. The batch procedure system requires an extensive manufacture chain to stop until the the group bottleneck is prepared for the following cycle. In the wine factory, batch squeezing delays the destemming, receival and pulverising stages. This can prompt temperature changes and extended oxidation. Screw squeezing permits the receival procedure motion to progress from a batch procedure to a continuous procedure. This can reduce bottlenecks all through the receival chain. Presses using screws have gradually been supplanted by press systems utilising membranes because of their capacity to diminish extraction of phenolic substances. Innovation in a screw press, for example, utilising bigger screw blades and moderating upheavals, can lessen quite a bit of this phenolic compound while optimising the speed of production.

## 6. Conclusion and Future Thrust

Winemakers are confronting increasing rivalry due to the enlarging gap between wine manufacture and wine utilisation, the trend for customer inclination far from fundamental ware wine to top quality wine and financial globalisation. Thus, there is a requirement for a global transformation in the realm of wine. One of the requirements is the usage of HACCP method in the beverages factory that has been of a colossal help. Despite the fact that alcoholic drinks are relatively more secure than different foods and beverages due to their high ethanol content, recognition of potential dangers and recommencement of inhibitory and restorative activities (at whatever point needed) are of essential significance. Foundation of basic control restrains in co-occurrence with suitable and viable checking methods completed by capable staff have figured out how to limit the episodes of occurrences that are perilous and malicious for human well-being. The way towards changing the wine factory from a manufacture to an oriented market industry is an increased reliance on biotechnological advancements and HACCP method. A great part of the procedure proficiency technology utilised in agro-industry factories is promptly accessible for use in the wine factory. A few wineries have executed this innovation while yet others are yet to do as such. Specifically, cross-flow filtration and computerisation-procedure efficiencies are yet being executed. These deferrals in usage are not because of the accessibility of innovation, but rather are expected partially to capital accessibility, the absence of information on proficient practices, or vulnerability regarding what the advantageous prices are for expanding the process productivity. In spite of a substantial number of results on wine, the characteristic course of the fermentation of wine is not completely investigated and fermentation procedures are not really completely managed to require the comprehension of the biochemical conduct of yeast and other organisms in the wine milieu. The overall spread of wine manufacture has prompted novel vineyards delivering quality wine by receiving tried systems, like centrifugation, filtration, stainless steel tanks for fermentation, monitoring of temperature and chosen yeasts, and so forth. A portion of the biotechnological developments, like specific yeasts, change of starters, treatment using enzymes, bioreactor planning and cell immobilisation are of key significance in wine production. The improvements in reactor innovation as for fruit wines other than grape are rare. Mastering ongoing technique efficiencies ought to be viewed as key to enhancing business productivities. Assuring these means can prompt noteworthy cost reserve funds for the business, especially with respect to assets, materials and manufacture rates. This will need an adjustment in worldview for some wineries, yet it is important to guarantee that they stay beneficial in the changing business condition.

## References


Amin, G. and Doelle, H.W. (1990). Production of high ethanol concentrations from glucose using *Zymomonas mobilis* entrapped in vertical rotating immobilised cell reactor. *Enzyme and Microbial Technology* 12: 443-446.

Andrews, G.F. (1988). Design of fixed film and fluidised bed bioreactors. pp. 765. *In*: L.E. Erickson and D.Y.C. Fung (Eds.). Handbook on Anaerobic Fermentations. Marcel Dekker, New York, USA.

Blade, W.H. and Boulton, R. (1988). Absorption of Protein by Bentonite in a Model Wine Solution. *American Journal of Enology and Viticulture* 39: 193-199.

Boulton, R., Singelton, V., Bisson and Kunkee, R. (1996). Principles and Practices of Winemaking. Chapman & Hall, New York.

Brodelius, P. and Vandamme, E.J. (1987). Immobilised cell systems. pp. 405. *In*: H. Rehm and G. Reed (Eds.). Biotechnology, VCH, Weinhein, Germany.

Cabras, P., Meloni, M., Pirisi, F.M., Farris, G.A.O. and Fatichenti, F. (1988). Yeast and pesticide interaction during aerobic fermentation. *Applied Microbiology and Biotechnology* 29: 298-301.

Cassidy, M.B., Lee, H. and Trevors, J.T. (1996). Environmental applications of immobilised microbial cells: A review. *Journal of Industrial Microbiology and Biotechnology* 16: 79.

Clement, T., Perez, M., Mouret, J.R., Sablayrolles, J.M. and Camarasa, C. (2011). Use of continuous multistage bioreactor to mimic winemaking fermentation. *International Journal of Food Microbiology and Immunology* 150: 42-49.

Clementi, F. (1990). Fluidbed dried immobilised cells of *Lactobacillus casei* used for L-malic acid degradation in must. *Italian Journal of Food Science* 1: 25.

Codex (1998). Report of the Thirtieth Session of the Codex Committee on Pesticide Residues. Alinorm 99/24. FAO/WHO. The Hague, 20-25 April 1998.

Colagrande, O., Silva, A. and Fumi, M.D. (1994). Recent applications of biotechnology in wine production. *Biotechnology Progress* 10: 2.

Constanti, M., Poblet, M., Arola, L., Mas, A. and Guillamon, J. (1997). Analysis of yeast population during alcoholic fermentation in a newly established winery. *American Journal of Enology and Viticulture* 48: 339-344.

Cooke, G.M. and Berg, H.W. (1984). A re-examination of varietal table wine processing practices in California, II: Clarification, stabilisation, aging and bottling. *American Journal of Enology and Viticulture* 35: 137-142.

Crapisi, A., Lante, A., Nuti, M.P., Zamorani, A. and Spettoli, P. (1990). Efficiency of bioreactors with bacterial cells immobilised on different matrixes for the control of malo-lactic fermentation in wine. pp. 349. *In*: P. Ribéreau-Gayon and A. Lonvaud (Eds.). Actualités oenologiques 89, Comptes rendus 4 Symp. International d' Oenologie, Bordeaux, 1989. Paris, France: Dunod.

Crapisi, A., Nuti, M.P., Zamorani, A. and Spettoli, P. (1987). Improved stability of immobilised Lactobacillus sp. cells for the control of malo-lactic fermentation in wine. *American Journal of Enology and Viticulture* 38: 310-312.

Crapisi, A., Spettoli, P., Nuti, M.P. and Zamorani, A. (1987). Comparative traits of *Lactobacillus brevis, Lactobacillus fructivorans* and *Leuconostoc oenos* immobilised cells for the control of malo-lactic fermentation in wine. *Journal of Applied Bacteriology* 63: 513-521.

Cysenski, G.R. and Wilke, C.R. (1978). Process design and economic studies of alternative fermentation methods for production of ethanol. *Biotechnology and Bioengineering* 20: 14-21.

Davis, C.R., Wibowo, D., Eschenbruch, R., Lee, T.H. and Fleet, G.H. (1985). Practical implications of malo-lactic fermentation: A review. *American Journal of Enology and Viticulture* 36: 290-301.

De-la-Presa-Owens, C. and Noble, A.C. (1997). Effect of Storage at Elevated Temperatures on Aroma of Chardonnay Wines. *American Journal of Enology and Viticulture* 48: 310-316.

Dempsey, M.J. (1990). Ethanol production by *Zymomonas mobilis* in a fluidised bed fermenter. pp. 137. *In*: J.A.M. de Bont, J. Visser, B. Mattiasson and J. Tramper (Eds.). Physiology of Immobilised Cells. Elsevier Science Publishers, Amsterdam, UK.

Deshusses, M.A., Chen, W., Mulchandani, A. and Dunn, I.J. (1997). Innovative bioreactors. *Current Opinion in Biotechnology* 8: 165.

Dibble, J.E. and Steinke, W.E. (1992). Principles and techniques of vine spraying. *In*: D.L. Flaherty, L.P. Christensen, W.T. Lanini, J.J. Marois, P.A. Phillips and L.T. Wilson (Eds.). Grape Pest Management. University of California. Division of Agriculture and Natural Resources, Oakland.

Diviès, C. (1988). Les Microorganisms immobilisés. Cas des bactéries lactiques et des levures en oenologie. pp. 449. *In*: G.H. Fleet (Ed.). Wine Microbiology and Biotechnology. Harwood Academic Publishers, Chur, Switzerland.

Diviès, C., Cachon, R., Cavin, J.-F. and Prévost, H. (1994). Immobilised cell technology in wine fermentation. *Critical Reviews in Biotechnology* 14: 135.

Dubourdieu, D. (1986). Wine technology: Current trends. *Experientia* 42: 914.

Ellison, P., Ash, G. and McDonald, C. (1998). An expert management system for the management of *Botrytis cinerea* in Australian vineyards. I. *Dev. Agric. Syst.* 56: 185-207.

Epchtein, J. (1984). Continuous process for the fermentation of must to produce wine or ethanol. US Patent No. 4487785.

Eschnauer, E. (1986). Lead in wine from tin-leaf capsules. *American Journal of Enology and Viticulture* 37: 158-162.

Farkas, J. (1984). Technology and Biochemistry of Wine. Gordon & Breach, New York.

Fatichenti, F., Farris, G.A., Deiana, P., Cabras, P., Meloni, M. and Pirisi, F.M. (1984). The effect of *Saccharomyces cerevisiae* on concentration of dicarboxymide and acylanilide fungicides and pyrethroid insecticides during fermentation. *Applied Microbiology and Biotechnology* 20: 419-421.

Fleet, G.H. (1994). Wine: Microbiology and Biotechnology. Harwood Academic Publishers, New York.

Forsythe, S.J. and Hayes, P.R. (1998). Food Hygiene, Microbiology and HACCP. Chapman & Hall, London.

Fugelsang, K. (1997). Wine Microbiology. Chapman & Hall, London.

Gao, C. and Fleet, G.H. (1994). Degradation of malic acid by high density cell suspensions of *Leuconostoc oenos*. *Journal of Applied Microbiology* 76: 632.

Genisheva, Z., Mota, A., Mussatto, S., Oliveira, J.M. and Teixeira, J.A. (2014). Integrated continuous winemaking process involving sequential alcoholic and malo-lactic fermentations with immobilised cells. *Process Biochemistry* 49: 1-9.

Gennari, M., Negre, M., Gerbi, V., Rainondo, E., Minati, J.L. and Gandini, A. (1992). Chlozolinate fates during vinification process. *Journal of Agricultural and Food Chemistry* 40: 898-900.

Ghose, T.K. and Bandyopadhyay, K.K. (1980). Rapid ethanol fermentation in immobilised cell reactor. *Biotechnology and Bioengineering* 22: 1489.

Gnaegi, F., Aerny, J., Bolay, A. and Crettenand, J. (1983). Influence des Traitement Viticoles Antifongiques sur la Vinification et la Qualite du vin, Revision Suisse de Viticulture. *Arboriculture et Horticulture* 15: 243-250.

Gnaegi, F. and Sozzi, T. (1983). Les Bacteriophages de Leuconostoc oenos et leur, Importance Oenologique. *Bulletin d'OIV* 56: 352-357.

Gôdia, C., Casas, C. and Solâ, C. (1987). A survey of continuous ethanol fermentation systems using immobilised cells. *Process Biochemistry* 22: 43.

Gôdia, F. and Solâ, C. (1995). Fluidised-bed Bioreactors. *Biotechnology Progress* 11: 479-497.

Goma, G. (1978). Optimisation de la mise en oeuvre des cultures anaérobies. pp. 171. *In*: L. Bolichon and G. Durand (Eds.). Utilisation Industrielle du Carbone d'Origine Végétale par voie Microbienne, Société Française de Microbiologie. Paris, France.

Grote, W., Lee, K.J. and Rogers, P.L. (1980). Continuous ethanol production by immobilised cells of *Zymomonas mobilis*. *Biotechnology Letters* 2: 481.

Guidoboni, G.E. (1984). Continuous fermentation systems for alcohol production. *Enzyme and Microbial Technology* 6: 194.

Gump, B. and Pruett, D. (1993). Beer and Wine Production, Analysis Characterisation and Technological Advances. American Chemistry Society, Washington DC.

Guzzo, J., Jobin, M.-P. and Divies, C. (1998). Increase of sulphite tolerance in *Oenococcus oeni* by means of acidic adaption. *FEMS Microbiology Letters* 160: 43-47.

Jackson, G. (1998). Practical HACCP in brewing industry. pp. 50-57. *In*: Monograph XXVI: European Brewery Convention. Stockholm.

Jackson, R.S. (1999). Grape-based fermentation products. pp. 583. *In*: V.K. Joshi and A. Pandey (Eds.). Biotechnology: Food Fermentation. Educational Publishers and Distributors, Ernakulam, India.

Keay, L., Eberhardt, J.J., Allen, B.R., Scott, C.D. and Davison, B.H. (1990). Improved production of ethanol and n-butanol in immobilised cell bioreactors. pp. 539. *In*: J.A.M. de Bont, J. Visser, B. Mattiasson and J. Tramper (Eds.). Physiology of Immobilised Cells. Elsevier Science Publishers, Amsterdam, UK.

Kourtis, L.K. and Arvanitoyannis, I.S. (2001). Implementation of hazard analysis critical control point (HACCP) system to the alcoholic beverages industry. *Food Reviews International* 17: 1-44.

Langhans, E. and Schlotter, H.A. (1985). Ursachen der Kupfer-Trung. *Deutse Weinband* 40: 530-536.

Lea, A. and Piggott, J. (1995). Fermented Beverage Production. Chapman & Hall, London.

Lichine, A. (1985). Alexis Lichine's Encyclopedia of Wines & Spirits. Cassell, London.

Maicas, S. (2001). The use of alternative technologies to develop malo-lactic fermentation in wine. Applied Microbiology and Biotechnology 56: 35.

Malfeito-Ferreira, M., Tareco, M. and Loureiro, V. (1997). Fatty acid profiling: A feasible typing system to trace yeast contamination in wine bottling plants. *International Journal of Food Microbiology* 38: 143-155.

Maner, P.J. and Stimmann, M.W. (1992). Pesticide safety. *In*: D.L. Flaherty, L.P. Christensen, W.T. Lanini, J.J. Marois, P.A. Phillips and L.T. Wilson (Eds.). Grape Pest Management. Publ. University of California. Division of Agriculture and Natural Resources, Oakland.

Marriott, N. (1994). Principles of Food Sanitation. Chapman & Hall, London.

Martinez, R.G., De-la-Serrana, H.L.G., Mir, M.V., Granados, J.Q. and Martinez, M.C.L. (1996). Influence of wood heat treatment, temperature and maceration time on vanillin, syringaldehyde and gallic acid contents in oak wood and wine spirit mixtures. *American Journal of Enology and Viticulture* 47: 441-446.

Maule, D.R. (1986). A century of fermenter design. *Journal of the Institute of Brewing* 92: 137.

Mehaia, M.A. and Cheryan, M. (1990). Membrane bioreactors: Enzyme processes. pp. 67. *In*: H.G. Schwartzberg and M.A. Rao (Eds.). Biotechnology and Food Process Engineering. Marcel Dekker, New York, USA.

Moresi, M. (1989). Fermenter design for alcoholic beverage production. pp. 107. *In*: C. Cantarelli and G. Lanzarini (Eds.). Biotechnology Applications in Beverage Production. Elsevier Applied Science, London, UK.

Mosedale, J.R. and Puech, J.L. (1998). Wood maturation of distilled beverages. *Trends in Food Science and Technology* 9: 95-101.

Nault, I., Gerbaux, V., Larpent, J.P. and Vayssier, Y. (1995). Influence of pre-culture conditions on the ability of *Leuconostoc oenos* to conduct malo-lactic fermentation in wine. *American Journal of Enology and Viticulture* 46: 357-362.

Neel, D. (1993). Advancements in processing Portuguese corks. *Australian Grapegrower and Winemaker* 353: 11-14.

Nielsen, J.C., Prahl, C. and Lonvaud-Funel, A. (1996). Malo-lactic fermentation in wine by direct inoculation with freeze-dried *Leuconostoc oenos* cultures. *American Journal of Enology and Viticulture* 47: 42-48.

Noble, A.C. (1988). Analysis of wine sensory properties. p. 9. *In*: H.E. Linskens and J.F. Jackson (Eds.). Wine Analysis – Modern Methods of Plant Analysis. Springer Verlag, Berlin.

Ogbonna, J.C., Amano, Y., Nakamura, K., Yokotsuka, K., Shimazu, Y., Watanabe, M. and Hara, S. (1989). A multistage bioreactor with replaceable bioplates for continuous wine fermentation. *American Journal of Enology and Viticulture* 40: 292.

OIV (1994). Pesticide Residue: Authorised Limits: Classification by Country, Classification by Pesticide. Office International de la Vigne et du Vin, Paris.

Oliva, J., Navarro, S., Barba, A. and Navarro, N. (1999). Determination of chlorpyrifos, penconazole, fenarimol, vinclozolin and metalaxyl in grapes, must and wine by on-line microextraction and gas chromatography. *Journal of Chromatography* A 833: 43-51.

Ough, C. (1992). Winemaking Basics. Food Products Press.

Parekh, S.R., Parekh, R.S. and Wayman, M. (1989). Ethanolic fermentation of wood-derived cellulose hydrolysates by *Zymomonas mobilis* in a continuous cynamic immobilised biocatalyst bioreactor. *Process Biochemistry* 24: 88.

Peyron, D. and Feuillat, M. (1985). Essais comparatifs de cuves d'automaceration en Bourgogne. *Revue d'Oenologie* 38: 7.

Pilone, G.J. (1986). Effect of Triadimenol Fungicide on Yeast Fermentation. *American Journal of Enology and Viticulture* 37: 304-305.

Popock, K.F., Strauss, C.R. and Somers, T.C. (1984). Ellagic acid deposition in white wines after bottling: A wood-derived instability. *Australian Grapegrower and Winemaker* 244: 87.

Quinn, M.K. and Singleton, V.L. (1985). Isolation and identification of ellagitannins from white oak wood and an estimation of their roles in wine. *American Journal of Enology and Viticulture* 35: 148-155.

Ranken, M.D., Kill, R.C. and Baker, C. (1997). Food Industries Manual. Blackie Academic & Professional, London.

Rapp, A. and Mandey, H. (1986). Wine aroma. *Experientia* 42: 873.

Ribereau-Cayon, P., Glories, Y., Maujean, A. and Dubourdieu, D. (1998). Traité d' Oenologie 2, Chimie du vin. Stabilisation et Traitements. Dunod, Paris.

Ribéreau-Gayon, P., Dubourdieu, D., Donèche, B. and Lonvaud, A. (2006). Handbook of enology. *In*: The Microbiology of Wine and Vinifications. John Wiley & Sons, Chichester.

Rosini, G. (1986). Winemaking by cell-recycle-batch fermentation process. *Applied Microbiology and Biotechnology* 24: 140.

Sala, C., Fort, F., Busto, O., Zamora, F., Alpola, L. and Guasch, J. (1996). Fate of some common pesticides during vinification process. *Journal of Agricultural and Food Chemistry* 44: 3668-3671.

Schreier, P. (1979). Flavour composition of wine: A review. *Critical Reviews in Food Science and Nutrition* 12: 59.

Shieh, Y.M. and Tsay, S.S. (1990). Malo-lactic fermentation by immobilised *Leuconostoc sp.* M-I. *Journal Chinese Agricultural and Chemical Society* 28: 246.

Shiotani, T. and Yamane, T. (1981). A horizontal packed bed bioreactor to reduce $CO_2$ gas hold up in the continuous production of ethanol by immobilised yeast cells. *European Journal of Applied Microbiology and Biotechnology* 13: 96.

Simpson, R.F. (1990). Cork taint in wine: A review of the causes. *Australian Grapegrower and Winemaker* 305: 286-296.

Simpson, R.F., Amon, J.M. and Daw, A.J. (1986). Off-flavour in wine caused by guaiacol. Food Technology Association of Australia, 38: 31-33.

Soufleros, E. (1997a). Winemaking. Thessaloniki, Greece.

Soufleros, E. (1997b). Winemaking. Thessaloniki, Greece.

Speck, L.M. (1984). Compendium of Methods for the Microbiological Examination of Foods. American Public Health Association, New York.

Sudraud, P. and Chauvet, S. (1985). Activite Antilevure de l'anhydride Sulfureux Moleculaire. *Connaissance de la Vigne et du Vin* 22: 251-260.

Towey, J.P. and Waterhouse, A.L. (1996). Barrel-to-Barrel variation of volatile oak extractives in barrel-fermented chardonnay. *American Journal of Enology and Viticulture* 47: 17-20.

Tucker, G.A. and Woods, L.F.J. (1996). Enzymes in Food Processing. Chapman & Hall, London.

Ubeda, J.F. and Briones, A.I. (1999). Microbiological quality of filtered and non-filtered wines. *Food Control* 10: 41-45.

Vaillant, H., Formysin, P. and Gerbaux, V. (1995). Malo-lactic fermentation of wine: Study of the influence of some physicochemical factors by experimental design assays. *Journal of Applied Bacteriology* 79: 640-650.

Van-Vuuren, H.J.J. and Jacobs, C.J. (1992). Killer yeasts in the wine industry: A review. *American Journal of Enology and Viticulture* 43: 119-128.

Verbelen, P., De-Schutter, D., Delvaux, F., Verstrepen, K.J. and Delvaux, F.R. (2006). Immobilised yeast cell systems for continuous fermentation applications. *Biotechnology Letters* 28: 1515-1525.

Viegas, M.C., Andrietta, S.R. and Andrietta, M.G.S. (2002). Use of tower reactors for continuous ethanol production. *Brazilian Journal of Chemical Engineering* 19: 167-173.

Vine, R., Harkness, E., Browing, T. and Wagner, C. (1997). Winemaking, From Grape Growing to Marketplace. Chapman & Hall, London.

Viriot, C., Scalbert, A., Lapierre, C. and Moutounet, M. (1993). Ellagitanins and lignins in aging of spirits in oak barrels. *Journal of Agricultural and Food Chemistry* 41: 1872-1879.

Vivas, N., Lonvaud-Funel, A. and Glories, Y. (1997). Effect of Phenolic acids and athocyanins on growth, viability and malo-lactic activity of a lactic acid bacterium. *Food Microbiol* 14: 291-300.

Webb, C., Dervakos, G.A. and Dean, J.F. (1990). Analysis of performance limitations in immobilised cell fermenters. pp. 589. *In*: W.E. Goldstein and H. Pederson (Eds.). Biochemical Engineering. Am. N.Y. Acad. Sci., USA.

Wood, B. (1998). Microbiology of Fermented Foods. Blackie A & P, London.

Zoecklein, B.W., Fugelsang, K.C., Gump, B.H. and Nury, F.S. (1994). Wine Analysis and Production. Chapman & Hall, New York.

# 20 Wine Maturation and Aging

**Hatice Kalkan Yıldırım**
Department of Food Engineering, Ege University, 35100 Bornova, Izmir, Turkey

## 1. Introduction

Many complex reactions take place during aging – special reactions involving phenolic compounds that affect the final sensory quality of the wine. The final taste, flavour, aroma and sensation of wine is determined by the interactions of the chemical components present in the wine (Sytger *et al.*, 2011). Aging can be examined in two stages:

- The first stage is called maturation, which includes the changes that occur during storage of wine between alcohol fermentation and bottling. It lasts from six to 24 months or sometimes may last for years. Maturation is also called 'oxidative aging' because it occurs in the presence of oxygen. The maturation process, including the chemical change that occurs in wine, is effected by many factors that affect the quality of wine. These factors include grape variety, fermentation (alcohol fermentation and malo-lactic fermentation) and storage conditions (Jacobson, 2006; Jackson, 2008).
- The second stage is the aging phase that begins after the bottling of the wine. The chemical reactions that occur at this stage are much slower. This stage is also called 'reducing aging' because it occurs in absence of oxygen in the bottle (Jacobson, 2006; Jackson, 2008).

Changes in wine during the maturation (storage of the wine in a tank or barrel) and aging phase (after bottling), together with determination of the boundary conditions, determine the composition of the final aged wine (Waterhouse, 2016).

## 2. Reactions Occurring during Maturation and Aging

The main compounds that are affected by aging are phenols. The phenolic compounds found in wine can be divided into two groups – flavonoids and non-flavonoids (Fig. 1).

The group of flavonoids includes anthocyanins, flavan-3 which contains flavonols, flavones, flavanones. The non-flavonoids include hydroxybenzoic acid, hydroxycinnamic acid and their derivatives, stilbenes (Jackson, 2008; Moreno-Arribas and Polo, 2009). Changes of phenolic compounds occurring at the beginning of wine production stages continues until aging. The new compounds formed during aging are usually with different organoleptic characters than their precursor compounds. In order to control the wine quality, the chemical changes and mechanisms that occur during wine production and aging must be well understood (Fig. 2) (Waterhouse and Ebeler, 1998).

Simultaneous presence of coloured phenolic compounds (anthocyanins) and colourless phenolic compounds (phenolic acids, tannins, flavanols, flavonols and flavanonols) in wine are responsible for the changes that occur during maturation and aging (Brouillard and Dangles, 1994). The amount of coloured pigments increases when the concentration of phenolic compounds decreases during aging in red wine (Dallas *et al.*, 1995). Change in colour properties of red wines within the first year of storage, reduction of anthocyanins and the formation of new stable pigments are influenced by many factors, such as temperature, pH, alcohol, $SO_2$ and oxygen content (Recameles *et al.*, 2006). Change in wine colour compounds during wine maturation is considered to be the fastest, especially during the first year. During wine maturation, interactions among anthocyanins and other phenolic compounds are followed by polymerisation reactions, which cause the formation of new pigments and colour changes in wine (Mateusand Freitas, 2001). Changes occurring in the wine phenols during aging are realised by the principal mechanisms that may take place. These complex reactions are divided into some groups:

- destruction of anthocyanins

E-mail: hatice.kalkan.yildirim@ege.edu.tr

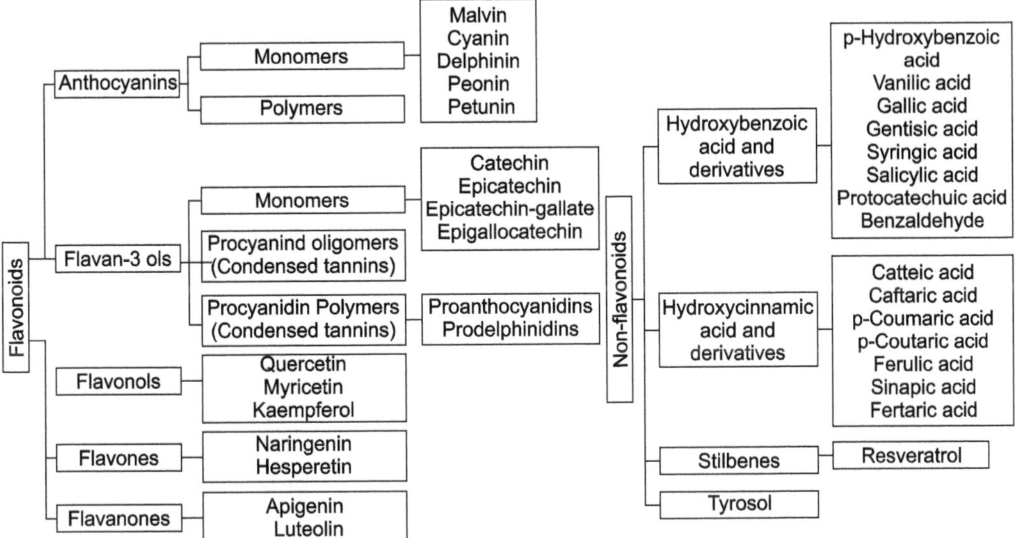


**Figure 1.** Wine phenolic compounds (Jackson, 2008; Moreno-Arribas and Polo, 2009)

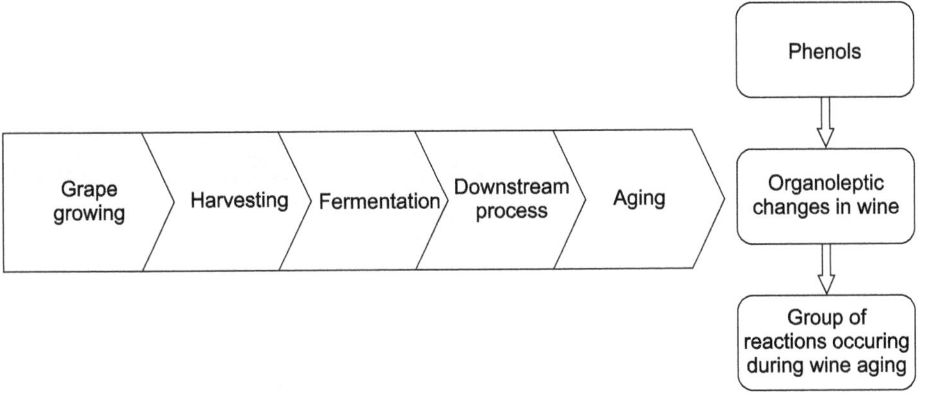


**Figure 2.** Factors related with aging

- reaction of anthocyanins with compounds containing polarised double bonds
- Co-pigment formation process of anthocyanins
- condensation reactions of anthocyanins with tannins
- reactions of tannins with proteins and polysaccharides, carbocation formation of procyanidins
- oxidation reactions of procyanidins, polymerisation reactions of procyanidins (Fig. 3). (Ribâereau-Gayon *et al.*, 2006)

## 2.1. Destructions of Anthocyanins

Main structure of anthocyanin is given in Fig. 4.

Anthocyanins, responsible for the colour of wine, are major in the changes that occur in the chemical structure of wine (Somers and Pocock, 1990). During maturation and aging, formation of new compounds takes place through oxidation, polymerisation and reduction reactions of anthocyanins, causing changes in the colour of the wine (de Freitas *et al.*, 2017). It is thought that 25 per cent of anthocyanins react with other flavonoid compounds during fermentation and this ratio can reach 40 per cent at the end of one year of aging. It has been reported that the amount of anthocyanin present in the monomeric form decreases in polymerisation reactions and the wine colour is retained in the polymeric pigments. However, the

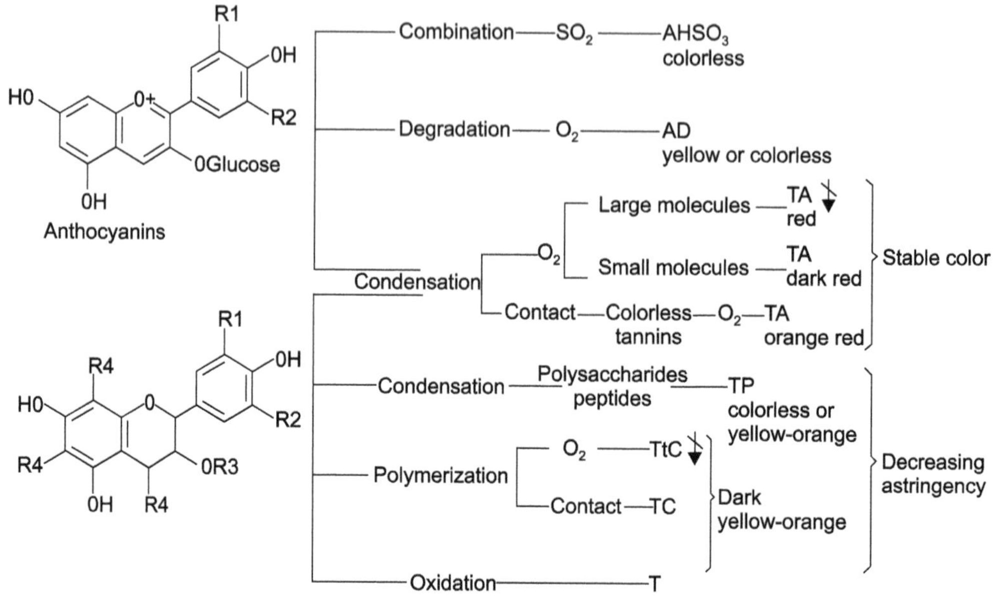

A = Anthocyanins T = Tannin P = Procyanidin C = Catechin Ttc = very condensed tannins

TC = Condensed tannins TP = Tannin polysaccharides ↓ = Precipitates AD = Degraded anthocyanins

**Figure 3.** Changes in phenolic compounds during wine aging (*Source*: Bruce *et al.*, 1995)

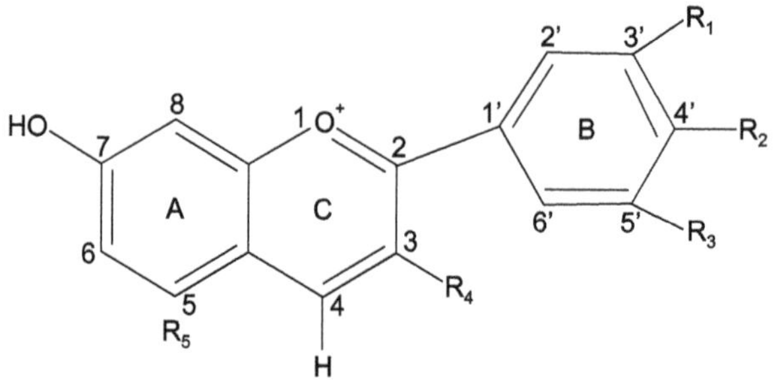

| Anthocyanins | $R_1$ | $R_2$ | $R_3$ | $R_4$ | $R_5$ |
|---|---|---|---|---|---|
| Pelargonidin 3-O-glucoside | H | OH | H | Glucoside | OH |
| Cyanidin 3-O-glucoside | OH | OH | H | Glucoside | OH |
| Delphinidin 3-O-glucoside | OH | OH | OH | Glucoside | OH |
| Peonidin 3-O-glucoside | $OCH_3$ | OH | H | Glucoside | OH |
| Malvidin 3-O-glucoside | $OCH_3$ | OH | $OCH_3$ | Glucoside | OH |
| Malvidin 3,5-O-glucoside | $OCH_3$ | OH | $OCH_3$ | Glucoside | Glucoside |
| Cyanidin 3-O-glucoside | OH | OH | H | Rutinoside | OH |

**Figure 4.** Structure of anthocyanins (Han and Xu, 2015)

colour density decreases over time due to the slow precipitation of the polymeric compounds and the drop in monomeric anthocyanins. In addition, the amount of co-pigmented anthocyanin decreases with time during aging (He *et al.*, 2012). Another reason for the destruction of anthocyanins with aging is connected to the structure of the coloured anthocyanins and tannin. The amount of tannins added to the coloured anthocyanins of aged wines is higher than that of young wines (McRae and Kennedy, 2011). Anthocyanins found in the grapes are in an aggregated form. During maturation, the clustered anthocyanins and free anthocyanin molecules may lose sugar and acyl groups. Such reactions are more susceptible to wine oxidation and browning reactions in the wine. These cases can be overcome by polymerisation reactions occurring among free anthocyanins and condensed tannins, proanthocyanidins and catechins. This will minimise the losses. As a result of the increase of the solubility by the polymerisation reactions, the losses due to the collapse are minimised (Jackson, 2016). Polymeric anthocyanins, which are formed as a result of various reactions, are more stable against oxidation and pH changes than monomeric anthocyanins (Fig. 5).

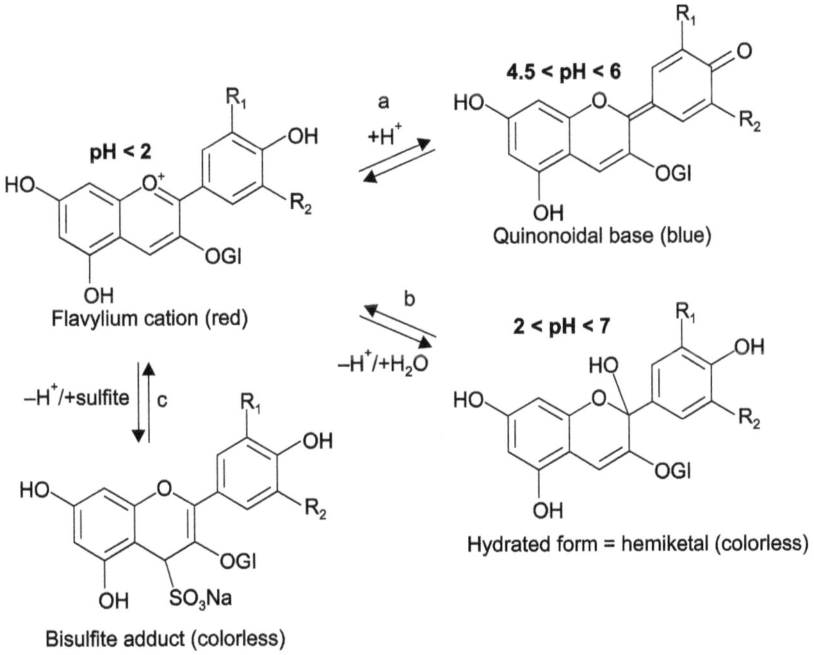

**Figure 5.** Reactions of anthocyanins in aqueous solutions and wine (R1, R2 = H, OH, $OCH_3$): proton transfer (a); hydration (b); and sulphite bleaching (c) (*Source*: Cheynier *et al.*, 2006)

Anthocyanins may act as nucleophiles and condense together with carbocations to form more polymeric pigments. The nucleophilic $C_6$ or $C_8$ position of the A-ring of flavanols performs an electrophilic substitution reaction with acetaldehyde, resulting in colourless condensed products. However, when the anthocyanins are added to this reaction, red or purple colour is observed. Consequently, as a result of chemical reactions in the anthocyanins during wine ripening and aging, various coloured or colourless pigments are formed (He *et al.*, 2008).

## 2.2. Reactions of Tannins with Proteins and Polysaccharides

Tannin, found in wines (Fig. 6), leads to astringency character of wines. Tannins form stable structures with proteins and polysaccharides. Tannins can combine with these components, depending on the electrical charge of the environment and ions concentration in the wine. Clarification agents are used to

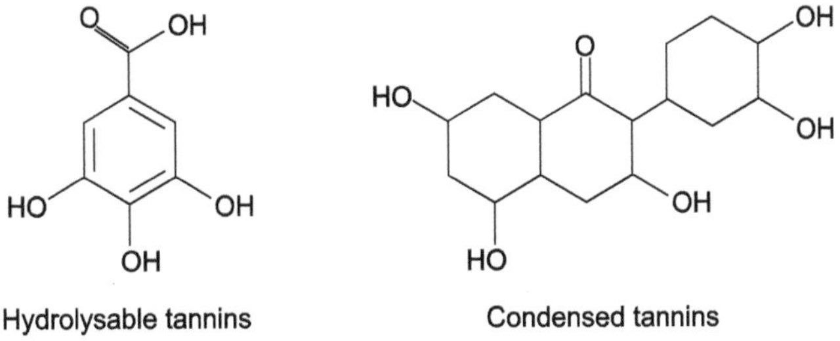

**Figure 6.** Structure of tannins (Anonymous, 2017)

remove these colloids when they are in excess. As a result, the content of tannin in the wine decreases and taste becomes softer (Markakis, 2012).

- The first stage of these reactions involves formation of planar heterocyclic amide bonds between the electron-rich phenol ring B-ring or galloylol ester of the tannin
- The second stage involves cross-linking reactions that occur by self-assocation between protein aggregates and bound tannins
- In the third stage, protein aggregates coalesce to form colloidal particles and thus the protein-tannin complex leads to precipitation (McRae and Kennedy, 2011)

The formation of the tannin-protein complex is directly related to various factors, such as pH, temperature, ionic strength, as well as protein type and molecular weight. It is easier to form complexes between proteins with high proline amino acid content and condensed tannins (Hagerman and Butler, 1980; Ribâereau-Gayon *et al.*, 2006). The astringency is caused by the proline content of the secreted salivary proteins by human saliva and their interactions with tannins (Haslam *et al.*, 1988). On the other hand, polysaccharides play the role as protective colloids in wine. They prevent the protein-tannin complex from precipitation, thus the leading to haze (Riou *et al.*, 2002; Cheynier, 2006; Del Barrio-Galán *et al.*, 2012).

## 2.3. Carbocation Formation of Procyanidins

Procyanidins are a sub-class of flavonoids. They are formed in the flavan units by a bond formed between two carbons ($C_4$, $C_6/C_8$). This bond is an unstable and is broken by acid action. Basically, the reaction leads to formation of carbocation ($C_4^+$), the first step in the reaction of the procyanidin B3 dimer with dissociation in acidic medium. After this step, various compounds may arise, depending on the effect of nucleophilic compounds or oxidation. It is known that the carbation is formed by nucleophilic compounds, such as thiols, as a result of the reaction of (+) -catechin 4-α-ethylthioflavan-3-ol. Carbocation can also lead to loss of the proton and electron from cyanide pigment. Thus lead to colour changes in wine (Haslam, 1997; Ribâereau-Gayon, 2006).

## 2.4. Oxidation Reactions of Procyanidins

The oxidation reactions include enzymatic, non-enzymatic and chemical-auto oxidation reactions occurring in wine. In oxidation reactions are involved primarily phenolic components, such as caftaric acid, catechin, gallic acid, anthocyanin and procyanidin. Metals ($Fe^{2+}$ and $Cu^+$) can act as catalysts in the oxidation. They react with molecular oxygen, leading to browning of wine; especially in white wines, catechins and procyanidins are the primary compounds causing oxidation. Due to the absence of maceration in the production stages of white wine, catechin and procyanidin compounds are present in small quantities. This is the main reason for the increase in oxidative browning in white wines. In red young wines, instability of these compounds is responsible for their marked differences (Butzke, 2010; Panda, 2011). In a previous study it has been reported that procyanidins are more easily oxidised than other phenolic compounds in must-like models containing more caffeoyl tartaric acid (Cheynier *et al.*, 1998).

## 2.5. Polymerisation Reactions of Procyanidins

Procyanidins can undergo oxidative and non-oxidative polymerisation reactions. Polymerisation reactions of procyanidins occur at the $C_4$-$C_8$ and $C_4$-$C_6$ interflavan bonds directly as a result of the wine pH values. These reactions among the anthocyanin and the procyanidin molecules may occur, both in the absence and presence of oxygen. The effect of oxygen in the medium oxidises ethanol to acetaldehyde, which increases the polymerisation reactions between the procyanidins and anthocyanins. These reactions affect grossly the taste of wine. Besides, these reactions also protect anthocyanins from oxidation and modifications that could occur with other compounds. Due to the increased solubility at the end of the polymerisation, the amount of precipitated tannin is minimised (Reynolds, 2010; Jackson, 2016).

### 2.6. Co-pigment Formation Process of Anthocyanins

Co-pigmentation is the result of interactions among the anthocyanins and the colourless compounds. Colourless compounds include flavonoids, such as flavonol and flavone subgroups (Boulton, 2001). Co-pigmentation causes significant changes in colour intensity (Castañeda-Ovando *et al.*, 2009). A large amount of pigments that emerge in young wines is obtained as a result of the pigmentation reactions of the pigments and their co-pigmentation co-factors (Neri and Boulton, 1966). This process is affected mainly by temperature and pH. At the pH value is 3.5, maximum co-pigmentation may occur, depending on the anthocyanins in the medium and their co-pigment co-factors (Sikorski, 2006). There is a sensible balance between the free anthocyanins and the co-pigments to result in co-pigmented anthocyanins. This balance is affected by competition between co-factors and reactions, such as oxidisation or hydrolysis of co-factors during aging. Therefore, the equilibrium is constantly changing and rebalanced. Co-pigmetation has an important contribution to wine quality since the co-pigmented anthocyanins give more colour than the free anthocyanins (Boulton, 2001).

### 2.7. Reactions of Anthocyanins with Compounds Containing Polarised Double Bonds

Anthocyanins react with molecules containing polar double bonds by cycloaddition reactions to their $C_4$/$C_5$ carbons. They tend to form new stable compounds by reacting with acetaldehyde and pyruvic acid, 4-vinylphenols, vinylflavanols which are known as pyranoanthocyanins (Bordiga, 2016). Basically, this reaction consists of a new cycloaddition of compounds between the flavylium molecule and polarised double bond coupled to the oxygen molecule held by the $C_4$/$C_5$ carbons of the double bonded anthocyanins. These new compounds occur at a very slow rate (Ribâereau-Gayon *et al.*, 2006).

### 2.8. Condensation Reactions of Anthocyanins with Tannins

Condensation reactions between anthocyanins and other phenolic compounds (procyanidins, catechins) are the most effective reactions causing changes in wine colour. These reactions occur among the $C_4$ carbon of anthocyanin molecule and the $C_6$ or $C_8$ carbon of the monomeric tannins, such as catechin, epicatechin or condense tannins (Du Toit *et al.*, 2017). In direct condensation reaction, the reaction involves the conversion of anthocyanins to flaven form. The electrophilic substitution between the anthocyanin flavylium form ($C_4$) and flavanols, such as procyanidins ($C_6$/$C_8$), involves the formation of a colourless flaven. Another direct condensation reaction involves the formation of a colourless compound as a result of the reactions between the carbocyclic ($C_4$) which occur after the interflavan bond breaks of procyanidin and the nucleophilic bonds ($C_6$/$C_8$) of anthocyanins (Ribâereau-Gayon *et al.*, 2006; González-Paramás *et al.*, 2006; Bakker and Clarke, 2011). Direct condensation reactions between anthocyanins and tannins or catechins lead to formation of anthocyanin-tannin or tannin-anthocyanin complexes which cause orange colour of wine (Jackson, 2016). There is an also polymeric structure formed due to acetaldehyde-mediated condensations between anthocyanins and flavanols. If this reaction occurs at acidic medium in presence of the catechin, then wine colour changes from reddish violet to orange (Ribâereau-Gayon *et al.*, 2006; González-Paramás *et al.*, 2006; Bakker and Clarke, 2011).

## 3. Effects of Wine Production Stages on Maturation and Aging

Many factors may influence the wine aging composition, such as grape variety, cultural processes, ecological factors and wine production techniques (Martin and Sun, 2013) as given in Fig. 7.

Wine composition varies with the grape variety, grape maturation and wine production techniques. Although the grape variety used in wine production reflects the composition of the wine, it can undergo considerable changes during the fermentation of grapes, production process and aging phase (Cheynier, 2010; Martin and Sun, 2013). The changes in the wine after fermentation can be explained by the fact that the wine is in a chemically dynamic state even during aging (Waterhouse, 2016). Specific changes occur in wine composition, contributing to the organoleptic properties of the wine (Morenoand Peinado, 2012). So, the aging stage can be considered as one of the important steps in wine production (Sun,

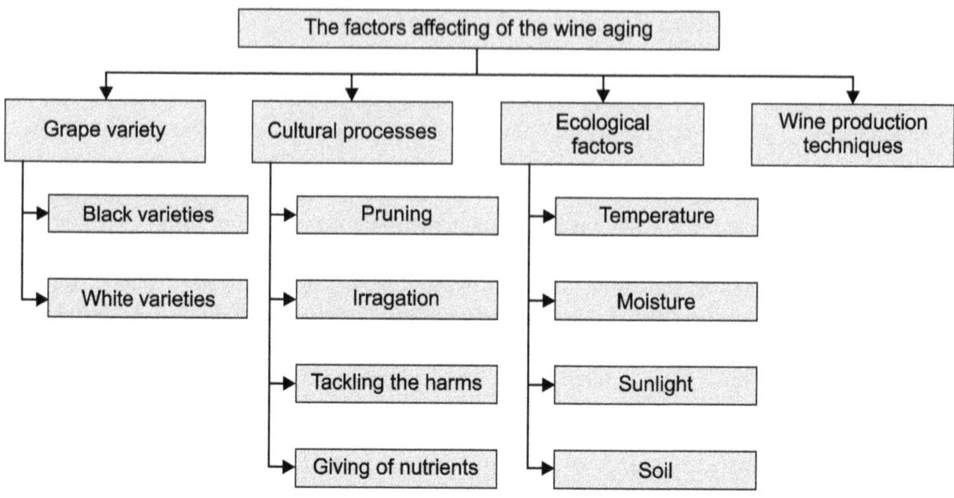


**Figure 7.** The factors affecting wine aging

2016). Different techniques used at each stage of wine production may contribute to the formation of high concentrations of pigments and tannins of wine, so that these compounds promise a potential for the complex taste and flavour to form during maturation and aging (Somers and Pocock, 1990).

### 3.1. The Effect of Grape Varieties

The aging period may vary, depending on the grape variety. Red wine produced from grapes of *Cabernet Sauvignon*, *Shiraz*, *Tempranillo*, *Nebbiolo* and *Pinot noir* can continue to retain its aroma after decades of aging for several years. White wines produced from grapes, such as *Riesling*, *Chardonnay*, *Sauvignon Blanc* and *Viura* have a high aging potential (Jackson, 2008). Some wine types (fresh, fruity whites, blushes, light reds and nouveau red wines) do not require prolonged maturation and aging because their quality peaks in a short time (Dharmadhikari, 2017). During aging, white wine generally loses itsr fruity aroma due to caramelisation and sherry-like odour formation while red wines develop bouquet due to reactions with tannins. This is explained by the fact that phenolic constituents in white wine are lower than that in red wines (Markakis, 2012). Additionally, aging time varies according to the grape varieties. White wines, such as *Riesling*, *Chardonnay*, *Sauvignon Blanc*, *Parellada*, *Semillon* and red wines, such as *Cabernet Sauvignon*, *Pinot noir*, *Syrah*, *Zinfandel* have a long aging period. White wines, such as *Trebbiano*, *Muscadet*, *Pinot Blanc*, *Kerner*, *Aligote* and red wines, such as *Gamay*, *Dolcetto*, *Carignan*, *Grenache*, *Grignolino* have a short aging period (Jackson, 2008).

### 3.2. The Effect of Grape Harvest Time

The grape harvesting time is closely related to the grape variety, climate, topography and vineyard management practices which directly affect chemical and sensory properties of the wine (Bindon *et al.*, 2013). The degree of grape ripening, the amount of acid and sugar and phenolic compounds are an important criterion for determining the harvesting date (Pérez-Magariño and González-San José, 2006). There are many studies in the literature that show that grape ripening and harvest date affect the phenolic compounds of wines (Pérez-Magariño and González-San José, 2006; Fang *et al.*, 2008). However, few studies have been found that examine the correlation between harvesting time and aging. Pérez-Magariño *et al.* investigated the influence of different grapes (Tino fino and Cabernet Sauvignon) and different harvest times (conventional, one week and two weeks later the conventional time) on wine aging. Wines matured in medium-fried American oak barrels for one year and then left to bottle aging for six months. It was found that the loss of free anthocyanin during the aging process was highest in Cabernet Sauvignon wines. However Tino fino (one week after conventional time) and Cabernet Sauvignon (two weeks after conventional time) were reported to have the highest levels of anthocyanin. The results demonstrated that the anthocyanins derived from the reduction of free anthocyanin and flavanols in wines increased and that these new pigments were also present in the two wine varieties made from more ripe grapes. However, in these wines, the differences between one week and two weeks after the conventional time were not statistically significant. As a result, they emphasised that the conventional harvest time will have better properties of wines with one or two week delays (Pérez-Magariño and González-San José, 2004).

### 3.3. The Effect of Maceration

In the stage of maceration, which is responsible for the wine colour, temperature, used method and time are important factors. At this state, significant changes occur in the phenolic compounds content of must; thus, high amounts of colour pigments and tannins can be formed causing colour stability in the wine (Sims and Bates, 1994; Alvarez *et al.*, 2006). Cejudo *et al.* investigated the effects of low toasted American oak chips (1 cm$^2$) added at 3 g/L for four days during maceration stage on Syrah wines, on phenolic component and colour properties of wine. The results demonstrated that must slightly increased the co-pigmentation reactions, including anthocyanin, flavan-3-ols and hydroxycinnamic acid. Additionally it was reported that this application increased the lightness and hue values of wines by 10 per cent and 70 per cent respectively. At last it was emphasised that the aging ability of wines could be improved by this method (Cejudo *et al.*, 2017). In another study done by Gómez-Plaza *et al.* the effect of different maceration times (four, five and 10 days) was investigated on phenolic compounds and colour properties on the one-year bottle-aged wine. It was found that 10 days of maceration caused high ionised anthocyanin, polymeric compound, chemical age and colour intensity values. On the other hand, four days of maceration caused a higher tint value with lower amounts of phenolic compounds. As a result, they emphasised that the increase in the amount of phenolic compounds extracted from grapes during a long maceration period gave the wines a high colour intensity and they retained the characteristics of the wines according to their short maceration times. Additionally, it has also been documented that 10-day maceration gave wine a higher sensory evaluation rank (Gómez-Plaza *et al.*, 2001).

### 3.4. The Effect of Press Process

Basically at this stage, fruit juice is obtained from crushed and pressed grapes. Usually, this process takes place after the must fermentation of red grapes but is directly produced if it is prepared from white grapes before the alcoholic fermentation (Robinson and Harding, 2015). The fruit juice obtained from grapes consists of 85-90 per cent of free-run juice while the remaining (10-15 per cent) portion is a juice obtained by pressing. This part may, however, depend on the grape variety, grape ripeness, pressing type, pressing time and pressing pressure (Amerine and Joslyn, 1970). The pressed juice includes more tannin, anthocyanin and total phenolic compounds than free-run juice. During winemaking, the producers can perform suitable blending procedures by mixing free-running and pressed juice compounds. Given the potential for low aging of free-run juice and the availability of low quality wines, the pressed juice, that enhances the potential of wine maturation and aging with higher tannin content, can be blended in various batches. Pressed juice ratio is determined according to the aging potential of wine (Ribéreau-Gayon *et al.*, 2006).

### 3.5. The Effect of Fermentation

Alcoholic fermentation is a process by which grape sugars (glucose, fructose), found at about 20-25 per cent, is converted to alcohol (10-15 per cent) and $CO_2$ by yeasts. During fermentation, there is a comprehensive change in wine due to the extraction of the grape components by yeast metabolism, the enzymes found in grape and in microorganisms. Changes that occur during fermentation include biochemical reactions, while changes occurring during aging include only chemical reactions (Cheynier *et al.*, 2010; Waterhouse *et al.*, 2016). During fermentation, different yeasts produce different metabolites including aroma compounds, such as ethyl esters, acetate esters, higher alcohols, fatty acids and aldehydes. Therefore, fermentation changes the composition of the wine by different yeast metabolites, but the final wine composition is determined by changes in the aging process (Ferreira *et al.*, 1995; Vilanova and Sieiro, 2006; Recamales *et al.*, 2011). Tomašević *et al.* (2017) investigated in Pošip the effect of maceration/non-maceration, different yeast strains (indigenous and commercial) and antioxidant additions (higher sulphur dioxide and glutathione) on wine aroma of one-year bottle-aged wine. They found that fermentation carried out by different yeasts (commercial and traditional yeast) both in maceration and non-maceration had an effect on the esters and higher alcohols that determine the aroma profile of the wine. In the samples of non-maceration and maceration, 12 months of bottle-aging showed a decrease in some important aroma compounds like terpenes, esters, thiols but the reduction was slower in wines with added sulphur dioxide and glutathione as antioxidants before bottling. (Tomašević *et al.*, 2017). Malo-lactic fermentation entailing conversion of malic acid to lactic acid in wine (Waterhouse *et al.*, 2016) has become a desirable step in the aging of red wine and even white wine, with increase in biological stability of wine and with deacidification that occurs during this fermentation (Boido *et al.*, 2009). Boido *et al.* (2009) investigated the effects of different culture strains used in malo-lactic fermentation and bottle-aging on volatile compounds of Tannat wines. The aging period in Tannat wines was affected by malo-lactic fermentation time and the strains used for this fermentation. As a result, decrease in some acetates and ethyl esters had an effect on the fruity aroma of wine (Boido *et al.*, 2009).

### 3.6. The Effect of Clarification

Occurring after wine maturation or before bottling, clarification of wine is provided by stabilisation, removing of unstable compounds and the haze of wine (de Freitas, 2017). Clarification agents used in this process stabilise besides clarifying wines. This process also affects the flavour of wine usually during fining (Henderson and Rex, 2012). Gelatine and egg are often used in combination to reduce the astringency. Bentonite is usually used for protein removal. Polyvinylpyrrolidone (PVP) used in combination with other agents removes polymerised brown pigments formed in wines (Blady, 1997). Bravo Haro *et al.* (1991) investigated the effect of different clarification agents (bentonite, gelatine and ovalbumin) on phenolic compounds in an aged red wine. They found that albumin had a greater effect on the phenol fractions, causing reduction in polymer forms. Bentonite significantly reduced the amount of free anthocyanin though it was not as effective as the polymer form. Gelatine was less effective than bentonite in reducing the amount of free anthocyanin (Bravo Haro *et al.*, 1991). Gómez-Plaza *et al.* (2000) conducted a study through different vinification techniques, such as maceration temperature (25°C and 10°C) and different fining agents (polyvinylpyrrolidone or bentonite and gelatine), different storage temperatures and time on phenolic compounds and colour parameters of Monastrell wine after bottling. The best colour properties of wine were obtained with low maceration temperature (10°C), the polyvinylpyrrolidone clarification agent used for the clarification and storage temperature of less than 20°C. Additionally, it was determined that the values of colour age and polymeric pigment colour increased during the storage period (zero-12 months) but hydroxyl-cinnamic acid and its esters decreased (Gómez-Plaza *et al.*, 2000).

### 3.7. The Effect of Storage Conditions

The storage of wines has an important influence on maturation. Oxidation reactions can occur when wine is placed in porous containers, such as wood and reduction reactions could occur when glass and stainless steel containers are used. These reactions can cause changes in the chemical structure of the wine and these may vary, depending on the type of containers used, the time of duration, wine quality and other storage conditions. The quality of the wine may increase with proper storage (Stevenson, 1997; Recamales *et al.*, 2006). Storage conditions, such as temperature, humidity, light are factors that have direct influence on wine quality (Robinson and Harding, 2015). Somers and Evans investigated the effects of maturation conditions under nitrogen and oxygen at two different temperatures (3°C and 25°C) on the colour composition of wines produced from Shiraz and Grenache grapes. Results demonstrated that reactions of polymeric pigments occur faster under nitrogen gas at 25°C, while oxygen increases the rate of change of colour composition of wines (Somers and Evans, 1986). A study conducted by Somers and Pocock demonstrated that storage conditions cause pigment degradation and affect the colour characteristics of wines (Somers and Pocock, 1990). Arapitasis *et al.* studied the influence of storage conditions in cellar (15-17°C) and house (20-27°C) on the phenolic compounds of Sangiovese wine for a period of two years. .The amount of anthocyanin in wine stored in the cellar was higher than that stored in the house (Arapitsas *et al.*, 2014). The effects of pH, temperature, alcohol content, storage temperature and duration in bottle on colour composition of Portuguese wines was determined.

The results demonstrated that the amount of anthocyanin decreased as storage time and temperature increased, but this loss decreased with increasing $SO_2$ content. The amount of coloured molecules that are attached to the polymeric pigment increases at high storage temperatures, but decreases with increasing pH and $SO_2$. Additionally, the anthocyanin concentration during storage was not affected by the pH and the alcohol content (Dallas and Laureano, 1994). Used cork varieties also affect the storage of wines. As an alternative to the natural cork that is still in use today, natural and technical cork, synthetic cork and metal screw cork have emerged. Some studies have emphasised that bottles in the vertical position during storage cause more oxygen input than those stored in the horizontal position (Linsenmeier *et al.*, 2010; Venturi *et al.*, 2017). However, the other study did not show any such effect (Lopes *et al.*, 2006). Reactions in bottled wines are slow, allowing limited contact with air (Boulton *et al.*, 1996). The rate of wine filling during bottling is also an important factor since the amount of oxygen in the bottle can lead to accelerated aging (Anonymous, 2017).

### 3.8. The Effect of Storage Time

The storage time affects the phenolic components, colour and sensory properties of wines. The effect of time on phenolic composition and colour parameters of Zalema wine was that the amount of total phenols decreased during the second month but increased during the four to six to eight months. Colour values (L*, a*, b*, croma, hue) fluctuated throughout the storage period. At the end of storage, L *, a *, b *, chroma values had the highest rates (Recamales *et al.*, 2006). The results further demonstrated that the anthocyanin monomers and total flavonols significantly reduced in three months' time and continued to decrease throughout the storage period (Marquez *et al.*, 2014). The chemometric properties of wines treated with oak chips at different toasting degrees were effective during the three months of aging and caused significant differences, whereas the effect of the toasting degrees was not observed in one-and-a-half month-aged wines (Dumitriu *et al.*, 2016).

## 4. Aging Techniques

### 4.1. Development in Existing Aging Systems

#### *4.1.1. Barrel System*

Barrel system is one of the most used types of aging in the wine industry (Fig. 8). In this system, the wine aroma is enriched through very slow oxidation. The most common type of oak used is *Quercusalba* from North America, *Quercusrobus* and *Quercussessilis* from France. The transfer of astringency-

**Figure 8.** Oak barrels in a winery

related phenolic compounds and oak as responsible aromatic compounds to wine are among the benefits of using oak barrels (Jackson, 2008). The composition and geographical origin of oak are important considerations in selecting the oak for barrelmaking. American oak barrels have high cis /trans-lactone ratios, while French oak barrels have a higher rate of oxygen transfer to barrels (Perez-Prieto *et al.*, 2003; Fernández de Simón *et al.*, 2008; Prida and Channet, 2010; Hernández-Orte *et al.*, 2014). At the same time, oak seasoning and toasting degree also affect the composition of the wine. Wines produced in light-toasted oak barrels have less aromatic compounds whereas medium-toasted oak barrels cause enrichment with phenolic compounds. Use of heavy toasted barrels generally leads to production of wines with high volatile phenols (Fernández de Simón *et al.*, 2008). The extraction of oak-related compounds is higher in 'new barrels' but the preservation effects of individual anthocyanins against oxidation is better in 'used barrels' due to release of lower content of ellagitannins, low molecule-weight phenolic, hydrolysable tannins and lower permeability of oxygen. One of the obstacles in use of old barrels (except their expensive care) is that they could be contaminated with *Brettanomyces* and *Dekkera* species (Perez-Prieto *et al.*, 2003; Prida and Channet, 2010).

There are some disadvantages in the existing methods of wine aging. In traditional methods, the use of the barrels is expensive because of the high-space occupation, the blocking of the pore and slow diffusion rate of oxygen. Furthermore, the development of unwanted microorganisms, such as *Brettanomyces* that affect the sensory properties of wine, may be adversely affected by the frequent use of old barrels (Suárez *et al.*, 2007; Tao *et al.*, 2014). Some studies related to barrel system on the chemical and sensory properties of wine are summarised in Table 1.

**Table 1.** The Effect of Barrel System on the Chemical and Sensory Properties of Wine

| | *Conditions* | *Types of wine* | *Effects* | *References* |
|---|---|---|---|---|
| Barrel system | New French and new Amerikan oak barrel for 6 months | Red wine | Chemical effects<br>• The total amount of anthocyanin is similar<br>• New American barrel aged wine HCl index higher than new French barrel aged wine<br>• New French barrel aged wine tannin content higher than new American oak barrel aged wine<br>• Sensory effects (rice wine)<br>• New American oak barrel aged wine woody aroma more than new French oak barrel aged wine.<br>• Astringency and acidity aromas are similar | Perez-Prieto *et al.*, 2003 |
| | New French oak barrel for 6 to 12 months | Red wine | Chemical effects<br>• Furanic compounds increase<br>• Lactones, eugenol, and vanillin content increase<br>• Furfural and 5-methylfurfural content decrease<br>Sensory effects<br>• Overall oak intensity increase<br>• Fruity intensity decrease<br>• Vanilla/pastry aroma increase | Prida and Channet, 2010 |

In recent years, new materials such, as Spanish oak (*Quercus robur*, *Quercus petraea*, *Quercus pyrenaica*) (Simon *et al.*, 2003; Simon *et al.*, 2008), acacia (*Robinia pseudoacacia*) (Kozlovic *et al.*, 2010), cherry (*Prunus avium*) (Rosso *et al.*, 2009), chestnut (Caldeira *et al.*, 2006; Gambuti *et al.*, 2010), mulberry (*Morusalba*) (Rosso *et al.*, 2009) emerged as an alternative to American and French oaks.

### *4.1.2. Use of Wood Fragments*

Wood fragments are used as an alternative aging system in the form of oak chips and oak staves. Oak chips and oak staves provide quick extraction of wood-related volatile compounds in the first three months of aging, but not later; therefore, it is appropriate only for short-term-aged wines. Wood fragment shapes (oak power, cubes, oak beans, granules and dominos) and geographical origin affect the composition of the wine (Fig. 9).

**Figure 9.** Some oak fragments

The use of wood fragments can reduce the aging time due to high extraction rate for a short time but colour evolution requires a longer time and small amounts of oxygen (Hernández-Orte *et al.*, 2014; del Barrio-Galán *et al.*, 2015). This is a new development as discussed in the subsequent section.

### *4.1.3. Microoxygenation*

Aerobic conditions could grossly influence the wine quality (Somers and Pocock, 1990). Oxygen affects the phenolic compound composition of the wine, leading to changes in sensory properties, such as colour and astringency of wine. A small amount of oxygen can improve the sensory properties of red wine (Picariello, 2017). Microoxygenation is an innovative aging method that is mainly applied with low controlled amounts of oxygen in the early stage of wine maturation. The critical point in this method is the amount of oxygen used and the time of application (Blaauw, 2009). This method is different from the passive oxygen uptake that occurs during barrel maturation or barrel aging. During microoxygenation, the amount of oxygen used to prevent dissolved oxygen from accumulating should be equal to/or lower than the amount of oxygen in the wine (Gómez-Plaza and Cano-López, 2011).

Microoxygenation can reproduce the benefits of barrel aging in a much shorter time. Microoxygenation application has many benefits, such as incensement of long-term oxidative stability, lowering of $SO_2$ requirements in winemaking, provision of more complex wine aroma and additional oxygen for yeast metabolism and decrease of sluggish fermentation (Blaauw, 2009; Gómez-Plaza and Cano-López, 2011). However, in this application there are some risks, such as formation of aldehyde/oxidided aromas and flavour development in wine, incensement of volatile acidity, colour losses caused by the precipitation of phenolics and microbial spoilage (acetic acid bacteria and Brettanomyces) (Pérez-Magariño et al., 2009; Gómez-Plaza and Cano-López, 2011). The effects of microoxygenation on the chemical and sensory properties of the wine are summarised in Table 2.

**Table 2.** The Effects of Microoxygenation on the Chemical and Sensory Properties of Wine

| | *Conditions* | *Types of wine* | *Effects* | *References* |
|---|---|---|---|---|
| Microoxygenation | 3 ml $L^{-1}$ $month^{-1}$ for 3 month | Red wine | Chemical effects<br>• Monomeric anthocyanin content decrease<br>• Polymeric peak increase<br>• Higher color intensity<br>• Phenolic and chromatic characteristics very similar to that of a 3 month oak aged wine. | López *et al.*, 2010 |
| Microoxygenation, Oak chips and oak staves | 1 mg $L^{-1}$ $month^{-1}$ for 6 month | Red wine | Chemical effects<br>• Color intensity increase<br>• Red color increase<br>• Similar effects to wood barrel<br>• aging (New American and French oak barrels)<br>Sensory effects<br>• Vanilla and woody aromas similar to American oak barrel (microoxygenation+ oak chips)<br>• Spicy aroma and sweet taste similar to French oak barrel (microoxygenation+ oak staves) | Oberholster *et al.*, 2015 |
| Microoxygenation and oak chips | 3 mg $L^{-1}$ $month^{-1}$ for 20 days. | Red wine | Oak chips<br>Chemical effects<br>• Highest concentration of alcohol, carbonyl compounds and lactones<br>• Lower concentration (–95%) of flavan-3-ol (respect to microoxygentaion treated ones)<br>Sensory effects<br>• Floral, fruity, herbaceous character decrease<br>• typical oak aroma become<br><br>Microoxygenation<br>Chemical effects<br>• Highest concentration of acids and esters<br>• Highest concentration (+306%) of anthocyanins (respect to oak treated ones)<br>Sensory effects<br>• Astingency and herbaceous character reduce<br>• Spicy and fruity aromas increase | Baiano *et al.*, 2016 |

#### 4.1.4. Aging on Lees

Aging with lees is a process carried out with yeast strains that develop a biofilm on the wine surface after fermentation. The yeast used in the aging process belongs to the genus *Saccharomyces* and is also called as flor yeast (Alexandre, 2013). The yeasts have the ability to form a biofilm at the liquid-air interface (Alexandre, 2013). The flour yeast consumes the remaining dissolved oxygen, creating a reducing environment. It affects the organoleptic properties of wine and creates special aromas (Moreno and Peinado, 2012). During the aging process with flour yeasts, some metabolites (mannoproteins, carbohydrates) combine with wines compounds (Leroy *et al.*, 1990).

Lees are used during aging of natural sparkling wines, white wines and sherry wine. Aging by using lees could improve the colour stability and modify the aromatic properties of wines. In aging by lees, the state of yeast lees, ethanol content, temperature, acidity, other compounds in wines which can be absorbed on yeast lees are important factors (del Barrio-Galán *et al.*, 2015; Juega *et al.*, 2015). The effects of lees on the chemical and sensory properties of the wine are shown in Table 3.

Application of non-*Saccharomyces* wine yeasts, lees aging combined with microoxygenation or wood fragments is the new development in this field.

### 4.2. New Wine Aging Systems

In recent years, alternative techniques have begun to be tried in wine aging with the idea that aging can be accelerated by new techniques, such as ultrasound, gamma radiation, electric field, high pressure (Suárez *et al.*, 2007; Tao *et al.*, 2014).

#### 4.2.1. Important Points of Ultrasound

The effect of ultrasound on wine aging process is related to acoustic cavitation, which is concerned with the formation, growth and collapse of microbubbles. The violent collapse of these bubbles produces high temperature and pressure (Martin and Sun, 2013; Zhang *et al.*, 2016; Delgado-González *et al.*, 2017). During the use of ultrasound treatment, frequency, exposure time and temperature are important parameters that must be considered. High-frequency ultrasound is not suitable; generally the best results are obtained in the range 20-100 kHz. (Martin and Sun, 2013; Zhang *et al.*, 2016; Delgado-González *et al.*, 2017). The effects of ultrasound treatment on the chemical and sensory properties of wine are shown in Table 4.

Application of different ultrasonic treatments by fixed or circulating wines and application of ultrasonic treatment combined with other aging techniques (aging in barrel, aging in bottles) is the new development in this field.

#### 4.2.2. Gamma Irradiation

Gamma radiation is one of the three types of natural radioactivity. The other two types of natural radioactivity are alpha and beta radiation, which are in the form of particles. Gamma radiation is a propagation of radioactive materials, known as electromagnetic quantum waves. Gamma radiation, as one of the types of ionising radiations, has higher photon energy and shorter wavelength than light (Wetherill, 1965; da Silva, 2012). Gamma radiation has emerged as an alternative to chemical preservatives of foods (Gupta *et al.*, 2015). The World Health Organisation (WHO) stated in 1981 that gamma irradiation technology could be used to increase protection and shelf- life of food products (Naresh *et al.*, 2015). The effects concerning wine are related to accelerating physical maturation. During use of gamma irradiation, gamma irradiation dosage (Gy), exposure time, wine type and composition are the important factors (Chang, 2003; Kondapalli *et al.*, 2014).

Wine treated with gamma irradiation (2400 Gy) had the lowest total anthocyanin content and colour intensity but highest colour age value. It was found that as the storage period increased, the amount of acetaldehyde increased, but after 18 months of storage, there was no significant difference in the amount of acetaldehyde in different doses of wine (Caldwell *et al.*, 1989). The application of gamma irradiation dose of 600 Gy reduced the pH value in wine. It was found that the colour values of the same dose L value

**Table 3.** The Effects of Less on the Chemical and Sensory Properties of Wine

| | *Conditions* | *Types of wine* | *Effects* | *References* |
|---|---|---|---|---|
| Lees | *Saccharomyces cerevisiae* 1 for 10, 20, 30, 40, 50 day | White wine | Chemical effects (20 day)<br>• Concentration of most esters and acetates increase<br>• Protein content decrease slightly<br>• Polymeric mannose concentration increase<br>• Lactone content decrease<br>Sensory effects (20 day)<br>• The best sensorial quality of the wine<br>• More aging than 20 days reduce sensorial quality | Juega *et al.*, 2015 |
| | Oak barrel aging + lees for 6 months | Red wine | Chemical effects<br>• Tartaric ester content decrease<br>• Total polyphenol concentration slightly decrease<br>• Flavonol content decrease<br>• Protein concentration decrease<br>Sensory effects<br>• Woody, astringency, fruity aroma decrease<br>• Acidity increase | Barrio-Galán *et al.*, 2011 |
| | Oak barrel aging + lees + microoxygenation (3 mg $L^{-1}$ $month^{-1}$) for 6 months | Red wine | Chemical effects<br>• Tartaric ester content not change<br>• Total polyphenol concentration slightly decrease<br>• Flavonol content increase<br>• Protein concentration increase<br>Sensory effects<br>• Woody, astringency, fruity aroma decrease | |
| | Oak barrel aging + oak chips for 6 months | Red wine | Chemical effects<br>• Tartaric ester content decrease<br>• Total polyphenol concentration slightly decrease<br>• Flavonol content increase<br>• Protein concentration increase<br>Sensory effects<br>• Woody, astringency, fruity aroma decrease<br>• Acidity increase | |

**Table 4.** The Effects of Gamma Radiation on the Chemical and Sensory Properties of Wine

| | *Conditions* | *Types of wine* | *Effects* | *References* |
|---|---|---|---|---|
| Gamma irradiation | 200, 400, 600, 800 Gy | Rice wine | Chemical effects<br>• Polyol concentration reduction<br>• Ethyl acetate content increase<br>• Acetaldehyde content increase<br>Sensory effects<br>• No change of wine colour<br>• Fruit fragrance slightly increase as gamma irradiation dosages increase<br>• Fragrance quality increase<br>• Greasy, rice-oil flavour decrease<br>• Astringent increase to some degree as gamma irradiation dosage increase | Chang, 2003 |
| | 200, 400, 600 and 800 Gy | Maize wine | Chemical effects<br>• Acetaldehyde content increase<br>• Polyol concentration reduction<br>Sensory effects<br>• Fruity fragrance slightly increase<br>• Rosy/flower fragrance dramatically increase<br>• greasy mouth feel decrease | Chang, 2004 |

**Table 5.** The Effects of Ultrasound Treatment on the Chemical and Sensory Properties of Wine

| | *Conditions* | *Types of wine* | *Effects* | *References* |
|---|---|---|---|---|
| Ultrasound | 20 kHz ultrasound for 1 week. | Rice wine, maize wine | Chemical effects<br>• Alcohol content reduction (rice wine)<br>• Alcohol content increase (maize wine)<br>• Acetaldehyde content decrease<br>• Ethyl acetate content increase (rice wine)<br>• Ethyl acetate content decrease (maize wine)<br>• Polyol concentration reduction<br>Sensory effects (rice wine)<br>• Overripe or sherry-like smell decrease<br>Sensory effects (maize wine)<br>• Tart, spicy, and unsmooth taste increase | Chang and Chen, 2002 |
| | 30 kHz ultrasound from 1.8 kPa to 20 kPa for 10 days | Red wine | Chemical effects<br>• Anthocyanin concentration increase<br>• Tannin concentration decrease | Masuzawa *et al.*, 2000 |

(*Contd.*)

**Table 5.** (*Contd.*)

| *Conditions* | *Types of wine* | *Effects* | *References* |
|---|---|---|---|
| | | • Lightness (L*) increase; redness (a*) and yellowness (b*) decrease<br>Sensory effects<br>• HCl index (index of maturation) decrease at 10 kPa and 20 kPa, increase 1.8 kPa and 5.8 kPa<br>• EtOH index( index of smoothness of wine) increase at 1.8 kPa, 5.8 kPa, 20 kPa, decrease at 10 kPa | |
| . | Greengage wine | Chemical effects<br>• Total acid content increase<br>• Total ester content increase<br>• Alcohol content reduction<br>• Aldehyde and ketone content decrease<br>Sensory effects<br>• Improve the sensory quality (highest sensory evaluation score at 45 kHz, 360 W for 30 min) | Zheng *et al.*, 2014 |
| <100 kHz multiple frequencies for 15 and 30 min. | Red wine, white wine | Chemical effects<br>• Accelerating chemical reactions<br>Sensory effects<br>• Extending the shelf life of wine | Leonhardt and Morabito, 2007 |

decreased, a* value increased, and b* value increased when compared with the control group (Harder *et al.*, 2013). The effects of gamma radiation on the chemical and sensory properties of wine are shown in Table 5. More work however, needs to be done to understand the effects of gamma irradiation on human health (Chang, 2003).

### 4.2.3. Electric Field

The application of AC electric field in food and bioengineering has been studied within the last 20 years. AC electric field treatment is also considered as an effective tool to accelerate wine aging. This technology has already been used in some Chinese wine factories. The high voltage electric field not only reduced the content of undesired compounds, such as aldehydes but also increased the contents of free amino acids and esters, which are linked to high-quality wines (Zeng *et al.*, 2008; Lu, 2013; Talele and Benseman, 2013). The effects of electric field treatment on the chemical and sensory properties of wine are summarised in Table 6. Studies on operational parameters should be carried out in future.

### 4.2.4. Pulsed Electric Field Technology

The pulsed electric field technology is based on the application of short-duration high-intensity electric field strengths that induce electroporation of the cell membranes. The pore formation provokes microbial inactivation and enhances the diffusion of the solutes through cell membranes. This technique is mainly used before fermentation with antimicrobial purpose and during maceration-fermentation with the purpose of extraction of phenolic compounds (Puértolas *et al.*, 2010a, b; El Darra *et al.*, 2013). This technique is mainly used before fermentation for the antimicrobial purpose and during maceration-fermentation with the purpose of extraction of phenolic compounds.

Application of pulsed electric field technology combined with other aging techniques (aging in barrel, aging in bottles) could be proposed for future work.

### 4.2.5. High Pressure

**Table 6.** The Effects of Electric Field Treatment on the Chemical and Sensory Properties of Wine

| | *Conditions* | *Types of wine* | *Effects* | *References* |
|---|---|---|---|---|
| Electric field | AC electric field 600 and 900 V/cm for 3-8 min | Red wine | Chemical effects<br>• Higher alcohols and aldehydes content decrease<br>• Esters and free amino acids content increase<br>Sensory effect (600 V/cm for 3 min)<br>• Harsh and pungent flavor decrease<br>• Harmonious and dainty flavor increase<br>• Fruit aroma decrease slightly<br>Sensory effect (900 V/cm for 8 min)<br>• Wine taste becomes burning and undrinkable<br>• Strange and unpleasant aroma increase | Zeng *et al.*, 2008 |

**Table 7.** The Effects of High Pressure Treatment on the Chemical and Sensory Properties of Wine

| | *Conditions* | *Types of wine* | *Effects* | *References* |
|---|---|---|---|---|
| High pressure | 100 MPa for 30 min | Red wine | Chemical effects<br>• Total phenol content increase<br>• Phenolic acid content increase<br>• Flavon-3 ols content reduction<br>Sensory effects (rice wine)<br>• Astringency and bitterness decrease | Sun *et al.*, 2016 |
| | 650 MPa for 2 hour | Red wine | Chemical effects<br>• Color intensity and phenolic compounds decrease<br>Sensory effects<br>• Sour and fruity odor decrease<br>• Sour, astringent, alcoholic, bitter taste increase<br>• Mouth-feel sensation slightly enhance | Tao *et al.*, 2012 |
| | 300-500 Mpa for 2 hour | Fresh claret (bordeaux wine) | Chemical effects<br>• Boiling point, relative density, redox potential, electrical conductivity, and total acidity change<br>Sensory effects<br>• Best taste and flavor at 300 Mpa<br>• Pressure higher above 500 MPa would break up claret styles | Li *et al.*, 2005 |

High pressure provides the activation energy to initiate chemical reactions in wines and consequently, shorten the aging time (Sun *et al.*, 2016).The application of high pressure during wine aging causes changes in boiling point, relative density, redox potential, electrical conductivity and total acidity of wines. Long-term studies are needed to further realise the full potential of this technique (Chen *et al.*, 2016; Sun *et al.*, 2016; Zhu *et al.*, 2016). The effects of high pressure treatment on the chemical and sensory properties of the wine are summarised in Table 7.

## 5. Conclusions

For production of high quality wines, aging is an important step in winemaking. Current technology in this regard that makes use of oak-barrel aging is an effective and reliable method and should not be completely abandoned. However, use of food fragments provides wood-related phenols and aromas to wines and could be used in combination with other technologies, Further, application of microoxygention and yeast lees improves physicochemical and sensory properties of wines and could also be used in combination with other technologies. Physical methods (ultrasonic waves, gamma rays, electric field, nano-gold photocatalysis, high pressure) provide drastic reduction in aging time but further studies on operational parameters and their effects should be carried out.

## References

Alexandre, H. (2013). Flor yeasts of *Saccharomyces cerevisiae* – Their ecology, genetics and metabolism. *International Journal of Food Microbiology* 167: 269-275.

Álvarez, I., Aleixandre, J.L., García, M.J. and Lizama, V. (2006). Impact of pre-fermentative maceration on the phenolic and volatile compounds in Monastrell red wines. *Analytica Chimica Acta* 563: 109-115.

Amerine, M.A. and Joslyn, M.A. (1970). Table Wines: The Technology of Their Production. University of California Press, Berkeley.

Anonymous (2017). Visual Clues and Aging Potential, http:// www.musingsonthevine. com/tips_age3.shtml, Accessed 22.07.2017

Anonymous (2017). What are Tannins in Wine? https://www.winecompass.com.au/blog/what-are-tannins-in-wine/, Accessed 22.07.2017

Arapitsas, P., Speri, G., Angeli, A., Perenzoni, D. and Mattivi, F. (2014). The influence of storage on the 'chemical age' of red wines. *Metabolomics* 10: 816-832.

Baiano, A., De Gianni, A., Mentana, A., Quinto, M., Centonze, D. and Del Nobile, M.A. (2016). Colour-related phenolics, volatile composition, and sensory profile of Nero di Troia wines treated with oak chips or by micro-oxygenation. *European Food Research and Technology* 242: 1631-1646.

Bakker, J. and Clarke, R.J. (2011). Wine: Flavour Chemistry. Blackwell Publishing, Oxford.

Bindon, K., Varela, C., Kennedy, J., Holt, H. and Herderich, M. (2013). Relationships between harvest time and wine composition in *Vitis vinifera* L. cv. Cabernet Sauvignon 1: Grape and wine chemistry. *Food Chemistry* 138: 1696-1705.

Blaauw, D.A. (2009). Micro-oxygenation in contemporary winemaking. Doctoral dissertation, Thesis. Cape Wine Academy, Stellenbosch, South Africa.

Blady, M.W. (1997). The University Wine Course. The wine appreciation guild, South San Francisco, CA, USA.

Boido, E., Medina, K., Fariña, L., Carrau, F., Versini, G. and Dellacassa, E. (2009). The effect of bacterial strain and aging on the secondary volatile metabolites produced during malo-lactic fermentation of Tannat red wine. *Journal of Agricultural and Food Chemistry* 57: 6271-6278.

Bordiga, M. (Ed.) (2016). Valorisation of Winemaking By-products. CRC Press, Boca Raton, FL, USA.

Boulton, R.B., Singleton, V.L., Bisson, L.F. and Kunkee, R.E. (1996). Principles and Practices of Winemaking. Chapman & Hall, New York.

Boulton, R. (2001). The co-pigmentation of anthocyanins and its role in the colour of red wine: A critical review. *American Journalof Enology and Viticulture* 52: 67-87.

Bravo Haro, S., Rivas Gonzalo, J.C. and Santos Buelga, C. (1991). Effect of some clarifying agents in phenolic and colour parameters in an aged red wine. *Revista de Agroquimica y Tecnologia de Alimentos* 31: 584-590.

Brouillard, R. and Dangles, O. (1994). Anthocyanin molecular interactions: The first step in the formation of new pigments during wine aging? *Food Chemistry* 51: 365-371.

Bruce, W.Z., Kenneth, C.F., Barry, H.G. and Fred, S.N. (1995). Wine Analysis and Production. Chapman & Hall, New York, USA.

Butzke, C.E. (Ed.) (2010). Winemaking Problems Solved. Woodhead Publishing Limited, Cambridge.

Caldeira, I., Clímaco, M.C., de Sousa, R.B. and Belchior, A.P. (2006). Volatile composition of oak and chestnut woods used in brandy ageing: Modification induced by heat treatment. *Journal of Food Engineering* 76: 202-211.

Caldwell, C.L. and Spayd, S.E. (1989). Effects of gamma irradiation on chemical and sensory evaluation of Cabernet Sauvignon wine. *American Chemical Society* 26: 337-345.

Cano-López, M., López-Roca, J.M., Pardo-Minguez, F. and Plaza, E.G. (2010). Oak barrel maturation vs. micro-oxygenation: Effect on the formation of anthocyanin-derived pigments and wine colour. *Food Chemistry* 119: 191-195.

Castañeda-Ovando, A., de Lourdes Pacheco-Hernández, M., Páez-Hernández, M.E., Rodríguez, J.A. and Galán-Vidal, C.A. (2009). Chemical studies of anthocyanins: A review. *Food Chemistry* 113: 859-871.

Cejudo-Bastante, M.J., Rivero-Granados, F.J. and Heredia, F.J. (2017). Improving the colour and aging aptitude of Syrah wines in warm climate by wood–grape mix maceration. *European Food Research and Technology* 243: 575-582.

Chang, A.C. (2003). The effects of gamma irradiation on rice wine maturation. *Food Chemistry* 83: 323-327.

Chang, A.C. (2004). The effects of different accelerating techniques on maize wine maturation. *Food Chemistry* 86: 61-68.

Chang, A.C. and Chen, F.C. (2002). The application of 20 kHz ultrasonic waves to accelerate the aging of different wines. *Food Chemistry* 79: 501-506.

Chen, X., Li, L., You, Y., Mao, B., Zhao, W. and Zhan, J. (2016). The effects of ultra-high pressure treatment on the phenolic composition of red wine. *South African Journal of Enology and Viticulture* 33: 203-213.

Cheynier, V., Dueñas-Paton, M., Salas, E., Maury, C., Souquet, J.M., Sarni-Manchado, P. and Fulcrand, H. (2006). Structure and properties of wine pigments and tannins. *American Journal of Enology and Viticulture* 57: 298-305.

Cheynier, V., Dueñas-Paton, M., Salas, E., Maury, C., Souquet, J.M., Sarni-Manchado, P. and Fulcrand, H. (2006). Structure and properties of wine pigments and tannins. *American Journal of Enology and Viticulture* 57: 298-305.

Cheynier, V., Owe, C. and Rigaud, J. (1988). Oxidation of grape juice phenolic compounds in model solutions. *Journal of Food Science* 53: 1729-1732.

Cheynier, V., Schneider, R., Salmon, J.M. and Fulcrand, H. (2010). Chemistry of wine. *Comprehensive Natural Products II: Chemistry and Biology* 3: 1119-1172.

Da Silva Aquino, K.A. (2012). Sterilisation by gamma irradiation. *In*: Gamma Radiation. Intech Press, Croatia.

Dallas, C. and Laureano, O. (1994). Effects of pH, sulphur dioxide, alcohol content, temperature and storage time on colour composition of a young Portuguese red table wine. *Journal of the Science of Food and Agriculture* 65: 477-485.

Dallas, C., Ricardo-da-Silva, J.M. and Laureano, O. (1995). Degradation of oligomeric procyanidins and anthocyanins in a Tinta Roriz red wine during maturation. *Vitis*, 34: 51-56.

De Freitas, V. (2017). Oenological perspective of red wine astringency. *OENO One* 51, http://oeno-one.eu/article/view/1816

De Freitas, V.A.P., Fernandes, A., Oliveira, J., Teixeira, N. and Mateus, N. (2017). A review of the current knowledge of red wine colour. *OENO One* 51, http://oeno-one.eu/article/view/1604

De Rosso, M., Panighel, A., Dalla Vedova, A., Stella, L. and Flamini, R. (2009). Changes in chemical composition of a red wine aged in acacia, cherry, chestnut, mulberry and oakwood barrels. *Journal of Agricultural and Food Chemistry* 57: 1915-1920.

Del Barrio-Galán, R., Medel-Marabolí, M. and Peña-Neira, Á. (2015). Effect of different aging techniques on the polysaccharide and phenolic composition and sensory characteristics of Syrah red wines fermented using different yeast strains. *Food Chemistry* 179: 116-126.

Del Barrio-Galán, R., Pérez-Magariño, S. and Ortega-Heras, M. (2012). Effect of the aging on lees and other alternative techniques on the low molecular weight phenols of Tempranillo red wine aged in oak barrels. *Analytica Chimica Acta* 732: 53-63.

Del Barrio-Galán, R., Pérez-Magariño, S. and Ortega-Heras, M. (2011). Techniques for improving or replacing ageing on lees of oak aged red wines: The effects on polysaccharides and the phenolic composition. *Food Chemistry* 127: 528-540.

Delgado-González, M.J., Sánchez-Guillén, M.M., García-Moreno, M.V., Rodríguez-Dodero, M.C., García-Barroso, C. and Guillén-Sánchez, D.A. (2017). Study of a laboratory-scaled new method for the accelerated continuous ageing of wine spirits by applying ultrasound energy. *Ultrasonics Sonochemistry* 36: 226-235.

Dharmadhikari, M. (2017). Wine Aging, http://www.extension.iastate.edu/wine/w-aging, Accessed 22.07.2017.

Du Toit, W.J., Marais, J., Pretorius, I.S. and Du Toit, M. (2017). Oxygen in must and wine: A review. *South African Journal of Enology and Viticulture* 27: 76-94.

Dumitriu, G.D., de Lerma, N.L., Cotea, V.V., Zamfir, C.I. and Peinado, R.A. (2016). Effect of aging time, dosage and toasting level of oak chips on the colour parameters, phenolic compounds and antioxidant activity of red wines (*var. Feteascăneagră*). *European Food Research and Technology* 242: 2171-2180.

El Darra, N., Grimi, N., Maroun, R.G., Louka, N. and Vorobiev, E. (2013). Pulsed electric field, ultrasound, and thermal pretreatments for better phenolic extraction during red fermentation. *European Food Research and Technology* 236: 47-56.

Fang, F., Li, J.M., Zhang, P., Tang, K., Wang, W., Pan, Q.H. and Huang, W.D. (2008). Effects of grape variety, harvest date, fermentation vessel and wine ageing on flavonoid concentration in red wines. *Food Research International* 41: 53-60.

Fernández de Simón, B., Cadahía, E. and Jalocha, J. (2003). Volatile compounds in a Spanish red wine aged in barrels made of Spanish, French, and American oakwood. *Journal of Agricultural and Food Chemistry* 51: 7671-7678.

Fernández de Simón, B., Cadahía, E., Sanz, M., Poveda, P., Perez-Magariño, S., Ortega-Heras, M. and González-Huerta, C. (2008). Volatile compounds and sensorial characterisation of wines from four Spanish denominations of origin, aged in Spanish Rebollo (*Quercus pyrenaica* Willd.) oakwood barrels. *Journal of Agricultural and Food Chemistry* 56: 9046-9055.

Ferreira, V., Fernández, P., Peña, C., Escudero, A. and Cacho, J.F. (1995). Investigation on the role played by fermentation esters in the aroma of young Spanish wines by multivariate analysis. *Journal of the Science of Food and Agriculture* 67: 381-392.

Gambuti, A., Capuano, R., Lisanti, M.T., Strollo, D. and Moio, L. (2010). Effect of aging in new oak, one-year-used oak, chestnut barrels and bottle on color, phenolics and gustative profile of three monovarietal red wines. *European Food Research and Technology* 231: 455-465.

Garde-Cerdán, T. and Ancín-Azpilicueta, C. (2006). Review of quality factors on wine ageing in oak barrels. *Trends in Food Science & Technology* 17: 438-447.

Gómez-Plaza, E. and Cano-López, M. (2011). A review on micro-oxygenation of red wines: Claims, benefits and the underlying chemistry. *Food Chemistry* 125: 1131-1140.

Gómez-Plaza, E., Gil-Muñoz, R., López-Roca, J.M. and Martínez, A. (2000). Colour and phenolic compounds of a young red wine: Influence of winemaking techniques, storage temperature, and length of storage time. *Journal of Agricultural and Food Chemistry* 48: 736-741.

Gómez-Plaza, E., Gil-Muñoz, R., López-Roca, J.M., Martínez-Cutillas, A. and Fernández-Fernández, J.I. (2001). Phenolic compounds and colour stability of red wines: Effect of skin maceration time. *American Journal of Enology and Viticulture* 52: 266-270.

González-Paramás, A.M., da Silva, F.L., Martín-López, P., Macz-Pop, G., González-Manzano, S., Alcalde-Eon, C., Pérez-Alonso, J.J., Escribano-Bailón, M.T., Rivas-Gonzalo, J.C. and Santos-Buelga, C. (2006). Flavanol-anthocyanin condensed pigments in plant extracts. *Food Chemistry* 94(3): 428-436.

Guerrero, E.D., Mejías, R.C., Marín, R.N., Bejarano, M.J.R., Dodero, M.C.R. and Barroso, C.G. (2011). Accelerated aging of a Sherry wine vinegar on an industrial scale employing microoxygenation and oak chips. *European Food Research and Technology* 232: 241-254.

Gupta, S., Padole, R., Variyar, P.S. and Sharma, A. (2015). Influence of radiation processing of grapes on wine quality. *Radiation Physics and Chemistry* 111: 46-56.

Hagerman, A.E. and Butler, L.G. (1980). Condensed tannin purification and characterisation of tannin-associated proteins. *Journal of Agricultural and Food Chemistry* 28: 947-952.

Han, F.L. and Xu, Y. (2015). Effect of the structure of seven anthocyanins on self-association and colour in an aqueous alcohol solution. *South African Journal of Enology and Viticulture* 36: 105-116.

Harder, M.N., Silva, L.A., Pires, J.A., Scanholato, M. and Arthur, V. (2013). Physical-chemical evaluation of wines subjected to gamma irradiation for aging. *Food Science and Technology* 1: 62-65.

Haslam, E. (1977). Symmetry and promiscuity in procyanidin biochemistry. *Phytochemistry* 16: 1625-1640.

Haslam, E., Lilley, T.H. and Butler, L.G. (1988). Natural astringency in foodstuffs – Amolecular interpretation. *Critical Reviews in Food Science & Nutrition* 27: 1-40.

He, F., Liang, N.N., Mu, L., Pan, Q.H., Wang, J., Reeves, M.J. and Duan, C.Q. (2012). Anthocyanins and their variation in red wines II. Anthocyanin derived pigments and their colour evolution. *Molecules* 17: 1483-1519.

He, F., Pan, Q.H., Shi, Y. and Duan, C.Q. (2008). Chemical synthesis of proanthocyanidins *in vitro* and their reactions in aging wines. *Molecules* 13: 3007-3032.

Henderson, J.P. and Rex, D. (2012). About Wine. Cengage Learning, Delmar, New York.

Hernández-Orte, P., Franco, E., Huerta, C.G., García, J.M., Cabellos, M., Suberviola, J., Orriols, I. and Cacho, J. (2014). Criteria to discriminate between wines aged in oak barrels and macerated with oak fragments. *Food Research International* 57: 234-241.

Jackson, R.S. (2008). Wine Science: Principles and Applications. Academic Press, San Diego, California.

Jackson, R.S. (2016). Wine Tasting: A Professional Handbook. AcademicPress, San Diego, California.

Jacobson, J.L. (2006). Introduction to Wine Laboratory Practices and Procedures. Springer Science & Business Media, New York.

Juega, M., Carrascosa, A.V. and Martinez-Rodriguez, A.J. (2015). Effect of short ageing on lees on the mannoprotein content, aromatic profile, and sensorial character of white wines. *Journal of Food Science* 80: 384-388.

Kondapalli, N., Sadineni, V., Variyar, P.S., Sharma, A. and Obulam, V.S.R. (2014). Impact of $\gamma$-irradiation on antioxidant capacity of mango (*Mangifera indica* L.) wine from eight Indian cultivars and the protection of mango wine against DNA damage caused by irradiation. *Process Biochemistry* 49: 1819-1830.

Kozlovic, G., Jeromel, A., Maslov, L., Pollnitz, A. and Orlić, S. (2010). Use of acacia barrique barrels – Influence on the quality of Malvazija from Istria wines. *Food Chemistry* 120: 698-702.

Leonhardt, C.G. and Morabito, J.A. (2007). Wine Aging Method and System. US Patent #7,220,439.

Leroy, M.J., Charpentier, M., Duteurtre, B., Feuillat, M. and Charpentier, C. (1990). Yeast autolysis during champagne aging. *American Journal of Enology and Viticulture* 41: 21-28.

Li, S.F., Duan, X.C., Liu, S.W. and Yang, G.M. (2005). Effects of ultra-high pressure treatment on physical properties of fresh claret. *Liquor Making Science and Technology* 8: 61.

Linsenmeier, A., Rauhut, D. and Sponholz, W.R. (2010). Ageing and flavour deterioration in wine. *In*: A.G. Reynolds (Eds.). Managing Wine Quality: Oenology and Wine Quality. Woodhead Publishing Limited, Cambridge.

Lopes, P., Saucier, C., Teissedre, P.L. and Glories, Y. (2006). Impact of storage position on oxygen ingress through different closures into wine bottles. *Journal of Agricultural and Food Chemistry* 54: 6741-6746.

Lu, X. (2013). The effects of alternating electric fields on wine. Doctoral dissertation. Auckland University of Technology, Auckland, New Zealand.

Markakis, P. (Ed.) (2012). Anthocyanins as Food Colours. Academic Press, New York.

Marquez, A., Serratosa, M.P. and Merida, J. (2014). Influence of bottle storage time on colour, phenolic composition and sensory properties of sweet red wines. *Food Chemistry* 146: 507-514.

Martín, J.F.G. and Sun, D.W. (2013). Ultrasound and electric fields as novel techniques for assisting the wine ageing process: The state-of-the-art research. *Trends in Food Science & Technology* 33: 40-53.

Martínez, J., Cadahía, E., Fernández de Simón, B., Ojeda, S. and Rubio, P. (2008). Effect of the seasoning method on the chemical composition of oak heartwood to cooperage. *Journal of Agricultural and Food Chemistry* 56: 3089-3096.

Masuzawa, N., Ohdaira, E. and Ide, M. (2000). Effects of ultrasonic irradiation on phenolic compounds in wine. *Japanese Journal of Applied Physics* 39: 2978.

Mateus, N. and de Freitas, V. (2001). Evolution and stability of anthocyanin-derived pigments during port wine aging. *Journal of Agricultural and Food Chemistry* 49: 5217-5222.

McRae, J.M. and Kennedy, J.A. (2011). Wine and grape tannin interactions with salivary proteins and their impact on astringency: A review of current research. *Molecules* 16: 2348-2364.

Moreno, J. and Peinado, R. (2012). *Enological Chemistry*. Academic Press, London.

Moreno-Arribas, M.V. and Polo, M.C. (2009). Wine Chemistry and Biochemistry, vol. 378. Springer, New York.

Naresh, K., Varakumar, S., Variyar, P.S., Sharma, A. and Reddy, O.V.S. (2015). Enhancing antioxidant activity, microbial and sensory quality of mango (*Mangifera indica* L.) juice by γ-irradiation and its *in vitro* radio-protective potential. *Journal of Food Science and Technology* 52: 4054-4065.

Neri, R. and Boulton, R.B. (1966). The assessment of co-pigmentation in red wines from the 1995 harvest. *In*: Forty-seventh Annual Meeting of the American Society for Enology and Viticulture, Reno.

Oberholster, A., Elmendorf, B.L., Lerno, L.A., King, E.S., Heymann, H., Brenneman, C.E. and Boulton, R.B. (2015). Barrel maturation, oak alternatives and micro-oxygenation: Influence on red wine aging and quality. *Food Chemistry* 173: 1250-1258.

Panda, H. (2011). The Complete Book on Wine Production. Niir Project Consultancy Services, Kamla Nagar, Delhi.

Pérez-Magariño, S. and González-San José, M.L. (2004). Evolution of flavanols, anthocyanins, and their derivatives during the aging of red wines elaborated from grapes harvested at different stages of ripening. *Journal of Agricultural and Food Chemistry* 52: 1181-1189.

Pérez-Magariño, S. and González-San José, M.L. (2006). Polyphenols and colour variability of red wines made from grapes harvested at different ripeness grade. *Food Chemistry* 96: 197-208.

Pérez-Magariño, S., Ortega-Heras, M., Cano-Mozo, E. and Gonzalez-Sanjose, M.L. (2009). The influence of oak wood chips, micro-oxygenation treatment, and grape variety on colour and anthocyanin and phenolic composition of red wines. *Journal of Food Composition and Analysis* 22: 204-211.

Perez-Prieto, L.J., la Hera-Orts, D., Luisa, M., López-Roca, J.M., Fernández, J.I. and Gómez-Plaza, E. (2003). Oak-matured wines: Influence of the characteristics of the barrel on wine colour and sensory characteristics. *Journal of the Science of Food and Agriculture* 83: 1445-1450.

Picariello, L., Gambuti, A., Picariello, B. and Moio, L. (2017). Evolution of pigments, tannins and acetaldehyde during forced oxidation of red wine: Effect of tannins addition. LWT – *Food Science and Technology* 77: 370-375.

Prida, A. and Chatonnet, P. (2010). Impact of oak-derived compounds on the olfactory perception of barrel-aged wines. *American Journal of Enology and Viticulture* 61: 408-413.

Puértolas, E., López, N., Condón, S., Álvarez, I. and Raso, J. (2010a). Potential applications of PEF to improve red wine quality. *Trends in Food Science & Technology* 21: 247-255.

Puértolas, E., Saldaña, G., Alvarez, I. and Raso, J. (2010b). Effect of pulsed electric field processing of red grapes on wine chromatic and phenolic characteristics during aging in oak barrels. *Journal of Agricultural and Food Chemistry* 58: 2351-2357.

Recamales, A.F., Gallo, V., Hernanz, D., González-Miret, M.L. and Heredia, F.J. (2011). Effect of time and storage conditions on major volatile compounds of Zalema white wine. *Journal of Food Quality* 34: 100-110.

Recamales, Á.F., Sayago, A., González-Miret, M.L. and Hernanz, D. (2006). The effect of time and storage conditions on the phenolic composition and colour of white wine. *Food Research International* 39: 220-229.

Reynolds, A.G. (Ed.) (2010). Managing Wine Quality: Viticulture and Wine Quality. Elsevier, Boca Raton, FL.

Ribâereau-Gayon, P., Glories, Y. and Maujean, A. (2006). Handbook of Enology: The Chemistry of Wine: Stabilisation and Treatments. John Wiley & Sons, Chichester.

Ribéreau-Gayon, P., Dubourdieu, D., Donèche, B. and Lonvaud, A. (Eds.) (2006). Handbook of Enology: The Microbiology of Wine and Vinifications, vol. 1. John Wiley & Sons, Chichester.

Riou, V., Vernhet, A., Doco, T. and Moutounet, M. (2002). Aggregation of grape seed tannins in model wine – Effect of wine polysaccharides. *Food Hydrocolloids* 16: 17-23.

Robinson, J. and Harding, J. (Eds.) (2015). The Oxford Companion to Wine. The Oxford University Press, Oxford.

Sikorski, Z.E. (Ed.) (2006). Chemical and Functional Properties of Food Components. CRC Press, Boca Raton, FL.

Sims, C.A. and Bates, R.P. (1994). Effects of skin fermentation time on the phenols, anthocyanins, ellagic acid sediment, and sensory characteristics of a red *Vitis rotundifolia* wine. *American Journal of Enology and Viticulture* 45: 56-62.

Somers, T.C. and Evans M.E. (1986). Evolution of red wines I. Ambient influences on colour composition during early maturation. *Vitis* 25: 31-39.

Somers, T. and Pocock, K. (1990). Evolution of red wines. III. Promotion of the maturation phase. *Vitis* 29: 109-121.

Stevenson, T. (1997). The New Sotheby's Wine Encyclopedia. Dorling Kindersley, New York.

Styger, G., Prior, B. and Bauer, F.F. (2011). Wine flavour and aroma. *Journal of Industrial Microbiology & Biotechnology* 38: 1145.

Suárez, R., Suárez-Lepe, J.A., Morata, A. and Calderón, F. (2007). The production of ethylphenols in wine by yeasts of the genera *Brettanomyces* and *Dekkera*: A review. *Food Chemistry* 102: 10-21.

Sun, X., Li, L., Ma, T., Zhao, F., Yu, D., Huang, W. and Zhan, J. (2016). High hydrostatic pressure treatment: An artificial accelerating aging method which did not change the region and variety non-colored phenolic characteristic of red wine. *Innovative Food Science & Emerging Technologies* 33: 123-134.

Sun, X., Li, L., Ma, T., Zhao, F., Yu, D., Huang, W. and Zhan, J. (2016). High hydrostatic pressure treatment: An artificial accelerating aging method which did not change the region and variety non-coloured phenolic characteristic of red wine. *Innovative Food Science & Emerging Technologies* 33: 123-134.

Talele, S. and Benseman, M. (2013). Wine maturation using high electric field. *In*: T. Sobh and K. Elleithy (Eds.). Emerging Trends in Computing, Informatics, Systems Sciences and Engineering. Springer, New York.

Tao, Y., García, J.F. and Sun, D.W. (2014). Advances in wine aging technologies for enhancing wine quality and accelerating wine aging process. *Critical Reviews in Food Science and Nutrition* 54: 817-835.

Tao, Y., Sun, D.W., Górecki, A., Błaszczak, W., Lamparski, G., Amarowicz, R., Fornal, J. and Jeliński, T. (2012). Effects of high hydrostatic pressure processing on the physicochemical and sensorial properties of a red wine. *Innovative Food Science & Emerging Technologies* 16: 409-416.

Tomašević, M., Gracin, L., Ćurko, N. and Ganić, K.K. (2017). Impact of pre-fermentative maceration and yeast strain along with glutathione and $SO_2$ additions on the aroma of *Vitis vinifera* L. Pošip wine and its evaluation during bottle aging. LWT – *Food Science and Technology* 81: 67-76.

Venturi, F., Sanmartin, C., Taglieri, I., Xiaoguo, Y., Quartacci, M.F., Sgherri, C., Andrich, G. and Zinnai, A. (2017). A kinetic approach to describe the time evolution of red wine as a function of packaging conditions adopted: Influence of closure and storage position. *Food Packaging and Shelf-Life* 13: 44-48.

Vilanova, M. and Sieiro, C. (2006). Contribution by *Saccharomyces cerevisiae* yeast to fermentative flavour compounds in wines from cv. *Albariño*. *Journal of Industrial Microbiology and Biotechnology* 33: 929-933.

Waterhouse, A.L. and Ebeler, S.E. (Eds.) (1998). Chemistry of Wine Flavour. ACS Symposium Series. American Chemical Society, Washington, DC.

Waterhouse, A.L., Sacks, G.L. and Jeffery, D.W. (2016). Understanding Wine Chemistry. John Wiley & Sons, Chichester.

Wetherill, J.M. (1965). Gamma Irradiation of Food. *Canadian Journal of Public Health/Revue Canadienne de Sante'e Publique* 56: 521-524.

Zeng, X.A., Yu, S.J., Zhang, L. and Chen, X.D. (2008). The effects of AC electric field on wine maturation. *Innovative Food Science & Emerging Technologies* 9: 463-468.

Zhang, Q.A., Shen, Y., Fan, X.H. and García Martín, J.F. (2016). Preliminary study of the effect of ultrasound on physicochemical properties of red wine. *CyTA – Journal of Food* 14: 55-64.

Zheng, X., Zhang, M., Fang, Z. and Liu, Y. (2014). Effects of low frequency ultrasonic treatment on the maturation of steeped greengage wine. *Food Chemistry* 162: 264-269.

Zhu, S.M., Xu, M.L., Ramaswamy, H.S., Yang, M.Y. and Yu, Y. (2016). Effect of high pressure treatment on the aging characteristics of Chinese liquor as evaluated by electronic nose and chemical analysis. *Scientific Reports* 6: 30273.

# Section 3
## Applied Aspects of Winemaking
### (A) Production of Wine and Brandy

# 21 Technology of Winemaking

**V.K. Joshi[1], Vikas Kumar[2*] and Jaspreet Kaur[2]**

[1] Department of Food Science and Technology, Dr Y.S. Parmar University of Horticulture & Forestry, Nauni, Solan – 173230, HP

[2] Food Technology and Nutrition, School of Agriculture, Lovely Professional University, Phagwara, Punjab – 144411, India

## 1. Introduction

No other beverage is discussed, adored or criticised in the same way as wine. To a few, it is something to be selected with the greatest care, laid down until optimum maturity, carefully prepared for serving, ritually tasted in the company of like-minded people and then, analysed in the manner of both the forensic scientist and literary critic. To many, it is simply the bottle bought from the supermarket according to the offer of the moment, drunk and perhaps enjoyed on the same day as purchased. To those favoured with living in wine-producing regions, it is often the beverage purchased from the local producers' co-operative from a dispenser resembling a petrol pump, taken home in a five or ten litre container and drunk with every meal. There is a wonderful diversity in the styles and quality of wines produced throughout the world, promoting discussion and disagreement amongst wine lovers (Grainger and Tattersall, 2005).

Wine has been extolled as a therapeutic agent. It is an important adjunct to the human diet, having polyphenols and other bioactive compounds that have antioxidant activities. In addition, the compounds bonded to insoluble plant compounds are released into the aqueous ethanolic solution during the winemaking process, which makes them more biologically available for absorption during consumption (Shahidi, 2009). Wines, because they are not distilled, have more nutrients, such as vitamins, minerals and sugars than distilled beverages, like brandy (Joshi *et al.*, 1999a), especially the polyphenolic compounds that act as antioxidants and antimicrobials. Moderate alcohol and/or wine consumption protects against the incidence of many diseases of modern society, like cardiovascular diseases, dietary cancers, ischemic stroke, peripheral vascular disease, diabetes, hypertension, peptic ulcers, kidney stones and macular degeneration, in addition to stimulating resistance to infection and retention of bone density (Jindal, 1990; Joshi *et al.*, 1999a; Stockley, 2011). With respect to their therapeutic value, the wines from non-grape fruits do not lag behind the grape wine.

The process of winemaking is multidisciplinary in its approach and nearly all the physical, chemical and biological sciences contribute to its production. Grapes are the main source of raw material for the production of wine by fermentation the world over, though the percentage contribution of grapes to total fruit production is only 15.53 per cent. The top six fruit producers, in declining order of importance, are China, India, Brazil, United States, Italy and Mexico. China, India and Brazil account for almost 30 per cent of the world's fruit supply. In the Southern Hemisphere, Chile, South Africa and New Zealand have become major suppliers to the international trade of fresh fruit commodities. Continued estimates of future growth for the wine industry in India and a growing middle class, there are considerable possibilities for opportunity in this market. With a rapidly growing export sector, expanding domestic consumer market and increasing industry support in major wine-producing states, the Indian wine industry has the potential to be a global market competitor (Grace, 2015). The amplified production of fruits leads to larger variety and availability of fruit-derived products, which could satisfy the global market and meet various requirements of consumers. Growth of the fruit processing industry is, thus, of utmost significance and production of wine is an integral component of this industry. Wine production, as an industry, is one of the options for value addition and waste minimisation including utilization of non-grape fruits (Joshi *et al.*, 2011). In view of the present scenario, unless the processing industry is linked with the horticultural industry, it is unlikely to achieve any worthwhile results, either for the farmers or for the consumers. A great advantage of production of fruit wines is that there are virtually no differences in the manufacturing plants required for the production of non-grape wines and grape wines, except for minor modifications

*Corresponding author: vkchoprafst@rediffmail.com

and the manufacturer can make use of the same facility. As a result, in Europe, one can find a significant production of non-grape fermentation products, where the United Kingdom and Germany represent attractive markets. The United Kingdom has a long tradition of fruit and other non-grape wines and is one of Europe's largest markets for fruit wine, with an annual production of 40-50 million litres a year, whereas the export of non-grape wine to other countries (mainly the United States) from Canada was recorded to be higher in comparison to grape wines. Further, development of global fruit-wine manufacturing is aligned with the current achievements in agriculture and global transportation, making raw materials available year-round, fresh, frozen or as a concentrate. Other reasons in favour of fruit wine production are based on the cost of fruit-wine production, which can be lower and the manufacturing process can be comparable with the grape winemaking. The higher space efficiency and higher potential profit can be achieved because of the consumer demand for such products (Kosseva *et al.*, 2017).

## 2. Basics of Wine Production

### 2.1. Grape Varieties

The grape variety or blend of grape varieties, from which a wine is made, is a vital factor in determining the design and style of the wine. However, it is not the only factor, although many a Chardonnay or Cabernet Sauvignon drinker believes otherwise. Wines made from a single variety are referred to as varietals. The name of the variety may or may not be stated on the label. Many top quality wines are made from a blend of two or more varieties, with each variety helping to make a harmonious and complex blend. Examples of well-known wines made from a blend of varieties include most red Bordeaux, which are usually made from two to five different varieties, and red Châteauneuf-du-Pape where up to 13 can be used. There are thousands of different grape varieties; the names of some for example Chardonnay, is very well known. Others are largely unknown. Some varieties are truly international, being planted in many parts of the world, others are found in just one country, or even in just one region within a country. A few varieties of grapes used for winemaking along with their characteristics are listed in Table 1. Physico-chemical characteristics of different fruit wines are given in Table 2.

**Table 1.** Varieties of Grapes Used for Winemaking along with Their Characteristics

| *Grape variety* | *Key points* |
|---|---|
| **Red Cultivars** | |
| • Barbera | Most extensively cultivated variety in Piedmont.<br>Is moderately high yielding, producing fruit.<br>Is intensely coloured, high in acidity and moderate in tannin content.<br>Is used to produce a distinctive, fruity wine and is often blended along with other cultivars for the sake of adding acidity and hence, high in pH. |
| • Cabernet Sauvignon | It is the best known variety which is small, acidic, seedy, tough and highly pigmented.<br>Is used for the production of red wine (Bordeaux) in Europe.<br>Is often combined with various wines produced from other grape varieties. |
| • Merlot | It has a higher tendency to undergo maturation than Cabernet Sauvignon and is therefore, used as a substitute of the same in wine production. |
| • Dolcetto | It is used for the production of light wine with bright colour and mild distinct odour. |
| • Gamay noir à jus blanć | This variety of grapes generally produces a light red wine after being crushed followed by fermentation.<br>A distinct fruity wine is also produced from this variety after being processed by carbonic maceration.<br>It is vulnerable to fungal grapevine diseases. |
| • Garnacha | This particular variety is used to produce rose wine or fortified wines.<br>It is also blended along with other varieties in order to speed up the maturation process. |

(*Contd.*)

| | |
|---|---|
| • Graciano | This variety is resistant to fungal diseases and also possesses good acidity and is therefore an important part of several Rioja wines. |
| • Nebbiolo | It is used for the production of red wines in north-western Italy.<br>It yields a wine with high amount of tannin and acidity with traditional vinification. |
| • Pinot noir | This variety of red grape is used for the production of flavourful wines, rose and sparkling wines.<br>Under optimal processing conditions, it yields distinctive aromatic wine while it yields non-distinctive wines under normal conditions. |
| • Sangiovese | It is used for the production of aromatic and distinctive light- to full-bodied wines as well as several red wines in Italy. |
| • Syrah | This variety is used to produce deep red-coloured and flavourful wines with a peppery finish. |
| • Tempranillo | It is one of the finest red grape-variety in Spain.<br>It produces a subtle wine under optimal conditions and ages well. |
| • Touriga National | It is commonly found in Portugese.<br>It is used to yield deep-coloured and flavourful red table wines. |
| • Zinfandel | Is commonly found in California.<br>Is used for the production of several flavourful wines ranging from ports to light blush wines which possess raspberry fragrance and rich berry flavours. |
| **White Cultivars** | |
| • Cheninblanc | It is extensively grown in France, Australia, California and South Africa.<br>Is used for the production of mild and fragrant wines including sweet and dry table wines as well as sparkling wines. |
| • Chardonnay | Is commonly grown white French cultivar.<br>Owing to its appealing fruity odour and its ability to grow well in wine-producing regions, it is used for the production of fine table wines and sparkling wines (Champagne). |
| • Ehrenfelser | Is commonly grown in Germany and Canada. |
| • Miiler-Thuragau | Is extensively grown in cold regions of Europe and New Zealand.<br>Is used to yield light wines with mild acidity and fruity odour. |
| • Mucatblanc | It is generally grown throughout the world.<br>It is used for the production of dessert wines owing to the characteristic features of this cultivar such as intense flavour and slight bitterness.<br>It is also used for the production of sparkling wines. |
| • Pinot gris and Pinot blanc | Both of these varieties extensively used for producing dry, fragrant, botrytised and sparkling wines. |
| • Riesling | It is a widely grown variety in Germany, California and Australia.<br>It is used for yielding flavourful, fragrant and well-aged wines varying from dry to sweet. |
| • Sèmillon | It is used to produce dry wine with distinctive flavour which develops on aging. |
| • Viura | It is chief white variety in Rioja.<br>It yields fresh wine with aromatic lemon flavour. |

*Source*: Jackson, 2008.

## 2.2. Composition of Grapes

Grapes are the principal fruit used in the preparation of various varieties of wine and the winemaker is directly concerned about the composition of berry at harvest. Grape berry contains a number of components, like water and other inorganic substances, carbohydrates, acids, phenolics, nitrogenous components, terpenoids, fats, volatile compounds, odourants, flavour compounds and vitamins which

**Table 2.** Physico-chemical Characteristics of Different Fruit Wines

| *Characteristics* | *Range* | | | | | | | | | | | |
|---|---|---|---|---|---|---|---|---|---|---|---|---|
| | *Dry white* | *Dry red* | *Grape wine* | *Orange wine* | *Pumpkin wine* | *Apple wine* | *Peach wine* | *Plum wine* | *Wild apricot wine* | *Mango wine* | *Strawberry wine* | *Banana wine* |
| Total soluble solids (°B) | - | - | - | - | 7.8 | 4.6-7.5 | 7.6-9.1 | 8.0-12.0 | 6.8 | - | 8.1-9.7 | 4.8 |
| Titratable Acidity (% as citric acid) | 0.586 | 0.649 | - | - | 0.36 | 0.37-0.41 | 0.61-0.80 | 0.62-0.68 | 0.75 | 0.6-0.8 | 0.63-0.73 | 0.85 |
| Volatile acidityic acid (% Acet) | 0.101 | 0.128 | - | 0.18-0.32 | 0.03 | 0.021-0.105 | 0.020-0.029 | 0.028-0.040 | 0.08 | 0.01-0.2 | 0.025-0.032 | 0.220 |
| Ethanol (% V/V) | 12.4 | 12.6 | 11.2 | 10.4-12.6 | 11.37 | 10.50-12.8 | 10.6-11.6 | 8.5-11.0 | 10.65 | 7-8.5 | 9.2-11.5 | 5.0 |
| Total esters (mg/litre) | - | - | - | - | 101 | 76-80 | 90.9-101.5 | 104-109 | - | 10-30 | 78.3-102.4 | - |
| Total phenols (mg/litre) | - | - | 0.3 | - | 899.1 | 124.50 | 206-278 | - | 240 | - | 126.8-144.7 | - |
| Sodium (mg /litre) | - | - | 51 | - | - | 18 | - | 20 | 43 | - | - | - |
| Potassium (mg/litre) | - | - | 803 | - | - | 1044 | - | 1008 | 2602 | - | - | - |
| Calcium (mg/litre) | - | - | 106 | - | - | 11 | - | 18 | 25 | - | - | - |
| Magnesium (mg/litre) | - | - | 88 | - | - | 144 | - | 82 | 94 | - | - | - |
| Copper (mg/litre) | - | - | 3.0 | - | - | 3.68 | - | 12.73 | 5.97 | - | - | - |
| Iron (mg/litre) | - | - | 0.13 | - | - | 0.21 | - | 0.20 | 0.50 | - | - | - |
| Manganese (mg/litre) | - | - | 0.66 | - | - | 0.76 | - | 1.04 | 2.69 | - | - | - |
| Zinc (mg/litre) | - | - | 0.70 | - | - | 0.84 | - | 0.95 | 0.99 | - | - | - |
| Reducing sugars (mg/litre) | 0.134 | 0.146 | - | - | 0.18 | - | - | - | - | - | 0.124-0.135 | - |
| Total sugars (mg/litre) | - | - | 0.3 | 3.8-70 | - | - | - | - | - | - | 0.6-1.7 | - |
| Glycerol (mg/litre) | 0.7019 | 0.6355 | - | - | 1.81 | - | - | - | - | - | - | - |

*Source*: Amerine *et al*., 1980; Joshi and Bhutani, 1990; Vyas and Joshi, 1982; Joshi *et al*., 1999c; Yang and Wiegand, 1949; Swami *et al*., 2014; Kosseva *et al*., 2017

are passed on to the resulting wine after the fermentation of grapes (Table 3). The major constituents of grape must and wine are several. There are different forms of phenols found in grapes, including forms of phenic acid, phenylic acid and oxybenzene. Phenolic acids in grape berries are located primarily in the skin and in grape pulp, where they are present at much lower concentrations than anthocyanins. The entire family of phenols, phenolics and tannins can be referred to as polyphenols. The chemistry of anthocyanins is complex as the pH of the medium affects the colour intensity. Phenolic compounds also contribute to sensory characteristics, particularly colour, astringency and bitterness, and are also involved in biochemical and pharmacological effects, including antimicrobial, anticarcinogenic and antioxidant properties.

**Table 3.** Approximate Composition of Grapes and Wine

| *Component/compound* | *% in grapes* | *% in wine* |
|---|---|---|
| Water | 75.0 | 86.0 |
| Sugars (fructose, glucose with minor levels of sucrose) | 22.0 | 0.3 |
| Alcohols (ethanol with trace levels of terpenes glycerol, higher alcohols) | 0.1 | 11.2 |
| Organic acids (tartaric, malic, with minor levels of lactic, succinic, oxalic acids etc.) | 0.9 | 0.6 |
| Minerals (potassium, calcium, with minor levels of sodium, magnesium, iron etc.) | 0.5 | 0.5 |
| Phenols (flavonoids such as colour pigments along with non flavonoids such as cinnamic acid and vanillin) | 0.3 | 0.3 |
| Nitrogenous compounds (protein, amino acids, humin, amides, ammonia etc.) | 0.2 | 0.1 |
| Flavour compounds (esters such as ethyl caproate, ethyl butyrate etc.) | Trace | Trace |

*Source*: Soni *et al*., 2011

## 2.3. Microorganisms Associated with Wine Fermentation

Various microorganisms are associated with winemaking and these are listed in Table 4.

## 2.4. Process of Winemaking

### *2.4.1. Fermentation*

The winemaking process includes two main steps, viz., alcoholic fermentation by yeasts (mainly *Saccharomyces cerevisiae* but include non-*Saccharomyces* yeasts also) followed by malo-lactic fermentation by lactic acid bacteria. Both types of microorganisms are present on grapes and on cellar equipment. Yeasts are, however, better adapted to grow in the grape must than lactic acid bacteria; that is why the alcoholic fermentation starts quickly. Throughout the alcoholic fermentation, a natural selection occurs and finally, *O. oeni* dominates due to interactions between yeasts and bacteria and between bacteria themselves. After bacterial growth, when the population is over $10^6$ CFU/ml, malo-lactic transformation is the obvious change in wine composition.

### *2.4.2. Transformation During Fermentation*

All these transformations greatly influence the sensory and hygienic quality of wine. Malic acid transformation is desirable as it results into deacidification of wine to make it palatable. Lactic acid is a product of malo-lactic acid fermentation in which malic acid is transformed into lactic acid, diacetyl and carbon dioxide. Citric acid is found in grapes only in trace amounts, but, is an important additive used by the winemakers. Grape contains various carbohydrates in the form of reducing and non-reducing sugars. Grape roots store sugars in the form of sucrose produced from photosynthesis. As maturation of the plant progresses, the sucrose is typically inverted to 1:1 ratio of D-glucose (dextrose) and D-fructose (levulose) as it is translocated to the grape berries. Fully ripened grapes may contain up to 1 per cent of sucrose which is inverted later by yeast-synthesised invertase during fermentation. Levulose is nearly twice as sweet as dextrose and, thus, for making sweet table wines, it would obviously be desirable to have high

**Table 4.** Microorganisms associated with wine making

| *Type of microorganism* | *Type of fermentation* | *Species involved* | *Benefits/Adverse effects* |
|---|---|---|---|
| FUNGI | NA | *Botrytis cinerea*<br>*Penicillium* spp.<br>*Aspergillus* spp. | It is considered as useful for the production of Botrytised wines.<br>Both are involved in the fruit spoilage<br>Besides, produces toxin |
| YEAST<br>*Saccharomyces cerevisiae* strains | Alcoholic fermentation | *Saccharomyces cerevisiae* var. *illipsoideus*<br>*Saccharomyces beticus*<br>*Saccharomyces Bayanus* | Conducts Alcoholic Fermentation leads to the production of alcohol in wines<br>Produces by-products especially flavor compounds<br>Produces fruit wines having desirable taste, colour and flavour. |
| YEASTS<br>Non-*Sacharomyces* spp | Mixed fermentation | Non-*Saccahromyces* Yeasts, *Candida*<br>Pichia | Increase titrable acidity, lower pH and inhibit the growth of spoilage microbes<br>Spoilage of wine<br>Imparts medicinal, phenolics and earthy aroma to wines. |
| | | *Zygosaccharomyces bacilli*<br>*Brettanomyces* sp.<br>*Schizosaccharomyces pombe*<br>*Torulaspora delbrueckii* | Removes fermentable source of carbon from wine and thus provides microbial stability from microorganisms.<br>Results into excessive volatile acidity, mousy taint and rancid isovaleric acid.<br>Produce higher amount of esters, estery taints, spoilage of wine.<br>Useful in deacidification of high acid must/wines |
| | | *Metschnikowia pulcherrima* | Sequential inoculation with *T. delbrueckii* and *S. cerevisiae* increases aroma intensity, including 'ripe red fruit' aroma, decreased intensity for vegetal attributes. |
| | | *Candida zemplinina/Candida* stellata | Wines of the grape obtained by sequential fermentation with *C. pulcherrima* and *S. cerevisiae* had higher quality scores than control wines (obtained by fermentation with *S. cerevisiae)*<br>Sequential inoculation with *C. zemplinina* and *S. cerevisiae* wine also showed a high 'ethyl acetate' aroma, had the highest concentrations of glycerol and succinic acid, and a lower concentration of ethanol. |

(*Contd.*)

| | | | |
|---|---|---|---|
| BACTERIA | | | |
| Lactic acid bacteria (LAB) | Malo-lactic fermentation | *Oenococcus oeni*, *Lactobacillus species* and *Pediococcus* species | Reduce acidity by degrading malic acid into lactic.<br>Increases pH of wine.<br>Impact the sensory characteristics of wine and imparts buttery-characters to it.<br>Enhances fruity aroma as well as flavours.<br>All kinds of red wine, a range of white and sparkling wines are produces by malolactic fermentation.<br>Production of excessive amount of exopoly-saccharides results in viscous texture of wine |
| Acetic acid bacteria | Acetic acid fermentation | *Acetobacter* spp<br>*Glucobacter* spp.<br>*Bacillus* | Results into excessive volatile acidity of wines.<br>It imparts vinegar taint to the wine.<br>Spoilage |
| ACTINOMYCETES | NA | *Streptomyces spp.* | Imparts earthy, corky taints. |

*Source*: Amerine et al., 1980; Joshi, 1997; Joshi et al., 1999.

levulose containing varieties and to retain residual sugar using a yeast strain, which ferments fructose at a slower pace. About one-third of the sugars can be attributed to various polysaccharides, which are complex forms of sugar unfermentable by yeasts, but, may be the substrates for malo-lactic fermentation. Several other sugars, like arabinose, rhamnose, ribose, xylose, maltose, mannose, melibiose, raffinose and stachyose have also been identified in the grapes.

### 2.4.3. Additives

A variety of enzymes including tannase, invertase, pectinase, ascorbase, catalase, dehydratase, esterase and proteases, and polyphenol oxidase are found in musts. Enzymatic oxidation by polyphenol oxidase is important for many wines particularly when made from the moldy grapes but the activity is completely inhibited by $SO_2$. Centrifuging musts can reduce the enzyme activity, which is mostly localised in the skins. In the production of wines, a number of additives such as pectinolytic enzymes, sulphur dioxide, bacteriocins etc. are used (Table 5). Further, application of methods to control the microflora during wine making as described in Table 5, is made. The addition of enzyme is made for the clarification of wine. Their role in wine production is also summarised in Table 5. Grape composition affects the sensory characteristics of wine, including flavour, colour and foam capacity. The must quality has been found to determine the foam capacity of base wine, and it also indicated that the maturation index (between 4 and 5.5) of grape berries increases the wine foam capacity.

### 2.4.4. Basic Principles of Vinification

The sugars contained in the pulp of grapes are fructose and glucose. During fermentation, enzymes from yeast convert the sugars into ethyl alcohol and carbon dioxide in approximately equal proportions and heat is liberated as:

$$C_6H_{12}O_6 \xrightarrow{\text{Yeast}} 2CH_3CH_2OH + 2CO_2 + \text{Heat}$$

Additionally, tiny amounts of other products are formed during the fermentation process, including glycerol, succinic acid, butylene glycol, acetic acid, lactic acid and higher alcohols. The winemaker has to control the fermentation process, aiming for a wine that is flavoursome, balanced and in the style required. The business of winemaking is however fraught with potential problems, including stuck fermentations (the premature stopping of fermentation whilst the wine still contains unfermented sugars), acetic spoilage or oxidation.

The amount of sugar in must, and the reducing sugars (glucose and fructose) in fermenting wine, can be determined by using a density hydrometer or a refractometer. Throughout the vinification process, it is essential to maintain accurate records, including temperature and specific gravity readings and total soluble solids (TSS). There are important differences in the making of red, white and rosé wines. For red wine, the colour must be extracted from the skins. For white wines, however, some winemakers choose to have a limited skin contact between the juice and skins because it can add a degree of complexity. Of course, white wine can be made from black grapes, commonly practised in the Champagne region. Rosé wine is usually made from black grapes whose juice has been in contact with the skins for a limited amount of time, e.g. 12 to 18 hours. Consequently, a little colour is leached into the juice. Wine is usually fermented in vats, although barrels are sometimes used, particularly for white wines. Traditionally, wine was made in either shallow stone tubs called 'lagars', in which the grapes would be trodden, the gentlest of crushing, or in open wooden vats. Lagars are sometimes still used in Portugal by some producers of Port wines. Cuboid vats made of concrete (béton) or cement (ciment) became very popular with producers in the early to mid-twentieth century and many still regard these as excellent fermentation vessels.

### 2.4.5. Fixing Colour

Steps may be taken to ensure that a red wine will hold its colour. The use of inner staves in a vat of wine is one way of achieving this, the tannins released by the wood, bind the colour to the wine. Historically, after fermentation, red wine often have a very lengthy macerating time on the skins. During this time, tannins would be absorbed, which would help to fix the colour and gives the wines structure, but the resulting wine would often be very firm in the mouth. During the past 20 years, many producers have

**Table 5.** Role of Different Additives and Physical Methods Used to Control Microflora during Wine Fermentation

| *Controlling method/ additives* | *Action of controlling agent/additives* | *References* |
|---|---|---|
| • Lysozyme | Inhibits the growth of lactic acid bacteria (LAB). Inhibits the growth of *Lactobacillus* species without effecting alcoholic fermentation. Reduces volatile acidity. Aroma of wine remains unaffected. | Gerbaux *et al.*, 1997; Bartowsky, 2003 |
| | Leads to bacterial cell lysis and hence, disrupts cell wall synthesis. | Gerbaux *et al.*, 1997 Bartowsky *et al.*, 2004 Bartowsky, 2008 |
| • Bacteriocins | These include nisin, pediocin and plantaricin. Causes cell lysis. | Bartowsky, 2008 |
| | | Bruno *et al.*, 1992 |
| • Phenolic compounds | Possess antimicrobial activity. | Vaquero *et al.*, 2007 |
| • Dimethyl dicarbonate | Inactivates cellular enzymes and inhibits the growth of spoilage causing microbes. | Daudt and Ough, 1980 |
| | Reacts with amino groups on active sites of enzymes in an irreversible manner. | Bartowsky, 2008 |
| • Sulphur dioxide | Inhibits the growth of spoilage causing microorganisms. | Bartowsky, 2008 |
| | Acts as an antioxidant. | Romano and Suzzi, 1993 |
| • Ultrahigh-pressure treatment | Delays microbial spoilage. | Hite, 1899 |
| | Inactivates microbes and enzymes. Doesn't affect the flavour, aroma or nutrient profile of wine. | Tauscher, 1995; Bartowsky, 2008 |
| | Causes cytoplasmic membrane lysis and hence, shows antimicrobial property. | Hoover *et al.*, 1989 |
| • High-power ultrasound | Helps to collapse microbial cell membranes and hence, results into their inactivation. | Fellows, 2000; Butz and Tauscher, 2002; Piyasena *et al.*, 2003; Jiranek *et al.*, 2008; Bartowsky, 2008 |
| • Pulsed electric field technology | This technology is known for wine sterilization without effecting the quality of wine. | Bartowsky, 2008 |
| | Prevents the formation of volatile compounds in grape must. Results into breakdown of bacterial cell membranes. | Grade-Cerdan *et al.*, 2008 Bartowsky, 2008 |
| • Ultraviolet irradiation | Inhibits the growth of bacteria and fungi. | Bartowsky, 2008 |

shortened this period, especially for wines that are not destined for lengthy bottle ageing. Moreover, it is now generally, accepted that during fermentation, there is little chance of the wine oxidising and indeed oxygen is necessary to maintain healthy yeast colonies. After fermentation, wines may benefit from absorbing tiny amounts of oxygen, for this will help polymerise the polyphenols and help fix the colour of the wine. There are three methods by which wine may be deliberately oxygenated, depending upon the purpose.

### 2.4.6. *Hyper-oxygenation*

The process includes addition of large amounts of oxygen to white must in pre-fermentation stage. The benefits include a reduction in volatile acidity and acetaldehyde.

#### 2.4.7. Macro-oxygenation

This is adding large amounts of oxygen during the fermentation. It may be undertaken as 'rack and splash', in which the wine is aerated when returned to the vat, or can be incorporated during pump overs. Micro-oxygenation is the continuous addition of tiny amounts of oxygen to wine in order to improve aromas, structure and texture. It may be carried out before or after the malo-lactic fermentation, with the former perhaps giving the greatest benefits.

#### 2.4.8. Removal of Excess Alcohol

Reverse osmosis is one relatively inexpensive method of removing excess alcohol from wines. An alternative method is to draw off a proportion of an over-alcoholic wine and send this to a specialist facility that utilises a spinning cone column (SCC). This is a distillation column through which wine is passed twice. The first distillation extracts the volatile aromas and the second removes the alcohol. The aromas are then, returned to the 'wine', which is returned to the winery and blended back into the bulk (Grainger and Tattersall, 2005).

## 3. Technology of Wine Production

For the preparation of wine from grapes, the technology is well standardised and wine production is an established industry in the grape-producing countries of the world. In the subsequent section of the chapter, methods to produce red, white and other wines have been described. The techniques used for the production of fruit wines are basically similar to those for the production of wines made from grapes (Amerine *et al.*, 1980; Joshi, 2009). However, differences arise from two facts: (1) it is quite difficult to extract the sugar and other soluble materials from the pulp of some fruits compared to grapes, and (2) the juices obtained from most fruits are lower in sugar content and higher in acids (Vyas and Gandhi, 1972; Joshi *et al.*, 1999b; Amerine *et al.*, 1980; Joshi, 2009; Swami *et al.*, 2014). The higher acidity in some fruits makes it more difficult to prepare wine of acceptable quality (Joshi *et al.*, 2011e), whereas in the case of citrus fruits, when the juice is extracted by a rack and cloth press or screw expeller, it contains so much essential oil from the peel that fermentation is reduced drastically. As a solution to these problems, the use of specialised equipment to thoroughly chop or disintegrate the fruit, such as berries, followed by pressing to extract the juice from the pulp, solves the first problem. The fermentation of some fruits is very slow or may even terminate before completion because of lack of certain nitrogenous compounds or other yeast growth factors in some fruit juices, such as pear. The addition of nitrogen source to such must have solved this problem (Joshi *et al.*, 1990b, 1991a; Amerine *et al.*, 1980). The second problem is solved by the addition of water to dilute the excess acid and the addition of sugar to balance the sugar deficiency. For reduction of acidity, the use of deacidifying yeast, like *Schizosaccharomyces pombe* (Vyas and Joshi, 1988; Joshi *et al.*, 1991b) or the malo-lactic acid bacteria have been successful (Bartowsky, 2011). The basic technique for production of fruit wines as stated earlier is essentially the same as that from grapes, involving routine alcoholic fermentation of the juice or pulp but modifications with respect to the physico-chemical characteristics, depending upon the type of wine to be prepared and the fruit used need to be made. In citrus fruits, bitterness of the juice is another problem, in addition to darkening of the wine. The bitterness of the juice is also carried on in the wine (Amerine *et al.*, 1980; Joshi and Thakur, 1994; Joshi *et al.*, 2012a), so to overcome the same, debittering with the XAD-16 adsorption technique has been successfully applied using kinnow juice (Joshi *et al.*, 1997). Like the grape wine industry, the waste from fruit wine production is a useful source of several components, which could be a source of value added products (Joshi *et al.*, 2011d; Kosseva *et al.*, 2016).

### 3.1. White Wines

White wines are made from juice, free from peels and stems. The quality of wine depends upon the variety of the grapes, its maturity and health. The grapes are crushed after de-stemming and then, pressed to obtain the free run juice. It is important to treat the must immediately with $SO_2$ in order to inactivate the oxidative enzymes and to prevent the growth of undesirable microorganisms. Oxidative enzymes can also be inactivated by high temperature short-time heating (85°C) of the must. The clarification of the must before fermentation is very important and is done by centrifugation or filtration or sedimentation.

Insufficiently clarified must generally produce crude wines with grassy-tastes. The lack of maceration in white wine however, is not an absolute factor and in some cases even short maceration of the skin is carried out. The brix in white grape juice generally ranges between 14-23°B. To achieve 12 per cent (v/v) alcohol level in white wine after fermentation, TSS of the juice needs to be adjusted to 22.5°B. The freshly pressed must frequently exhibits a disharmonious sugar-acid ratio which must be regulated. The addition of sugar is called chaptalisation. When the grapes are excessively high in acidity, both sugar and water are added, referred to as amelioration. Though, both the adjustments have historically been controversial in fine wine circles. The new techniques, like reverse-osmosis, cryo-extraction and entropie concentration can be used to increase the sugar concentration without the addition of sugar. If the acidity level is undesirably low (>5g/L), acids such as tartaric or citric can be added. In case of high acid musts, deacidification can be done by using calcium carbonate or the calcium 'double salt' precipitation of malic and tartaric acid. The inoculation with pure wine yeast culture is preferred to eliminate the risk of contamination. The fermentation temperature for most of the white wines is in the range of 18-24°C. A maximum of aromatic substances, i.e. acetates of higher alcohols and esters of fatty acids are formed when the fermentation is conducted at 15-20°C. Fermentations conducted at higher temperature however, leads to a loss of aroma and wine quality. Different methods used for production of white wines of different styles are depicted in Fig. 1.

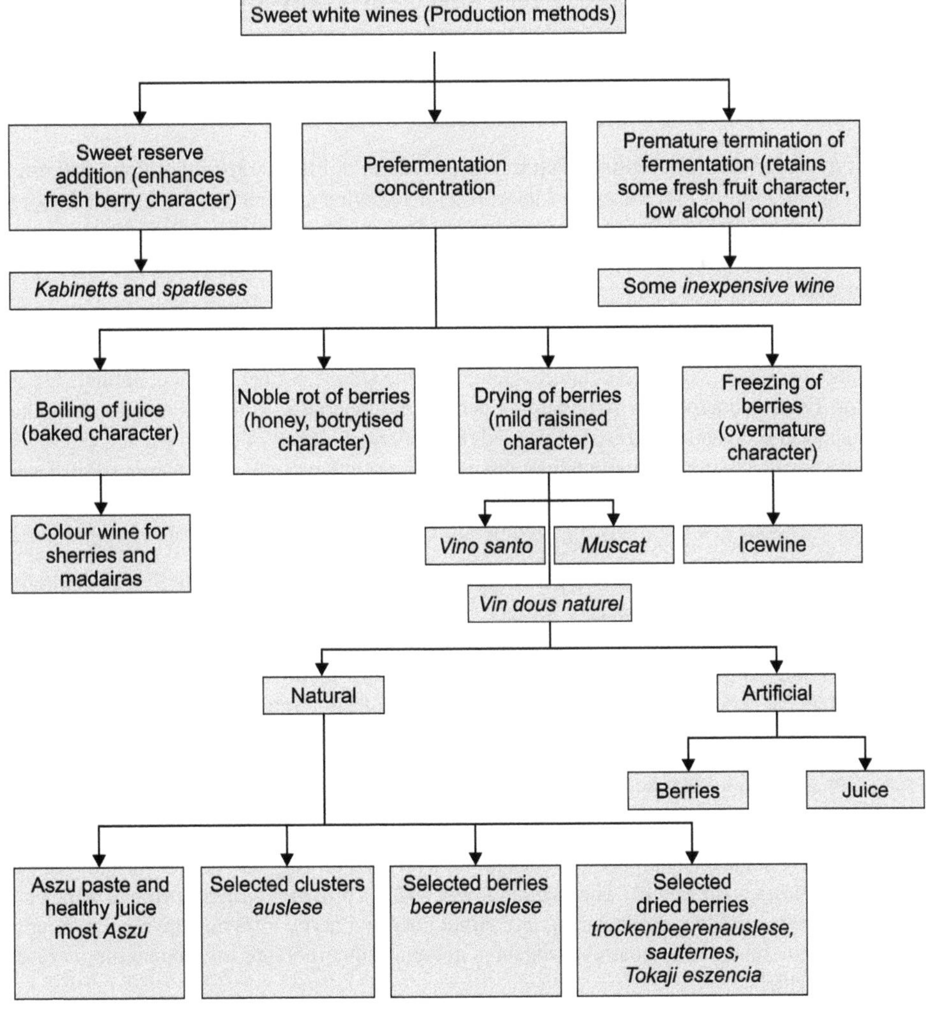


**Figure 1.** Classification of sweet wines based on production methods
*Source:* Based on Jackson, 2004

## 3.2. Red Wines

Red wine is a wine made from fermentation of macerated red or black grape juice along with their skins and seeds. The longer skin maceration time produce wines with higher colour intensity, phenolics compounds and higher sensory ratings. During storage, they also preserve their characteristics better than wines made with shorter maceration time. Red wine fermentation is always conducted at higher temperatures (24-27°C) than white wines. Red wines fermented at lower temperatures tend to be lighter in colour and display the fruitier range of esters. Co-pigmentation of anthocyanins accounts for 30-50 per cent of the colour in young wines and is primarily influenced by the levels of several specific, non-coloured phenolic components or co-factors. The formation of maximum colour by mild fermentation and then, a loss of some of it in the days and weeks following the end of fermentation, is a pattern that has been known for long. The loss has generally been attributed to a fall in the anthocyanin content, although the loss in colour is more than that of anthocyanin content, which is probably the result of break up of the co-pigmentation stacks formed earlier in fermentation. The presence of particulate matter in the must or residual cloudiness has a substantial effect on both the kinetics of fermentation and formation of by-products. The addition of montmorillonite clay bentonite to the fermentation medium has been recommended to facilitate even fermentation of juice low in solids and ensure rapid classification at the end of fermentation. Red winemaking differs from white winemaking, not only in terms of skin-juice contact and temperature, but also because some exposure to oxygen, barrel maturation or at least cask aging, etc. is also more common for red wines than white wines (Singh *et al.*, 2011). Rest of the procedure is similar to that of white wine.

### 3.2.1. Different Types of Maceration (Red Wine)

Red wine is a wine made from fermentation of macerated red or black grape juice along with their skins and seeds. The duration of maceration produces rose or red wine and this can be regulated accordingly. The longer skin maceration time produces wines with higher colour intensity, phenolic compounds and higher sensory ratings. During storage, they also preserve their characteristics better than wines made with shorter maceration time.

*Traditional maceration*: In traditional maceration, the crushed grapes are transferred to the fermentation containers and inoculated with yeast culture. Generally, the stainless steel or oak containers are used for fermentation. The maceration and fermentation occur simultaneously in this conventional method. The length of the maceration and fermentation depends entirely on the level of tannin and the extent of colour required in the final wine. Short fermentations are suitable for light wines, while longer times are used for wines destined with intense colour. Generally, the total fermentation of grape sugars is carried out for five to eight days. It is done at a relatively high temperature of 25-30°C, since it enhances colour extraction.

*Traditional French pigeage*: It is still used in more traditional wineries. Pigeage or punching the cap down consists of stomping the grapes with feet. Two wooden beams are placed across the top of the vat and workers jump on the cap to break it and force it down. Once the cap has softened, pigeage is continued by sitting on the edge of a fermenting vat with a pole having a round disk on the end. The disk is used to plunge the mass of grape skins below the surface of the fermenting juice.

*Submerged cap or heading down system*: In this system, the must is placed in a tank below a stainless steel screen or a slotted lid that keeps the skins below the surface of juice.

*Pumping over technique*: In the technique, the juice is pumped out of the bottom of the vat and sprayed on to the top of the skincap. The juice extracts colour as it falls through the cap. A variation of pumping over is called 'délestage' in French. The pumping over usually provides a predetermined juice volume to the cap which permeates the cap, displacing interstitial juice and partly lowering the cap temperature. The cooling of the juice by external heat exchangers is generally incorporated into the pump-over operation in the large fermenters.

*Carbonic maceration*: Another method of increasing red wine fruit flavour is the carbonic maceration and sometimes also called 'whole grape fermentation'. It is a very ancient and technically simple process. The pre-fermentation treatment involves holding the intact grapes in $CO_2$ environment at elevated temperature (20-35°C) for eight to 10 days. The anaerobic life of the intact grapes is the maceration phase of the process. This is an enzymatic process that occurs in whole grapes and is quite distinct from microbial fermentation. In this process, the intact grapes are allowed to respire and to have partial fermentation by glycolytic enzymes present in the grapes. This self-fermentation generates unique flavours associated with carbonic maceration.

*Thermovinification*: Thermovinification lines were developed with the goal of automating winemaking. The process is often employed with the grapes that are poor in pigmentation, caused by either very warm or very cool climate conditions. It is the process of heating of crushed must or whole grapes to 60-80°C for a period of 20-30 minutes. This process enhances the colour extraction from skins, inactivation of juice and mold-derived enzymes. The use of temperature above 60°C leads to a more complete but usually unacceptable level of phenol extraction, although such conditions are used in pigment recovery process. The method is usually integrated with traditional winemaking to increase the concentration of phenolic compounds, especially anthocyanins. This also leads to decrease in the time in winemaking process and labour cost.

*Other methods of maceration*: The use of pectinase-based enzyme preparations added to the must prior to inoculation to extract the colour has also been made. This causes a relatively extensive breakdown of berry and skin lamellae or cell walls to allow a greater extraction of phenols in the aqueous phase. The use of enzymes may be beneficial where wines have limited residence time in red wine fermenters, like rotary fermenters. The other method of maceration involves the freezing of whole berries at -4°C instead of crushing, a process called as freeze extraction or supra extraction. The ice formed causes the cells to rupture and the skins to split, so that when the berries are warmed to about 10°C and pressed, the extraction of phenols is increased. Flash expansion is the recent technology under investigation for colour extraction in red wines where the grapes are heated to 85°C for 5 minutes and then, subjected to flash expansion under a vacuum of 60 mbar. Here, some of the water in the cells will flash to the vapour phase at this temperature and pressure. In Fig. 2, different methods used to make red and white wines are depicted .

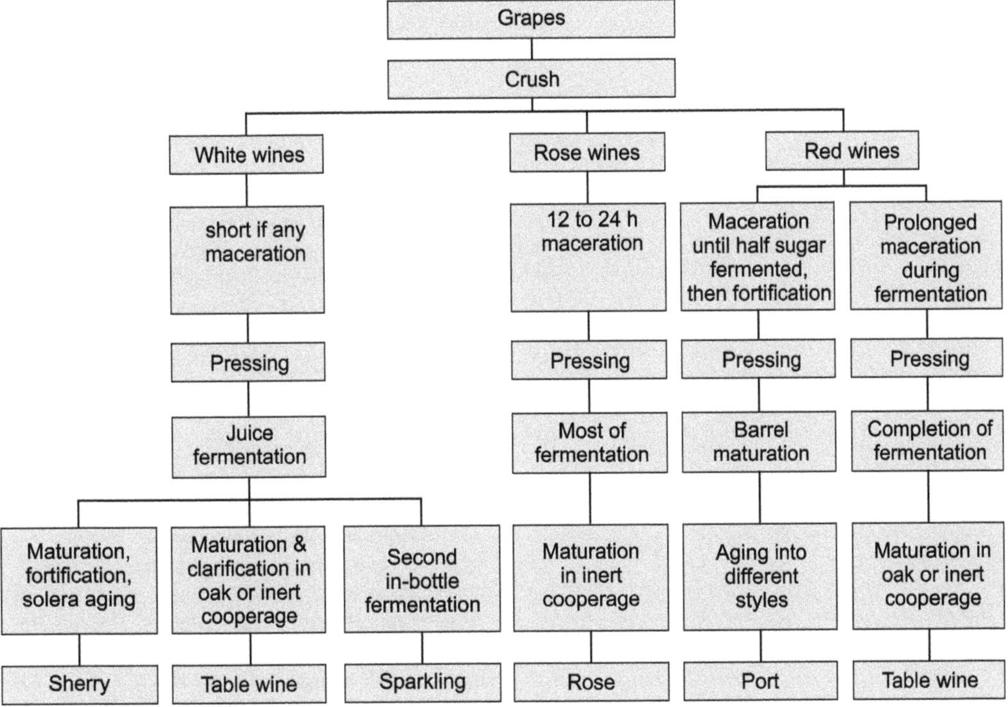


**Figure 2.** Flow chart of different steps for wine (white, red, rose and wines of other styles) production (Jackson, 2004)

Some of the new findings on wines including red and white are also summarised in Table 6.

**Table 6.** Some of the New Findings on Wines including Grape (Red and White)

| *Wine* | *Yeast* | *Key remarks* | *References* |
|---|---|---|---|
| Grape (*Vitis vinifera*) wine | *Saccharomyces cerevisiae* strain isolated from Palm wine | Fermentation of fruit must was carried out in the study.<br>pH of the fruit must was found to decrease after fermentation.<br>Volatile acidity was found to be increased.<br>Study revealed that during the fermentation of fruits, low pH inhibits the growth of spoilage microorganisms and also proves to be advantageous for the growth of desirable microorganisms. | Yabaya *et al.*, 2016 |
| Grape wine | *Saccharomyces cerevisiae* produced from rotten grapes | Fermentation of rotten undamaged grapes was carried out.<br>pH was found to be decreased whereas, titrable acidity tend to increase during the process of fermentation. | Guleria, 2014 |
| Red wine | *Schizosaccharomyces pombe* (*Sc. pombe*) | All types of fermentation of fruit (*Vitis vinifera*) must was carried out at a constant temperature of 25°C.<br>Fermentation was carried out with different strains of microorganisms including *Schiz. pombe* and *Saccharomyces cerevisiae*.<br>It was found that *Schiz. pombe* is a poor producer of alcohol during the production of wine.<br>Wine produced by fermentation with *Schiz. pombe* was found to be organoleptically more acceptable in terms of alcohol levels, total malic acid content, glycerol production and urea degradation; as compared to the wine produced with *Saccharomyces cerevisiae*. | Benito *et al.*, 2014 |
| Grape wine | *Schizosaccharomyces pombe* | This study aimed to prove the potential role of *Schiz. pombe* in the effective management of malic acid deacidification as well as reduction in the levels of biogenic amines and ethyl carbamate.<br>Wine produced using *Schiz. pombe* strain has an appropriate volatile aroma profile alongwith low levels of malic acid, acetic acid, ethyl carbamate and biogenic amines. | Benito *et al.*, 2016 |

### 3.3. Maturation of Wine

Maturation and/or aging denotes "the reactions and changes that occur after the first racking that lead to improvement at some stage rather than spoilage, but are separated from other post-fermentation processes, such as clarification, tartarate stabilisation and malo-lactic fermentation." Wine aging can be sub-categorised into two major stages, i.e. bulk storage and bottle storage or aging. During aging/maturation changes are caused by three types of reactions occurring simultaneously and continually in the barrel: i) extraction of complex wood constituents by the liquid, ii) oxidation of components originally in the liquid and of material extracted from the wood, iii) reactions between the various organic substances present in the liquid, leading to the formation of new congeners.

#### *3.3.1. Metal Cooperage*

Metal containers contribute iron to wine, ferrous or ferric ions which catalyse undesirable reactions in wine. Brass and copper contribute $Cu^{+}$ ions which cause unwanted quality characteristics in wine even at a low concentration and thus, leaving wooden and stainless steel cooperage for modern wine storage and maturation.

#### *3.3.2. Plastic Cooperage*

These types of containers are made up of different types of materials, such as fibre glass polystyrene tanks, light pass through tanks that can affect the wine detrimentally and may contribute monomers of plastic itself or extractable plasticisers.

#### *3.3.3. Wooden Cooperage*

Oak barrel is the oldest storage container, used for the maturation and aging of wine. In France, two species are used for the construction of oak barrels: *Quercus robur* and *Q. petraea*. Barrels used for the aging of wines and brandies are constructed from oak heartwood which are the most abundant constituents that are potentially extracted into the wine during aging. The distribution of extractives in the tissue and factor(s) that influence the permeability of wood are likely to affect the movement of these extractive liquids and gases through the staves. Moreover, anatomical features may correlate with chemical properties that affect maturation.

#### *3.3.4. Wood Segments for Enhanced Fermentation*

In order to achieve barrel-aged-style wines, the use of wood chips, beans, segments, etc. using stainless steel tanks has been suggested due to many advantages, such as lower long-term storage, greater protection against oxygen exposure, better utilisation of cellar space and smaller evaporative losses (Bhushan and Sharma, 2011).

### 3.4 Production of Fruit Wines

A number of fruits can be used to make wine called fruit wines. The method is similar to wine preparation from grapes but modifications with respect to amelioration i.e. correction of acidity and tannins etc. need to be made to make it a palatable wine (Joshi *et al*., 2017). Methods for production of different wines have been developed and described earlier (Kossesva *et al.*, 2017). Significant research findings made on wine production including different fruit wines are summarized in Table 7.

## 4. Technology of Sparkling Wine Production

Grape harvesting occurs earlier for sparkling wine production than is typical for still wines. This provides fruits higher in total acidity and lower in °Brix. The additional acidity provides the freshness typical of sparkling wines, whereas the lower sugar content generates wines with an alcohol content between 9-10.5 per cent. The reduced alcohol content facilitates initiation of the second, sparkle-producing fermentation. Fruit immaturity also reduces varietal aroma, thereby minimising interference with the subtle processing bouquet so essential to most the sparkling wines. The latter comes from prolonged exposure to lees

**Table 7.** Significant Findings on Wine Production including Different Fruit Wines

| *Research findings* | *References* |
|---|---|
| **A. Fruit wines** | |
| Wines, having more than 14 per cent alcohol, are prepared by fortification with wine spirit or alcohol and are called dessert wines. Vermouths with 15-20 per cent alcohol are produced by fortification with herbs or spice mixtures and these may be white or red and sweet or dry. Sparkling wine contains 2-6 per cent alcohol and excess of $CO_2$, giving an effervescence like a carbonated beverage. | Vyas and Chakravorthy, 1971 |
| **B. Screening of suitable variety** | |
| Thompson Seedless and Mendeline Anguine were suitable for the production of white wines and Beauty Seedless and Rubired for red wines. | Amda Dan *et al*., 1972 |
| Beauty Seedless and Carignane produced acceptable quality red wines in Haryana. | Vyas and Gandhi, 1972; Bardiya *et al*., 1980 |
| Early Muscat, Champion and Perlette were found suitable for white wine production in Haryana. | Vyas and Gandhi, 1972; Kundu *et al*., 1980 |
| Bangalore Blue, Bangalore Purple and Beauty Seedless used for quality red wine production and Anab-e-Shahi and Selection-7 for the production of low alcoholic beverages. | Sreekantiah and Johar, 1968; Subba Rao, 1972 |
| Screening of exotic grape varieties for winemaking revealed that Chenin Blanc and Bayan Sherei for white wine; Black Cornichon and Feteasca Niagra for red wine and Convent Large Black and Black Champa for dessert wines were suitable. | Negi *et al*., 1972; Suresh and Negi, 1975 |
| Arkavati variety for good quality dry white table wine, Arka Shyam for dry red table and Arka Shyam and Arka Kanchan for dessert wine were found suitable. | Suresh *et al*.,1985 |
| Suitability of eight peach cultivars (Sunhaven, Redhaven, Kateroo, J.H. Hale, Flavourcrest, July Elberta, Stark Early Giant and Rich-haven) for preparation of table wine revealed that all cultivars were suitable but Redhaven, Sunhaven, J.H. Hale, Flavourcrest and July Elberta were adjudged as better than others. | Joshi *et al*., 2005 |
| Studies revealed that a wine with organoleptic properties comparable to grape wine could be prepared from custard apple (*Anonas quamosa*). | Kotecha *et al*., 1995 |
| Mango varieties were screened and Fazri, Langra and Chausa produced good quality wines. Sweet wines from Dashehari had a characteristic fruity flavour. | Kulkarni *et al*., 1980 |
| Custard apple (*Anonas quamosa*) wine was prepared with different dilutions (1:2, 1:3 and 1:4 dilution with and without DAHP). Among all the dilutions 1:4 dilution with DAHP had higher value for physico-chemical characteristics, sensory scores and acceptability in comparison to other dilutions. | Kumar *et al*., 2011 |
| It was found that Banginapalli, Banglora and Alphonso varieties were most suitable for mango wine production. | Reddy and Reddy, 2005 |
| **C. Harvest maturity** | |
| Wines prepared from hot-season crop were better than those from cold season. The changes in carbohydrates, nitrogenous compounds, organic and inorganic constituents revealed distinct variations in Bangalore Blue, Black Champa, Malvasia Bianca and Thompson Seedless due to maturity and season. | Chikkasubbana, 1982 |
| Grapes harvested at early or mid-mature stage produced good quality wines from Thompson Seedless, Arkavati, Arka Shyam and Arka Kanchan. | Suresh and Ethiraj, 1987 |
| Changes in amino acid in the must and wine revealed that proline was present in largest amounts, both in grapes and wines due to failure of *Saccharomyces cerevisiae* to metabolise this amino acid during fermentation. | Ethiraj and Suresh, 1982 |

(*Contd.*)

| | |
|---|---|
| **D. Improving composition and sensory quality** | |
| A blend proportion in 2:1 and 3:1 ratios of light-coloured Gulabi with deep coloured Baily Alicante or Rubired grapes produced wines with desirable colour, whereas, commercial varieties like Thompson Seedless and Anab-e-Shahi were blended with Rubired in different proportions for the production of table and dessert wines to impart sufficient intensity of colour. | Suresh *et al.*, 1983 |
| The flavour profiling characterised the strawberry wines of different treatments successfully. Those fermented on the skin were specific for intensity of phenolic, higher alcoholic and strawberry-like, while control wine was peculiar for vegetative, alcoholic and sweety notes. The wines fermented with carbonic maceration were characteristic for all-spices, bitterness, astringency and sour descriptors. However, among cultivars 'Camarosa' and 'Chandler' had significantly higher flavour intensities than 'Doughlas'. | Sharma and Joshi, 2004 |
| Application of Principal Component Analysis (PCA) to the mean of flavour scores generated from flavour profiling, weakly separated and characterised the apple wines fermented by different sources of fermentation but did not differentiate the wines fermented with or without nitrogen source. The descriptors developed could characterise apple wine of different quality attributes. | Joshi *et al.*, 2002 |
| Production of cider and brandy from Indian apples has been reported. | Patel *et al.*, 1977 |
| Cider containing 4.5-8.23 per cent alcohol was produced from apple varieties: Red Delicious, Rich-a-Red, Kesari, Golden Delicious and Maharaji. Further, Golden Delicious and Red Delicious apples were suitable for cider production | Singh *et al.*, 1976 |
| The studies of cider during storage revealed that alcohol content continued to increase but tannins and reducing sugars decreased during storage of cider at two different temperatures. | Singh *et al.*, 1976 |
| Studies were also made on the production and acceptability of cider from Himachal Pradesh apple. | Rana *et al.*, 1986 |
| Preparation of cider from scabbed apple fruit. The processed apple juice from scabbed fruits was found to be free from patulin, aflatoxin and microfloras. The cider prepared from scabbed fruit juice, too, was comparable in all the respects, including the fermentation behaviour with that of normal juice. | Azad *et al.*, 1987 |
| Studies on cider preparation from apple juice concentrate revealed that fortification with DAHP as nitrogen source was essential for rapid fermentation and the must prepared from diluting the concentrate fermented faster than the amelioration done with sugar, while prepared cider showed a high acceptability. | Joshi *et al.*, 1991 |
| A technological profile for the production of cider is reported with suitable cultivar and method employed. | Joshi, 1998 |
| Apple juice concentrate diluted directly to the desired level of sugar gave higher fermentability and more ester content in the wine. | Joshi and Sandhu, 1994 |
| The addition of pectolytic enzyme increased the rate of fermentation, alcohol content and K, Na, Ca, Zn, Cu and Fe while Mg and Mn levels remained unaffected with improvement in different sensory quality attributes. | Joshi and Bhutani, 1991 |
| Apple honey wine made by blending apple juice with honey was adjudged to be the best wine amongst other fruit. | Joshi *et al.*, 1990 |
| With increase in initial sugar content, increase in ethanol, titrable acidity, colour of wine took place, while aldehyde level decreased. Initial sugar concentration of 20 and 24°B was found to be optimum for preparation of cider and apple wine, respectively. | Joshi and Sandhu, 1997 |
| Use of increasing concentration of juice decreased the alcohol and acidity but increased the TSS and total sugar in apple wine. A product with 20 per cent juice was adjudged as the best cider. | Joshi and Sandhu, 2000 |

(*Contd.*)

**Table 7.** (*Contd.*)

| *Research findings* | *References* |
|---|---|
| Antimicrobial activity was shown against *E.coli, Staphylococcus, Aspergillus, Candida and Bacillus* by adding extract of garlic, hops, honey, etc. in apple wine. | Joshi and Sibby John, 2002 |
| Wines made from guava juice were found to be highly acceptable due to low tannin content, optimum colour and flavour. Pectinase treatment of guava pulp prior to fermentation gave about 18 per cent increase in wine yield. | Bardiya *et al.*, 1974 |
| Fermentation of banana pulp diluted in 1:1, 1:2 or 1:3 ratio produced acceptable quality wines but the wine flavour was lost after six months of storage. The wine recovery ranged between 60-76 per cent in different varieties. | Kundu *et al.*, 1976 |
| Muskmelon (*C.melo*), unfit for table purpose was converted into alcoholic beverages with 6.5 per cent (w/v) alcohol and exhibited a very good sensory quality. | Teotia *et al.*, 1991 |
| Jamun var., viz. Pharenda, Jamun and Kathjamun were used for wine preparation by diluting of whole fruit in the ratio of 1:1 with water and treatment with pectinase enzyme fruit was adjudged to be the best. | Shukla *et al.*, 1991 |
| Sapota wine can be made either from clarified or non-clarified sapota juice by raising TSS to 25°B, addition of 0.7 per cent citric acid, 30 ppm $SO_2$, heating to 80-85°C for 10 min., pressing of pulp for four days, 0.1 per cent pectinase addition, siphoning, etc. | Gautam and Chundawat, 1998 |
| Among different methods (control, thermo-vinification, fermenting on the skin and carbonic maceration), must from fruits fermented on the skin gave the highest rate of fermentation and ethanol content. The carbonic macerated wines had a higher amount of alcohol, higher pH, lower acidity, lesser higher alcohol and volatile acidity while that fermented on the skin-treated wines were typical for higher amounts of anthocyanin, lower reducing sugar, total sugar than control wines. Thermo-vinification imparted more desirable quality characteristics to the wine than the control or the other treatments tried. Thermo-vinified wines, irrespective of cultivars, scored the highest sensory quality attributes. Wines from Camarosa cultivar registered many desirable characteristics, like more esters, optimum acidity, better colour and best sensory quality. | Joshi *et al.*, 2005 |
| Wild apricot (*Prunus armeniaca*) wine has been prepared by dilution of fruit pulp in 1:1 ratio with water, raising TSS to 24°B, addition of DAHP @ 0.1 per cent, 0.5 per cent pectinase followed by ameliorating to 12°B with sugar. | Joshi *et al.*, 1990 |
| Out of cyclo-dextrin and Amberlite X AD-16, Amberlite-16 was considered better with respect to the extent of bitterness and changes in composition in kinnow juice for winemaking and concluded that debitterness of juice either prior to or during fermentation improved the sensory quality. | Joshi *et al.*, 1997 |
| De-acidification of high-acid musts of eight grape varieties fermented with *Saccharomyces pombe* (a malic acid-metabolising yeast) showed a considerable reduction in acidity. | Ethiraj and Suresh, 1978a |
| Production of acceptable quality wines from plums by reducing the acidity by use of *Schizo. pombe* has been reported. | Vyas and Joshi, 1988 |
| De-acidification activity of yeast was rapid at pH 3-4.5 but adversely affected when pH was lowered to 2.5 in plum must. It was also influenced by higher concentration of ethanol, $SO_2$ and addition or not of nitrogen source. | Vyas and Joshi, 1988 |
| Standardisation of plum wine production was earlier done with 1:1 ratio of water and whole fruit pulp which reduced the acidity by dilution and produced the wine of acceptable quality. | Vyas and Joshi, 1982 |

(*Contd.*)

| | |
|---|---|
| Strawberry (*Fragaria X ananassa* Duchensne) wine is prepared with different cultivars. Wine from cultivar Caramosa had higher value for physico-chemical characteristics, sensory scores and acceptability in comparison to other two cultivars (Doughlas and Chandler). | Joshi *et al.*, 2005 |
| *Schizosaccharomyces pombe* was employed to produce plum wine, whose acidity can be adjusted as a dry wine. This technique may serve as an alternative to dilution of plum must, practiced at present to produce wine of desirable acidity. | Joshi *et al.*, 2002 |

following the second fermentation. Early picking also avoids late-season infections and their associated off-odours. It also reduces the need for sulphur dioxide addition. Finally, pigmentation of red grapes is less well developed, making it easier to obtain pale coloured juice for producing white sparkling wines.

Traditionally, pressing involves whole clusters, which are especially important to minimise pigment release from red grapes. Pressing uncrushed grapes also limits the release of aroma compounds, grape solids, polyphenoloxidases and potassium. Because pressing is slow, readily oxidised phenolics usually combine with oxygen and precipitate during clarification or fermentation. This protects the wine against subsequent in-bottle browning. Because the grapes are pressed unstemmed, the stalks provide channels through which the juice can escape. This minimises the pressure required to extract the juice. In addition, the large surface area of the traditional shallow presses in Champagne, further reduces the pressure required.

However, in many regions, the traditional press is now largely illustrative; the more efficient, but equally effective, pneumatic press having replaced it. The first press fractions (~ 50 litres per 100 kg) are usually combined for clarification and vinification. The second major fraction (~12.5 l/100 kg) possesses more flavour, suspended solids and phenolics. It, correspondingly, requires extra clarification and is vinified separately. Further press fractions are seldom used in producing sparkling wines.

Depending on the Brix, sugar may be added to the juice. Bentonite, activated carbon, pectinase and sulphur dioxide may also be added to remove glucans, pigments, or minimise the activity of laccase released from the diseased fruit. If the grapes are crushed before pressing, removal of the extra solids (2-4 per cent *versus* 0.5 per cent) is facilitated by bentonite addition, followed by centrifugation or filtration. Under standard conditions, however, clarification occurs spontaneously; typically lasting 12-15 hours at 10°C.

The initial fermentation (to make the base wine) employs procedures typical of most modern white wines. For example, fermentation typically occurs between 15-18°C. Lower temperatures are considered to generate grassy odours, whereas higher temperatures yield wines lacking in finesse. Inoculation with an aromatically neutral yeast strain is almost universal. It minimises the production of fermentation odours that could interfere with the subtle toasty character of wine. If the pH of juice is below 3.0, malo-lactic deacidification is encouraged. Some producers also believe that it provides an additional desired complexity. In addition, by reducing excess acidity, a greater proportion of the wine can be used to produce the popular dry (brut) style of sparkling wines. If encouraged, malo-lactic fermentation must come to completion, early-preparation of the cuvée often occurring in the spring. The bacterial sediment generated cannot be effectively removed after the second, in-bottle fermentation. If producers wish to avoid malo-lactic fermentation (higher acidity being considered to donate greater aging potential), the base wines are sterile-filtered. Maturation of the base wines usually lasts only a few months, but can last for years. The bottles will be stored horizontally for the wine to mature, including the development of autolitic (yeasty) character. For Champagne, the minimum legal maturation period is 15 months and for Vintage Champagnes, three years. From time to time, the bottles may undergo poignettage – shaking and restacking – to prevent the yeasty sediment from sticking to the glass (Grainger and Tattersall, 2005). Traditionally, the bottles are placed into wooden vessels called 'pupitres', each side of which contains 60 slanted holes as shown in Fig. 3. All the large Champagne houses (and other wineries that make sparkling wines by the traditional method) now use an automated process of rémuage. The most commonly used system is that of gyropalettes, as shown in Fig. 4.

Aging typically happens in stainless steel tanks, but can occasionally occur in oak cooperage. This can supply an additional complexity to a portion of the base wines. Following the maturation process,

**Figure 3.** Pupitres

**Figure 4.** Gyropalettes

the wines are ready for selection and inclusion in one or more of the blends (cuvies) used in preparing various brands.

Several methods are available to transform the base wine into a sparkling wine. These are: the method champenoise, the Cremants from France and Luxembourg, the methode traditionnelle (formerly the methode champenoise), e.g. used for Cavas, the transfer method, the methode ancestrale (Limoux, Gaillac), also including the Dioise method and the bulk method (Cuvee Close).

### 4.1. Champenoise Method

Elaboration of sparkling wines from France and Luxembourg uses the champenoise method for both the base wine and the *prise de mousse*. Differences occur with the Champagne winemaking, especially when considering the separation of juices after pressing the whole grapes. The second phase, that is, second fermentation and aging in the bottle, is very close to that of Champagne wines except for the fact that Cremants only have nine months aging in contact with lees. Cava elaboration theoretically, uses the *methode traditionnelle* (formerly the *methode champenoise*), but the base wine is rarely obtained after pressing whole grapes, which differs from the Champagne method. The most commonly used technique for making the base wine is rather that of a traditional white wine. The second phase is perfectly controlled as for Champagne or Cremants. Aging on lees usually takes nine months for Cava wines.

### 4.2. The Transfer Method

In the transfer method, the wine (obtained by traditional white winemaking) is fermented and aged on lees in the bottle (as for the *methode traditionnelle*) but there are no constraints of riddling (a crucial, expensive and lengthy operation) and disgorging. After bottle fermentation and proper aging, the bottles are automatically emptied into a steel tank without degassing since the wine is maintained at an isobarometric pressure. At this stage, the dosage can be directly added in the tank (scheme 1) but winemakers generally prefer adding the dosage in another steel tank after having filtered the wine (scheme 2). After standing several days, the wine is filtered and bottled (scheme 1) or just bottled (scheme 2). All operations are carried out under a carbon dioxide atmosphere, using isobarometric bottling. Advantages of the transfer method are the suppression of the lengthy operation of remuage and the fact that the dosage is more uniform. However, this method is expensive, energy-consuming and a risk of oxidation of the wine does really exist.

### 4.3. Methode Ancestrale

The methode ancestrale, which is an elaborating method but very difficult to control, has been developed in the vineyards of Limoux and Gaillac. The base wine is made with whole grapes (mainly of the Mauzac variety) or through a traditional white winemaking, using a semi-fermented wine. In fact, sugars are used for both the primary alcoholic fermentation and the second fermentation. At different steps of the elaboration process, it is essential to stop fermentation each time it starts to accelerate. In this way, refrigeration (until 0°C), sulphiting, depletion of yeast nutrients (using settling, fining, filtration or centrifugation) are repeated as many times as necessary to regulate or stop the activity of yeast. The wine is then, filtered and kept at 0°C until spring time. At this step, the second fermentation takes place (two to three months) in the bottle (at rigorously controlled temperatures) with yeast and remaining sugars of the semi-fermented wine. Riddling and disgorging (without dosage) then, occur, but the wine can also be sold with a slight yeast deposit at the bottom of the bottle. These wines receive the Blanquette Methode Ancestrale Appellation of Controlled Origin. For the Dioise method, the same principles of winemaking as for the ancestrale method are used, that is, elaboration of a semi-fermented wine from the *Muscat à petits* grains variety, filtration and refrigeration at 0°C, second fermentation in the bottle (using sugars remaining after the first fermentation). To increase the extraction of aroma compounds, pectinolytic enzymes are added to Muscat grape berries in the crusher. This imposes fining treatments since the must obtained presents high turbitidies ranging from 1000-1500 NTU. The flotation technique is used for clarification of the must. After the second fermentation is stopped by refrigerating the cellar, bottles are emptied into a steel tank maintained at an isobarometric pressure by $CO_2$ to avoid degassing (as for the transfer method). After filtration, the wine is bottled using isobarometric bottling. The final alcoholic content of the Clairette de Die is of approximately 7.5° with 40-50 g/L residual sugars. The elaboration of Asti spumante follows the same process and is made from the same grape variety.

### 4.4. Bulk Method

The bulk method (cuvee close) is a simpler and more cost-effective technique that has been developed to obtain ordinary wines of low prices. It is also called as 'tank method' or 'charmat process'. The second fermentation does not take place in the bottle, but the base wine is sent to a reinforced steel fermentation tank able to contain several hundreds hectolitres of wine. Yeast and sugars are added and the wine is maintained at a temperature of 20-25°C. The *prise de mousse* duration does not exceed 10 days. The second fermentation is stopped by a light sulphiting and by refrigerating the wine at –2°C. After having been cold-stabilised at –5°C for several days, the wine is filtered at a low temperature and then, bottled using isobarometric bottling. One disadvantage of this method is that there is no aging of the wine as the wine contact with lees is insufficient (Jeandet *et al.*, 2011).

## 5. Technology of Rose Wine Production

Still rosé wines are generally classified by their colour, regardless of their vinification technique. Normally, they are produced by similar techniques of white winemaking. To achieve the desired rosé colour, the grape skins are removed from the juice shortly after fermentation has begun. Thus, the uptake of substances that contribute red wines their sensory characteristics is limited. Very few rosé wines age well and most are made to have a sweet finish. Rosé wines are mainly produced in France, followed by Spain, Portugal, United States (Californian .blush. wines), South Africa, Chile, Argentina, Australia, Morocco (Boulaouane) and Greece. Three types of still rosé wines can be distinguished according to the vinification technique as detailed below:

(a) Pale rosé wines from red grapes pressed immediately to extract juice with very little colour, sometimes called 'grey' wines (*Vins Gris*) or *Blancs de Noirs*.
(b) Rosé wines made by saignée technique. These wines are produced from red grape juice left in contact with skins from two to 20 hours and then, pressed and fermented like white wines.
(c) Rosé wines derived by blending red and white wines. The use of this technique is actually limited.

### 5.1. Methods Used for Producing Wines

*Blending*: The blending of red and white wine together is now permitted in the European Union. Historically, this method could not be used, except in the Champagne region where it has long been the main method for the production of rosé Champagne.

*Skin contact*: This style of rosé is made from red grapes or white grapes with heavily pigmented skins, such as Pinot Gris, using the white grape method of winemaking. Some grapes are crushed and strained; others are pressed. The free run juice together with the lightly pressed juice is then, fermented without the skins. Pigments from the skins give some degree of colour to the juice. Quite pale rosés often result from this method and are sometimes labelled as 'vin gris' (grey wine). Many rosé wines from the Loire are made in this way, similar to California's 'Blush' wines. If extra colour is needed, a little red wine may be blended in.

*Saignée*: This French term means 'bleeding'. The wine may be made by the method involving short (six to 12 hours) fermentation on the grape skins; after which the juice is drained off. It is also possible to produce a rosé by-product, by draining part of a vat for a rosé and leaving behind a great concentration of skins to produce a full-bodied red wine. Alternatively, red grapes can be chilled and lightly crushed just sufficiently to release their juice. They are then allowed to macerate for between 12 and 24 hours, during which time the colour pigments and flavourings are allowed to 'bleed' into the juice. The juice is then, drained off and fermentation begins. This method can give quite good depth of colour, especially if the grapes used have deep coloured skins, e.g. Cabernet Sauvignon. This style of wine is rose, often show some tannins in it (Grainger and Tattersall, 2005).

## 6. Technology of Fortified Wine Production

Two main properties, the high alcohol content and the addition of alcohol or wine spirits at some stage of production characterise the large group of fortified wines. Such wines are produced in many parts of the world, particularly in Spain, France, United States, Portugal, South Africa, Australia, Italy and Greece. Fortified wines can be made in a wide range of styles i.e. white or red, dry or sweet, with or without added flavours. Because of their more alcohol content and their flavour intensity and complexity, they are not intended to accompany a meal. Instead, they serve as aperitifs, dessert wines or cocktail beverages. Dry fortified wines such as Fino-style sherry and dry vermouth are generally considered to be perfect aperitifs, especially before a meal at which fine wines are to be served. More commonly, fortified wines possess a sweet character, such as ports, Oloroso-style sherry, madeiras and vinsdoux naturels. These wines are consumed after meals or as a dessert. Classification of fortified wines may be done according to their sugar content, with or without added flavours or colour. However, the distinctive types of fortified wines, developed during the last centuries in southern Europe, are the base for the most appropriate classification.

A glimpse of the research carried out on wines from different fruits is presented in Table 7.

## 7. Technology of Vermouth Production

### 7.1. Vermouth Preparation Technology

Vermouth is prepared traditionally from grape by making the base wine, extracting the herbs and spices in wine and brandy mixture, blending the extract with base wine, fortifying the base wine to the desired level of alcohol and finally maturing the vermouth.

The base wine is prepared from grape juice or a concentrate as per the routine method. The essential requirements of a base wine for conversion into vermouth are that the wine should be sound, neutral and cheap. For example, among Italian vermouths, wine from Emilia district is popular as it is a fairly neutral wine with 10-11 per cent (v/v) alcohol and a low acidity of 0.5-0.6 per cent. It is prepared largely from Ugni Blanc grapes. Many Italian producers use refined beet sugar for the preparation of vermouth although in France, mistelas (fortified grape wine) are preferred. Caramel is an important constituent vermouth and is carefully prepared. American vermouth is produced if a wine of natural higher acidity is used. It should be fortified with neutral high-proof brandy. The extract can be prepared by direct extraction method in which the calculated amount is placed in the base wine till the wine has absorbed the desired flavour and aromas. The wine may be heated during the extraction process up to 60°C and the container should be covered to minimise the loss of aromas. However, details of the extraction processes differ among various manufacturers (Amerine *et al.*, 1980). Some companies use a type of fractional blending system to maintain consistency in the composition of the botanical extract. It is believed that the more important botanical ingredients used are wormwood, coriander, cloves, chamomile, dittany of Crete, orris and quassia. Brandy or alcohol extracts of spices and herbs are also available for flavouring the vermouth. Use of brandy in the extraction of flavour can dispense with the need of heating usually carried out in the process with wine. Dry vermouth should not be aged very long; rather it should be finished and bottled very young. Sweet or dry vermouth should be low in pH and preserved with $SO_2$ to prevent any spoilage by *Lactobacillus trichodes*. It may also be flash pasteurised or hot-bottled. Further, the quality and type of vermouth depend upon the quality and nature of base wine and on the type, quality and amounts of various herbs used. In general, the viscosity of wine that correlates with the body of the wine, can be affected by parameters like ethanol and dry extract concentration of wine. It has also been observed that ethanol and dry extract of wine are the constituents that mainly affect its viscosity, while glycerol has a negligible effect due to its low concentration. The flow sheet of entire process is shown in Fig. 5.

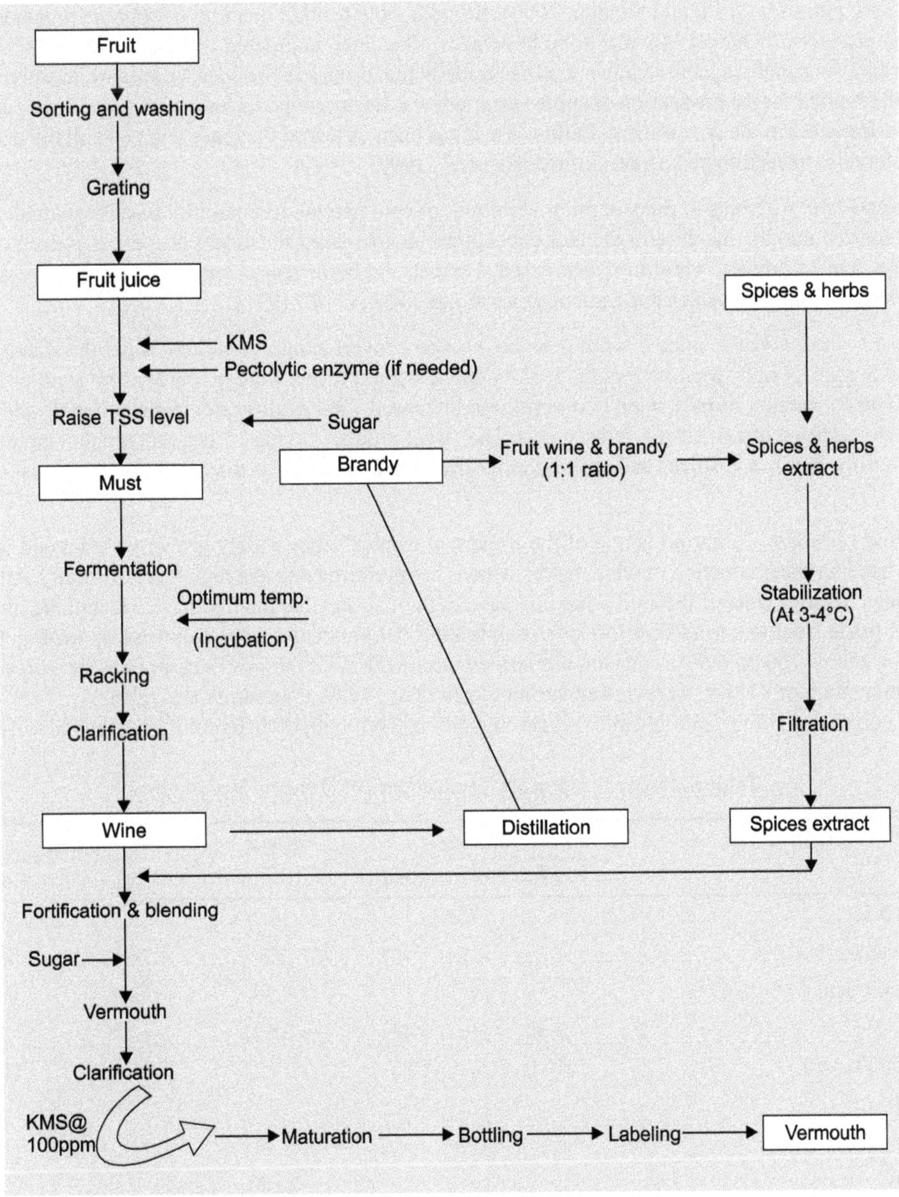


**Figure 5.** Basic process involved in the preparation of vermouth (Panesar *et al.*, 2009)

## 7.2. Preparation of Vermouth from Non-grape Fruits

The method for the preparation of vermouth from grapes is well established but quality and type of vermouth depend upon the base wine and the type, quality and amounts of various herbs/spices used. Besides grapes, other fruit juices, such as that of apple, mango, plum and sand pear have shown their potential in vermouth production (Joshi *et al.*, 1999, 2011, 2017).

*Mango vermouth*: The production of aromatic wine from mango, known as mango vermouth, has been carried out successfully. The base wine was made from cv. 'Banganpalli', raising TSS to 22°Brix, adding 100 ppm $SO_2$, 0.5 per cent pectinol enzyme and carrying out fermentation at 22+1°C using the Montrachet strain 522 of *Saccharomyces cerevisiae*. The composition of mango vermouth in respect of pH, total acidity, alcohol, aldehydes and total phenols was comparable to the values reported for vermouths prepared from grapes (Reddy *et al.*, 2014).

*Apple vermouth*: Apple is produced and relished all-over the world and used both for dessert and processing purposes (Joshi and Sandhu, 2000). In India, only a small quantity of apple is processed into various products, including low alcoholic beverages, like cider, compared to other advanced countries. Spices like *amla* and ginger are known to possess medicinal properties besides antimicrobial activity. The method reported for the production of apple vermouth was the modified technique of grape vermouth, due to the differences in the type of fruit. In this case it has been observed that base wine had all the desirable characteristics needed to make vermouth (Joshi *et al.*, 1999).

*Plum vermouth*: Attempts to prepare plum vermouth of commercial acceptability have been made. It has been reported that the increase in alcohol concentrations increased the aldehydes, ester, phenol content and TSS, but acidity and vitamin C decreased. Further, the herbs/spices extract addition increased the total phenols, aldehydes and ester content of vermouth (Joshi *et al.*, 1991).

*Sand pear vermouth*: The juice of sand pear can also be converted into vermouth as per the standardised methodology. Dry and sweet vermouths with variable alcohol levels were prepared from sand-pear base wine. The conversion of base wine into vermouth increased TSS, acidity, aldehydes, phenols and esters due to the addition of extracts of herbs/spices. The sweet product having 15 per cent alcohol gave higher acceptability. Herbs and spices and their quantity used were the same as that used for plum (Joshi *et al.*, 1991).

*Tamarind vermouth*: Tamarind is one of the important tropical trees widely grown in India and its fruit can be used for the production of wine, but its wine is not preferred due to a high level of acidity. Attempts have been made to convert the fruit wine into vermouth of acceptable quality. In this technique, the base wine is made from tamarind fruit (50 g/l), maintaining 0.9 per cent acidity followed by raising TSS to 23°Brix, adding 150 ppm $SO_2$, carrying out fermentation with *Saccharomyces cerevisiae* var *ellipsoideus* at a temperature of 27±1°C as reviewed earlier (Joshi *et al.*, 1999; Panesar *et al.*, 2009).

A comparison of physico-chemical characteristics of vermouth of different fruits is made in Table 8.

**Table 8.** Physico-chemical Characteristics of Different Vermouths

| *Characteristics* | *Mango* | | *Apple* | *Plum* | |
|---|---|---|---|---|---|
| | *Dry* | *Sweet* | | *Dry* | *Sweet* |
| TSS °Brix | | | 16.2 | | |
| Total sugars (%) | | | 7.8 | ND | 4.8 |
| Titratable acidity (% MA) | | | 0.39 | 0.81 | 0.9 |
| pH | 3.40 | 3.42 | 3.29 | | |
| Ethanol (% v/v) | 17.0 | 17.2 | 15.2 | 15.0 | 14.5 |
| Colour: | | | | | |
| Red | | | 2.92 | | |
| Yellow | | | 20.00 | | |

*(Contd.)*

| | | | | | |
|---|---|---|---|---|---|
| Apparent viscosity (flow) | | | 1.95 | | |
| Free aldehyde (mg/l) | | | 46 | | |
| Total esters (mg/l) | | | 181.0 | 204 | 219 |
| Volatile acidity (%) (as acetic acid) | | | 0.040 | 0.03 | 0.04 |
| Total tannins (mg/l) | | | 524 | | |
| Total phenolics (%) | 0.055 | 0.070 | | 417 mg/l | 390 mg/l |

*Source:* Based on Reddy *et al.*, 2014; Joshi and Sandhu, 2000; Joshi *et al.*, 1991

## 8. Technology of Production of Reduced Alcoholic Wines

During the last three decades, a fundamental shift in the attitude of consumers to alcohol (ethanol) consumption has arisen. Changes in legal blood alcohol limits for driving and increased awareness amongst consumers of the adverse health effects of excessive alcohol intake, have resulted in decreasing wine consumption in traditional market segments. The emergence of a perceived increased consumer demand for reduced alcoholic strength beverages has resulted in several manufacturers producing wine and wine products with lower than normal alcohol content. The term 'low alcohol' may be applied to beverages derived from fermented grape juice that contain less than 1.15 per cent v/v ethanol and 'non-intoxicating' implies the beverage contains less than 0.5 per cent v/v ethanol.

### 8.1. Production Techniques

Production techniques for manufacturing low or reduced alcoholic-strength beverages have been developed over the last 15-20 years in order to satisfy the consumer demand for healthier alcoholic products. The production and sale of alcohol-reduced wines and the lowering of ethanol concentration in wines with alcohol levels greater than acceptable for a specific wine style however, pose a number of technical and marketing challenges. Several engineering solutions and wine production strategies that focus upon pre- or post-fermentation technologies have been described and patented for production of wines with lower ethanol concentrations than would naturally arise through normal fermentation and wine production techniques. The production methods that are of most relevance to the wine industry, though not exhaustive, are summarised in Table 9 and given below.

**Table 9.** Technologies for Reducing Ethanol Concentration in Wine and Fermented Beverages

| *Stage of wine production* | *Principle* | *Technology* |
|---|---|---|
| Pre-fermentation | Reduced fermentable sugars | Early fruit harvest |
| | | Juice dilution |
| | | Glucose oxidase enzyme |
| Concurrent with fermentation | Reduced alcohol production | Modified yeast strains |
| | | Arrested fermentation |
| Post-fermentation | Alcohol removal | Spinning cone column |
| | | Reverse osmosis |
| | | Solvent extraction |
| | | Ion exchange |

*Source:* Schmidtke *et al.*, 2011

### 8.2. Pre-fermentation Technologies for Limiting Alcohol Production

Limiting alcohol production during fermentation by reducing fermentable sugars in juice through early grape harvest, juice dilution or arresting fermentation while significant levels of unfermented sugars remain in the wine are some of the options that enable wines with reduced alcohol levels to be produced. Early grape harvest may result in wines that are sensorily undeveloped due to reduced flavour precursor development in grapes prior to harvest, high acidity levels and lack of yeast-contributing flavour compounds (Schmidtke *et al.*, 2011).

## 9. Mead

Mead, or honey wine, is regarded as the ancestor of all fermented drinks, being the oldest alcoholic drink known to man. The invention of making mead has been considered as ante-dating the cultivation of soil and as a marker of the passage from nature to culture. Traditional mead is made by fermenting a mixture of honey and water. However, a number of variations in the protocol are being followed throughout the world. Depending on local traditions and specific recipes, it may be brewed with spices, fruits, or grainmash. It may be still carbonated, or sparkling; it may be dry, semi-sweet, or sweet. The amount of alcohol in mead varies from 10-18 per cent, depending upon the alcohol-tolerance capacity of yeast culture being used for fermentation.

### 9.1. Method for Mead Production

Honey is diluted with water (approximately 1 kg with 2.5-3.0 litres of water). To create the fruit-containing mead, 10-20 per cent fruit juice or purees are added to the honey-water mixture (Joshi *et al.*, 1990). Citric acid is added up to 0.4 per cent acidity. This step balances the flavour and acid sugar blend of the product. The mixture may be heated to kill the natural microflora associated with honey. The must is cooled and yeast nutrition including nitrogen [yeast assimilable nitrogen (YAN)], vitamins (thiamine) and mineral salts (Mg, Zn) are added. Managing nutrient requirements regulates fermentations and enhances sensory quality. Thiamin is a vitamin used as a coenzyme in the alcoholic fermentation pathway. Minerals are components of the yeast cell membrane and help maintain fermentation metabolism activities. Nitrogen is metabolised by yeast to synthesise proteins. The mixture should incubate at a temperature between 16-25°C. Lower than this range will inhibit the fermentation, but higher will affect the quality of the product. The mixture should ferment for about two weeks and then siphoned into a clean glass container. Almost any spice or herb can be added to mead, either as an extract or directly, at almost any time during the mead-making process. Blends of two or more spices and herbs are commonly used. Adding hops to mead adds a distinctive flavour, clarifies the mead and preserves its freshness. Addition of tannins increases astringency and helps in brewing and clarification (http://www.honey.com/images/downloads/makingmead.pdf; Joshi *et al.*, 1990). When making still mead, the addition of potassium sorbate or wine stabiliser prevents the onset of second fermentation by killing remaining yeast cells (http://www.honey.com/images/downloads/makingmead.pdf). The mead is stored at 16°C for a further two months, racked again after two months. By this time, yeast and the other heavy impurities will settle down and the mead can be bottled or stored at a low temperature for future bottling. The table given below describes the physico-chemical characteristics of mead.

**Table 10.** Physico-chemical Characteristics of Mead

| *Compound* | *Minimum* | *Maximum* |
|---|---|---|
| Alcohol (%) | 12.2 | 20.8 |
| pH | 2.9 | 3.75 |
| Total acidity (g L-l) | 2.20 (36.636 mEq L-l) | 7.08 (117.902 mEq L-l) |
| Volatile acidity (g L-l) | 0.14 (2.331 mEq L-l) | 0.779 (12.973 mEq L-l) |
| Residual sugar (%) | 2.5 | 27.8 |
| Acetaldehyde (mg L-l) | 18.2 | 125.5 |
| Ashes (%) | 0.046 | 0.520 |
| Calcium (%) | 0.41 | 5.11 |
| Magnesium (%) | 0.43 | 2.03 |
| Potassium (%) | 8.62 | 74.19 |
| Sodium (%) | 1.24 | 14.02 |

*Source:* Based on Steinkraus and Morse, 1966; Steinkraus and Morse, 1973; Steinkraus, 1983

### 9.2. Factors Affecting the Quality of Mead

Some of the methods affecting the quality of Mead are summarized here:

- Quality of raw materials affects the final product. Honey, the main raw material for mead production, is made with *Apis mellifera*. The physico-chemical characteristics of honey affect the quality of mead. Different treatments, storage temperature and period affect the quality of honey (Gupta *et al*., 1992; Kaushik *et al*., 1996). The flavour and colour of mead depends upon the age of the honey – if the honey is old and dark, the product will be dark in colour with a strong flavour. The flavour of the product will depend upon the floral source of honey.
- The yeast strain used for fermentation is important since different strains exhibit different resistance to acidity and alcohol concentration.
- The initial sugar concentration in the must determines the final alcoholic concentration and the type of mead (dry, medium dry, or sweet). Other ingredients, such as water, acid (malic or tartaric), spices and herbs, type of fruit and sulphites also affect the quality of mead.
- Fermentation temperature between 16-25°C is optimal, while temperatures as high as 32.0-38.0°C reduce the quality of the product (Morse and Steinkraus, 1975).
- Pasteurisation of must at 80-85°C preserves the delicate flavour of honey; therefore boiling should be avoided to preserve the flavour (Steinkraus, 1983; Steinkraus and Morse, 1966).
- To avoid heating at all, sulphiting (60-70 ppm $SO_2$) is the best option.
- Repeated rackings improve the clarity of the mead (Steinkraus and Morse, 1973).
- The equipment, containers and bottles must be thoroughly rinsed with chlorinated water to ensure the long shelf-life of the product. Most off-flavours are indicative of poor sanitation practices (Steinkraus and Morse, 1973; Kosseva *et al*., 2016).

## 10. Herbal Wine

An herbal purple sweet potato wine can be prepared from purple-fleshed sweet potato rich in anthocyanin pigment and medicinal plant parts, by fermenting with wine yeast, *S. cerevisiae.* The herbal purple sweet potato wine is a novel product with ethanol content of 8.61 per cent (v/v) and a rich source of antioxidant anthocyanin, which offers remedies for colds, coughs, skin diseases and dysentery (Panda *et al*., 2012). To make pumpkin pulp is diluted with water in the ratio of 1:2 by weight. The initial TSS was raised to 24°Bx with honey and acidity was adjusted with pomegranate extract to 0.25 per cent. Further, pectinol and DAHP were added at the rate of 1.0 per cent and 0.1 per cent, respectively. Various herbs (ginger, *tulsi*, *mulhati*) and their combinations were added in different concentration (1 per cent, 2 per cent, and 4 per cent) during the must preparation. To each treatment, 50 ppm $SO_2$ in the form of KMS (potassium metabisulfite) and 5 per cent yeast inoculum each were added. Each fermentation was carried out at 27°C and monitored for reduction in sugar and alcohol production. Finally, the filled bottles were pasteurised at 62-65°C for 20 min. The pumpkin-based herbal wine with *tulsi* (1 per cent) had many desirable characteristics, like higher amounts of alcohol, phenols and esters; higher sensory scores and other comparable characteristics, like acidity, sugars, higher alcohols and volatile acidity compared to other treatments. The presence of flavonoids, saponins, tannins and reducing sugars and absence of alkaloids, phlobatannins and cardiac glycosides were observed. Further, it can be concluded that pumpkin-wine with herbal extract has healthful components.

## 11. Future Prospects

The advancement in the social, educational and economical status of consumers has led to the shift from other alcoholic beverages to wine due to the numerous health benefits that wine offers, especially red wine. Challenges still however, remain in the industry with regards to consumer's culture, importing and state regulations which vary according to the country and even state, and are very complicated to understand. Besides these, ban on advertising alcohol, multiple tax authorities and lack of adequate infrastructure are some other challenges that the wine industries face. There are different fields of wines which need to be explored such as zero per cent alcohol wines, health wines, medicinal/herbal wines and vermouth from different fruits and finally, their commercialization.

## References


Agouridis, N., Bekatorou, A., Nigam, P. and Kanellaki, M. (2005). Malo-lactic fermentation in wine with Lactobacillus casei cells immobilised on delignified cellulosic material. *Journal of Agricultural and Food Chemistry* 53(7): 2546-2551.

Almela, L., Lazaro, I. and Lopez Roca, J.M. (1991). Effect of yeasts on the content of phenolic acids in wines. *Rev Agrog Technol Aliment* 31(3): 393-399.

Amda, Dan, Anand, J.C. and Yadav, I.S. (1972). Preliminary evaluation of grape varieties grown in Delhi region for winemaking. Proc. Symp. on Alcoholic Beverage Industries in India: Present Status and Future Prospects, Mysore, pp. 28-30.

Amerine, M.A. (1953). The composition of wines. *The Scientific Monthly* 77(5): 250-254.

Amerine, M.A., Kunkee, K.E., Ough, C.S., Singleton, V.L. and Webb, A.D. (1980). The Technology of Wine Making, fourth edn. AVI Publishing Co. Inc., Westport, CT.

Azad, K.C., Vyas, K.K. and Joshi, V.K. (1987). Some observations on the properties of juice and cider made from scabbed apples. *Indian Food Packer* 41(1): 56-61.

Bardiya, M.C., Kundu, B.S. and Tauro, P. (1974). Studies on fruit wines – Guava wine. *Haryana J. Hort. Sci.* 3: 140.

Bardiya, M.C., Kundu, B.S., Daulta, B.S. and Tauro, P. (1980). Evaluation of exotic grapes grown in Haryana for red wine production. *J. Res. Haryana Agric. Univ.* 10: 374-379.

Bartowsky, E.J. and Henschke, P.A. (2008). Acetic acid bacteria spoilage of bottled red wine – A review. *International Journal of Food Microbiology* 125(1): 60-70.

Bartowsky, E.J. and Pretorius, I.S. (2009). Microbial formation and modification of flavour and off-flavour compounds in wine. pp. 209-231. *In*: Helmut König, Gottfried Unden, Jürgen Fröhlich (Eds.). Biology of Microorganisms on Grapes, in Must and in Wine. Springer, Berlin, Heidelberg.

Bartowsky, E.J. (2009). Bacterial spoilage of wine and approaches to minimise it. *Letters in Applied Microbiology* 48: 149-156.

Bartowsky, E.J., Xia, D., Gibson, R.L., Fleet, G.H. and Henschke, P.A. (2003). Spoilage of bottled red wine by acetic acid bacteria. *Letters in Applied Microbiology* 36(5): 307-314.

Benito, S., Palomero, F., Gálvez, L., Morata, A., Calderón, F., Palmero, D. and Suárez-Lepe, J.A. (2014). Quality and composition of red wine fermented with *Schizosaccharomyces pombe* as sole fermentative yeast, and in mixed and sequential fermentations with *Saccharomyces cerevisiae*. *Food Technology and Biotechnology* 52(3): 376-382.

Bhushan, S. and Sharma, S. (2011). Maturation of Wines and Brandies. pp. 861-900. *In*: Joshi, V.K. (Ed.). Handbook of Enology: Principles, Practices and Recent Innovations, Vol 2. Asia Tech Publisher and Distributors, New Delhi.

Bruno, M.E., Kaiser, A.L.A.N. and Montville, T.J. (1992). Depletion of proton motive force by rise in Listeria monocytogenes cells. *Applied and Environmental Microbiology* 58(7): 2255-2259.

Butz, P. and Tauscher, B. (2002). Emerging technologies: Chemical aspects. *Food Research International* 35(2-3): 279-284.

Cherubini Alves, A., Carneiro Zen, A. and Domingus Padula, A. (2011). Routines, capabilities and innovation in the Brazilian wine industry. *Journal of Technology Management & Innovation* 6(2): 128-144.

Chikkasubbanna, V. (1982). Effect of variety and maturity on the composition of musts and quality of grapes wines. Ph.D. thesis. University of Agricultural Sciences, Bangalore.

Chilaka, C.A., Uchechukwu, N., Obidiegwu, J.E. and Akpor, O.B. (2010). Evaluation of the efficiency of yeast isolates from palm wine in diverse fruit wine production. *African Journal of Food Science* 4(12): 764-774.

Costello, P.J. and Henschke, P.A. (2002). Mousy off-flavour of wine: Precursors and biosynthesis of the causative N-heterocycles 2-ethyltetrahydropyridine, 2-acetyltetrahydropyridine and 2-acetyl-1-pyrroline by Lactobacillus hilgardii DSM 20176. *Journal of Agricultural and Food Chemistry* 50(24): 7079-7087.

Costello, P.J., Lee, T.H. and Henschke, P. (2001). Ability of lactic acid bacteria to produce N-heterocycles causing mousy off-flavour in wine. *Australian Journal of Grape and Wine Research* 7(3): 160-167.

Daudt, C.E. and Ough, C.S. (1980). Action of dimethyldicarbonate on various yeasts. *American Journal of Enology and Viticulture* 31(1): 21-23.

Deak, T. and Beuchat, L.R. (1996). Yeasts in specific types of foods. *In*: Handbook of Food Spoilage Yeasts 61-96.

Duarte, W.F., Dias, D.R., Oliveira, J.M., Teixeira, J.A., e Silva, J.B.D.A. and Schwan, R.F. (2010). Characterisation of different fruit wines made from cacao, cupuassu, gabiroba, jaboticaba and umbu. *LWT – Food Science and Technology* 43(10): 1564-1572.

Ethiraj, S. and Suresh, E.R. (1978a) De-acidification of high acid grape musts and wine making with *Schizosaccharomyces pombe*. *J. Food Sci. Technol* 15: 111-113.

Ethiraj, S. and Suresh, E.R. (1982). The proline content of some experimental wines made in India. *Amer. J. Enol. Vitic.* 33: 231-232.

Fellows, P.J. (2009). Food Processing Technology: Principles and Practice. Elsevier.

Fleet, G.H. (1993). Wine Microbiology and Biotechnology. Harwood Academic Publishers, Singapore.

Fowles, G. (1989). The complete home wine-maker. *New Scientist*, 123(1680): 38-43.

Fugelsang, K.C. and Edwards, C.G. (2006). Wine Microbiology: Practical Applications and Procedures. Springer Science & Business Media, Springer US.

Fugelsang, K.C., Osborn, M.M. and Muller, C.J. (1993). Brettanomyces and Dekkera: Implications in winemaking. pp. 110-129. *In*: Gump, B.H. and Pruett, D.J. (Eds.). Beer and Wine Production: Analysis, Characterization, and Technological Advances. Vol. 536. American Chemical Society, Washington, DC.

Garde-Cerdán, T., Marsellés-Fontanet, A.R., Arias-Gil, M., Ancín-Azpilicueta, C. and Martín-Belloso, O. (2008). Influence of $SO_2$ on the evolution of volatile compounds through alcoholic fermentation of must stabilised by pulsed electric fields. *European Food Research and Technology* 227(2): 401-408.

Gautam, S.K. and Chundawat, B.S. (1998). Standardisation of technology of Sapota winemaking. *Indian Food Packer* 17-21.

Gerbaux, V., Villa, A., Monamy, C. and Bertrand, A. (1997). Use of lysozyme to inhibit malo-lactic fermentation and to stabilise wine after malo-lactic fermentation. *American Journal of Enology and Viticulture* 48(1): 49-54.

Gnekow, B. and Ough, C.S. (1976). Methanol in wines and musts: Source and amounts. *American Journal of Enology and Viticulture* 27(1): 1-6.

Grace Tate (2015). Investigating in India's Emerging Wine Industry, http://www.india-briefing.com/news/india-wine-9761.html/

Grainger, K. and Tattersall, H. (2005). Wine production: Vine to bottle. Blackwell Pub., UK.

Gupta, J.K., Kaushik, R. and Joshi, V.K. (1992). Influence of different treatments, storage temperature and period on some physico-chemical characteristics and sensory qualities of Indian honey. *Journal of Food Science and Technology*, Mysore 29(2): 84-87.

Hite, B.H. (1899). The Effect of Pressure in the Preservation of Milk: A Preliminary Report, vol. 58. West Virginia Agricultural Experiment Station.

Hoover, D.G. (1989). Biological effects of high hydrostatic pressure on food microorganisms. *Food Technol.* 43: 99-107.

Houtman, A.C., Manns, J. and Du Pihssis, C. (1980). Factors affecting the reproducibility of fermentation of grape juice and of the aroma composition of wines l. Grape maturity, sugar, inoculum concentration, aeration. *Vitis* 19: 37-54.

Jackson, R. (1994). Wine Science: Principle and Application. Academic Press, San Diego.

Jackson, R.S. (2000). Wine Science – Principles, Practices, Perception, second ed. Academic Press, San Diego.

Jackson, R.S. (2003). Wines: Types of table wine. *In*: Caballero, B., Trugo, L., Figlas, P.N. (Eds.). Encyclopedia of Food Sciences and Nutrition, second ed. Elsevier Science, UK.

Jackson, R.S. (2011). Red and White Wines. *In*: Joshi, V.K. (Ed.) *Handbook of Enology: Principles, Practices and Recent Innovations* 3: 981-1020. Asia Tech Publisher, New Delhi.

Jeandet, P., Vasserot, Y., Liger-Belair, G. and Marchal, R. (2011). Sparkling wine production. pp. 1064-1115. *In*: Joshi, V.K. (Ed.). *Handbook of Enology: Principles, Practices and Recent Innovations*. Asiatech Publisher Inc., New Delhi.

Jindal, P.C. (1990). Grape. p. 85. *In*: Bose, T.K. and Mitra, S.K. (Eds.). Fruits, Tropical and Sub-tropical. Naya Prakashan, Calcutta.

Jiranek, V., Grbin, P., Yap, A., Barnes, M. and Bates, D. (2008). High-power ultrasonics as a novel tool offering new opportunities for managing wine microbiology. *Biotechnology Letters* 30(1): 1-6.

Joshi, V.K. (1997). Fruit Wines. Dr. Y.S. Parmar University of Horticulture and Foresty, Nauni, Solan (HP), p. 155.

Joshi, V.K. (1998). Cider: Technological profile: Process. *Food Ind.* 1(9): 10-13.

Joshi, V.K. (2009). Production of wines from non-grape fruit. *In*: Natural Product Radiance, special issue, July-August. NISCARE, New Delhi.

Joshi, V.K. and Sandhu, D.K. (1994). Influence of juice contents on quality of apple wine prepared from apple juice concentrate. *Res. Ind.* 39(4): 250-252.

Joshi, V.K. and Sandhu, D.K. (1997). Effect of different concentrations of initial soluble solids on physico-chemical and sensory qualities of apple wine. *Indian Journal of Horticulture* 54(2): 116-123.

Joshi, V.K. and Sandhu, D.K. (2000). Studies on preparation and evaluation of apple cider. *Ind. J. Hort.* 57(1): 42-46.

Joshi, V.K. and Sibby John (2002). Antimicrobial activity of apple wine against some pathogenic and microbes of public health and significance. *Alimentaria* November: 67-72.

Joshi, V.K. and Bhutani, V.P. (1991). The influence of enzymatic clarification on fermentation behaviour and qualities of apple wine. *Sci. Des Aliments* 11(3): 491.

Joshi, V.K. and Kumar, V. (2011). Importance, nutritive value, role, present status and future strategies in fruit wines in India. pp. 39-62. *In*: Panesar, P.S., Sharma, H.K. and Sarkar, B.C. (Eds.). Bio-processing of Foods. Asiatech Publishers Inc, New Delhi.

Joshi, V.K. and Thakur, N.K. (1994). Preparation and evaluation of Citrus Wines. p. 149. *In*: 24th *International Horticultre Congress*, held in Japan.

Joshi, V.K., Attri, B.L., Gupta, J.K. and Chopra, S.K. (1990). Comparative fermentation behaviour, physico-chemical characterstics of fruit honey wines. *Indian J. Hort.* 47(1): 49-54.

Joshi, V.K., Sandhu, D.K., Attri, B.L. and Walia, R.K. (1991). Cider preparation from apple juice concentrate and its consumer acceptability. *Indian J. Hort.* 48(4): 321-327.

Joshi, V.K., Attri, D., Singh, T.K. and Abrol, G.S. (2011a). Fruit Wines: Production Technology. *In*: Joshi, V.K. (Ed.) *Handbook of Enology: Principles, Practices and Recent Innovations*, Asiatech Publishers Inc.

Joshi, V.K., Bhutani, V.P. and Sharma, R.C. (1990). The effect of dilution and addition of nitrogen source on chemical, mineral, and sensory qualities of wild apricot wine. *American Journal of Enology and Viticulture* 41(3): 229-231.

Joshi, V.K. and Bhutani, V.P. (1990). Evaluation of plum cultivars for wine preparation. *In*: XXIII Int. Hort. Congress, held at Italy, Abst. 3336.

Joshi, V.K., Bhutani, V.P. and Thakur, N.K. (1999a). Composition and nutrition of fermented products. pp. 259-320. *In*: Joshi, V.K. and Pandey, A. (Eds.). Biotechnology: Food Fermentation, vol. I. Educational Publishers and Distributors, New Delhi.

Joshi, V.K., Sandhu, D.K. and Thakur, N.S. (1999b). Fruit-based alcoholic beverages. pp. 647-744. *In*: Joshi, V.K. and Ashok, P. (Eds.). Biotechnology: Food Fermentation, vol. II, Educational Publishers and Distributors, New Delhi.

Joshi, V.K., Rakesh, S. and Ghanshyam, A. (2011). Stone fruit: Wine and brandy. pp. 273-304. *In*: Hue *et al.* (Eds.). Handbook of Food and Beverage Fermentation Technology. CRC Press, Florida.

Joshi, V.K., Sandhu, D.K., Thakur, N.S. and Walia, R.K. (2002). Effect of different sources of fermentation on flavour profile of apple wine by Descriptive Analysis Technique. *Acta Alimentaria* 31(3): 211-226.

Joshi, V.K., Shah, P.K. and Kumar, K. (2005). Evaluation of different peach cultivars for wine preparation. *J. Food Sci. Technol.* 42(1): 83.

Joshi, V.K., Sharma, P.C. and Attri, B.L. (1991). A note on deacidification activity of *Schizosaccharomyces pombe* in plum must of variable compositions. *Journal of Applied Microbiology* 70(5): 385-390.

Joshi, V.K., Sharma, S. and Thakur, A.D. (2017). Wines: White, red, sparkling, fortified and cider. pp. 353-406. *In*: Ashok Pandey, Maria Ángeles Sanromán, Guocheng Du, Carlos Ricardo Soccol and Claude-Gilles Dussap (Eds.). Current Developments in Biotechnology and Bioengineering. Elsevier, Netherlands.

Joshi, V.K. and Sharma, S.K. (1993). Effect of method of must preparation and initial sugar levels on the quality of apricot wine. *Research on Industry* 39(4): 255-257.

Joshi, V.K. and Sharma, S.K. (1995). Comparative fermentation behaviour and physico-chemical sensory characteristics of wine as affected by the type of preservatives. *Chemie Microbiologie Der Labensmittl* 17: 45-53.

Joshi, V.K., Sharma, S., Bhushan, S. and Attri, D. (2004). Fruit-based alcoholic beverages. pp. 335-345. *In*: Pandey, A. (Ed.). Concise Encyclopedia of Bioresource Technology. Haworth Inc., New York.

Joshi, V.K., Sharma, S. and Bhushan, S. (2005). Effect of method of preparation and cultivar on the quality of strawberry wine. *Acta Alimentaria* 34(3): 339-353.

Joshi, V.K., Thakur, N.K. and Lal Kaushal, B.B. (1997). Effect of debittering of kinnow juice on physico-chemical and sensory quality of kinnow wine. *Indian Food Packer* 51: 5-10.

Joshi, V.K., Thakur, N.K. and Lal, B.B. (1997). Effect of cyclodextrin addition on fermentation behaviour: Physico-chemical characteristics and sensory quality of kinnow wine. *Indian Food Packer* 4: 5-9.

Joshi, V.K., Thakur, N.S., Bhatt, A. and Chayanika, G. (2011). Wine and brandy. pp 1-45. *In*: V.K. Joshi (Ed.). Handbook of Enology: Principles, Practices and Recent Innovations, 3 volume set. Asia Tech, New Delhi.

Karagiannis, S.D. (2011). Classification and characteristics of wine and brandies. pp. 46-65. *In*: V.K. Joshi (Ed.). Handbook of Enology: Principles, Practices and Recent Innovations. New Delhi: Asia Tech Publishers.

Kosseva, M., Joshi, V.K. and Panesar, P.S. (Eds.) (2017). Science and Technology of Fruit Wine Production. Academic Press, United Kingdom.

Kosseva, M.R. (2017). Chemical engineering aspects of fruit wine production. pp. 253-293. *In*: Science and Technology of Fruit Wine Production. Academic Press, United Kingdom.

Kotecha, P.M., Adsule, R.N. and Kadam, S.S. (1995). Processing of custard apple: Preparation of ready-to-serve beverage and wine. *Indian Food Packer* 5-7.

Kulkarni, J.H., Harmail Singh and Chadha, K.L. (1980). Preliminary screening of mango varieties for wine making. *J. Food Sci. Technol.* 17: 218-220.

Kumar, K., Kaur, R. and Sharma, S.D. (2011). Fruit cultivars for winemaking. pp. 237-265. *In*: Joshi, V.K. (Ed.). Handbook of Enology: Principles Practices, and Recent innovations vol. 1. Asia Tech Publisher, New Delhi.

Kumar, Vikas, Veerana, P., Goud, Babu, J. Dilip and Reddy, R.S. (2011). Preparation and evaluation of custard apple wine: Effect of dilutions on physico-chemical and sensory quality characteristics, Abstract (BQS-15). *In*: National Conference on New Horizon in Bio-Processing of Foods (NHBF-2011), pp. 51.

Kundu, B.S., Bardiya, M.C. and Tauro, P. (1976). Studies on fruit wines: 1 banana wine. *Haryana J. Hort. Sci.* 5: 160-163.

Kundu, B.S., Bardiya, M.C., Daulta, B.S. and Tauro, P. (1980). Evaluation of exotic grapes grown in Haryana for white table wines. *J. Food Sci. Technol.* 17: 221-224.

Kunkee, R.E. and Bisson, L.F. (1993). Winemaking yeasts. pp. 69-127. *In*: H. Rose and J.S. Harrison (Eds). The Yeasts, second ed. Vol. 5. Academic Press, London.

Kunkee, R.E. and Goswell, R.W. (1977). Table wines. *Alcoholic Beverages* 315.

Landete, J.M., Ferrer, S. and Pardo, I. (2005). Which lactic acid bacteria are responsible for histamine production in wine? *Journal of Applied Microbiology* 99(3): 580-586.

Landete, J.M., Ferrer, S. and Pardo, I. (2007). Biogenic amine production by lactic acid bacteria, acetic bacteria and yeast isolated from wine. *Food Control* 18(12): 1569-1574.

Li, H. and Förstermann, U. (2012). Red wine and cardiovascular health. *Circulation Research* 111(8): 959-961.

Loureiro, V. and Malfeito-Ferreira, M. (2003). Spoilage yeasts in the wine industry. *International Journal of Food Microbiology* 86(1-2): 23-50.

Manes-Lazaro, R., Ferrer, S., Rossello-Mora, R. and Pardo, I. (2009). *Lactobacillus oeni* sp. nov. from wine. *International Journal of Systematic and Evolutionary Microbiology* 59(8): 2010-2014.

Morse, R.A., Steinkraus, K.H. and Paterson, P.D. (1975). Wines from the fermentation of honey. *In*: Crane, E. (Ed.). Honey. Crane, Russak & Company, Inc., New York.

Mylona, A.E., Del Fresno, J.M., Palomero, F., Loira, I., Bañuelos, M.A., Morata, A., Calderón, F., Benito, S. and Suárez-Lepe, J.A. (2016). Use of *Schizosaccharomyces* strains for wine fermentation – Effect on the wine composition and food safety. *International Journal of Food Microbiology* 232: 63-72.

Nanda, Kasabe (2015). Maharashtra's Wine industry Cheers New Liquor Policy, http://www.financialexpress.com/article/markets/commodities/maharashtras-wine-industry-cheers-new-liquor-policy/133870/

Negi, S.S., Suresh, E.R. and Randhawa, G.S. (1972). Raw materials recruited for wine. Proc. Symp. on Alcoholic Beverage Industries in India, Present Status and Future Prospects, Mysore, pp. 12-14.

Özhan, D., Anli, R.E., Vural, N. and Bayram, M. (2009). Determination of chloroanisoles and chlorophenols in cork and wine by using HS-SPME and GC-ECD detection. *Journal of the Institute of Brewing* 115(1): 71-77.

Pando Bedrinana, R., Lastra Queipo, A. and Suarez Valles, B. (2012). Screening of enzymatic activities in non-*Saccharomyces* cider yeasts. *Journal of Food Biochemistry* 36(6): 683-689.

Panesar, P.S., Kumar, N., Marwaha, S.S. and Joshi, V.K. (2009). Vermouth Production Technology – An Overview. *Natural Product Radiance* 8(4): 334-344.

Panesar, P.S., Panesar, R. and Singh, B. (2009). Application of response surface methodology in the optimisation of process parameters for the production of kinnow wine. *Natural Product Radiance* 8: 366-373.

Patel, J.D., Venkataramu, K. and Subba Rao, M.S. (1977). Studies on the preparation of cider and brandy from some varieties of Indian apples. *Indian Food Packers* 31(6): 5-8.

Piyasena, P., Mohareb, E. and McKellar, R.C. (2003). Inactivation of microbes using ultrasound: A review. *International Journal of Food Microbiology* 87(3): 207-216.

Pretorius, I.S. (2000). Tailoring wine yeast for the new millennium: Novel approaches to the ancient art of winemaking. *Yeast* 16(8): 675-729.

Rana, R.S., Vyas, K.K. and Joshi, V.K. (1986). Studies on the production and acceptability of cider from Himachal apples. *Indian Food Packer* Nov.-Dec. 48-55.

Reddy, L.V.A. and Reddy, O.V.S. (2005). Production and characterisation of wine from mango fruit (*Mangifera indica* L.). *World Journal of Microbiology and Biotechnology* 21(8-9): 1345-1350.

Reddy, L.V.A. and Reddy, O.V.S. (2011). Effect of fermentation conditions on yeast growth and volatile composition of wine produced from mango (*Mangifera indica* L.) fruit juice. *Food and Bioproducts Processing* 89(4): 487-491.

Ribereau-Gayon, P. (1978). Wine flavour. pp. 355-380. *In*: Charalambous, G. and Inglett, G.E. (Eds.). Flavour of Foods and Beverages. Academic Press, New York, USA.

Romano, A., Fischer, L., Herbig, J., Campbell-Sills, H., Coulon, J., Lucas, P., Cappellin, L. and Biasioli, F. (2014). Wine analysis by Fast GC proton-transfer reaction-time-of-flight-mass spectrometry. *International Journal of Mass Spectrometry* 369: 81-86.

Romano, P. and Suzzi, G. (1993). Higher alcohol and acetoin production by *Zygosaccharomyces* wine yeasts. *Journal of Applied Bacteriology* 75(6): 541-545.

Romano, P. and Suzzi, G. (1993). Sulphur Dioxide and Wine Microorganisms. pp. 373-393. *In:* Fleet, G.H. (Ed.). Wine Microbiology and Biotechnology. Harwood Academic Publishers, London.

Sánchez-Hernández, J.L., Aparicio-Amador, J. and Alonso-Santos, J.L. (2010). The shift between worlds of production as an innovative process in the wine industry in Castile and Leon (Spain). *Geoforum* 41(3): 469-478.

Sandhu, D.K., Joshi, V.K. (1995). Technology, quality and scope of fruit wines with special reference to apple. *Indian Food Industry* 14(1): 24-34.

Schmidtke, L.M., Delves, T. and Agboola, S. (2011). Technology of Production of Reduced Alcoholic Wines. pp. 1152-1176. *In:* V.K. Joshi (Ed.). Handbook of Enology: Principles, Practices and Recent Innovations 3: Asia Tech Publisher, New Delhi.

Sevda, S.B. and Rodrigues, L. (2011). Fermentative behaviour of *Saccharomyces* strains during guava (*Psidium guajava* L.) must fermentation and optimisation of guava wine production. *J. Food Process Technol.* 2(118): 2.

Shah, P.K. and Joshi, V.K. (1999). Influence of different sugar sources and addition of wood chips on physicochemical and sensory quality of peach brandy. *J. Sci. Ind. Res.* 58: 995-1004.

Shahidi, F. (2009). Nutraceuticals and functional foods: Whole versus processed foods. *Trends in Food Science and Technology* 20: 376-387.

Sharma, Somesh and Joshi, V.K. (2004). Flavour profiling of strawberry wine by quantitative descriptive analysis technique. *J.Food Sci. Technol.* 41(1): 22-26.

Shukla, K.G., Joshi, M.C., Sarswati, Y. and Bisht, N.S. (1991). Jambal winemaking: Standardisation of methodology and screening of cultivars. *J. Food Sci. Technol.* 28(3): 142- 144.

Singh Nagi, H.P.P. and Manjrekar, S.P. (1976). Studies in the preparation of cider from North Indian apples: II. Storage studies. *Indian Food Packer* 30(1): 12-15.

Singh, R.S., Sooch, B.S. and Attri, D. (2011). Bioreactor technology in wine production. pp. 802-860. *In*: Joshi, V.K. (Ed.). Handbook of Enology: Principles, Practices and Recent Innovations, Vol. 1. Asia Tech Publisher and Distributors, New Delhi.

Soni, S.K., Marwaha, S.S., Marwaha, U. and Soni, R. (2011). Composition and nutritive value of wine. pp. 89-145. *In*: Joshi, V.K. (Ed.). Handbook of Enology: Principles, Practices and Recent Innovations, Vol. 1. Asia Tech Publisher and Distributors, New Delhi.

Sreekantiah, K.R. and Johar, D.S. (1968). Processing of grapes. *In*: The Grape. Andhra Pradesh Grape Growers Association. Hyderabad, India.

Steinkraus, K.H. and Morse, R.A. (1973). Chemical analysis of honey wines. *Journal of Apicultural Research Cardiff* 12(3): 191-195.

Stockley, S. (2011). Therapeutic value of wine. pp. 146-208. *In*: Joshi, V.K. (Ed.). Handbook of Enology, Principles,Practices and Recent Innovations vol. 1. Asia Tech Publication, New Delhi.

Subba Rao, M.S. (1972). Wine technology. Proc. Symp. on Alcoholic Beverage Industries in India, Present Status and Future Prospects, Mysore, pp. 35-38.

Sun, B., Neves, A.C., Fernandes, T.A., Fernandes, A.L., Mateus, N., De Freitas, V., Leandro, C. and Spranger, M.I. (2011). Evolution of phenolic composition of red wine during vinification and storage and its contribution to wine sensory properties and antioxidant activity. *Journal of Agricultural and Food Chemistry* 59(12): 6550-6557.

Suresh, E.R., Ethiraj, S. and Negi, S.S. (1985). Evaluation of new grape cultivars for wine. *J. Food Sci. Technol.* 22: 211-212.

Suresh, E.R., Ethiraj, S. and Onkarayya, H. (1983). Blending of grape musts for production of red wines. *J. Food Sci. Technol.* 20: 313-315.

Suresh, E.R. and Ethiraj, S. (1987). Effect of grape maturity on the composition and quality of wines made in India. *Amer. J. Enol. Vitic.* 38: 329-331.

Suresh, E.R. and Negi, S.S. (1975). Evaluation of some grape varieties for wine quality. *J. Food Sci. Technol*. 12: 79-81.

Swami, S.B., Thakor, N.J. and Divate, A.D. (2014). Fruit wine production: A review. *Journal of Food Research and Technology* 2: 93-94.

Tauscher, B. (1995). Pasteurisation of food by hydrostatic high pressure: Chemical aspects. *ZeitschriftfürLebensmittel-Untersuchung und Forschung* 200(1): 3-13.

Teotia, M.S., Manan, J.K., Berry, S.K. and Sehgal, R.C. (1991). Beverage development from fermented (*S. cerevisiae*) muskmelon (*C. melo*) juice. *Indian Food Packer* (4): 49.

Towantakavanit, K., Park, Y.S. and Gorinstein, S. (2011). Quality properties of wine from Korean kiwifruit new cultivars. *Food Research International* 44(5): 1364-1372.

Unwin, T. (1991). Wine and the Vine: An Historical Geography of Viticulture and the Wine Trading. Routledge, London.

Vaquero, M.R., Alberto, M.R. and De Nadra, M.M. (2007). Antibacterial effect of phenolic compounds from different wines. *Food Control* 18(2): 93-101.

Vyas, K.K. and Joshi, V.K. (1982). Plum winemaking: Standardisation of a methodology. *Indian Food Packer* 36: 80-86.

Vyas, K.K. and Joshi, V.K. (1988). Deacidification activity of *Schizosaccharomyces pombe* in plum musts. *Journal of Food Science and Technology* 25(5): 306-307.

Vyas, S.R. and Chakravorthy, S.C. (1971). Wines from Indian Grapes. Haryana Agril. Univ. Hisar, pp. 1-69.

Vyas, S.R. and Gandhi, R.C. (1972). Enological qualities of various grape varieties grown in India. Proc. Symp. on Alcoholic Beverage Industries in India, Present Status and Future Prospects, Mysore, pp. 9-11.

Wade, M.E., Strickland, M.T., Osborne, J.P. and Edwards, C.G. (2018). Role of *Pediococcus* in winemaking. *Australian Journal of Grape and Wine Research*, Wade. doi: 10.1111/ ajgw.12366

Wang, D., Xu, Y., Hu, J. and Zhao, G. (2004). Fermentation kinetics of different sugars by apple wine yeast *Saccharomyces cerevisiae*. *Journal of the Institute of Brewing* 110(4): 340-346.

Wildenradt, H.A. and Caputi, A. (1977). Wine analysis collaborative study, 1975. *American Journal of Enology and Viticulture* 28(3): 145-148.

Yabaya, A., Bobai, M. and Adebayo, L.R. (2016). Production of wine from fermentation of *Vitis vinifera* (grape) juice using *Saccharomyces cerevisiae* strain isolated from palm wine. *International Journal of Information Research and Review* 3: 2834-2840.

Yang, H.Y. and Weigand, E.H. (1949). Production of fruit wines in the Pacific North West. *Fruits Product Journal* 29(8-12): 27-29.

Zoecklein, B.W., Fugelsang, K.C., Gump, B.H. and Nury, F.S. (1995). Volatile acidity. pp. 192-198. *In*: Wine Analysis and Production. Springer, Boston, MA.

# 22 Cider: The Production Technology

**V.K. Joshi[1], Somesh Sharma[1*] and Vikas Kumar[2]**

[1] School of Bioengineering and Food Technology, Shoolini University, Solan, Himachal Pradesh – 173229

[2] Food Technology and Nutrition, School of Agriculture, Lovely Professional University, Phagwara, Punjab – 144411, India

## 1. Introduction

One of the most important temperate fruit crops of the world is apple (*Malus domestica* Borkh.). The major apple-producing countries of the world are China, USA, India, Germany, France, Italy and Turkey (FAO, 2018). In India, apple is commercially cultivated in the states of Jammu & Kashmir, Himachal Pradesh, Uttarakhand and Arunachal Pradesh. It is used for both dessert and processing purposes, such as apple juice concentrate, vinegar, apple sauce, juice, butter, preserve, candy, jam, jellies, canned and alcoholic beverages (cider, wine, vermouth and brandy).

Out of these beverages, cider is a low alcoholic drink produced from apple, a popular beverage especially in those countries where grapevine cultivation is not practiced due to the prevailing agro-climatic conditions. Worldwide, cider varies in alcohol content from less than 1.2 per cent alcohol by volume (ABV), as found in French cidredoux, to 8.5 per cent ABV in traditional English ciders. In the United States of America, the legal definition of cider for tax purposes specifies 7 per cent or lower ABV; anything above 7 per cent ABV falls into a different tax category – it can still be called cider, but is taxed at a different rate (Bureau of Alcohol, Tobacco and Firearms 1998). France is the world's largest cider-producing country. Normandy and Brittany in Northern France is the main apple cider-producing region and is famous for its traditional sweet *cidre.* Some restaurants even substitute a bottle of cider for the usual free bottle of wine (Herrero *et al.*, 2001). The United Kingdom leads the world in hard cider production and consumption, though the United States of America is catching up. The quantity of cider produced is second only to the wine produced from grapes. Out of the two, cider has become an increasingly important commercial product in recent years (Jarvis *et al.*, 1995). It was a common drink at the time of Roman invasion of England in 55 BC and was drunk throughout Europe in the third century AD. Commercial cider production started during the 19th century. Today, Bullmers alone makes 65 per cent of the 5 million hectolitre cider produced in UK.

Recently, low alcoholic beverages have gained importance in preventing cardiovascular diseases. Wine consumption prevents the formation of LDL and increases HDL levels (having protective effects against heart diseases) (Joshi *et al.*, 1999). The use of hops in cider and spices has enhanced the antimicrobial activity and is considered significant (Joshi and Siby, 2002). Here, in this chapter, the word 'cider' is used to describe the fermented juice of apple. Details of its production and its quality and factors effecting it have been described.

## 2. History

The first record of cider was documented in 1205. Description of cidermaking in the Mediterranean basin is found in the works of the Roman writer Pliny during the first century AD (Pliny, 1967). Thereafter, its production appears to have moved toward north, so that cidermaking was well established in France by the time of Charlemagne (9th century) and probably was introduced in England from Normandy well before Duke William's conquest in 1066 (Revier, 1985; Roach, 1985). Cider was established in the Basque country well before the 12th century and by the 11th and 12th centuries, it was also being made in Contentin, the area around Caenand in the Paysd' Auge. During the 13th century, it was found in Southeast England and by the end of 15th, and the beginning of 16th centuries, its production had

*Corresponding author: someshsharma@shooliniuniversity.com

spread to Eastern Normandy, as well as to Brittany, but it remained a drink mainly of poorer classes. The continuing preference of the people of Brittany for wine during the 16th century is well attested by the large quantities of wine imported through the Breton ports. By the 17th and 18th centuries, its production in England had reached something of an art form and had become the subject matter of a number of learned discourses. In the aftermath of the Napoleonic wars, corn growing and cattle rearing became more profitable. Naturally, orchards and consequently, cidermaking received less skilled attention. Fallen trees were replaced with worthless seedlings and the enormous volume of cider produced was of low or indifferent quality, resulting in depressed prices. Even much of the trade was given to cider merchants, who bought everything wholesale at their own price; thus, the quality of cider started deteriorating at that time. During the 19th century, imports of continental and American apples began to the market in Britain. Cider apples were grafted over the table type and eventually, led to the growth of the specialised table apple industry. During the 19th century, the popularity and quality of cider declined, until it became a little more than a cheap source of alcohol for itinerant farm workers and acquired its unfortunate 'scrumpy' image. Increasingly, since 1900, however, cider has prospered in new markets and the last decade has witnessed an increase in sales against a generally static or declining consumption pattern for many other types of alcoholic drinks. During 19th and 20th centuries, cider and perry were produced and consumed throughout the world, for more details, see Unwin (1980).

## 3. Definition and Characteristics of Cider

Definition and characteristics of cider are given in Table 1. Cider is an alternative term used for cyder, though both the terms have been used since at least 1631, including that in Australia. Nevertheless, distinction between the two is also made. Cider is an alcoholic beverage, whereas cyder is usually apple juice or a non-alcoholic beverage (apple juice). The fermented juice, called cider in England, is known as 'hard cider' in the USA. In Europe, fermented apple juice is known as cider (France), Sidre (Italy), 'Sidra' (Spain), 'Applewein' (Germany and Switzerland) where the name for the corresponding unfermented product is clearly distinguished as apple juice. Figure 1 shows the various classes of cider. Depending upon the alcohol content, cider could be a soft cider (1-5 per cent) or hard cider (6-7 per cent) (Downing, 1989) and may be made from fresh juice or juice from single cultivar, and may be classified as vintage cider or white ciders which is made from decolourised apple juice or pale coloured juice (Jarvis, 1993). The sweet cider has residual sugar from fermentation or is sweetened after fermentation, and still cider is

**Table 1.** Definition of Cider

**1. Definition**

*Cider*: It is a beverage obtained by the complete or partial fermentation of the juice of fresh apples or a mixture of the juice of fresh apples and fresh pears, with or without the addition of drinking water.

**2. Characteristics of Cider**

| | | *Pure juice cider* | *Other cider* |
|---|---|---|---|
| Actual potential alcohol by volume | Minimum | 5 per cent | 4 per cent |
| Total dry extract | Minimum | 16 g/l | 13 g/l |
| Without sugar | Minimum | 1 g/l | - |
| Volatile acid | Maximum | 1.4 g/l as acetic acid | 1.4 g/l as acetic acid |
| Iron | Maximum | 150 mg /l | 150 mg /l |

**3. Forbidden Treatments or Practices**

Use of colouring matter except caramel, any other practice intended to alter the composition with a view to deceiving the purchaser of the true nature or origin of the product or to disguise its deterioration is forbidden.

Draft regulation on cider, Resolution 7, Council of Europe, Strasbrough, 30th November 1962 (A.S./Agr / V.Sp (14)10 Rev)
*Source*: Beech and Carr, 1977

with low sugar and without carbon dioxide. Dry cider is without sugar and with an alcohol content of 6-7 per cent, whereas ciders having alcohol content not more than 1.2 per cent alcohol by volume are made by removing the alcohol from strong cider by thermal evaporation, by reverse osmosis or generally by adding apple juice to it (Jarvis, 1993). The cider produced by the 'Methode Champenoise' is called champagne cider (Downing, 1989). Sparkling sweet cider is produced by fermenting apple juice containing not more than 1 per cent alcohol (v/v) and the natural $CO_2$ formed during fermentation is retained. Sparkling cider has lower sugar and higher alcohol content of 3.5 per cent but with partial retention of $CO_2$ formed during fermentation. Carbonated cider is charged with commercial $CO_2$ to produce effervescent.

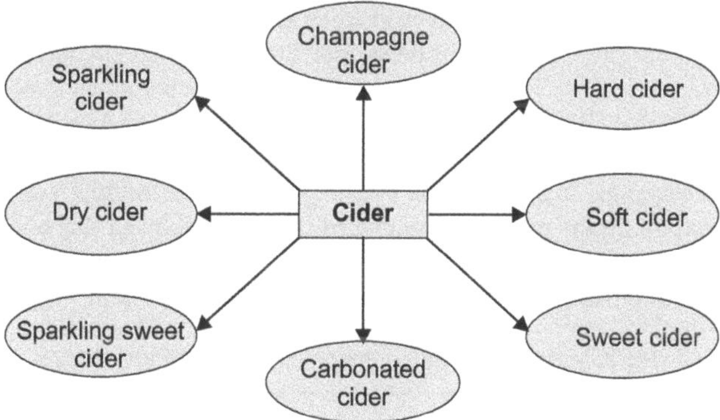


**Figure 1.** Classification of cider

Usually, the sparkling cider is produced through two different methods:

1. Sparkling cider made from concentrated apple juice and/or fresh apple must with addition of sugars and carbon dioxide being permitted, as well as the use of different stabilisation processes.
2. 'Natural cider' is made according to traditional methods, which imply, among other practices (such as the prohibition of addition of sugars and $CO_2$), the exclusive use of juices obtained from the pressing of cider apples (Picinelli *et al.*, 2000).

## 4. Factors Affecting Fermentation

The quality of cider depends upon various factors, such as cultivars, composition of the fruit, method of preparation and condition of fruit fermentation.

### 4.1. Cultivars

Cider makers know from repeated observations that some cultivars have good, stable pressing properties, while others give lower yields or yields which decrease considerably upon maturation (Guillermina *et al.*, 2006). Theoretically, cider can be prepared from any apple but choice of right cultivar is one of the important factors influencing the quality of cider (Downing, 1989). Broadly, the cider apples are classified into four categories (Cider Advisory Committee, 1956):

1. *Bittersweet* – high in tannin, low in acid
2. *Bittersharp* – high in tannin, high in acid
3. *Sharp* – low in tannin, high in acid
4. *Sweet* – low in tannin, low in acid

Further, different varieties of apple suitable for cidermaking have also been recommended (Table 2). The type of apple grown in countries like France and England for fermentation into cider are usually unsuitable for other products because of their very high content of pectin-estrase and total phenol content than that of dessert apples (Burroughs, 1973). The phenolic compounds increase the tendency to enzymatic browning during juice extraction and the polymers of the catechins and leucoanthocyanins, which contribute huge bitterness and astringency to the cider. It has been used for many years as a criterion for classification of juice for suitability for fermentation into cider. However, to specify which apple variety makes the best cider is difficult (Beech and Carr, 1977). Still the use of apple juice concentrate for cider production has increased considerably due to several advantages offered by it, such as price stabilisation, quality maintenance and storage of concentrate for a long time without spoilage (Downing, 1989; Joshi *et al*., 1991), though it leads to a loss of development of specific cultivars for cidermaking (Beech and Carr, 1977; Labelle, 1979; Jarvis *et al.*, 1995). Since more than the cultivars, fermentation conditions effect the quality of cider, the varietal effect is not given much importance these days. In India, cider production is in infancy and the suitability of Indian varieties for cider production has not been adequately worked out, though Ambri-Kashmiri, Red Delicious, Golden Pippin, Maharaji apples and crab apples, Golden Delicious, Red delicious and Rus Pippin have been found suitable for cider making (Joshi *et al*., 1991; Joshi *et al*., 1994). A comparative study of scabbed fruits with normal fruits showed that fruits with less than 15 spots did not affect the fermentation behaviour or the physico-chemical and sensory qualities of cider produced (Azad *et al*., 1987). Guillermin *et al*. (2006) compared technological and rheological properties of two cider apple cultivars, Avrolles and Doucecoetligne. Cvr Avrolles had higher juice yield and rheological characters than to Doucecoetligne. Use of cultivar and pasteurisation also has effect on formation of alcoholic beverage. In a study done by Hang and Woodams (2009) on the influence of apple cultivar juice pasteurisation on the methanol content of the wines showed that among the four cultivars used, crispin apple yielded more methanol in hard cider than Empire, Jonagold or Pacific rose apples. However, pasteurisation of crispin apple juice reduced the methanol content.

**Table 2.** Varieties Suitable for Apple Wine and Cidermaking

| *Varieties suitability* | *Quality characteristics* |
|---|---|
| • Delicious, Cortland, Cider, Rome Beauty | Low acid group |
| • Jonathan, Winesap, Stayman, Cider Stayman, Northern spy, Rhode Island, Greening Wealthy, Newton, Wayne | Higher acid group |
| • McIntosh, Gravenstein, Golden Russet, Cox's Orange, Roxbury Russet, Wealthy | Aromatic group |
| • Dabinett, Michelin, Chisel Jersey, Harry Masters Jersey, Yarlington Mill, Viberie, Medaille, Bedan, Kermerrein | Bitter sweet |
| • Breakwells seedling, Backwell Red, Brown's Apple, Crimson King, Stoke Red Frederick, Kingston Black | Sharp/Bitter sharp |
| • Sweet Copin, Sweet Alford, Northwood | Sweet |

*Source*: Labelle, 1980; Pourlx and Nicholas, 1980; Rana *et al*., 1986; Vyas and Kochhar, 1986; Jarvis *et al*., 1995

## 4.2. Composition of Fruits

It is certainly an important factor influencing the quality of wine and cider. Carbohydrates are the principal food constituents in apple (Table 3) but is a poor source of protein. Among the minerals, potassium, phosphorus and calcium are present in significant amounts in apple fruit. There are large variations for various physico-chemical and flavour characteristics amongst the various cultivars used to make cider (Jarvis *et al*., 1995). The amino acid composition of cider apple includes asparagine, aspartic acid, glutamic, serine and alanine to be present while other are just in traces (Table 4). Phenolic compounds constitute a significant component of apple, due to its antioxidant role (Table 5). Different cultivars, like bittersweet cultivars (particularly French and English) have relatively high concentrations of polyphenols, conferring bitterness and astringency to the finished beverage. In addition to the procyanidins, other classes of polyphenols are also present, such as phenolic acids (chlorogenic and p-coumaroylquinic), together with phloretin glucoside (phloridzin) and the xylo-glucoside (Lea, 1978; Lea, 1982; Lea, 1984). Levels of all these components in bittersweet cider cultivars may be tenfold higher than in the dessert apples. Since the polyphenols make a major contribution to flavour, colour, pressability and also have same antimicrobial properties, therefore their retention in cider is considered useful.

**Table 3.** Composition of Apple Fruit

| *Constituents* | *Average range* |
|---|---|
| *Physico-chemical characteristics* | |
| Calories (k cal/100 g) | 37-46 |
| Water (g) | 84.3-85.6 |
| Fiber (g) | 2.0-2.4 |
| Total Nitrogen (g/100 g) | 0.04-0.05 |
| Protein (%) | 0.19 |
| Lipid (%) | 0.36 |
| Sugar (per 100 g flesh) | 9.2-11.8 |
| Total sugars (%) | 10.65-13.23 |
| Reducing sugars (%) | 7.05-10.67 |
| Sucrose (%) | 1.95-5.02 |
| Pectin (%) | 0.32-0.75 |
| Phenolic compounds (%) | 0.15-2.4 |
| Vitamin C (mg/100 g) | 3.15-5.7 |

*Source*: Gebhardt *et al.*, 1982; Mitra, 1991; Sharma and Joshi, 2005; Upshaw *et al.*, 1978

**Table 4.** Amino Acids in Cider Apple Juices

| *Amino acid* | *Concentration range (mg nitrogen /100 ml)* |
|---|---|
| Asparagine | 0.19-17.5 |
| Aspartic acid | 0.35-1.5 |
| Glutamic acid | 0.21-0.31 |
| Serine | 0.04-0.65 |
| Alpha-Alanine | 0.03-0.18 |
| Gamma-Aminobutyric acid | Trace |
| Valine | Trace |
| Leucine+isoleucine | Trace |

*Source:* Beech and Carr, 1977

**Table 5.** Phenolic Compounds Present in Cider Apple Juices

| *Compound* | *Example* |
|---|---|
| Phenolic acids | Chlorogenic acid |
| Phloretin derivatives | Phloridzin |
| Simple catechins | (-) Epicatechin |
| Condensed procyanidins | B2 |
| Flavonol glycosides | Anthocyanins |

*Source*: Gebhardt *et al.*, 1982; Mitra, 1991; Sharma and Joshi, 2005; Upshaw *et al.*, 1978

### 4.3. Yeast Strains for Fermentation

In traditional fermentation, where no yeast is added and no sulphite is used, the first few days are dominated by the non-*Saccharomyces* species (*Candida pulcherrima*, species of *Pichia*, *Torulopsis*, *Hansenula* and *Kloeckera apiculata*) which multiply quickly to produce a rapid evolution of gas and alcohol. They also generate a distinctive range of flavours, characterised by ethyl acetate, butyrate and related esters. As the alcohol level rises (2-4 per cent), these initial fermenters begin to die out and the microbial succession is taken over by *Saccharomyces uvarum* which completes the conversion of sugar to alcohol and generation of a more wine-like flavour. The yeast cells become sub-lethally damaged by the increasing concentration of alcohol, but death does not occur until the concentration exceeds 9 per cent alcohol (v/v). The formation of higher alcohols and esters during alcoholic fermentation is related to a particular yeast strains used (Castelli, 1973). The desirable characteristics for yeast used in cidermaking include production of polygalacturonase to breakdown soluble pectins; produce rapid onset of fermentation; relatively resistant to $SO_2$, low pH value and high ethanol level; low requirements for vitamins, fatty acids and oxygen; ferments to 'dryness' (i.e. no residual fermentable sugar); does not produce excessive foam; efficient utilisation of sugars; minimal production of $SO_2$; non-producer of $H_2S$ and acetic acid; produces required aroma components, organic acids and glycerol (Beveridge, 1986; Beech 1993). The effect of different yeast starter cultures on the overall quality of pumpkin wine reveals that a mixed culture of *Saccharomyces cerevisiae*, *Torulaspora delbrueckii* and *Zygosaccharomyces rouxii* produced a wine with better quality in comparison to individual culture (Sharma *et al.*, 2018). The number and type of yeast in the apple juice are also influenced by the method of juice extraction (Table 6). Consequently, the quality of cider is also affected.

**Table 6.** Comparison of Yeast Composition of Apple Juice Extracted by the Two Methods of Pressing

| | *Number of yeasts/ml* | |
|---|---|---|
| *Species* | *Screw press* | *Hydraulic press* |
| *Kloeckera apiculata* | 360.000 | 160,000 |
| *Saccharomyces* spp. | - | 23,000 |
| *Pichia* spp. | 5,000 | 23,000 |
| *Torulopsis* spp. | - | 2,030,000 |
| *Candida pulcherrima* | - | 23,000 |
| Carotenoid-containing yeasts | - | 23,000 |

*Source*: Beech, 1977

### 4.4. Nitrogen Source

Nitrogen containing compounds of must are important to the growth of yeast and hence, to fermentation rate and aroma compounds production. Supplementation with nitrogen source is also essential as in its absence, the yeast uses the amino acids of the must resulting in the formation of higher alcohols (Amerine *et al.*, 1980). DAHP @ 0.1 per cent has been used as a yeast food in alcoholic fermentation. Addition of nitrogen source to the musts made from apple juice concentrate and that from pear for alcoholic fermentation has also been reported (Joshi and Sandhu, 1994). The effect of nitrogen on the quality of persimmon wine has similarly been documented (Mahant *et al.*, 2017).

### 4.5. Must Clarification

Clarification of cider must is an indispensable factor affecting the overall quality. Pectic enzymes are often added to the juice to hasten and improve the clarification of cider during and after fermentation. Enzymatically clarified apple wines were rated better in terms of colour, appearance, body and flavour than non-clarified apple wines (Joshi and Bhutani, 1991). The addition of increasing levels of insoluble solids to the apple juice leads to the production of undesirable physico-chemical characteristics on the apple wine (Joshi *et al.*, 2013). Thus, to prepare quality apple wine, juice without insoluble solids by pre-settling and clarification using pectolytic enzyme should be practiced.

### 4.6. Sulphur Dioxide

The treatment of juice with $SO_2$ before fermentation is the most common means for controlling undesirable micro-organisms as well as to prevent enzymatic and non-enzymatic browning reactions and is a well-established practice in winemaking (Beech and Carr, 1977; Rana *et al.*, 1986). Table 7 shows the effect of $SO_2$ and temperature in cider. If $SO_2$ is added immediately after pressing, nearly all the colour (chemically and visually) is reduced as the sulphite binds to the quinoidal forms. If the sulphite is added at the later stage, less reduction in colour will take place; presumably the quinones become more tightly cross-linked and less susceptible to nucleophilic addition and reduction. Sulphur dioxide has also clarifying action and reduces volatile acidity while exerting solvent effect on anthocyanin pigments (Amerine *et al.*, 1980). Usually, 100-200 ppm $SO_2$ is added to the musts for cider production.

**Table 7.** Effect of $SO_2$ and Temperature in Cider and Wine Preparation

| *Concentration* | *Effects* | *Optimum temperature* | *Low temperature* | *High temperature* |
|---|---|---|---|---|
| $SO_2$ (50-200 ppm) | Controls undesirable micro organisms | 15-18°C | Less activity of bacteria and wild yeast in the must | Enhanced growth of thermophillic organisms |
| | Prevents enzymatic browning of the juice | | Less loss of volatile aromatic principles | More loss of volatile compounds |
| | Has clarifying action, i.e. it neutralizes negatively charged colloids | | Greater alcohol yield | Slowing down of fermentation |
| | $SO_2$ has solvent effect on anthocyanin pigments | | More residual carbon dioxide production | Less carbon dioxide production |
| | Reduces volatile acidity | | | |
| | Increases glycerol production | | | |

*Source*: Amerine *et al.*, 1980; Frazier and Westhoff, 1995; Jarvis, 1993

### 4.7. Temperature

The temperature affects the rate of fermentation and nature of metabolites formed. It takes 3 to 4 weeks to attenuate cider fermentation at temperatures within the range of 20-25°C. Although temperature of 15-18°C is preferred for flavour development in Germany and France, the optimum temperature for cider fermentation was found to be 15-18°C (Jarvis *et al.*, 1995). Higher temperature increases the rate of fermentation but enhances the chances of contamination with undesirable thermophillic microorganisms (Frazier and Westhoff, 1995). The changes in viable cell count, ethanol, glucose, fructose and sucrose during cider fermentation at 20°C with *Saccharomyces cerevisiae* have been noted.

### 4.8. Factors Affecting Flavour of Cider

Flavour is one of the important sensory attribute that determines the overall quality of cider. The various factors that influence the flavour of cider are described in Table 8.

**Table 8.** Factors Influencing the Flavour of Cider

| | |
|---|---|
| Apple juice | Specific Variety of fruit(s); Maturity and condition of apple fruit at pressing |
| Other ingredients of raw material | Fresh juice or concentrate<br>Condition of concentrate<br>Type of chaptilization sugar(s)<br>Quantity of $SO_2$<br>Yeast nutrients, amelioration of pH by addition of acid |
| Yeast | Natural or Inoculated fermentation<br>Strain of yeast(s)<br>Condition of yeast(s) when inoculated |
| Fermentation | Temperature, time of fermentation |
| Fermenter design and operation | Hydrostatic pressure; operational pH/acidity level |
| Secondary fermentation | Natural or induced malo-lactic fermentation secondary yeast fermentation<br>Spoilage organisms |
| Maturation | Chemical and enzymatic changes |
| Processing factors | Decolourisation of juice or final cider<br>Dealcoholisation |
| Final product make-up | Carbonation |

*Source*: Jarvis *et al.*, 1995

## 5. Cider Production Technology

### 5.1. Methods of Cidermaking

The cider production process that combines two successive biological fermentations: the first is the classical alcoholic fermentation of sugar into alcohol conducted by yeast strains, like *Saccharomyces cerevisiae* species and the second is malo-lactic fermentation that occurs during maturation process by lactic acid bacteria. The latter is an important manufacturing step to reduce the acidity of cider and stabilises it with respect to microbial spoilage through the bacteriostatic effect of the lactic acid produced. The different methods used to make cider have been reviewed in a systematic way earlier (Amerine *et al.*, 1980) and are summarised in Table 9. There are some reports on the studies of cider preparation from India also (Kerni and Shant, 1984).The yeast strain affect the formation of flavour compounds also in cider (Jarvis *et al.*, 1995). A method for sweet cider making has been developed. Cider with 5 per cent alcohol, TSS/acid ratio of 25 was found to be the most favoured at laboratory and consumer survey scales (Rana *et al.*, 1986).

### 5.2. Raw Materials

Geographic location may have a greater effect upon the finished product than the cultivar itself. Americans interested in cultivating English and European cider apple cultivars may not grow them in their particular location; hence different cultivars are recommended in different areas of even small countries. In traditional orchards, fruits are generally allowed to fall naturally or are shaken from the trees using long poles (lugs), and are then, picked up either by hand or by machine, but in intensive bush orchards, mechanically shaking of the tree permits fruits to fall which are collected and washed immediately after falling and transferred to the mill (Jarvis, 1993). Generally, apples for cider production are different from culinary apples as they have a higher tannin and sugar content, but are lower in acid. Dessert and culinary apples lose more body and flavour during fermentation (Smock and Neubert, 1950) than do the cider apples.

**Table 9.** Summary of Methods Used in Cider Preparation

| *Type of method* | *Fruit* | *Juice* | *Parameters/additives* | *Fermentation* | *Maturation* | *Others* |
|---|---|---|---|---|---|---|
| **European** | | | | | | |
| Method-1 | (a) Some stored for 3-4 days and others macerated | Extracted as usual, cold stabilised at 0-7.8°C | $SO_2$ 50-100 mg/1 | Temperature 4.4-10°C, pure yeast in some, mixed in others | Secondary fermentation in casks for several months | Malo-lactic fermentation, produces $CO_2$ in bottles |
| | (b) – | – | Pectic enzymes for clarification | | | |
| Method -2 | Lower sugar, higher acidity | Juice extracted, no maceration, juice centrifuged for bacteria and yeast removal | Lactic acid added to increase acidity, if needed | Pure yeast such as Steinberg added | – | – |
| Method -3 | Sound fruits separated by flotation | Juice extracted in hydraulic press | – | Natural fermentation from 1.008-1.005 sp. gravity | Storage in concrete tanks lined with wooden coatings | Before delivery, cider is sweetened with syrup |
| Method -4 | – | – | – | Fermentation allowed up to TSS of 5-7.5°B filtered or centrifuged | Stored in wooden casks | Carbonated and bottled |
| **American** | | | | | | |
| Method -1 | (a) Sound apples are used for cider making | Juice is extracted in usual press after crushing in a mill | Sulphur dioxide 100-125 ppm added, glucose added to give 13 per cent alcohol | Spontaneous fermentation may begin during settling | – | Clarified by bentonite treatment, blended to give 10°B, filtered, bottled and pasteurised |
| | (b) Juice extraction instead of apple juice concentrate used | Juice made from concentrate | Sweetened | Yeast Champagne, 24.4°C temperature was the best | – | – |

*Source*: Amerine *et al*., 1980

Over the years, cider apples have been classified into various categories (see earlier section on cultivars) based on the properties of the juice they yield. A recent classification is based on the level of flavour, acids and tannins (Beech and Carr, 1977). Apples for cidermaking should be mature and free from starch. Blending has always been an important step in controlling uniformity of the finished product and suggestions for blending have also been made (Pourlx and Nicholas, 1980). The preferred procedure is to use fermented stock for blending because the effect of fermentation on the fresh juice cannot always be accurately predicted.

Juice from apples in the sweet group is generally, considered good for blending with strong flavoured juices, while that in the bittersweet group gives cider a tangy sensation. Juice for making sparkling sweet cider should neither be sweet nor too heavy in the body. Astringency is considered less important than the correct sugar/acid ratio. Campo *et al.* (2007) showed that in the apple musts with high content of acid and phenols, having malo-lactic fermentation first followed by alcoholic fermentation, had comparatively low production of acetic acid.

The sweet, low-acid cultivars, such as Delecious, Cortland, Ben Davis and Rome beauty are recommended for basic juice, while those of Jonathan, Stayman Winesap, Northern Spy, Rhode Island Greening, Wayne, Newtown possess higher acid levels and add tartness to the cider. MacIntosh, Gravenstein, Ribston Pippin, Golden Russet, Delecious are aromatic and add flavour and bouquet to the cider. The body and flavour can be improved by using astringent apples, such as Red Astrakhan, Lindel, and crab apples. A good thumb rule is to add less than 10 per cent of astringent cider to an acidic cider and not more than 20 per cent should be added to any blend.

## 5.3. Milling and Pressing

Fresh and ripe apple fruits here used to extract the juice are generally stored for a few weeks after harvest so that all the starch can be converted into sugar. Apples selected for juice processing are then, washed and inspected for the presence of any foreign materials and decayed fruits, which have adverse effects on microbiological status and ultimate cider quality. Earlier practice was to empty bulk truckloads or bins of apples on a de-leafing screen into a tank of water. A circulating pump was used to direct the apples to an elevating conveyor which discharged to an inspection belt. At this point, inspectors removed any damaged or decayed fruit and extraneous material. Rinsing with clean water was accomplished at the scrubber or after inspection. The routine replacement of holding water was necessary. Fruits were also transferred into the mill using a water flume, which provided an additional advantage of washing of fruits (Jarvis, 1993).

In preparation for pressing, the apples are ground to a mash using either a hammer or grating mill and even slicers are required for difficult extraction. A recent development in the production of apple juice for concentrate is by diffusion extraction (Bump, 1989; Downing, 1989) and the juice can be used for cidermaking. Several types of equipment have been developed for pressing and extracting apples. These include hydraulic presses, screw presses, basket presses, belt presses and pneumatic presses. The factors involved in the choice of equipment are production capacity, product yield, ease of cleaning and sterilisation and length of production season (Bump, 1989). For many years, rack and frame press has been used in the apple juice industry which consists of a frame containing a slatted board covered by a cloth into which a measured amount of mash is transferred and corners of the cloth are folded over to form an envelope. The frame is removed and next slatted board added, together with the frame and another cloth; the procedure is repeated 10-20 times. Then, it is pressed hydraulically to remove the juice, under a ram pressure of 15 Mpa giving a recovery of 80 per cent of juice (Jarvis, 1993).

The advantage of this type of press is that no press aids are required and the apple juice produced has a low level of solid particles, though it involves high labour cost alongwith the need for great care in cleaning and sterilising the cloths and racks is required. The other presses used are the stoll press, the bucher-guyer press, bullmer continuous belt press, atlas-pacific press for the apple juice extraction. The use of screw presses for the recovery of additional juice for hydraulically pressed pomace in France has been made for additional recovery of juice from centrifuged apple pulp containing cellulose fibres (Lowe *et al.*, 1964). Using press aids, the Zenith and Jones presses were among the first continuous presses used for apple juice extraction in the United States of America. The amount of apple solids in the juice from the screw presses is much greater than that in juice from hydraulic rack and cloth presses (Bump, 1989).

A screw press that is well adapted to apple juice production is the Reitz press system. Electroacoustic dewatering process is also one of the recent methods that employs passing of electric current through the pulp prior to pressing has been claimed to release a higher yield of juice. Immediately after pressing, the juice is treated with sulphur dioxide, which acts both as antioxidant to prevent browning of the juice and a preservative by destroying wild yeast and bacteria (Bump, 1989). From an economic standpoint, maximum recovery of juice is most important for cheap cider production. Cider can also be produced from the apple concentrate directly after diluting it to a desired Brix level.

## 5.4. Controlling Microorganisms Before Fermentation

In earlier times, natural fermentation of apple juice was the common practice. However, for proper fermentation the microflora of the juice must be controlled before inoculation with yeast to avoid off-flavour or similar defects in the cider (Beech and Carr, 1977). There are several methods to accomplish this (Beech and Carr, 1977). In northern France, centrifuging or fining of the juice with gelatine and tannin followed by filtration to reduce the rate of fermentation is practiced. Another approach is to treat the juice with pectin hydrolysing enzymes and filter before adding yeast. The danger of bacterial spoilage is still present with these methods. Use of sulphur dioxide is made extensively for this purpose. The natural fermentation of apple juice depends upon the ability of naturally occurring yeasts in the juice to convert the fruit sugars to ethyl alcohol. These yeasts are native to fruit or normal contaminants on the pressing equipment. Liang *et al.* (2004) reported that Pulsed Electric Field (PEF) can also be used for inactivation of spoilage microorganisms. Further, the use of PEF along with clove oil showed additional reduction in the microbial count.

The Swiss reportedly dilute sterile fourfold concentrate to a specific gravity of 1.050, treat the juice with 35-40 ppm $SO_2$ and then, pitch inoculate with yeast two days later. However, sterile juices or diluted concentrates with little or no sulphur dioxide need to be kept in sterile containers to prevent contamination with spoilage microorganisms.

The treatment of juice with $SO_2$ before fermentation is undoubtedly the most common means for controlling undesirable microorganisms (Table 10) but the amount required depends on the pH of the juice as well as on the concentrations of the sulphite-binding compounds that are present in the juice (Table 11). Since the effectiveness of $SO_2$ is pH-dependent undissociated form (so called molecular $SO_2$), both $SO_2$ and lysozyme prevent the development of undesirable bacterial fermentations. Addition of lysozyme and oenological tannis during alcoholic fermentation could represent a promising alternative to the use of $SO_2$ and for the production of wines with reduced content of $SO_2$ (Sonni *et al.*, 2009). The cider apple juices should always be brought below a pH of 3.8 by the addition of malic acid before $SO_2$ addition. The $SO_2$-binding compounds produced by acetic acid bacteria are present in greater quantities in the poor quality fruit. The binding of $SO_2$ is dependent upon the nature and origin of carbonyl group. Compounds, such as glucose, xylose, and xylosone present in the juice bind with $SO_2$. Such juices will require addition of higher amounts of $SO_2$ to control the microorganisms effectively. Consequently, all the additions must be completed immediately after pressing the juice, provided the initial fermentation is inhibited and further addition of free $SO_2$ can be made during the following 24 hours (Jarvis, 1993).

**Table 10.** Typical Microorganisms of Freshly Pressed Apple Juice

| *Microorganisms type* | *Typical species* |
|---|---|
| Yeast | *Saccharomyces cerevisiae, S. uvarum, Saccharomycodes ludwigii, Kloeckera apiculata, Candida pulcheriima, Pichia* spp., *Torulopsis famata, Aureobasidium pullulans, Rhodotorula* spp. |
| Bacteria | *Acetobacter xylinum, Pseudomonas* spp., *Escherichia coli, Salmonella* spp., *Micrococcus* spp., *Staphylococcus* spp., *Bacillus* spp., *Clostridium* sp. |

*Source*: Beech and Carr, 1977

The composition of musts may also control the bacterial population. In fact, a pH lower than 3.5 is recommended for initial musts (Jarvis *et al.*, 1995) and apple varieties rich in phenolic compounds are usual in cidermaking (Guyot *et al.*, 1998). Juices of pH >3.8 should be brought down to this value by blending or acid addition and then, 150 ppm $SO_2$ is added. However, it has been noted that juices with pH 3.8 could not be satisfactorily treated within the legal limit of 200 ppm $SO_2$ (Beech, 1972; Burroughs, 1973). After sulphiting, the juice should be allowed to equilibrate for a minimum of six hours before free $SO_2$ is determined. In countries where legal limits of sulphur dioxide lower than 200 ppm are prescribed, the best approach would be to use only good microbial quality fruit, maintain good plant sanitation and monitor the pH of the raw material. In Table 11, the amount of sulphur dioxide to be added to the juice of particular pH is given.

**Table 11.** Amount of Sulfur Dioxide to be Added to Juice Based on pH

| *pH* | *Concentration of sulphur dioxide (ppm) (Sodium metabisulphite solution)* |
|---|---|
| 3.0-3.3 | 75 |
| 3.3-3.5 | 100 |
| 3.5-3.8 | 150 |

*Source*: Beech, 1972; Burroughs, 1973

## 5.5. Amelioration

In alcoholic beverage-production terminology, correction of raw material to make a product of consistent quality is referred to as amelioration, i.e. adjustment of the sugar and/or acid content of the juice, as regulated by the respective standards. Controlling the sugar content of apple juice is required to maintain the proper final alcohol content. It is achieved by the addition of water, juice from the second pressing of the pomace, sugar, or concentrated juice. Initial sugar concentration (ISC) influences the quality of the cider and its value of 20°B was found optimum (Joshi and Sandhu, 1997). Fortification of apple juice after dilution from its concentrate with diammonium hydrogen phosphate is essential for rapid fermentation, as discussed earlier. The must prepared by direct dilution of the concentrate reportedly ferments faster than that ameliorated with sugar (Joshi *et al*., 1991; Joshi and Sandhu, 1994).

## 5.6. Inoculation

The traditional method of cidermaking does not employ any external source of yeast. The indigenous micro-flora of apple in the order of $5\times10^4$ cells per gram of stored fruits carries out spontaneous fermentation (Lea, 1995). After sulphiting the juice, it is inoculated with the desired yeast culture in case of inoculated fermentation, wherever employed. The growth of yeasts, acetic acid and lactic acid bacteria can be excluded by washing and sorting of apples before milling and pressing. High counts of bacteria (including lactic acid bacteria) were observed during alcoholic fermentation and storage of cider. The levels of LAB found in musts fermented in small vessels, using acid-washed apples, were low. However, the must fermented by using unwashed apples but blended with different varieties had a limited number of microorganisms only. The growth of microorganisms could have been limited by fermentation and storage temperature of 10°C together with low pH (Ribereau-Gayon *et al*., 1975). A yeast strain making a clean-tasting beverage, with a minimum of yeasty flavour and a maximum fermentation rate along with other desirable characteristics, is selected for the preparation of cider. The use of a mixed inoculum of *S. uvarum* and *S. bayanus* is a widespread practice, on the grounds that the first yeast provides a speedy start, but the second will cope up better with the fermentation to dryness to produce high alcohol level.

These dried yeasts require no pre-propagation and are simply hydrated in warm water before pitching directly into the juice. A small quantity of heat-sterilised juice is inoculated with a dry culture or liquid-nitrogen-frozen culture and after fermentation, the inoculum is added to a larger volume of sulphited juice. The procedure is continued until a final inoculum of 1 per cent or greater by volume is obtained. It may be added that not many different cultures are actually used in cidermaking (Beech and Carr, 1977), such as A.W. Y.350r (*Saccharomyces uvarum*) (Australian Wine Research Institute, Aldaide), Champagne strains, Champagne Epernay, Geisenheim GE 1 (Institute for Microbiology and Biochemistry at Geisenheim),

Pinnacle (No. 729), Montrachet (UCD 522) or Champagne A.Y.D. (Australian Wine Research Institute, Aldaide Universal Foods Corporation, USA).

A comparison of two methods of cidermaking has also been made with respect to evolution of microbial population and malo-lactic fermentation (Deunas *et al.*, 1994). The two methods were: the traditional method where unwashed apples of different varieties were used and in the other, a sole acidic varieties of apple with temperature control during fermentation was used. The frequency distribution (%) of yeast species isolated during cidermaking is summarised in Table 12. The occurrence of malo-lactic fermentation together with alcoholic fermentation is not considered desirable in French and English ciders (Salih *et al.*, 1990) and degradation of malic acid occurs after alcoholic fermentation. However, it does not occur until the population of lactic acid bacteria reaches $10^6$ CFU/ml (Deunas *et al.*, 1994). Interestingly alcoholic fermentation was carried out by *Kloeckera apiculata* and *Saccharomyces cerevisiae* and the distribution was found similar in both the methods. In the traditional method, the malo-lactic fermentation proceeded at the same time as alcoholic fermentation but in the modified method, no malo-lactic fermentation occurred, but produced cider with lower volatile acid. Controlled malo-lactic fermentation in cider using *Oenococcus oeni* immobilised in alginate beads has been made as a starter culture (Herrero *et al.*, 2001). Malic acid degradation was similar to that with free cells of *Oenococcus*. Immobilised cells synthesised less ethanoic acid and ethyl ethanoate but the profile of evolution of pyruvic acid, shikmic acid and succinic acid was similar. The immobilised cells produced more ethanol during earlier four days but it declined during the later periods. The results are promising with respect to production of better quality cider but may need more research work (Nedovic *et al.*, 2000).

Another interesting approach for continuous production of cider was attempted with the use of Ca-alginate material to co-immobilised *Saccharomyces bayanus* and *Leuconostoc oenos* in one integrated biocatalyst system which permitted much faster fermentation than traditional cidermaking with better flavour fermentation. After completion of fermentation, D-lactate was produced while L-lactate progress of lactic acid bacteria in cidermaking also took place.

**Table 12.** Frequency (%) of Yeast Species Isolated During Cidermaking

| | | *Period of fermentation*[a] | | |
|---|---|---|---|---|
| *Species* | | *A* | *B* | *C* |
| *Sacchromyces cerevisiae* | 1[b] | 12 | 42 | 78 |
| *Zygosaccharomyces cidri* | 1 | - | 2 | - |
| *Zygosaccharomyces florentius* | 1 | 2 | 2 | - |
| *Kloeckera apiculata* | 2 | 68 | 52 | 12 |
| *Sacchromycodes ludwigii* | 2 | 4 | - | 2 |
| *Candida pulcherrima* | 3 | 6 | - | - |
| *Rhodotorula rubra* | 3 | 2 | - | 2 |
| *Torulasporadel brueckii* | 3 | 2 | 2 | - |
| *Candida vini* | 3 | 2 | - | 2 |
| *Pichiamem braaefaciens* | 3 | 2 | - | 4 |

*Source*: Deunas *et al.*, 1994

[a] A: After barrel filling, at beginning of alcoholic fermentation; B: Active fermentation (density at 20°C between 1.0351.005); C: at the end of alcoholic fermentation (density at 20°C below 1.005).

[b] 1: Yeasts with strong fermentative metabolism; 2: *apiculata* species with low fermentative activity; 3: species with mainly an oxidative metabolism.

## 5.7. Fermentation

Mostly stainless steel tanks are used these days for fermentation of cider (Downing, 1989) though traditionally barrels of oak have been used for this purpose. Wooden barrels or vats of mild steel, with a ceramic or resin lining, bitumen lined, concrete vats and, more recently, stainless steel and even lined fibre glass-resin vats or tanks have been employed commonly for cider fermentation. The former may still be employed but today most companies use vertical stainless steel tanks while other cidermakers use conico-cylindrical vats. These tanks may be equipped with temperature controlled systems, level indicators and carbon dioxide venting and blanketing systems. Correct sulphiting of the juice and proper cleaning of all the equipment will help ensure a good start to fermentation.

The best procedure for assuring a good fermentation is to employ larger inocula as all yeast strains perform in the same manner at the same concentrations. Either juice or cider, if exposed to air during fermentation, will usually develop a surface film of acetic acid bacteria or yeast. This aerobic spoilage can be prevented by excluding the air from the vats properly by sulphiting the juice. Treatment of juice before inoculation influences the fermentation. Heated juice ferments faster than unheated juice but sulphited juices ferment slower than those not treated with sulphur dioxide. The availability of soluble nitrogen in the juice affects the rate of fermentation of cider. Fermentation with *Schizosaccharomyces pombe* reduced the malic acid in several fruits, including apple (Azad *et al.*, 1986), though with low rate of alcohol production (Parkand John, 1980). *Leuconostoc* has also been employed to reduce the acidity of the fermented product. Simultaneous inoculation of apple juice with *Saccharomyces cerevisiae* and *Schizosaccharomyces pombe* produced cider with acceptable levels of alcohol and acidity. Ion exchange sponge with tailored surface charge for immobilisation of *Saccharomyces cerevisiae* encouraged yeast growth but reduced fermentation (O'Reilly and Scott, 1993). In most of the ciders studied, the malo-lactic fermentation and the alcoholic fermentation started at the same time.

The best flavoured cider is generally produced by a slow fermentation process. The rate of fermentation may be controlled by maintaining the temperature around 16°C, by reducing the yeast population by racking or by the addition of sulphur dioxide. However, if the cider fermentation is too slow, it may be susceptible to cider sickness, imparting milky white appearance to the cider with a sweet pungent odour. This condition is encouraged at the elevated temperatures but is reduced in ciders with 0.5 per cent malic acid. Occasionally, fermentation may be slowed down or even stalled. Aeration has been found useful for restoring yeast activity in such cases (Burroughs and Pollard, 1954). Stuck fermentations could be restarted if the temperature could be 12-13°C, some fermentable sugar remained and at least 10,000 yeast cells per ml are present in the fermenting musts (Whiting, 1961).

A process was developed, based on alcoholic fermentation of the available carbohydrates present in ciders. The impact of inhibitors at different pH, size and reuse of inocula and different nutrient supplementation on the ethanol yield were evaluated. The use of a 0.5 g/l yeast inoculum and corn steep water as the nutrient source allowed depletion of the sugars in less than 48 hours, which increased the content of ethanol to more than 70 g/l (Seluy *et al.*, 2018).

## 5.8. Clarification

Juice and cider can also be clarified and one of the clarification treatments consists in adding gelatine. This treatment can be used either on the juice before fermentation or on cider before bottling. By removing selectively high DP procyanidins, this treatment modifies both the total tannin content and the profile of the residual tannins, and thus, may change the composition and the taste. However, the most common clarifying method in French cidermaking uses endogenous pectin methyl esterase as a clarifying agent. Calcium is added to induce a formation of a calcium pectate gel that includes all particles of the cloudy juice. This gel is then, separated by 'natural' flotation due to $CO_2$ bubbles of the beginning fermentation. This process produces a pectin-free clarified juice with a reduced nitrogen amount and results in slower fermentation and better stabilisation (Que´re *et al.*, 2006).

After completion of fermentation, the cider is left on the lees for few a days to facilitate the yeast to autolyse, thereby adding enzymes and amino acids to the cider. The cider will be separated from the lees and transferred after clarification into the storage vats or storage tanks (Jarvis, 1993). Initial clarification may be performed by the natural settling of a well-flocculated yeast, by centrifugation, by fining, or by a combination of all the three. Typical fining agents are bentonite, gelatine, isinglass or chitosan. Gelatine forms a block with negatively charged tannins in the cider and brings down other suspended materials by entrapment, and can also be used together with bentonite for similar effect.

## 5.9. Aging/Maturation and Secondary Fermentation

After clarification, cider may be bulk-stored or bottled. Extreme care in the sanitation of storage vessels however, is necessary to prevent contamination with undesirable microorganisms. Stored cider should be cultured periodically and removed from storage for special processing if unwanted growth occurs. Storage temperature can be as low as 4°C, but not higher than 10°C. Sparkling or charged ciders have to be stored in pressure tanks to avoid any loss of $CO_2$. Uncharged cider should be kept in a full, closed tank with an air trap or under a blanket of carbon dioxide or nitrogen or a mixture of the two (Cant, 1960). If air is not excluded from the tanks, acetic acid bacteria will produce acetic acid taints. Film yeasts, which may develop, also produce volatile acids in cider. A temperature of 4°C for bulk storage of cider is desirable. After fermentation, the cider is racked and filtered. Maturation is an important step in cidermaking during which most of the suspended material settles down, leaving rest of the liquid clear which may be clarified with bentonite, casein, gelatine followed by filtration.

Enumeration, isolation and identification of lactic acid bacteria in processing and storage of Australian cider revealed *Leuconostoc oenos* as the predominant bacteria (Salih *et al.*, 1990). During maturation, growth of LAB cultures can occur extensively, especially if wooden vats are used (Table 13). This growth results in malo-lactic fermentation. Such fermentation would convert malic acid into lactic acid and reduce acidity, impart subtle flavour, which generally improves the flavour of the product. However, in certain circumstances, metabolites of LAB cultures damage the cider flavour by excessive production of diacetyl, the butter scotch-like taste (Jarvis, 1993). Since malic acid is a predominant acid in the apple, so reduction in acidity due to malo-lactic acid fermentation might be detrimental to the quality of cider (Salih *et al.*, 1990).

**Table 13.** Progress of Lactic Acid Bacteria During Cidermaking

| *Period* | *During active fermentation* | | *During malo-lactic fermentation* | | *During storage* | |
|---|---|---|---|---|---|---|
| Sample | Total LAB count (CFU/ml) | Species | Total LAB count (CFU/ml) | Species | Total LAB count (CFU/ml) | Species |
| A11 | $1.2 \times 10^5$ | *L. oenos* | $7.4 \times 10^6$ | *L. oenos* | $7.5 \times 10^6$ | *L. oenos* |
| A2 | $1.2 \times 10^5$ | *L. oenos* | $1.2 \times 10^7$ | *L. oenos* | $1.2 \times 10^7$ | *L. oenos* |
| B1 | $5.8 \times 10^5$ | *L. oenos* | $6.8 \times 10^6$ | *L. oenos* *Tediococcub* sp. | $1.0 \times 10^5$ | *L. oenos* |
| B2 | $9.3 \times 10^6$ | *L. oenos* | $1.2 \times 10^7$ | *L. oenos* | $1.5 \times 10^6$ | *L. brevis* |
| | | *L. mesenteroides* | $1.3 \times 10^6$ | *L. oenos* | $6.3 \times 10^6$ | *L. brevis* |
| C1 | $1.8 \times 10^5$ | *L. oenos* | | *L. brevis* | | *L. oenos* |
| C2 | $8.6 \times 10^4$ | *L. oenos* | $7.9 \times 10^6$ | *L. oenos* | $7.6 \times 10^5$ | *L. brevis* |

LAB = Lactic acid bacteria
*Source*: Salih *et al.*, 1990

## 5.10. Biochemical Changes During Aging

Production of aldehydes as one of the flavouring compounds takes place as a result of auto-oxidation of polyphenolic compounds and oxidation of ethanol by direct chemical reaction with air (Wildenradt and Singleton, 1974). Alcohols in wine react with organic acids, like tartaric, malic, succinic and lactic acid to form their respective esters, which have been reported to increase with the aging of wine (Amerine *et al.*, 1980). The concentration of total volatile compounds also increases during fermentation as well as in storage. Higher alcohols formation has been found to be closely related to the aroma and taste of wine. During maturation, a decrease in the tannins due to their complexing with protein and polymerisation and subsequent, precipitation takes place (Amerine *et al.*, 1980).

### 5.11. Final Treatment and Packaging

The desired product determines the final treatments that cider receives. Different batches of cider, generally made from a mixture of different juices, are blended to give a specific flavour. To make cider with no haze, it is desirable to treat the raw cider with fining agents organic or inorganic agents, such as bentonite, gelatin or chitin, silica solution, albumen, casein, isinglass and tannin, and filtered. A number of invisible components, such as polymeric carbohydrates, proteins and polyphenols are present in the wine. If the negative charge in wine is altered by adding positively-charged particles, it becomes neutralised. Particles combine and are flocculated or coagulated and clear wine can be recovered after removal of these sediments. The use of cross-flow microfiltration systems has also been used to obviate the need for fining and reducing the processing time and labour requirement (Jarvis, 1993). Cider can be sold as a still or sparkling beverage with varying degrees of sweetness and clarity. The amount of carbonation ranges from saturation for ciders in jars to 2-2.5 volumes of $CO_2$ in most bottled ciders, and up to 5 volumes in Champagne cider. Carbonation pressure ranges between 2.5-3.5 bar, higher pressure is being used in case of PET bottles (Jarvis, 1993). The sweetening may be from unfermented juice sugars, added juice or concentrate, sweetening agents, depending upon the appropriate regulations. In terms of clarity, ciders range from turbid farm cider to brilliantly clear ciders. Majority of commercial cider is filled into kegs, bottles or cans. Keg cider is carbonated and pasteurised in-line and filled into stainless steel kegs in a plant which rinses, washes and sterilises the kegs prior to filling. It can also be filled in glass bottles that may be carbonated and pasteurised after filling. Common container closures are crown caps and roll-on or plastic stoppers, which have replaced corks. It is then, pasteurised at 60°C for 20-30 min. or preserved with $SO_2$ as the best practical approach. Sulphiting the cider after finishing the alcoholic and MLF is a solution which is sometimes employed in order to eliminate the bacteria that cause undesirable alterations. However, in the elaboration of natural cider in the Basque Country, this practice is rarely used (Dueñas *et al.*, 2002). It is preferable to keep the addition of chemicals to a minimum in order to maintain the sensory qualities of the final product.

Results indicate that UV light is effective for reducing pathogen, like *E. coli* in cider. However, with the dosages used, additional reduction measures are necessary to achieve the required 5-log reduction (Wright *et al.*, 2000). The effect of pulsed electric field (PEF) in inactivating naturally occurring microorganisms (yeast and molds) in freshly squeezed apple cider in a continuous flow system was investigated by Liang *et al.* (2006) and reported that the microbial count decreased with an increase of applied pulses (17.6-58.7 total) and treatment temperature (45-50°C), and a decrease of flow rate (3-10 l/h). At field strength of 27-33 kV/cm (3 mm electrode gap in a concentric chamber), 200 pulses/s, 3 l/h flow rate, and 50°C process temperature, there was a 3.10 log reduction in microbial counts (Table 14).

**Table 14.** Effect of PEF Using a Continuous Flow System on Microbial Inactivation in Apple Cider

| *Treatment* | *Field strength (kV/cm)* | *Flow rate (l/h)* | *Pulses applied* | *Pulse rate (pulse/s)* | *Number of microorganisms* | | | |
|---|---|---|---|---|---|---|---|---|
| | | | | | 45°C | | 50°C | |
| | | | | | $\log_{10}N_0$ | $\log_{10}N_0/N$ | $\log_{10}N_0$ | $\log_{10}N_0/N$ |
| PEF | 33 | 3 | 58.7 | 200 | 4.20±0.73 | 2.65±0.17a | 6.67±0.18 | 3.1±0.10 |
| | | 6 | 29.3 | 200 | 7.52±0.11 | 1.90±0.17b | 7.44±0.16 | 2.20±0.19 |
| | 27 | 10 | 17.6 | 200 | — | — | 4.52±0.16 | 1.12±0.12c |
| PEF + nisin-lyso (1:3) | 27 | 10 | 17.6 | 200 | — | — | 5.44±0.09 | 1.78±0.20b (0.02±0.00)[a] |
| PEF +clove oil, 3 ml/100 ml | 27 | 10 | 17.6 | 200 | — | — | 5.26±0.21 | 2.88±0.36a (0.06±0.02) b |
| PEF + clove oil, 5 ml/100 ml | 27 | 10 | 17.6 | 200 | — | — | 4.36±0.10 | 3.11±0.23a (0.07±0.03) b |

*Source:* Liang *et al.*, 2006

Aluminium-spotted caps are not satisfactory for bottling of cider due to excessive corrosion but vinyl spotted caps are satisfactory for this purpose. Inert gases like $CO_2$, $N_2$ or their mixture can also be used for storage of cider (Cant, 1960). Locally sold, still cider may be sold in plastic containers used for or dispensed from a refrigerated bulk container. When refrigerated, cider remains stable for a week. Carbonated cider is either sterile-filtered or flash-pasteurised before packaging. Various requirements to be fulfilled for labelling bottled cider as select cider (Champagne) process are enumerated in Table 15.

**Table 15.** Requirements for Labelling Bottled Cider as Select Cider (Champagne) Process

- To be made from clean sound cider apples only or from a blend (including pears).
- No sweeteners other than cane sugar or beet can be added.
- No additions of concentrate or other fruit juice, foreign acids, artificial essences or artificial carbonation be done.
- To the undiluted juice or battery diffusion juice may be added not more than 25 per cent of its own volume of syrup made from pure cane or beet sugar. The original specific gravity of the pure juice and cider to which the syrup was added must not be less than 1.040 at 15°C.
- No preservative or colouring matter prohibited by the Public Health Regulations be added.
- Acetic acid should not be discernible on the palate and volatile acidity not to exceed 0.15 per cent as acetic acid.
- To be free of disorders.
- The last stages of fermentation must take place in the bottle and the deposit removed by disgorging

# 6. Quality of Cider

## 6.1. Chemical Composition of Cider

The most important compounds formed during fermentation which are considered key products effecting sensory profile of cider are higher alcohols, esters, organic acids, carbonyl compounds, sugars and tannins (Table 16). Except for extensive hydrolysis by pectolytic and cellulolytic enzymes, the composition of fermented products, especially the flavour components, remained almost similar in the products obtained by mechanical or mild enzymatic extraction process (Poll, 1993).

### *6.1.1. Ethyl Alcohol*

Different types of ciders are classified according to their ethanol content, which varies from 0.05-13.6 per cent (Amerine *et al.*, 1980; Jarvis *et al.*, 1995).

### *6.1.2. Acids*

The acids are important in maintaining the pH low enough so as to inhibit the growth of many undesirable bacteria. Like apple, must cider contains a variety of organic acids and their concentration depends on the maturity and fermentation conditions (Beechand Carr, 1977; Labelle, 1980). Sweet cider could have less than 0.45 gm acid. In dry ciders made by traditional method of fermentation, i.e. in which the apples are not washed, have a high amount of volatile acidity (1 g/l) than the ciders made after washing and blending of apples. In storage, the acetic acid is generally increased. In traditional method of fermentation, malic acid in must is low (3-3.8 g/l) but in the must made by modern fermentation high concentration (4.8 g/l) is observed because of acidic apples. The complete degradation of L-malic acid was carried out rapidly by LAB in all the musts, except that made by modern methods of fermentation, where no MLF occurred.

### *6.1.3. Higher Alcohols*

The formation of higher alcohol is an important criterion to determine the quality of the alcoholic beverages but varies from strain to strain of yeast, cultivars used and fermentation conditions employed (Amerine *et al.*, 1980). The biosynthesis of higher alcohols is generally linked with amino

**Table 16.** Analysis of 15 Commercial Ciders

| *Parameters* | *Range* | *Mean* |
|---|---|---|
| TSS (°B) | 4.60-7.40 | 6.8 |
| Volatile acidity (g/100 ml) | 0.060-0.105 | 0.926 |
| Tannins (mg/L) | 45-100 | 68 |
| Esters (mg/100 ml) | 17.60-28.72 | 21.76 |
| Acidity (% malic acid) | 0.40-0.69 | 0.55 |
| Sweetness (%, w/v) | 1.56-5.58 | 2.80 |
| Alcohol (%, w/v) | 3.2-6.6 | 4.71 |
| Tannins (%, w/v) | 0.028-0.17 | 10.10 |
| Total sulphur dioxide ( ppm) | 64-189 | 130 |
| Total nitrogen (ppm) | 18-63 | 42 |
| Thiamin (μg/ml) | all<0.005 | <0.005 |
| Nicotinic acid (μg/ml) | 0.03-0.33 | 0.16 |
| Pantothenate (μg/ml) | 0.10-0.80 | 0.38 |
| Riboflavin (μg/ml) | 0.41-4.7 | 1.35 |
| n-Propyl alcohol (ppm) | 4-27 | 12 |
| Isopropyl alcohol (ppm) | 24-82 | 45 |
| n-Butyl alcohol (ppm) | 3-6 | 5 |
| 2- and 3-Methyl butyl alcohol (ppm) | 113-176 | 150 |
| n-Hexyl alcohol (ppm) | 2-29 | 10 |
| 2-Phenethanol (ppm) | 51-160 | 79 |
| Magnesium (ppm) | 8-41 | 27 |
| Chloride (ppm) | 33-146 | 112 |
| Phosphate (ppm) | 20-195 | 100 |
| Sulphate (ppm) | 120-380 | 227 |
| Sodium (ppm) | 30-275 | 123 |
| Potassium (ppm) | 415-1420 | 722 |
| Iron (ppm) | 0.95-6.73 | 3.7 |
| Carbon (ppm) | 0.10-1.05 | 0.42 |
| Zinc (ppm) | 0.21-1.77 | 0.56 |

*Source*: Jarvis *et al.*, 1995; Vyas and Kochhar, 1993

acid metabolism. Higher alcohols are formed as by-products of both anabolic (Genevois pathway) and catabolic metabolism (Ehrlich pathway) and allow the re-equilibrium of the redox balance involving $NAD^+$/NADH cofactors (Hammond, 1986). Therefore, they may appear *via* biosynthesis route using the amino acid biosynthesis pathway of the yeast or by the deamination and decarboxylation of amino acids present in the substrate. It is also known that higher fusel alcohols are generated from cloudy rather than clear juice fermentation (Beech, 1993). Table 17 shows the amounts of certain higher alcohols present before and after fermentation in different varieties of apple.

### *6.1.4. Tannins*

Tannins enhance the sensory qualities of wines by affecting their astringency level, which vary in cider from 50-100 mg/100 ml (Azad *et al.*, 1987). Polyphenols play an important role in the cider quality as they are related to the colour, bitterness and astringency, whose balance defines the overall mouthfeel of the beverage (Guyot *et al.*, 1998; Alonso-Salces *et al.*, 2001; Lea and Drilleau, 2003; Alonso-Salces, *et al.*, 2004). They may be involved in providing the cider aroma, and as inhibitors of the microorganism development, controlling the fermentation rates and avoiding some faults that can develop in cider from the

**Table 17.** The Major Higher Alcohols in Apple Juices and Ciders

| *Type of higher alcohols* | *Content (ppm)* | | | | | | | |
|---|---|---|---|---|---|---|---|---|
| | *Yarlington Mill* | | *Sweet Coppin* | | *Kingston Black* | | *Bramley's seedling* | |
| | *Juice* | *Cider* | *Juice* | *Cider* | *Juice* | *Cider* | *Juice* | *Cider* |
| n-Propanol | 0 | 52 | 0.6 | 46 | 0 | 34 | 3 | 44 |
| n-Butanol | 35 | 2 | 16 | 1 | 40 | 34 | 8 | 6 |
| Isobutanol | 0 | 6 | 1 | 6 | 0.5 | 6 | 2 | 25 |
| n-Pentanol | 0.05 | 0.01 | 0.2 | tr. | 0.5 | 0.3 | 0.1 | 0.1 |
| 2-and 3-Methyl butanol | 2 | 96 | 1 | 107 | 9 | 105 | 8 | 90 |
| Hexanol | 4 | 4 | 3 | 1 | 11 | 9 | 6 | 4 |
| 2-Hexanol | 1 | 0.05 | 1 | 0.05 | 0.1 | 0.2 | 0.3 | 0.1 |
| 2-Phenethanol | 0 | 19 | 0 | 34 | 0 | 30 | 0 | 19 |

action of lactic acid bacteria, such as acidification, mannitol taint, 'framboisé', bitterness (Alonso-Salces *et al.*, 2004). Furthermore, the phenolic compounds participate in the formation of sediments during cider storage due to their colloidal interaction with the proteins through the van der Waals forces (Siebert *et al.*, 1996; Kawamoto and Nakatsubo, 1997). They can also inhibit the pre-fermentative clarification enzymes (Cowan, 1999). The cider-making steps mainly responsible for the extraction and content of the phenols in the final product are maceration, pressing, enzymatic clarification of the must prior to fermentation, centrifugation, filtration and fining. During the maceration and pressing time, intensive oxidation of the polyphenols takes place, due to the activity of the polyphenoloxidase (PPO) and the subsequent coupled oxidation reactions with other polyphenols. In addition, a large proportion of the procyanidins from the fruits remain in the pomace after the pressing step because of their adsorption onto the cell-wall matrix (Renard *et al.*, 2001; Guyot *et al.*, 2003). These lead to musts with lower phenolic content (Siebert *et al.*, 1996). It has been proved that must oxidation was higher when it was in contact with the apple pulp. The enzymatic clarification, centrifugation, filtration and fining of the French ciders lead to partial elimination of procyanidins due to their ability to precipitate proteins and to interact with cell wall polysaccharides (Alonso-Salce *et al.*, 2004).

The type of polyphenols or tannins found in the bittersweet English cider are listed in Table 18. No significant change is seen in the content during fermentation although the chlorogenic, caffeic and p-coumaryl acids may be reduced to dihydroshikimic acid and ethyl catechol (Jarvis, 1993). The chlorogenic and caffeic acid in apple juice cultivars and ciders correlated very well with total phenols. The chlorogenic acid constitutes 6.2-10.7 per cent of total phenols and the involvement of these acids is responsible for non-enzymic auto-oxidative browning reaction (Cilliers *et al.*, 1989; Cilliers *et al.*, 1990).

**Table 18.** Polyphenols in Bitter Sweet English Cider

| *Polyphenol type* | *Conc. in cider (mg/100 ml)* |
|---|---|
| Chlorogenic acid | 98 |
| Epicatechin | 38 |
| Dimericprocyanidin | 79 |
| Trimericprocyanidin | 26 |
| Tetramericprocyanidin | 21 |

*Source*: Jarvis, 1993

### 6.1.5. Carbonyl Compounds

The most important carbonyl compounds formed in cider fermentation are acetaldehydes, diacetyl and 2,3-pentanedione. Aldehydes, having very low flavour thresholds, tend to be considered as off-flavour (green leaf-like flavour). As intermediates in the formation of ethanol and higher alcohols from amino acids and sugars, the conditions favouring alcohols production also generate the formation of small quantities of aldehydes. They are excreted and then, reduced to ethanol during the later stage of fermentation. Diacetyl makes an important contribution to the flavour of cider and its presence is considered essential for correct flavour.

Aldehyde is a by-product of alcoholic fermentation and its low amounts are considered responsible for the flavour and taste of wine. The yeast strains affect the formation of flavour compounds in wine (Jarvis *et al*., 1995). Diacetyl and acetaldehyde may also be produced if the process is inhibited by excess sulphite and/or if controlled lactic fermentation occurs. Methanol is also produced in small quantities (10-100 ppm) as a result of demethylation of pectin juice (Jarvis, 1993).

### *6.1.6. Total Esters*

Generally, esters are present in smaller concentrations than alcohols (Table 19), with the notable exception of ethyl acetate and 2- and 3-methyl butyl acetates, which in Yarlington Mill juice, increase 200-fold during fermentation. Esters constitute a major group of desirable compounds. Among the esters that can be formed, the most significant in fermented beverages are ethyl acetate (fruity), isoamyl acetate (pear drops), isobutyl acetate (banana like), ethyl hexanoate (apple like), and 2-phenyl acetate (honey, fruity, flowery). These esters are formed by yeast during fermentation in a reaction between alcohols, fatty acids, co-enzyme A (COASH) and an esters synthesising enzyme (Nedovic *et al*., 2000). The supplementation of amino acids was also found responsible for the production of esters in ciders. The addition of aspartate, asparagine and glutamate positively influenced the production of esters in the cider models. In addition, the combination of aspartate and glutamate predicted a higher production. The optimal suggested concentrations were 43.4 per cent of aspartate and 56.6 per cent of glutamate for 120 mg/L of total nitrogen. The apple must supplemented with these two amino acids resulted in production of four times more esters than in the same cider without supplementation (Eleutério dos Santos, 2016).

**Table 19.** Concentration of Major Esters in Apple Juices and Ciders of Different Varieties

| *Major esters* | *Content (ppm)* | | | | | | | |
|---|---|---|---|---|---|---|---|---|
| | *Yarlington Mill* | | *Sweet Coppin* | | *Kingston Black* | | *Bramley's seedling* | |
| | *Juice* | *Cider* | *Juice* | *Cider* | *Juice* | *Cider* | *Juice* | *Cider* |
| Ethyl acetate | 2 | 35 | 1 | 20 | 1 | 17 | 2 | 15 |
| Isobutyl acetate | - | 0.2 | - | 0.003 | - | 0.1 | 0.03 | 0.3 |
| Ethyl butyrate | 0.3 | - | 0.01 | 0.01 | 0.2 | 0.4 | 0.3 | 0.1 |
| 2- and 3-Methyl butyl acetates | 0.15 | 30 | 0.02 | 3 | 0.2 | 4 | 0.1 | 0.9 |
| Ethyl-methyl butyrate | 0.006 | - | - | - | 0.01 | - | 0.04 | - |
| Hexyl acetate | 0.6 | 6 | 0.3 | 0.1 | 0. 2 | 1.5 | 0.3 | 0.7 |
| Ethyhexanoate | - | 2 | - | 0.02 | - | 0.6 | - | 0.9 |
| 2- and 3-Methyl butyl octanoates | - | 4 | - | 0.1 | - | 0.7 | - | 0.01 |

## 6.2. Sensory Qualities

Appearance, colour, aroma, taste and subtle taste factors, such as flavour constitute the quality of cider. The aroma and taste are very complex and depend on a number of factors, such as varieties, agricultural land, vinification practices, fermentation and maturation (Gayon, 1978). The taste of cider is determined more by the apple composition whereas cider odour is governed by the technological factors and yeast employed than the apple varieties used to make cider. Cider with higher juice content was preferred to that with lower juice content in various sensory quality characteristics (Joshi *et al.*, 1991). The effect of addition of insoluble solids, pectolytic enzyme and strains of wine yeast has been evaluated using various descriptors (Joshi *et al.*, 2013). Influence of cider-making technology on low boiling-point compounds can be clearly seen in preparation of semi-sweet cider (Mangas *et al.*, 1993; Mangas *et al.*,1993). A comparison of the concentration of volatile compounds produced in cider made by batch and continuous fermentation has also been made. The cider flavour is assessed using both by subjective and objective approaches. In human beings, flavour sensation by taste is limited to sweetness, sourness, bitterness and astringency together with such tastes as metallic and pungency (Piggott, 1988). Quantitative descriptive analysis (QDA) has also been applied to cider to profile their flavour analytically (Williams, 1975). Out of 86 descriptors used, 33 descriptors had greatest meaning to characterise the cider aroma and perry essence. The development of a cider flavour wheel (Fig. 2) like that for wine and beer flavour profile is employed in cider industry. In cider, bitterness and astringency are due to the polyphenols, especially procyanidins which are polymers of catechins. The degree of polymerisation (DP) is the main factor that influence the ratio between the two sensations: small procyanidins (up to DPn 4) are rather bitter and higher (DPn 5-9) are rather astringent but both sensations are usually associated. There is often an interaction with other constituents of the beverage: alcohol and polysaccharides reduce astringency, while pH can increase it without changing the bitterness. Sugars are also known to reduce acidity and bitterness (Quére *et al.*, 2006).

At a simple level, a number of general descriptors can be used, such as fruity, acidity, sweetness, astringency, alcohol, body, bitterness and sulphury, but at the analytical level, the number of descriptors

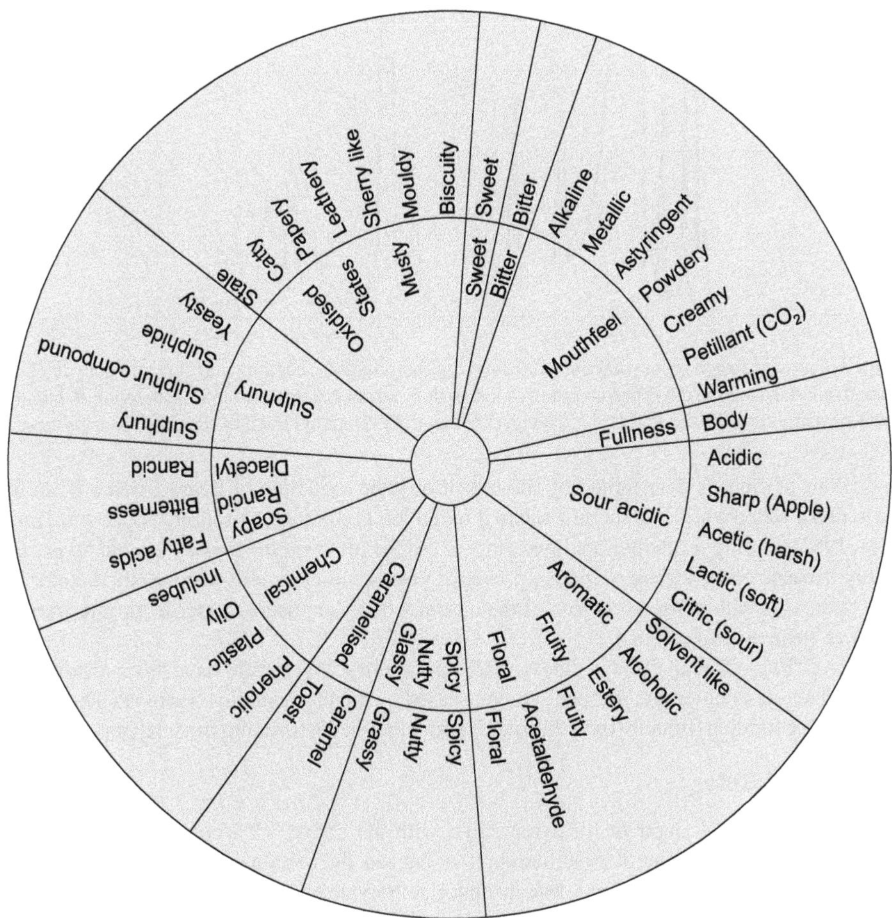


**Figure 2.** Cider flavour wheel (*Source*: National Association of Cider Maker, 1994)

are kept large to differentiate the ciders of different types. Another approach which has been applied to the flavour profiling is sniff analysis, where the effluent from GC is assessed by specially-trained judges. Capillary gas chromatography (GC) on head-space samples of cider has been made to characterise the aroma compounds (Fig. 3) (Jarvis *et al.*, 1995). As many as 200 compounds reportedly contribute to the flavour of ciders, but the key compounds are alcohols, acids, aldehydes, esters and sulphur compounds. The spicy, aromatic and apple-like are the notes which differentiate the cider from other fermented beverages.

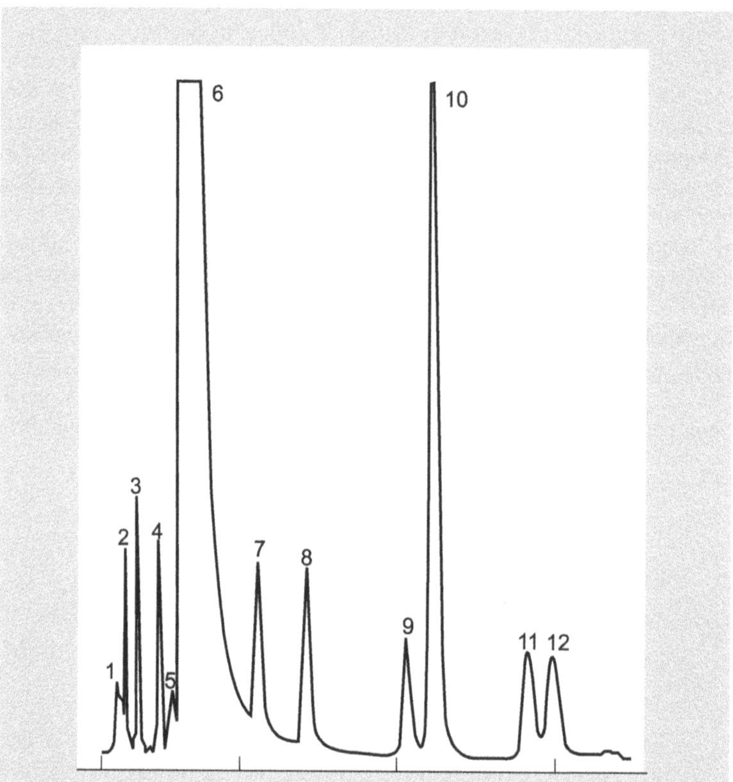


**Figure 3.** Chromatograph of a synthetic solution of low-boiling point components: 1. acetaldehyde; 2. ethyl formate; 3. ethyl acetate; 4. ethyl propionate; 5. methanol; 6. ethanol; 7. propanol; 8. isobutanol; 9. butanol; 10. 4-methyl 2-pentanol (internal standard); 11. 2-methyl butanol; 12. 3-methyl butanol (*Source*: Mangas *et al.*, 1993a)

The colour of cider is determined by the extent of juice oxidation or degradation and, in fact, it is possible to make water-white high tannin ciders if oxidation is completely inhibited (Lea and Timberlake, 1978; Lea, 1982). During fermentation, however, the initial colour diminishes by around 50 per cent. It is presumably because of the strong reductive power of yeasts, which readily reduces the keto or carbonyl groups to hydroxyls with consequent loss of the chromophore (exposure to sterile air after fermentation will slowly re-generate the colour).

Traditional English and French ciders made from bittersweet fruit have been distinguished by relatively high levels of bitterness and astringency caused by the procyanidins (tannins). The cultivars and juice-processing condition (notably oxidation) also play a part in determining the final non-volatile flavour.

## 6.3. Spoilage of Cider

Some ciders with residual sugar or the sweet cider with pH above 3.8 stored at ambient temperature develops a defect called ropiness or oiliness. It is caused by certain strains of lactic acid bacteria (*Lactobacillus* and *Leuconostoc* spp.) that produce a polymeric glucan (Carr, 1983, 1987), which thickens its consistency and when poured, it appears oily in texture with a detectable sheen. At higher concentration of glucan, the texture thickens so that the cider moves as a slimy ‘rope’ when poured from a bottle. Properly sulphited juice with a low pH can correct the defect (Beech and Carr, 1977). A much simpler approach would be to pasteurise the affected juice. However, the treated cider would need blending before use. Another defect is referred to as mousiness (Tucknott and Williams, 1973). The exact cause of this defect is not known but it occurs in unsulphited cider with a high pH that has necessarily been exposed to air during fermentation. The growth of film-forming yeasts, such as *Brettanomyces* spp., *Pichia membranefaciens*, *Candida mycoderma* also produce 'mousy' flavour (1,4,5,6-tetrahydro-2-acetopyridine). *Sacchromycodes ludwigii* is often resistant to $SO_2$ levels (1000-1500 ppm). These can grow slowly during fermentation and maturation, resulting in production of butyric flavour and formation of flaky particles, which spoil the appearance of cider. Contamination of final product with *Saccharomyces cerevisiae*, *S. bailli* and *S. uvarum* increases the concentration of $CO_2$.

Ciders low in acidity, tannin and nitrogen but high in mineral matter occasionally develop an olive green colour; the fermentation ceases and starch is deposited. If iron in the ciders combines with tannins, a black or greenish black colour develops. Bottled cider stored at high temperature sometimes produces a sediment called *casse*. The action of peroxidase on tannins causes this defect, which can be prevented by the addition of $SO_2$ after fermentation.

The classical microbiological disorder of stored bulk ciders is known as 'cider sickness' or '*Framboise*' in French (Beech and Carr, 1977; Carr, 1987) which is caused by the bacterium *Zymomonas anaerobia* which ferments sugar in bulk sweet ciders stored at pH values greater than 3.7. The features of cider sickness are a renewed and 'almost explosive fermentation' accompanied by a raspberry or banana-peel aroma and a dense white turbidity in the beverage due to production of acetaldehyde at high levels by *Zymomonas*. The acetaldehyde reacts with the 'tannins' to produce an insoluble aldehyde-phenol complex and consequently, turbidity.

Flavour taints in ciders may arise from the presence of naphthalene and related hydrocarbons where tarred rope had been stored adjacent to a cider keg. A new taint in ciders is caused by indole and is derived from tryptophan breakdown (Wilkins, 1975) at levels in excess of 200 ppb where its odour becomes increasingly faecal and unpleasant.

### 6.4. Waste from Cider Industry

Wastes and wastewater generated during the cider-making process were identified as potential sources to obtain value added products, such as ethanol, via alcoholic fermentation mediated by yeast. This wastewater comprises the purges from the fermentation process, cider losses during transfers, products discarded due to quality policies and products that have returned from the market past the expiration date. The wastewater also exhibit a high Chemical Oxygen Demand (COD), with values greater than 170,000 mg $O_2$/l, due to its elevated sugar and ethanol content, and usually represents approximately 10 per cent of the volume of cider produced. Therefore, it must be treated prior to discharge into the environment.

## References

Alonso-Salces, R.M., Barranco, A., Abad, B., Berrueta, L.A., Gallo, B. and Vicente, F. (2004). Polyphenolic profiles of basque cider apple cultivars and their technological properties. *Journal of Agricultural and Food Chemistry* 52: 2938-2952.

Alonso-Salces, R.M., Korta, E., Barranco, A., Berrueta, L.A., Gallo, B. and Vicente, F. (2001). Determination of polyphenolic profiles of Basque cider apple varieties using accelerated solvent extraction. *Journal of Agricultural and Food Chemistry* 49: 3761-3767.

Amerine, M.A., Kunkee, R.E., Ough, C.S., Singleton, V.L. and Webb, A.D. (1980). Technology of Wine Making. AVI Publ. Co., Westport, Connecticut.

Azad, K.C., Vyas, K.K., Joshi, V.K. and Srivastava, M.P. (1986). Deacidification of fruit juices for alcoholic fermentation. *Abst. ICOFOST-86*, 53.

Azad, K.C., Vyas, K.K. Joshi, V.K. and Sharma, R.P. (1987). Observations on juice and cider made from scabbed apple fruit. *Indian Food Packer* 41(1): 56-61.

Babsky, N.E., Toribio, J.L. and Lozano, J.E. (1986). Influence of storage on the composition of clarified apple juice concentrate. *J. Food Sci.* 51: 564.

Beech, F.W. (1972). Quick determination of adequate juice sulphiting. *J. Inst. Brew*. 78: 477.

Beech, F.W. and Carr, J.G. (1977). Cider and perry. p. 139. *In*: A.H. Rose (Ed.). Economic Microbiology, vol. VI: Alcoholic Beverages. Academic Press, London.

Beech, F.W. (1993). Yeasts in cidermaking. p. 169. *In*: A.H. Rose and J.S. Harrison (Eds.). The Yeasts, second ed., vol. 5: Yeast Technology. Academic Press, London.

Beveridge, T., Franz, K. and Harrison, J.E. (1986). Clarified natural apple juice: Product storage stability of juice and concentrate. *J. Food. Sci*. 51: 411.

Bump, V.L. (1989). Apple processing and juice extraction. pp. 53-82. *In*: Donald, L. Downing (Ed.). Processed Apple Products. AVI Publishing Co., New York.

Bureau of Alcohol, Tobacco and Firearms (1998). Implementation of Public Law 105-34, Sections 908, 910 and 1415, related to hard cider, semi-generic wine designations, and wholesale liquor dealers' signs (97-2523). US Department of the Treasury. *Federal Register* 63(162): 44819-44820, http://www.atf.gov/regulations-rulings/rulemakings/notices/notice-859.html

Burroughs, L.F. and Pollard, J.P. (1954). Annual Report of the Agricultural and Horticultural Research Station. Long Ashton, 1953, p. 184, University of Bristol, England.

Burroughs, L.F. (1973). Report. Long Ashton, 1972, p. 124, University of Bristol, England.

Campo, G.D., Berregi, I., Santos, J.I., Duenas, M. and Gud Irastorza, A. (2008). Development of alcoholic and malo-lactic fermentations in highly acidic and phenolic apple musts. *Bioresource Technology* 99: 2857-2863.

Cant, R.R. (1960). The effect of nitrogen and carbon dioxide treatment of wine on dissolved oxygen levels. *Am. J. Enol. Vitic.* 11: 164.

Carr, J.G. (1983). Microbes I have known. *J. Appl. Bacteriol.* 55: 383.

Carr, J.G. (1987). Microbiology of wines and ciders. p. 291. *In*: J.R. Norris and G.L. Pettipher (Eds.). Essays in Agricultural and Food Microbiology. John Wiley, London.

Castelli, T. (1973). Lecologie Des Levures. *Collogue Inter. D's Conolgie Arcensenas, Vignasvins Mai* 19: 25.

Cilliers, J.J.L., Singleton, V.L. and Lamuela-Raventos (1990). Total polyphenols in apples and ciders: Correlation with chlorogenic acid. *J. of Food Sci.* 55(5):1458-1459.

Cilliers, J.J.L. and Singleton, V.L. (1991). Non-enzymic autoxidative phenolic browning reactions in a caffeic acid model system. *J. Agric. Food Chem.* 37: 1298-1303.

Cowan, M.M (1999). Plant products as antimicrobial agents. *Clinical Microbiology Reviews* 12: 564-582.

Dinsdale, M.G., Lloyd, D. and Jarvis, B. (1994). Membrane potential studies of *Saccharomyces cerevisiae* during cider fermentation. *In*: Biochemical Society Transactions. 650th Meeting, Cardiff, p. 325.

Downing, Donald, L. (1989). Processed Apple Products. AVI Publishing Company, New York.

Dueñas, M., Irastorza, A., Fernandez, A.B. and Huerta, A. (1994). Microbial populations and malo-lactic fermentation of apple cider using traditional and modified methods. *J. Food Sci.* 59(5): 1060-1064.

Dueñas, M., Irastorza, A., Munduate, A., Santos, J.I., Berregi, I. and del Campo, G. (2002). Influence of enzymatic clarification with a pectin methylesterase on cider fermentation. *J. Inst. Brew*. 108: 243-247.

Eleutério dos Santos, M.C., Alberti, A., Arruda Moura Pietrowski, G. de, Ferreira Zielinski, A.A., Wosiacki, G., Nogueira, Alessandro and Matos Jorge, R.M. (2016). Supplementation of amino acids in apple must for the standardisation of volatile compounds in ciders. *J. Inst. Brew*, DOI 10.1002/jib.318

FAO (2018). Production Year Book. Food and Agriculture Organisation, Rome, www.fao.org

Frazier, W.C. and Westhoff, D.C. (1995). Food Microbiology. Tata McGraw-Hill Publishing Co. Ltd, New Delhi.

Garcia, Y.D., Valles, B.S. and Lobo, A.P. (2009). Phenolic and antioxidant composition of by-products from the cider industry: Apple pomace. *Food Chemistry* 117: 731-738.

Gary, L.M., Renee, T.T. and Justin, R.M. (2007). Reduction of malic acid in wine using natural and genetically enhanced microorganisms. *Am. J. Enol. Vitic*. 58: 341-345.

Gayon, P.R. (1978).Wine flavour. p. 335. *In*: G. Charlambous and G.E. Inglett (Eds.). Flavour of Food and Beverage Chemistry and Technology. Academic Press, Inc, New York, London.

Guillermin, P., Dupont, N., LeMorvan, C., Le Quéré, J.-M., Langlais, C. and Mauget, J.C. (2006). Rheological and technological properties of two cider apple cultivars. *LWT* 39: 995-1000.

Gullon, B., Yanez, R., Alonso, J.L. and Parajo, J.C. (2008). L-Lactic acid production from apple pomace by sequential hydrolysis and fermention. *Bioresource Technology* 99: 308-319.

Guyot, S., Marnet, N., Laraba, D., Sanoner, P. and Drilleau, J.F. (1998). Reversed-phase HPLC following thiolysis for quantitative estimation and characterisation of the four main classes of phenolic compounds in different tissue zones of a French cider apple variety. *J. Agric. Food Chem*., 46: 1698-1705.

Guyot, S., Marnet, N., Sanoner, P., Drilleau, J.-F. (2003). Variability of the polyphenolic composition of cider apple (*Malus domestica*) fruits and juices. *Journal of Agricultural and Food Chemistry* 51: 6240-6247.

Gebhardt, S.E., Cutrufelli, R. and Mathews, R.H. (1982). Composition of Foods. *Agric. Handbook* 8-9, US Dept of Agri., Washington.

Hang, Y.D. and Woodams, E.E. (2010). Influence of apple cultivar and juice pasteurisation on hard cider and ean-de-vie methanol content. *Bioresource Technology* 101: 1396-1398.

Hammond, J. (1986). The contribution of yeast to beer flavor. *Brew. Guardian* 115: 27-33.

Herrero, M., Laca, A., Garcia, L.A. and Diaz, M. (2001). Controlled malo-lactic fermentation in cider using *Oenococcus oeni* immobilised in alginate beads and comparison with free cell fermentation. *Enzy. Microb.Technol*. 28: 35-41.

Jarvis, B. (1993). Chemistry and microbiology of cidermaking. *In*: Encyclopedia: Food Science and Nutrition, first ed. Coleraine Campus, Cromore Road, Coleraine, BT52ISA, Belfast Coleraine Jordans town, Magee.

Jarvis, B. (1993). Cider: Hard cider. *In*: Encyclopedia: Food and Nutrition, first ed., Coleraine Campus, Cromore Road, Coleraine, BT52ISA, Belfast Coleraine Jordans town, Magee.

Jarvis, B., Foster, M.J. and Kinsella, W.P. (1995). Factors affecting the development of cider flavour. *J. Appl. Bacteriol*. Symp. supp., 79: 55.

Joshi, V.K. and Bhutani, V.P. (1991). The influence of enzymatic clarification on fermentation behaviour and qualities of apple wine. *Sci. Des Aliments*. 11(3): 491-496.

Joshi, V.K., Sandhu, D.K., Attri, B.L. and Walia, R.K. (1991). Cider preparation from apple juice concentrate and its consumer acceptability. *Indian J. Hort.* 48(4): 321-327.

Joshi, V.K. and Sandhu, D.K. (1994). Influence of juice contents on quality of apple wine prepared from apple juice concentrate. *Res. Ind.* 39(4): 250-252.

Joshi, V.K. and Sandhu, D.K. (1997). Effect of different concentrations of initial soluble solids on physicochemical and sensory qualities of apple wine. *Indian J. Hort*. 54(2): 116-123.

Joshi, V.K., Sandhu, D.K. and Thakur, N.S. (1999). Fruit-based alcoholic beverages. pp. 647-744. *In*: V.K. Joshi and Ashok Pandey (Eds.). Biotechnology: Food Fermentation, vol. II. Educational Publishers and Distributors, New Delhi.

Joshi, V.K. and Siby John (2002). Antimicrobial activity of apple wine against some pathogenic and microbes of public health significance. *Alimentaria* November: 31(2): 67-69.

Joshi, V.K., Sandhu, D.K. and Kumar Vikas (2013). Influence of addition of insoluble solids, wine yeast strains and pectinolytic enzymes on the flavour profile of apple wine. *International Journal of Food and Fermentation Technology* 3(1): 57-66.

Kawamoto, H. and Nakatsubo, F. (1997). Effects of environmental factors in two-stage tannin-protein co-precipitation. *Phytochemistry* 46: 379-483.

Kerni, P.N. and Shant, P.S. (1984). Commercial Kashmir apple for quality cider. *Indian Food Packer* 38(1): 78.

Labelle, R.L. (1979). The many faces of (hard) cider. *N.Y. State Agric. Exp. Stn. Spec. Rep*. 32: 1.

Labelle, R.L. (1980). Apple cultivars tested as naturally fermented cider at Geneva. Memo, N.Y. *State Agric. Exp. Stn.* Geneva, New York.

Lea, A.G.H. (1978). Phenolics of cider – Procyanidins. *J. Sci. Food Agric*. 29: 484-492.

Lea, A.G.H. (1982). Analysis of phenolics in oxidising apple juice by HPLC using a pH shift method. *J. Chromatog.* 238: 253.

Lea, A.G.H. (1984). Colour and tannins in English cider apples. *Flussiges Obst.* 51: 356.

Lea, A.G.H. (1995). Cidermaking. p. 66. *In*: A.G.H. Lee and J.R. Piggott (Eds.). Fermented Beverage Production. Blackie Academic and Professional, London, UK.

Lea, A.G.H. and Timberlake, C.F. (1974). Phenolics of Ciders – The effect of processing. *J. Sci. Fd. Agric.* 25: 1537-1545.

Lea, A. and Drilleau, J.-F. (2003). Cidermaking. pp. 59-87. *In*: Lea, A. (Ed.). Fermented Beverage Production. London.

Liang Ziwei, Cheng, Z. and Mittal, G.S. (2006). Inactivation of spoilage microorganisms in apple cider using a continuous flow-pulsed electric field system. *LWT* 39: 350-356.

Lowe, E., Durkee, E.L., Hamilton, W.E. and Moyan, A.I. (1964). Bitter apple juice dejuicing through thick cake extraction. *Food Eng.* 36(12): 48-50.

Mahant, K., Sharma, S., Sharma, S. and Thakur, A.D. (2017). Effect of nitrogen source and citric acid addition on wine preparation from Japanese persimmon. *Journal of the Institute of Brewing* 123(1): 144-150.

Mangas, J., Paz-Gonzallez, M. and Blanco, D. (1993a). Influence of cider-making technology on low-boiling point volatile compounds. *Zest Schrift fur labensmittelutersuchung Forschung*, 197(6): 522-524.

Mangas, J.J., Moreno, J., Cabranes, C., Dapana, E. and Blanco, D. (1993b). Study of semi-sweet cider. *Alimentaria* 243: 85.

Mitra, S.K. (1991). Apples. p. 122. *In*: S.K. Mitra, T.K. Bose and D.S. Rathore (Eds.). Temperate Fruits. Horticulture and Applied Publ., Calcutta.

Moulton, G.A., Miles, C.A., King and Zimmerman, J.A. (2010). Hard Cider Production & Orchard Management in the Pacific Northwest. A Pacific Northwest Extension Publication. Washington State University Extension, Oregon State University Extension Service, University of Idaho Cooperative Extension System and the US Department of Agriculture Co-operative.

Nedovic, V.A., Durieux, A., Van Nedervelde, L., Rossels, P., Vandegans, J., Plainsant, A.M. and Simon, J.F. (2000). Continuous cider fermentation by co-immobilised yeast and *Leuconostoc oenos* cells. *Enz. Micro. Technol.* 26: 834.

Nicolini, G., Román, T., Carlin, S., Malacarne, M., Nardin, T., Bertoldi, D. and Larcher, R. (2017). Characterisation of single variety still ciders produced with dessert apples in the Italian Alps. *J. Inst. Brew.*, DOI 10.1002/jib.510

Nogueira, A., Guyot, S., Marnet, N., Lequere, J.M., Drilleav, J.F. and Wosiacki, G. (2008). Effect of alcoholic fermentation in the content of phenolic compound in cider processing. *Brazilian Archives of Biology and Technology* 51(5): 1025-1032.

O' Reilly, A. and Scott, J.A. (1993). Use of an ion-exchange sponge to immobilise yeast in high gravity apple-based cider alcoholic fermentation. *Biotechnol.Letters* 15(10): 1061-1069.

Park, Y.J. and John, C.B. (1980). Decomposition of acid in wine by yeast. *Res. Rep. Agric. Sci. Technol.* Chungnam Nat, Univ., Daejeon, S. Korea, 7(2): 176.

Picinelli, A., Suarez, B., Moreno, J., Rodrıguez, R., Caso-Garcıa, L.M. and Mangas, J.J. (2000). Chemical characterisation of Asturian cider. *Journal of Agricultural and Food Chemistry* 48: 3997-4002.

Piggott, J.R. (1988). Sensory Analysis of Foods, second ed. Elsevier Applied Science, London, New York.

Pliny (the Elder), Reissue (1967). Natural History. Loeb Parallel text ed., Heinemann, London.

Poll, L. (1993). The effect of pulp holding time and pectolytic enzyme treatment on the acid content of apple juice. *Food Chem.* 47(1): 73-75.

Pourlx, A. and Nicholas, L. (1980). Sweet and Hard Cider. Gardenway Publishing Co., Charlotte, V.T.

Quéré Jean-Michel Le, Husson Franc-ois, Catherine, Renard, M.G.C. and Primault, Jo (2006). French cider characterisation by sensory, technological and chemical evaluations. *LWT*, 39: 1033-1044.

Rana, R.S., Vyas, K.K. and Joshi, V.K. (1986). Studies on production and acceptability of cider from Himachal Pradesh apples. *Indian Food Packer* 40(6): 48-56.

Renard, C.M.G.C., Baron, A., Guyot, S. and Drilleau, J.F. (2001). Interactions between apple cell walls and native apple polyphenols: Quantification and some consequences. *International Journal of Biological Macromolecules* 29: 115-125.

Revier, M. (Ed.) (1985). Le cidre-heiretaujourd'hui. La Nouvelle Libraire, Paris.

Reedy, D., Mcclactchey, W.C., Smith, C., Lan, Y.H. and Bridges, K.W. (2009). A monthful of diversity: Knowledge of cider apple cultivars in the United Kingdom and northwest United States. *Economic Botany* 63(1): 2-15.

Ribereau-Gayon, J., Penaud, E., Ribereau-Gayon, P. and Sudraud (1975). Traited'oenologie. *Sciences ettechniquies du vin*, vol. 2, Dunod, Paris.

Roach, F.A. (1985). Cultivated Fruits in Britain – Their Origin and History. Basil Blackwell, Oxford.

Salih, A.G., Le Quere, J.M., Drilleau, J.F. and Fernandez, J.M. (1990). Lactic acid bacteria and malo-lactic fermentation in manufacture of Spanish cidermaking. *J. Inst. Brew*. 96: 369-372.

Seluy, L.G., Comelli, R.N., Benzzo, M.T. and Isla, M.A. (2018). Feasibility of bioethanol production from cider waste. *J. Microbiol. Biotechnol*. 28(9): 1493-1501.

Sharma, R.C. and Joshi, V.K. (2005). Apple processing technology. pp. 445-498. *In*: K.L. Chadha and R.P. Awasthi (Eds.). The Apple. Malhotra Publish. House, New Delhi.

Sharma, S., Thakur, A.D., Sharma, S. and Attenasova, M. (2018). Effect of different yeast species on the production of pumpkin wine making. *Journal of the Institute of Brewing* 124(2): 187-193.

Siebert, K.J., Carrasco, A. and Lynn, P.Y. (1996). Formation of protein-polyphenol haze in beverages. *Journal of Agricultural and Food Chemistry* 44: 1997-2005.

Smock, R.M. and Neubert, A.M. (1950). Apple and Apple Products. Interscience Publishers, New York.

Sonni, F., Cejudo, B., Maria, J., Chinnici, F., Natali, Nadia and Riponi Claudio (2009). Replacement of sulphur dioxide by lysozyme and oenological tannins during fermentation: Influence on volatile composition of white wines. *Science Food Agri.* 89(4): 688-696.

Tucknott, O.G. and Williams A.A. (1973). Report. Long Ashton, 1972, p. 150, University of Bristol, England.

Unwin, T. (1991). Wine and the Vine: An Historical Geography of Viticulture and the Wine Trading. Routledge, London.

Upshaw, S.C., Lopez, A. and Williams, H.L. (1978). Essential elements in apples and canned apple sauce. *J. Fd. Sci.* 43(2): 449-456.

Vyas, K.K. and Kochhar, A.P.S. (1993). Studies on cider and wine from culled apple fruit available in Himachal Pradesh. *Indian Food Packer* 47(4): 15. 63-69.

Whiting, G.C. (1961). Annual Report of the Agricultural and Horticultural Research Station. Long Ashton, 1960, p. 135, University of Bristol, England.

Wildenradt, H.L. and Singleton, V.L. (1974). The production of aldehydes as a result of oxidation of polyphenolic compounds and its relation to wine aging. *Am. J. Enol. Vitic*. 25: 119-126.

Wenlai, F., Yan, Xu and Yu, Aimei (2006). Influence of oak chips geographical origin, toast level, dosage and aging time on volatile compounds of apple cider. *J. Inst. Brew.* 112(3): 255-263.

Wilkins, C.K. (1990). Analysis of indole and skatol in porcine gut contents. *Inter. J. Fd. Sci. Technol*. 25: 313-317.

Williams, A.A. (1975). The development of vocabulary and profile and profile assessment method for evaluating the flavour contribution of cider and perry aroma constituents. *J. Sci. Fd. Agric*., 26: 567-582.

Wong, M. and Stanton, D.W. (1993). Effect of removal of amino acid and phenolic compounds on non enzymatic browning of stored kiwifruit juice concentrate. *Lebensmittel-Wissenching und technologie* 26: 138.

Wright, J.R., Summer, S.S., Hackney, C.R., Pierson, M.D. and Zoechlein, B.W. (2000). Efficacy of ultraviolet light for reducing *Escherichia coli* O157:H7 in unpasteurised apple cider. *Food Protection* 63: 563-567.

Wrolstad, R.E., Spanos, G.A. and Durst, R.W. (1990). Changes in phenolics and amino acid profiles of apple juice concentration during processing and storage. *Berichte Intenational Fruchtsaft-union, Wissenchaftlich-Technische Kommission* 103. www.google.com. www.nhb.org (Hard Honey Cider).

# 23 Brandy Production: Fundamentals and Recent Developments

**Francisco López**

Departament d'Enginyeria Química, Facultat d'Enologia, Universitat Rovira i Virgili Av. Països Catalans 26, 43007 Tarragona

## 1. Introduction

Brandy is a spirit obtained through the distillation of wine and generally contains 35-60 per cent v/v of ethanol. If the name of the brandy is not associated with the type of raw material originating from this spirit (fruit brandy, grain brandy, pomace brandy, etc.), it is understood that it is made exclusively from grape wine. The origin of the word 'brandy' comes from the Dutch *brandewijn*, whose meaning is 'wine burned'. Some brandies are aged in wooden barrels, some are coloured with caramel to emulate the effect of aging and/or homogenise the final product, and some brandies are produced using a combination of aging and colouration (Amerine *et al.*, 1989; Christoph and Bauer-Christoph, 2007).

Different types of brandy are made all over the world from wine. The best known are produced in France under the appellation of Cognac and Armagnac. In other countries, different types of brandies are made from wine, for example, brandy of Jerez and brandy of Penedès in Spain, Italian brandy produced from regional wine grapes and distilled by column stills, although there are also a number of low-scale producers, which employ pot stills. German brandy, which is called *weinbrand* (burnt wine), is made usually from imported wine. The most known South American brandy is Pisco. In Peru, it is made mainly from Muscat grapes. In Chile, it is made from different grape varieties and is distilled in pot stills. It is obvious that, worldwide, there are various legal definitions according to the national traditions and commercial interests (Tsakiris *et al.*, 2014).

The use of the word 'brandy' to define a spirit obtained from distillation of wine leads to confusion, since it is sometimes used as a generic of a product and sometimes to define a specific type of distillate. European legislation distinguishes wine spirits and brandies (European Union 2008). Among the first are Cognac and Armagnac. Table 1 shows the characteristics of these two types of spirit drinks.

**Table 1.** Characteristics of Wine Spirit and Brandy According to European Legislation

| *Spirit drink* | *Wine spirit* | *Brandy or weinbrand* |
|---|---|---|
| Maximum distillation alcoholic strength | 86% v/v | 94.8% v/v |
| Minimum alcoholic strength | 37.5% v/v | 36% v/v |
| Maximum commercial alcoholic strength | Not indicated | 50% v/v |
| Maturation time | Not indicated, but if it is matured as a brandy can be labelled as 'wine spirit' | 1 year oak receptacle<br>6 months oak casks <1000 L |
| Volatile substances | >125 g/hL p.a. | >125 g/hL p.a. |
| Methanol content | <200 g/hL p.a. | <200 g/hL p.a. |
| Flavouring | Not, except traditional production methods | Not, except traditional production methods |
| Caramel addition | Only to adapt colour | Only to adapt colour |
| Addition of alcohol | No | No |

E-mail: francisco.lopez@urv.cat

In United States, brandies are defined according to the US Code of Federal Regulations, Title 27: Alcohol, Tobacco and Firearms PART 5, Sub-part C, Standards of Identity for Distilled Spirits, as class 4: 'Brandy' is an alcoholic distillate made from fermented juice, mash, or wine of fruit, or from the residue thereof, produced at less than 190°C, in such a manner that the distillate possesses the taste, aroma and characteristics generally attributed to the product, and bottled at not less than 80°C. Brandy, or mixtures thereof, not conforming to any of the standards in paragraphs (d) (1) through (8) of this section shall be designated as 'brandy', and such designation shall be immediately followed by a truthful and adequate statement of composition. The fruit brandy, derived from grapes, shall be designated as 'grape brandy' or 'brandy', except that in the case of brandy (other than neutral brandy, pomace brandy, marc brandy or grappa brandy) distilled from the fermented juice, mash, or wine of grapes, or the residue thereof, which has been stored in oak containers for less than two years, the statement of class and type shall be immediately preceded, in the same size and kind of type, by the word 'immature'. Fruit brandy, other than grape brandy, derived from one variety of fruit, shall be designated by the word 'brandy' qualified by the name of such fruit (for example, 'peach brandy'), except that 'apple brandy' may be designated 'applejack'. Fruit brandy derived from more than one variety of fruit shall be designated as a 'fruit brandy' qualified by a truthful and adequate statement of composition.

## 2. History of Brandy and Economical Aspects

### 2.1. Some Historical Aspects

Distillation was a technique already used by ancient cultures in China (3000 years BC), India (2500 years BC), Egypt (2000 BC), Greece (1000 years BC) and Rome (200 BC). Initially, all these cultures produced a distilled liquid, later called alcohol by the Arabs, for the preparation of medicines and perfumes. In the 7th century, the Arabs invaded Europe, introducing the technique of distillation there. Later, in Christian Europe, the doctor and theologian Arnau de Vilanova (Valencia, 1238?-1311) published the book *Liber Aqua Vitae*, a treatise on wines and spirits, which was a manual of the period for the production of distillates (Lopez *et al.*, 2017).

Until the end of the 15th century, distilled wine (*aqua vitae*) was a product used as a medicine and doctors and apothecaries controlled the distillation technique. From the authorisation to distil to vinegarmakers by Louis XII and later to the winegrowers by Francisco I, in France, its elaboration like product of consumption in the 16th century increased. In the wine trade with Holland, to reduce costs, the wine was transported as distillate, which was then reconstituted with water. However, this wine concentrate was consumed as such, being called '*Brandewijn*', whose name evolved to brandy. Although, the Dutch had provided much of the original capital together with the stills and the technology necessary for distillation, local producers had also invested in the production of brandy and they were, therefore, forced to seek new markets for their products. Elsewhere throughout France, but particularly in Armagnac, other wine producers had also begun to distill their wines, especially those that were of poor quality. However, brandy gradually came to be made where the raw material was available. Later, it spread to southern wine-growing lands, such as Andalusia, Catalonia and Languedoc (Dhiman and Attri, 2011).

### 2.2. Economical Aspects of Brandy

The Wiseguy report (2018) indicates that despite the slowdown in global economic growth, the spirits industry has also suffered a certain impact, but still maintained a relatively optimistic growth in the past four years. Spirits market maintain the average annual growth rate of 1.86 per cent from 218,200 million $ in 2014 to 230,600 million $ in 2017, the analysts believe that in the next few years, spirits' market size will be further expanded. They expect that by 2022, the market size of the spirits will reach 239,200 million $. Figure 1 shows the brand values of spirits worldwide in 2017 (Statista, 2018), with 8.0 per cent of spirits brand value.

The top markets for brandy are India, Philippines, Russia, Brazil and Germany. Grape-based brandy-producing countries have a significant-sized wine industry that provides the base wine for brandy production. Brandy production and consumption tends to be highly regionalised, with the exception of

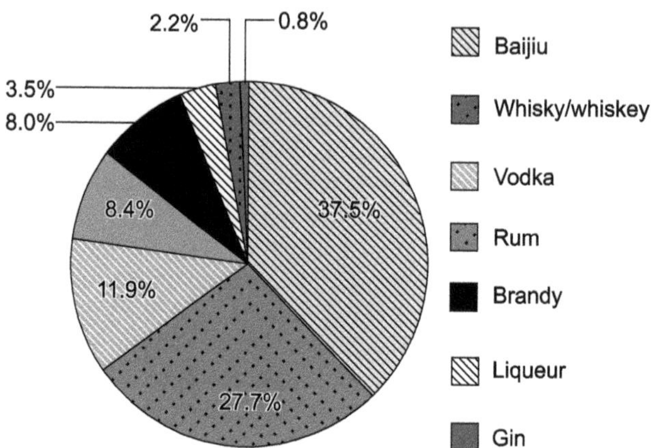


**Figure 1.** Brand values share of spirits worldwide in 2017, by spirit type (Statista, 2018)

Cognac, that is known and consumed globally. The production methods thus differ significantly (Bougas, 2014; Lambrecths *et al*., 2016).

However, in recent years the trend towards the consumption of artisanal distillates has increased and the consequent growth in the number of distilleries, due to a greater perception of new consumers of these artisanal products which are of a higher quality (Kohlmann, 2016).

## 3. Techniques Used in Brandy Elaboration

### 3.1. Wine Characteristics

For the production of one litre of brandy, around 4.5 litres of wine are needed and during the distillation, the aromas of the wine are concentrated. These aromas are produced mostly during fermentation, which is why grape varieties with a lower aromatic potential and higher productivity are usually used to make the wines that are then distilled (Tsakiris *et al*., 2014). The characteristics of the wines elaborated for their distillation must minimise the presence of sulphur dioxide and be distilled as soon as possible. The presence of phenolic compounds should be minimised; therefore, continuous presses or the use of pectolytic enzymes is not recommended, as they can increase the presence of methanol (Tsakiris *et al*., 2016). Wines with lees affect the aromatic profile and their complexity; so sometimes in different types of brandies or wine distillates, the regulation authorises its presence in wine. Another important aspect is the preservation of wines before fermentation, since they can pass periods of up to five months between obtaining the wine (case of Cognac and Armagnac), is to control the presence of bacteria that can impart strange aromas to wine, which can be increased in the distillate (Amerine *et al*., 1989; Tsakiris *et al*., 2014).

### 3.2. Distillation Systems

For the distillation of wine, different distillation systems are used throughout the world. These systems range from the simple alembic (batch distillation system) to continuous distillation systems with distillation columns. Usually batch distillation produces more complex products, since during the distillation, the distillate varies in a wide range of ethanol concentrations, dragging more variety of minor compounds to the final distillate, while the continuous distillations are carried out at a fixed and higher alcoholic strength, so that the final distillate has fewer congeners.

The traditional alembic (Fig. 2), which is formed by a kettle with deposits of the wine to distill. Over the kettle is placed the hat, in which a small rectification of distillate is produced, which allows obtaining products with an alcoholic strength higher than that expected by an ideal and simple differential distillation. The hat is connected to the total condenser by the swan neck, in which the distillate is still rectified gently. Before the distillate passes to the total condenser, it can go through a preheater of wines, so that part of the energy supplied in the kettle is used to preheat the wine for a second batch of distillation.

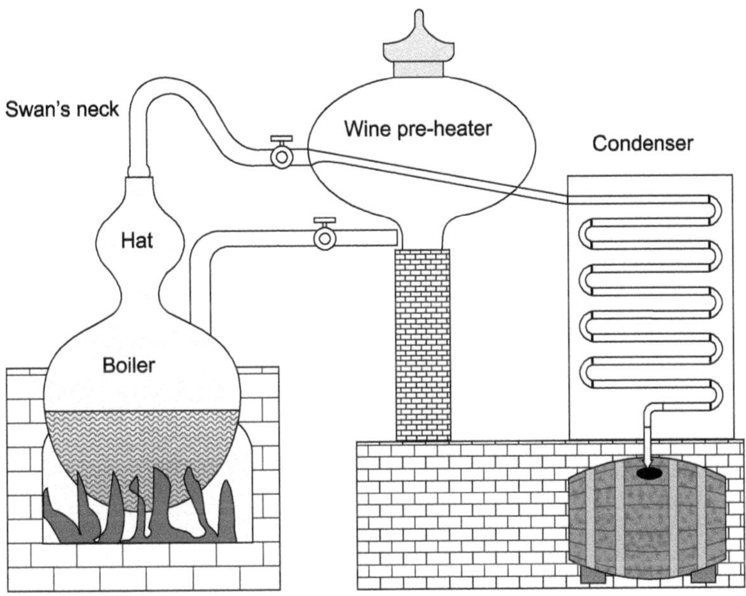


**Figure 2.** Schema of traditional alembic (Charantais type) (Adapted from Bureau National Interprofessionel de Cognac, http://www.cognac.fr/cognac/_fr/2_cognac/index.aspx?page=distillation)

In the total condenser, the vapours of the distillate are finally condensed and cooled to a low temperature to collect finally the distillate.

The hat shape can affect the characteristics of the distillate, since the rectification produced will depend on it surface to generate more or less reflux inside the equipment. Hats with a large surface area will produce less aromatic distillates and contain fewer compounds with more body, such as the longer chain fatty acids (Bougass *et al.*, 2014). For example, the hats used in making cognacs are onion-shaped and therefore small; so these products are more complex and aromatic. Another effect of the lower reflux generated with small hats is that it minimises the re-concentration of water in the kettle, causing the concentration of ethanol in the waste to be higher and, therefore, the recovery will be lower and the final product will have a minor alcoholic strength.

The distillation time will also affect the characteristics of the distillate, since during boiling, it can produce different reactions of formation or degradation of compounds (Cantagrel, 2003; Bougass *et al.*, 2014), such as equations 1.1-1.4.

| | | |
|---|---|---|
| Esterification/Hydrolysis: | Acid + Alcohol $\leftrightarrow$ Ester + Water | [1.1] |
| Acetal formation: | Aldehyde + Alcohol $\leftrightarrow$ Acetal + Water | [1.2] |
| Maillard's reaction: | Sugar + Amino acid $\rightarrow$ Pyrazine, Furans | [1.3] |
| Strecker degradation: | $\alpha$-Amino acid $\rightarrow$ Aldehydes $\rightarrow$ Acetals | [1.4] |

The operation of batch distillation systems consists in the separation of a first fraction, called 'heads', in which negative volatile compounds, such as ethyl acetate and acetaldehyde, are concentrated. So this fraction is rejected. Next, a second fraction, called 'heart', is collected in which most of the positive compounds in the distillate are concentrated. Finally, a third fraction, called 'tails', is collected, in which a significant amount of alcohol remains, as well as less compounds, which in general are organoleptically negative in the final distillation. This fraction and the head fraction are usually reintroduced in subsequent distillations in order to recover the maximum possible ethanol. The residue left in the kettle is rejected.

With this discontinuous system, the distillate obtained has a variable alcoholic strength between the beginning and the end that favours the complexity of the final distillate, as previously mentioned. This discontinuous system is applied to beverages, such as cognac and Peruvian Pisco, using different strategies.

The other classic alternative is the column distillation, operating discontinuous or continuous, as in the case of Armagnac. Figure 3 presents a scheme of this distillation system.

The distillation strategy operating in discontinuous is similar to the traditional alembic, separating the three distillation fractions (head, heart and tail). In the column top can be placed a partial condenser or dephlegmator to generate the reflux necessary to operate the column, which can also be external. In this case, the reflux generated is reintroduced through the upper part of the column. According to this distillation system, beverages, such as Armagnac and Chilean Pisco, are obtained with some design modifications and/or distillation strategies.

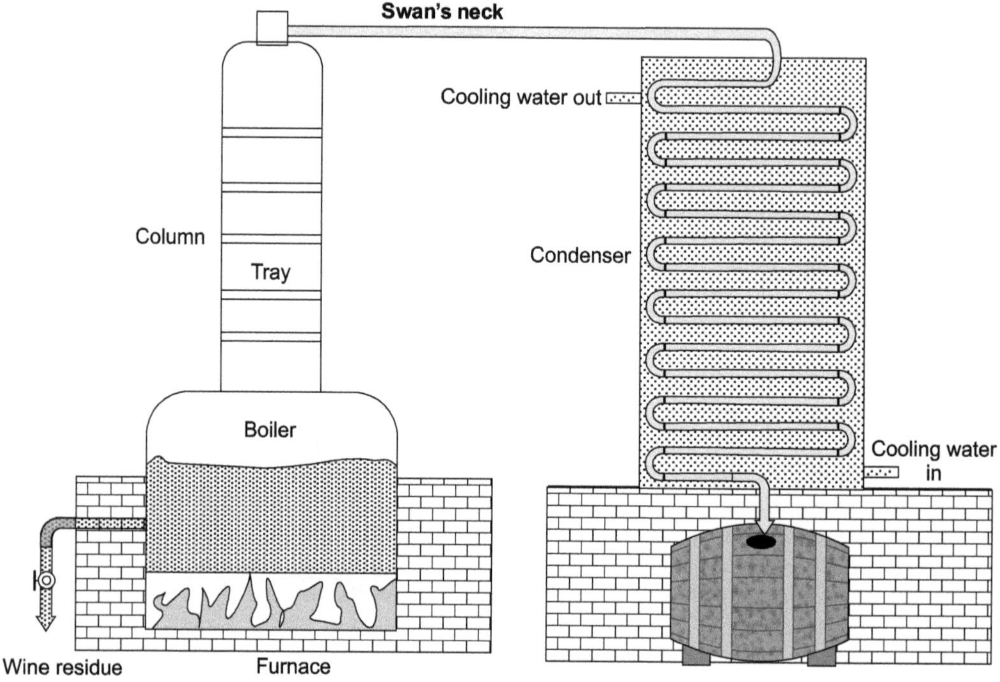


**Figure 3.** Batch column distillation (Adapted from Bureau National Interprofessionel de l'Armagnac, http://www.armagnac.fr/distillation-et-vieillissement)

In other areas, brandies are distilled in continuous still by means of distillation columns (Fig. 4). In this case, columns usually have the distillate output before the upper part, since the last trays are used to concentrate the more volatile compounds, such as acetaldehyde, ethyl acetate and others, corresponding to the head fraction in discontinuous distillation. At the bottom of the column, the residue is extracted, in which the water would be concentrated with the less volatile compounds. In this column, lateral extractions could be carried out, in which the concentration of higher alcohols is higher, so that their presence in the final distillate can be minimised. This, however, has the disadvantage that the quantity of first quality brandy obtained is reduced; in addition, this practice complicates the operation of the column (Guymon, 1974).

These distillation columns usually have a large number of trays with the goal of producing distillates at 95-96 per cent v/v, when only three or four trays are required to obtain distillates of 85 per cent v/v. It is also possible to concentrate other undesired compounds, which are eliminated at the top of the column, while the distillate is extracted in two or three trays below. In general, 60-70 per cent of the trays are located above the feed tray, while the rest are below. The source of heat is usually by open steam, although some brandy producers use indirect heating, such as reboilers (Guymon, 1974).

Another alternative is the employment of split column units as shown in Fig. 5, or systems with three or four columns (Fig. 6) in which one of them concentrates the heads (aldehydes) up to 20 times and the ethanol, devoid of aldehydes and other components of low point of boiling, is recycled to the main unit (Bertrand, 2003).

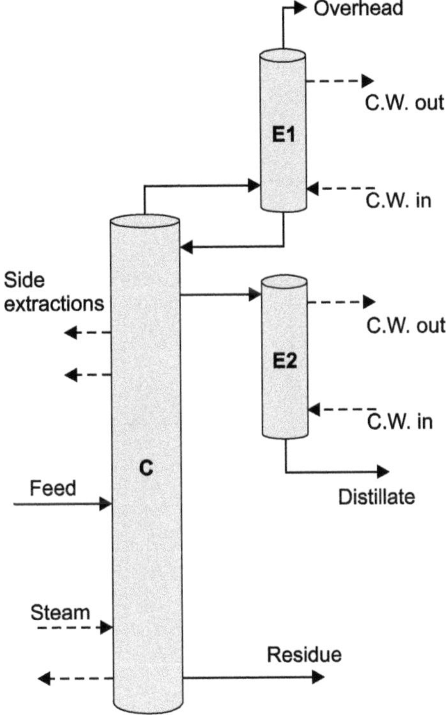


**Figure 4.** Continuous distillation column

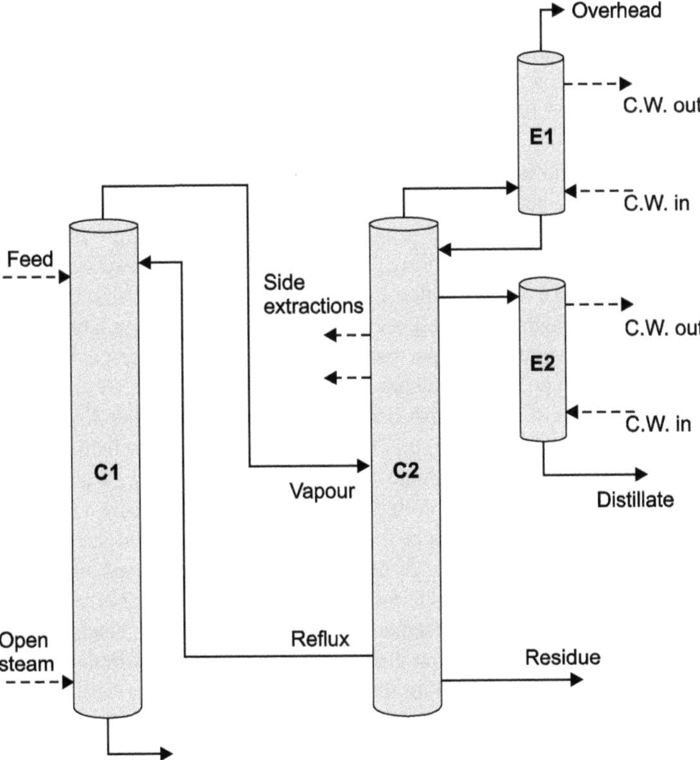


**Figure 5.** Double column distillation

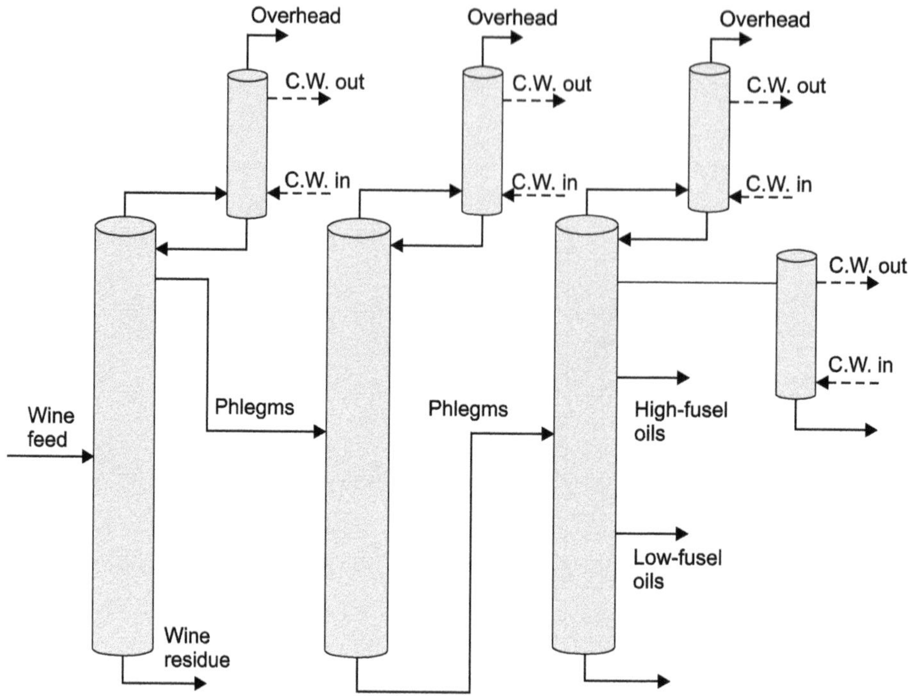


**Figure 6.** Split column with a head concentrating column

### 3.3. Distillation Strategies

In the elaboration of brandies and distillates of the wine, different distillation strategies are carried out to obtain a final distillate, which together with the subsequent ripening or aging treatments will return the commercial product. Given below are some of those currently used in the production of different brandies.

The first strategy is that in a single distillation in a still with a column or not, the head and tail fractions are separated as described in Section 1.3.2. The obtained heart is the distillate to be used for the elaboration of the final product. This strategy is applied, for example, in the production of Peruvian Pisco.

Another strategy consists in the realisation of a double distillation, known as the 'charantais' method, which is used in France to make cognac with a traditional alembic. In the first distillation, a small head fraction with an alcoholic strength around 60 per cent v/v is separated; then a heart fraction is collected with an average alcoholic strength of 28 per cent v/v (range of 60-5 per cent v/v in ethanol), called *brouillis*, and finally a tail with an average alcoholic strength of 3 per cent v/v (range of 5-0 per cent v/v in ethanol). The heads and tails of the first distillation are reintroduced in the distillation of later batches, mixing with the wine to be distilled. Then, this first heart (heart 1) is re-distilled where the heads are separated (approximately 1 per cent by volume from the initial load and an alcoholic strength around 75 per cent v/v), in which the distillate is collected between 75-60 per cent v/v, obtaining an average alcoholic strength of 70 per cent v/v. This is followed by heart 2 (called *secondes*) with an average alcoholic strength of 30 per cent v/v (range of 60-5 per cent v/v) and finally the tails with an average alcoholic strength of 3 per cent v/v (range from 5-0 per cent v/v). The heart 2 is redistilled and mixed with the *brouillis* from the first distillation in subsequent batches. Likewise, the heads and tails of this second distillation, with the heads and tails of the first distillation, are usually reintroduced in the first successive distillations, mixing with the initial wine. With this strategy at the end of a distillation campaign, almost all the alcohol from the wine distilled initially is recovered (Léauté, 1990).

Another strategy is to use a still with a column, as is the case of the Armagnac (Fig. 6), which involves introducing the wine continuously into the upper part of the column and this is stripped of alcohol as it goes down the column to the boiler. The vapour generated in the boiler produces the bubbling in the trays of the column so that it is rectified as it rises through it. The vapour, once outside the column, condenses in the condenser, using the wine that is continuously fed as a cooling fluid, which in turn warms up before entering the column. Operating in this way, the energy consumption in the boiler is reduced. The obtained distillate has an alcoholic strength of 50-54 per cent v/v. At the same time, exhausted wine (wine residue) must be extracted from the bottom intermittently to operate in a semi-continuous manner (Fig. 7).

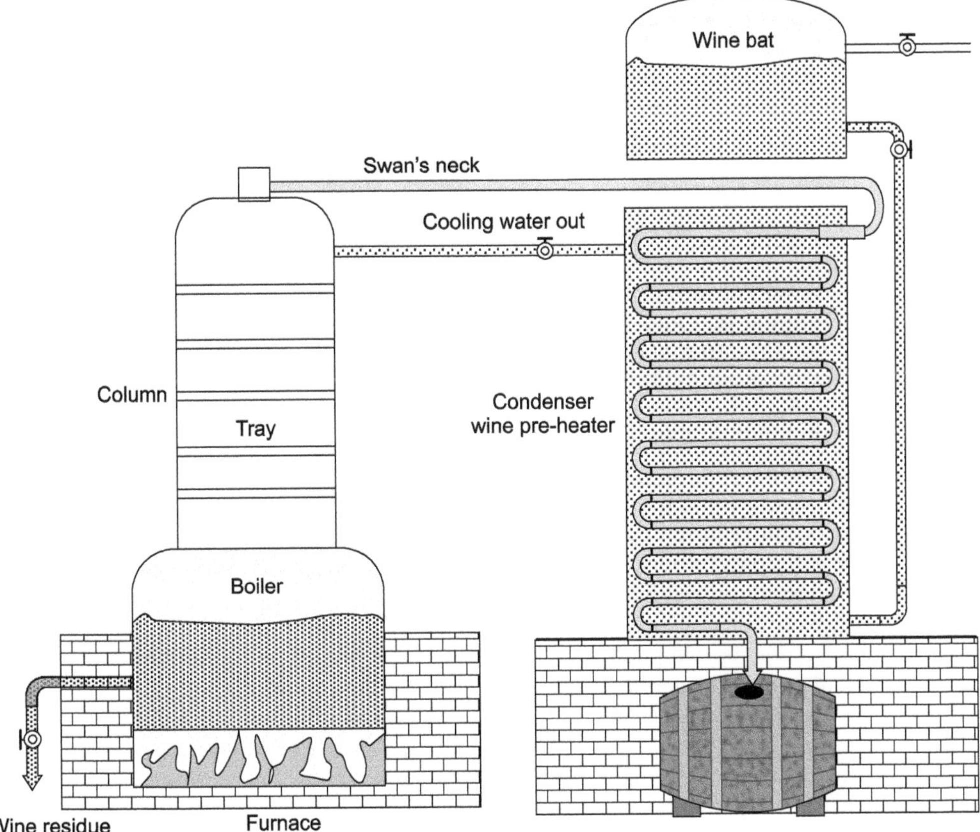


**Figure 7.** Armagnac column distillation operating in continuous (Adapted from Bureau National Interprofessionel de l'Armagnac, http://www.armagnac.fr/distillation-et-vieillissement)

On the other hand, with similar equipment in the elaboration of Chilean Pisco (Fig. 8), the distillation is usually performed discontinuously, working in devices with larger boilers, in which the wine is loaded and distilled in a discontinuous manner, separating the head, heart and tail fractions. With this strategy, the product obtained is a mixture of distillate in a range of concentrations in ethanol from 70-40 per cent v/v, being therefore, the characteristic of the final product more complex aromatically than Armagnac.

## 3.4. Aging Systems

The aging of brandy is one of the most important operations to be carried out in wooden barrels of oak. This process may last from several months to several years. Relatively new distillates are placed in barrels with a maximum capacity of 225-1000 L, while more aged, distillates can be stored in 5000 L barrels. During this period, there is a loss of 2-3.5 per cent of the distillate per year due to evaporation (Tsakiris *et al.*, 2014). The sensory attributes of the distillates, such as colour, flavour and taste are greatly influenced by the botanical species of the wood, the different heat treatments applied to wooden barrels, the times that the wooden barrel has been used and the ageing time. This treatment will largely define the quality of the final product, due to the physico-chemical processes that involve the distilled alcohol and the wood of the barrels. During aging, the brandy is oxidised slowly. Its acidity grows by oxidation of alcohol into volatile acids and by dissolution of the acid substances in the wood. Moreover, acetals are formed and their odours are softer than those of aldehydes. The odour characteristic of young brandy becomes blurred and eventually disappears, to be replaced by a vanilla odour induced by vanillic aldehyde together with other phenolic aldehydes and acids arising from the alcoholysis of lignin in oak wood (Bertrand, 2003).

An oak is a tree or shrub in the genus *Quercus*. There are around 600 species of *Quercus* worldwide, of which about 20 are economically important. Historically, the species most used in cooperage are *Q. alba* (American white oak) and two European species, *Q. robur* (pedunculate oak) and *Q. petraea* (sessile oak). In practice, they can often serve the same purpose but there are differences in wood extractives; most notably the fact that American oak wood contains lower ellagitannin content than the European species, and its wood is denser and more coarse-grained than European oak (Hornsey, 2016). Figure 9 presents the composition of oak heartwood.

**Figure 8.** Pisco distillation column

Cernîsev (2017) suggested a simplified scheme for lignin degradation during spirit maturation. Lignin consists mainly of coniferyl, p-coumaryl and sinapyl alcohols. These alcohols, relatively easy, quickly convert to coniferaldehyde, p-coumaryl aldehyde and sinapaldehyde, which then are oxidised to respective acids. Theoretically, a part of the formed phenolic acids can be esterified in the presence of alcohol. Ferulic and protocatechuic acids are also present in aged spirits: the first one is formed as result of gentle oxidation of coniferaldehyde to vanillin. Formation of protocatechuic acid can be explained by conversion via demethylation of vanillic acid or oxidation of p-hydroxybenzoic acid. Aromatic acids can be also converted to volatile phenols, such as 4-methylguaiacol, 4-ethylguaiacol, eugenol, guaiacol, syringol, 4-methylsyringol and 4-ethylphenol. Thus, the concentrations of ethyl esters of fatty acids increase during ageing, but the concentrations of esters of other alcohols, such as 3-methylbutyl acetate, decrease by transesterification (Christoph and Bauer-Christoph, 2007). The colour becomes brown by dissolution of tannin. Moreover, the taste softens with the appearance of sugars arising from the hydrolysis of wood hemicelluloses. There is also the appearance of a rancid taste by oxidation of fatty acids (Bertrand, 2003).

Among the many species of the genus *Quercus*, only a few are of major technological interest for ageing: American white oak (*Quercus alba*), sessile oak (*Quercus petraea*) and pedunculate oak (*Quercus robur*). While European pedunculate oak has high quantities of extractable ellagic tannins, sessile oak releases much smaller quantities of polyphenols and white oak even less. The American species have a greater aromatic potential than European oak due to their high content of *cis/trans* isomers of β-methyl-γ-octalactone. American white oak is easily identified by the low quantity of extractable polyphenols, the high methyl-octalactone content and the presence of two isomers of 3-oxo-retro-α-ionol. European sessile oak and American white oak are ideal for aging fine wine. Pedunculate oak, with its low aromatic potential and high ellagitannin content, is best suited to aging spirits. Proper control of toasting operations in barrelmaking could facilitate the use of this type of oak, by modelling the release of volatile and odourous substances from wood (Chatonnet and Dubourdieu, 1998; Prida and Puech, 2006). The extraction of tannins will depend on the size of the grains with lesser for fine grains than for coarse grains. One of the most important treatments is toasting, since it will affect both the aroma and the colour of the final product. For example, the hydrolysable tannins extracted in the brandy will cause different reactions, such as the formation of esters from alcohols and acids. Lignin is a wood material that releases aromatic aldehydes, such as vanillin, coniferaldehyde and syringaldehyde. Vanillin is a characteristic

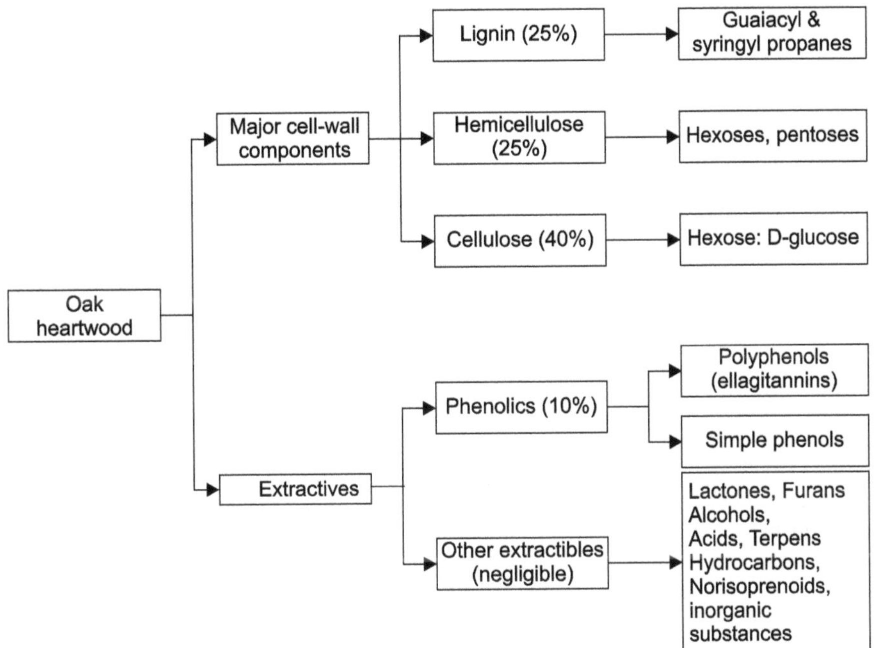


**Figure 9.** Composition of oak heartwood, where the percentages indicate the percentage of wood mass made up by the constituent (Adapted from Hornsey, 2016; Mosedale and Puech, 2003)

aroma of cognacs and other alcoholic beverages aged in oak, as it is found in large concentrations after toasting. The sweetness of the brandies is attributed to glycerol, xylose, arabinose and glucose extracted from wood. Other typical aromas due to the aging of brandies in oak are the oak lactones, since at very low concentrations they contribute aromas of coconut – an aroma associated with alcoholic beverages aged in oak. Semi-volatile and non-volatile compounds of wood change the colour of the distillate and contribute to an up-rounded flavour. The wooden barrels that are permeable allow air to pass in and cause ethanol to evaporate; thus, the ethanol content decreases and the aroma gets more intense, complex and concentrated. It has been proved that a decrease in the oxygen transfer rate is governed mainly by the advance of the moisture front in the wood (wood impregnation), in contrast to the hypothesis which attributes it to the oxygen consumption by the soluble ellagitannins of the wood (del Alamo-Sanza and Nevares, 2017). In addition, harsher aroma constituents are removed and the spirit changes to mellow. The period of maturation depends on the size of the casks used, the alcoholic strength, as well as the temperature and humidity in the warehouse, which leads to a smoother flavour (Christoph and Bauer-Christoph, 2007). Another factor to consider is the aging temperature, since it affects the aging rate as well as a faster loss of ethanol by evaporation.

Globally, the aging process is characterised by changes in flavour, aroma and colour of the brandies, as well as a decrease in the amount of product and the alcoholic strength. The aging time in oak barrels varies according to the country (Louw and Lambrechts, 2012). The European regulations fix that brandies must be aged for at least six months in oak barrels, while in United States brandies must be labelled with the word 'immature' if the period of aging is less than two years and in South Africa, the minimum time is three years.

Regarding the aging time, it will obviously affect the amount of compounds extracted from the wood, as well as the reactions that occur. Therefore, longer aging time means higher compounds extraction. According to Bertrand (2003), a barrel can yield substances to the cognac for about 40 years though several stages in aging may be distinguished:

- By 1.5-5 years, the main process is dissolution of substances in the wood.
- By 5-10 years, astringency decreases and the brandy becomes rounder.
- By 10-35 years, a rancid taste appears.
- After 40 years, one should no longer keep brandies out of the barrel.

Caldeira *et al.* (2016) studied the kinetics of the odourant compounds of a wine brandy during two years of ageing in two ageing systems on an industrial scale. The odourant compounds in the analysed brandies changed significantly over the time, but with different evolution patterns. With regards to the odourant compounds proceeding from the distillate, namely alcohols (isobutanol, 2+3-methyl-1-butanol,trans-2-hexen-1-ol, linalool, 2-phenylethanol) and acids (butanoic, isovaleric, hexanoic and dodecanoic acid) the tendency was to decrease over two years of ageing. Regarding the esters, also derived from the distillate, an inverse tendency for two esters was noticed (diethyl malate and ethyl octanoate), which increased over the time, while the other four esters (isobutyl acetate, ethyl butyrate, isoamyl acetate and ethyl hexanoate) were not affected by the ageing time. The kinetics of the majority of wood-related odourant compounds (acetic acid, furanic aldehydes, volatile phenols, vanillin, acetovanillone and cis, trans-b-methyl-c-octalactone) followed a hyperbolic pattern with a major increment at the beginning of the ageing period, along with the diffusion of the compounds from the wood into brandies. However, for some of the wood–related compounds, such as vanillin, acetovanillone, guaiacol, eugenol, 4-methylguaiacol and trans-b-methyl-c-octalactone, the initial hyperbolic increase was followed by a linear enrichment, suggesting their formation during ageing.

Cernîsev (2017), in his study, performed on 24 wine distillates with ageing time from one to 50 years and containing 69.1-43.4 per cent v/v alcohol observed that concentration of unsaturated aromatic compounds (sinapaldehyde, coniferaldehyde) decreases with increasing maturing time of the distillates (Fig. 10), probably due to their oxidation during long maturation, while the concentration of other substances, such as syringaldehyde, vanillin, syringic and vanillic acids, increases. It is also interesting to note that ratio syringaldehyde/vanillin lies between 1.80 and 2.21 almost for all distillates. The fact that the ratio of syringaldehyde/vanillin is constant ($2.0 \pm 0.2$) for most of aged distillates indicates a possible correlation between the transformations of these aldehydes.

The use of oakwood fragments, as an alternative to the traditional barrels, is a rapid and economical method of ageing treatment. Nowadays, there is no legislation applied to the ageing of spirits in contact with oakwood fragments, and as a result, research on the accelerated ageing of distillates is scarce (van Jaarsveld and Hattingh, 2012; Rodriguez-Solana *et al.*, 2017). Schwarz *et al.* (2014) carried out accelerated aging on a laboratory scale of Brandy de Jerez, employing oak chips and ultrasound as extraction method and achieved in one-month sensorial characteristics and acceptability which are similar to those of brandies whose average aging time was between six and 18 months.

Delgado-González *et al.* (2017) developed an accelerated aging process consisting of circulation of the distillate through packed oak chips and the application of ultrasound. The best conditions were obtained with wine distillate of 65 per cent (v/v), operating at room temperature (25°C) and equivalent doses of 5 g/L oak chips, using longer than fragmented ultrasound pulses, since the brandy presents

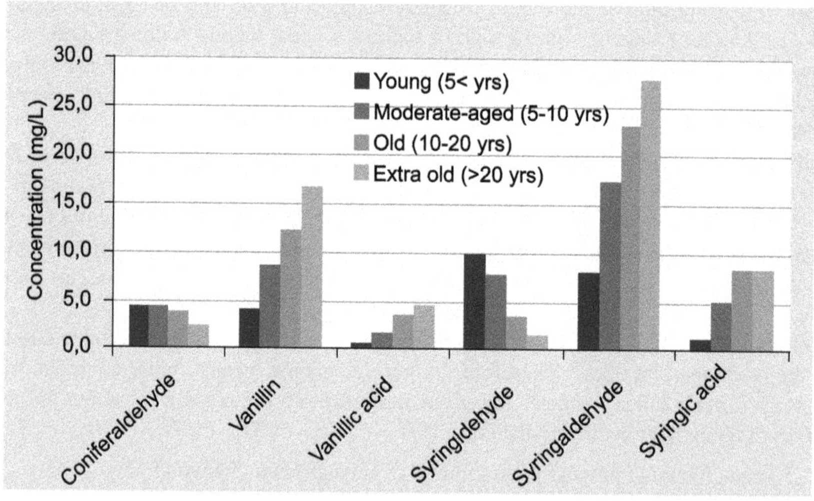


**Figure 10.** Mean concentrations of ageing markers in wine distillates from different age groups (Adapted from Cernîsev, 2017)

higher intensity of colour, TPI and extraction of furfuraldehyde, syringic acid, vanillin, syringaldehyde and aromatic intensities and visual impressions. On the other hand, the aeration is positive, since the TPI and the extraction of tartaric acid, syringaldehyde and syringic acid increases. The sensory analysis showed that the distillates aged by five distillates of different grape varieties used in the Jerez area allow brandies with characteristics of spirits aged by traditional methods in only three days.

Nevertheless, Rodriguez-Solana *et al.* (2017) for a pomace brandy conclude that in general, high toast-oak fragments provide greater colour intensity in the accelerated aged spirits but reduce the antioxidant capacity of the corresponding beverage. The best results are obtained with smaller fragment size (granular) from *Quercus petraea* with medium toast level. The contact time did not significantly influence the parameters evaluated. Caldeira *et al.* (2017) compared the influence of brandy aging in 650-L wooden barrels and in 3000-L stainless steel tanks. Both had Limousin oak (*Quercus robur*) and Portuguese chestnut staves (*Castaneas ativa*). The brandy samples were profiled by descriptive sensory analysis during the ageing period. The brandies aged in stainless steel tanks with staves presented higher intensity of attributes, such as topaz, coffee, caramel and unctuous and lower intensity of golden, woody and green attributes, and higher overall quality than the same wine distillate aged in wooden barrels. Regardless of these differences, it is not possible to clearly distinguish the brandy samples proceeding from different ageing systems by multidimensional analysis (MDS) of data. This effect seems to contain the influence of a cooperage heating process, which imparts a different volatile wood composition to the staves and wood barrels. Consequently, the brandies produced in wooden barrels are associated with higher amounts of acetic acid, 5-methylfurfural, eugenol, acetic acid, cis-b-methyl-c-octalactone and HMF, while brandies from stainless steel tanks with staves are linked to higher amounts of volatile phenols (4-methylsyringol, syringol, guaiacol, 4-methylguaiacol) and acetovanillone (Caldeira *et al.*, 2016).

Zang *et al.* (2013a) studied the effects of applying an electric field (EF) treatment with 1 kV/cm (50 Hz) on brandy stored in 5-L and 2-L oak barrels, respectively, for over 14 months to simulate the natural aging process and it was compared with brandy sample naturally aged in 225-L oak barrels. Results demonstrated that the content of phenol compounds in brandy, such as tannins, total phenols as well as volatile phenols, significantly increased after treatment by EF, with tannin concentration of the brandy treated with EF being 54.4 per cent and 43.9 per cent higher than those of the control brandy after 14 months of maturation in 5-L and 2-L barrels, respectively. It was also demonstrated that the EF-treated samples in 5-L barrels for seven months and in 2-L barrels for five months exceeded the content of tannins of those naturally aged for 12 months in 225-L oak barrels used in the brandy industry. The kinetic model of oak phenol compounds extracted by brandy demonstrated that EF treatment played a positive role in accelerating the extraction of phenol compounds and its effect was more significant than the size of the barrel. Zang *et al.* (2013b) demonstrated that the application of an EF, the ester concentration, in the brandy was higher than in the untreated samples. On the other hand, the concentration of some unpleasant compounds apparently decreased; thus, the aroma of the brandy improved.

## 4. Brandy Styles in the World

The elaboration of brandies varied considerably in the world and in many places, it has a protected geographical indication (GI) under the corresponding regulations. In some zones, practices, such as the addition of wine, must, caramel and flavouring extracts based on dried plums, raisins, walnuts and almond shells, are carried out. These spirits are normally aged in wooden casks (usually oak), a process by which colour, mouthfeel and flavour are significantly changed (Christoph and Bauer-Christoph, 2007). The characteristics of different types of brandies are described in the following section.

### 4.1. France

According to European regulations (European Union, 2008) in France, the geographical indications of Cognac and Armagnac are recognised in the category of wine spirits, and the brandy Français in the category of brandies. Cognac and Armagnac are made from wines from a limited geographical area and are not characterised by a single specific flavour, but their quality depends on factors, such as the grape varieties as well as the distillation system, aging and blending (Christoph and Bauer-Christoph, 2007).

Cognac is produced in the region of Cognac (France), located north of Bordeaux. In 1909, the French government established by means of a law that Cognac can be denominated only in a well-defined zone that surrounds the city of Cognac. The Cognac production area is divided into six zones: Grande Champagne, Petite Champagne, Borderies, Fins Bois, Bons Bois and Bois à Terroirs.

The grape varieties used are Ugni Blanc, Colombard and Folle Blanche, which have a minimum of 90 per cent of the wine destined for distillation. Although other varieties, such as Semillon, Folignan, Juraçon Blanc, Montils, Sélect and Meslier St-François are authorised but with a proportion of less than 10 per cent in the production of the wine destined for distillation.

The grape harvest takes place in the second half of October and is pressed in horizontal presses (continuous presses are prohibited), with the use of sulphur dioxide. The distillation of wine is with lees and chaptalisation is not allowed. The alcoholic strength is generally relatively low (8-10 per cent v/v). The wines are too acidic for direct consumption at pH 3 or even less. This acidity makes it possible to some extent to compensate for the absence of sulphur dioxide. However, according to regulations, the wines must be distilled before the end of March (Bertrand, 2003; Owens and Dikty, 2009).

Distillation is usually done with a copper Charantais traditional alembic (Fig. 1) and the procedure used is double distillation (Fig. 11). The volume of the boiler is 30 hL maximum. The first distillation is called '*chauffe du brouillis*' and the second distillation '*bonne chauffe*'. The distillation strategy is described in Section 1.3.2. Usually the second distillation is performed at a lower temperature for obtaining better rectification of the spirit.

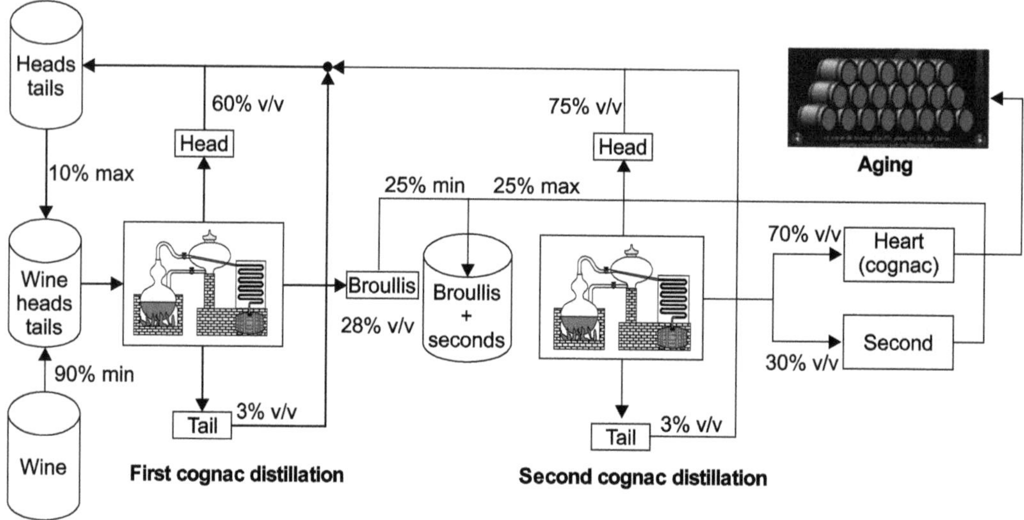


**Figure 11.** Cognac distillation process (Adapted from Leauté, 1989)

There are also variants of these two methods. Varying the intensity of heating is important according to the alcoholic strength of *brouillis* required. Slow distillation gives good rectification. An odourous, fine brandy is obtained but with dryness that may be detected on tasting due to the lack of certain products of tail distillation (e.g. ethyl lactate, diethyl succinate). In contrast, fast heating involves the formation of a marrowy brandy with little bouquet. Excessive heating results in a heavy taste (Bertrand, 2003).

The cognac is aged in oak barrels (200-600 L) from the forests of Tronçais, Allier, Limousin and the Vosges in France. During aging, there is a loss of volume (around 3 per cent on an average per year) and a 1 per cent v/v reduction in the alcoholic strength. During conservation, there is little evolution of the volatile substances. The alcohol concentrate, the esters, are slightly hydrolysed and the unsaturated fatty acids oxidise, giving a rancid taste. The Cognac ages by slow oxidation in barrels. In bottles, there is no further evolution (Bertrand, 2003).

During aging, once a year, the cognac is racked and all the barrels of the same production are mixed together. The alcoholic strength is gradually diminished by adding demineralised water to finally obtain an alcoholic strength of 40.0 per cent (v/v) in commercialised bottles (never less). During aging, it is common to mix spirits of different origins, quality and age in order to obtain the same odour and taste for a given brand name.

Accordingly the ageing period can be defined for different Cognac categories, as:

- Very special (V.S.) or three stars (***) where the youngest brandy is at least two years old
- Very Superior Old Pale (V.S.O.P) or *Réserve* where the youngest brandy is at least four years old
- Napoléon, X.O or *Hors d'âge* where the youngest brandy is at least six years old.

The colour can be adjusted with the addition of caramel. The taste can be adjusted with the addition of woody water extract obtained from small pieces of oakwood to give more body to the spirit, more astringency and a little bitterness. On the other hand, excessive hardness can be diminished by adding sucrose syrup, generally less than 8 g/L. Generally, all these procedures should be carried out at least two or three months before bottling.

Armagnac is produced in the region of Gascoigne (France), located east of Bordeaux. The Armagnac production area is divided into three zones: Haute Armagnac, Bas Armagnac and Tenareze. The main grape varieties used are Ugni Blanc, Colombard, Folle Blanche and Baco 22 (Owens and Dikty, 2009). The distillation system used is described in Section 1.3.2 (Fig. 6). The wines to be distilled can be made from musts obtained with continuous press and distilled with their lees. It is forbidden to add sulphur dioxide. Wines are required to be distilled between the end of the harvest and 31 March of the following year. According to the regulation, the maximum and minimum alcoholic strength allowed are 72 per cent (v/v) and 52 per cent (v/v), respectively.

The distillation by means of stills with columns allows regulation of the alcoholic strength of the distillate varying in wine flow and heating power. Therefore, lowering the heating or increasing the wine flow brings down the temperature at the head of the column and results in a higher alcoholic strength. In this case, higher alcohol and ester concentrations remain constant; on the other hand, the amount of substances called tailings, of which there is usually a surplus in Armagnac, decreases exponentially when the percentage of alcohol increases. For prolonged aging, a large quantity of tailings is an advantage because of the 'winey' character of their molecules; but if the Armagnac is to be marketed soon, it is preferable to make a high-proof distillate to limit the amount of such substances (Bertrand, 2003).

Armagnac is aged in oak casks and the coarse-grained wood is preferred (Gascony or Limousin) to fine-grained wood, as it is slightly more permeable to oxygen and yields more tannin. Although there are a variety of aging methods, spirits are usually kept in new casks for six months to one year before being transferred to old casks. Armagnac contains vanillin, syringaldehyde, coniferaldehyde and sinapaldehyde, but only vanillin is detectable on tasting. Prior to being marketed, several wine spirits are blended and the alcoholic strength of the blend is reduced to a minimum of 40 per cent (v/v) with distilled water. The naturally golden-yellow colour can be enhanced with caramel. Sometimes, infusions or decoctions made from oak shavings are added to make the Armagnac more astringent and to give it more body; however, these preparations must be at least of the same age as the youngest spirit used for the commercial designation of the final product. Sugar solutions are sometimes added to attenuate the 'burn' of the alcohol (about 6 g/L).

Finally, before being bottled, the spirits are cold processed (usually one week at 5°C) and passed through a cellulose filter to eliminate any possible cloudiness due to excess of calcium or fatty acids.

Armagnac comprises several hundreds of substances, but the main features of old Armagnac are its aroma of prunes, its rancid taste, its complexity; it is vigorous and even rough, with a long-lasting palate.

The blend of Armagnacs from different years of aging is authorised to obtain a more homogeneous product in quality, but the age of the blend is that of the youngest Armagnac. The Bureau National Interprofessionnel de l'Armagnac (BNIA) has harmonised the Armagnac since 2010 into four categories:

- Three Stars (***) or Very Special (V.S.) where the youngest brandy is at least one year old
- Very Superior Old Pale (V.S.O.P) or *Réserve*, where the youngest brandy is at least four years old Napoléon or Extra Old (X.O.) where the youngest brandy is at least six years old
- *Hors d'Âge*, where the youngest brandy is at least 10 years old

Lastly, the *Millésimes* (Vintage), 10 years of minimum of ageing in wood, specific to Armagnac, corresponds solely to the year of harvest declared on the label.

It has been found (Bertrand, 2003) that there are about four times less ethyl esters of fatty acids with 8, 10 and 12 carbon atoms in Armagnac than in Cognac. Some brandies contain noticeably lower quantities of esters than others. This can be attributed to the utilisation of wine distillate, resulting in poorer distillates containing volatile substances (Tsakiris *et al.*, 2014). Cognac can then be differentiated from Armagnac because this spirit contains highest contents of furan derivatives, such as furfural, 5-methylfurfural, furfuryl ethyl ether, and 2-acetylfuran due to the effect of double distillation (Ledauphin *et al.*, 2010).

## 4.2. Spain

In Spain, there are two brandies with geographical indication in accordance with the European regulation of spirits (European Union, 2008): Sherry brandies (*Brandy de Jerez*) and Penedès brandy (*Brandy del Penedès*).

Sherry brandy is produced in the area between the municipalities of Jerez de la Frontera, Sanlúcar de Barrameda and El Puerto de Santamaría, located in the province of Cádiz (Spain). The regulation on the preparation of these brandies (RDE Brandy de Jerez, 2005) allows the use of wine alcohol without specifying its origin; however, most of the alcohol used in its preparation is distilled in the region of La Mancha, located in the centre of Spain, using wines from the Airén variety.

The traditional distillation system is by means of *alquitaras*, a system similar to the charantais alembic, in which distillates obtained are denominated *holandas* with a graduation between 40-70 per cent (v/v) and content in volatile substances of 200-600 g/hL p.a. This distillation system is used to elaborate the highest quality Sherry brandies. It is also used in more modern and efficient distillation columns in which the wine is fed continuously. There are two types of columns – the so-called low-grade ones in which spirits are obtained between 70-86 per cent (v/v) and a content in volatile substances between 130-400 g/hL p.a., and the high-grade ones in which the distillates are with a graduation between 86-94.8 per cent (v/v) and a content of volatile substances less than 50 g/hL p.a. In traditional distillation with *alquitara*, the *holandas* are obtained in a single distillation unlike the Cognac, since the wines distilled have a greater alcoholic strength, conserving better the characteristics of the wines.

According to its specific regulation, sherry brandy is matured through the traditional criaderas and soleras system (Fig. 12), during which the spirit extracts its principal components from the interior surface of the oak casks, although the vintage system is also used. The barrels used for aging usually are of American oak (*Quercus alba*) with a capacity less than 1000 L, which have previously been used in the production of Sherry wines.

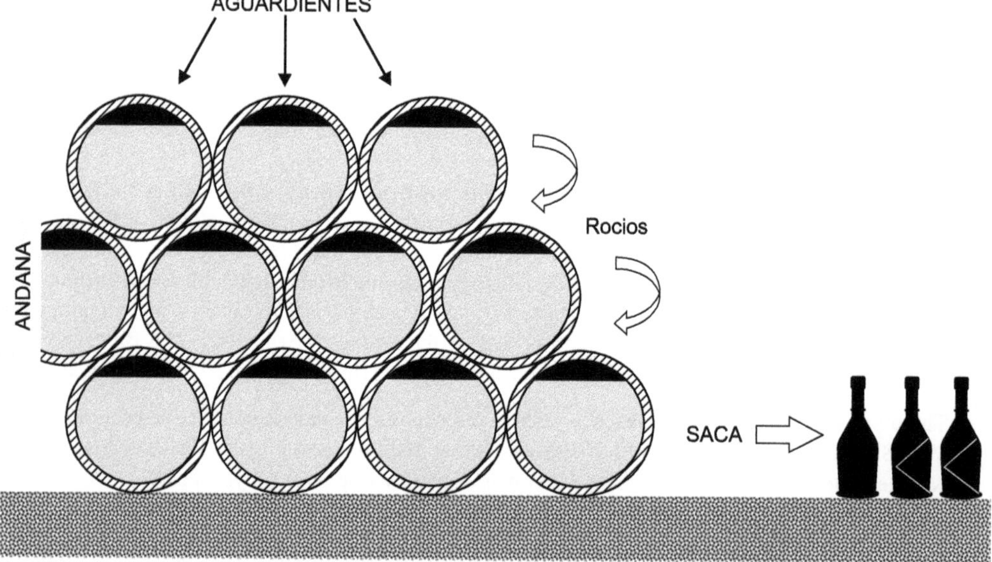


**Figure 12.** Sherry brandy traditional aging system of criaderas and soleras

According to the characteristics of the spirits and distillates of wine, and the process of elaboration and aging, three types of 'Brandy de Jerez' are distinguished: *Solera*, *Solera Reserva* and *Solera Gran Reserva*. Table 2 indicates the main characteristics of 'Brandy de Jerez'.

**Table 2.** Characteristics of 'Brandy de Jerez'

| *Brandy de Jerez type* | *Aging system* | *Aging time* | *Barrel size (L)* | *Volatile substances (g/hL p.a.)* |
|---|---|---|---|---|
| *Solera* | criaderas and solera | > 6 mon | 1000 | 150 |
| *Solera Reserva* | criaderas and solera | > 1 yr | 1000 | 200 |
| *Solera Gran Reserva* | criaderas and solera | > 3 yr | 1000 | 250 |

The *Solera* sherry brandy is the youngest and fruity, with an average aging of one year, while the *Reserva* usually has an average aging time of three years and finally the brandy *Gran Reserva* has an average of 10 years. However, in practice, the best reserves and large reserves are aged 12-15 years. With this procedure, it is possible to maintain the characteristics (flavour, aroma and colour), identical quality and peculiarities of each brand. It is usually aged at a graduation between 50-60 per cent (v/v), that allows the processes of maturation or extraction of characteristics of the wood to be more intense. In prolonged aging, the graduation is reduced to a certain extent, mainly due to the evaporation that in Jerez is intense – 7 per cent annually, due to its climate.

The taste of sherry brandies is sweet of ripe fruit, not only due to aging, but also due to the elaboration process that authorises the addition of different substances, called *Cabeceo*, in which small amounts of extracts and fruit macerations are used. According to the legislation, they can also use concentrates of grapes and plums, oak essences, almond pericarp, vanilla pods and green walnuts and can be also sweetened with natural sugars and caramel to adjust the colour. The alcohol content must be between 36-45 per cent (v/v), although the usual commercial value is 38 per cent v/v and the highest quality products are 39 per cent (v/v).

Penedès brandies are made in Catalonia, in the northeast of Spain, near Barcelona. The grape varieties used are usually Macabeo and Parellada, as well as the Ugni Blanc variety used in the elaboration of Cognac. Penedès brandy production is currently limited to two producers (Owens and Dikty, 2009). In accordance with Penedès brandy regulation (Reglament IGP Brandy del *Penedès*, 2017), they can be obtained by continuous (tray column) or discontinuous distillation (double distillation as in Cognac). The spirits suitable for the production of 'Brandy del Penedès', obtained by distilling wines, must preserve the organoleptic characteristics and volatile components of the raw material.

The types of distillates obtained are wine 'holandas' with alcohol content less than 70 per cent (v/v), wine spirits with an alcohol content between 70-86 per cent (v/v) and wine distillates wines with an alcohol content between 86-94.8 per cent v/v. In the final product, the sum of *holandas* and wine spirits must represent more than 50 per cent alcoholic strength. In the case of double distillation, the maximum degree of the second distillation will be 70 per cent (v/v).

Aging is performed both in static and dynamic systems in the area of Protected Geographical Indication Brandy del Penedès. After distillation, the *holandas* are introduced in oak containers for aging for a minimum of six months. In the case of dynamic aging, the process can be started in containers of more than 1000 L, but it must be finished in barrels less than 500 L. For static aging, it must be in barrels of less than 500 L.

For dynamic aging, there are the same categories as for Jerez brandies: Solera, Solera Reserva and Solera Gran Reserva; for static aging or vintages, they are distinguished: a) 'Reserva', when aging is higher than one year, b) 'Gran Reserva', when aging is greater than three years and c) 'Extra Or', when aging is greater than five years with a minimum alcoholic strength of 40 per cent (v/v). 'Extra Or' category must be elaborated by double distillation with a final product lower than 65 per cent (v/v).

The Penedès brandies must have a graduation higher than 36 per cent (v/v) and a low sugar content (less than 20 g/L), being a dry product. The colour varies from amber to topaz; aromatically they are intense with traces of nuts and notes of species, such as cinnamon, vanilla and nutmeg. The oak manifests itself in the mouth with caramel and roasted notes.

## 4.3. Other European Countries

In Italy, there is a tradition to make brandies from grapes, but they do not have defined geographical indications. Normally, they are elaborated from native grapes and the distillation system used is distillation columns and subsequent minimum aging of one to two years in oak barrels, up to ages of eight years. They are characterised as being light and delicate with a soft touch of sweetness. In Germany, brandies have been distilled since the middle ages. They are called *weinbrand* and have traditionally been made from imported wines. Most of them are made in alembics and are aged in oak barrels for a minimum of six months. Those aged for a minimum of one year are called old (in German *uralt* or *alter*). Brandies tend to be softer and lighter than Cognacs and have a touch of sweetness. In Greece, distillates are made from grapes using traditional alembics. Greek brandy is distilled from Muscat wine. Mature distillates are made from sun-dried Savatiano, Sultana and Black Corinth grape varieties blended with an aged Muscat wine. One of the best known is called Metaxa, which is flavoured with muscatel wine, anise and other species. There are many varieties: centennial, golden, great reserve, Rhodes, etc. The number of stars indicates the years of aging: three means a minimum of three years of aging; five stars, five years; seven stars, seven. 'Private reserve' means at least 20 years (Dhiman and Attri, 2011; Owens and Dikty, 2009).

## 4.4. America

In the United States, the production of grape brandy was confined to California due to the tradition of the old Spanish missions. At present, after the vicissitudes due to phylloxera, the dry law, after the Second World War in California, the so-called California-style was developed under the influence of the University of California at Davis (Owens and Dikty, 2009; Amerine *et al.*, 1989). The brandies made in this style are clean on the palate, light and with aromas that make them suitable for consumption. Brandies are made by distillation columns from table grapes. The most commonly used varieties of grapes are the Thompson seedless and the Flame Tokay, although grapes with little varietal character are currently used, such as Chenin blanc, Folle blanche, French Colombard and Palomino (Owens and Dikty, 2009). The California brandies use the grape variety which is of lesser importance than clean low-temperature fermentations; also use fresh wine of immature grapes, good distillation practice and proper aging (Amerine *et al.*, 1989). These brandies are aged from two to 12 years in used American oak barrels. Some producers use the aging method of sherry brandy. The more expensive brandies usually contain a certain percentage of very aged brandies distilled in alembic.

In Latin America, in countries like Mexico, elaborated wine is used practically to obtain brandy. The distillation methods include both traditional alembics and distillation columns. The grape varieties most used are Thompson Seedless, Palomino and Ugni Blanc. The method of aging most used is that of soleras.

Brandy is also distilled from grapes in Chile and Peru to make a drink called Pisco. The most commonly used grape variety is muscatel in both countries and the distillation procedures are traditional alembic mainly in Peru and still with column in Chile. While in Peru the product does not age, in Chile the trend is also to make Pisco age in oak barrels. Pisco is a very aromatic drink with a remarkable presence of terpenic aromas.

According to Chilean legislation, Pisco is defined as a brandy produced and bottled in units of consumption in the Atacama and Coquimbo regions through the distillation of genuine wine, from *Vitis vinifera* varieties specified by law and planted in these regions (Bordeu *et al.*, 2012). The Pisco is a non-aged or slightly aged brandy. The distillation is mainly done by using batch column alembics, separating head, heart and tail fractions. The average alcoholic strength of heart fractions is around 60 per cent (v/v). After the maturation or aging, Pisco is adjusted to the final consumption alcohol content (35-50 per cent v/v), clarified and bottled. The product ranges from completely colourless to light amber and is characterised by fruit Muscat aromas combined in different degrees with aromas coming from aging and oak.

The commercial categories are:

- Artisanal Pisco, produced by small distillers with wines elaborated with skin contact during winemaking. The distillate is more complex and needs some maturation or aging.
- White Pisco is young, distillates with less than six months aging, in general with no contact with wood, frequently come from single varietal wines and can be triple distilled.

- Mature Pisco, aged between six and 12 months, usually comes from a mixture of grape varieties. Aromatically it is a mixture of fruity aromas and wooden notes.
- Aged Pisco, aged around one year, comes in young American oak barrels and aromatically have an important contribution of wood-derived aromas combined with fruity aromas. It is elaborated from multi-varietal wines and have an amber colour.

In general, Chilean Pisco is characterised by Muscat aroma, like geraniol, linalool, raisin and honey. Even if Pisco is not aged for long periods in wood, descriptors like vanilla, oak and toasted are important in describing products with some aging (Bordeu *et al.*, 2004).

Peruvian Pisco is a grape brandy distilled from must of fresh grape in alembics, which do not rectify the final product. Thus, it is transparent, slightly yellowish with an alcohol content of about 42 per cent v/v. According to the Peruvian legislation (Indecopi, 2006), Pisco is the product obtained exclusively by distillation of fresh musts of freshly fermented Pisco grapes, using methods that maintain the principles of quality. It is produced on the coast of the specified departments. The Pisco varieties can be aromatics (Italy, Muscat, Albilla and Torontel) or non-aromatics (Quebranta, Negracriolla, Mollar and Uvina). The distillation system could be a traditional alembic called 'falca' and the Charantais alembic with and without wine pre-heater. Rectification is not authorised and the addition of water is not allowed to adjust the alcohol content of the final product. Pisco must mature for three months in containers that do not give aromas or flavour. The different types of Pisco recognised are:

- Pure Pisco (*Pisco puro*), obtained from a single grape variety
- Pisco green must (*Pisco mosto verde*), obtained from the distillation of fresh musts with incomplete fermentation
- *Pisco acholado*, which is obtained from the mixture of aromatic and non-aromatic grapes, musts, wines or Piscos

The main aromatic descriptors associated with Peruvian Pisco for the Quebranta variety are fresh fruit, dried fruit, citrus, chocolate, herbaceous, syrup, alcohol, chemical, acetic and empyreumatic; for the variety Italia. they are the same in addition to aromatic grass, floral and sulphurous (Cacho *et al.*, 2012).

### 4.5. Other Regions

In South Africa, there is also a tradition in the production of brandies since the 17th century with the arrival of the Dutch. The introduction of new regulations and modern processing techniques at the beginning of the 20th century allowed the development and improvement of traditional products. The brandies are usually made with grape varieties – Ugni Blanc, Colombard, Chenin Blanc and Palomino in pot stills and in column stills. (Owens and Dikty, 2009; Bougass, 2014; Bougass *et al.*, 2014).

According to South African Liquor Products Act No. 60 of 1989, Sections 10, 12, 13 and 14 define the requirements for grape spirit, pot still brandy, brandy and vintage brandy, respectively. If the grape spirit is distilled in a continuous still, the alcohol content will be higher than 75 per cent (v/v), while if it is distilled by pot still, the alcohol content will be lower than 75 per cent (v/v). The minimum alcohol content of commercial product will be 43 per cent (v/v). The pot still brandy is distilled below 75 per cent (v/v) and matured by storage for a period of at least three years in oak casks, with a capacity of not more than 340 L. The alcohol content of the commercial product is a minimum of 38 per cent (v/v) and it accepts a maximum of 10 per cent of wine spirits. Brandy consists of a mixture of not less than 30 per cent of pot still brandy without grape spirit, wine spirit, spirit or a mixture thereof added in terms of regulation. Brandy shall have an alcohol content of at least 43 per cent (v/v). Vintage brandy shall be produced in such a manner that at least 90 per cent of the volume thereof is brandy according to the regulation. The pot still portion requires an additional period of maturation of at least five years in oaken casks with a capacity of not more than 1000 L; and the other portion must be matured for at least eight years in oaken casks. The alcohol content of the commercial product is at least 38 per cent (v/v).

Pot still brandy is by far the more complex brandy of the three styles and is considered the richest, fruitiest and most layered in style (Bougass *et al.*, 2014). South African brandies are distinguished by their overt fragrances of stone fruit, like apricots, peaches and pears (Owens and Dikty, 2009).

### 4.6. Fruit Brandies

Fruit distillates, normally called brandies, are produced exclusively through alcoholic fermentation and distillation of fleshy fruit or the must of berry, or vegetable, with or without stones and distilled at less than 86 per cent (v/v). According to the raw material, three main types of brandies can be distinguished: distillates obtained with pome fruits, of which apples and pears are the most common; those obtained with stone fruits, mainly sweet cherries, sour cherries, plums, apricots, and peaches; and finally the distillates obtained from berries. While the distillates with pome fruits are those traditionally obtained from pears and apples, new brandies with other fruits are also being developed to valorise surpluses (Lopez *et al.*, 2017).

## 5. Brandy Composition and Sensorial Characteristics

Brandy is a wine distillate, so its initial chemical composition will depend on the volatile compounds in the wine. Although ethyl alcohol and water are the two main components of any distilled spirit, the character of aroma and flavour depend on a multitude of minor compounds, which are generally referred to as congeners or congenerics. After distillation, brandy does not contain the non-distillable wine's organic and inorganic compounds, and this absence of non-volatile organic acids in the distillate affects the balance of the taste. The brandy's high alcohol content, about 40 per cent v/v, 'burns' the mouth and simultaneously augments the sweet taste (Tsakiris *et al.*, 2014). Therefore, the aging process produces a complexity of minor components that generally improve palatability and improves aroma and flavour.

More than 500 substances have been detected in spirits (Cantagrel, 2003); they belong to a large number of chemical classes: alcohols (ethanol, methanol, higher alcohols, etc.), aldehydes, esters, volatile acids (acetic acid, fatty acids, etc.), ketones, acetals, nitrogen-, oxygen- and sulphur-containing heterocyclic compounds, phenolic acids, and aldehydes. These different aromatic constituents are contributed by:

- the grape (primary aromas)
- winemaking (fermentation aromas)
- distillation (specific aromas produced by the heating process)
- aging (aromas imparted by the oak wood)

The volatile composition of brandies is also characterised by high contents of low molecular weight ethyl esters and some phenols, aldehydes, and acetals. However, these major components, due to their high concentrations in all brandies, may not be helpful in the discrimination of samples. The volatile compositions of brandies, such as Mirabelle, Calvados, Cognac and Armagnac are qualitatively rather similar. However, their organoleptic characteristics are really different, due to slight differences in the concentrations of volatiles (Ledauphin *et al.*, 2010). Wine distillates' authentication, mainly in terms of varieties and regions of geographical origin, is rather difficult. However, differentiation can be realised by the utilisation of statistical methods, such as principal component analysis (Tsakiris *et al.*, 2014).

### 5.1. Alcohols

Alcohols possessing more than two carbon atoms are known as higher alcohols or fusel alcohols and are by-products of yeast fermentation during the elaboration of base wine and form an important part of the volatile composition of brandy (Louw and Lambrechts, 2012), being quantitatively the largest group of volatile flavour compounds. Table 3 summarises the odour quality, odour threshold value in water and/ or ethanol solution, and concentration range of single volatile compounds in distilled spirits produced during alcoholic fermentation from carbohydrates by yeasts and other microorganisms.

The term 'fusel alcohols' refers to their malty and burnt flavour, with the exception of 2-phenethyl alcohol, which has a rose-like odour. The higher alcohol content remains almost unaffected before distillation. The most important alcohols are 1-propanol, 2-methyl-1-propanol, 2-methylbutanol, 3-methylbutanol, and the aromatic alcohol 2-phenylethanol. Most straight-chain alcohols and their esters have a strong pungent smell. At high concentration levels, higher alcohols are characterised by pungent odours, which mask the aromatic finesse. They reach concentration in the order of 2.5-5.0 g/Lp.a and their recovery is about 90 per cent with the exception of 2-phenyl-ethanol which is recovered 10 per

**Table 3.** Odor Quality, Odour Threshold Value in Water and/or Ethanol Solution and Concentration Range of Main Alcohols in Distilled Spirits

| *Compound* | *Flavour quality* | *Threshold (mg/L water)** | *Threshold (mg/ L ethanol solution)** | *Typical concentration (g/L p.a.)* |
|---|---|---|---|---|
| Methanol | Alcoholic | - | 668 | 0.30–0.70 |
| 1-butanol | Alcoholic | 0.5; 1.3 | 820 | 0.0025-0.200 |
| 2-butanol | Alcoholic | - | 1,000 | 0.001-0.8 |
| 1-propanol | Stupefying | 500 | 830 | 0.1-2 |
| 2-Methyl-1-propanol | Alcoholic | - | 40; 75 | 0.1-1 |
| 2-Methyl-1-butanol | Malty | 0.32 | 7; 30 | 0.02-1.8 |
| 3-Methyl-1-butanol | Malty | 1 | 7; 30 | 0.01-3 |
| Allyl alcohol | Unpleasant | 19 | - | 0.01-0.13 |
| Phenethyl alcohol | Rose-like | 1 | 7.5; 10 | 0.01-0.08 |

*Minimum and maximum threshold values cited from the literature

cent (Christoph and Bauer-Christoph, 2007; Tsakiris *et al.*, 2014). At lower concentration levels, higher alcohols may add complexity to brandy due to synergistic interactions with other compounds. In addition, higher alcohols may be esterified during the aging process, resulting in esters that impart more pleasant aromas (Louw and Lambrechts, 2012).

Methanol, 1-butanol and 2-butanol are not compounds of alcoholic fermentation and their threshold values are rather high (668, 820 and 1000 mg/L p.a., respectively) and therefore, they do not contribute significantly to the flavour (Christoph and Bauer-Christoph, 2007). Methanol is always present in very small quantities of 40-60 mg/L in wine. However, in distillates and brandy, it is found in higher concentrations of 0.30–0.70 g/L p.a. with a recovery of 90 per cent. Its taste is similar to ethanol and it does not affect the organoleptic quality of the spirits. However, it affects the safety of brandy because its toxicity is well known. European Union legislation requires a limit lower than 2.0 g $L^{-1}$ of pure alcohol. Methanol is reduced during ageing in barrels (Tsakiris *et al.*, 2014); 2-Butanol levels higher than 0.5 g/L p.a. indicate a bacterial spoilage of raw materials or mash (Christoph and Bauer-Christoph, 2007).

The herbaceous odour of brandies is due to grape-derived carbonyl compounds with six carbon atoms. Unripe grapes and continuous presses may induce herbaceous tastes by liberating compounds, such as hexanols (hexanol-1 and hexanol-2) and hexenols (cis-3-hexene-1-ol, trans-2-hexen-1-ol, cis-2-hexen-1-ol). 1-Octen-3-ol is characterised by a mushroom odour and is produced in grapes infected by *Botrytis cinerea* (Tsakiris *et al.*, 2014).

## 5.2. Terpenes

The Muscat flavor is produced by terpenes. The chemical compounds responsible for this specific aroma are mainly geraniol, nerol, α-terpineol, linalool and ß-citronellol. These monoterpenes are part of a large family of molecules obtained by association of isoprene units. The total free terpene concentration varies between 0.6-1.5 mg/L, depending on the grape variety and cultural parameters, with important variations in the respective proportions of terpenes. The olfactory perception thresholds of these compounds are rather low (a few hundred micrograms per litre). In Muscat wines, terpenes are found either free or bound to sugars as glycosides; the latter are also called aroma precursors, as they are unable to express their aromatic character. The sensorial characteristics are described as rose flower (geraniol, nerol), rosewood (linalool) and geranium flower (geraniol, citronellol, terpineol) (Tsakiris *et al.*, 2014; Colonna-Ceccaldi, 2010).

## 5.3. Volatile Acids

Acetic acid can be produced during and/or after fermentation by oxidation of ethanol under aerobic conditions by the acetic acid bacteria Acetobacter; acetic acid levels should not be higher than 1 g/L p.a.

in distilled spirits, since higher levels may contribute to a typical vinegar-like off-flavor (Christoph and Bauer-Christoph, 2007).

In wine, it is found in concentrations ranging between 300-700 mg/L, while in distillates the concentration ranges from 0.20-1.0 g/L p.a. Recovery is as low as 2-5 per cent due to the removal of the distillation 'heads'. However, distillates produced by continuous distillation may contain higher amounts of acetic acid due to the absence of the removal of the distillation heads. During wine maturation in barrels, a small quantity of acetic acid can be produced from ethanol oxidation or it can be extracted from wood hemicelluloses (Tsakiris *et al.*, 2014).

Other carboxylic acids, such as propionic acid and butyric acids, may also be present and they are associated with bacterial activity. Butyric acid is characterised by an unpleasant, buttery and cheesy aroma and its concentration increases during ageing. Hexanoic, octanoic, decanoic, dodecanoic, myristic (14 carbon atoms), palmitic (16 carbon atoms) and stearic (18 carbon atoms) acids are formed by yeasts (Tsakiris *et al.*, 2014). In distilled spirits the concentration of free fatty acids is low owing to the esterification and separation by distillation; thus, the concentration in wine distillates, like Cognac, is in the range of 50 mg 0.1 $L^{-1}$ p.a. (Christoph and Bauer-Christoph, 2007).

## 5.4. Ethyl Esters

Esters are condensation products of the carboxyl group of an organic acid and the hydroxyl group of an alcohol or a phenol. Yeasts produce esters after cell division has ceased. Table 4 summarises the odour quality, odour threshold value in water and/or ethanol solution, and concentration range of main esters in distilled spirits.

**Table 4.** Odour Quality, Odor Threshold Value in Water and/or Ethanol Solution and Concentration Range of Main Alcohols in Distilled Spirits

| *Compound* | *Flavour quality* | *Threshold (mg/L water)** | *Threshold (mg/ L ethanol solution)** | *Typical concentration (g/L p.a.)* |
|---|---|---|---|---|
| Ethyl acetate | Solvent-like, nail polish | 17.6 | 7.5 | 0.01–2 |
| Ethyl butanoate | Fruity, floral | 0.001 | 0.02 | <0.00025–0.008 |
| Methyl butyl acetate | Fruity, banana, pear | 0.3 | 0.03 | 0.003–0.03 |
| 2-Phenethyl acetate | Rose, honey, fruity | 0.02 | 0.25 | 0.01–0.03 |
| Ethyl hexanoate | Apple, banana, violet | 0.005 | 0.005 | 0.001–0.008 |
| Ethyl octanoate | Pineapple, pear | 0.07 | 0.26; 0.002 | 0.01–0.05 |
| Ethyl decanoate | Floral, fatty | 0.5 | - | 0.01–0.09 |
| Ethyl dodecanoate | Floral | - | - | 0.004–0.08 |
| Diethyl succinate | - | - | 100 | 0.005–0.03 |
| Ethyl lactate | - | - | 100 | <0.025-1 |

*Minimum and maximum threshold values cited from the literature

Ethyl acetate, mainly produced as a result of esterification of acetic acid, is the main ester to occur in fermented products and their distillates; it contributes significantly to a solvent-like nail polish off-flavour at levels higher than 400 mg /100 mL p.a. (Christoph and Bauer-Christoph, 2007) and the range is about 0.4 to 0.8 g $L^{-1}$ of pure alcohol (Tsakiris *et al.*, 2014). Ethyl acetate has a recovery of 100 per cent in continuous distillation and 60 per cent in batch distillation (Tsakiris *et al.*, 2014). Ethyl acetate is mostly found in the head fraction, and the ethyl lactate in the tail fraction; therefore, the distillation cuts can influence the presence of these compounds in the final brandy. Ethyl lactate has a negative influence on brandy quality, and usually is accompanied by diethyl succinate, which is regarded as a spoilage compound that can be removed along with the tails (Louw and Lambrechts, 2012). Ethyl butyrate concentration increases with ageing, with the absence of antioxidant and antimicrobial agents, and mainly with increased temperature (Tsakiris *et al.*, 2014).

Longer-chain ethyl esters also contribute to the total ester content of brandy. These esters are generally associated with fruity aromas and contribute to the overall brandy quality. Their quantities and mutual proportions are of great importance for the perceived flavour of a spirit drink since their concentrations are generally above the sensory threshold values. Especially the low-boiling ethyl esters, like ethyl 2-methylbutanoate, ethyl hexanoate, and ethyl octanoate, and the acetates like ethyl acetate, isoamyl acetate, isobutyl acetate, hexyl acetate, and 2-phenethyl acetate are of great importance for the flavour of distilled spirits (Christoph and Bauer-Christoph, 2007). Over 160 esters have been identified in wines and most of them are also present in brandies. Almost all esters are low-boiling-point compounds and are distilled in the beginning. Thus, the point at which the heads are cut to remove undesirable compounds must be controlled in such a way no to remove too many desirable esters. Similarly, the tails must not be cut too early in the distillation to preserve more esters. Recovery varies between 40-60 per cent, depending on the distillation technique (Louw and Lambrechts, 2012; Tsakiris *et al.*, 2014).

The flavour fraction with the lowest volatility is composed of C14-C18 fatty acid esters; these esters as well as the long-chain fatty alcohols may contribute to the stearine-like smell that is characteristic of Scotch malt whisky, in particular. Malo-lactic fermentation also has an influence on the concentration of these compounds; distillates show a loss of fruitiness and aroma intensity with decreasing levels of ethyl hexanoate, hexyl acetate, 2-phenethyl acetate, and with increasing levels of ethyl lactate, acetic acid, and diethyl succinate (Christoph and Bauer-Christoph, 2007). In general, wines pressed by continuous presses contain higher amounts of ethyl esters of long-chain fatty acids (14-18) because they contain relatively lower oxygen, which affects their synthesis by yeasts. Distilling wine in the presence of yeasts contained in fermentation lees leads to enrichment in fatty acid esters, such as ethyl octanoate (fruity, floral, pineapple, apple and pear) and ethyl decanoate (fruity, pear, wine, etc.). Aromatic synergy between those esters strengthens their olfactory impact. Fermentation lees also supply the wine spirit with fatty acids ($C_8$-$C_{18}$, saturated and unsaturated) which are at the origin of aromatic derivatives formed by their oxidation into aldehydes or ketones during the ageing process (Lurton *et al.*, 2012). Recent research has shown that some brandies contain naturally rare ethyl esters, which may have some impact on their aroma. Such compounds are ethyl 2-, 3- and 4-methylpentanoate and ethyl cyclohexanoate, which exhibit pleasant strawberry–liquorices-like odours (Tsakiris *et al.*, 2014). The concentration of ethyl pentanoate is higher in German brandies than in Cognac and similar to some Spanish and French brandies (Uselmann and Schieberle, 2015).

The ethyl ester content of brandies increases during ageing because of the slow esterification of different organic acids with ethanol. As brandy matures, ethyl esters become less flavour-active due to an increase in their solubility in aqueous ethanol by the wood-extracted materials. Preserving wines before distillation in the presence of lees has been connected with increased content of ethyl esters, since they are generally retained within yeast cells rather than being released into the fermenting must during fermentation (Tsakiris *et al.*, 2014). Duran-Guerrero *et al.* (2011) found that ethyl esters were the main family of volatile compounds responsible for the differentiation among the three different categories of sherry brandies. The concentration of ethyl esters increased during the aging process, appearing in higher amounts in those sherry brandies submitted to longer aging in wood.

## 5.5. Higher Alcohol Acetates

The formation of esters between acetic acid and higher alcohols is also important since they may provide a fruity character, for example, isoamyl acetate, which has a characteristic banana odour, positively influences brandy's aroma. Low fermentation temperatures favour synthesis of fruity esters, such as isoamyl, isobutyl and hexyl acetates, while higher temperatures favour the production of higher molecular-weight esters. Both low $SO_2$ levels and juice clarification favour ester synthesis and retention. The absence of oxygen during yeast fermentation enhances further ester formation (Tsakiris *et al.*, 2014).

Malo-lactic fermentation also has an influence on the concentration of these compounds; distillates show a loss of fruitiness and aroma intensity with decreasing levels of ethyl hexanoate, hexyl acetate, 2-phenethyl acetate and with increasing levels of ethyl lactate, acetic acid, and diethyl succinate (Christoph and Bauer-Christoph, 2007).

Higher alcohol acetate synthesis as well hydrolytic breakdown continue enzymatically during ageing based on the chemical composition and storage conditions of the brandy, due to formation of acetic acid from the xylans extracted from the wood (Louw and Lambrechts, 2012; Tsakiris *et al.*, 2014)

## 5.6. Aldehydes and Ketones

Acetaldehyde (ethanal) is the major important carbonyl compound of alcoholic fermentation and is formed as an intermediate compound by degradation of pyruvate; its production by yeasts depends on the pyruvate decarboxylase activity of the yeast (Christoph and Bauer-Christoph, 2007). The toxicity associated with acetaldehyde is well known and its presence in the alcoholic beverages is quite often related to nausea and vomiting. In distillates and brandies, it is found in concentrations ranging between 0.20-0.25 g $L^{-1}$ p.a. (Tsakiris *et al.*, 2014) and in wine distillates, higher concentrations do not affect the quality owing to an odour threshold of about 100 mg/L (Christoph and Bauer-Christoph, 2007). Since acetaldehyde is one of the most volatile compounds, the highest levels are in the 'head cut' of the distillation and thus can be separated from the 'heart cut' (Christoph and Bauer-Christoph, 2007). Acrolein (2-propenal) has a peppery, horseradish-like smell and is formed either by dehydration of glycerol during distillation in the presence of acids on hot metallic surfaces or especially by bacteria during fermentation of spoiled raw materials. The biochemical pathway of the formation of acrolein is initiated by a bacterial dehydratase enzyme, which converts glycerol to 3-hydroxypropionaldehyde (Christoph and Bauer-Christoph, 2007). Acetaldehyde as well as acrolein react with ethanol to form the acetals 1,1-diethoxyethane and 1,1,3-triethoxypropane, respectively, leading to a reduction in the pungent odour of the aldehydes; the equilibrium concentration of the 1,1-diethoxyethane formed is close to 10 per cent in relation to the amount of acetaldehyde present. Higher aldehydes and their acetals can be found at concentrations less than 0.1 mg/L (Christoph and Bauer-Christoph, 2007).

Isobutanal, at concentrations higher than 25 mg $L^{-1}$, could give an herbaceous character to the brandy. However, during ageing, its content declines due to acetylysation and selective evaporation. Trans-Nonenal is characterised by a paper-like sense, while octanal contributes to the aroma complexity by adding an orange flavour. Although concentrations of both the above aldehydes is increased during ageing, it generally remains below the perception threshold. However, significant statistical correlations have been obtained among the herbaceous odour and aldehyde concentration. β-Damascenone is an isoprenoid ketone, which is present in grapes. Since it is a highly odoriferous compound with a powerful and pleasant fragrance, it is an important compound in the perfume and flavouring industries. As it has a very low sensory threshold, β-damascenone is considered a key odour compound in brandy, imparting a stewed apple, fruity–flowery and honey-like character (Tsakiris *et al.*, 2014). The concentration of β-damascenone is quite high in Cognac than in German brandies, as well as in Spanish and some French brandies. The ratio of the β-damascenone concentration to that of ethyl pentanoate allows differentiating the German and Spanish brandies from the Cognacs (Uselmann and Schieberle, 2015).

Diacetyl (2,3-dioxobutane) is a ketone produced during wine fermentation through oxidation of acetoin, a degradation product of citric acid. It has an important sensory influence on brandy since its odour is characterised as sweet, buttery or butterscotch-like (Tsakiris *et al.*, 2014; Christoph and Bauer-Christoph, 2007). Other aldehydes, which may be found in brandies, are formaldehyde, 5-hydroxymethylfyrfural, propionaldehyde, butyraldeyde, benzaldehyde, isovaleraldeyde and n-valeraldeyde (Tsakiris *et al.*, 2014).

## 5.7. Furfural and Furanic Compounds

Another aldehyde having a sensory impact like that of 'baked' in brandy, is furfural (0.5-82.5 mg $L^{-1}$ of pure alcohol). Its synthesis involves sugar oxidation and is activated by heat. It is mainly produced during distillation from the remaining pentose content of the lees and consequently is highly influenced by the distillation system employed and double distillation enhances the amount of all furanic species. For this reason, the concentration of furfural in brandy shows high fluctuation. Furfural and its derivatives may also be derived from both the wooden cask and the possible addition of caramel. Concentration of furanic compounds, such as furfuryl ethyl ether, furfural, 2-acetylfuran and 5-methylfurfural, varies according to the type of cask and ageing time. Furanic aldehydes are derived from thermal degradation of wood polysaccharides of wooden casks (Tsakiris *et al.*, 2014).

## 5.8. Volatile Compounds from Casks

It was found that significant differences in vanilla, woody, caramel, burned/toasted, green, tails and rubbery aromas existed between brandies aged in barrels from different wood origins. By storing distillates in wooden casks, volatile aroma compounds, like cis-β-methyl-γ-octalactone and trans-β-methyl-γ-octalactone (oak lactone), furfural, 4-hydroxy-2-butenoic acid lactone, hexanoic acid, vanillin, guaiacol, eugenol, cresols and other phenolic compounds migrate from the toasted wood into the distillate. These compounds are responsible for the characteristic oak wood and vanilla-like flavour (Tsakiris *et al.*, 2014; Christoph and Bauer-Christoph, 2007; Louw and Lambrechts, 2012).

The toasting process can modify strongly the volatile composition of the different types of wood, particularly the levels of furanic aldehydes (furfural, 5-methylfurfural with toasted almond aromas, 5-hydroxylmethyl furfural), volatile phenols (syringol and 4-allyl-syringol), propanoic acid, 4-hydroxy-2-butenoic acid lactone and vanillin (Tsakiris *et al.*, 2014). In contrast, non-toasted oak wood had a small quantity of volatile phenols, mainly eugenol and traces of phenolic aldehydes (Zhang *et al.*, 2015). After three consecutive uses (three years), the wood will no longer liberate fatty acids, coniferaldehyde, furfural, 5-methylfurfural and 5-hydroxymethyl-furfural into the wine spirits (Tsakiris *et al.*, 2014).

## 5.9. Non-volatile Compounds from Casks

The ageing of spirits in oak barrels is a complex process. Direct extraction of wood components or degradation products of macromolecules of the wood may occur, as well as reactions between the components of the distillate itself and/or those originating from the oak wood (polymerisations, esterifications, acetylysations, and hydrolysis), in addition to major oxidation processes. Apart from ellagitannins, oak releases a certain number of other compounds into brandies, mainly lignins. Depending on conditions, oak may also release polysaccharides, mostly consisting of hemicelluloses that contribute to spirit flavour (Tsakiris *et al.*, 2014).

## 5.10. Other Compounds

Brandy may contain extremely low concentrations of different unpleasant (rotten eggs, garlic) volatile sulphur compounds, such as hydrogen sulphide, carbonyl sulphide, sulphur dioxide, thiols, sulphides, polysulfides and thiosterols. The determination of these compounds in wines and spirits is difficult because of their volatility and their very low concentrations, which require the use of highly sensitive detectors (Tsakiris *et al.*, 2014). Ethyl carbamate has been detected in several types of fermented foods and beverages, such as alcoholic beverages, particularly stone fruit spirits. Ethyl carbamate is typically formed from various precursors and one of the most important precursors is urea, which might be formed during the degradation of arginine by yeast; another source is cyanate from the oxidation of cyanide. The precursors react with ethanol to form ethyl carbamate and this occurs even during storage after distillation, which explains why the concentration of the chemical in spirits is high (Tsakiris *et al.*, 2014; Pang *et al.*, 2017). Ethyl carbamate is a potential carcinogenic compound and its presence is strictly monitored in wines and spirits. Ethyl carbamate cannot be completely avoided because it occurs naturally in fermented foods and beverages and the maximum levels for ethyl carbamate must be established based on risk assessment and the levels of this chemical in spirits must be periodically monitored by producers and supervisors (Tsakiris *et al.*, 2014; Pang *et al.*, 2017).

## 5.11. Caramel

According to the European legislation, wine brandies may only contain added caramel as a means to acquire colour. Nevertheless, there is no legal limit for the concentration of this additive, neither specifications for the product and the technological procedure have been prescribed. The addition of caramel is quite common in the production of aged spirit beverages since it gives them an amber colouration, that is attractive to the consumer. The chemical composition of caramel is complex due to the large number of substances produced because of pyrolysis of carbohydrates, such as sucrose, glucose or starch. However, furanic compounds are also present, such as furfural or 5-hydroxylmethyl furfural, of which the latter is found in much higher concentrations (Tsakiris *et al.*, 2014; Canas and Belchior, 2013). For quality control purposes, the correlation analysis between the caramel concentration and the characteristics of the brandies reveals that caramel mainly influences HMF content, total phenolic index and coordinate a*, since the correlations are positive and significant for all brandies as well as for the lightness which are negative and are very significant for all of them. It is also shown that the ratio of furfural/HMF is a useful tool to detect the addition of caramel in aged wine brandies (Canas and Belchior, 2013).

## References


Amerine, M.A., Berg, H.W., Kunkee, R.E., Ough, C.S., Singleton, V.L. and Webb, A.D. (1989). The Technology of Wine Making. AVI Publishing, Westport, CT, USA, pp. 582-642.

Bertrand, A. (2003). Brandy and cognac: Armagnac, brandy, and cognac and their manufacture. pp. 584-601. *In*: Caballero, B., Trugo, I.C. and Finglas, P.M. (Eds.). Encyclopedia of Food Sciences and Nutrition, second ed. Academic Press, Oxford, UK.

Bordeu, E., Agosin, E. and Casaubon, G. (2012). Pisco: Production, flavour chemistry, sensory analysis and product development. pp. 331-347. *In*: Piggott, J. (Ed.). Sensory Evaluation and Consumer Research. Woodhead Publishing Ltd., Oxford (UK).

Bordeu, E., Formas, G. and Agosin, E. (2004). Proposal for standardised set of sensory terms for Pisco, a young Muscat wine distillate. *American Journal of Enology and Viticulture* 55: 104-107.

Bougass, N.V., Van Rensburg, P., Snyman, C.L.C. and Lambrechts, M.G. (2014). Brandy, cognac and armagnac. pp. 248-276. *In*: Bamforth, C.W. and Ward, R.E. (Eds.). Food Fermentations. Oxford University Press, New York, USA.

Bougass, N.V. (2014). Factors influencing the style of brandy. Ph.D. thesis. Stellenbosch University.

Cacho, J., Moncayo, L., Palma, J., Ferreira, V. and Culleré, L. (2012). Characterisation of the aromatic profile of the Italian variety of Peruvian Pisco by gas chromatography – olfactometry and gas chromatography coupled with flame ionisation and mass spectrometry detection systems. *Food Research International* 49: 117-125.

Caldeira, I., Santos, R., Ricardo-da-Silva, J.M., Anjos, O., Mira, H., Belchior, A.P. and Canas, S. (2016). Kinetics of odourant compounds in wine brandies aged in different systems. *Food Chemistry* 211: 937-946.

Caldeira, I., Anjos, O., Belchior, A.P. and Canas, S. (2017). Sensory impact of alternative ageing technology for the production of wine brandies. *Ciência e Técnica Vitivinícola* 32: 12-22.

Canas, S. and Belchior, A.P. (2013). Effects of caramel addition on the characteristics of wine brandies. *Ciência e Técnica Vitivinícola* 28: 51-58.

Canas, S., Caldeira, I., Anjos, O., Lino, J., Soares, A. and Belchior, A.P. (2016). Physicochemical and sensory evaluation of wine brandies aged using oak and chestnut wood simultaneously in wooden barrels and in stainless steel tanks with staves. *International Journal of Food Science and Technology* 51: 2537-2545.

Cantagrel, R. (2003). Chemical composition and analysis of cognac. pp. 601-606. *In*: Caballero, B., Trugo, I.C. and Fingla, P.M. (Eds.). Encyclopedia of Food Science and Nutrition, second ed. Academic Press, Oxford (UK).

Cernîsev, S. (2017). Analysis of lignin-derived phenolic compounds and their transformations in aged wine distillates. *Food Control* 73: 281-290.

Chatonnet, P. and Dubourdieu, D. (1998). Comparative study of the characteristics of American white oak (*Quercus alba*) and European oak (*Quercus petraea* and *Q. robur*) for production of barrels used in barrel aging of wines. *American Journal of Enology and Viticulture* 49: 79-85.

Christoph, N. and Bauer-Christoph, C. (2007). Flavour of spirit drinks: Raw materials, fermentation, distillation, and ageing. pp. 219-239. *In*: Berger, R.G. (Ed.). Flavours and Fragrances. Springer, Berlin, Heidelberg, Germany.

Colonna-Ceccaldi, B. (2010). Use of terpene-producing yeasts in brandy production. pp. 63-67. *In*: Walker, G.M. and Hughes, P.S. (Eds.). Distilled Spirits, New Horizons: Energy, Environmental and Enlightenment. Nottingham University Press, Nottingham, United Kingdom.

del Alamo-Sanza, M. and Nevares, I. (2017). Oak wine barrel as an active vessel: A critical review of past and current knowledge. *Critical Reviews in Food Science and Nutrition*, DOI: 10.1080/10408398.2017.1330250

Delgado-González, M.J., Sánchez-Guillén, M.M. García-Moreno, M.V., Rodríguez-Dodero, M.C., García-Barroso, C. and Guillén-Sánchez, D.A. (2017). Study of a laboratory-scaled new method for the accelerated continuous ageing of wine spirits by applying ultrasound energy. *Ultrasonics Sonochemistry* 36: 226-235.

Dhiman, A.K. and Attri, S. (2011). Production of Brandy (Chapter 35). pp. 1-60. *In*: Joshi, V.K. (Ed.). Handbook of Enology: Principles, Practices and Recent Innovations, vol. III. Asiatech Publisher, INC, New Delhi.

Duran-Guerrero, E., Cejudo-Bastante, M.J., Castro-Mejías, R., Natera, R. and García-Barroso, C. (2011). Differences in the volatile compositions of French labelled brandies (Armagnac Calvados Cognac and Mirabelle) using GC-MS and PLS-DA. *Journal of Agricultural and Food Chemistry* 59: 2410-2415.

European Union (2008). Regulation (EC) No. 110/2008 of the European Parliament and of the Council of 15 January 2008. *Official Journal of the European Union L* 39: 30-31.

Guymon, J.F. (1974). Chemical aspects of distilling wines into brandy. pp. 232-253. *In*: Webb, A.D. (Ed.). Advances in Chemistry, vol. 137. Chemistry of Winemaking. American Chemical Society, Washington, DC, USA.

Hornsey, I. (2016). Way through the wood. *Barrel Biochemistry for Brewers and Distillers*, Brewer and Distiller International, April: 48-55.

Indecopi (2006). Reglamento de la Denominación de Origen Pisco. INDECOPI. NTP 211.001.2006, Bebidas Alcohólicas, Pisco, Requisitos, Norma Técnica, Peru, 2006: 1-11.

Kohlmann, H. (2016). Future of the US spirit is market in light of accelerated fragmentation. *Global Drinks Forum*, October 10, Berlin, Germany.

Lambrecthts, M., van Velden, D., Louw, L. and van Rensburg (2016). Brandy and Cognac: Consumption, Sensory and Health Effects. pp. 456-461. *In*: Caballero, B., Finglas, P.M. and Toldrá, F. (Eds.). Encyclopedia in Health. Academic Press, Oxford, UK.

Léauté, R. (1990). Distillation in Alembic. *American Journal of Enology and Viticulture* 41: 90-103.

Ledauphin, J., Milbeau, C., Barillier, D. and Hennequin, D. (2010). Differences in the volatile compositions of French labelled brandies (Armagnac Calvados Cognac and Mirabelle) using GC-MS and PLS-DA. *Journal of Agricultural and Food Chemistry* 58: 7782-7793.

López, F., Rodríguez-Bencomo, J.J., Orriols, I. and Pérez-Correa, J.R. (2017). Fruit brandies. pp. 531-556. *In*: Kosseva, M.R., Joshi, V. and Panesar, P. (Eds.). Science and Technology of Fruit Wine Production. Academic Press.

Louw, L. and Lambrechts, M.G. (2012). Grape-based brandies: Production, sensory properties and sensory evaluation. pp. 281-298. *In*: Piggot, J. (Ed.). Alcoholic Beverages: Sensory Evaluation and Consumer Research. Woodhead Publishing Ltd., Cambridge, UK.

Lurton, L., Ferrari, G. and Snakkers, G. (2012). Cognac: Production and aromatic characteristics. pp. 242-266. *In*: Piggott, J. (Ed.). Sensory Evaluation and Consumer Research. Woodhead Publishing Ltd., Oxford (UK).

Mosedale, J.R. and Puech, J.L. (2003). Barrels, wines, spirits and other beverages. pp.393-403. *In*: Caballero, B., Trugo, I.C. and Finglas, P.M. (Ed.). Encyclopedia of Food Sciences and Nutrition, second ed. Academic Press, Oxford, UK.

Owens, B. and Dikty, A. (2009). The art of distilling whiskey and other spirits. Query Books, Beverley, Massachusetts (USA).

Pang, X.N., Li, Z.J., Chen, J.Y., Gao, L.J. and Han, B.Z. (2017). Comprehensive review of spirit drink safety standards and regulations from an international perspective. *Journal of Food Protection* 80: 431-442.

Prida, A. and Puech, J.L. (2006). Influence of geographical origin and botanical species on the content of extractives in American, French and East European oak woods. *Journal of Agricultural and Food Chemistry* 54: 8115-8126.

RDE Brandy de Jerez (2005). Reglamento de la Denominación Específica 'Brandy de Jerez' y su Consejo Regulador. *Boletín Oficial de la Junta de Andalucía* No. 122, Spain, 2005.

Reglament IGP Brandy del Penedès (2017). ORDRE ARP/246/2017, de 27 d'octubre, per la qual s'aprova el reconeixement transitori i el Reglament de la Indicació Geogràfica Protegida Brandy del Penedès, i es reconeix el seu Consell Regulador provisional. *Diari Oficial de la Generalitat de Catalunya Núm*, 7488, Spain, 2017.

Rodriguez-Solana, R., Rodriguez-Feijedo, S., Salgado, J.M., Dominguez, J.M. and Cortes-Dieguez, S. (2017). Optimisation of accelerated ageing of grape marc distillate on a micro-scale process using a Box–Benhken design: Influence of oak origin, fragment size and toast level on the composition of the final product. *Australian Journal of Grape and Wine Research* 23: 5-14.

Statista (2018). https://www.statista.com/statistics/693822/global-market-share-spirits/

Schwarz, M., Rodríguez, M.C., Sánchez, M., Guillén, D.A. and Barroso, C.G. (2014). Development of an accelerated aging method for Brandy. *LWT – Food Science and Technology* 59: 108-114.

Tsakiris, A., Kallithraka, S. and Kourkoutas, Y. (2014). Grape brandy production, composition and sensory evaluation. *Journal of the Science of Food and Agriculture* 94: 404-414.

U.S. Code of Federal Regulations, Title 27: Alcohol, Tobacco and Firearms, PART 5, Subpart C. *Standards of Identity for Distilled Spirits* (2013).

Uselmann, V. and Schieberle, O. (2015). Decoding the combinatorial aroma code of a commercial cognac by application of the sensomics concept and first insights into differences from a German brandy. *Journal of Agricultural and Food Chemistry* 63: 1948-1956.

van Jaarsveld, F.P. and Hattingh, S. (2012). Rapid induction of ageing character in brandy products: Ageing and general overview. *South African Journal for Enology and Viticulture* 33: 225-252.

Wiseguy (2018). Global Spirits Market Report (2018). WISEGUY RESEARCH CONSULTANTS PVT LTD. https://www.wiseguyreports.com/reports/2941237-global-spirits-market-report-2018

Zhang, B., Cai, J., Duan, C.Q., Reeves, M.J. and He, F. (2015). A review of polyphenolics in oak woods. *International Journal of Molecular Sciences* 16: 6978-7014.

Zhang, B., Zeng, X.A., Lin, W.T., Sun, D.W. and Cai, J.L. (2013). Effects of electric field treatments on phenol compounds of brandy aging in oak barrels. *Innovative Food Science and Emerging Technologies* 20: 106-114.

Zhang, B., Zeng, X.A., Sun, D.W., Yu, S.J., Yang, M.F. and Ma, S. (2013). Effect of electric field treatments on brandy aging in oak barrels. *Food Bioprocess Technology* 6: 1635-1643.

# 24 Biovalorisation of Winery Wastes

**Konstantinos V. Kotsanopoulos[1*], Ramesh C. Ray[2] and Sudhanshu S. Behera[2,3]**

[1] School of Agricultural Sciences, Department of Agriculture Ichthyology and Aquatic Environment, University of Thessaly, Fytokou Str., Nea Ionia Magnessias, 38446 Volos, Hellas, Greece

[2] Centre of Food and Environment Studies, 1071/17 Jagamohan Nagar, PO: Khandagiri, Bhubaneswar – 751030, Odisha, India

[3] Department of Fisheries and Animal Resource Development, Government of Odisha, India

## 1. Introduction

Wine is primarily produced by fermentation of grape (*Vitis vinifera* L.) juice. Recently, fruits such as mango, pine apple, avocado, litchi and many other tropical as well as temperate fruits are processed for winemaking. Winemaking industries generate huge amounts of different types of residues/wastes derived from the de-stemmed grapes, sediments obtained during clarification, bagasse from pressing, lees, exhausted yeast and loaded wastewater which are obtained after different decanting steps (Lin *et al.*, 2014; Barba *et al.*, 2016; Devesa-Rey *et al.*, 2011) (Fig. 1). Jozinović *et al.* (2014) reported that the major residues from winemaking are organic wastes (grape pomace, seeds, pulp and skins, grape stems and grape leaves), wastewater, emission of greenhouse gases ($CO_2$, volatile organic compounds, etc.) and inorganic wastes (diatomaceous earth, bentonite clay and perlite). Sometimes environmental-friendly technologies have been proposed for the valorisation of winery waste products. Specifically, winery wastes can be an alternative source of added value products (e.g. polyphenols, bioethanol, lactic acid, tartaric acid, enzymes, etc.) and are considered safer in comparison to synthetic counterparts (i.e. fertilisers, synthetic antioxidants, etc.) (Arvanitoyannis *et al.*, 2006). According to Cuervas-mons *et al.* (2013), world wine industry transforms 10-25 per cent of raw grapes into residues, mainly represented by grape marc, lees, seeds and stems. These by-products are a rich source of polyphenols and can therefore, be used to produce new added-value products. Grape pomace has a vast array of applications in food industries, such as,

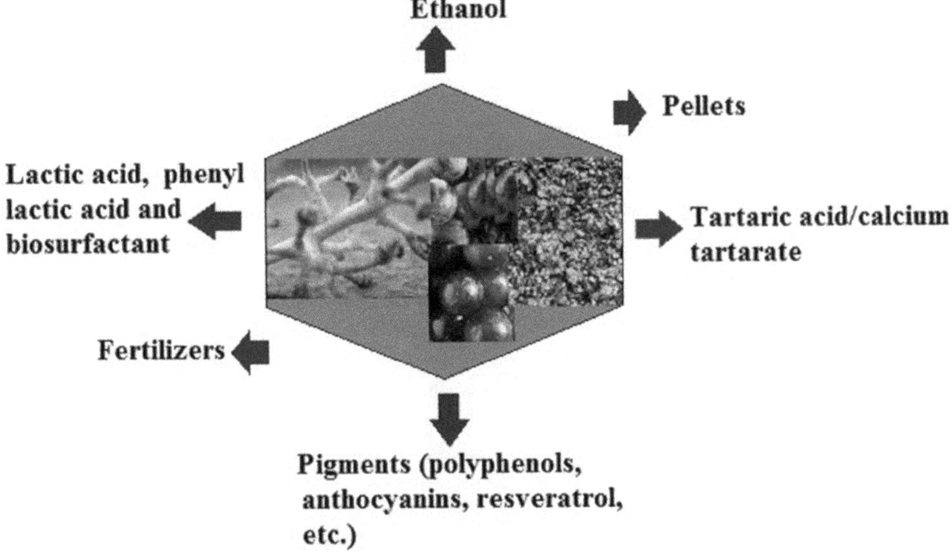


**Figure 1.** Different types of wastes generated during winemaking

*Corresponding author: kostaskot@yahoo.gr

**Abbreviations**

**BOD:** Biological oxygen demand
**COD:** Chemical Oxygen demand
**DPPH:** α, α-diphenyl-β-picrylhydrazyl
**DSC:** Differential scanning calorimetry
**ESI-MS/MS:** Electrospray ionisation mass spectrometry
**FTIR:** Fourier-transform infrared spectroscopy
**GC-MS:** Gas chromatography mass spectrometryW
**HPLC:** High performance liquid chromatography
**NMR Spectroscopy:** Nuclear magnetic resonance spectroscopy
**XRD:** X-ray diffraction

functional foods (dietary fibre + polyphenols), food processing (biosurfactants), cosmetics (grape seeds oil + antioxidants), biomedical/pharmaceutical (pullulan) and food supplements (grape pomace powder). To date, there has been no assessment of the market potential of grape pomace by-products (Dwyer *et al.*, 2014; Rondeau *et al.*, 2013). A sustainable winemaking process consists of maximising resources and decreasing greenhouse gas emissions that are generated during the production process (Castillo-Vergara *et al.*, 2015). On a global scale, the wine sector is responsible for around 0.3 per cent of annual greenhouse gas emissions (Amienyo *et al.*, 2014). In this regards, the problem of food and winery wastage has attracted considerable public attention in recent years and was considered by the European Parliament on January 19$^{th}$, 2012 (2011/2175(INI)) (Barba *et al.*, 2016). The recovery and use of wine processing by-products are thus, practical procedures that can lessen the waste disposal problems while increasing the limited resources (Barba *et al.*, 2016). All these aspects have found a place in this chapter.

## 2. Wine and Vine By-products

Currently, up to 210 million tons of grapes are produced annually, with 15 per cent of these being used in the winemaking industry (Teixeira *et al.*, 2014). This commercial activity generates huge amounts of solid waste (up to 30 per cent w/w). Winery wastes include biodegradable solids such as stems, skins, seeds, lees and waste water (Teixeira *et al.*, 2014). The following sections review the different by-products of the wine-industry, the technologies used for the valorisation process, the main substrates that can be extracted and used and the added-value products that can be obtained. A list of the main wine and vine waste products, the resulting added-value products and the operative conditions of the relevant process are given in Table 1.

### 2.1. Grape Stalks and Vine Prunings

Grape stalks can be an interesting source of solid biomass to cover energy needs. In comparison to other agricultural residues (e.g. wheat straw or corncobs), grape stalks contain relatively low levels of ash (2.90 per cent), which is, however, 10 times higher than the level contained in softwoods (e.g. spruce or pine) (González-Centeno *et al.*, 2010; Prozil *et al.*, 2012). The chemical composition and the structure of macromolecular components of grape stalks from red grape pomaces were evaluated by Prozil *et al.* (2012). These compounds were mainly composed of cellulose (30.3 per cent), hemicelluloses (21.0 per cent), lignin (17.4 per cent), tannins (15.9 per cent) and proteins (6.1 per cent). Ping *et al.* (2011) evaluated the composition of grape stalk (i.e. 36 per cent cellulose, 34 per cent lignin, 24 per cent hemicellulose and 6 per cent tannins). The tannins, analysed by using a solution-state 13C NMR were mainly of procyanidin type and the dichloromethane extractives fraction, characterised by GC-MS, was principally composed of fatty acids. Vine shoots are wastes that are produced in quantities of 1.4 and 2.0 tons/ha/year (Sánchez-Gómez *et al.*, 2016). The vine has been a traditional cultivar in Greece for many years. Every year after February, pruning leads to large quantities of vine prunnings that remain in the field as by-products. The average amount of prunings per year reaches ~5 t/ha, which is higher than the average yield of forests in temperate zones (Ntalos and Grigoriou, 2002). Vine-shoots have been shown to possess antioxidant, antifeedant and phytotoxic activities and could therefore be considered for applications in the agri-food industry (Sánchez -Gómez *et al.*, 2016).

**Table 1.** Global Vision of Wine and Vine Waste, Operative Conditions, and Value-added Products Produced

| *Wine and vine waste* | *Operative conditions* | *Value-added products* | *References* |
|---|---|---|---|
| Wine grape marc | SFE; 20-60°C | Polyphenols; 20.2 mg GAE/g DM | Vatai *et al.*, 2009 |
| Pomace of *Palomino fino* grapes | SC-$CO_2$, 100,400 bar, 35-55 °C, 5% (v/v) EtOH | Resveratrol; 0.39-0.68 mg/100 g fresh weight | Casas *et al.*, 2010 |
| Grape pomace extracts | HVED, 20-60°C, 1h, 30% (v/v) EtOH | Polyphenols (2.8 ± 0.4 gGAE/100 gDM) | Boussetta *et al.*, 2011 |
| Grape pomace | 3 mL of 20% (w/v) $Na_2CO_3$; 40°C; 20 min | TPC of RWGP (21.4–26.7 mg GAE/g DM); TPC of WWGP (11.6–15.8 mg GAE/g DM) | Deng *et al.*, 2011 |
| Grape seeds and/or skin | MAE; mixed extractant (EtOH 75%, HCl 1% in water) in an 1:10 (w/v) ratio | TPC, 36.8% (expressed as GAE) | Pérez-Serradilla and De Castro, 2011 |
| Grape marcs | 2 h of extraction, 75% EtOH liquid mixture at pH 2 | Polyphenols | Librán *et al.*, 2013 |
| SdWPP | Methanol/HCl (97:3) for 24 h | TPC (untreated: 42.72±0.79; UV treated: 38.59±0.67; thermally treated: 41.66±0.34) | García-Lomillo *et al.*, 2014 |
| Grape pomace | Conventional (mechanical stirring, 200 rpm) and acoustic (55±5 kHz, 435±5 W/L); 20-50°C | Polyphenols; 770.9±77.5 mgGAE/100 gDM | González-Centeno *et al.*, 2015 |
| Grape skins and defatted seeds | 10MPa, 80-120°C, 2-5 mL/min, 2h | Polyphenols; 124±1 mg/g (grape seeds); 77±3 mg/g (skins) | Duba *et al.*, 2015 |
| Grape seeds | 80 bar, 6kg/h $CO_2$ flow rate, 20% co-solvent | >1000 mg catechin/100 g DM) and FII >800 mg catechin/100 g DM | Da Porto and Natolino, 2017 |
| Grape seeds | Temperature 313-333 K, pressure up to 40.0MPa | Linolenic acid (67 %), and oleic acid (20%) | Coelho *et al.*, 2017 |
| Vine pruning | Fluidized bed gasifier, (70 kg/h), and operates at 800°C | Gasification gas (in terms of its CO and $CO_2$ contents) raw-material | Brito *et al.*, 2014 |
| Lee, grape marc | Yeast (*Saccharomyces cerevisiae*) induced fermentation | Acetic acid | Vilela-Moura *et al.*, 2011 |
| Vinasses and grape marc | Fermentation with *Trichoderma viride* | Lactic acid, biosurfactants, xylitol, ethanol and other compounds | Devesa-Rey *et al.*, 2011 |
| Vinasse | Alkali treatment (8%), microwave power (700 W) 24h fermentation | Lactic acid (17.5 g/L) | Liu *et al.*, 2010 |

(*Contd.*)

**Table 1.** (*Contd.*)

| *Wine and vine waste* | *Operative conditions* | *Value-added products* | *References* |
|---|---|---|---|
| Lee, grape marc | Yeast (*S.cerevisiae* and *Candida parapsilosis*) induced fermentation; pH (3.0-5.0); 28-36°C | Animal feed supplement and SCP | Silva *et al.*, 2011 |
| Vinasse | Fermentation; pH5.0, 30°C | Protein rich fungal biomass (45.55% crude protein) | Nitayavardhana & Khanal, 2010 |
| Vinasse and grape marc | Fermentation with *Trichoderma viride* WEBL0703 ($6.65 \times 10^9$ CFU/g) | Biocontrol agent | Zhihui *et al.*, 2008 |
| Vinasse | Fermentation (100 mL, 100 rpm and 31°C for 194 h), xylose (55 g/L) | Tartaric acid and calcium tartrate | Salgado *et al.*, 2010 |
| Grape pomace | SsF ($4.5 \times 10^8$ spores/g solid substrate); *Aspergillus awamori* (0.2 mL), | Exo-PG, xylanase and cellullase | Díaz *et al.*, 2012 |
| Grape +sugarbeet pomaces | SsF (sugar, 16.5%; pH 4.5; humidity 68%; $10^8$ cells/g), 28°C, 96 h | Bioethanol | Rodríguez *et al.*, 2010 |
| Grape marc | Fermentation with *Lactoacillus plantarum* | Anti-allergen | Tominaga *et al.*, 2010 |

**Abbreviations:**

TPC: Total phenol content; HVED: High-voltage electric discharge; PEF: Pulsed electric fields; US: Ultrasounds; GAE: Gallic acid equivalents; DM: Dry matter; SFE: Supercritical fluid extractions; SC-$CO_2$: Supercritical carbon dioxide; TCC: Total catechin content; SdWPP: Seed wine pomace product; MAE: microwave-assisted extraction; SCP: Single cell protein; SsF: Solid state fermentation; Exo-PG: Exo-polygalacturonase

## 2.2. Grape Pomace

Grape pomace consists mainly of seeds, peels (skins) and stems and accounts for about 20-25 per cent of the weight of the grape crushed for wine production. Grape seed is rich in extractable phenolic antioxidants, such as phenolic acid, procyanidins, flavonoids and resveratrol, while grape skins contain abundant anthocyanins. The health benefits of grape pomace polyphenols have been the centre of interest of researchers, the food and the nutraceutical industries. In addition to phenolic antioxidants, grape pomace also contains significant amounts of lipid, proteins, non-digestible fibre and minerals (Yu and Ahmedna, 2013). The polyphenol composition of grape pomace has been well characterised and its biological and functional properties are also intensively studied. These properties of polyphenols are linked to chemoprevention, lower risk of cardiovascular disease and other diseases and have been revealed by researchers working over the world. Therefore, grape pomace has a great potential to serve as a source of an ingredient for functional foods (Yu and Ahmedna, 2013). Grape pomace is also a source of antioxidant dietary fibre. Apart from promoting human health, grape pomace as an antioxidant dietary fibre plays an important role as an antioxidant and antimicrobial agent, extending the shelf-life of food products. For instance, grape pomace powder has been added into minced fish and chicken breast to delay the lipid oxidation. Also, grape pomace extract exhibited antimicrobial activity when added into beef patties. Tseng and Zhao (2013) investigated the feasibility of fortifying foods (yoghurt and salad dressing) with grape pomace, to enhance their dietary fibres and polyphenols content.

## 2.3. Grape Seeds

Grape seeds (5 per cent of the fruit weight) are another by-product (38-52 per cent of dry matter of pomace) of winemaking (Ovcharova *et al.*, 2016). These seeds contain lipid, protein, carbohydrates, and 5-8 per cent polyphenols, depending on the variety. Kim *et al.* (2006) reported that grape seeds have a complex composition containing approximately 40 per cent fibre, 16 per cent oil, 14 per cent and 7 per cent phenols besides sugar, minerals, salts etc. Mironeasa *et al.* (2010) reported that the grape seeds contain 28 per cent cellulose, 4-6 per cent tannins, 10-25 per cent oil and 2-4 per cent minerals. The most abundant phenolic compounds isolated from grape seeds are catechins, epicatechin, procyanidin and some dimmers and trimers (Shi *et al.*, 2003). Grape seed extract is a powerful antioxidant that seems to protect the body from premature aging, disease and decay. Scientific studies have shown that the antioxidant power of proanthocyanidins is 20 times greater than that of vitamin E and 50 times greater than that of vitamin C (Shi *et al.*, 2003). Kim *et al.* (2006) evaluated the effect of heating (50-200°C for 10-40 min.) on grape seeds and the antioxidant activity of their extracts. Based on the GC-MS analysis, several low-molecular weight phenolic compounds, such as azelaic acid, 3,4-dihydroxy benzoic acid and o-cinnamic acid were formed in the grape seed extracts (Perumalla and Hettiarachchy, 2011). Grape seeds are also a source of beneficial fatty acids and dietary fibres. Some unsaturated fatty acids contained in grape seed oil, such as $\alpha$-linolenic acid ($\omega$-3) and $\gamma$-linolenic acid ($\omega$-6), are considered essential fatty acids because the human body cannot produce them and they therefore need to be taken through the diet (Hussein and Abdrabba, 2015). Consumption of unsaturated fatty acids has been linked to a reduction in the rate of cardiovascular diseases, cancer, hypertension and immune disorders (Hussein and Abdrabba, 2015). A high portion of the composition of grape seeds and peels (about 80 per cent of their dry weight) is composed of dietary fibres, which have been linked to lower risks of heart disease, obesity, diabetes and colon cancer (Hussein and Abdrabba, 2015). Ovcharova *et al.* (2016) investigated the oil yield (16.63 per cent) obtained from grape seeds. It was proved that a high percentage of the fatty acid composition of the seeds of red grape was polyunsaturated fatty acids, especially linoleic acid (55.30 per cent) followed by oleic acid (25.81 per cent), while palmitic acid was the dominant saturated fatty acid (11.87 per cent) (Ovcharova *et al.*, 2016). This material is therefore, a good source of polyunsaturated fatty acids, vitamins and antioxidants. For this reason, it is used for the prevention of a variety of diseases, such as thrombosis, cardiovascular diseases, reduction of cholesterol in serum, dilation of blood vessels, cancer reduction and regulation of autonomic nerves (Yi *et al.*, 2009; Fontana *et al.*, 2013). The interest in grape seed oil as a functional food product has increased, mainly because of its high levels of hydrophilic constituents, such as phenolic compounds and lipophilic constituents, such as vitamin E, unsaturated fatty acids and phytosterols (Karaman *et al.*, 2015). The oil content of grape seeds varies between 8-20 per cent, depending on grape variety and agricultural conditions (Garavaglia *et al.*, 2016).

### 2.4. Wine Lees

Lees are the waste generated during the fermentation and aging process of different industrial activities related to alcoholic drinks, such as wine, cider and beer. Wine lees constitute approximately 2-6 per cent of the initial grape used in winemaking. It has been reported that approximately 0.42-1.26 million tons of wine lees are generated annually in Europe. Wine lees are currently used for commercial production of platform chemicals, such as calcium tartarate, ethanol, lactic acid and xylitol (Pérez-Bibbins *et al.*, 2015; Dimou *et al.*, 2015; Bordiga *et al.*, 2015). Dimou *et al.* (2015) demonstrated the development of a novel wine lees-based integrated biorefinery for the production of several added-value products. Wine lees were initially fractionated for the production of antioxidants, tartarate and ethanol and the remaining stream was converted into a fermentation nutrient supplement for poly (3-hydroxybutyrate) production (30.1 g/L), using the strain *Cupriavidus necator* DSM 7237.

## 3. Treatment Methods in Volarisation of Added-value Products

Various pre-treatment methods, such as low temperature pyrolysis, enzymatic extraction, mechanical process, etc. are carried out for valorisation of winery wastes.

### 3.1. Low-temperature Pyrolysis

Biomass pyrolysis is the process of breaking chemical bonds in biomass macromolecules, using heat energy under inert atmosphere and can convert biomass into low-polymerisation products or even small molecular compounds through a series of complex reactions, such as depolymerisation, ring opening and cleavage. Biomass pyrolysis products can be classified into gas, tar and char according to their phases at room temperature. The chemical reactions are extremely complicated during biomass pyrolysis owing to the diverse distribution of biomass components (Bennadji *et al.*, 2013; Wang and Luo, 2017). Dried grape can be converted to carbon products by low-temperature pyrolysis. The calorific value of these products was determined and compared with that of commercial barbeque briquet (Walter and Sherman, 1976). The gross heat of combustion of grape charcoal briquets was approximately 90 per cent of that of the commercial briquet, while the dried press cake contained approximately 65 per cent. The thermolytic reactions generally augmented the fuel value of the dried press-cakes by 37-45 per cent.

### 3.2. Enzymatic Extraction

Pectic enzymes can be used for a more efficient extraction of desirable red grape pigments and other compounds which are bound in plant cells (Munoz *et al.*, 2004; Saigal and Ray, 2006). When the enzymatic extraction of anthocyanin pigments from the grape pigments of three varieties of grape from central Chile was considered, it was shown that the best results of extraction of anthocyanins could be obtained with Vinozym EC using skin grape Ribier after two hours of treatment (Muñoz *et al.*, 2004). In the study of Rodriguez-Rodriguez *et al.* (2012), grape pomace derived from winemaking extracted by an enzymatic process and its composition of polyphenols was evaluated by HPLC and ESI-MS/MS. Kaempferol, catechin, quercetin and procyanidins as well as trace levels of resveratrol and traces of gallocatechin and anthocyanidins were detected. The extraction of phenolic compounds from grapes usually involves the use of organic solvents, which implies a health risk as well as a potential environmental contamination potential. Various methods have been designed in order to recover a larger amount of phenolic compounds with minimal use of solvents (Xia *et al.*, 2010). An enzymatic extraction method gives rise to a new water-soluble product from grape pomace; a grape pomace enzymatic extract, which provides a significant amount of phenolic compounds is among the other components (Rodriguez-Rodriguez *et al.*, 2012).

### 3.3. Mechanical Technologies

Several studies have employed a wide range of mechanical technologies for extracting valuable products (especially antioxidants) from winery by-products. The methodologies used are detailed in these sections here. Monrad *et al.* (2012) reported that grape pomace contains appreciable amounts of polyphenolic compounds, such as anthocyanins and procyanidins, which can be recovered for use as food supplements. The extraction of these polyphenols from the pomace is usually accomplished at slightly elevated temperatures, frequently employing hydroethanolic solvents. Traditional extraction methods, including liquid-liquid and solid-liquid extraction, can involve long extraction times accompanied by organic solvent uptake into the remaining pomaces. These conditions can produce an extract containing 139 mg/100 g dry weight (DW) of anthocyanins and 2077 mg/100 g DW (dry weight) of procyanidins. Fontana *et al.* (2013) characterised high contents of phenolics in grape pomaces due to incomplete extraction during the winemaking process. Monrad *et al.* (2012) reported that due to governmental regulations and the cost involved in using ethanol as a solvent, as well as the loss in polyphenolics due to thermal degradation, improved extraction techniques are required. A semi-continuous extraction apparatus employing only water was developed to maximise the recovery of anthocyanins and procyanidins from red grape pomace. Water was preheated prior to its entry to the extraction cell containing the grape pomace sample, where it was then, allowed to flow continuously through the unheated extraction vessel before its collection at ambient conditions. Extraction variables that impacted the polyphenolic recovery included pomace moisture content (crude or dried), sample mass, water flow rate and extraction temperature.

### 3.4. Anaerobic Digestion

Anaerobic digestion is widely used for wastewater treatment, especially in the food industry (Moletta, 2005). After the anaerobic treatment, there is generally an aerobic post-treatment in order to return the treated water to the environment (Devesa-Rey *et al.*, 2011). Waste waters are typically characterised by exceptionally high levels of chemical oxygen demand (COD), both particulate and soluble, and high biodegradability. Semi-solid wastes, like lees and vinasses are often treated in anaerobic stirred reactors to recover renewable energy (Moletta, 2005). Several technologies are applied for winery wastewater treatment. These technologies employ free cells or flocs (anaerobic contact digesters, anaerobic sequencing batch reactors and anaerobic lagoons), anaerobic granules (Upflow Anaerobic Sludge Blanket), biofilms on fixed support (anaerobic filter) or on mobile support as with the fluidised bed. Some technologies include two strategies, e.g. a sludge bed with anaerobic filter, as in the hybrid digester (Moletta, 2005). With winery wastewaters (as for vinasses from distilleries) the removal yield of anaerobic digestion is very high (up to 90-95 per cent COD removal). The organic loads are between 5-15 kg COD/m$^3$ of digester/day. The biogas production is between 400-600 L per kg COD removed with 60-70 per cent methane content. For anaerobic and aerobic post-treatment of vinasses in the Cognac region, the REVICO Company has reported 99.7 per cent COD removal at a cost of 0.52 Euro/m$^3$ of vinasses (Moletta, 2005). Rodríguez *et al.* (2007) examined the use of anaerobic batch reactors, treating winery wastewater combined with waste activated sludge in different proportions under mesophilic conditions. It was shown that for anaerobic digestion of winery wastewater alone, the methane production rate was lower than the rates achieved when winery wastewater and waste activated sludge were treated together. A simplified anaerobic model was used to determine the main kinetic parameters, such as the maximum COD reduction rate ($q_{DA}$) and maximum methane generation rate ($k_{max}$). The maximum values of $q_{DA}$ and $k_{max}$ were 16.50 kg COD/d and 14.34 kg $CH_4$/d, respectively. On a pilot scale, the anaerobic co-digestion of wine lees together with waste activated sludge in mesophilic and thermophilic conditions was tested by Da Ros *et al.* (2014). Three organic loading rates (OLRs 2.8, 3.3 and 4.5 kg COD/m$^3$/d) and hydraulic retention times (HRTs 21,19 and 16 days) were applied to the reactors in order to evaluate the best operational conditions for the maximisation of the biogas yields. The addition of lee to sludge determined a higher biogas production; the best yield obtained was 0.40 $Nm^3_{biogas}$/kg $COD_{fed}$. Recently, the use of solar photo-Fenton oxidation (stimulated solar light) for treating the concentrate with the reverse osmosis process proved to be a successful combined process for the integrated treatment of winery effluents (Lofrano and Meric, 2016).

### 3.5. Composting

Composting of winery waste is an alternative to the traditional disposal of residues and involves a commitment to reducing the production of waste products (Bertran *et al.*, 2004). Diaz *et al.* (2002)

investigated an incubation process of binary mixture of grape marc/vinasse for an optimum composting process. Mixtures with increasing amounts of vinasse (0-40 per cent) were incubated in a laboratory-scale reactor under aerobic conditions at 55°C for 43 days. The results of the incubation experiment indicated that the composting process of vinasse and grape marc was technically suitable; however, a moderate amount of vinasse (10-20 per cent) would be the best compromise to optimise the process and obtain high quality compost. Composting can also be a successful strategy for sustainable and complete recycling of grape marc. Moldes *et al.* (2007) reported that the wine industry generates a large amount of wastes, including grape marc and vinification lees. These substances can be used to produce enzymes or other food additives. The mesophilic biodegradation of grape marc (several ratios of skin, seed and stem) during 60 days under microaerobic conditions was studied. The presence of *Penicillium* spp. was detected at the beginning of composting. Biodegraded grape marc with stem showed the best organic matter properties (C/N ration of 14 and N content of 37 g/kg) and a germination index of 155 per cent for the growth of ray grass seeds. The results suggested that the biodegradable grape marc could be used as a fertiliser, especially for ray grass crops. An experiment was conducted by Carmona *et al.* (2012) to study the potential of compost derived from de-alcoholised grapevine marc and grape stalk as growing medium components in the plug seedlings production of lettuce, tomato, pepper and melon. The compost was found to be an effective component of a fertiliser that could be used for plug production of vegetable seedlings. Paradelo *et al.* (2012) prepared five grape marc composts following different procedures (composting and vermicompost at several scales) and tested them as potential components of plant-growth media. The five composts had a high organic matter content (>90 per cent), low electric conductivity (<1 dS/ m) and a pH of 7 to 8. The results showed that four out of the five composts were suitable for promoting plant growth and increasing the productivity of barley (*Hordeum vulgare* L.).

Zhang and Sun (2016) currently studied the conditions and the efficiency of two-staged composting of green waste derived from sugarcane bagasse (at 0, 15, and 25 per cent) and/or exhausted grape marc (at 0, 10, and 20 per cent). The combined addition of sugarcane bagasse and exhausted grape marc improved the composting conditions and the quality of the compost product in terms of temperature, water-holding capacity, particle-size distribution, coarseness index, pH, electric conductivity, microbial counts, etc. The optimal two-stage composting and the best quality compost were obtained with the combined addition of 15 per cent sugarcane bagasse and 20 per cent exhausted grape marc. The two-staged green waste-compost needed only 21 days to mature instead of 90-270 days required for traditional composting.

## 4. Value-added Products

Turning vinification wastes into valuable products (Fig. 2) is becoming an essential part of good winemaking practices, further reducing concerns of waste disposal and cutting costs for partly imported wine additives (e.g. tartaric acid, ethanol, lactic acid, etc.) (Rivas *et al.*, 2006).

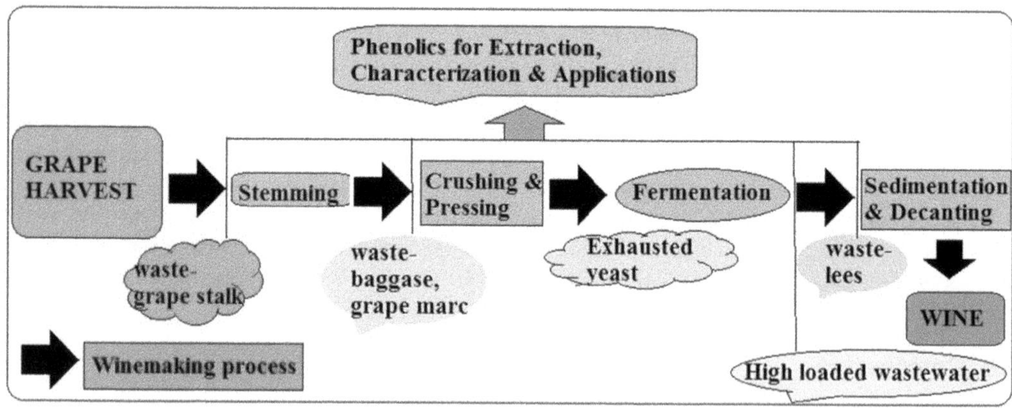


**Figure 2.** Value-added products obtained from winery wastes

## 4.1. Tartaric Acid/Calcium Tartarate

Wine waste lees are currently partly exploited for tartaric acid production (Kontogiannopoulos *et al.*, 2017). Rivas *et al.* (2006) recovered tartaric acid from distilled vinification lees of white and red winemaking technology and further optimised the process parameters using response surface methodology and Statistica 5.0 software. The sequential treatment of dissolving tartaric acid and the additional calcium tartarate precipitation can be used to recover up to 92.4 per cent of initial tartaric acid concentration. Moreover, the remaining lees can be used as cost-effective nutrients for lactic acid production from trimming vine shoot hydrolysates using *Lactobacillus pentosus* CECT-4023. In the study of Yalcin *et al.* (2008), tartaric acid-containing waste samples obtained from the wine- and grape-juice industries were characterised by using DSC, HPLC, FTIR, and XRD. HPLC to determine the tartaric acid content of samples. The decomposition temperatures of waste samples were found to be relatively higher compared with that of pure tartaric acid. This difference in decomposition temperatures was attributed to the presence of potassium tartarate. According to Salgado *et al.* (2010), vinasses, the main liquid wastes from the distillation process of grape marc and wine lees, are acidic effluents with high organic content, including acids, carbohydrates, phenols and unsaturated compounds with high COD, BOD and solid concentrations. These wastes can be revalued to provide additional benefits when they are employed as feedstock of compounds, including tartaric acid and cost-effective nutrients for elaboration of fermentable broths. Kontogiannopoulos *et al.* (2016) developed a novel cost-effective and environment-friendly process, using cation exchange resin for recovering tartaric acid and polyphenolic compounds from wine lees. An experimental design was carried out, based on central composite design with Response Surface Methodology to evaluate the effects of process parameters and their interaction in order to determine the optimum conditions. A set of optimum values of the three main variables was determined at pH = 3.0, water dosage 10 ml/g dry lees and resin dosage 3.5 g/g dry lees. Under these optimum conditions, the predicted tartaric acid recovery could be as high as 74.9 per cent. In an another report, Kontogiannopoulos *et al.* (2017) investigated an integrated environment-friendly process, using mild conditions to recover tartaric acid and simultaneously exploit the total polyphenols content of wine lees. Several ultrafiltration and nano-filtration membranes were assessed in bench-scale filtration tests for their efficiency in separating the two main products (i.e. tartaric acid; 44.2 g/L and total polyphenols; 323.3 mg/L) from the main stream (wine lees).

## 4.2. Ethanol

The feasibility of fermenting grapes to produce bioethanol fuel in the European Community was assessed in the study of Scrase *et al.* (1993). The net energy balances, costs and environmental impacts were considered for a range of management scenario. Typical wine-producing vineyards cannot produce ethanol with a positive net energy balance and costs are four to six times as great as those using wheat or sugar beet as raw materials. Wine production in the European Community exceeds demand by 20-40 per cent. Producing ethanol for fuel from surplus wine in the European Community is a drain on energy and financial resources. It is possible to improve the performance of grapes as an energy crop, principally by raising yields. It was also calculated that annual crops, such as wheat and sugar beet, remain preferable to grapes as raw materials. However, the surplus land under grape vines is often steeply sloping and has thin, dry soils, which would be subject to considerable soil erosion if planted with arable crops; it is considered that perennial energy crops are more environmentally acceptable than annual arable crops. Grape pomace was also used as a substrate for the production of ethanol under solid-state fermentation conditions in the study of Hang *et al.* (1986). The yield of ethanol amounted to greater than 80 per cent of the theoretical value, based on the fermentable sugar consumed. The grape pomace used in this study contained 13.7 per cent sugar as glucose and had a moisture content of 64.4 per cent, while the pH was 3.6. Rodríguez *et al.* (2010) described the production of ethanol through solid state fermentation, using grape pomace and sugar beet pomaces as substrates. This work reported the use of laboratory-scale solid state fermentation to obtain alcohol from grape pomace and sugar beet pomaces, using *Saccharomyces cerevisiae* yeasts. The initial conditions of the culture media were sugar 16.5 per cent (p/p), pH 4.5, and humidity 68 per cent (p/p). The cultures were inoculated with $10^8$ cells/g of pomace and incubated in an anaerobic environment, at 28°C, for 96 hours. Solid state fermentation showed ethanol maximum concentrations at 48 hours and ethanol yield on sugars consumed more than 82 per cent.

### 4.3. Lactic acid, Phenyl Lactic Acid and Biosurfactant

Trimming of vine shoots and vinasses means agricultural wastes of little use, but have the potential as an alternative cost-effective media for lactic acid and cell-bound biosurfactant production (Rodríguez *et al.*, 2010). Bustos *et al.* (2004) hydrolysed vine shoots with dilute sulphuric acid (1-5 per cent) in order to obtain sugar solutions suitable as fermentation media. The operational conditions for hydrolysis were selected on the basis of both the generation of hemicellulosic sugars (mainly xylose) and glucose and the concentrations of reaction byproducts (furfural, hydroxymethyl furfural and acetic acid) affecting fermentation. Hemicellulosic hydrolysates were supplemented with nutrients and fermented with *Lactobacillus pentosus*, without any previous detoxification stage, to produce lactic acid. Under the best operational conditions assayed (3 per cent $H_2SO_4$ at 15 min.), 21.8 g lactic acid /L was produced ($Q_p$=0.844 g/L/ h, $Y_{p/s}$ = 0.77 g/g), which represents a theoretical yield of 99.6 per cent. Acetic acid was the primary byproduct formed from xylose, at about 25 per cent of the lactic acid level. Similarly, an effective process for the chemical-biotechnological utilisation of trimming wastes of vines roots was reported in the study of Bustos *et al.* (2005). Initial treatment with sulphuric acid (prehydrolysis) allowed the solubilisation of hemicelluloses to xylose and glucose-containing liquors (suitable for the production of fermentation media for lactic acid production with *Lb. pentosus*) and a solid phase containing cellulose and lignin. The solid residues from prehydrolysis were treated with NaOH in order to increase their cellulase digestibility. In the alkaline treatments, the effects of temperature (in the range of 50-130°C), retention time (30-120 min.) and NaOH concentration (4-12wt, per cent of solution) on the composition and susceptibility to enzymatic hydrolysis of solid residues were assessed by means of an experimental plan with factorial structure. The lignin content decreased, whereas the susceptibility towards enzymatic hydrolysis increased with temperature, reaction time and NaOH concentration within the tested range. Using the cellulosic residues under harsher conditions, favourable fermentation kinetics during simultaneous saccharification for lactic acid production carried out by *Lactobacillus rhamnosus* were observed. Rivera *et al.* (2007) evaluated the sugar-containing liquors obtained from hydrolysates of distilled grape marc as the media for lactic acid and biosurfactants production. In order to obtain the best operational conditions for hydrolysis, the variables temperature, reaction time and $H_2SO_4$ concentration were studied, using factorial design. Selected operational conditions were chosen to carry out fermentation by *Lb. pentosus*. The hydrolysis (30 min.) at 130°C with 3.3 per cent $H_2SO_4$ was most suitable in order to carry out the fermentation, using non-detoxified hydrolysates and providing the best results in terms of lactic acid and biosurfactants production. *Lactococcus lactis* is an interesting microorganism with several industrial applications, particularly in the food industry. As well as being a probiotic species, *Lb. lactis* produces several metabolites with interesting properties, such as lactic acid and biosurfactants. The potential of *Lb. lactis* CECT-4434 as a lactic acid and biosurfactant producer was studied. The financial cost of *Lb. lactis* cultures could be reduced by replacing the MRS (De Man, Rogosa and Sharpe) medium with the use of two waste materials: trimming vine shoots as C source, and 20 g/L distilled wine lees (vinasses) as N, P and micronutrient sources.

From the hemicellulosic fraction, 14.3 g/L lactic acid and 1.7 mg/L surfactin equivalent was achieved after 74 hours (surface tension reduction of 14.4 N/m); meanwhile, a simultaneous saccharification and fermentation process allowed the generation of 10.8 g/L lactic acid and 1.5 mg/L surfactin equivalent after 74 hours, reducing the surface tension by 12.1 units at the end of fermentation (Rodríguez *et al.*, 2010). Co-culture fermentations show advantages for producing food additives from agro-industrial wastes, considering that different specified microbial strains are combined to improve the consumption of mixed sugars obtained by hydrolysis. Rodríguez-Pazo *et al.* (2013) developed a profitable technology for the use of both hemicellulosic and cellulosic fractions of trimming wastes by co-culture of *Lactobacillus plantarum* and *Lb. pentosus*. Different bioactive compounds, such as lactic acid, phenyl lactic acid and biosurfactants) were analysed in the exhausted culture media. The highest lactic acid and phenyl lactic acid concentrations, 43.0 g/L and 1.58 m/M, respectively, were obtained after 144 hours during the fermentation of hemicellulosic sugars and simultaneous saccharification and fermentation carried out by co-cultures of *Lb. plantarum* and *Lb. pentosus*.

### 4.4. Antioxidants and Pigments

It has been demonstrated that wine and other products derived from grapes have a high antioxidant capability and as a result, they may offer potential health benefits. Solid by-products obtained from the white and red wine industry were subjected to evaluation in the study of Makris *et al.* (2007) as potential sources of antioxidant phytochemicals on the basis of their content in phenols and *in vitro* antioxidant activity. The results showed that extracts from grape seeds (either white or red) contain exceptionally high amounts of total polyphenols (10.3-11.1 per cent on dry weight basis), a great part of which is composed of flavanols.

Polyphenols, the well-known naturally occurring antioxidants, are the most abundant secondary metabolites in grape wastes. Casazza *et al.* (2010) investigated several non-conventional extraction methods *vs.* classic solid-liquid extraction to obtain phenolic compounds from grape seeds and skins. The several non-conventional extraction methods, such as ultrasound-assisted extraction, microwave-assisted extraction and high pressure and temperature extraction, were investigated and compared with solid-liquid extraction and extracts were defined on the basis of extraction yield and antioxidant power. Quali-quantitative analyses were performed using colorimetric and HPLC methods. The highest content in total polyphenols, ortho-diphenols and flavonoids, both for seeds and skins, was obtained with a high pressure and temperature extraction method, while the highest antiradical power was determined in seed extracts obtained by using a microwave-assisted extraction (78.6$\mu l_{extract}$ $\mu g/_{DPPH}$). In another study, Casazza *et al.* (2012) investigated a non-conventional extraction technology in which a high-pressure high-temperature reactor was employed to extract polyphenols from grape skins. The extraction time (15-330 min.) and temperature (30-150°C) were selected as independent variables and their effects were studied. The total polyphenol and total flavonoid yields, as well as the antiradical power of the extracts, were analysed. The use of high-pressure high-temperature technology resulted in extracts rich in polyphenols with high antiradical power. The highest total polyphenol (60.7 mg GAE (gallic acid equivalent) g/DW) and total flavonoids (15.1 mgCE g/DW) yields were obtained at 150°C for 270 min. and 150°C for 15 min., respectively. Antiradical power values were found between 8.45 and 52.17 $\mu l_{extract}$ $\mu g$/DPPH. Furthermore, Librán *et al.* (2013) aimed at determining the best process conditions (treatment time, per cent ethanol, pH of the solvent) for the solid-solid extraction of polyphenols from grape marcs and analysed the effects of these conditions on several extraction yields, namely on total phenolics, flavonoids, phenolic acids and anthocyanins and also on the antioxidant power of the extracts. Among all the polyphenols extracted, anthocyanins were the most abundant, representing over 40 per cent of the total polyphenol content. The best process conditions were obtained after two hours of extraction in a 75 per cent ethanol- liquid mixture at pH 2. In the study of Duba *et al.* (2015), polyphenols were extracted from grape skins and defatted grape seeds (cultivar: Pinot Nero) under a constant pressure of 10 MPa and a flow rate of 2-5 mL/min and under three operating temperatures, namely, 80, 100 and 120°C. For both skins and defatted seeds, total polyphenol yield significantly increased with temperature: for skin from 44.3±0.4 to 77±3 mg/g, while for defatted seeds from 44±2 to 124±1 mg/g when temperature increased from 80 to 120°C. González-Centeno *et al.* (2015) evaluated the kinetics of both conventional (mechanical stirring, 200 rpm) and acoustic (55±5 kHz, 435±5 W/L) aqueous extraction of total phenolic content and antioxidant capacity from grape pomace by-products and modelled them at different extraction temperatures (20, 35 and 50°C). A gradual and significant increase of total phenolic content (770.9±77.5 mg GAE/100g DM) and antioxidant capacity (722.4±41.0 mg TE/100 g DM) of the extracts was observed as the temperature increased (maximum at 50°C), and the highest values were reported in the case of the acoustically-assisted extraction. As observed, the acoustic process required less extraction time and lower operating temperatures to obtain extracts with similar phenolic and antioxidant characteristics than those resulting from conventional extraction. More recently, Da Porto and Natolino (2017) used Response Surface Methodology for the optimisation of supercritical extraction of total polyphenols and proanthocyanidins from grape seeds in order to evaluate the effects of pressure, co-solvent percentages and $CO_2$ flow rate, through a Box-Behnken design. A two-steps supercritical carbon dioxide extraction was conducted, with the first step using supercritical carbon dioxide to remove the non-polar components from the grape seeds, and the second using supercritical carbon dioxide added with a co-solvent, finally leading to recovery from the defatted grape seeds. The statistically generated optimum conditions to obtain the highest total polyphenols concentration were a pressure of 80 bar, $CO_2$ flow rate of 6 kg/h and 20 per cent (w/w) co-solvent. Coelho *et al.* (2017) studied the supercritical extraction of oil from grape seed samples from a Portuguese industry. The process was carried out at temperatures of 313-333 K, pressures up to 40.0 MPa and different supercritical carbon dioxide flow rates. Higher oil yield was achieved using supercritical extraction (in the range of 12.0-12.7 per cent) as compared to 12.3 per cent obtained by conventional n-hexane extraction. The main fatty acids present in the supercritical carbon dioxide oil extracts were linolenic and oleic acids, with an average percentage of 67 and 20 per cent, respectively.

### 4.5. Pellets Solid Fuel

Prozil *et al.* (2014) evaluated pelletised solid fuel from grape stalks and compared it with fuel produced from softwood. It was found that the specific energy consumption for pelletising grape stalks was approximately 25 per cent lower when compared to that of softwood sawdust. The bulk density of the produced grape stalk pellets (670 $kg/m^3$) was similar to that of pellets produced from softwoods (660 $kg/m^3$), whereas the particle density was slightly higher in the case of grape stalks than in the case of softwood pellets (1129 against 1098 $kg/m^3$). The durability of pellets from grape stalks and softwood was practically the same – 95.8 per cent and 95.6 per cent, respectively. The grape stalks showed a higher heating value of 16.7 MJ/kg, which is slightly lower than that obtained for softwood (18.27 MJ/kg).

The wine industry, an important economic activity in Portugal, particularly in the Alto-Alentejo region, generates large amounts of residues, especially in the pruning of vines. Various technologies, including energy and agricultural applications, have been considered for the economical valorisation of these residues. Brito *et al.* (2014) assessed the potential use of biomass energy available in the Norte Alentero region and studied the technical feasibility of energy recovery from wastes resulting from the wine industry by using thermal gasification technologies. The study was conducted in a pilot thermal gasification plant and was based on fluidied bed technology, with a processing capacity of 70 kg/h and operating at 800°C. The gasification tests were performed continuously for several days, using the pellets of mixtures with different mass rations of vine pruning and wood pellets, in order to optimise the heat value and the composition of the produced gas and obtained condensate. The thermodynamic model developed in this work seems to perform well as an estimator of the gasification gas composition (in terms of its CO and $CO_2$ contents). An increase in the gasification temperature improved the gasification gas Net Heat Value. The study of Benetto *et al.* (2015) analysed the production chain of grape marc pellets and evaluated the overall environmental performance of the use of grape marc pellets for heat production, also comparing this performance with alternative fossil and renewable energy resources. A Life Cycle Assessment based on primary data from field experiments was used. A sensitivity analysis was carried out concerning the type of fuel used for drying, the methodological approach to solve multi-functionality, as well as the influence of the water content of grape marc. The combustion and drying of pellets were found to be the main contributors to the environmental impacts, although a crucial influence of methodological choices on co-products management was observed.

### 4.6. Fertilisers

Vinasse is utilised in agriculture as a cheap nutrient source, for ameliorating agents and animal feeds (Vadivel *et al.*, 2014). The optimised dose of vinasse application has significance over soil properties, crop qualities and yield improvements. It also contributes a substantial amount of phosphorous, sulphur, calcium and micronutrients to crops Vadivel *et al.*, 2014).

## 5. Future Research Focus and Conclusion

The use of trimming waste, grape marc and wine lees for the production of value-added products is a promising method of reducing the total price of biotechnological processes, but is also an environmentally friendlier method of removing these waste products, which may cause damage to the environment. Companies must therefore, invest in new technologies to decrease the impact of agro-industrial residues on the environment and establish new processes that will provide additional sources of income. Another way of valoraising winery waste is by composting, which in most cases generates compost and fertiliser of high agronomic value. Nevertheless, the health benefit of winery waste is currently undervalued and its use is limited to alcohol production by fermentation and distillation as well as to the manufacture of animal feed. Further research and practical experimentation is necessary since, in the case of winery waste, limited studies have been conducted. Moreover, the life cycle analysis regarding full economic costing of the use of wine waste as a resource, is needed.

It appears from this review that bioconversion of winery waste and their products to value-added products are economically viable. Although different applications have been assayed, including the recovery of ethanol by distillation, extraction of polyphenolic compounds or salts, use as compost and fertilisers, use as raw material for L-lactic acid production, or even for the production of biogas, winery waste appears as an undervalued by-product up to now. The currently available results of biotechnological/treatment techniques considered nowadays for the use of winery waste are a promising alternative that is focusing on the waste remediation and treatment, rather than on resource recovery. However, winery industry by-products, including grape seeds, grape pomace and stems are very rich sources of antioxidant polyphenols compared to other agri-food solid wastes and therefore, their exploitation as a source of added-value products may be more cost-effective and merits a profounder investigation.

## References

Aliakbarian, B., Fathi, A., Perego, P. and Dehghani, F. (2012). Extraction of antioxidants from winery wastes using subcritical water. *The Journal of Supercritical Fluids* 65: 18-24.

Amienyo, D., Camilleri, C. and Azapagic, A. (2014). Environmental impacts of consumption of Australian red wine in the UK. *Journal of Cleaner Production* 72: 110-119.

Arvanitoyannis, I.S. and Kotsanopoulos, K.V. (2016). Food waste generation and bio-valorisation. *Fermented Foods, Part I: Biochemistry and Biotechnology* 349.

Arvanitoyannis, I.S., Ladas, D. and Mavromatis, A. (2006). Potential uses and applications of treated wine waste: A review. *International Journal of Food Science and Technology* 41(5): 475-487.

Barba, F.J., Zhu, Z., Koubaa, M., Sant'Ana, A.S. and Orlien, V. (2016). Green alternative methods for the extraction of antioxidant bioactive compounds from winery wastes and by-products: A review. *Trends in Food Science and Technology* 49: 96-109.

Benetto, E., Jury, C., Kneip, G., Vázquez-Rowe, I., Huck, V. and Minette, F. (2015). Life cycle assessment of heat production from grape marc pellets. *Journal of Cleaner Production* 87: 149-158.

Bennadji, H., Smith, K., Shabangu, S. and Fisher, E.M. (2013). Low-temperature pyrolysis of woody biomass in the thermally thick regime. *Energy and Fuels* 27(3): 1453-1459.

Beres, C., Costa, G.N., Cabezudo, I., da Silva-James, N.K., Teles, A.S., Cruz, A.P., Mellinger-Silva, C., Tonon, R.V., Cabral, L.M. and Freitas, S.P. (2017). Towards integral utilisation of grape pomace from winemaking process: A review. *Waste Management* 68: 581-594.

Bertran, E., Sort, X., Soliva, M. and Trillas, I. (2004). Composting winery waste: Sludges and grape stalks. *Bioresource Technology* 95(2): 203-208.

Bordiga, M., Travaglia, F., Locatelli, M., Arlorio, M. and Coïsson, J.D. (2015). Spent grape pomace as a still potential by-product. *International Journal of Food Science and Technology* 50(9): 2022-2031.

Borja, R., Martin, A., Maestro, R., Luque, M. and Durán, M.M. (1993). Enhancement of the anaerobic digestion of wine distillery wastewater by the removal of phenolic inhibitors. *Bioresource Technology* 45(2): 99-104.

Boussetta, N., Vorobiev, E., Deloison, V., Pochez, F., Falcimaigne-Cordin, A. and Lanoisellé, J.L. (2011). Valorisation of grape pomace by the extraction of phenolic antioxidants: Application of high voltage electrical discharges. *Food Chemistry* 128(2): 364-370.

Brito, P.S., Oliveira, A.S. and Rodrigues, L.F. (2014). Energy valorisation of solid vines pruning by thermal gasification in a pilot plant. *Waste and Biomass Valorisation* 5(2): 181-187.

Bustamante, M.A., Paredes, C., Moral, R., Moreno-Caselles, J., Pérez-Espinosa, A. and Pérez-Murcia, M.D. (2005). Uses of winery and distillery effluents in agriculture: Characterisation of nutrient and hazardous components. *Water Science and Technology* 51(1): 145-151.

Bustamante, M.A., Paredes, C., Morales, J., Mayoral, A.M. and Moral, R. (2009). Study of the composting process of winery and distillery wastes using multivariate techniques. *Bioresource Technology* 100(20): 4766-4772.

Bustamante, M.A., Said-Pullicino, D., Agulló, E., Andreu, J., Paredes, C. and Moral, R. (2011). Application of winery and distillery waste composts to a Jumilla (SE Spain) vineyard: Effects on the characteristics of a calcareous sandy-loam soil. *Agriculture, Ecosystems and Environment* 140(1): 80-87.

Bustos, G., De la Torre, N., Moldes, A.B., Cruz, J.M. and Domínguez, J.M. (2007). Revalorisation of hemicellulosic trimming vine shoots hydrolyzates trough continuous production of lactic acid and biosurfactants by *Lactobacillus pentosus*. *Journal of Food Engineering* 78(2): 405-412.

Bustos, G., Moldes, A.B., Cruz, J.M. and Domínguez, J.M. (2004). Production of fermentable media from vine-trimming wastes and bioconversion into lactic acid by *Lactobacillus pentosus*. *Journal of the Science of Food and Agriculture* 84(15): 2105-2112.

Bustos, G., Moldes, A.B., Cruz, J.M. and Domínguez, J.M. (2005). Production of lactic acid from vine-trimming wastes and viticulture lees using a simultaneous saccharification fermentation method. *Journal of the Science of Food and Agriculture* 85(3): 466-472.

Carmona, E., Moreno, M.T., Avilés, M. and Ordovás, J. (2012). Use of grape marc compost as substrate for vegetable seedlings. *Scientia Horticulturae* 137: 69-74.

Casas, L., Mantell, C., Rodríguez, M., de la Ossa, E.M., Roldán, A., De Ory, I., Caro, I. and Blandino, A. (2010). Extraction of resveratrol from the pomace of Palomino fino grapes by supercritical carbon dioxide. *Journal of Food Engineering* 96(2): 304-308.

Casazza, A.A., Aliakbarian, B., Mantegna, S., Cravotto, G. and Perego, P. (2010). Extraction of phenolics from *Vitis vinifera* wastes using non-conventional techniques. *Journal of Food Engineering* 100(1): 50-55.

Casazza, A.A., Aliakbarian, B., De Faveri, D., Fiori, L. and Perego, P. (2012). Antioxidants from winemaking wastes: A study on extraction parameters using response surface methodology. *Journal of Food Biochemistry* 36(1): 28-37.

Casazza, A.A., Aliakbarian, B., Sannita, E. and Perego, P. (2012). High-pressure high-temperature extraction of phenolic compounds from grape skins. *International Journal of Food Science and Technology* 47(2): 399-405.

Castillo-Vergara, M., Alvarez-Marin, A., Carvajal-Cortes, S. and Salinas-Flores, S. (2015). Implementation of a cleaner production agreement and impact analysis in the grape brandy (pisco) industry in Chile. *Journal of Cleaner Production* 96: 110-117.

Coelho, J.P., Filipe, R.M., Robalo, M.P. and Stateva, R.P. (2017). Recovering value from organic waste materials: Supercritical fluid extraction of oil from industrial grape seeds. *The Journal of Supercritical Fluids* 120: 102-112.

Cortés-Camargo, S., Pérez-Rodríguez, N., de Souza Oliveira, R.P., Huerta, B.E.B. and Domínguez, J.M. (2016). Production of biosurfactants from vine-trimming shoots using the halotolerant strain *Bacillus tequilensis* ZSB10. *Industrial Crops and Products* 79: 258-266.

Cuervas-mons, C.M.L., López, L.M., Castelló, E.M.G. and Brotons, D.J.V. (2013). Polyphenol extraction from grape wastes: Solvent and pH effect. *Agricultural Sciences* 4: 56-62.

Da Porto, C. and Natolino, A. (2017). Supercritical fluid extraction of polyphenols from grape seed (*Vitis vinifera*): Study on process variables and kinetics. *The Journal of Supercritical Fluids* 130: 239-245.

Da Ros, C., Cavinato, C., Cecchi, F. and Bolzonella, D. (2014). Anaerobic co-digestion of winery waste and waste activated sludge: Assessment of process feasibility. *Water Science and Technology* 69(2): 269-277.

Daffonchio, D., Colombo, M., Origgi, G., Sorlini, C. and Andreoni, V. (1998). Anaerobic digestion of winery wastewaters derived from different winemaking processes. *Journal of Environmental Science and Health, Part A* 33(8): 1753-1770.

Deng, Q., Penner, M.H. and Zhao, Y. (2011). Chemical composition of dietary fibre and polyphenols of five different varieties of wine grape pomace skins. *Food Research International* 44(9): 2712-2720.

Devesa-Rey, R., Vecino, X., Varela-Alende, J.L., Barral, M.T., Cruz, J.M. and Moldes, A.B. (2011). Valorisation of winery waste vs. the costs of not recycling. *Waste Management* 31(11): 2327-2335.

Díaz, A.B., de Ory, I., Caro, I. and Blandino, A. (2012). Enhance hydrolytic enzymes production by *Aspergillus awamori* on supplemented grape pomace. *Food and Bioproducts Processing* 90(1): 72-78.

Diaz, M.J., Madejon, E., Lopez, F., Lopez, R. and Cabrera, F. (2002). Optimisation of the rate vinasse/ grape marc for co-composting process. *Process Biochemistry* 37(10): 1143-1150.

Dimou, C., Kopsahelis, N., Papadaki, A., Papanikolaou, S., Kookos, I.K., Mandala, I. and Koutinas, A.A. (2015). Wine lees valorisation: Biorefinery development including production of a generic fermentation feedstock employed for poly (3-hydroxybutyrate) synthesis. *Food Research International* 73: 81-87.

Duba, K.S., Casazza, A.A., Mohamed, H.B., Perego, P. and Fiori, L. (2015). Extraction of polyphenols from grape skins and defatted grape seeds using subcritical water: Experiments and modeling. *Food and Bioproducts Processing* 94: 29-38.

Dwyer, K., Hosseinian, F. and Rod, M. (2014). The market potential of grape waste alternatives. *Journal of Food Research* 3(2): 91.

Elagamey, A.A., Abdel-Wahab, M.A., Shimaa, M.M.E. and Abdel-Mogib, M. (2013). Comparative study of morphological characteristics and chemical constituents for seeds of some grape table varieties. *Journal of American Science* 9(1): 447-454.

Fiori, L., De Faveri, D., Casazza, A.A. and Perego, P. (2009). Grape by-products: Extraction of polyphenolic compounds using supercritical $CO_2$ and liquid organic solvent – A preliminary investigation; Subproductos de la uva: Extracción de compuestos polifenólicos usando $CO_2$ supercrítico y disolventes orgánicos líquidos – Una investigación preliminary. *Cyta-Journal of Food* 7(3): 163-171.

Fontana, A.R., Antoniolli, A. and Bottini, R. (2013). Grape pomace as a sustainable source of bioactive compounds: Extraction, characterisation and biotechnological applications of phenolics. *Journal of Agricultural and Food Chemistry* 61(38): 8987-9003.

Garavaglia, J., Markoski, M.M., Oliveira, A. and Marcadenti, A. (2016). Grape seed oil compounds: Biological and chemical actions for health. *Nutrition and Metabolic Insights* 9: 59.

García-Lomillo, J., González-SanJosé, M.L., Del Pino-García, R., Rivero-Pérez, M.D. and Muñiz-Rodríguez, P. (2014). Antioxidant and antimicrobial properties of wine by-products and their potential uses in the food industry. *Journal of Agricultural and Food Chemistry* 62(52): 12595-12602.

Gardiman, M., Giust, M., Flamini, R. and Dalla Vedova, A. (2013). New uses for old grapevine germplasm: Agronomic evaluation of hybrids for the production of biomass to energy use. *In*: International Symposium on Fruit Culture and Its Traditional Knowledge along Silk Road Countries 1032: 43-48.

González-Centeno, M.R., Comas-Serra, F., Femenia, A., Rosselló, C. and Simal, S. (2015). Effect of power ultrasound application on aqueous extraction of phenolic compounds and antioxidant capacity from grape pomace (*Vitis vinifera* L.): Experimental kinetics and modelling. *Ultrasonics Sonochemistry* 22: 506-514.

González-Centeno, M.R., Rosselló, C., Simal, S., Garau, M.C., López, F. and Femenia, A. (2010). Physico-chemical properties of cell wall materials obtained from ten grape varieties and their by-products: Grape pomaces and stems. *LWT – Food Science and Technology* 43(10): 1580-1586.

Hang, Y.D., Lee, C.Y. and Woodams, E.E. (1986). Solid-state fermentation of grape pomace for ethanol production. *Biotechnology Letters* 8(1): 53-56.

Hussein, S. and Abdrabba, S. (2015). Physico chemical characteristics, fatty acid, composition of grape seed oil and phenolic compounds of whole seeds, seeds and leaves of red grape in Libya. *International Journal of Applied Science and Mathematics* 2(5): 175-181.

Ioannou, L.A., Puma, G.L. and Fatta-Kassinos, D. (2015). Treatment of winery wastewater by physicochemical, biological and advanced processes: A review. *Journal of Hazardous Materials* 286: 343-368.

Jozinović, A., Šubarić, D., Aćkar, Đ., Miličević, B., Babić, J., Jašić, M. and Valek Lendić, K. (2014). Food industry by-products as raw materials in functional food production. *Hrana u Zdravlju i Bolesti* 3(1): 22-30.

Kamel, B.S., Dawson, H. and Kakuda, Y. (1985). Characteristics and composition of melon and grape seed oils and cakes. *Journal of the American Oil Chemists' Society* 62(5): 881-883.

Karaman, S., Karasu, S., Tornuk, F., Toker, O.S., Geçgel, U., Sagdic, O., Ozcan, N. and Gül, O., (2015). Recovery potential of cold press by-products obtained from the edible oil industry: Physicochemical, bioactive, and antimicrobial properties. *Journal of Agricultural and Food Chemistry* 63(8): 2305-2313.

Kim, S.Y., Jeong, S.M., Park, W.P., Nam, K.C., Ahn, D.U. and Lee, S.C. (2006). Effect of heating conditions of grape seeds on the antioxidant activity of grape seed extracts. *Food Chemistry* 97(3): 472-479.

Kiran, E.U., Trzcinski, A.P., Ng, W.J. and Liu, Y. (2014). Bioconversion of food waste to energy: A review. *Fuel* 134: 389-399.

Kontogiannopoulos, K.N., Patsios, S.I. and Karabelas, A.J. (2016). Tartaric acid recovery from winery lees using cation exchange resin: Optimisation by response surface methodology. *Separation and Purification Technology* 165: 32-41.

Kontogiannopoulos, K.N., Patsios, S.I., Mitrouli, S.T. and Karabelas, A.J. (2017). Tartaric acid and polyphenols recovery from winery waste lees using membrane separation processes. *Journal of Chemical Technology and Biotechnology* 92(12): 2934-2943.

Librán, C.M., Mayor, L., Garcia-Castello, E.M. and Vidal-Brotons, D. (2013). Polyphenol extraction from grape wastes: Solvent and pH effect. *Agricultural Sciences* 4(09): 56.

Lin, C.S.K., Koutinas, A.A., Stamatelatou, K., Mubofu, E.B., Matharu, A.S., Kopsahelis, N., Pfaltzgraff, L.A., Clark, J.H., Papanikolaou, S., Kwan, T.H. and Luque, R. (2014). Current and future trends in food waste valorisation for the production of chemicals, materials and fuels: A global perspective. *Biofuels, Bioproducts and Biorefining* 8(5): 686-715.

Liu, J.G., Wang, Q.H., Ma, H.Z. and Wang, S. (2010). Effect of pretreatment methods on L-lactic acid production from vinasse fermentation. pp. 1302-1305. *In*: Advanced Materials Research, vol. 113. Trans Tech Publications.

Lofrano, G. and Meric, S. (2016). A comprehensive approach to winery wastewater treatment: A review of the state-of the-art. *Desalination and Water Treatment* 57(7): 3011-3028.

Makris, D.P., Boskou, G. and Andrikopoulos, N.K. (2007). Polyphenolic content and *in vitro* antioxidant characteristics of wine industry and other agri-food solid waste extracts. *Journal of Food Composition and Analysis* 20(2): 125-132.

Mateo, J.J. and Maicas, S. (2015). Valorisation of winery and oil mill wastes by microbial technologies. *Food Research International* 73: 13-25.

Mironeasa, S., Leahu, A., Codina, G.G., Stroe, S.G. and Mironeasa, C. (2010). Grape seed: Physico-chemical, structural characteristic and oil content. *Journal of Agroalimentary Process and Technologies* 16: 1-6.

Moldes, A.B., Vázquez, M., Domínguez, J.M., Díaz-Fierros, F. and Barral, M.T. (2007). Evaluation of mesophilic biodegraded grape marc as soil fertiliser. *Applied Biochemistry and Biotechnology* 141(1): 27-36.

Moletta, R. (2005). Winery and distillery wastewater treatment by anaerobic digestion. *Water Science and Technology* 51(1): 137-144.

Monrad, J.K., Srinivas, K., Howard, L.R. and King, J.W. (2012). Design and optimisation of a semicontinuous hot-cold extraction of polyphenols from grape pomace. *Journal of Agricultural and Food Chemistry* 60(22): 5571-5582.

Munoz, O., Sepulveda, M. and Schwartz, M. (2004). Effects of enzymatic treatment on anthocyanic pigments from grapes skin from Chilean wine. *Food Chemistry* 87(4): 487-490.

Nicolle, P., Marcotte, C., Angers, P. and Pedneault, K. (2018). Co-fermentation of red grapes and white pomace: A natural and economical process to modulate hybrid wine composition. *Food Chemistry* 242: 481-490.

Nitayavardhana, S. and Khanal, S.K. (2010). Innovative biorefinery concept for sugar-based ethanol industries: Production of protein-rich fungal biomass on vinasse as an aquaculture feed ingredient. *Bioresource Technology* 101(23): 9078-9085.

Ntalos, G.A. and Grigoriou, A.H. (2002). Characterisation and utilisation of vine prunings as a wood substitute for particleboard production. *Industrial Crops and Products* 16(1): 59-68.

Ovcharova, T., Zlatanov, M. and Dimitrova, R. (2016). Chemical composition of seeds of four Bulgarian grape varieties. *Ciência e Técnica Vitivinícola* 31(1): 31-40.

Paradelo, R., Moldes, A.B., González, D. and Barral, M.T. (2012). Plant tests for determining the suitability of grape marc composts as components of plant growth media. *Waste Management and Research* 30(10): 1059-1065.

Parajó, J.C., Dominguez, H. and Dominguez, J.M. (1995). Production of xylitol from raw wood hydrolysates by *Debaryomyces hansenii* NRRL Y-7426. *Bioprocess and Biosystems Engineering* 13(3): 125-131.

Pedretti, E.F., Duca, D., Toscano, G., Riva, G., Pizzi, A., Rossini, G. and Flamini, R. (2014). Sustainability of grape-ethanol energy chain. *Journal of Agricultural Engineering* 45(3): 119-124.

Pérez-Bibbins, B., Torrado-Agrasar, A., Pérez-Rodríguez, N., Aguilar-Uscanga, M.G. and Domínguez, J.M. (2015). Evaluation of the liquid, solid and total fractions of beer, cider and wine lees as economic nutrient for xylitol production. *Journal of Chemical Technology and Biotechnology* 90(6): 1027-1039.

Perez-Bibbins, B., Torrado-Agrasar, A., Salgado, J.M., de Souza Oliveira, R.P. and Dominguez, J.M. (2015). Potential of lees from wine, beer and cider manufacturing as a source of economic nutrients: An overview. *Waste Management* 40: 72-81.

Pérez-Serradilla, J.A. and De Castro, M.L. (2011). Microwave-assisted extraction of phenolic compounds from wine lees and spray-drying of the extract. *Food Chemistry* 124(4): 1652-1659.

Perumalla, A.V.S. and Hettiarachchy, N.S. (2011). Green tea and grape seed extracts – Potential applications in food safety and quality. *Food Research International* 44(4): 827-839.

Pinelo, M., Ruiz-Rodríguez, A., Sineiro, J., Señoráns, F.J., Reglero, G. and Núñez, M.J. (2007). Supercritical fluid and solid-liquid extraction of phenolic antioxidants from grape pomace: A comparative study. *European Food Research and Technology* 226(1-2): 199-205.

Ping, L., Brosse, N., Sannigrahi, P. and Ragauskas, A. (2011). Evaluation of grape stalks as a bioresource. *Industrial Crops and Products* 33(1): 200-204.

Prozil, S.O., Evtuguin, D.V. and Lopes, L.P.C. (2012). Chemical composition of grape stalks of *Vitis vinifera* L. from red grape pomaces. *Industrial Crops and Products* 35(1): 178-184.

Prozil, S.O., Evtuguin, D.V., Lopes, S.M., Lopes, L.C., Arshanitsa, A.S., Solodovnik, V.P. and Telysheva, G.M. (2014). Evaluation of grape stalks as a feedstock for pellets production. *In*: 13th European Workshop on Lignocellulosics and Pulp, *EWLP* 2014: 24-27.

Rajha, H.N., Boussetta, N., Louka, N., Maroun, R.G. and Vorobiev, E. (2015). Effect of alternative physical pretreatments (pulsed electric field, high voltage electrical discharges and ultrasound) on the dead-end ultrafiltration of vine-shoot extracts. *Separation and Purification Technology* 146: 243-251.

Riva, G., Pedretti, E.F., Toscano, G., Duca, D., Pizzi, A., Saltari, M. and Flamini, R. (2013). Sustainability of grape-ethanol energy chain. *Journal of Agricultural Engineering* 44(2s).

Rivas, B., Torrado, A., Moldes, A.B. and Domínguez, J.M. (2006). Tartaric acid recovery from distilled lees and use of the residual solid as an economic nutrient for *Lactobacillus*. *Journal of Agricultural and Food Chemistry* 54(20): 7904-7911.

Rivas, B., Torrado, A., Rivas, S., Moldes, A.B. and Domínguez, J.M. (2007). Simultaneous lactic acid and xylitol production from vine trimming wastes. *Journal of the Science of Food and Agriculture* 87(8): 1603-1612.

Rivera, O.M.P., Moldes, A.B., Torrado, A.M. and Domínguez, J.M. (2007). Lactic acid and biosurfactants production from hydrolysed distilled grape marc. *Process Biochemistry* 42(6): 1010-1020.

Rockenbach, I.I., Gonzaga, L.V., Rizelio, V.M., Gonçalves, A.E.D.S.S., Genovese, M.I. and Fett, R. (2011). Phenolic compounds and antioxidant activity of seed and skin extracts of red grape (*Vitis vinifera* and *Vitis labrusca*) pomace from Brazilian winemaking. *Food Research International* 44(4): 897-901.

Rodríguez, L.A., Toro, M.E., Vazquez, F., Correa-Daneri, M.L., Gouiric, S.C. and Vallejo, M.D. (2010). Bioethanol production from grape and sugar beet pomaces by solid-state fermentation. *International Journal of Hydrogen Energy* 35(11): 5914-5917.

Rodriguez, L., Villasenor, J., Buendia, I.M. and Fernandez, F.J. (2007). Re-use of winery wastewaters for biological nutrient removal. *Water Science and Technology* 56(2): 95-102.

Rodríguez, L., Villasenor, J., Fernández, F.J. and Buendía, I.M. (2007). Anaerobic co-digestion of winery wastewater. *Water Science And Technology* 56(2): 49-54.

Rodríguez, N., Salgado, J.M., Max, B., Torrado, A., Cortés, S. and Domínguez, J.M. (2010). Trimming vine shoots and vinasses as alternative economical media for lactic acid and cell-bound biosurfactants production by *Lactococcus lactis*. *Journal of Biotechnology* 150: 320.

Rodríguez-Pazo, N., Salgado, J.M., Cortés-Diéguez, S. and Domínguez, J.M. (2013). Biotechnological production of phenyllactic acid and biosurfactants from trimming vine shoot hydrolysates by microbial coculture fermentation. *Applied Biochemistry and Biotechnology* 169(7): 2175-2188.

Rodriguez-Rodriguez, R., Justo, M.L., Claro, C.M., Vila, E., Parrado, J., Herrera, M.D. and de Sotomayor, M.A. (2012). Endothelium-dependent vasodilator and antioxidant properties of a novel enzymatic extract of grape pomace from wine industrial waste. *Food Chemistry* 135(3): 1044-1051.

Rondeau, P., Gambier, F., Jolibert, F. and Brosse, N. (2013). Compositions and chemical variability of grape pomaces from French vineyard. *Industrial Crops and Products* 43: 251-254.

Ruggieri, L., Cadena, E., Martínez-Blanco, J., Gasol, C., Rieradevall, J., Gabarrell, X., Gea, T., Sort, X. and Sánchez, A. (2009). Recovery of organic wastes in the Spanish wine industry: Technical, economic and environmental analyses of the composting process. *Journal of Cleaner Production* 17(9): 830-838.

Saigal, D. and Ray, R.C. (2007). Winemaking: Microbiology, biochemistry and biotechnology. pp. 1-33. *In*: Ramesh C. Ray and O.P. Ward (Eds.). Microbial Biotechnology in Horticulture, vol. 3. Science Publishers, New Hampshire, USA.

Salgado, J.M., Carballo, E.M., Max, B. and Domínguez, J.M. (2010). Characterisation of vinasses from five certified brands of origin (CBO) and use as economic nutrient for the xylitol production by *Debaryomyces hansenii*. *Bioresource Technology* 101(7): 2379-2388.

Salgado, J.M., Rodríguez, N., Cortés, S. and Domínguez, J.M. (2010). Improving downstream processes to recover tartaric acid, tartrate and nutrients from vinasses and formulation of inexpensive fermentative broths for xylitol production. *Journal of the Science of Food and Agriculture* 90(13): 2168-2177.

Sánchez-Gómez, R., Zalacain, A., Pardo, F., Alonso, G.L. and Salinas, M.R. (2016). An innovative use of vine-shoots residues and their 'feedback' effect on wine quality. *Innovative Food Science and Emerging Technologies* 37: 18-26.

Scram, J.I., Hall, D.O. and Stuckey, D.C. (1993). Bioethanol from grapes in the European community. *Biomass and Bioenergy* 5(5): 347-358.

Serrano, L., De la Varga, D., Ruiz, I. and Soto, M. (2011). Winery wastewater treatment in a hybrid constructed wetland. *Ecological Engineering* 37(5): 744-753.

Shi, J., Yu, J., Pohorly, J.E. and Kakuda, Y. (2003). Polyphenolics in grape seeds: Biochemistry and functionality. *Journal of Medicinal Food* 6(4): 291-299.

Silva, C.F., Arcuri, S.L., Campos, C.R., Vilela, D.M., Alves, J.G. and Schwan, R.F. (2011). Using the residue of spirit production and bio-ethanol for protein production by yeasts. *Waste Management* 31(1): 108-114.

Šťavíková, L., Polovka, M., Hohnová, B., Karásek, P. and Roth, M. (2011). Antioxidant activity of grape skin aqueous extracts from pressurised hot water extraction combined with electron paramagnetic resonance spectroscopy. *Talanta* 85(4): 2233-2240.

Teixeira, A., Baenas, N., Dominguez-Perles, R., Barros, A., Rosa, E., Moreno, D.A. and Garcia-Viguera, C. (2014). Natural bioactive compounds from winery by-products as health promoters: A review. *International Journal of Molecular Sciences* 15(9): 15638-15678.

Tominaga, T., Kawaguchi, K., Kanesaka, M., Kawauchi, H., Jirillo, E. and Kumazawa, Y. (2010). Suppression of type-I allergic responses by oral administration of grape marc fermented with *Lactobacillus plantarum*. *Immunopharmacology and Immunotoxicology* 32(4): 593-599.

Tseng, A. and Zhao, Y. (2013). Wine grape pomace as antioxidant dietary fibre for enhancing nutritional value and improving storability of yogurt and salad dressing. *Food Chemistry* 138(1): 356-365.

Vadivel, R., Minhas, P.S., Kumar, S., Singh, Y., DVK, N.R. and Nirmale, A. (2014). Significance of vinasses waste management in agriculture and environmental quality-Review. *African Journal of Agricultural Research* 9(38): 2862-2873.

Vatai, T., Škerget, M. and Knez, Ž. (2009). Extraction of phenolic compounds from elder berry and different grape marc varieties using organic solvents and/or supercritical carbon dioxide. *Journal of Food Engineering* 90(2): 246-254.

Versari, A., Castellari, M., Spinabelli, U. and Galassi, S. (2001). Recovery of tartaric acid from industrial enological wastes. *Journal of Chemical Technology and Biotechnology* 76(5): 485-488.

Vilela-Moura, A., Schuller, D., Mendes-Faia, A., Silva, R.D., Chaves, S.R., Sousa, M.J. and Côrte-Real, M. (2011). The impact of acetate metabolism on yeast fermentative performance and wine quality: Reduction of volatile acidity of grape musts and wines. *Applied Microbiology and Biotechnology* 89(2): 271-280.

Walter, R.H. and Sherman, R.M. (1976). Fuel value of grape and apple processing wastes. *Journal of Agricultural and Food Chemistry* 24(6): 1244-1245.

Wang, S., Dai, G., Yang, H. and Luo, Z. (2017). Lignocellulosic biomass pyrolysis mechanism: A state-of-the-art review. *Progress in Energy and Combustion Science* 62: 33-86.

Xia, E.Q., Deng, G.F., Guo, Y.J. and Li, H.B. (2010). Biological activities of polyphenols from grapes. *International Journal of Molecular Sciences* 11(2): 622-646.

Yalcin, D., Ozcalik, O., Altiok, E. and Bayraktar, O. (2008). Characterisation and recovery of tartaric acid from wastes of wine and grape juice industries. *Journal of Thermal Analysis and Calorimetry* 94(3): 767-771.

Yi, C., Shi, J., Kramer, J., Xue, S., Jiang, Y., Zhang, M., Ma, Y. and Pohorly, J. (2009). Fatty acid composition and phenolic antioxidants of winemaking pomace powder. *Food Chemistry* 114(2): 570-576.

Yu, J. and Ahmedna, M. (2013). Functional components of grape pomace: Their composition, biological properties and potential applications. *International Journal of Food Science and Technology* 48(2): 221-237.

Zacharof, M.P. (2017). Grape winery waste as feedstock for bioconversions: Applying the biorefinery concept. *Waste and Biomass Valorization* 8(4): 1011-1025.

Zhang, L. and Sun, X. (2016). Improving green waste composting by addition of sugarcane bagasse and exhausted grape marc. *Bioresource Technology* 218: 335-343.

Zhihui, B.A.I., Bo, J.I.N., Yuejie, L.I., Jian, C.H.E.N. and Zuming, L.I. (2008). Utilisation of winery wastes for *Trichoderma viride* biocontrol agent production by solid state fermentation. *Journal of Environmental Sciences* 20(3): 353-358.

# Section 3
# Applied Aspects of Winemaking
## (B) Evaluation of Wine

# 25 Analytical Techniques in Oenology

**Disney Ribeiro Dias[1], Leonardo de Figueiredo Vilela[2] and Rosane Freitas Schwan[2*]**

[1] Department of Food Science, Food Microbiology Sector, Federal University of Lavras, Campus Universitario, Lavras, MG, Brazil, 37.200-000

[2] Department of Biology, Microbiology Sector. Federal University of Lavras, Campus Universitario, Lavras, MG, Brazil, 37.200-000

## 1. Introduction

Since the times of Pasteur, wine production has been developing with strong research in the areas of viticulture and enology. Enology has long been said to be a science, while winemaking is an art. Viticulture is related to the study of grapes and their cultivation, while oenology is concerned with post-harvesting of grapes and the elaboration of wines. It is known that the quality of the grapes varies from year, region, date of harvest and even for each type of vineyard.

There is a tendency to seek, every day, new technologies that bring, besides greater productivity, an improvement in the final product quality. Winemaking processes have been a good example of this progress. The alcoholic fermentation of grape must, which was once completely empirical, has become one of the strongest areas of agro-industrial research. Within this context, viticulture and oenology study several aspects for the improvement of production, ranging from the selection of the best cultivars and vine varieties and the microorganism, and system for fermentation, until the care to obtain the final beverage, its stabilisation, bottling and sale. The yeasts used in the fermentation process are extremely important for the final product obtained. Yeasts will transform the sugars into ethanol and their metabolism generates the other aroma compounds that characterise, together with varietals aroma-forming compounds, the quality of the wine.

The elaboration of quality beverages requires care, ranging from the ideal cultivation of the vine (climatic and soil factors), handling and hygiene in post-harvest processing and within the industry. From berry to bottling, there are several stages of wine processing. At each stage, analytical methods are involved to a greater or lesser extent, always necessary for the maintenance and guarantee of the chemical, microbiological and sensorial quality of wine, as well as its authenticity and safety. Analytical methodologies are essential in winemaking and enology, since all commercial wines, from a wide range of geographical origins, are ruled by specific standards that determine several parameters, such as density, pH, acidity, ashes, amounts of sugars, ethanol, methanol, higher alcohols, esters and minerals.

Several methodologies and techniques are employed to analyse the various parameters in enology, some with greater or lesser operational complexity and equipment infrastructure. In this chapter, we will present an overview of the most advanced techniques in wine analysis, addressing physico-chemical, chromatographic, sensory, microbiological and wine analysis (Fig. 1).

## 2. Advances in Wine Physico-chemical Analysis

During the processes of wine production, considering the growing of the grapes until the beverage is obtained, it is necessary to evaluate certain chemical components that parameterise the quality of the wine. Wines are made up of a myriad of chemical compounds, which may be varietal, originating from the grape itself, synthesised or transformed by microbial metabolism or can be formed during storage (Fleet, 2007; Polášková *et al.*, 2008; Ebeler and Thorngate, 2009; Puertas *et al.*, 2018). It is certain that evaluating all the compounds present in wine would be a laborious, expensive task. To avoid unnecessary spending of time and money on wine analysis, but guaranteeing the quality and safety of wines, the regulatory bodies of each country, as well as international organisations, such as the OIV, standardised

*Corresponding author: rschwan@ufla.br

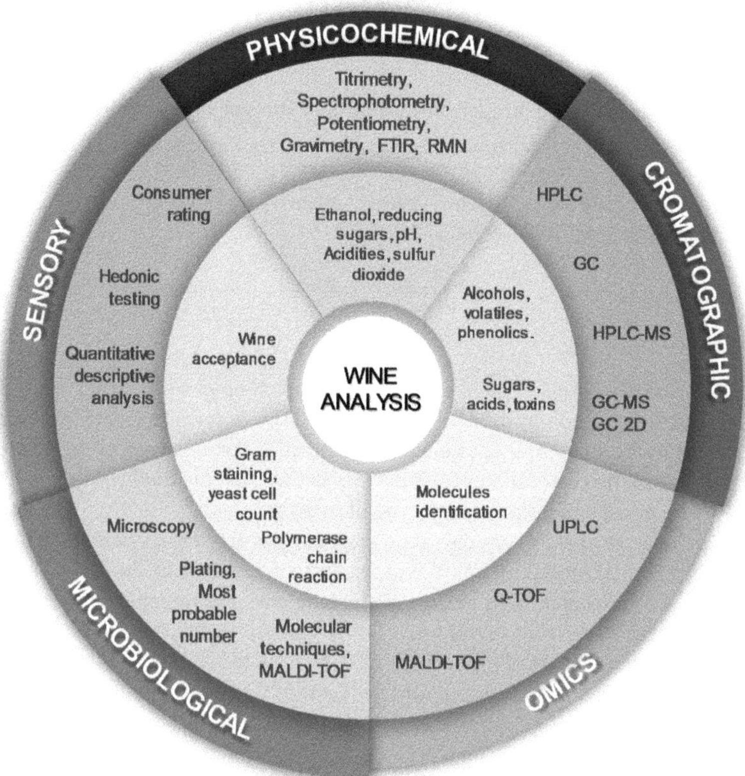


**Figure 1.** Techniques used in wine analysis

Color version at the end of the book

some of the minimum parameters of quality wine. For the determination of these parameters, the physico-chemical analysis are indispensable. The physico-chemical analysis comprises the determination of important wine quality parameters and evaluates the control of the fermentative process.

## 2.1. Classical Techniques in Wine Analysis

The main physico-chemical parameters analysed in wines around the world comprise pH, total acidity, volatile acidity, ethanol, alkalinity of ash, sulphur dioxide, reducing sugars, alcohol strength and density, among others. To achieve the values of these parameters, several classical techniques, which use robust and low-priced apparatus, are employed, such as titrimetry, potentiometry, spectrophotometry and spectrometry (Amerine and Ough, 1980; OIV, 2015; Dias *et al.*, 2017).

Titrimetry, also known as titration, is one of the oldest classic techniques and it is used in enological research laboratories and wineries to determine acidity (total and volatile), alkalinity of ash, carbon dioxide, hydroxymethyl furfural, sulphates, and sulphur dioxide, for example (Johansson, 1988; Dias *et al.*, 2017). Potentiometry is an electroanalytical technique used to measure variation in electric potential in an electrochemical cell. Potentiometry is a common and reliable technique used in wine analysis and pH meter is the most common and useful potentiometric-based laboratory apparatus (Zoski, 2006; Ribéreau-Gayon *et al.*, 2006; Dias *et al.*, 2017). The spectrophotometric analysis is based on the light propagation capacity and its absorption/reflection in solutions, generating spectral responses and correlating these responses with the concentration of a certain compound. Such an analysis is important in the quantification of several compounds in wines, among them are organic acids, reducing sugars, and phenolic compounds (Aleixandre-Tudo *et al.*, 2017).

## 2.2. Recent Techniques in Wine Analysis

Although the classic techniques mentioned above are still used in many laboratories and small wineries, the development of more complex, robust and automated equipment, capable of analysing several physico-chemical parameters and compounds in wine samples simultaneously, has been developed in the last few decades (Schindler *et al.*, 1998; Serban *et al.*, 2004; Lima and Reis, 2017). Such equipment, often referred to as 'automatic wine analyzers' are highly versatile, performing numerous tests on several samples over a much shorter period than would normally occur using standard equipment such as pH meters, titrators, distillers or a spectrophotometers.

Noticeably, all these benefits come at a cost, and the lab, the winery, or the industry needs to assess whether the cost acquisition and maintenance of the automatic analyzers is worth the investment Table 1 presents the pros and cons of the use of an automatic analyser as well as the main analyses performed on commercially available analysers, manufactured by companies, such as Bruker (ALPHA II Wine Analyser), Foss Analytical (WineScan™), and Skalar (San$^{++}$ Automated Wet Chemistry Analyser).

**Table 1.** Merits and Demerits of the Use of Wine Analyser and the Main Parameters Analysed by this Kind of Equipment

| *Pros* | *Cons* |
|---|---|
| • Several compounds analysed per sample<br>• Larger number of samples analysed in less time<br>• Reduced number of reagents per analysis<br>• Improves accuracy of data acquisition due to lower incidence of operator errors | • Acquisition and maintenance costs<br>• Cost of reagents and other consumables<br>• Requires trained personnel or supervision of a qualified technician |
| **Main automated analyses of wines and musts** | |
| Free and total sulphur dioxide, ethanol, glucose/fructose, malic acid, volatile acid, total acid, pH, tartaric acid, brix, density, acetaldehyde, diacetyl, total alkalinity, ascorbic acid, sodium, potassium. | |

# 3. Chromatographic Analysis

The chemistry of flavour of grape wine has been the main focus of many researches due to the complexity of the volatile and non-volatile compounds that contribute to the flavor of different grape varieties (Polášková *et al.*, 2008; Ebeler and Thorngate, 2009). The spectroscopic methods applied for wine and grape analyses include techniques as spanning atomic spectroscopic methods, such as atomic absorption spectroscopy (AAS), inductively coupled plasma (ICP) and some molecular spectroscopic methods, such as infrared and ultraviolet/visible spectrophotometry, nuclear magnetic resonance (NMR) spectroscopy and mass spectrometry (MS) (Grindlay *et al.*, 2011). Particularly, ultraviolet-visible (UV/Vis) spectrophotometry and infrared (IR) spectrometry offer good features that make them ideal for a very large volume of the routine grape and wine analyses (McGoverin *et al.*, 2010). Despite the power of other techniques, as spectroscopy techniques, for the high throughput analysis of a wide variety of compounds in wine samples, many applications in grape and wine analysis require individual separation of compounds, as complex organic fractions, such as the volatile compounds, phenolics and influential trace-level constituents. The most common chromatographic methods used for wine analysis are gas chromatography (GC) and high-performance liquid chromatography (HPLC) and more recently, techniques more advanced, as gas chromatography-mass spectrometry (GC-MS), liquid-chromatography-mass spectrometry (LC-MS) (De Villiers *et al.*, 2012).

The main compounds found in grape wines that contribute to flavour are carbohydrates, such as glucose and fructose, the major sugars present in grapes and juices (Eyduran *et al.*, 2015). The malic, lactic and tartaric acids influence directly important parameters, as the taste balance, chemical stability and pH of the beverage (Ali *et al.*, 2010; Silva *et al.*, 2015). Proanthocyanidins (tannins), terpenoids (monoterpenoids, sesquiterpenoids and C13-norisoprenoids) and various precursors of aromatic aldehydes, esters and thiols are detectable in finished wines (Lund *et al.*, 2006; Coelho *et al.*, 2017).

### 3.1. High Performance Liquid Chromatography (HPLC)

The high-performance liquid chromatography (HPLC) has been one of the more studied analytic techniques, which is widely spread and used in food and beverage analysis. In grape wines analysis, the HPLC technique has been widely used for determination of sugars, alcohols, organic acids, phenolic compounds among others. Sugar analysis is performed using different separation methods (columns) and may be successfully used with different detection systems (detectors). The chromatographic columns commonly used in separation of sugars in foods and beverages are the $NH_2$ stationary phase and cation exchange resins. The mobile phase, water, mixtures of acetonitrile and water (80:20:75:25) or acid solutions of sulfuric acid, orthophosphoric acid and perchloric acid are used under different conditions of temperature and flow. Organic acids analysis in grape wine has also been performed, in most cases, employing ion exchange columns, acidified mobile phases and detectors based on the use of ultraviolet light as shown in Fig. 2.

The organic acids affect the flavour, enhance colour stability, limit oxidation and together with ethanol, are largely responsible for the microbial and physico-chemical stability of table wines (Waterhouse *et al.*, 1997; Jackson, 2000). The carbohydrates, as glucose and fructose, the major hexoses present in grapes

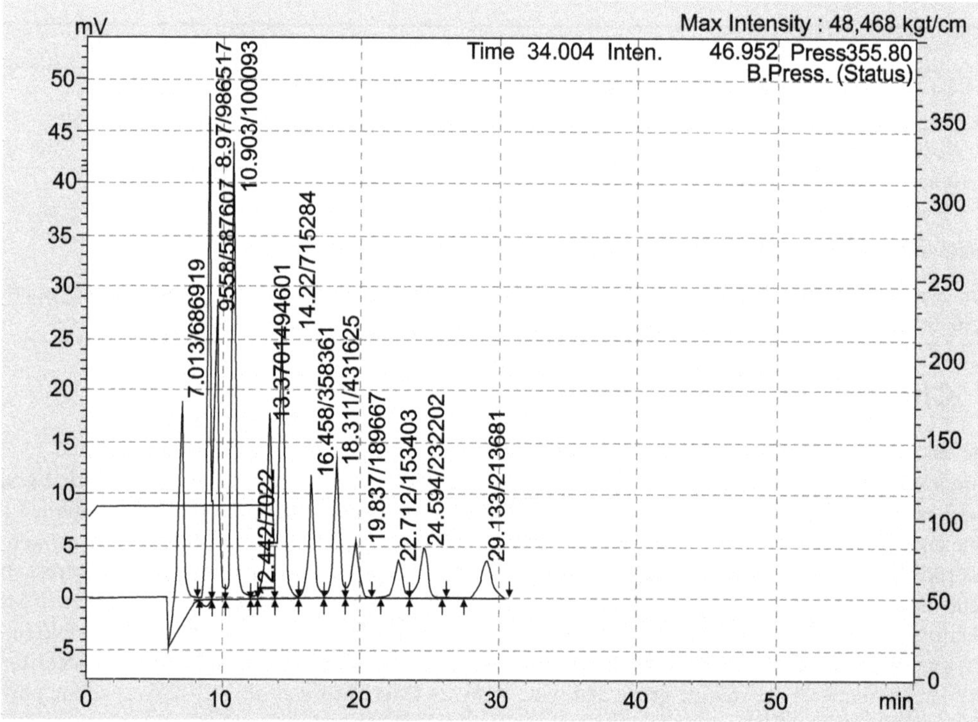


**Figure 2.** Chromatogram from organic acids (mix of standards). 9.55 min = citric acid; 10.80 min = tartaric acid; 12.44 min = malic acid; 18.30 min = acetic acid. Conditions: UV detector 210 nm; SCR 101H column; 100 mM perchloric acid 0.6 ml of flow rate; temperature 50°C.

are responsible for ethanol formation as well formation of secondary metabolites that determine the fermentation endpoint (De Villiers *et al.*, 2012).

Phenolic compounds, identified by HPLC, play an important role in the wines ageing and influence the wine flavour, affecting the organoleptic properties through their contribution to astringency, bitterness and colour (Armstrong *et al.*, 2001). The anthocyanins are phenolic compounds responsible for the colour of red grapes and wine and are important to the health benefits. The chlorophylls and carotenoids are photosynthetic pigments and important as precursors to produce isoprenoids, which are known to be significant contributors to wine aroma (De Villiers *et al.*, 2012).

## 3.2. Gas Chromatography

In many studies, the gas-chromatograph with flame ionisation detectors (FID), GC-FID, and mass spectrometers (MS), GC-MS, are used for volatile compounds determination in grape wine; an example of the chromatogram generated by such equipment is shown in Fig. 3.

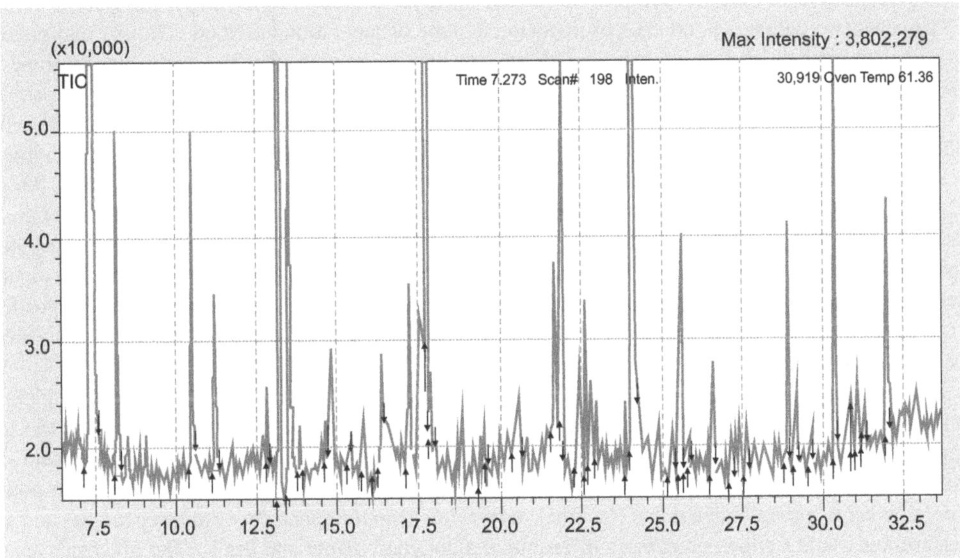


**Figure 3.** Example of the chromatogram generated by GC-MS analysis

The analyses are preceded by extraction steps aiming at the sample clean up and analytes concentration. The volatile compounds extraction is performed by Liquid-Liquid Extraction (LLE) and Solid Phase Micro Extraction (SPME). For grape wine, volatile compounds are grouped as majority and minority. The major compounds are present in higher concentrations, while the minorities are the compounds presents in minor concentration (De Villiers *et al.*, 2012; Panighel *et al.*, 2014; Mencarelli *et al.*, 2018).

According to Flamini *et al.* (2014), the extraction of volatile compounds is performed by solid phase micro extraction. SPME was developed in the 1990s by Pawliszyn *et al.* (1990) and many other papers describe different aspects of this approach and applications in different fields (Bojko *et al.*, 2012). This extraction technique was demonstrated to be rapid, simple and reproducible, without solvent use and needs a small sample volume (Yu *et al.*, 2012; Harmon *et al.*, 1997). For these reasons it has been used to study the volatile profile of many beverages, including grapes and wine (Vas *et al.*, 2004; Castro *et al.*, 2008; Flamini *et al.*, 2010). An interesting alternative to SPME, Stir Bar Sorptive Extraction (SBSE) has been recently developed (Baltussen *et al.*,1999; Sandra *et al.*, 2001; Zalacain *et al.*, 2007). In this technique, a magnetic stir bar coated with a polymeric sorbent (polymethylsiloxane, PDMS), is placed in the sample and stirred for a defined time to extract nonpolar analytes from the sample into the polymeric coating. After extraction, the stir bar is placed in a thermal desorption unit coupled online to a GC, usually equipped with an MS detector. The apparent advantage of SBSE is the relatively high content of polymeric sorbent (about 50-250 times the amount present on a SPME fiber) available for extraction of analytes, making it about 50-250 times more sensitive than SPME. Various kinds of fibres are commercially available, as PDMS, PDMS/CAR (Polydimethylsiloxane/Carboxen) and PDMS/DVB/CAR (Polydimethylsiloxane/Divinylbenzene/Carboxen). The competition for sorptive sites on the fibre can occur so that small changes in the matrix composition can significantly changes the quantitative extraction of the analytes of interest. The time of fibre exposure, sample temperature and in the case of liquid samples, the pH, ionic strength and type of solvent or matrix composition (i.e. water and ethanol solvents in the case of grape wines) that may be present are parameters that determine the success of analysis (Murray *et al.*, 2001; Howard *et al.*, 2005; Polášková *et al.*, 2008).

With the use of capillary GC, the number and quantification of compounds in a single analysis improve significantly. The phases for separation of grape wine volatiles are polyethylene glycol (PEG) or 'WAX' and nitroterephtalic acid-modified PEG phases (free fatty acid phases, FFAP). Non-polar phases, such as polydimethylsiloxane (PDMS), are used for the analysis of apolar compounds, such as terpenoids and volatile phenols (Pawliszyn *et al.*, 2006). The bi-dimensional gas chromatography (GCXGC) improves separation of highly complex mixtures, such as encountered in grape wine samples. The GCXGC uses two columns to create a bi-imensional plane of separation based on volatility and polarity, thus demonstrating a high-resolution power identifying new compounds that may contribute to grape and wine aroma (Ryan *et al.*, 2005). The volatile compounds are identified on the basis of comparisons of the mass spectra and GC retention indices (RI) to synthesised standards (Polášková *et al.*, 2008). Many of the peaks detected by the GC do not actually contribute to our perception of flavours or fragrances because they are present below our thresholds for detecting. Therefore, the technique, GC-olfactometry (GC-O) was a landmark development in flavour, aroma and fragrance research, as it provides valuable information on compounds and a powerful tool for identifying important odourants that contribute to grape wine aroma and for relating the contributions of individual odourants to the differences among different wines samples. In the technique GC-O, a human assessor sniffs the effluent as it emerges from the GC column and the aroma quality; the time at which the aroma is sensed and in some cases, the aroma intensity is recorded (Polášková *et al.*, 2008).

Grape aroma is composed of about 800 compounds and the volatile profile of wine also includes compounds produced from the fermentative process and which are the largest percentage of the total aroma composition of wine. Volatile organic compounds are responsible for the wine 'bouquet' and vital to wine quality, determining their aroma characteristics. The main groups of aroma and flavour compounds are organic acids, proanthocyanidins (tannins), terpenoids (monoterpenoids, sesquiterpenoids, and C13 norisoprenoids) and various precursors of aromatic aldehydes, esters and thiols. The glycosidases and peptidases being water-soluble, play a vital role in wine flavour and aroma (González-Barreiro *et al.*, 2015).

## 4. Current Techniques for Microbiological Analysis of Wines

Microorganisms belonging to several groups, including filamentous fungi, yeasts and bacteria are present in the grape since its cultivation. These microorganisms, especially yeasts, can remain during the fermentation process and are important for obtaining quality wines. In spontaneous fermentations of healthy grapes, yeasts predominate during fermentation, with a prevalence of *Saccharomyces cerevisiae* species (Fleet *et al.*, 1984). Other yeast species, called non-*Saccharomyces*, are also present at the beginning of the fermentation and may contribute to the quality of the wine, by the formation of aroma-forming compounds (Ciani and Comitini, 2011). In addition to yeast, lactic acid bacteria (LAB) contribute to the improvement of wine quality, reducing the acidity of the wine during malo-lactic fermentation (Davis *et al.*, 1985; Lasik, 2013). Filamentous fungi and acetic acid bacteria are associated with deterioration or contamination of the wine. Due to the characteristics of the microorganisms and their importance in winemaking, species identification is essential during and after vinification to control and guarantee the wine quality (Bartowsky and Henschke, 2008).

### 4.1. Overview of Detection and Counting of Microorganisms

The counting and detection of microorganisms or the products of their metabolism, using classical techniques and some techniques of molecular biology is established by OIV (OIV 2015). The classical microbiological methods generally involve the use of an appropriate pre-enrichment and enrichment culture media, isolation in selective media followed by morphological and/or biochemical tests. All this is laborious, obtaining results can take days or weeks and, additionally, they present low sensitivity. In addition to this, it has been demonstrated that some microbial cells can enter a viable but not cultivable state (unculturable cells), due to the processing to which the wine is subjected, making the use of culture methods limited. A series of alternative, rapid and sensitive molecular methods for the detection, identification and quantification of microorganisms have been developed to overcome these drawbacks. The best way to solve the problem of microbial identification is the use of a polyphasic approach, combining classical and molecular techniques (Ivey and Phister, 2011; Fröhlich *et al.*, 2017). As mentioned in previous topics for physico-chemical and chromatographic techniques, several techniques in microbiology have also been developed or improved, with emphasis on molecular and spectrometric techniques, to identify microorganisms in an accurate and fast way (Fig. 4).

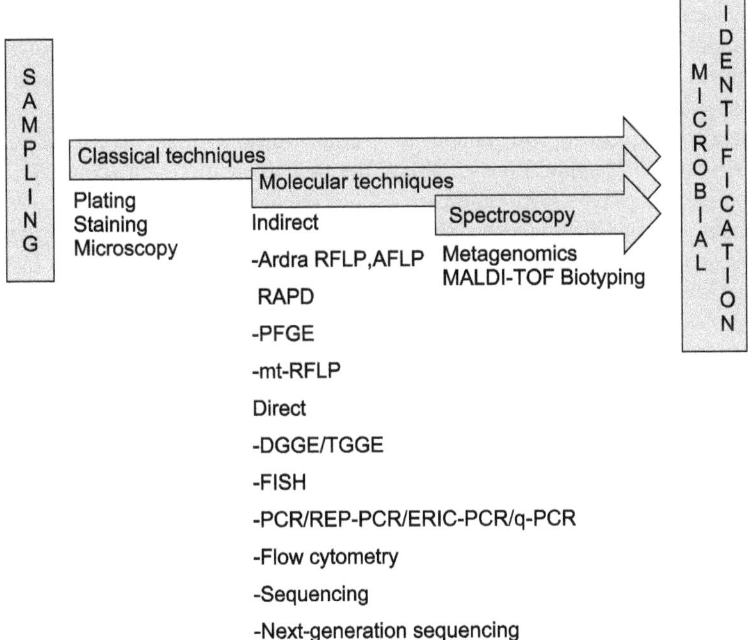


**Figure 4.** Evolution of microbial identification techniques

## 4.2. Microbial Identification Using Molecular Techniques

One of the great limitations of classical methods of microbial identification is the difficulty in describing the ecological interactions and the identification of the microbial diversity present in wine and other foods or a complex sample. In this sense, the use of indirect, for the identification of cultured microorganisms, and direct, used to profile entire microbiota or identify specific microbes in a mixed population (Fig. 4), molecular methods are of great value in the evaluation and identification of the microbiota present in the winemaking processes (Giraffa and Carminati, 2008; Cocolin *et al.*, 2011; Ivey and Phister, 2011).

The main advances in the tests of detection of microorganisms in food, based on nucleic acids, occurred in 1990. The first molecular identification methods were DNA-DNA hybridisation, 16S rDNA sequence analysis, hybridisation with a specific probe and RFLP (restriction fragment length polymorphism) or ribotyping analysis. In contrast to physiological and biochemical characteristics, molecular identification is based on the constitutive composition of nucleic acids rather than the products of their expression (Querol and Ramon, 1996; Kurtzman, 2011).

The use of indirect molecular techniques (PFGE, pulse field gel electrophoresis; RAPD, polymorphism random amplification; AFLP, amplified fragment length polymorphism; RFLP, restriction fragment length polymorphism) allows differentiation of detection and strains, identification of species and analysis of genetic similarity even in the absence of nucleotide sequence information, has been widely used successfully in the characterisation of yeasts in different wine environments. Among the direct molecular methods, PCR (polymerase chain reaction) and associated techniques, such as rep-PCR (repetitive extragenic palindromic), ERIC-PCR (enterobacterial repetitive intergenic consensus) and q-PCR (quantitative) are recognised by simplicity in operation. The DGGE (denaturing gradient gel electrophoresis) and TGGE (temperature gradient gel electrophoresis) techniques have been increasingly used for analysis of the microbial diversity, to provide information on the diversity of microorganism populations in wines (Querol and Ramon, 1996; Cocolin *et al.*, 2011; Ivey and Phister, 2011; Cappello *et al.*, 2014; Lonvaud-Funel, 2015; Liu *et al.*, 2017; Fröhlich *et al.*, 2017).

## 4.3. Next Generation Sequencing (NGS)

DNA sequencers are devices that read a DNA sample and generate an electronic file with symbols representing the sequence of nitrogenous bases – A, C, G, T contained in the sample. The first popular DNA sequencing methods were the chemical method of base degradation (Gilbert's method) and the dideoxy method or fragment termination (Sanger's sequencing). Both methods are based on the production of a set of single strands of DNA that are separated by electrophoresis. Due to its high efficiency and low radioactivity, Sanger's sequencing was adopted as the core technology in the 'first generation' of research and commercial sequencing applications. In 1987, the first automatic DNA sequencer, ABI 370, was launched by Applied Biosystems. With automation it was possible to carry out large sequencing projects, such as the human genome, the mouse and others (Liu *et al.*, 2012). This technology was the base for complete sequencing of *Saccharomyces cerevisiae* genome (Goffeau *et al.*, 1996). After the human genome project (Collins *et al.*, 2003), which encouraged the development of new sequencing technologies (Next Generation Sequencing), the 454 company launched the 454 in 2005. In 2006, Solexa launched the Genome Analyser, followed by Agencourt launching the SOLiD (sequencing by oligo ligation detection), which are three more typical sequencing systems in next generation sequencing (NGS). In the year 2006, Agencort was purchased by Applied Biosystems. In 2007, the 454 was purchased by Roche and Solexa was purchased by Illumina.

Next Generation Sequencing (NGS) techniques, in comparison to Sanger sequencing, use different analytical methodologies and provide high-throughput and high resolution, producing thousands or even millions of sequences at the same time. These sequences allow precise identification of the microbial rate, including non-cultivable organisms and those present in small numbers. In specific applications, NGS provides a complete inventory of all operons and microbial genes present or expressed under different study conditions. NGS techniques are revolutionising the field of microbial ecology and have recently been used to examine various food ecosystems (Liu *et al.*, 2012; Mayo *et al.*, 2014). After years of evolution, the three systems, Roche 454, AB SOLiD and Illumina, exhibit better performance and their own advantages in terms of read length, accuracy, consumables, hand-power requirements and computer infrastructure (Mergulies *et al.*, 2005; Schuster, 2007; Liu *et al.*, 2012; Liu *et al.*, 2017). NGS has been used, for example, for the identification of microbial diversity and dynamics during wine fermentation (del Carmen Portillo and Mas, 2016; del Carmen Portillo *et al.*, 2016), microbial biogeography of wine grapes (Bokulich *et al.*, 2014), wine grapes microbial *terroir* (Gilbert *et al.*, 2014) and bacterial diversity in botrytised wine (Bokulich *et al.*, 2012).

## 4.4. MALDI-TOF Identification of Wine Microbiota

MALDI-TOF (Matrix Associated Laser Desorption-Ionisation – Time of Flight) is a omics approach technique within the mass spectrometry area used for the identification of microorganisms in the most varied samples. The system consists of irradiation of a laser on the fresh sample of microbial colony in the presence of polymeric organic matrices (e.g. 2,5-dihydroxybenzoic acid (DHB), α-Cyano-4-hydroxycinnamic acid (α-CHCA). The material irradiated with the laser is vapourised, generating ionisation of the molecules, which fly, under vacuum, through a tube to a detector. The characteristics of the ionised molecules, especially regarding their mass to charge ratio (m/z), are responsible for the differences in the time of arrival to the detector (time of light). The detector begins to acquire data and generates spectra for each sample, which are compared with databases, in the equipment itself, leading to the rapid identification of the microorganism (Tanaka *et al.*, 1988). The microbial identification is based on analyses of ribosomal proteins patterns, which are synthesised under all microbial growth conditions and are the most abundant cellular proteins (Ryzhov and Fenselau, 2001).

MALDI-TOF is an emerging technique that can contribute significantly to fast microbial identification at a low cost of analysis. However, the diffusion of this technique for routine analyses in laboratories of industrial, environmental and food microbiology and the analysis of fermented beverages, such as wine, still faces some barriers, mainly the lack of information, or database for microorganisms not common to medical clinical area (Rahi *et al.*, 2016). Another barrier is the value of the equipment and the database, with the main equipment available on the market being the Bruker-Biotyper (Bruker Daltonics, Germany), SARAMIS AnagnosTec (acquired by bioMérieux, France and restructured as Vitek-MS), and Axima Assurance (Shimadzu, Japan). These platforms use very similar analytical methodologies as they are mainly related to the sample preparation, spectra acquisition and comparison of the generated spectra with the reference spectra of the database (Cassagne *et al.*, 2016).

This technique of microbial identification was initially developed for application in clinical samples, especially for the identification of pathogenic bacteria and later for the identification of human pathogenic fungi (Tanaka *et al.*, 1988; Cain *et al.*, 1994; Holland *et al.*, 1996). Nowadays MALDI-TOF is used in the identification of food-contaminating fungi (Lima and Santos, 2017), bacteria and yeasts present in cocoa fermentation (Miguel *et al.*, 2016), vinegar (Viana *et al.*, 2017) and microorganisms present in wines (Moothoo-Padayachie *et al.*, 2013; Andrés-Barrao *et al.*, 2013; Usbeck *et al.*, 2014; Gutiérrez *et al.*, 2017).

## 5. Wine Sensory Analysis

Wine is made up of hundreds of compounds with varied chemical structures that, due to their fixed or volatile nature, will jointly define the sensory characteristics of the beverage. These compounds aoriginate from the grape variety itself and are produced during the fermentation process, by the microorganisms and in stages after fermentation (Puertas *et al.*, 2018). The presence of these molecules in solution, responsible for the generation of aroma and flavour, added to the colour and the density of the drink, constitute the basis of the sensorial attributes of wine. And that, to a large extent, defines consumer acceptance of wine. For these reasons, sensorial analysis is a fundamental step in wine production (Lesschaeve, 2007).

Sensory analysis of wines can be performed by using several methods with specific objectives, which are selected according to the purpose of the analysis, such as sensitivity methods to select or trained judges (trained panel), or affective methods to verifying the acceptability of the consumer market (Amerine *et al.*, 1980; Jackson, 2000; Joshi, 2006; Narasimhan and Stephen, 2011). There are several sensory tests used in wine analysis, such as discriminative tests (triangular, duo-trio, ordering, paired comparison and multiple comparison), descriptive tests (taste profile, texture profile and quantitative descriptive analysis) and affective tests (preference, hedonic acceptance, ideal scale acceptance and purchase intention).

The quality of wine is usually evaluated by winemakers, who are highly trained and able to detect quality attributes or possible defects (Lesschaeve, 2007), but this does not dispense the need for consumers' affective tests (Lockshin and Corsi, 2012; Francis and Williamson, 2015).

Among the several methodologies in sensory analysis, some of them have been used more frequently in wine evaluation. In the case of the analysis by untrained tasters, the main methodologies are the affective tests of hedonic scale, consumer's intention of purchase test and the Check-All-That-Apply (CATA) test. When considering evaluations from trained panels, methodologies such as Quantitative Descriptive Analysis (QDA) and Temporal Dominance of Sensations (TDS) are unusually applied (Murray *et al.*, 2001; Varela and Gámbaro, 2006; Cadot *et al.*, 2010; Schlich, 2017).

### 5.1. Electronic Sensory Analysis of Wines

The advancement of analytical technologies in recent years has contributed to the development of new techniques for identifying compounds in wines, which tend to be cheaper, reproducible, accurate and faster. In this sense, in addition to the evolution in the development of electronic and chemical sensors, new instruments of sensorial analysis have been improved aiming to analyse wines, like the sensorial analysis by human trained panels. Among the sensors developed are the electronic nose and electronic tongue, which are capable of accurately characterising the wine sensory quality in a short time and at a low cost, when compared to the last generation chromatographs or spectrometers (Ebeler and Thorngate, 2009; Ferreira, 2010; Sáenz-Navajas *et al.*, 2012; Smyth and Cozzolino, 2013).

The electronic nose consists of a chemo-electronic sensor array capable of interacting with volatile compounds and mimicking the human olfactory system, generating an electrical response. This response is recognised by a computerised system, or through an artificial neural network, generating a result (Fig. 5). Similar to the human nose, and unlike gas chromatographic systems, the electronic nose recognises the sample from a set of volatile compounds present in it, not only from one compound pattern (Röck *et al.*, 2008; Brattoli *et al.*, 2011; Sáenz-Navajas *et al.*, 2012).

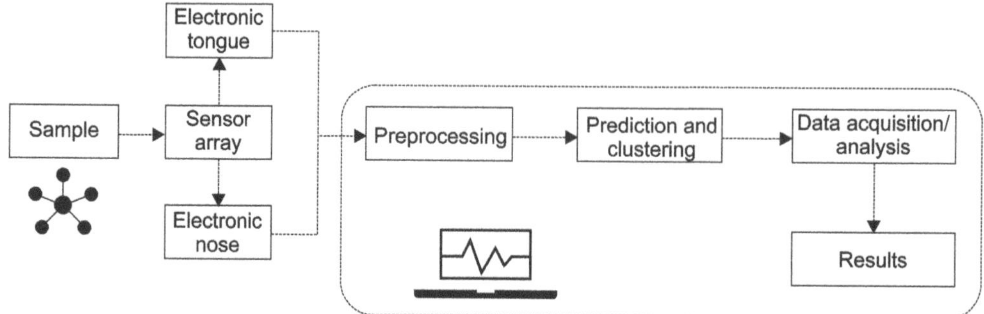


**Figure 5.** General scheme of electronic mimicking of human sensory

The electronic tongue is also composed of a sensor array capable of mimicking the human gustatory system and recognise the different types of flavour (umami, sweet, salt, bitter and sour). However, the sensors most commonly employed in the electronic tongue are based on electrochemical methods, such as potentiometry, amperometry and cyclic voltammetry (Winquist *et al.*, 2000; Vlasov *et al.*, 2005; Riul Jr. *et al.*, 2010), while the electronic nose uses sensors of a chemical nature, such as conducting polymers, electrochemical cells, piezoelectric devices, metal oxide sensors, and metal-insulator semiconductor field effect transistors (Smyth and Cozzolino, 2013). Like the electronic nose, the electronic tongue is also based on the ability of global selectivity, recognising a set of information associated with taste and generating a response related to quality (Riul Jr. *et al.*, 2010; Ha *et al.*, 2015). Both the electronic nose and the electronic tongue have been used in sensory analysis of wines, wine deterioration and wine discrimination, among others. Table 2 summarises some papers regarding the application of electronic nose and tongue in wine analyses.

## 6. Omics Approach in Wine Analysis

The term metabolomics refers to the complete (qualitative and quantitative) analysis of all low molecular weight (less than 1.5 kDa) metabolites present in, and around, the cells at a given time of their growth or in a sample (Dunn and Ellis, 2005; Alañón *et al.*, 2015). Koek *et al.*, 2011). Regarding its application in wine analysis, metabolomic studies allow the identification of many compounds, generating data capable of guaranteeing a wine unbiased discrimination in relation to quality assurance, authenticity, variety, vintage, origin and deterioration (Alañon *et al.*, 2015; Cozzolino, 2016).

A wide array of modern analytical technologies has been carried out in the metabolomic analysis of wines, such as nuclear magnetic resonance (NMR), vibration spectroscopy (MIR; near infrared, NIR; Raman), Fourier transform (FT), capillary electrophoresis (EC). Some of these technologies are used coupled to single mass spectrometers or tandem MS (MS/MS), accelerating and increasing the capacity of molecule identification, either qualitatively or quantitatively. Recently, the development of modern high-resolution mass spectrometers, mainly time-of-flight (TOF), quadrupole TOF (Q-TOF), ion-trap TOF (IT-TOF) and Orbitrap analysers, have significantly increased accuracy and the sensitivity of compound identification (Hong, 2011; Alañon *et al.*, 2015; Ebeler, 2015; Markley *et al.*, 2017).

The choice of the technology to be employed in metabolomic studies will depend on the compounds to be identified, their concentration on the sample and the physico-chemical characteristics of the sample. The metabolomic approach can be classified as targeted, or profiling and untargeted, or fingerprinting. Target metabolomics refer to the analysis of previously defined compounds that are expected to be found and quantified in the sample. Untargeted metabolomic refers to the approach of the largest possible number of compounds and their quantification in a sample, which is one of the most applied in the metabolomic analysis of an extremely complex samples such as wines (Cozzolino, 2016; Gallo and Ferranti, 2016). Figure 6 shows a ranking of the most frequent analytical techniques and the approach, targeted or untargeted, applied in wine metabolomic in the last 10 years, published in research papers.

**Table 2.** Application of Electronic Nose and Electronic Tongue in Wine Analyses

| *Sample* | *Electronic sensor, additional technique* | *Application* | *References* |
|---|---|---|---|
| Red wine | Nose<br>Standard chemical analytical approach | Recognition of the vineyard | Di Natale *et al.*, 1996 |
| Red wines | Tongue | Discrimination between wines | Legin, 2003. |
| Barbera wine | Nose<br>Tongue | Characterization and classification | Buratti *et al.*, 2004 |
| Red wines | Tongue | Wine adulterations | Parra *et al.*, 2006 |
| Red wine | Nose | Wine discrimination | García *et al.*, 2006 |
| Red and white wines | Nose<br>Artificial neural networks | Typical aroma components | Lozano *et al.*, 2006 |
| Red wine | Nose<br>Tongue<br>Spectrometry | Prediction of sensorial descriptors | Buratti *et al.*, 2007 |
| Port wine | Tongue<br>Chemical analyses | Prediction of wine aging | Rudnitskaya *et al.*, 2007 |
| Red wine | Nose | Wine aging | Lozano *et al.*, 2008 |
| Red wine | Nose (metal oxide-based and MS-based) | Wine spoilage | Berna *et al.*, 2008 |
| Red wine | Nose<br>Human sensory panel | Threshold of aromatic compounds | Santos el al., 2010 |
| Red wine | Tongue<br>Chemical analyses | Bitter taste | Rudnitskaya *et al.*, 2010 |
| Madeira wine | Tongue<br>HPLC | Prediction of wine aging<br>Organic acids and phenolics quantification | Rudnitskaya *et al.*, 2010 |
| Red and White wines | Nose<br>Tongue<br>Titratable total acidity | Wine deterioration | Gil-Sánchez *et al.*, 2011 |
| Red wine | Nose<br>Tongue<br>Near infrared and mid infrared spectroscopies | Monitoring of alcoholic fermentation | Buratti *et al.*, 2011 |
| Grape | Nose<br>GC-MS | Off-vine dehydration time | De Lerma *et al.*, 2012 |
| Red and white wines | Nose<br>Artificial neural networks | Prediction and classification of wines | Aguilera *et al.*, 2012 |
| Red wines | Tongue<br>Trained sensory panel | Discrimination based on the maturing in barrel. Prediction of the global scores | Cetó *et al.*, 2017 |

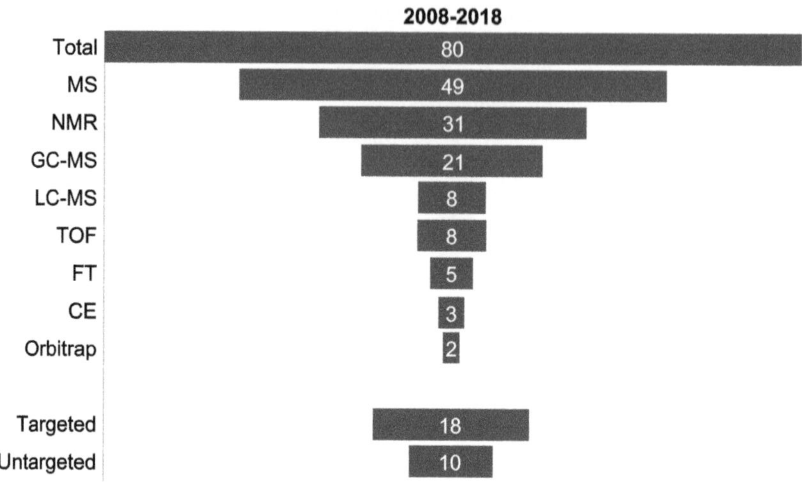


**Figure 6.** Web of Science/Clarivate Analytics search result for published papers in wine metabolomic from 2008 to 2018. Search report (26 March 2018): Citation report for 80 results from Web of Science Core Collection between 2008 and 2018, searched for: TOPIC: (metabolomic*) AND TITLE: (wine*), Refined by: DOCUMENT TYPES: (ARTICLE OR REVIEW). Indexes: SCI-EXPANDED, SSCI, A&HCI, CPCI-S, CPCI-SSH, ESCI. Based on total results and on 80 papers, we performed searches using the following keywords: MS, NMR, GC-MS, LC-MS, TOF, FT, CE, Orbitrap, Targeted, Untargeted

## 7. Future Prospects

Analytical techniques have evolved immensely in the last decade, generating new equipments and methodologies capable of identifying and quantifying hundreds of chemical compounds as well as microbial diversity, in a matrix as complex as wine, fast and accurately at a relatively low cost per analysis. However, the cost of many state-of-the-art equipment for the analysis of volatile compounds, in a metabolomic approach, for the identification of the microbiota (MALDI-TOF or NGS), even for automation of physico-chemical analyses, is still far from reality of many winemakers. The trend, however, is that over time, these technologies will decrease the cost, allowing the acquisition and use of them, favouring the analysis and guaranteeing the wine quality.

In the coming years, it is expected that the advances in analytical techniques will make possible identification and quantification of compounds present in wine not yet chemically described.

## References

Aguilera, T., Lozano, J., Paredes, J.A., Alvarez, F.J. and Suárez, J.I. (2012). Electronic nose based on independent component analysis combined with partial least squares and artificial neural networks for wine prediction. *Sensors* 12(6): 8055-8072.

Alañón, M.E., Pérez-Coello, M.S. and Marina, M.L. (2015). Wine science in the metabolomics era. *TrAC Trends in Analytical Chemistry* 74: 1-20.

Aleixandre-Tudo, J.L., Buica, A., Nieuwoudt, H., Aleixandre, J.L. and du Toit, W. (2017). Spectrophotometric analysis of phenolic compounds in grapes and wines. *Journal of Agricultural and Food Chemistry* 65(20): 4009-4026.

Ali, K., Maltese, F., Choi, Y.H. and Verpoorte, R. (2010). Metabolic constituents of grapevine and grape-derived products. *Phytochemistry Reviews* 9: 357-378.

Amerine, M.A. and Ough, C.S. (1980). Methods for Analysis of Musts and Wines. John Wiley and Sons, New York.

Andrés-Barrao, C., Benagli, C., Chappuis, M., Pérez, R.O., Tonolla, M. and Barja, F. (2013). Rapid identification of acetic acid bacteria using MALDI-TOF mass spectrometry fingerprinting. *Systematic and Applied Microbiology* 36(2): 75-81.

Armstrong, G.O., Lambrechts, M.G., Mansvelt, E.P.G., van Velden, D.P. and Pretorius, I.S. (2001). Wine and health. *South African Journal of Science* 97: 279-282.

Baltussen, E., Sandra, P., David, F. and Cramers, C. (1999) Stir bar sorptive extraction (SBSE), a novel extraction technique for aqueous samples: Theory and principles. *Journal of Microcolumn Separations* 11: 737.

Bartowsky, E.J. and Henschke, P.A. (2008). Acetic acid bacteria spoilage of bottled red wine – A review. *International Journal of Food Microbiology* 125(1): 60-70.

Berna, A.Z., Trowell, S., Cynkar, W. and Cozzolino, D. (2008). Comparison of metal oxide-based electronic nose and mass spectrometry-based electronic nose for the prediction of red wine spoilage. *Journal of Agricultural and Food Chemistry* 56(9): 3238-3244.

Bojko, B., Cudjoe, E., Gómez-Ríos, G.A., Gorynski, K., Jiang, R., Reyes-Garcés, N., Risticevic, S., Silva, É.A.S., Togunde, O. and Vuckovic, D. (2012). SPME-Quo vadis? *Anal. Chimica. Acta* 750: 132-151.

Bokulich, N.A., Joseph, C.L., Allen, G., Benson, A.K. and Mills, D.A. (2012). Next-generation sequencing reveals significant bacterial diversity of botrytised wine. *PloS One* 7(5): e36357.

Bokulich, N.A., Thorngate, J.H., Richardson, P.M. and Mills, D.A. (2014). Microbial biogeography of wine grapes is conditioned by cultivar, vintage and climate. *Proceedings of the National Academy of Sciences* 111(1): E139-E148.

Boulton, R. (1980). The relationships between total acidity, titratable acidity and pH in wine. *American Journal of Enology and Viticulture* 31(1): 76-80.

Brattoli, M., De Gennaro, G., De Pinto, V., Demarinis Loiotile, A., Lovascio, S. and Penza, M. (2011). Odour detection methods: Olfactometry and chemical sensors. *Sensors* 11(5): 5290-5322.

Buratti, S., Ballabio, D., Benedetti, S. and Cosio, M.S. (2007). Prediction of Italian red wine sensorial descriptors from electronic nose, electronic tongue and spectrophotometric measurements by means of Genetic Algorithm regression models. *Food Chemistry* 100(1): 211-218.

Buratti, S., Ballabio, D., Giovanelli, G., Dominguez, C.Z., Moles, A., Benedetti, S. and Sinelli, N. (2011). Monitoring of alcoholic fermentation using near infrared and mid infrared spectroscopies combined with electronic nose and electronic tongue. *Analytica Chimica Acta* 697(1-2): 67-74.

Buratti, S., Benedetti, S., Scampicchio, M. and Pangerod, E.C. (2004). Characterisation and classification of Italian Barbera wines by using an electronic nose and an amperometric electronic tongue. *Analytica Chimica Acta* 525(1): 133-139.

Cadot, Y., Caillé, S., Samson, A., Barbeau, G. and Cheynier, V. (2010). Sensory dimension of wine typicality related to a terroir by Quantitative Descriptive Analysis, just about right analysis and typicality assessment. *Analytica Chimica Acta* 660(1-2): 53-62.

Cain, T.C., Lubman, D.M., Weber, W.J. and Vertes, A. (1994). Differentiation of bacteria using protein profiles from matrix-assisted laser desorption/ionisation time-of-flight mass spectrometry. *Rapid Communications in Mass Spectrometry* 8(12): 1026-1030.

Cappello, M.S., De Domenico, S., Logrieco, A. and Zapparoli, G. (2014). Bio-molecular characterisation of indigenous *Oenococcus oeni* strains from Negroamaro wine. *Food Microbiology* 42: 142-148.

Cassagne, C., Normand, A.C., L'ollivier, C., Ranque, S. and Piarroux, R. (2016). Performance of MALDI-TOF MS platforms for fungal identification. *Mycoses* 59(11): 678-690.

Castro, R., Natera, R., Durán, E. and García-Barroso, C. (2008). Application of solid phase extraction techniques to analyse volatile compounds in wines and other enological products. *European. Food Research Technology* 228: 1-18.

Cetó, X., González-Calabuig, A., Crespo, N., Pérez, S., Capdevila, J., Puig-Pujol, A. and del Valle, M. (2017). Electronic tongues to assess wine sensory descriptors. *Talanta* 162: 218-224.

Ciani, M. and Comitini, F. (2011). Non-*Saccharomyces* wine yeasts have a promising role in biotechnological approaches to winemaking. *Annals of Microbiology* 61(1): 25-32.

Cocolin, L., Bisson, L.F. and Mills, D.A. (2000). Direct profiling of the yeast dynamics in wine fermentations. *FEMS Microbiology Letters* 189(1): 81-87.

Cocolin, L., Campolongo, S., Alessandria, V., Dolci, P. and Rantsiou, K. (2011). Culture independent analyses and wine fermentation: An overview of achievements 10 years after first application. *Annals of Microbiology* 61(1): 17-23.

Coelho, E.M., Padilha, C.V.S., Miskinisa, G.A., de Sá, A.G.B., Pereira, G.E., Azevêdo, L.C. and Lima, M.S. (2018). Simultaneous analysis of sugars and organic acids in wine and grape juices by HPLC: Method validation and characterization of products from northeast Brazil. *Journal of Food Composition and Analysis* 66(3): 160-167.

Collins, F.S., Morgan, M. and Patrinos, A. (2003). The Human Genome Project: Lessons from large-scale biology. *Science* 300(5617): 286-290.

Davis, C.R., Wibowo, D., Eschenbruch, R., Lee, T.H. and Fleet, G.H. (1985). Practical implications of malolactic fermentation: A review. *American Journal of Enology and Viticulture* 36(4): 290-301.

De Lerma, N.L., Bellincontro, A., Mencarelli, F., Moreno, J. and Peinado, R.A. (2012). Use of electronic nose, validated by GC-MS, to establish the optimum off-vine dehydration time of wine grapes. *Food Chemistry* 130(2): 447-452.

de Villiers, A., Alberts, P., Tredoux, A.G.J. and Nieuwoudt, H.H. (2012). Analytical techniques for wine analysis: An African perspective: A review. *Analytica Chimica Acta.* 730: 2-23.

del Carmen Portillo, M. and Mas, A. (2016). Analysis of microbial diversity and dynamics during wine fermentation of Grenache grape variety by high-throughput barcoding sequencing. *LWT – Food Science and Technology* 72: 317-321.

del Carmen Portillo, M., Franquès, J., Araque, I., Reguant, C. and Bordons, A. (2016). Bacterial diversity of Grenache and Carignan grape surface from different vineyards at Priorat wine region (Catalonia, Spain). *International Journal of Food Microbiology* 219: 56-63.

Di Natale, C., Davide, F.A., D'Amico, A., Nelli, P., Groppelli, S. and Sberveglieri, G. (1996). An electronic nose for the recognition of the vineyard of a red wine. *Sensors and Actuators B: Chemical* 33(1-3): 83-88.

Dias, D.R., Duarte, W.F. and Schwan, R.F. (2017). Methods of evaluation of fruit wines. pp. 227-252. *In*: Kosseva, M.R., Joshi, V. and Panesar, P. (Eds.). Science and Technology of Fruit Wine Production. Academic Press.

Dunn, W.B. and Ellis, D.I. (2005). Metabolomics: Current analytical platforms and methodologies. *TrAC Trends in Analytical Chemistry* 24(4): 285-294.

Ebeler, S.E. (2015). Analysis of grapes and wines: An overview of new approaches and analytical tools. pp. 3-12. *In*: Advances in Wine Research. American Chemical Society.

Ebeler, S.E. and Thorngate, J.H. (2009). Wine chemistry and flavour: Looking into the crystal glass. *Journal of Agricultural and Food Chemistry* 57(18): 8098-8108.

Eyduran, S.P., Akin, M., Ercisli, S., Eyduran, E. and Maghradze, D. (2015). Sugars, organic acids, and phenolic compounds of ancient grape cultivars (*Vitis Vinifera* L.) from Igdir province of eastern Turkey. *Biology Research* 48: 1-8.

Ferreira, V. (2010). Viticulture and wine quality. *In*: Reynolds, A.G. (Eds.). Managing Wine Quality, vol. 1. CRC Press, Boca Raton.

Flamini, R. (2010). Volatile and aroma compounds in wines. pp. 117-162. *In*: Flamini, R. and Traldi, P. (Eds.). Mass Spectrometry in Grape and Wine Chemistry. John Wiley and Sons, Inc., Hoboken, NJ, USA.

Fleet, G.H. (2007). Yeasts in foods and beverages: Impact on product quality and safety. *Current Opinion in Biotechnology* 18(2): 170-175.

Fleet, G.H., Lafon-Lafourcade, S. and Ribéreau-Gayon, P. (1984). Evolution of yeasts and lactic acid bacteria during fermentation and storage of Bordeaux wines. *Applied and Environmental Microbiology* 48(5): 1034-1038.

Francis, I.L. and Williamson, P.O. (2015). Application of consumer sensory science in wine research. *Australian Journal of Grape and Wine Research* 21(S1): 554-567.

Fröhlich, J., König, H. and Claus, H. (2017). Molecular methods for identification of wine microorganisms and yeast development. pp. 517-547. *In*: Biology of Microorganisms on Grapes, in Must and in Wine. Springer, Cham.

Gallo, M. and Ferranti, P. (2016). The evolution of analytical chemistry methods in foodomics. *Journal of Chromatography A* 1428: 3-15.

García, M., Aleixandre, M., Gutiérrez, J. and Horrillo, M.C. (2006). Electronic nose for wine discrimination. *Sensors and Actuators B: Chemical* 113(2): 911-916.

Gilbert, J.A., van der Lelie, D. and Zarraonaindia, I. (2014). Microbial terroir for wine grapes. *Proceedings of the National Academy of Sciences* 111(1): 5-6.

Gil-Sánchez, L., Soto, J., Martínez-Máñez, R., Garcia-Breijo, E., Ibáñez, J. and Llobet, E. (2011). A novel humid electronic nose combined with an electronic tongue for assessing deterioration of wine. *Sensors and Actuators A: Physical* 171(2): 152-158.

Giraffa, G. and Carminati, D. (2008). Molecular techniques in food fermentation: Principles and applications. pp. 1-30. *In*: Molecular Techniques in the Microbial Ecology of Fermented Foods. Springer, New York, NY.

Goffeau, A., Barrell, B.G., Bussey, H., Davis, R.W., Dujon, B., Feldmann, H. and Louis, E.J. (1996). Life with 6000 genes. *Science* 274(5287): 546-567.

González-Barreiro, C., Rial-Otero, R., Cancho-Grande, B. and Simal-Gándara, J. (2015). Wine aroma compounds in grapes: A critical review. *Critical Reviews in Food Science and Nutrition* 55(2): 202-218.

Grindlay, G., Mora, J., Gras, L. and de Loos-Vollebregt, M.T.C. (2011). Atomic spectrometry methods for wine analysis: A critical evaluation and discussion of recent applications – Review. *Analytica Chimica Acta.* 691: 18-32.

Gutiérrez, C., Gómez-Flechoso, M.Á., Belda, I., Ruiz, J., Kayali, N., Polo, L. and Santos, A. (2017). Wine yeasts identification by MALDI-TOF MS: Optimisation of the pre-analytical steps and development of an extensible open-source platform for processing and analysis of an in-house MS database. *International Journal of Food Microbiology* 254: 1-10.

Ha, D., Sun, Q., Su, K., Wan, H., Li, H., Xu, N., Sun, F., Zhuang, L., Hu, N. and Wang, P. (2015). Recent achievements in electronic tongue 4 and bioelectronic tongue as taste sensors. *Sensors and Actuators B: Chemical*, 207: 1136-1146.

Harmon, A.D. (1997). Solid-phase microextraction for the analysis of flavours. pp. 81-112. *In*: Marsili, R. (Ed.). Techniques for Analysing Food Aroma. Marcel Decker, Inc., New York, NY, USA.

Holland, R.D., Wilkes, J.G., Rafii, F., Sutherland, J.B., Persons, C.C., Voorhees, K.J. and Lay, J.O. (1996). Rapid identification of intact whole bacteria based on spectral patterns using matrix-assisted laser desorption/ionization with time-of-flight mass spectrometry. *Rapid Communications in Mass Spectrometry* 10(10): 1227-1232.

Hong, Y.S. (2011). NMR-based metabolomics in wine science. *Magnetic Resonance in Chemistry* 49(S1).

Howard, K.L., Mike, J.H. and Riesen, R. (2005). Validation of a solid-phase microextraction method for headspace analysis of wine aroma components. *Am. J. Enology & Viticulture* 56: 37-45.

Ivey, M.L. and Phister, T.G. (2011). Detection and identification of microorganisms in wine: A review of molecular techniques. *Journal of Industrial Microbiology and Biotechnology* 38(10): 1619-1634.

Jackson, R.S. (2000). Wine Science – Principles, Practice, Perception. Academic Press, New York.

Johansson, A. (1988). The development of the titration methods: Some historical annotations. *Analytica Chimica Acta* 206: 97-109.

Joshi, V.K. (2006). Sensory Science: Principles and Application in Food Evaluation. Agro-Tech Academy, Udaipur, pp. 527.

Joshi, V.K., Thakur, N.S., Bhat, A. and Garg, C. (2011). Wine and brandy: A perspective. pp. 1-45. *In*: V.K. Joshi (Ed.). Handbook of Enology, vol. 3. Asia Tech Publisher, New Delhi.

Koek, M.M., Jellema, R.H., van der Greef, J., Tas, A.C. and Hankemeier, T. (2011). Quantitative metabolomics based on gas chromatography mass spectrometry: Status and perspectives. *Metabolomics* 7(3): 307-328.

Kurtzman, C., Fell, J.W. and Boekhout, T. (Eds.). (2011). The Yeasts: A Taxonomic Study. Elsevier.

Lasik, M. (2013). The application of malo-lactic fermentation process to create good-quality grape wine produced in cool-climate countries: A review. *European Food Research and Technology* 237(6): 843-850.

Legin, A., Rudnitskaya, A., Lvova, L., Vlasov, Y., Di Natale, C. and D'amico, A. (2003). Evaluation of Italian wine by the electronic tongue: Recognition, quantitative analysis and correlation with human sensory perception. *Analytica Chimica Acta* 484(1): 33-44.

Lesschaeve, I. (2007). Sensory evaluation of wine and commercial realities: Review of current practices and perspectives. *American Journal of Enology and Viticulture* 58(2): 252-258.

Lima, M.J. and Reis, B.F. (2017). Fully automated photometric titration procedure employing a multi-commuted flow analysis setup for acidity determination in fruit juice, vinegar and wine. *Microchemical Journal* 135: 207-212.

Lima, N. and Santos, C. (2017). MALDI-TOF MS for identification of food spoilage filamentous fungi. *Current Opinion in Food Science* 13: 26-30.

Liu, L., Li, Y., Li, S., Hu, N., He, Y., Pong, R., Lin, D., Lu, L. and Law, M. (2012). Comparison of next-generation sequencing systems. *Bio Med. Research International* 251364, pp. 11.

Liu, Y., Rousseaux, S., Tourdot-Maréchal, R., Sadoudi, M., Gougeon, R., Schmitt-Kopplin, P. and Alexandre, H. (2017). Wine microbiome: A dynamic world of microbial interactions. *Critical Reviews in Food Science and Nutrition* 57(4): 856-873.

Lockshin, L. and Corsi, A.M. (2012). Consumer behaviour for wine 2.0: A review since 2003 and future directions. *Wine Economics and Policy* 1(1): 2-23.

Lonvaud, A. (2011). Microbial spoilage of wine. pp. 1367-1391. *In*: V.K. Joshi (Ed.). Handbook of Enology, vol. 3. Asia-Tech Publisher, New Delhi.

Lonvaud-Funel, A. (2015). Lactic acid bacteria and malo-lactic fermentation in wine. pp. 231-247. *In*: F. Mozzi, R.R. Raya and G.M. Vignolo (Eds.). Biotechnology of Lactic Acid Bacteria. John Wiley & Sons, Chichester, UK.

Lozano, J., Arroyo, T., Santos, J.P., Cabellos, J.M. and Horrillo, M.C. (2008). Electronic nose for wine ageing detection. *Sensors and Actuators B: Chemical* 133(1): 180-186.

Lozano, J., Santos, J.P., Aleixandre, M., Sayago, I., Gutierrez, J. and Horrillo, M.C. (2006). Identification of typical wine aromas by means of an electronic nose. *IEEE Sensors Journal* 6(1): 173-178.

Lund, S.T. and Bohlmann, J. (2006). The molecular basis for wine grape quality – A volatile subject. *Science* 311.

Margulies, M., Egholm, M., Altman, W.E., Attiya, S., Bader, J.S., Bemben, L.A. and Dewell, S.B. (2005). Genome sequencing in microfabricated high-density picolitre reactors. *Nature* 437(7057): 376.

Markley, J.L., Brüschweiler, R., Edison, A.S., Eghbalnia, H.R., Powers, R., Raftery, D. and Wishart, D.S. (2017). The future of NMR-based metabolomics. *Current Opinion in Biotechnology* 43: 34-40.

Mato, I., Suárez-Luque, S. and Huidobro, J.F. (2005). A review of the analytical methods to determine organic acids in grape juices and wines. *Food Research International* 38(10): 1175-1188.

Mayo, B., Rachid, C.T., Alegría, Á., Leite, A. Mo, Peixoto, R.S. and Delgado, S. (2014). Impact of next generation sequencing techniques in food microbiology. *Current Genomics* 15(4): 293-309._

McGoverin, C.M., Weeranantanaphan, J., Downey, G. and Manley, M. (2010). The application of near infrared spectroscopy to the measurement of bioactive compounds in food commodities. *J. Near Infrared Spectroscopy* 18: 87-111.

Mencarelli, F. and Bellincontro, A. (2018). Recent advances in postharvest technology of the wine grape to improve the wine aroma. *J. Sci. Food Agric.* doi:10.1002/jsfa.8910

Miguel, M.G.D.C.P., de Castro Reis, L.V., Efraim, P., Santos, C., Lima, N. and Schwan, R.F. (2017). Cocoa fermentation: Microbial identification by MALDI-TOF MS, and sensory evaluation of produced chocolate. *LWT – Food Science and Technology* 77: 362-369.

Mills, D.A., Phister, T., Neeley, E. and Johannsen, E. (2008). Wine fermentation. pp. 162-192. *In*: Cocolin and Ercolini (Eds.). Molecular Techniques in the Microbial Ecology of Fermented Foods. Springer New York.

Moothoo-Padayachie, A., Kandappa, H.R., Krishna, S.B.N., Maier, T. and Govender, P. (2013). Biotyping *Saccharomyces cerevisiae* strains using matrix-assisted laser desorption/ionisation time-of-flight mass spectrometry (MALDI-TOF MS). *European Food Research and Technology* 236(2): 351-364.

Murray, J.M., Delahunty, C.M. and Baxter, I.A. (2001). Descriptive sensory analysis: Past, present and future. *Food Research International* 34(6): 461-471.

Murray, R.A. (2001). Limitations to the use of solid-phase microextraction for quantitation of mixtures of volatile organic sulphur compounds. *Anal. Chemistry* 7: 16-46.

Narasimhan, S. and Stephen, N.S. (2011). Sensory evaluation of wine and brandy. pp. 1331-1366. *In*: V.K. Joshi (Ed.). Handbook of Enology, vol 3. Asia-Tech Publisher, New Delhi.

OIV (2015). Compendium of international methods of wine and must analysis. *Organisation Internationale de la vigne et du vin (OIV)*, edition 2015, vols. 1 and 2, OIV, Paris.

Panighel, A. and Flamini, R. (2014). Applications of solid-phase microextraction and gas chromatography/mass spectrometry (SPME-GC/MS) in the study of grape and wine volatile compounds. *Molecules* 19: 21291-21309.

Parra, V., Arrieta, Á.A., Fernández-Escudero, J.A., Rodríguez-Méndez, M.L. and De Saja, J.A. (2006). Electronic tongue based on chemically modified electrodes and voltammetry for the detection of adulterations in wines. *Sensors and Actuators B: Chemical* 118(1-2): 448-453.

Pawliszyn, J. and Pedersen-Bjergaard, S. (2006). Analytical microextraction: Current status and future trends. *J Chromatogr Sci*. 6: 291-307.

Pawliszyn, J. (1990). Solid-phase microextraction with thermal desorption using silica optical fibres. *Anal. Chem*. 62: 2145-2148.

Polášková, P., Herszage, J. and Ebeler, S.E. (2008). Wine flavour: Chemistry in a glass. *Chemical Society Reviews* 37(11): 2478-2489.

Puertas, B., Jimenez-Hierro, M.J., Cantos-Villar, E., Marrufo, A., Carbú, M., Cuevas, F.J. and Ruiz-Moreno, M.J. (2018). The influence of yeast on chemical composition and sensory properties of dry white wines. *Food Chemistry* 253: 227-235.

Querol, A. and Ramon, D. (1996). The application of molecular techniques in wine microbiology. *Trends in Food Science and Technology* 7(3): 73-78.

Rahi, P., Prakash, O. and Shouche, Y.S. (2016). Matrix-assisted laser desorption/ionisation time-of-flight mass-spectrometry (MALDI-TOF MS) based microbial identifications: Challenges and scopes for microbial ecologists. *Frontiers in Microbiology* 7: 1359.

Ribéreau-Gayon, P., Glories, Y., Maujean, A. and Dubourdieu, D. (2006). Handbook of Enology: The Chemistry of Wine Stabilisation and Treatment, vol. 2. John Wiley and Sons.

Riul Jr., A., Dantas, C.A., Miyazaki, C.M. and Oliveira Jr., O.N. (2010). Recent advances in electronic tongues. *Analyst* 135(10): 2481-2495.

Röck, F., Barsan, N. and Weimar, U. (2008). Electronic nose: Current status and future trends. *Chemical Reviews* 108(2): 705-725.

Rossouw, D.R. and Bauer, F.F. (2016). Wine science in the omics era: The impact of systems biology on the future of wine research. *South African Journal of Enology and Viticulture* 30(2): 101-109.

Rudnitskaya, A., Delgadillo, I., Legin, A., Rocha, S.M., Costa, A.M. and Simões, T. (2007). Prediction of the Port wine age using an electronic tongue. *Chemometrics and Intelligent Laboratory Systems* 88(1): 125-131.

Rudnitskaya, A., Nieuwoudt, H.H., Muller, N., Legin, A., du Toit, M. and Bauer, F.F. (2010). Instrumental measurement of bitter taste in red wine using an electronic tongue. *Analytical and Bioanalytical Chemistry* 397(7): 3051-3060.

Rudnitskaya, A., Rocha, S.M., Legin, A., Pereira, V. and Marques, J.C. (2010). Evaluation of the feasibility of the electronic tongue as a rapid analytical tool for wine age prediction and quantification of the organic acids and phenolic compounds: The case-study of Madeira wine. *Analytica Chimica Acta* 662(1): 82-89.

Ryan, D., Watkins, P., Smith, J., Allen, M. and Marriott, P. (2005). Analysis of methoxypyrazines in wine using headspace solid phase microextraction with isotope dilution and comprehensive two-dimensional gas chromatography. *Journal Separation Science* 9-10: 1075-1082.

Ryzhov, V. and Fenselau, C. (2001). Characterisation of the protein subset desorbed by MALDI from whole bacterial cells. *Analytical Chemistry* 73(4): 746-750.

Sáenz-Navajas, M.P., Fernández-Zurbano, P. and Ferreira, V. (2012). Contribution of nonvolatile composition to wine flavour. *Food Reviews International* 28(4): 389-411.

Sandra, P., Tienport, B., Vercammen, J., Tredoux, A., Sandra, T. and David, F. (2001). Stir bar sorptive extraction applied to the determination of dicarboximide fungicides in wine. *Journal Chromatograph* 928: 117.

Santos, J.P., Lozano, J., Aleixandre, M., Arroyo, T., Cabellos, J.M., Gil, M. and del Carmen Horrillo, M. (2010). Threshold detection of aromatic compounds in wine with an electronic nose and a human sensory panel. *Talanta* 80(5): 1899-1906.

Schindler, R., Vonach, R., Lendl, B. and Kellner, R. (1998). A rapid automated method for wine analysis based upon sequential injection (SI)-FTIR spectrometry. *Fresenius' Journal of Analytical Chemistry* 362(1): 130-136.

Schlich, P. (2017). Temporal Dominance of Sensations (TDS): A new deal for temporal sensory analysis. *Current Opinion in Food Science* 15: 38-42.

Schuster, S.C. (2007). Next-generation sequencing transforms today's biology. *Nature Methods* 5(1): 16.

Serban, S., Danet, A.F. and El Murr, N. (2004). Rapid and sensitive automated method for glucose monitoring in wine processing. *Journal of Agricultural and Food Chemistry* 52(18): 5588-5592.

Silva, F.L.N., Schmidt, E.M., Messias, C.L., Eberlin, M.N., Frankland, Helena and Sawaya, A.C. (2015). Quantitation of organic acids in wine and grapes by direct infusion electrospray ionisation mass spectrometry. *Anal. Methods* 7: 53-62.

Tanaka, K., Waki, H., Ido, Y., Akita, S., Yoshida, Y., Yoshida, T. and Matsuo, T. (1988). Protein and polymer analyses up to m/z 100 000 by laser ionisation time-of-flight mass spectrometry. *Rapid Communications in Mass Spectrometry* 2(8): 151-153.

Usbeck, J.C., Wilde, C., Bertrand, D., Behr, J. and Vogel, R.F. (2014). Wine yeast typing by MALDI-TOF MS. *Applied Microbiology and Biotechnology* 98(8): 3737-3752.

Varela, P. and Gámbaro, A. (2006). Sensory descriptive analysis of Uruguayan Tannat wine: Correlation to quality assessment. *Journal of Sensory Studies* 21(2): 203-217.

Vas, G. and Vékey, K. (2004). Solid-phase microextraction: A powerful sample preparation tool prior to mass spectrometric analysis. *J. Mass Spectrum* 39: 233-254.

Viana, R.O., Magalhães-Guedes, K.T., Braga Jr., R.A., Dias, D.R. and Schwan, R.F. (2017). Fermentation process for production of apple-based kefir vinegar: Microbiological, chemical and sensory analysis. *Brazilian Journal of Microbiology* 48(3): 592-601.

Vlasov, Y., Legin, A., Rudnitskaya, A., Di Natale, C. and D'amico, A. (2005). Nonspecific sensor arrays ('electronic tongue') for chemical analysis of liquids (IUPAC Technical Report). *Pure and Applied Chemistry* 77(11): 1965-1983.

Waterhouse, A.L. and Ebeler, S.E. (1997). Chemistry of wine flavor. *In*: Proceedings of a Symposium at the 213th National Meeting of the American Chemical Society, held 13-17 April 1997, in San Francisco, California.

Winquist, F., Holmin, S., Krantz-Rülcker, C., Wide, P. and Lundström, I. (2000). A hybrid electronic tongue. *Analytica Chimica Acta* 406(2): 147-157.

Yu, Y.-J., Lu, Z.-M., Yu, N.-H., Xu, W., Li, G.-Q., Shi, J.-S. and Xu, Z.-H. (2012). HS-SPME/GC-MS and chemometrics for volatile composition of Chinese traditional aromatic vinegar in the Zhenjiang region. *Journal Institute Brewing* 118: 133-141.

Zalacain, A., Marín, J., Alonso, G.L. and Salinas, M.R. (2007). Analysis of wine primary aroma compounds by stir bar sorptive extraction. *Journal Talanta* 4: 1610-1615.

Zhang, X., Quinn, K., Cruickshank-Quinn, C., Reisdorph, R. and Reisdorph, N. (2018). The application of ion mobility mass spectrometry to metabolomics. *Current Opinion in Chemical Biology* 42: 60-66.

Zoecklein, B.W., Fugelsang, K.C. and Gump, B.H. (2011). Analytical techniques in wine and distillates. pp. 1287-1330. *In*: V.K. Joshi (Ed.). Handbook of Enology, vol. 3. Asia-Tech Publisher, New Delhi.

Zoski, C.G. (Ed.) (2007). Handbook of Electrochemistry. Elsevier. Oxford, UK.

# 26 Advances in Analytical Techniques: Determination of Toxic Components, Microelements, Compounds of Aroma and Therapeutic Significance

**Simona Guerrini[1,2], Silvia Mangani[2], Giovanna Fia[1] and Lisa Granchi[1*]**

[1] Department of Agriculture, Food, Environment and Forestry (DAGRI), University of Florence, Piazzale delle Cascine, 18, Florence, Italy

[2] Food Micro Team s.r.l, Academic Spin-Off of University of Florence, via Santo Spirito 14, I-50125 – Florence (Italy)

## 1. Analytical Chemistry Instrumental Methods

### 1.1 Introduction

Continuous development in analytical chemistry instrumentation and methods has resulted in the increased application of advanced chromatographic and spectroscopic methods to grape and wine analysis. Innovations in Instrumentation continues to be used to provide more detailed chemical information, especially using hyphenated techniques such as gas chromatography mass spectrometry (GC-MS), liquid chromatography mass spectrometry (HPLC-MS) and advanced spectroscopic detection systems such as tandem mass instruments, etc. As instrumentation become more sensitive, detection limits are driven down, thus permitting the identification of new compounds as well as to the quantification of trace compounds.

Significant advances in instrumentation and chromatographic column technology led to the introduction of ultra-performance liquid chromatography (UPLC) which are coupled to columns packed with sub-2 micron fully porous particles and allowed achieving dramatic increases in resolution, speed and sensitivity in liquid chromatography. Modern developments in gas-phase separation technologies such as the progression from wide-bore packed to capillary columns have played a crucial role in the expansion of analytical possibilities for wine analysis. Further, important developments in sample pre-treatment procedures and more sensitive and selective GC detectors have been influential in extending the application of GC for analysis of wine volatiles.

Sample preparation represents an especially important step in the analytical processes. Effective extraction and pre-concentration of the analyte from the hydro-alcoholic wine matrix is essential for the accurate quantitative analysis. The choice of the sample pre-treatment technique depends on the goals of the analysis. Solid phase extraction (SPE) is regarded as the most popular technique for extraction of natural and anthropogenic compounds from wine. It is a sample preparation technique designed to extract, partition, and/or adsorb one or more components from a liquid phase using a suitable stationary phase. The adsorbed substances can be removed from the adsorbent by step-wise increase of elution strength of the eluent. The relatively low breakthrough volumes of SPE sorbents for wine matrix can be compensated by further combination with dispersive liquid-liquid micro extraction, aiming not only to increase the obtained enrichment factors, but also to remove some undesired species from the primary SPE extract (Rodríguez-Cabo *et al.*, 2016).

Sorptive extraction techniques such as solid phase micro-extraction (SPME) have been shown by several Authors to be advantageous for the extraction of analytes from complex matrices such as wine. SPME is a solvent-free adsorption/desorption technique, combining extraction, concentration and

*Corresponding author: lisa.granchi@unifi.it

chromatographic injection into one-step, dramatically reducing labor, materials and waste disposal. It consists of coated fibers that are used to isolate and concentrate analytes into a range of coating materials. After extraction, fibers are transferred to an analytical instrument for separation and quantification of the target analytes. A breakthrough in SPME for application in food matrices involves the development of molecularly-imprinted polymer (MIP) coatings to enhance method sensitivity and selectivity. The utilisation of task-specific MIPs as SPME coatings has demonstrated outstanding capabilities to overcome the limitations associated with low sorbent volumes, rendering enhanced selectivity and improved method selectivity of SPME (Souza-Silva *et al.*, 2015).

Derivatisation is a commonly used technique to augment chromatographic analysis and it is employed to permit analysis of compounds with inadequate volatility or stability as well as to improve chromatographic behavior or detectability. Derivatisation of wine constituents for example, is often used to modify non-volatile or highly polar chemical compounds not otherwise amenable to GC analysis.

Finally, spectroscopic methods applied for wine and grape analyses include a wide range of techniques, spanning atomic spectrometry methods such as atomic absorption spectroscopy (AAS) and inductively coupled plasma (ICP) and several molecular spectroscopy methods such as nuclear magnetic resonance spectrometry (NMR) and mass spectrometry (MS).

The focus of the first part of this chapter is specifically on the application of advanced instrumental techniques, including chromatography and spectroscopy, to the analysis of toxic components, microelements, aroma compounds and constituents of therapeutic significance in grape and wine.

## 1.2. Biogenic Amines

Biogenic amines (BAs) are nitrogenous low molecular weight organic bases that can have an aliphatic, aromatic or a heterocyclic structure and are receiving much attention in wine science because of their potential implication for human health (Silla Santos, 1996). The main BAs present in wines are histamine, tyramine, putrescine, cadaverine, 2-phenethylamine, agmatine and tryptamine mostly originating from the decarboxylation of their respective free precursor amino acids, through the action of substrate-specific microbial decarboxylases. Other amines, possibly present in wines, include the aliphatic volatile amines (methylamine, ethylamine and isoamylamine), that can be formed by the amination of non-nitrogen compounds, such as aldehydes and ketones (Bauza *et al.*, 1995), and the polyamines, spermine and spermidine, that can be produced from putrescine (1,4-diaminobutane), through methylation reactions involving *S*-adenosyl-methionine (Vincenzini *et al.*, 2016).

BA quantification in wine is still problematic due to their low concentration, the lack of chromophores of most BAs, the complexity of the sample matrix and the presence of potentially interfering substances. Many analytical methods have been developed to quantify these compounds in wines, including gas chromatography (Fernandes and Ferreira, 2000), capillary electrophoresis (Herrero *et al.*, 2010), enzymatic methods and immunoassays (Lange and Wittmann, 2002). Nowadays, HPLC is by far the most frequently used technique, due to its high resolution and sensitivity, especially when coupled to a fluorescence detector. As BAs do not show satisfactory absorption in the visible and ultraviolet range nor do they show fluorescence, chemical derivatisation is considered a necessary analytical step for their detection technology. The derivatisation methods can be mainly divided into two categories: pre-column (the derivatisation is carried out before the chromatographic separation) or post-column (the derivatisation is carried out after the chromatographic separation) derivatisation methods. The pre-column derivatisation technique is used more frequently than the post-column derivatisation because of providing more sensitive detection (Önal *et al.*, 2013). Several derivatisation reactions have been employed, as for example those using orto-phthaldialdehyde (OPA) (Vidal-Carou *et al.*, 2003; Kelly *et al.*, 2010; Arrieta and Prats-Moya, 2012), 9-fluorenylmethylchloroformate (FMOC), 2,4,6-trinitrobenzenesulfonic acid (TNBS), diethyl ethoxymethylenemalonate (DEMM), 4-fluoro-3-dinitro-fluomethylbenzene, dansylchloride (Dns-Cl) (Zotou *et al.*, 2003; Hernández-Borges *et al.*, 2007; Proestos *et al.*, 2008), dansyl chloride (Dabs-Cl) (Romero *et al.*, 2000), benzoyl chloride (Bnz-Cl) (Özdestan and Üren 2009), 1,2-naphtoquinone-4-sulphonate (NQS) (Hlabangana *et al.*, 2006), 6-aminoquinolyl-N-hydroxysuccinimidyl carbamate (Hernández-Orte *et al.*, 2006), 1-fluoro-2-nitro-1-(trifluoromethyl) benzene (Jastrzębska *et al.*, 2016). Currently, the International Organization of Vine and Wine (OIV) proposes derivatisation with OPA and

DEMM to determine BAs in musts and wines by HPLC using fluorimetric (OIV-MA-AS315-18) and spectrophotometric detection methods (OIV-MA-AS315-26).

Recent liquid chromatography combined with tandem mass spectrometry and ultra-performance liquid chromatography coupled to quadrupole-time of flight mass spectrometry have been shown to be very powerful techniques to increase the performance of BAs analysis (García-Villar *et al.*, 2009; Jia *et al.*, 2012), also without derivatisation (Millán *et al.*, 2007).

A fast and reliable HPLC method for the determination of 11 BAs in beverages has been performed by Preti *et al.*, 2015. After pre-column derivatisation with Dns-Cla C18 core-shell particle column has been employed and the BAs were identified and quantified in a total run time of 13 minutes with ultraviolet (UV) or fluorescence detector.

## 1.3. Ethyl Carbamate

Ethyl carbamate (EC), commonly called urethane, is the ethyl ester of carbamic acid. It is genotoxic and a multi-site carcinogen in experimental animals and probably carcinogen to humans (EFSA, 2007). The EC occurs in wine where it is thought to be formed from non-enzymatic reaction between ethanol and a compound containing a carbamoyl group, such as urea (produced from arginine breakdown by yeasts), citrulline and carbamoylphosphate (produced from arginine breakdown by lactic acid bacteria) (Vincenzini *et al.*, 2016). There are currently no harmonised maximum EC levels for table wine in the EU, but Canada and USA recommend maximum values of 30 and 15 $\mu gL^{-1}$, respectively (EFSA, 2007).

The determination presents problems owing to the low concentration of EC and to matrix interferences, The former problem can be solved by application of MS techniques, which provided sensitive determination at trace levels, while the latter can be solved by various sample clean-up procedures (Zhang and Zhang, 2008; Jiao *et al.*, 2014). Many methodological publications regarding EC analysis have summarised the precise quantification of EC concentrations and detailed general pretreatment steps, such as extraction and cleanup steps, detection systems, general analytical methods, and proficiency tests (Jiao *et al.*, 2014). Regarding the extraction steps, the most traditional method is liquid-liquid extraction, which requires a large amount of chlorinated toxic solvent (usually dichloromethane) and the use of intensive labor effort and prolonged time during the concentration step (Nobrega *et al.*, 2015). As an improved measure for extracting EC, solid-phase extraction provides considerable advantages over liquid-liquid extraction and it is applied in the standard method for EC determination in wine (AOAC method 994.07, AOAC, 2006), also adopted by OIV (method MA-AS315-04, OIV, 2013), and as reference method in the European Union (commission Regulation 1999). The AOAC method involves analysis by gas-chromatography coupled to mass spectrometry in selected ion monitoring (GC-MS-SIM) after a sample preparation procedure that implies addition of propyl carbamate as internal standard, cleanup through diatomaceous earth SPE columns, EC extraction by dichloromethane, and eluate concentration. Jagerdeo *et al.* (2002) described a SPE method for multidimensional GC-MS (MDGC-MS) determination of EC levels in wines that eliminates the use of dichloromethane. The utilisation of SPE apparently decreases the amount of solvent and labor required but is far from satisfactory, as it is always accompanied by background interference, poor reproducibility between cartridges, and high economic cost due to the inability to reutilise cartridges during extraction (Zhang and Zhang, 2008). Another powerful technique for extracting EC is solid-phase microextraction (SPME) that combines extraction, concentration and chromatographic injection into one-step with a carbowax/divinylbenzene (CW/DVB) fiber (Whiton and Zoecklein, 2002, Zhang and Zhang, 2008). This technique requires only a small amount of sample and shortens the time required for extraction, lowers the economic cost, and preserves the natural content of the sample (Jiao *et al.*, 2014).

Although these alternative preparations have advantages over the standard procedure, they have not been extensively adopted for EC analysis in wine and are not without problems. For instance, the CW/DVB fiber is no longer commercially available (Liu *et al.*, 2012), and the alcohol part in the sample may influence the SPME extraction yield (Lachenmeyer *et al.*, 2006). Furthermore, the method proposed by Jagerdeo *et al.* (2002) involves a previous time-consuming step for ethanol removal from wine by vacuum.

An improved sample preparation procedure by GC/MS/SIM was proposed by Nobrega *et al.* (2015). This method differs from AOAC reference procedure by the use as internal standard of deuterated ethyl carbamate (EC-d5), more similar to ethyl EC than propyl carbamate, the extraction by diethyl ether instead of more toxic dichloromethane and the concentration by vacuum automated parallel evaporation. Applicability of the method was tested by analysis of 5 wine samples. EC concentration ranged from 5.2 ± 0.2 to 29.4 ± 1.5 $\mu gL^{-1}$.

## 1.4. N-nitrosamines

N-nitrososamines (NAms) are highly active carcinogen that have been detected in red wine. The presence of these compounds could be attributed to the nitrosable compounds present in red wine and grape juice, such as phenolic compounds, biogenic amines and flavonoids among others. There is controversial information about the presence or the absence of these compounds in wine, which could be related to the analytical method used as well as on the origin and type of red wine analyzed (Lona-Ramirez *et al.*, 2016).

Recent techniques have been used to detect and quantify NAms in different types of matrices, but none of these techniques was applied in wine. Jurado-Sanchez and co-workers (2007) proposed a semi automatic method for the determination of seven NAms by gas chromatography with nitrogen-phosphorus detection following automatic solid-phase extraction, but no NAms were detected in red wine. Recently the head space solid phase micro-extraction technique coupled to gas chromatography-mass spectrometry (HS-SPME-GC-MS) was applied to quantify four NAms, N-nitrosodimethylamine (NDMA), N-nitrosodiethylamine (NDEA), N-nitrosopiperazine (NPIP), N-nitrosodibutylamine (NDBA), in red wines from different types of grapes, from different countries and having undergone different aging processes (Lona-Ramirez *et al.*, 2016) using a polydimethylsiloxane-divinylbenzene fiber. The mass detector was equipped with a positive pole ion, single quadrupole with electron impact ionisation (EI) source. The quantification was carried out with the application of the standard addition technique. The method was validated by calculating the linearity, limit of detection and quantification. Two of the four NAms analyzed, NDMA and NDBA, were found to be present in red wines.

## 1.5. Resveratrol and Other Phenolic Compounds

Resveratrol is a member of the stilbene family of phenolic compounds. Two isomers (*trans*-resveratrol and *cis*-resveratrol), and their glucosylated derivatives (*trans*-piceid and *cis*-piceid, respectively) have been detected in wine (Hashim *et al.*, 2013, Rodríguez-Cabo *et al.*, 2014).

Phenolic compounds are molecules consisting of a phenyl ring backbone with a hydroxyl group or other substituents. They are the molecules which are naturally derived from grapes and microbes. In wine, phenolic compounds are classified as non-flavonoid, such as hydroxybenzoic acids, hydroxycinnamic acids, and stilbenes, and flavonoid compounds, such as anthocyanins, flavan-3-ols and flavonols. Despite a not negligible presence of non-flavonoid compounds, flavonoids make up a significant portion of phenols in wines (Texeira *et al.*, 2013). In wines, flavonoid contributes significantly to sensorial quality. Indeed, they are the main compounds responsible of colour, flavour, texture and astringency. Phenols contained within the skin, and flesh of grapes are extracted into wine during the vinification processes. The amount of phenols extracted during vinification is influenced by many factors, including temperature, length of skin contact, mixing, type of fermentation vessel, ethanol concentrations, $SO_2$, yeast strain, pH, and pectolytic enzymes. Extraction is ultimately limited by the amount present in the fruit, and this varies with cultivar, vintage, macro-and micro-climatic conditions, and vinification process (Romboli *et al.*, 2017). Besides, anthocyanins fingerprints have been widely utilised for the classification and differentiation of grape cultivars and monovarietal wines (Mangani *et al.*, 2011).

A recent interest in these substances has been stimulated by abundant evidence of their beneficial role in human health, such as anticarcinogenic, anti-inflammatory, and antimicrobial activity, and many of these biological functions have been attributed to their free radical scavenging and antioxidant activity. Reverse-phase HPLC with detection by UV-vis absorption with a diode array detector (DAD) represents the most popular technique. Enhancing selectivity and sensitivity for the determination of certain wine phenolics requires the application of different detection techniques, such as fluorometry, electrochemistry, chemiluminescence, and/or MS coupled with different techniques: electrospray ionisation (ESI), matrix-assisted laser desorption/ionisation (MALDI), and atmospheric pressure chemical ionisation (APCI) (Medić-Šarić 2011). Due to the complexity of wine phenolics, extensive pre-fractionation is also often employed (Lorrain 2013). In the last few years, influential developments in HPLC, in terms of the use of smaller particle-packed columns, elevated temperature and multidimensional separations, have also been exploited for these compounds (Kalili and de Villiers, 2011; de Villiers, 2012; Lorrain *et al.*, 2013).

Silva *et al.* (2011) described an ultra-fast, efficient and high throughput analytical method based on UPLC equipped with a PDA detection system using a 50 mm column packed with 1.7 μm particles. After a polyphenol extraction from wines by SPE on a new hydrophilic-lipophilic balanced sorbent (N-vinylpyrrolidone-divinylbenzene copolymer), the method managed to separate and analyzed, in five minutes, fifteen bioactive phenolic compounds mainly belonging to flavonols, flavan-3-ols and phenolic acids, including *trans*-resveratrol.

The stilbene content of wine has received extensive attention in literature due to the beneficial biological activity ascribed to this class of compounds (Di Donna *et al.*, 2017; Hashim *et al.*, 2013; Rodríguez-Cabo *et al.*, 2014; Cacho *et al.*, 2013; Airado-Rodríguez *et al.*, 2010). Hashim *et al.* (2013) reported the application of reusable molecularly imprinted Polymers (MIPs), synthetic materials designed to have a predetermined selectivity for defined molecular targets, for the selective and robust SPE and rapid analysis of *trans*-resveratrol. Optimisation of the molecularly imprinted solid-phase extraction (MISPE) protocol resulted in the significant enrichment in *trans*-resveratrol and several structurally related polyphenols. The metabolites were subsequently identified by capillary RP-HPLC and electrospray ionisation mass spectrometry (ESI-MS/MS) and μLC-ESI ion trap MS/MS methods.

Regarding anthocyanin, the International Organization of Vine and Wine (OIV) proposes analysis by direct separation by RP-HPLC using spectrophotometric detection methods (OIV-MA-AS315-11).

## 1.6. Mycotoxins

Mycotoxins are toxins produced by fungi as secondary metabolites that may occasionally contaminate bottled wines (Bolton *et al.*, 2017). They represent a chronic health risk since prolonged exposure to them through diet has been linked to a range of adverse health effects (Hussein and Brasel 2001).They are colourless, odorless, and tasteless, and though primary produced in the vineyard on grapes, they have the ability to remain toxic throughout the winemaking process. More than 300 mycotoxins have been reported; currently, OchratoxinA (OTA), is the most common mycotoxin found in wine. It is produced on grapes by *Aspergillus carbonarius* and, to a lesser extent, *Aspergillus niger*. The European Union has regulated the maximum limit of OTA in wine at 2.0 μg/L (EC 123/2005, 26 Jan 2005).

As low as ppb's concentrations are usually involved, very sensitive analytical methods for mycotoxins are needed. Conventional analytical methods are mainly based on separative instrumental techniques (CE, HPLC, GC) (Almeda *et al.*, 2008; Soleas *et al.*, 2001), immunological methods (ELISAs) (Soares *et al.*, 2014; Vidal *et al.*, 2013) and sensor-based systems (Turner *et al.*, 2015). Among these above-mentioned methods, the most commonly used technique is based on HPLC coupled with FLD, the native fluorescence of OTA favouring the development of very sensitive methods. Before the chromatographic separation, sample preparation step involving extraction, purification and concentration of the extract must be carried out to remove the major interferences present in the sample and to pre-concentrate the analytes in order to achieve the desired sensitivity (Lee *et al.*, 2012).

The most widely used clean-up and pre-concentration methods for OTA determination are liquid-liquid extraction (LLE) or solid-phase extraction (SPE) (Lee *et al.*, 2012). A number of SPE columns are commercially available; the monoclonal antibody base immunoaffinity columns (IACs) are the most commonly used (Lee *et al.*, 2012; Fabiani *et al.*, 2010). The main advantage of these columns is that OTA is bound specifically to the antibody and the matrix interferences can be removed nearly completely. Furthermore, IACs give an optimal performance in terms of precision and accuracy within a wide range of concentrations and they reduce the use of dangerous solvents. However, these columns present several problems such as the relatively high cost; limited capacity cannot be reused, limited lifetime and in some cases, lack of specificity was observed due to cross–reaction with ochratoxin C (Aresta *et al.*, 2006). Attempts to replace the biological recognition element by a synthetic counterpart e.g. by OTA molecularly imprinted polymers (MIP), have been proposed (Lee *et al.*, 2012; Cao *et al.*, 2013).

Currently, the method recommended for OTA determination in wines and beer (European Standard prEN 14133, OIV MA-AS315-10, OIV 2016) uses IACs columns to clean-up OTA after dilution of the samples in aqueous solution of polyethylene glycol and $NaHCO_3$ and the samples are analyzed by HPLC with fluorescent detection (excitation wavelength at 333 nm, emitting wavelength at 460 nm). A fast separation of OTA in red wines was obtained by Mao *et al.* (2013) utilising a core-shell column (C18, 2.6 µm, 100 Å). Under optimised condition, OTA was separated in less than 5 minutes by HPLC-FLD in isocratic conditions with mobile phase constituted by acetonitrile/water (50/50 v/v, both acidified with 1% volume formic acid) and a flow rate of 0.66 mL/min after IAC cleanup.

A fast multi method was developed and optimised by Pizzutti *et al.* (2014) for the analysis of 36 mycotoxins in wine, based on an acetonitrile extraction, followed by a partitioning and a subsequent drying step with magnesium sulfate, and detection using UPLC-MS/MS (ESI positive mode). No cleanup was necessary, because matrix effects were kept at an acceptable level.

## 1.7. Microelements (Especially Cu, As, Zn, Fe)

Wine typically contain macro-elements such as K, Ca, Na and Mg (concentration >10 mg/L), micro-elements such as Fe, Cu, Zn Mn, Pb (concentration >10 µg/L), and ultra micro-elements such as Cr, As, Cd, and Ni (concentration <10 µg/L) (Geana *et al.*, 2013; Grindlay *et al.*, 2011). Elements in wine can be classified into two groups: endogenous and exogenous. Endogenous elements are related with the type of soil on which vines are grown, the grape variety and the climatic conditions during their growth. Exogenous elements derive from external impurities that reach wine during growth of grapes or at different stages during winemaking (from harvesting to bottling and cellaring) (Grindlay *et al.*, 2011; Pohl, 2007).

The micro-element composition of grapes and wines is important from several standpoints. Firstly, some elements are regulated, with most countries following the maximum legal limits established by the International Organization of Vine and Wine (OIV-MA-C1-01). Examples for these are As, B, Br, Cd, Pb, and Zn, for which maximum legal limits in the upper parts per million (mg/L) to mid parts per billion (µg/L) have been defined. Secondly, many elements influence the quality of vine and wine, as they are macro-, micro-, and trace nutrients for the vine plant or are used in agrochemicals (Hopfer *et al.*, Spectroscopy 2014). Additionally, some trace elements play a relevant role in winemaking, for example, Zn is essential at low concentrations for the correct development of alcoholic fermentation, while Cu, Fe and Mn have organoleptic effects at increased levels (Cozzolino, 2008; Ronkainen, 2016); besides some elements have detrimental effects on wine stability and need to be closely monitored during wine making (Pohl, 2007). Examples for this, are Cu and Fe, which can act as oxidation catalysts, and can also cause haze formation in wines, similar to Zn or Al. Lastly, elemental fingerprints can provide important information on the geographical origin in which the grapes were grown due to the direct relationship with soil composition as well as wine processing and storage conditions (Ebeler Chapter 1, Pyrzynska, 2007). For the determination of mineral content of wines, atomic spectrometric techniques are the most often used. Application of flame atomic absorption spectroscopy (AAS) and electrothermal AAS for metal analysis in wine have been reported (de Villiers *et al.*, 2012; Pohl, 2007). Recently inductively coupled plasma mass spectrometry (ICPMS) and inductively coupled plasma atomic emission spectrometry (ICPAES) have gained popularity as rapid and sensitive approaches for simultaneous screening of a large number of elements (Grindlay *et al.*, 2011). Hoepfer *et al.* (2015) determined a total of 63 elements with ICP-MS in 65 red wines from grapes harvested in five different vineyards within 40 miles of each other and processed in at least two different wineries. Based on the multi-elemental pattern, they were able to classify the wines according to their vineyard origin, their processing winery, as well as the combined effect of both origin and processing (Hoepfer *et al.*, 2015).

Metals may exist in wines as free ions, as complexes with organic acids as well as with large molecules of pectic polysaccharides, peptides, proteins and polyphenols (Pyrzynska, 2007). The chemical speciation and fractionation analysis of metals are gaining interest because metals bioavailability and toxicity depend on the chemical forms in which they are present. Pyrzynska provided (2007) a review of the main developments in chemical speciation and fractionation of metals in wine samples.

## 1.8. Aroma Components

The volatile fraction of wines determines largely its aroma, which is one of the most important characteristics influencing wine quality and the consumer acceptance. The flavour of a wine is extremely complex, and is due to the presence of several classes of compounds, such as alcohols, terpenes, hydrocarbons, ketones, esters, acids, aldehydes ethers, sulfur nitrogen compounds and lactones. More than 1000 aroma compounds have been identified, covering a wide range of polarities and volatilities and spanning few orders of magnitude in concentrations. All these compounds are responsible for the so-called "bouquet" of the wine on sniffing the head space from a glass, and the odour/aroma component of the overall flavour perceived on drinking (Cincotta *et al.*, 2015). Several factors influence the wine aroma: grape variety, grape ripeness, climate, soil, fermentation conditions, yeast and bacteria strains, production process, and aging (Barros *et al.*, 2012). Furthermore, many aroma compounds are chemically very unstable and can be easily oxidised or thermo degraded (Andujar-Ortiz *et al.*, 2009).

Analysis of the base aroma compounds comprising the so-called major volatiles, which include the principal fermentation derived esters, alcohols and acids is routinely performed using generic GC methods combined with flame ionisation detector (FID) or MS (de Villiers *et al.*, 2012). On the other hand, the analysis of specific minor volatile compounds requires dedicated methods with selective extraction and pre-concentration steps due to the complexity of wine matrix and relatively low concentrations and selective detection strategies.

For example, haloanisoles (e.g, (2,4,6-trichloroanisole, 2,3,4,6-tetrachloroanisole, 2,3,4,5,6-pentachloanisole, 2,4,6-tribromoanisole) are responsible for musty off-aromas of wines described as cork-taint. Hjelmeland *et al.* (2012) presented a method for the simultaneous analysis of four haloanisoles in wine by HS-SPME coupled to a GC-triple quadrupole MS. The method, fully automated, required no sample preparation other than the addition of internal standards, and was high throughput, with a 10 minutes extraction time and a 5 minutes incubation prior to extraction. Limits of detection and quantification were mainly in the sub-ng/L range.

The volatile phenols 4-ethyl phenol, 4-ethyl guaiacol,4-vinyl phenol an 4-vinyl guaiacol are mainly associated to *Brettanomyces* spoilage. These compounds are one of the most significant problems in modern winemaking, as they can give the wine "off-flavours", described as phenolic, medicinal, pharmaceutical smoky and clove-like flavours (Nicolini *et al.*, 2007). HPLC is a frequently used analytical technique; as an example a rapid method was established by Nicolini *et al.* (2007) using HPLC coupled with a fluorimeter detector. This method did not require sample preparation and it carried out chromatographic separation in less than 5 minutes with a detection limit of 4 $\mu gL^{-1}$. However, the most frequent approach to measuring volatile phenolsis GC-MS. In a recent work, Zhou *et al.* (2015) developed an ethylene glycol-polymethyl siloxane based stir bar sorptive extraction (EG/PDMS) coupled with GC-MS method, with quantification limits lower than 3 µg/L. Stir bar sorptive extraction employs a magnetic stir bar coated with a thick layer polymer for volatile extraction which increase phase volume and minimise the absorptive competition.

An automated HS-SPME combined with GC-ion trap/MS was developed by Barros *et al.* (2012) in order to quantify a large number of volatile compounds such as alcohols, esters, norisoprenoids, and terpenes. The procedure was optimised for SPME fiber selection, pre-incubation temperature and time, extraction temperature and time, and salt addition. The method allowed the identification of 64 volatile compounds, besides for 20 compounds considered as important aromatic contributors for the aroma of white wines calibration and validation were also performed.

Furthermore, the volatilome has been used to derive classification models for the identification of individual cultivars (de Villiers *et al.*, 2012; Villano *et al.*, 2017). In particular, univariate and multivariate principal component analysis-discriminant analysis statistics applied to the combined SPME-GC and $^{1}H$ NMR data allowed a chemometric discrimination of 270 wines from Galicia (Spain) according to the type of grape and identifying, in part, the geographical subzone of origin (Martin-Pastor *et al.*, 2016).

Considering that, the release of some impact compounds in aroma wine, depends onthe action of simple enzymatic steps promoted by microorganism, the quantification of such enzyme activities can be a useful tool. Indeed, very efficient fluorimetric methods have been developed to perform assays of esterase and β-glucosidase activities from yeasts and lactic acid bacteria of enological interest (Fia *et al.*, 2005; Rosi *et al.*, 2007). Commercial enzyme preparations, widely used in winemaking, were studied by the same methods (Fia *et al.*, 2016; Fia *et al.*, 2014). Esterase activity was assayed by measuring the amount of 5(6)-carboxyfluorescein released from 5(6)-carboxyfluorescein diacetate used as substrate after incubation at 37°C for 5 min., while β-glucosidase activity was assayed by measuring the amount of

4-Methylumbelliferone liberated from 4-Methylumbelliferyl-β-D-glucuronide after incubation at 37°C for 5 min.. Enzyme assays were carried out using black 96-well microtitre plates with flat transparent bottom by their exposition to the long-wavelength UV light of a transilluminator. The images were acquired with Gel Doc 2000 System and analysed by Quantity One v.4.3.0 (Bio-Rad) software.

## 2. Microbial and Biomolecular Analysis

### 2.1. Introduction

The yeast and bacterial microbiota occurring in grape must and during alcoholic or malolactic fermentation have been investigated by several Authors (Baleiras Couto *et al.*, 2005; Fernandez *et al.*, 1999; Gangaand Martinez, 2004; Gonzalez *et al.*, 2007; Hierro *et al.*, 2006b; Lopandic *et al.*, 2008; Spano and Torriani, 2017) and different techniques have been used with this purpose (Ivey and Phister, 2011; Zott *et al.*, 2010; Longin *et al.*, 2017). Before the advent of molecular biology, microbial population size and diversity were analyzed using methods mainly based on growth of cultivable microbiota on nutrient media (Lafon-Lafourcade and Joyeux, 1979). Molecular biology has brought forth significant new advances in microbiological analysis of grape must and wine (Mils *et al.*, 2008). In particular, traditional methods for microbial identification (morphological and/or physiological criteria) have been almost completely replaced by ribosomal DNA – based methods. Moreover, in the past few years, successful culture-independent methods, such as Polymerase Chain Reaction (PCR) real Time, PCR-DGGE, in situ hybridisation, flow cytometry with fluorescent antibodies, or microbial genome sequences have been used to describe the microbial ecology of musts and wines (Capece *et al.*, 2003; Doare-Lebrun *et al.*, 2006; Rodriguez and Thornton, 2008; Xufreetal, 2006; Longin *et al.*, 2017). The following paragraphs describe not only the various methods to quantify, identify and characterise the wine microorganisms, but also to detect their metabolic capabilities of technological and healthy interest.

### 2.2. Microbial Techniques to Quantify Wine Microorganisms

Monitoring of the microbial changes occurring during the winemaking is fundamental to carry out a process under control. In fact, this information is necessary to not only promote and guide the growth of yeasts during alcoholic fermentation and of lactic acid bacteria during malolactic fermentation, but also to ensure the proper aging and stability of the wine before bottling and storage (Delfini and Formica, 2001). Microbial enumeration in wine can involve two kind of methods: “indirect” and “direct”. The indirect methods, such as plate count or most probable number, do not enumerate the original cells in the sample but their progeny, as enriched in a specific medium. On the contrary, the direct methods allow the counting of microbial cells directly into must and wine using microscopy techniques or the more complex flow cytometry. The following paragraphs briefly describe the main methods for counting microorganisms in wine.

#### *2.2.1. Microscopic Count, Plate Count, Most Probable Number*

Classical microbiological techniques (microscopic technique, plate count, most probable number) to monitor wine fermentation and to detect undesirable microorganisms are well described in the resolution OIV-Oeno 206/2010. The resolution also reports, for each technique, a description of the principles, reagents and materials, installations and equipment, sampling procedures, and finally quality tests. In Table 1, a schematic description of various techniques shown in the resolution 206/2010, is reported.

#### *2.2.2. Direct Epifluorescence Technique (DEFT)*

Direct epifluorescence technique (DEFT) is a particularly successful direct analysis when applied to aging wines or to control filtered wines after bottling. This technique exploits the microbial-based cleavage of a fluorescent substrate, which enables direct counting of viable cells through a fluorescent microscope. Therefore, DEFT enables to quantify viable cells of both bacteria and yeast in aging wine, including those viable but non-cultivable (VBNC) that cannot be quantified by the culture dependent methods (Millet and Lonvoud-Funel, 2000; du Oit *et al.*, 2005; Divol and Lonvoud-Funel, 2005). To analyze aging wine,

**Table 1.** Schematic Description and Aim of Various Techniques Reported in the Resolution 206/2010

| *Technique* | *Description* | *Aim* |
|---|---|---|
| Microscopic techniques (culture independent technique) | Microscopic examination of liquids or deposits | To detect and differentiate the yeasts from the bacteria in terms of their size and shape. Microscopic observation cannot distinguish between viable and non-viable microorganisms. |
| | Gram staining for the differentiation of bacteria isolated from colonies | To differentiate between lactic acid bacteria (Gram positive)and acetic bacteria (Gram negative) and also to observe their morphology |
| | Catalase Test for the differentiation of bacteria isolated from colonies | To differentiate between acetic and lactic acid bacteria. Acetic bacteria have a positive reaction. Lactic acid bacteria give a negative response |
| | Yeast cell count by haemocytometry | To determinate yeast cell population in fermenting musts, wines, and active dry yeasts (starter cultures) |
| | Yeast cell count after methylene blue staining of yeast cells | To allow a rapid estimation of the percentage of viable yeast cells,which are not stained, while dead cells are blue-stained |
| Plate count (culture dependent technique) | Enumeration of viable and cultivable yeasts, moulds and lactic or acetic bacteria in musts, concentrated musts, partially fermented musts, wines (including sparkling wines) during their manufacture and after bottling, by counting the colonies grown on solid differential and/or selective media after suitable incubation | To control the winemaking process and prevent microbial spoilage of musts or wines |
| Most probable number (culture dependent technique) | Estimation of the number of viable microorganisms in selective liquid media, starting from the principle of its normal distribution in the sample | To evaluate the number of viable and cultivable microorganisms in wines having high contents of solid particles in suspension and/or high incidence of plugging |

DEFT methods provides that a homogenised sample is filtered through a 0.2 μm pore polycarbonate filter where the wine microorganisms are concentrated and remained on. Then the microorganisms are stained with fluorochromes and the filter examined under a fluorescence microscope to count viable yeast and/or bacteria cells. Siegrist *et al.* (2015) report an exhaustive list of fluorochromes for rapid detection not only viable, but also damage and death microbial cells. DEFT is particularly useful for rapid detection of spoilage yeasts such as *Brettanomyces* spp. which may negatively affect wine quality producing off-flavours and which is more difficult to be revealed by traditional plate count method because of long time of incubation and the possible presence of cells in VBNC state (Granchi *et al.*, 2006). DEFT, with some modifications, can be also used to quantify viable yeasts during alcoholic fermentation. This procedure, named Thoma-Epifluorescence-Microscopy-Technique (TEMT), is based on the combined use of Thoma-counting chamber and epifluorescence microscopy (Granchi *et al.*, 2006). TEMT is a suitable tool for monitoring yeast populations during the winemaking process. Moreover, the rapid assessment of inadequate *Saccharomyces cerevisiae* cell concentrations during alcoholic fermentation, allowing timely interventions, can contribute in preventing stuck fermentations.

### 2.2.3. Flow Cytometry

Flow cytometry (FCM) can allow the rapid acquisition of multi-parametric data, such as cell numbers, as well as shape, size, and cell viability regarding individual cells suspended in a fluid stream (Díaz *et al.*, 2010; Longobardi-Givan, 2001; O'Neill *et al.*, 2013; Longin *et al.*, 2017). A flow cytometer is mainly composed of a fluidic, an optical, and an electronic part (Longin *et al.*, 2017). The fluidic component is composed of a flow chamber that separates and aligns the cells by passing them through the light source with the refraction or scattering of light (laser, arc lamp or light emitting diode) at all angles. The light emitted from the cells after they are irradiated in the flow chamber is directed to appropriate detectors. The magnitude of "forward scatter" (FSC: the amount of light scattered in the forward direction when the light strikes the cell) is roughly proportional to the size of the cell. In addition to this, a system of focusing lens is used to direct the light rays to a set of filters separating the various wavelengths present. If the cells are labelled with fluorochromes, they generate a fluorescence signal crossing the path of light. The fluorescent light is then directed to an appropriate detector. Finally, the light, which passes through an optical system, will be converted into electronic signals generated by photodiodes and photomultiplier tubes and collected to enable data acquisition and analysis (Longin *et al.*, 2017).

Most flow cytometry protocols applied in enology concern detection (presence/absence) and/or enumeration of viable microorganisms. Using special dyes coupled to flow cytometry, physiological analysis of wine yeasts and bacteria can be also performed. Even if only viable cells of yeasts and bacteria are able to perform alcoholic and malolactic fermentations respectively, they could have a weak metabolic activity that can negatively affect the fermentation rate. For this reason it could be important to measure also the microbial "vitality", which reflects metabolic activity of the cells. Vitality dyes, such as Fluorescein Di-Acetate (FDA), point out life essential functions such as enzymatic activity. In particular, FDA staining highlights the metabolic activity of cells through esterase activity. This dye has provided excellent results by measuring the cell vitality of *S. cerevisiae* during alcoholic fermentation in synthetic wine, in grape must, in red and white wines (Malacrino *et al.*, 2001; Salma *et al.*, 2013; Malacrino *et al.*, 2001; Gerbaux and Thomas, 2009). On the contrary, viability dyes determine if cells are in a physiological state sufficient to ensure their survival,for example measuring the membrane integrity in terms of permeability. Propidium Iodide (PI) is a dye frequently used with this aim (Delobel *et al.*, 2012). Finally, comparing the results of cell viability obtained with FCM and culture dependent methods such as plate counts, it is possible quantify for difference the VBNC cells of yeasts (Andorra *et al.*, 2011; Divol and Lonvaud-Funel, 2005; Serpaggi *et al.*, 2012; Salma *et al.*, 2013) and bacteria (Herrerp *et al.*, 2006; Quiròs *et al.*, 2009; Oliver, 2005) in wine. Table 2 reports the dyes used in FCM to monitor enological microorganisms.

## 2.3. Molecular Techniques to Identify and Characterise Wine Microorganisms

This section will focus on molecular techniques used to identify and characterise yeasts, lactic acid bacteria and acetic bacteria obtained from must and wine after an enrichment phase in plates containing specific growth media. These molecular methods can be grouped in three categories: those that are aimed to identify species, those that are suitable to differentiate strains of the same species, and finally those that highlight special enzymatic abilities of the strains assayed.

### 2.3.1. Species Identification and Strain Characterisation

*Yeasts*

One of the most significant systems to identify yeast species is the sequencing of the ribosomal genes, followed by the comparison of the sequences experimentally obtained with those deposited in special database (Mills *et al.*, 2008). The 5.8S, 18S, and 26S ribosomal genes are grouped in tandem to form transcription units that are repeated 100-200 times throughout the genome of the yeasts. Each transcription unit contains the internal transcribed spacer (ITS), while non-transcribed spacers (NTSs) separate the coding regions. The sequences of 5.8S, 18S, and 26S ribosomal genes and the ITS and NTS spacers can be

**Table 2.** Dyes Used in FCM and Related Function to Monitor Enological Microorganisms (modified from Longin *et al.*, 2017)

| *Dyes* | *Function* | *Monitored microorganisms* | *References* |
|---|---|---|---|
| Fluorescein diacetate (FDA) | VBNC State (Flow cytometry versus plate count) | Spoilage yeasts | Serpaggi *et al.*, 2012 |
| | Vitality | Alcoholic fermentation microorganisms | Malacrinò *et al.*, 2001<br>Salma *et al.*, 2013<br>Gerbaux and Thomas, 2009 |
| | | Malolactic fermentation microorganisms | Boiux and Ghorbal, 2013<br>Salma *et al.*, 2012<br>Malacrinò *et al.*, 2001 |
| Carboxyfluorescein diacetate (cFDA) | | Alcoholic fermentation microorganisms | Bouix and Leveau, 2001; Bouchez *et al.*, 2004 |
| 2-chloro-4-(2,3-dihydro-3-methyl-(benzo-1,3-thiazol-2-yl)-methylidene)-1-phenylquinolinium iodide (FUN-1) | | Alcoholic fermentation microorganisms | Salma *et al.*, 2013 |
| Propidium iodide (PI) | Viability (membrane integrity) | Microorganisms of sediments | Gerbaux and Thomas, 2009 |
| | | Alcoholic fermentation microorganisms | Landolfo *et al.*, 2008<br>Mannazzu *et al.*, 2008<br>Branco *et al.*, 2012<br>Delobel *et al.*, 2012<br>Chaney *et al.*, 2006<br>Farthing *et al.*, 2007 |
| | | Malolactic fermentation microorganisms | Bouix and Ghorbal, 2013<br>Salma *et al.*, 2012 |
| Chemchrom V6 and Propidium iodide (CV6/PI) | VBNC State (Flow cytometry versus plate count) | Malolactic fermentation microorganisms | Herrero *et al.*, 2006 |
| | Vitality and Viability | Alcoholic fermentation microorganisms | Herrero *et al.*, 2006 |
| | | Malolactic fermentation microorganisms | Da Silveira *et al.*, 2002<br>Bouix and Ghorbal, 2013 |
| Carboxyfluorescein diacetate and Propidium iodide (cFDA/PI) | | Alcoholic fermentation microorganisms | Monthèard *et al.*, 2012 |

used to identify yeasts species because of their conservation and concerted evolution (Fernàndez-Espinar *et al.*, 2011). In other words, the sequence similarity between repeated transcription units within a given species is greater than between units belonging to different species (Fernàndez-Espinar *et al.*, 2011). This sequence similarity within the species makes the ribosomal genes a powerful tool with which to identify yeasts. In particular, the D1 and D2 regions at the 5′ end of the genes encoding the 26S (Kurtzman and Robnett 1998) and 18S (James *et al.*, 1997) ribosomal subunits are the two most commonly used regions to identify yeasts also thanks to a wide availability of deposited sequences in DNA databases. The sequencing of D1/D2 region allows identifying the yeasts species when the homology of the sequences experimentally obtained with those deposited in the database is greater than 99% (Kurtzman and Robnett, 1998). The sequence homology is performed by using the program WU-BLAST2 (http://www.ebi.ac.uk/Blas2/index.html).

Unfortunately, sequencing of ribosomal genes (or portions thereof) are expensive and time consuming, two aspects that make this method inadequate for large-scale ecological studies (Mils *et al.*, 2008). The rRNA gene sequence analysis combined with PCR methods have enabled the rapid identification of species yeast isolated from wine. A common approach is to isolate yeasts based on colony or microscopic morphology, and then to perform the identification, by rRNA gene sequencing, of only the isolates representatives of each morphology.

Another economical method to identify yeasts is the ITS-Restriction Fragment Length Polymorphism (RFLP) (Mils *et al.*, 2008). This method consists in PCR amplification of rDNA regions followed by restriction analysis of the amplified products (Guillamon *et al.*, 1998; Esteve-Zarzoso *et al.*, 1999; Granchi *et al.*, 1999). The fragments generated are separated by electrophoresis in agarose gel and their size is compared to databases occurring in literature to recognise the species. Despite rITS-RFLP is not as discriminatory as 26S rRNA gene sequences (Arias *et al.*, 2002), several authors successfully used this method to describe the yeast ecology during wine fermentations (Fernandez *et al.*, 1999; Granchi *et al.*, 1999; Pramateftaki *et al.*, 2000; Esteve-Zarziso *et al.*, 2001; Jemec *et al.*, 2001; Raspor *et al.*, 2002; Romancino *et al.*, 2008).

The most commonly used molecular methods to characterise yeast strains are the restriction analysis of mitochondrial DNA (mtDNA) and Polymerase Chain Reaction (PCR)-based methods such as Random Amplification of Polymorphic DNA (RAPD), PCR-Analysis of Repetitive Genomic DNA (Microsatellites and Minisatellites), Amplification of d Sequences, and Amplified Fragment Length Polymorphism (AFLP). (Fernàndez-Espinar *et al.*, 2011). Restriction analysis of mtDNA exploit the high polymorphism of mtDNA to differentiate *S. cerevisiae* strains. The DNA is extracted and digested by restriction enzymes providing specific patterns for each strain.Several Authors used this technique to describe wine yeast biodiversity and ecology (Fernàndez-Gonzàlez *et al.*, 2001; Torija *et al.*, 2001; Beltràn *et al.*, 2002; Lòpes *et al.*, 2002; Sabatè *et al.*, 2002; Granchi *et al.*, 2003; Torija *et al.*, 2003; Esteve-Zarzoso *et al.*, 2001; Martìnez *et al.*, 2004; Lòpes *et al.*, 2006; Gonzàlez *et al.*, 2007; Lòpes *et al.*, 2007; Capece *et al.*, 2016).With the same purpose, PCR-based methods have been developed to detect DNA polymorphisms without using restriction enzymes. All of these techniques use oligonucleotide primers, which bind to target sequences of the DNA. The target sequences are amplified by PCR and the amplification products are visualised in agarose gels. Thanks to the strain specific nature of these amplification products, different strains show different profiles. The two most frequently PCR-based techniques used to differentiate yeast strains are the RAPD (Lopandic *et al.*, 2008; Urso *et al.*, 2008; Tofalo *et al.*, 2009) and microsatellite analysis (Caruso *et al.*, 2002; Capece *et al.*, 2003; Howell *et al.*, 2004; Ayoub *et al.*, 2006; Capece *et al.*, 2016). Finally, the amplification of d sequences and intron splice sites have been used to differentiate between wine strains belonging to the species *S. cerevisiae* (Pramateftaki *et al.*, 2000; Fernàndez-Espinar *et al.*, 2001; Lòpes *et al.*, 2002; Ciani *et al.*, 2004; Le Jeune *et al.*, 2006). To conclude, another approach have been used to differentiate wine yeast strains: whole or sub-genomic analysis trough pulse field gel electrophoresis (PFGE) In PFGE, two transverse electric fields are alternated forcing the chromosomes to continually change the direction of their migration. Consequently, these large fragments of DNA can be separated in the agarose gel matrix. This method of karyotype analysis has been demonstrated to be highly efficient for the differentiation of *S. cerevisiae* strains and numerous studies have used karyotype analysis to characterise wine strains of *S. cerevisiae* (Marinez *et al.*, 2004; Rodriguèz *et al.*, 2004; Schuller *et al.*, 2004; Naumov *et al.*, 2002).

*Bacteria*

Various molecular techniques are available to identify bacteria (lactic acid bacteria and acetic bacteria) from wine after an enrichment phase in plate containing selective media. As already shown for yeasts, also for bacteria a common approach is to randomly withdraw from the plates a significant number of isolates and identify them by molecular methods. The most significant method is certainly the direct sequencing of the 16S and 23S rDNA gene (Cloe *et al.*, 2005; Yamada and Yukphan, 2008), or parts of it, through comparison to existing database. As previously underlined, sequencing of ribosomal genes are expensive and time consuming. Therefore, rRNA gene RFLP approaches have been used also to identify bacteria of musts and wines. In particular, the most rapid and reliable method is the Amplified 16S rDNA restriction analysis (ARDRA) which provides for the amplification of the 16S rRNA gene with appropriate primers and the cut of amplicon with restriction enzymes. The digestion patterns obtained are compared with literature (Rodas *et al.*, 2003; Ventura *et al.*, 2000; Poblet *et al.*, 2000; Ruiz *et al.*, 2000; Ruiz *et al.*, 2000; Gonzàlez *et al.*, 2006b; Guillamòn and Mas, 2011).

Among the lactic acid bacteria (LAB) present in wine, *Oenococcus oeni* is the main species associated with malolactic fermentation, therefore PCR methods have been developed for rapid detection and identification of this species using specific primers (Zapparoli *et al.*, 1998) or multiplex RAPD-PCR (Reguant and Bordons, 2003). The RAPD analysis is also useful for distinguishing between different strains of the same species (Guerrini *et al.*, 2003; Rodas *et al.*, 2005; Solieri *et al.*, 2010; Marques *et al.*, 2011), as well as PFGE (Larisica *et al.*, 2008; Lòpez *et al.*, 2008; Pramateftak *et al.*, 2012; Zapparoli *et al.*, 2012). Although less frequently, other molecular techniques are used for typing strains of *O. oeni*: AFLP (Cappello *et al.*, 2014), Variable Number of Tandem Repeats (VNTR) (Garofalo *et al.*, 2015; Claisse and Lonvaud-Funel, 2014), and Multilocus Sequence Typing (MLST) (Garofalo *et al.*, 2015; Bridier *et al.*, 2010). Croz-Pio *et al.* (2017) demonstrated that the use of molecular fingerprints (RAPD and VNTR) is in many cases enough to discriminate *O. oeni* strains and to quantify diversity even if some isolates sharing the same genomic profiles can have different fermentative profiles, and *vice versa*. Consequently, the polyphasic approach, combining phenotypic (such as carbohydrates degradation) and genotypic profiles (molecular fingerprint), provides an optimum typing of *O. oeni* strains.

*Metabolic Capabilities of Microorganisms*

A reliable tool to identify microbial populations able to produce unwanted substances in wine is to use PCR screenings after an enrichment phase in plate containing selective media. This approach allow understanding the ecological distribution of specific genes (Mills *et al.*, 2008). Genes responsible for ropiness (Gindreau *et al.*, 2001), acrolein taint (Claisse and Lonvoud-Funel, 2001) and biogenic amines (BA) production (Landete *et al.*, 2005; Costantini *et al.*, 2006) have been identified, sequenced and used as target of PCR with properly designed primers. The gene occurs in the assayed isolate when an amplification product is obtained. This approach is particularly important in order to assess for example the potential risk of BA accumulation in wine. PCR screens to detect BA producing LAB are described in RESOLUTION OIV-OENO 449-2012. The methods described in this document consist in detecting LAB that have the genes of amino acids decarboxylases and/or agmatine deiminase using the primers listed in Table 3. Obviously, the results obtained with these methods are not able to predict the final BA concentrations in wine, but identify the risk of BA formation due to the presence of the decarboxylases and agmatine deiminase genes in the LAB population (Lucas *et al.*, 2008). Assessing the potential risk of a BA accumulation in wine at an early stage of the winemaking, these methods can assist in managing the fermentation process in order to reduce the BA formation (Vincenzini *et al.*, 2017).

## 2.4. Molecular Techniques for Direct Microbial Species or Strain Detection

The molecular methods previously described to identify or characterise yeasts and bacteria in wine involve enrichment techniques. Unfortunately, enrichment procedures are not only time consuming, but also underestimate the viable but not cultivable cells, which are unable to grow on plates but are still metabolically active. Alternatively, various molecular techniques have been developed for microbial species or strain detection directly in must and wine. The following paragraphs briefly describe these methods.

**Table 3.** Oligonucleotide Primers for the Detection of BA Producing LAB in Wine reported by RESOLUTION OIV-OENO 449-2012 (Vincenzini *et al.*, 2017)

| *Gene* | *Primers 5'→3' sequence* | *References* |
|---|---|---|
| Histamine decarboxylases | HDC3: GATGGTATTGTTTCKTATGA<br>HDC4: CAAACACCAGCATCTTC | Coton and Coton, 2005 |
| Tyramine decarboxylases | 41: CAYGTNGAYGCNGCNTAYGGNGG<br>42: AYRTANCCCATYTTRTGNGGRTC<br>TD5: CAAATGGAAGAAGAAGTAGG<br>TD2: ACATAGTCAACCATRTTGAA | Marcobal *et al.*, 2005<br>Coton *et al.*, 2004 |
| Putrescine decarboxylases | 4: ATNGARTTNAGTTCRCAYTTYTCNGG<br>15: GGTAYTGTTYGAYCGGAAWAAWCAYAA<br>OdF: CATCAAGGTGGACAATATTCCG<br>OdR: CCGTTCAACAACTTGTTTGGCA | Marcobal *et al.*, 2005<br>Granchi *et al.*, 2006 |

### *2.4.1. PCR – DGGE and TGGE*

Muyzer *et al.* (1993) have introduced for the first time denaturing gradient gel electrophoresis (DGGE) of PCR products to study microbial ecology. Unlike simple gel electrophoresis above reported, DGGE allows to separate DNA amplicons of the same length based on sequence differences. In fact, the separation of these DNA fragments is based on the different electrophoretic mobility of a partially melted double stranded DNA molecule in polyacrylamide gels containing a linear gradient of denaturing agents (usually a mixture of urea and formamide). DNA migration is retarded when the DNA strands dissociate and the strand dissociation at a specific concentration of denaturing agent depends on the basis sequence. A similar technique is the temperature gradient gel electrophoresis (TGGE), which is based on a linear temperature gradient for separation of DNA molecules. DNA bands in DGGE and TGGE methods can be visualised using ethidium bromide or SYBR Green I. Moreover, PCR fragments can be extracted directly from the gel and sequenced for species identification.

Both methods have been used for yeast identification in wine fermentations (Andorra *et al.*, 2008; Cocolin *et al.*, 2000; di Maro *et al.*, 2007; Prakitchaiwattana *et al.*, 2004; Renouf *et al.*, 2007; Stringini *et al.*, 2009; Urso *et al.*, 2008). In particular, Cocolin *et al.* (2000) developed primers that amplified a portion of D1-D2 region of rDNA. Using these primers in DGGE-PCR, the Authors described the ecology of various wine yeasts species (both *Saccharomyces* and non-*Saccharomyces* spp.) during alcoholic fermentation. PCR-DGGE is also useful to detect spoilage yeast such as *Brettanomyces bruxellensis* (Renouf *et al.*, 2006). Moreover, some Authors concluded that PCR-DGGE analysis is less sensitive than agar culture at least for determining the yeast ecology of grapes (Andorra *et al.*, 2008).

As far as LAB are concerned, PCR-DGGE in wine is complicated by inherent problems associated with primer specificity (Mills *et al.*, 2008). Indeed, several Authors observed how primers, commonly used for bacterial PCR-DGGE in other foods, amplify incorrectly DNAs of yeast, moulds, or plants (Lopez *et al.*, 2003; Dent *et al.*, 2004). Anyway, recently some authors have successfully used this technique to describe the ecology of *O. oeni* in wine (González-Arenzana *et al.*, 2017).

### *2.4.2. Direct PCR Approaches and Real Time PCR*

In recent years, Authors reported the use of real-time quantitative PCR (Q-PCR) to detect and quantify micro-organisms in different foods (Blackstone *et al.*, 2003; Bleve *et al.*, 2003; Hein *et al.*,2001). The main advantage of this method is the low detection level, theoretically as low as 1 cell/mL (Zot *et al.*, 2010). In Q-PCR the logarithmic amplification of a DNA target sequence is linked to the fluorescence of reporter molecules. These molecules binds double stand DNA and, when linked, they have excitation and emission at accurate wavelengths. This fluorescence, which is measured after each cycle of DNA amplification, may either be compared to an external standard curve (absolute quantification) or to an internal or external control sample (relative quantification). Relative quantification is primarily used to follow gene expression, while absolute quantification is the most common type of Q-PCR employed in wine ecology (Mills *et al.*, 2008). Several different reporter molecules exist, but the most common for detection of wine microorganisms is SYBER Green, which has an excitation wavelength of about 250 nm and an emission wavelength of about 497 nm.

In recent years, Q-PCR has been applied in many aspects of wine microbial ecology (Hierro *et al.*, 2007). It was used to quantify the total yeast population (Hierro *et al.*, 2006a; Salinas *et al.*, 2009), *Saccharomyces cerevisiae* (Martorell *et al.*, 2005), *Zygosaccharomyces bailii* (Rawsthorne and Phister, 2006), *Brettanomyces bruxellensis/Dekkera bruxellensis* (Delaherche *et al.*, 2004; Phister and Mills, 2003; Tessonniere *et al.*, 2009), lactic acid bacteria (Neely *et al.*, 2005), *O. oeni* (Pinzani *et al.*, 2004), acid acetic bacteria (Gonzalez *et al.*, 2006).

At present, the current use of Q-PCR to describe wine microbial ecology is limited (Mills *et al.*, 2008). Despite the large number of Q-PCR assays developed and the diffusion of these to service laboratories, few wineries use this approach to keep under control the wine fermentations. It is an exception the quantification of *Brettanomyces bruxellensis/Dekkera bruxellensis* by Q-PCR. Indeed, the enumerations of this yeast required from 5 to 10 days with conventional analysis by plate counts (Phister and Mills, 2003), while only 1-2 hours with Q-PCR (Mills *et al.*, 2008). This time difference provides winemakers the possibility of intervening before the damage caused by this yeast occurs. Another exception is represented by the rapid detection and quantification of BA producing LAB in wine with Q-PCR. Indeed, in the RESOLUTION OIV-OENO 449-2012, in order to assess the potential risk of BA accumulation in wine, Q-PCR methods to detect LAB that have the genes of amino acids decarboxylases by targeting the suitable genes are described.

### *2.4.3. Oligonucleotide Probes and Specific Antibodies*

Another approach to detect microorganisms in wine is fluorescence in situ hybridisation (FISH). FISH allows the direct detection and identification of microorganisms combining fluorescence microscopy with the reliability of molecular methods. Indeed, in this technique, a set of specific probes are used to detect different microorganisms directly in the sample. The probes can be labelled with different fluorophores, thereby allowing detection of several species simultaneously. When FISH probes are designed to target ribosomal RNA, only living cells are detected (Bottari *et al.*, 2006). In wine, this technique has been used for the rapid monitoring of LAB (Sohier and Lonvaud-Funel1998; Blasco *et al.*, 2003), *Brettanomyces bruxellensis/Dekkera bruxellensis* (Stender *et al.*, 2001), *S. cerevisiae* and various non-*Saccharomyces* yeasts (Xufre *et al.*, 2006). The D1/D2 domains at the 5′ end of the 26S rRNA subunit provide an excellent basis to develop species-specific FISH probes for yeasts, while the 16S rRNA subunit is suitable for LAB. FISHmay be coupled with FCM to specifically quantify wine microorganisms (Longin *et al.*, 2017).Thanks to the species-specific properties of the FISH probes, the FISH-FCM technique has been used to monitor fermentation directly in wine quantifying *S. cerevisiae, H. guilliermondii, H. uvarum, Starmerella bacillaris, B. bruxellensis* (Andorra *et al.*, 2011; Branco *et al.*, 2012; Wang *et al.*, 2014b, Serpaggi *et al.*, 2010). FCM may be also coupled with polyclonal antibodies. Rodriguez and Thornton (2008) developed a method to distinguish and quantify *S. cerevisiae* from other yeasts such as the genus *Hanseniaspora* in must fermentation. First, they incubated grape must with an anti-*Saccharomyces* polyclonal antibody and then a second incubation was performed with a secondary antibody coupled with Alexa Fluor® 488 before FCM analysis. These Authors developed also a green fluorescent polyclonal antibody for *O. oeni* to monitor malolactic fermentation. Recently, an immune-cytometric test has been developed also to detect and quantify *B. bruxellensis* (Chaillet *et al.*, 2014). This method consists in using an anti-*Brettanomyces* polyclonal antibody conjugated with a fluorochrome to specifically distinguish and quantify this yeast among other yeast species with a greater efficiency and rapidity than plate count methods.

## 2.5. Next-generation DNA Sequencing Applied to Wine Microbiota

Recently, the study of genetic material recovered directly from environmental samples (metagenomic approach) has been applied to the study of microbial communities in ecosystems, providing a great insight into the processes responsible for microbial diversity. In this contest, DNA sequence represents a single format onto which a broad range of biological phenomena can be projected for high throughput data collection. Over the past ten years, DNA sequencing platforms have become widely available, reducing considerably the cost of DNA sequencing. These new technologies are rapidly evolving, and near-term challenges include the development of robust protocols for generating sequencing libraries and building effective new approaches to data-analysis (Shendure and Ji, 2008). In particular, High-Throughput Sequencing (HTS) technologies such as the 454 pyrosequencing of amplicons (Shendure and Ji, 2008), can be used to characterise more precisely the microbial diversity of complex environmental ecosystems, including food samples (Ercolini, 2013; Galimberti *et al.*, 2015; Solieri, *et al.*, 2013). In the 454 pyrosequencing, libraries may be constructed by any method that gives rise to a mixture of short, adaptor-flanked fragments. Clonal sequencing features are generated by emulsion PCR (Dressman *et al.*, 2003), with amplicons captured to the surface of 28-μm beads. After breaking the emulsion, beads are treated with denaturant to remove untethered strands, and then subjected to a hybridisation-based enrichment for amplicon bearing beads. A sequencing primer is hybridised to the universal adaptor at the appropriate position and orientation, that is, immediately adjacent to the start of unknown sequence. Finally, sequencing is performed by the pyrosequencing method (Ronaghi *et al.*, 1996).

HTS technology has recently been used to determine the bacterial diversity of botrytised wines (Bokulich *et al.*, 2012), to monitor seasonal changes in winery-resident microbiota (Richardson and Mills, 2013) or to analyze the microbial biogeography of grapes from a Californian region (Bokulich *et al.*, 2014). Using HTS techniques, Bokulich *et al.* (2014) showed that the microbial population in wine fermentation is strongly relatedto climatic conditions, grape variety, and vineyard environmental conditions. In other words, the authors confirmed what was already known by using other molecular methods: there is a unique microbial pattern that influences the wine quality and asserts the existence of non-random "microbial terroir". Likewise De Filippis et al. (2017) monitored yeast and mould populations involved in spontaneous fermentations of Aglianico and Greco di Tufo grape must by high-throughput sequencing (HTS) of 18S rRNA gene amplicons. As expected, the Authors found a complex microbiota at the beginning of the fermentation, mainly characterised by non-*Saccharomyces* yeasts and several moulds, with differences between the two types of grapes.

To conclude, HTS techniques can be used for monitoring microbial changes in wine fermentations and winemakers could exploit this information to drive fermentation, but the too expensive instrumentation and bioinformatic knowledge needed to use these techniques confine, at least in the near future, this method to research laboratories.

## References

Airado- Rodríguez, D., Galeano-Díaz, T. and Durán-Merás, I. (2010). Determination of trans-resveratrol in red wine by adsorptive stripping square-wave voltammetry with medium exchange. *Food Chemistry* 122: 1320-1326.

Almeda, S., Arce, L. and Valcárel, M. (2008). Combined use of supported liquid membrane and solid-phase extraction to enhance selectivity and sensitivity in capillary electrophoresis for the determination of ochratoxin A in wine. *Electrophoresis* 29: 1573-1581.

Andorrà I., Landi, S., Mas, A., Guillamom, J.M. and Esteve-Zarzoso, B. (2008). Effect of oenological practices on microbial populations using culture-independent techniques. *Food Microbiology* 25: 849-856.

Andorrà, I., Monteiro, M., Esteve-Zarzoso, B., Albergaria, H. and Mas, A. (2011). Analysis and direct enumeration of *Saccharomyces cerevisiae* and *Hanseniaspora guilliermondii* populations during alcoholic fermentation by fluorescence in situ hybridization, flow cytometry and quantitative PCR. *Food Microbiology* 28: 1483-1491.

Andujar-Ortiz, L., Moreno-Arribas, M.V., Martín-Álvarez, P.J. and Pozo-Bayón, M.A. (2009). Analytical performance of three commonly used extraction methods for the gas chromatography–mass spectrometry analysis of wine volatile compounds. *Journal of Chromatography A* 1216: 7351-7357.

Aresta, A., Vatinno, R., Palmisano, F. and Zambonin, C.G. (2006). Determination of Ochratoxin A in wine at sub ng/mL levels by solid-phase microextraction coupled to liquid chromatography with fluorescence detection. *Journal of Chromatography A* 1115: 196-201.

Arias, C.R., Burns, J.K., Friedrich, L.M., Goodrich, R.M. and Parish. M.E. (2002). Yeast species associated with orange juice: Evaluation of different identification methods. *Applied and Environmental Microbiology* 68: 1955-1961.

Arrieta, M.P. and Prats-Moya, M.S. (2012). Free amino acids and biogenic amines in Alicante Monastrell wines. *Food Chemistry* 135: 1511-1519.

Association of Official Analytical Chemists (1997). AOAC official method 994.07: Ethyl carbamate in alcoholic beverages and soy sauce. pp. 14-15. *In*: AOAC Official Methods of Analysis. 17th edition. *Journal of AOAC International*, Gaithersburg.

Ayoub, M.J., Legras, J.L., Saliba, R. and Gaillardin, C. (2006). Application of multi locus sequences typing to the analysis of the biodiversity of indigenous *Saccharomyces cerevisiae* wine yeasts from Lebanon. *Journal of Applied Microbiology* 100: 699-711.

Baleiras-Couto, M.M., Reizinho, R.G. and Duarte, F.L. (2005). Partial 26S rDNA restriction analysis as a tool to characterise non-*Saccharomyces yeasts* present during red wine fermentations. *International Journal of Food Microbiology* 102: 49-56.

Barros, E.P., Moreira, N., Pereira, G.E., Leite, S.G.F., Rezende, C.M. and de Pinho, P.G. (2012). Development and validation of automatic HS-SPME with a gas chromatography-ion trap/mass spectrometry method for analysis of volatiles in wines. *Talanta* 101: 177-186.

Bauza, T., Blaise, A., Daumas, F. and Cabanis, J.C. (1995). Determination of biogenic amines and their precursor amino acids in wines of the Vallée du Rhône by high-performance liquid chromatography with precolumn derivatization and fluorimetric detection. *Journal of Chromatography A* 707: 373-379.

Beltràn, G., Torija, M.J., Novo, M., Ferrer, N., Poblet, M. and Guillamòn, J.M. (2002). Analysis of yeast populations during alcoholic fermentation: Six year follow-up study. *Systematic and Applied Microbiology* 25: 287-293.

Blackstone, G.M., Nordstrom, J.L., Vickery, M.C.L., Bowen, M.D., Meyer, R.F. and DePaola, A. (2003). Detection of pathogenic *Vibrio parahaemolyticus* in oyster enrichments by real time PCR. *Journal of Microbiology Methods* 53: 149-155.

Blasco, L., Ferrer, S. and Pardo, I. (2003). Development of specific fluorescent oligonucleotide probes for in situ identification of lactic acid bacteria. *FEMS Microbiology Letters* 225: 115-123.

Bleve, G., Rizzotti, L., Dellaglio, F. and Torriani, S. (2003). Development of reverse transcription (RT)-PCR and real-time RT-PCR assays for rapid detection and quantification of viable yeasts and molds contaminating yogurts and pasteurized food products. *Applied and Environmental Microbiology* 69: 4116-4122.

Bokulich, N.A., Joseph, C.M.L., Allen, G., Benson, A.K. and Mills, D.A. (2012). Next generation sequencing reveals significant bacterial diversity of botrytized wine. *PLoS One* 7: 36357. http://dx.doi.org/10.1371/journal.pone.0036357.

Bokulich, N.A., Ohta, M., Richardson, P.M. and Mills, D.A. (2013). Monitoring seasonal changes in winery-resident microbiota. *PLoS One* 8: 66437. http://dx.doi.org/10.1371/journal.pone.0066437.

Bokulich, N.A., Swadener, M., Sakamoto, K., Mills, D.A. and Bisson, L.F. (2014). Sulfur dioxide treatment alters wine microbial diversity and fermentation progression in a dose-dependent fashion. *American Journal of Enology and Viticulture* 66: 73-79.

Bolton, S.L., Mitchell, T., Brannen, P.M. and Glenn, A.E. (2017). Assessment of Mycotoxins in *Vitis vinifera* Wines of the Southeastern United States. *American Journal of Enology and Viticulture* 68: 336-343.

Bottari, B., Ercolini, D., Gatti, M. and Neviani, E. (2006). Application of FISH technology for microbiological analysis: Current state and prospects. *Applied Microbiology and Biotechnology* 73: 485-494.

Bouchez, J.C., Cornu, M., Danzart, M., Leveau, J.Y., Duchiron, F. and Bouix, M. (2004). Physiological significance of the cytometric distribution of fluorescent yeasts after viability staining. *Biotechnology and Bioengineering* 86: 520-530.

Bouix, M. and Leveau, J.-Y. (2001). Rapid assessment of yeast viability and yeast vitality during alcoholic fermentation. *Journal of the Institute of Brewing* 107: 217-225.

Bouix, M. and Ghorbal, S. (2013). Rapid enumeration of *Oenococcus oeni* during malolactic fermentation by flow cytometry. *Journal of Applied Microbiology* 114: 1075-1081.

Branco, P., Monteiro, P.M., Moura and Albergaria, H. (2012). Survival rate of wine-related yeasts during alcoholic fermentation assessed by direct live/dead staining combined with fluorescence in situ hybridization. *International Journal of Food Microbiology* 158: 49-57.

Bridier, J., Claisse, O., Coton, M., Coton, E. and Lonvaud-Funel, A. (2010). Evidence of distinct populations and specific subpopulations within the species *Oenococcus oeni*. *Applied and Environmental Microbiology* 76: 7754-7764.

Cacho, J.I., Campillo, N., Viñas, P. and Hernández-Córdoba, M. (2013). Stir bar sorptive extraction with gas chromatography–mass spectrometry for the determination of resveratrol, piceatannol and oxyresveratrol isomers in wines. *Journal of Chromatography A* 1315: 21-27.

Cao, J., Kong, W., Zhou, S., Yin, L., Wan, L. and Yang, M. (2013). Molecularly imprinted polymer-based solid phase clean-up for analysis of ochratoxin A in beer, red wine, and grape juice. *Journal of Separation Science* 36: 1291-1297.

Capece, A., Salzano, G. and Romano, P. (2003). Molecular typing techniques as a tool to differentiate non-*Saccharomyces* wine species. *International Journal of Food Microbiology* 84: 33-39.

Capece, A., Granchi, L., Guerrini, S., Mangani, S., Romaniello, R., Vincenzini, M. and Romano, P. (2016). Diversity of *Saccharomyces cerevisiae* strains isolated from two Italian wine-producing regions. *Frontiers in Microbiology* 7: 1-11.

Cappello, M.S., De Domenico, S., Logrieco, A. and Zapparoli, G. (2014). Bio-molecular characterisation of indigenous *Oenococcus oeni* strains from Negroamaro wine. *Food Microbiology* 42: 142-148.

Caruso, M., Capece, A., Salzano, G. and Romano, P. (2002). Typing of *Saccharomyces cerevisiae* and *Kloeckera apiculate* strains from Aglianico wine. *Letters in Applied Microbiology* 34: 323-328.

Chaillet, L., Martin, G. and Genty, V. (2014). Mise au point d'une mèthode de dètection des *Brettanomyces* par immunocytomètrie. Personal Communication in SFI-AFC Congress.

Chaney, D., Rodriguez, S., Fugelsang, K. and Thornton, R. (2006). Managing high-density commercial scale wine fermentations. *Journal of Applied Microbiology* 100: 689-698.

Ciani, M., Mannazzu, I., Marinangeli, P., Clementi, F. and Martini, A. (2004). Contribution of winery-resident *Saccharomyces cerevisiae* strains to spontaneous grape must fermentation. *Antonie van Leeuwenhoek* 85: 159-164.

Cincotta, F., Verzera, A., Tripodi, G. and Condurso, C. (2015). Determination of Sesquiterpenes in Wines by HS-SPME Coupled with GC-MS. *Chromatography* 2: 410-421.

Claisse, O. and Lonvaud-Funel, A. (2001). Primers and specific DNA probe for detecting lactic acid bacteria producing 3-hydroxypropioaldheyde from glycerol in spoiled ciders. *Journal of Food Protection* 64: 883-887.

Claisse, O. and Lonvaud-Funel, A. (2014). Multiplex variable number of tandem repeats for *Oenococcus oeni* and applications. *Food Microbiology* 38: 80-86.

Cocolin, L., Bisson, L.F. and Mills, D.A. (2000). Direct profiling of the yeast dynamics in wine fermentations. *FEMS Microbiology Letters* 189: 81-87.

Costantini, A., Cersosimo, M., del Prete, V. and Garcia-Moruno, E. (2006). Production of biogenic amine by lactic acid bacteria, screening by PCR, thin layer chromatography, and high-performance liquid chromatography of strains isolated from wine and must. *Journal of Food Protection* 69: 391-396.

Coton, E. and Coton, M. (2005). Multiplex PCR for colony direct detection of gram-positive histamine and tyramine-producing bacteria. *Journal of Microbiology Methods* 63: 296-304.

Cozzolino, D. (2015). Elemental composition in grapes and wine: Role, analytical methods and their use. pp. 473-487. *In*: M. de la Guardia and S. Garrigues (Eds.). Handbook of Mineral Elements in Food. J. Wiley & Sons, Ltd, Chichester.

Cruz-Pio, L.E., Poveda, M., Alberto, M.R., Ferrer, S. and Pardo, I. (2017). Exploring the biodiversity of two groups of *Oenococcus oeni* isolated from grape musts and wines: Are they equally diverse? *Systematic and Applied Microbiology* 40: 1-10.

Da Silveira, M.G., Romao, M.V.S., Loureiro-Dias, M.C., Rombouts, F.M. and Abee, T. (2002). Flow cytometric assessment of membrane integrity of ethanol-stressed *Oenococcus oeni* cells. *Applied and Environmental Microbiology* 68: 6087-6093.

De Filippis, F., La Storia, A. and Blaiotta, G. (2017). Monitoring the mycobiota during Greco di Tufo and Aglianico wine fermentation by 18S rRNA gene sequencing. *Food Microbiology*, 63: 117-122.

de Villiers, A., Cabooter, D., Lynen, F., Desmet, G. and Sandra, P. (2009). High performance liquid chromatography analysis of wine anthocyanins revisited: Effect of particle size and temperature. *Journal of Chromatography A* 1216: 3270-3279.

de Villiers, A., Alberts, P., Tredoux, A.G.J. and Nieuwoudt, H.H. (2012). Analytical techniques for wine analysis: An African perspective: A review. *Analytica Chimica Acta* 730: 2-23.

Delaherche, A., Claisse, O. and Lonvaud-Funel, A. (2004). Detection and quantification of *Brettanomyces bruxellensis* and 'ropy' *Pediococcus damnosus* strains in wine by real-time polymerase chain reaction. *Journal of Applied Microbiology* 97: 910-915.

Delfini, C. and Formica, J.V. (2001). Wine Microbiology: Science and Technology. Marcel Dekker, New York.

Delobel, P., Pradal, M., Blondin, B. and Tesniere, C. (2012). A "fragile cell" sub-population revealed during cytometric assessment of *Saccharomyces cerevisiae* viability in lipid-limited alcoholic fermentation. *Letters in Applied Microbiology* 55: 338-344.

Dent, K.C., Stephen, J.R. and Finch-Savage, W.E. (2004). Molecular profiling of microbial communities associated with seeds of *Beta vulgaris* subsp *vulgaris* (sugar beet). *Journal of Microbial Methods* 56: 17-26.

Di Donna, L., Taverna, D., Indelicato, S., Napoli, A., Sindona, G. and Fabio Mazzotti (2017). Rapid assay of resveratrol in red wine by paper spray tandem mass spectrometry and isotope dilution. *Food Chemistry* 229: 354-357.

Di Maro, E., Ercolini, D. and Coppola, S. (2007). Yeast dynamics during spontaneous wine fermentation of the Catalanesca grape. *International Journal of Food Microbiology* 117: 201-210.

Díaz, M., Herrero, M., García, L.A. and Quiròs, C. (2010). Application of flow cytometry to industrial microbial bioprocesses. *Biochemical Engineering Journal* 48: 385-407. Invited Review Issue 2010.

Divol, B. and Lonvaud-Funel, A. (2005). Evidence for viable but non culturable yeasts in botrytis-affected wine. *Journal of Applied Microbiology* 99: 85-93.

Doare-Lebrun, E., ElArbi, A., Charlet, M., Guerin, L., Pernelle, J.J., Ogier, J.C. and Bouix, M. (2006). Analysis of fungal diversity of grapes by application of temporal temperature gradient gel electrophoresis – potentialities and limits of the method. *Journal of Applied Microbiology* 101: 1340-1350.

Dressman, D., Yan, H., Traverso, G.K.W., Kinzler and Vogelstein, B. (2003). Transforming single DNA molecules into fluorescent magnetic particles for detection and enumeration of genetic variations. *Proceedings of the National Academy of Sciences of the United States of America* 100: 8817-8822.

Du Toit, W.J., Pretorius, I.J. and Lonvaud-Funel, A. (2005). The effect of sulphur dioxide and oxygen on the viability and culturability of a strain of *Acetobacter pasteurianus* and a strain of *Brettanomyces bruxellensis* isolated from wine. *Journal of Applied Microbiology* 98: 862-871.

Eder, R. (2001). Pigments. pp. 825-880. *In*: L.M.L. Nollet (Ed.). Food Analysis by HPLC. M. Dekker Inc., Basel.

Ercolini, D. (2013). High-throughput sequencing and metagenomics: Moving forward in the culture-independent analysis of food microbial ecology. *Applied and Environmental Microbiology* 79: 3148-3155.

Esteve-Zarzoso, B., Belloch, C., Uruburu, F. and Querol, A. (1999). Identification of yeast by RFLP analysis of the 5.8S rRNA gene and the two ribosomal internal transcribed spacers. *International Journal of Systematic Bacteriology* 49: 329-337.

Esteve-Zarzoso, B., Peris-Toran, M.J., García-Maiquez, E., Uruburu, F. and Querol, A. (2001). Yeast population dynamics during the fermentation and biological aging of sherry wines. *Applied and Environmental Microbiology* 67: 2056-2061.

European Commission (EC) (2006). Commission regulation (EC) n◦1881/2006 of 19 December 2006 setting maximum levels for certain contaminants in foodstuffs. *Official Journal of the European Union* 49: 5-24.

European Food Safety Authority (EFSA) (2007). Scientific opinion on ethyl carbamate and hydrocyanic acid in food and beverages. *The EFSA Journal* 551: 1-44.

Fabiani, A., Corzani, C. and Arfelli, G. (2010). Correlation between different clean-up methods and analytical techniques performances to detect Ochratoxin A in wine. *Talanta* 83: 281-285.

Farthing, J.B., Rodriguez, S.B. and Thornton, R.J. (2007). Flow cytometric analysis of *Saccharomyces cerevisiae* populations in high-sugar Chardonnay fermentations. *Journal of the Science of Food and Agriculture* 87: 527-533.

Fernandes, I., Pérez-Gregorio, R., Soares, S., Mateus, N. and de Freitas, V. (2017). Wine Flavonoids in Health and Disease Prevention. *Molecules* 22: 1-30.

Fernandes, J.O. and Ferreira, M.A. (2000). Combined ion-pair extraction and gas chromatography–mass spectrometry for the simultaneous determination of diamines, polyamines and aromatic amines in Port wine and grape juice. *Journal of Chromatography* A 886: 183-195.

Fernandez, M.T., Ubbeda, J.F. and Briones, A.I. (1999). Comparative study of non-*Saccharomyces* microflora of musts in fermentation, by physiological and molecular methods. *FEMS Microbiology Letter* 173: 223-229.

Fernàndez-Espinar, M.T., Llopis, S., Querol, A. and Barrio, E. (2011). Molecular identification and characterization of wine yeasts. *Molecular Wine Microbiology* 5: 111-141.

Fernàndez-Espinar, M.T., Lòpez, V., Ramòn, D., Bartra, E. and Querol, A. (2001). Study of the authenticity of commercial wine yeast strains by molecular techniques. *International Journal of Food Microbiology* 70: 1-10.

Fernàndez-Gonzàlez, M., Espinosa, J.C., Ubeda, J.F. and Briones, A.I. (2001). Yeast present during wine fermentation: Comparative analysis of conventional plating and PCR-TTGE. *Systematic and Applied Microbiology* 24: 634-638.

Fia, G., Giovani, G. and Rosi, I. (2005). Study of β-glucosidase production by wine-related yeasts during alcoholic fermentation. A new rapid fluorimetric method to determine enzymatic activity. *Journal of Applied Microbiology* 99: 509-517.

Fia, G., Canuti, V. and Rosi, I. (2014). Evaluation of potential side activities of commercial enzyme preparations used in winemaking. *International Journal of Food Science and Technology*, 49: 1902-1911.

Fia, G., Olivier, V., Cavaglioni, A., Canuti, V. and Zanoni, B. (2016). Side activities of commercial enzyme preparations and their influence on the hydroxycinnamic acids, volatile compounds and nitrogenous components of white wine. *Australian Journal of Grape and Wine Research* 22: 366-375.

Galimberti, A., Bruno, A., Mezzasalma, V., De Mattia, F., Bruni, I. and Labra, M. (2015). Emerging DNA-based technologies to characterize food ecosystems. *Food Research International* 69: 424-433.

Ganga, M.A. and Martinez, C. (2004). Effect of wine yeast monoculture practice on the biodiversity of non-*Saccharomyces* yeasts. *Journal of Applied Microbiology* 96: 76-83.

García-Villar, N., Hernández-Cassou, S. and Saurina, J. (2009). Determination of biogenic amines in wines by pre-column derivatization and high-performance liquid chromatography coupled to mass spectrometry. *Journal of Chromatography A* 1216: 6387-6393.

Garofalo, C., El-Khoury, M., Lucas, P., Bely, M., Russo, P., Spano, G. and Capozzi, V. (2015). Autochthonous starter cultures and indigenous grape variety for regional wine production. *Journal of Applied Microbiology* 118: 1395-1408.

Geana, I., Iordache, A., Ionete, R., Marinescu, A., Ranca, A. and Culea, M. (2013). Geographical origin identification of Romanian wines by ICP-MS elemental analysis. *Food Chemistry* 138: 1125-1134.

Gerbaux, V. and Thomas, J. (2009). Utilisations pratiques de ma cytometrie de flux pour le suivi des levures en oenologie. *Rev. Fr. Oenologie* 8-13.

Gindreau, E., Wailling, E. and Lonvaud-Funel, A. (2001). Direct polymerase chain reaction detection of ropy *Pediococcus damnosus* strains in wine. *Journal of Applied Microbiology* 90: 535-542.

Gonzàlez, A., Guillamòn, J.M., Mas, A. and Poblet, M. (2006). Application of molecular methods for routine identification of acetic acid bacteria. *International Journal of Food Microbiology* 108: 141-146.

Gonzalez, S.S., Barrio, E. and Querol, A. (2007). Molecular identification and characterization of wine yeasts isolated from Tenerife (Canary Island, Spain). *Journal of Applied Microbiology* 102: 1018-1025.

González-Arenzana, L., Santamaría, P., Gutiérrez, A.R., López, R. and López-Alfaro, I. (2017). Lactic acid bacteria communities in must, alcoholic and malolactic Tempranillo wine fermentations, by

culture-dependent and culture-independent methods. *European Food Research and Technology* 243: 41-48.
Granchi, L., Bosco, M., Messini, A. and Vincenzini, M. (1999). Rapid detection and quantification of yeast species during spontaneous wine fermentation by PCR-RFLP analysis of the rDNA ITS region. *Journal of Applied Microbiology* 87: 949-956.
Granchi, L., Ganucci, D., Viti, C., Giovannetti, L. and Vincenzini, M. (2003). *Saccharomyces cerevisiae* biodiversity in spontaneous commercial fermentations of grape musts with "inadequate" assimilable-nitrogen content. *Letters in Applied Microbiology* 36: 54-58.
Granchi, L., Talini, D., Rigacci, S., Guerrini, S., Berti, A. and Vicenzini, M. (2006). Detection of putrescine producer *Oenococcus oeni* strains by PCR. *In*: 8th Symposium on Lactic Acid Bacteria. The Netherlands.
Granchi, L., Carobbi, M., Guerrini, S. and Vincenzini, M. (2006a). Rapid enumeration of yeast during wine fermentation by the combined use of Thoma Chamber and epifluorescence microscopy. Proceedings "XXIXth World Congress of Vine and Wine" 1-5.
Grindlay, G., Mora, J., Gras, L. and de Loos-Vollebregt, M.T.C. (2011). Atomic spectrometry methods for wine analysis: A critical evaluation and discussion of recent applications. *Analytica Chimica Acta* 691: 18-32.
Guerrini, S., Bastianini, A., Blaiotta, G., Granchi, L., Moschetti, G., Coppola, S., Romano, P. and Vincenzini, M. (2003). Phenotypic and genotypic characterization of *Oenococcus oeni* strains isolated from Italian wines. *International Journal of Food Microbiology* 85: 1-14.
Guillamòn, J.M., and Mas, A. (2011). Acetic acid bacteria. *Molecular Wine Microbiology* 9: 227-252.
Guillamòn, J.M., Sabate, J., Barrio, E., Cano, J. and Querol, A. (1998). Rapid identification of wine yeast species based on RFLP analysis of the ribosomal internal transcribed spacer ITS region. *Archives of Microbiology* 169: 387-392.
Hashim, S.N.N.S., Schwarz, L.J., Boysen, R.I., Yang, Y., Danylec, B. and Hearn, M.T.W. (2013). Rapid solid-phase extraction and analysis of resveratrol and other polyphenols in red wine. *Journal of Chromatography A* 1313: 284-290.
He, F., Mu, L., Yan, G.-L., Liang, N.-N., Pan, Q.-H., Wang, J., Reeves, M.J. and Duan, C.-Q. (2010). Biosynthesis of anthocyanins and their regulation in colored grapes. *Molecules* 15: 1141-1153.
Hernández-Borges, J., D'Orazio, G., Aturki, Z. and Fanali, S. (2007). Nano-liquid chromatography analysis of dansylated biogenic amines in wines. *Journal of Chromatography A* 1147: 192-199.
Hernández-Orte, P., Peña-Gallego, A., Ibarz, M.J., Cacho, J. and Ferreira, V. (2006). Determination of the biogenic amines in musts and wines before and after malolactic fermentation using 6-aminoquinolyl-N-hydroxysuccinimidyl carbamate as the derivatizing agent. *Journal of Chromatography A* 1129: 160-164.
Herrero, M., Quiròs, C., García, L.A. and Díaz, M. (2006). Use of flow cytometry to follow the physiological states of microorganisms in cider fermentation processes. *Applied Environmental Microbiology* 72: 6725-6733.
Herrero, M., Garcia-Cañas, V., Simo, C. and Cifuentes, A. (2010). Recent advances in the application of capillary electromigration methods for food analysis and Foodomics. *Electrophoresis* 31: 205-228.
Hierro, N., Gonzalez, A., Mas, A. and Guillamon, J.M. (2006). Diversity and evolution of non-*Saccharomyces* yeast populations during wine fermentation: Effect of grape ripeness and cold maceration. *FEMS Yeast Research* 6: 102-111.
Hierro, N., Esteve-Zarzoso, B., Gonzalez, A., Mas, A. and Guillamon, J.M. (2006). Real-time quantitative PCR (QPCR) and reverse transcription – QPCR for detection and enumeration of total yeasts in wine. *Applied and Environmental Microbiology* 72: 7148-7155.
Hierro, N., Esteve-Zarzoso, B., Mas, A. and Guillamon, J.M. (2007). Monitoring of *Saccharomyces* and *Hanseniaspora* populations during alcoholic fermentation by real-time quantitative PCR. *FEMS Yeast Research* 7: 1340-1349.
Hjelmeland, A.K., Collins, T.S., Miles, J.L., Wylie, P.L., Mitchell, A.E. and Ebeler, S.E. (2012). High-throughput, sub ng/L analysis of haloanisoles in wines using HS-SPME with GC-triple quadrupole MS. *American Journal of Enology and Viticulture* 63: 494-499.

Hlabangana, L., Hernández-Cassou, S. and Saurina, J. (2006). Determination of biogenic amines in wines by ion-pair liquid chromatography and post-column derivatization with 1,2-naphthoquinone-4-sulphonate. *Journal of Chromatography A* 1130: 130-136.

Hopfer, H., Nelson, J., Collins, T.S., Heymann, H. and Ebeler, S.E. (2015). The combined impact of vineyard origin and processing winery on the elemental profile of red wines. *Food Chemistry* 172: 486-496.

Hussein, H.S. and Brasel, J.M. (2001). Toxicity, metabolism, and impact of mycotoxins on humans and animals. *Toxicology* 167: 101-134.

Ivey, M.L. and Phister, T.G. (2011). Detection and identification of microorganisms in wine: A review of molecular techniques. *Journal of Industrial Microbiology and Biotechnology* 38: 1619-1634.

Jagerdeo, E., Dugar, S., Foster, G.D. and Schenck, H. (2002). Analysis of ethyl carbamate in wines using solid-phase extraction and multidimensional gas chromatography/mass spectrometry. *Food Chemistry* 50: 5797-5802.

James, S.A., Cai, J., Roberts, I.N. and Collins, M.D. (1997). Phylogenetic analysis of the genus *Saccharomyces* based on 18S rRNA gene sequences: Description of *Saccharomyces kunashirensis* sp. nov., and *Saccharomyces martiniae* sp. nov. *International Journal of Systematic and Evolutionary Microbiology* 47: 453-460.

Jastrzębska, A., Piasta, A., Kowalska, S., Krzemiński, M. and Szłyk, E. (2016). A new derivatization reagent for determination of biogenic amines in wines. *Journal of Food Composition and Analysis* 48: 111-119.

Jemec, K.P., Cadez, N., Zagore, T., Bubic, V., Zupec, A. and Raspor, P. (2001). Yeast population dynamics in five spontaneous fermentations of Malvasia must. *Food Microbiology* 18: 247-259.

Jia, S., Kang, Y.P., Park, J.H., Lee, J. and Kwon, S.W. (2012). Determination of biogenic amines in Bokbunja (*Rubus coreanus* Miq.) wines using a novel ultra-performance liquid chromatography coupled with quadrupole-time of flight mass spectrometry. *Food Chemistry* 132: 1185-1190.

Jiao, Z., Dong, Y. and Chen, Q. (2014). Ethyl carbamate in fermented beverages: Presence, analytical chemistry, formation mechanism, and mitigation proposals. *Comprehensive Reviews in Food Science and Food Safety*, 13: 611-626.

Jurado-Sánchez, B., Ballesteros, E. and Gallego, M. (2007). Gas Chromatographic determination of N-nitrosamines in beverages following automatic solid-phase extraction. *Journal of Agricultural and Food Chemistry* 55: 9758–9763.

Kalili, K.M. and de Villiers, A. (2011). Recent developments in the HPLC separation of phenolic compounds. *Journal of Separation Science* 34: 854–876.

Kelly, M.T., Blaise, A. and Larroqueb, M. (2010). Rapid automated high performance liquid chromatography method for simultaneous determination of amino acids and biogenic amines in wine, fruit and honey. *Journal of Chromatography A* 1217: 7385-7392.

Kurtzman, C.P. and Robnett, C.J. (1998). Identification and phylogeny of ascomycetous yeasts from analysis of nuclear large subunit 26S ribosomal DNA partial sequences. *Antonie van Leeuwenhoek* 73: 331-371.

Lachenmeier, D.W., Nehrlich, U. and Kuballa, T. (2006). Automated determination of ethyl carbamate in stone-fruit spirits using headspace solid-phase microextraction and gas chromatography–tandem mass spectrometry. *Journal of Chromatography A* 1108: 116-120.

Lafon-Lafourcade, S. and Joyeux, A. (1979). Techniques simplifiées pour le dénombrement et l'identification des microorganismes vivants dans les moûts et les vins. *Conn. Vigne Vin* 13: 295-310.

Landete, J.M., Ferrer, S. and Pardo, I. (2005). Which lactic acid bacteria are responsible for histamine production in wine? *Journal of Applied Microbiology* 99: 580-586.

Landolfo, S., Politi, H., Angelozzi, D. and Mannazzu, I. (2008). ROS accumulation and oxidative damage to cell structures in *Saccharomyces cerevisiae* wine strains during fermentation of high-sugar-containing medium. *Biochimica et Biophysica Acta (BBA) – General Subjects* 1780: 892-898.

Lange, J. and Wittmann, C. (2002). Enzyme sensor array for the determination of biogenic amines in food samples. *Analytical and Bioanalytical Chemistry* 372: 276-283.

Larisika, M., Claus, H. and Konig, H. (2008). Pulsed-field gel electrophoresis for the discrimination of *Oenococcus oeni* isolates from different wine-growing regions in Germany. *International Journal of Food Microbiology* 123: 171-176.

Le Jeune, C., Erny, C., Demuyter, C. and Lollier, M. (2006). Evolution of the population of *Saccharomyces cerevisiae* from grape to wine in a spontaneous fermentation. *Food Microbiology* 23: 709-716.

Lee, T.P., Saada, B., Khayoona, W.S. and Salleh, B. (2012). Molecularly imprinted polymer as sorbent in micro-solid phase extraction of ochratoxin A in coffee, grape juice and urine. *Talanta* 88: 129-135.

Liu, J., Xu, Y. and Zhao, G. (2012). Rapid determination of ethyl carbamate in Chinese rice wine using headspace solid-phase microextraction and gas chromatography–mass spectrometry. *Journal of the Institute of Brewing* 118: 217-222.

Lona-Ramirez, F.J., Gonzalez-Alatorre, G., Rico-Ramírez, V., Perez-Perez, M.C.I. and Castrejón-González, E.O. (2016). Gas chromatography/mass spectrometry for the determination of nitrosamines in red wine. *Food Chemistry* 196: 1131-1136.

Longin, C., Petitgonnet, C., Guilloux-Benatier, M., Rousseaux, S. and Alexandre, H. (2017). Application of flow cytometry to wine microorganisms. *Food Microbiology* 62: 221-231.

Lopandic, K., Tiefenbrunner, W., Gangl, H., Mandl, K., Berger, S., Leitner, G., Abd-Ellah, G.A., Querol, A., Gardner, R.C., Sterflinger, K. and Prillinger, H. (2008). Molecular profiling of yeasts isolated during spontaneous fermentations of Austrian wines. *FEMS Yeast Research* 8: 1063-1075.

Lòpes, C.A., Rodrìguez, M.E., Sangorrìn, M., Querol, A. and Caballero, A.C. (2007). Patagonia wines: Implantation of an indigenous strain of *Saccharomyces cerevisiae* in fermentations conducted in traditional and modern cellars. *Journal of Industrial Microbiology & Biotechnology* 34: 139-149.

Lòpes, C.A., van Broock, M., Querol, A. and Caballero, A.C. (2002). *Saccharomyces cerevisiae* wine yeast populations in a cold region in Argentinean Patagonia: A study at different fermentation scales. *Journal of Applied Microbiology* 93: 608-615.

Lòpes, C.A., Lavalle, T.L., Querol, A. and Caballero, A.C. (2006). Combined use of killer biotype and mtDNARFLP patterns in a Patagonian wine *Saccharomyces cerevisiae* diversity study. *Leeuwenhoek* 89: 147-156.

Lopez, I., Ruiz-Larrea, F., Cocolin, L., Orr, E., Phister, T., Marshall, M., Vander Gheynst, J. and Mills, D.A. (2003). Design and evaluation of PCR primers for analysis of bacterial populations in wine by denaturing gradient gel electrophoresis. *Applied and Environmental Microbiology* 69: 6801-6807.

Lòpez, I., Torres, C. and Ruiz-Larrea, F. (2008). Genetic typification by pulsed-field gel electrophoresis (PFGE) and randomly amplified polymorphic DNA (RAPD) of wild *Lactobacillus plantarum* and *Oenococcus oeni* wine strains. *European Food Research and Technology* 227: 547-555.

Lòpez, V., Fernàndez-Espinar, M.T., Barrio, E., Ramòn, D. and Querol, A. (2002). A new PCR-based method for monitoring inoculated wine fermentations. *International Journal of Food Microbiology* 81: 63-71.

Lorrain, B., Ky, I., Pechamat, L. and Teissedre, P.-L. (2013). Evolution of analysis of polyhenols from grapes, wines, and extracts. *Molecules* 18: 1076-1100.

Lucas, P.M., Claisse, O. and Lonvaud-Funel, A. (2008). High frequency of histamine-producing bacteria in the enological environment and instability of the histidine decarboxylase production phenotype. *Applied and Environmental Microbiology* 74: 811-817.

Malacrinò, P., Zapparoli, G., Torriani, S. and Dellaglio, F. (2001). Rapid detection of viable yeasts and bacteria in wine by flow cytometry. *Journal of Microbiology Methods* 45: 127-134.

Mangani, S., Buscioni, G., Collina, L., Bocci, E. and Vincenzini, M. (2011). Effects of microbial populations on anthocyanin profile of Sangiovese wines produced in Tuscany, Italy. *American Journal of Enology and Viticulture* 62: 487-493.

Mannazzu, I., Angelozzi, D., Belviso, S., Budroni, M., Farris, G.A., Goffrini, P., Lodi, T., Marzona, M. and Bardi, L. (2008). Behaviour of *Saccharomyces cerevisiae* wine strains during adaptation to unfavourable conditions of fermentation on synthetic medium: Cell lipid composition, membrane integrity, viability and fermentative activity. *International Journal of Food Microbiology* 121: 84-91.

Mao, J., Lei, S., Yang, X. and Xiao, D. (2013). Quantification of ochratoxin A in red wines by conventional HPLC-FLD using a column packed with core-shell particles. *Food Control* 32: 505-511.

Marcobal, A., de las Rivas, B., Moreno-Arribas, M.V. and Muñoz, R. (2005). Multiplex-PCR method for the simultaneous detection of lactic acid bacteria producing histamine, tyramine and putrescine, three major biogenic amines. *Journal of Food Protection* 68: 874-878.

Marques, A.P., Duarte, A.J., Chambel, L., Teixeira, M.F., Romão, M.V.S. and Tenreiro, R. (2011). Genomic diversity of *Oenococcus oeni* from different winemaking regions of Portugal. *International Microbiology* 14: 155-162.

Martìnez, C., Gac, S., Lavin, A. and Ganga, M. (2004). Genomic characterization of *Saccharomyces cerevisiae* strains isolated from wine-producing areas of South America. *Journal of Applied Microbiology* 96: 1161-1168.

Martin-Pastor, M., Guitian, E. and Riguera, R. (2016). Joint NMR and solid-phase microextraction–gas chromatography chemometric approach for very complex mixtures: Grape and zone identification in wines. *Analytical Chemistry* 88: 6239-6246.

Martorell, P., Querol, A. and Fernandez-Espinar, M.T. (2005). Rapid identification and enumeration of *Saccharomyces cerevisiae* cells in wine by real-time PCR. *Applied and Environmental Microbiology* 71: 6823-6830.

Mattivi, F., Guzzon, R., Vrhovsek, U., Stefanini, M. and Velasco, R. (2006). Metabolite profiling of grape: Flavonols and anthocyanins. *Journal of Agricultural and Food Chemistry* 54: 7692-7702.

Medić-Šarić, M., Rastija, V. and Bojić, M. (2011). Recent advances in the application of high performance liquid chromatography in the analysis of polyphenols in wine and propolis. *Journal of the Association of Official Agricultural Chemists International* 94: 32-42.

Millán, S., Sampedro, M.C., Unceta, N., Goicolea, M.A. and Barrio, R.J. (2007). Simple and rapid determination of biogenic amines in wine by liquid chromatography–electrospray ionization ion trap mass spectrometry. *Analytica Chimica Acta* 584: 145-152.

Millet, V. and Lonvaud-Funel, A. (2000). The viable but non-culturable state of wine micro-organisms during storage. *Letters in Applied Microbiology* 30: 136-141.

Mills, D.A., Phister, T., Neeley, E. and Johannsen, E. (2008). Wine fermentation. pp. 161-192. *In*: Cocolin, L. and Ercolini, D. (Eds.). Molecular Techniques in the Microbial Ecology of Fermented Foods. Springer, New York.

Monthèard, J., Garcier, S., Lombard, E., Cameleyre, X., Guillouet, S., Molina-Jouve, C., Alfenore, S. (2012). Assessment of Candida shehatae viability by flow cytometry and fluorescent probes. *Journal of Microbiology Methods* 91: 8-13.

Muyzer, G., de Waal, E.C. and Uitterlinden, A.G. (1993). Profiling of complex microbial populations by denaturing gradient gel electrophoresis analysis of polymerase chain reaction-amplified genes encoding for 16S rRNA. *Applied Environmental Microbiology* 59: 695-700.

Naumov, G.I., Naumova, E.S., Antunovics, Z. and Sipiczki, M. (2002). *Saccharomyces bayanus* var. *uvarum* in Tokaj wine-making of Slovakia and Hungary. *Applied Microbiology and Biotechnology* 59: 727-730.

Neeley, E.T., Phister, T.G. and Mills, D.A. (2005). Differential real-time PCR assay for enumeration of lactic acid bacteria in wine. *Applied and Environmental Microbiology* 12 : 8954-8957.

Nicolini, G., Larcher, R., Bertoldi, D., Puecher, C. and Magno, F. (2007). Rapid quantification of 4-ethylphenol in wine using high-performance liquid chromatography with a fluorimetric detector. *Vitis* 46: 202-206.

Nóbrega, I.C.C., Pereira, G.E., Silva, M., Pereira, E.V.S., Medeiros, M.M., Telles, D.L., Albuquerque Jr., E.C., Oliveira, J.B. and Lachenmeier, D.W. (2015). Improved sample preparation for GC–MS–SIM analysis of ethyl carbamate in wine. *Food Chemistry* 177: 23-28.

O'Neill, K., Aghaeepour, N., Spidlen, J. and Brinkman, R. (2013). Flow cytometry bioinformatics. *PLOS Computational Biology* 9, https://doi.org/10.1371/journal.pcbi.1003365

Oliver, J.D. (2005). The viable but non-culturable state in bacteria. *Journal of Microbiology* 43: 93-100.

Önal, A., Tekkeli, S.E.K. and Önal, C. (2013). A review of the liquid chromatographic methods for the determination of biogenic amines in foods. *Food Chemistry* 138: 509-515.

Özdestan, Ö and Üren, A. (2009). A method for benzoyl chloride derivatization of biogenic amines for high performance liquid chromatography. *Talanta* 78: 1321-1326.

Phister, T.G. and Mills, D.A. (2003). Real-time PCR assay for detection and enumeration of *Dekkerab ruxellensis* in wine. *Applied and Environmental Microbiology* 69:7430-7434.

Pinzani, P., Bonciani, L., Pazzagli, M., Orlando, C., Guerrini, S. and Granchi, L. (2004). Rapid detection of *Oenococcus oeni* in wine by real-time quantitative PCR. *Letters in Applied Microbiology* 38: 118-124.

Pizzutti, I.R., de Kok, A., Scholten, J., Righi, L.W., Cardoso, C.D., Necchi Rohers, G. and da Silva, R.C. (2014). Development, optimization and validation of a multimethod for the determination of 36 mycotoxins in wines by liquid chromatography–tandem mass spectrometry. *Talanta* 129: 352-363.

Poblet, M., Rozès, N., Guillamòn, J.M. and Mas, A. (2000). Identification of acetic acid bacteria by restriction fragment length polymorphism analysis of a PCR-amplified fragment of the gene coding for 16S rRNA. *Letters in Applied Microbiology*, 31: 63-67.

Pohl, P. (2007). What do metals tell us about wine? *Trends in Analytical Chemistry* 26: 941-949.

Prakitchaiwattana, C.J., Fleet, G.H. and Heard, G.M. (2004). Application and evaluation of denaturing gradient gel electrophoresis to analyse the yeast ecology of wine grapes. *FEMS Yeast Res* 4: 865-877.

Pramateftaki, P.V., Lanaridis, P. and Typas, M.A. (2000). Molecular identification of wine yeasts at species or strains level: A case study with strains from two vine growing areas of Greece. *Journal of Applied Microbiology* 89: 236-248.

Pramateftaki, P.V., Metafa, M., Karapetrou, G. and Marmaras, G. (2012). Assessment of the genetic polymorphism and biogenic amine production of indigenous *Oenococcus oeni* strains isolated from Greek red wines. *Food Microbiology* 29: 113-120.

Preti, R., Antonelli, M.L., Bernacchia, R. and Vinci, G. (2015). Fast determination of biogenic amines in beverages by a core-shell particle column. *Food Chemistry* 187: 555-562.

Proestos, C., Loukatos, P. and Komaitis, M. (2008). Determination of biogenic amines in wines by HPLC with precolumn dansylation and fluorimetric detection. *Food Chemistry* 106: 1218-1224.

Pyrzynska, K, (2007). Chemical speciation and fractionation of metals in wine. *Chemical Speciation and Bioavailability* 19: 1-8.

Quiròs, C., Herrero, M., García, L.A. and Díaz, M. (2009). Quantitative approach to determining the contribution of viable-but-non culturable subpopulations to malolactic fermentation processes. *Applied Environmental Microbiology* 75: 2977-2981.

Raspor, P., Cus, F., Jemec, K.P., Zagorc, T., Cadez, N. and Nemanic, J. (2002). Yeast population dynamics in spontaneous and inoculated alcoholic fermentation of Zametovka must. *Food Technology and Biotechnology* 40: 95-102.

Rawsthorne, H. and Phister, T.G. (2006). Areal-time PCR assay for the enumeration and detection of *Zygosaccharomyces bailii* from wine and fruit juices. *International Journal of Food Microbiology* 112: 1-7.

Reguant, C. and Bordons, A. (2003). Typification of *Oenococcus oeni* strains by multiplex RAPD-PCR and study of population dynamics during malolactic fermentation. *Journal of Appllied Microbiology* 95: 344-353.

Renouf, V., Falcou, M., Miot-Sertier, C., Perello, M.C., De Revel, G. and Lonvaud-Funel, A. (2006). Interactions between *Brettanomyces bruxellensis* and other yeast species during the initial stages of winemaking. *Journal of Applied Microbiology* 100: 1208-1219.

Rodas, A.M., Ferrer, S. and Pardo, I. (2005). Polyphasic study of wine *Lactobacillus* strains: Taxonomic implications. *International Journal of Systematic and Evolutionary Microbiology* 55: 197-207.

Rodrìguez, M.E., Lopes, C.A., van Broock, M., Valles, S., Ramon, D. and Caballero, A.C. (2004). Screening and typing of Patagonia wine yeasts for glycosidase activities. *Journal of Applied Microbiology* 96: 84-95.

Rodriguez, S.B. and Thornton, R.J. (2008). Use of flow cytometry with fluorescent anti-bodies in real-time monitoring of simultaneously inoculated alcoholic-malo-lactic fermentation of Chardonnay. *Letters in Applied Microbiology* 46: 38-42.

Rodríguez-Cabo, T., Rodríguez, I., Ramil, M., Silva, A. and Cela, R. (2016). Multiclass semi-volatile compounds determination in wine by gas-chromatography accurate time-of-flight mass spectrometry. *Journal of Chromatography A* 1447: 107-117.

Rodríguez-Cabo, T., Rodríguez, I., López, P., Ramil, M. and Cela, R. (2014). Investigation of liquid chromatography quadrupole time-of-flight mass spectrometry performance for identification and determination of hydroxylated stilbene antioxidants in wine. *Journal of Chromatography A* 1337: 162-170.

Romancino, D.P., di Maio, S., Muriella, R. and Oliva, D. (2008). Analysis of non-*Saccharomyces* yeast populations isolated from grape musts from Sicily (Italy). *Journal of Applied Microbiology* 105: 2248-2254.

Romboli, Y., Di Gennaro, S.F., Mangani, S., Buscioni, G., Matese, A., Genesio, L. and Vincenzini, M. (2017). Vine vigour modulates bunch microclimate and affects the composition of grape and wine flavonoids: An unmanned aerial vehicle approach in a Sangiovese vineyard in Tuscany. *Australian Journal of Grape and Wine Research* 23: 1-10.

Romero, R., Gázquez, D., Bagur, M.G. and Sánchez-Viñas, M. (2000). Optimization of chromatographic parameters for the determination of biogenic amines in wines by reversed-phase high-performance liquid chromatography. *Journal of Chromatography A* 871: 75-83.

Ronaghi, M., Karamohamed, S., Pettersson, B., Uhlen, M. and Nyren, P. (1996). Real-time DNA sequencing using detection of pyrophosphate release. *Analytical Biochemistry* 242: 84-89.

Ronkainen, N. (2016). Determination of trace elements in wine by atomic spectroscopy and electroanalytical methods. pp. 417-439. *In*: Morata, A. (Ed.). Grape and Wine Biotechnology. IntechOpen.

Rosi, I., Fia, G., Millarini, V. and Nannelli, F. (2007). Characterization of *Oenococcus oeni* strains for the production of beta-glucosidase and esterase activity. Oeno 2007, 8th International Enology Symposium, Bordeaux, France, June, 25-27, 2007.

Ruiz, A., Poblet, M., Mas, A. and Guillamon, J.M. (2000). Identification of acetic acid bacteria by RFLP of PCR amplified 16S rDNA and 16S-23S rDNA intergenic spacer. *International Journal of Systematic and Evolutionary Microbiology* 50: 1981-1987.

Sabatè, J., Cano, J., Esteve-Zarzoso, B. and Guillamòn, J.M. (2002). Isolation and identification of yeasts associated with vineyard and winery by RFLP analysis of ribosomal genes and mitochondrial DNA. *Microbiology Research* 157: 267-274.

Salinas, F., Garrido, D., Ganga, A., Veliz, G. and Martinez, C. (2009). Taqman real-time PCR for the detection and enumeration of *Saccharomyces cerevisiae* in wine. *Food Microbiology* 26: 328-332.

Salma, M., Rousseaux, S., Sequeira-Le Grand, A., Divol, B. and Alexandre, H. (2013). Characterization of the viable but nonculturable (VBNC) state in *Saccharomyces cerevisiae*. *PLoS One* 8, https://doi.org/10.1371/journal.pone.0077600.

Schuller, D., Valero, E., Dequin, S. and Casal, M. (2004). Survey of molecular methods for the typing of wine yeast strains. *FEMS Microbiology Letters* 231: 19-26.

Serpaggi, V., Remize, F., Sequeira-Le Grand, A. and Alexandre, H. (2010). Specific identification and enumeration of the spoilage microorganism *Brettanomyces* in wine by flow cytometry: A useful tool for winemakers. *Cytometry* 77A: 497-499.

Serpaggi, V., Remize, F., Recorbet, G., Gaudot-Dumas, E., Sequeira-Le Grand, A. and Alexandre, H. (2012). Characterization of the "viable but nonculturable" (VBNC) state in the wine spoilage yeast *Brettanomyces*. *Food Microbiology* 30: 438-447.

Shendure, J. and Ji, H. (2008). Next-generation DNA sequencing. *Nature Biotechnology* 26: 1135-1145.

Siegrist, J., Kohlstock, M., Merx, K. and Vetter, K. (2015). Cap. 14 - Rapid detection and identification of spoilage bacteria in beer. pp. 287-318. *In*: Brewing Microbiology. Managing Microbes, Ensuring Quality and Valorising Waste Woodhead Publishing. Series in Food Science, Technology and Nutrition.

Silla Santos, M.H. (1996). Biogenic amines: Their importance in foods. *International Journal of Food Microbiology* 29: 213-231.

Silva, C., Pereira, J., Wouter, V.G., Giró, C. and Câmara, J.S. (2011). A fast method using a new hydrophilic–lipophilic balanced sorbent in combination with ultra-high performance liquid chromatography for quantification of significant bioactive metabolites in wines. *Talanta* 86: 82-90.

Soares, R.R.G., Novo, P., Azevedo, A.M., Fernandes, P., Chu, V., Conde, J.P. and Aires-Barros, M.R. (2014). Aqueous two-phase systems for enhancing immunoassay sensitivity: Simultaneous concentration of mycotoxins and neutralization of matrix interference. *Journal of Chromatography A* 1361: 67-76.

Sohier, D. and Lonvaud-Funel, A. (1998). Rapid and sensitive in situ hybridization method for detecting and identifying lactic acid bacteria in wine. *Food Microbiology* 15: 391-397.

Soleas, G.J., Yan, J. and Goldberg, D.M. (2001). Assay of Ochratoxin A in wine and beer by high-pressure liquid chromatography photodiode array and gas chromatography mass selective detection. *Journal of Agricultural and Food Chemistry* 49: 2733-2740.

Solieri, L., Genova, F., De Paola, M. and Giudici, P. (2010). Characterization and technological properties of *Oenococcus oeni* strains from wine spontaneous malolactic fermentations: A framework for selection of new starter cultures. *Journal of Applied Microbiology* 108: 285-298.

Solieri, L., Dakal, T.C. and Giudici, P. (2013). Next-generation sequencing and its potential impact on food microbial genomics. *Annals of Microbiology* 63: 21-37.

Souza-Silva, É.A., Gionfriddo, E. and Pawliszyn, J. (2015). A critical review of the state of the art of solid-phase microextraction of complex matrices II. *Food Analysis* 71: 236-248.

Spano, G. and Torriani, S. (2016). Editorial: Microbiota of grapes: Positive and negative role on wine quality. *Frontiers Microbiology* 7: 2036, doi: 10.3389/fmicb.2016.02036.

Stender, H.C., Kurzman, C., Hylding-Nielsen, J.J., Soresen, D., Broomer, A., Oliveira, K., Perry O'Keefe, H., Sage, A., Young, B. and Coull, J. (2001). Identification of *Dekkera bruxellensis* (*Brettanomyces*) from wine by fluorescence in situ hybridation using peptide nucleic acid probes. *Applied and Environmental Microbiology* 67: 938-941.

Tessonniere, H., Vidal, S., Barnavon, L., Alexandre, H. and Remize, F. (2009). Design and performance test of a real-time PCR assay for sensitive and reliable direct quantification of *Brettanomyces* in wine. *International Journal of Food Microbiology* 129: 237-243.

Tofalo, R., Chaves-Lòpez, C., di Fabio, F., Schirone, M., Felis, G.E. and Torriani, S. (2009). Molecular identification and osmotolerant profile of wine yeasts that ferment a high sugar grape must. *International Journal of Food Microbiology* 130: 179-187.

Torija, M.J., Rozes, N., Poblet, M., Guillamòn, J.M. and Mas, A. (2001). Yeast population dynamics in spontaneous fermentations: Comparison between two different wine-producing areas over a period of three years. *Antonie van Leeuwenhoek* 79: 345-352.

Torija, M.J., Rozes, N., Poblet, M., Guillamòn, J.M. and Mas, A. (2003). Effects of fermentation temperature on the strain population *Saccharomyces cerevisiae*. *International Journal of Food Microbiology* 80: 47-53.

Turner, N.W., Bramhmbhatt, H., Szabo-Vezse, M., Poma, A., Coker, R. and Piletsky, S.A. (2015). Analytical methods for determination of mycotoxins: An update (2009-2014). *Analytica Chimica Acta* 901: 12-33.

Urso, R., Rantsiou, K., Dolci, P., Rolle, L., Comi, G. and Cocollin, L. (2008). Yeast biodiversity and dynamics during sweet wine production as determined by molecular methods. *FESM Yeast Research* 8: 1053-1062.

Ventura, M., Casas, I.A., Morelli, L. and Callegari, M.L. (2000). Rapid amplified ribosomal DNA restriction analysis (ARDRA) identification of *Lactobacillus* spp. Isolated from fecal and vaginal samples. *Systematic and Applied Microbiology* 23: 504-509.

Vidal, J.C., Bonel, L., Ezquerra, A., Hernández, S., Bertolín, J.R., Cubel, C. and Castillo, J.R. (2013). Electrochemical affinity biosensors for detection of mycotoxins: A review. *Biosensors and Bioelectronics* 49: 146-158.

Vidal-Carou, M.C., Lahoz-Portolés, F., Bover-Cid, S. and Mariné-Font, A. (2003). Ion-pair high-performance liquid chromatographic determination of biogenic amines and polyamines in wine and other alcoholic beverages. *Journal of Chromatography A* 998: 235-241.

Villano, C., Lisanti, M.T., Gambuti, A., Vecchio, R., Moio, L., Frusciante, L., Aversano, R. and Carputo, D. (2017). Wine varietal authentication based on phenolics, volatiles and DNA markers: State of the art, perspectives and drawbacks. *Food Control* 80: 1-10.

Vincenzini, M., Guerrini, S., Mangani, S. and Granchi, L. (2017). Amino Acid Metabolisms and Production of Biogenic Amines and Ethyl Carbamate. pp. 231-253. *In*: König, H., G. Unden and J. Fröhlich (Eds.). Biology of Microorganisms on Grapes, in Must and in Wine. Springer International Publishing AG, Cham.

Wang, C., Esteve-Zarzoso, B. and Mas, A. (2014). Monitoring of *Saccharomyces cerevisiae*, *Hanseniaspora uvarum* and *Starmerella bacillaris* (synonym *Candida zemplinina*) populations during alcoholic fermentation by fluorescence in situ hybridization. *International Journal of Food Microbiology* 191: 1-9.

Whiton, R.S. and Zoecklein, B.W. (2002). Determination of ethyl carbamate in wine by solid-phase microextraction and gas chromatography/mass spectrometry. *American Journal of Enology and Viticulture* 53: 60-63.

Xufre, A., Albergaria, H., Inácio, J., Spencer-Martins, I. and Gírio, F. (2006). Application of fluorescence in situ hybridization (FISH) to the analysis of yeast population dynamics in winery and laboratory grape must fermentations. *International Journal of Food Microbiology* 108: 376-384.

Yamada, Y. and Yukphan, P. (2008). Genera and species in acetic acid bacteria. *International Journal of Food Microbiology* 125: 15-24.

Zapparoli, G., Fracchetti, F., Stefanelli, E. and Torriani, S. (2012). Genetic and pheno-typic strain heterogeneity within a natural population of *Oenococcus oeni* from Amarone wine. *Journal of Applied Microbiology* 113: 1087-1096.

Zapparoli, G., Torriani, S., Pesente, P. and Dellaglio, F. (1998). Design and evaluation of malolactic enzyme gene targeted primers for rapid identification and detection of *Oenococcus oeni* in wine. *Letters in Applied Microbiology* 27: 243-246.

Zhang, Y. and Zhang, J. (2008). Optimization of headspace solid-phase microextraction for analysis of ethyl carbamate in alcoholic beverages using a face-centered cube central composite design. *Analytica Chimica Acta* 627: 212-218.

Zhou, Q., Qian, Y. and Qian, M.C. (2015). Analysis of volatile phenols in alcoholic beverage by ethylene glycol-polydimethylsiloxane based stir bar sorptive extraction and gas chromatography–mass spectrometry. *Journal of Chromatography A* 1390: 22-27.

Zotou, A., Laukou, Z., Soufleros, E. and Stratis, I. (2003). Determination of biogenic amines in wines and beers by high performance liquid chromatography with pre-column dansylation and ultraviolet detection. *Chromatography* 57: 429-439.

Zott, K., Claisse, O., Lucas, P., Coulon, J., Lonvaud-Funel, A. and Masneuf-Pomarede, I. (2010). Characterization of the yeast ecosystem in grape must and wine using real-time PCR. *Food Microbiology* 27: 559-567.

# 27 Astringency and Colour of Wine: Role, Significance, Mechanism and Methods of Evaluation

**M. Teresa Escribano-Bailón***, **Alba M. Ramos-Pineda and Ignacio García-Estévez**

Polyphenols Research Group, Department of Analytical Chemistry, Nutrition and Food Science, Faculty of Pharmacy, University of Salamanca, Campus Miguel de Unamuno, E-37007, Salamanca, Spain

## 1. Introduction

Alcoholic beverages including wines and brandies have been consumed and enjoyed by the mankind for centuries (Joshi *et al.*, 2011). Wine is the fermented juice of the grape. Fruit wines, derived from fruits other than grapes, include cider from apple, perry from pears, plum wine, cherry wine and others from various berries (Joshi *et al.*, 1999; Kosseva *et al.*, 2017). With the advances in the technology, horticultural practices or newer sources of wine, brandy or other exotic drinks are developed, testing of their acceptance by consumers has become very important (Narasimhan and Stephen, 2011). Consumer acceptance is defined as the reaction of the user to a product determined by personal factors as age, like, dislike, familiarity, economics etc. This is an "affective" response and the major focus is on a population, its behaviour/response. Sensory analysis on the other hand, is testing of product under bias free conditions, using human beings as tools.

Sensory evaluation is a scientific discipline used to evoke, measure, analyze and interpret the reactions to characteristics of food/beverage or materials as perceived by senses of sight, smell, taste, touch and hearing (Narasimhan and Rajalakshmi, 1999). The measurements are not simple physical quantities but involved interpretations of the sensations evoked and perceived at a psychological level, in response to various physical and chemical stimuli. It is certainly different from casual examination of products.

In evaluation of wine, colour and astringency are very important characteristics. The acceptance of a quality wine depends greatly on how its colour is perceived or liked while the astringency influences the acceptance of the taste of a wine. Regarding astringency, recent findings on the molecular mechanisms of astringency and the instrumental analysis performed to assess the astringency from a molecular point of view are discussed. As for colour, the available oenological tools to enhance colour extraction and to improve wine colour stability are also addressed. The consequences of global climate change have several environmental impacts including the warming of the Northern Hemisphere and the alteration of the rainfall patterns. These changes are expected to have high impact in the viticulture and oenology of the Southern European wine-producing regions. Indeed, the effects have already been observed especially in warm vintages: an important gap between the technological maturity (coming earlier) and the phenolic and aromatic maturity (which come later), resulting in wines with unbalanced astringency or with irregular, poor or unstable colour. This chapter is focused on the astringency and colour of red wine with the aim to understand these two sensory properties.

## 2. Astringency

Astringency has been defined as "the complex of sensations due to shrinking, drawing or puckering of the epithelium as a result of exposure to substances such as alums or tannins" by the America Society for Testing Materials (ASTM, 2004). Although the bases of the astringency mechanism are not well understood yet, it is known to be engendered by different classes of astringent compounds, including salt of multivalent metallic cations (particularly aluminum salts), dehydrating agents (ethanol and acetone), mineral and organic acids and polyphenols (Bajec and Pickering, 2008; Joslyn and Goldstein, 1964).

*Corresponding author: escriban@usal.es

## 2.1. Mechanisms for Astringency

The word astringency is derived from the Latin word, *ad stringem*, meaning 'to bind', showing the basis of this primary chemical process. The first mechanism for astringency was proposed by Bate-Smith (1954) indicating the main reaction whereby astringency develops is *via* the precipitation of proteins and mucopolysaccharides in the mucous secretions (Bate-Smith, 1973). This view is still broadly accepted, although widely different opinions exist around the astringency process.

Lee and Lawless (1991) proposed that astringency can be broken down into multiple sub-qualities. Their data suggested that the tactile attributes of drying and roughing were the most closely associated with astringency, implying changes in the texture of the oral mucosa. Green (1993) also pointed to a tactile origin of the astringency sensation, mainly caused by the precipitation of salivary proteins and possibly cross-linking of proteins in the mucosa. Since some wine polyphenols are able to bind salivary proteins, they can form insoluble tannin-protein precipitates in the mouth, causing a loss of lubrication and increased friction in the oral cavity, which would explain its astringency (Baxter *et al.*, 1997). The most accepted mechanisms to explain these facts was proposed by Siebert *et al.* (1996). Regarding this mechanism (Fig. 1), a protein has a fixed number of sites to which tannin can bind while each polyphenol also has fixed number of binding sites. When the number of binding sites in the polyphenol equals the number of binding sites in the protein, the largest network and the maximum protein precipitation will be produced. Depending on the ratio of protein or tannin used, different protein-polyphenol complexes will be formed (Soares *et al.*, 2012).

Moreover, the interaction process between polyphenols and peptides has been divided into three stages (Charlton *et al.*, 2002). Initially, reversible associations between the hydrophobic face of the aromatic rings of the polyphenol and the pyrrolidine ring of the proline residues of the protein give a soluble complex. In general, several molecules can bind to the same peptide. In the second stage, two peptides are cross-linked by the addition of more polyphenols, which can bind to the peptide acting as a linker between two peptides, by cooperative weak intermolecular binding interactions, leading to a bigger and insoluble complex which starts to precipitate. Finally, the complex aggregates into larger or smaller

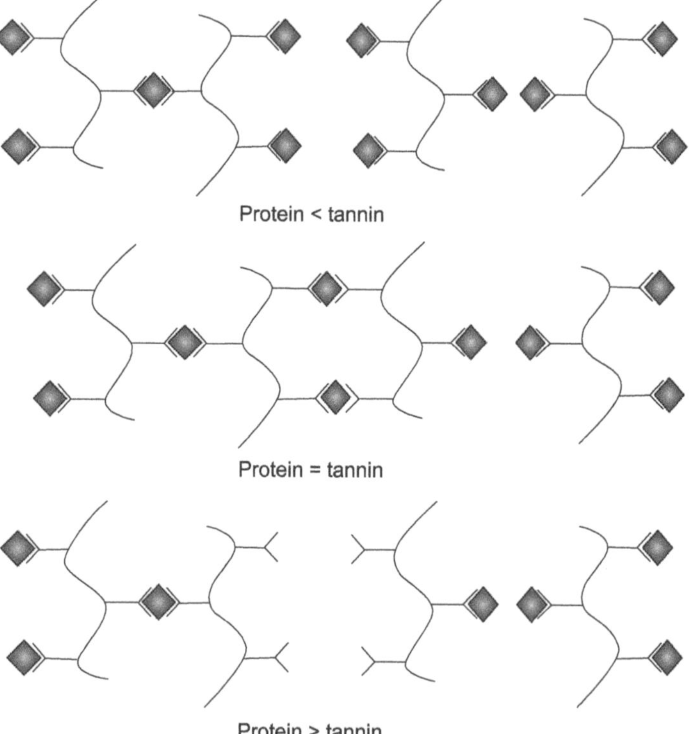


**Figure 1.** Model for protein/polyphenol interaction (Adapted from Siebert *et al.*, 1996)

particles, seen as a phase separation process. The 3-stage model was later confirmed and expanded by Jöbstl *et al.* (2004).

Protein-polyphenol aggregates have been described as both soluble and insoluble, and its stability depends on several variables like protein: polyphenol ratios, pH, temperature and ionic strength of solution, but also on the types of polyphenol and protein used (Bajec and Pickering, 2008).Furthermore, the existence of synergistic effect between phenolic compounds has also recently been suggested. It has been proposed that it could be due to a cooperative behaviour between phenolic compounds when binding proteins, and it could explain why astringency of wine is more influenced by the qualitative phenolic composition than by the total concentration (Ferrer-Gallego *et al.*, 2014; Ramos-Pineda *et al.*, 2017). Even more, the existence of a synergistic effect of the coexistence of different salivary protein families on the interaction with wine flavanols has been recently proposed (Ramos-Pineda *et al.*, 2019).

Although precipitation of salivary proteins, namely PRPs (proline-rich proteins), is one of the most accepted mechanisms, not all astringents cause salivary protein precipitation, evidencing there must be other mechanisms implicated in astringency development. For instance, Lee and Vickers (2012) studied the influence of loss of salivary lubricity on the development of the astringency sensation. They determined both sensory and instrumentally if different astringent compounds could induce changes in saliva lubricity, evincing that precipitation of PRPs or mucins is not a requirement to the development of astringency. Considering these results, they proposed that changes in friction or lubricity are not necessary conditions for astringency and they stipulated that direct tissue effects are related with tannin astringency while acid astringency could be related with coating disruption (Lee *et al.*, 2012).

Other approaches have suggested that astringency could be engendered by activation of specific taste receptors (Tachibana *et al.*, 2004) or even by direct interactions between tannins and oral epithelial cells (Payne *et al.*, 2009). In this study, a cell-based assay was developed with which demonstrated that interactions between grape seed tannins occur (Payne *et al.*, 2009). Gibbins and Carpenter (2013) proposed that the complex sensation of astringency could involve multiple mechanisms occurring simultaneously (Fig. 2): aggregation of salivary proteins, salivary film disruption, decrease in salivary lubrication, receptors exposure and mechanoreceptors stimulation in the oral mucosal epithelium.

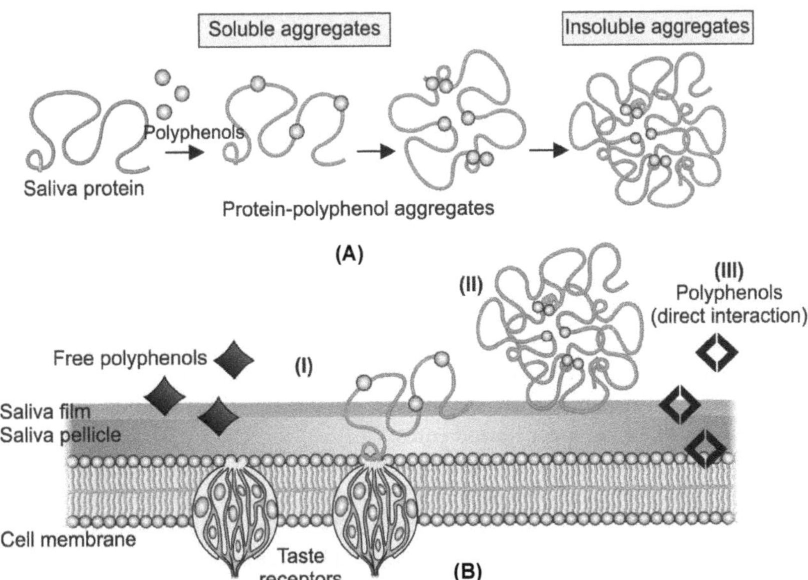


**Figure 2.** Proposed astringency mechanisms: (A) Protein aggregation and complex formation by the interaction between polyphenols and salivary proteins; (B) Astringency development: (I) Free polyphenols and soluble polyphenol–protein aggregates could disrupt the salivary film and reach the pellicle or even activate specific taste receptors. (II) Insoluble aggregates are rejected against salivary film, causing a loss of lubrication and increased friction in the oral cavity. (III) Direct interaction between polyphenols an oral epithelium (Adapted from Ma *et al.*, 2014)

Color version at the end of the book

Nevertheless, since there are no conclusive results, the scientific community is still discussing the different proposed mechanisms that explain this complex phenomenon. Although the literature suggests several theories related to astringency, we could differentiate two main hypotheses: astringency as a tactile or taste sensation.

## 2.2. The Theory of Astringency as a Tactile Sensation

According to this theory, astringency is a feeling not a taste, attributing its tactile nature to the increased friction between oral surfaces after the loss of lubrication. Many authors have furthered the tactile theory of astringency, suggesting that protein precipitation could lead to oral tissue constriction due to a loss of oral lubrication, perceived as an increase in oral friction (Clifford, 1997). Other authors have suggested that astringent compounds could cause a sensation of roughness when changing the oral epithelium (Jellinek, 1985). Breslin *et al.* (1993) contribute with this theory presenting additional evidence that astringency is a tactile sensation resulting from the stimulation of mechanoreceptors during movement of the oral mucosa, consistent with Bate-Smith's (1954) previous speculation. Moreover, Green (1993) explained not only how dehydration can lead to sensations of dryness, but also how the cross-linking of proteins may contribute to the development of astringency. The main reason probably lies in the effects that astringents have on the lubricating capacity of saliva. Cross-linking mucoproteins could induce astringents to precipitate, leaving a losing of viscosity and lubrication in the oral fluid. Additionally, free precipitated proteins are able to adhere to the mucosa and dentition, where they can form a sticky residue. Both effects may increase the coefficient of friction between the surfaces of the oral cavity, changing the tactile sensation perceived (Green, 1993).

Although most studies have assumed that the oral perception of astringency is related to salivary protein precipitation, mainly PRPs, other authors have also proposed the possibility of free astringent stimulus that interact directly with oral tissue through receptors (Schwarz and Hofmann 2008, Lee *et al.*, 2012). Moreover, several studies have shown that the precipitation of PRPs is not required for astringency development, but they could play a protective role. PRP-binding reaction could prevent astringent compounds to interact with the oral mucosa (Horne *et al.*, 2002).

As discussed above, the interaction between salivary proteins and wine polyphenols can lead to the formation of insoluble aggregates, decreasing lubricity in the oral cavity. The increased friction stimulates mechanoreceptors in the oral mucosa, which are responsible to pressure, touch, vibration, tension and stretch responses. These mechanoreceptors are both superficial slow-adapting (SA) and rapidly adapting (RA) receptor units (Van Aken, 2010) and could be the most susceptible gustatory receptors to stimulation by this mechanism (Breslin *et al.*, 1993). However, Breslin *et al.* (1993) and Lim and Lawless (2005) revealed direct physiological data to explain astringency as a tactile sensation that is mediated by non-gustatory mechanisms. In these studies, they tested some astringent compounds in an area of the mouth exempt of taste receptors, finding that they could elicit the sensation of astringency. Based on these results, these authors considered the possibility that taste receptors were not responsible for astringency, so the presence of these receptors was not an essential factor. On the contrary, Green (1993) suggested that "the mechanoreceptors responsible for astringency may be RA receptor units that have been identified in the chorda tympani and lingual nerve". In order to prove the tactile origin of astringency, Breslin *et al.* (1993) confirmed by an experimental study that astringency was evoked on non-gustatory surface between the upper lip and gums.

Another explanation against the taste hypothesis has been proposed, since it seems that perceived astringency intensifies with repeated sampling. Guinard *et al.* (1986) and Lyman and Green (1990) demonstrated that the astringency of wine and beer increased over three and five exposures, respectively, and that the rate of ingestion affected the rate of increase, a typical feature of trigeminal, but not taste, sensations.

Finally, many astringent compounds have shown its ability to precipitate mucins, large salivary glycoproteins produced by epithelial tissues. However, published research studies focused on the role of mucins in the development of astringency are scarce, although it is broadly assumed its importance in oral lubrication. It has even been proposed that mucosa pellicle could be more important than the salivary film in the perception of astringency (Nayak and Carpenter 2008, Lee *et al.*, 2012). In relation with the tactile theory of astringency, it has been suggested that polyphenols compounds could disrupt the salivary mucosal pellicle, causing an increased friction in the oral surface, which could stimulate mechanoreceptors as it has been previously explained (Gibbins and Carpenter, 2013). More recently, some authors have explored sensory astringency using "oral tribology" approaches. Tribology is the science of adhesion, friction, and lubrication of interaction surfaces that are in relative motion (Upadhyay *et al.*, 2016). The latest studies using this approach have found a relationship between sensory and friction, but no conclusive results have been obtained clarifying astringency perception, despite being a potential tool (Brossard *et al.*, 2016, Laguna *et al.*, 2017).

## 2.3. The Theory of Astringency as a Taste Sensation

Taste is firstly recognised on the tongue, where epithelial receptors cells are located, organised into taste buds in papillae. Gustatory papillae can be divided into three types: fungiform papillae, foliate papillae and circumvallate papillae (see Fig. 3). Mammalian taste receptors cells transmit action potentials to neurons of the gustatory fibers that innervate taste buds. Fungiform papillae are placed in the anterior two-thirds of the tongue, and receive innervation from the chorda tympani branch of the facial nerve (cranial nerve VII). The facial nerve connects gustatory papillae of this area of the tongue, ending on the lingual branch of the trigeminal nerve (CN V). Foliate papillae are found on the lateral edges whereas circumvallate papillae are found in the posterior one-third of the tongue. These papillae are innervated by the glossopharyngeal nerve (CN IX) (Scott 2005, Bajec and Pickering 2008). Those three nerves (facial nerve, glossopharyngeal nerve and trigeminal nerve) provide innervation to the oral cavity (Matthews 2001), receiving taste information that is transmitted to the thalamus and finally to the gustatory areas of the cortex (Scott 2005).

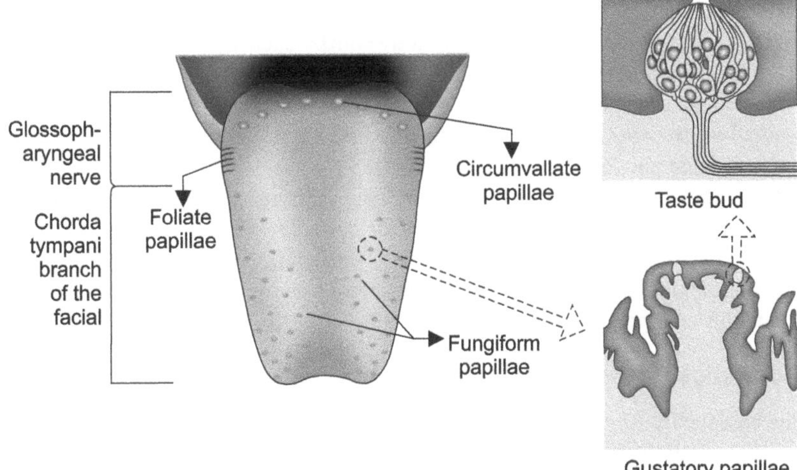


**Figure 3**. Gustatory papillae location

Nowadays, astringency is not considered one of the five basic taste modalities, including salty, sour, umami, sweet and bitter. These taste modalities are sensed by taste buds on the tongue and the information is transmitted to the brain through taste nerves, as explained above. It is not clear if any receptor exists that recognise astringents, or the signaling cascades downstream (Jiang 2014).However, several authors have supported a possible interpretation of astringency as a taste sensation. Kawamura *et al.* (1969) demonstrated that tannic acids do not interact directly with mechanoreceptors. Subsequently, Schiffman *et al.* (1992) attribute the tactile, thermal, and pain response to the lingual nerve, although it is also known that this nerve is responsible to chemical stimulation (Wang *et al.*, 1993). Moreover, they established that the chorda tympani branch of the facial nerve (which innervates taste cells) was sensitive to astringent compounds, whereas the lingual branch of the trigeminal nerve was not. They conclude that mechanoreceptors could not be responsible for the perception of astringency due to the lack of stimulation of the lingual nerve by the astringent compounds. However, these studies did not provide a definitive proof of the gustatory basis of astringency. The studies by Simon *et al.* (1992) also supported this theory. These authors published electrophysiological investigations showing that some astringent compounds affect ion transport through $Na^+$ channels in the lingual epithelium. In the same way, the ability of some astringent compounds to change the membrane potential of a lipid taste sensor has been presented as additional evidences supporting the theory of astringency as a taste sensation (Iiyama *et al.*, 1995). More recently, several studies have shown that polyphenols could directly activate functional transient receptor potential channels in oral epithelial cells, producing intracellular calcium concentration changes when they are opened (Kurogi *et al.*, 2012, Wang *et al.*, 2011). Nevertheless, Carpenter (2013) found that most human oral epithelial cells did not respond to either the transient receptor potential agonist or a black tea solution containing polyphenols.

On the other hand, the cellular effects of astringent substances in cortical signaling have been studied. Critchley and Rolls (1996) investigated the cortical representation of the taste of tannic acid, which produces the taste of astringency. They suggested a sub-population of neuron specific for tannic acid and proposed that it's astringent taste should be considered as a distinct taste quality apart from the five basic taste modalities (Critchley and Rolls, 1996).

Moreover, in the last few years, some research has focused on the study of receptors that could be involved in oral sensing of astringent compounds. Studies in cancer cell lines have identified a receptor for epigallocatechin-gallate as the mammalian 67 kDa laminin receptor (Tachibana *et al.*, 2004), a protein also identified as an extracellular matrix receptor in the oral mucosa (Hakkinen *et al.*, 2000). Other studies have suggested that several astringent wine phenols activate bitter taste receptors (Soares *et al.*, 2013).

Deepening in the neural basis of this complex sensation, Shovel *et al.* (2014) proposed that human astringency perception was mediated by trigeminal nerves through the activation of a G-protein and adenylyl cyclase, and analyze the downstream signaling holding this response. They proved in human subjects that only when trigeminal and taste nerves were both blocked, they lost their astringency sensation, while this sensation was not affected when only blocking taste nerves, indicating that astringency is more likely a trigeminal sensation for humans. Moreover, they tested whether astringents could directly activate trigeminal *neurons in vitro*, suggesting the existence of a trigeminal G protein–coupled receptor for galloylated astringent phenols (Schöbel *et al.*, 2014).

It has been demonstrated that the mechanism of wine astringency goes much beyond the traditional view focused on mechanosensation. The theories discussed here are not exclusive, so a possible explanation of the mechanism for astringency lies in the sum of several mechanisms working together. In this way, caution should be taken when interpreting results and further investigation will be needed to better understand this enigmatic sensation.

## 2.4. Methodologies for Astringency Determination

Because of the complexity of astringency sensation, it is not easy to find an objective methodology for measuring and characterizing it. That is way, nowadays, sensory analysis is still the method regularly used for the determination of wine astringency. However, new approaches using different instrumental techniques have been developed for solving the several drawbacks related to sensory analysis. These new methodologies aim to unravel the astringency mechanisms and/or to predict the astringency sensations elicited by different compounds, which may allow providing an objective explanation for the astringency of different food and beverages, such as wines.

### *2.4.1. Sensory Analysis*

Sensory analysis requires a group of testers (comprised of 8-20 people) that have been previously trained using a set of reference compounds and descriptors, to familiarise them with the astringency sensation and terminology and to standardise the criteria used for evaluating (quantitatively and qualitatively) astringency. Different scales have been employed for quantifying astringency in wines, from the simplest linear magnitude estimations to more complex alternatives involving non-linear spaces, the later usually providing better results in astringency quantification (Green *et al.*, 1996). Among these alternatives, the Labeled Magnitude Scale (LMS) was introduced by Green and co-workers for rating perceptual magnitudes related to taste or aroma (Green *et al.*, 1993). This scale is a quasilogarithmic scale with verbal label descriptors ranging between "barely detectable" to "strongest imaginable" that is not affected by the ceiling effect, thus improving other scales (Pickering *et al.*, 2004).

From a qualitative point of view, to divide astringency into different sub-qualities is helpful for characterizing it. The terminology traditionally employed for describing wine astringency (astringent, puckering, roughing, drying…) can turn out to be quite general and insufficient and, that is why, Gawel and co-workers proposed a structured vocabulary for assisting tasters in the interpretation of the mouth-feel sensations elicited by red wines (Gawel *et al.*, 2000). These authors suggested the terms *paticulate*, *drying*, *harsh* and *unripe* for grouping the negative sub-qualities of astringency whereas the terms *surface smoothness*, *complex* and *dynamic* would group the positive ones (Gawel *et al.*, 2000). Thus, it seems that sensory analysis allowed a comprehensive description of astringency. In fact, it has been possible to develop predictive models for wine astringency from sensory results (Cliff *et al.*, 2002). However, sensory analysis shows important drawbacks: it is time-consuming, expensive and it usually leads to important standard deviations in the determination even when trained panelist are involved, due to the fact that astringency perception is highly subjective (Simoes Costa *et al.*, 2015). The use of instrumental techniques for determining astringency mainly tries to solve the latter issue (García-Estévez *et al.*, 2018).

### 2.4.2. Instrumental Analysis

Most of the instrumental analyses employed for assessing astringency in wines are based on the premise that the key mechanism for astringency development is the interaction between salivary proteins (namely PRPs) and different compounds (mainly tannins and other phenolic compounds). For this reason, the simplest and most used approaches for assessing the astringency properties of phenolic compounds studied its ability for interacting with proteins. This is the basis of the gelatin index, which measures the extent of precipitation of phenolic compounds by means of a gelatin solution. The main limitation of this method is related to the high variability of the results obtained, which is mainly due to the fact that gelatin is a complex and non-standardised mixture of proteins (Llaudy *et al.*, 2004). Moreover, it has been established that this method does not provide good results for high levels of tannins (Goldner and Zamora, 2010). For this reason, other commercial proteins showing more similarities to salivary proteins are used for assessing astringency. Among them, the most utilised are bovine serum albumin (BSA) and α-amylase obtained from porcine pancreas, the latter showing a high degree of homology to salivary α-amylase (Soares *et al.*, 2009). The use of these proteins for assessing the astringency of different wine compounds has provided successful results from a quantitative point of view (Hofmann *et al.*, 2006, Llaudy *et al.*, 2004). However, to be close to the real conditions in the astringency development, recent studies have used purified salivary proteins (SP) for assessing astringency (Silva *et al.*, 2017, Soares *et al.*, 2018).

Moreover, the precipitation of salivary proteins has been studied by means of different techniques, such as SDS-PAGE (Sodium Dodecyl Sulfate PolyAcrylamide *Gel* Electrophoresis), which has been used for assessing the changes in the salivary protein profile after the interaction with different phenolic compounds. As a result, the reactivity of the different salivary proteins (SPs) and the ability of different wine phenolic compounds for interacting with SPs can be determined (Gambuti *et al.*, 2006, Rinaldi *et al.*, 2010, Sarni-Manchado *et al.*, 1999). A similar approach but using liquid chromatography (LC) has been used, allowing the determination of not only the most reactive families of SPs but also the formation of soluble aggregates (Kallithraka *et al.*, 1998; Quijada-Morín *et al.*, 2016). LC coupled to mass spectrometry has been used for proteomics studies about the effect on salivary profile of different astringent stimuli, thus providing qualitative and quantitative information about the interaction (Delius *et al.*, 2017). Moreover, mass spectrometry, namely MALDI-TOF (Matrix Assisted Laser Desorption/Ionisation-Time of Flight), has also been employed for assessing the identity of wine phenolic compound-salivary proteins soluble aggregates (Ferrer-Gallego *et al.*, 2015a; García-Estévez *et al.*, 2017; Pérez-Gregorio *et al.*, 2014).

Other recent approaches have studied the interaction process using techniques further than precipitation of proteins, such as infrared spectroscopy, electronic tongues, fluorescence, nephelometry, dynamic light scattering (DLS), small angle X-ray scattering (SAXS), circular dichroism (CD), nuclear magnetic resonance (NMR) or isothermal titration calorimetry (ITC). Several of these techniques allow studying the interaction even if it does not lead to protein precipitation. For instance, middle infrared spectroscopy (MIR) has been used as a predictive tool for wine astringency. The studies using MIR assess the most important wavelengths for estimating astringency and they build predictive models by means of chemometric tools, such as Partial Least Square regression (PLS) (Simoes Costa *et al.*, 2015; Vera *et al.*, 2010). PLS was also used to build calibration models from near infrared (NIR) spectroscopy that results to predict the sensory attributes related to astringency of grape skin and seed, pointing out the potential of infrared spectroscopy to predict different astringency parameters (Ferrer-Gallego *et al.*, 2013). Similarly, electronic tongues based on spectroscopic, potentiometric and/or electrochemical sensors can be used to estimate astringency. Those allows making measurements that, when calibrated, have been used trying to simulate sensory analysis (Diako *et al.*, 2016). The results obtained by using electronic tongues seems promising, allowing the discrimination and classification of wines (Gay *et al.*, 2010). However, although these methodologies usually show good correlations to sensory analysis, they do not provide any information about the astringency process.

Fluorescence quenching measurements study the reduction in the intrinsic fluorescence of proteins (mainly due to the tryptophan residues (Soares *et al.*, 2009)), as a result of the interaction between proteins and phenolic compounds. The study of the fluorescence quenching allows measuring the extent of the interaction. Results usually showed that the higher concentration of wine phenolic compounds is assayed, the higher quenching effect is observed (Jauregi *et al.*, 2016; Yao *et al.*, 2010). However, Ferrer-Gallego *et al.* (2012) have pointed out that higher phenolic contents do not involve larger affinity towards proteins since the structural features of the wine phenolic compounds (molecular size or galloylation in the structure) could modify the affinity, thus affecting quenching results.

Compound astringency can also be assessed by means of nephelometry, which studies the formation of phenolic compound-protein aggregates by measuring the scattered light when a beam of light is passed through a solution containing suspended particles. Studies carried out by using this technique showed a direct relationship between the level of phenolic compounds and the nephelometric values (Ferrer-Gallego *et al.*, 2012; Monteleone *et al.*, 2004), which, in turn, are also related to the ability of wine phenolic compounds to induce astringency in sensorial analysis (Monteleone *et al.*, 2004). Nephelometry has also been used to assess the astringency mechanism when an agent for modulating astringency of wines is involved (Brandão *et al.*, 2017). However, nephelometry measurements are affected by several factors, mainly by the size of aggregates, since all the particles should be small and of identical size (Monteleone *et al.*, 2004). Thus, to avoid the formation of larger aggregates, nephelometry measurements should be done after a short reaction time. Dynamic light scattering (DLS) could help for solving this problem since it measures the relaxation rate of particles in a solution that scatter light and, thereby, permits an estimation of their diameter (Charlton *et al.*, 2002) providing size distribution of aggregates. This technique provides good results, showing a direct relationship between the size of aggregates and the concentration of wine phenolic compound assayed, suggesting the formation of complexes or metastable aggregates (Ferrer-Gallego *et al.*, 2016; Jauregi *et al.*, 2016; Ramos-Pineda *et al.*, 2018). The size distribution of the aggregates for a large range of particle sizes can also be studied by SAXS. This technique measured the scattered radiation (X-rays in this case) by a solution containing the aggregates when it is irradiated with an X-ray collimated beam. Measurements are done very close to the primary beam ("small angles") and, depending on the angle, different ranges of aggregate sizes can be studied (Petoukhov and Svergun, 2013). Moreover, this technique allows obtaining quantitative information about the strength of the interaction (Pascal *et al.*, 2008).

Nuclear magnetic resonance (NMR) allows a deeper study of the interaction, since it could provide not only quantitative information, measuring the strength of the interaction, but also qualitative, by providing information about the number and the nature of binding sites (Cala *et al.*, 2010a, b). Experiments usually involves a phenolic compound titration, maintaining the protein concentration constant throughout the entire titration. The chemical shifts of protein protons obtained from this experiments are used for obtaining information about the binding sites and the type of aggregates (soluble or insoluble) (Faurie *et al.*, 2016). The association constants can be obtained both from the chemical shifts (Cala *et al.*, 2010b) and from saturation transference difference (STD) experiments (Ferrer-Gallego *et al.*, 2015b). In the latter, the subtraction of the *on-resonance* spectrum (in which protein was selectively saturated by irradiating at a region of the spectrum in which protein protons appear) from the *off-resonance* spectrum (that recorded without protein saturation) is done. In the difference spectrum, only those protons of the wine phenolic compounds that are close to protein *via* binding will appear, since they could receive saturation transfer from the protein (Viegas *et al.*, 2011). This technique also allows the determination of the binding epitope of the phenolic compound, *i.e.* the protons of the astringent compound that are closer to the protein upon binding (Viegas *et al.*, 2011). For instance, recent studies support that some procyanidins could be multidentate ligands and that the first epitopes involved in the interaction between these wine tannins and salivary proteins depends on the structure of the procyanidins, being rings B and E in the case of non-galloylated procyanidins dimers and galloyl ring for galloylated dimers (Soares *et al.*, 2018). Moreover, NMR diffusion experiments can be useful for detecting the formation of small aggregates between phenolic compounds and proteins, allowing the determination of the number of binding sites and the association constants by following the changes in the diffusion of the protein throughout the titration (Cala *et al.*, 2010a). Structural information about the aggregates (size, binding epitopes in protein and phenolic compound, etc.) can be obtained from two-dimensional NMR experiments, such as TOCSY (Total Correlation Spectroscopy), NOESY (Nuclear Overhauser Spectroscopy), ROESY (Rotating-frame Overhauser Spectroscopy), HSQC (Heteronuclear Single Quantum Correlation) and DOSY (Diffusion Ordered Spectroscopy) (Faurie *et al.*, 2016; Pascal *et al.*, 2009; Simon *et al.*, 2003). Qualitative information about protein-phenolic compounds aggregates can also be obtained from circular dichroism (CD) spectroscopy. CD studies the difference in the absorption of the left and right circularly polarised light of the solution containing the protein and/or aggregates, providing information about the bonds and structures responsible for this chirality (Rodger *et al.*, 2005). As for studies about wine astringency, it is usually employed the far-ultraviolet (190-370 nm) spectra for obtaining information about the conformational changes in the protein structure because of interaction (Pascal *et al.*, 2009, Simon *et al.*, 2003).

As for the mechanism of aggregation, ITC (Isothermal Titration Calorimetry) could be a helpful tool for establishing the main forces driving the interaction. In fact, from the titration curves the change in energies (enthalpy ($\Delta H$), Gibbs free energy ($\Delta G$) and entropy ($\Delta S$)) can be determined. Titration at different temperatures helps for distinguishing among the different forces driven the interaction (Kilmister *et al.*, 2016, McRae *et al.*, 2010). From the values of energy changes, it could be ascertained the type of forces involved in the interaction: important $\Delta H$ negative values; i.e. enthalpy drives the interaction, are related to exothermic hydrogen bonding between protein and phenolic compounds. On the contrary, hydrophobic interactions can be considered as the main force of the interaction when the process is entropy-driven, i.e. when positive values for $\Delta H$ and for $\Delta S$ are obtained (Kilmister *et al.*, 2016, McRae *et al.*, 2010, Ramos-Pineda *et al.*, 2017). Moreover, from ITC result, the stoichiometry and the binding constants of association can also be obtained, thus helping to assess the whole interaction process (García-Estévez *et al.*, 2018). As for wine tannins, it has been pointed out that the situation is not so well dichotomised and that both hydrophobic and hydrophilic interactions can play complementary roles in the network formation depending on the protein and procyanidin structures (Soares *et al.*, 2018). However, this study also reports that some proteins such as aPRPs, which show in their structure an N-terminal with acidic residues, seem to preferentially establish H-bonds in the interaction with wine procyanidins, mainly when these compounds present a galloyl group in the structure.

In addition to these experimental techniques, molecular dynamics (MD) simulations have been employed to establish theories about the strength, the mechanisms and the number of molecules involved in the interaction. Calculations are done by using model peptides (Cala *et al.*, 2012; Ferrer-Gallego *et al.*, 2017) or salivary peptides (Soares *et al.*, 2018) to simulate the interaction with one or several molecules of ligand.

Moreover, it has been stated that quantifying only the extend of protein binding could not be enough for explaining sensory perception and that it is necessary to also consider other *free* astringent stimulus in the saliva liquid (Schwarz and Hofmann, 2008). For this reason, new methodologies, based on salivary rheology and oral tribology, are being used with the aim to explain astringency not only based on protein complexation but also in a wider sense. These approaches, which study the modifications in friction and lubrication of saliva when mixture with astringent compounds for explaining astringency (Brossard *et al.*, 2016) have shown a great potential for establishing relationships to the perceived texture and the mouthfeel attributes of different foods, including wine (Upadhyay *et al.*, 2016).

# 3. Colour

Colour is an important sensory property for evaluating the quality of red wine and it is one of the most influential factors when consumers make their choice. The colour of a red wine is determined mainly by the composition and concentration of anthocyanins and anthocyanin derived pigments. Furthermore, colourless or poor coloured compounds like catechins, proanthocyanidins or flavonols, have been found to play an important role in the protection of wine colour since they can act as anthocyanin co-pigments and also contribute to the development of anthocyanin derived pigments during wine aging.

## 3.1. Chemical Stabilisation of Coloured Pigments

The chemical stabilisation of pigments in wine is one of the main research areas of enology. The stability and quality of wine colour is related to its phenolic composition. The anthocyanin stabilisation by co-pigmentation and polymerisation mechanisms is highly dependent on the concentration and nature of other colourless phenols also extracted from grape skins and seeds during maceration. Several strategies have been tried to enhance red wine colour and stability. One of them is the incorporation of extra co-pigments in the wine. This can be achieved by:

- the external addition of phenolic extracts or oenological tannins, that may consist of condensed or hydrolysable tannins or a mixture of them,
- co-vinification of different grape varieties, each contributing with additional cofactors, and
- addition of winery by-products like seeds or skins from white grapes, in order to provide supplementary sources for the extraction of phenolics.

### *3.1.1. External Addition of Oenological Tannins to Improve Wine Colour*

The objective of this practice is mainly to compensate wine quality deficiencies. Oenological tannins could supply compounds to the wine that can take part directly or indirectly in reactions with anthocyanins favouring the synthesis of derivative pigments or can provide compounds that participate in co-pigmentation interactions or that protect them from oxidation. The usefulness of the addition of enological tannin, containing condensed and hydrolysable tannins (ellagitannins), in chemical stabilisation of the colouring matter of the wines has been demonstrated (Alcalde-Eon, 2014a). This chemical stabilisation implies the increase of more stable anthocyanin-derived pigments in relation to control wines (Alcalde-Eon, 2014b), which can be formed by condensation reactions favoured by the presence of ellagitannins and other compounds that can promote the formation of reactive oxygen species (Garcia-Estevez *et al.*, 2015). Furthermore, greater and faster extraction of the anthocyanins in wines treated with oenological tannin has been observed (Boulton, 2001; Darias-Martín *et al.*, 2001; González-Manzano *et al.*, 2009). Although the effect of this oenological practice depends on the characteristics of the grapes at harvest, which in turn may change from one vintage to another (Bautista-Ortín *et al.*, 2007). Also, it has been observed that the addition after alcoholic fermentation has better effect on the phenolic composition than the addition before alcoholic fermentation (Neves *et al.*, 2010).

### *3.1.2. Co-vinification of Different Grape Varieties, Each Contributing with Additional Cofactors*

The extraction and retention of colour in red wines is mainly influenced by the content of cofactors in it. Since not all grape varieties have the same quantities and percentages of individual anthocyanins and other polyphenols, that could act as colour cofactors, co-maceration of different grape varieties could compensate cofactor deficiencies favouring and increasing the content of anthocyanins and contribute to an increase in the co-pigmentation process (García-Marino *et al.*, 2010). It has been observed that blends from different wines could lead to wines with a more balanced anthocyanin/flavanol ratio (Monagas *et al.*, 2006; Garcia-Marino *et al.*, 2013). Pigment extraction and retention in Tempranillo wines seems to be increased by the incorporation of the Graciano variety during the pre-fermentative maceration step and may be linked to the fact that the flavanols from grape skins of the Graciano variety are better co-pigments than those of Tempranillo (García-Marino *et al.*, 2010, 2013). When Monastrell grapes were co-fermented in the presence of Cabernet Sauvignon and Merlot grapes, an increase in the phenolic extraction was observed which influences the colour of the finished wine (Lorenzo *et al.*, 2005).

### *3.1.3. Addition of Winery By-products like Seeds or Skins from White Grapes*

Grape-skins and seeds from white grapes are considered a good source of phenolic compounds (catechin, procyanidins, quercetin glycosides, etc.), which have enological interest to be used in co-fermentations with red grape musts, especially when red grapes do not present a good balance between the concentrations of anthocyanins and co-pigmentation cofactors (Gordillo *et al.*, 2014). It has been observed that the addition of white-grape pomace during wine fermentation increases the extraction of phenolic compounds from grapes. Therefore, it has a positive effect on co-pigmentation and on the colour, although the effect of the addition depends on the doses applied (Gordillo *et al.*, 2014). Recently, Rivero and coworkers observed that the addition of overripe seeds from white grapes during the fermentative stage of red winemaking leads to wines with higher pigment extraction, darker colours and with bluish tones (Rivero *et al.*, 2017). This underlines the potential of the use of by-products for enhancing the stability of red wine colour.

### **3.2. Addition of Mannoproteins to Stabilise Red Wine Colour**

Mannoproteins are glycoproteins of *Saccharomyces cerevisiae* cell wall that can be excreted to the wine during alcoholic fermentation or be released to the wine during yeast autolysis (Ribereau-Gayon, 2000). They are considered as protective colloids that protect wine from protein haze (Dupin *et al.*, 2000) and that can prevent tartrate precipitation (Moine-Ledoux and Dubourdieu, 2002). It has also been reported that mannoproteins can improve colour stability (Escot *et al.*, 2001), probably due to stabilisation effect of the colloidal colouring matter (Alcalde-Eon *et al.*, 2014) and that could increase the monomeric anthocyanin content (Ghanem, 2017). Nevertheless, some other studies have shown that mannoproteins do not affect wine colour (Guadalupe *et al.*, 2010) and even could provoke losses of colour (Guadalupe *et al.*, 2008). These discrepancy shows that it maybe due to the yeast strain used to obtain the mannoprotein or the technique used for obtaining it (acidic hydrolysis, enzymatic hydrolysis, type of enzyme utilised) that could lead to differences in mannoprotein composition. In fact, Fernández and coworkers observed that wines treated with mannoproteins obtained by acidic and enzymatic treatment showed higher intensity of colour than those treated with mannoproteins obtained only by acidic hydrolysis (Fernández *et al.*, 2011).

## 4. Summary and Conclusion

As a consequence of the global warming, the viticulture and the oenology have different challenges to address. The gap between technological and phenolic maturity that leads to wines with low astringency quality and problems of colour stability during the wine ageing is one of the concerns of oenologists in the last years. This chapter is focused on astringency and on colour of red wines with the aim to understand these two sensory properties. Regarding astringency, despite its importance on the quality of red wines, and therefore, its economic importance in the winemaking industry, the mechanisms of astringency are not well understood, neither the possibility of an objective determination. Several aspects that have been outlined in this chapter such as the utilisation of new methodologies as molecular dynamics simulations or tribology and the application of instrumental techniques to unravel the astringency mechanisms and/or to predict the astringency sensations elicited by different compounds, may allow providing an objective explanation for the astringency of wines. This knowledge is essential, for example, to successfully deal, in a non-empirical way, with processes of modulation of astringency and stabilisation of colour in wineries, for example using biopolymers. The addition of oenological products based on cell wall material, obtained from winemaking by-products, could be a suitable and sustainable strategy that worth to be further investigated to modulate wines with unbalanced astringency and to enhance their colour stability.

## Acknowledgments

The authors thank FEDER-Interreg España-Portugal Programme (Project ref. 0377_IBERPHENOL_6_E) and the Spanish MINECO (Project ref. AGL2017-84793-C2-1-R co-funded by FEDER) for the financial support. IGE thanks University of Salamanca for the postdoctoral contract and AMRP thanks MINECO for the FPI scholarship.

## References

Alcalde-Eon, C., García-Estévez, I., Puente, V., Rivas-Gonzalo, J.C. and Escribano-Bailón, M.T. (2014a). Color stabilization of red wines. A chemical and colloidal approach. *Journal of Agricultural and Food Chemistry* 62: 6984-6994.

Alcalde-Eon, C., García-Estévez, I., Ferreras-Charro, R., Rivas-Gonzalo, J.C., Ferrer-Gallego, R. and Escribano-Bailón, M.T. (2014b). Adding oenological tannin vs. overripe grapes: Effect on the phenolic composition of red wines. *Journal of Food Composition and Analysis* 34: 99-113.

ASTM (2004). Standard Definitions of Terms Relating to Sensory Evaluation of Materials and Products. America Society for Testing and Materials, Philadelphia, PA, USA, 2004.

Bajec, M.R. and Pickering, G.J. (2008). Astringency: Mechanisms and perception. *Critical Reviews in Food Science and Nutrition*, 48: 858-875.

Bate-Smith, E.C. (1954). Astringency in foods. *Food* 23: 124-127.

Bate-Smith, E.C. (1973). Haemanalysis of tannins: The concept of relative astringency. *Phytochemistry* 12: 907-912.

Bautista-Ortín, A.B., Fernández-Fernández, J.I., López-Roca, J.M. and Gómez-Plaza, E. (2007). The effects of enological practices in anthocyanins, phenolic compounds and wine colour and their dependence on grape characteristics. *Journal of Food Composition and Analysis* 20: 546-552.

Baxter, N.J., Lilley, T.H., Haslam, E. and Williamson, M.P. (1997). Multiple interactions between polyphenols and a salivary proline-rich protein repeat result in complexation and precipitation. *Biochemistry* 36: 5566-5577.

Boulton, R. (2001). The copigmentation of anthocyanins and its role in the colour of red wine: A critical review. *American Journal of Enology and Viticulture* 52: 67-87.

Brandão, E., Silva, M.S., García-Estévez, I., Williams, P., Mateus, N., Doco, T., de Freitas, V. and Soares, S. (2017). The role of wine polysaccharides on salivary protein-tannin interaction: A molecular approach. *Carbohydrate Polymers* 177: 77-85.

Breslin, P.A.S., Gilmore, M.M., Beauchamp, G.K. and Green, B.G. (1993). Psychophysical evidence that oral astringency is a tactile sensation. *Chemical Senses* 18: 405-417.

Brossard, N., Cai, H., Osorio, F., Bordeu, E. and Chen, J. (2016). "Oral" tribological study on the astringency sensation of red wines. *Journal of Texture Studies* 47: 392-402

Cala, O., Dufourc, E.J., Fouquet, E., Manigand, C., Laguerre, M. and Pianet, I. (2012). The colloidal state of tannins impacts the nature of their interaction with proteins: The case of salivary proline-rich protein/procyanidins binding. *Langmuir* 28: 17410-17418.

Cala, O., Fabre, S., Fouquet, E., Dufourc, E.J. and Planet, I. (2010a). NMR of human saliva protein/wine tannin complexes. Towards deciphering astringency with physico-chemical tools. *Comptes Rendus Chimie* 13: 449-452.

Cala, O., Pinaud, N., Simon, C., Fouquet, E., Laguerre, M., Dufourc, E.J. and Pianet, I. (2010b). NMR and molecular modeling of wine tannins binding to saliva proteins: Revisiting astringency from molecular and colloidal prospects. *Faseb J.* 24: 4281-4290.

Carpenter, G.H. (2013). Do transient receptor protein (TRP) channels play a role in oral astringency? *Journal of Texture Studies* 44: 334-337.

Charlton, A.J., Baxter, N.J., Khan, M.L., Moir, A.J.G., Haslam, E., Davies, A.P. and Williamson, M.P. (2002). Polyphenol/peptide binding and precipitation. *Journal of Agricultural and Food Chemistry* 50: 1593-1601.

Cliff, M.A., Brau, N., King, M.C. and Mazza, G. (2002). Development of predictive models for astringency from anthocyanin, phenolic and color analyses of British Columbia red wines. *Journal of International Science Vigne Vin* 36: 21-30.

Clifford, M. (1997). Astringency. Proceedings of the Phytochemical Society of Europe 41: 87-107.0

Critchley, H.D. and Rolls, E.T. (1996). Responses of primate taste cortex neurons to the astringent tastant tannic acid. *Chemical Senses* 21: 135-145.

Darias-Martín, J., Carrillo, M., Díaz, E. and Boulton, R.B. (2001). Enhancement of red wine color by pre-fermentation addition of copigments. *Food Chemistry* 73: 217-220.

Delius, J., Medard, G., Kuster, B. and Hoffmann, T. (2017). Effect of astringent stimuli on salivary protein interactions elucidated by complementary proteomics approaches. *Journal of Agricultural and Food Chemistry* 65: 2147-2154.

Diako, C., McMahon, K., Mattinson, S., Evans, M. and Ross, C. (2016). Alcohol, tannins, and mannoprotein and their interactions influence the sensory properties of selected commercial merlot wines: A preliminary study. *Journal of Food Science* 81: S2039-S2048.

Dupin, I.V.S., Stockdale, V.J., Williams, P.J., Jones, G.P., Markides, A.J. and Waters, E.J. (2000). *Saccharomyces cerevisiae* mannoproteins that protect wine from protein haze: Evaluation of extraction methods and immunolocalization. *Journal of Agricultural and Food Chemistry* 48: 1086-1095.

Escot, S., Feuillat, M., Dulau, L. and Charpentier, C. (2001). Release of polysaccharides by yeasts and the influence of released polysaccharides on color stability and wine astringency. *Australian Journal of Grape and Wine Research* 7: 153-159.

Faurie, B., Dufourc, E.J., Laguerre, M. and Pianet, I. (2016). Monitoring the interactions of a ternary complex using NMR spectroscopy: The case of sugars, polyphenols, and proteins. *Analytical Chemistry* 88: 12470-12478.

Fernández, O., Martínez, O., Hernández, Z., Guadalupe, Z. and Ayestarán, B. (2011). Effect of the presence of lysated lees on polysaccharides, color and main phenolic compounds of red wine during barrel ageing. *Food Research International* 44: 84-91.

Ferrer-Gallego, R., Brás, N.F., García-Estévez, I., Mateus, N., Rivas-Gonzalo, J.C., de Freitas, V. and Escribano-Bailón, M.T. (2016). Effect of flavonols on wine astringency and their interaction with human saliva. *Food Chemistry* 209: 358-364.

Ferrer-Gallego, R., Gonçalves, R., Rivas-Gonzalo, J.C., Escribano-Bailon, M.T. and de Freitas, V. (2012). Interaction of phenolic compounds with Bovine Serum Albumin (BSA) and alpha-amylase and their relationship to astringency perception. *Food Chemistry* 135: 651-658.

Ferrer-Gallego, R., Hernández-Hierro, J.M., Brás, N.F., Vale, N., Gomes, P., Mateus, N., de Freitas, V., Heredia, F.J. and Escribano-Bailón, M.T. (2017). Interaction between wine phenolic acids and salivary proteins by Saturation-Transfer Difference Nuclear Magnetic Resonance Spectroscopy (STD-NMR) and Molecular Dynamics Simulations. *Journal of Agricultural and Food Chemistry* 65: 6434-6441.

Ferrer-Gallego, R., Hernández-Hierro, J.M., Rivas-Gonzalo, J.C. and Escribano-Bailon, M.T. (2013). Evaluation of sensory parameters of grapes using near infrared spectroscopy. *Journal of Food Engineering* 118: 333-339.

Ferrer-Gallego, R., Hernández-Hierro, J.M., Rivas-Gonzalo, J.C. and Escribano-Bailón, M.T. (2014). Sensory evaluation of bitterness and astringency sub-qualities of wine phenolic compounds: Synergistic effect and modulation by aromas. *Food Research International* 62: 1100-1107.

Ferrer-Gallego, R., Soares, S., Mateus, N., Rivas-Gonzalo, J.C., Escribano-Bailon, M.T. and de Freitas, V. (2015a). New anthocyanin-human salivary protein complexes. *Langmuir* 31: 8392-8401.

Ferrer-Gallego, R., Quijada-Morín, N., Brás, N.F., Gomes, P., de Freitas, V., Rivas-Gonzalo, J.C. and Escribano-Bailón, M.T. (2015b). Characterization of sensory properties of flavanols – A molecular dynamic approach. Chemical Senses 40: 381-390.

Gambuti, A., Rinaldi, A., Pessina, R. and Moio, L. (2006). Evaluation of aglianico rape skin and seed polyphenol astringency by SDS-PAGE electrophoresis of salivary proteins after the binding reaction. *Food Chemistry* 97: 614-620.

García-Estévez, I., Alcalde-Eon, C., Le Grottaglie, L., Rivas-Gonzalo, J.C. and Escribano-Bailón, M.T. (2015). Understanding the ellagitannin extraction process from oak wood. *Tetrahedron* 71: 3089-3094.

García-Estévez, I., Cruz, L., Oliveira, J., Mateus, N., de Freitas, V. and Soares, S. (2017). First evidences of interaction between pyranoanthocyanins and salivary proline-rich proteins. Food Chemistry 228: 574-581.

García-Estévez, I., Ramos-Pineda, A.M. and Escribano-Bailón, M.T. (2018). Interactions between wine phenolic compounds and human saliva in astringency perception. *Food and Function* 9: 1294-1309.

García-Marino, M., Escudero-Gilete, M.L., Heredia, F.J., Escribano-Bailón, M.T. and Rivas-Gonzalo, J.C. (2013). Color-copigmentation study by tristimulus colorimetry (CIELAB) in red wines obtained from Tempranillo and Graciano varieties. *Food Research International* 51: 123-131.

García-Marino, M., Hernández-Hierro, J.M., Rivas-Gonzalo, J.C. and Escribano-Bailón, M.T. (2010). Color and pigment composition of red wines obtained from co-maceration of Tempranillo and Graciano varieties. *Analytica Chimica Acta* 660: 134-142.

Gawel, R., Oberholster, A. and Leigh Francis, I. (2000). A 'Mouth-feel Wheel': Terminology for communicating the mouth-feel characteristics of red wine. *Australian Journal of Grape and Wine Research* 6: 203-207.

Gay, M., Apetrei, C., Nevares, I., del Álamo, M., Zurro, J., Prieto, N., de Saja, J.A. and Rodríguez-Mendez, M.L. (2010). Application of an electronic tongue to study the effect of the use of pieces of wood and micro-oxygenation in the aging of red wine. *Electrochimica Acta* 55: 6782-6788.

Ghanem, C., Taillandier, P., Rizk, M., Rizk, Z., Nehme, N., Souchard, J.P. and El Rayess, Y. (2017). Analysis of the impact of fining agents types, oenological tannins and mannoproteins and their concentrations on the phenolic composition of red wine. *LWT – Food Science and Technology* 83: 101-109.

Gibbins, H.L. and Carpenter, G.H. (2013). Alternative mechanisms of astringency –What is the role of saliva? *Journal of Texture Studies* 44: 364-375.

Goldner, M.C. and Zamora, M.C. (2010). Effect of polyphenol concentrations on astringency perception and its correlation with gelatin index of red wine. *Journal of Sensory Studies* 25: 761-777.

González-Manzano, S., Dueñas, M., Rivas-Gonzalo, J.C., Escribano-Bailón, M.T. and Santos-Buelga, C. (2009). Studies on the copigmentation between anthocyanins and flavan-3-ols and their influence in the color expression of red wine. *Food Chemistry* 114: 649-656.

Gordillo, B., Cejudo-Bastante, M.J., Rodríguez-Pulido, F.J., Jara-Palacios, M.J., Ramírez-Pérez, P., González-Miret, M.L. and Heredia, F.J. (2014). Impact of adding white pomace to red grapes on the phenolic composition and color stability of Syrah wines from a warm climate. *Journal of Agricultural and Food Chemistry* 62: 2663-2671.

Green, B.G. (1993). Oral astringency: A tactile component of flavor. *Acta Physiologica* 84: 119-125.

Green, B.G., Dalton, P., Cowart, B., Shaffer, G., Rankin, K. and Higgins, J. (1996). Evaluating the 'labeled magnitude scale' for measuring sensations of taste and smell. *Chemical Senses* 21: 323-334.

Green, B.G., Shaffer, G.S. and Gilmore, M.M. (1993). Derivation and evaluation of a semantic scale of oral sensation magnitude with apparent ratio properties. *Chemical Senses* 18: 683-702.

Guadalupe, Z. and Ayestarán, B. (2008). Effect of commercial mannoprotein addition on polysaccharide, polyphenolic, and color composition in red wines. *Journal of Agricultural and Food Chemistry* 5: 9022-9029.

Guadalupe, Z., Martínez, L. and Ayestarán, B. (2010) Yeast mannoproteins in red winemaking: Effect on polysaccharide, polyphenolic, and color composition. *American Journal of Enology and Viticulture* 61: 191-200.

Guinard, J.X., Pangborn, R.M. and Lewis, M.J. (1986). Preliminary studies on acidity vs. astringency interactions in model solutions and wines. *Journal of the Science of Food and Agriculture* 37: 811-817.

Häkkinen, L., Uitto, V.J. and Larjava, H. (2000). Cell biology of gingival wound healing. *Periodontology* 24: 127-152.

Hofmann, T., Glabasnia, A., Schwarz, B., Wisman, K.N., Gangwer, K.A. and Hagerman, A.E. (2006). Protein binding and astringent taste of a polymeric procyanidin, 1,2,3,4,6-penta-O-galloyl-beta-D-glucopyranose, castalagin, and grandinin. *Journal of Agricultural and Food Chemistry* 54: 9503-9509.

Horne, J., Hayes, J. and Lawless, H.T. (2002). Turbidity as a measure of salivary protein reactions with astringent substances. *Chemical Senses* 27: 653-659.

Iiyama, S., Ezaki, S., Toko, K., Matsuno, T. and Yamafuji, K. (1995). Study of astringency and pungency with multichannel taste sensor made of lipid membranes. *Sensors and Actuators B: Chemical* 24: 75-79.

Jauregi, P., Olatujoye, J.B., Cabezudo, I., Frazier, R.A. and Gordon, M.H. (2016). Astringency reduction in red wine by whey proteins. *Food Chemistry* 199: 547-555.

Jellinek, G. (1985). Sensory Evaluation of Food Theory and Practice. Deerfield Beach, FL: VCH Publishers.

Jiang, Y., Gong, N.N. and Matsunami, H. (2014). Astringency: A more stringent definition. *Chemical Senses* 39: 467-469.

Jöbstl, E., O'Connell, J., Fairclough, J.P.A. and Williamson, M.P. (2004). Molecular model for astringency produced by polyphenol/protein interactions. *Biomacromolecules* 5: 942-949.

Joslyn, M.A. and Goldstein, J.L. (1964). Astringency of fruits and fruit products in relation to phenolic content. *Advances in Food Research* 13: 179-217.

Joshi, V.K., Sandhu, D.K. and Thakur, N.S. 1999. Fruit based alcoholic beverages. pp. 647-744. *In*: V.K. Joshi and Ashok Pandey (Eds.). Biotechnology: Food Fermentation, vol. II. Educational Publishers and Distributors, New Delhi.

Joshi, V.K., Attri Devender, Singh, Tuhin Kumar and Abrol Ghanshyam (2011). Fruit wines: Production technology. pp. 1177-1221. *In*: V.K. Joshi (Ed.). Handbook of Enology, vol. II1. Asia Tech Publishers, INC. New Delhi.

Kallithraka, S., Bakker, J. and Clifford, M.N. (1998). Evidence that salivary proteins are involved in astringency. *Journal of Sensory Studies* 13: 29-43.

Kawamura, Y., Funakoshi, M., Kasahara, Y. and Yamamoto, T. (1969). A neurophysiological study on astringent taste. *The Japanese Journal of Physiology* 19: 851-865.

Kosseva, M.R., Joshi, V.K. and Panesar, P.S. (Eds.) (2017). Science and Technology of Fruit Wine Production. Academic Press is an imprint of Elsevier, London, United Kingdom, pp. 705.

Kilmister, R.L., Faulkner, P., Downey, M.O., Darby, S.J. and Falconer, R.J. (2016). The complexity of condensed tannin binding to bovine serum albumin – An isothermal titration calorimetry study. *Food Chemistry* 190: 173-178.

Kurogi, M., Miyashita, M., Emoto, Y., Kubo, Y. and Saitoh, O. (2011). Green tea polyphenol epigallocatechin gallate activates TRPA1 in an intestinal enteroendocrine cell line, STC-1. *Chemical Senses*, 37: 167-177.

Laguna, L. and Sarkar, A. (2017). Oral tribology: update on the relevance to study astringency in wines. *Tribology – Materials, Surfaces and Interfaces* 11: 116-123.

Lee, C.A. and Vickers, Z.M. (2012). Astringency of foods may not be directly related to salivary lubricity. *Journal of Food Science* 77: S302-S306.

Lee, C.A., Ismail, B. and Vickers, Z.M. (2012). The role of salivary proteins in the mechanism of astringency. *Journal of Food Science* 77: C381-C387.

Lee, C.B. and Lawless, H.T. (1991). Time-course of astringent sensations. *Chemical Senses* 16: 225-238.

Lim, J. and Lawless, H.T. (2005). Oral sensations from iron and copper sulfate. *Physiology and Behaviour* 85: 308-313.

Llaudy, M.C., Canals, R., Canals, J.M., Rozes, N., Arola, L. and Zamora, F. (2004). New method for evaluating astringency in red wine. *Journal of Agricultural and Food Chemistry* 52: 742-746.

Lorenzo, C., Pardo, F., Zalacain, A., Alonso, G.L. and Salinas, M.R. (2005). Effect of red grapes co-winemaking in polyphenols and color of wines. *Journal of Agricultural and Food Chemistry* 53: 7609-7616.

Lyman, B.J. and Green, B.G. (1990). Oral astringency: Effects of repeated exposure and interactions with sweeteners. *Chemical Senses* 15: 151-164.

Ma, W., Guo, A., Zhang, Y., Wang, H., Liu, Y. and Li, H. (2014). A review on astringency and bitterness perception of tannins in wine. *Trends in Food Science and Technology* 40: 6-19.

Matthews, G.G. (2000). Neurobiology: Molecules, cells and systems. 2nd ed., Malden MA: Blackwell Science.

McRae, J.M., Falconer, R.J. and Kennedy, J.A. (2010). Thermodynamics of grape and wine tannin interaction with polyproline: Implications for red wine astringency. *Journal of Agricultural and Food Chemistry* 58: 12510-12518.

Moine-Ledoux, V. and Dubourdieu, D. (2002). Rôle des mannoprotéines de levures vis-à-vis de la stabilisation tartrique des vins. *Bulletin de l'O.I.V* 75: 471-482.

Monagas, M., Bartolomé, B. and Gómez-Cordovés, C. (2006). Effect of the modifier (Graciano vs. Cabernet Sauvignon) on blends of Tempranillo wine during ageing in the bottle. I. Anthocyanins, pyranoanthocyanins and non-anthocyanin phenolics. *LWT – Food Science and Technology* 39: 1133-1142.

Monteleone, E., Condelli, N., Dinnella, C. and Bertuccioli, M. (2004). Prediction of perceived astringency induced by phenolic compounds. *Food Quality and Preference* 15: 761-769.

Narasimhan, S. and Rajalakshmi, D. (1999). Sensory evaluation of fermented foods. Chapter 7. pp. 345-372. *In*: V.K. Joshi and Ashok Pandey (Eds.). Biotechnology: Food Fermentation, vol. I. Educational Publishers & Distributors, Ernakulam, New Delhi, India.

Narasimhan, S. and Stephen, N. Samuel (2011). Sensory evaluation of wine and brandy. pp. 1331-1366. *In*: V.K. Joshi (Ed.). Handbook of Enology, vol. 3. Asia Tech. Publisher and Distributors, New Delhi.

Nayak, A. and Carpenter, G.H. (2008). A physiological model of tea-induced astringency. *Physiology and Behaviour* 95: 290-294.

Neves, A.C., Spranger, M.I., Zhao, Y., Leandro, M.C. and Sun, B. (2010). Effect of addition of commercial grape seed tannins on phenolic composition, chromatic characteristics, and antioxidant activity of red wine. *Journal of Agricultural and Food Chemistry* 58: 11775-11782.

Pascal, C., Pate, F., Cheynier, V. and Delsuc, M.A. (2009). Study of the interactions between a proline-rich protein and a flavan-3-ol by NMR: Residual structures in the natively unfolded protein provides anchorage points for the ligands. *Biopolymers* 91: 745-756.

Pascal, C., Poncet-Legrand, C., Cabane, B. and Vernhet, A. (2008). Aggregation of a proline-rich protein induced by epigallocatechin gallate and condensed tannins: Effect of protein glycosylation. *Journal of Agricultural and Food Chemistry* 56: 6724-6732.

Payne, C., Bowyer, P.K., Herderich, M. and Bastian, S.E. (2009). Interaction of astringent grape seed procyanidins with oral epithelial cells. *Food Chemistry* 115: 551-557.

Pérez-Gregorio, M.R., Mateus, N. and de Freitas, V. (2014). Rapid screening and identification of new soluble tannin-salivary protein aggregates in saliva by mass spectrometry (MALDI-TOF-TOF and FIA-ESI-MS). *Langmuir* 30: 8528-8537.

Petoukhov, M.V. and Svergun, D.I. (2013). Applications of small-angle X-ray scattering to biomacromolecular solutions. *The International Journal of Biochemistry and Cell Biology* 45: 429-437.

Pickering, G.J., Simunkova, K. and DiBattista, D. (2004). Intensity of taste and astringency sensations elicited by red wines is associated with sensitivity to PROP (6-n-propylthiouracil). *Food Quality and Preference* 15: 147-154.

Quijada-Morín, N., Crespo-Expósito, C., Rivas-Gonzalo, J.C., García-Estévez, I. and Escribano-Bailón, M.T. (2016). Effect of the addition of flavan-3-ols on the HPLC-DAD salivary-protein profile. *Food Chemistry* 207: 272-278.

Ramos-Pineda, A.M., García-Estévez, I., Brás, N.F., Martín del Valle, E.M., Dueñas, M. and Escribano-Bailón, M.T. (2017). Molecular approach to the synergistic effect on astringency elicited by mixtures of flavanols. *Journal of Agricultural and Food Chemistry* 65: 6425-6433.

Ramos-Pineda, A.M., García-Estévez, I., Dueñas, M. and Escribano Bailón, M.T. (2018). Effect of the addition of mannoproteins on the interaction between wine flavonols and salivary proteins. *Food Chemistry* 264: 226-232.

Ramos-Pineda, A.M., García-Estévez, I., Soares, S., de Freitas, V., Dueñas, M. and Escribano Bailón, M.T. (2019). Synergistic effect of mixture of two proline-rich-protein salivary families (aPRP and bPRP) on the interaction with wine flavanols. *Food Chemistry* 272: 210-215.

Ribéreau-Gayon, P., Dubourdieu, D., Donèche, B. and Lonvaud, A. (2000). Handbook of Enology. Vol. 1: The Microbiology of Wine and Vinifications, 1st ed. Wiley, Chichester, UK, pp. 454.

Rinaldi, A., Gambuti, A., Moine-Ledoux, V. and Moio, L. (2010). Evaluation of the astringency of commercial tannins by means of the SDS-PAGE-based method. *Food Chemistry* 122: 951-956.

Rivero, F.J., Gordillo, B., Jara-Palacios, M.J., González-Miret, M.L. and Heredia, F.J. (2017). Effect of addition of overripe seeds from white grape by-products during red wine fermentation on wine color and phenolic composition. *LWT – Food Science and Technology* 84: 544-550.

Rodger, A., Marrington, R., Roper, D. and Windsor, S. (2005). Circular dichroism spectroscopy for the study of protein-ligand interactions. *Methods in Molecular Biology* 305: 343-364.

Sarni-Manchado, P., Cheynier, V. and Moutounet, M. (1999). Interactions of grape seed tannins with salivary proteins. *Journal of Agricultural and Food Chemistry* 47: 42-47.

Schiffman, S.S., Suggs, M.S., Sostman, L. and Simon, S.A. (1992). Chorda tympani and lingual nerve responses to astringent compounds in rodents. *Physiology and Behaviour* 51: 55-63.

Schöbel, N., Radtke, D., Kyereme, J., Wollmann, N., Cichy, A., Obst, K., Kallweit K., Kletke O., Minovi, A., Dazert, S., Wetzel, C.H., Vogt-Eisele, A., Gisselman, G., Ley, J.P., Bartoshuk, L.M., Spehr, J., Hofmann, T. and Hatt, H. (2014). Astringency is a trigeminal sensation that involves the activation of G protein –coupled signaling by phenolic compounds. *Chemical Senses* 39: 471-487.

Schwarz, B. and Hofmann, T. (2008). Is there a direct relationship between oral astringency and human salivary protein binding? *European Food Research and Technology* 227: 1693-1698.

Scott, K. (2005). Taste recognition: Food for thought. *Neuron* 48: 455-464.

Siebert, K.J., Troukhanova, N.V. and Lynn, P.Y. (1996). Nature of polyphenol– protein interactions. *Journal of Agricultural and Food Chemistry* 44: 80-85.

Silva, M.S., García-Estévez, I., Brandão, E., Mateus, N., de Freitas, V. and Soares, S. (2017). Molecular interaction between salivary proteins and food tannins. *Journal of Agricultural and Food Chemistry* 65: 6415-6424.

Simoes Costa, A.M., Costa Sobral, M.M., Delgadillo, I., Cerdeira, A. and Rudnitskaya, A. (2015). Astringency quantification in wine: Comparison of the electronic tongue and FT-MIR spectroscopy. *Sensors and Actuators B: Chemical* 207: 1095-1103.

Simon, C., Barathieu, K., Laguerre, M., Schmitter, J.M., Fouquet, E., Pianet, I. and Dufourc, E.J. (2003). Three-dimensional structure and dynamics of wine tannin-saliva protein complexes: A multitechnique approach. *Biochemistry* 42: 10385-10395.

Simon, S.A., Hall, W.L. and Schiffman, S.S. (1992). Astringent-tasting compounds alter ion transport across isolated canine lingual epithelia. *Pharmacology Biochemistry and Behaviour* 43: 271-283.

Soares, S., García-Estévez, I., Ferrer-Gallego, R., Brás, N.F., Brandão, E., Silva, M., Teixeira, N., Fonseca, F., Sousa, S.F., Ferreira-da-Silva, F., Mateus, N. and de Freitas, V. (2018). Study of human salivary proline-rich proteins interaction with food tannins. *Food Chemistry* 243: 175-185.

Soares, S., Gonçalves, R.M., Fernandes, I., Mateus, N. and de Freitas, V. (2009). Mechanistic approach by which polysaccharides inhibit alpha-amylase/procyanidin aggregation. *Journal of Agricultural and Food Chemistry* 57: 4352-4358.

Soares, S., Kohl, S., Thalmann, S., Mateus, N., Meyerhof, W. and de Freitas, V. (2013). Different phenolic compounds activate distinct human bitter taste receptors. *Journal of Agricultural and Food Chemistry* 61: 1525-1533.

Soares, S., Mateus, N. and de Freitas, V. (2012). Interaction of different classes of salivary proteins with food tannins. *Food Research International* 49: 807-813.

Tachibana, H., Koga, K., Fujimura, Y. and Yamada, K. (2004). A receptor for green tea polyphenol EGCG. *Nature Structural and Molecular Biology* 11: 380-381.

Upadhyay, R., Brossard, N. and Chen, J. (2016). Mechanisms underlying astringency: Introduction to an oral tribology approach. *Journal of Physics D: Applied Physics* 49: 104003 (11 pp).

Van Aken, G.A. (2010). Modelling texture perception by soft epithelial surfaces. *Soft Matter* 6: 826-834.

Vera, L., Acena, L., Boque, R., Guasch, J., Mestres, M. and Busto, O. (2010). Application of an electronic tongue based on FT-MIR to emulate the gustative mouthfeel "tannin amount" in red wines. *Analytical and Bioanalytical Chemistry* 397: 3043-3049.

Viegas, A., Manso, J., Nobrega, F.L. and Cabrita, E.J. (2011). Saturation-Transfer Difference (STD) NMR: A simple and fast method for ligand screening and characterization of protein binding. *Journal of Chemical Education* 88: 990-994.

Wang, B., Danjo, A., Kajiya, H., Okabe, K. and Kido, M.A. (2011). Oral epithelial cells are activated via TRP channels. *Journal of Dental Research* 90: 163-167.

Wang, Y., Erickson, R.P. and Simon, S.A. (1993). Selectivity of lingual nerve fibers to chemical stimuli. *The Journal of General Physiology* 101: 843-866.

Yao, J.W., Lin, C.J., Chen, G.Y., Lin, F. and Tao, T. (2010). The interactions of epigallocatechin-3-gallate with human whole saliva and parotid saliva. *Archives of Oral Biology* 55: 470-478.

# Index

# Color Section

**Chapter 1:** Fig. 5, p. 17

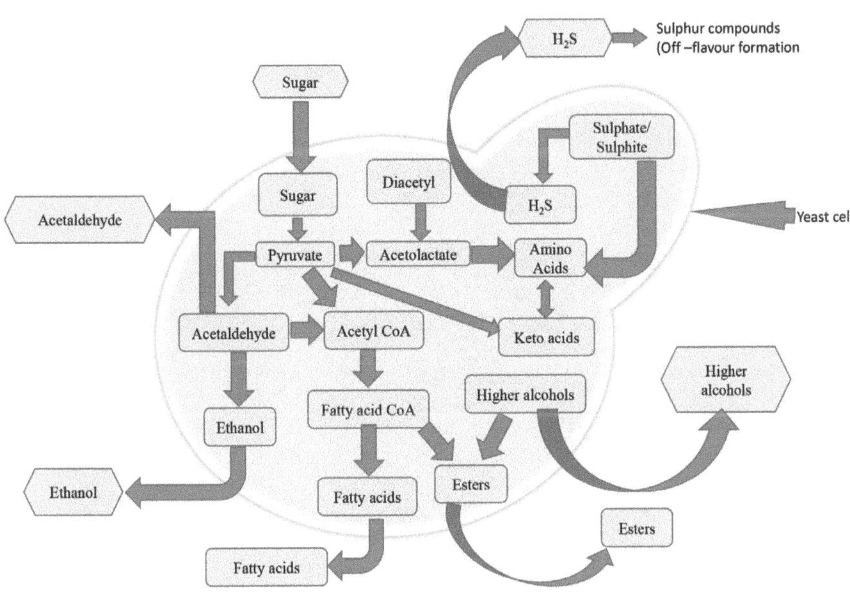


**Figure 5.** A schematic representation of the biochemical mechanisms of yeast metabolism during alcoholic fermentation (*Source*: Adapted from Pretorius (2000))

**Chapter 2:** Fig. 1, p. 38

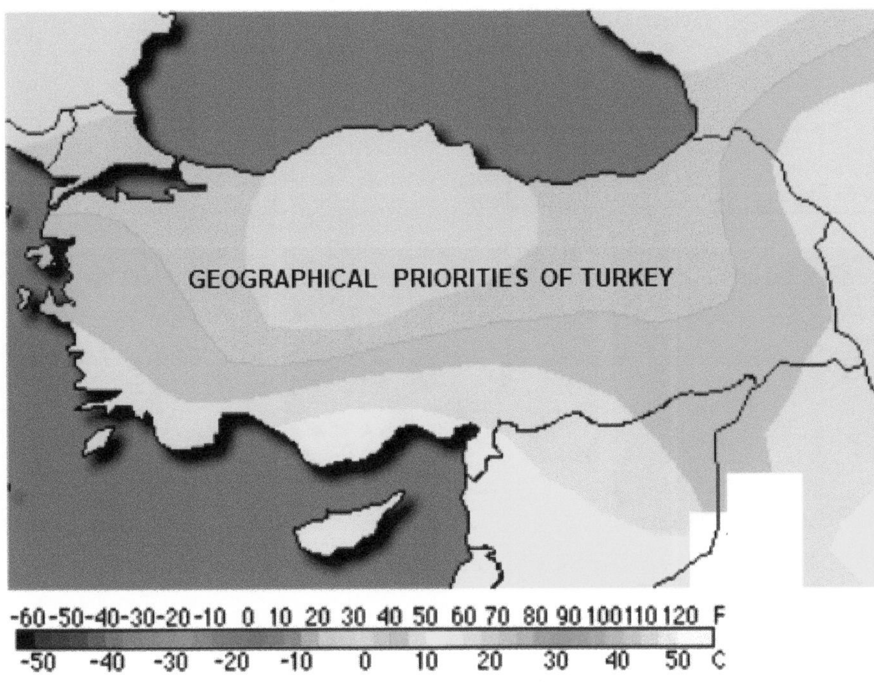


**Figure 1.** Anatolia (east of Turkey)

**Chapter 4:** Fig. 1, p. 102

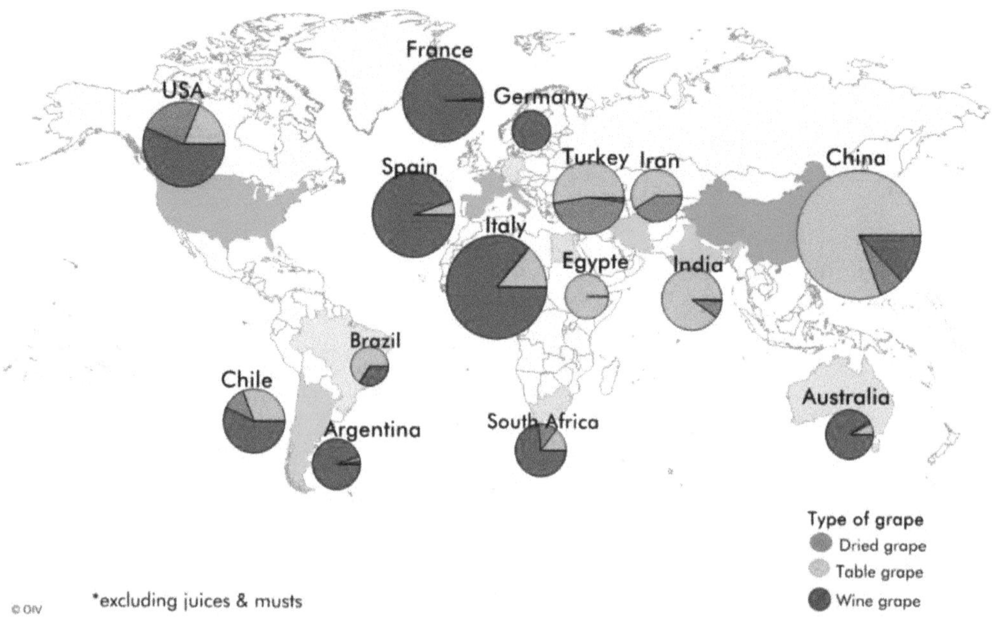


**Figure 1.** World's major grape producers
(*Source*: OIV statistical report on world vitiviniculture, 2016)

**Chapter 5:** Fig. 3, p. 123

**Figure 3.** Evolution of berry colour during ripening (green phase, veraison, maturity)

**Chapter 8:** Fig. 2, p. 199

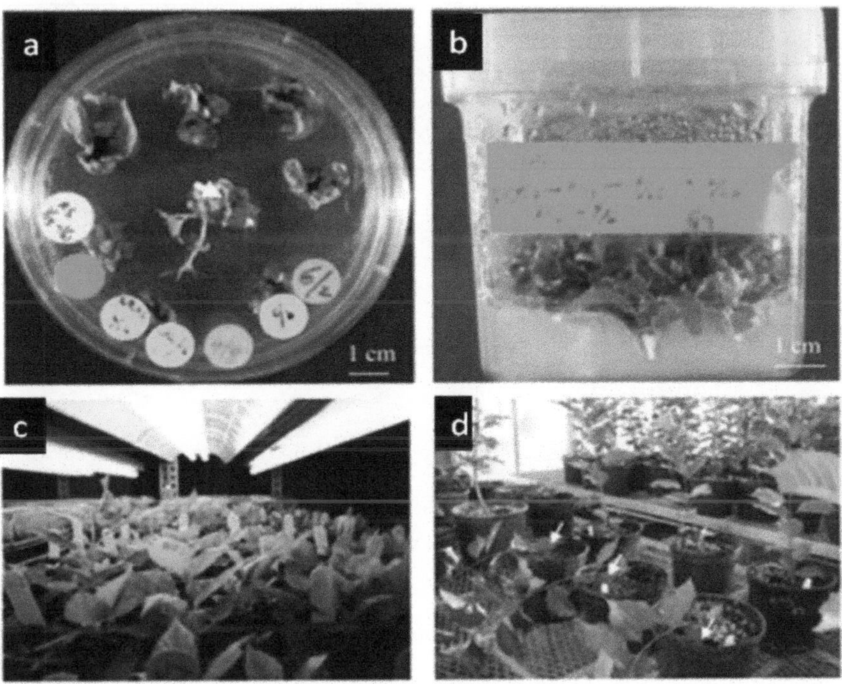


**Figure 2.** Transformation of cherry rootstocks (a) shoot regeneration from leaf explants; (b) proliferation of a putative transgenic; (c) rooting of transgenic plants; (d) growing of transgenic plants. The white arrows are showing three procumbent plants derived from one transgenic event of 'Gisela 6' (*Source*: Song *et al.*, 2013)

**Chapter 9:** Fig. 1, p. 227

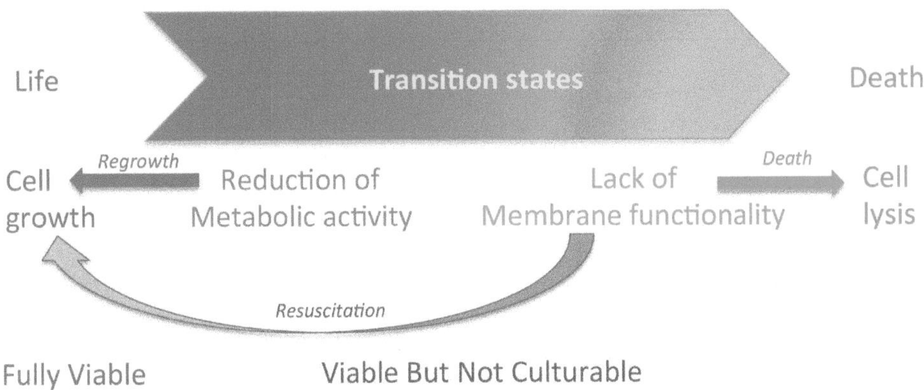


**Figure 1:** Cell statuses during alcoholic fermentation

**Chapter 9:** Fig. 2, p. 231

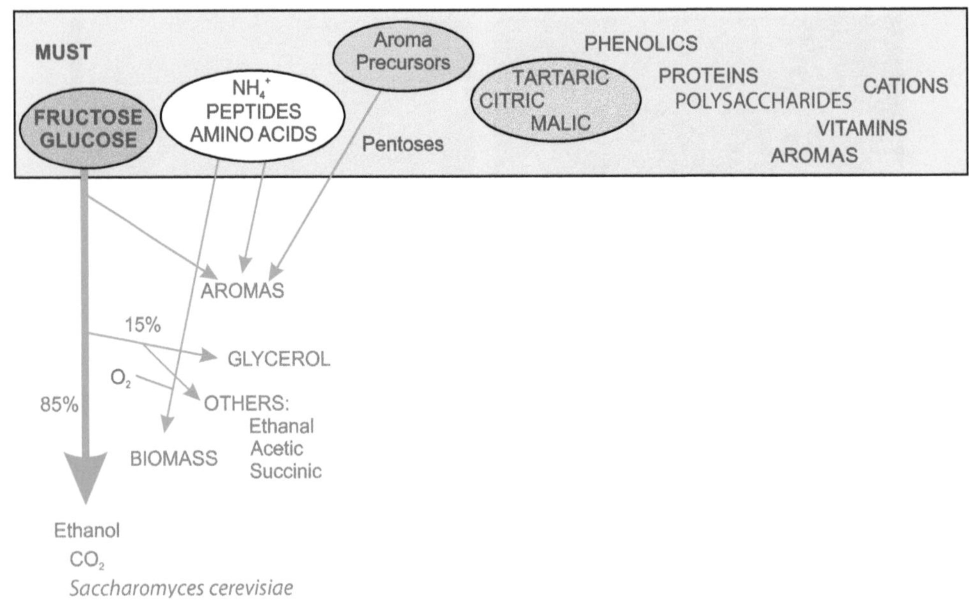


**Figure 2:** Alcoholic Fermentation

**Chapter 9:** Fig. 3, p. 236

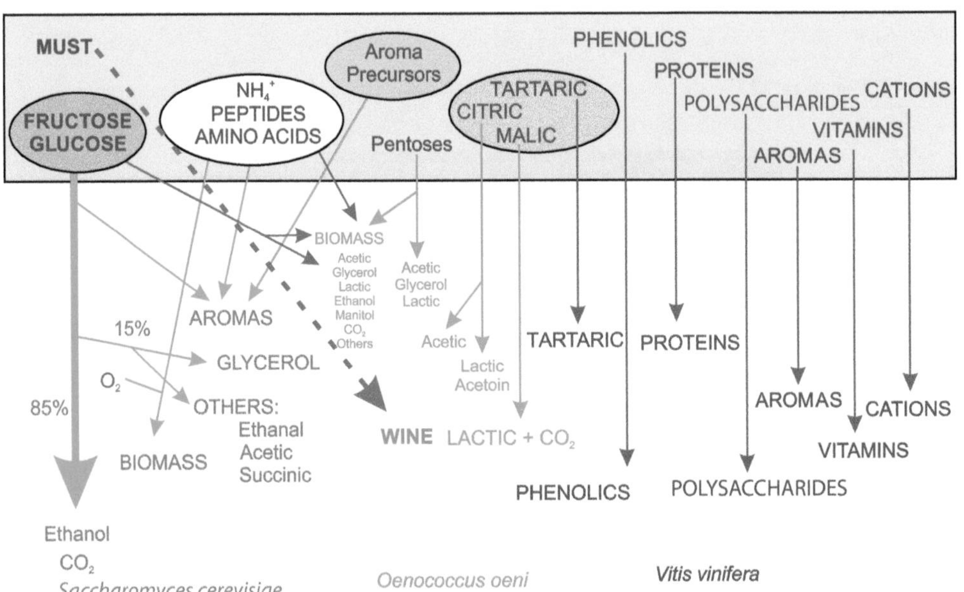


**Figure 3:** The conversion of must into wine

**Chapter 11:** Fig. 1, p. 277

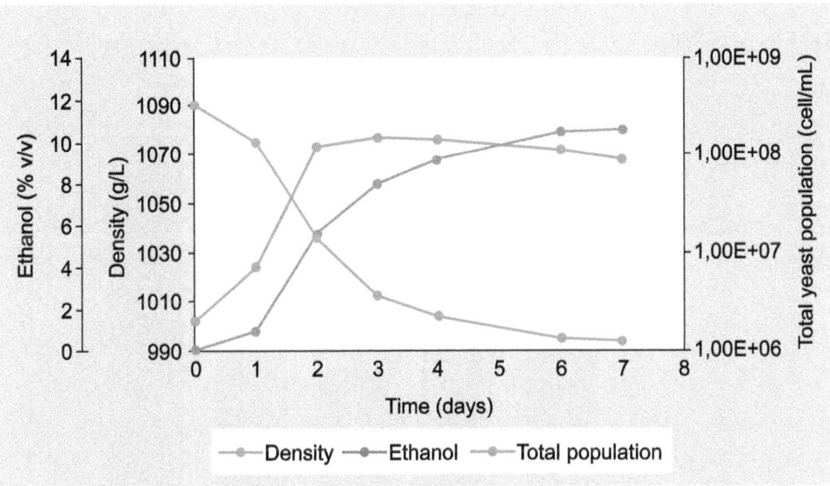


**Figure 1.** Evolution of alcoholic fermentation with the indication of the total yeast population, decrease of density (a typical measure of the fermentable sugars) and increase of ethanol

**Chapter 11:** Fig. 2, p. 278

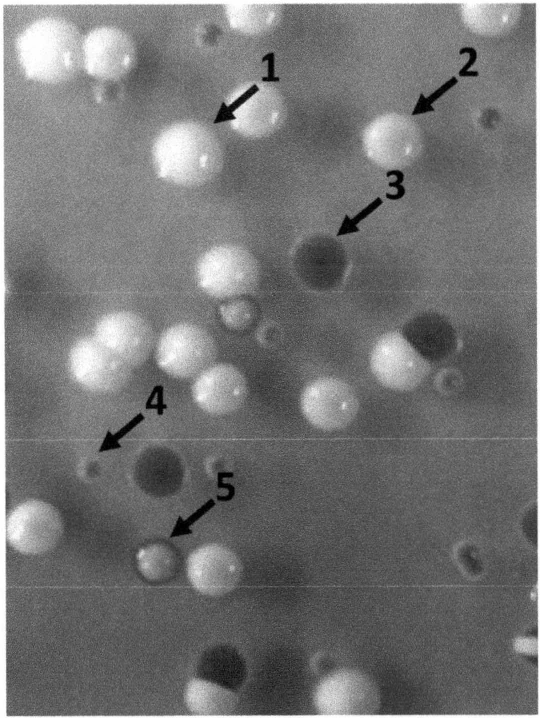


**Figure 2.** Different wine yeast morphologies in Wallenstein medium (WLN medium): 1. *S. cerevisiae*; 2. *T. delbrueckii*; 3. *H. uvarum*; 4. *St. bacillaris*; 5. *M. pulcherrima*

**Chapter 11:** Fig. 3, p. 280

A)

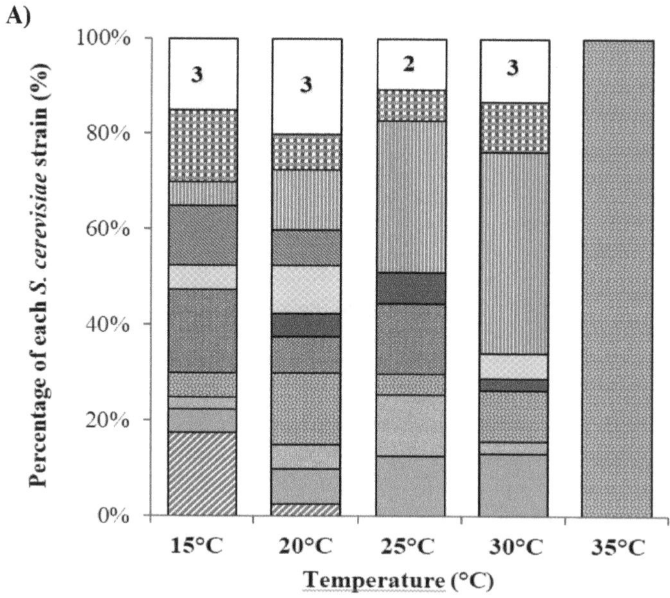


B)

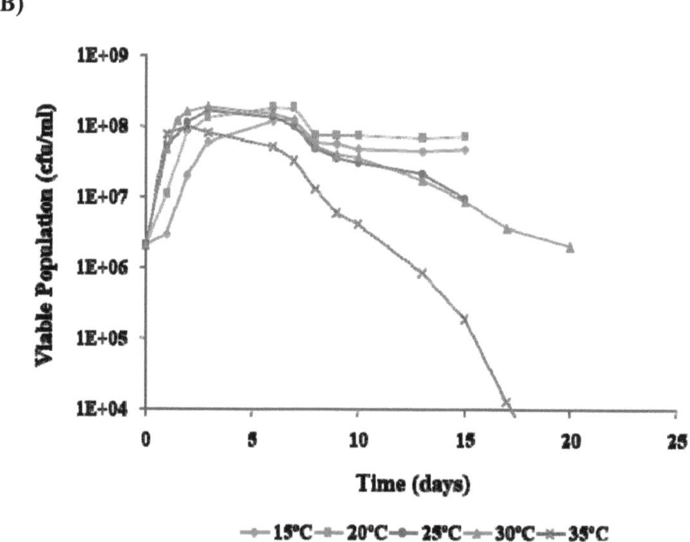


**Figure 3.** Effect of fermentation temperature on *S. cerevisiae* strain diversity at the end of mixed alcoholic fermentations (A) and on yeast viability during alcoholic fermentation (B). Each frame indicates a different *S. cerevisiae* strain. The numbers on the top of the bars indicate the number of different minority strains (<10 per cent) (Adapted from Torija *et al.*, 2003a)

**Chapter 11:** Fig. 4, p. 285

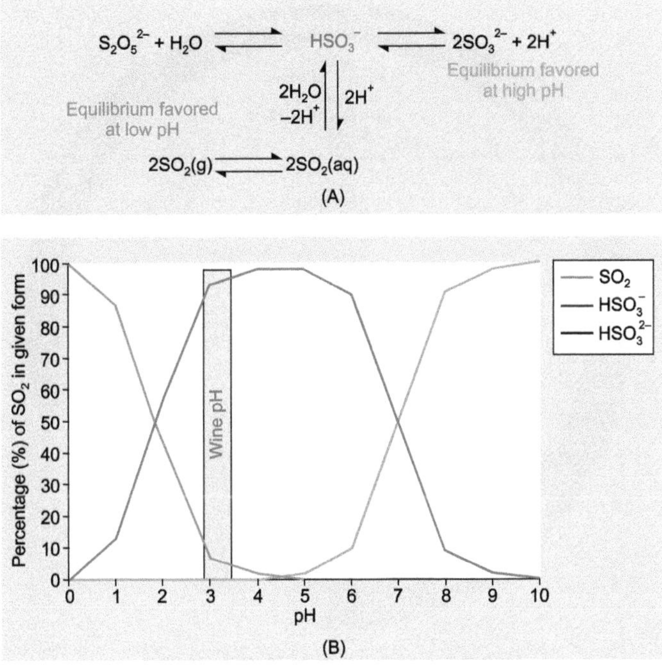


**Figure 4.** Equilibrium between the different sulphur dioxide forms (molecular, bisulphite and sulphite) in aqueous solution after supplementation in the form of metabisulphite (A) and percentage of each form in relation to the pH (B). The grey box indicates the range of pHs usually found in wines

**Chapter 12:** Fig. 4, p. 311

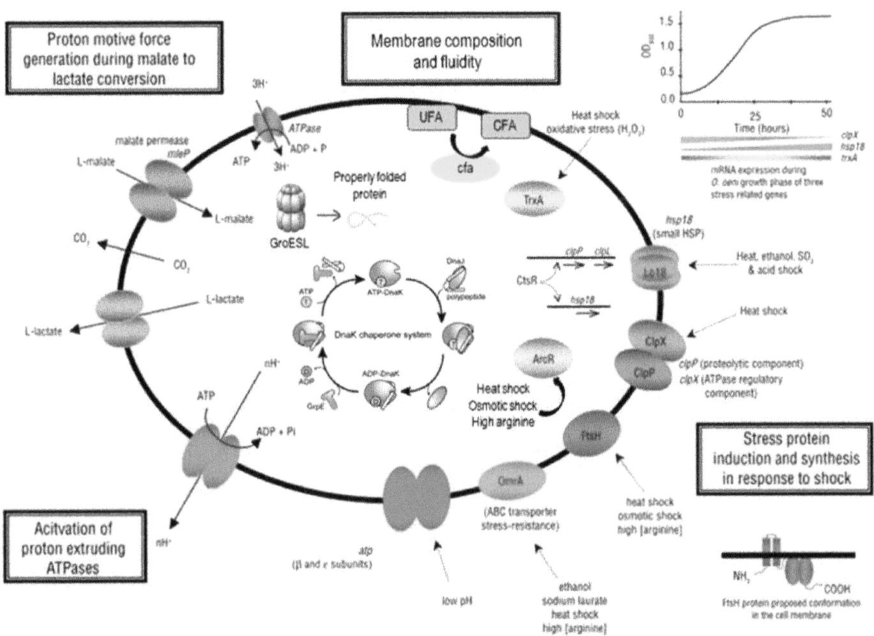


**Figure 4.** Mechanisms that are important in conferring, in *O. oeni*, the ability to survive in wine: the membrane composition, the proton motive force generated by malic acid metabolism, the activation of proton-extruding ATPase, and the stress protein induction and synthesis in response to shock (Adapted from Bartowsky, 2005)

**Chapter 25:** Fig. 1, p. 656

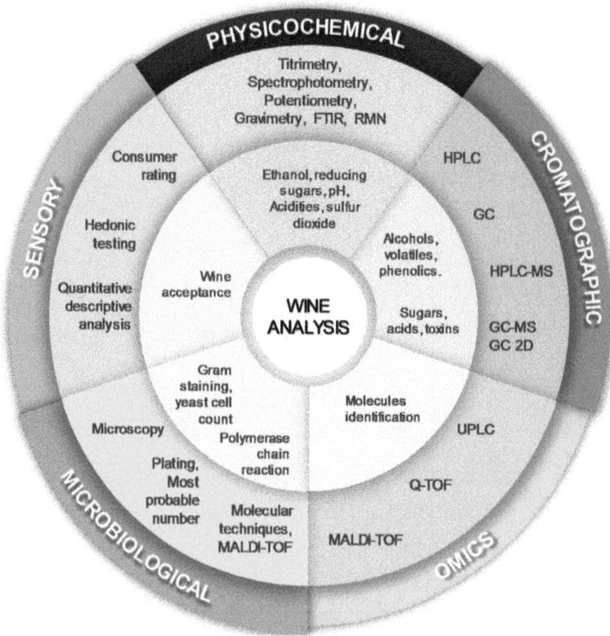


**Figure 1.** Techniques used in wine analysis

**Chapter 27:** Fig. 2, p. 703

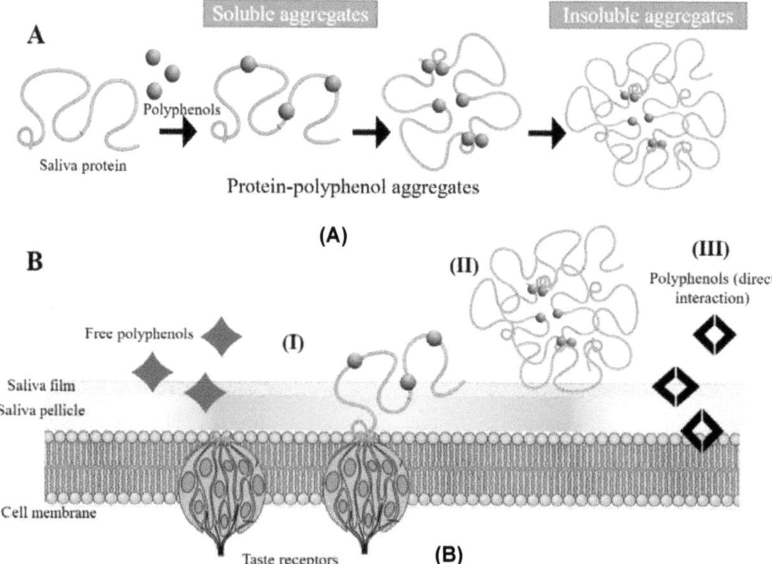


**Figure 2.** Proposed astringency mechanisms: (A) Protein aggregation and complex formation by the interaction between polyphenols and salivary proteins; (B) Astringency development: (I) Free polyphenols and soluble polyphenol–protein aggregates could disrupt the salivary film and reach the pellicle or even activate specific taste receptors; (II) Insoluble aggregates are rejected against salivary film, causing a loss of lubrication and increased friction in the oral cavity. (III) Direct interaction between polyphenols an oral epithelium. Adapted from Ma *et al.* (2014).

Milton Keynes UK
Ingram Content Group UK Ltd.
UKHW050448071024
449327UK00014B/293